POLYMER MORPHOLOGY

POLYMER MORPHOLOGY

Principles, Characterization, and Processing

Edited by

QIPENG GUO

WILEY

Published by John Wiley & Sons, Inc., Hoboken, New Jersey

Published simultaneously in Canada

Library of Congress Cataloging-in-Publication Data

Names: Guo, Qipeng, editor.
Title: Polymer morphology : principles, characterization, and processing / edited by Qipeng Guo.
Description: Hoboken, New Jersey : John Wiley & Sons, Inc., [2016] | Includes bibliographical references and index.
Identifiers: LCCN 2015038844 (print) | LCCN 2015040343 (ebook) | ISBN 9781118452158 (cloth) | ISBN 9781118892763 (ePub) | ISBN 9781118892770 (Adobe PDF)
Subjects: LCSH: Polymers. | Polymerization. | Crystalline polymers. | Surface chemistry.
Classification: LCC QD381.9.S97 P625 2016 (print) | LCC QD381.9.S97 (ebook) | DDC 547/.7–dc23
LC record available at http://lccn.loc.gov/2015038844

Typeset in 10/12pt TimesLTStd by SPi Global, Chennai, India

Printed in the United States of America

10 9 8 7 6 5 4 3 2 1

CONTENTS

PREFACE

Polymer morphology refers to the order within macromolecular solids and the structural constitution on the levels from nanoscale to a submicron and micron scale. The morphology plays an important role in the processing and application of polymeric materials. It embraces the processes of formation such as crystallization, deformation, and so on and the consequences for material properties as well as product performance.

Many books have been written on different particular aspects of polymer morphology. However, this book includes the basic principles and methods for morphological characterization, followed by a comprehensive understanding of polymer morphology, properties, and processing in a range of polymeric materials. It is composed of two parts: I. Principles and Methods of Characterization and II. Morphology, Properties, and Processing. Part I starts with an overview and prospects of polymer morphology, and then presents a detailed account of the principles and methods that are most used for investigation of polymer morphology. Morphology–property relationships are critical in the processing and applications of polymeric materials, which are addressed in Part II. The way in which the material is processed is a key determinant of the morphology and resulting properties. Understanding the morphology and the processes of its formation is essential in order to achieve the desired properties of a polymeric material in a certain application. A sampling of work in this area is also provided in Part II. I hope that it will not only serve as a useful textbook for advanced undergraduate and postgraduate students but also as a concise handbook for researchers in academia and engineers in related industries.

I express my appreciation and respects to all the contributors for their commitment, patience, and cooperation. Finally, I wish to express my sincere gratitude to the staff of John Wiley & Sons, Inc., especially Jonathan T. Rose who invited me to edit this comprehensive book and provided helpful support, and also Purvi Patel and Amanda Amanullah for her assistance during the edition.

Qipeng Guo
2015
Australia

LIST OF CONTRIBUTORS

Ruth Cardinaels, Department of Chemical Engineering, KU Leuven, 3001 Leuven, Belgium

Pierre J. Carreau, Chemical Engineering Department, Ecole Polytechnique de Montreal, Montreal H3T 1J4, Canada

K. L. A. Chan, Department of Chemical Engineering, Imperial College London, London SW7 2AZ, UK

Peng Chen, Department of Materials, School of Chemistry and Chemical Engineering, Anhui University, Hefei, 230039, China

Sossio Cimmino, Istituto per i Polimeri, Compositi e Biomateriali, Consiglio Nazionale delle Ricerche, 80078, Pozzuoli, Italy

M.-B. Coltelli, Department of Civil and Industrial Engineering, University of Pisa, 56126, Pisa, Italy

Duraccio Donatella, Istituto per i Polimeri, Compositi e Biomateriali, Consiglio Nazionale delle Ricerche, 80078, Pozzuoli, Italy; Istituto per le Macchine Agricole e Movimento Terra (IMAMOTER) – Consiglio Nazionale delle Ricerche, Torino, 10135, Italy

So Fujinami, RIKEN SPring-8 Center, RIKEN, Hyogo 679-5148, Japan

Abbas Ghanbari, Chemical Engineering Department, Ecole Polytechnique de Montreal, Montreal H3T 1J4, Canada

G. Groeninckx, Department of Chemistry, Division of Molecular and Nanomaterials, Laboratory of Macromolecular Structure Chemistry, Katholieke Universiteit Leuven, 3001 Heverlee, Belgium

Jean-Michel Guenet, Institut Charles Sadron, CNRS-Université de Strasbourg, 67034 Strasbourg, France

Qipeng Guo, Institute for Frontier Materials, Deakin University, Geelong, Victoria, 3220, Australia

Yanchun Han, State Key Laboratory of Polymer Physics and Chemistry, Changchun Institute of Applied Chemistry, Chinese Academy of Sciences, Changchun, 130022, China

Masanori Hara, Department of Chemical and Biochemical Engineering, Rutgers University, Piscataway, NJ 08854-8058, USA

Charef Harrats, Laboratoire de Chimie Appliquée (LAC), DGRSDT, Institut dcs Scicnccs ct Tcchnologic, Ctr Univ Belhadj Bouchaib, Ain Temouchent 46000, Algeria

Xuehao He, Department of Chemistry, School of Science, Tianjin University, Tianjin, 300072, China

Marie-Claude Heuzey, Ecole Polytechnique de Montreal, Montreal H3T 1J4, Canada

Wenbing Hu, Department of Polymer Science and Engineering, School of Chemistry and Chemical Engineering, Nanjing University, Nanjing 210093, China

S. G. Kazarian, Department of Chemical Engineering, Imperial College London, London SW7 2AZ, UK

Werner Lebek, Institute of Physics and Institute of Polymeric Materials, Martin Luther University Halle-Wittenberg, 06099 Halle, Germany

Xuejin Li, Division of Applied Mathematics, Brown University, Providence, RI 02912, USA

Hao Liu, Zhengzhou University, Zhengzhou, 450001, China

Jiangang Liu, State Key Laboratory of Polymer Physics and Chemistry, Changchun Institute of Applied Chemistry, Chinese Academy of Sciences, Changchun, 130022, China

Arnaldo T. Lorenzo, The Dow Chemical Company, Performance Plastics R&D, Freeport, TX 77541, USA

Antonella Marra, Istituto per i Polimeri, Compositi e Biomateriali, Consiglio Nazionale delle Ricerche, 80078, Pozzuoli, Italy

Hironori Marubayashi, Department of Organic and Polymeric Materials, Graduate School of Science and Engineering, Tokyo Institute of Technology, Tokyo 152-8552, Japan

Rose Mary Michell, Grupo de Polímeros USB, Departamento de Ciencia de los Materiales, Universidad Simón Bolívar, Apartado 89000, Caracas 1080-A, Venezuela

Goerg H. Michler, Institute of Physics and Institute of Polymeric Materials, Martin Luther University Halle-Wittenberg, 06099 Halle, Germany

Paula Moldenaers, Department of Chemical Engineering, KU Leuven, 3001 Leuven, Belgium

Alejandro J. Müller, POLYMAT and Polymer Science and Technology Department, Faculty of Chemistry, University of the Basque Country UPV/EHU, Paseo Manuel de Lardizabal 3, 20018. Donostia-San Sebastián, Spain; IKERBASQUE, Basque Foundation for Science, 48013 Bilbao, Spain; Grupo de Polímeros USB, Departamento de Ciencia de los Materiales, Universidad Simón Bolívar, Apartado 89000, Caracas 1080-A, Venezuela

N. Sanjeeva Murthy, New Jersey Center for Biomaterials, Rutgers University, Piscataway, NJ 08854, USA

Ken Nakajima, Department of Organic and Polymeric Materials, Graduate School of Science and Engineering, Tokyo Institute of Technology, Tokyo 152-8550, Japan

Toshio Nishi, International Division, Tokyo Institute of Technology, Tokyo 152-8550, Japan

Shuichi Nojima, Department of Organic and Polymeric Materials, Graduate School of Science and Engineering, Tokyo Institute of Technology, Tokyo 152-8552, Japan

Anbazhagan Palanisamy, Institute for Frontier Materials, Deakin University, Geelong, Victoria, 3220, Australia

Marilena Pezzuto, Istituto per i Polimeri, Compositi e Biomateriali, Consiglio Nazionale delle Ricerche, 80078, Pozzuoli, Italy

Holger Schönherr, Department of Chemistry – Biology, Physical Chemistry I, University of Siegen, 57076 Siegen, Germany

Jerold M. Schultz, Department of Chemical Engineering, University of Delaware, Newark, DE 19716-3110, USA

Anne Seidlitz, Department of Physics, Martin-Luther University Halle-Wittenberg, 06120 Halle, Germany

Clara Silvestre, Istituto per i Polimeri, Compositi e Biomateriali, Consiglio Nazionale delle Ricerche, 80078, Pozzuoli, Italy

Xiaoli Sun, State Key Laboratory of Chemical Resource Engineering and Beijing Key Laboratory of Membrane Science and Technology, Beijing University of Chemical Technology, Beijing 100029, China

Thomas Thurn-Albrecht, Department of Physics, Martin-Luther University Halle-Wittenberg, 06120 Halle, Germany

Minh-Tan Ton-That, Automotive and Surface Transport, National Research Council Canada, Ottawa, K1V 1S2, Canada

Hiroki Uehara, Department of Chemistry and Chemical Biology, Gunma University, Gunma 376-8515, Japan

Dong Wang, State Key Laboratory of Organic-Inorganic Composites, College of Materials Science and Engineering, Beijing University of Chemical Technology, Beijing 100029, China

Longjian Xue, Dynamic Biointerfaces, Max Planck Institute for Polymer Research, Mainz, 55128, Germany

Takeshi Yamanobe, Department of Chemistry and Chemical Biology, Gunma University, Gunma 376-8515, Japan

Shouke Yan, State Key Laboratory of Chemical Resource Engineering and Beijing Key Laboratory of Membrane Science and Technology, Beijing University of Chemical Technology, Beijing 100029, China

Xinhong Yu, State Key Laboratory of Polymer Physics and Chemistry, Changchun Institute of Applied Chemistry, Chinese Academy of Sciences, Changchun, 130022, China

Liyun Zha, Department of Polymer Science and Engineering, School of Chemistry and Chemical Engineering, Nanjing University, Nanjing 210093, China

PART I

PRINCIPLES AND METHODS OF CHARACTERIZATION

1

OVERVIEW AND PROSPECTS OF POLYMER MORPHOLOGY

JEROLD M. SCHULTZ
Department of Chemical Engineering, University of Delaware, Newark, DE, USA

1.1 INTRODUCTORY REMARKS

Why are we interested in the morphology of polymers? I would like to say that our interest is in the inherent beauty and intriguing complexity of the patterns. Two examples are microphase-separated block copolymers and homopolymers crystallized from the melt. Figure 1.1 shows the range of morphologies typical to simple AB copolymers [1]. These structures are defined by the composition of the diblock copolymer: alternating plates for approximately equal amounts of components A and B, proceeding through double gyroid, rod, and sphere morphologies as the composition is made increasingly unbalanced (see Chapters 10, 14, and 15). More complicated repetitive morphologies are found in more complex block systems [1,2]. Importantly, the repetition scales of these morphologies are of the same order as the dimensions of the molecular coils of the blocks – typically, tens of nanometers. Homopolymers crystallized from the melt display morphological features from the micrometer range to tens of nanometers. Figure 1.2 is an optical micrograph of spherulites of poly(ether ketone ketone) crystallizing from the melt. Seen in the figure are arms radiating from a central point, the arms then branching at small angles, to fill all space between the arms. Higher resolution images reveal that the arms are composed of stacks of long, ribbonlike crystals, with the molecules running in the thin direction of the crystals, as sketched in Figure 1.3 [3]. Why and how did such long-chain molecules, very highly intertangled in the melt, disentangle themselves to form this spherically symmetric array of bundles of lamellar crystals?

Certainly, the beauty of and the desire to understand the complexity of systems such as block copolymers and melt-crystallized polymers have played a role in driving the study of morphology. But most of the research in this area is funded and executed because properties of the polymeric materials are tied to the morphological detail. For instance, the fine-scale repetitive morphology of block copolymers makes them well suited for photonics [4], as well as for photovoltaic [5,6] and battery [7] applications. Figure 1.4 shows a recent result for a block copolymer used as a photovoltaic system. The block copolymer is poly(3-hexylthiophene)–*block*–poly((9,9-dioctylfluorene)-2,7-diyl-alt-[4,7-bis(thiophen-5-yl)-2,1,3-benzothiadiazole]-2′,2″-diyl) (P3HT-*b*-PFTBT), with a composition of 56 wt% P3HT. With nearly equivalent volume fractions of each block, the system has an alternating plate morphology, as shown in Figure 1.4a. The *I*–*V* curve shown at the right demonstrates an efficiency of about 3%. While this efficiency is not competitive with current commercial photovoltaics, it is encouraging for the earliest stages of a new approach.

While the study of the morphology of polymers has been an occasional topic for over a century, it became a field of study in its own right with the advent of commercial transmission electron microscopes some 60 years ago. It was only then that the fine structures unique to polymers could be resolved and directly observed. But the electron beam in electron microscopies typically destroys the specimen in a few tens of seconds, precluding much in the way of following the evolution of fine morphological detail. Studies of morphological evolution were based on less direct (but

Polymer Morphology: Principles, Characterization, and Processing, First Edition. Edited by Qipeng Guo.

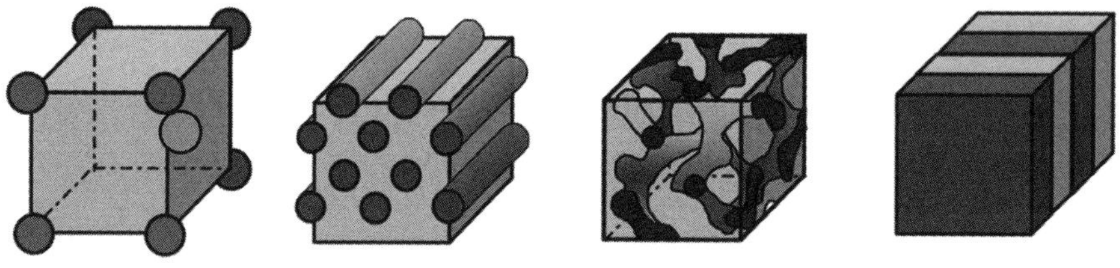

Figure 1.1 Morphology of AB diblock copolymers. From (a–d), in increasing composition from 0 to 50 vol%, spheres arranged on a body-centered cubic lattice, hexagonally packed cylinders, gyroid, and lamellae. Balsara and Hahn [1]. Reproduced with permission of World Scientific.

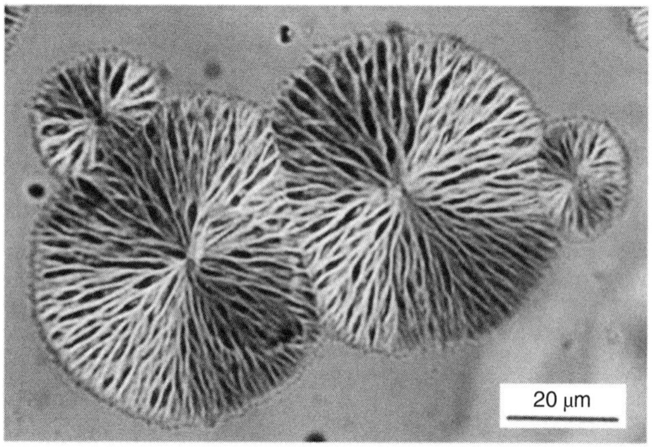

Figure 1.2 Spherulites growing into a melt of poly(ether ketone ketone) (PEKK 70/30, a copolymer of 70% terephthalate and 30% isophthalate moieties) at 280 °C.

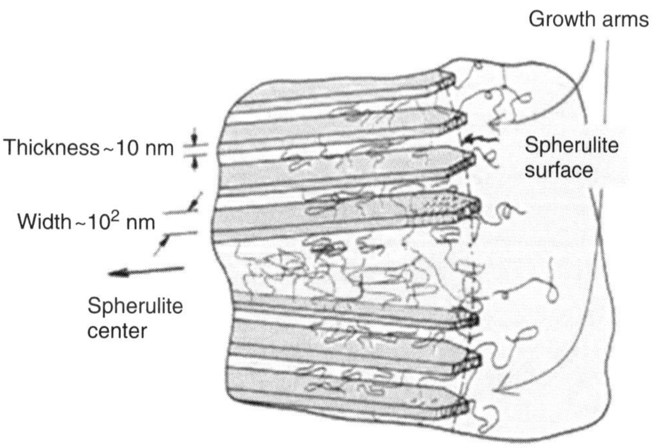

Figure 1.3 Sketch of a growing spherulite, showing crystalline lamellae and growth arms (stacks of lamellae). Schultz [3]. Reproduced with permission of American Chemical Society.

nonetheless useful) scattering, diffraction, spectroscopic, and calorimetric methods, in which local structure was educed from bulk behavior. This situation changed in the late 1990s with the advent of scanning probe microscopies and, somewhat more recently, with imaging based on spectroscopies. The current state of structural tools is detailed in Chapters 2–9 of this book.

1.2 EXPERIMENTAL AVENUES OF MORPHOLOGICAL RESEARCH

There are broadly three avenues of investigation of morphology. One is the characterization of the morphological state of a polymeric material. All of Part 1 and parts of Part 2 of this book deal specifically with characterization. As mentioned, there would be no need for a science of polymer morphology, were the morphology unimportant in establishing properties. Morphology–property relationships are then a second important area of study. A sampling of recent morphology–property research is given in Chapters 18 and 21, with examples included in other chapters. The third avenue is the study of how processing controls morphological detail, and hence also defines the behavior of the product. A sampling of work in this area is provided in Chapters 11–16, 19, 20, and 22. The three avenues of research are treated in the following subsections.

1.2.1 Morphological Characterization: The Enabling of *in situ* Measurements

Because so much of this compilation is already devoted to characterization, we concentrate here on only two aspects: rapid measurements and combined techniques.

One of the most interesting developments over the few decades of morphological study has been the development of tools for following morphological development *in situ* during processing operations. Many of these *in situ* methods awaited the development of fast measurement tools. Synchrotron radiation has provided X-ray and infrared (IR) intensities orders of magnitude higher than had been possible in laboratory-scale instruments. This beam intensity, plus the creation of detectors capable of capturing an entire spectrum of data in parallel, has reduced scan times for individual measurements from the order of an hour to the order of milliseconds. The technologies that enabled such work were the development of one- [7, 8] and two-dimensional [9, 10] position-sensitive wire detectors in the 1970s and of polymer-oriented beamlines at synchrotrons, beginning in the mid-1980s [11]. In parallel, more recently, Chase and Rabolt have similarly provided a rapid advance for infrared spectroscopy, developing a parallel capture system [12–14]. Another pair of important breakthroughs in the first decade

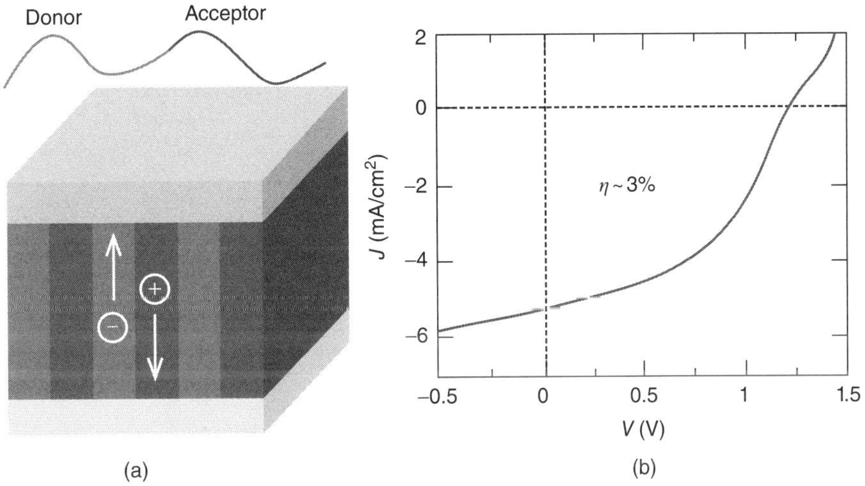

Figure 1.4 (a) Sketch of alternating lamellar morphology of a photovoltaic device made from the block copolymer poly(3-hexylthiophene)–block–poly((9,9-dioctylfluorene)-2,7-diyl-alt-[4,7-bis(thiophen-5-yl)-2,1,3-benzothiadiazole]-2′,2″-diyl) (P3HT-*b*-PFTBT), with a composition of 56 wt% P3HT. (b) *I–V* curve for the device. Guo et al. [6]. Reproduced with permission of American Chemical Society.

of the 2000s were the recognition that morphological development from the melt could be followed at high resolution by atomic force microscopy (AFM) [15–17], and the subsequent development of a very fast method of obtaining AFM images [18–21]. Using AFM, the same area can be probed many times at high resolution, in contrast to the situation for electron microscopies.

Another area of recent advances is in imaging using signals other than light, electrons, or neutrons. Scanning microscopies have enabled the use of any of a wide variety of signals, among which are surface friction (AFM phase mode), near-field optics, time-of-flight secondary ion mass spectroscopy (ToF-SIMS), and infrared and Raman absorption. An example of ToF-SIMS mapping across a spherulite, from Sun et al. [22], is shown in Figure 1.5. Seen is a map of the positions from which molecular fragments representing poly(ethylene oxide) (PEO) and poly(L-lactic acid) (PLLA) occur in a ring-banded spherulite of a 50/50 blend of PEO and PLLA. This image shows a radially periodic alternation of the two components. The periodic alternation is as yet unexplained. Figure 1.6, from Cong et al. [23], shows IR images taken at three different times during the growth from the melt of an isotactic polypropylene (iPP) spherulite at 142 °C. The band at 1303 cm^{-1} represents crystalline iPP; the band at 998 cm^{-1} represents ordered sequences of iPP in the melt. The spatial disposition of the 998 cm^{-1} band demonstrates the preordering of iPP chains in the melt ahead of the spherulite front. IR imaging is described in detail in Chapter 7.

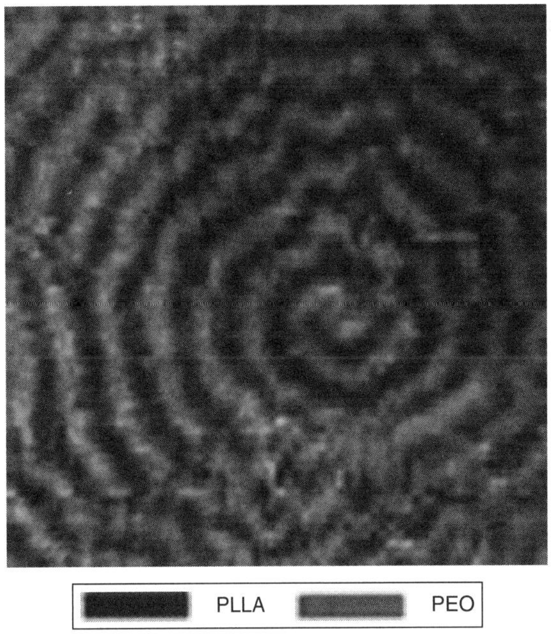

Figure 1.5 An ion map of a banded spherulite that formed in a 3-μm-thick film of a PLLA/PEO (50/50) blend crystallized between a silicon wafer and a Kapton cover for 5 h at 125 °C. Image obtained after removal of the Kapton. Sun et al. [22]. Reproduced with permission of Elsevier.

1.2.2 Morphology–Property Investigation

An ongoing example of correlating morphology and properties is that of the mechanical behavior of engineering polymers. In the case of tensile deformation, one ideally follows changes in structure at the intermolecular, lamellar, and spherulite levels. Optical microscopy and small-angle light scattering can be used to follow spherulite-level changes during deformation, while wide-angle and small-angle X-ray scattering are available for intermolecular and lamellar study, respectively. Infrared absorption can be used to follow molecular-level changes. Until relatively recently, only light scattering has been fast enough to follow events *in situ* during deformation.

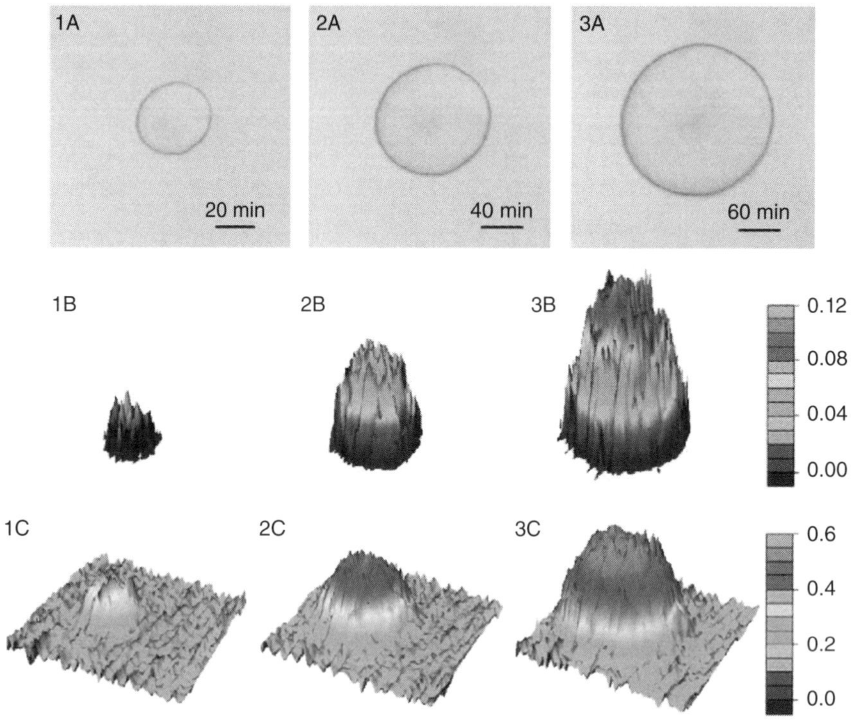

Figure 1.6 *In situ* optical microscope images of a single spherulite (set A) and the corresponding 3D images of intensity distribution of different conformational bands. Sets B and C refer to 1303 and 998 cm^{-1} bands, respectively. 1–3 are collected at different times during isothermal crystallization at 142 °C. The scale bar is 50 μm. Cong et al. [22]. Reproduced with permission of American Chemical Society.

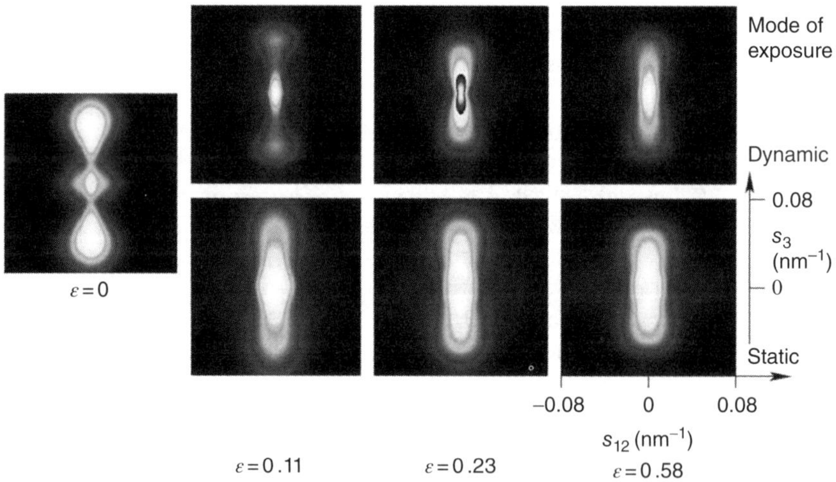

Figure 1.7 SAXS patterns obtained during tensile testing of oriented iPP films. Straining direction is vertical. Top row: during continuous straining at 10^{-3} s^{-1}. Bottom row: stretch-hold method. Stribeck et al. [31]. Reproduced with permission of John Wiley and Sons.

Optical microscopy [24, 25] and light scattering [26] studies in the 1960s showed clearly that spherulites underwent large-scale deformation as entities. In this earlier work, a specimen was extended a certain amount and then held there until a measurement was made (over several minutes) and then extended to the next level (*stretch-hold* investigation). But polymer spherulites are composed of ribbonlike crystals, with intervening layers of uncrystallized material. How does this finer structure deform so as to allow spherulites to change shape, and how are these mechanisms reflected in stress–strain behavior? To address these questions, tensile deformation devices were built onto laboratory-scale X-ray systems, and small- and wide-angle X-ray scattering (SAXS and WAXS) patterns sampled during interruptions of deformation, again each pattern requiring several minutes [27, 28]. A good deal was learned about intraspherulitic deformation mechanisms: sequentially, deformation of the amorphous layer, shear of lamellae along the chain axis, and destruction of lamellae,

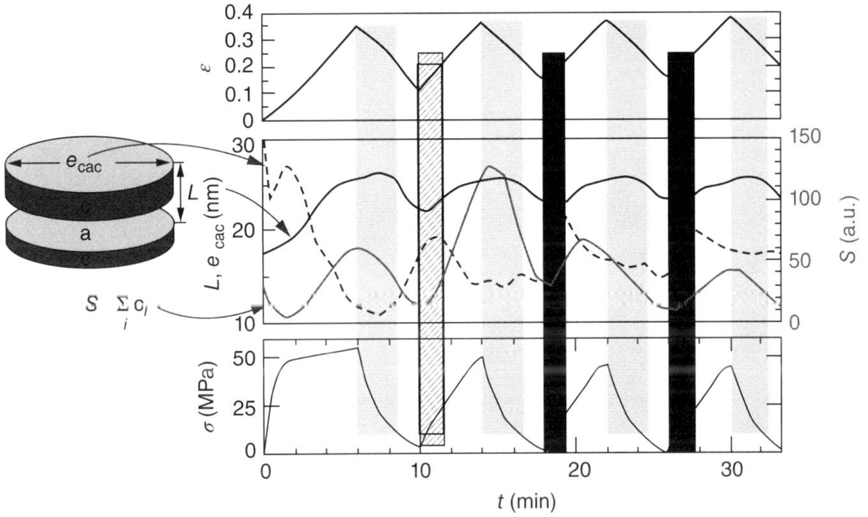

Figure 1.8 Dynamic load-reversal mechanical test of hard-elastic iPP film at a strain rate of $\dot{\varepsilon} \approx 10^{-3}\ s^{-1}$. The following are shown as a function of elapsed time *t*: top – elongation; middle – long period *L* (solid line), lateral extension of a sandwich made of two crystalline lamellae (broken line), and strength of the chord distribution function (dotted line); bottom – tensile stress. Vertical bars indicate zones of strain-induced crystallization (black) and relaxation-induced melting (gray). Stribeck et al. [32]. Reproduced with permission of John Wiley and Sons.

followed later by the creation of fibrils [29] along the stress direction (see Ref. [30] for a review of the earlier work).

While this older work provided glimpses of mechanisms, the studies were problematic in two major ways. First, there was considerable relaxation of the system during the time in which measurements were taken. This has been brought home recently in parallel synchrotron-based studies of the same polymer deformed according to the older stretch-hold procedure and by continuous drawing [31]. Results are shown in Figure 1.7. The difference between the stretch-hold SAXS patterns (top) and the continuous deformation patterns (bottom) are large, and show broadly that the perfection of the lamellar stacks increased considerably during the hold period. In most studies reported in the past 15–20 years, SAXS and WAXS data was collected during continuous deformation. The results relate to intermolecular strain (WAXS) and lamellar deformation (SAXS). An example of such results for cyclic loading and unloading of a hard elastic iPP film is shown in Figure 1.8 [32]. It is interesting that the strength of the SAXS signal *S* rises and falls with the applied strain, apparently showing alternating crystallization and melting. Excellent descriptions of experimental methods and devices for synchrotron studies of mechanical behavior are found in Refs [33] and [34].

Fairly recently, IR measurements have been combined with mechanical testing and synchrotron X-rays, providing additional information on changes within the chain [35]. Finally, microbeam synchrotron X-ray investigation has permitted the study of structural variations at different positions within a specimen, for instance, skin versus core in fibers [36].

It should be mentioned that the spherulite size itself can be important in the failure behavior of spherulitic polymers. Some early literature in this area can be found in Refs [37] and [38].

An interesting finding is reported by Sandt [39]. In this work, it was shown that the mode of crack propagation gradually changed from interspherulitic to intraspherulitic with increasing rate of deformation, as shown in Figure 1.9.

1.2.3 Morphology Development

1.2.3.1 Flow-Induced Crystallization In most polymer processing, the material is highly stretched in the melt while it crystallizes. The most extreme cases of such processing are in fiber spinning and film blowing. In the case of the spinning of poly(ethylene terephthalate) (PET) fibers from the melt, stretch rates of the order of kilometers per minute are typical, and crystallization rates can be increased by some five orders of magnitude in melt-oriented fibers, relative to the quiescent state [40]. The final state of melt-spun PET fibers is that of row structures: stacks (rows) of thin lamellar crystals with the chain axis aligned along the thin dimension of the lamellae and along the fiber axis [41]. It had long been conjectured that these lamellae nucleated on very fine precursor fibrils, and it was suspected that the fibrillar precursors were either present only ephemerally or were hidden by the prolific overgrowth of lamellar crystals. Ephemeral precursory fibrils (with diameters of a few nanometers) were identified, using transmission electron microscopy (TEM) [42, 43] and also by performing radial distribution function analysis [42] (two very difficult experiments!) in fibers crystallized in the solid state after quenching to an oriented, noncrystalline structure. But the study of the early stages of crystallization during actual spinning operations awaited the construction of the spinning apparatus at synchrotron sites. The early stages are now well documented, including the initial precursor fibrils and the subsequent growth of lamellae. This work has been reviewed by Somani et al. [44].

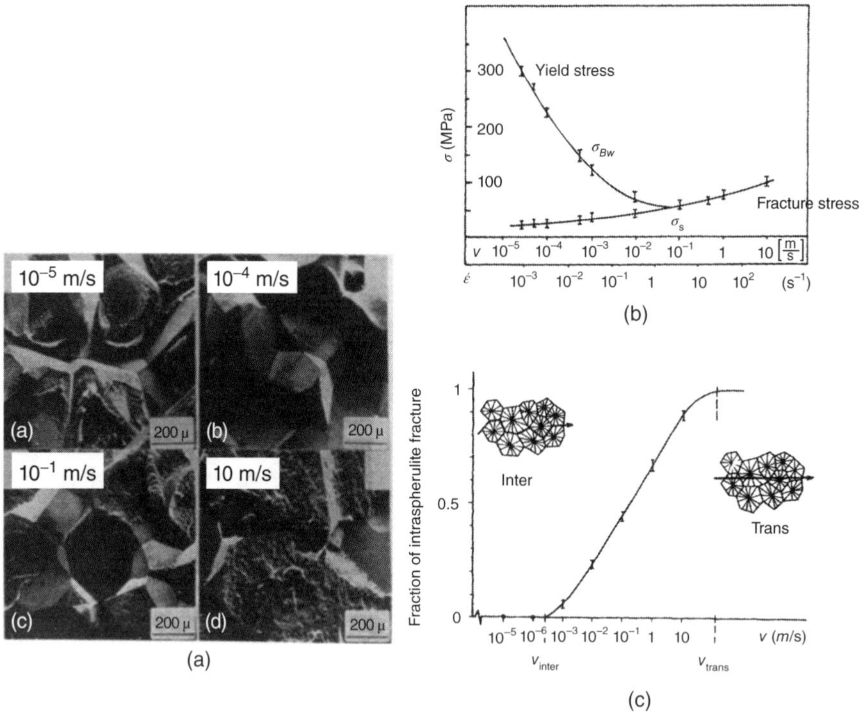

Figure 1.9 Effect of strain rate on the fracture of spherulitic iPP: (a) Scanning electron micrographs of the fracture surface, in which fracture along interspherulite boundaries is seen at the three lowest strain rates, while intraspherulite fracture is seen at 10 m/s; (b) yield stress and fracture stress versus strain rate; (c) fraction of intraspherulite fracture versus strain rate. Sandt [39]. Reproduced with permission of Ruhr-Universitaet Bochum.

It is relatively easy to build heaters and other ancillary equipment in the spacious and accessible specimen area at synchrotron beamlines. Consequently, the synchrotron-based study of morphological development during crystallization from the melt is now nearly routine. Similarly, deformation devices, with heaters attached, are found at synchrotron polymer beamlines, and numerous synchrotron-based reports of morphological development during drawing can be found.

1.2.3.2 Spherulite Formation It has been recognized for almost 60 years that polymer spherulites are composed of ribbonlike lamellar crystals in which the chains lie parallel to the thin direction and must reenter each crystal numerous times (see, e.g., Ref. [45], for an excellent early review). The most interesting phenomena occur at the level of individual lamellae, and until relatively recently sufficiently high resolution has been available only in electron microscopy. Because of the low specimen life in an electron beam, and because of the usual use of irrecoverable staining or etching, electron microscope study of the fine structure of spherulites has been limited to "post mortem" observation, at room temperature, of fully crystallized material. Some of the most detailed and careful TEM studies have been performed by Bassett and coworkers, using etching methodologies. The earlier work in this laboratory is reviewed in Refs [46] and [47]. In this work, done primarily on polyethylene and polypropylene, it was found that spherulites develop through the growth first of individual primary lamellae, with backfilling by secondary lamellae. It was also shown that, at least some of the time, lamellar twist occurred discretely, rather than continuously. More recent work, using AFM, has been able to follow the fine-scale development of spherulites *in situ* (see Ref. [48] for a recent review).

While it is the time-lapse sequences that are unique, "still" shots during transformations are also instructive. Examples are shown in Figures 1.10 and 1.11. Figure 1.10, taken during the early stages of growth of a spherulite of PBA-C8 (a poly(bisphenol A octane ether)) shows the transition from a single lamella to a spherulite, via small-angle branching [50]. Figure 1.11 shows the growth front of a poly(caprolactone) spherulite. Evident are the spade-like tips of the lamellae and the cloning of lamellae in a stack by growth about giant screw dislocations [49].

1.3 MODELING AND SIMULATION

In the past two to three decades, quantitative modeling of morphogenesis has shifted significantly from analytical theory to numerical simulation. Although the facile use of analytical expressions in engineering application is valuable, the ability of numerical simulation to predict details of morphological development has opened new paths of understanding. Two areas in which our understanding has been moved forward are (i) numerical simulations of polymer phase separation and

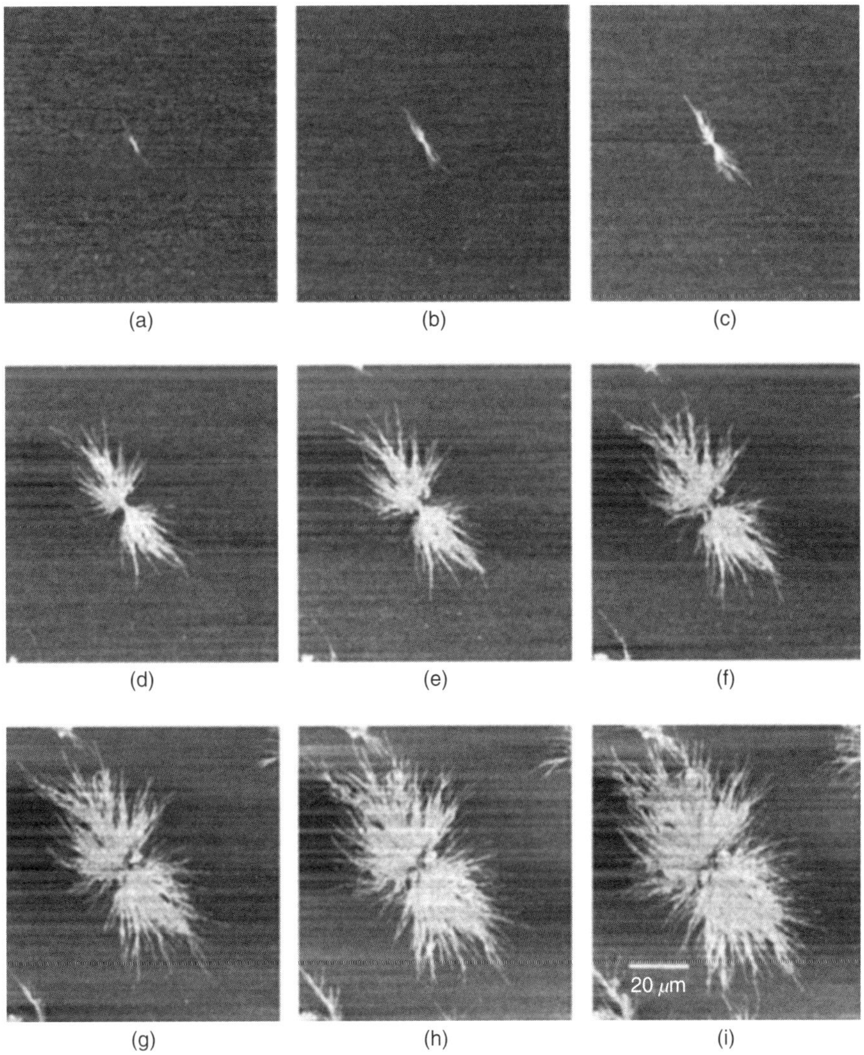

Figure 1.10 AFM phase images of spherulite development in PBA-C8. The temperature was 30 °C and the overall time was 167 min. Lei et al. [50]. Reproduced with permission of Elsevier.

crystallization and (ii) analyses of the roles of self-generated fields in polymer crystallization. Simulations are nicely covered in Chapters 13 and15 and are not treated further here.

1.3.1 Self-Generated Fields

As a crystallization front progresses into a melt, three kinds of fields are formed in the melt ahead of the front. First, because of the density difference between crystal and melt and the slow stress relaxation in the viscous melt, a pressure or stress field is set up, with a gradient of negative pressure decreasing in magnitude with the distance from the interface. Second, the latent heat of fusion is continuously released at the crystallization front, resulting in a thermal field highest at the interface and decreasing into the melt. If there exist in the melt molecular species that cannot crystallize, or crystallize slowly in the propagating crystals, these species are excluded from the crystal and must diffuse away, down a composition gradient, into the melt. This "solute" buildup and its gradient constitute a third type of self-generated field, a compositional field. These fields are characterized by a diffusion length ö. δ is the distance from the interface at which the field strength (temperature, pressure, or impurity composition) has dropped from its value f_i at the solid/melt interface to a level $(f_i - f_o)/e$ above the far-field value f_o (i.e., the far-field value plus $1/e$ of the difference between the built-up value at the interface and the far-field value). For thermal and compositional fields, ö $= D/V$, where D is the thermal or mass diffusivity and V is the velocity of propagation of the solid/melt interface. For a pressure field, $\delta_P = \frac{D_{self}}{V} = \frac{\beta}{\eta V}$, where D_{self} is the self-diffusivity of chains in the melt, η is the melt viscosity, and β is the Einstein equation coefficient, relating diffusivity and viscosity. High levels of temperature, pressure, or impurities in the melt at the interface act to slow the velocity of the interface. A very small diffusion length indicates that the field level is high at the interface. The interface propagation velocity V is set by the undercooling $T_m^o - T_c$, where T_m^o is the equilibrium melting point and T_c is the crystallization

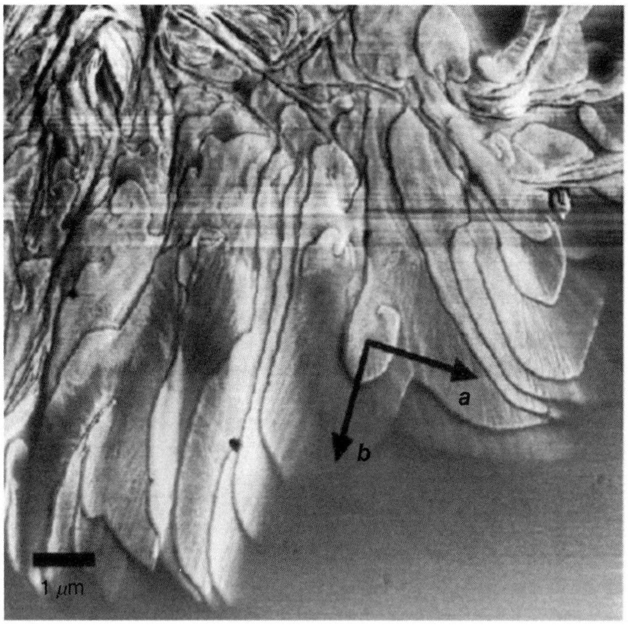

Figure 1.11 AFM image of a larger PCL structure grown at 57 °C from a molten film. Spiral growths from giant screw dislocations are abundant in views that are intermediate between flat-on and edge-on. Beekmans and Vancso [49]. Reproduced with permission of Elsevier.

temperature. The interface velocity increases steeply with the undercooling. The buildup of stress, heat, or uncrystallizable material (solute) at the interface all act to decrease the equilibrium melting point, thereby decreasing the V. For a smooth, continuous interface (a plane or a sphere), buildup continues as the front propagates, slowing the motion of the front continuously. The velocity decreases with time t, proportionally to $t^{-1/2}$ [51]. Thermal diffusivity is relatively very high, and consequently the thermal diffusion length is almost always very large, rendering heat buildup at the interface negligible. Thermal effects can be ignored. (The only possible exception is for the very high values of interface velocity associated with the spinning or heat treating of melt spun fiber [52, 53].)

The reasons that self-generated pressure or compositional fields are morphologically important is that the buildup of stress or impurities is greatly diminished if growth occurs as a fine needle or ribbon, because the stress or impurities at the tip can then be dissipated in three dimensions, as opposed to one-dimensional flow for a large flat or spherical surface. The growing body then adjusts its fineness to match the rate of dissipation with the rate of growth. In this context, then, the width of crystal lamellae and the size of stacks of lamellae could be governed by stress or compositional fields, and indeed finite-element [54] and analytical [55] modeling show this to be the case. A finite element result representing the growth-front propagation in a 50/50 blend of syndiotactic (crystallizable) and atactic (not crystallizable) polystyrene (iPS and aPS) is shown in Figure 1.12. In this example, a stack of ribbonlike lamellae of small thickness and infinite width is made to propagate forward in fine growth steps. At each growth step, the diffusion equation is solved locally in front of each lamella, and also the stack is made to increase by one new crystal (simulating spawning by a giant screw dislocation). The lamellar crystal grows at the velocity dictated by the instantaneous composition in the melt at the growth front. Shown in Figure 1.12 are the front positions after 20, 40, 60, 80, and 100 growth increments. One sees the formation of a narrow growth arm propagating linearly with time into the melt. Most of the uncrystallizable polymer has diffused laterally and produces a large interface composition in front of the lamellae at the sides of the growth arm. The growth front of these lamellae at either side propagates at a velocity proportional to $t^{-1/2}$. For these "secondary," trailing lamellae, the solute or pressure can be dissipated only normal to the front, whereas at the growth arm tip these can be dissipated laterally, as well as forward. This change in the dimensionality of the field is important and allows a constant growth velocity for the arm. The absolute growth arm diameters given by this method are in broad agreement with measurement for the iPS/aPS system [54, 55].

Another area in which self-generated field studies has been useful is in the crystallization from miscible blends in

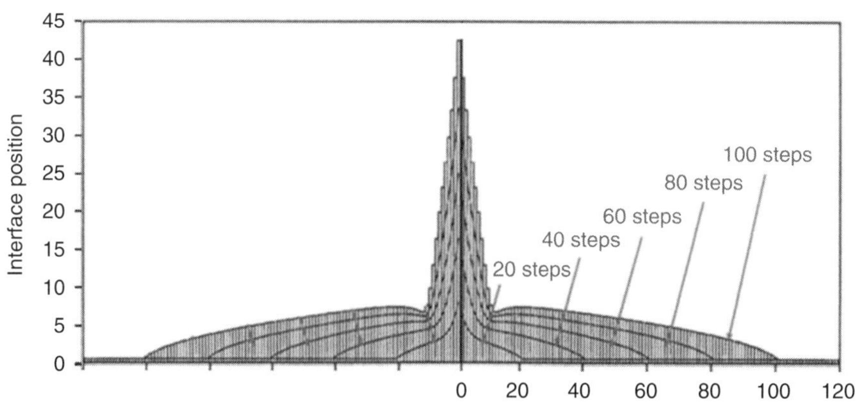

Figure 1.12 Computed growth front after 20, 40, 60, 80, and 100 steps. Parameters match a 50/50 iPS/aPS blend crystallizing into the melt at 260 °C. Kit and Schultz [54]. Reproduced with permission of John Wiley and Sons.

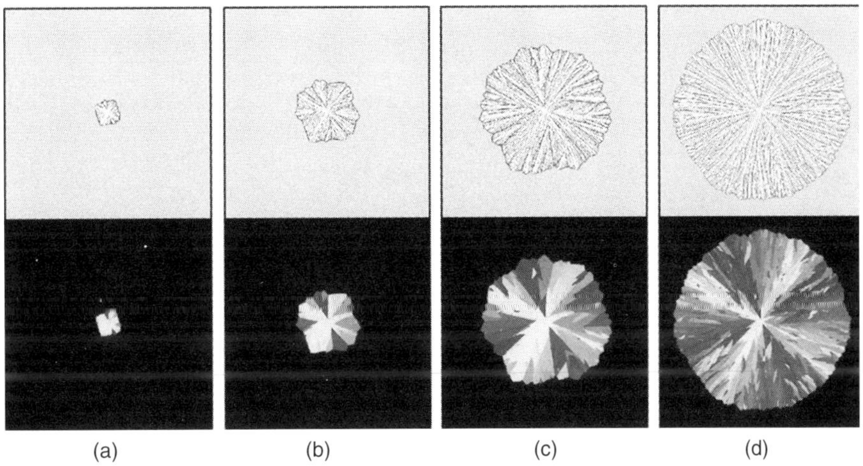

Figure 1.13 Phase field simulations of the growth of a spherulite. Upper row: composition maps. A grayscale map was used to increase the contrast. Lower row: orientation map. Schultz [56]. Reproduced with permission of American Physical Society. (*See color plate section for the color representation of this figure*)

which only one component crystallizes (or crystallizes first). In these systems, the noncrystallizing component can reside finally between crystalline lamellae, between growth arms, or between spherulites. Selection from these possibilities is governed by the magnitude of the diffusion length. A review of this area can be found in Ref. [56].

This topic has also been approached by phase-field modeling. Phase-field modeling is an elegant approach for following the propagation of a front at a fine scale, in the presence of self-generated fields. In this method, equations representing the local evolution of order (crystallinity) and the local transfer of heat or chemical species are coupled. The order parameter is allowed to vary continuously from a crystalline phase (value 1) to a noncrystalline phase (value 0), over a very short distance (the "interface thickness"). The order parameter is related to the local entropy, and use is made of equilibrium and irreversible thermodynamics formalisms. Phase-field modeling applied to two-dimensional spherulitic polymer crystallization produces time-lapse images of the growing spherulite and can replicate known features of spherulitic growth [57–59]. Figure 1.13 is an example of the phase-field simulation of the growth of a spherulite in a blend. Most of the important features are captured, but the relationships to real-life conditions (particularly temperature) are not available. The current weakness in the phase-field modeling of polymer crystallization is that the coupling of growth kinetics and front propagation does not yet have the steep temperature dependence of front propagation which is experimentally known, the argument being made that only a more simplified coupling can allow simulation to be carried out in reasonable time.

1.4 WISHFUL THINKING

I have no crystal ball and am not endowed with second sight; I cannot predict the future. I comment here only on what I perceive to be useful. Certainly the ultimate aim of all polymer morphology research is to produce useful products. In order to engineer such products, one would like to schedule a sequence of processes whereby a morphology is created suitable to the end use of the product. In polymer processing, this is almost always done by "feel," with the connections of processing to morphology and morphology to property not fully made. In other areas of materials science – think, for instance, of solid-state devices and of steels and cast irons – these connections have been made, and processing to achieve specific end uses is done with significantly better quantitative understanding. In the case of polymers, much greater effort has been made in characterizing morphology than in putting processing–morphology and morphology–property relations on a strong, quantitative footing. Advances in measurement technology and simulation methods have now put us in position to perform the studies that will make possible the detailed, quantitative processing–morphology and morphology–property connections. It is hoped that more effort will be placed in these areas, in both academia and industry, with profitable interactions between the two.

1.5 SUMMARY

The ultimate goal of morphology research is in the end product. Work in this area can be divided into the areas of morphology characterization, processing–morphology relationships, and morphology–property relationships. Recent advances in characterization tools have allowed measurements, at appropriate accuracy, to be made orders of magnitude faster than was previously possible. Further, microscopic mapping of properties such as IR absorption and viscosity are now available. These tools are enabling *in situ* studies of the development of morphology and the correlation of morphology with properties. Examples of synchrotron X-ray investigation of morphological development in fiber spinning and of the changes of morphology with mechanical behavior are given. It is hoped that such studies will place the engineering of polymer products on an improved quantitative footing.

REFERENCES

1. Balsara N, Hahn H. Block copolymers in nanotechnology. In: Yong P, editor. *The Chemistry of Nanostructured Materials*. Singapore: World Scientific Publishing Co; 2003. p 317–327.
2. Abetz V, Simon PFW. Phase behaviour and morphologies of block copolymers. Adv Polym Sci 2005;189:125–212.
3. Schultz JM. Self-generated fields and polymer crystallization. Macromolecules 2012;45:6299–6323.
4. Darling SB. Block copolymers for photovoltaics. Energy Environ Sci 2009;2:1266–1273.
5. Orilall MC, Wiesner U. Block copolymer based composition and morphology control in nanostructured hybrid materials for energy conversion and storage: Solar cells, batteries, and fuel cells. Chem Soc Rev 2011;40:520–535.
6. Guo C, Lin Y-H, Witman MD, Smith KA, Wang C, Hexemer A, Strzalka J, Gomez ED, Verduzco R. Conjugated block copolymer photovoltaics with near 3% efficiency through microphase separation. Nano Lett 2013;13:2957–2963.
7. Dupont Y, Gabriel A, Chabre M, Bulik-Krzywicki T, Schecter E. Use of a new detector for X-ray diffraction and kinetics of the ordering of the lipids in *E. coli* membranes and models. Nature 1972;238:331–333.
8. Gabriel A, Dupont Y. A position sensitive proportional detector for X-ray crystallography. Rev Sci Instrum 1972;43:1600–1602.
9. Borkowski CJ, Kopp MK. Some applications and properties of one- and two-dimensional position-sensitive proportional counters. IEEE Trans Nucl Sci 1970;17:340–349.
10. Borkowski CJ, Kopp MK. Proportional counter photon camera. IEEE Trans Nucl Sci 1972;19:161–168.
11. Elsner G, Riekel C, Zachmann HG. Synchrotron radiation in polymer science. Adv Polym Sci 1985;67:1–58.
12. Elmore D, Tsao M, Chase B, Rabolt J. Planar array infrared (PAIR) spectrograph that operates in the 3400 to 2000 cm^{-1} region. Appl Spectrosc 2002;56:145–149.
13. Pellerin C, Snively C, Liu Y, Chase DB, Rabolt JF. Performance and application of a new planar array infrared spectrograph operating in the mid-infrared (2000–975 cm^{-1}) fingerprint region. Appl Spectrosc 2004;58:639–646.
14. Pellerin C, Frisk S, Rabolt JF, Chase DB. A faster approach to infrared rheo-optics using a planar array infrared spectrograph. Appl Spectrosc 2004;58:799–803.
15. Pearce R, Vancso GJ. Imaging of melting and crystallization of poly(ethylene oxide) in real-time by hot-stage atomic force microscopy. Macromolecules 1997;30:5843–5848.
16. Pearce R, Vancso GJ. Real-time imaging of melting and crystallization of poly(ethylene oxide) by atomic force microscopy. Polymer 1998;39:1237–1243.
17. Schultz JM, Miles MJ. AFM study of morphological development during the melt-crystallization of poly(ethylene oxide). J Polym Sci Polym Phys Ed 1998;36:2311–2325.
18. McMaster TJ, Brayshaw DJ, Miles MJ, Walsby AE, Dunton PG. A new ultra high speed AFM technique for biophysics: 3-dimensional imaging of surfaces, molecules and processes with true millisecond resolution. Biophys J 2005;88:541A.
19. Hobbs JK, Vasilev C, Humphris ADL. Real time observation of crystallization in polyethylene oxide with video rate atomic force microscopy. Polymer 2005;46:10226–10236.
20. Picco LM, Ulcinas A, Engledew DJ, Antognozzi M, Miles MJ. Video rate AFM for the real time characterisation of polymers and biological systems. Nano-molecular analysis for emerging technologies II, NMAETII; 2006 Oct 17–18; Teddington, UK; (2006).
21. Hobbs JK, Vasilev C, Humphris ADL. Video AFM – A new tool for high speed surface analysis. Analyst 2006;131:251–256.
22. Sun G, Weng L-T, Schultz JM, Chan C-M. Formation of banded and non-banded poly(L-lactic acid) spherulites during crystallization of films of poly(L-lactic acid)/poly(ethylene oxide) blends. Polymer 2014;55:1829–1836.
23. Cong Y, Hong Z, Qi Z, Zhou W, Li H, Liu H, Chen W, Wang X, Li L. Conformational ordering in growing spherulites of isotactic polypropylene. Macromolecules 2010;43:9859–9864.
24. Hay IL, Keller A. Polymer deformation in terms of spherulites. Kolloid-Z u Z Polymere 1965;204:43–74.
25. Kargin VA, Tsarevskaya IY. Deformation of crystalline polybutylene. Polym Sci U.S.S.R. 1996;13:1601–1607.
26. Erhardt P, Stein RS. A technique for the study of spherulite deformation: Light scattering movies. J Polym Sci 1965;3B: 553–555.
27. Cowking A, Rider JG, Hay IL, Keller A. A study on the orientation effects in polyethylene in the light of crystalline texture. J Mater Sci 1968;3:646–654.
28. Cowking A, Rider JG. On molecular and textural reorientations in polyethylene caused by applied stress. J Mater Sci 1969;4:1051–1058.
29. Sakaoku A, Peterlin A. Bond rupture in highly oriented crystalline polymers. J Polym Sci 1969;7A-2:1151–1163.
30. Schultz JM. *Polymer Materials Science*. Englewood Cliffs, NJ: Prentice-Hall; 1974. , Chapter 11.
31. Stribeck N, Nöchel U, Funari SS, Schubert T. Tensile tests of polypropylene monitored by SAXS: Comparing the stretch-hold technique to the dynamic technique. J Polym Sci Polym Phys 2008;46:721–726.
32. Stribeck N, Nöchel U, Funari SS, Schubert T, Timmann A. Nanostructure of polypropylene in dynamic mechanical tests. Polym Chem Phys 2008;209:1992–2002.
33. Stribeck N. Deformation behavior of nanocomposites studied by X-ray scattering: Instrumentation and methodology. In: Karger-Kocsis J, Fakirov S, editors. *Nano- and Micro-Mechanics of Polymer Blends and Composites*. Munich, Germany: Carl Hanser Verlag; 2009. p 269–300.
34. Davies RJ, Zafeiropoulos NE, Schneider K, Roth SV, Burghammer M, Riekel C, Kotek JC, Stamm M. The use of synchrotron X-ray scattering coupled with in situ mechanical testing for studying deformation and structural change in isotactic polypropylene. Colloid Polym Sci 2004;282:854–866.
35. Davies RJ, Burghammer M, Riekel C. Simultaneous microfocus Raman and microfocus XRD: Probing the deformation of a single high-performance fiber. Macromolecules 2006;39: 4834–4840.

36. Davies RJ, Koenig C, Burghammer M, Riekel C. On-axis microbeam wide- and small-angle scattering experiments of a sectioned poly(*p*-phenylene terephthalamide) fiber. Appl Phys Lett 2008;92:1019011–10190133.
37. Remaly LS, Schultz JM. Time-dependent effect of spherulite size on the tensile behavior of polypropylene. J Appl Polym Sci 1970;14:1871–1877.
38. Kleiner LW, Radloff MR, Chou T-W, Schultz JM. Spherulite size effects in linear polyethylene. J Poly Sci Polym Phys Ed 1974;12:819–821.
39. Sandt A. *Rissausbildung in teilkristallinen polymeren*. Doktorarbeit: Universität Bochum; 1981.
40. Schultz JM. Structure development in polyesters. In: Schultz JM, Fakirov S, editors. *Solid State Behavior of Linear Polyesters and Polyamides*. Englewood Cliffs, NJ: Prentice Hall; 1990. p 75–130.
41. Keller A. Unusual orientation phenomena in polyethylene interpreted in terms of the morphology. J Polym Sci 1955;15: 31–49.
42. Hristov HA, Schultz JM. Thermal response and structure of PET fibers. J Poly Sci Polym Phys Ed 1990;28:1647–1663.
43. Chang H, Lee K-G, Schultz JM. Structure development of polyethylene terephthalate (PET) fibers during post-spinning annealing. J Macromol Sci Phys 1994;B33:105–127.
44. Somani RH, Yang L, Zhu L, Hsiao BS. Flow-induced shish kebab precursor structures in entangled polymer melts. Polymer 2005;46:8587–8623.
45. Geil PH. *Polymer Single Crystals*. New York: John Wiley and Sons, Inc.; 1963.
46. Bassett DC. *Principles of Polymer Morphology*. Cambridge, UK: Cambridge University Press; 1981.
47. Bassett DC. Etching and microstructure of crystalline polymers. In: Allen GC, Bevington JC, editors. *Comprehensive Polymer Science, Volume 1, Polymer Characterization*. Oxford, UK: Pergamon Press; 1989. p 841–866.
48. Crist B, Schultz JM. Atomic force microscopy studies of polymer crystals: Nucleation, growth, annealing, and melting. In: Kyu T, editor. *Encyclopedia of Polymers and Composites*. Heidelberg, Germany: Springer; 2014.
49. Beekmans LGM, Vancso GJ. Real-time crystallization study of poly(ε-caprolactone) by hot-stage atomic force microscopy. Polymer 2000;41:8975–8981.
50. Lei YG, Chan CM, Wang Y, Ng KM, Jiang Y, Li L. Growth processes of homogeneously and heterogeneously nucleated spherulites as observed by atomic force microscopy. Polymer 2003;44:4673–4679.
51. Carslaw HS, Jaeger JC. *Conduction of Heat in Solids*. 2nd ed. Oxford, UK: Clarendon Press; 1986, Chapter XI.
52. Tiller WA, Schultz JM. Crystallization of polymers under high tension: A dendrite model. J Polym Sci Polym Phys Ed 1984;22:143–161.
53. Schultz JM. Theory of crystallization in high-speed spinning. Polym Eng Sci 1991;31:661–666.
54. Kit KM, Schultz JM. Study of bundle formation during crystallization in polymer blends. J Polym Sci Polym Phys Ed 1998;36:873–888.
55. Balijepalli S, Schultz JM. Modeling of crystallization in a blend containing at least one crystallizable component: An analogy from eutectic systems. Macromolecules 2006;39:7407–7414.
56. Schultz JM. The crystallization and morphology of melt-miscible polymer blends. Front Chem China 2010;5:262–276.
57. Mehta R, Kyu T. Dynamics of spherulite growth in blends of polypropylene isomers. J Polym Sci Polym Phys Ed 2004;42:2892–2899.
58. Xu X, Keawwattana W, Kyu T. Effect of thermal transport on spatiotemporal emergence of lamellar branching morphology during polymer spherulitic growth. J Chem Phys 2005;123: 124908.
59. Gránásy L, Pusztai T, Tegze G, Warren JA, Douglas JF. Growth and form of spherulites. Phys Rev E 2005;72:011605.

2

X-RAY DIFFRACTION FROM POLYMERS

N. Sanjeeva Murthy
New Jersey Center for Biomaterials, Rutgers University, Piscataway, NJ, USA

2.1 INTRODUCTION

Polymers can be broadly classified as either amorphous or semicrystalline. In amorphous polymers, crystallization of polymer chains can be inhibited by steric hindrances (e.g., polycarbonates and polymethacrylates), by the presence of cross-links (e.g., epoxies), or by entanglements (e.g., elastomers). Categories of semicrystalline polymers include thermoplastics (e.g., polyethylene, PE; and poly(ethylene terephthalate), PET) thermoplastic elastomers (e.g., ethylene-vinyl acetate polymers, EVA), and liquid crystalline polymers (LCPs, e.g., aromatic polyamides). In all these different classes of polymers, the structure (spatial arrangement of the atoms) and morphology (size, distribution, and orientation of crystallites and lamellae), which are determined by processing conditions, affect the performance of the polymer. For instance, crystallization kinetics determines the morphology, and thus affects the surface finish and influence subsequent coating processes such as electrodeposition and painting. Orientation, crystallinity, and the distribution of the crystalline and amorphous domains are determined by the linkages between various domains that are determined by the processing conditions. These characteristics influence the mechanical properties, such as tenacity, modulus, and elongation to break and fatigue behavior, and the barrier properties by controlling the diffusion of water and gas molecules. Morphology also plays an important role in controlling the dimensional stability, which is important in engineering materials and biomaterials where tight tolerances are desired. This chapter discusses the measurement of some of the characteristics of polymer structure and morphology that impact the polymer properties.

Several different techniques are required to examine the structural characteristics in polymers at several length scales (Fig. 2.1). This chapter describes two X-ray diffraction (XRD) techniques that are commonly used in the characterization of the structure and morphology in polymers and polymer composites: wide-angle X-ray scattering (WAXS) for the atomic and molecular structures, and small-angle X-ray scattering (SAXS) for the mesoscale structures [1–4]. The use of WAXS for determining crystal structure, disorder, and orientation is discussed. SAXS methods are discussed in the context of analyzing the fibrillar and lamellar morphology in semicrystalline polymers. Other techniques highlighted are combination measurements such as simultaneous XRD and elongation, microbeam technique to map the spatial variation in structure, and grazing incidence diffraction (GID) to determine the variation in structure with depth.

2.2 BASIC PRINCIPLES

X-rays are electromagnetic radiations. X-rays of wavelength from ~0.1 to 2.5 Å (ca. 100 to 5 keV) are typically used in scattering or diffraction experiments. Although diffraction refers to the interference of radiations scattered from structural features of the sample, the terms scattering pattern and diffraction are often used interchangeably. The wavelength of X-rays is such that the diffraction data permits analysis of structures at sub nanometer to sub micrometer length scales. Short wavelength or hard X-rays penetrate through thick samples, and permits sample cells to be made of metals such as thermal analysis pans and high-pressure cells. Long wavelengths or

Polymer Morphology: Principles, Characterization, and Processing, First Edition. Edited by Qipeng Guo.

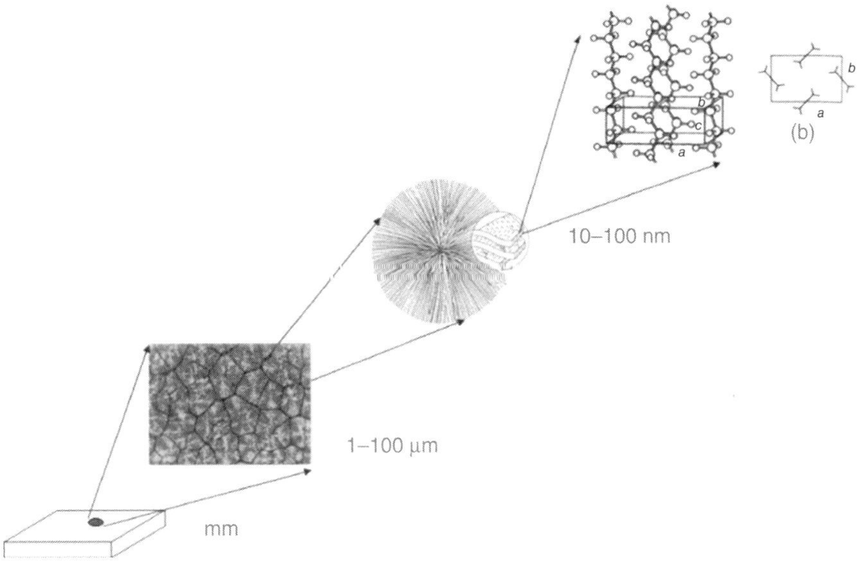

Figure 2.1 Schematic of the hierarchical structure in crystallizable polymers.

soft X-rays make it convenient to examine structures at longer length scale.

When X-rays pass through matter, depending on the absorption coefficient of the material and the wavelength of the X-rays, a fraction of their energy is absorbed by the atomic electrons in the material. The fraction of the X-rays that is not absorbed is scattered. A large fraction of this scattering occurs without any change in the wavelength. This scattering, called coherent scattering, is used in structural analysis. Scattered radiation whose wavelength is different from that of the incident radiation, called incoherent scattering, contributes to the diffuse background to the observed diffraction pattern. This diffuse incoherent scattering needs to be calculated and subtracted from the observed diffraction pattern. This is especially important in the quantitative analysis of the diffraction patterns from polymers that are more often highly disordered, and thus the coherent scattering from these disordered domains also appears as diffuse background (see Sections 2.5 and 2.6).

Even though the coherently scattered X-ray from each electron is in phase with the incident X-rays, because of the electrons are spread over an appreciable volume around the nucleus, the X-rays scattered by the various electrons within one atom are more or less out of phase with each other depending on the scattering angle 2θ. The coherent scattering power thus decreases with 2θ. This is expressed in the scattering factor or the form factor (f) of the atom. The coherent scattered intensity is a product of two factors, the form factor f, and the structure factor F. F is the structure, the organization of the atoms and molecules that is being sought. Established crystallographic methods for determining and describing the polymer structure at the atomic and molecular level can be found in well-written textbooks [5, 6] and handbooks [7]. Therefore, rather than describing these methods, this chapter focuses on the aspects unique to polymeric materials.

The diffracted intensities are recorded as a function of the scattering vector (or momentum transfer) q, which is related to the scattering angle 2θ by the relation $q = (4\pi \sin\theta)/\lambda$. When coherently scattered X-rays from atoms in neighboring planes interfere with each other, then Bragg's law (Eq. 2.1) holds:

$$n\lambda = 2d \sin\theta \quad \text{or} \quad d = 2\pi/q \tag{2.1}$$

where d is the distance between the atomic planes (d-spacing) and λ is the wavelength. n is an integer; $n = 1$ corresponds to the reflection of the first order for a given set of atomic planes, and second and higher orders of reflections occur at larger values of 2θ. This is the key relation in relating the structure to the observed peaks in the diffraction pattern. Bragg's law applies to materials with repeating discrete reflections that occur from the crystallographic planes. However, the inverse relationship between the real space (d) and the reciprocal space (q) is often used to interpret the positions of diffuse halos present in amorphous materials in terms on interchain distances even though no clear crystallographic planes are present.

Scattering at $q > 0.5$ Å^{-1}, labeled wide-angle X-ray diffraction or scattering (WAXD or WAXS), reveals the structure at $d = 0.1$–10 Å, at the level of atoms and molecules. Scattering at $0.005 < q < 0.5$ Å^{-1}, labeled SAXS, reveals the structure at $d \sim 10$–100 nm, the mesoscale structure arising from the aggregates of atoms and molecules. Scattering at $q < 0.005$ Å^{-1}, often called ultrasmall-angle X-ray scattering (USAXS), is used to analyze the structure at $d \sim 0.5$ μm corresponding to colloidal structures that are also seen in electron microscopy. The smallest length scale that can be analyzed is $\lambda/2$, ~0.5 Å, the diffraction limit. In theory, there is no upper limit to the length scale of the structures that

TABLE 2.1 The Structural Parameters That Can be Derived from WAXS Data from Polymers in the Solid State

Structural Feature	Characteristic Parameter	Feature in the WAXD Pattern
Crystal structure	Unit cell dimensions	Positions of the crystalline reflections
	Atomic positions	Positions and the intensities of the crystalline reflections
Crystallites	Total weight fraction	Area under the crystalline peak relative to the scattered intensity
	Size	Width of the Bragg peaks
	Disorder	Increase in the widths of the peak with scattering angle
	Orientation	Azimuthal spread of the intensity in the crystalline reflections
Polymorphs/composition	Relative fractions	Relative intensities of the corresponding reflections
Amorphous domains	Interchain distance	Position of the amorphous halo
	Orientation	Azimuthal spread of the amorphous halo
	Oriented or intermediate phase	Ratio of the areas due to the oriented and unoriented components

TABLE 2.2 The Structural Parameters That Can be Derived from SAS Data from Polymers in the Solid State

Structural Feature	Characteristic Parameter	Feature in the SAS Pattern
Voids	Size and distribution	Characteristics of the decay in central diffuse scattering
	Height (length) when voids are elongated	Width of the diffuse scattering along the meridian extrapolated from a series of meridional slices
	Misorientation	Azimuthal intensity distribution of diffuse streak
Fibrillar structure	Spacing between the fibrils	Position of the diffuse interference peak in the equatorial streak
	Height (length) of the fibrils in oriented structures are elongated	Width of the diffuse scattering along the meridian after accounting for the void contribution
	Misorientation	Azimuthal intensity distribution of diffuse streak after accounting for the void component
Lamellar structures or domains in oriented samples	Lamellar or domain spacing	Position of the interference peak along the meridian or the minor axis of the elliptical trajectory of the lamellar reflection
	Lamellar or domain thickness and diameter	Extent of the reflection in meridional and equatorial directions
	Tilt angle of the lamellae	Angular separation of the lamellar reflections across the meridian in 4-point patterns
	Misorientation	Variation in the meridional width with distance from the meridional axis

can be analyzed by diffraction, but the practical limits of collimation sets the current limits at ~5 μm [8]. Tables 2.1 and 2.2 show some of the parameters that are discussed in the following sections.

2.3 INSTRUMENTATION

X-ray scattering equipment typically have the following components: an X-ray source, collimation, sample positioning system, detector and data-analysis software. Although synchrotron sources, which provide 1000 times more flux than the in-house sources, are now readily available, in-house X-ray facilities using sealed and rotating anodes are still widely used because of ease of use, flexibility, and accessibility. Synchrotron radiation makes it possible to carry out specialized investigations such as spatial mapping with micron-size beams, real-time measurement during thermal cycling in a differential scanning calorimeter, and *in situ* measurement during mechanical deformation.

The most basic X-ray measurement device is a powder diffractometer that includes a collimator, a sample holder, and a counter that scans along the 2θ arc (Fig. 2.2a). The apparatus of the type shown in the figure is optimized for collecting the data in reflection or Bragg–Brentano geometry, the most common mode for powder diffraction. Most polymers are weak absorbers of X-rays, a common exception being fluorine- and chlorine-containing polymers such as poly(tetrafluoro ethylene) and poly(vinyl chloride). Therefore, XRD data from polymeric materials are commonly obtained in transmission geometry where X-rays pass through the sample. For this purpose, equipment of the type shown in Figure 2.2a can be fitted with a transmission stage. It is also possible to carry out microbeam XRD using capillary optics. With special collimators and sample positioning devices, this apparatus can also be used for reflectivity and grazing incidence mode to study the surface structure.

An example of a SAXS apparatus is shown in Figure 2.2b. In these experiments, to obtain data close to the primary

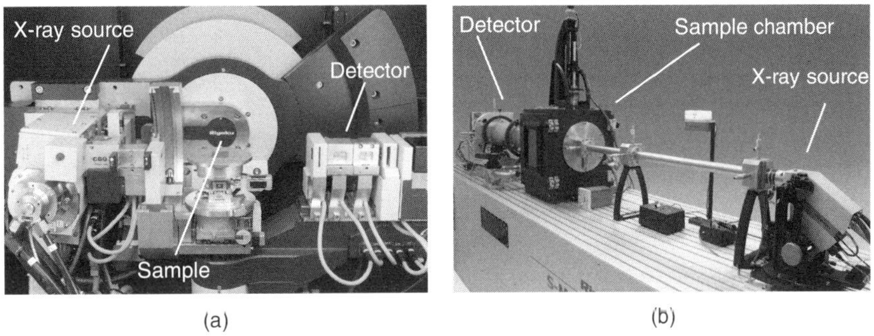

Figure 2.2 Instrumentation for (a) WAXD and (b) SAXS. Images courtesy of Rigaku Americas Corporation.

beam, a highly collimated beam obtained with slits separated by large distances between the source and the sample, large flight path between the sample and the detector, and evacuated beam paths are used.

The intensity of the scattered X-rays is measured using point detectors (scintillation, proportional or solid-state detectors) that offer resolution at the expense of speed. Point detectors typically collect data from 5° to 120° 2θ ($q \sim 0.4$–7 Å^{-1} with Cu Kα). 1D and 2D detectors offer speed with some loss of resolution, and can be expensive. Figure 2.3 shows a combined SAXS–WAXS pattern that can be obtained with a large 2D detector.

Data analyses software are often provided by the manufacturers of the diffraction equipment. However, special attention needs to be exercised when analyzing data from nonroutine samples.

Figure 2.3 A combined SAXS and WAXS pattern from a drawn PA6 fiber. The small-angle pattern near the origin covers about 2.5° 2θ. The d-spacing of the SAXS reflection is ~100 Å. The d-spacings of the two wide-angle reflections along the equator are 4.4 and 3.8 Å [9]. (*See color plate section for the color representation of this figure.*)

2.4 STRUCTURE DETERMINATION

Methods for determining crystal structures of small molecules and proteins using single crystals are highly developed. But, these methods are not helpful in solving polymer structures. This is because of the comparatively low degree of order that can be attained in many polymers. Many polymers do not crystallize. And even in those that crystallize, a significant fraction of the polymer can be amorphous. Furthermore, small crystallites (~100 Å) and crystalline disorder due, among other things, to the linkages between polymer chains in the amorphous and the crystalline region broaden the few observable reflections. Overlap of the reflections due to this broadening is further compounded by the additional overlap that occurs when the samples are not single crystals, but at best fibers or films, where only one or two axes are distinct. Therefore, even though crystal structures of polymers can, in principle, be derived using standard crystallographic techniques, alternative methods based on structural models are invariably used. There have been attempts to use direct methods for polymer structure determination as illustrated in the work on polyethylene using powder diffraction data, and on poly(ε-caprolactone) using fiber diffraction data [10]. Details of the techniques to prepare the samples, and collecting the diffraction data for structure analysis can be found in textbooks devoted to polymer structure analysis by Tadokoro [11], and Alexander [1]. A combination of techniques, especially electron diffraction, is useful in determining the structures [12, 13].

2.4.1 Lattice Dimensions

Unit cell and the lattice dimensions, while necessary for crystal structure determination, are by themselves valuable in polymer characterization. This is because in contrast to the lattice dimensions in crystals of small molecules, which remain essentially unchanged, these dimensions in polymeric materials vary over a relatively large range. The polymer needs to be obtained in at least a fiber or a biaxially oriented film in the initial identification of the unit cells, and in assigning the reflections to appropriate lattice planes. Once the reflections are indexed, cell dimensions can be calculated even from unoriented polymers, and used in identifying the effect of processing on the packing of the polymer chains in the unit cell [14], and in relating these changes to the properties of the material. These dimensions change when the crystal is under internal stress, when the polymer is subjected to external strain [9, 15], and in conducting polymers where the charge transfer affects the chain-axis dimensions [16]. Some lattice dimensions are less variable than others, for example, the dimension along the chain axis varies less than those in the lateral directions. The variations in the unit cell dimensions often serve as a proxy to the more subtle changes in the morphology of the polymer. These unit cell changes are often reflected in thermal characteristics such as the melting point and the crystallization temperature, and in mechanical properties such as the impact strength and the extensibility of a polymer.

2.4.2 Molecular Modeling

Polymer structures are often determined by trial-and-error in which structures are first modeled, and then refined by comparing the calculated and observed diffraction patterns. Models typically use the constraints of standard bond distances, bond angles, and internal rotational angles. These parameters that define bonding geometry of an isolated chain, called the secondary structure in the literature on biological macromolecules, can often be obtained by spectroscopic methods, Raman, infrared (IR) and nuclear magnetic resonance (NMR). In some instances, oligomers may provide a template to model the polymer chain [17]. This is based on the observation, first made by Staudinger, that packing of chains is independent of the degree of polymerization so that the crystal structure of the polymer can be represented by the structure of a few monomer units. In using oligomers as a template, proper choice of the end groups and the number of monomer units that takes part in the crystal packing are important concerns.

An example of the structure determined by modeling is shown in Figure 2.4. This structure of poly(ethylene terephthalate) was simulated using modeling software [18], and is in agreement with the structure determined much earlier from photographic data [19]. In recent times, structures have been determined by energy minimization using commercially available molecular modeling software [20]. There have been efforts to derive the structure from *ab initio* quantum chemical methods [21–23]. These structures are in good agreement with the structure derived from the XRD data. In addition to the determination of the crystal structures, computational modeling is use to calculate the physical properties, determination of the effects of processing, and to understand the effect of chemical and structural imperfections [18].

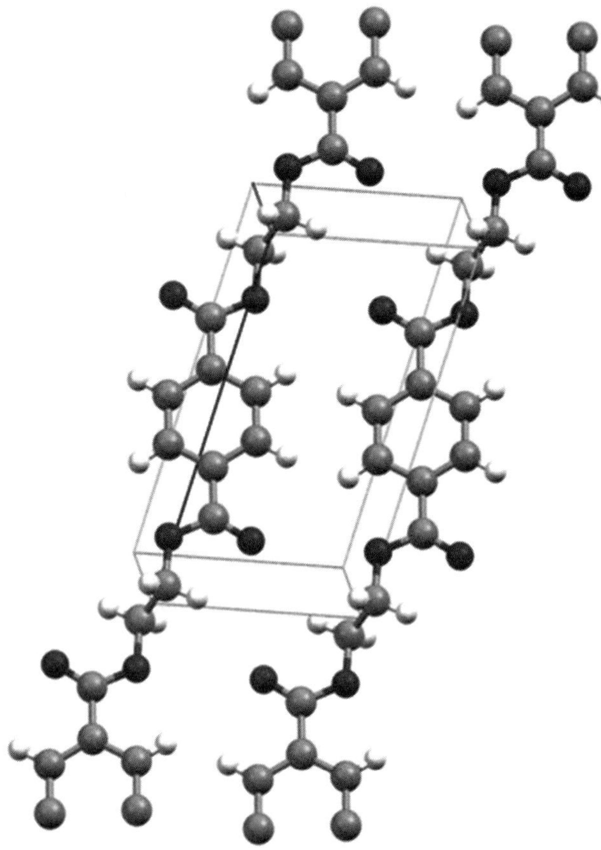

Figure 2.4 Crystal structure of poly(ethylene terephthalate). Carbon atoms are shown in gray, oxygen atoms in dark and hydrogen atoms in white colors.

2.4.3 Rietveld Method

Another tool useful in discriminating between several possible structural models is Rietveld refinement [24, 25]. This technique was developed for the analysis of structures of small molecules, particularly inorganic materials. This is essentially a curve-fitting procedure in which the unit cell parameters and the molecular structure are refined using the whole powder diffraction pattern. The technique is useful when only powder diffraction data is available, and is applicable when the polymer is sufficiently ordered and crystalline so that ~20 or more diffraction peaks are observed up to high q values (~4 Å^{-1}, $2\theta = 60°$ with Cu Kα radiation). There has been some success in the use of this method in polymers in refining the structures such as of polythiophene and poly(hydroxy butarate) [24, 25], in identifying structures of intermediate order [23], and in characterizing the microstructure [26].

2.4.4 Pair Distribution Functions

Amorphous polymers, which do not have any crystalline reflections, are not amenable to structure analysis by the methods described in Sections 2.4.2 and 2.4.3. In such instances, the structure can be assessed using atomic pair distribution function (PDF) [27–29]. PDF, $g(r)$, is the probability of finding two atoms separated by a distance r. It can be obtained by a numerical Fourier transform of the measured coherent scattered intensity, $I_{coh}(q)$, by the using equations 2.2 and 2.3.

$$G(r) = 4\pi r \rho_0 [g(r) - 1] = \frac{2}{\pi}\int_0^{\infty} q[S(q) - 1] \sin(qr)\, dq \qquad (2.2)$$

$$S(q) = \frac{I_{coh}(q) - \sum_i c_i f_i^2(q)}{\left[\sum_i c_i f_i(q)\right]^2} \qquad (2.3)$$

The function $G(r)$ given in equation 2.2 is widely used in the literature and gives information about the number of atoms in a spherical shell of unit thickness at a distance r from a reference atom. The function $S(q)$ in the equations is called the structure function. In equation 2.3, c_i and f_i are the atomic concentration and X-ray atomic form factor, respectively, for the atomic species of type i. A drawback with $g(r)$ is that it emphasizes the atom-atom correlations, whereas interchain correlations are of more importance in polymeric materials. For this reason, the structure function $S(q)$, which emphasizes the structure at larger distances, is often preferred in analyzing polymer structure. The PDFs calculated from the structural models obtained from molecular dynamics (MD) simulations are compared with those measured by XRD to arrive at the most plausible structural representation of the polymer.

Structures of polymers such as poly(methyl methacrylate) [30, 31], poly(dialkyl siloxanes) [32], polyethylene [33, 34], and polylactides [35] have been determined using PDFs or structure functions. The observed and the calculated $S(q)$s for poly(methyl methacrylate) are compared in Figure 2.5 [31]. The $S(q)$ curves can be transformed into PDF, and the different peaks in the PDF correspond to the interatomic distances such as O–O, C–C, C–O. The data shown in Figure 2.5 are significant for two reasons [30]. First, modeling was able to show that the first peak at $q \sim 1$ Å^{-1} is a composite of intermolecular distance of 6.6 Å and an intramolecular distance of 5.2 Å. Other peaks are primarily intramolecular in character. Second, even with powder diffraction data, it was possible to show that the splitting of the peak at 2.5 Å^{-1} into two peaks corresponding to two distances, 2.15 and 2.85 Å, a consequence of the curvature of the molecule as shown in Figure 2.5 due mainly to the unequal backbone bond angles. Similarly, analysis in polyethylene was used to show that structure of molten polyethylene consists of random chains with no substantial orientation correlation between adjacent chains, whereas chains in amorphous polyethylene are considerably more ordered with regions of parallel chains [33]. PDF has been found to useful in validating the computationally derived models, in following the structural changes that occur during processing, including hydration [35], and in understanding the small structural differences between closely related product types [36].

2.5 PHASE ANALYSIS

Both synthetic and natural polymers are rarely fully crystalline. Even when polymers are crystallizable, only a fraction of the crystallizable chains is incorporated into crystalline domains. Depending on the polymer and the conditions of crystallizations, a significant fraction remains amorphous. Furthermore, polymers usually crystallize in more than one form. These different polymorphs can be identified, and their relative amounts and the fractions of the chains that remain amorphous can be unambiguously assessed from their XRD patterns.

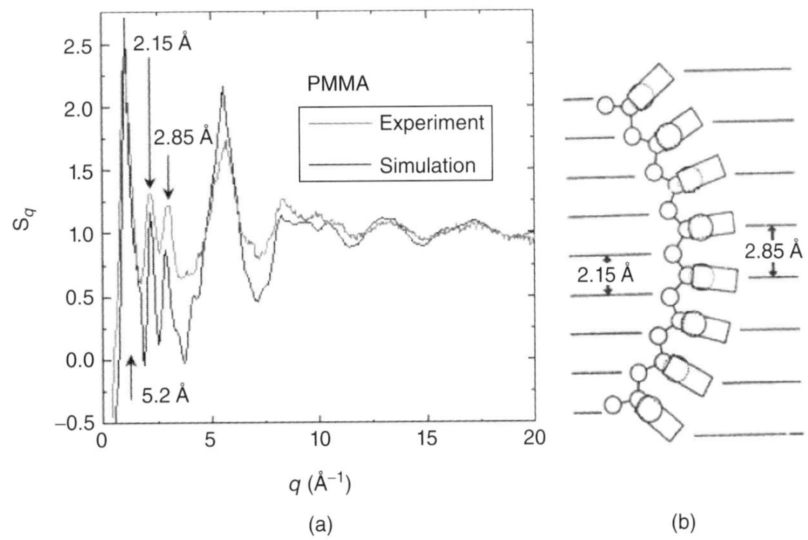

Figure 2.5 (a) Structure functions, $S(q)$s, for poly(methyl methacrylate) (PMMA). The observed data from an X-ray diffraction measurement and that simulated using molecular modeling are shown. Li et al. [31]. Reproduced with permission of American Chemical Society. (b) A curvature in the chain to account for the splitting of the peak at 2.5 Å^{-1}.

2.5.1 Crystallinity Determination

Polymers usually have a continuum of structures, from crystallites that are almost defect-free single crystals, to crystals with dislocations and other defects, chain folded and micellar crystals, 1D and 2D ordered aggregates, and finally liquid-like amorphous domains. Despite this complexity, it has been found to be useful to treat the diffraction patterns from semicrystalline polymers as arising from two phases, crystalline and amorphous. Within this framework, Bragg reflections with a Scherrer crystallite size >30 Å are attributed to crystalline phases, and the broad diffuse halos that resemble thermal diffuse scattering to the amorphous domains. To account for the range of order that can be present between these two extremes, a third phase, variously called as intermediate phase, mesomorphic structures, rigid amorphous phase, dense amorphous phase, is sometimes introduced especially in oriented polymers (see Section 2.7.4) [37–42]. Assuming a two-phase system with crystalline and amorphous segments, the total crystallinity can be determined by comparing the area under the crystalline peaks to the total scattering area [43–46].

The scattered intensity is used to calculate the volume integral of the diffracted peak.

$$V_c \propto \int_0^\infty I_c(q)\,dV_Q = \int_0^\infty q^2 I_c(q)\,dq \tag{2.4}$$

Weight fraction of crystallinity ϕ_c is then given by the ratio of the integral of the diffraction intensity scattered by the crystalline fraction to the total coherent scattered intensity:

$$\varphi_c = \frac{\int_0^\infty q^2 I_c(q)\,dq}{\int_0 q^2\,[I(q) - I_{Compton}(q)]\,dq} \tag{2.5}$$

Alternatively, areas A_a and A_c of the amorphous and the crystalline peaks, respectively, are used to calculate a crystalline index (CI) defined as

$$CI = 100 \times A_c/(A_a + A_c) \tag{2.6}$$

A diffraction scan over a q range of 0.5–3 Å^{-1} ($2\theta = 10$–$50°$, Cu Kα) is used. The areas A_a and A_c are determined by resolving the diffractometer scan into amorphous and crystalline peaks (Fig. 2.6a) [49].

The accuracy and reliability of crystallinity determination can be considerably improved by carefully addressing three issues: background subtraction, identification of the crystalline peaks – especially when they are broad and weak, and, most importantly, the choice of profile for the amorphous halo. Background subtraction requires drawing a proper baseline between the minima in the scattering curve, or, more appropriately, by collecting empty cell scattering. Crystalline peaks are sometimes easily mistaken to be amorphous scattering, thus leading to a severe underestimation of the crystallinity. This can be alleviated by identifying the broad crystalline peaks by suitable peak sharpening algorithms [50]. In drawing the amorphous halo, a scan from a molten polymer provides upper limits for the widths of the amorphous halo, and lower limits for their positions. The most versatile method to derive an amorphous template is to obtain the scan of a fully amorphous polymer or to extract it from a well-crystallized specimen in which the crystalline peaks can be easily stripped [45, 47]. Determining the crystallinity using this approach is illustrated with data from polyamide 6 (PA6, also called as nylon 6) (Fig. 2.6a). The amorphous profile was derived from a scan of highly crystalline (64%), unoriented film of PA6, in which the crystalline peaks could be easily identified and removed to leave behind the amorphous halo. The parameters of the amorphous halo were constrained while the scan was profile fitted with the four additional crystalline peaks shown in the figure.

Crystallinity determination requires that the XRD scan be obtained from an unoriented sample. If the samples are oriented, then numerous scans from different orientations have to be collected and properly weighted to obtain a scan equivalent to that of an unoriented sample [51]. Such randomization can be readily accomplished for uniaxially oriented samples than for samples with biaxial or more complex orientation [52]. A measure of the intensity of the crystalline peak in just a single scan, for example, equatorial

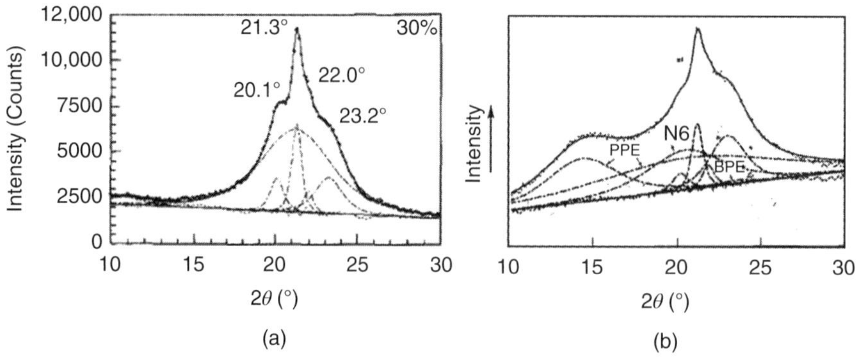

Figure 2.6 (a) Profile-fitted data from poorly crystallized polymer film (PA6). The scan is resolved into an amorphous halo, and crystalline peaks from the two crystalline polymorphs, α and γ. Murthy and Minor [47]. Reproduced with permission of Elsevier. (b) Profile-fitted scan from a blend of PA6 and poly(2,6-dimethyl-p-phenylene ether) (PPE). Murthy et al. [48]. Reproduced with permission of John Wiley and Sons.

scan, can be sufficient in some instances such as when the goal is a qualitative comparison of the crystallinities in fibers processed under different conditions.

2.5.2 Composition Analysis

The methods used for the crystallinity determination can also be used to determine the ratios of the different polymorphs that are often present in some polymers. In the case of PA6 shown in Figure 2.6a, the scan is resolved into the contribution from its two polymorphs, α and γ, along with the amorphous halo [47]. Relative areas of the various peaks are used to calculate the relative amounts of the α and γ components as well as the total crystallinity. The method can also be extended to determine when more than one polymer is present in the sample, such as polymer blends [48, 53]. Figure 2.6b shows an example of a mixture of amorphous poly(2,6-dimethyl-*p*-phenylene ether) (PPE) and PA6. The amorphous templates of PPE and PA6 were obtained from the scans of the homopolymers as discussed in Section 2.5.1. These templates were used as constraints in least squares fitting the data from the blend. Such analyses were useful in demonstrating that crystallinity and crystallite sizes of the PA6 were smaller in an alloy of the two polymers than in a blend [48]. Similar analyses have been carried out in a blend of two crystalline polymers, polyethylene and polypropylene [53].

2.6 CRYSTALLITE SIZE AND DISORDER

In addition with the level of crystallinity in a polymer, characteristics of the crystals that make up the crystalline regions also determine the performance of polymers. These characteristics include changes in the unit cell dimensions, crystallite size, and crystalline defects. These characteristics can be evaluated from the positions of the diffraction peaks, from the width of the diffraction peaks, and from the increase in the width of the diffraction peaks with q, respectively, as discussed below.

The width of the crystalline peaks is a measure of the size (L) of the crystals, which can be determined from the Scherrer equation:

$$L = \frac{K\lambda}{\beta_o \cos\theta} = \frac{2\pi K}{\Delta q} \tag{2.7}$$

where K is a constant that depends on how the width is calculated (0.9, if full width at half maximum, and 1 for integral breadth) [54], λ is the wavelength of the radiation, and β_o and Δq are the breadth of the peak corrected for instrumental broadening. XRD line broadening is not only affected by the crystallite size but also by the crystalline disorder. The contribution of this disorder needs to be taken into account as described below to calculate the actual size of the crystallites.

Two kinds of crystalline disorder can be identified in XRD data. The disorder of the first kind refers to the imperfections that, like thermal vibrations, alter the position of the atoms locally around the lattice points; examples are vacancies, interstitials, frozen-in displacements, substitutions, all of which leave the long-range order unchanged. This disorder reduces the intensity of the Bragg peak, without altering their widths. The missing intensity appears as diffuse scattering underneath the Bragg peak. This diffuse scattering can be evaluated by global fitting, as shown in Figure 2.7. This component measures the degree of lattice distortions of the first kind, and has to be taken into account in evaluating absolute crystallinities [49]. The crystalline disorders of the second kind manifest in the form of increase in the breadth of the crystalline peaks with q. The characteristic of the disorder of the second kind is that the lattice order is not preserved over a long range (>20 unit cells).

Two lattice disorder models are used to capture the increase in the widths of the diffraction peaks with q. One is the paracrystalline model of Hosemann [55, 56], and the other is lattice strain or the microstrain model of Warren and Averbach [57, 58], which is generally implemented in practice using the method developed by Williams and Hall [59, 60]. In the paracrystalline model, the unit cell vectors, rather than being fixed over the entire crystal as in the ideal lattice, fluctuate around a mean value. These fluctuations are cumulative, and atomic positions lose register over long distances. Thus, the lattice shows only a short-range order. In the micro strain model, the increase in the peak width is attributed to lattice strain due to internal or external stresses. It is possible to distinguish between these two types of disorders. Figure 2.8 shows one such example from a PA6 fiber. In the figure, the widths of a series of *0 k 0* reflections are plotted as a function of the order of the reflection k [62]. After correcting for various artefacts, it was shown that in this particular instance the increase in the width is due to lattice strain, and not to paracrystalline disorder. In a plot of the type

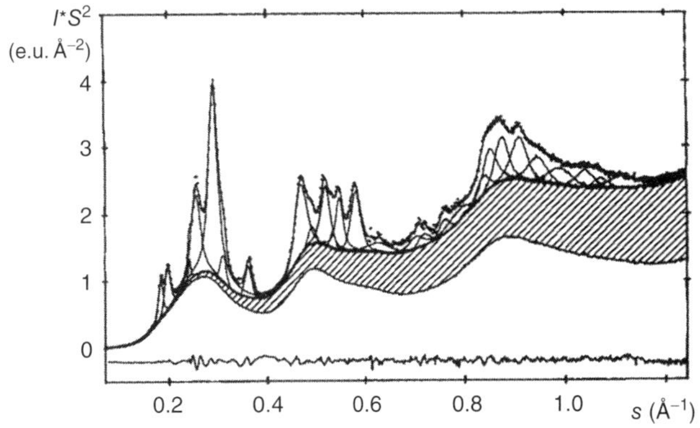

Figure 2.7 Global fitting of the data from a highly crystalline PET sample. The scan is resolved into crystalline peaks (sharp peaks), and two broad scattering, one due to the amorphous phase (the lower background line) and the other due to crystalline disorder (hatched area). Polizzi et al. [49]. Reproduced with permission of John Wiley and Sons.

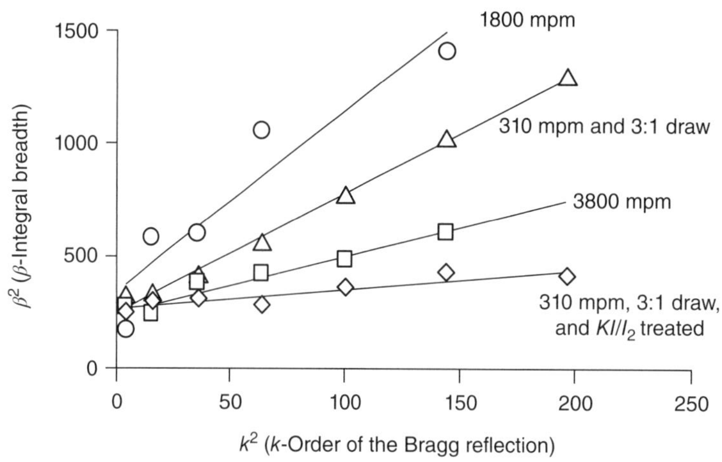

Figure 2.8 Evaluation of the disorder in polymer crystals by measuring the changes in the widths of a series of reflections. The data are from PA6 fibers drawn at different speeds given in meters per minute, mpm [61].

shown in the figure, the intercept can be used to calculate the crystallite size, and the slope of the lines to calculate the degree of crystalline disorder.

2.7 ORIENTATION ANALYSIS

Orientation of a polymer strongly influences many of its properties, and therefore is one the most commonly measured characteristics of the polymer. For instance, modulus of fibers can be directly correlated to the degree of crystalline orientation of the fiber. Because oriented amorphous segments are not entropically favored, these segments tend to become randomized when exposed to heat, causing the fibers to shrink. More generally, oriented amorphous segments contribute to the dimensional stability of the molded plastic parts. Therefore, a description of orientation in polymers requires a proper accounting of the orientation of the chain segments in both the crystalline and amorphous domains.

2.7.1 Crystalline Orientation

The orientation of the crystals is described in terms of the orientations of the normals to the selected crystallographic planes (*h k l*), more specifically of the poles, which are points at which the plane normals intersect a sphere that surrounds the crystal (Fig. 2.9). The angular dispositions of the various poles in 3D are often presented in 2D in the form of pole figures by means of stereographic projections, one for each reflection (Fig. 2.10) [1]. In this procedure, lines are drawn from the points of intersection of the plane normals with the sphere, h k l poles, to the south pole of the sphere. The locus of the intersection of the isointensity lines from each h k l pole with the equatorial plane (the plane of projection) forms the pole figure for that particular reflection. Pole figures are measured typically by rotating the sample and the detector in diffractometer along a programmed path so as to explore the required part of the reciprocal space.

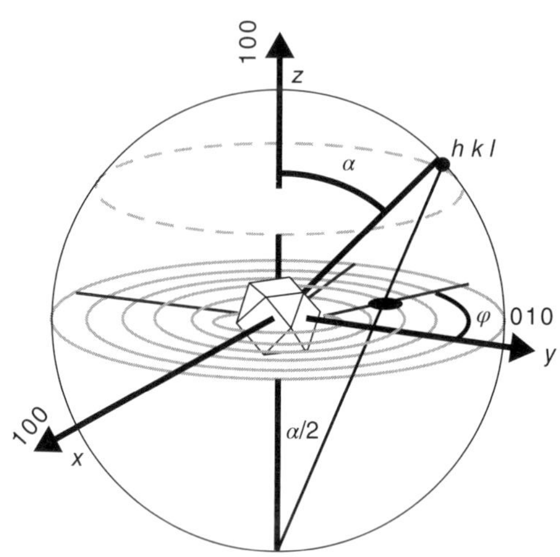

Figure 2.9 Stereographic projections of the crystallographic planes.

As an alternative, or in addition, to pole figures, orientation distribution function (ODF), which describes the orientation of one of the crystal axis with a reference axis, is often used. The orientation of the amorphous segments, which meander from one crystalline region to another and exist outside the crystalline segments, can be measured in terms of a different set of orientation functions. ODF in polymers is usually broad, the ODFs of the amorphous domains being significantly broader than those of the crystalline domains. XRD is perhaps the only technique that can provide a complete description of ODF. Roe and Krigbaum have described a method for calculating the crystallite orientation in terms of pole figures, which are the orientation distribution of the normals to the different crystallographic planes [63–65].

2.7.2 Uniaxial Orientation

In uniaxially drawn specimens (cylindrical symmetry) the ODF, $\rho(\alpha)$, for any given axis (e.g., chain axis) inclined at an

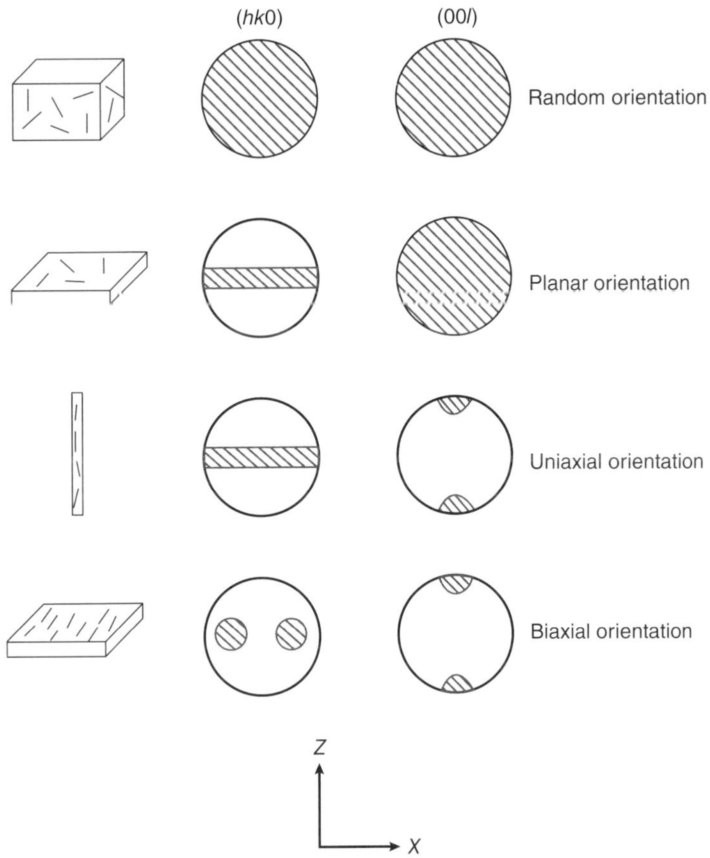

Figure 2.10 Examples of pole figure plots of four common types of orientation. The pole figures vary depending on how the sample is mounted with respect to the X-ray beam.

angle α with respect to an external axis (e.g., draw direction) is be expressed as a Legendre polynomial of order n:

$$\langle P_n \rangle = \langle P_n(\cos\alpha)\rangle = \int_0^{\pi} \rho(\alpha)\sin(\alpha)P_n(\cos\alpha)d\alpha \qquad (2.8)$$

When the distribution function is replaced by variation in XRD intensity, we have

$$\langle P_n \rangle = \frac{\int_0^{\pi} I(\alpha)\,\sin(\alpha)P_n(\cos\alpha)\,d\alpha}{\int_0^{\pi} I(\alpha)\,\sin(\alpha)d\alpha} \qquad (2.9)$$

For $n = 2$, the Legendre polynomial reduces to Hermans orientation function, commonly referred to as P_2

$$\langle P_2 \rangle = \frac{3\langle\cos^2\alpha - 1\rangle}{2} \qquad (2.10)$$

P_2 is widely used to describe the orientation in uniaxially drawn samples, such as fibers. Generalized spherical harmonics, of which Legendre polynomials are a subset, are required to describe orientation in samples without cylindrical symmetry [66].

In samples with cylindrical symmetry, $<P_2>$ is obtained from the XRD patterns from the azimuthal intensity distributions $I(\alpha)$ of suitably chosen reflections, where α is the angle between a crystal axis and the deformation direction (external reference). For example, a meridional reflection can be used to calculate the orientation of the chain axis with respect to the draw direction. A degree of orientation, f, of a crystal axis with respect to the fiber axis is calculated from the intensity distribution by first calculating the mean $\cos^2\phi$ from the expression

$$\langle\cos^2\phi\rangle = \frac{\int_0^{\pi/2} I(\phi)\ \sin(\phi)\cos^2(\phi)d\phi}{\int_0^{\pi/2} I(\phi)\sin(\phi)\ d\phi} \qquad (2.11)$$

In practice, f can be evaluated from the full width at half-maximum (FWHM, $\Delta\phi$) of the intensity distribution, assuming the distribution to be Gaussian:

$$I(\varphi) = e^{(-4\varphi^2 \ln 2/\Delta\varphi^2)} \qquad (2.12)$$

The following expression is then used to obtain f:

$$f = \frac{3\langle\cos^2\phi\rangle - 1}{2} \qquad (2.13)$$

This expression looks similar to the definition of P_2 (Eq. 2.10). If $\phi = 0°$ is the deformation direction, then the value of f varies from 0 in a randomly oriented system to 1 in a perfectly oriented system, and is −0.5 when the chosen crystal axis is perpendicular to the fiber axis. The number of Hermans' orientation functions that need to be evaluated for a complete description of orientation depends on the unit cell: two for an orthorhombic and up to five for a triclinic unit cell [1]. Figure 2.11 shows an example of the evaluation of the degree of orientation in uniaxially oriented samples. Figure 2.11b is the azimuthal trace of one of the crystalline reflections in the diffraction pattern (Fig. 2.3a) [40, 68]. A practical measure of parallelism or orientation given by expression 2.14 is sometimes useful:

$$\text{Degree of orientation} = \frac{180° - \Delta\phi(\text{in}\,°)}{180°} \tag{2.14}$$

This simple intuitive measure of orientation is especially useful in evaluating poorly oriented structures such as amorphous domains (see Section 2.7.3) [2, 52].

2.7.3 Biaxial Orientation

A large fraction of the polymers that are commercially produced are fabricated into films. Films are produced by stretching the polymer along orthogonal axes, machine direction (MD), and in the transverse direction (TD), resulting in biaxial orientation. XRD is often used to characterize this biaxial orientation. There have been a few attempts to systematically characterize this biaxial orientation. The one proposed by White and Spruiell [69] can be readily used in practice. In the previous section, it was shown that uniaxial orientation can be described by one parameter, the degree of orientation of the chain axis with respect to the fiber axis.

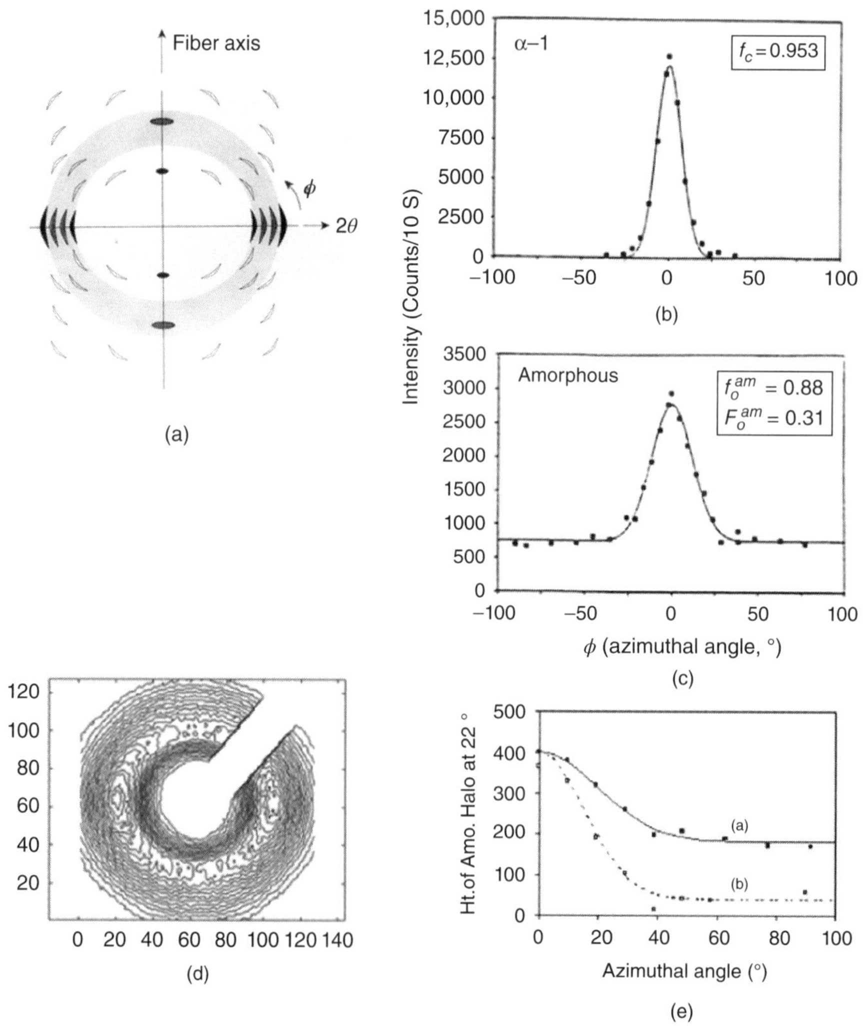

Figure 2.11 (a) 2D pattern of an oriented semicrystalline polymer (PA6). (b) Plot of the intensity in one of the crystalline reflections, the first peak in (a), as a function of the azimuthal angle. (c) Plot of the intensity of the amorphous halo in (a). (d) 2D diffraction data from a partially oriented amorphous poly(ethylene terephthalate) (PET) fibers. (e) Intensity as a function of the azimuthal angle of one of the amorphous halos in a PET fiber. Full line – undrawn fiber with a low degree of orientation. Dotted line – drawn fiber with a high degree of orientation [41, 67].

Similarly, it can be shown that a minimal description of biaxial orientation requires at least two parameters shown below:

$$f_z^B = 2\langle \cos^2\varphi_{1z}\rangle + \langle \cos^2\varphi_{1y}\rangle - 1$$
$$f_y^B = 2\langle \cos^2\varphi_{1y}\rangle + \langle \cos^2\varphi_{1z}\rangle - 1 \quad (2.15)$$

The angle brackets represent averaged values as given in Equation 2.11. x, y, and z are the coordinate axes given in Figure 2.9 and the subscript 1 refers to the direction of the $(h\,k\,l)$ vector in the figure.

2.7.4 Amorphous Orientation

When polymers are deformed and as the crystal axes tilt, the chains in the amorphous domains also become oriented. This orientation results in concentration of the intensity of the amorphous halo into an arc centered on the equator, as shown in Figure 2.11d from a drawn PET fiber [67]. The oriented chains in the amorphous domains become more densely packed than when they are not oriented. These domains are sometimes considered as intermediate phases or rigid amorphous phases [38, 70]. The nature of the oriented amorphous phase influences numerous polymer properties such as diffusion, strength and modulus, and shrinkage and dimensional stability. XRD patterns can be used to determine the fraction of this phase, the degree of lateral and orientational order that is present in these ordered but noncrystalline domains [41].

The methods described in the previous section for determining the crystalline orientation can be adapted to the orientation of the chains in the amorphous regions [39–41, 71, 72]. By comparing the azimuthal intensity distribution data in Figure 2.11b and c that while the intensity of the crystalline peaks is zero away from the peak maximum, the baseline intensity of the amorphous halo is not zero. Because the amorphous components in polymers can be both isotropic (unoriented) and anisotropic (oriented), the nonzero intensity can be attributed to the unoriented amorphous segments in the specimen. The width of the amorphous azimuthal intensity distribution, as in the case of the crystalline peak, is a measure of the degree of orientation of the oriented amorphous region f_o^{am}. The fraction of the intensity above the baseline represents the fraction of the amorphous segments that are oriented, or fractional orientation, F_o^{am}:

$$F_o^{am} = \frac{A_p}{A_p + A_b} \quad (2.16)$$

where A_p is the area in the peak above the baseline and A_b is the area under the baseline. These parameters can be conveniently obtained from 2D fitting of the diffraction pattern [73]. Figure 2.11e shows that as the PET fiber is drawn, both f_o^{am} and F_o^{am} increase as indicated by the decrease in the peak width and height of the baseline. It should be noted that the degree of orientation of the amorphous regions lags behind that of the crystalline regions, and is always less that the crystalline orientation [40].

2.8 SMALL-ANGLE SCATTERING

The intensity scattered at small angles (<5° 2θ, Cu Kα) contains information about the structure at large length scales (50–1000 Å). Such structures, often called mesoscale structures, are invariably present in polymers. The contrast that produces SAXS arises from the density heterogeneities that produce differences in electron density distribution at these length scales. A closely related technique, small-angle neutron scattering (SANS), relies on the scattering contrast that arises from the differences in the strength of interaction between the neutrons and the atomic nuclei. The basic theories and models used to interpret both the SAXS and SANS data are quite similar. Some of the applications of SAS, such as determination of the chain dimension, are discussed in Chapter 4. This chapter describes the use of these two techniques, collectively referred to as small-angle scattering (SAS), to probe structural inhomogeneities or fluctuations at the nanoscale. Table 2.2 lists some of the parameters that can be derived from a detailed analysis of the SAXS pattern [74].

Any phase separation that segregates molecules of different chemical composition invariably results in electron density fluctuation. The SAXS that results from such structures can be usefully analyzed to characterize the phase behavior in polymers. For instance, while copolymerization inhibits micron-length scale phase separation, di-, tri-, and multiblocks as well as graft copolymers show interesting phase behavior at length scales ~100 Å. [75]. Examples include phase separation in polyurethane-segmented block copolymers [76–78]. The peak position in these patterns corresponds to the spacing between the low T_g "soft" blocks that are covalently linked to "hard" blocks. The factors that govern phase behavior of these hard and soft blocks, and, in general, the formation of microdomains can be investigated using SAS. Such phase separation gives rise to either or both a central diffuse scattering (CDS) at low-q values (<0.05 Å^{-1}) and discrete reflections at slightly higher q values (~0.05 Å^{-1}). This diffuse scattering, when not arising from phase separated domains, is usually labeled as void scattering, although it arises for the most part not from the voids themselves but from the polymer–air interfaces in the voids, along with contributions from surface refraction [79], and a weaker contribution from structural entities such as fibrils [80–82].

Semicrystalline polymers show additional SAXS features from the structural heterogeneities at 10–100 nm that arise from the natural segregation of crystalline and amorphous regions into separate domains. Such phase separation manifests as three basic types of morphologies (Fig. 2.12): fringed micelles (e.g., cellulose), fibrillar structures (e.g., silk and many liquid crystalline fibers), and lamellar structures in polymers that form folded chain crystallites (e.g., polyethylene, aliphatic polyamides, poly(ethylene terephthalate), polypropylene) [9, 81]. In fringed micelles, a single polymer chain with sequences that are long enough to crystallize will be incorporated into different crystals that serve

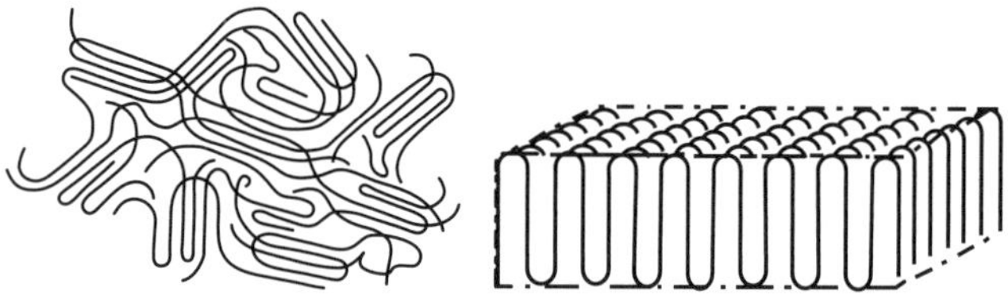

Figure 2.12 Schematic representation of the two basic morphologies that arise in the course of crystallization in polymers. (a) Micelles creating an overall connectedness that leads to a network where the crystals are the junctions. The molecular detail is not to be taken literally [83]. (b) Folded chain lamellae. The figure shows regular and adjacently reentrant polymer chains, whereas in practice random reentry is often observed.

as multifunctional junctions to form a network in a matrix of noncrystalline chain segments. This model was proposed by Frey-Wyssling and Hermans to explain the properties and the diffraction patterns of cellulose fibers [84, 85]. The fringed micelle concept is particularly relevant to polymer chains with limited crystallizable sequence lengths, such as homopolymers with crystallizable stereoregular sequences like poly(vinyl chloride), ethylene-vinyl acetate copolymers, and segmented block copolymers such as thermoplastic polyurethanes [86], and polymers such as polyacrylonitrile that have crystalline fragments. The discrete reflection that results from the separation of the crystalline domains can be used to discern the characteristics of the micelle [87]. Often, as the degree of order increases, for instance, by a decrease in the noncrystallizable sequences in a copolymer, these micelles transform into lamellae [88].

2.8.1 Central Diffuse Scattering

There is typically an intense CDS near the primary beam ($q < 0.05$ Å^{-1}) from polymeric samples (Fig. 2.13). Such scattering is usually diamond shaped or shaped like a bow-tie or two-bladed propeller [89]. There are at least two features in the CDS that can be easily recognized in the scattering from oriented polymers such as fibers [74]. One of the components in this scattering is a nearly isotropic inner scattering and the other is an extended equatorial streak. The inner isotropic CDS is attributed to small spherical voids (100–1000 Å) [1, 90–92], and is usually called void scattering. Voids >0.1 μm scatter at too small an angle to be observed in SAXS. This isotropic scattering can also occur from solid particulate matter embedded in the polymer matrix. These entities do not become aligned or elongated when polymer is stretched. The equatorial streak is due to objects that become aligned when the polymer is stretched. These objects include internal phase boundaries [93], elongated voids, and surfaces of macroscopic entities such as fibers themselves [79].

The two components of the CDS can be analyzed to determine the sizes of the entities that give rise to these features. The isotropic scattering can yield an estimate of the size of voids or particulate defects. The equatorial scattering can be analyzed to determine the orientation distribution of the scattering entities [93] and, when appropriate, the size and length of the elongated scattering entities such as microfibrils

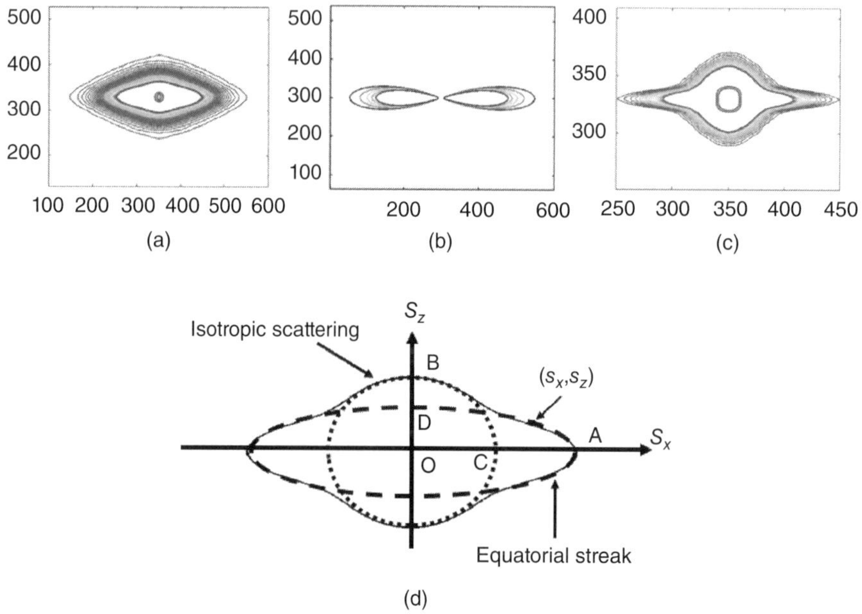

Figure 2.13 (a) Diamond-shaped SAXS pattern from solution-spun PAN. (b) Propeller-shaped SAXS pattern from gel-spun PAN fibers. (c) Solid outline is a schematic of an iso-intensity contour from propeller-shaped scattering pattern. (d) The dotted outline is a circle and dashed outline is an ellipse. The propeller-shaped contour coincides with the circle and ellipse in the vicinity of point A and B, respectively [89].

[80, 94–96]. Other methods of analysis based on Guinier law [74, 97] invariants [97], and Porod law [93] have also been used.

In addition to the isotropic scattering and the elongated equatorial streak, there is a third much weaker feature within the CDS in the form of an interference peak that is sometimes present along the equator [80, 98]. This feature is often not easy to distinguish from the more intense isotropic and streak scattering. On the basis of additional data obtained from electron micrographs, it has been shown that the interference peaks are due to the scattering from ordered crystalline fibrils embedded in the amorphous matrix, the diameter of these fibrils, deduced from the position of the peak, is ~50 Å. If the fibrils are not sufficiently ordered, then the peak degenerates into a diffuse scattering from which, using a Guinier analysis ($\log(I(q))$ vs q^2 plot), it is possible to deduce the diameter of the fibrils [82]. Alignment of the scattering entities such as fibrils, and the lateral and longitudinal size of these entities are usually derived from a detailed analysis of the equatorial streak [80, 93–97].

2.8.2 Discrete Reflections from Lamellar Structures

Polymer chains in semicrystalline polymers such as polyolefines, aliphatic polyamides (nylons), and aliphatic polyesters crystallize in the form of lamellae of folded chains that are organized to form lamellar stacks (Fig. 2.14a). The spaces between the lamellae and between the stacks are occupied by

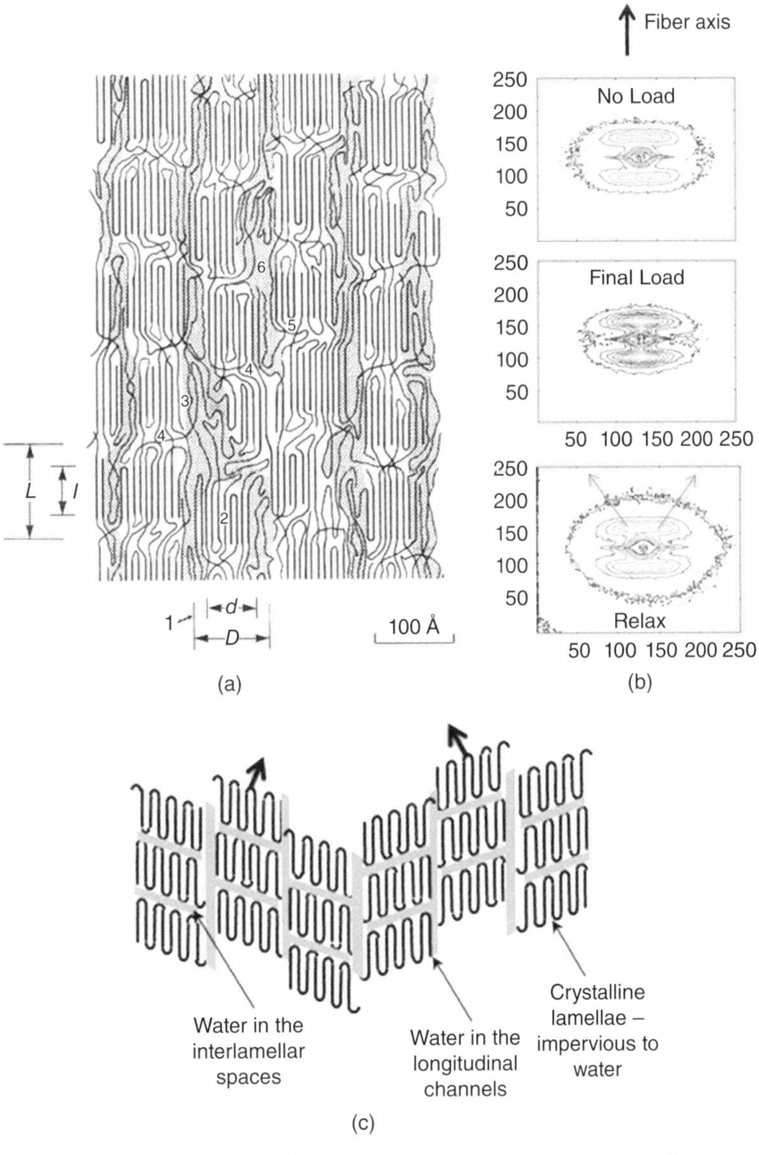

Figure 2.14 (a) SAXS scattering from a PA6 fiber. (1) Fibrils. (2) Lamellae. (3) Partially extended chains in the interfibrillar regions. (4) Tie molecules in the interlamellar regions. (5) Free chain ends. (6) Amorphous chain segments with large free volume. (7) Fusion of adjacent fibrils. Shaded areas represent the interfibrillar amorphous regions that form longitudinal channels for the diffusion of water along the fiber axis [82]. (b) A model to interpret the features in the SAXS and SANS pattern in terms of folded chain lamellae that form fibrils. (c) Model for the tilted lamellae. Also illustrated is the greater permeability of the amorphous domains to solvents.

amorphous segments. Separation of the lamellae along the fiber axis (lamellar spacing) gives rise to peaks close to the draw direction (meridional peaks). The details of the arrangements of the lamellae determine the nature of the resulting diffraction pattern. Three commonly observed diffraction patterns, the 2-point, bar-type, and 4-point patterns, are shown in Figure 2.14b. The 2-point patterns are due to 1D repeating arrangement of lamellae (crystals) along a microfibril. The positions of the crystals in adjacent microfibrils are uncorrelated. The 4-point pattern usually occurs when the polymers are stretched, such as in the fiber-forming process, and is due to the tilt of the lamellar normal away from the draw direction as shown in Figure 2.14c. As the fiber is stretched, a combination of chain slip and lamellar slip can give rise to the tilted lamellae. This lamellar tilt is due to the mismatch in the cross-sectional area between the tightly packed crystalline stems and the loosely packed amorphous chains at the lamellar surface. The resulting 4-point reflection that is tilted so as to give either an eyebrow (tilted in) or a butterfly (tilted out) pattern. This tilt angle can be measured from the azimuthal-angular separation of the reflection on either side of the meridional axis. The bar-type pattern could arise from the small lateral width of the diffracting element, a narrow microfibril, or more likely, from the coexistence of 2-point and 4-point structures.

The changes in the lamellar spacings with processing parameters such as crystallization temperatures, annealing, and drawing are most frequently followed and easily analyzed. Figure 2.15, for example, shows that long period increases with annealing temperature and draw ratio [80, 99]. The increase in lamellar spacing produced by stretching in combination with heat is accompanied by an increase in the tilt angle [80, 81]. However, when the fibers are annealed under no constraints, the lamellar spacing increases and the tilt angle decreases [80]. These and other changes can be accurately measured by analyzing the diffraction patterns in an elliptical coordinate system [100, 101]. The lateral spread of these reflections, that is, along a straight line perpendicular the meridian, is due to the limited width of the fibrils or stacks that are within the fibril. Misorientation of the fibrils would spread the lamellar reflections along a circular arc. However, the lamellar reflections spread neither along a straight line nor along a circular arc. In fact, no combination of the straight line and a circular arc is able to give the observed loci of the lamellar peak. It has been shown that the lamellar reflections, specifically the loci of the maxima of the traces, fall neither on a Cartesian (x,y) nor a polar grid (r,θ) but on an elliptical (u,v) grid (Fig. 2.16a). An elliptical contour can be constructed through all the intensity maxima of each vertical slice. The equation for such an ellipse with semi-axes a and b is

$$\frac{X^2}{a^2} + \frac{Z^2}{b^2} = 1 \qquad (2.17)$$

A plot of Z^2 versus X^2 is a straight line with slope of $(-b^2/a^2)$ and an intercept of b^2 as given by the relation

$$Z^2 = \frac{X^2}{\beta^2} + b^2 \qquad (2.18)$$

where β, a measure of anisotropy, is the ratio of the semi-major axis (a) to the semi-minor axis (b) [89, 102]. Alternatively, Equation 2.17 can be written as

$$\frac{1}{Z^2} = \frac{1}{b^2} + \frac{1}{a^2}\tan^2\phi \qquad (2.19)$$

where $\phi = \tan^{-1}\left|\frac{X}{Z}\right|$. A plot of $1/Z^2$ versus $\tan^2\phi$ would be a straight line (Fig. 2.16c) with intercept related to the meridional spacing ($1/b$). Because these analyses use the complete azimuthal data to determine the least-squares-fitted values of the lamellar spacing and the tilt angle, the values are more reliable than those obtained from the positions of the center of the reflections.

The elliptical trace corresponds to the deformation of the lamellar lattice, and the disposition of the reflections along this trace reflects the orientation of the lamellae in this lattice [103]. One possibility is that as the lamellar spacing increases, the lateral distance between the adjacent lamellar columns decreases. Absence of any correlation between adjacent lamellar columns results in an eyebrow pattern (Fig. 2.16b). However, if the tilt

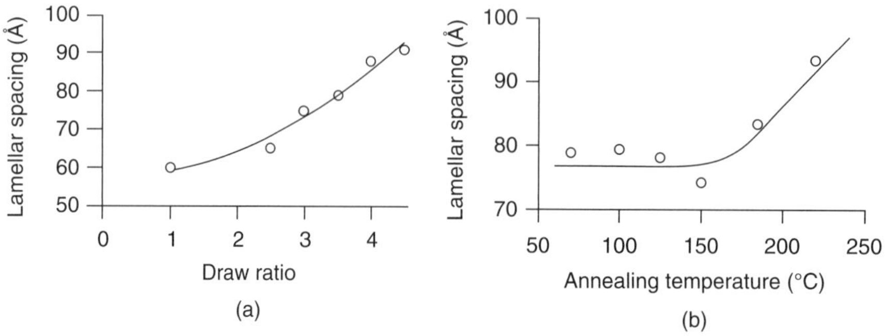

Figure 2.15 Changes in the long spacing in PA6 fibers with (a) draw ratio and (b) with annealing temperature [80, 99].

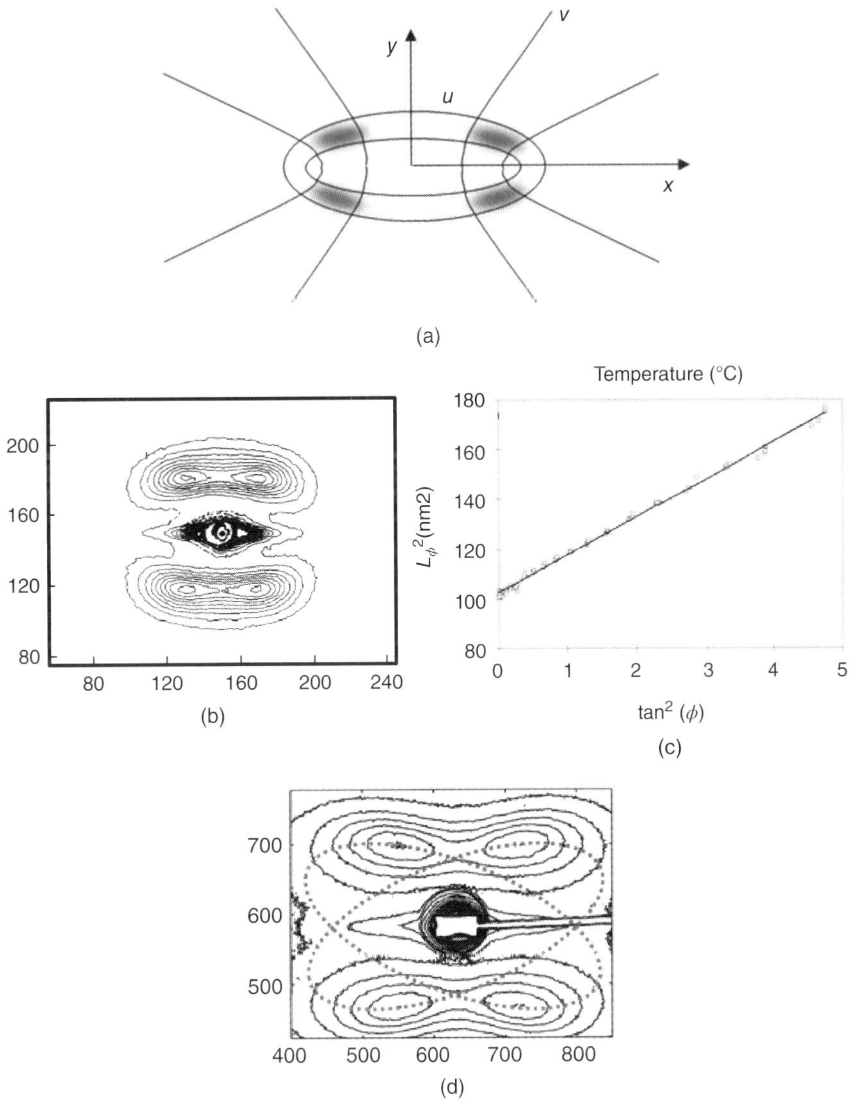

Figure 2.16 (a) SAXS pattern shown in an elliptical coordinate system. Cartesian coordinates (x–y) are shown through the origin. A pair of lines with constant u and v values is shown near the reflections. (b) Two-dimensional SAXS pattern from an oriented PA6 fiber. (c) Plot of L_ϕ^2 versus $\tan^2 \phi$. (d) A butterfly SAXS pattern interpreted as due to two overlapping ellipses that are overlaid on the observed pattern.

of the lamellae in the neighboring columns is correlated, then a butterfly pattern is obtained (Fig. 2.16d). Such butterfly patterns are attributed to clusters of lamellae within which the lamellae are tilted either to the right or to the left [101]. Simulation of the patterns in an elliptical coordinate system can provide a means to fit the observed SAXS pattern from a lamellar structure with the fewest parameters [104].

2.8.3 Small-Angle Neutron Scattering and Solvent Diffusion

Because of the similarities in wavelengths of X-rays and neutrons used in scattering experiments, the sizes of structures accessible to X-rays and neutron are similar. In polymeric materials, the substitution of deuterium for hydrogen in a structure dramatically changes its neutron scattering pattern, while the X-ray scattering remains unchanged because the deuteration in most instances does not appreciably alter the structure. Therefore, neutron scattering has emerged as a powerful tool for studying the molecular structure by following the distribution of deuterium-labeled molecules in polymers. Some of the SANS results that rely on the deuterium-labeling techniques are discussed in Chapter 4. This section describes the studies that take advantage of the contrast that arises from the preferential diffusion solvents (deuterated versions) into the more accessible regions of the polymer. This reveals the inhomogeneity at nanolength scales.

Figure 2.17a shows an example of the neutron scattering that results when D_2O is preferentially absorbed by one segment in a polymer. These data were obtained from a polymer in which poly(ethylene glycol), PEG, segments are randomly inserted along the polymer chains composed mostly of hydrophobic groups, tyrosine-derived polycarbonates

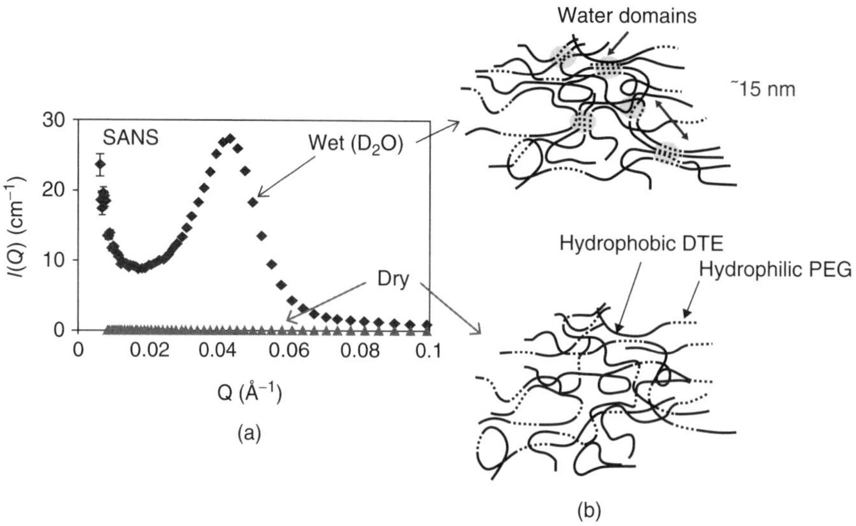

Figure 2.17 (a) Comparison of the SANS scans from dry and hydrated polymers (poly(desaminotyrosyl-tyrosine ethyl ester carbonate-co-poly(ethylene glycol) carbonate). (b) Model for phase separation upon hydration [105, 106]. The top figure shows hydrated PEG-rich domains after hydration. The bottom figure shows uniformly distributed PEG segments in the dry samples.

[105, 106]. The interference peak seen in the SANS data shows that D_2O is preferentially incorporated into certain domains in the polymer. SAXS data obtained from the same polymer show that the PEG segments are homogeneously distributed throughout in the dry sample. However, when hydrated, the SAXS pattern that results is very similar to that shown in Figure 2.17a. Because SAXS captures the distribution of PEG domains, and SANS the distribution of water, in combination the two data show that PEG segments get segregated when the polymer is hydrated, and that D_2O reside in these PEG domains. Thus, the peak in Figure 2.17a is a signature of phase segregation of hydrated polar segments and the nonhydrated hydrophobic polymers that typically occurs as water is absorbed by amphiphilic polymers (Fig. 2.17b).

Because water preferentially diffuses into the amorphous and porous regions in a polymer, SANS can be used to deduce the distribution, and thus the interconnectivity of the amorphous segments and void spaces by using D_2O to generate the contrast. SANS patterns from drawn semicrystalline PA6 fibers show meridional peaks that arise from the presence of D_2O in the interlamellar regions of the fiber and the diffuse equatorial streak near the center that is attributed to D_2O present along the fiber axis in the spaces between the fibrils [96]. These two pathways are schematically shown in Figure 2.14c that shows a typical lamellar structure of a semicrystalline polymer in which the stacks of folded lamellae form fibrils that are oriented along the fiber axis [82].

2.9 SPECIALIZED MEASUREMENTS

Availability of high-brilliance synchrotron-based X-ray sources, progress in X-ray focusing optics, and fast 2D large-area detectors have made it possible to carry out measurements at submicron spatial resolutions and at microsecond time scale. It is now possible to routinely carry out X-ray measurements in combination with other techniques such as thermal analysis, rheology, and spectroscopy, and also to follow the development of structure during processing such as such as in fiber spinning and drawing operations. Three such applications are described here.

2.9.1 In situ Experiments

Study of the changes in the structure and morphology during fabrication of plastic is important from both practical and academic perspectives. For example, experiments carried out as a function of heat flow, relative humidity, and external stress yield, respectively, fundamental insights into the crystallization kinetics [107], hydration [79], and deformation [9, 81] behavior of polymer chains.

Figure 2.18A shows an example of the *in situ* studies of the changes in the lamellar structure during stretching and relaxation of a PET fiber using SAXS data [81]. The figures in the inset are the 2D pattern of the starting fiber, the fiber while stretched, and after it was relaxed. These figures show qualitatively that the 4-point pattern in the starting fibers transitions into a 2-point pattern when the fiber is stretched, but reverts to the initial 4-point pattern. The plots in the figure are the azimuthal traces through the lamellar reflections that show these changes in further detail. Whereas the starting and the relaxed fibers both show clear 4-point patterns, the fiber under stress is a composite of the 2- and 4-point patterns. More importantly, note that the 4-point pattern is weaker in the stretched fiber indicating that the 2-point pattern appears at the expense of the 4-point pattern. This transition from a 4-point to a 2-point pattern is reversible. This shows that lamellae reversibly transform from a tilted to an untilted state under strain. This is shown schematically in Figure 2.18B. WAXS data obtained simultaneously with these SAXS data

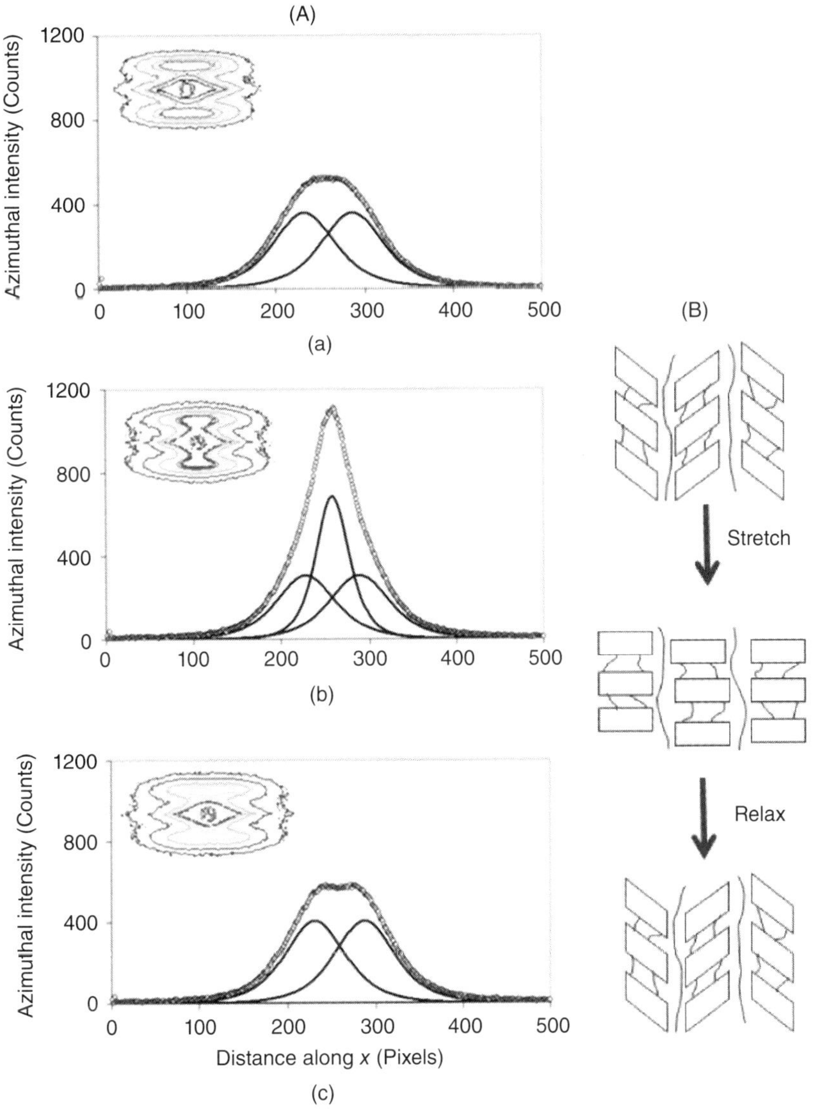

Figure 2.18 (A) SAXS data observed during various stages of deformation of a PET fiber. The inset in each figure shows the 2D data during deformation. The plots are the azimuthal intensity distribution through the lamellar peak maxima: (a) before loading, (b) at intermediate loading (12% strain, about half the maximum strain), and (c) after relaxation of the fiber. Circles represent the measured intensities. The curves under these circles are the resolved components of the four- and two-point patterns. The sum of these components almost completely overlaps the observed data and therefore is not shown [81]. (B) A model for the reversible changes in the lamellar tilt observed during stretching and relaxing of the fiber.

show that there are corresponding changes in the crystallite dimensions with stress [81].

2.9.2 Microbeam Diffraction

Polymeric materials, usually considered homogeneous, can be inhomogeneous at mesoscopic length scales (0.1–10 μm). Such inhomogeneities are the result of temperature and stress gradients that are present during fabrication of fibers, films, and plastic parts. Scanning micro X-ray diffraction (μ-XRD) techniques [108, 109] can be used to investigate these inhomogeneities by mapping grain orientations, crystalline phase distribution, and full strain/stress tensors with micrometer spatial resolutions. Microbeam diffraction with a 0.4 μm beam has also been used to understand some of the features in SAXS, especially in identifying that the equatorial streak in central scattering has contributions from surface refraction [79]. Diffraction patterns obtained from Kevlar® with a 3 μm beam were able to show structural gradients that occur in a fiber [110]. The data show clear evidence of a skin-core effect that is commonly present in polymers [111]. While in some materials this is associated with changes in crystallinity, in the case of fibers, this manifests as a higher degree of orientation at the skin due to large shear stresses at the spinneret. This orientational gradient is accompanied by a decrease in the modulus from skin to core.

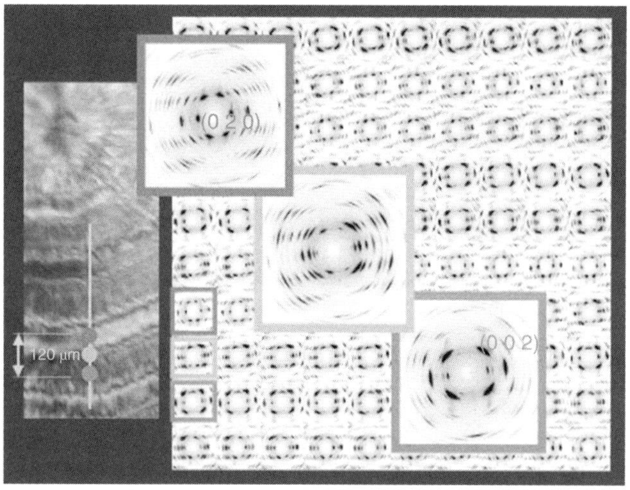

Figure 2.19 A series of wide-angle X-ray diffraction photographs from crystals located along the vertical line within a spherulite of poly(hydroxy butarate) shown in the left inset. The enlarged diffractograms are from three areas separated by 60 μm as shown in the optical micrograph. Courtesy of C. Riekel, ESRF. (*See color plate section for the color representation of this figure.*)

Figure 2.19 shows an example of the use of microbeam X-rays to follow the changes in the structure from one micron-size crystal to the next within a spherulite [112, 113]. The figure shows a series of hundreds of WAXS patterns recorded from a single spherulite of poly(hydroxyl butarate). These patterns show the changes in the orientation and the rotation of the crystals as the beam was stepped in 3 μm increments along the vertical line drawn in the polarized optical micrograph shown on the left. The three photographs in the foreground show the differences in the texture of three crystals 60 μm apart. Microbeam diffraction has been used in general to identify the disordered structures between microfibrils and the defects inside the crystallites in cellulose [114], phase transformation between different polymorphs [115], and in analyzing the morphology transcrystalline regions [116].

2.9.3 Grazing Incidence Diffraction

Multilayer films and films deposited onto substrates are now commonly used in many commercial applications. The structures of such μm-thick layers are important in evaluating these films. Furthermore, the arrangement of polymer chains near the surface is often different from that in the bulk because of thermal gradients [111, 117], and because the air–polymer interface places restrictions on the conformation of the long chain molecules. These structures cannot be investigated using the techniques described thus far. The surface structures can be investigated using GID. In GID, X-rays are incident at a very shallow angle so that the resulting large path length will attenuate the X-ray intensity within a depth of several nanometers up to a few tens of micrometers, depending on the incident angle and the absorption coefficient of the surface layers. Thus, the diffracted X-rays represent the structure within the top surface layers. A depth profile (z-profile) of the structure can be generated by inverting a sequence of scans at a series of incidence angles (τ-profiles) [118]. GID experiments are carried out using an experimental arrangement similar to those of a related technique, reflectometry. Whereas the reflectometry measures the electron density variations normal to the surface from which parameters such as the surface thickness and the arrangement of different layers in multilayered surfaces can be examined [119, 120], GID can be used to analyze the atomic and molecular structure near the surface.

Figure 2.20 shows the use of GID to study the different polymer layers, PE and PA6 [121]. In this figure, the depth of penetration was increased from a fraction of a μm to ~100 μm by increasing the incident angle (α) from 0.25° to 3°. The diffraction pattern at $\alpha = 0.25°$ is due only to the top PE layer at the film–air interface. At $\alpha = 0.5°$, a new peak appeared at 22.8° and it was attributed to the bulk PE in the top PE layer near the N6–PE interface. At higher incidence angles, the contribution due to the surface PE becomes weaker relative to

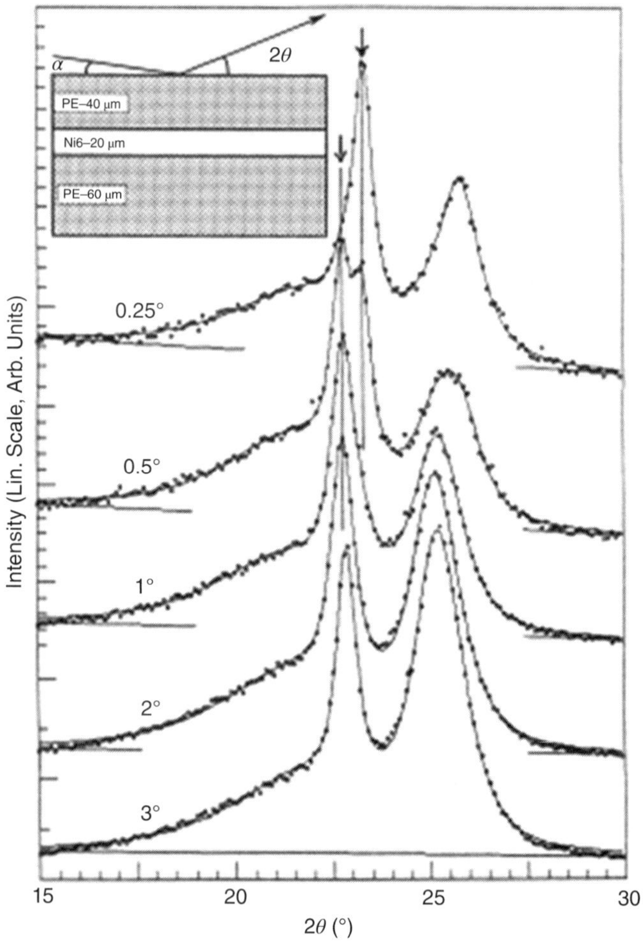

Figure 2.20 Examples of 2θ scans from polyethylene/PA6/polyethylene laminate at five incident angles (α). The depths of penetration at these angles are 11, 21, 42, 80, and 114 μm in PE and 7, 13, 26, 50, and 72 μm in PA6.

that from the bulk PE and the middle biaxially oriented PA6 layer that shows only the peak at 25°. These measurements show that GID is useful in studying multilayer polymer films; multicomponent ultrathin films [122]; skin-core effects common to injection-molded plastics; nanostructured surface coatings; the orientation, conformation, and packing modes of chains near the surface [123, 124]; as well as the structure at interfaces and adhesives.

2.10 SUMMARY

XRD patterns from polymers can be analyzed at various levels, based on the quality of the data and the intended use of the results. Earlier efforts were focused on determining the crystal structure, lamellar morphology, and in improving the methods to determine the crystallinity and orientation. Later efforts were focused on determining the nature of the lattice distortions and the nature of the noncrystalline phases. With the availability of synchrotrons, fast 2D detectors, software tools to analyze the data, future effort will more likely be toward understanding the evolution of these structures with time and temperature. While the analysis and interpretation of WAXS data is now straightforward, there is opportunity to improve the analysis and interpretation of the SAXS data.

REFERENCES

1. Alexander LE. *X-Ray Diffraction Methods in Polymer Science*. NY: Wiley-Interscience; 1969. p 582.
2. Kakudo M, Kasai N. *X-Ray Diffraction by Polymers*. Tokyo: Kodansha; 1972.
3. Spruiell JE, Clark ES. Unit cell and crystallinity. Methods Exp Phys 1980;16:1–127.
4. Wang JI, Harrison IR. Crystallite size and lamellar thickness by X-ray methods. Methods Exp Phys 1980;16:128–184.
5. Giacovazzo C. *Fundamentals of Crystallography*. Vol. 7. USA: Oxford University Press; 2002.
6. Woolfson MM. *An Introduction to X-Ray Crystallography*. Cambridge University Press; 1997.
7. Prince E, Wilson AJC, Hahn T, Shmueli U. *International Tables for Crystallography*. International Union of Crystallography; 1999.
8. Wanakule NS, Nedoma AJ, Robertson ML, Fang Z, Jackson A, Garetz BA, Balsara NP. Characterization of micron-sized periodic structures in multicomponent polymer blends by ultra-small-angle neutron scattering and optical microscopy. Macromolecules 2008;41:471–477.
9. Murthy NS, Grubb DT. Deformation of lamellar structures: Simultaneous small- and wide-angle X-ray scattering studies of polyamide-6. J Polym Sci Polym Phys 2002;40:691–705.
10. Dorset DL. Direct determination of polymer crystal structures from fibre and powder X-ray data. Polymer 1997;38:247–253.
11. Tadokoro H. *Structure of Crystalline Polymers*. New York: Wiley; 1979.
12. Yang J, Sidoti G, Liu J, Geil P, Li C, Cheng S. Morphology and crystal structure of CTFMP and bulk polymerized poly(trimethylene terephthalate). Polymer 2001;42:7181–7195.
13. Wang B, Li CY, Hanzlicek J, Cheng SZD, Geil PH, Grebowicz J, Ho R-M. Poly(trimethylene teraphthalate) crystal structure and morphology in different length scales. Polymer 2001;42:7171–7180.
14. Murthy NS, Minor H. A discussion of the structural changes in nylon 6 and comments on Salem and Weigmann's paper. Polym Commun 1991;32:297–300.
15. Matsuo M, Sawatari C. Elastic modulus of polyethylene in the crystal chain direction as measured by X-ray diffraction. Macromolecules 1986;19:2036–2040.
16. Murthy NS, Shacklette LW, Baughman RH. Effect of charge transfer on chain dimension in trans-polyacetylene. J Chem Phys 1987;87:2346.
17. Enkelmann V. Crystal structure. In: Müllen K, Wegner G, editors. *Electronic Materials: The Oligomer Approach*. Wiley Online Library; 2007. p 295–344.
18. Liu J, Geil P. Crystal structure and morphology of poly(ethylene terephthalate) single crystals prepared by melt polymerization. J Macromol Sci Phys 1997;36:61–85.
19. Daubeny RP, Bunn CW, Brown CJ. The crystal structure of polyethylene terephthalate. Proc R Soc Lond A Math Phys Sci 1954;A226:531–542.
20. Pazur RJ, Hocking PJ, Raymond S, Marchessault RH. Crystal structure of syndiotactic poly(β-hydroxybutyrate) from X-ray fiber and powder diffraction analyses and molecular modeling. Macromolecules 1998;31:6585–6592.
21. Dasgupta S, Hammond WB, Goddard WA III. Crystal structures and properties of nylon polymers from theory. J Am Chem Soc 1996;118:12291–12301.
22. Li Y, Goddard WA III, Murthy NS. Crystal structure and properties of N6/AMCC copolymer from theory and fiber XRD. Macromolecules 2003;36:900–907.
23. Baker AME, Windle AH. Evidence for a partially ordered component in polyethylene from wide-angle X-ray diffraction. Polymer 2001;42:667–680.
24. Bruckner S, Meille SV, Malpezzi L, Cesaro A, Navarini L, Tombolini R. The structure of poly(d-(−)-. beta.-hydroxybutyrate). A refinement based on the Rietveld method. Macromolecules 1988;21:967–972.
25. Brückner S, Porzio W. The structure of neutral polythiophene. An application of the Rietveld method. Die Makromol Chem 1988;189:961–967.
26. Dupont O, Jonas AM, Legras R. Adaptation of the Rietveld method for the characterization of the lamellar microstructure of polymers. J Appl Crystallogr 1997;30:921–931.
27. Chupas PJ, Qiu X, Hanson JC, Lee PL, Grey CP, Billinge SJL. Rapid-acquisition pair distribution function (RA-PDF) analysis. J Appl Crystallogr 2003;36:1342–1347.
28. Mitchell GR, Rosi-Schwartz B, Ward DJ, Warner M. Local order in polymer glasses and melts. Philos Trans: Phys Sci Eng 1994;348:97–115.
29. Petkov V, Billinge SJL, Shastri SD, Himmel B. High-resolution atomic distribution functions of disordered materials by high-energy X-ray diffraction. J Non-Cryst Solids 2001;293–295:726–730.
30. Lovell R, Windle AH. Determination of the local conformation of PMMA from wide-angle X-ray scattering. Polymer 1981;22:175–184.

31. Li X, Murthy NS, Latour RA. Construction and validation of all-atom bulk-phase models of amorphous polymers using the TIGER2/TIGER3 empirical sampling method. Macromolecules 2011;44:5452–5464.
32. Habenschuss A, Tsige M, Curro JG, Grest GS, Nath SK. Structure of poly(dialkylsiloxane) melts: Comparisons of wide-angle X-ray scattering, molecular dynamics simulations, and integral equation theory. Macromolecules 2007;40:7036–7043.
33. Mitchell GR, Lovell R, Windle AH. The local structure of molten polyethylene. Polymer 1982;23:1273–1285.
34. Yoda O, Kuriyama I, Odajima A. Interchain ordering in amorphous solid polyethylene. Appl Phys Lett 1978;32:18–20.
35. Li X, Murthy NS, Latour RA. Structure of hydrated poly(D,L-lactic acid) studied with X-ray diffraction and molecular simulation methods. Macromolecules 2012;45:4896–4906.
36. Rosenfield H, Barton R Jr. Pair-density function of nano-scale morphology in oriented polymer fibers: Application to nomex aramid. In: Gilfrich JV, Cev Noyan I, Jenkins R, Huang TC, Snyder RL, Smith DK, Zaitz MA, Predecki PK, editors. *Advances in X-Ray Analysis*. Springer; 1998. p 523–533.
37. Chen W, Fu Y, Wunderlich B, Cheng J. The morphology of gel-spun polyethylene fibers, investigated by solid-state 13C NMR. J Polym Sci B 1994;32:2661–2666.
38. Fu Y, Chen W, Pyda M, Londono D, Annis B, Boller A, Habenschuss A, Cheng J, Wunderlich B. Structure-property analysis for gel-spun, ultrahigh molecular mass polyethylene fibers. J Macromol Sci Phys 1996;35:37–87.
39. Murthy NS, Correale ST, Minor H. Structure of the amorphous phase in crystallizable polymers: Polyethylene terephthalate. Macromolecules 1991;24:1185–1189.
40. Murthy NS, Correale ST, Moore RAF. Characterization of the amorphous phase in nylon 6 fibers by X-ray diffraction. J Appl Polym Sci 1991;47:185–197.
41. Murthy NS, Minor H, Bednarczyk C, Krimm S. Structure of the amorphous phase in oriented polymers. Macromolecules 1993;26:1712–1721.
42. Wunderlich B. Reversible crystallization and the rigid-amorphous phase in semicrystalline macromolecules. Prog Polym Sci 2003;28:383–450.
43. Hermans PH, Weidinger A. X-ray studies on the crystallinity of cellulose. J Polym Sci 1949;4:135–144.
44. Ruland W. X-ray determination of crystallinity and diffuse disorder scattering. Acta Crystallogr 1961;14:1180–1185.
45. Murthy NS, Minor H. General procedure for evaluating amorphous scattering and crystallinity from X-ray diffraction scans of semicrystalline polymers. Polymer 1990;31:996–1002.
46. Krimm S, Tobolsky AV. Quantitative X-ray studies of order in amorphous and crystalline polymers. Quantitative X-ray determination of crystallinity in polyethylene. J Polym Sci 1951;7:57–76.
47. Murthy NS, Minor H. Analysis of poorly crystallized polymers using resolution enhanced X-ray diffraction scans. Polymer 1995;36:2499–2504.
48. Murthy NS, Minor H, Akkapeddi MK, Buskirk BV. Characterization of polymer blends and alloys by constrained profile-analysis of X-ray diffraction scans. J Appl Polym Sci 1990;41:2265–2272.
49. Polizzi S, Fagherazzi G, Benedetti A, Battagliarin M, Asano T. A fitting method for the determination of crystallinity by means of X-ray diffraction. J Appl Crystallogr 1990;23:359–365.
50. Murthy NS, Zero K, Minor H. Resolution enhancement of polymer X-ray diffraction scans using maximum entropy methods: Poly(ethylene terephthalate). Macromolecules 1994;27:1484–1488.
51. Desper CR, Stein RS. Randomization of orientation of films and fibers. J Polym Sci B 1967;5:893–900.
52. Murthy NS, Barton R Jr. Polymer industry. In: Chung FH, Smith DK, editors. *Industrial Applications of X-Ray Diffraction*. New York: Marcel Dekker; 2000.
53. Trottier AM, Zwanziger JW, Murthy NS. Amorphous orientation and its relationship to processing conditions of blended polypropylene/polyethylene fibers. J Appl Polym Sci 2008;108:4047–4057.
54. Guinier A. *X-Ray Diffraction in Crystals, Imperfect Crystals and Amorphous Bodies*. San Francisco: Freeman and Co.; 1963.
55. Hosemann R, Bagchi S. *Direct Analysis of Diffraction by Matter*. Amsterdam: North-Holland Publishing Company; 1962.
56. Bonart R, Hosemann R, McCullough RL. The influence of particle size and distortions upon the X-ray diffraction patterns of polymers. Polymer 1963;4:199–211.
57. Warren BE, Averbach BL. The effect of cold-work distortion on X-ray patterns. J Appl Phys 1950;21:595–599.
58. Warren BE, Averbach BL. The separation of cold-work distortion and particle size broadening in X-ray patterns. J Appl Phys 1952;23:497.
59. Williamson GK, Hall WH. X-ray line broadening from filed aluminium and wolfram. Acta Metall 1953;1:22–31.
60. Ungár T, Revesz A, Borbély A. Dislocations and grain size in electrodeposited nanocrystalline Ni determined by the modified Williamson–Hall and Warren–Averbach procedures. J Appl Crystallogr 1998;31:554–558.
61. Murthy NS, Aharoni SM, Szollosi AB. Stability of the γ form and the development of the α form in nylon 6. J Polym Sci B 1985;23:2549–2565.
62. Murthy NS. Analysis of meridional X-ray diffraction pattern of the γ form of nylon 6 and comparison of paracrystalline and microstrain models of lattice disorder. J Polym Sci B 1986;24:549–561.
63. Roe RJ, Krigbaum WR. Description of crystallite orientation in polycrystalline materials having fiber texture. J Chem Phys 1964;40:2608.
64. Roe RJ, Krigbaum WR. Crystallite orientation in materials having fiber texture. II. A study of strained samples of crosslinked polyethylene. J Chem Phys 1964;41:737.
65. Roe RJ, Krigbaum WR. Description of crystallite orientation in polycrystalline materials. III. General solution to pole figure inversion. J Appl Phys 1965;36:2024–2031.
66. Butler JH, Wapp SM, Chambon FH. Quantitative pole figure analysis of oriented polyethylene films. Adv X-Ray Anal 2000;43:141–150.
67. Murthy NS, Zero K. Amorphous orientation in polymers determined using two-dimensional X-ray diffraction data and its significance. In: *ANTEC 1999*. Society of Plastic Engineers; 1999.

68. Murthy NS, Bray RG, Correale ST, Moore RAF. Drawing and annealing of nylon-6 fibres: Studies of crystal growth, orientation of amorphous and crystalline domains and their influence on properties. Polymer 1995;36:3863–3873.
69. White JL, Spruiell JE. Specification of biaxial orientation in amorphous and crystalline polymers. Polym Eng Sci 1981;21:859–868.
70. Fu Y, Busing WR, Jin Y, Affholter KA, Wunderlich B. Structure analysis of the noncrystalline material in poly(ethylene terephthalate) fibers. Macromol Chem Phys 1994;195:803–822.
71. Biangardi HJ. Determination of the orientation distribution function of amorphous polymers by wide angle X-ray scattering measurement. Makromol Chem 1982;183:1785–1802.
72. Harget PJ, Oswald HJ. Birefringence in amorphous poly(ethylene terephthalate) fibers. J Polym Sci B 1978;17: 531–534.
73. Murthy NS, Zero K. Full-pattern parameterization of two-dimensional wide-angle diffraction data from oriented polymers. Polymer 1997;38:2277–2280.
74. Murthy NS. Small-angle scattering. In: Salem DR, editor. *Structure Formation in Polymeric Fibers*. Munich: Hanser Gardner Publications; 2001.
75. Castillo RV, Müller AJ. Crystallization and morphology of biodegradable or biostable single and double crystalline block copolymers. Prog Polym Sci 2009;34:516–560.
76. Lamba NMK, Woodhouse KA, Cooper SL. *Polyurethanes in Biomedical Applications*. CRC Press; 1998.
77. Lee D, Seung-Heon L, Sangcheol K, Kookheon C, Jae HP, Bae YH. Micro-phase-separation behavior of amphiphilic polyurethanes involving poly(ethylene oxide) and poly(tetramethylene oxide). J Polym Sci B 2003;41:2365–2374.
78. da Silva GR, da Cunha AS Jr, Ayres E, Oréfice RL. Effect of the macromolecular architecture of biodegradable polyurethanes on the controlled delivery of ocular drugs. J Mater Sci 2009;20:481–487.
79. Grubb DT, Murthy NS. Real-time X-ray study of nylon-6 fibers during dehydration: Equatorial small-angle scattering is due to surface refraction. Macromolecules 2009;43:1016–1027.
80. Murthy NS, Bednarczyk C, Moore RAF, Grubb DT. Analysis of small-angle X-ray scattering from fibers: Structural changes in nylon 6 upon drawing and annealing. J Polym Sci B 1996;34:821–835.
81. Murthy NS, Grubb DT. Study of the deformation in lamellar and crystalline structures: In-situ simultaneous SAXS and WAXD measurement on poly(ethylene terephthalate) fibers. J Polym Sci B 2003;41:1538–1553.
82. Murthy NS, Reimschuessel AC, Kramer V. Changes in void content and free volume in fibers during heat setting and their influence on dye diffusion and mechanical properties. J Appl Polym Sci 1990;40:249–262.
83. Keller A. Crystalline polymers: An introduction. Faraday Discuss Chem Soc 1979;68:145–166.
84. Keller A. Morphology of crystalline polymers. Makromol Chem 1959;34:1–28.
85. Hearle JWS. Fine structure of fibers and crystalline polymers. I. Fringed fibril structure. J Appl Polym Sci 1963;7:1175–1192.
86. Bensason S, Stepanov EV, Chum S, Hiltner A, Baer E. Deformation of elastomeric ethylene-octene copolymers. Macromolecules 1997;30:2436–2444.
87. Li L, de Jeu WH. Shear-induced smectic ordering as a precursor of crystallization in isotactic polypropylene. Macromolecules 2003;36:4862–4867.
88. Bensason S, Minick J, Moet A, Chum S, Hiltner A, Baer E. Classification of homogeneous ethylene-octene copolymers based on comonomer content. J Polym Sci B 1996;34:1301–1315.
89. Wang W, Murthy NS, Grubb DT. Central small-angle diffuse scattering from fibers is made of two components. J Polym Sci B 2012;50:797–804.
90. Hermans PH, Heikens D, Weidinger A. A quantitative investigation on the X-ray small angle scattering of cellulose fibers. Part II. The scattering power of various cellulose fibers. J Polym Sci A 1959;35:145–165.
91. Kaburagi M, Bin Y, Zhu D, Xu C, Matsuo M. Small angle X-ray scattering from voids within fibers during the stabilization and carbonization stages. Carbon 2003;41:915–926.
92. Thünemann AF, Ruland W. Microvoids in polyacrylonitrile fibers: A small-angle X-ray scattering study. Macromolecules 2000;33:1848–1852.
93. Wang W, Murthy NS, Chae HG, Kumar S. Small-angle X-ray scattering investigation of carbon nanotube-reinforced polyacrylonitrile fibers during deformation. J Polym Sci B 2009;47:2394–2409.
94. Grubb DT, Prasad K. High-modulus polyethylene fiber structure as shown by X-ray diffraction. Macromolecules 1992;25:4575–4582.
95. Grubb DT, Prasad K, Adams W. Small-angle X-ray diffraction of Kevlar using synchrotron radiation. Polymer 1991;32:1167–1172.
96. Murthy NS, Orts WJ. Hydration in semicrystalline polymers: Small-angle neutron scattering studies of the effect of drawing in nylon 6 fibers. J Polym Sci B 1994;32:2695–2703.
97. Astley OM, Donald AM. A small-angle X-ray scattering study of the effect of hydration on the microstructure of flax fibers. Biomacromolecules 2001;2:672–680.
98. Murthy NS, Grubb DT, Zero K, Nelson CJ, Chen G. Lamellar structure and properties in poly(ethylene terephthalate) fibers. J Appl Polym Sci 1998;70:2527–2538.
99. Murthy NS, Minor H, Latif RA. Effect of annealing on the structure and morphology of nylon 6 fibers. J Macromol Sci B 1987;26:427–446.
100. Murthy NS, Grubb DT, Zero K. Structural implications of the elliptical form of small-angle reflections in oriented semicrystalline polymers. Macromolecules 2000;33:1012–1021.
101. Wang W, Murthy NS, Grubb DT. 'Butterfly' small-angle X-ray scattering patterns in semicrystalline polymers are double-elliptical. Polymer 2007;48:3393–3399.
102. Brandt M, Ruland W. SAXS studies on the deformation of macrolattices in block copolymers. Acta Polym 1996;47:498–506.
103. Murthy NS, Grubb DT. Tilted lamellae in an affinely deformed macrolattice and elliptical features in small-angle scattering. J Polym Sci B 2006;44:1277–1286.
104. Murthy NS, Zero K, Grubb DT. Full-pattern analysis of two-dimensional small-angle scattering data from oriented polymers using elliptical coordinates. Polymer 1997;38:1021–1028.

105. Luk A, Murthy NS, Wang W, Rojas R, Kohn J. Study of nanoscale structures in hydrated biomaterials using small-angle neutron scattering. Acta Biomater 2012;8:1459–1468.
106. Murthy NS, Wang W, Kohn J. Microphase separation in copolymers of hydrophilic PEG blocks and hydrophobic tyrosine-derived segments using simultaneous SAXS/WAXS. Polymer 2010;51:3978–3988.
107. Chu B, Hsiao BS. Small-angle X-ray scattering of polymers. In: Michl J, editor. *Chemical Reviews*. American Chemical Society; 2001. p 1727–1761.
108. Riekel C, Davies RJ. Applications of synchrotron radiation micro-focus techniques to the study of polymer and biopolymer fibers. Curr Opin Colloid Interface Sci 2005;9:396–403.
109. Uelcer Y, Cakmak M. Effect of annealing on the structure of injection-molded PEN. J Appl Polym Sci 1998;62:1661–1678.
110. Riekel C, Dieing T, Engström P, Vincze L, Martin C, Mahendrasingam A. X-ray microdiffraction study of chain orientation in poly(*p*-phenylene terephthalamide). Macromolecules 1999;32:7859–7865.
111. Murthy NS, Kagan VA, Bray RG. Effect of melt temperature and skin-core morphology on the mechanical performance of nylon 6. Polym Eng Sci 2002;42:940–950.
112. Gazzano M, Focarete ML, Riekel C, Ripamonti A, Scandola M. Structural investigation of poly(3-hydroxybutyrate) spherulites by microfocus X-ray diffraction. Macromol Chem Phys 2001;202:1405–1409.
113. Mahendrasingam A, Martin C, Fuller W, Blundell D, MacKerron D, Rule R, Oldman R, Liggat J, Riekel C, Engstrom P. Microfocus X-ray diffraction of spherulites of poly-3-hydroxybutyrate. J Synchrotron Radiat 1995;2: 308–312.
114. Muller M, Czihak C, Burghammer M, Riekel C. Combined X-ray microbeam small-angle scattering and fibre diffraction experiments on single native cellulose fibres. J Appl Crystallogr 2000;33:817–819.
115. Schoeck J, Davies R, Martel A, Riekel C. Na-cellulose formation in a single cotton fiber studied by synchrotron radiation microdiffraction. Biomacromolecules 2007;8: 602–610.
116. Assouline E, Wachtel E, Grigull S, Lustiger A, Wagner H, Marom G. Lamellar twisting in α isotactic polypropylene transcrystallinity investigated by synchrotron microbeam X-ray diffraction. Polymer 2001;42:6231–6237.
117. Wenig W, Herzog F. Injection molded polypropylene: X-ray investigation of the skin-core morphology. J Appl Polym Sci 1993;50:2163–2171.
118. Zhu X, Ballard B, Predecki P. Determination of *Z*-profiles of diffraction data from profiles using a numerical linear inversion method. Adv X-Ray Anal 1995;38:255–262.
119. Lu JR, Lee EM, Thomas RK. The analysis and interpretation of neutron and X-ray specular reflection. Acta Crystallogr A: Found Crystallogr 1996;52:11–41.
120. Parratt LG. Surface studies of solids by total reflection of X-rays. Phys Rev 1954;95:359.
121. Murthy NS, Bednarczyk C, Minor H. Depth-profiles of structure in single-and multilayered commercial polymer films using grazing-incidence X-ray diffraction. Polymer 2000;41:277–284.
122. Gutmann JS, Müller-Buschbaum P, Wolkenhauer M, Hermsdorf N, Stamm M. X-ray and neutron grazing incidence scattering study of a ternary ultra-thin polymer blend film. J Macromol Sci B 2005;43:207–217.
123. Factor BJ, Russell TP, Toney MF. Grazing incidence X-ray scattering studies of thin films of an aromatic polyimide. Macromolecules 1993;26:2847–2859.
124. Kuhl TL, Majewski J, Howes PB, Kjaer K, von Nahmen A, Lee KYC, Ocko B, Israelachvili JN, Smith GS. Packing stress relaxation in polymer-lipid monolayers at the air–water interface: An X-ray grazing-incidence diffraction and reflectivity study. J Am Chem Soc 1999;121:7682–7688.

3

ELECTRON MICROSCOPY OF POLYMERS

GOERG H. MICHLER AND WERNER LEBEK

Institute of Physics and Institute of Polymeric Materials, Martin Luther University Halle-Wittenberg, Halle, Germany

3.1 INTRODUCTION

The structures and morphologies of polymers have been under investigation by means of electron microscopy for more than 60 years. With developing microscopic techniques, all structural details from macroscopic sizes down to about 0.1 nm ($=10^{-10}$ m) can be revealed (see Table 3.1). Alternative techniques are scattering methods (SALS – small-angle light scattering; SAXS, WAXS – small- and wide-angle X-ray scattering, SANS – small-angle neutron scattering), because scattering and microscopic techniques often yield additional results.

Recent advances in techniques used, in particular availability of high-resolution transmission electron microscopes, as well as interpreting and processing the images, have allowed resolutions of the order of atomic structures in inorganic crystal (below 0.1 nm). However, the best resolution achieved in polymers is, in practice, poorer than this because of polymer-specific problems with high electron irradiation sensitivity and low contrast.

Electron microscopy (EM) can be divided into the techniques of *transmission electron microscopy* (*TEM*) and *scanning electron microscopy* (*SEM*). A step in study structures of materials came with the development of *scanning probe microscopy*, in particular with a modified, for polymers interesting technique of *atomic force microscopy* (*AFM*). In general, all of the different types of microscopes can be classified according to whether imaging is achieved by irradiation of the object with a "lamp" or to feeling the surface with a "finger" or "needle" (see Fig. 3.1) [1, 2]:

Type 1 – Transmission: A fixed beam of light or electrons is transmitted through the thin specimen in the transmission mode of the optical microscope and in transmission electron microscopes; in this case, thin (optical microscopy, OM) and ultrathin samples (EM) must be used.

Type 2 – Reflection: A stationary beam is reflected off the specimen surface in the reflection mode of the optical microscopes or – for inorganic material only – in electron mirror microscopes; here, bulk samples can be used.

Type 3 – Scanning beam: A focused beam (laser light or electron beam) is scanned across the specimen, resulting in a reflected beam from the surface (as in confocal laser scanning microscopy) or in secondary or backscattered electrons (BSE; in scanning electron microscopes).

Type 4 – A focused scanning beam is passed through the thin specimen (scanning transmission electron microscopes, STEMs).

Type 5 – Scanning tip: A mechanical tip is scanned across the specimen in order to make use of different physical properties in AFMs (or tunneling microscopes for conductive samples).

Many reviews discuss the details of optical [3, 4], electron microscopy [1, 5–10], and AFM [1, 11, 12].

Here, the techniques of SEM and TEM are discussed with particular consideration of contrast development and preparation of polymeric materials. Representative examples are shown, which are discussed in more detail in Refs [1, 2].

3.2 MICROSCOPIC TECHNIQUES

3.2.1 Scanning Electron Microscopy (SEM)

SEM is currently the most popular of the microscopic techniques, due to the user-friendliness of the apparatus, the ease

Polymer Morphology: Principles, Characterization, and Processing, First Edition. Edited by Qipeng Guo.

TABLE 3.1 Sizes of Structural Details of Polymers, Resolutions, and Magnifications Attainable with the Different Microscopic Techniques and Scattering Methods

Structure:	*macroscopic*	*mesoscopic* phases, spherlites, domains...,	*microscopic* crystal lattice, molecules...	*atomar*

Size	1 mm	100 µm	10 µm	1 µm	100 nm	10 nm	1 nm	0.1 nm

Technique

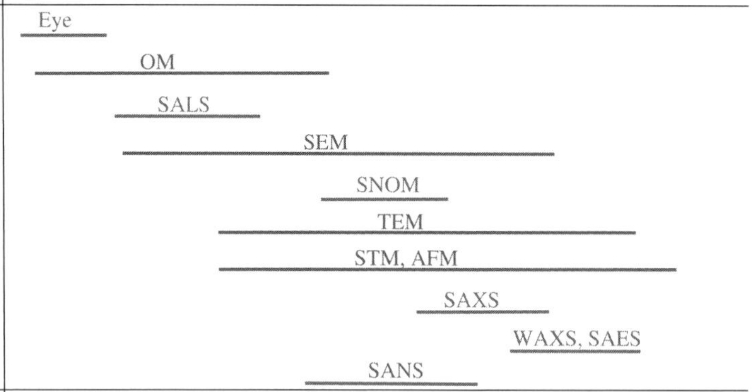

Magnification	1×	10×	100×	1,000×	100,000×	10,000,000×

OM, optical microscopy; SALS, small-angle light scattering; SEM, scanning electron microscopy; SNOM, scanning near-field optical microscopy; TEM, transmission electron microscopy; STM, scanning tunneling microscopy; AFM, atomic force microscopy; SAXS, small-angle X-ray scattering; WAXS, wide-angle X-ray scattering; SAES, small angle electron scattering.

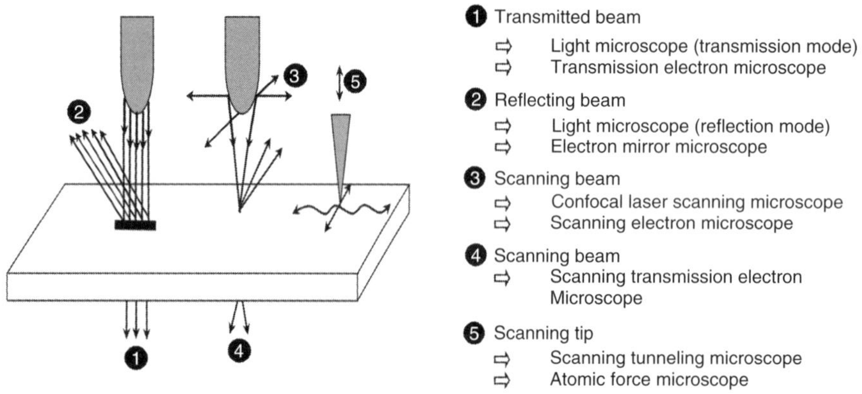

Figure 3.1 Schematic representation of the principles of different types of microscopes [1, 2].

of specimen preparation, and the general simplicity of image interpretation. The obvious limitation is that only surface features are easily accessible.

In a SEM, a focused electron beam (energies between 1 and 50 keV) scans line by line over the specimen surface in the evacuated microscope column and forms signals based on the interactions between the beam and the sample, which are electronically detected and amplified (Fig. 3.2). This response signal is displayed as a brightness modulation on a cathode ray tube (display) or, nowadays, by digital computer techniques. As the area of the displayed image remains unchanged, the magnification of the image is determined by the dimension of the scanned sample area (i.e., ratio of the lateral length of the image displayed to that of the scanned area) and can reach several 100,000 times.

The following interaction types between the electron beam and the specimen surface may occur (Fig. 3.3):

- Primary beam electrons interact with electrons within the specimen atoms, knocking them free ("*secondary electrons*," SE, low energy of usually 0–200 eV).
- Electrons are "backscattered" as a consequence of the electrostatic attraction with the positively charged atomic nucleus within the specimen ("*backscattered electrons*," BSE).
- After a secondary electron has been removed from an inner shell of an atom, an electron from an outer shell (from a less tightly bound state) falls into the inner shell, with the emission of an X-ray photon ($h{\cdot}\upsilon$) ("*characteristic X-rays*," because the energetic jump of

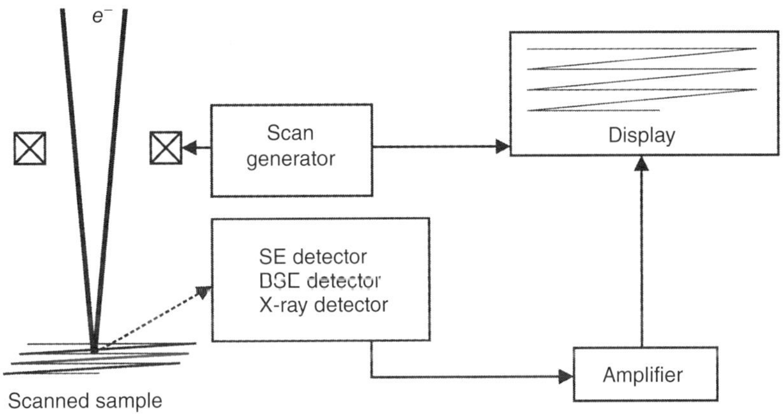

Figure 3.2 Scheme showing the principle of SEM.

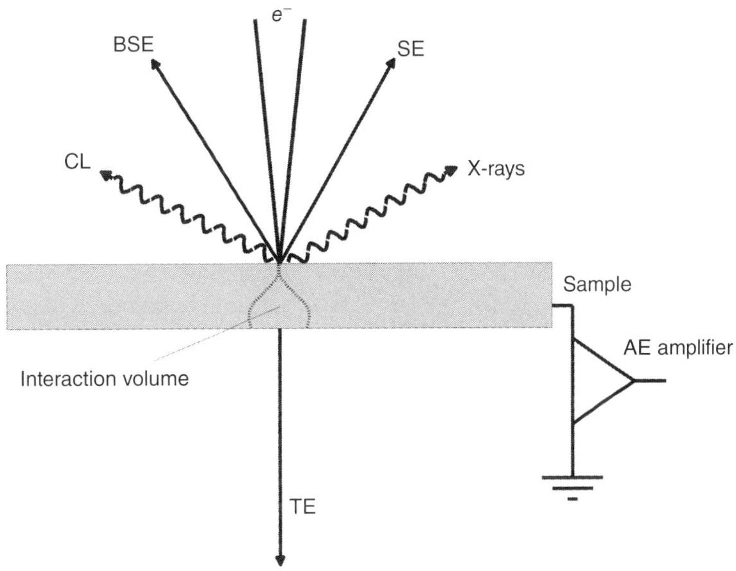

Figure 3.3 Elastic and inelastic interactions between primary electrons and the sample.

the falling electron has a characteristic X-ray emission energy for each atom).

- The remaining charges in the specimen, called absorbed electrons (AE) and – in the case of thin samples – the transmitted electrons (TE) are further signals resulting from the interaction process.
- At least some samples can be excited to cathodoluminescence (CL), and also potential differences and differences in electronic states can be detected in SEM.

The number of secondary electrons produced per primary electron (PE) depends on the atomic number of the specimen and on the angle between the primary beam and the surface (Fig. 3.4). If the incident electron beam impinges on a flat site or on a cavity of the sample, only few secondary electrons reach the detector (registering a dark region on the image). If the primary beam impinges at the side of a step or on an asperity, a large number of secondary electrons can leave the sample surface (producing a bright area in the image). This so-called edge effect gives an excellent contrast of the surface topography. Figure 3.5 shows as an example of a defect-initiated secondary crack in a PBTP/PC blend in overview (a) and in larger magnification of the crack initiation at a morphological defect (b).

The yield of reflected BSEs increases with increasing atomic number. Elements having higher atomic number offer a higher probability of backscattered elements, including an additional atomic number contrast ("*material contrast*") to the surface topography contrast (Fig. 3.6).

A SEM is usually equipped with an X-ray detector (either *energy dispersive*, EDX, or *wavelength dispersive*, WDX). Thus, the characteristic signal X-ray emission of a particular element can be displayed on the imaging screen and the distribution of different elements on a surface can be readily obtained.

Besides the integral analysis, the SEM provides the ability to determine the elemental distribution (elemental mapping) over a sample surface. To do this, in an EDX system, one or

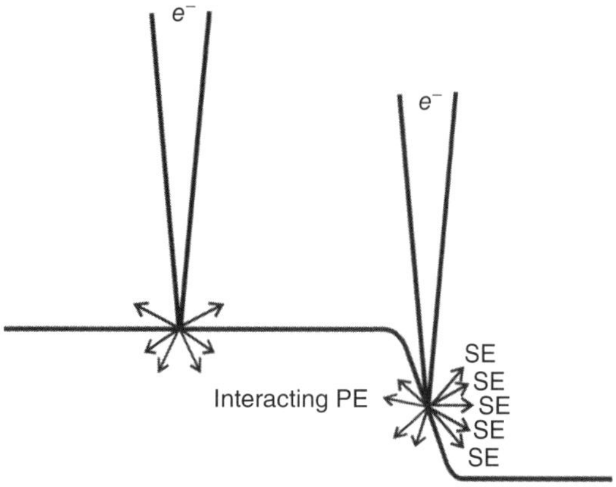

Figure 3.4 Contrast formation in secondary electron mode due to surface topography (edge effect):– At a smooth surface, electrons are collected by the specimen and fail to reach the detector (dark region in the image). – At an edge, electrons are easily emitted and are drawn toward the positively charged detector (bright region in the image, located to the right of the diagram).

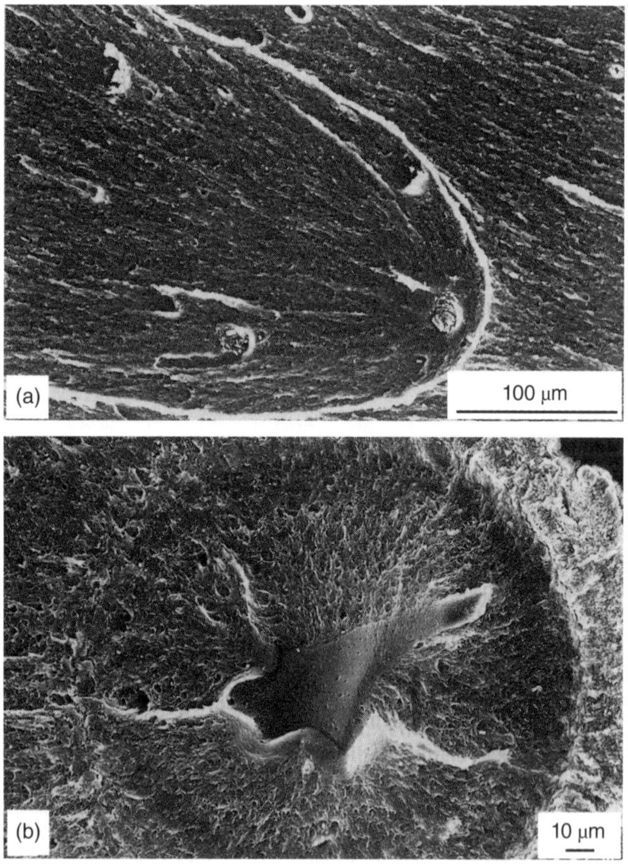

Figure 3.5 Secondary crack in a PBTP/PC blend (from [2]): (a) formation of a parabola-like pattern due to superposition of the main crack front (from right) with a secondary crack and (b) origin of the secondary crack.

more peaks from an interesting element are selected from the spectrum by so-called regions of interest. While scanning the electron beam, it is possible to record the X-ray counts of these regions of interest correlating to points on the specimen surface. By accumulating X-ray counts per pixel, it is possible to create an elemental distribution image – X-ray mapping, see Figure 3.7.

If bulk material with a multiphase morphology is studied, three-dimensional information about the structure can be gained by simply varying the electron energy in the SEM in combination with image processing [13].

There is a major problem in the observation of polymers by means of SEM. Polymers are poor conductors of electricity and a charge rapidly builds up on the specimen when it is bombarded with the electron beam. A conductive coating must be applied to solve this problem in a conventional SEM.

A *low-vacuum SEM* or *environmental scanning electron microscopy* (abbreviated as ESEM) is a modified SEM with a lower vacuum of about 1 mbar in the specimen chamber, which is achieved using a special pumping system. The term "environmental" refers to the possibility of imaging wet samples such as biological materials when using the SEM ("ESEM" is a trademark of the company FEI). In a usual SEM, incident primary electrons cause negative charging of the surfaces of electrically nonconducting samples. Under low-vacuum conditions, a part of the residual gas molecules are ionized by the primary electron beam. The positive ions near the surface are attracted by the negatively charged sample surface, so that the surface is compensated for by the impacting ions. In addition, the patented gaseous secondary electron detector (GSED) of the ESEM generates further gas ions and accelerates it into the surface of the specimen. Therefore, under low-vacuum conditions, the surface charges can be ideally so compensated that there is no need for a conductive coating. In addition, wet samples (i.e., in biological, medical applications) must not be dried before investigation and they are not dried during the investigation, thus avoiding artefacts. Hence, this technique offers new applications and advantages over the conventional SEM. The principles of these techniques with specific applications to polymers are described in detail in Ref. [1].

The relatively easy preparation of polymer surfaces makes SEM a very valuable tool to study larger and even smaller polymer structures. Often, fracture surfaces of polymers give information on large structural details and structural defects, which are the source of damage and fracture of polymeric materials (compare Fig. 3.5). After selective etching of a smooth surface, many structures are visible in SEM. A well-developed spherulite with a central part of parallel-packed lamellae and bended lamellae around the so-called eyes of a spherulite (left and right of the central stack of lamellae) is shown in Figure 3.8.

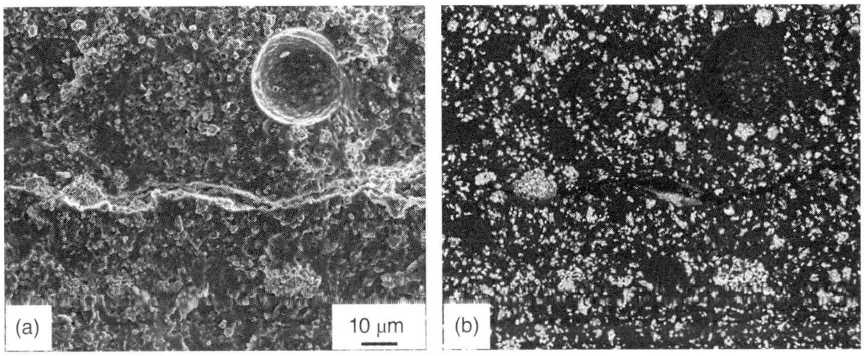

Figure 3.6 Comparison of secondary electron (SE, a) and backscattered electron (BSE, b) imaging of a polymeric concrete sample consisting of epoxy, quartz and calcite.

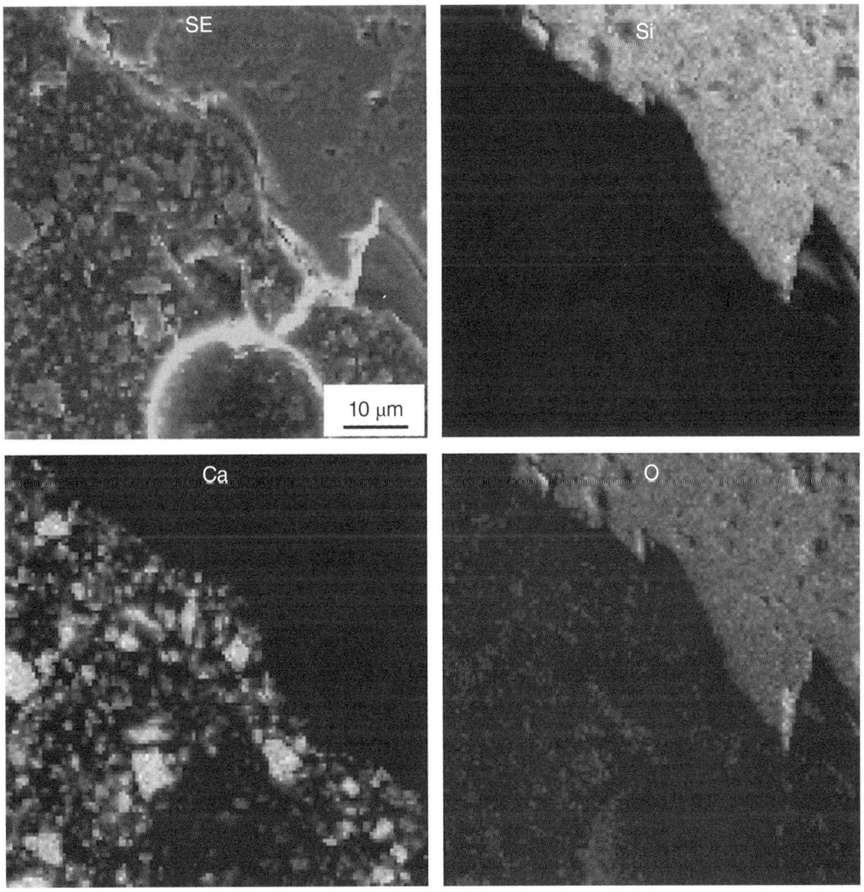

Figure 3.7 Secondary electron (SE) imaging and X-ray mapping of the elements silicon (Si), calcium (Ca), and oxygen (O) of polymeric concrete sample consisting of epoxy, quartz (SiO_2), calcite ($CaCO_3$).

Additional details of the structure of semicrystalline polymers can be determined on such selectively etched surfaces. SEM micrographs in Figure 3.9 show the two main types of isotactic polypropylene (PP): Micrograph (a) shows the most common α-modification with the so-called cross-hatched arrangement of the crystalline lamellae; the main lamellae are growing radially from an initial site (center of the spherulite), whereas the "secondary lamellae" are formed by an epitaxial growth onto them [14]. In β-modified PP, the lamellae form a more sheaf-like superstructure with a

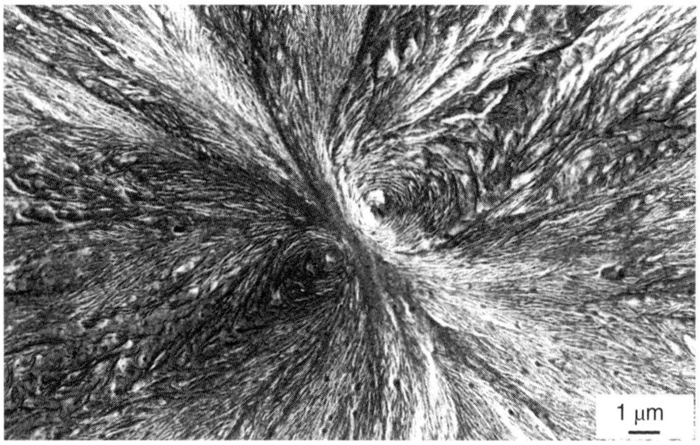

Figure 3.8 Structure of a spherulite in HDPE (high-density polyethylene) with radial oriented and branched lamellae in a SEM micrograph of a selectively etched surface (from [2]).

parallel arrangement of bundles of lamellae, as shown in micrograph (b).

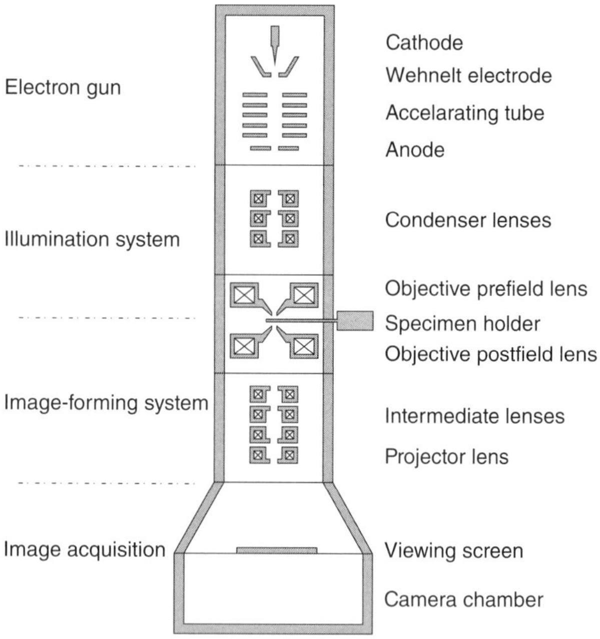

Figure 3.10 Basic components of a TEM.

3.2.2 Transmission Electron Microscopy (TEM)

There are two types of TEM, the conventional transmission electron microscopy (CTEM, type 1 in Fig. 3.1) and the scanning TEM (STEM type 4 in Fig. 3.1). The CTEM is similar to the light microscope operated in transmission with an electron beam instead of light. A major difference follows from the ready interaction of electrons with matter, including gas, so that the path from source to detector must remain in vacuum. The principle and basic components of a TEM are illustrated in Figure 3.10. Electrons emitted from a thermionic or a field-emission source are accelerated in the gun by a high voltage produced via a high-voltage generator. The electron beam is formed with the aid of condenser lenses, a condenser lens aperture, a condenser lens stigmator and beam tilt and translation coils for alignment, and then it enters the objective lens and strikes the specimen. The specimen is placed via the specimen holder in the goniometer stage within the objective lens between the upper and the lower pole pieces. After passing through the specimen, the electrons form an image through the action of the objective lens and an objective aperture in the back focal plan of the lens. The image is corrected by an objective stigmator and enlarged by an image-forming system consisting of a series of intermediate and projector lenses and alignment units. Finally, a highly magnified image becomes visible on the screen or it can be recorded by a camera or an electron-sensitive film.

Local elemental distributions in a specimen can be investigated in TEM and STEM using the analytical techniques electron energy-loss spectroscopy (EELS) and X-ray microanalysis. X-ray microanalysis in TEM and STEM corresponds to EDX spectroscopy in SEM. However, the thin

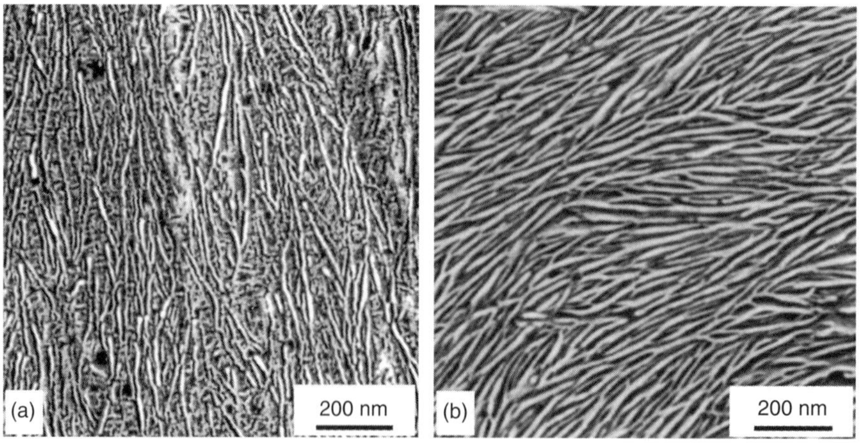

Figure 3.9 SEM micrographs of PP surfaces after permanganic etching, showing (a) the typical morphology of α – iPP with the typical cross-hatched arrangement of the lamellae and (b) the typical morphology of β – iPP with the sheaf-like lamellar arrangement.

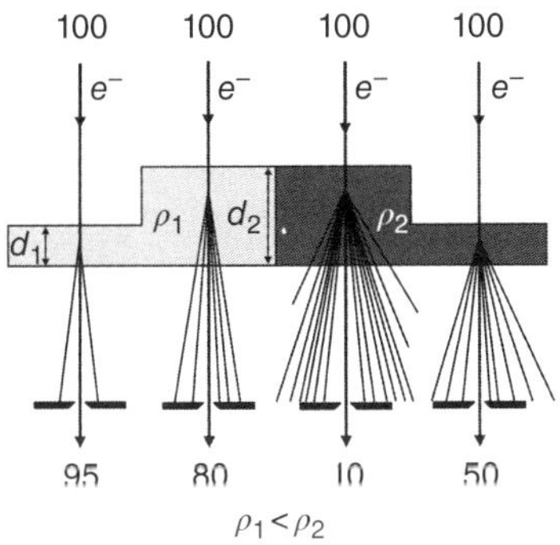

Figure 3.11 Illustration of the mass-thickness contrast in TEM: The influence of locally increased specimen thickness and density on scattering of electrons is shown; thicker specimen and more density material scatter more electrons, resulting in darker area on the micrograph.

samples of light element polymers in TEM only emit a few X-ray quanta, therefore, X-ray microanalysis of polymers is usually used in conjunction with SEM investigations that generally yield better signal-to-noise ratios.

In practice, most TEM investigations of polymers make use of mass-thickness contrast, that is, specimen parts containing elements with a higher atomic number and/or which are thicker scatter electrons stronger, but not contributing to image formation. Therefore, such parts appear darker than the surroundings (see Fig. 3.11).

To increase the density of particular structures in polymers, usually local chemical staining is applied. As an example, Figure 3.12 shows TEM micrographs of thin sections of PP. The amorphous parts and particularly the boundaries of the lamellae are chemically stained, appearing dark, whereas the crystalline lamellae are bright. Comparison of Figure 3.12 of TEM micrographs with SEM micrographs in Figure 3.9 reveals the differences in the two techniques.

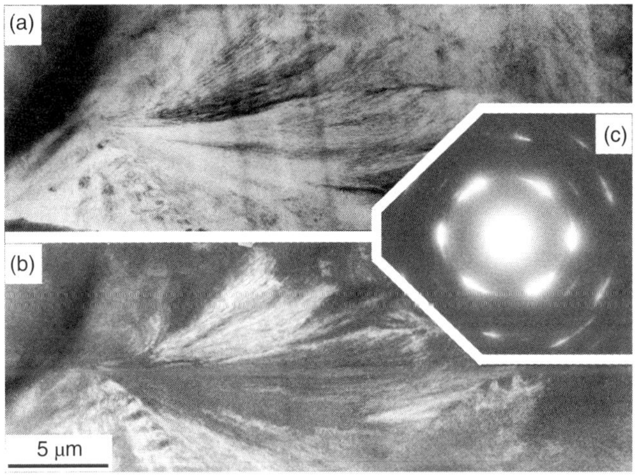

Figure 3.13 Images of a sheaflike structure in LDPE obtained by different operating modes of the TEM, cryo-thin section in 1000 kV HEM (a) bright/field image, (b) dark-field image, and (c) electron diffraction diagram.

Crystalline samples can be studied using diffraction contrast by diffraction patterns and dark field and bright field contrast. An example of these different types of contrast is given in Figure 3.13 with a sheaflike morphology of a low-density polyethylene (LDPE) material. Precondition to realize such pictures is working at low electron beam intensities and using high-sensitive registration procedure ("low-dose" technique).[1]

A crystalline dark-field image is formed if only diffracted electrons are allowed to contribute to image formation and the crystalline regions appear as bright objects on a dark background. Dark-field imaging can be used to suppress sample artefacts (e.g., cutting striations, scratches, shatter marks) because mass-thickness effects do not contribute to the image contrast (see Fig. 3.14).

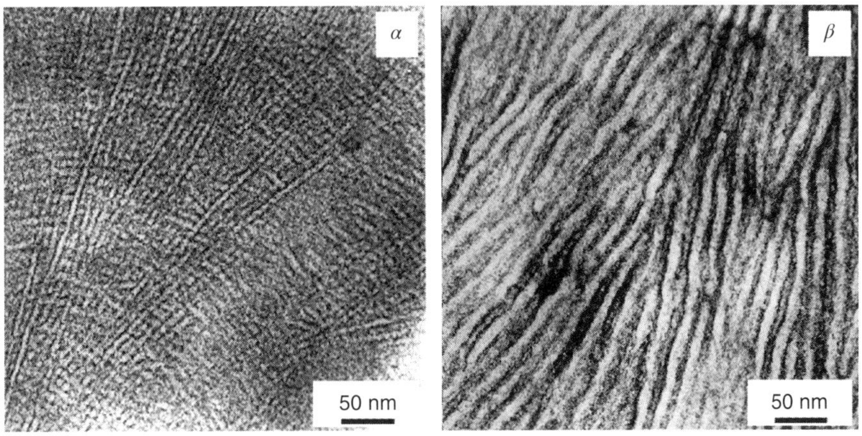

(a) (b)

Figure 3.12 Mass-thickness contrast in TEM micrographs of isotactic PP ultrathin sections due to selectively chemical staining of the amorphous regions and the boundaries between the lamellae; (a) α – iPP with the typical cross-hatched structure of lamellae and (b) β – iPP with a bundle of parallel lamellae (compare also with Fig. 3.9).

[1] In Ref. [1, Chapter 8, pp. 175–183].

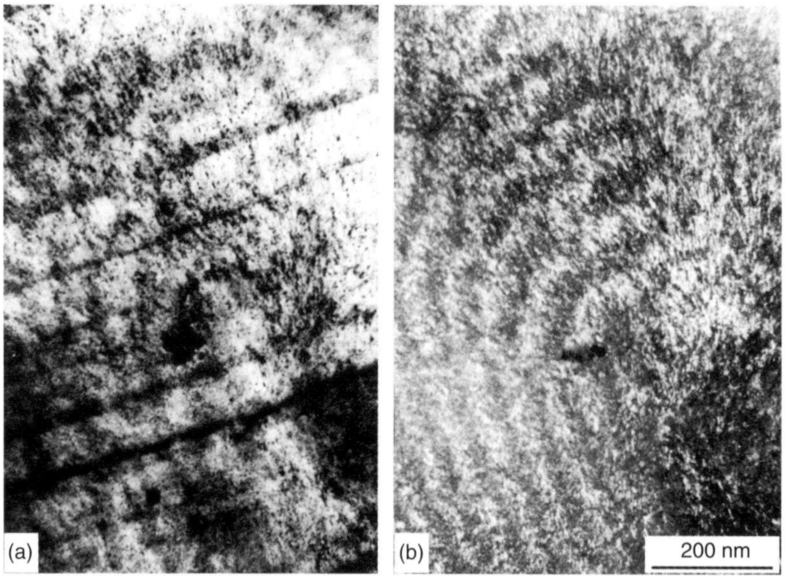

Figure 3.14 TEM micrographs of the same area of a HDPE with a banded spherulite in HEM [15] (a) bright-field image with visible cutting artefacts and (b) dark-field image which do not show these cutting artefacts.

The transmission electron microscope provides detailed structural information at levels down to atomic dimensions. However, such high-resolution examination is seldom possible in case of polymers. Nevertheless, it is possible to obtain information within the range 1–10 nm with varying degrees of difficulty. This range is beyond that of optical microscopy, and the TEM can provide information that can rarely be obtained by any other means.

Generally, polymeric materials are composed of only low-atomic-number elements; therefore, significant contrast due to variations of the local density in the specimen cannot be expected. Usually, a chemical staining is used, that is, treatment with heavy metal compounds, for example, OsO_4 or RuO_4, and deposition in different regions of the polymer (e.g., amorphous, crystalline) (see Fig. 3.12).

The second and main disadvantage of the TEM is that it can be used only on thin samples, less than 1 μm thick and preferably 50–100 nm thick. These ultrathin samples can be cut from bulk material using (cryo-)ultramicrotomy with special glass or diamond knives.

A third disadvantage of studying polymers with electron microscopy is the radiation damage due to the electron beam. There are several primary and secondary irradiation effects, which can, on one side, damage the polymer, but on the other hand, contribute to a contrast enhancement, for example, if in polymer blends the thickness or the density in one part of the specimen is reduced by evaporation of volatile fractions of polymer chains, as in PVC/SAN blends (see Fig. 3.15) or if secondary cross-linking effects in semicrystalline polymers are stronger in the amorphous regions than in the crystalline ones, as in PEs [1, 15, 16].

The accelerating voltage of a TEM is usually in the range between 100 and 300 kV. High-voltage electron microscopes (HEM) work at a voltage of 1 MV or even higher. In a HEM, the reduction of the electron wavelength can generally improve resolution, but – and this is more important for materials science – reduces the scattering cross section in the specimen, improving the penetration power of the imaging electron beam. In addition, the lifetime of the macromolecular structure and particularly the crystallinity in semicrystalline polymers is increased as compared to 100 kV TEM [17]. The main goal of employing HEM at a voltage of 1 MV or even higher for polymers is the possibility of using specimen thicknesses that are about 3–10 times larger than possible with a conventional TEM [7, 15, 18].

On the other hand, the image contrast decreases with increasing energy of the electrons. Unlike the conventional voltage values around 100 kV, a voltage of the order of units of kV is exploited in low-voltage transmission electron microscopy (LVTEM) [19]. At those voltages, the instrument is able to provide an enhanced imaging contrast nearly 20 times higher than for 100 kV, but with a much lower penetration depth (much lower sample thickness necessary) and resolution.

Analysis of the energy distribution of electrons that have passed through the specimen (EELS) is used in an energy-filtered transmission electron microscopy (EFTEM). EFTEM offers a powerful tool for the chemical analysis of materials at the nanometer scale. On the other hand, the mode of zero-loss filtering, where only the unscattered and elastically scattered electrons contribute to the image, enhance the contrast and improve the resolution by avoiding

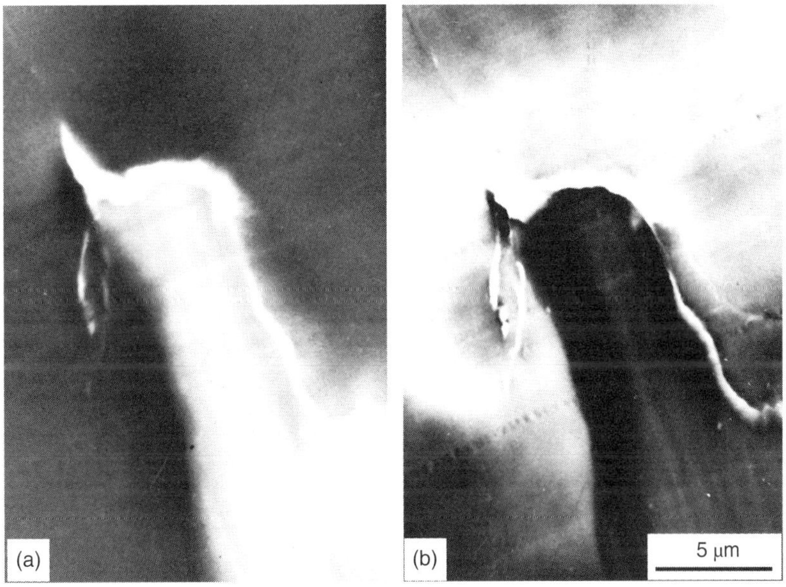

Figure 3.15 Differentiation of the two polymer phases in a PVC/SAN blend due to different densities and mass loss after electron irradiation (semi-thin section, 1000 kV HEM, from [15]) (a) at the start of irradiation: PVC darker than SAN and (b) after intense irradiation: PVC brighter than SAN.

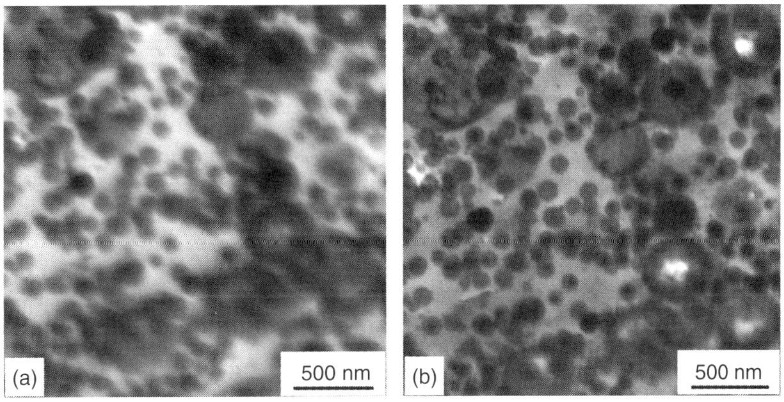

Figure 3.16 TEM images of a stained 400-nm-thick ABS section produced by (a) global mode and (b) zero-loss filtering [5].

chromatic aberration even in thicker specimens. This is demonstrated in Figure 3.16 by a stained 400-nm-thick ABS (acrylonitrile-butadiene-styrene) polymer semi-thin section imaged in an EFTEM with Ω-filter used in the global mode (a) and by zero-loss filtering (b).

3.2.3 Comparison of Different Microscopic Techniques

With the different techniques, the surface of bulk polymeric materials or its interior is the target of the microscopic investigations (see Fig. 3.17 at the top). The surface can be studied directly with SEM or ESEM, AFM and, indirectly after replication, with TEM. Ultra- and semi-thin sections from the interior can be used for TEM and thicker sections for SEM or AFM. To compare imaging using TEM, SEM, and AFM, Figure 3.17 compares the morphology of a commercial iPP in the α-form with the so-called cross-hatched arrangement consisting of primary lamellae radiating from a spherulite center and smaller secondary (or "daughter") lamellae perpendicular to the primary ones (at the bottom, compare also with Figs. 3.9 and 3.12). Several preparation techniques, necessary for the different microscopic techniques, are mentioned in Section 3.3.

In addition to the study of the structure of materials, the influence of several parameters on morphology can be studied, in particular the influence of mechanical loading. There are several methods to investigate deformation and fracture processes, which are discussed in detail in [1, 2, 5]. The direct imaging techniques of microscopy allow very direct determination of the morphology and

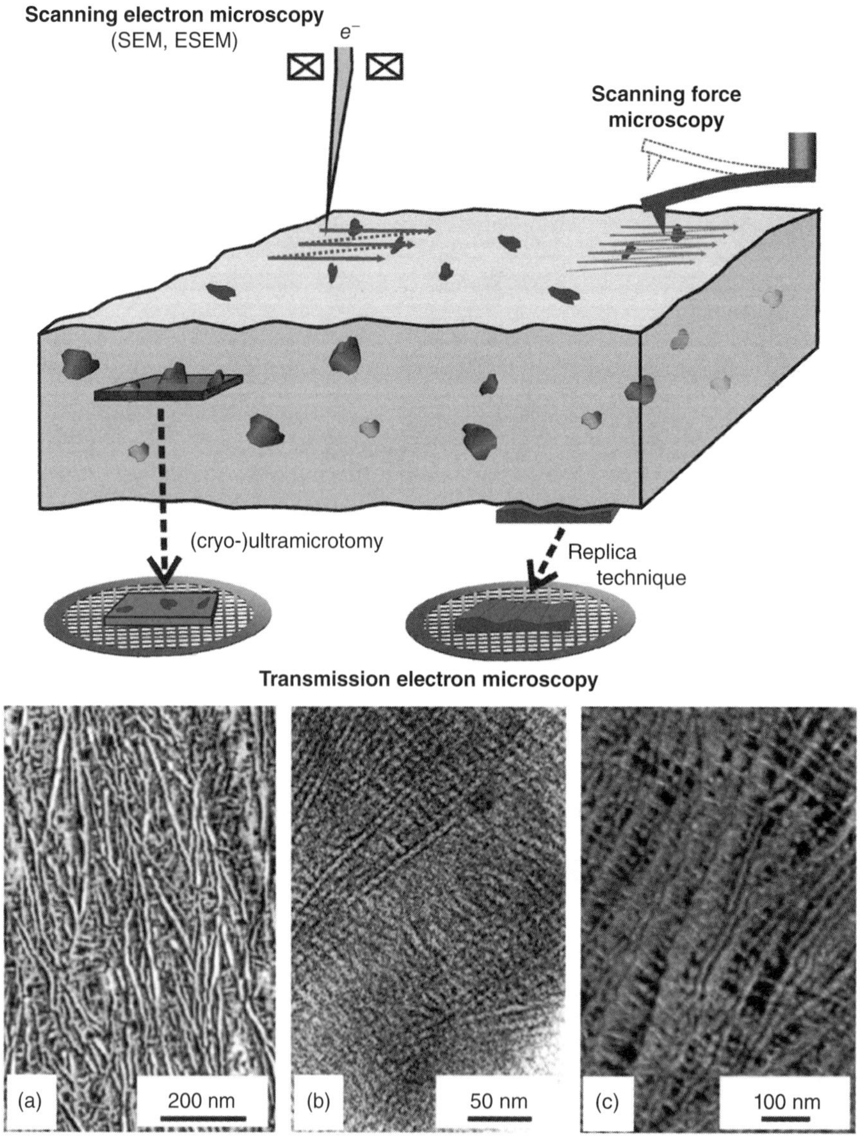

Figure 3.17 Top: Application of different microscopic techniques to study the surface and interior of a bulk polymeric material. Bottom: Comparison of the results from different microscopic techniques applied to α – iPP: (a) SEM image after permanganic etching, (b) TEM image of chemically stained ultrathin section, and (c) AFM tapping mode image.

nano- or micromechanical processes in the same material area and, therefore, offering a very direct way to determine structure–property correlations.

3.2.4 Image Processing and Image Analysis

The microstructures of modern materials are becoming more and more complex and nonuniform. Therefore, it is very important to evaluate microscopy images in detail, objectively and quantitatively. Today, the use of computer-assisted methods for this analysis and evaluation is inevitable. The initial task of image processing is to enhance the quality of digital images for further analysis. This optimization comprises the use of grayscale, contrast, shading correction, specific filtering methods (e.g., sharpness, high pass, low pass), arithmetic operations, as well as methods to determine relative composition, particle size, interparticle distances, intensity profile, and so on in heterogeneous polymers. Of particular interest are analysis procedures for the determination of orientations and stereoscopic imaging. Using various filter operations, it is possible to highlight or attenuate specific structural details. After applying a "low-pass filter" to an image, it will appear less sharp, with the edges of the gray areas becoming blurred and noise diminishing. In contrast, a "high-pass filter" makes the image appear sharper, highlighting the finer structural details (such as edges) and removing the heterogeneous regions of an image (see Fig. 3.18).

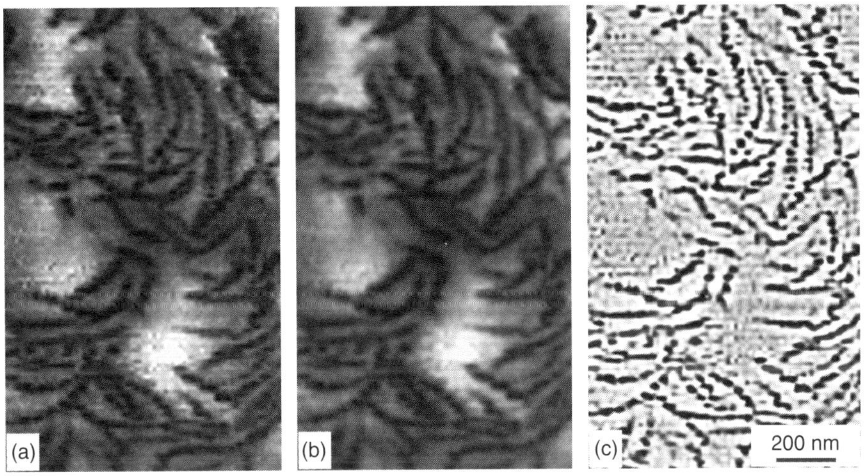

Figure 3.18 Examples of filter operations leading to changes in image detail (a) AFM phase images of a HDPE/VLDPE blend, (b) after low-pass filter, and (c) after high-pass filter.

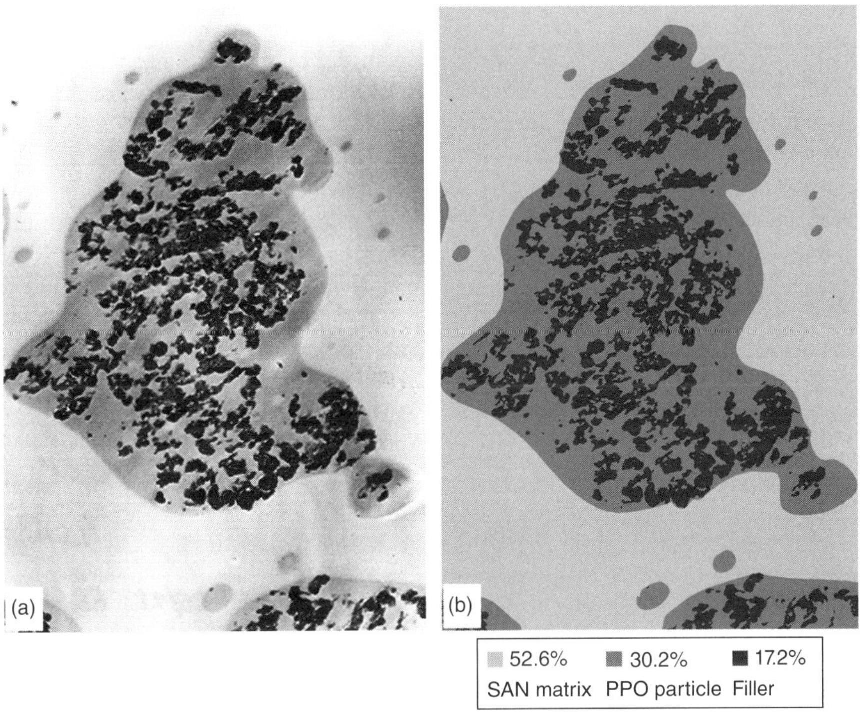

Figure 3.19 (a) TEM micrograph of a filled SAN/PPO blend and (b) determining the volume fractions of the constituents by assigning defined grayscale values to individual components.

The fundamental possibilities offered by image processing for the quantitative evaluation of structural details are demonstrated by determining the volume fraction of components in polymer blends. Figure 3.19 illustrates a particle-filled SAN/PPO polymer blend by defining the grayscale regions of the image that one can easily calculate the relative compositions of the different phases in this multicomponent system.[2] Special image analysis procedures based on the so-called fast Fourier transformation (FFT) allow to determine periodicities in micrographs, such as orientation of structures, long periods, and domain thickness.

[2]In Ref. [1, Chapter 7, pp. 161–171].

3.3 SAMPLE PREPARATION

The polymers that are studied by electron microscopy can take various shapes, forms, and sizes. Examples include powders obtained directly from macromolecular synthesis and granules from compaction on a first extrusion step, and injection- and extrusion-molded test parts in the centimeter range or large pieces for different applications. In general, direct investigations of polymers by electron microscopy involve three problems:

1. The usual preparation techniques applied to inorganic samples cannot be applied, and the preparation of ultrathin specimens from bulk polymers is often difficult.
2. Polymers as inorganic substances are particularly sensitive to electron beam irradiation.
3. The contrast between structural details is often very low because polymers usually consist of the same light elements (C, H, O, and others) that interact only weakly with the electron beam.

Several preparation and investigation techniques have been developed to overcome these difficulties. Three methods for investigating the morphologies of polymers are generally available (see Fig. 3.20) [1, 2, 5]:

1. The preparation of special surfaces (brittle fracture surfaces, smooth and selectively etched surfaces, etc.) that yield information on the internal structure of the material. These surfaces are investigated by means of replicas in the TEM or directly in the SEM or AFM.
2. The preparation of thin sections by ultramicrotomy, generally after special fixation and staining procedures has been performed. Investigations are carried out by conventional TEM, HEM, or AFM.
3. The preparation of special thin films (solution-cast films, focused ion beam (FIB) sections) with an additional staining treatment (as used for ultramicrotomy) and then studying by TEM or AFM.

The mostly used and most successful method to prepare polymers for TEM inspection is ultramicrotomy (including cryoultramicrotomy) in combination with chemical staining. It is the standard method for the preparation of ultrathin and semi-thin sections as well as very flat surfaces of plastics and biological and biomedical objects for various microscopic investigations. Improvements in preparation techniques over the past few decades have demonstrated that thin sections of different materials that are free from artefacts can be successfully prepared for electron microscopic investigations.

Successful sectioning today depends primarily on the experience of the experimentalist rather than on the instrumentation used. In order to avoid sectioning-induced errors and to fully exploit the capability of an ultramicrotome, one must master the optimum specimen preparation and sectioning technique; for details, see [5].[3]

An (ultra)microtome is a sensitive instrument for thin sectioning and is equipped with a knife made of glass or diamond. The microtome sample holder is moved toward the knife a given distance during every cycle. Figure 3.21 shows the principle of an (ultra)microtome. During the downward motion of the sample holder (A → B, cutting step), a section is produced with a thickness specified by the forward motion (difference A − B). The thickness of the resulting section is determined not only by the amount of forward motion but also by other parameters such as the type, quality, and inclination of the knife as well as the properties of the material and the size of the surface being sectioned. The backward motion of the sample and the sample holder takes place at a higher speed.

The sections are generally allowed to swim on a liquid placed on the knife truf, which are then transferred to microscope grids and are ready for microscopic investigations.

Soft polymeric materials or polymers containing soft components need to be fixed as well as hardened before undergoing sectioning. An alternative method is sectioning at cryogenic temperatures, that is, below the glass transition temperature of the material. If cryoultramicrotomy is required, the instrument must be equipped with a cryo-compartment. The latter is cooled by means of liquid nitrogen, and sectioning is generally carried out at temperatures of up to −185 °C. Any type of knife can be used including special cryo-diamond knives [5].

Because of the usual low contrast of polymers, different methods are employed for the selective staining procedure, either on the bulk material before sectioning or on the thin sections obtained after cutting.

Chemical treatment is the method most commonly employed for the fixation or hardening of polymers. Following this route, the specimen is treated with one or several chemicals that preferentially react with or diffuse into specific polymer phases. The component that reacts with the chemical becomes hard or even brittle and thus contributes to the staining effect. Today, the staining agents most frequently used for polymers are osmium tetroxide and ruthenium tetroxide. An overview of the staining agents that have been successfully employed in polymer microscopy is provided in Table 3.2 and illustrated together with application examples in Refs [1, 5]. The effect of chemical staining may be enhanced by performing a special chemical pretreatment (activation) of the material before staining. Reactants often used for pretreatment include chlorosulfonic acid, tungsten-phosphoric acid, formalin, and others.

Besides chemical staining, contrast enhancement of polymeric materials can also be achieved through some physical effects. This is of particular interest if the polymers do not possess reactive groups needed for chemical staining to work. Two techniques are generally used: irradiation-induced effects due to γ- or electron irradiation and straining-induced effects. Two examples illustrate these effects. Figure 3.22 shows spherulites with the typical concentric rings of LDPE (in contrast to spherulites in HDPE; Fig. 3.8). During γ- or electron irradiation, PE is known to show a stronger tendency to cross-link than to undergo chain scission, yielding a fixation (hardening) and staining effect. The lamellae appear as bright

[3] In Ref. [1, Chapters 10 and 12, pp. 199–217 and 231–240].

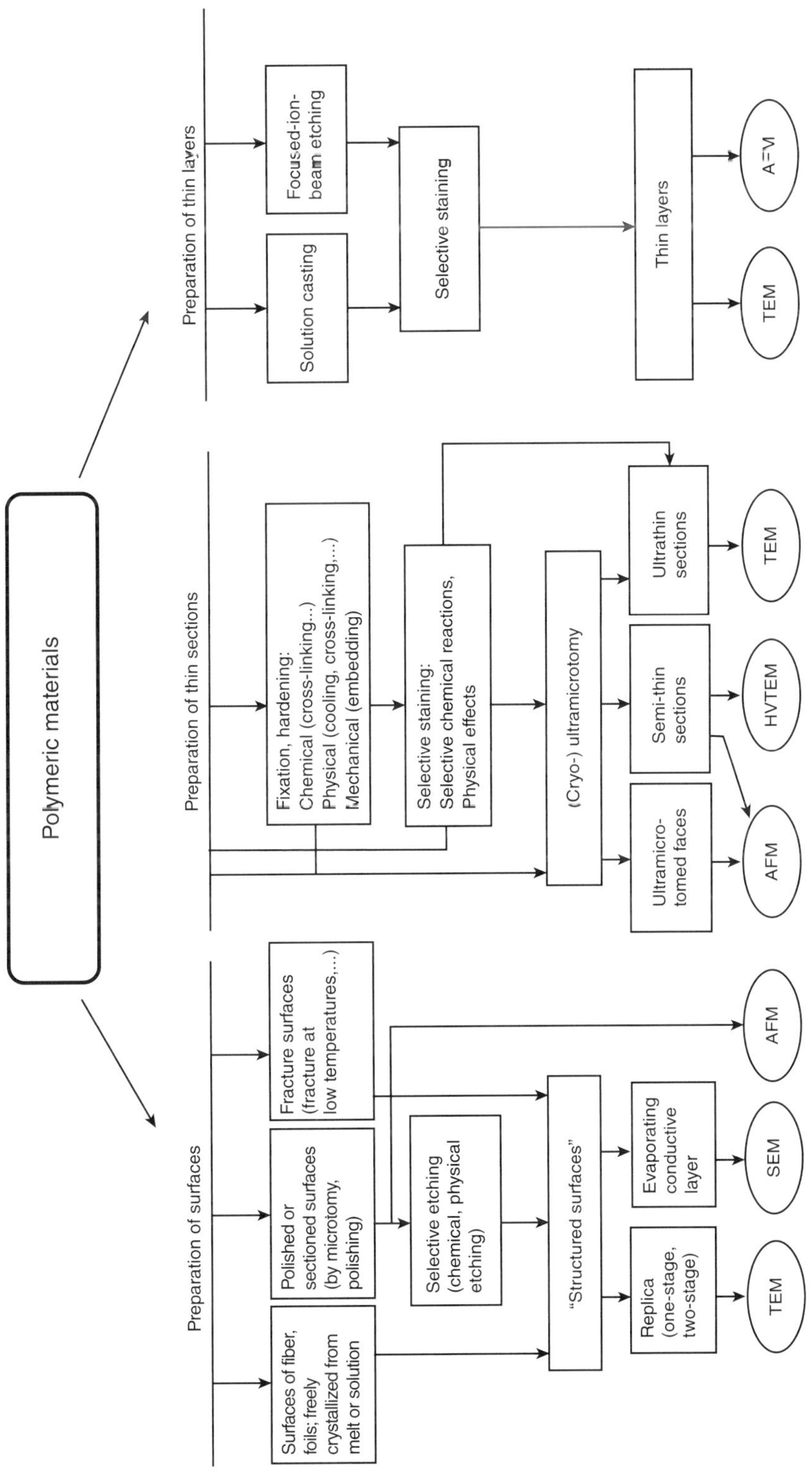

Figure 3.20 Overview of the preparation techniques and electron microscopic methods successfully used to investigate the morphologies of bulk polymers.

TABLE 3.2 Chemicals Successfully Employed as Staining Agents in Polymer Research

Polymers	Staining Agents
Polyolefines (e.g., PE, PP)	Chlorosulfonic acid/osmium tetroxide Chlorosulfonic acid/uranyl acetate Ruthenium tetroxide
Polyamides (nylons)	Formalin/osmium tetroxide Tungsten phosphoric acid/osmium tetroxide Ruthenium tetroxide
Polyacrylates	Hydrazine/osmium tetroxide Chlorosulfonic acid/osmium tetroxide Ruthenium tetroxide
Polystyrene, styrene copolymers	Ruthenium tetroxide
Polyurethanes	Chlorosulfonic acid/osmium tetroxide Ruthenium tetroxide
Polyvinyl chloride	Chlorosulfonic acid/osmium tetroxide
Polymers with double bonds (such as PB, PI, HIPS, ABS)	Osmium tetroxide; Bromine solution
Polymers with OH groups	Osmium tetroxide

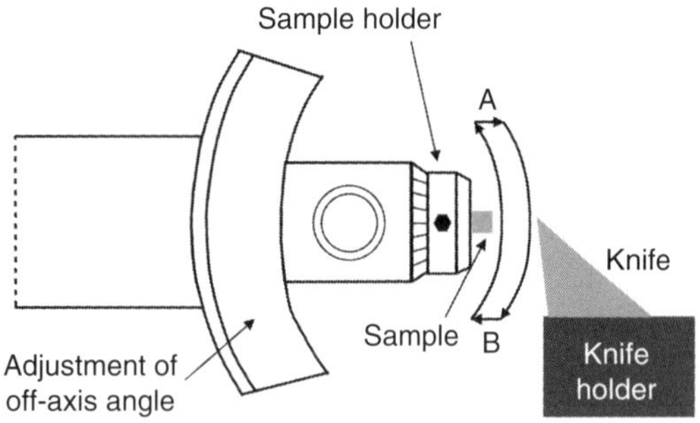

Figure 3.21 Scheme showing the movement of the sample relative to the knife during the sectioning process; A forward motion; A → B cutting; B backward motion; B → A upward movement; section thickness = length A − length B.

lines in a surrounding darker amorphous area. The thicknesses of the lamellae are about 10–15 nm, which corresponds to the values for lamellae measured in chemically stained sections. In chemically stained sections, the amorphous regions appear dark as a consequence of absorption of heavy atoms; here, the amorphous zones are darker due to irradiation-induced cross-linking and contraction (see Fig. 3.12). Figure 3.23 shows a domain-like structure in a SAN copolymer in a deformed zone in front of a crack tip. The small domains are somewhat weaker than the surroundings and appear under load as brighter particles. This method of straining-induced contrast enhancement is the only technique to make these small domains in SAN visible.[4] The difference in density and material composition is too small for using chemical staining to enhance the contrast.

[4]In Ref. [15, Section 7.4, p. 161].

3.4 IN SITU MICROSCOPY

Due to the variety of different structural details that can occur in polymers, there are also a wide variety of nano-/micromechanical processes that can appear under load. These include changes to individual macromolecular segments (on a nanometer scale), localized plastic yielding in the form of crazes or shear bands (at the micrometer scale), up to crack propagation and macroscopic fracture (at the millimeter scale). Therefore, different techniques for studying these processes are required, which are shown in Figure 3.24.

Three main methods exist:

1. Investigation of fracture surfaces (from tensile or impact test, events of damage of materials, etc.) directly in the SEM (or in TEM using – as in the past – replicas).
2. Deformation of bulk samples (from the first elongations up to fracture) is followed by investigating the changes at the surfaces by SEM (or after replication by TEM). Changes inside the bulk material are studied by preparing semi- or ultrathin sections using an ultramicrotome (occasionally after cooling or chemical staining) and investigating by TEM or AFM.
3. Deformation of thin films or semi-/ultrathin sections in a tensile device and investigation after deformation or *in situ* by TEM, HEM, ESEM, or AFM.

When going from method 1 to methods 2 and 3, more and more details of the micromechanical processes can be revealed and investigated in their dependence on the real morphology. Analysis of fracture surfaces by SEM (method 1 – microfractography) yields information mainly about the processes of crack initiation and crack propagation up to the final fracture. Particularly, the influence of structural

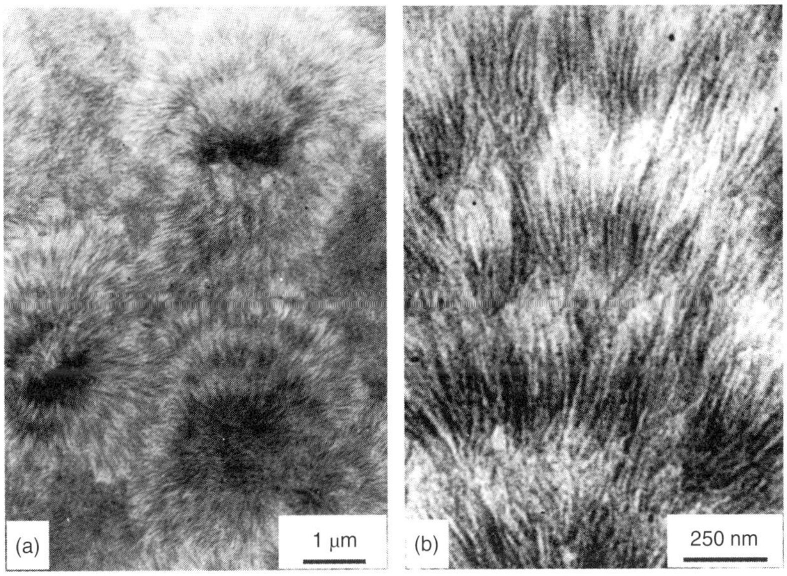

Figure 3.22 Contrast enhancement by the physical effect of γ-irradiation in semi-crystalline LDPE (γ-irradiation with a dosage of 20 MGy) [15] (a) banded spherulites and (b) lamellae inside the concentric bands.

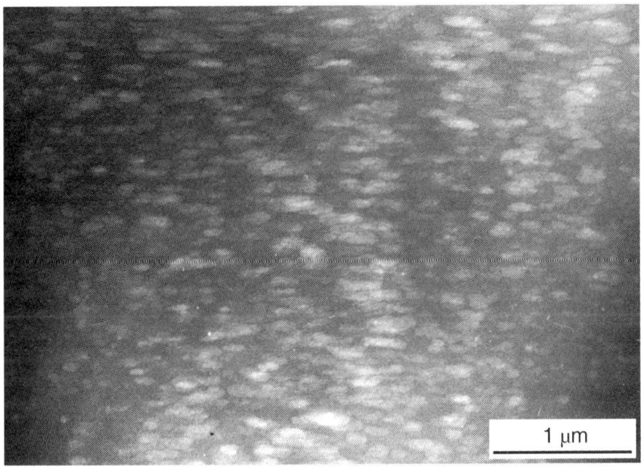

Figure 3.23 SAN copolymer with small domains, which are visible due to a straining-induced contrast enhancement; area in front of a crack tip, deformation direction horizontal, semi-thin section in HEM.

heterogeneities ("defects") on initiation and propagation of cracks, and the results of phase separation or low interfacial strength in polymer combinations can be studied.

Ductile materials show large amounts of plastic deformation before fracture occurs, making identification of the whole process on the final fracture surface difficult or impossible. Such ductile mechanisms and, in general, processes that occur before fracture, such as crack initiation, microvoid formation, local fibrillation, and crazing, can be seen with method 2.

Method 3 enables the morphology and deformation processes to be investigated with high resolution. As the sample size and thickness decreases, the resolution usually increases. The use of several tensile stages applicable for SEM, TEM, and AFM enables to deform a sample with a "representative thickness," containing the characteristic morphological features of the material of interest. The study of mechanically loaded samples and the changes under deformation in a microscope is an example of the so-called *in situ* techniques. The term "*in situ*" is Latin and means "in its place." There are several types of *in situ* experiments that can be performed: mechanical deformation tests, heating or cooling, electron irradiation, application of electric or magnetic fields, and application of different ambient atmospheres. One example of an *in situ* deformation test of a rubber-toughened polymer (high impact polystyrene, HIPS) is demonstrated in Figure 3.25: Deformation of HIPS is connected with an opaque whitening of the stress area; visible in low microscopic magnifications by lots of whitening bands (the so-called crazes) perpendicular to the loading direction. The larger magnification Figure 3.25b reveals some details: crazes are initiated at zones of stress concentration at the rubber particles, that is, in the equatorial zones of the matrix perpendicular to the loading direction. A general problem of these methods, however, is associated with radiation damage (see above), since material properties and particularly (micro)mechanical properties are drastically changed by electron irradiation.

Such tests not only provide us with information about the "macroscopic" properties of the material but also a deeper understanding of its deformation behavior at the micro- and nanoscopic scales. *In situ* tools allow the local morphology and the deformation in a material that is subjected to local mechanical stresses to be mapped out. Therefore, these studies open up a very direct way to determine structure–property correlations of polymers and to explain them on the basis of nano- and micromechanical mechanisms in dependence on real morphology [2].

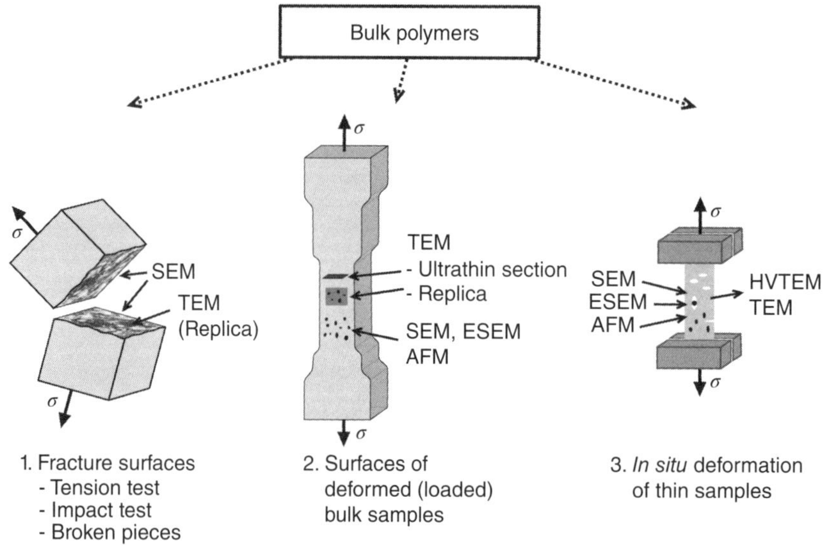

Figure 3.24 Survey of electron microscopic methods for investigating nano-/micromechanical processes in polymers (from [1, 2]).

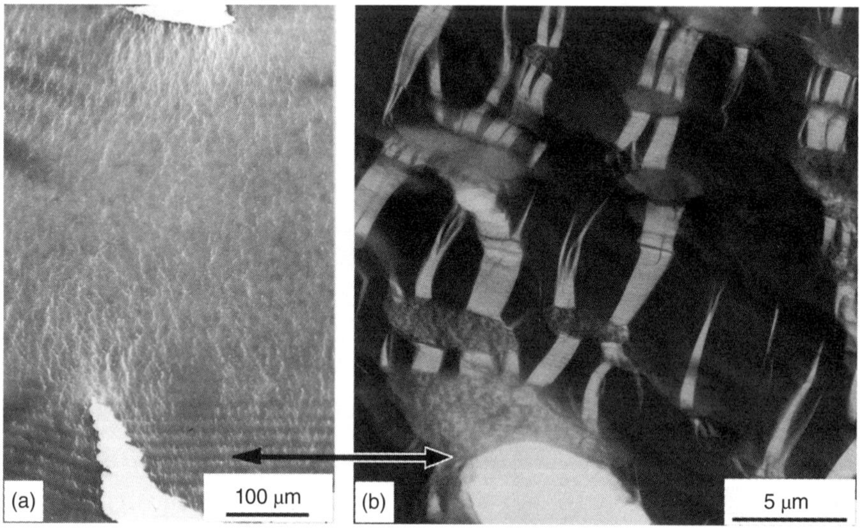

Figure 3.25 *In situ* deformation of a rubber-toughened polymer (HIPS – high impact polystyrene) (a) overview of the area under load and (b) area in front of a crack tip (in micrograph (a) at the bottom) with rubber particles (gray) in a matrix (black) with crazes (bright) 2-μm-thick deformed section, deformation direction see arrow, in 1000 kV HEM.

REFERENCES

1. Michler GH. *Electron Microscopy of Polymers*. Berlin, Heidelberg: Springer-Verlag; 2008.
2. a) Michler GH, Balta-Calleja FJ. *Nano- and Micromechanics of Polymers: Structure Modification and Improvement of Properties*. München: Carl Hanser Verlag; 2012.
   b) Michler GH, *Atlas of Polymer Strucures: Morphology, Deformation and Fracture Structures*. München: Carl Hanser Verlag; 2016.
3. Kern M, Trempler J. *Observation and Measurement Microscopy in Material Science*. Berlin Heidelberg: Brünne-Verlag; 2008.
4. Wu J, Chan C-M, Mai Y-W. In: Shonaike GO, Simon GP, editors Chapter 18. *Polymer Blends and Alloys*. 1964. p 505–548 New York: Marcel Dekker Inc.; 1999.
5. Michler GH, Lebek W. *Ultramikrotomie in der Materialforschung*. München: Carl Hanser Verlag; 2004.
6. Michler GH. New possibilities of electron microscopical determination of the morphology of polymers. Ultramicroscopy 1984;15:81–100.
7. Bethge H, Heydenreich J., editors. *Electron Microscopy in Solid State Physics*. Amsterdam: Elsevier Science Publisher; 1987.
8. Goodhew PJ, Humphreys FJ, Beanland R. *Electron Microscopy and Analysis*. 3rd ed. London: Taylor & Francis; 2000.
9. Zhang X-F, Zhang Z, editors. *Progress in Transmission Electron Microscopy, 1: Concepts and Techniques*. Berlin: Springer; 2001.
10. Li ZR, editor. *Industrial Application of Electron Microscopy*. New York: Marcel Dekker Inc.; 2003.

11. Magonov SN. In: Meyers RA, editor. *Encyclopedia of Analytical Chemistry*. Chichester, UK: Wiley; 2000. p 7432.

12. Schönherr H, Vancso GJ. *Scanning Force Microscopy of Polymers*. Berlin, Heidelberg: Springer-Verlag; 2010.

13. Poelt P, Ingolic E, Gahleitner M, Bernreitner K, Geymayer W. Characterization of modified polypropylene by scanning electron microscopy. J Appl Polym Sci 2000;78:1152–1161.

14. Karger-Kocsis J. *Polypropylene: Structure, Blends and Composites, Vol. 1: Structure and Morphology*. London: Chapman & Hall; 1995.

15. Michler GH. *Kunststoff-Mikromechanik: Morphologie, Deformations- und Bruchmechanismen*. München: Hanser-Verlag; 1992.

16. Michler GH. Electron microscopic investigations of morphology and structure formation of polymers. J Macromol Sci Phys 1996;B35:329–355.

17. Michler GH, Dietzsch C. Investigation of crystalline polymeric solids at reduced beam damage in the high voltage electron microscope. Cryst Res Technol 1982;17:1241–1254.

18. Michler GH. Determination of the morphology and mechanical microprocesses in polymer combinations by electron microscopy. Polymer 1986;27:323–328.

19. Delong A, Hladil K, Kolarik V. A Low Voltage Transmission Electron Microscope. Eur Microsc Anal 1994;27:13–15.

4

CHARACTERIZATION OF POLYMER MORPHOLOGY BY SCATTERING TECHNIQUES

JEAN-MICHEL GUENET

Institut Charles Sadron, CNRS-Université de Strasbourg, Strasbourg, France

4.1 INTRODUCTION

Since the bold hypothesis by Leucippus and Democritus that matter ought to be discontinuous and made up with indivisible particles, Democritus named atoms; man has pursued his quest for visualizing these entities and, correspondingly, for unveiling the structure of matter. The invention of optical microscopy did not allow one to achieve this goal as resolution using light is still not good enough. Eventually, more than 20 centuries had to elapse since Democritus and Leucippus' hypothesis before the discovery of X-rays by Roentgen [1], which was to provide scientists with the very first tool capable of probing the most intimate scale of matter. The later development of quantum mechanics and the prediction by de Broglie [2] that particles belonging to the quantum world should possess a wavelength and so should behave as X-rays still further opened up new horizons soon to be enlightened with the development of the electron microscope by Knoll and Ruska [3]. The existence of neutrons by Chadwick gave physicists a new particle for the same purposes as was shown later by Wollan and Shull [4], who performed the very first neutron diffraction experiments. Yet, the credit for using neutron scattering in studying polymers ought to be given to Jannink, Benoit, and coworkers [5], who carried out the very first investigations on these systems.

In the mid-20th century, Guinier [6], Porod [7], Kratky [8], and others realized that X-ray small-angle scattering, also termed forward scattering, could bring useful information, in particular in the studies of metallic alloys. In this range, the intensity is in many cases monotonously decreasing with angle, and, despite the absence of peaks as occur in X-ray diffraction, allows one to determine the shape of large nanometric structures. Such giant structures also occur with high-molecular-weight polymer chains, but most of them could hardly be evidenced by X-rays due to contrast problems. For instance, discriminating one chain among similar chains was insuperable in the absence of proper tagging. The design of small-angle neutron scattering set up in the early 1970s allowed one to start exploring these systems thanks to the difference in scattering properties of hydrogen and deuterium.

Since these early days of pioneering works, large-scale facilities that provide scientists with very high flux of X-rays and neutrons have been built throughout the world. While neutrons have always required large-scale facilities (a nuclear reactor or a spallation source), those X-ray sources from synchrotron radiation have revolutionized this technique. The flux and the brilliance of X-ray photons is so intense that diffraction experiments can now be performed in a few seconds or even tenth of seconds as compared to hours with devices available in usual laboratories in spite of progress in technologies such as rotating anodes. Time-resolved investigations are now feasible, which give access to new perspectives for studying polymer systems. Also, a broader range of photon wavelengths are easily accessible, which has promoted the development of techniques such as EXAFS (extended X-ray absorption fine structure) or XANES (X-ray absorption near-edge structure).

Thanks to the newly built neutron large-scale facilities, successes in polymer science were immediate in the mid-1970s [9]. Neutron scattering experiments gave the first direct confirmation of the Gaussian behavior of polymer chains in the solid state as was predicted earlier by Flory [10]. They further revealed the chain conformation in semi-dilute solutions, which confirmed the relevancy of extending scaling

Polymer Morphology: Principles, Characterization, and Processing, First Edition. Edited by Qipeng Guo.

concepts to polymer physics as advocated by de Gennes [11]. Other systems such as semicrystalline polymers were then studied, providing information about the way chains fold in the bulk state or in single crystals [12, 13]. Nowadays, these scattering techniques are routinely used in the field of soft matter to which belong polymer systems.

A monograph centered on polymer morphology should undoubtedly include a chapter on these techniques. This chapter is therefore focused on the possibilities offered by X-ray and neutron scattering/diffraction for determining the structure of polymer systems. These techniques are often complementary as X-ray photons and neutrons do not "see" matter the same way. The difference of neutrons versus X-rays lies in the way they interact with atoms: X-rays interact with the electron cloud, while neutrons interact with the nucleus. Neutrons have an unquestionable advantage over X-rays when polymers are at stake thanks to the difference in scattering amplitude between hydrogen and deuterium, as is detailed below.

The field of radiation scattering studies of polymers is a vast one, and cannot be treated in only one chapter. Consequently, this chapter is restricted to specific examples that are intended to present the potentialities of the scattering techniques. It is divided into three main sections: a short presentation of the minimum theory to understand the concepts, the discussion about the notion of contrast factor, and finally a series of relevant examples taken from various systems.

4.2 A SHORT THEORETICAL PRESENTATION

4.2.1 General Expressions

The general expression for the intensity scattered by a binary system composed of molecules a and b is written as

$$S(q) = \overline{A}^2(q)S_a(q) + \overline{B}^2(q)S_b(q) + 2\overline{A}(q)\overline{B}(q)S_{ab}(q) \quad (4.1)$$

where $\overline{A}^2(q)$ and $\overline{B}^2(q)$ are the coherent scattering amplitudes of molecules a and b, and $S_a(q)$, $S_b(q)$, and $S_{ab}(q)$ are the scattering factors that characterize the spatial correlations between molecules a, molecules b, and molecules a and b. q is the momentum transfer, which reads:

$$q = 2\pi s = \frac{4\pi}{\lambda}\sin\theta/2$$

where λ is the radiation wavelength and θ the scattering angle.

The expression of such a scattering factor is

$$S_a(q) = \sum_i \sum_j \langle \exp iqr_{ij} \rangle \quad (4.2)$$

where r_{ij} is the distance between two molecules a. Similar expressions hold for $S_b(q)$ and $S_{ab}(q)$. The scattering amplitude, which characterizes the correlations within one molecule, is written as

$$\overline{A}(q) = \sum_k a_k \langle \exp iq\rho_k \rangle \quad (4.3)$$

where a_k is the scattering factor of atom k (or scattering length for neutron) and ρ_k its distance from the center of mass of the molecule.

It is customary to consider two main q-ranges depending on the desired type of information: the *small-angle scattering* range and the *diffraction* range.

The *small-angle scattering range* is such that $q\rho_k < 1$ so that $\overline{A}(q)$ reduces to a constant:

$$\overline{A}(q) = \overline{A} = \sum_k a_k \quad (4.4)$$

The value and the expression of this constant depend on the radiation type, as discussed below. It particularly differs whether one uses X-rays or neutrons. This range is therefore located at low q-values, that is, at a resolution typically below $q \approx 2$–$3\ \mathrm{nm}^{-1}$. Under these conditions, *the system can be considered a continuum.*[1] This further leads one to assume that any point of the system contains a molecule a or b [14]. This is the so-called incompressibility hypothesis [15], which leads to the following expression for $S(q)$:

$$S(q) = [\overline{A} - \overline{B}]^2 S_a(q) = [\overline{B} - \overline{A}]^2 S_b(q) \quad (4.5)$$

The cross-term $S_{ab}(q)$ has vanished; only correlations between molecules a, on the one hand, and molecules b, on the other hand, remain. The term $[\overline{A} - \overline{B}]^2 = [\overline{B} - \overline{A}]^2$ is the contrast factor.

That the contrast factor is the square of the difference of the scattering amplitudes is equivalent to Babinet's principle in optics, which states that the diffraction pattern from an opaque body is identical to that from a hole of the same size and shape. For instance, a circular hole scatters like a circular disc of the same diameter.

Note that in the case of polymer systems, $S(q)$ reads:

$$S(q) \sim \left(\overline{A} - \overline{B}\right)^2 \left[CP(q) + C^2 Q(q)\right] \quad (4.6)$$

where $P(q)$ is the labeled species form factor (intramolecular interaction), C their concentration, and $Q(q)$ stands for the intermolecular interactions between labeled species.

In the *diffraction range*, the resolution is much higher because larger q-values are involved, so that the *discontinuous* character of matter cannot be ignored any longer. As a result, Equation 4.1 holds, which particularly means that the cross-term $S_{ab}(q)$ must be taken into account.

In the case of semicrystalline polymers, we should distinguish between two cases: classical *nonsolvated crystals*, on the one hand, and *solvated crystals* (*crystallosolvates*), on the other hand. In the case of *solvated crystals*, changing the labeling type of the solvent or of the polymer (the case of neutrons, for instance) alters the diffraction pattern as some peaks

[1] A way of understanding this assumption simply is to look at newspaper pictures. At a distance they look continuous, while a closer look reveals that they are made up of dots.

may disappear or appear. This arises from the presence of the cross-term arising from the intimate organization between solvent and polymer.

In the case of *nonsolvated crystals* surrounded by a liquid phase, the cross-term $S_{ab}(q)$ can be dropped provided crystals are larger than sizes of about q^{-1}. The intensity then reads:

$$S(q) = \overline{A(q)}^2 S_c(q) + \overline{B(q)}^2 S_l(q) \tag{4.7}$$

where subscripts c and l correspond to the crystal phase and to the liquid phase, respectively.

Therefore, changing the labeling type of the solvent will have no effect on the diffraction pattern of the polymer.

Another case of interest is the "dilute regime" where conditions are such that one may consider "single particle scattering" conditions, that is, intermolecular terms between the different particles can be neglected. The scattered intensity is simply expressed as

$$I(q)_{C=0} = KCMP(q) \tag{4.8}$$

in which K is a constant (which includes the calibration constant of the apparatus used for the measurements and the contrast factor of the scattering species $[\overline{A} - \overline{B}]^2$) and C and M are the concentration and the molecular weight of the scattering species. As is suggested by the subscript $(C = 0)$, relation 4.8 is strictly valid for infinite dilution.

Yet, if the particles are randomly dispersed with no specific interactions, relation 4.8 is likely to hold for relatively high concentrations. If there are specific interactions, then a first-order corrective term must be taken into account (often referred to as the virial coefficient), and experiments at different concentrations must be carried out in order to evaluate this term.

Relation 4.8 can also be valid for $q\xi > 1$, where ξ is the average distance between particles and provided that $\xi > R$ where R is the largest dimension of the particle. If $\xi \approx R$, then the relation can still hold for $qR > 1$.

When studying particles of unknown size, it is always highly recommended to investigate different concentrations.

The case of ternary systems may be commonly encountered in polymers. For instance, one may deal with a system containing two types of particles embedded in a solvent (gels, for instance). The scattering intensity is then [16]:

$$\begin{aligned} S(q) &= (\overline{A_{p1}} - \overline{A_s})^2 S_{p1}(q) + (\overline{A_{p2}} - \overline{A_s})^2 S_{p2}(q) \\ &\quad + 2(\overline{A_{p1}} - \overline{A_s}) \times (\overline{A_{p2}} - \overline{A_s}) S_{p1p2}(q) \end{aligned} \tag{4.9}$$

The term related to the solvent scattering due to density fluctuations, which is q-independent, is not included. This term is usually subtracted by measuring the intensity scattered by a pure solvent sample.

Interestingly, if one can achieve the conditions where $\overline{A_{p1}} = \overline{A_s}$ or $\overline{A_{p2}} = \overline{A_s}$, then one can determine independently the structure of either particles $p2$ or particles $p1$, respectively.

4.2.2 The Form Factor

As shown in relation 4.8, an important parameter is the *particle form factor*, which is the Fourier transform of the pair-correlation function. A few models that are relevant to situations encountered in "solid" systems such as bulk polymers, gels, and so on, are discussed subsequently. As is customary, two ranges can be distinguished in small-angle scattering for the form factor: the *Guinier range* [6], which is defined for $qR < 1$, where R is the radius of gyration of the particles, and the *intermediate range*, for $qR > 1$, where the short-range structure of the particle is observed (expressions of the radius of gyration for some objects are given in Table 4.1).

In the *Guinier range*, irrespective of the particles' shape, the intensity always reduces to

$$I(q) \sim \left[1 - \frac{q^2R^2}{3}\right] \quad \text{or} \quad \exp\left(-\frac{q^2R^2}{3}\right) \tag{4.10}$$

Plotting the intensity $I(q)$ versus q^2 or $\log I(q)$ versus q^2 yields the radius of gyration. Zimm representation is another way of processing the data, especially when several concentrations, C, are used [17]. Zimm equation reads:

$$I^{-1}(q) \sim \left[1 + \frac{q^2R^2}{3}\right](1 + 2A_2C) \tag{4.11}$$

Therefore, plotting $I^{-1}(q)$ versus q^2 gives both the radius of gyration and the second virial coefficient A_2. This coefficient is nothing but the intermolecular interactions between labeled species, and is therefore related to $Q(q)$ in Equation 4.6. Strong interactions, namely, segregation, give negative values, and strong repulsions as in good solvents yield positive values.

In the intermediate range, no universal expression can be derived as intensity depends strongly on the particle's shape. Some examples for which analytical calculations are available are given below. In other cases, no analytical expressions are obtainable and so simulation must come into play.

In polymers, Debye [18] has derived an analytical expression in the case of the *Gaussian chains*:

$$P_G(q) = \frac{2}{q^2R_G^4}\left[\exp\left(-q^2R_G^2\right) + q^2R_G^2 - 1\right] \tag{4.12}$$

where R_G is the chain radius of gyration. Note that for $qR_G < 1$, the Guinier expression is retrieved from Equation 4.11.

In the intermediate range, $P_G(q)$ reads:

$$P_G(q) = \frac{2}{q^2R_G^2} \tag{4.13}$$

The Debye's Gaussian chain is rather ideal as the statistical segment is considered negligible with respect to R_G. More realistic models include either larger statistical segments (freely rotating rods), and wormlike chains [19, 20].

Before discussing these models, it ought to be mentioned that chains may depart from their Gaussian behavior in a good

TABLE 4.1 Some Expressions of the Radius of Gyration as a Function of Dimensional Parameters of the Objects

Object	R_G^2
Gaussian chain of contour length L and segment length $2l_p$ with $L \gg l_p$	$Llp/3$
Wormlike chain of contour length L and persistence length l_p [a]	$\frac{l_p^2}{3}(e^{-L/l_p} - 1 + L/l_p)$
Excluded volume chain of contour length L	$L^{6/5}$
Infinitely thin rod of length L	$\frac{L^2}{12}$
Oblong solid cylinder of length L and radius r	$\frac{L^2}{12} + \frac{r^2}{2}$
Oblong hollow cylinder of length L, outer radius r_{out}, and inner radius γr_{out}	$\frac{L^2}{12} + \frac{(1-\gamma^2)r^2}{2}$
Flat disc of radius r and thickness l	$\frac{r^2}{2} + \frac{l^2}{12}$
Slab of length L, width l, and thickness δ	$\frac{L^2}{12} + \frac{l^2}{12} + \frac{\delta^2}{12}$
Sphere of radius r	$\frac{3}{5}r^2$
Hollow sphere of outer radius r and inner radius γr	$\frac{3}{5}(r^2 - \gamma r^2)$

[a] Note that for $L/l_p \to \infty$ one retrieves the expression for the Gaussian chain and for $L/l_p \to 0$ the value for a rod.

solvent. In many cases, a fractal approach can be used so that the scattering in the intermediate range reads [16]:

$$P(q) \propto \frac{1}{(qR_G)^{D_F}} \tag{4.14}$$

where D_F is the fractal dimension of the chain. Excluded volume corresponds to a fractal dimension $D_F = 5/3$. For a fractal dimension $D_F = 1$, that is a rod, relation 4.14 yields the same scattering behavior as that calculated by another approach (see below).

Freely rotating rods consist of an assembly of N rods linked in series to one another with no spatial orientation with respect to one another. The form factor is then written as [21]:

$$P_{fr}(q) = \frac{1}{N}\left[2\Lambda(\beta) - \frac{4}{\beta^2}\sin^2\frac{\beta}{2}\right] + \frac{2}{N^2}\Lambda^2(\beta) \times \left[\frac{N(1-\nu) - (1-\nu^N)}{(1-\nu)^2}\right] \tag{4.15}$$

where $\beta = ql_r$, l_r being the length of the rods, $\nu = \sin\beta/\beta$, and with:

$$\Lambda(\beta) = \frac{1}{\beta}\int \frac{\sin t}{t}dt = \frac{1}{\beta}Si(\beta) \tag{4.16}$$

For $ql_r > 1$, Equation 4.15 reads [22]:

$$P_{wl}(q) \propto \frac{\pi}{ql_p} + \frac{8(N-1)(\pi^2-4)}{q^2 l_r} \tag{4.17}$$

Unlike freely rotating rods, *wormlike chains* are characterized by a so-called persistence length l_p and the following relations: for $ql_p < 1$, the chains display a Gaussian behavior, while for $ql_p > 1$ the chains are rodlike. The term *wormlike* is a speaking image of this conformation. No analytical calculation can be derived for these chains but only two types of asymptotic behavior for $qR_G > 1$ [19, 22–24]:

For $ql_p < 1$

$$P_{wl}(q) \propto \frac{6}{q^2 l_p^2} \tag{4.18}$$

For $ql_p > 1$

$$P_{wl}(q) \propto \frac{\pi}{ql_p} + \frac{2}{3q^2 l_p^2} \tag{4.19}$$

In a $q^2I(q)$ *versus* q representation, one should therefore observe two asymptotes: a flat asymptote at low-q and a straight line at high-q. The intersection between these two regimes yields q^*, which reads:

$$q^* = \frac{16}{3\pi l_p} \tag{4.20}$$

For an *infinitely thin rod* of length L, the form factor reads [23]:

$$P_{wl}(q) \propto \frac{\pi}{qL} + \frac{2}{q^2L^2} \tag{4.21}$$

an expression closely reminiscent of the asymptotic behavior of the wormlike chain for $ql_p > 1$, except for a slightly differing $1/q^2$ term. Note that absolute calibration gives Equation 4.8 so that in this range $\mu_L = M/L$, namely, *the mass per unit length* can be determined. This parameter may be of interest, as will be highlighted in the case of thermoreversible gels.

These equations give the asymptotic behavior, but cannot be of any use for a total fit of the scattering curve. Yoshizaki and Yamakawa [24] have developed a pseudoanalytical approach by establishing several equations for covering the entire q-range.

If the cross section, r_c, of the rod is not negligible; with $L \gg r_c$, then one is dealing with an *elongated solid cylinder*, and the relation reads [25]:

$$P_{wl}(q) = \frac{\pi}{qL} \times \frac{4J_1^2(qr_c)}{q^2 r_c^2}\left[1 + \frac{2}{\pi qL}\right] \propto \frac{\pi}{qL} \times \frac{4J_1^2(qr_c)}{q^2 r_c^2} \tag{4.22}$$

where J_1 is the *Bessel function* of the first kind and the first order. For $qr_c < 1$, the *Bessel function* can be developed, which eventually gives:

$$P_{wl}(q) \propto \frac{\pi}{qL} \times \left[1 - \frac{q^2 r_c^2}{2}\right] \propto \frac{\pi}{qL} \times \exp\left(-\frac{q^2 r_c^2}{2}\right) \tag{4.23}$$

As a result, a plot of the type log $qI(q)$ versus q^2, often designated as a Porod plot, allows determination of r_c.

The general equation for the scattering by *any type of cylinder* has been derived by Fournet [26]:

$$P_c(q) = \int_0^{\pi/2} \frac{\sin^2(qL\cos\theta)}{q^2L^2\cos^2\theta} \times \frac{4J_1^2(qr_c\sin\theta)}{q^2 r_c^2 \sin^2\theta} \sin\theta \, d\theta \tag{4.24}$$

Thanks to personal computer performances, such an equation can nowadays be rapidly calculated for all types of cylinders.

As with very long rods, an approximation for a *flat disc* ($r_d \gg L_d$) gives simpler expressions [25]:

For $qL_d > 1$ with $qr_d < 1$,

$$P_d(q) \approx \frac{8}{(qr_d)^2} \exp\left(-\frac{q^2 L_d^2}{12}\right) \tag{4.25}$$

For $qL_d > 1$ and $qr_d > 1$

$$P_d(q) \approx \frac{8}{(qr_d)^2} \times \frac{\sin^2(qL_d/2)}{(qL_d/2)^2} \tag{4.26}$$

Other very long cylinders are of interest: *hollow cylinders*. The equation derived by Mittelbach and Porod [27] for $qL_{hc} > 1$ is as follows:

$$P_{hc}(q) = \frac{\pi}{qL_{hc}} \left\{ \frac{2}{\left(1-\gamma^2\right) qr_{hc}} \times \left[J_1(qr_{hc}) - \gamma J_1(q\gamma r_{hc})\right] \right\}^2 \tag{4.27}$$

where r_{hc} is the outer radius and γr_{hc} the inner radius. For $\gamma = 0$, one retrieves relation 4.22 while for $\gamma \approx 0$ the scattering by a *thin-walled sleeve* of cross-radius r_S for $qL_S > 1$ is obtained:

$$P_s(q) = \frac{\pi}{qL} J_o^2(qr_S) \tag{4.28}$$

As a rule, any kind of cross section can be considered for very long straight objects. The general relation is simple [25]:

$$P(q) = \frac{\pi}{qL} \phi(q\sigma) \tag{4.29}$$

where $\phi(q\sigma)$ is the cylindrical Fourier transform of the cross-section shape, and is written as

$$\phi(q\sigma) = \int_\sigma \rho(\sigma) J_o(q\sigma) 2\pi\sigma \, d\sigma \bigg/ \int_\sigma \rho(\sigma)\, 2\pi\sigma d\sigma \tag{4.30}$$

In many cases, cylinders display cross-section polydispersity. Guenet [28] has derived a calculation that takes into account the *effect of a polydispersity* by means of a distribution function of the type $w(r) \sim r^{-\lambda}$ with two cutoff radii r_{min} and r_{max}. In the range $r_{max} > q^{-1} > r_{min}$, the form factor is written as:

$$\frac{q^4 I_A(q)}{C} = 4\pi^2 \rho \left[A(\lambda) q^\lambda - \frac{1}{\lambda r_{max}^\lambda}\right] \bigg/ \int_{r_{min}}^{r_{max}} w(r) dr \tag{4.31}$$

with:

$$A(\lambda) = \frac{\Gamma(\lambda)\Gamma\left(\frac{3-\lambda}{2}\right)}{2^\lambda \Gamma\left(\frac{\lambda+1}{2}\right)\Gamma\left(\frac{\lambda+3}{2}\right)\Gamma\left(\frac{\lambda+1}{2}\right)} \tag{4.32}$$

where Γ is the gamma function.

For $qr_{min} > 1$, the function possesses the following asymptote:

$$\frac{q^4 I_A(q)}{C} = 4\pi\rho \int_{r_{min}}^{r_{max}} \left[1 + \frac{3}{8q^2r^2} + \cdots\right] w(r) dr \bigg/ \int_{r_{min}}^{r_{max}} w(r) dr \tag{4.33}$$

Cross-section polydispersity entails, in most cases, the disappearance of the oscillations due to the Bessel functions. A monotonously decreasing curve is then observed. Note that in the case where $\lambda = 1$, the intersection with the q-axis, q_o is $q_o = 2/\pi r_{max}$, while the intersection between the two types of behavior given by Equations 4.31 and 4.33 occurs at q^*, whose value is $q^* = 2/\pi r_{min}$.

It is equally worth emphasizing here that low-resolution continuous or discontinuous helices scatter like cylinders in the small-angle range for $q < 2\pi/P$, where P is the helix pitch [29].

Another case worth mentioning is that of a cylinder composed of a *hollow cylinder fitted with an inner solid cylinder* of differing scattering amplitudes. The cross-section form factor $\phi(qr)$ is then:

$$\phi(qr_{hc}) = \left[\int_0^{\gamma r_{hc}} A_{sc} J_o(qr) 2\pi r \, dr + \int_{\gamma r_{hc}}^{r_{hc}} A_{hc} J_o(qr) 2\pi r \, dr / A\right]^2 \tag{4.34}$$

where A is

$$A = \int_0^{\gamma r_{hc}} A_{sc} 2\pi r\, dr + \int_{\gamma r_{hc}}^{r_{hc}} A_{hc} 2\pi r\, dr \tag{4.35}$$

where A_{sc} and A_{hc} are the scattering amplitudes of the solid cylinder and the hollow cylinder, respectively.

Introducing:

$$A_m = \gamma^2 A_{sc} + (1-\gamma^2) A_{hc} = A/\pi r_{hc}^2 \tag{4.36}$$

$\phi(qr_{hc})$ is finally written as:

$$\phi(qr_{hc}) = \left[\frac{2A_{sc}\gamma}{A_m q r_{hc}} J_1\left(q\gamma r_{hc}\right) + \frac{2A_{hc}}{A_m q r_{hc}}\left(J_1(qr_{hc}) - \gamma J_1(q\gamma r_{hc})\right)\right]^2 \tag{4.37}$$

In the q-range where $qr_{hc} < 1$, Equation 4.37 is approximated to

$$\phi(qr_{hc}) = 1 - \frac{q^2}{4}\left[\frac{r_{hc}^2}{A_m}\left(A_{sc}\gamma^4 + A_{hc}\left(1-\gamma^4\right)\right)\right] \tag{4.38}$$

Unless $A_{sc} = A_{hc}$, *an apparent square cross-section radius* will be measured. As shown below, the scattering pattern can be drastically altered by toying with the scattering amplitudes. Under certain circumstances, the bracketed term of Equation 4.38 can be negative so that a "*negative apparent square radius*" can even be observed [16]!

These calculations can be easily extended to cases where a tube is made up with several concentric cylinders of differing contrast [30].

Another case of interest is a *slab* of length L_s, width l_s, and thickness δ_s. The form factor can be derived from the relation calculated by Mittelbach and Porod for elongated objects with a rectangular cross section [27]:

$$P(q) \approx \frac{\pi}{qL_s}\frac{2}{\pi}\int_0^{\pi/2}\left[\frac{\sin ql_s/2\cos\theta}{ql_s/2\cos\theta} \times \frac{\sin q\delta_s/2\sin\theta}{q\delta_s/2\sin\theta}\right]^2 \sin\theta\, d\theta \tag{4.39}$$

Three domains can be considered:

For $qL_c > 1$ with $ql_c < 1$,

$$P(q) \approx \frac{\pi}{qL_s}\exp\left(-\frac{q^2 l_s^2}{24}\right) \tag{4.40}$$

For $qL_c > 1$, $ql_c > 1$ and $q\delta_c < 1$

$$P(q) \approx \frac{2\pi}{q^2 l_s L_s} \tag{4.41}$$

Interestingly, in a Kratky plot ($q^2 I(q)$ vs q), the scattering vector q^* at which the asymptotic behavior of Equation 4.40 ($1/q$) meets the plateau of Equation 4.41 gives $l_s = 2/q^*$.

For $qL_c > 1$, $ql_c > 1$ and $q\delta_c < 1$:

$$P(q) \approx \frac{2\pi}{q^2 l_s L_s} \times \exp\left(-\frac{q^2\delta_s^2}{12}\right) \tag{4.42}$$

In many instances, cylinders can be bunched together in a parallel fashion. The intensity of such a *collection of n parallel cylinders* has been calculated by Oster and Riley [31]:

$$P(q) = \frac{\pi \mu_L}{qn^2}\phi(q\sigma)\sum_{j=1}^{n}\sum_{k=1}^{n} J_o(qr_{jk}) \tag{4.43}$$

r_{jk} stands for the distances between center of mass of the cylinders, and $\phi(q\sigma)$ is the function related to the cylinder cross section.

Spheres may be encountered in polymer systems when dealing with spherical inclusions or at the early stage of a liquid–liquid phase separation. The equation for sphere scattering reads [32]:

$$I(q) = \left[3\frac{\sin qr - qr\cos qr}{q^2 r^2}\right]^2 = \frac{9\pi}{2}\left[\frac{J_{\frac{3}{2}}(qr)}{(qr)^{3/2}}\right]^2 \tag{4.44}$$

where $J_{3/2}$ is the Bessel function of the first kind of order 3/2.

Similarly, a hollow sphere where the ratio between the inner and the outer radii is γ and is written [32, 33] as:

$$I(q) = \left[3\frac{(\sin qr - qr\cos qr) - (\sin q\gamma r - q\gamma r\cos q\gamma r)}{(1-\gamma^2)q^2 r^2}\right]^2 \tag{4.45}$$

It is worth noting that in all cases involving objects of density different from that of the surrounding medium, the asymptote of the scattering behavior is always proportional to $1/q^4$, which is known as Porod's law [34]:

$$I(q) \sim \frac{2\pi S}{Vq^4} \tag{4.46}$$

where S and V are the surface and volume of the particle.

Note that Porod's relation is valid only for a smooth surface. If the surface of the particle displays some roughness of amplitude of the order of $1/q$, then the scattering law is $I(q) \sim q^{-6+D_s}$, where D_s is the fractal dimension of the surface's particle [16].

Finally, a singular case that deserves to be mentioned deals with *partially labeled diblock copolymers*. The result is rather unexpected, as it gives a maximum that must not be confused with a peak due to order. This predictable effect was actually evidenced by neutron scattering thanks to the deuterium-labeling technique. For a symmetric copolymer where half of the system is labeled, Benoit derived the following relation [16]:

$$I(q) \sim (b_D - b_H)^2 (P_{DD} - P_{DH}) \tag{4.47}$$

where P_{DD} is the form factor of the deuterium-labeled part of the copolymer and P_{DH} the interference term between each branch of the copolymer. Since for $q = 0$ $P_{DD} = P_{DH} = 1$, the intensity $I(q) = 0$ for $q = 0$. Further, the intensity must be a monotonously decreasing function for larger q, which implies the occurrence of a maximum related to the radius of gyration of the chains. More sophisticated calculations show that the position of this maximum and its magnitude also depend on the interaction parameter between the two moieties of the copolymer [35].

4.3 EXPERIMENTAL ASPECTS

The choice of the most appropriate scattering technique depends upon two main requirements: the *contrast factor* and/or the need for *time-resolved* experiments. Presently, time-resolved experiments can be essentially carried out with X-rays thanks to the extremely high flux of photons delivered by synchrotron radiation facilities. In some high flux neutron facilities, such as ILL in Grenoble, some apparatuses allow one to do near-time-resolved experiments under certain experimental setup conditions (low sample-detector distances).

The contrast factor is by far the characteristic difference between neutrons and X-rays. As was said in Section 4.1, these particles do not "see" matter the same way.

4.3.1 The Contrast Factor

Neutrons interact only with the nucleus of atoms and are not perturbed by the electronic cloud. It is therefore a point-like interaction whatever the q-range explored so that the scattering amplitude of a molecule is simply the sum of the scattering amplitude of the atoms it is made up with. The main advantage of neutrons, particularly when organic systems containing many hydrogen atoms such as polymers are dealt with, lies in the conspicuous difference in the scattering length of hydrogen ($a_H = -0.375 \times 10^{-12}$ cm) and that of deuterium ($a_D = 0.675 \times 10^{-12}$ cm) [36]. Their signs are opposite,[2] which clearly enhances the contrast between hydrogenous and deuterated species, as the contrast is the difference between the scattering amplitudes (relation 4.5).

Another item of interest is the low variability in scattering amplitudes of most of the atoms. These amplitudes do not depend on the atomic number and are rather random. Values are, in most cases, somewhere between $0.2 < a < 1$ (10^{-12} cm), with a few negative values often found for isotopes (Table 4.2 shows some of the values for atoms often found with polymers). Unlike X-rays, heavy atoms do not scatter very differently from lighter atoms. Note that scattering lengths are only determined from experiments.

The calculation of the contrast factor must take into account molar volumes of each species, as the scattered intensity is always by unit of volume. The contrast factor K_N for an experiment intended to give absolute unit reads for objects of scattering amplitude A embedded in a medium of scattering amplitude B is written as:

$$K_N = \frac{N_A}{m_o^2}\left(A - \frac{v_o}{v_m}B\right)^2 \tag{4.48}$$

where v_o is the molar volume of species A and v_m the molar volume of species B, N_A is Avogadro's number, and m_o the molecular weight of the basic constituent of the object (a monomer unit for instance).

While the high contrast available by deuterium labeling is a decisive advantage of neutrons over X-rays in the field of polymer physics, neutron scattering presents a major drawback: hydrogen atoms also give off a high *incoherent* signal (see Table 4.2). As a result, the ratio signal/noise is not necessarily favorable. One way for reducing this effect consists, instead, in using the deuterated species as the main component of the system under study. Yet, a restrictive factor is all too often the price of the deuterated species. This incoherent signal must be determined separately or calculated, and then subtracted from the sample spectrum to extract the coherent scattering. This procedure becomes exceedingly delicate when the ratio signal/noise is low (typically lower than 1.1). Improper subtraction may yield an utterly irrelevant result.

Also, the procedures depend on whether the major component is hydrogenous or deuterated.

If the major component is deuterated (matrix), that is, the species under study are hydrogenous, then the intensity reads after subsequent rescaling by transmission and thickness:

$$I_h(q) = \frac{1}{K}[I_s(q) - I_{inc} - (1 - C_p) \times I_d(q)] \tag{4.49}$$

where $I_s(q)$ is the sample's normalized intensity, $I_d(q)$ the deuterated matrix normalized intensity, I_{inc} the incoherent intensity scattered by the hydrogenous species, C_p the concentration of hydrogenous species, and K is the contrast

TABLE 4.2 Values of Scattering Lengths a_c and Incoherent Scattering Cross Sections ($\sigma_{inc} = 4\pi a_{inc}^2$) for Various Atoms Often Encountered in Polymer Systems

Atom	a_c (10^{-12} cm)	σ_{inc} (barn)
H	−0.375	80
D	0.670	2.2
C	0.662	0.01
N	0.940	0.4
O	0.575	0.04
F	0.574	0.2
S	0.285	0.01
Cl	0.958	2.8
Br	0.685	0.4

Note that C and Br have virtually identical neutron scattering lengths, whereas they differ drastically as to the number of electrons (6 against 35). This means that X-ray intensity scattered by Br atoms is about 36 times larger than that by C atoms.

[2] Negative scattering length values can be experimentally measured as they are obtained from the extrapolation to zero of the neutron wave function for distances larger than the atomic nucleus radius.

factor. I_{inc} is far from being negligible and is not so easy to measure. Several methods exist, yet the best way is to calculate it from an equation determined experimentally, as has been achieved by Fazel et al. for polymers [37].

Now, if the major component is hydrogenous (hydrogenous matrix), and therefore deuterated species are the labeled species, then subtraction is subtler. If the hydrogenous sample displays a transmission value T, then $1 - T$ fraction of the neutrons flux is scattered *incoherently* over the entire solid angle 4π. So the incoherent intensity per unit angle is $I_{inc} = (1 - T)/4\pi$. As a result, the normalized intensity of the deuterated species, $I_d(q)$, is derived from the normalized intensity of the sample, $I_s(q)$, and the normalized, incoherent intensity scattered by a hydrogenous matrix, $I_h(q)$, through the following equation:

$$I_d(q) = \frac{1}{K}\left[I_s(q) - (1 - C_p) \times \frac{(1 - T_s)}{T_s} \times \frac{T_h}{(1 - T_h)} \times I_h(q)\right] \tag{4.50}$$

where T_s and T_h are the transmission of the sample and of the hydrogenous matrix, respectively, and C_p is now the concentration of deuterated species.

Other corrections taking into account the dynamics of the detector that are useful in case of low-ratio signal/noise can also be applied. Further details about data reduction can be found in the book edited by Zemb and Lindner [38].

Note that intensity normalization is achieved by dividing the spectra by the intensity scattered by a light water sample, whose incoherent scattering can be calculated and which coherent scattering is negligible. This procedure, which also allows for correction of detector efficiency, is carried out on all samples prior to any background subtraction.

Another noticeable drawback is related to the perturbation of the thermodynamic properties of the system when using mixtures of hydrogenous and of deuterated materials. In some cases, significant alterations of the properties are observed. Benoit et al. have shown the effect of deuteriation on the θ-temperature of polymer/solvent binary solutions [39]. Blends of hydrogenous and deuterated polystyrene will exhibit demixing for molecular weights above 2×10^6 [40]. In the case of polyethylene, phase separation has been reported in the crystallization process [41]. Therefore, one has to be rather careful when using such mixtures.

X-ray photons interact only with the electronic clouds of atoms. As a result, the contrast K_X depends on the number of electrons, and reads in the case of small-angle experiments where relation 4.5 applies [42]:

$$K_X = \frac{N_A e^4}{m_p^2 \mu^2 c^4}\left(Z_p - \frac{v_p}{v_m} Z_m\right)^2 \tag{4.51}$$

where e and μ are the charge and the mass of one electron, c the speed of light, Z_p, v_p and Z_m, v_m the number of electrons per basic unit of species p and m, and their molar volumes, respectively. N_A is again Avogadro's number and m_o the molecular weight of the basic constituent of the polymer object p.

The main disadvantage of X-ray over neutrons is the virtual impossibility of studying the single chain behavior in bulk systems or the difficulties in distinguishing one component with respect to another. Indeed, in many instances the only way to create enough contrast would be the use of heavier atoms, which would result in a drastic alteration of both the thermodynamic properties of the system and the molecular structure.

Unlike neutrons, the incoherent background is much lower and is usually negligible in the small-angle range. As a result, in some cases the contrast can be high enough to yield a good ratio signal/noise, as shown below. This has been recently improved by using very narrow X-ray beams, which reduce considerably the noise for the smallest angles. Once transmission corrections have been made, subtraction of the appropriate background (solvent, for instance) can be performed.

Another current advantage is the possibility of performing time-resolved experiments thanks to the high flux available at synchrotron radiation facilities.

4.3.2 Experimental Setup

The production of X-rays and neutrons is not detailed here as numerous monographs are available on the subject. This short section is a reminder of how these particles are detected and the signal processed.

A typical experimental setup used in most cases for both radiations is shown in Figure 4.1. For neutrons produced from a reactor, cold neutrons obtained by crossing a heavy water tank then fly through a rotating drum possessing helical grooves along its surface, whose effect is to select those neutrons that have the desired speed, and correspondingly the appropriate wavelength, to be capable to cross the drum without being absorbed. With this technique, resolution is about 10% (full width at half maximum, FWHM). For a small angle, this resolution is good enough. For diffraction experiments, far better resolutions must be achieved. Wavelength selection is obtained by reflection onto a crystal positioned at a Bragg angle with respect to the incident beam, which provides FWHM of about 1% (details can be found in [43]). Other facilities, such as spallation sources, work on another principle. Neutrons are produced by a proton beam hitting a target made up with depleted uranium or tantalum. The resulting neutrons possess a wide wavelength distribution. The data collected onto the 2D detector are then analyzed by a *time-of-flight* procedure, namely, a given cell of the detector may correspond to different values of q.

X-ray photons, produced by rotating anodes or by synchrotron radiation, are focused by means of bending mirrors and monochromated by means of a crystal oriented in such a way as to fulfill Bragg conditions. The monochromatic particles, after being scattered by the sample, are collected onto a two-dimensional detector. In the case of X-ray, the detector consists of a pixelated CMOS readout circuit connected to a semiconductor sensor. The photoelectric charge created by the absorption of an X-ray photon in the sensor is detected by

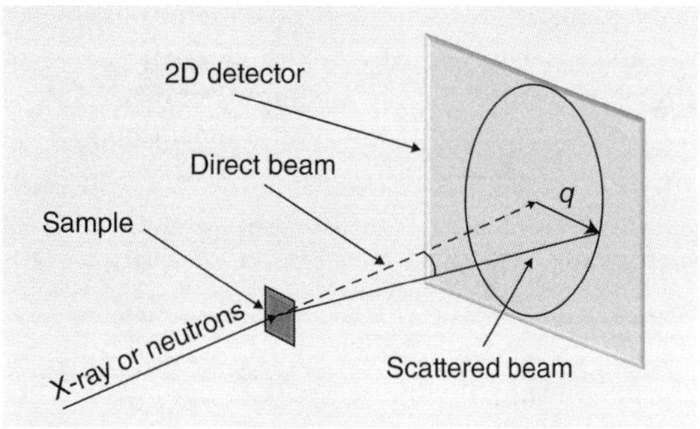

Figure 4.1 Schematic representation of a scattering experimental setup with data collection on a 2D detector.

the pixel closest to the absorption point. A map of the counted X-ray photons is thus obtained (details can be found in [44]).

In the case of neutrons, the detector is a proportional counter consisting of a container filled with ^3He gas and an array of wires. When neutrons hit ^3He atoms, electrons are released and their positions are detected by the array of wires, which gives both the position of the scattered neutrons and their number [16].

The signal processing differs depending on whether isotropic or oriented systems are investigated. For isotropic systems, all the cells located on the same circle correspond to the same value of q and their intensities are simply summed. For anisotropic systems, specific areas of the detector can be selected so as to determine the intensity for a given orientation of the sample. *Currently, the lowest accessible values of q are about* $q \approx 5 \times 10^{-3}\ nm^{-1}$*, which corresponds to objects of size lower than 200 nm.*

4.4 TYPICAL RESULTS

A few examples of the recent results gained by X-ray and neutron scattering/diffraction are detailed in this section. These examples are solely intended for illustrating the potentialities of both techniques, but are in no way a review of the many experiments reported so far, especially those on large-scale facilities. Most of the cases discussed here are taken from results published by the author of this chapter.

Important enough to be mentioned here is the way scientists plot their scattering data. Unless diffraction peaks occur, virtually nobody can plot data by means of $I(q)$ versus q, as this representation does not provide useful information. Depending on the types of systems under study, two main ways of plotting data are $\log I(q)$ versus $\log q$ and/or $q^n I(q)$ versus q with $n = 1$, 2, 4 in most cases (referred to as Kratky plot). The first representation is often used with systems possessing fractal dimensions, for which $I(q) \sim q^{\nu}$, while the second case is used for more complex situations such as cylinders. Other types of plots are also worth mentioning such as $I^{-1}(q)$ versus q^2 (Zimm plot) for determining radii of gyration or $\log qI(q)$ versus q^2 (Porod plot) for determining cross-section radii of oblong objects.

4.4.1 Neutrons Experiments: A Contrast Variation Story

4.4.1.1 The "single-chain" Investigation The very first advantage of neutrons is beyond doubt the possibility of investigating the chain conformation in a concentrated state. A few labeled chains randomly dispersed stand in a dilute situation so that the effect of the intermolecular term $Q(q)$ (relation 4.6) can be either determined or got rid of [16]. Undoubtedly, neutron scattering is the only technique permitting such investigations (see Fig. 4.2 for a schematic illustration).

Significant breakthroughs have been obtained by the collaborative group between Saclay (L. Brillouin Laboratory) and Strasbourg (C. Sadron Institute) for determining the chain conformation in the polymer bulk state [9]. These scientists were then able to confirm the assumption by Flory that the chains should be described by a Gaussian statistics under

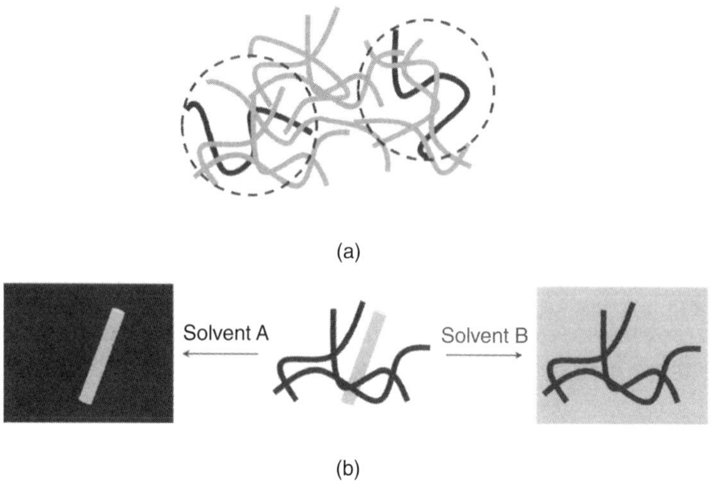

Figure 4.2 (a) Labeled chains (black) are here sufficiently spaced so as to consider a "single-chain" behavior provided the q-range explored is larger than d^{-1}, where d is the diameter of a sphere encompassing the chain (dotted line here). (b) Case where two structures are present with differing labeling. By using an appropriate solvent, only the coherent scattering of one structure is observed (contrast matching).

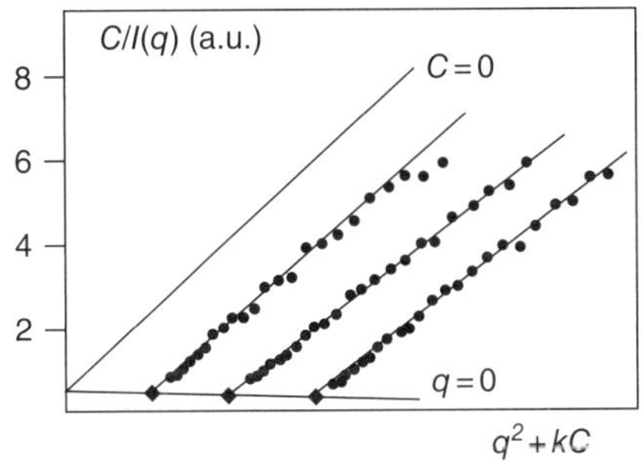

Figure 4.3 A typical Zimm plot obtained on 1%, 2%, and 3% of labeled chain in a gel of total polymer concentration 15% (iPS/*trans*-decalin gels). Extrapolation to $C=0$ (radius of gyration) and $q=0$ (second virial coefficient) are shown. Replotted from [49].

these conditions. To be sure, these experiments triggered the interest toward neutron scattering in the polymer community. Investigations of the chain trajectory in semicrystalline polymers followed, with the aim of testing whether chain folding proceeded in crystalline lamellae via adjacent reentry or simply at random [12, 13, 45, 46]. Very few reliable results were obtained in this case, particularly for polyethylene due to isotopic segregation. Significant results were obtained for isotactic polystyrene, which have shown that adjacent reentry was the main effect yet its amount depends very much on the crystallization conditions [47, 48].

Similarly, the chain trajectory in thermoreversible gel of isotactic polystyrene (iPS) has been determined. Zimm plots can be obtained (Fig. 4.3) that, after the usual proper extrapolation both to $C=0$ and $q=0$, have given values of radii of gyration R_g about twice larger than its value in the polymer bulk state [49].

A study in the range $qR_g>1$ does reveal a wormlike chain behavior (relations 4.18 and 4.19) consistent with a persistence length of about 4 nm as obtained after a theoretical fit (Fig. 4.4). Chains in iPS thermoreversible gels are thus locally extended with a persistence length about 4 times larger than in their usual random state, but are otherwise Brownian over large distances. Interestingly, it was also shown that this conformation did not vanish at the gel melting temperature. Guenet et al. concluded that chains are too rigid locally to allow for folding, this process requiring a persistence length of about 1 nm, and the only way to organize is then to "bunch," which results in the formation of fibrillar structures instead of lamellae [50].

Clearly, only neutrons could provide one with such a result, which gave the key for understanding why iPS forms a gel in some solvents rather than the usual chain-folded crystals.

Similar results [51] were obtained with syndiotactic polystyrene (sPS) solutions prior to gelation (Fig. 4.5). A persistence length of about $l_p=9$ nm was derived from a fit with Yamakawa pseudoanalytical approach. The broader q-range explored with respect to iPS experiments allowed one to fit the departure from linearity with Equation 4.22, which yielded a cross-section radius, $r_H=0.62\pm0.1$ nm, and a mass per unit length, $\mu_L=530$ g/nm mol, values consistent with

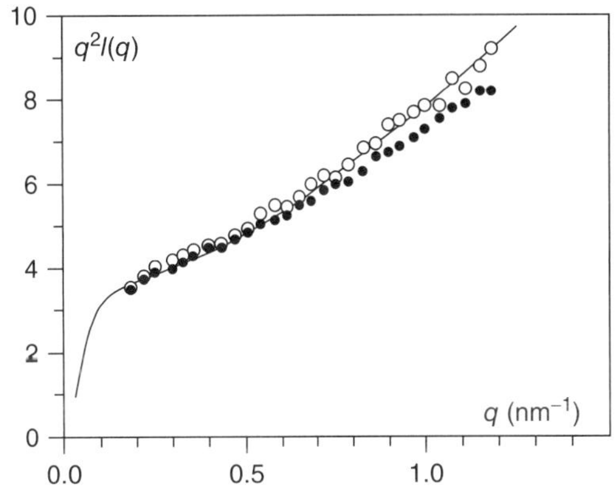

Figure 4.4 Kratky plot ($q^2I(q)$ vs q) for 1% labeled chains in a 15% iPS gel prepared in *trans*-decalin (open circle), and in the same system after gel melting (filled circle). The full line is a fit with Yamakawa's pseudoanalytical scattering curve. Replotted from [50].

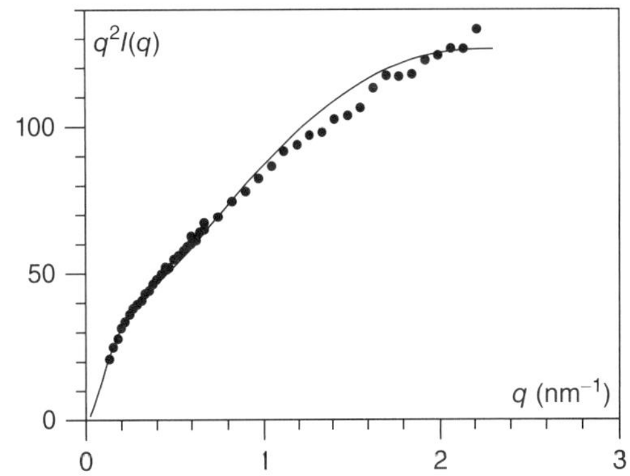

Figure 4.5 Scattering curve plotted by means of a Kratky plot for syndiotactic polystyrene in the SOL state. The fit is achieved with Yamakawa's pseudoanalytical equations and by considering a finite chain cross section, hence the departure from a linear variation at large q. Here, $l_p=9$ nm and $r_H=0.62\pm0.1$ nm. Replotted from [50].

the 2_1 helical form observed in the gel state for this polymer. This means that the chains are again locally rigid and already possess a conformation in solutions close to the helical form they will have in the gel state. As has been proposed earlier for iPS gels, this is a clear indication of the involvement of the solvent in the helical for stabilization process.

Another field where neutron scattering has provided scientists with valuable information concerns semicrystalline polymers. More specifically, one has attempted to determine the chain trajectory in lamellae either located in spherulites or under the form of single crystals in order to find out the extent of adjacent chain folding, which was a matter of controversy in the 1970s. Earlier investigations with polyethylene proved to be mainly unsuccessful due to a strong isotopic segregation [46]. It turned out that this segregation phenomenon was chiefly due to the differing degrees of branching of the hydrogenous and the deuterated samples [45]. Experiments on isotactic polystyrene were more successful as this segregation

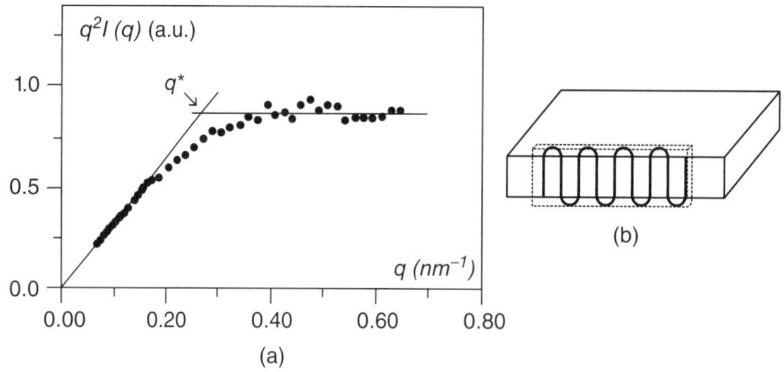

Figure 4.6 (a) Scattering curve obtained on iPS single crystals grown from a dilute solution where 1% of the iPS chains are deuterium labeled. (b) A sketch of regularly folded chain in a single crystal with the equivalent slab (dotted lines). Replotted from [13].

effect was absent. Basically, it has been shown that the amount of adjacent folding depends on the relaxation time of the chains versus the crystal growth rate. When chains have time to rearrange before being frozen in by the crystal growth front, a large amount of adjacent chain folding occurs, and vice versa.

The case of chains in iPS single crystals, where a few chains are deuterated, is presented here as an example of *slab* scattering (Fig. 4.6). Indeed, two regimes can be clearly indentified: a $1/q$ regime followed by a $1/q^2$ regime, which corresponds to the behavior described by Equations 4.40 and 4.41 [13, 52]. This is so because regular adjacent folding produces a conformation that can be regarded as a *slab* provided $qd_f < 1$, where d_f is the distance between two adjacent stems (see inset in Fig. 4.6).

4.4.1.2 Systems Involving Three Labeled Components

Some systems that are *partially labeled* may show differing scattering curves whether deuterated or hydrogenous solvents, or a mixture of them are used. A typical example is given by a polyelectrolyte/surfactant complex (Fig. 4.7).

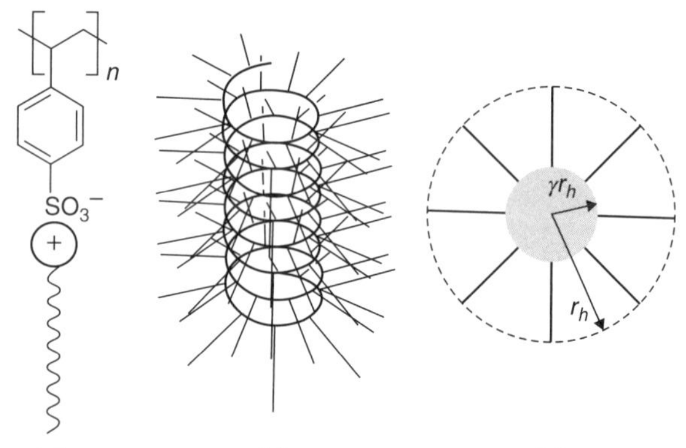

Figure 4.7 (a) Chemical structure of polystyrene sulfonate and its interaction with CTAB. (b) The bottle-brush structure. (c) The cross section of the bottle-brush structure, where r_h is the outside radius including the tails of the surfactant molecules pointing outward, and γr_h the inner radius corresponding to the chain core.

These complexes form in some solvents and are highly stable. One can then prepare a complex where the polymer chain is deuterated, while the surfactant is hydrogenous. In Figure 4.8 are shown the scattering curve of an $iPS_D/CTAB_H$ complex in deuterated *n*-butanol and hydrogenous *n*-butanol [53]. The scattering curve for $iPS_D/CTAB_H/n\text{-}butanol_D$ can be clearly fitted with an oblong cylinder (Eq. 4.22). Although the structure is rather reminiscent of a bottle brush by virtue of a helical arrangement of the polymer together with the CTAB moiety pointing outward, the fit with Equation 4.22 is relevant as cylinders and helices scatter the same at this level of resolution [29]. Thus, one can define an internal radius (the polymer chain) and an external radius (the CTAB molecules pointing outward). Actually, the authors of this investigation chose a more sophisticated model by contemplating a structure with three radii as they considered that a helix of the hollow type could not be discarded.

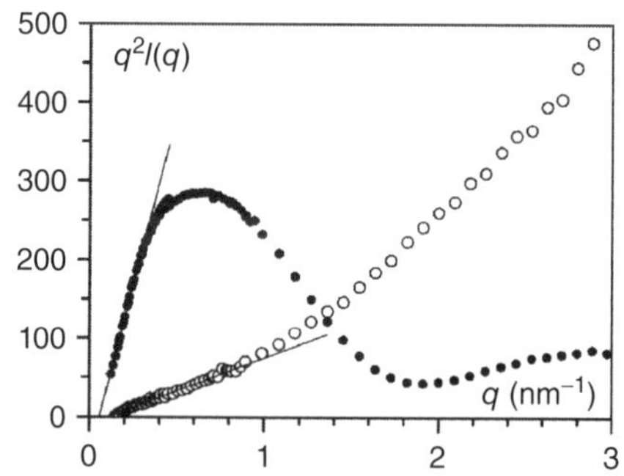

Figure 4.8 Scattering curves for $iPS_D/CTAB_H/n\text{-}butanol_D$ (filled circle), and $iPS_D/CTAB_H/n\text{-}butanol_H$ (open circle). Linear domains are highlighted with solid lines. Replotted from [53].

Schematically, linear behavior typical of this structure can be observed at small q followed by one oscillation at larger q for the $iPS_D/CTAB_H/n\text{-}butanol_D$ system. Conversely, the curve for $iPS_D/CTAB_H/n\text{-}butanol_H$ is utterly different just by altering the solvent labeling. In $iPS_D/CTAB_H/n\text{-}butanol_D$, the solution's departure from linearity occurs through a decrease in the scattering curve, while the reverse situation is seen for $iPS_D/CTAB_H/n\text{-}butanol_H$. These outcomes illustrate perfectly the occurrence of an *apparent negative mean square radius* due to a contrast effect as is highlighted by Equation 4.38.

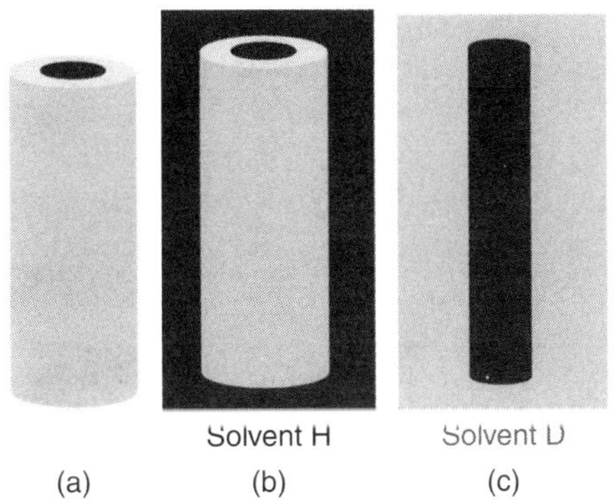

Figure 4.9 (a) A nanocomposite material made up with two concentric cylinders of different scattering amplitudes (schematized by "gray" or "black" colors). In a "black" solvent (b) the cylinder is of the hollow cylinder type, while in a "gray" solvent (c) it is of the solid cylinder type.

Toying with the solvent labeling can also be interest for determining the structure of the different *components in a nanocomposite material*. As a result, one can match the coherent intensity of one component, making it "invisible" to neutrons while keeping the other component "visible" as is highlighted in relation 4.9. Nanomaterials involving fibrillar thermoreversible gels, where one component stands for the core and the other component encapsulates or sheathes this core, are typical examples of such a situation (see Fig. 4.9). Here, two cases are described.

In one case, gel bicopper complex filaments shown in Figure 4.10 are encapsulated within those fibrils of a polymer thermoreversible gel. These filaments form in organic solutions through the piling of bicopper complex molecules in one dimension [54], and so contain only one molecule per filament cross section (Fig. 4.10). The process to achieve this type of encapsulated structure is the *heterogeneous nucleation* of polymer fibrils by bicopper complex filaments. The encapsulation is carried out in a common solvent to the polymer and the bicopper complex, namely, *trans*-decalin in the case studied. The filaments grow at temperatures above the gelation threshold of the polymer, and so act as heterogeneous nuclei [55]. The neutron scattering experiments were performed with deuterated polymer sample (deuterated isotactic polystyrene), a hydrogenous bicopper complex. Whether one expects to "see" the polymer conformation or the bicopper complex structure, highly hydrogenous solvent (92% hydrogenous *trans*-decalin + 8% deuterated *trans*-decalin) was used to match the coherent signal of the bicopper complex, or highly deuterated solvent (91% deuterated *trans*-decalin + 9% hydrogenous *trans*-decalin) to match the coherent signal of the polymer. The results reported in Figure 4.11 show the two different curves obtained while altering the solvent labeling. When only the bicopper complex is "seen," the scattering curve is identical to that observed for only the filaments. The mass per unit length and the cross-section radius are the same within experimental uncertainties. When only the polymer is "seen," the scattering curve can be interpreted with a model where 4 polymer chains stand around a hole whose dimensions correspond to the diameter of the bicopper filaments. The corresponding equation is [55]:

$$q^2 I(q) = \pi \mu_L C_p q \times \frac{4J_1^2(qr_H)}{q^2 r_H^2} \times \left[1 + \exp\left(-q^2 l_o^2\right) + 2\exp(-q^2 l_o^2/2)\right] \quad (4.52)$$

where the part in brackets corresponds to the positioning of the polymer chains spaced by a mean distance l_o, while the left terms correspond to cylinder scattering.

These results are clearly consistent with filaments encapsulated within polymer fibrils, which could be only demonstrated directly by neutron scattering thanks to the contrast variation.

In the second case, a reverse situation occurs as self-assembled nanotubules sheathe gel fibrils. These nanotubules are produced from the molecules shown in Figure 4.12 (named BHPB-10 for short) when the ribbons they form eventually warp with a "pitch" of about 150 nm [57]. The warping process [56] is clearly seen from the atomic force microscopy (AFM) picture (Fig. 4.12). Neutron scattering allows one to demonstrate beyond doubt these are hollow cylinders by using relation 4.27 (Fig. 4.12). The inner (r_{in} = 9.44 nm) and outer

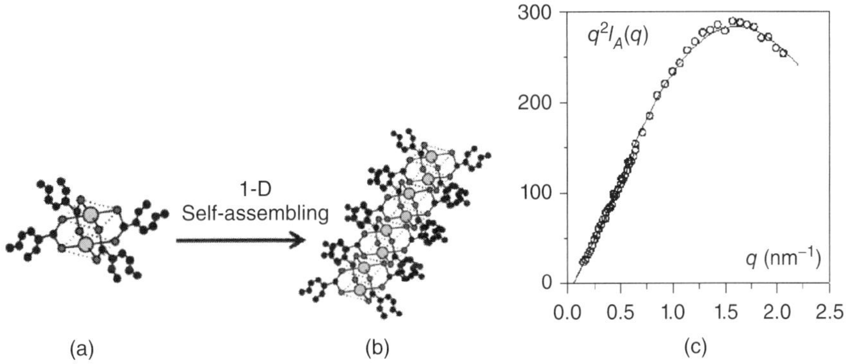

Figure 4.10 Filaments (b) obtained by the piling in one dimension of a bicopper complex (a). Gray balls = copper atoms, each surrounded by four oxygens, black balls = carbon atoms. (c) Scattering curve for organic solutions of these filaments with a fit by Equation 4.22 with r_h = 0.4 nm and μ_L = 1515 ± 160 g/mol nm. Replotted from [54].

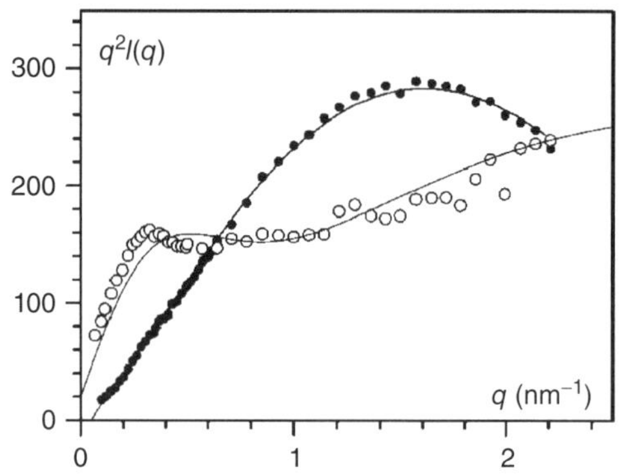

Figure 4.11 Kratky plot for iPS/CuS8/*trans*-decalin systems. (filled circle) = iPSD/CuS8/solventD; (open circle) = iPSD/CuS8/solventH (see text for details). The full lines fit with Equation 4.51. Replotted from [55].

(r_{out} = 12.75 nm) radii of these nanotubules are determined with great precision thanks to the many oscillations observed. Radius polydispersity is therefore extremely low; otherwise, a rapid damping of these oscillations would be noticed.

In order to prove that the sheathing of polymer fibrils (from isotactic polystyrene gels) has occurred, neutron scattering is again the appropriate tool. Whether one investigates ternary systems with different labeled species, namely, iPS_D/BHPB-10_H/*trans*-$decalin_D$ and iPS_H/BHPB-10_H/*trans*-$decalin_D$, the scattering curves differ markedly (Fig. 4.13). In particular, when only the BHPB-10 is "visible," one observes oscillations, and the scattering curve is identical with that recorded in the binary system BHPB-10_H/*trans*-$decalin_D$. When both the polymer and the BHPB-10 molecules are "visible," these oscillations have vanished [56].

This can be accounted for by examining the cartoons in the inset of Figure 4.13. The observation of oscillations when only BHPB-10 molecules are "visible" clearly means that the hollow cylinder structure is there. Yet, there are two issues: either nanotubules do sheathe the fibrils or they just stand outside in the fibrils' surrounding phase. The fact that oscillations vanish when both systems are hydrogenous, and therefore both "visible," proves that they do sheathe the fibrils, but only those for which the cross section is in register with the inner diameter of the nanotubules. Under these conditions, the object *sheathing nanotubule + fibril* cannot be distinguished from any other fibrils; they are all solid cylinders. Because there is a large cross-section polydispersity of the fibrils, oscillations are simply damped. Equations 4.31 and 4.33 are thus characterizing the system. If those nanotubules were in the fibrils' environment but without sheathing any of them, then they would still scatter as they do in binary systems so that oscillations would be again detected.

4.4.1.3 Neutron Diffraction As is said above, neutron diffraction experiments can be used as a complementary technique to X-ray diffraction for the determination of crystallographic structures. It is particularly useful for distinguishing between "anhydrous" crystals and solvated crystals. The technique is of great interest in the case of polymers that contain many hydrogen atoms that can be precisely positioned in the crystalline lattice by using deuterated species. It has been used in the case of polymer-solvent molecular compounds. The difference in the diffraction profile of a sample whether one uses differing labeled species allows one to draw relevant conclusions as to the existence or not of solvated crystals. In Figure 4.14, two case studies are displayed.

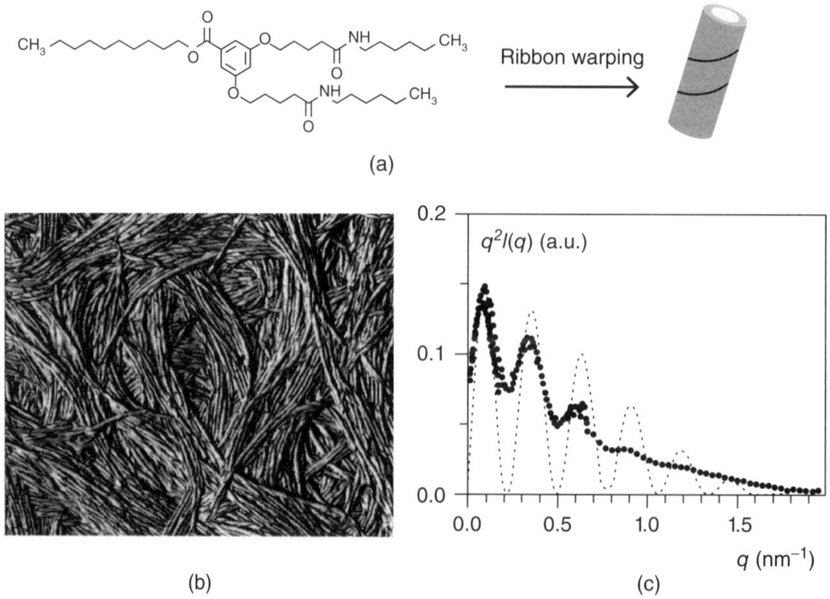

Figure 4.12 (a) Structure of BHPB-10, and cartoon of how the nanotubules are formed through ribbon warping. (b) AFM picture of the nanotubules. (c) Neutron scattering curve of nanotubules formed in 1% *trans*-decalin solutions; dotted line = fit by means of Equation 4.27 with r_{hc} = 12.8 nm and γ = 0.74. Curve replotted from [56].

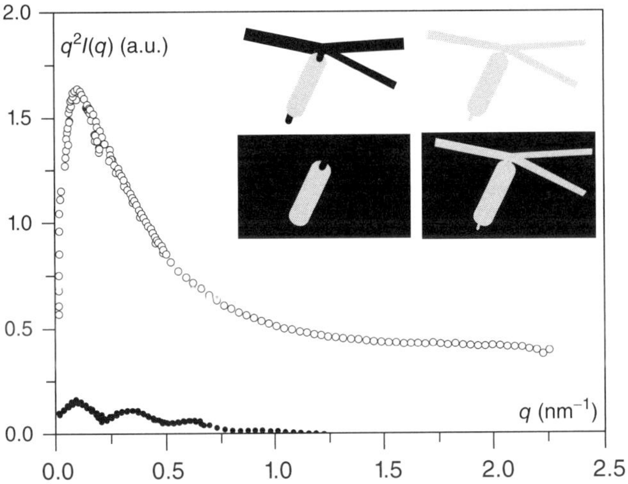

Figure 4.13 Scattering curves for iPS_D/BHPB-10_H/*trans*-$decalin_D$ (filled circle) (case corresponding to the left cartoons); iPS_H/BHPB-10_H/*trans*-$decalin_D$ (open circle) corresponding to the right cartoon. Curve replotted from [56].

Figure 4.14 shows the diffraction on assemblies of spherulites prepared from a solution of deuterated syndiotactic polystyrene (sPSD) in hydrogenous *trans*-decalin and deuterated *trans*-decalin. Syndiotactic polystyrene is here under the β form, that is, absence of crystal solvation. As can be seen, the intensities of the peaks do not change with respect to one another [58]. There is only a significant upturn for sPSD/*trans*-$decalin_H$ due to scattering. This situation corresponds to Equation 4.7. Conversely, the diffraction patterns by gels prepared from solutions in toluene show conspicuous differences [51]: peaks appear or disappear whether one uses deuterated toluene or hydrogenous toluene (in particular, those at $q = 7.26\,nm^{-1}$ and $q = 9.5\,nm^{-1}$). This is due to crystal solvation, and here the pattern is consistent with the δ form. This situation corresponds to Equation 4.1.

4.4.2 X-Ray Experiments: A Time-Resolved Story

X-ray scattering can be used in systems where the contrast is not an insuperable issue. Under these conditions, the main advantage of the technique is the possibility of performing nowadays time-resolved experiments in large-scale facilities. Otherwise, X-ray diffraction still remains the best technique for studying crystalline arrangements in polymers.

Systems where the contrast is high enough are, for example, *gels of agarose* in water or water/organic solvents. Ramzi et al. have studied the gel structure for different mixtures of aqueous solvents [59]. A typical example is given in Figure 4.15. The scattering curve can be interpreted with Equations 4.31 and 4.32. In both cases, in the intermediate q-range ($q < q^*$), the intensity can be fitted by considering a cross-section distribution of the type $w(r) \sim r^{-\lambda}$ with $\lambda = 1$ with $r_{max} = 7 \pm 1$ nm. Note that the intermediate behavior extends over one decade for the *water50/DMF50* sample. The Porod domain ($I(q) \sim 1/q^4$) is observable for *water70/DMF30* samples, which yield $r_{min} = 0.9 \pm 0.1$ from the value of q^* ($q^* = 2/\pi r_{min}$ see relation 4.33). In the second case, *water50/DMF50*, this behavior is absent. Yet, in both cases, a strong upturn is seen at very large q. Ramzi et al. have assumed that this upturn arises from pendent and/or free chains that scatter as $I(q) \sim 1/q^2$. Therefore, at larger q, this signal obliterates the $I(q) \sim 1/q^4$ signal scattered by the fibrils.

The *nanotubules* already described above represent a typical system that can be studied by both X-ray and neutron scattering (Fig. 4.16). Interestingly, the scattering curves differ conspicuously at large q values [56]. As was stressed in the introduction section, this is so because neutrons and X-ray do not see matter the same way. In the X-ray scattering curve, there is an additional term to that of the hollow cylinder, which arises from the molecular arrangement of the molecules in the nanotubules. This term exists because the electron density is not uniform in the BHPB-10 molecule: it is far higher in the vicinity of the benzene ring than anywhere else. As a result, the contrast is high enough to produce a diffraction peak. Conversely, as far as neutrons are concerned, contrast is much uniform within the molecular arrangement, and hence the absence of a diffraction peak. This may further suggest that molecules pile up head-to-tail onto one another, which eventually produces contrast uniformity. So far the molecular arrangement remains unknown for these nanotubules.

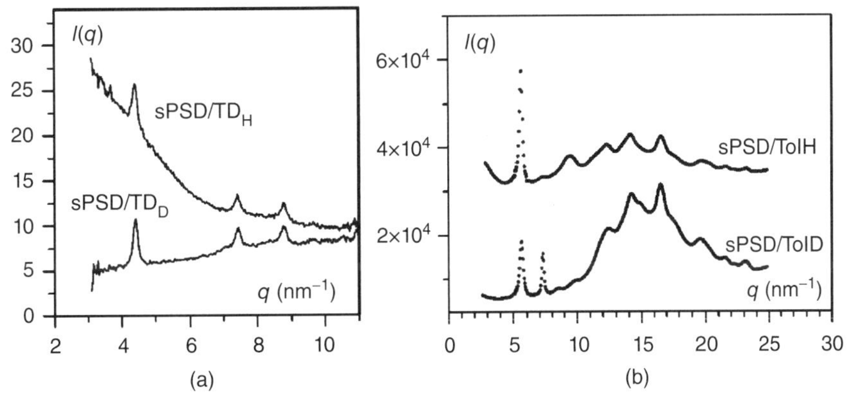

Figure 4.14 (a) Diffraction patterns for sPS_D/*trans*-$decalin_H$ and sPS_D/*trans*-$decalin_D$. sPS is here under the β form, that is, a nonsolvated form. Diffraction patterns for sPS_D/$Toluene_H$ and sPS_D/$Toluene_D$. Here, sPS is under the δ form, a solvated form. Replotted from [58].

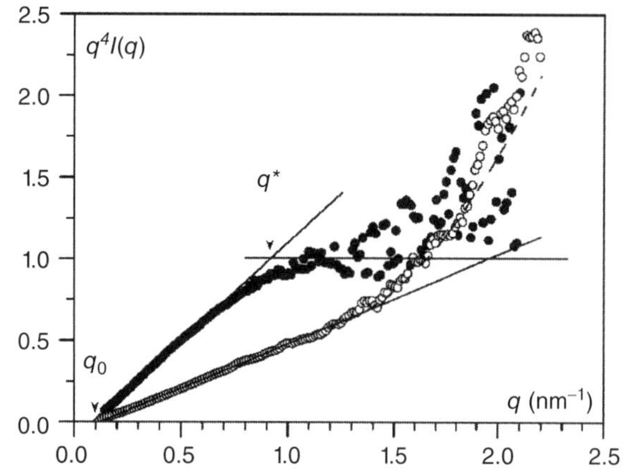

Figure 4.15 Scattering curves for agarose/water/DMF gels at room temperature. $C_{agarose} = 5$ g/L; (filled circle) = 70/30 (vol/vol), (open circle) = 50/50 (vol/vol). See text for further explanations. Replotted from [59].

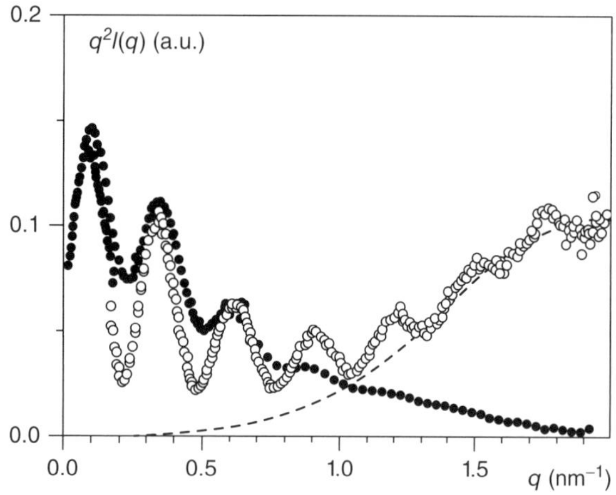

Figure 4.16 Scattering curves for 1% nanotubules in *trans*-decalin by X-ray (open circe) and by neutrons (filled circle). The additional term in X-ray scattering is highlighted by a dashed line and stands for the molecular arrangement within the nanotubules. Replotted from [56].

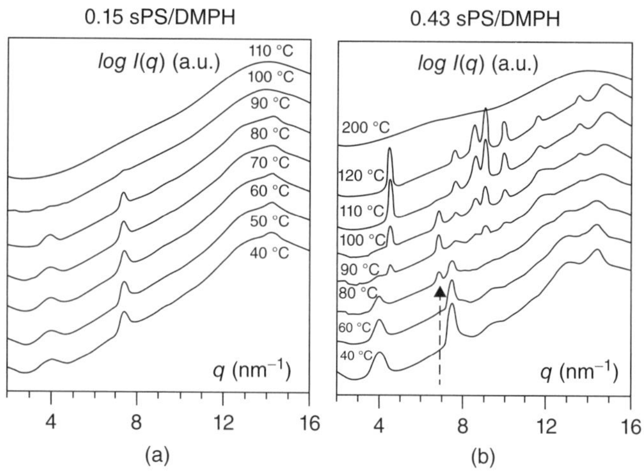

Figure 4.17 (a) Diffraction patterns obtained as a function of temperature for a heating rate ~2 °C/min. Polymer fractions as indicated. (b) The arrow indicates the new phase formed between 80 and 100 °C. Replotted from [60].

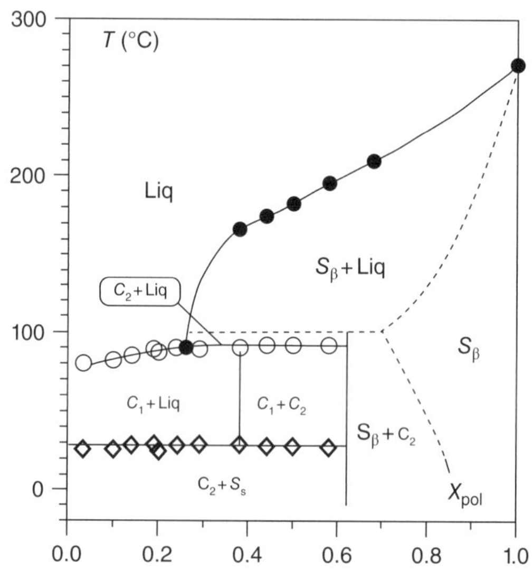

Figure 4.18 Temperature–concentration phase diagram for sPS/diphenyl methane. C_1 and C_2 stand for molecular compounds of differing stoichiometry; S_s and S_β for solid phases (the former crystallized solvent, the latter to the β form of sPS). C_2 is the form evidenced by X-ray diffraction and is most probably a mesophase of the nematic type. Replotted from [60].

As emphasized in the above title, one of the main interests in using X-ray scattering and/or diffraction on large-scale facilities lies in the possibility of performing time-resolved experiments. Coupling this with differential scanning calorimetry (DSC) experiments, the latter being always carried out at finite cooling or heating rate, allows one to clearly identify the occurrence of phases and therefore to confirm and/or complete the temperature-concentration phase diagrams mapped out through this calorimetric technique.

Typical examples displayed in Figure 4.17 present diffraction patterns obtained at a heating rate of 2 °C/min for syndiotactic polystyrene/diphenyl methane (sPS/DPM) in two concentration domains for which different thermal events occur [60]. This binary system is known to form intercalates, namely, crystallosolvates arising from the intercalation of solvent molecules in the crystal lattice of sPS. It turns out that Gibbs phase rules can be applied for mapping out the temperature–concentration phase diagram of these systems (Fig. 4.18).

Fulfilling these rules, particularly the variance rule which states that a maximum of three phases can coexist in a binary mixture at constant pressure, requires inventing an additional phase domain in a relatively large concentration range yet in a very narrow temperature range (designated as C_2+*Liq* in Fig. 4.18).

As is customary, the upper limit of this hypothetical phase domain is highlighted with a dashed line. This expected phase domain, hardly detectable by means of DSC experiments chiefly because of the overlap of endotherms, appears in the X-ray patterns by two distinctive peaks as shown in Figure 4.17b. Conversely, the molecular compound occurring in this phase domain is no longer present at lower concentration, as shown by the diffraction pattern of Figure 4.17a, where its characteristic peaks are absent.

Time resolved diffraction is thus a precious tool not only for determining molecular structures but also for investigating the thermodynamic properties of semicrystalline polymer/solvent binary systems.

4.5 CONCLUDING REMARKS

X-ray and neutron scattering techniques are highly important tools in the field of materials science, especially for polymers, thanks to their specificities as well as their nondestructive character. They give access to the meso and nanoscopic structures while keeping them virtually unaltered.

When it comes to writing some concluding remarks, it seems appropriate to emphasize their advantages while not hiding their shortcomings and drawbacks.

Neutron scattering is undoubtedly a powerful technique in polymer and colloid sciences thanks to the possibility of labeling a few molecules by means of deuteriation. This clearly allows one to study one object (a chain, an aggregate, etc.) dispersed amidst "brother" objects. This has been a decisive breakthrough in the study of amorphous or semicrystalline polymers. Also, the possibility of partial labeling has given access to the structure of parts of complex systems such as copolymers, nanocomposite materials, and the like.

The major drawback of this technique lies, however, in the limited compatibility of hydrogenous and deuterated "brother" objects. This can still be checked by carefully measuring molecular weight whenever possible or detecting it on the scattering curve. The thermodynamic properties, such as θ-temperatures, are also slightly altered, and these have to be taken into account in interpreting the data.

X-ray scattering is just fine as long as global structures of binary systems are investigated, provided some subsequent electronic contrast exists. As has been shown, this is typically the case of physical gels or phase-separated systems. Unlike neutrons, no object can be "singled out" among "brother" objects as labeling would require the use of heavy atoms, which in turn would totally modify the system. Conversely, the currently most interesting advantage is the opportunity of carrying out time-resolved experiments, at least in large-scale facilities, and thereby studying the evolution of the structures under formation.[3]

[3] Actually, time-resolved experiments can be achieved by neutron scattering at ILL for short sample-detector distances (D11 and D22).

One of the shortcomings of both techniques is the need for theoretical models to interpret the data, particularly when one is dealing with monotonously decreasing intensities. Two differing structures may have close scattering curves. For instance, flat cylinders or Gaussian chains possess scattering functions of the type $1/q^2$. The choice of the model therefore depends very much on other pieces of information gathered by other techniques. It is always recommended to cross-check conclusions drawn from these techniques with those derived from other techniques, even destructive techniques such as AFM, electron microscopy, and so on,

REFERENCES

1. Roentgen W. (1895) Sitzung Berichte der Würzburger Physik.medic. Gesellschaft, 3.
2. de Broglie L. (1925) Recherches Sur La Théorie Des Quanta. Ann Phys, 10e Série, 3.
3. Knoll M, Ruska E. The electron microscope. Z Phys 1932;78: 318–339.
4. Shull CG, Wollan EO. Applications of neutron diffraction to solid state problems. Solid State Phys: Adv Res Appl 1956;2: 137–217.
5. Benoit H, Duplessix R, Ober R, Daoud M, Cotton JP, Farnoux B, Jannink G. Scattered intensity by a polymer-chain in a sample with external constraints. Macromolecules 1975;8:451–453.
6. Guinier A. *Théorie et Technique de la Radiocristallographie*. Paris: Dunod Eds; 1956.
7. Kratky O, Porod G. Diffuse small-angle scattering of X-rays in colloid systems. J Colloid Sci 1949;4:35–70.
8. Porod G. Die Rontgenkleinwinkelstreuung Von Dichtgepackten Kolloiden. 1 Systemen. Kol-Z Z Polym 1951;124:83–114.
9. Cotton JP, Picot C, Farnoux B, Jannink G, Mons J. Observation on conformation of polymeric chain in amorphous solid by small-angle neutron-scattering. CR Acad Sci C 1972; 275:175–177.
10. Flory PJ. *Principles of Polymer Chemistry*. New York: Cornell University Press; 1953.
11. Daoud M, Cotton JP, Farnoux B, Jannink G, Sarma G, Benoit H, Duplessix R, Picot C, Gennes PGD. Solutions of flexible polymers – Neutron experiments and interpretation. Macromolecules 1975;8:804–818.
12. Spells SJ, Sadler DM, Keller A. Chain trajectory in solution grown polyethylene crystals – Correlation between infrared-spectroscopy and small-angle neutron-scattering. Polymer 1980;21:1121–1128.
13. Guenet JM. A neutron-scattering study of the chain trajectory in isotactic polystyrene single-crystals. Macromolecules 1980; 13:387–391.
14. Sadler DM, Spells SJ, Keller A, Guenet JM. Wide-angle neutron-scattering from isotactic polystyrene – The fold arrangement in solution grown crystals. Polym Commun 1984;25:290–293.
15. Egelstaff PA. *An Introduction to the Liquid State*. London: Academic Press; 1967.

16. Higgins JS, Benoit HC. *Polymer and Neutron Scattering*. Oxford: Clarendon Press; 1994.
17. Zimm BH. Apparatus and methods for measurement and interpretation of the angular variation of light scattering – Preliminary results on polystyrene solutions. J Chem Phys 1948;16:1099–1116.
18. Debye P. Light scattering in solutions. J Appl Phys 1944; 15:338–342.
19. a Kuhn W. Concerning the shape of thread shapes molecules in solution. Kol Z 1934;68:2–15;b Descloizeaux J. Form-factor of an infinite Kratky–Porod chain. Macromolecules 1973;6: 403–407.
20. Jannink G, Descloizeaux J. Polymers in solution. J Phys 1990; 2:1–24.
21. Hermans J, Hermans JJ. Light scattering by random zig-zag and by worm-like chains. J Phys Chem 1959;62:1543–1546.
22. Luzzati V, Benoit H. Diffusion Centrale Des Rayons X Par Des Particules Filiformes. Acta Crystallogr 1961;14:297–304.
23. Neugebauer T. Calculation of the light diffusion of filament chain solutions. Ann Phys 1942;42:509–533.
24. Yoshizaki T, Yamakawa H. Scattering functions of wormlike and helical wormlike chains. Macromolecules 1980;13:1518–1525.
25. Mittelbach P. Zur Röntgenkleinwinkelstreuung Verdunnter Kolloider Systeme. Acta Phys Austriaca 1964;19:53–57.
26. Fournet G. Fonctions de diffusion pour des formes géométriques. B Soc Fr Min Crist 1951;74:39.
27. Mittelbach P, Porod G. Zur Röntgenkleinwinkelstreuung verdünnter kolloiden Systeme. Acta Phys Austriaca 1961;14:185–211.
28. Guenet JM. Scattering by a network of prolate, cross-section-polydispersed cylinders applicable to fibrillar thermoreversible gels. J Phys II 1994;4:1077–1082.
29. a Schmidt PW. Small angle X-ray scattering from helical filaments. J Appl Crystallogr 1970;3:257–265;
b Pringle OA, Schmidt PW. Small-angle X-ray scattering from helical macromolecules. J Appl Crystallogr 1971;4:290–295.
30. Saiani A, Guenet JM. On the helical form in syndiotactic poly (methyl methacrylate) thermoreversible gels as revealed by small-angle neutron scattering. Macromolecules 1997; 30:966–972.
31. Oster G, Riley DP. Scattering from cylindrically symmetric systems. Acta Crystallogr 1952;5:272–276.
32. Rayleigh L. On the diffraction of light by spheres of small relative on the index. Proc R Soc 1914;90:219–225.
33. Oster G, Riley DP. Scattering from isotropic colloidal and macromolecular systems. Acta Crystallogr 1952;5:1–6.
34. Porod G. Die Rontgenkleinwinkelstreuung Von Dichtgepackten Kolloiden. 2. Systemen. Kol-Z Z Polym 1951;125:51–57.
35. Leibler L. Theory of microphase separation in block co-polymers. Macromolecules 1980;13:1602–1617.
36. Bacon GE. Coherent neutron-scattering amplitudes. Acta Crystallogr A 1972;A28:357–370.
37. Fazel N, Brulet A, Guenet JM. Molecular-structure and thermal-behavior of poly(methyl methacrylate) thermoreversible gels and aggregates. Macromolecules 1994;27:3836–3842.
38. Zemb T, Lindner P, editors. *Neutron, X-Ray and Light: Scattering Methods Applied to Soft Condensed Matter*. Amsterdam: North Holland; 2002.
39. Strazielle C, Benoit HC. Some thermodynamic properties of polymer-solvent systems – Comparison between deuterated and undeuterated systems. Macromolecules 1975;8:203–205.
40. Bates FS, Wignall GD, Koehler WC. Critical-behavior of binary-liquid mixtures of deuterated and protonated polymers. Phys Rev Lett 1985;55:2425–2428.
41. Sadler DM, Keller A. Trajectory of polyethylene chains in single-crystals by low-angle neutron-scattering. Polymer 1976; 17:37–40.
42. Kirste RG, Wunderlich W. Determination of radius of gyration and 2nd viral coefficient of polymers in solution from small angle radiography. Z Phys Chem 1968;58:133–140.
43. http://www.ill.eu/. Accessed 2015 Oct 30.
44. http://www.esrf.eu/. Accessed 2015 Oct 30.
45. a Crist B, Graessley WW, Wignall GD. Chain dimensions of crystallizable polymers in the solid and melt states. Polymer 1982;23:1561–1567;
b Guenet JM. Chain dimensions of crystallizable polymers: An alternative interpretation of the experimental-data – Comments. Polym Commun 1983;24:323–325.
46. Schmatz W, Schelten J, Wignall GD, Ballard DGH. Neutron small-angle scattering from deuterium-tagged chain molecules in crystalline polyethylene. Acta Crystallogr A 1975;31:S162.
47. Guenet JM, Picot C. Conformation of isotactic polystyrene (Ips) in the bulk crystallized state as revealed by small-angle neutron-scattering. Polymer 1979;20:1483–1491.
48. Guenet JM. Neutron-scattering investigations on the effect of crystallization temperature and thermal-treatment on the chain trajectory in bulk-crystallized isotactic polystyrene. Polymer 1981;22:313–320.
49. Guenet JM. Neutron-scattering investigation of the chain trajectory in thermoreversible gels. Macromolecules 1987; 20:2874–2878.
50. Klein M, Brulet A, Guenet JM. Molecular-structures in isotactic polystyrene thermoreversible gels. Macromolecules 1990; 23:540–548.
51. Daniel C, Menelle A, Brulet A, Guenet JM. Thermoreversible gelation of syndiotactic polystyrene in toluene and chloroform. Polymer 1997;38:4193–4199.
52. Guenet JM, Picot C. Bulk crystallization of isotactic polystyrene near its melting-point – A neutron-scattering study of the chain trajectory. Macromolecules 1983;16:205–210.
53. Ray B, El Hasri S, Guenet JM. Effect of polymer tacticity on the molecular structure of polyelectrolyte/surfactant stoichiometric complexes in solutions and gels. Eur Phys J E 2003;11:315–323.
54. (a) Terech P, Schaffhauser V, Maldivi P, Guenet JM. Rheological and neutron-scattering investigations of the jelly state of binuclear copper-complexes in cyclohexane. Europhys Lett 1992;17:515–521; (b) Lopez D, Guenet JM. Encapsulation of self-assembled bicopper complex filaments in thermoreversible gel fibrils: Effect of the solvent isomer. J Phys Chem B 2002;106:2160–2165.

55. Lopez D, Guenet JM. Encapsulation of filaments of a self-assembling bicopper complex in polymer nanowires. Eur Phys J B 1999;12:405–411.

56. Dasgupta D, Kamar Z, Rochas C, Dahmani M, Mesini P, Guenet JM. Design of hybrid networks by sheathing polymer fibrils with self-assembled nanotubules. Soft Matter 2010;6:3573–3581.

57. Diaz N, Simon FX, Schmutz M, Rawiso M, Decher G, Jestin J, Mesini PJ. Self-assembled diamide nanotubes in organic solvents. Angew Chem Int Ed 2005;44:3260–3264.

58. Malik S, Rochas C, Schmutz M, Guenet JM. Syndiotactic polystyrene intercalates from naphthalene derivatives. Macromolecules 2005;38:6024–6030.

59. Ramzi M, Rochas C, Guenet JM. Structure–properties relation for agarose thermoreversible gels in binary solvents. Macromolecules 1998;31:6106–6111.

60. Malik S, Rochas C, Guenet JM. Thermodynamic and structural investigations on the different forms of syndiotactic polystyrene intercalates. Macromolecules 2006;39:1000–1007.

5

DIFFERENTIAL SCANNING CALORIMETRY OF POLYMERS

ALEJANDRO J. MÜLLER[1,2,3] AND ROSE MARY MICHELL[3]

[1]*POLYMAT and Polymer Science and Technology Department, Faculty of Chemistry, University of the Basque Country UPV/EHU, Paseo Manuel de Lardizabal 3, 20018. Donostia-San Sebastián, Spain*
[2]*IKERBASQUE, Basque Foundation for Science, Bilbao, Spain*
[3]*Grupo de Polímeros USB, Departamento de Ciencia de los Materiales, Universidad Simón Bolívar, Apartado 89000, Caracas 1080-A, Venezuela*

In this chapter, the basics of differential scanning calorimetry (DSC) analysis and its correlation to polymer morphology for semicrystalline polymeric materials are presented. After a brief review of fundamental concepts, the utility of the technique is illustrated by a series of practical applications.

5.1 INTRODUCTION TO DIFFERENTIAL SCANNING CALORIMETRY. BASIC PRINCIPLES AND TYPES OF DSC EQUIPMENT

The differential scanning calorimetry (DSC) technique is probably the most widely used thermal analysis technique in polymer science. It determines in a quantitative way the amount of heat absorbed or evolved by a sample as compared to a reference during a specific thermal transition (which could be a first-order or second-order transition). The changes experienced by the thermal properties of the samples can be determined isothermally as a function of time or at a constant heating or cooling rate as a function of temperature [1, 2].

Figure 5.1 shows a schematic that represents the two basic types of DSC. In Figure 5.1a a power compensation equipment is represented. It consists of two separate adiabatic ovens where the sample and the reference are placed within aluminum pans. The typical reference material is air (i.e., an empty pan). Each oven is equiped with a resistance and a thermocouple. Power is supplied to both ovens at the same time, so that the temperature is increased linearly with time. The thermocouples are employed to monitor the temperatures. The equipment operates by keeping the difference in temperature between the sample and the reference always at zero. This is done by compensating the power supplied to the sample whenever it undergoes a phase change or a thermal transition that requires either more or less power (when heat is absorbed or released from the sample). In this way, this power compensation is determined, which is directly proportional to the amount of heat that was transfered to the sample. Figure 5.1c shows a photograph of a cross section of a typical power compensation cell, and the left oven was sectioned for illustration purposes.

Figure 5.1b shows the configuration of a heat flux DSC. In this case, both pans, containing the reference and the sample, are placed inside a single oven and the temperatures are carefully determined as power is supplied to produce a linear temperature increase of a desired constant heating rate. The temperature difference generated whenever a thermal transition takes place in the sample is proportional to the heat transferred to or from the sample [1].

In both cases, suitable calibration and operation allowed calorimetric quantities to be measured, such as heat capacity or a quantity proportional to the heat capacity [1].

Top-notch power compensation calorimeters can be very sensitive and are an excellent tool for academic studies. However, they can be noisy, not so easy to calibrate, and frequently present baseline stability problems. Heat flux calorimeters tend to be more robust, with better baseline

Polymer Morphology: Principles, Characterization, and Processing, First Edition. Edited by Qipeng Guo.

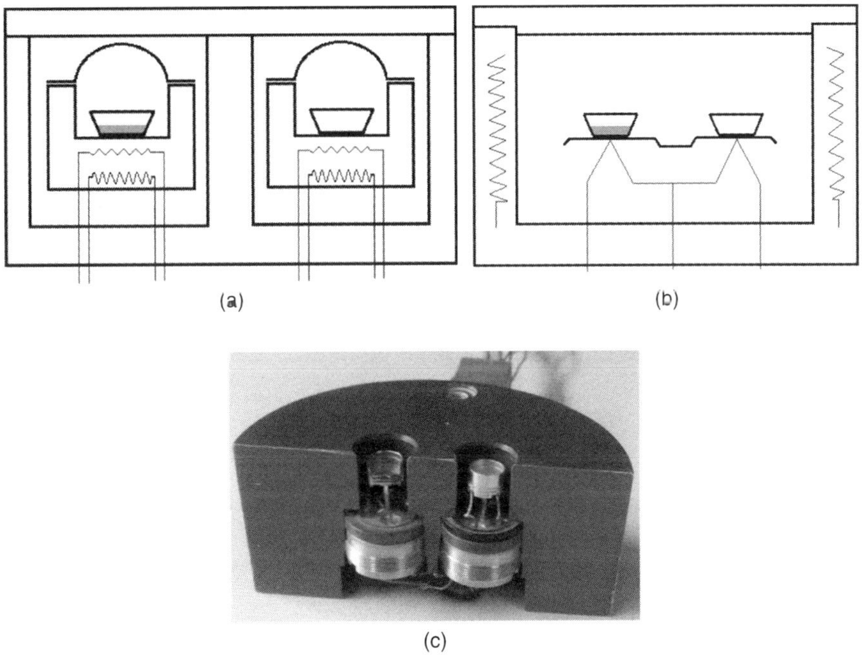

Figure 5.1 Schematic representation of the basic types of DSC equipments: (a) power compensation DSC, (b) heat flux DSC, (c) photograph of transversal cut of a Perkin Elmer® cell for a DSC 7 calorimeter.

TABLE 5.1 Typical Applications of DSC [28]

Melting Ranges	Release of Strains
Rates of crystallization and reaction	Purity determination
Degree of crystallinity	Quality control
Glass-transition phenomena	Phase diagrams
Heat capacity	Thermal and oxidative stability
Enthalpy of transitions	Hazards evaluation
Identification	Mesophase transitions
Thermal conductivity	Nucleation phenomena
Analysis of copolymers and blends	Catalysis

stability, and the simplest models can be suitable for quality control routine operations. The most sophisticated models can match the sensitivity of the power compensation calorimeters.

A great effort has been recently made to build faster calorimeters [3–13]. They can perform analysis at heating or cooling rates that range from 500 to 30,000 °C/min. The faster end of the heating/cooling rate range has been achieved with the so-called fast chip calorimeters [14]. In these devices, the measuring cell is a chip where the sample is placed directly on its surface. To avoid heat transfer problems, the mass has to be reduced to the micrograms or even the picograms range. More details can be found in several recent publications by the group of Schick (see Refs [14–27]).

DSC has many diverse applications for polymeric materials, some of which are summarized in Table 5.1.

In this chapter, we concentrate on the following applications that are important for the study of polymer thermal properties, nucleation, and crystallization and their relationship with morphology:

- Standard tests. These are the most fundamental DSC tests performed by heating or cooling samples at a constant rate. These tests must be performed to have an idea of the basic thermal transitions of the samples under study.
- Self-nucleation (SN) tests. In these specialized tests, the self-nucleation of a semicrystalline polymer can be studied. Also, with its results, the efficiency of nucleating agents (NAs) or fillers can be quantitatively determined.
- Thermal fractionation by successive self-nucleation and annealing (SSA) technique. Amongst the thermal fractionation methods, the SSA technique is the most efficient and sensitive. It is mostly employed for ethylene/α-olefin copolymers in order to evaluate in a quick way their short-chain branching (SCB) distribution. However, it can be employed to analyze the effect that any defect along the chain can produce in the distribution of linear crystallizable sequences.

Another important application for DSC is in isothermal tests. They are performed at constant temperature and are useful to determine overall crystallization kinetics, but can also be used to determine curing kinetics or monitoring isothermal polymerization. Chapter 11 details the use of DSC in isothermal mode, so it is not treated in this chapter.

5.2 DETECTION OF FIRST-ORDER AND SECOND-ORDER TRANSITIONS BY DSC. APPLICATIONS OF STANDARD DSC EXPERIMENTS TO THE DETERMINATION OF THE GLASS TRANSITION TEMPERATURE AND THE MELTING TEMPERATURE OF POLYMERIC MATERIALS

Standard DSC tests at constant scanning rates can be employed to extract the following data from amorphous or semicrystalline polymers:

1. Glass transition temperature (T_g);
2. Crystallization temperature (T_c);
3. Melting temperature (T_m);
4. Latent heat of fusion or enthalpy of fusion (ΔH_f);
5. Latent heat of crystallization or crystallization enthalpy (ΔH_c);
6. Crystallinity degree (X_c).

In order to understand the origin of the transitions detected by DSC, it is useful to understand the differences between first- and second-order transitions.

Phase transitions are generally classified by the criteria developed by Ehrenfest. It is considered that first-order transitions are those at which the free energy as a function of a given state variable (volume (V), pressure (p), and temperature (T)) is continuous, while the first derivative is discontinuous or steplike. First-order thermodynamic quantities are those that can be expressed as the first derivative of Gibbs free energy, such as

$$\left(\frac{\partial G}{\partial T}\right)_p = -S \tag{5.1}$$

$$\left(\frac{\partial G}{\partial T}\right)_T = V \tag{5.2}$$

$$\left(\frac{\partial\left(\frac{G}{T}\right)}{\partial\left(\frac{1}{T}\right)}\right)_p = H \tag{5.3}$$

Hence, the entropy (S), the volume (V), and the enthalpy (H) are all first-order thermodynamic quantities. If a first-order thermodynamic quantity is monitored as a function of temperature, such quantities experience a discontinuity or jump whenever a first-order transition is encountered, such as melting or freezing (see Fig. 5.2b).

Second-order thermodynamic quantities are those that can be expressed as a second derivative of free energy, such as

$$\frac{\partial}{\partial T}\left[\frac{\partial\left(\frac{G}{T}\right)}{\partial\left(\frac{1}{T}\right)}\right]_p = \left(\frac{\partial H}{\partial T}\right)_p = C_p \tag{5.4}$$

where C_p is the heat capacity of a sample at constant pressure. In fact, the aim of any quantitative calorimetry experiment performed at constant pressure is to obtain the differential amount of heat dQ required to increment the temperature of the sample by a differential amount dT, which can be expressed as

$$C_p = \frac{dQ}{dT} = \left(\frac{\partial H}{\partial T}\right)_p \tag{5.5}$$

During a first-order transition, like crystallization or melting, the heat capacity (a second-order quantity) will experience a peak. These are the peaks that can be seen when standard DSC scans are performed (Fig. 5.3b). The area under such endothermic or exothermic peaks will then be the associated enthalpy change with melting or crystallization.

A second-order thermodynamic transition is a transition that experiences only a change in slope whenever a first-order quantity is being monitored as a function of temperature (see Fig. 5.2a). On the other hand, when a second-order quantity like C_p is being monitored as a function of temperature like in a DSC run, a step or discontinuity will be visible at the second-order transition. The glass transition temperature (T_g) behaves like a second-order transition in this sense (as shown in Fig. 5.3a). However, it has a kinetic nature, because the exact value of the T_g depends on the rate at which it is measured. Therefore, it cannot be considered a

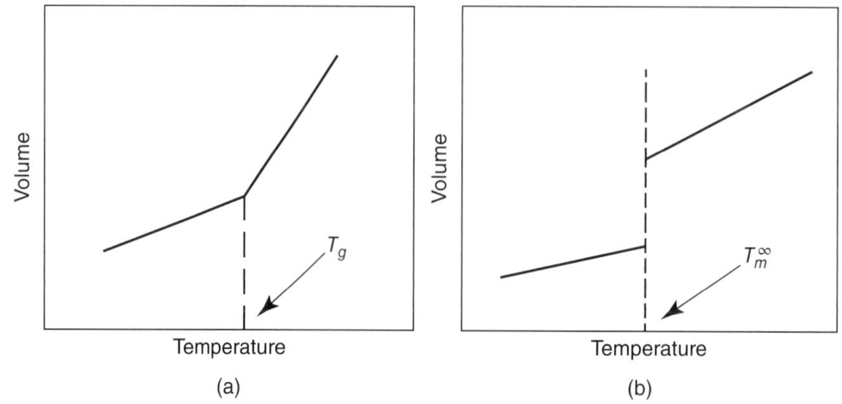

Figure 5.2 Schematic representation of monitoring volume as a function of temperature during (a) a second-order transition and (b) a first-order transition. Hiemenz [29]. Reproduced with permission of Taylor and Francis.

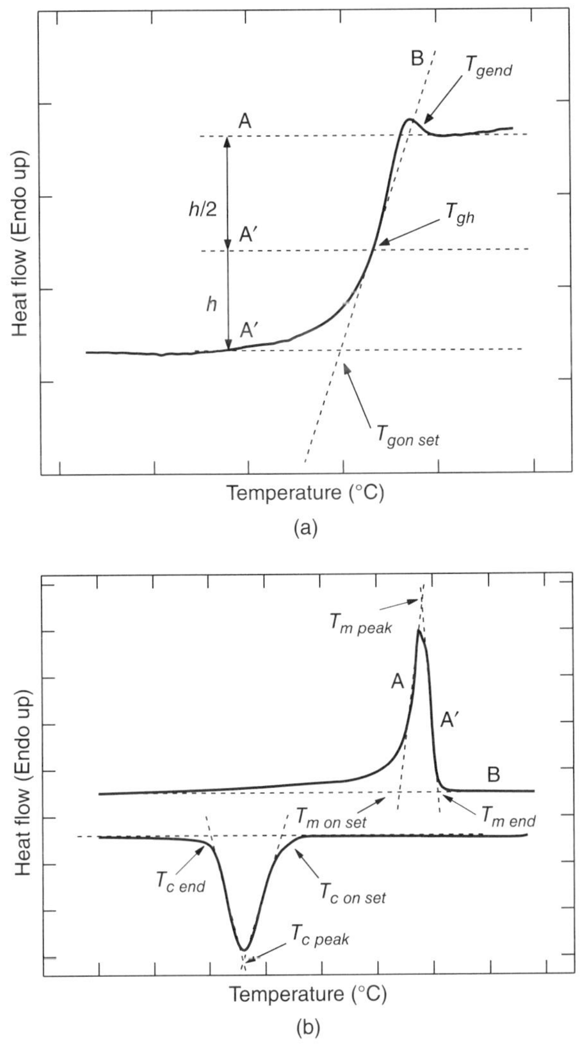

Figure 5.3 Schematic representation of thermal transition characteristic values: (a) determination of characteristic glass transition temperatures, (b) determination of characteristic melting and crystallization temperatures.

true thermodynamic transition, but a pseudothermodynamic second-order transition.

As is well known, the thermal transitions like T_g, T_c, or T_m do not occur at a single temperature but at a temperature range. Therefore, certain criteria had to be applied to determine these transitions and report a single average or meaningful value. One possibility is to follow the ASTM D3418 standard, whose guidelines are illustrated in Figure 5.3.

Figure 5.3a illustrates a material undergoing its glass transition, as an endothermal step, as the temperature is increased. The T_g can be expressed as the onset, midpoint, or end value of the temperature range involved in the transition. One of the values most frequently used is the midpoint value. The scanning rate should always be reported when values of T_g are given.

Figure 5.3b shows cooling and heating scans where a sample is undergoing crystallization and melting. The values most frequently employed as characteristic temperatures representing T_c and T_m are the peak values.

5.3 SELF-NUCLEATION

Self-nucleation is a technique to produce self-seeds or self-nuclei within a polymer melt, so that its nucleation can be greatly enhanced. In principle, the best nucleating agent for any polymer should be its own crystallographically ideal crystal fragments or chain segments with a residual crystal memory [30–33].

The self-nucleation (SN) technique was originally applied by Keller et al. [32] to create in a controlled fashion single crystals from solution. Fillon et al. devised a thermal protocol to induce and study self-nucleation by DSC [30].

The SN procedure is schematically shown in Figure 5.4 and can be described as follows [30, 33]:

(a) Erasure of previous thermal history and crystalline memory. The sample is kept in the melt for 3 min at a high enough temperature (typically, 25–30 °C above its peak melting temperature). All thermally sensitive nuclei must be destroyed in this first step, leaving only temperature-resistant heterogeneous nuclei of unknown nature (catalyst residues, impurities, or any other heterogeneities).

(b) Creation of the initial "standard" semicrystalline state. The sample is cooled from the melt at a constant rate (typically, 10 °C/min) down to a minimum temperature that should be low enough to allow the sample to crystallize until saturation. The peak crystallization temperature recorded during this cooling scan is the "standard" crystallization temperature (or standard T_c), because it is a function of the density of thermally stable nuclei of the sample. The sample is held at the minimum temperature for 3 min.

(c) Thermal conditioning at a specific temperature (where the sample could melt, self-nucleate, or self-nucleate and

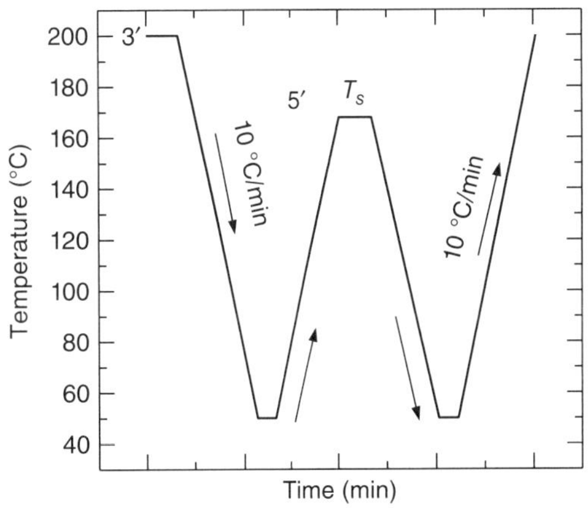

Figure 5.4 Schematic representation of a self-nucleation treatment.

anneal depending on the temperature). The sample is now heated at a constant rate (the same rate employed in step (b)) from the chosen minimum temperature up to a selected self-seeding or self-nucleation temperature (commonly denoted as T_s), and then the sample is held at this T_s temperature for 5 min.

(d) Subsequent cooling at a constant rate from T_s down to the minimum temperature chosen in step (b).

(e) Final melting. Subsequent heating at a constant rate (once again, it is recommended to use the same rate as in steps (b) through (d), i.e., 10 °C/min) from the minimum temperature chosen in (b) up to the maximum melting temperature established in step (a).

The most important parameters during SN are (i) the heating and cooling rates employed, (ii) the T_s temperature, and (iii) the time spent at T_s.

Figure 5.5 presents the self-nucleation behavior of an isotactic polypropylene (PP). Figure 5.5a shows the cooling runs after thermal conditioning at the indicated T_s temperatures and Figure 5.5b shows the subsequent heating scans.

Fillon et al. [30] have defined the so-called domains of self-nucleation for isotactic polypropylene (PP) and we will follow their definitions:

Domain I (*complete melting domain*). The polymer is under *Domain I* when complete melting occurs and the crystalline history of the material is completely erased. For the PP employed in the example provided in Figure 5.5, complete melting was found to persist down to 169 °C, because no change was detected in the peak crystallization temperature (T_c), as compared to the standard crystallization temperature obtained with a T_s temperature of 200 °C. The PP of Figure 5.5 is under *Domain I* at any T_s temperature larger or equal than 169 °C.

Domain II (*exclusive self-nucleation domain*). Thermal conditioning at 168 °C for 5 min was able to self-nucleate the PP, and T_c (the peak crystallization temperature) was shifted to higher temperatures while the subsequent melting (see Fig. 5.5b) did not reveal any traces of annealing. This is the characteristic behavior of *Domain II* or exclusive self-nucleation *domain*. The exact nature of the self-seeds that are causing self-nucleation is controversial. According to Fillon et al. [30] the temperature in *Domain II* is high enough to melt almost all of the polymer crystals, but low enough to leave "small" crystal fragments that can act as self-nuclei. This is probably true for the lowest temperatures of *Domain II* (i.e., 166 °C). Lorenzo et al. have also considered another possibility to explain the increase in nucleation density detected within the higher temperature range of *Domain II*. They argue that 5 min at those high T_s temperatures are not enough to erase the crystalline memory of PP or to relax the residual segmental orientation in the melt that can remain just after melting occurs. In fact, the actual presence of crystal remains is not needed to produce self-nucleation, as demonstrated in the work of Lorenzo et al. [31].

Regardless of the origin of the self-seeds, the experimental fact remains that when the sample is within *Domain II*, an increase in nucleation density is obtained as indicated by the increase in T_c. This increase in nucleation density can be exponential for polymers

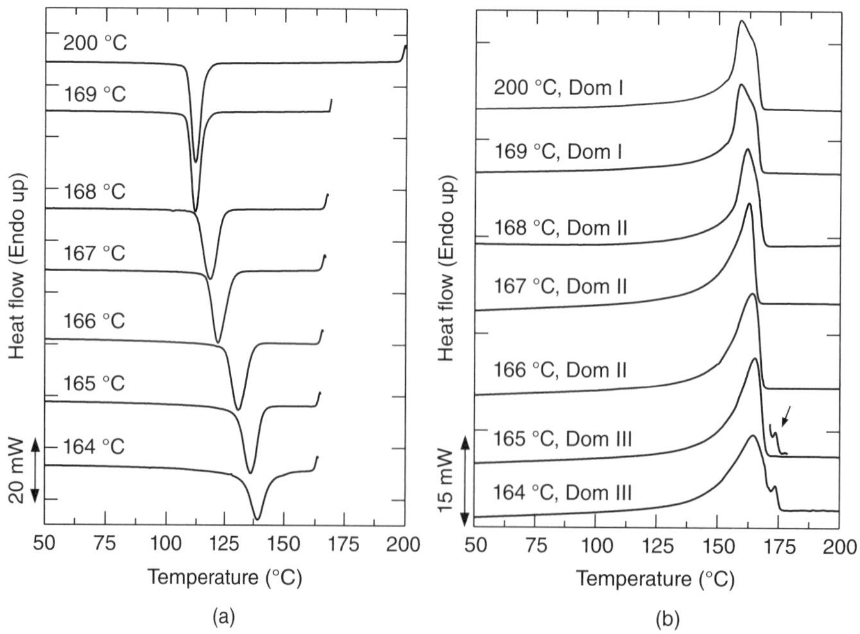

Figure 5.5 (a) DSC cooling scans (at 10 °C/min) for PP after 5 min at the indicated T_s. (b) Subsequent heating scans (at 10 °C/min) after the cooling runs shown in (a). Lorenzo et al. [[31], Figure 1]. Reproduced with permission of John Wiley and Sons.

with intrinsically low nucleation densities. In the case of PP, as the T_s temperature decreases from 169 to 166 °C, the nucleation density approximately increases from 10^6 nuclei/cm^3 in *Domain I* to 10^{12} nuclei/cm^3. The nucleation density can change more than an order of magnitude per degree in *Domain II* for PP [30]. Figure 5.5 shows that the minimum T_s temperature within *Domain II* for the PP examined was 166 °C, because the sample was self-nucleated without any annealing. The minimum T_s temperature within *Domain II* is defined as the "ideal self-nucleation temperature," because it is the temperature that causes the maximum self-nucleation (maximum increase in T_c) without producing any annealing.

Domain III (*self-nucleation and annealing domain*). When the T_s temperature employed is too low, partial melting is produced and the unmelted crystals can experience annealing during the 5-min thermal conditioning at T_s. In Figure 5.5, that at a T_s value of 165 °C, even though the crystallization exotherm does not reveal any unusual signs to indicate that the sample has crossed over to *Domain III*, the subsequent melting endotherm in Fig. 5.5b shows a small high-temperature sharp peak that is due to the melting of the annealed crystals at that T_s temperature. At a T_s value of 164 °C, the crystallization exotherm exhibits almost immediate crystallization upon cooling (in the form of a high-temperature tail), which is also a revealing sign that the sample is in *Domain III*.

Figure 5.6 provides a schematic diagram to illustrate at a molecular level the different self-nucleation *domains*. In *Domain I*, after the thermal conditioning at T_s, the material melts completely and the chains adopt the random coil conformation characteristic of the isotropic melt. In *Domain II*, two possibilities have been considered. At the high-temperature end of *Domain II*, the melt just retains some residual segmental orientation that constitutes its crystalline memory which allows the polymer to exhibit self-nucleation. At the low-temperature end of *Domain II*, small crystal fragments survive melting and constitute the self-seeds. Finally, in *Domain III*, partial melting and annealing of unmolten crystals coexist.

5.3.1 Quantification of the Nucleation Efficiency

Self-nucleation can be used as the basis of comparison to determine the nucleation power of additives, nanofillers, or nucleating agents. This is a very useful application in order to quantify the efficiency of nucleation of a foreign additive into any polymer. The idea was proposed by Fillon et al. [34]. The first step would be to self-nucleate the polymer of interest, just like in Figure 5.4, to determine its self-nucleation domains.

Employing the PP of Figure 5.5 as an example, the crystallization temperature of PP under standard crystallization conditions is determined, that is, T_c or the peak crystallization temperature corresponding to Domain I (in Fig. 5.5a that would be the peak T_c value corresponding to any cooling scan at T_s values equal or larger than 169 °C). Then the ideal self-nucleation temperature must be determined, as indicated above. For the PP of Figure 5.5a such ideal T_s temperature is 166 °C, or the minimum temperature within *Domain II*. The peak crystallization temperature upon cooling from the ideal self-nucleation T_s is then determined and labeled $T_{c,Max}$, because this would be the maximum crystallization

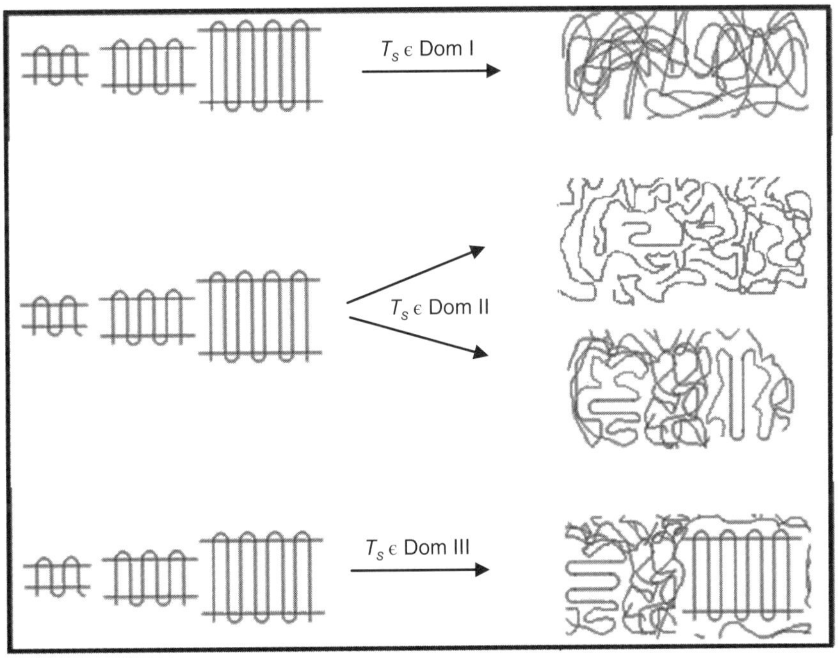

Figure 5.6 Schematic molecular representation of the different self-nucleation *domains*.

temperature of the self-nucleated sample (without any annealing).

We now consider the case where an nucleating agent (NA) is added to the polymer of interest. The sample should be treated as the parent polymer without the additive and after erasing its thermal history, a cooling scan from the melt should reveal its peak crystallization temperature, which is labeled $T_{c,NA}$. Then Equation 5.6 will yield the nucleation efficiency (NE) in percentage [34]:

$$NE = \frac{T_{C,NA} - T_C}{T_{C,Max} - T_C} \times 100 \tag{5.6}$$

This equation represents a simple way to quantify the nucleating action of an additive in relative terms to its self-nucleation behavior.

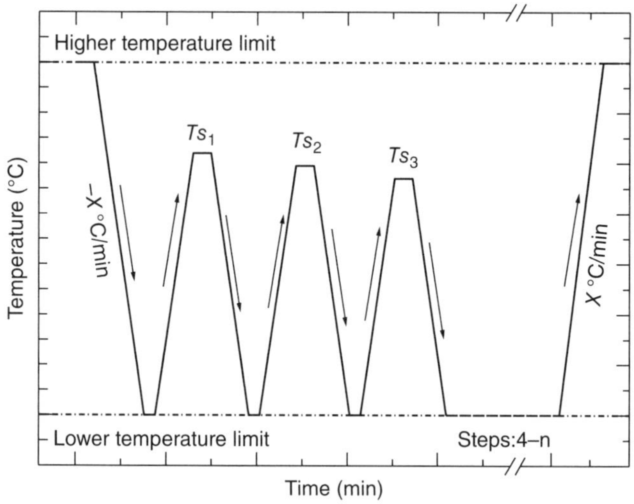

Figure 5.7 SSA thermal protocol schematic representation. Cooling and heating scans are performed at a constant scanning rate.

5.4 THERMAL FRACTIONATION

The SCB distribution produced by copolymerization of ethylene with α-olefins determines many of the polymer physical and mechanical properties. The comonomer distribution can be ascertained by temperature rising elution fractionation (TREF). This technique produces separation by the elution of fractions at increasingly rising temperatures of a given polymer that has been previously crystallized from solution on an inert support during multiple steps or very slow cooling [35, 36]. Slow crystallization from solution favors molecular segregation by SCB content and distribution with very limited influence of molecular weight. Hence, the more linear chains precipitate at higher temperatures, and those with higher comonomer content will do so at lower temperatures [36, 37]. A related technique is CRYSTAF (crystallization analysis fractionation) [35, 36]. CRYSTAF monitors the concentration of the polymer in solution during the crystallization stage. Therefore, analysis times in CRYSTAF are shorter but still significant and both TREF and CRYSTAF employ solvents and involve costly equipment investment. On the other hand, thermal fractionation techniques that can be performed with a DSC do not involve solvents, but the physical separation of fractions is not possible.

Thermal fractionation employs a temperature routine (either programmed step cooling or a series of heating and cooling cycles) that is designed to produce a distribution of lamellar crystals whose sizes reflect the distribution of methyl sequence length (MSL) that are present in the ethylene/α-olefin copolymer [33]. The experiment is performed in a conventional DSC.

Step crystallization (SC), where a programmed step cooling is applied [33, 38–40], and SSA, where a series of heating and cooling cycles are employed [33, 41–48], are the two most frequently employed thermal fractionation techniques, and their comparative advantages and shortcomings have been reviewed [33]. The polymer chains are never physically separated during thermal fractionation and therefore the technique is sensitive to linear and uninterrupted chain sequences, that is, MSL. This implies that thermal fractionation is performed both intramolecularly and intermolecularly. The important unit is the linear chain length in between branch points, or the MSL.

Müller et al. designed and implemented SSA [33, 41, 42, 46–48]. After thermal conditioning of the sample, a final DSC heating run reveals the distribution of melting points induced by SSA as a result of the heterogeneous nature of the chain structure of the polymer under analysis. SSA is performed at substantially shorter times than SC and with better resolution [33, 41–48].

Figure 5.7 shows a schematic representation of the thermal protocol involved in SSA:

The first two steps are identical to the SN protocol described above (Fig. 5.4)

(a) Erasure of previous thermal history and crystalline memory.
(b) Creation of the initial "standard" semicrystalline state.
(c) The sample is heated at a constant rate (the same rate employed in step (b)) from the lower temperature limit to the ideal self-nucleation temperature that must be determined in separate SN experiments. In Figure 5.7, this first T_s temperature (which must be the ideal self-nucleation temperature) is denoted as T_{s1}.
(d) The sample is held at the ideal T_s for 5 min. This is the isothermal fractionation time and it should be kept constant. The time of 5 min is usually enough to get well-separated thermal fractions. This isothermal treatment at T_{s1} results in maximum self-nucleation without annealing, because the sample is at the lowest temperature of *Domain II*.
(e) Cooling from T_{s1}. When the sample is cooled at a constant rate from T_{s1} to the lower temperature limit,

the polymer will crystallize, having been ideally self-nucleated.

(f) The sample is heated at a constant rate (the same rate employed in step (b)) from the lower temperature limit to T_{s2}. The difference in temperature between T_{s1} and T_{s2} is usually set at 5 °C. This is the fractionation window (ΔT_f) because it determines the size of the thermal fraction and it should be kept constant throughout the SSA experiment. T_{s2} is within *Domain III*, and hence it will produce annealing of unmolten crystals and self-nucleation of the molten polymer when the sample is cooled down. This is how the first fraction will be generated.

(g) Steps "e," "f," and "g" are repeated at progressively lower T_s. The number of repetitions (cycles) can be chosen to cover the entire melting range of the sample with a "standard" thermal history or a shorter range.

(h) Final melting. The sample is heated at the chosen constant rate up to the melt state. In this step, the fractionation will be revealed.

The effects of changing ΔT_f, the fractionation time, and the initial T_s have been previously discussed in the literature [33].

Figure 5.8 shows DSC heating scans before (Fig. 5.8a) and after SSA fractionation (Fig. 5.8b) of two polyethylene samples. A commercial ethylene/α-olefin copolymer denoted 11U4 is a linear low-density polyethylene (LLDPE) synthesized by Ziegler–Natta-type catalysis and therefore the material exhibits a bimodal distribution of melting points (therefore of lamellar thicknesses) after SSA that reflects its bimodal distribution of SCB [33, 41, 42]. On the contrary, a model hydrogenated polybutadiene (HPB) exhibits a monomodal distribution of SCB as expected from the approximately random incorporation of 1, 2 units in the polybutadiene precursor during anionic polymerization (see more details about the samples in Ref. [46]).

The fractionated HPB by SSA shows a series of melting peaks corresponding to the melting of crystallites with different mean lamellar thickness formed and annealed at each T_s. The SSA final DSC heating scan shows the effects of the accumulation of seven self-nucleation and annealing steps using T_s values ranging from 102 to 72 °C, every 5 °C. Because 102 °C is the ideal self-nucleation temperature for this HPB [33, 46], the thermal treatment at this T_s does not cause any annealing. Therefore, only six steps of the SSA procedure were able to produce annealing. Figure 5.8b shows the six sharp melting peaks of HPB obtained after SSA plus an additional broad melting peak (at around 65 °C) corresponding to crystals formed during cooling from the lowest T_s employed. The observed distribution of melting peaks is unimodal and reflects the random distribution of 1, 2 units in the PB precursor, that is, the distribution of the resulting ethyl branches along the polyethylene backbone.

For 11U4, 13 thermal cycles were applied starting from the ideal self-nucleation temperature, that is, 124 °C. Therefore, 12 sharp meeting peaks are appreciated in Figure 5.8b corresponding to those T_s temperatures that produced annealing and thermal fractions.

SSA has been typically performed employing a constant sample mass (approximately 10 mg) and various heating rates (5, 10, and 20 °C/min) in previous works [41, 42]. However, Pijpers et al. [3] introduced high-speed calorimetry concepts

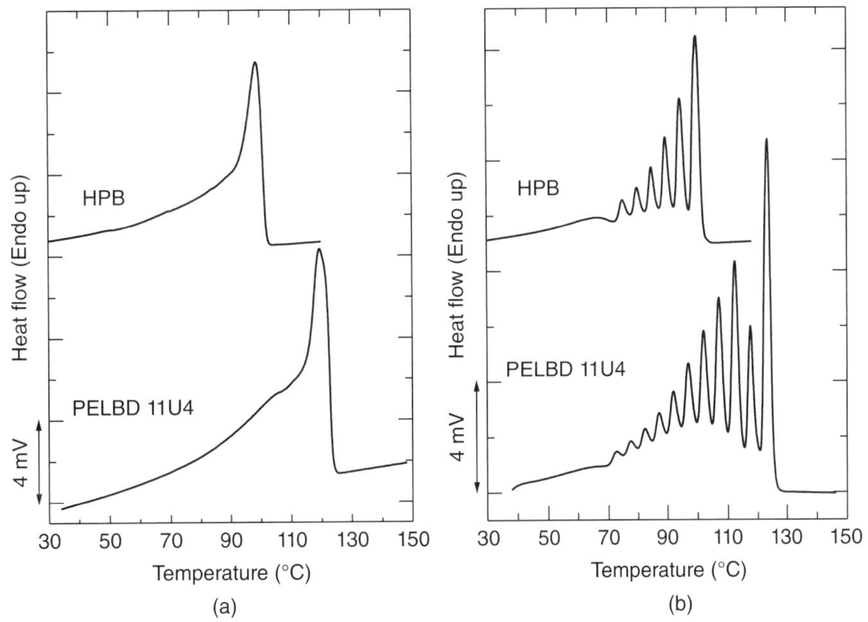

Figure 5.8 DSC heating scans (10 °C/min) for HPB and 11U4 before (a) and after (b) SSA fractionation. Lorenzo et al. [46]. Reproduced with permission of John Wiley and Sons.

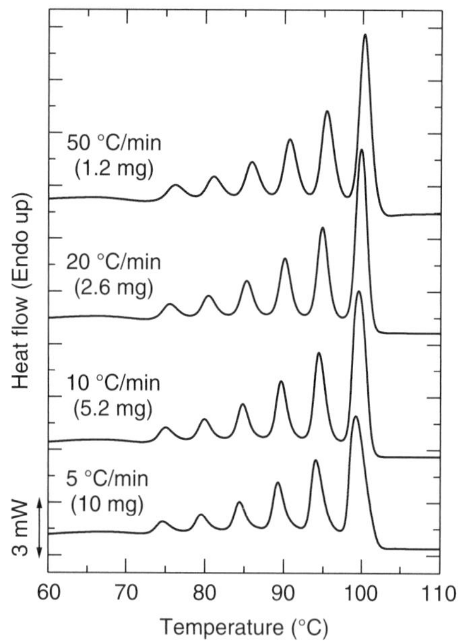

Figure 5.9 Final heating scans after SSA fractionation of HPB performed at the indicated rates and sample masses. At 50 °C/min, the quality of the fractionation is still very good and the fractionation time has been reduced to only 78 min. Lorenzo et al. [46]. Reproduced with permission of John Wiley and Sons.

that were advantageously applied to thermal fractionation experiments [33, 46, 47].

Following Pijpers et al.'s [3] recommendations, Müller et al. [33, 46, 47] performed SSA at faster heating rates while reducing sample mass. Figure 5.9 shows that SSA can be performed employing rates as high as 50 °C/min (in principle, much faster rates could be used with faster DSC equipments) using a conventional PerkinElmer PyrisTM 1 DSC equipped with an intracooler device. The resolution at high- and low-scanning rates is almost identical, according to Figure 5.9, when the mass has been conveniently reduced to avoid any superheating effects. A slight shift in the melting point of each fraction to higher temperatures is noticeable, although it is less than 1.1 °C when the scan rate is increased 10 times (from 5 to 50 °C/min) [46].

The fractionation of HPB was achieved in merely 78 min when all the heating and cooling cycles employed during SSA were set at 50 °C/min. This yields an approximate 70% time gain per self-nucleation and annealing cycle (where one cycle comprises heating from −20 to 120 °C at a specific scanning rate, a holding time of 5 min at 120 °C, and then cooling from 120 to −20 °C) if 50 °C/min is used instead of 10 °C/min. The time of 78 min is the fastest thermal fractionation time ever reported in the literature [33] (compared with commonly used values like 12 or 24 h [33, 38–40] or even 1 week) and with resolution equivalent to usual conditions involving 10 °C/min.

Table 5.2 reports typical thermal fractionation times employing SSA and step crystallization (SC). Step crystallization is based on the sequential isothermal crystallization of the sample at progressively lower temperatures encompassing the entire crystallization range of the polymer. There are no intermediate heating steps applied in SC as in SSA. SSA is by far faster than SC. If high heating and cooling rates are employed (i.e., at least 50 °C/min), the SSA time advantage is even greater, because the isothermal steps in SSA are just 5 min long. Even higher heating and cooling rates are possible; however, special equipment and/or cooling devices may be needed.

TABLE 5.2 Comparison between SC and SSA Typical Fractionation Times [46]

Method	Description	Total Time (min)
SC[a]	50 min	348
	6 h	2,520
	24 h	10,080
SSA[b]	5 °C/min	354
	10 °C/min	210
	20 °C/min	138
	50 °C/min	78

[a]The SC protocol consisted in seven steps from 102 to 72 °C using a 5 °C fractionation window. The holding time within each step was varied as indicated.

[b]The SSA protocol included seven isothermal steps between 102 and 72 °C employing a 5 °C fractionation window. In SSA, each isothermal step was applied during 5 min. The heating and cooling rates of all thermal cycles including the final heating scans were kept constant at the indicated values.

As indicated in previous references [33, 41–54], SSA can yield quantitative information when compared with other techniques; and SCB distributions, dispersity of the SCB distribution, and distributions of lamellar thickness can all be derived.

The SSA technique has been used to characterize the products of functionalization, because any chemical group that is inserted along the chain by grafting reactions will interrupt the MSL sequences [55–57]. With the same principle, cross-linked and irradiated polyethylenes have also been studied by SSA [58–63].

Another SSA application has been the assessment of miscibility in polyethylene blends [64, 65]. SSA has also been valuable in studying the effects of confinement into micro- and nanophases or in nanocomposites [66–68] as well as in characterizing the products of biodegradation [69] and oxidative degradation [70–73], among other applications [74–90].

The SSA technique is now being used rather frequently for all the applications quoted above. Unfortunately, in many recent references, the authors do not perform a previous self-nucleation study to determine the ideal self-nucleation temperature. Therefore, they start their SSA protocol at an arbitrary T_s value that will have an important impact on the fractionation profile. Any attempt to quantify the data and calculate SCB or MSL distributions will not be correct unless the first T_s value employed to perform SSA is the ideal self-nucleation temperature.

5.5 MULTIPHASIC MATERIALS: POLYMER BLENDS AND BLOCK COPOLYMERS. FRACTIONATED CRYSTALLIZATION AND CONFINEMENT EFFECTS

The confinement of semicrystalline polymers into micro- or nanophases is an attractive subject that has been reviewed recently several times [91–100].

The dispersion of a polymer into isolated phases can be achieved by different means:

1. One possibility is to disperse the polymer into droplets of different sizes [101–110]. Droplets can be obtained by dewetting a thin film deposited on an incompatible substrate [111–118], by annealing films composed of immiscible micro or nanolayers [119–128], by miniemulsions and water dispersions, among other methods [129–134].
2. Immiscible blends can be prepared at compositions where one of the polymeric phases is dispersed in a matrix of the second component [91, 92, 100, 135–145].
3. Block copolymers are one of the most versatile ways of producing nanoscopic dispersions [92–99]. Block copolymers can exhibit a wide range of morphologies depending on their segregation strength, given by the product χN, where χ is the Flory–Huggins interaction parameter between the block components and N is the degree of polymerization. If we simply consider a linear diblock copolymer where one block is amorphous and the second can crystallize, then the situations illustrated in Table 5.3 will arise depending on the relative values of the order–disorder transition (T_{ODT}), the crystallization temperature of the crystallizable block (T_c), and the glass transition temperature of the amorphous block (T_g) [92, 96, 146–150]. The most common isolated dispersed domains encountered in block copolymers are spheres and cylinders [92].
4. Nanostructures can also be formed by self-assembly from solution [134, 151–189].
5. Inorganic templates, such as anodized aluminum oxide (AAO) can be prepared with tailored nanopores and can be used to infiltrate polymers [190–204].

5.5.1 Blends and Fractionated Crystallization

In confined polymers, one of two situations can arise during freezing from the melt:

(a) Crystallization occurs in a single crystallization exotherm at much lower temperatures than the usual crystallization temperature (T_c) of the bulk polymer. This is found in systems characterized by heterogeneity free domains, when the number of domains is several orders of magnitude higher than the number of heterogeneities present in the bulk polymer.

(b) The crystallization occurs in several steps well spaced in temperature, that is, the so-called fractionated crystallization [91, 92]. This case occurs when the number of active heterogeneities is of the same order of magnitude than the number of domains, so that a significant population of domains still contains some type of heterogeneity.

In case (a), the nucleation typically changes from heterogeneous (present in the bulk polymer) to either homogeneous or surface nucleation of the domains (i.e., initiated at the surface of the domains or at the interface between the crystallizable domains and the matrix surrounding them). Surface nucleation is frequently encountered, although not often recognized in the literature [99], because it requires a lower free energy as compared with homogeneous nucleation.

TABLE 5.3 Possible Morphologies of a Linear Diblock Copolymer with One Crystallizable Block Depending on the Segregation Strength and T_{ODT}, T_g, and T_c Values

Segregation Level	$T_{ODT}/T_g/T_c$	Morphology in the Solid State
Homogeneous melt	$T_{ODT} < T_c > T_g$	Crystalline lamellae surrounded by amorphous material
Weakly segregated systems (low χN values)	$T_{ODT} > T_c > T_g$	The crystallization destroys previous melt structure (by breakout) and crystalline lamellae are formed
Medium segregated systems (medium χN values)	$T_{ODT} > T_c > T_g$	Quenching: The melt-segregated morphology is preserved Slow cooling: Breakout and crystalline lamellae are formed
Strongly segregated systems (high χN values)	$T_{ODT} > T_c > T_g$	The crystallization can be confined within the MDs dispersed in a rubbery block matrix (soft confinement)
	$T_{ODT} > T_g > T_c$	A strictly confined crystallization within the MDs dispersed in a glassy matrix (hard confinement)

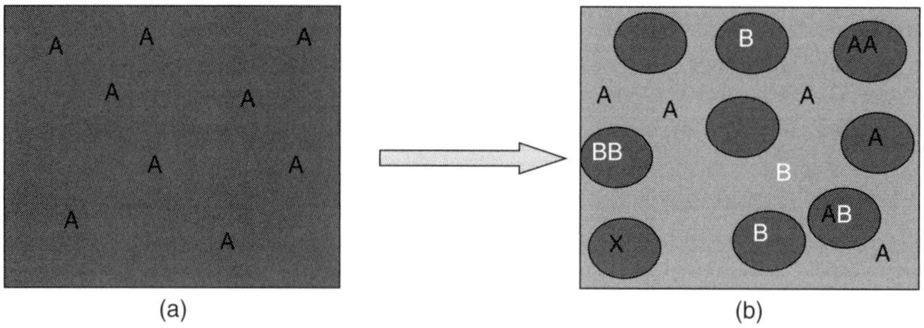

Figure 5.10 Schematic illustration of two different types of heterogeneities in bulk isotactic polypropylene (iPP) (a) and how they are distributed, once the material is dispersed into a polystyrene amorphous phase in an 80/20 PS/iPP blend (b).

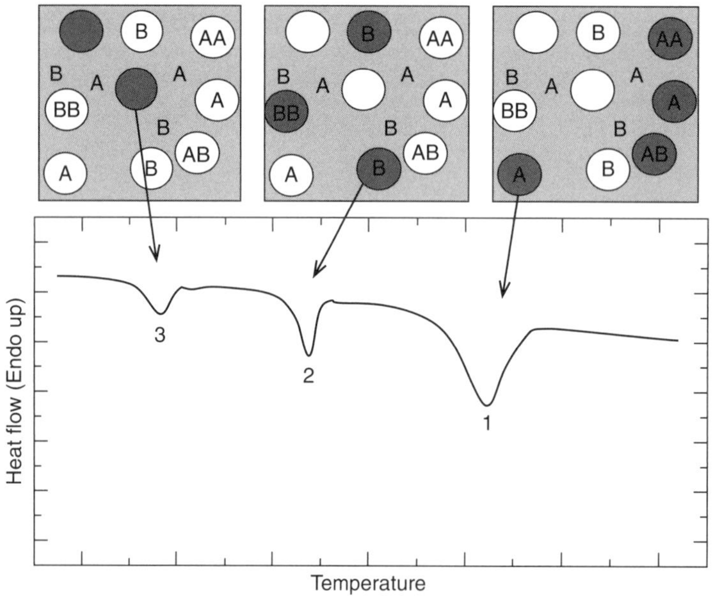

Figure 5.11 A schematic illustration of the fractionated crystallization of polymer droplets dispersed in an immiscible polymer matrix.

The fractionated crystallization phenomenon described in case (b) arises from the existence of different nucleation events, where each crystallization exotherm starts from a distinct nucleation event. This behavior is produced as a result of the much larger number of isolated microdomains (MDs), in comparison to the number of highly active heterogeneities (e.g., type A) usually present in an equivalent bulk sample of a crystallizable polymer. A cartoon illustrating this effect is provided in Figure 5.10.

Figure 5.11 shows a schematic DSC cooling scan where the crystallization does not occur in a single step but in a fractionated crystallization fashion. Three crystallization exotherms labeled "1" (higher temperature crystallization exotherm) to "3" (lowest temperature crystallization exotherm) are depicted.

When a crystallizable component is finely dispersed in a matrix (the number density of the isolated phase must be at least of the same order of magnitude, than the concentration of heterogeneities present in the bulk crystallizable polymer), as in an immiscible polymer blend (where the crystallizable polymer is the minority component), the number of heterogeneities present in the bulk crystallizable polymer will be distributed among the isolated phases or migrate to the matrix component. Then, the situation shown in the cartoons of Figure 5.11 could be envisaged. Starting at a high temperature, some microdomains (MDs) or droplets with the most effective heterogeneity (type "A") inside them will be able to crystallize at the highest crystallization temperature, that is, at the same T_c at which the bulk polymer crystallizes. The MDs with the second most active heterogeneity (i.e., type "B" in Figs. 5.10 and 5.11) will need a larger supercooling in order to nucleate and crystallize. Finally, the lowest temperature exothermic signal is originated by the crystallization of clean MDs. This particular crystallization exotherm can occur at the lowest possible temperature or maximum supercooling and could be originated from (i) surface or interfacial nucleation events (due to the MDs' interface) or (ii) from bulk or volume homogeneous nucleation events.

Immiscible blends of iPP and PS were studied by Arnal et al. [100]. Figure 5.12a shows a scanning electron

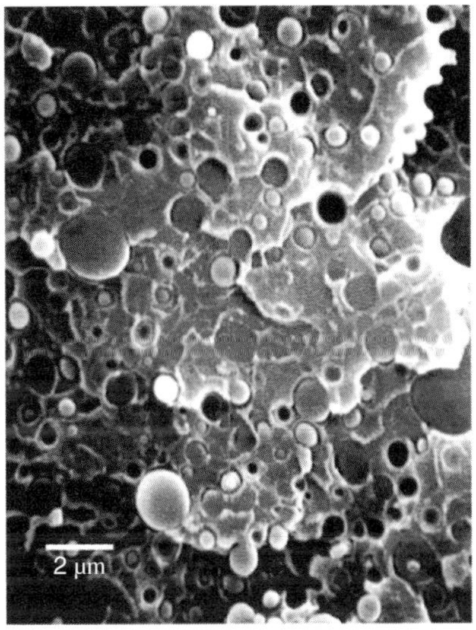

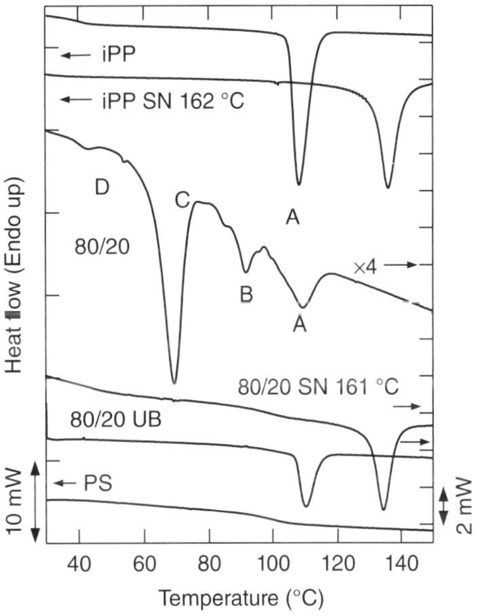

Figure 5.12 (a) SEM micrograph of the PS/iPP 80/20 wt/wt blend and (b) cooling scans at 10 °C/min for the 80/20 PS/iPP blend and the corresponding homopolymers, the iPP and the 80/20 PS/iPP blend before and after self-seeding. Arnal et al. [[100], Figures 1, 2, 6]. Reproduced with permission of John Wiley and Sons.

microscopy (SEM) micrograph revealing a dispersion of iPP droplets in the glassy amorphous PS matrix. The cooling DSC scans of PS, iPP, and their 80/20 blends are shown in Figure 5.12b. A cooling scan of an "unmixed blend" is also shown. This was prepared using the same weight proportions of PS and iPP as in the corresponding melt-mixed blend, but placing both polymers in a DSC pan separated by aluminum foil, so that no contact between the two polymers is made [205].

PP exhibits its usual bulk behavior, displaying an exotherm that peaks at 111 °C. The cooling DSC scan of the atactic PS only shows the T_g at approximately 100 °C. A superposition of these two typical scans can be seen in the 80/20 PS/iPP unmixed blend (UB), where only the exotherm corresponding to the crystallization of the iPP can be clearly seen because the T_g of the PS matrix occurs within the same temperature range.

The melt-mixed blend displays a quite different behavior in Figure 5.12b, where the heterogeneous nucleation of the iPP component is nearly completely suppressed as indicated by the marked decrease in the crystallization enthalpy in the temperature region where the iPP crystallizes in bulk, that is, at 109–111 °C. The number of droplets for this composition is known (by SEM observations) to be of the order of 10^{11} particles/cm^3 and polarized optical microscopy experiments have shown that the iPP contains approximately 9×10^6 heterogeneities/cm^3. The concomitant fractionated crystallization of the iPP droplets follows in such a way that iPP crystallization occurs in at least four distinct steps, which we shall distinguished by the letters A, B, C, and D in decreasing temperature order.

When iPP is dispersed into droplets, the content of heterogeneities of each droplet can be different. iPP may contain different types of heterogeneities that activate at different supercoolings depending on their specific interfacial energy differences with the polymer melt [91]. In bulk iPP, the heterogeneity with the lowest specific interfacial energy difference will be activated at lower supercoolings and will dominate the crystallization of the polymer via secondary nucleation at the created crystals. Figure 5.12b shows that bulk iPP crystallization occurs in only one exotherm, located at a temperature range A, where a certain heterogeneity, which we shall also term A (or type A), will be activated during cooling from the melt. In bulk iPP, there is no chance for other heterogeneities that are active at higher supercoolings to cause any nucleation because the polymer crystallizes at higher temperatures (it is this effect that can be inhibited if the original volume of the material is divided into many small droplets).

In the case of the 80/20 PS/iPP blend, only a certain number of droplets contain heterogeneities of type A, while others contain type B, and so on. A number of exotherms will be generated depending on the relative supercoolings needed to activate each dominating heterogeneity within a certain droplet population. It is also possible that some droplets contain no heterogeneities at all, and in that case, homogeneous or surface nucleation could well be the origin of the lowest temperature exotherm observed (exotherm D). Therefore, in the 80/20 PS/iPP case, exotherms A, B, and C are probably due to the presence of three different kinds of heterogeneities in the different crystallizing groups of droplets, while exotherm D may be due to the crystallization of clean droplets.

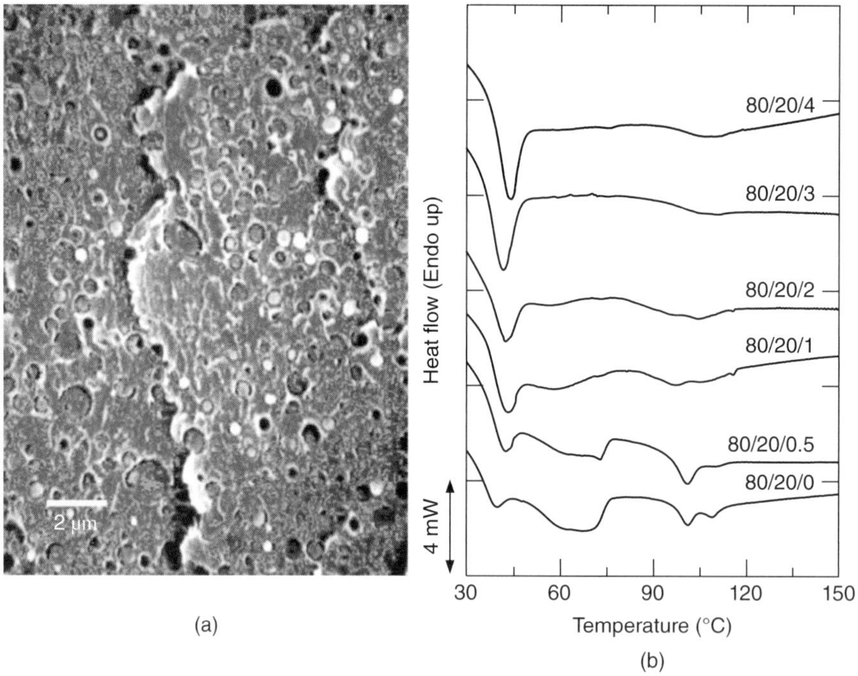

Figure 5.13 (a) SEM micrograph of the PS/iPP/SEP 80/20/4 wt/wt blend and (b) DSC cooling scans at 10 °C/min for the 80/20 PS/iPP blend with different contents of SEP. Arnal et al. [[100], Figures 1, 2, 6]. Reproduced with permission of John Wiley and Sons.

The definite demonstration that the fractionated crystallization behavior of the 80/20 PS/iPP blend is due to the lack of highly active nuclei in every droplet is provided by self-nucleation experiments and/or by the addition of a nucleating agent [135–137]. Figure 5.12b shows the DSC cooling scan of iPP in the bulk after the self-nucleation process at the ideal self-seeding temperature, which, in this case, was a T_s of 162 °C [30]. The self-nucleation process substantially increases the number of nuclei, such that bulk iPP crystallizes at 136.2 °C after self-nucleation; this means that an increase of 28 °C in T_c has been achieved. In order to produce an equivalent self-nucleation of the iPP component in the 80/20 PS/iPP blend, a T_s temperature of 161 °C had to be employed. This fact alone indicates that because the polymer is confined into droplets, it is experiencing more difficulties in becoming self-nucleated. Nevertheless, after SN, the DSC cooling scan from T_s shows clearly in Figure 5.12b that almost every iPP droplet can now crystallize at much higher temperatures, that is, at 134.5 °C.

Figure 5.13 shows the effect of adding the compatibilizer SEP (polystyrene-*b*-poly(ethylene-*ran*-propylene)) on the morphology and thermal behavior of 80/20 PS/iPP/SEP blends. In this case, the same neat polymers as in the blends described in Figure 5.12b were used for the uncompatibilized blend, but a different extruder was employed for melt mixing (see Ref. [100]). This is the reason the 80/20/0 blend of Figure 5.13b does not present a cooling behavior identical to that of the same composition presented in Figure 5.12b (see Ref. [100] for more details).

The addition of the SEP diblock copolymer results in a reduction in mean volume particle size from 1.56 μm for the unmodified blend to 0.87 μm for the blend with the maximum amount of interfacial modifier. The difference in particle sizes is apparent from a comparison of the SEM micrographs of Figures 5.12a and 5.13a.

Addition of SEP compatibilizer to the 80/20/0 PS/iPP/SEP blend in amounts higher than 1% by total weight of the sample produced significant changes in the PS/iPP/SEP blend crystallization because all high temperature exotherms (A and C) tend to disappear, while the 45 °C exotherm becomes more important. In a previous publication [206], we reported the determination of the interfacial tension by the breaking thread method in these blends. A decrease in the interfacial tension was observed upon increasing the amount of SEP. The mean number of particles increased by a factor of 3 upon addition of the maximum amount of SEP used here. The higher number of small dispersed particles lowers the probability of finding heterogeneities that usually crystallize the polymer in bulk (heterogeneity type A) in a significant number of iPP droplets, thereby favoring the possible activation of other types of heterogeneities and/or surface or homogeneous nucleation processes at much lower temperatures. We have determined the approximate number of nuclei present in bulk iPP by polarizing optical microscopy, and the average value is in the order of 9×10^6 nuclei/cm^3. On the other hand, Ref. [100] shows that the calculated average number of droplets derived from SEM examinations for the blend with the maximum amount of SEP used (PS/iPP/SEP 80/20/4) is in the order of 21.9×10^{11} particles/cm^3. Therefore, we have at least six orders of magnitude more droplets than heterogeneities available in bulk iPP, hence the increase probability on encountering clean droplets that can only crystallize in exotherm D.

5.5.2 Copolymers

We present as an example of segregated block copolymers two analog sets (in wt% composition) of PCL-containing copolymers of different molecular architectures.

The first set is composed by $(PCL)_2$-*b*-$(PS)_2$ miktoarm star block copolymers (i.e., stars with two arms of PS and two arms of PCL radiating from the same center) have χN values in the range between 7.1 and 25.6 (at room temperature) that indicate a behavior in between total miscibility of both blocks (χN lower than 10) and a weakly segregated regime (especially for $\chi N = 25.6$).

In the second set, the linear PCL-*b*-PS diblock copolymers have slightly higher χN values between 8.4 and 38.4 (at 25 °C). Figure 5.14 shows selected transmission electron microscopy (TEM) for PCL-*b*-PS linear diblock copolymers and for (PCL_2)-*b*-(PS_2) miktoarm star block copolymers. For the linear diblock copolymers (Fig. 5.14a and c). the morphologies obtained conform to the theoretical expectations on the basis of their composition (i.e., lamellae for the diblock with 41% PCL and PCL cylinders for that with 20% PCL) for segregated diblock copolymers [92, 94–96, 150].

The miktoarm star block copolymers (Fig. 5.14b and d) exhibited morphologies quite different from those observed for their linear analogs. The *TEM* for the $(PCL_2)_{39}$-*b*-$(PS_2)_{61}{}^{62}$ miktoarm copolymer (where the subscripts outside the brackets indicate the composition in wt%, while the superscript denotes the number average molecular weight of the entire block copolymer), see Figure 5.14b, exhibits a cylindrical

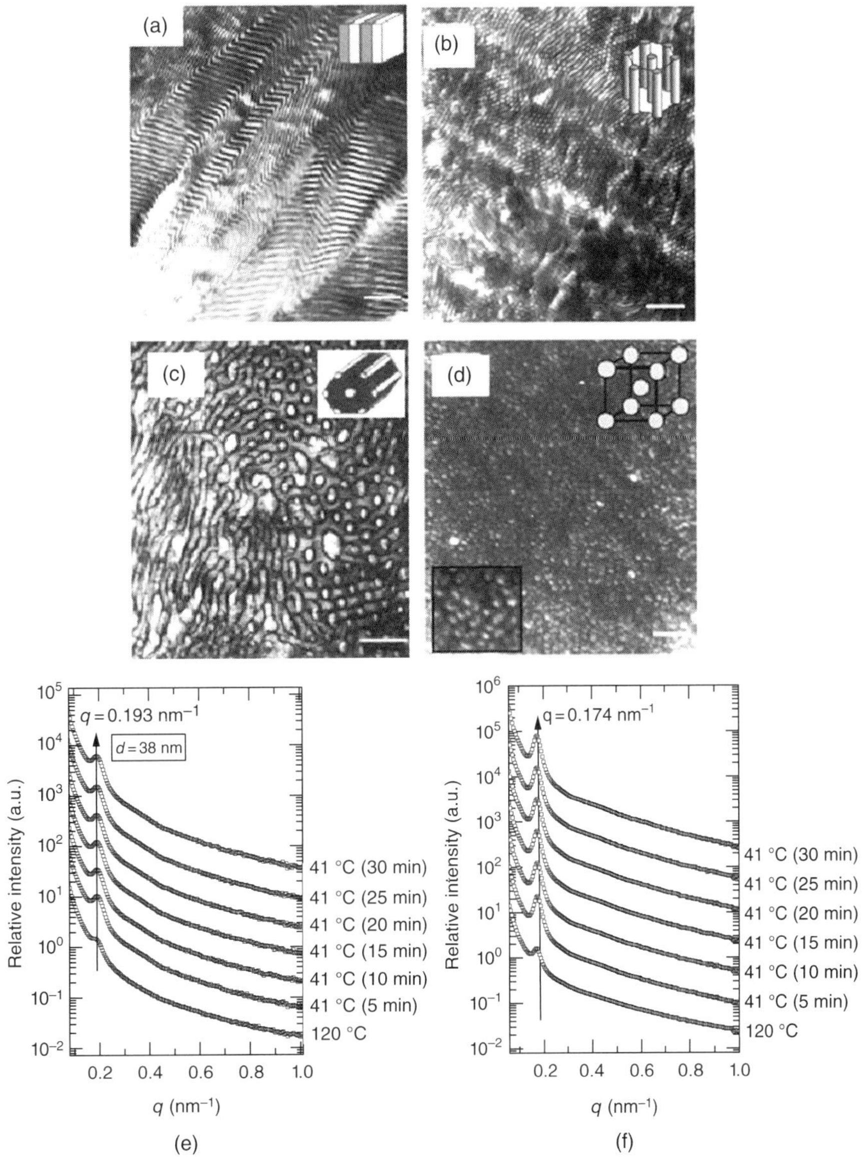

Figure 5.14 *TEM* of (a) PCL_{41}-*b*-$PS_{59}{}^{73}$ (scale bar: 100 nm), (b) $(PCL_2)_{39}$-*b*-$(PS_2)_{61}{}^{62}$ (scale bar: 200 nm), (c) PCL_{20}-*b*-$PS_{80}{}^{153}$ (scale bar: 100 nm), (d) $(PCL_2)_{27}$-*b*-$(PS_2)_{73}{}^{100}$ (scale bar: 200 nm) and *SAXS* scattering profiles of (e) $(PCL_2)_{39}$-*b*-$(PS_2)_{61}{}^{62}$ and (f) $(PCL_2)_{27}$-*b*-$(PS_2)_{73}{}^{100}$. The copolymers were stained with RuO_4 (PS component is gray). Lorenzo et al. [207, 208].

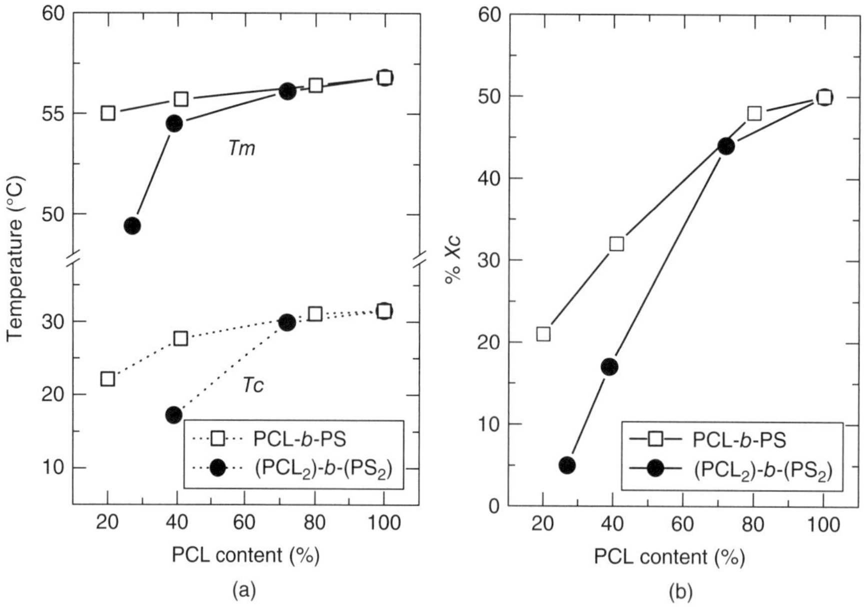

Figure 5.15 (a) Crystallization (T_c) and melting peak temperatures (T_m) and (b) crystallinity degree for linear PCL-*b*-PS diblock copolymers and (PCL_2)-*b*-(PS_2) miktoarm star copolymers as function of PCL content. Lorenzo et al. [208].

morphology with a clear hexagonal array of PCL cylinders. In addition, SAXS patterns for this miktoarm star copolymer are displayed in Figure 5.14e. A clear scattering peak at 120 °C (when the sample is in the melt) confirms that the copolymer undergoes phase segregation in the melt state. The form factor position allows the calculation of the diameter of the cylindrical microdomains, yielding approximately 38 nm, which closely agrees with that estimated by *TEM* (41 nm).

Figure 5.14 demonstrates that the microdomain morphologies are more confined for the PCL component within the miktoarm star copolymers than expected on the basis of their composition as compared to the linear diblock copolymer analogs. Therefore, the segregation strength within the miktoarm star block copolymers may be stronger than that exhibited by the corresponding linear diblocks at similar PCL contents. The observation of a clear hexagonally packed cylindrical morphology by TEM in the $(PCL_2)_{39}$-*b*-$(PS_2)_{61}{}^{62}$ miktoarm star block copolymer (Fig. 5.14a), in spite of its 0.39 volume fraction of PCL, is not expected because for such a composition, lamellar microdomains (Fig. 5.14a) are observed for its linear diblock copolymer analog. This behavior was explained thanks to previous theoretical predictions for star copolymers [209, 210]. These previous theoretical works have argued that different morphologies could be expected for the stars, because of the higher resistance of the arms to be stretched in a miktoarm star copolymer as compared to a linear one, and because of the relative strong dependence of the phase boundaries on the number of each type of arm [207–210].

DSC cooling and heating scans of the nanostructured block copolymers presented here can be found elsewhere [207, 208]. Figure 5.15 shows the peak melting and crystallization temperatures (T_m and T_c) and the crystallinity degree ($\%X_c$) for the two sets of PCL-containing copolymers (linear diblocks and miktoarm stars). A PCL homopolymer sample was employed for comparison purposes (with a M_n of 29 kg/mol) and it exhibited a T_c and T_m of 31.5 and 56.8 °C, respectively. Figure 5.15a demonstrates how the values of T_c and T_m decrease as morphological confinement increases (i.e., as the PS content increases) [207, 208]. The decrease in the melting temperature and in the degree of crystallinity (X_c) (see Fig. 5.15b) signals that the previous crystallization process was greatly affected by the composition of each copolymer. As confinement increased, crystallization needed higher supercoolings to be produced, and thinner lamellar crystals that melt at lower temperatures were generated. The decreased MD dimensions and larger surface area of the MDs (on going from lamellae to cylinders and finally to spheres) produced increased topological confinements in the PCL block chains.

Figure 5.15 also compares the effect of molecular architecture as a function of PCL content. All the parameters under consideration (T_m, T_c, and X_c) are smaller for the PCL phase within the miktoarm stars than within the linear diblock copolymers at equivalent compositions (same content of PCL), thereby indicating a higher confinement degree for the miktoarm star copolymers.

The effect of confinement caused by the miktoarm molecular architecture is so important, that even comparing similar morphologies with their linear counterparts (at different compositions), a greater depression in T_c, T_m, and X_c is obtained. For example, if the two cylinder-forming samples are compared, the values of T_c, T_m, and X_c are all much smaller for the miktoarm star copolymer $(PCL_2)_{39}$-*b*-$(PS_2)_{61}{}^{62}$ than for the linear diblock copolymer sample PCL_{20}-*b*-$PS_{80}{}^{153}$, even though the first one contains 39% PCL and the second one only 20%. This is a result of the larger confinement experienced by the PCL chains in the miktoarm versus the linear diblock copolymers.

5.5.3 Copolymers Versus Blends

In this section, we have chosen an interesting example, where DSC and morphology interplay in an illustrative way, when blends and block copolymers of poly(L-lactic acid) (PLLA) and poly(ε-caprolactone) are examined.

PLLA and PCL blends were prepared by dissolution in a common solvent followed by solvent evaporation. Figure 5.16 contains DSC data and polarized light optical microscopy images taken at two different T_c temperatures. The blends exhibit two crystallization and melting temperatures that are almost identical to those obtained for the parent homopolymers, a clear sign of immiscibility. The micrographs of Figure 5.16b corroborate the immiscibility of the blends. At 120 °C, only the PLLA component can crystallize. The PLLA component forms clear negative spherulites that surround dark isotropic regions (one of which has been marked by a white ellipse) where the molten PCL component is segregated. These regions without birefringence are clearly filled with small PCL spherulites when the sample is quenched to 35 °C, a temperature at which PCL can grow spherulites.

Figure 5.17 shows DSC cooling runs for some PLLA-*b*-PCL diblock copolymers and similar molecular weight PCL. Both block crystallize during cooling from the melt at lower temperatures than the corresponding homopolymer. In the composition range presented in Figure 5.17, the lowering of the PCL block crystallization peak is clearly seen.

Figure 5.18A shows how the T_c and T_m values of the PLLA homopolymer and PLLA block within the diblock copolymers depend on composition. In general, the crystallization and melting temperature of the PLLA block within the diblock copolymers is depressed when the PCL content increases. These PLLA-*b*-PCL diblock copolymer systems are partially miscible or weakly segregated. Hence, because of the weak segregation state, a certain level of thermodynamic interactions between the blocks is expected and that allows the PCL to act as a plasticizer for PLLA.

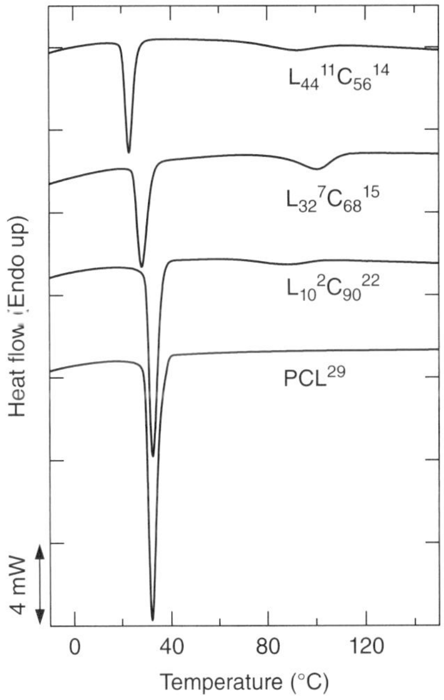

Figure 5.17 DSC cooling scans at 10 °C/min of PCL and the indicated PLLA-*b*-PCL copolymers, after melting at 190 °C for 3 min. Castillo et al. [212]. Reproduced with permission of American Chemical Society.

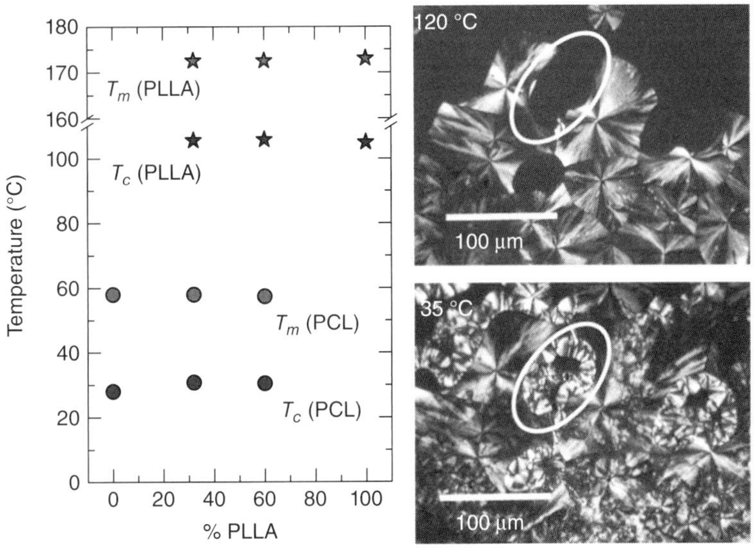

Figure 5.16 (a) Crystallization and melting temperatures (obtained from the data reported in Ref. [211]) for PLLA and PCL blends as a function of PLLA content. (b) Photomicrograph obtained by PLOM for the 32/68 PLLA/PCL blend. Castillo et al. [212]. Reproduced with permission of American Chemical Society. (*See color plate section for the color representation of this figure.*)

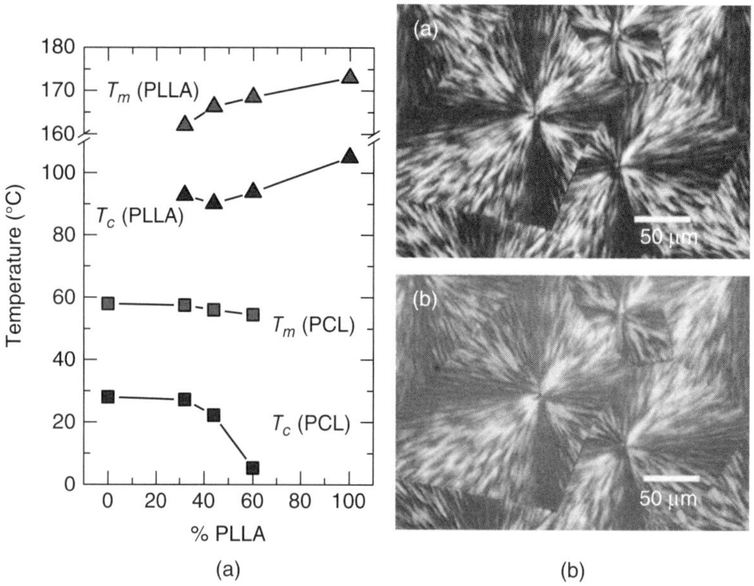

Figure 5.18 (A) Crystallization and melting temperatures (obtained from the data reported in Ref. [211]) for the PLLA and PCL block within all copolymers versus PLLA content. (B) Photomicrograph obtained by PLOM for the (a) $L_{32}{}^{7}C_{68}{}^{15}$ after 30 min at 122 °C. (b) $L_{32}{}^{7}C_{68}{}^{15}$ after 15 min at 42 °C. Castillo et al. [212]. Reproduced with permission of American Chemical Society. (*See color plate section for the color representation of this figure.*)

Figure 5.16 showed that PLLA and PCL form immiscible blends and separate spherulitic structures. However, when block copolymers are prepared, PLLA and PCL chains are covalently bonded in every chain of the linear diblock copolymers. This effect is enough to produce partial miscibility as demonstrated by the thermal properties shown in Figure 5.18A, which should be contrasted with those of the blends in Figure 5.16.

Furthermore, when the superstructural morphology of the PLLA-*b*-PCL diblock copolymers was examined, just one type of spherulite was seen, that is, a mixed spherulite. An example is provided in Figure 5.18B. The diblock copolymer $L_{32}{}^{7}C_{68}{}^{15}$ (subscripts indicate the composition in weight percent and superscripts the number average molecular weight in kg/mol of each block) was crystallized at 122 °C, a temperature at which the PCL block cannot crystallize. The PLOM field was filled with PLLA spherulites, as shown in the top micrograph of Figure 5.18. These spherulites contain in the amorphous interlamellar regions the molten PCL block chains mixed together with those PLLA block chains that remained amorphous. Upon cooling to 42 °C, the PCL block crystallizes within the interlamellar regions, simply filling the template spherulitic structure provided by the PLLA block previous crystallization. Such a mixed spherulite (Fig. 5.18b) is formed by interdigitized PLLA/PCL lamellae.

Figure 5.19 shows more examples of the remarkable ability of the PLLA block (and PLLA) to form spherulites at temperatures where the PCL block is molten, thus templating the spherulitic morphology that will characterize the copolymer upon cooling [211, 213, 214]. PLLA crystallized in spherulitic morphology regardless of composition, as already reported [211–214]. Figures 5.18B and 5.19 show well-defined negative PLLA spherulites, although the Maltese cross-extinction pattern tends to get blurry with increasing PCL content (see also Fig. 11 from Ref. [211]).

In the case of the $L_{10}C_{90}{}^{24}$ block copolymer, axialite-like superstructural crystal aggregates are formed during isothermal crystallization at 100 °C (above the melting point of PCL) of the PLLA block chains, even though the PLLA block can only achieve a 50% crystallinity (but the sample only contains 10% of PLLA) and therefore 95% of the sample remains in the melt (i.e., 50% of the 10% PLLA block plus 90% of the PCL block). These two-dimensional superstructural features (i.e., axialites) are in agreement with the Avrami index values close to 2 obtained by DSC experiments (see Ref. [212]). A morphology referred to as concentric spherulites was reported by Wang et al. [215] for a two-arm star-shaped PLLA-*b*-PCL block copolymer with 28 wt% PLLA.

Albuerne et al. [216] reported dramatic spherulitic morphology changes for poly(*p*-dioxanone)-*b*-poly(*ε*-caprolactone) (PPDX-*b*-PCL) diblock copolymers. The differences are surely due to a higher thermodynamic repulsion between the blocks in this case, although the copolymers are still in the weak segregation regime. At higher crystallization temperatures where the PCL block is molten, granular structures were formed if PCL was the major component. At temperatures where both blocks crystallized, banded spherulites similar to those formed by PPDX homopolymer were clearly observed.

5.5.4 The Crystallization of Polymers and Copolymers within Nanoporous Templates

Masuda and Fukuda were the first to prepare AAO templates by a double anodization process [217]. After the first

Figure 5.19 Polarized light optical micrographs during isothermal crystallization: (a) PLLA[24], after 8 °min at 140 °C. (b) $L_{81}C_{19}$[21], after 10 min at 140 °C. (c) $L_{60}C_{40}$[21], after 30 min at 140 °C. (d) $L_{10}C_{90}$[24] after 10 min at 100 °C. Castillo et al. [212]. Reproduced with permission of American Chemical Society. (*See color plate section for the color representation of this figure.*)

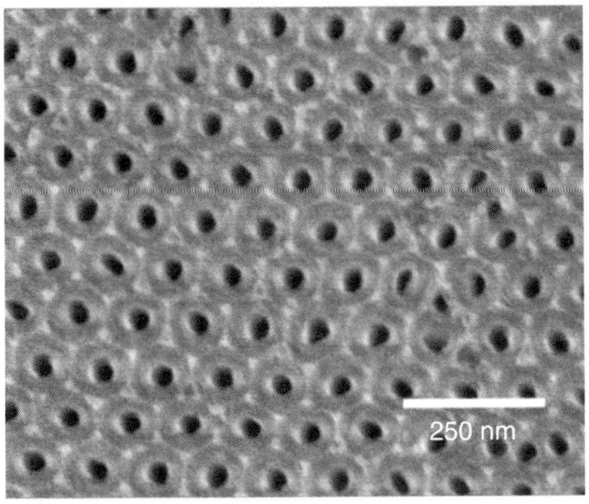

Figure 5.20 SEM (top view) of an AAO template with mean pore diameter of 35 nm. Adapted from Michell et al. [99]. Reproduced with permission of Elsevier.

anodization, an irregular AAO layer is formed. This layer is removed and in the second anodization a hexagonally porous array of AAO is formed, as illustrated in the examples provided in Figure 5.20. Different pore sizes can be produced by varying the conditions employed during template preparation (Refs [203, 218]).

The infiltration of a polymeric material into the nanopores of the AAO templates can be achieved by different methods. Melt infiltration has been extensively employed, where a polymer film is placed on top of the pores and heated above its melting point. The molten polymer fills the pores after some time has elapsed (for detailed information on the filling process, see Ref. [218]). This infiltration process is also known as nanomolding [219–223]. An interconnecting polymer layer may remain on the top of the template. It is very important that this layer should be removed in order to obtain a truly confined behavior of the nanodomains within the nanopores of the templates.

The crystallization of infiltrated polymers within AAO templates has been recently studied [190–204, 224–226]. We can extract some general trends from a review of the literature:

1. The crystallization temperature (T_c) has been found to decrease with pore size. Woo et al. [197] infiltrated linear polyethylene (PE) in AAO templates. Several pore sizes were studied, from 15 to 110 nm, and even though a clear decreasing trend with pore size diameter was found, the crystallization temperatures found were too high to correspond to homogeneous nucleation of PE (see [92, 97, 100, 112, 200]). The nucleation of the confined PE nanocylinders must have been produced by surface nucleation. Similar results of decreasing T_c with pore size have been reported for infiltrated PP [190], PEO [227, 200, 202], and PE-*b*-PS [200] diblock copolymers.
2. Confinement within AAO nanopores produces lower degrees of crystallinity than those obtained for neat polymers in the bulk state [190, 191, 202]. Similar results have been obtained for the confined crystallization of micro- and nanodomains within block copolymers and polymer blends (see [92, 97] and references therein).
3. The presence of traces of the polymer film employed during the infiltration process may lead to the presence

of an additional exotherm at the same undercooling as in the neat polymer, in view of the percolating path connecting a series of nanopores. Once the layer is completely removed, only the crystallization peak at maximum supercooling is observed [99].

4. The melting temperature has also been found to decrease with pore diameter. This behavior is especially important at small pore sizes, where crystalline lamellae are forced to be thinner due to a reduction in the available space for crystallization [99].
5. In the case of strongly segregated block copolymers, double confinement effects can cause further reductions in crystallization temperature. This novel effect can be observed in Figure 5.21 for PS-*b*-PE infiltrated within AAO nanopores. The chosen diblock copolymer was E_{26}-*b*-S_{74}^{105}, whose morphology is that of a PS matrix with hexagonally packed PE cylinders (Refs [66, 200]). The already confined PE cylinders within a glassy PS matrix diblock copolymer was infiltrated into AAO nanopores. Figure 5.21 shows the different behavior upon cooling from the uninfiltrated diblock and that inside the 60-nm-diameter AAO template. A double confinement effect was experienced by the PE block chains (reducing its crystallization temperature), because the PE block feels the confinement of the PS vitreous block covalently bonded to it plus that caused by the inorganic AAO walls (further details can be found in Ref. [200]). The PE block (as well as the PE homopolymer infiltrated for comparison purposes) crystallization started by surface nucleation [200].

Figure 5.22 presents examples of the thermal properties of PEO and PBd-*b*-PEO polybutadiene-*b*-poly(ethylene oxide) before and after infiltration in AAO templates.

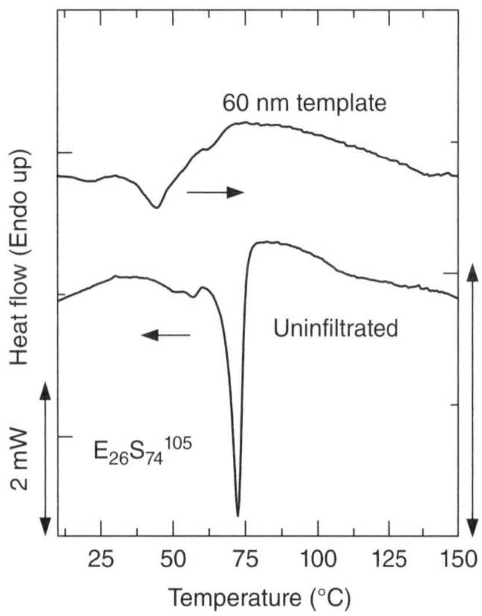

Figure 5.21 DSC cooling scans at 30 °C/min from the melt for the indicated samples. Michell et al. [200].

TABLE 5.4 Transition Temperatures and Enthalpies Obtained by DSC Standard Heating and Cooling Scans

Sample	T_c (°C)	ΔH_c (J/g)	T_m (°C)	ΔH_m (J/g)
EO_{100}^{100}	42.7	−127	65.5	135
EO_{100}^{100}/AAO 60 nm	−22.1	−0.7[a]	58.3	1.6[a]
EO_{100}^{100}/AAO 35 nm	−22.3	−0.3[a]	59.7	0.8[a]
Bd_{21}-*b*-EO_{79}^{257}	46.2	−111	66.1	122
Bd_{21}-*b*-EO_{79}^{257}/AAO 35 nm	−26.0	−0.5[a]	50.8	0.6[a]

[a]Nonnormalized, because the exact weight of the polymer inside the nanopores is unknown.

The results shown in Figure 5.22 for the neat uninfiltrated PEO are typical of the homopolymer in the bulk. Both melting and crystallization temperatures occur at shallow supercoolings and are within the expected range for a PEO with M_n value of 100 kg/mol (see Table 5.4). The Bd_{21}-*b*-EO_{79}^{257} is a strongly segregated linear diblock copolymer that exhibits a morphology of PB cylinders inside a PEO matrix [228]. Hence, the matrix component is formed by percolated PEO chains. The PEO block within the diblock copolymer contains a narrow molecular weight distribution, while the commercial PEO homopolymer used for comparison purposes is polydisperse. As a result, the values of T_c differ by 3 °C (see Table 5.4).

Confinement within the AAO templates caused a dramatic reduction of the crystallization temperature (see Fig. 5.22 and Table 5.4). For the PEO homopolymer infiltrated within a 35-nm template, the very large reduction in T_c value of 65.0 °C was obtained, in agreement with a previous work by us with a different PEO sample [200]. A similar behavior has been reported for PEO and also for PP, PE, sPS, and PVDF infiltrated in naopores [190–200].

The crystallization temperature of the PEO block within Bd_{21}-*b*-EO_{79}^{257} changed from 46.2 °C in the bulk to −26.0 °C in the template, that is, a reduction of 72.2 °C. These low-crystallization temperatures for PEO chains are characteristic of heterogeneity-free nanodomains (considering the volume of the PEO nanocylinders inside the AAO nanopores).

The number of nanopores in the templates is approximately 10 orders of magnitude higher than the number of active heterogeneities in bulk PEO; therefore, statistically speaking, most nanopores are free from heterogeneities.

The crystallization could have been started by homogeneous nuclei within the volume of the nanopores or by the AAO surface in contact with the polymer. If homogeneous nucleation were to be the origin of the crystallization process, the crystallization temperature should occur at very large supercoolings (close to vitrification) and should be a function of the nanopore volume [92, 115, 200, 229].

Müller et al. [92] gathered data from various PEO-confined systems where homogeneous nucleation was most likely. From these data, an empirical relationship between the volume of the microdomains and the crystallization temperature of PEO was

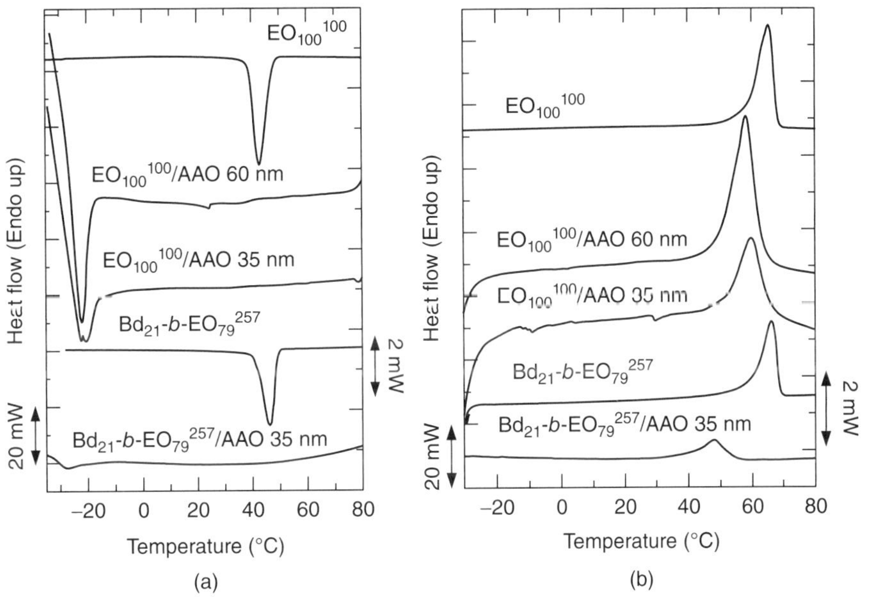

Figure 5.22 (a) DSC cooling and (b) heating scans for the indicated samples. The y-axis scale on the left of the plots (i.e., 20 mW) corresponds to the uninfiltrated neat materials, while that on the right (2 mW) corresponds to the AAO-infiltrated materials. Michell et al. [99]. Reproduced with permission of Elsevier.

proposed as follows:

$$T_c = -41.8 + 2.89\log(v_d) \tag{5.7}$$

where T_c is the peak crystallization temperature of the PEO microdomains and v_d their average volume. Using Equation 5.7, the predicted crystallization temperatures for the PEO nanocylinders within the AAO templates are −19 and −17 °C for the templates with 35 and 60 nm, respectively. These values can be compared with those experimentally found by DSC tests (−22 °C for both cases); hence, a homogeneous nucleation process is highly likely in these two cases.

The crystallization temperature reported in Table 5.4 for the PEO block within the infiltrated block copolymer is lower (−26° C) than for neat infiltrated PEO. It is not possible to determine exactly the phase volume of PEO in the block copolymer case, because the morphology probably changed with infiltration, as reported for other block copolymers in the literature [230–232]. However, in any case, the volume occupied by the PEO chains should be lower than the size of the nanopore. Therefore, the crystallization temperature for the block copolymer case should be lower than the crystallization temperature of the infiltrated PEO homopolymer, exactly as obtained. Hence, the nucleation of the PEO block nanodomains is also homogeneous.

Two different template diameters were employed in the PEO case. As already stated, an increase in the crystallization temperature with pore diameter has been reported for several infiltrated polymers [190, 196–200]. However, the difference in pore volume employed here was most probably not large enough to generate significant differences in crystallization temperature.

5.6 SELF-NUCLEATION AND THE EFFICIENCY SCALE TO EVALUATE NUCLEATION POWER

In this section, two examples of the use of self-nucleation are presented for block copolymers and nanocomposites. The first example shows how self-nucleation can help separate the coincident crystallization of a double crystalline diblock copolymer. In the second example, the efficiency scale based on self-nucleation helps to define supernucleation.

Figure 5.23 presents DSC cooling and subsequent heating scans for PPDX-*b*-PCL diblock copolymers in a wide composition range. These copolymers are in the weak segregation regime. The cooling scans from the melt indicate that regardless of the composition, only one crystallization exotherm is observed. However, when the subsequent melting run is carried out, the separate melting of each component is clearly seen. This behavior was examined [216] by time-resolved WAXS and isothermal crystallization kinetics and it was concluded that a coincident crystallization process takes place.

Upon cooling from the melt, the PCL molten block at high temperatures slows down the crystallization of the PPDX block. By the time the crystallization of the PPDX block starts at lower temperatures, PPDX crystallizes in the same temperature range as the PCL block, and hence the term coincident crystallization. In fact, once the crystallization of the PPDX block starts, it is quickly followed by the crystallization of the PCL block, which happens to be nucleated by the PPDX block. As a result, the crystallization of both blocks occurs in the same temperature range and in quick succession, leading to an overlap of the exothermic signals during cooling.

The DSC heating scans in Figure 5.23b show that each phase melts separately, in a temperature range close to

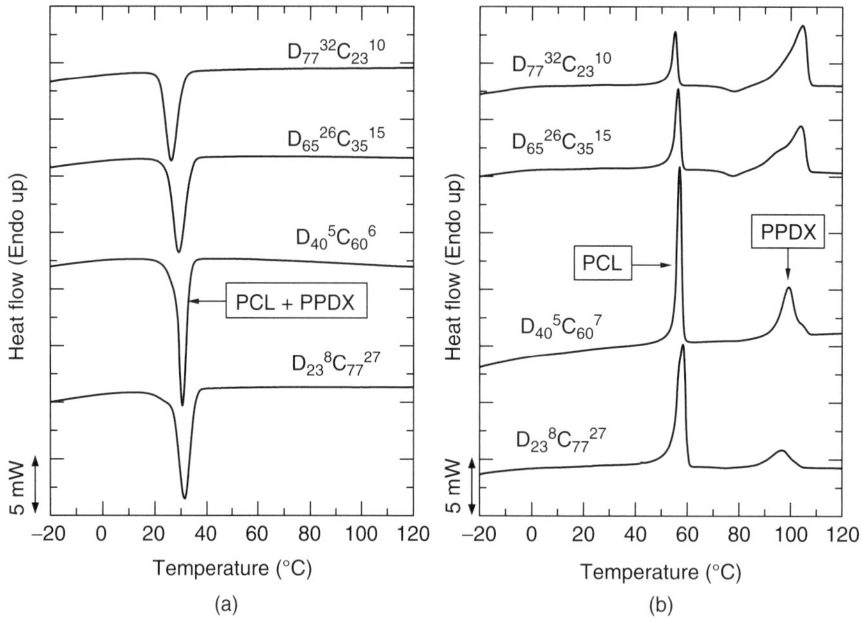

Figure 5.23 (a) DSC cooling scans (10 °C/min) for PCL-*b*-PPDX, diblock copolymers. (b) Subsequent heating scans (10 °C/min). Adapted from Müller et al. [233] with permission from The Royal Society of Chemistry.

that exhibited by the equivalent homopolymers (see Ref. [233]). The ability of PPDX to reorganize during heating has been reported previously [234] and cold crystallization exotherms can be detected in some of the heating DSC scans in Figure 5.23b.

The self-nucleation technique was employed in order to explore if such coincidences of crystallization temperature could be overriden. Details of the experimental procedure can be found in previous works [235].

Figure 5.24a shows DSC cooling scans from T_s for $D_{55}{}^{7}C_{45}{}^{6}$ diblock copolymer. At 125 °C, the T_s is too high and the sample is in *Domain I* or complete melting domain. The DSC cooling scan exhibits a single crystallization exotherm where both PPDX and PCL blocks crystallize

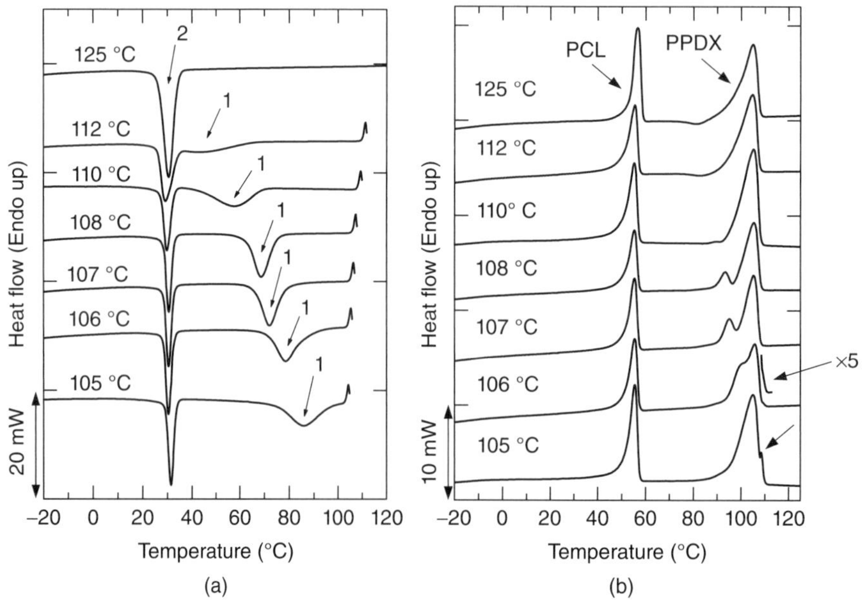

Figure 5.24 (a) DSC cooling scans (10 °C/min) after self-nucleation for 3 min at the indicated T_s temperatures for $D_{55}{}^{7}C_{45}{}^{6}$. (b) Subsequent melting scans. Müller et al. [236]. Reproduced with permission of Wiley-VCH.

coincidently [216]. Figure 5.24b shows the subsequent heating scans after the cooling runs shown in Figure 5.24a. The two melting endotherms are due to the separate melting of the PCL component (at lower temperatures) and the PPDX component (at higher temperatures).

An interesting result is shown in Figure 5.24a, thanks to the self-nucleation technique. When a T_s of 112 °C is used, the PPDX block crosses over into *Domain II*. Only the PPDX block is self-nucleated (because the PCL block is molten at that temperature, see Fig. 5.24b) and upon cooling from 112 °C a new high temperature exotherm, labeled 1 in Figure 5.24a, starts to develop (corresponding to the self-nucleated PPDX block), while the PCL component still crystallizes at lower temperatures (exotherm 2). Lowering T_s leads to a clear separation of the crystallization of the PPDX block and the PCL block.

Domain III is easily detected by observing the subsequent heating scans of Figure 5.24b [235, 237]. At a T_s temperature of 106 °C, a very small high-temperature melting peak signaled with an arrow that is the trademark of the first signs of annealing can be seen. The PPDX block of $D_{55}{}^{7}C_{45}{}^{6}$ at 106 °C is in *Domain III* or self-nucleation and annealing domain.

5.6.1 Supernucleation

In general terms, the efficiency of a nucleating agent calculated according to Equation 5.6, is normally not larger than 100%. However, Müller et al. have recently reported [238–241] nucleation efficiencies in excess of 100% when carbon nanotubes are incorporated into PE, PEO, and PCL matrices by different means, and this effect has been termed supernucleation. Precisely, an example of supernucleation is presented in this section.

Figure 5.25 plots the nucleation efficiency, calculated with Equation 5.6, for melt-mixed blends of PCL and multiwall carbon nanotubes (MWCNT) [241]. The nucleation efficiency rapidly increases with MWCNT content and saturates at values close to 200% efficiency. This means that the MWCNT are being more efficient than self-nuclei in promoting PCL nucleation.

TEM of selected multiwall nanotube (MWNT)-containing materials are shown in Figure 5.26. In Figure 5.26a, a distinct core–shell structure with MWNTs at the center and a thin layer around the MWNTs with varying thickness (average 25 nm) for the MWNT-*g*-PEO sample can be observed [240].

Figure 5.26b, shows a micrograph of the $PE_{41}M_{26}A_{33}$ nanocomposite, where a thin coating of HDPE can be seen surrounding the CNTs [67, 239]. The subscripts indicate the wt% composition for these *in situ* polymerized carbon nanotubes. M stands for MWCNT, while A indicates residual alumina content, a by-product of the metallocene catalytic reaction (see Refs [67, 239]). Figure 5.26c shows the lamellar morphology surrounding the MWNTs in $PE_{83}M_6A_{11}$, after a prolonged isothermal crystallization (1 week at 124 °C). The lamellae grew from the CNT surface in a direction roughly perpendicular to the CNT main axis, creating a bottlebrush- or shish-kebab-like morphology [67, 239, 242, 243]. This bottlebrush morphology is derived from the outstanding nucleation effect that the MWNT create on the PE, especially because of the *in situ* polymerization procedure (i.e., the PE chains grew from the surface of the CNT and are therefore ideally positioned to become nucleated by its surface). Evidences of this nucleation effect are presented below.

A summary of the thermal properties exhibited by *in situ* polymerized HDPE/CNT nanocomposite samples is presented in Figure 5.27. Different types of nanotubes were employed to perform the *in situ* polymerization: single (S), double (D), and multiwall (M) carbon nanotubes. The T_c determined during cooling from the melt at 10 °C/min, the T_m recorded during a subsequent heating DSC scan at 10 °C/min, and the degree of crystallinity (% X_c) obtained after cooling the sample from the melt are presented. The results of Figure 5.27 clearly show that:

i) Regardless of the CNT nature (single-wall nanotubes (SWNT), double-wall nanotubes (DWNT), or MWNT), a strong nucleation effect is produced on the HDPE according to the increase in T_c.
ii) The higher T_m values indicate that crystals produced upon nucleation on CNTs are thermodynamically more stable as compared to neat HDPE.
iii) The degree of crystallinity decreases as the amount of CNT increases, especially at high CNT contents. The presence of high contents of CNT probably interferes with crystal growth, especially when the extremely high number of nucleating sites they provide, in view of their enormous surface, are taken into account.
iv) When CNT content increases, the amount of Al_2O_3 residues also increases. Accordingly, the HDPE/CNT nanocomposites behave as materials with only 30–50 wt% PE and a very high load of fillers (both CNT and alumina); thus, a confinement effect is produced.

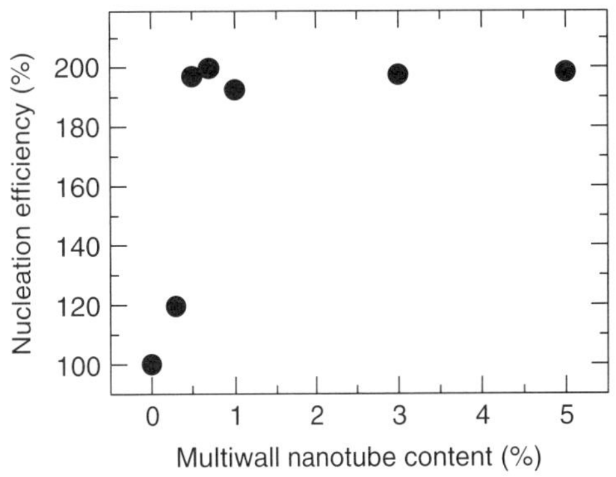

Figure 5.25 Nucleation efficiency (calculated with Eq. 5.6) as a function of MWCNT content for PCL/MWCNT nanocomposites. Adapted from Trujillo et al. [241]. Reproduced with permission of Elsevier.

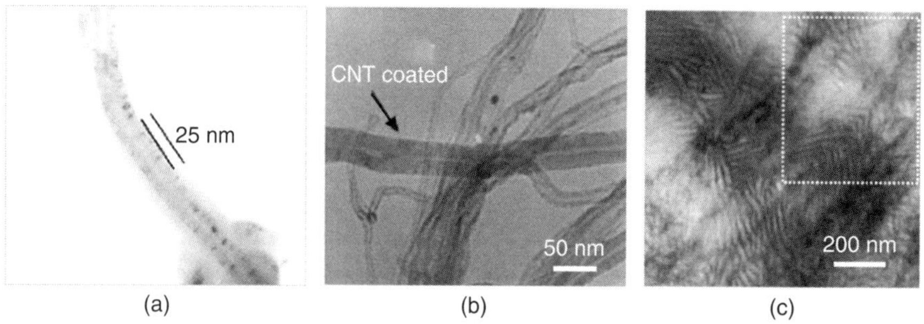

Figure 5.26 *TEM* images of (a) MWNT-*g*-PEO; (b) MWNT coated with PE for $PE_{41}M_{26}A_{33}$; (c) $PE_{83}M_6A_{11}$ after isothermal crystallization for 1 week at 124 °C. Adapted from Trujillo et al. [239]; Priftis et al. [240]; Trujillo et al. [67].

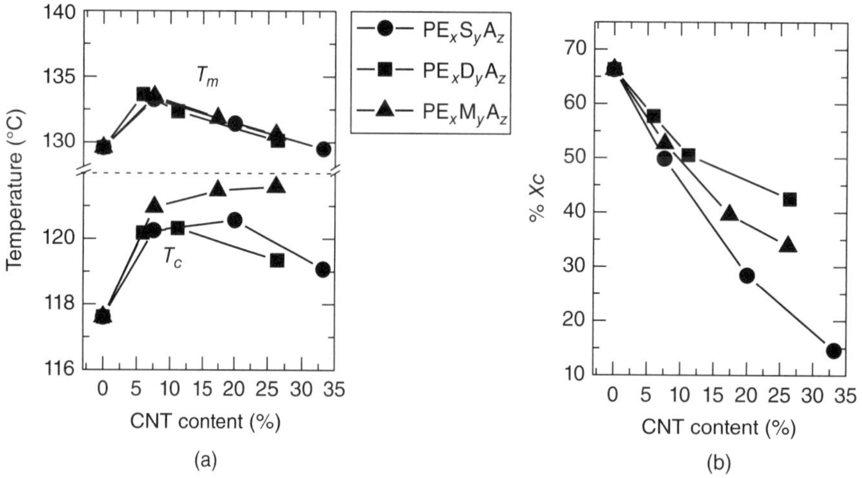

Figure 5.27 (a) Crystallization (T_c) and melting peak temperatures (T_m) and (b) crystallinity degree, for the HDPE/CNT nanocomposites as a function of the CNT content. Adapted from Trujillo et al. [67, 239].

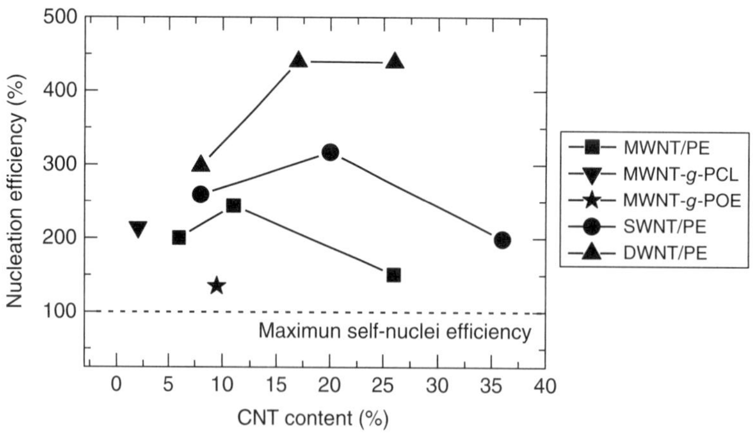

Figure 5.28 Efficiency of carbon nanotubes as a nucleating agent for several matrices. The estimated experimental error of each data point is less than 3%. Müller et al. [238]. Reproduced with permission of Elsevier.

The trend observed in Figure 5.27a can be interpreted as a competition between the supernucleation effect caused by the CNTs (which, in this case, fluctuates between 200% and 400% depending on the type and content of CNTs; see Ref. [238]) and the confinement effect they produced on the HDPE matrix. At low CNT contents, the nucleation effect dominates and a sharp increase in T_c and T_m is observed. However, as the content of CNTs is larger than about 7%, the confinement effect dominates and a maximum in the plot of Figure 5.27a can be observed in most cases. The confinement effect will depend on the interaction between the CNT surface and the HDPE chains, and it is expected that the larger the surface area of CNT the higher the confinement effect. Figure 5.27b illustrates this effect, because at higher CNT contents the crystallinity is a function of the CNT type employed. The confinement effect, taken in terms of crystallinity reduction, increases in the following order: MWNT, DWNT, and SWNT. At the same CNT content, the surface area of CNTs should increase in that same order.

The CNT nucleation efficiency for three different nanocomposites employing CNTs is summarized in Figure 5.28. The efficiency calculated with Equation 5.6 yields values above 100% in all cases, indicating that CNTs are much better nucleating agents than the homopolymer self-nuclei. Because of this reason, this nucleating effect has been termed "supernucleation."

The trends regarding the type of nanocomposite and the type of CNT employed are not clear (see Ref. [238]). A maximum in efficiency is observed in some cases in Figure 5.28. The trend would have to be validated with more data; however, if the trend is confirmed, it would be compatible with the mentioned competence between supernucleation and confinement effects at large CNT contents.

The most important result to be highlighted from this section is the occurrence in all cases of the supernucleation effect (i.e., nucleating efficiencies larger than 100%). Such an effect is probably due to the fact that the polymer chains in all these cases have a good contact with the surface of the CNTs and are well dispersed. Nevertheless, the nature of the interactions between CNTs and polymer chains and the reason why in some cases supernucleation can be achieved and in others it cannot is the subject of ongoing investigations.

5.7 DETERMINATION OF OVERALL ISOTHERMAL CRYSTALLIZATION BY DSC

Because there is a specific chapter in this book (Chapter 11) that deals with isothermal crystallization kinetics, the interested reader of this important topic is referred to that chapter.

5.8 CONCLUSIONS

In this chapter, we have described briefly the fundamentals of DSC and provided several examples on a wide range of uses for the technique while highlighting its correspondence with polymer morphology. Through the selected examples and the vast literature on the subject, we can conclude that DSC is one of the most powerful techniques to correlate thermal properties and morphology, although it must always be complemented by other techniques, such as PLOM, SEM, TEM, SAXS, and WAXS, among others depending on the objective of the ongoing study.

ACKNOWLEDGMENT

We acknowledge fruitful collaborations and discussions with Julio Albuerne, María Luisa Arnal, R. Verónica Castillo, Hsin-Lung Chen, Philippe Dubois, Nikos Hadjichristidis, A. Tomas Lorenzo, Mariselis Trujillo, and Carmen Mijangos.

REFERENCES

1. Ehrenstein GW, Riedel G, Trawiel P. *Thermal Analysis of Plastics*. Munich: Hanser; 2004.
2. Hatakeyama T, Quinn FX. *Thermal Analysis Fundamentals and Applications to Polymer Science*. Chichester: Wiley; 1999.
3. Pijpers TFJ, Mathot VBF, Goderis B, Scherrenberg RL, van der Vegte EW. Macromolecules 2002;35:3601.
4. Vanden Poel G, Mathot VBF. Thermochim Acta 2007;461:107.
5. Vanden Poel G, Mathot VBF. Thermochim Acta 2006;446:41.
6. Mathot VBF, Pyda M, Pijpers T, Poele GV, van de Kerkhof E, van Herwaarden S, van Herwaarden F, Leenaers A. Thermochim Acta 2011;522:36.
7. Tol RT, Minakov AA, Adamovsky SA, Mathot VBF, Schick C. Polymer 2006;47:2172.
8. Luruli N, Pijpers T, Brüll R, Grumel V, Pasch H, Mathot VBF. J Polym Sci Pol Phys 2007;45:2956.
9. Vanden Poel G, Istrate D, Magon A, Mathot VBF. J Therm Anal Calorim 2012;110:1533.
10. Van Herwaarden S, Iervolino E, van Herwaarden F, Wijffels T, Leenaers A, Mathot VBF. Thermochim Acta 2011;522:46.
11. Mathot VBF, Vanden Poel G, Pijpers TFJ. Benefits and potentials of high performance differential scanning calorimetry (HPer DSC). In: Brown ME, Gallagher PK, editors. *The Handbook of Thermal Analysis and Calorimetry. Volume 5: Further Advances, Techniques and Applications*. Oxford: Elsevier; 2008. p 269–298.
12. Pijpers MFJ, Mathot VBF. J Therm Anal Calorim 2008;93:319.
13. Herwaarden AW. Thermochim Acta 2005;432:192.
14. Schick C. Anal Bioanal Chem 2009;395:1589.
15. Yang B, Abyzov AS, Zhuravlev E, Gao Y, Schmelzer JWP, Schick C. J Chem Phys 2013;138:054501.
16. Mollova A, Androsch R, Mileva D, Schick C, Benhamida A. Macromolecules 2013;46:828.
17. Mileva D, Androsch R, Zhuravlev E, Schick C. Polymer 2012;53:3994.
18. Mileva D, Androsch R, Zhuravlev E, Schick C, Wunderlich B. Mol Cryst Liq Cryst 2012;556:74.

19. Wurm A, Zhuravlev E, Eckstein K, Jehnichen D, Pospiech D, Androsch R, Wunderlich B, Schick C. Macromolecules 2012;45:3816.
20. Huth H, Minakov AA, Schick C. J Polym Sci Pol Phys 2006;44:2996.
21. Mileva D, Androsch R, Zhuravlev E, Schick C, Wunderlich B. Polymer 2012;53:277.
22. Kolesov I, Mileva D, Androsch R, Schick C. Polymer 2011;52:5156.
23. Zhuravlev E, Schmelzer JWP, Wunderlich B, Schick C. Polymer 2011;52:1983.
24. Zhuravlev E, Schick C. Thermochim Acta 2010;505:1.
25. Chen W, Zhou D, Xue G, Schick C. Front Chem China 2009;4:229.
26. Mileva D, Androsch R, Zhuravlev E, Schick C. Thermochim Acta 2009;492:67.
27. De Santis F, Adamovsky S, Titomanlio G, Schick C. Macromolecules 2007;40:9026.
28. Chartoff RP, Sircar AK. Thermal analysis of polymers. In: Mark HF, editor. *Encyclopedia of Polymer Science and Technology*. New York: John Wiley & Sons; 2005. p 1–85.
29. Hiemenz PC. *Polymer Chemistry – The Basic Concepts*. New York: Marcel Dekker; 1984.
30. Fillon B, Wittmann J, Lotz B, Thierry A. J Polym Sci Pol Phys 1993;31:1383.
31. Lorenzo AT, Arnal ML, Sánchez JJ, Müller AJ. J Polym Sci Pol Phys 2006;44:1738.
32. Blundell DJ, Keller A, Kovacs AJ. J Polym Sci B Polym Lett 1966;4:481.
33. Müller AJ, Arnal ML. Prog Polym Sci 2005;30:559.
34. Fillon B, Lotz B, Thierry A, Wittmann JC. J Polym Sci Pol Phys 1993;31:1395.
35. Wild L. Adv Polym Sci 1991;98:1.
36. Anantawaraskul S, Soares JBP, Wood-Adams PM. Adv Polym Sci 2005;182:1.
37. Monrabal B. J Appl Polym Sci 1994;52:491.
38. Keating MY, McCord EF. Thermochim Acta 1994;243:129.
39. Starck P. Polym Int 1996;40:111.
40. Shanks RA, Amarasinghe G. Polymer 2000;41:4579.
41. Müller AJ, Hernández ZH, Arnal ML, Sánchez JJ. Polym Bull 1997;39:465.
42. Arnal ML, Balsamo V, Ronca G, Sánchez A, Müller AJ, Cañizales E, Urbina de Navarro C. J Therm Anal Cal 2000;59:451.
43. Zhang M, Lynch DT, Wanke SE. Polymer 2001;42:3067.
44. Starck P, Rajanen K, Löfgren B. Thermochim Acta 2003;395:169.
45. Zhang M, Wanke S. Polym Eng Sci 2003;43:1878.
46. Lorenzo AT, Arnal ML, Müller AJ, Boschetti de Fierro A, Abetz V. Macromol Chem Phys 2006;207:39.
47. Müller AJ, Lorenzo AT, Arnal ML. Macromol Symp 2009;277:207.
48. Lorenzo AT, Arnal ML, Müller AJ, Lin MC, Chen HL. Macromol Chem Phys 2011;212:2009.
49. Shen G, Shen H, Xie B, Yang W, Yang M. J Appl Polym Sci 2013;129:2103.
50. Gai JG, Cao Y. J Appl Polym Sci 2013;129:354.
51. Zhang S, Cheng R, Dong Q, He X, Wang Q, Tang Y, Yu Y, Xie K, Da J, Terano M, Liu B. Macromol Reac Eng 2013;7:254.
52. Matsko MA, Vanina MP, Echevskaya LG, Zakharov VA. J Therm Anal Calorim 2012;113:923.
53. Porcello JMB, Cardozo NSM, Forte MMC, Wolf CR, de Camargo M. Int J Polym Anal Charact 2011;16:95.
54. Wang Y, Yan W. Chin Sci Bull 2007;52:736.
55. Márquez L, Rivero I, Müller AJ. Macromol Chem Phys 1999;200:330.
56. Ding S, Liu M, Fang H. J Appl Polym Sci 2008;108:423.
57. Rojas De Gáscue B, López J, Prin JL, Hernández G, Reyes Y, Marcano LM, Carrasquero FL, Puig C, Müller AJ. Interciencia 2005;30:388.
58. Paolini Y, Ronca G, Feijoo JL, Da Silva E, Ramírez J, Müller AJ. Macromol Chem Phys 2001;202:1539.
59. Rojas de Gáscue B, Prin JL, Hernández G, Vallés EM, Lorenzo AT, Müller AJ. J Therm Anal Calorim 2011;103:669.
60. Perez CJ, Villarreal N, Pastor JM, Failla MD, Valles EM, Carella JM. Polym Degrad Stab 2009;94:1639.
61. Gai JG, Zhu SP, Kang J, Cao Y, Xiang M. J Macromol Sci B 2012;51:1322.
62. Perera R, Albano C, Sánchez Y, Karam A, Silva P, Pastor JM. J Appl Polym Sci 2012;124:1106.
63. Cardenas MA, Villarreal N, Gobernado-Mitre I, Merino JC, Pastor JM. J Polym Sci Pol Phys 2007;45:2432.
64. Arnal ML, Sánchez JJ, Müller AJ. Polymer 2001;42:6877.
65. Arnal ML, Cañizales E, Müller AJ. Polym Eng Sci 2002;42:2048.
66. Lorenzo AT, Arnal ML, Müller AJ, Boschetti de Fierro A, Abetz V. Eur Polym J 2006;42:516.
67. Trujillo M, Arnal ML, Müller AJ, Bredeau S, Bonduel D, Dubois P, Hamley IW, Castelletto V. Macromolecules 2008;41:2087.
68. Zhang Q, Song S, Feng J, Wu P. J Mater Chem 2012;22:24776.
69. Sabino MA, Feijoo JL, Müller AJ. Polym Degrad Stab 2001;73:541.
70. Benítez A, Sánchez JJ, Arnal ML, Müller AJ, Rodríguez O, Morales G. Polym Degrad Stab 2013;98:490.
71. Pérez CJ, Failla MD, Carella JM. Polym Degrad Stab 2013;98:177.
72. Gauthier E, Laycock B, Cuoq FJJ-M, Halley PJ, George KA. Polym Degrad Stab 2013;98:425.
73. Jin J, Chen S, Zhang J. J Polym Res 2010;17:827.
74. Michell RM, Müller AJ, Deshayes G, Dubois P. Eur Polym J 2010;46:1334.
75. Tenía R, Karam A, Yañez JD, Prin JL, Rojas de Gáscue BMAJ. Rev Iber Polímeros 2011;12:352.
76. Chang H, Ren S, Dang X, Zhang L, Li H, Hu Y. J Appl Polym Sci 2013;129:1026.
77. Denisova YI, Shandryuk GA, Krentsel LB, Blagodatskikh IV, Peregudov AS, Litmanovich AD, Kudryavtsev YV. Polym Sci Ser A 2013;55:385.
78. Kang J, Li J, Chen S, Zhu S, Li H, Cao Y, Yang F, Xiang M. J Appl Polym Sci 2013;130:25.

79. Luo J, Lin ZL, Tan SW, Chen ZJ, Yuan F, Liang YL, Lin H, Wang HJ, Liu DR, Tu KH, Jiang HL, Yu FS, Wang LQ. Acta Polym Sin 2013;0:263.
80. Liu L, Tong ZZ, Xu JT, Fan ZQ, Yu KM. J Therm Anal Calorim 2013;114:573.
81. Tong ZZ, Xu JT, Xia SJ, Fan ZQ. Polym Int 2013;62:228.
82. Fan J, Feng J. Ind Eng Chem Res 2013;52:761.
83. Mazzola N, Cáceres CA, França MP, Canevarolo SV. Polym Test 2012;31:870.
84. Luo J, Huo F, Lin H, Lin Z, Chen Z, Yuan F, Jiang H, Wang H, Tu K, Liu D, Tan S, Wang LQ. J Polym Sci Pol Phys 2012;50:1277.
85. Tian Z, Gu XP, Wu GL, Feng LF, Fan ZQ, Hu GH. Ind Eng Chem Res 2012;51:2257.
86. Zhang C, Shangguan Y, Chen R, Zheng Q. J Appl Polym Sci 2011;119:1560.
87. Franco-Urquiza E, Santana OO, Gámez-Pérez J, Martínez AB, Maspoch ML. Express Polym Lett 2010;4:153.
88. Ouhib F, Hiorns RC, de Bettignies R, Bailly S, Desbrières J, Dagron-Lartigau C. Thin Solid Films 2008;516:7199.
89. Sabino MA. Polym Degrad Stab 2007;92:986.
90. Seraidaris T, Puranen A, Karesoja M, Löfgren B, Repo T, Leskelä M, Seppälä J. J Polym Sci Polym Chem 2006;44:4743.
91. Frensch H, Harnischfeger P, Jungnickel BJ. Fractionated crystallization in incompatible polymer blends. In: Utracky L, Weiss R, editors. *Multiphase Polymers: Blends and Ionomers*. Toronto: ACS Symp Series 395; 1989.
92. Müller AJ, Balsamo V, Arnal ML. Adv Polym Sci 2005;190:1.
93. Hamley IW. Adv Polym Sci 1999;148:113.
94. Loo YL, Register AR. Crystallization within block copolymer mesophases. In: Hamley IW, editor. *Developments in Block Copolymer Science and Technology*. New York: Wiley; 2004.
95. Castillo RV, Müller AJ. Prog Polym Sci 2009;34:519.
96. Müller AJ, Balsamo V, Arnal ML. Crystallization in block copolymers with more than one crystallizable block. In: Reiter G, Strobl G, editors. *Lecture Notes in Physics: Progress in Understanding of Polymer Crystallization*. Vol. 714. Berlin: Springer; 2007. p 229–259.
97. Müller AJ, Arnal ML, Lorenzo AT. Crystallization in nano-confined polymeric systems. In: Piorkowska E, Rutledge G, editors. *Handbook of Polymer Crystallization*. New York: Wiley; 2013.
98. Nadan B, Hsy JY, Chen HL. Polymer Rev 2006;46:143.
99. Michell RM, Blaszczy-Lezak I, Mijangos C, Müller AJ. Polymer 2013;54:4059.
100. Arnal ML, Matos ME, Morales RA, Santana OO, Müller AJ. Macromol Chem Phys 1998;199:2275.
101. Price FP. Nucleation in polymer crystallization. In: Zettlemoyer A, editor. *Nucleation*. New York: Marcel Dekker; 1969.
102. Vonnegut B. J Colloid Sci 1948;3:563.
103. Turnbull D, Cech RC. J Appl Phys 1950;21:804.
104. Pound GM, La Mer VK. J Am Chem Soc 1952;74:2323.
105. Turnbull D. J Chem Phys 1952;20:411.
106. Turnbull D, Cormia RL. J Chem Phys 1961;34:820.
107. Cormia RL, Price FP, Turnbull D. J Chem Phys 1962;37:1333.
108. Burns JR, Turnbull D. J Appl Phys 1966;37:4021.
109. Koutsky JA, Walton AG, Baer E. J Appl Phys 1967;38:1832.
110. Gornick F, Ross GS, Frolen LJ. J Polym Sci Pol Sym 1967;18:79.
111. Kailas L, Vasilev C, Audinot J, Migeon H, Hobbs J. Macromolecules 2007;40:7223.
112. Carvalho JL, Dalnoki-Veress K. Eur Phys J E 2011;34:1.
113. Carvalho JL, Massa M, Dalnoki-Veress K. J Polym Sci Pol Phys 2006;44:3448.
114. Carvalho JL, Dalnoki-Veress K. Phys Rev Lett 2010;105:237801.
115. Massa MV, Carvalho JL, Dalnoki-Veress K. Eur Phys J 2003;12:111.
116. Massa MV, Dalnoki-Veress K. Phys Rev Lett 2004;92:255509.
117. Massa MV, Carvalho JL, Dalnoki-Veress K. Phys Rev Lett 2006;97:247802.
118. Massa MV, Lee MSM, Dalnoki-Veress K. J Polym Sci Pol Phys 2005;43:3438.
119. Jin Y, Rogunova M, Hiltner A, Baer E, Nowacki R, Galeski A, Piorkowska E. J Polym Sci Pol Phys 2004;42:3380.
120. Bernal-Lara TE, Liu R, Hiltner A, Baer E. Polymer 2005;46:3043.
121. Jin Y, Hiltner A, Baer E, Masirek R, Piorkowska E, Galeski A. J Polym Sci Pol Phys 2006;44:1795.
122. Jin Y, Hiltnet A, Baer E. J Polym Sci Pol Phys 2007;45:1138.
123. Jin Y, Hiltner A, Baer E. J Polym Sci Pol Phys 2007;45:1788.
124. Jin Y, Hiltner A, Baer E. J Appl Polym Sci 2007;105:3260.
125. Bernal-Lara TE, Masirek R, Hiltner A, Baer E, Piorkowska E, Galeski A. J Appl Polym Sci 2006;99:597.
126. Puente Orench I, Stribeck N, Ania F, Baer E, Hiltner A, Baltá Calleja F. Polymer 2009;50:2680.
127. Wang H, Keum JK, Hiltner A, Baer E, Freeman B, Rozanski A, Galeski A. Science 2009;323:757.
128. Pethe VV, Wang HP, Hiltner A, Baer E, Freeman BD. J Appl Polym Sci 2008;110:1411.
129. Montenegro R, Landfester K. Langmuir 2003;19:5996.
130. Tongcher O, Sigel R, Landfester K. Langmuir 2006;22:4504.
131. Taden A, Landfester K. Macromolecules 2003;36:4037.
132. Ibarretxe Uriguen J, Bremer L, Mathot VBF, Groeninckx G. Polymer 2004;45:5961.
133. Ibarretxe J, Groeninckx G, Bremer L, Mathot VBF. Polymer 2009;50:4584.
134. Weber CHM, Chiche A, Krausch G, Rosenfeldt S, Ballauf M, Harnau L, Göttker-Schnetmann I, Tong Q, Mecking S. Nano Lett 2007;7:2024.
135. Santana OO, Müller AJ. Polym Bull 1994;32:471.
136. Morales RA, Arnal ML, Müller AJ. Polym Bull 1995;35:379.
137. Manaure AC, Morales RA, Sánchez JJ, Müller AJ. J App Polym Sci 1997;66:2481.
138. Arnal ML, Müller AJ, Maiti P, Hikosaka M. Macromol Chem Phys 2000;201:2493.
139. Tol RT, Mathot VBF, Groeninckx G. Polymer 2005;46:369.
140. Tol RT, Mathot VBF, Groeninckx G. Polymer 2005;46:383.
141. Tol RT, Mathot VBF, Groeninckx G. Polymer 2005;46:2966.
142. Tol RT, Mathot VBF, Groeninckx G. Polymer 2005;46:2955.
143. Yordanov C, Minkova L. Eur Polym J 2005;41:527.

144. Arnal ML, Müller AJ. Macrom Chem Phys 1999;200:2559.
145. Manaure A, Müller AJ. Macrom Chem Phys 2000;201:958.
146. Brown RA, Masters AJ, Price C, Yuan X. Chain segregation in block copolymers. In: Aggarwal S, editor. *Comprehensive Polymer Science – The Synthesis, Characterizations, Reaction & Applications of Polymers*. 1 ed. Vol. 7. London: Pergamon Press; 1989.
147. Hadjichristidis N, Pispas S, Floudas G. *Block Copolymers: Synthetic Strategies, Physical Properties, and Applications*. 1st ed. New Jersey: Wiley-Interscience; 2002.
148. Abetz V. Block copolymers. In: Mark HF, editor. *Encyclopedia of Polymer Science and Engineering*. New York: John Wiley & Sons; 2001.
149. Abetz V. Assemblies in complex block copolymer systems. In: Ciferri A, editor. *Supramolecular Polymers*. New York: Marcel Dekker Inc.; 2000.
150. Hamley IW. *The Physics of Block Copolymers*. London: Oxford University Press; 1998.
151. Li ZY, Liu R, Mai BY, Wang WJ, Wu Q, Liang GD, Gao HY, Zhu FM. Polymer 2013;54:1663.
152. Yin LG, Hillmyer MA. Macromolecules 2011;44:3021.
153. Qi F, Guerin G, Cambridge G, Xu W, Manners I, Winnik M. Macromolecules 2011;44:6136.
154. Gilroy JB, Gadt T, Whittell GR, Chabanne L, Mitchels JM, Richardson RM. Nat Chem 2010;2:566.
155. Lopez-Quintela MA, Lazzari M. Macromol Rapid Commun 2009;30(1785).
156. Gao L, Zhang K, Chen YM. Polymer 2011;52:3681.
157. Zh S, Lu HJ, Chen ZC, Cheng RS, Chen DZ. Polymer 2012;53:359.
158. She MS, Ho RM. Polymer 2012;53:2628.
159. Petzetakis N, Walker D, Dove AP, O'Reilly RK. Soft Matter 2012;8:7408.
160. Li ZC, Ono RJ, Wu ZQ, Bielawski CW. Chem Commun 2011;47:197.
161. Qin S, Li H, Yuan WZ, Zhang YM. Polymer 2011;52:1191.
162. Bian QQ, Xiao Y, Lang MD. Polymer 2012;53:1684.
163. Xu LM, Zhang XK, Yang H, Li X, Li CS, Zhang SX. Polymer 2010;51:3808.
164. Fu J, Luan B, Yu X, Cong Y, Li J, Pan CY, Han Y, Yang Y, Li B. Macromolecules 2004;37:976.
165. Chen YF, Zhang FB, Xie XM, Yuan JY. Polymer 2007;48:2755.
166. Yu CH, Chuang YH, Tung SH. Polymer 2011;52:3994.
167. Yi FP, Yu RT, Zheng SX, Li XH. Polymer 2011;52:5669.
168. Mihut AM, Chiche A, Drechsler M, Schmalz H, Di Cola E, Krausch G. Soft Matter 2009;5:208.
169. Mihut AM, Drechsler M, Moller M, Ballauff M. Macromol Rapid Commun 2010;31:449.
170. Du ZX, Xu JT, Fan ZQ. Macromol Rapid Commun 2008;29:467.
171. Zheng JX, Xiong HM, Chen WY, Lee KM, Horn RMV, Quirk RP, Lotz B, Thomas EL, Shi A-C, Cheng SZD. Macromolecules 2006;39:641.
172. Petzetakis N, Dove AP, Reilly RKO. Chem Sci 2011;2:955.
173. Zhao Y, Shi XB, Gao HY, Zhang L, Zhu FM, Wu Q. J Mater Chem 2012;22:5737.
174. Schmalz H, Schmelz J, Drechsler M, Yuan JY, Walther A, Schweimer K, Mihut AM. Macromolecules 2008;41:3235.
175. Schmelz J, Karg M, Hellweg T, Schmalz H. ACS Nano 2012;5:9523.
176. Massey JA, Temple K, Cao L, Rharbi Y, Raez J, Winnik MA, Manners I. J Am Chem Soc 2000;122:11577.
177. Raez J, Manners I, Winnik MA. Langmuir 2002;18:7229.
178. Guerin G, Raez J, Manners I, Winnik MA. Macromolecules 2005;38:7819.
179. Raez J, Manners I, Winnik MA. J Am Chem Soc 2002;124:10381.
180. Mihut AM, Crassous JJ, Schmalz H, Ballauff M. Colloid Polym Sci 2010;288:573.
181. Mihut AM, Crassous JJ, Schmalz H, Drechsler M, Ballauff M. Soft Matter 2012;8:3163.
182. Wang XS, Wang H, Frankowski DJ, Lam PG, Welch PM, Winnik MA, Hartmann J, Manners I, Spontak RJ. Adv Mater 2007;19:2279.
183. Wang JD, Horton JH, Liu GJ, Lee SY, Shea KJ. Polymer 2007;48:4123.
184. Lin EK, Gast AP. Macromolecules 1996;29:4432.
185. Richter D, Schneiders D, Monkenbusch M, Willner L, Fetters L, Huang J, Lin M, Mortensen K, Farago B. Macromolecules 1997;30:1053.
186. Ramzi A, Prager M, Richter D, Efstratiadis V, Hadjichristidis N, Young R, Allgaier J. Macromolecules 1997;30:7171.
187. Radulescu A, Mathers RT, Coates GW, Richter D, Fetters LJ. Macromolecules 2004;37:6962.
188. Li ZY, Liu R, Mai BY, Feng S, Wu Q, Liang GD, Gao HY, Zhu FM. Polym Chem 2013;4:954.
189. Rochette CN, Rosenfeldt S, Henzler K, Polzer F, Ballauff M, Tong Q, Mecking S, Drechsler M, Narayanan T, Harnau L. Macromolecules 2011;44:4845.
190. Duran H, Steinhart M, Hans-Jürgen B, Floudas G. Nano Lett 2011;11:1671.
191. Shin K, Woo E, Jeong YG, Kim C, Huh J, Kim KW. Macromolecules 2007;40:6617.
192. Garcia-Gutiérrez MC, Linares A, Hernández JJ, Rueda DR, Ezquerra TA, Poza P, Davies R. Nano Lett 2010;10:1472.
193. Hui W, Wei W, Huang Y, Wang C, Su Z. Macromolecules 2008;41:7755.
194. Lutkenhaus JL, McEnnis K, Serghei A, Russell TP. Macromolecules 2010;43:3844.
195. Steinhart M, Haissam PG, Prabhukaran M, Gösele U. Phys Rev Lett 2006;97:027801.
196. Steinhart M, Senz S, Wehrspohn RB, Gösele U, Wendorff JH. Macromolecules 2003;36:3646.
197. Woo E, Huh J, Jeong YG, Shin K. Phys Rev Lett 2007;98:136103.
198. Wu H, Wang W, Huang Y, Su Z. Macromol Rapid Commun 2009;30:194.
199. Wu H, Wang W, Yang H, Su Z. Macromolecules 2007;40:4244.
200. Michell RM, Lorenzo AT, Müller AJ, Lin MC, Blaszczyk-Lezak I, Martín J, Mijangos C. Macromolecules 2012;45:1517.
201. Wu H, Su Z, Takahara A. Soft Matter 2012;8:3180.

202. Suzuki Y, Duran H, Steinhart M, Butt H-J, Floudas G. Soft Matter 2013;9:2621.
203. Martín J, Maiz J, Sacristan J, Mijangos C. Polymer 2012;53:1149.
204. Maiz J, Schäfer H, Rengarajan GT, Hartmann-Azanza B, Eickmeier H, Haase M, Mijangos C, Steinhart M. Macromolecules 2013;46:403.
205. Müller AJ, Balsamo V. Thermal characterization of linear low density and low density polyethylene blends. In: Finlayson K, editor. *Advances in Polymer Blends and Alloys Technology*. Vol. V. Technomic Pub. Co.; 1994. p 1–21.
206. M. E. Matos, C. Rosales, A. J. Müller and B. D. Favis, Polystyrene/polypropylene blends: Morphology/interphase relationship, rheological and calorimetric behavior. SPE-ANTEC 97, Toronto; 1997.
207. Lorenzo AT, Müller AJ, Priftis D, Pitsikalis M, Hadjichristidis N. J Polym Sci Pol Chem 2007;45:5387.
208. Lorenzo AT, Müller AJ, Lin MC, Chen HL, Jeng US, Priftis D, Pitsikalis M, Hadjichristidis N. Macromolecules 2009;42:8353.
209. Olvera de la Cruz M, Sanchez IC. Macromolecules 1986;19:2501.
210. Milner ST. Macromolecules 1994;27:2333.
211. Hamley IW, Parras P, Castelletto V, Castillo RV, Müller AJ, Pollet E, Dubois P, Martin CM. Macromol Chem Phys 2006;207:941.
212. Castillo RV, Müller AJ, Raquez JM, Dubois P. Macromolecules 2010;43:4149.
213. Hamley IW, Castelletto V, Castillo RV, Müller AJ, Martin CM, Pollet E, Dubois P. Macromolecules 2005;38:463.
214. Laredo E, Prutsky N, Bello A, Grimau M, Castillo RV, Müller AJ, Dubois P. Eur Phys J E 2007;23:295.
215. Wang Y, Funar SS, Mano JF. Macromol Chem Phys 2006;207:1262.
216. Albuerne J, Marquez L, Müller AJ, Raquez JM, Degee P, Dubois PCV, Hamley I. Macromolecules 2003;36:1633.
217. Masuda H, Fukuda K. Science 1995;268:1466.
218. Steinhart M. Adv Polym Sci 2008;220:123.
219. Martin CR. Science 1994;266:1961.
220. Martin CR. Acc Chem Res 1995;28:61.
221. Martín J, Mijangos C. Langmuir 2009;25:1181.
222. Steinhart M, Wendorff JH, Greiner A, Wherspohn RB, Nielsch K, Schilling J, Choi J, Gösele U. Science 2002; 296:1997.
223. Zhang M, Dobriyal P, Chen JT, Russell TP, Olmo J, Merry A. Nano Lett 2006;6:1075.
224. Shin K, Xiang H, Moon SI, Kim T, McCarthy TJ, Russell TP. Science 2004;306:76.
225. Michell RM, Blaszczyk-Lezak I, Mijangos C, Müller AJ. Macromol Symp 2014;337:109.
226. Michell RM, Blaszczyk-Lezak I, Mijangos C, Müller AJ. J Polym Sci Polym Phys 2014;52:1179.
227. Maiz J, Martin J, Mijangos C. Langmuir 2012;28:12296.
228. Castillo RV, Arnal ML, Müller AJ, Hamley IW, Castelletto V, Schmalz H, Abetz V. Macromolecules 2008;41:879.
229. Chen HL, Wu JC, Lin TL, Lin JS. Macromolecules 2001;34:6936.
230. Xiang H, Shin K, Kim T, Moon SI, McCarthy TJ, Russell TP. Macromolecules 2005;38:1055.
231. Dobriyal P, Xiang H, Kazuy M, Russell TP. Macromolecules 2009;42:9082.
232. Sevink GJA, Zvelindovsky AV, Fraaije JGE, Huinink HP. J Chem Phys 2001;115:8226.
233. Müller AJ, Albuerne JML, Raquez JM, Degée P, Dubois P, Hobbs J, Hamley I. Faraday Discuss 2005;128:231.
234. Sabino MA, Feijoo JL, Müller AJ. Macromol Chem Phys 2000;201:2687.
235. Müller AJ, Balsamo V, Arnal ML, Jakob T, Schmalz H, Abetz V. Macromolecules 2002;35:3048.
236. Müller AJ, Albuerne J, Esteves LM, Márquez L, Raquez JM, Degée P, Dubois P, Collings S, Hamley I. Macromol Symp 2004;215:369.
237. Balsamo V, Paolini Y, Ronca G, Müller AJ. Macromol Chem Phys 2000;201:2711.
238. Müller AJ, Arnal ML, Trujillo M, Lorenzo AT. Eur Polym J 2011;47:614.
239. Trujillo M, Arnal ML, Müller AJ, Laredo E, Bredeau S, Bonduel D, Dubois P. Macromolecules 2007;40:6268.
240. Priftis D, Sakellariou G, Hadjichristidis N, Penott E, Lorenzo AT, Müller AJ. J Polym Sci Pol Chem 2009;47:4379.
241. Trujillo M, Arnal ML, Müller AJ, Mujica MA, Urbina de Navarro C, Ruelle B, Dubois P. Polymer 2012;53:832.
242. Haggenmueller R, Fisher JE, Winey KI. Macromolecules 2006;39:2964.
243. Minus ML, Chae HG, Kumar S. Polymer 2006;47:3705.

6

IMAGING POLYMER MORPHOLOGY USING ATOMIC FORCE MICROSCOPY

HOLGER SCHÖNHERR

Department of Chemistry – Biology, Physical Chemistry I, University of Siegen, Siegen, Germany

6.1 INTRODUCTION

Atomic force microscopy (AFM) [1] belongs to the family of scanning proximity probe microscopy techniques and has developed in the past decades into a versatile and widely used microscopy technique that complements the more traditional techniques of optical microscopy [2] and electron microscopy in studies of polymers [3]. In addition to direct quantitative imaging of micro- and nanostructures (metrology) at surfaces covering literally all length scales of polymer morphologies, from molecular-scale crystal structures, single polymer chains, lamellae up to spherulites, polymeric superstructures, and microscale morphologies, different AFM imaging modes reveal additional information [4]. Prime among these is qualitative compositional mapping in semicrystalline and heterogeneous systems. These qualitative and, in many cases, quantitative data obtained also include, for example, maps of polymeric (micro)phases, identification of interfaces and interphases, as well as maps of mechanical properties, thermal properties, and chemical composition. Furthermore, *in situ* experiments are feasible focusing on mechanical deformation or polymer crystallization from the melt (compare Chapter 1), among others.

AFM complements, as mentioned, other more established microscopy methods. The choice of the most applicable method is naturally guided by a critical analysis of the imaging physics, required information, and the advantages and disadvantages of the corresponding microscopy methods. Unlike scanning electron microscopy (SEM) and transmission electron microscopy (TEM) (Chapter 3), AFM can be applied essentially independent of the surrounding medium and in a broad range of experimental conditions, for example, in ambient, in vacuum, at elevated temperatures, as well as in liquid media, which renders AFM applicable for *in situ* studies of, for example, biomaterials under realistic physiological conditions. In terms of ultimate resolution, tomography capabilities and access to chemical information, AFM cannot rival electron microscopy in general. Without discussing the choice of methods at length, the reader is referred to textbooks on polymer microscopy in this context [5–7] and specifically to Chapter 3, where the different areas of application are summarized in a concise manner.

The wide range of surface structures and properties that can be analyzed by AFM is a consequence of (i) the exquisite nano- and micropositioning devices developed and constantly refined in the past decades, (ii) the microfabrication of suitable nanoprobe tips as local sensors, and (iii) the general applicability of *forces* as the basis for the underlying imaging process and the contrast mechanisms. Unlike in scanning tunneling microscopy (STM), in which only conducting samples can be probed by a very sharp conductive nanoprobe tip, exploiting electrons tunneling via the classically forbidden gap between tip and sample, AFM is generally applicable to condensed matter [7]. Even drops of viscous liquids [8], fragile supramolecular structures such as lipid vesicles [9] or enclosed gas in liquid media [10] can be interrogated. Interatomic forces between the atoms located at the apex of the tip and the sample surface lend themselves as a distinct measure of tip–sample separation and hence as a parameter to track the surface profile under minimally invasive conditions [11, 12]. In addition, these interaction forces provide, in some cases, directly useful information of polymer properties, including adhesion (adherence), friction, elastic properties, and so on [13]. Common essentially to all AFM modes is

Polymer Morphology: Principles, Characterization, and Processing, First Edition. Edited by Qipeng Guo.

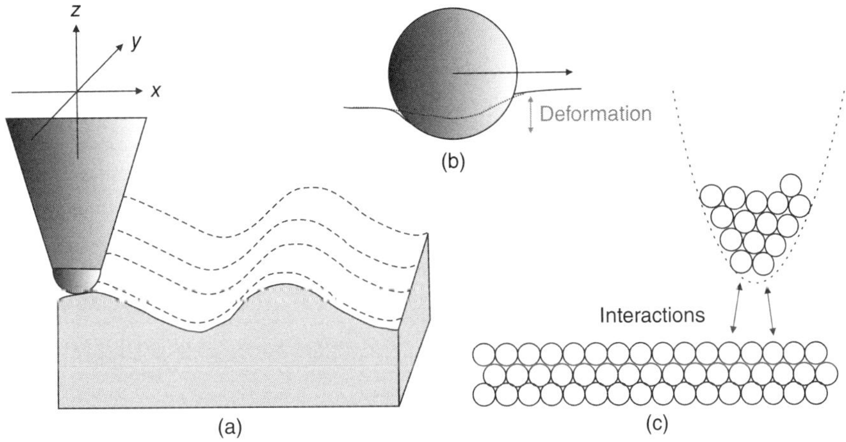

Figure 6.1 Schematics of (a) an AFM tip scanning across a polymer surface to assess the topography of the underlying sample by following accurately the surface contour, (b) the probe tip (symbolized by a sphere) may deform the sample locally (in the surface normal direction and also laterally, if shear forces are present), and (c) the interatomic and intermolecular forces that determine, in addition to the tip size, tip shape and deformation in the tip–sample contact area the resolution.

the nanoprobe tip, which is brought into close proximity or contact with the sample surface. By mutual displacement, for example, a common raster scan, the surface of the sample is probed in each position and, depending on the range of the underlying forces, the tip size and the local deformation involved, data with high spatial resolution are being captured (Figure 6.1).

AFM is by definition a surface analytical tool and polymer morphology can hence be analyzed at free surfaces of neat samples or, following suitable sample preparation established mainly for electron microscopy, surfaces of exposed bulk material specimens (compare Chapter 3). Among the broad range of scanning proximity microscopy probe techniques utilizing a solid nanoprobe tip to interrogate different properties of polymers and the large number of different AFM experiments proposed in the literature to date, a number of key applications of AFM in analyzing polymers can be identified, which have become routine approaches or lend themselves to broad applicability. More specialized techniques, however, may not be generally applicable, and require specialized equipment or, in order to be conducted and analyzed correctly, a more elaborate treatment, which is beyond the scope of this chapter. A good illustration is the determination of local tribological [14] or mechanical properties, such as elastic moduli, treated in Chapter 17, which requires, in addition to elaborate calibration, an appropriate use of adequate contact mechanics theories. Therefore, we focus in this chapter on *imaging of polymer morphologies* and selected *spatially resolved analyses of some polymer properties*.

AFM suffers like all microscopy and other analytical imaging techniques of artifacts that have their roots in the imaging physics or signal and data processing. If AFM data are captured, for instance, with inadequate imaging parameters (such as too high imaging forces), the results may be completely useless. Likewise, the interpretation of AFM height data without knowledge of the imaging mode and parameters may lead to erroneous conclusions. To help address these vital issues and to appreciate the opportunities and limitations of AFM for imaging of polymer morphology, the central imaging modes are discussed first, illustrated by a selected example each. The essential features and elements of an atomic force microscope are introduced in the following first section on contact mode AFM. Subsequent to the sections on the various AFM modes, the imaging of polymer morphology is illustrated for all levels of structural hierarchy.

6.2 FUNDAMENTAL AFM TECHNIQUES

For any AFM experiment, and also the interpretation of AFM data, an appreciation of the underlying fundamental AFM techniques and their opportunities and limitations is required. This statement refers in the context of this chapter to the rudimentary imaging mechanisms involved. More detailed treatments on specific aspects, including the underlying physics, have been published and the reader is referred to these reviews [13], handbooks [6], and the exact specifications of the AFM manufacturers.

6.2.1 Contact Mode AFM

In contact mode AFM, which is the originally reported AFM mode [1] and may serve as the general example of a typical AFM concerning its features, a sharp nanoprobe tip, mounted to a flexible cantilever (typically made of Si_3N_4 by chemical vapor deposition or Si by silicon etching microfabrication technology) is brought into contact with the sample surface. The sample (or, likewise, the tip) can be positioned in all three directions independently by appropriate transducers. The tip, characterized by its radius of curvature R_c, may be as sharp as several nanometers; typical values are $R_c \approx 5$–25 nm.

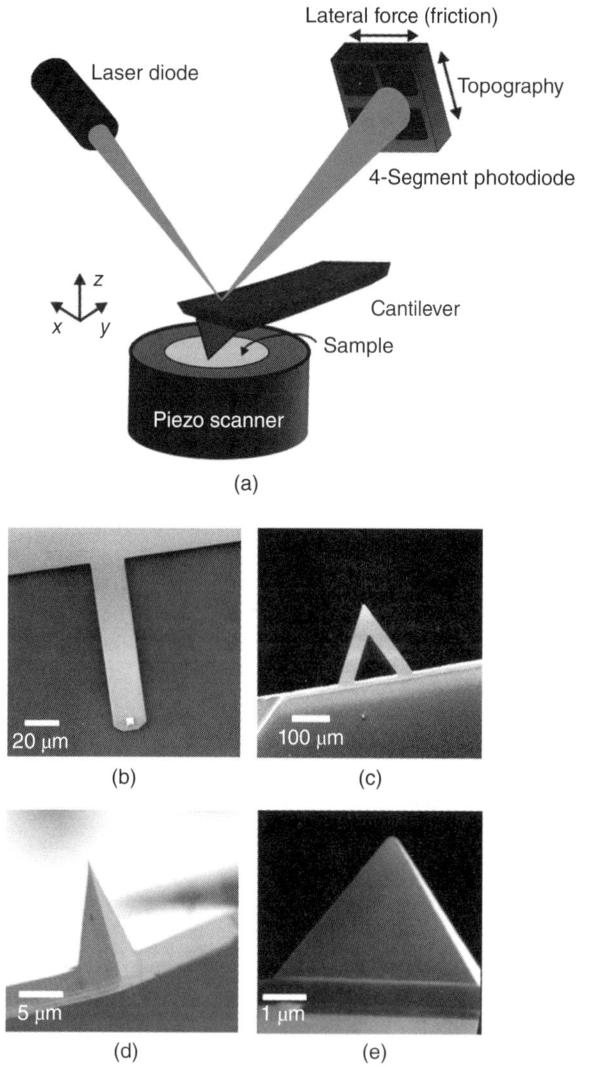

Figure 6.2 (a) Schematic of contact mode AFM with optical beam deflection detection. Any deflection of the cantilever results in a change of the inclination angle, which is detected on a four-segment photodiode. By comparing the signals recorded in the upper versus the lower sectors the vertical displacement is measured; likewise, the comparison of the left and right sectors provides information about the cantilever torsion. Schönherr et al. [14]. Reproduced with permission of Springer. (b)–(d) SEM images of typical (b, c) AFM cantilevers and (d, e) AFM tips. Courtesy: C.J. Padberg, G. J. Vancso, H. Schönherr, unpublished data.

The cantilever has a spring constant k_c of approximately 0.05–1.0 N/m and deflects up or down, if repulsive or attractive forces, respectively, are experienced by the tip. The lever behaves as a Hookean spring and by measuring the cantilever deflection Z_C (equal to its vertical displacement Δz), for example, exploiting the widely used optical beam deflection technique, tip–sample interaction forces F ($F = k_c\,\Delta z$) can be measured (Figs. 6.2 and 6.3).

6.2.1.1 ***Force–Displacement Curve*** In the so-called force–displacement experiment, the sample is brought periodically

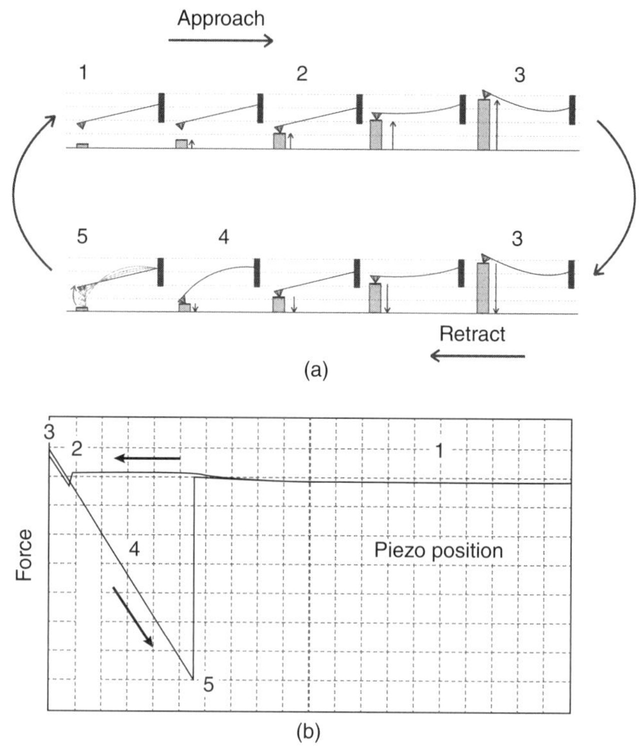

Figure 6.3 (a) Schematic (side view) of a piezo tube, sample, and cantilever during one cycle of a force–displacement curve. (b) Measured force–displacement curve ("displacement" and piezo position Z_P are identical and are used interchangeably). Reproduced with permission from [15].

in and out of contact with the tip at a fixed position (x, y). This can be achieved by moving the sample up and down, as shown in the scheme, or by lowering the AFM head comprising the tip down and up. In Figure 6.3, the movements of the sample (positioned via a piezo tube scanner) and the corresponding cantilever deflection as well as the resulting force–displacement curve are shown. The sample is approaching the tip (1, top); at some distance, the gradient of the force overcomes the cantilever spring constant and the tip jumps into contact (2); further movement up causes a deflection of the cantilever (3); during retraction, the tip sticks usually much longer (4); and snaps off when the spring constant overcomes the force gradient (5, bottom). The pull-off or pull-out force (the sudden jump of force when the tip breaks free from the surface) characterizes the adhesive interactions.

6.2.1.2 ***Constant Force Imaging*** For imaging of nano- and microscale surface structures, the tip is scanned with respect to the sample surface, for example, in a raster fashion, using suitable displacement elements (Fig. 6.2). The displacement in x and y, which must be accurate and precise down to (sub)nanometer scale, is afforded by piezoelectric tube scanners or similar high precision displacement elements. By compensating any upward or downward deflection of the tip with respect to a defined set point bending (which equals a set point force) using a vertical z displacement element

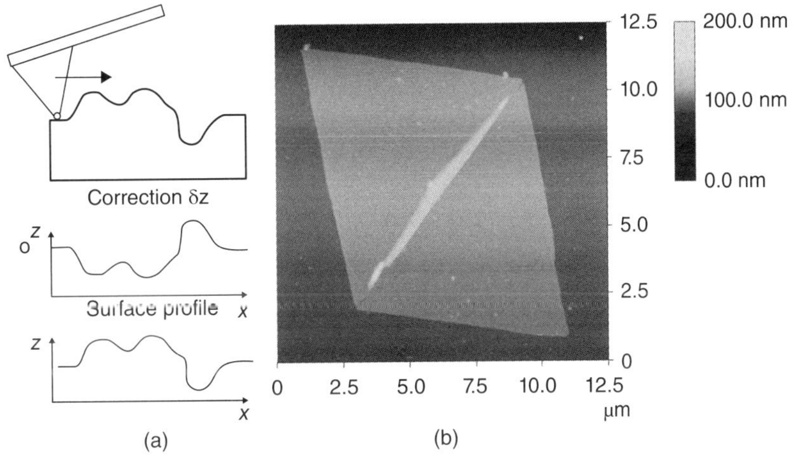

Figure 6.4 (a) Schematic illustration of the relation of true profile and correction δz (of feedback loop applied to keep the cantilever bending/applied force constant) used for the profile reconstruction. (b) AFM contact mode height image of lamellar crystal of polyethylene grown in solution, imaged in ambient conditions on a mica substrate. Schönherr and Vancso [16]. Reproduced with permission of American Chemical Society.

controlled by a feedback loop, the topography $z = z(x, y)$ of the sample surface can be reconstructed from the recorded (i.e., known) compensation δz for each pixel (x, y) (Fig. 6.4a). The feedback loop settings and the scan velocity determine how well the surface is being tracked. In each experiment, these settings must be optimized. Assuming that the tip interacts in all positions and in a similar manner with the sample surface, *both* in terms of surface deformation *and* adhesive properties of the contact, a quantitative height profile is thus obtained in this *constant force imaging mode* (Fig. 6.4). For the example shown in Figure 6.4b, this means that the height profile encoded in the color scale is correct only if the polyethylene is deformed to an extent similar to that of the underlying mica substrate by the force applied by the AFM tip. If the imaging forces are minimized, this assumption is, to a first approximation, valid. In addition to the constant force "height" image, the error signal of the feedback loop (the actual residual deflection of the cantilever) may provide information on surface structures.

Constant force contact mode AFM hence affords a quantitative 3D map of the surface topography. On a neat material with one phase, this works adequately, unless lateral forces deform the sample (see below). On multiphase systems, however, such as semicrystalline polymers comprising an amorphous and a crystalline phase, or on polymer blends with two phases with widely different mechanical properties, this may lead to height artifacts. Namely, if the force exerted by the tip leads to different local deformation of phase A compared to phase B, the softer phase will appear too low. Likewise, if different adhesive forces are present between the tip and the constituent phases, the force applied in the surface normal direction onto the sample (the so-called normal or imaging force, which is the sum of the applied force by cantilever bending and the adhesive force between tip and sample) varies locally. Consequently, the image is, strictly speaking, *not* acquired at constant force imaging conditions and the increased normal force in areas with higher adhesion may result in a height underestimate compared to areas with lower adhesion. In practice, experiments with different imaging forces can be conducted to assess potential errors in profile assessment. Minimum set point forces help to minimize this potential source of errors.

While contact mode AFM lends itself to the analysis of a broad range of materials, it is no longer the central AFM imaging mode applied to polymers. Due to (i) the repulsive contact of tip and sample and (ii) the mutual lateral displacement, (lateral) shear forces may occur that can be significant and may severely deform the polymeric sample or even damage the area to be analyzed. This is primarily due to the very high torsional spring constant of AFM cantilevers, which is often orders of magnitude above the spring constant for normal deflection.

In friction force microscopy (also called lateral force microscopy, LFM), on the contrary, these lateral forces are measured and may yield important insight into friction forces, their dependence on particular phases, and orientation of the underlying polymer or environmental conditions. Due to the difficulties in obtaining truly quantitative friction force data and the dependence of friction forces on load and scan velocity (due to the time–temperature superposition principle) [14], which all require tedious experimental procedures, friction force microscopy is not widely used in the analysis of polymer morphologies.

The resolution of contact mode AFM is, as is typical for AFM in general, limited by the tip size, tip shape, and deformation in the tip–sample contact area. The range and strength of the interatomic and intermolecular forces between tip and sample may become significant, if the imaging force is very small. For operation in air, however, capillary condensation may occur in the tip–sample contact [12], which results in significant attractive forces that cannot be counteracted in a stable manner. Imaging in liquid (e.g., water or ethanol) can be applied to circumvent this problem, if the sample is stable and unaltered under these conditions [6].

6.2.2 Intermittent Contact (Tapping) Mode AFM

The high lateral forces and concomitant drawbacks of contact mode are circumvented in intermittent contact (also called tapping) mode AFM (Fig. 6.5). This mode utilizes an oscillating tip–cantilever assembly and relies on a feedback from the amplitude (constant amplitude imaging). A typical cantilever for operation in air is much stiffer than a contact mode cantilever (10–100 N/m) and is excited to resonance or near resonance. The forced oscillator is damped upon interaction of the tip with the sample surface (Fig. 6.5b and c). If the cantilever spring constant or the amplitude is too low, the energy in the forced oscillator is not sufficient to overcome the adhesive interactions and the tip remains trapped in contact and is consequently dragged across the surface.

6.2.2.1 Constant Amplitude Imaging The sample topography $z = z(x, y)$ of the sample surface can be reconstructed from the recorded (i.e., known) compensation δz for each pixel (x, y) by scanning the sample with respect to the oscillating tip/cantilever assembly, while compensating any increase or decrease of the cantilever's *rms* amplitude with respect to a defined set point amplitude (typically, the amplitude of the freely oscillating lever) using the vertical z displacement element controlled by the feedback loop. Assuming that the tip interacts in all positions in a similar manner with the sample surface in terms of amplitude damping (energy dissipation), a quantitative height profile is thus obtained in this *constant amplitude imaging mode* (Fig. 6.5). Similar to the contact mode, the feedback loop settings and scan velocity must be optimized to ensure that the amplitude is indeed constant. The residual changes in amplitude (deviation of the set point value) may be displayed as an error signal (amplitude channel).

The acquisition of an accurate surface profile is based, as mentioned, on the assumption of identical energy dissipation in all areas of the surface imaged. If the forced oscillation of the cantilever–tip assembly is damped in different areas to a different extent, the correction δz does not match exactly the "true" surface profile. If, for instance, phase A in a heterogeneous phase-separated structure damps the oscillation more efficiently compared to phase B, the more dissipative phase A will appear too high compared to phase B. The example shown

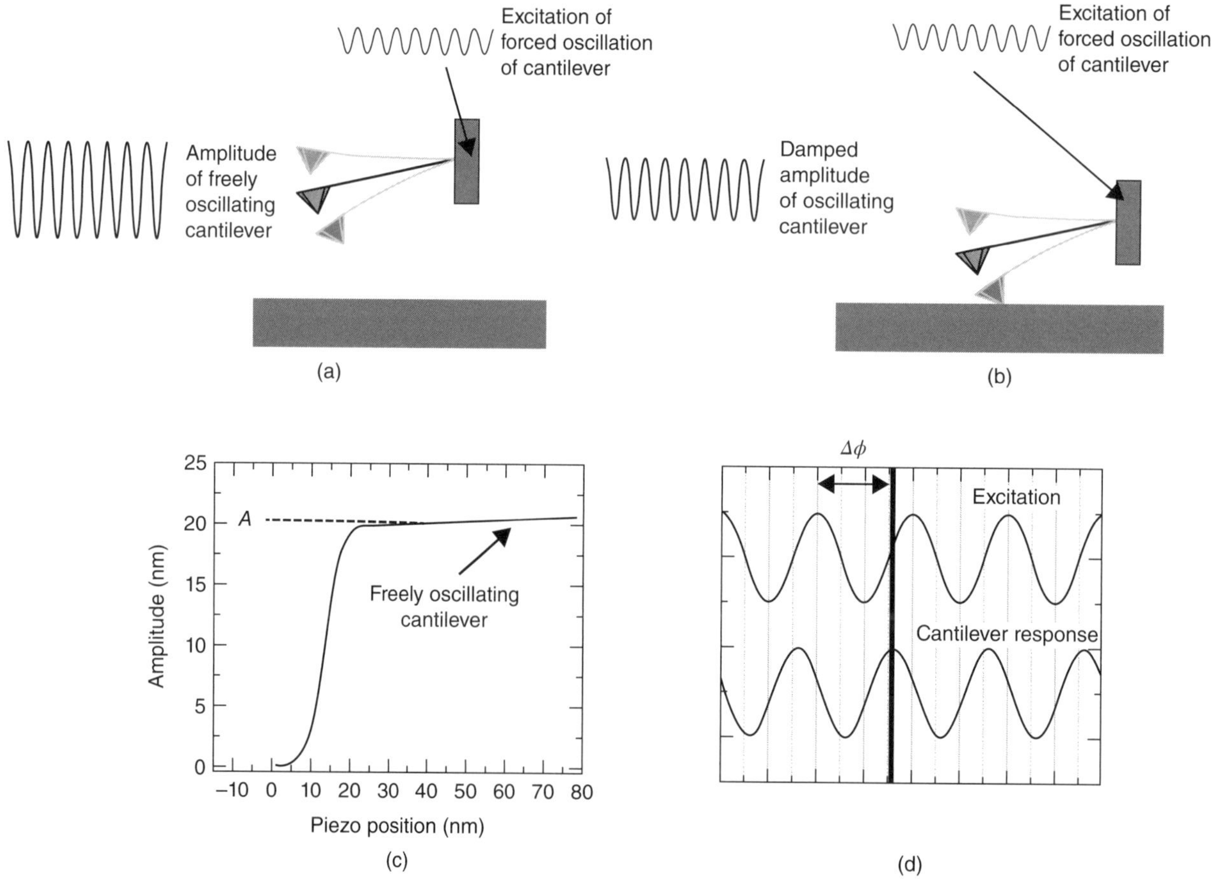

Figure 6.5 Schematic of oscillating tip/cantilever assembly oscillating (a) in air far away from the sample surface, (b) in close proximity to the sample surface; (c) schematic amplitude versus piezo position curve; (d) phase shift between excitation oscillation and cantilever response. Reproduced with permission from [4]. Copyright 2004. John Wiley & Sons, New York.

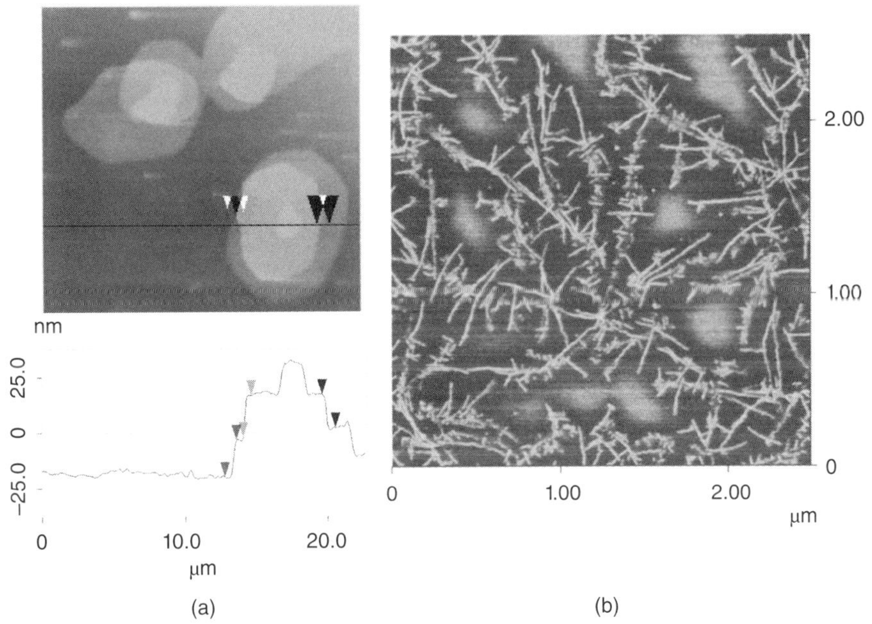

Figure 6.6 (a) Intermittent contact mode AFM height image (top) and cross-sectional analysis along the horizontal line (bottom) of melt-crystallized poly(ethylene oxide). The sample was crystallized from a 46-nm-thin spin-coated film on oxidized silicon at a constant crystallization temperature of 58 °C. The vertical distances between the markers are 16.5, 14.9, and 14.9 nm. Schönherr and Frank [17]. Reproduced with permission of American Chemical Society. (b) Intermittent contact mode AFM phase image of a fraction of elastomeric polypropylene crystallized in a thin film on silicon. The bright features correspond to lamellar crystals, which are embedded in an amorphous matrix. Courtesy: H. Schönherr, R. M. Waymouth, C. W. Frank, unpublished data.

in Figure 6.6a is devoid of this potential artifact, because the poly(ethylene oxide) lamellae reside on a poly(ethylene oxide) film.

6.2.2.2 Intermittent Contact Mode Phase Imaging It is fair to conclude that intermittent contact mode AFM has been the prime mode for imaging polymer morphologies. This is largely due to the useful mapping capability of the phase angle shift ("phase signal"). In addition to (constant amplitude) height data that is devoid of lateral force induced effects, the phase angle shift, which is calculated from the relation of the forced sinusoidal oscillation of the cantilever to the excitation oscillation, provides useful compositional information on the analysis of polymer morphology (Fig. 6.5d). The phase shift is related to energy dissipation in the tip–sample contact and hence serves as an excellent means to differentiate areas of different composition in blends or microphase-separated block copolymers, among others.

The interpretation of conventional phase shifts is not trivial (the reader is referred to the monograph by Garcia [18]) and, depending on the instrument make, the displayed phase angle shift may possess different meanings. As a rule of thumb, a highly dissipative (polymer) phase shows a larger phase lag compared to phases with low dissipation. For instance, the amorphous and crystalline phases of semicrystalline polymers can be differentiated easily (Fig. 6.6b). Filler particles can also be easily differentiated from the surrounding polymer matrix based on the phase signal. However, because the phase shift depends on various parameters, including the cantilever amplitude, the quality factor and the amplitude-set point ratio, the data remain semi-quantitative in most application-oriented scenarios.

6.2.3 Further Dynamic AFM Modes

For imaging of polymer morphology, many additional AFM imaging and analysis modes have been developed. We limit our brief discussion here to two related *imaging* modes, which utilize intermittent tip–sample contacts and address not only topography imaging but also allow one to capture simultaneously the data that characterize the adhesive and mechanical properties of the contact. Both the "pulsed force mode" and the "peak force tapping (PFT) mode" utilize the continuous capture of force–displacement curves; they differ, however, in the way the feedback circuit ensures a consistent tracking of the surface.

6.2.3.1 Pulsed Force Mode AFM In pulsed force mode AFM, a conventional contact mode AFM scan with feedback on the cantilever deflection is carried out. Simultaneously, the sample is modulated in the z direction sinusoidally (with variable frequency, e.g., 1 kHz) such that the tip breaks free from the surface periodically. The resulting plot of cantilever deflection versus time is shown in Figure 6.7. In the figure, it can be observed that the tip snaps to contact and then indents the sample. Upon retraction, the tip breaks free from the surface (pull-off) and the free cantilever rings down in air. Hence, instead of a slow acquisition of force–displacement data in

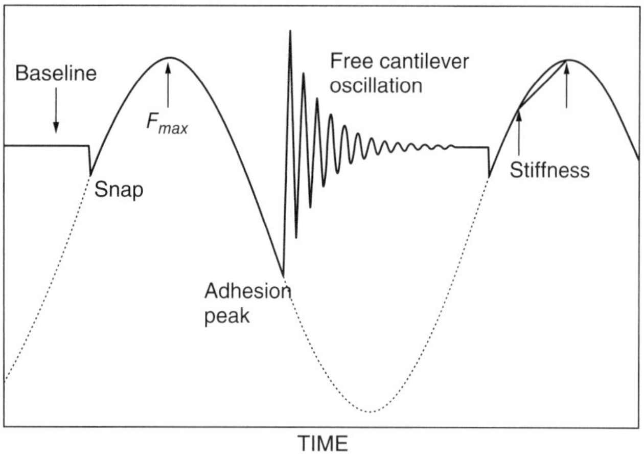

Figure 6.7 Schematic of the force signal in pulsed force mode AFM. Reprinted from [19]. Copyright 2004. Elsevier.

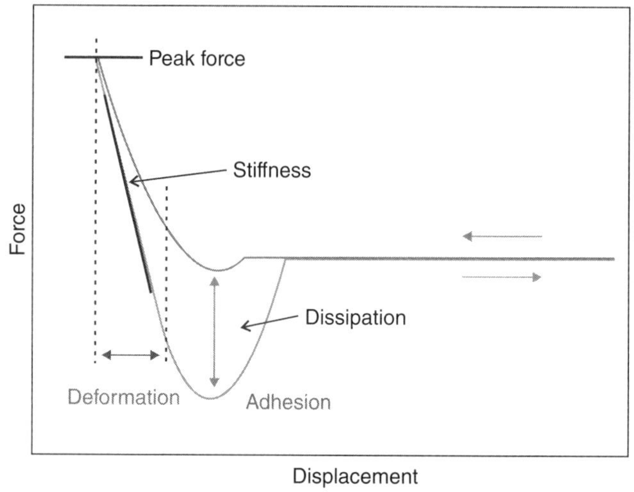

Figure 6.8 Schematic force–displacement curve obtained in PFT mode AFM.

a point-by-point fashion [13], the tip experiences the same process as during a contact mode scan, albeit at an increased rate. Instead of recording the entire force–displacement curve, only four essential points are being captured in the original pulsed force mode: the baseline, the maximum repulsive force, the maximum adhesion force, and a second point on the repulsive part of the force–displacement curve to define the slope in the loading regime as a measure of stiffness. Therefore, the pulsed force mode affords the simultaneous mapping of sample topography, pull-off forces, and stiffness at useful scan rates.

6.2.3.2 PFT Mode AFM Related to this dynamic pulsed force mode is the so-called PFT mode. Using a cantilever with intermediate spring constant ($\approx$1 N/m), a somewhat altered intermittent contact mode experiment is carried out during which the entire force–displacement curve is being captured. Unlike the intermittent contact mode, however, in which the amplitude is used as the feedback parameter, in this mode the force is controlled directly by using the maximum exerted force, the "peak force," as the feedback parameter. This mode allows exquisite control of the force exerted with the tip/cantilever and provides information on sample deformation, stiffness, and adhesive forces from an analysis of the force–displacement curve (Fig. 6.8).

In the PFT mode, as mentioned, the AFM tip is brought intermittently into contact with the sample surface with an adjustable amplitude $\geq$0.1 nm. Similar to the intermittent contact mode, this minimizes the lateral forces as the lateral displacement is slow compared to the residence time of the tip in contact, even though the tip (or the sample) is oscillated out of resonance at a frequency of 1–10 kHz, which is lower than in standard TM in air and also in liquid, and even though the imaging speeds in both modes are comparable.

In each oscillation cycle, the force–displacement curve is recorded and analyzed. Initially, the force is zero (baseline force), which corresponds to the deflection of the undisturbed cantilever with the tip being far away from the sample surface. The value of the peak force threshold is held constant by the feedback loop, so that at each point on the sample scanned the force exerted on the sample is nominally the same. Provided that the lateral differences in adhesive forces are negligible, this affords constant normal force imaging conditions.

Far from the sample, the force is by definition zero. Close to the sample, the tip may experience van der Waals attraction, snap into contact, and subsequently bend due to increasing repulsive forces up to the peak force threshold. Upon retraction, the tip may stick to the sample and the cantilever bends down due to adhesive forces (the force drops below the zero force line), until the nanoprobe tip eventually snaps off the surface and the deflection returns to the base line deflection position. This procedure is repeated pixel by pixel.

"Adhesion" (pull-off force) data are calculated as the largest negative force detected during the retraction curve. In addition to directly extracted data on the maximum adhesive force, further data may be calculated from the force–displacement curve. The area enclosed between the approach force curve and the retract force curve accounts for the dissipation of the energy per oscillation cycle. Finally, the maximum deformation of the sample is calculated as the difference in the piezo-displacement between the points of maximum and zero force, measured along the approach curve, and corrected for the change in the deflection of the cantilever. The calculated value includes both elastic and plastic contributions and reaches its maximum at the peak force.

An illustrative example is shown in Figure 6.9, where a phase-separated polyurethane synthesized form of 4,4′-methylenbis (phenyl isocyanate) (MDI), 1,4-butanediol (BD), and poly(tetrahydrofurane) polyether polyol (PTHF) was analyzed by PFT AFM. The data confirm that stiff and soft segments of the phase-separated polyurethane samples can be resolved at nanoscale resolution. Compared to intermittent contact mode phase imaging, these data reveal contrast directly due to the elastic moduli of the segments [20].

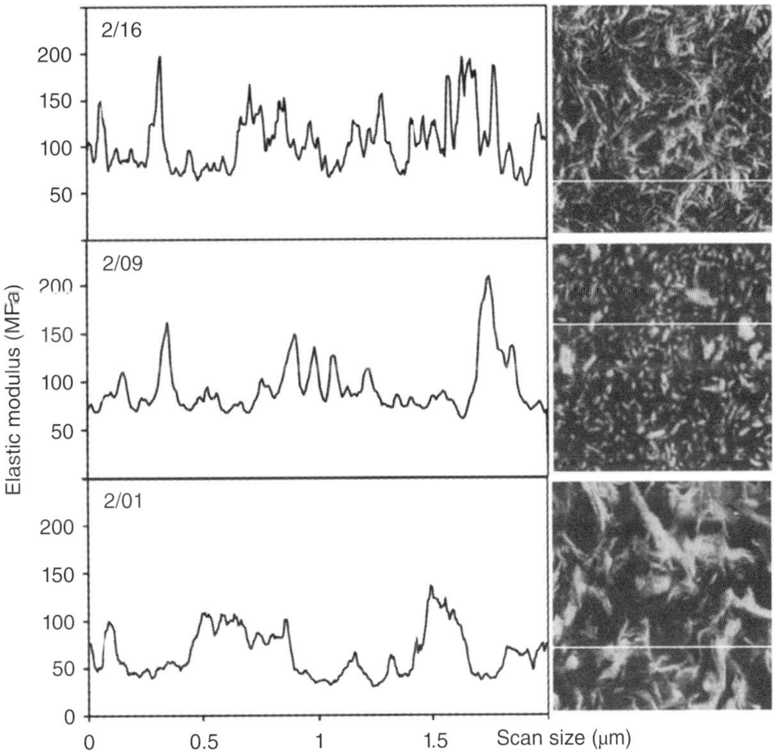

Figure 6.9 DMT modulus images with corresponding representative cross sections of linear-phase-separated polyurethane sample obtained by PFT mode AFM. The labels denote the ratio of the isocyanate to hydroxyl groups (NCO/OH), which changed between 0.940 and 1.150. The –OH functional group ratio of polyol/total diol (poly-(OH)/OH ratio) was kept constant at 0.4 in all experiments. Reprinted from [20]. Copyright 2010. Elsevier.

6.3 IMAGING OF POLYMER MORPHOLOGY

The AFM techniques briefly introduced above provide direct real-space access on structures and phases on all relevant length scales of hierarchical ordering of polymers. The subsequent sections shall not encompass all possible structures and examples of polymer morphology, but rather focus on selected examples that illustrate the rich information extracted from AFM. These examples include classic morphologies, as well as more recently reported polymeric nano- and microstructures.

6.3.1 Single Polymer Chains

Individual (bio)polymer molecules, in particular if they are long and shape persistent, can be easily imaged by intermittent contact or PFT mode AFM. Noteworthy is the difference in step heights measured with these two related mode. While intermittent contact mode AFM shows systematically too low heights for DNA (the example shown in Fig. 6.10a is plasmid DNA on highly oriented pyrolytic graphite), PFT mode AFM shows step height values that coincide with the structural model and diffraction data (Fig. 6.10b). Hoogenboom et al. could even reconstruct the oligonucleotide secondary structure from their AFM data [21].

Molecular level detail was also discernible in intermittent contact mode images of PMMA molecular brushes with poly(*n*-butyl acrylate) side chains (Fig. 6.11) [22]. The polymer backbone and the grafted side chains can be differentiated in these individual molecules, giving rise to a nonensemble average characterization of macromolecular architectures that cannot be done with nonmicroscopic techniques. This is also true for the analysis of branching (Fig. 6.12) [23]. Sheiko and coworkers demonstrated that molecularly resolved AFM images afford quantitative information about branching topology including length and distribution of branches.

6.3.2 Crystal Structures

The periodic arrangement of polymer chains in crystallites of semicrystalline polymers has been in the focus of early AFM investigations, in which known, but also unknown, crystal structures have been reported. Due to the local character of the experiment, ensemble averaging is avoided, unlike in conventional X-ray crystallographic methods. It is, however, important to note that most data published in the literature is contact mode AFM data, which represent, in most cases, lattice resolution only. One prominent example is the structure of poly(oxy methylene) unveiled by Snétivy and Vancso (Fig. 6.13) [24]. The information on the periodic arrangement of molecules is hence averaged over the length scale of the tip–sample contact. Consequently, point defects and other deviations in the periodic structure cannot be resolved.

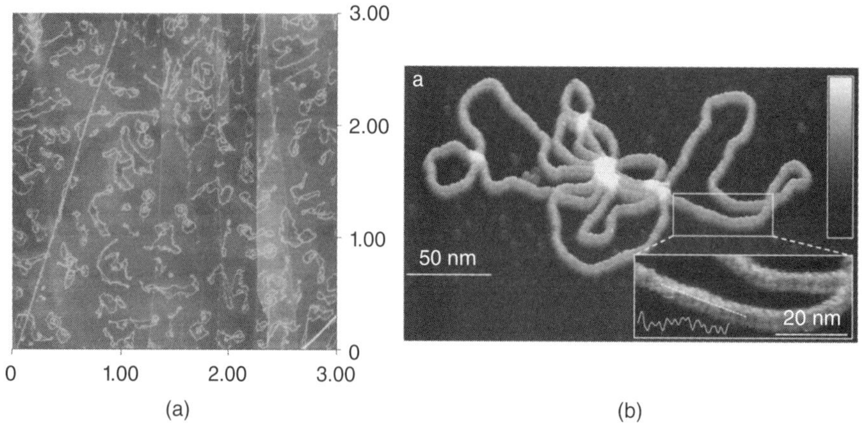

Figure 6.10 (a) Low-resolution intermittent contact mode AFM height image of plasmid DNA on HOPG obtained in ambient conditions. Courtesy: B. Song, H. Schönherr, G. Nöll, unpublished data. (b) PFT AFM height image of DNA plasmid on mica acquired at a peak force of 49 pN in buffer showing the double-helix corrugation and height of the DNA plasmid. Reproduced unaltered with permission from [21]. Copyright 2014. The Authors. The Figure is covered by the CC-BY license (http://creativecommons.org/licenses/by/4.0/). Published by Wiley-VCH Verlag GmbH & Co. KGaA, Weinheim.

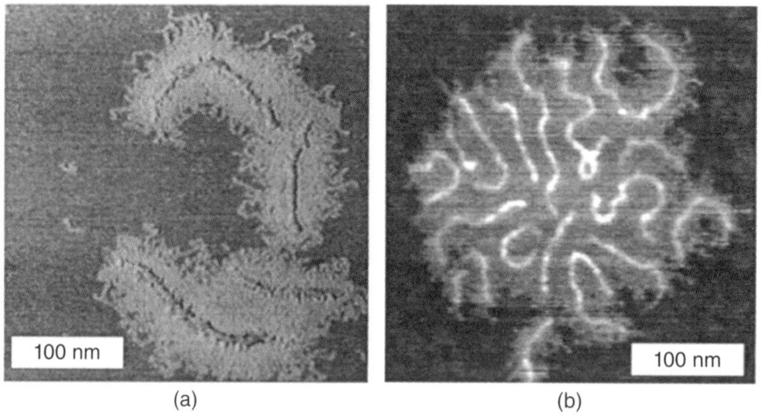

Figure 6.11 PMMA molecular brushes with poly(n-butyl acrylate) side chains of different degrees of polymerization: (a) $n = 52$ and (b) $n = 10$. Reproduced with permission from [22]. Copyright 2001. American Chemical Society.

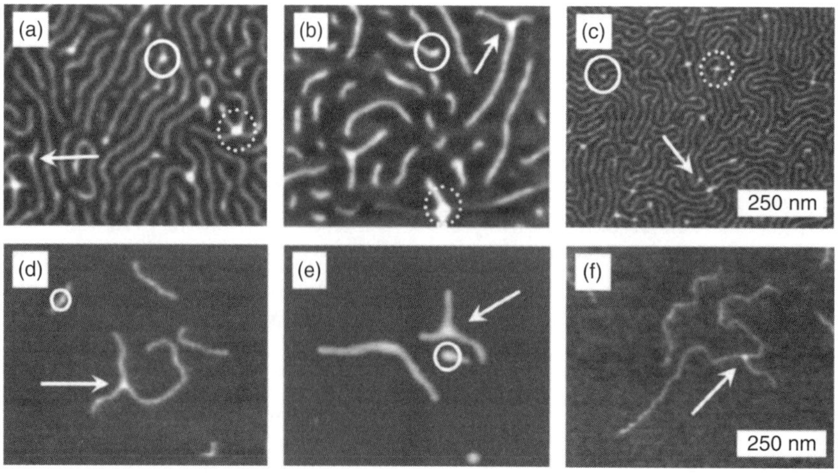

Figure 6.12 AFM images of poly(n-butyl acrylate) (brushes made from poly(alkyl acrylate) and poly(alkyl methacrylate) backbones. The branching points are highlighted by arrows in images of dense films. Further evidence for the presence of branched macromolecules is provided by the AFM images of each brush polymers embedded in linear poly(n-butyl acrylate) matrix (white arrows pointing to branch junctions) in panels d–f. Reproduced with permission from [23]. Copyright 2011. American Chemical Society.

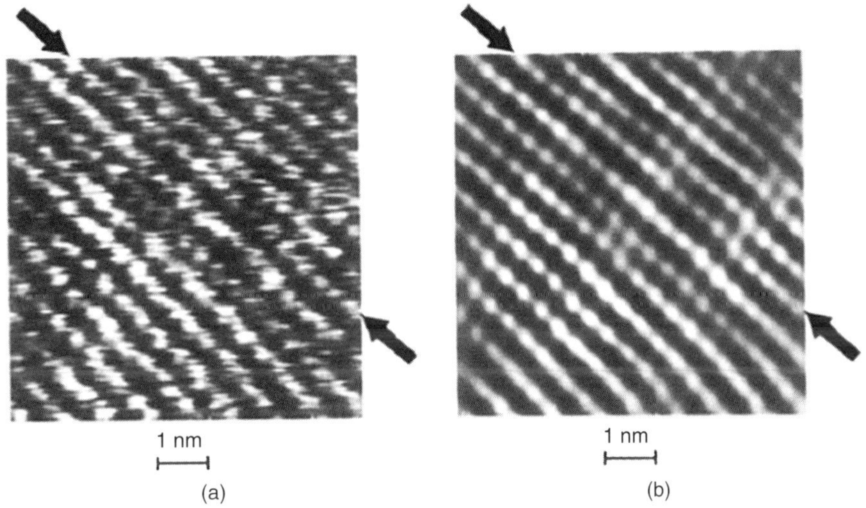

Figure 6.13 Contact mode AFM height image of POM with molecular (lattice) resolution: (a) raw data; (b) image obtained from Fourier reconstruction. The arrow indicates the polymer chain direction (image size 7×7 nm^2). Snétivy and Vancso [24]. Reproduced with permission of American Chemical Society.

6.3.3 Lamellar Crystals

Lamellar crystals obtained by crystallization from solution were successfully imaged by contact mode AFM (compare Fig. 6.2). These lamellae, including polyethylene or poly(oxy methylene) are mechanically robust enough to retain their shape and to suffer no noticeable damage. When scanned under an angle of 90° with respect to the long axis of the cantilever, friction forces can be recorded simultaneously. The polyethylene specimen shown in Figure 6.2 reveals in friction mode (Fig. 6.14) a clearly discernible sectorization, which coincides with the direction of folding at the fold surface.

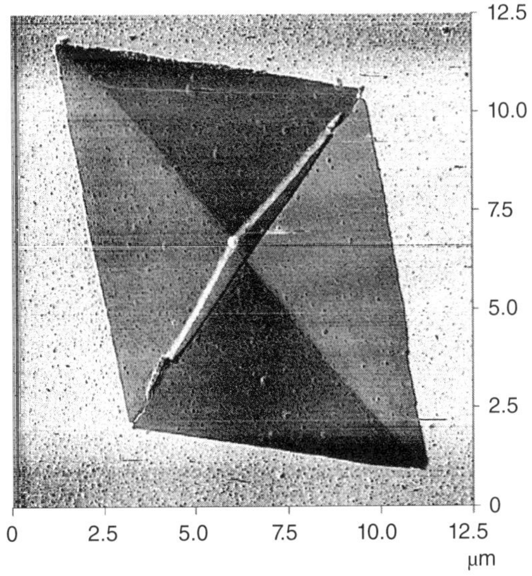

Figure 6.14 Friction force microscopy image of a solution grown polyethylene lamellar crystal obtained in air. Schönherr and Vancso [16]. Reproduced with permission of American Chemical Society.

Lamellar habit is also known for crystallization from the melt under isothermal and nonisothermal conditions. Typical views of lamellae in isotactic polypropylene or isotactic sequences in elastomeric polypropylene are shown in Figure 6.15, also for comparison with data discussed in Chapter 3. The typical cross-hatched morphology with angles of 80° agrees well with the theoretically predicted and experimentally confirmed value [26]. Independent of the AFM mode (contact mode, intermittent contact mode, pulsed force mode) consistent observations were made.

6.3.4 Spherulites

Many polymers crystallize in the form of spherulites. Hedrites, as early stages of development of spherulites, may reveal information about the growth and development of spherulites in time. As shown below in Figure 6.16, the morphological analysis is often feasible at the free surface of melt-crystallized films; however, permanganate-etched films of PP can also be studied to reveal information in the depth of a specimen [27].

6.3.5 Multiphase Systems

Constituent components in multiphase systems, such as phase-separated blends and microphase-separated block copolymers can be conveniently identified by various AFM imaging modes. Compared to the qualitative or semi-quantitative intermittent contact mode phase images, which requires careful analysis in order to assign the phases [28], pulsed force or PFT modes possess the advantage that they afford pull-off force *and* stiffness data from the captured force–displacement curves, which are acquired parallel to sample topography at useful scan rates. An example of PFT was already introduced in Figure 6.9. Here, we show an early

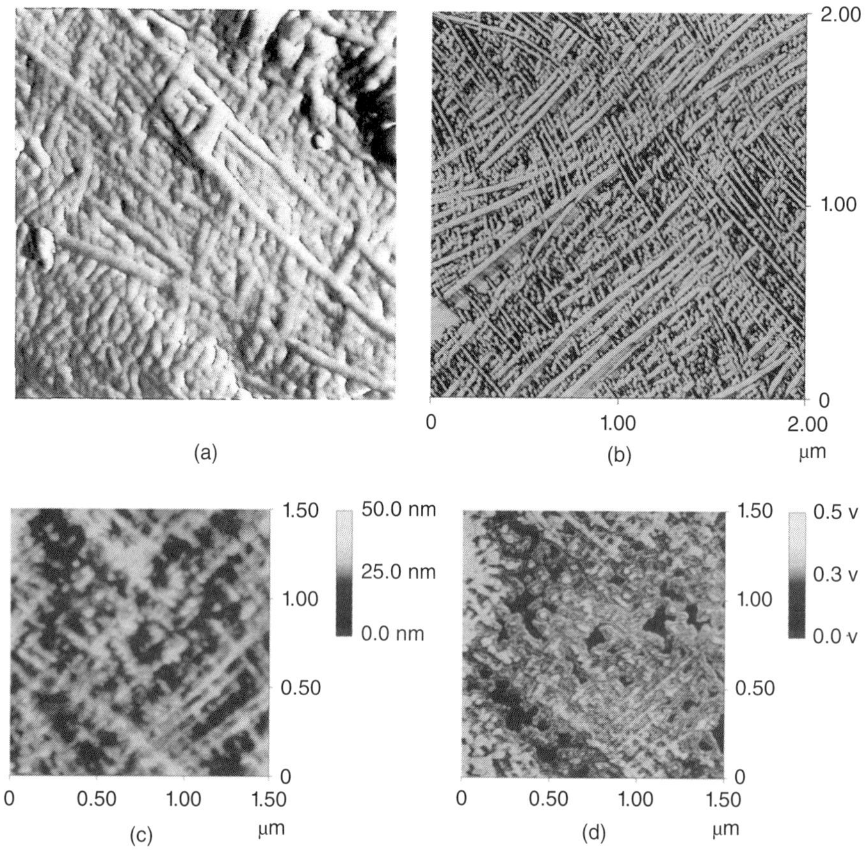

Figure 6.15 (a) Contact mode AFM deflection image of cross-hatched morphology found at the surface of melt-crystallized isotactic PP (image size 1.0 μm^2). Cross-hatching in α-m spherulite, $T_C = 110\,°C$, AFM scan, 1.0 μm × 1.0 μm, force mode, NanoTip. Reproduced with kind permission from Springer Science and Business Media from [25]. Copyright 1993. Springer-Verlag. (b) Intermittent contact mode AFM phase image of isotactic PP in film of the heptane insoluble fraction of elastomeric PP. Courtesy: H. Schönherr, R. M. Waymouth, C. W. Frank, unpublished data. (c) Pulsed force mode AFM height image and (d) pulsed force mode stiffness image (right) of cross-hatched lamellae of iPP. Schönherr et al. [6]. Reproduced with permission of Springer.

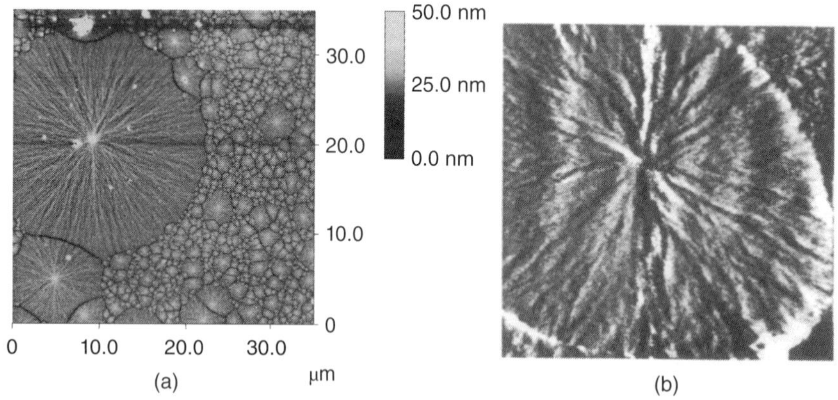

Figure 6.16 (a) Intermittent contact mode AFM height image of spin-coated iPP film on silicon after nonisothermal crystallization (film thickness: 40 nm). Courtesy: H. Schönherr, R. M. Waymouth, C. W. Frank, unpublished data. (b) Contact mode AFM deflection image of spherulite found at the surface of isothermally melt-crystallized iPP film (image sizes, 15 μm × 15 μm). α-m spherulite, AFM scan, force mode, 15 μm × 15 μm, NanoTip. Schönherr et al. [25]. Reproduced with permission of Springer.

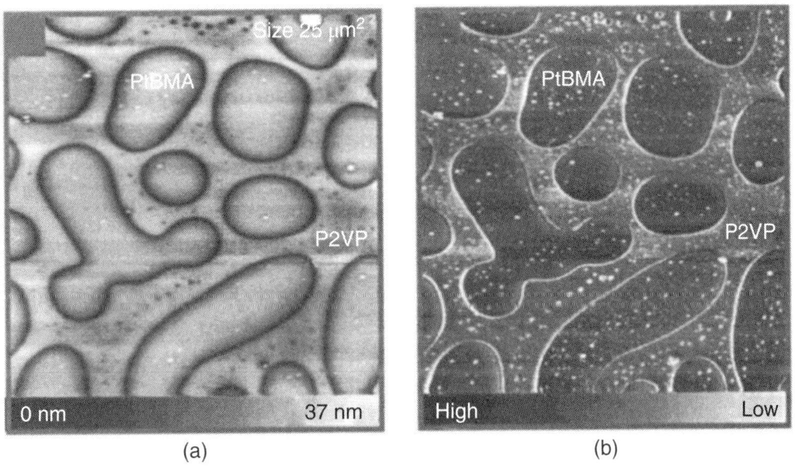

Figure 6.17 Pulsed force mode measurement on a blend of P2VP and PtBMA spin coated on silicon. The pulsed force mode AFM images clearly distinguish the PtBMA islands from the surrounding P2VP in both topography (a) and in adhesion (b). Reproduced with permission from [29]. Copyright 2000. AIP Publishing LLC.

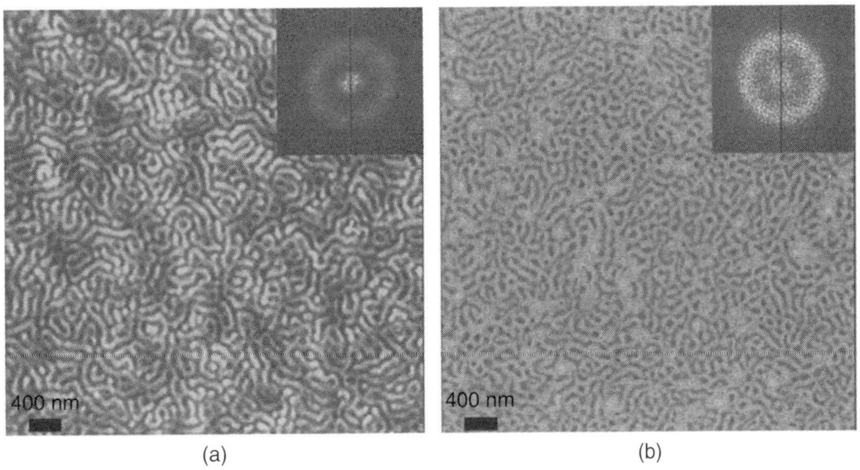

Figure 6.18 Intermittent contact mode AFM height images of thin films of PS_{2048}-*b*-P*t*BA_{1152} prior to hydrolysis and after hydrolysis in trifluoro acetic acid, respectively. Courtesy: D. Tranchida, H. Schönherr, unpublished data.

example of blend analysis by pulsed force mode AFM, in which a phase-separated blend of P2VP and PtBMA spin coated on silicon was interrogated (Fig. 6.17).

Figure 6.18a and b shows intermittent contact mode AFM height images of the morphology observed on thin-film samples of the diblock copolymer PS_{2048}-*b*-P*t*BA_{1152} prior to hydrolysis and after hydrolysis in trifluoro acetic acid, respectively. The two phases can be clearly distinguished, despite the fact that "height" data are being displayed. As discussed by Pickering and Vancso, the contrast in intermittent contact height and phase images depends sensitively on the cantilever amplitude and the amplitude set point ratio [28]. From the fast Fourier transforms (FFT) (see insets in Fig. 6.18) and the power spectral density, the periodicities of the phases can be estimated and thereby it can be checked whether the hydrolysis altered the morphology.

6.3.6 Polymeric Nanostructures

With its proven capabilities to image the structure and nanoscale composition of polymers, AFM can be applied to a sheer unlimited range of problems. Below we show a concise collection of examples that illustrate the power of AMF to analyze essentially nano- or microstructured polymers and architectures "as is" in ambient conditions.

6.3.6.1 Nanofibers The first example is a study on the organization of lamellae in electrospun fibers of a semicrystalline polymer. Poly(ε-caprolactone) has been spun into submicron fibers by electrospinning [30, 31]. The fiber morphology can be imaged successfully by intermittent contact mode AFM, as shown in Figure 6.19a, where a rather dense web of electrospun poly(ε-caprolactone) fibers can

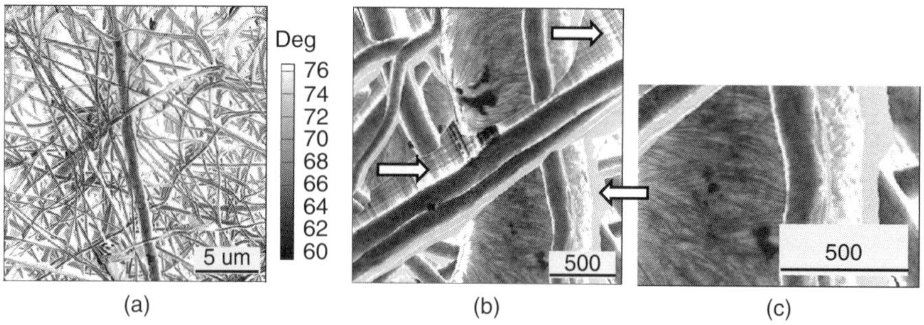

Figure 6.19 Intermittent contact mode AFM phase images of electrospun poly(ε-caprolactone) fibers. The white arrows highlight parts of the fiber that could not be imaged due to sidewall contact of the tip. Reproduced with permission from [32]. Copyright 2014. The Royal Society of Chemistry.

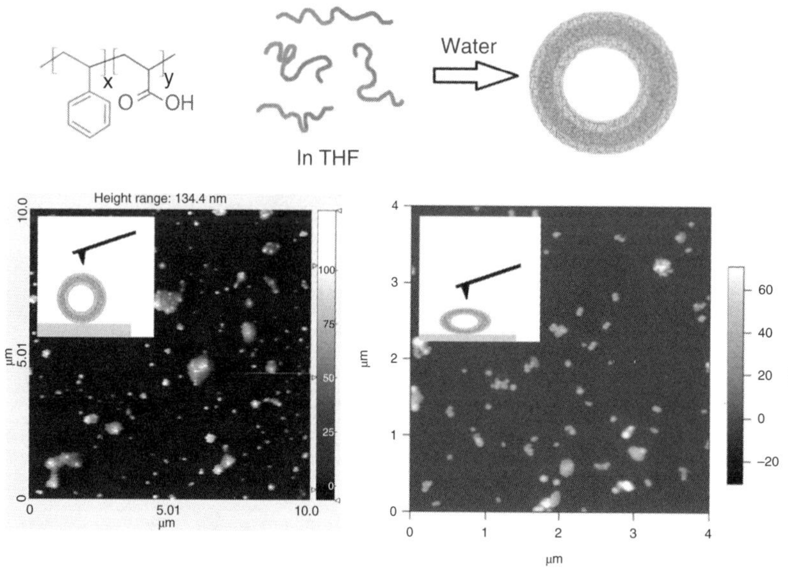

Figure 6.20 (a) Schematic of polymersome formation of poly(styrene)-block-poly(acrylic acid) and intermittent contact mode AFM height image of (b) PS_{115}-*b*-PAA_{15} vesicles and (c) PEG_{114}-*b*-PLA_{167} vesicles deposited on glass substrate. Courtesy: D. Wesner, S. Handschuh-Wang, H. Schönherr, unpublished data.

be recognized. In the close up in Figure 6.19b, however, it become apparent that the steep sidewalls cannot be imaged by the AFM probe tip, as the tip–sample contact is at the sidewalls not defined by the point of the tip, but rather by the sides of the pyramid. Despite this shortcoming, the phase images reveal the lamellar organization, which was found to be better aligned at 90° to the fiber axis for thinner fibers compared to the thicker fiber (Fig. 6.19c).

6.3.6.2 ***Polymeric Vesicles*** Polymeric vesicles (polymersomes) formed by self-organization of amphiphilic block copolymers are interesting hollow containers for encapsulation and controlled release strategies. They are assembled in liquid medium and may encapsulate water-soluble molecules in their aqueous interior. The wall is formed by the amphiphilic polymers, as shown in the scheme in Figure 6.20. Using intermittent contact mode AFM, the structure of the dried vesicles was unraveled on a glass substrate. By comparing the measured vesicle height to the dimensions determined by dynamic light scattering or SEM, the degree of deformation of the vesicles on the substrate, that is, the deviation of the shape from the spherical shape adopted by the vesicles in solution, can be determined. While poly(styrene)-block-poly(acrylic acid) vesicles deform only slightly (Fig. 6.20a), softer polymers may result in a collapsed wall, as has been observed for block copolymers of polyethylene and poly(lactic acid) (Fig. 6.20b).

6.3.6.3 ***Micropatterned Polymer Brushes*** Micropatterned polymer brushes are interesting interfacial polymer architectures for the control of interactions at biointerfaces. They can be conveniently synthesized by surface-initiated polymerization starting from micropatterned initiator monolayers using suitable monomers [33]. Depending on the type of monomer, protein-resistant or even stimulus-responsive brushes are obtained. A prominent example are brushes of poly(oligo ethylene glycol methacrylate), which have been shown to impart excellent protein resistance to surfaces. The characterization of the brush height is often carried out by ellipsometry or related methods in the dry state. In buffer

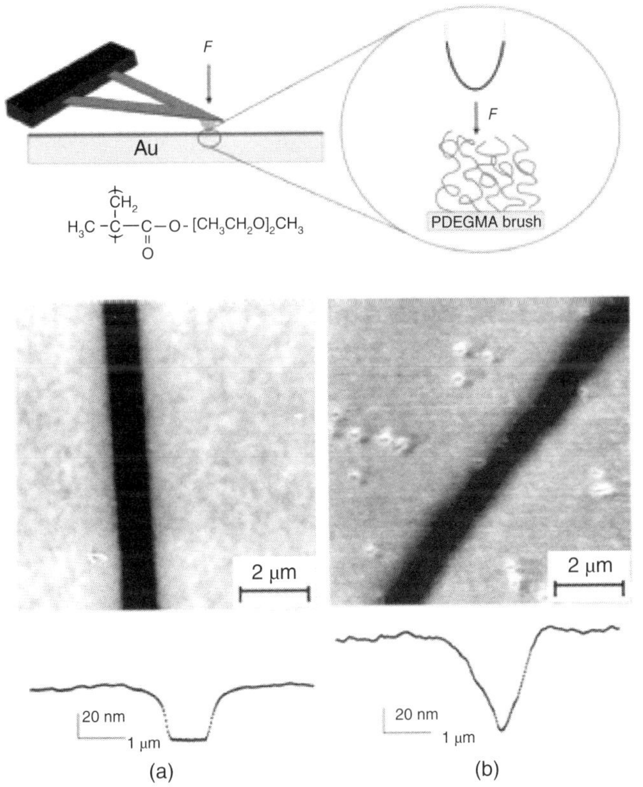

Figure 6.21 Schematic and intermittent contact mode AFM height images acquired on a patterned PDEGMA brush (a) in air and (b) in buffer. The apparent thickness increases from 41 to 75 nm due to swelling of the brush. Tranchida et al. [34]. Reproduced with permission of American Chemical Society.

or aqueous medium, the brushes may swell considerably, as shown in Figure 6.21 in intermittent contact mode height images of a micropatterned poly(diethylene glycol methacrylate) (PDEGMA) brush that was imaged first in ambient air and then after equilibration in the liquid cell in phosphate buffer. Force–displacement measurements may also unravel the mechanical properties [34].

6.4 PROPERTY MAPPING

In addition to its undoubted imaging capabilities that give access to structures and thus quantitative height profiles, AFM has also its benefits in mapping surface properties. This unique aspect is discussed only briefly using relevant examples for polymers.

6.4.1 Nanomechanical Properties

Because AFM is based on forces between the nanoprobe tip and the surface, the method is inherently suited to analyze mechanical properties of polymers. This has been realized experimentally very early [35]. For polymers, the determination of nanomechanical response of polymers, measurements of viscoelasticity by contact resonance force microscopy [36] and AFM nanorheology [37] have been reported. These more specialized modes are not further discussed as they are not considered to be standard methods.

Elastic properties can be assessed quantitatively by AFM in point-by-point force–displacement measurements (nanoindentation). The rigorous quantification of elastic moduli requires a thorough characterization of the indenter geometry and size as well as the use of an appropriate model to describe the contact between tip and sample. Previous work, for example, by Tranchida et al. [38] and Cappella and coworkers [39] has shown that excellent agreement between data from conventional mechanical testing and AFM nanoindentation can be obtained (see also Chapter 17).

A lateral resolution of 6 nm for mechanical property mapping has been recently reported in a comparative tapping force–displacement AFM study of thin-film blends of polybutadiene/polystyrene on glass by Cappella and coworkers [40] (Fig. 6.22). The measurement of force–distance curves gave access to quantitative Young's moduli and adhesion data that were useful to unveil the aging behavior of the blend exposed to air. The aging resulted in an increase of the Young's modulus due to cross-linking and dewetting processes.

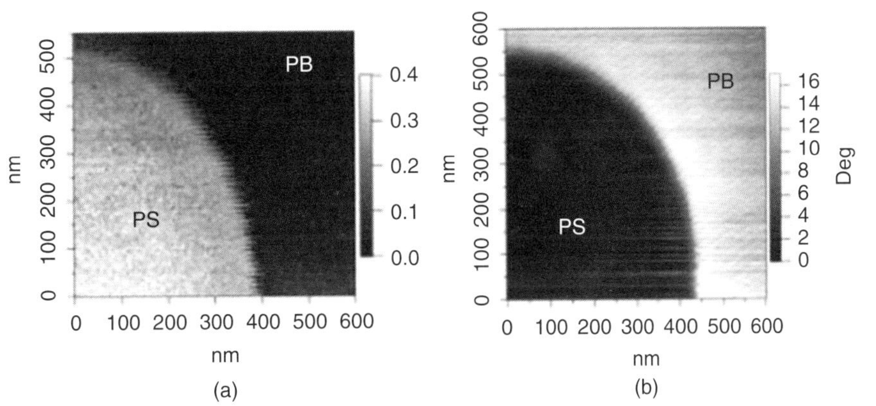

Figure 6.22 Compositional map of phase separated blend of polybutadiene/polystyrene on glass in (a) f–d curve mode and (b) intermittent contact mode phase imaging. Krämer et al. [40]. Reproduced with permission of Elsevier.

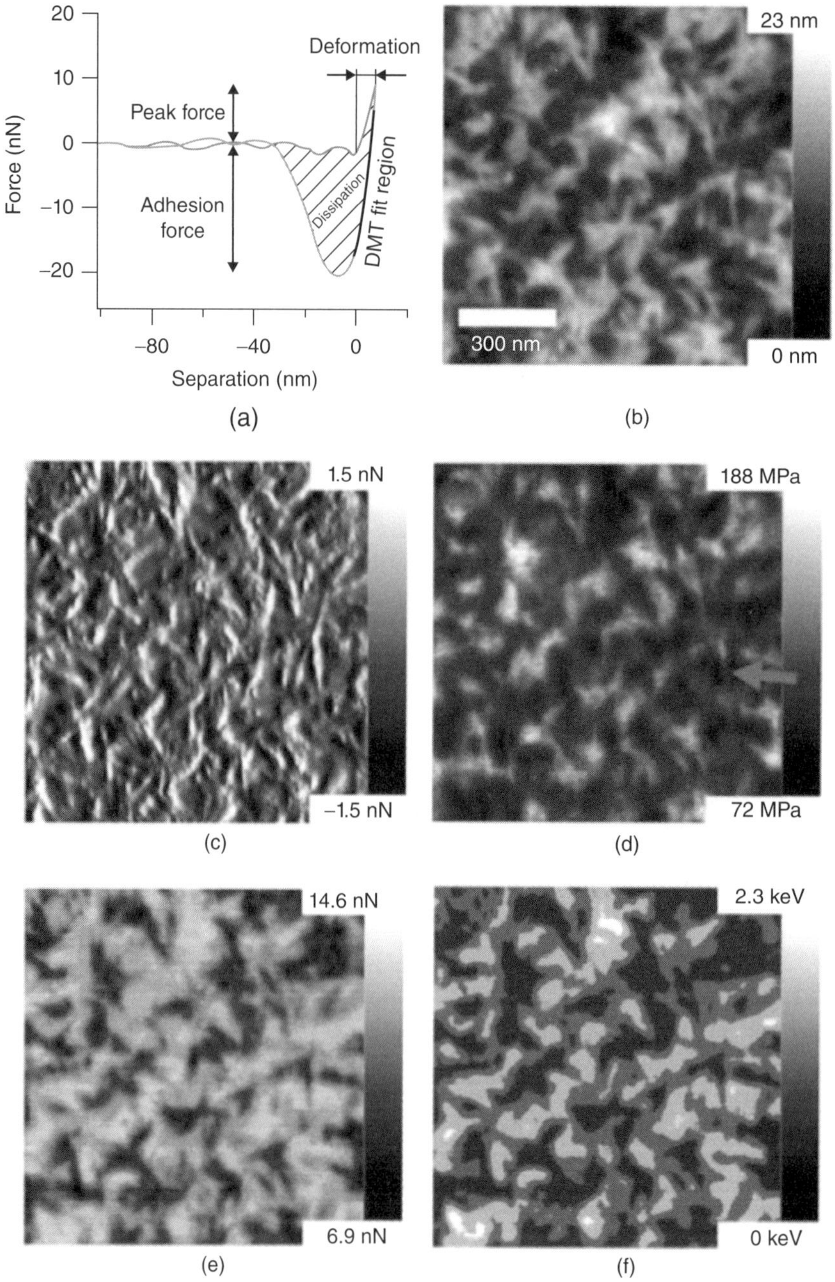

Figure 6.23 Mechanical surface properties of an unetched ePP sample measured via quantitative nanomechanical mapping: (a) Approach and retract curves of a single force-versus-separation measurement. The derived physical quantities are highlighted in the graph. (b) Topography image. The bright areas correspond to the crystalline regions of the polymer, whereas the dark areas correspond to the amorphous regions. (c) Corresponding error map for the feedback loop maintaining a constant peak force of 8 nN during imaging. (d) Elasticity map of ePP derived from a DMT model fit. The crystals appear bright due to their high stiffness. (e) Adhesion map of ePP. Please note the inverted contrast. The force necessary to separate the tip from the amorphous regions is greater than the adhesion force on crystalline regions. (f) Map of the energy dissipated between the tip and the sample surface during one oscillation cycle. The determined cantilever/tip properties are $k = 6.6$ N/m and $R = 8$ nm. Voss et al. [41]. Reproduced with permission of American Chemical Society.

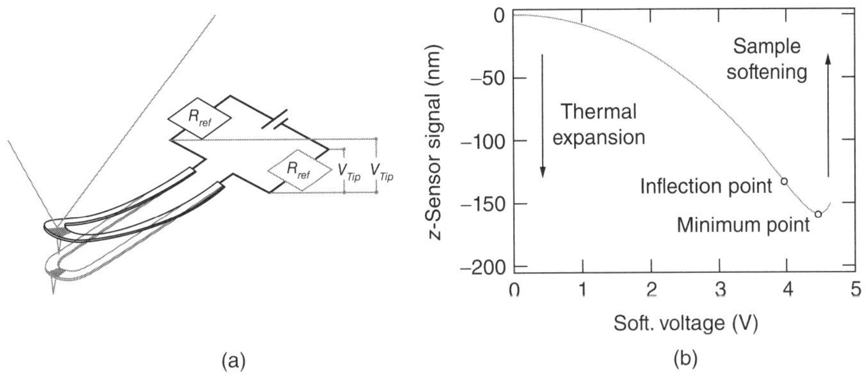

Figure 6.24 (a) The basic experimental configuration for local thermal analysis (LTA). (b) z-Sensor signal as a function of applied voltage. Fischinger et al. [42]. Reproduced with permission of American Chemical Society.

Using the newer dynamic modes, such as PFT, a more rapid assessment of the mechanical response of polymers can be recorded. If continuum contact mechanics models, such as the Hertz model, the Johnson–Kendall–Roberts (JKR) model, or the Derjaguin–Muller–Toporov (DMT) model are applicable, the data can be rapidly analyzed (see Chapter 17). This is illustrated in the work by Voss and coworkers, who analyzed the difference between the mechanical properties of elastomeric polypropylene at the surface and in the bulk by AFM in the PFT mode (Fig. 6.23) [41]. Via wet-chemical ablation access to the bulk was afforded in a layer-by-layer fashion. On top of the as-prepared samples, little difference in mechanical properties was observed for the amorphous and the crystalline regions due to an approximately 20 nm thin amorphous top layer. In the bulk, that is, when this top layer was removed, the modulus of crystalline regions was found to be greater than that of amorphous regions. Figure 6.23 shows the simultaneously captured maps of topography and mechanical surface properties of a pristine sample of elastomeric polypropylene.

6.4.2 Scanning Thermal Microscopy

The localized measurement of thermal properties is feasible using heated AFM probes. Initially, relatively large Wollaston wire probes were utilized in scanning thermal probe microscopy; more recently microfabricated probes have become available. Using such probe tips, thermal transitions, such as glass transition temperatures, can be analyzed locally. For instance, the effect of film thickness on T_g can be determined by measuring the surface softening temperature of corresponding polymer films using a heated AFM-like tip via the change in cantilever deflection induced by the onset of the transition.

In such an experiment, the thermal probe is engaged on the polymer and the force is kept constant by the feedback loop. Using the thermal probe, it is possible to control and measure the tip temperature by varying the voltage applied to the thermal probe. The response of a closed-loop feedback loop can be assessed from the z-sensor signal, as shown in Figure 6.24 [42]. Noteworthy are the thermal expansion of the film and the inflection point of the curve, which signifies the onset of tip penetration into the softened polymer.

The combination of AFM with other techniques enhances obviously the capabilities of the isolated techniques for the study of polymers and polymer morphologies. Particularly interesting is the combination of AFM with spectroscopic techniques, such as (confocal) Raman or fluorescence microscopy and/or spectroscopy. These diffraction-limited techniques cannot rival the AFM in terms of spatial resolution. However, the added information (vibrational fingerprint information or emission of embedded fluorescent molecules) goes well beyond the capabilities of conventional AFM.

More recently, the implementation of newly developed AFM approaches that beat the diffraction limit have garnered increasing attention. In particular AFM-infrared (IR) spectroscopy appears to be useful for the label-free identification of local chemical composition. The AFM-IR technique exploits a pulsed, tunable IR laser that irradiates the sample. Upon absorption of light of a suitable wavelength due to the excitation of molecular vibrations in the polymer, a rapid thermal expansion in the sample is induced, which is detected by a resonant AFM probe. Spectral information and position (plus all additional AFM information) is thereby captured simultaneously [43, 44]. Similarly, optical microscopy including fluorescence spectroscopy can be operated in parallel to the AFM operation in inverted optic microscopes, giving rise to simultaneous data acquisition. These combined techniques possess again countless applications that go well beyond mere imaging of morphologies, but allow one to interrogate and analyze advanced polymer materials at the micro- and nanoscale.

REFERENCES

1. Binnig G, Quate CF. Atomic force microscope. Phys Rev Lett 1986;56:930.
2. Hemsley DA. *Applied Polymer Light Microscopy*. London: Elsevier Applied Science; 1989.
3. Michler GH. *Electron Microscopy of Polymers*. Springer; 2010.

4. Schönherr H. Scanning force microscopy. In: Mark HF, Bikales N, Overberger CG, Menges G, Kroschwitz JI, editors. *Encyclopedia of Polymer Science and Technology*. New York: John Wiley & Sons; 2004.
5. Sawyer LC, Grubb DT, Meyers GF. *Polymer Microscopy: Characterization and Evaluation of Materials*. 3rd ed. Berlin: Springer; 2008.
6. Schönherr H, Vancso GJ. *Scanning Force Microscopy of Polymers*. Springer; 2010.
7. Magonov SN, Whangbo M-H. *Surface Analysis with STM and AFM*. VCH: Weinheim; 1996.
8. Mugele F, Becker T, Nikopoulos R, Kohonen M, Herminghaus S. Capillarity at the nanoscale: an AFM view. J Adh SciTechnol 2002;16:951.
9. Schönherr H, Johnson JM, Lenz P, Frank CW, Boxer SG. Vesicle adsorption and lipid bilayer formation on glass studied by atomic force microscopy. Langmuir 2004;20:11600.
10. Walczyk W, Hain N, Schönherr H. Hydrodynamic effects of the tip movement on surface nanobubbles: a combined tapping mode, lift mode and force volume mode AFM study. Soft Matter 2014;10:5945.
11. Israelachvili JN. *Intermolecular and Surface Forces*. 3rd ed. Academic Press; 2011.
12. Butt H-J, Graf K, Kappl M. *Physics and Chemistry of Interfaces*. 3rd ed. Wiley-VCH; 2013.
13. Vancso GJ, Hillborg H, Schönherr H. Chemical composition of polymer surfaces imaged by atomic force microscopy and complementary approaches. Adv Polym Sci 2008;182:55.
14. Schönherr H, Tocha E, Vancso GJ. Friction and surface dynamics of polymers on the nanoscale by AFM. In: Samori P, editor. *STM and AFM Studies on (Bio)molecular Systems: Unravelling the Nanoworld*. Topics in Current Chemistry. Vol. 285. Springer; 2008. p 103.
15. H. Schönherr, From functional group ensembles to single molecules: scanning force microscopy of supramolecular and polymeric systems. Ph. D. Thesis, University of Twente. 1999.
16. Schönherr H, Vancso GJ. In: Tsukruk VV, Wahl KJ, editors. *Microstructure and Microtribology of Polymer Surfaces*. Molecular alignment and nanotribology of polymeric solids studied by lateral force microscopy. ACS Symposium Series. Vol. 741. 1999. p 317.
17. Schönherr H, Frank CW. Ultrathin films of poly(ethylene oxides) on oxidized silicon. 2. In situ study of crystallization and melting by hot stage AFM. Macromolecules 2003;36:1199.
18. Garcia R. *Amplitude Modulation Atomic Force Microscopy*. Wiley-VCH; 2010.
19. Kresz N, Kokavecz J, Smausz T, Hopp B, Csete M, Hild S, Marti O. Thin. Investigation of pulsed laser deposited crystalline PTFE thin layer with pulsed force mode AFM. Solid Films 2004;453–454:239.
20. Schön P, Bagdi K, Molnár K, Markus P, Pukánszky B, Vancso GJ. Quantitative mapping of elastic moduli at the nanoscale in phase separated polyurethanes by AFM. Europ Poly J 2011;47:692.
21. Pyne A, Thompson R, Leung C, Roy D, Hoogenboom BW. Single-molecule reconstruction of oligonucleotide secondary structure by atomic force microscopy. Small 2014;10:3257.
22. Sheiko SS, Möller M. Visualization of macromolecules – a first step to manipulation and controlled response. Chem Rev 2001;101:4099.
23. Yu-Su SY, Sun FC, Sheiko SS, Konkolewicz D, Lee H-I, Matyjaszewski K. Molecular imaging and analysis of branching topology in polyacrylates by atomic force microscopy. Macromolecules 2011;44:5928.
24. Snétivy D, Vancso GJ. Selective visualization of atoms in extended-chain crystals of oriented polyoxymethylene by atomic force microscopy. Macromolecules 1992;25:3320.
25. Schönherr H, Snétivy D, Vancso GJ. A nanoscopic view at the spherulitic morphology of isotactic polypropylene by atomic force microscopy. Polym Bull 1993;30:567.
26. Lotz B, Wittmann JC. The molecular origin of lamellar branching in the α (monoclinic) form of isotactic polypropylene. J Polym Sci Polym Phys 1986;24:1541.
27. Olley R. Selective etching of polymeric materials. Sci Progr Oxf 1986;70:17.
28. Pickering JP, Vancso GJ. On the formation of oriented nanometer scale patterns on amorphous polymer surfaces studied by atomic force microscopy. Appl Surf Sci 1999;148:147.
29. Krotil HU, Stifter T, Marti O. Concurrent measurement of adhesive and elastic surface properties with a new modulation technique for scanning force microscopy. Rev Sci Instrum 2000;71:2765.
30. Reneker DH, Chun I. Nanometre diameter fibres of polymer, produced by electrospinning. Nanotechnology 1996;7:216.
31. Jaeger R, Bergshoef MM, Martín i Batlle C, Schönherr H, Vancso GJ. Electrospinning of ultra-thin polymer fibers. Macromol Symp 1998;127:141.
32. Goonoo N, Bhaw-Luximon A, Rodriguez IA, Wesner D, Schönherr H, Bowlin GL, Jhurry D. Poly(ester-ether)s: III. assessment of cell behaviour on nanofibrous scaffolds of PCL, PLLA and PDX blended with amorphous PMeDX. J Mater Chem B 2015;3: 673–687.
33. Ducker R, Garcia A, Zhang JM, Chen T, Zauscher S. Polymeric and biomacromolecular brush nanostructures: progress in synthesis, patterning and characterization. Soft Matter 2008;4:1774.
34. Tranchida D, Sperotto E, Staedler T, Jiang X, Schönherr H. Nanomechanical properties of oligo(Ethylene Glycol Methacrylate) polymer brush-based biointerfaces. Adv Eng Mater 2011;13:B369.
35. Burnham N, Colton R. Measuring the nanomechanical properties and surface forces of materials using an atomic force microscope. J Vac Sci Technol A 1989;7:2906.
36. Killgore JP, Yablon DG, Tsou AH, Gannepalli A, Yuya PA, Turner JA, Proksch R, Hurley DC. Viscoelastic property mapping with contact resonance force microscopy. Langmuir 2011;27:13983.
37. Yablon DG, Proksch R, Gannepalli A, Tsou AH. Mapping storage modulus and loss modulus of polyolefin/polystyrene blends with atomic force microscopy. Rubber Chem Technol 2012;85:559.
38. Tranchida D, Piccarolo S, Soliman M. Nanoscale mechanical characterization of polymers by AFM nanoindentations: Critical approach to the elastic characterization. Macromolecules 2006;39:4547.

39. Griepentrog M, Krämer G, Cappella B. Comparison of nanoindentation and AFM methods for the determination of mechanical properties of polymers. Polym Testing 2013;32:455.
40. Krämer G, Griepentrog M, Bonaccurso E, Cappella B. Study of morphology and mechanical properties of polystyrene – polybutadiene blends with nanometre resolution using AFM and force-distance curves. Euro Polym J 2014;55:123.
41. Voss A, Stark RW, Dietz C. Surface versus volume properties on the nanoscale: Elastomeric polypropylene. Macromolecules 2014;47:5236.
42. Fischinger TJ, Laher M, Hild S. Direct measurement of acceptor group localization on donor-acceptor polymers using resonant Auger spectroscopy. J Phys Chem B 2014;118:5570.
43. Marcott C, Lo M, Kjoller K, Prater C, Noda I. Spatial differentiation of sub-micrometer domains in a poly(hydroxyalkanoate) copolymer using instrumentation that combines atomic force microscopy (AFM) and infrared (IR) spectroscopy. Appl Spectr 2011;65:1145.
44. Awatani T, Midorikawa H, Kojima N, Ye J, Marcott C. Morphology of water transport channels and hydrophobic clusters in Nafion from high spatial resolution AFM-IR spectroscopy and imaging. Electrochemistry Communications 2013;30:5.

7

FTIR IMAGING OF POLYMERIC MATERIALS

S. G. Kazarian and K. L. A. Chan
Department of Chemical Engineering, Imperial College London, London, UK

7.1 INTRODUCTION

In polymer processing, whether it is a blending, annealing or diffusion process, it is vital to characterize the chemical composition and spatial distribution of the materials involved. Fourier transform infrared (FTIR) spectroscopy has traditionally been used as a powerful nondestructive material identification method, as well as an important tool to study intermolecular interactions and different polymer morphologies. With the introduction of infrared (IR) microscopes and array detectors, chemical imaging using FTIR spectroscopy has become an established tool in polymer analysis.

In this chapter, we introduce the principles of the FTIR imaging method and some of the recent applications of FTIR imaging in the study of polymeric materials. This includes an introduction to the general approach of FTIR image acquisition, the transmission and attenuated total reflection (ATR) modes of measurement, the different approaches of measuring liquid and solid samples, the considerations needed to be taken in order to obtain a reliable and accurate imaging measurement and also some of the new opportunities that exist for FTIR imaging.

7.2 PRINCIPLES OF FTIR IMAGING

FTIR imaging is the measurement of a large number of spatially resolved IR spectra, usually arranged in an array format, which can then be used to form images that represent the spatial distribution of various materials in a heterogeneous sample. FTIR images can be collected using a mapping approach (point-by-point measurement with the aid of an aperture to restrict the area on the sample where IR light is measured) or array detectors (linear arrays or focal plane arrays (FPA) where no aperture is used), illustrated in Figure 7.1. Point-by-point mapping, in comparison to the FPA detector imaging approach, is a few orders of magnitude slower and not as efficient. This is because the aperture used for mapping greatly reduces the throughput of light, especially when small apertures are used to try to obtain higher spatial resolution images. In contrast, the FPA approach is fast. For example, a 64×64 pixel FPA detector can be used to simultaneously collect 4096 spectra in a few minutes with noise level of 0.001 absorbance unit (a.u.). However, FPA detectors are more expensive than the single-element MCT detectors and the detection range is limited to above $900\,cm^{-1}$. The linear array detector method lies between these two approaches. A linear array detector (usually in a 1×16 pixel format) can be used to collect 16 spectra in one measurement and an image is collected by rastering across the sampling area. No aperture is needed and thus this system gives a similar throughput of light as a system that employs an FPA detector. The speed of measurements is also comparable to the systems that utilize an FPA detector. The measured spectral range is larger than the FPA approach (down to $720\,cm^{-1}$), but because images are collected by rastering, not all spectra are measured simultaneously.

Each of the measured spectra becomes a pixel which, when placed together, forms the final image (see Fig. 7.1). The spectral imaging data is therefore a three-dimensional data cube: the *X* and *Y* axis of the array of pixels and the wavenumbers in the spectrum. A polymer that contains different functional groups will produce specific spectral bands, which can then be used as a marker for that polymer species. The spatial distribution of the various polymer

Polymer Morphology: Principles, Characterization, and Processing, First Edition. Edited by Qipeng Guo.

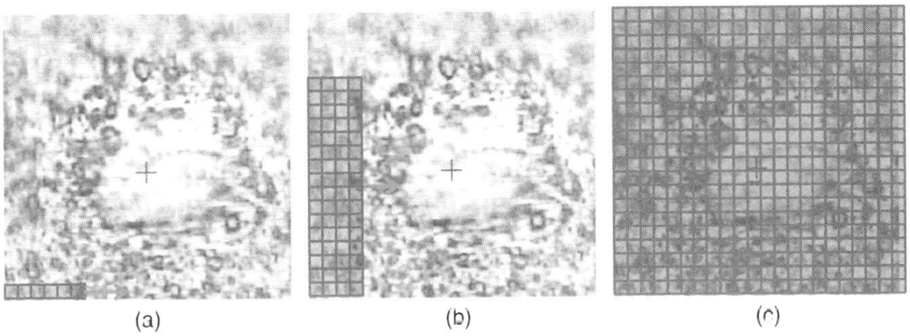

Figure 7.1 A schematic showing the difference between (a) single-element mapping where each pixel on the image is collected individually and consecutively, (b) linear array mapping where a line of 16 pixels are collected simultaneously while each line is collected consecutively, and (c) FPA imaging where all 64 × 64 or 128 × 128 pixels are collected simultaneously.

species can be presented as 2D maps by plotting the integrated area under the specific band that characterizes each of the components in the mixture across all the pixels measured. As an example, earlier work has demonstrated that chemical images showing the spatial distribution of polyamide (PA), silicon oil, and polytetrafluoroethylene (PTFE) in a blend can be obtained on the basis of distinct spectral bands of the three components [1]. The characteristic spectral bands used in that study are the 1640 cm^{-1} band representing the PA, the 1260 cm^{-1} band representing the silicon oil, and the 1155 cm^{-1} band representing the PTFE. Chemical images representing the spatial distribution of the concentration of the component in the sample have been generated using these three bands. The chemical images revealed that the PA is the major component in the PA–PTFE blend, while the PTFE formed small domains (<50 μm) and the silicon oil is present next to the PTFE clusters (see Fig.7.2). In comparison to the visible image, the contrast between the different polymers in the blend is greatly enhanced in the FTIR chemical images.

The different chemical images generated in Figure 7.2 are obtained from a single imaging measurement which makes FTIR imaging a powerful tool to quantitatively track multiple components simultaneously. Because absorbance, A, is proportional to concentration, c, according to the Beer–Lambert law (ε is the molar absorptivity and l is the path length, Eq. 7.1), these maps can also be presented as concentration maps when a suitable calibration curve is made. The colors on the map can then be assigned to a concentration according to the calibration curve.

$$A = \varepsilon c l \tag{7.1}$$

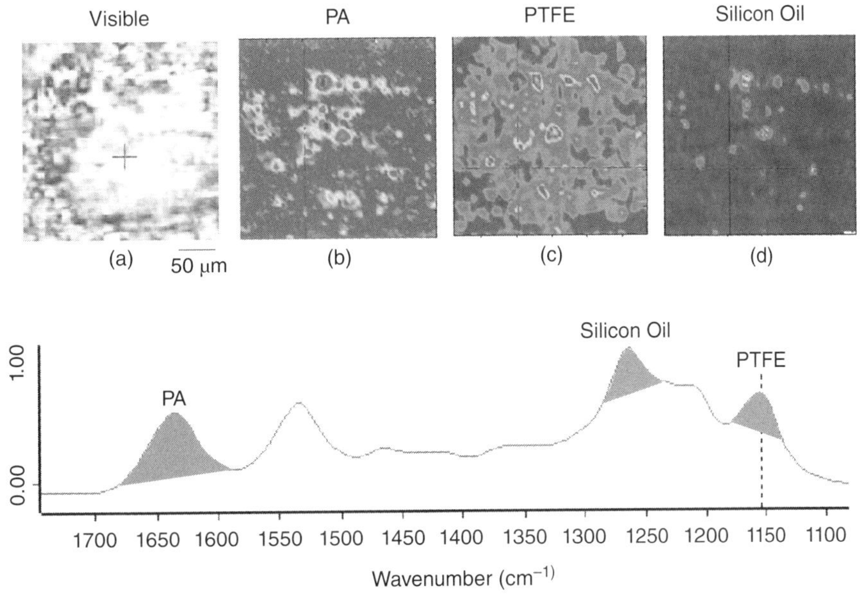

Figure 7.2 Transmission FTIR images of PA, silicone oil, and PTFE in the polymer blend. Image size is 266 × 266 μm^2. (a) Visible image of the polymer blend. FTIR images created by integrating, (b) the polyamide band between 1677 and 1597 cm^{-1}, (c) the silicone oil band between 1290 and 1240 cm^{-1}, and (d) the PTFE band between 1174 and 1122 cm^{-1}. Red regions represent a high concentration, while blue regions represent a low concentration. (See eBooks for color version of this figure).

It is important to note that FTIR images provide not only an enhanced image contrast, compared to a visible image, but also provides the identification of the different polymer species, the concentration, and the molecular state of the polymers at localized areas.

7.3 SAMPLING METHODS

7.3.1 Transmission Mode

In a transmission measurement, the IR light passes through the sample and the absorbance of light is calculated by comparing the measured intensity to a reference measurement. The sample can be measured in a microscope for high spatial resolution imaging or in a large sample compartment (also called the macro sample chamber) to image with a larger field of view. Samples should be flat and relatively thin (i.e., a film) and the film thickness, which determines the path length, should be uniform. A nonuniform film thickness could produce complications in the result, where the absorbance can be strongly influenced by the sample thickness as well as the composition (see Eq. 7.1). In this case, the band ratio should be adapted to remove contribution of absorbance variation due to heterogeneity in film thickness [2]. The result of this ratio will be

$$\frac{A_1}{A_2} = \frac{\varepsilon_1 c_1}{\varepsilon_2 c_2} \tag{7.2}$$

where the subscript "1" and "2" represent the two components that produced the different absorbance used in the band ratio. The sample can be measured as a free-standing film (in which case, air should be measured as the background) or on an IR transparent substrate such as CaF_2, BaF_2, or ZnSe where the clean substrate should be used as the background.

7.3.1.1 ***Liquid Cell*** A liquid cell with a known path length may be used to produce a liquid sample film of uniform thickness for the imaging measurement. The assembly of the major part of a liquid cell is illustrated in Figure 7.3. The path length is governed by the thickness of the spacer in between the two windows. A liquid cell is also useful in the study of solvent interactions with polymer films. Early examples include the studies of solvent diffusion into a polymer and the dissolution of a polymer in organic solvents [3–6]. The interaction between water and polymer can also be studied using a liquid cell, such as the study of polymer dissolution in water in transmission mode. However, studying a system containing water in transmission is a challenging task because water has strong absorbance in the 3700–3000 cm^{-1} region (O–H stretching band) as well as in the 1700–1600 cm^{-1} region (O–H bending band) and below 1000 cm^{-1} (water librations) [7]. When the spectral region of interest is masked by the absorption of water band, deuterated water (D_2O) may be used where the O–H bands are replaced with O–D bands. The

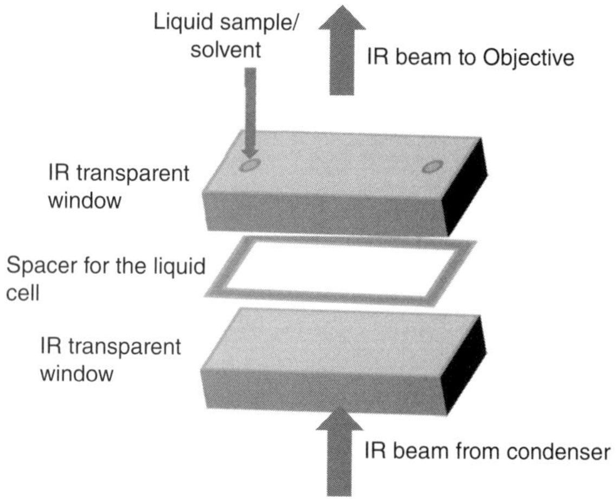

Figure 7.3 Schematic showing the major components of a liquid FTIR cell. The cell is shown as assembled to display the various components (liquid cell holder not shown).

strong absorbance of D_2O is in the 2750–2100 cm^{-1} region (O–D stretching band) as well as in the 1250–1150 cm^{-1} region (O–D bending) [8]. Water may be replaced with D_2O without complications, provided that the polymer does not undergo H–D isotopic exchange or the exchange is insignificant within the time frame of the experiment. The experiment must also be designed to prevent isotopic exchange with atmospheric water. Otherwise, the formation of HDO can complicate the spectrum by introducing an absorbance band at around 1460 cm^{-1} (HDO bending mode) and other new bands associated with the H–D exchange. On the other hand, the isotopic exchange between H and D can be used as an indicator for the study of diffusion of D_2O into a polymer [9].

7.3.1.2 ***Refraction Effects*** While the liquid cell is commonly used in single-spectrum measurements, refraction of light at the IR windows of the liquid cell may have an undesirable effect on the spatially resolved spectra in an imaging measurement [10–12]. For most FTIR imaging measurements, IR light is focused by optics onto the sample and collected by a second set of optics before being directed to the detector. IR light, therefore, approaches the surface of the liquid cell at an angle and refraction occurs as a result of the change in refractive index between the window of the liquid cell and air. The refraction is greater when the optics used have higher focusing power, for instance in an IR microscope. The degree of refraction depends on the refractive index of the window. Because the refractive index of most materials used for IR windows (e.g., CaF_2) is a function of wavenumbers, chromatic dispersion of light occurs when the light enters and exits the window. As a result of this dispersion, the IR light at different wavenumbers cannot be focused to the same point such that, when the IR light of smaller wavenumber is in focus, the light with larger wavenumber will be out of focus and vice versa [12]. Strategies to reduce the dispersion of light through a transmission cell include using thin CaF_2 windows [13] or thin diamond windows. Diamond is advantageous because of its relatively constant refractive index in the mid-IR region [14]. Another method to completely eliminate the dispersion through the liquid cell using a window of ordinary thickness (~4 mm) has been recently introduced [12]. This method involves the use of a lens, which creates a hemisphere above the sample when placed on the window. The lens, which is made of the same material as the windows, has shown to be able to remove the refraction such that the chromatic dispersion effect is eliminated when light passes through the transmission cell.

7.3.2 Attenuated Total Reflection (ATR) Mode

The ATR sampling mode is suitable for samples when microtoming is not possible (or not desirable) or when a small path length is needed for the measurement of materials with strongly absorbing IR bands. When measuring in the ATR mode, in contrast to the transmission mode, IR light does not pass through the whole sample. Instead, the IR light enters an IR transparent ATR element with a high refractive index (e.g., diamond, $ZnSe_2$, or germanium) and undergoes total internal reflection at the interface between the ATR element and the sample before exiting and being directed to the IR detector. At the region where IR light is internally reflected, the IR light penetrates into the sample as an evanescent wave and the absorption of the IR light from the sample will attenuate the light (see Fig. 7.4).

This attenuation of light is then recorded as absorbance in the spectrum. The light penetrates into the sample at a depth given by the following equation [15]:

$$d_p = \frac{\lambda}{2\pi n_1 \left(\sin^2\theta - \left(\frac{n_2}{n_1}\right)^2 \right)^{0.5}} \tag{7.3}$$

d_p corresponds to the depth of penetration where the strength of the evanescent wave decays accordingly to e^{-1} of the field strength at the interface. Typical values of d_p for ranges between 0.2 and 5 μm depending on the wavelength of light, λ, angle of incidence θ (which should be equal or greater than the critical angle) and the refractive index of the ATR element, n_1 and sample, n_2. Note that d_p is not the same as the effective path length (d_e) in an ATR measurement. However, d_e is of a magnitude similar to that of d_p and is given by [15, 16]

$$\frac{d_e}{\lambda} = \frac{\frac{n_2}{n_1}\cos\theta \left[3\sin^2\theta - 2\left(\frac{n_2}{n_1}\right)^2 + \left(\frac{n_2}{n_1}\right)^2 \sin^2\theta\right]}{2\pi\left(1 - \left(\frac{n_2}{n_1}\right)^2\right)\left[\left(1 + \left(\frac{n_2}{n_1}\right)^2\right)\sin^2\theta - \left(\frac{n_2}{n_1}\right)^2\right] \times \left(\sin^2\theta - \left(\frac{n_2}{n_1}\right)^2\right)^{0.5}} \tag{7.4}$$

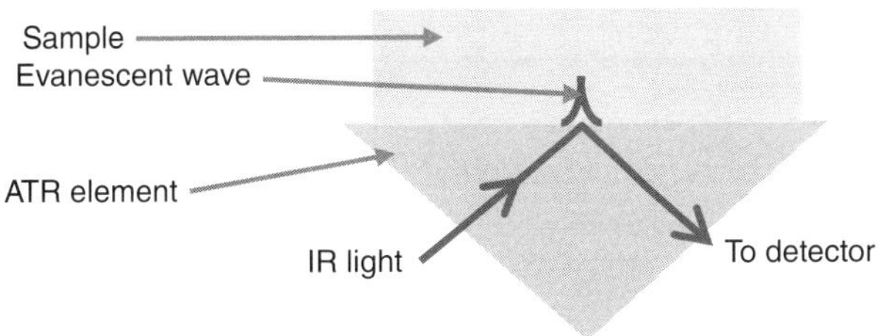

Figure 7.4 Schematic of a typical ATR–FTIR measurement.

For thick samples, both d_p and d_e are independent of sample thickness; hence, an ATR measurement often involves no microtoming and minimal sample preparation. The relatively small path length generated in an ATR measurement allows the study of polymer dissolution in water without being significantly obscured by the strong water absorbance [17]. The small depth of penetration means that the measurement in ATR mode requires the sample to be in intimate contact with the ATR element for the measurement to be reproducible and reliable. A uniform contact is especially important for ATR imaging because the absorbance will not follow the Beer–Lambert law (Eq. 7.1) when the sample is not in good contact with the ATR element. The image generated on the basis of the variation in absorbance across the different points on the map will be dominated by the variation in the quality of contact across the imaged area rather than the composition. To ensure a good contact is achieved, samples can be simply deposited directly onto the ATR element when in a liquid form or may need to be pressed if a solid. Polymers with low melting points can also be melted onto the ATR surface to ensure a good uniform contact is made. Because only one side of the sample is required to be in contact with the ATR element, the sample can be exposed to various environments including high-pressure gases and temperatures providing an opportunity for studying polymer processing *in situ*. We have demonstrated this by studying the effects of temperature and CO_2 pressure on the speed of diffusion between two mixable polymers [18] and phase separation between a poly(vinylmethylether) (PVME) and a polystyrene (PS) blend [19]. The PVME–PS blend is a mixture that exhibits a low critical solution temperature (LCST) such that phase separation occurs when the temperature is increased to above the LCST. However, phase separation of the blend can also be induced by subjecting the blend to high-pressure CO_2. We have studied this process by applying *in situ* ATR–FTIR imaging to monitor the chemical distribution of the blend before and during exposure to high-pressure CO_2 (see Fig. 7.5) [19]. The results showed that the blend phase separated when subjected to high-pressure CO_2 without a change in the temperature, demonstrating that high-pressure CO_2 can mimic the effect of high temperature on polymer blends with an LCST.

As d_p is dependent on the angle of incidence, an opportunity exists to change the angle of incidence for depth profiling

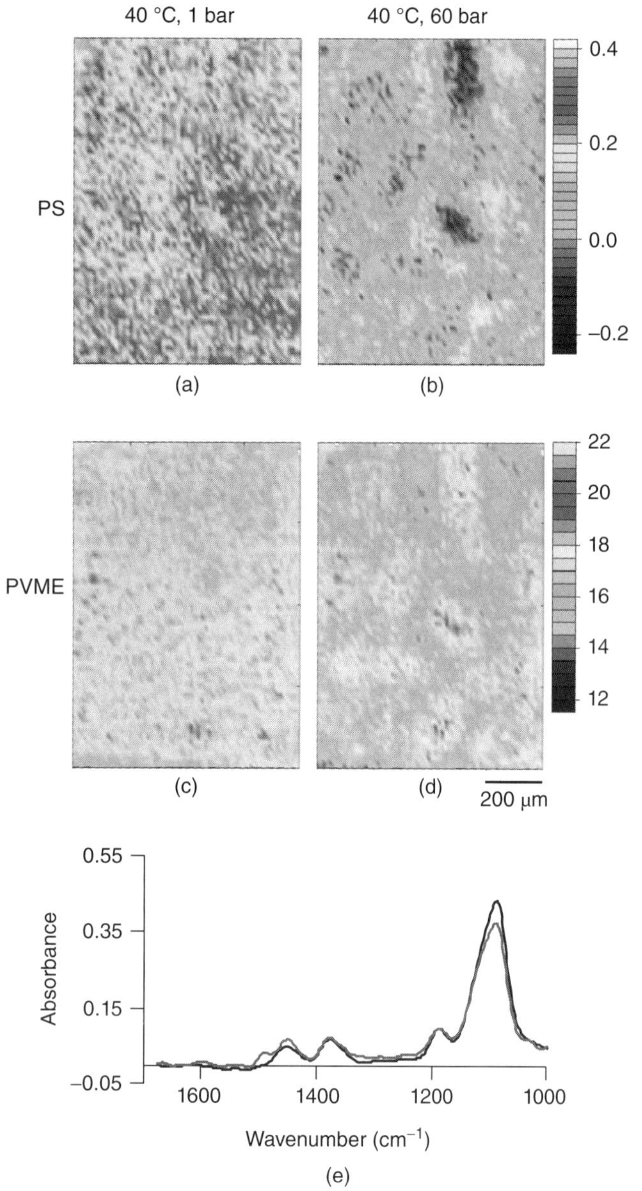

Figure 7.5 ATR–FTIR images of a PS-PVME blend before (a and c) and during exposure to 60 bar of CO_2 (b and d). The images (a and b) are based on the spectral band of PS, while images c and d are based on the spectral band of PVME. (e) Spectra extracted from PVME-rich (blue) and PS-rich (red) domains. Kazarian and Chan [19]. Reproduced with permission of the American Chemical Society. *(See color plate section for the color representation of this figure.)*

purposes. Depth profiling, by changing the angle of incidence, is an established technique in ATR–FTIR studies. However, only recently, the combination of variable angle of incidence with imaging has been demonstrated [20]. We have shown that it is possible to resolve polymer films of 240–400 nm thickness, embedded in a polymer laminar using the variable angle method in ATR–FTIR imaging [21]. Two laminates of four polymer films were created. The main difference between the two laminate was in the order of the polymer layers. The first laminate, S1, consisted of PS – poly(butyl methacrylate) (PBMA) – polycarbonate (PC) – poly(dimethylsiloxane) (PDMS) layers, while the second laminate, S2, consisted of PS–PC–PBMA–PDMS. The thickness of PC and PBMA is on the order of 240–400 nm. Experimental results have shown that the variable angle imaging technique can clearly distinguish between the two laminates and the order of the polymer films can be accurately determined (see Fig. 7.6) [21].

It is important to remember that d_e is also a function of the angle of incidence; therefore, when using the ATR–FTIR imaging approach, it is important to ensure that the angle of incidence across the imaging area is uniform. A previous report has shown that when using a specific ATR accessory, which has limited room for alignment, a gradient of absorbance of a pure liquid across the imaged area is observed (the so-called gradient effect) [22]. This gradient effect can be minimized or removed by ensuring that the alignment of the ATR accessory is optimized. It is, therefore, recommended that testing of the alignment should include a measurement of a pure liquid, such as paraffin oil, to ensure that the measurement produces a uniform absorbance of the liquid across the imaged area [16], which is especially important when quantitative analysis is needed.

7.4 SPATIAL RESOLUTION

In an FTIR image, whether it is collected via mapping, FPA, or linear array detectors, each spectrum is collected from a specific area of a sample which forms one of the pixels of the image. The size of that area is dependent on a number of factors, such as the aperture size (if images were collected via point-by-point mapping) and the magnification of the optical system (or projected pixel size). The size of this area where spectral information is collected in a single pixel is determined by the spatial resolution of the system, which is limited by the diffraction effect for imaging systems with far-field optics. It is important to stress the significance of this because there has been a misconception that spatial resolution is the same as the projected pixel size. The Rayleigh criterion can be used to estimate the expected resolving power of an optical system, which is given as an Equation 7.4 below:

$$r = \frac{0.61\lambda}{n \sin\theta} \quad (7.4)$$

where n is the refractive index of the medium where light approaches the sample, θ is the angle of the marginal ray collected by the objective. Together, $n\sin\theta$ is the numerical aperture (NA) of the system, which is often indicated as a reference for the resolving power. λ is the wavelength of light and r is the distance required to have a minimum contrast of 26.4% between two nearby objects. When a diffraction limited image is achieved, the spatial resolution will not be increased by using a smaller aperture or higher magnifying optics without increasing the NA of the system. A recent report has indicated that diffraction-limited images can be achieved when the projected pixel size is at least $1/4^{th}$ of ref. [23].

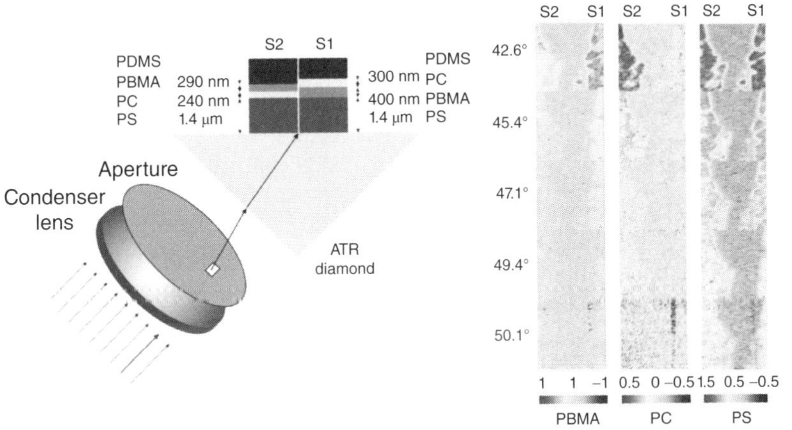

Figure 7.6 Variable angle ATR–FTIR imaging of two polymer laminate samples. S1 consists of PS–PBMA–PC–PDMA layers, while S2 consists of PS–PC–PBMA–PDMS layers. The setup of the experiment is shown on the schematic on the left and the imaging results are shown on the right. A selection of apertures can be introduced to allow only part of the IR beam with restricted angle of incidence to reach the sample. The position of the aperture determines the resultant angle of incidence of the measurement. The angle of incidence is indicated on the left of the imaging result. The d_p has an inverse relationship to the angle of incidence, that is, 50.1° represents a shallow d_p, while 42.6° represents a larger d_p. Frosch et al. [21]. Reproduced with permission of the American Chemical Society.

7.4.1 Transmission FTIR Imaging

The large sample compartment usually consists of 1:1 magnification optics with low NAs (<0.1) and does not involve a motorized stage. The spatial resolution is therefore on the order of 40 μm, which is similar to the physical and projected pixel size. Using an IR microscope in transmission mode, IR light from the interferometer is focused to the sample by a condenser where the light passes through and is absorbed by the sample before being collected by the objective. Typically, a 15× objective with an NA of 0.4 is used, providing a spatial resolution of ~15 μm depending on the wavelength of light used. To resolve smaller domains, an objective with a higher NA will be required. Previous work has shown that a 74× objective with an NA of 0.65 can be used in transmission mode for high spatial resolution imaging. However, with such high-power magnification, a multibeam synchrotron source was needed to improve the spectral quality [23]. We have recently demonstrated that by introducing a CaF_2 hemisphere on top of a sample in a transmission FTIR imaging, the NA of the system is increased by ~1.4 times (the refractive index of CaF_2), from 0.4 to 0.56, without any modification to the existing FTIR imaging system. An increase in sharpness of the image and magnification can be clearly observed (see Fig. 7.7) [12].

7.4.2 ATR–FTIR Spectroscopic Imaging

Similar to the transmission measurement, samples can be measured using the microscope or the large sample compartment in ATR mode. The large sample compartment is

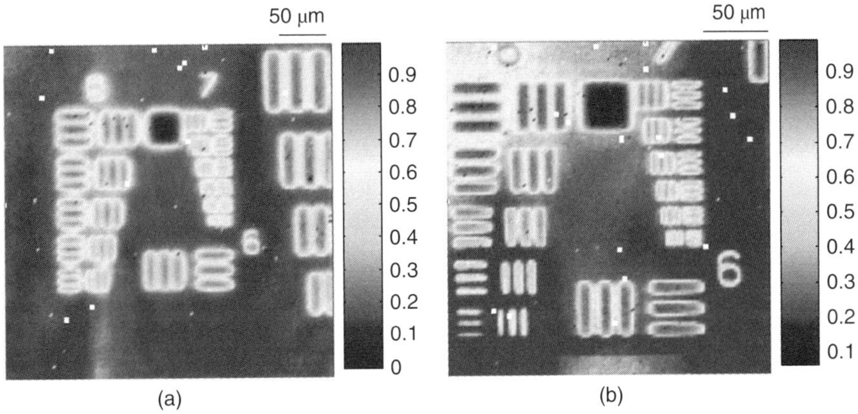

Figure 7.7 Transmission FTIR image of a USAF spatial resolution target (a) without lens (imaging size of 341 μm × 341 μm) and (b) with lens (imaging size of 240 μm × 240 μm). Chan and Kazarian [12]. Reproduced with permission of the American Chemical Society. *(See color plate section for the color representation of this figure.)*

designed to accommodate larger samples or accessories, for example, high-pressure ATR cells that are normally not easy to fit under a microscope objective. It provides a larger field of view but with lower spatial resolution. A diamond ATR accessory that provides an intermediate spatial resolution (~15 μm) is also available [24]. This accessory employs a small diamond (an inverted prism that gives an imaging area of ~0.6 mm × 0.6 mm) as the ATR element and uses optics to focus and collect light through the small diamond. However, it has been shown that it is crucial to ensure that good alignment is achieved with this type of measurement [16] to avoid the gradient effect (a small but noticeable change in absorbance across the image from a uniform sample) as a result of a spread of the angle of incidence across the imaging area in the ATR imaging measurement [22]. A small aperture can be inserted to limit the spread of the angle of incidence but this will result in a reduced throughput of light [25]. On the other hand, the introduction of the aperture provides a new opportunity to control the angle of incidence for the depth profiling of a sample with imaging (see Fig. 7.6) [25]. A larger imaging area can be obtained by using expanded optics, which is useful in high-throughput studies and sample deposition [26] but comes at the expense of poorer spatial resolution. A solution to this is to combine imaging with mapping, similar to the linear array approach, achieving a larger field of view without a reduction in spatial resolution. However, in this way, not all pixels on the image are measured simultaneously and therefore can only be suitable for static system or very slow dynamic system [27, 28].

Micro-ATR imaging is an ATR–FTIR imaging measurement using an infrared microscope. Apart from the fact that the optics employed in the microscope have higher magnification and NAs, one of the main differences between micro- and macro-ATR is the shape of the ATR element. In macro-ATR imaging, the ATR element is often a prism and therefore infrared light refracts as it enters the ATR element. Because $\mathrm{NA} = n\sin\theta$, the refraction reduces the $\sin\theta$ component that compensated for the increase in n for the light approaching the samples in a high-refractive-index ATR element. The overall NA, therefore, remains unchanged through the ATR prism. Fortunately, the NAs in macro-ATR mode are often relatively small (<0.1) or a diamond is used as the ATR element, and the effect of dispersion from the refraction is less severe when compared to imaging through a liquid cell in a microscope in transmission (Section 7.3.1.1). In micro-ATR, the ATR element is a hemisphere or a section of a hemisphere. IR light enters the hemisphere of high refractive index element without refraction. The NA is increased as a result of the increase in the value of n, which results in an increase in resolving power. A fourfold improvement in spatial resolution, when compared to images measured in transmission mode with the same microscope, has been demonstrated using a Ge hemisphere [24]. Images with spatial resolution of 2–4 μm can be achieved, which helps in resolving small domains in, for example, a polymer blend [1, 29].

7.5 RECENT APPLICATIONS

7.5.1 Polymer Blends

FTIR imaging has shown to be a valuable tool in the study of polymer blends [30]. One of the important products where polymer blends are often used is for compacted pharmaceutical tablets. Pharmaceutical tablets are formulated with various polymer blends to achieve the desirable dissolution characteristics. We have applied ATR–FTIR imaging and X-ray microtomography to assess the effect of compaction and composition on the distribution of a model drug, caffeine, in a matrix of hydroxypropyl methylcellulose (HPMC). While the ATR–FTIR imaging approach can be used to characterize the drug and the different components in the polymer blend, due to the small depth of penetration of the ATR measurement, this technique can only measure the outer layer of the intact tablet. X-ray tomography, on the other hand can be used to obtain a 3D image of the tablet but the resulting image is based on density which, in contrast, is not chemical specific. The combined method allows the best of both techniques to be realized: ATR–FTIR imaging for the high chemical specificity and X-ray microtomography for the nondestructive 3D imaging capability (see Fig. 7.8) [31]. The X-ray microtomograph has shown localized domains of high density of 100–150 μm, but without providing the chemical information for these domains. Imaging over the same area of the tablet with the ATR–FTIR method has revealed that these localized domains of 100–150 μm domains are caffeine. Overlaying the two images has shown a similar pattern (caffeine appeared to be more dense than HPMC). The two images, however, did not overlay exactly because of the difference in sampling volume by the two imaging methods.

By blending the natural polymers cellulose and bacterial-synthesized biopolymer (PHBV) using an ionic liquid as the processing solvent, the effect of polymer composition on the final spatial distribution has been shown by both micro-ATR–FTIR imaging and scanning electron microscopy [32]. In this study, the two polymers were clearly distinguished by the spectral features of the polymers (the biopolymer contains a strong carbonyl band at 1730 cm^{-1} and the cellulose contain a strong C–O–C band at 1030 cm^{-1}). Images (Fig. 7.9) showing their relative concentration over the imaged area were generated and shown to be complementary, consistent with the fact that it is a binary mixture.

In addition to determining the spatial distribution, it is also possible to monitor molecular interactions between the polymers in the blend. A shift and change in the ratio of the absorbance of carbonyl bands as a function of blend composition has been used as an indicator of a change in the hydrogen bonding between the polymers[32]. Apart from studying blends, FTIR imaging has been used to verify the homogeneity in samples [33] including novel polymerization processes of the conductive polymer polyaniline [34], estimation of the size of liquid crystal inclusions in a polymer

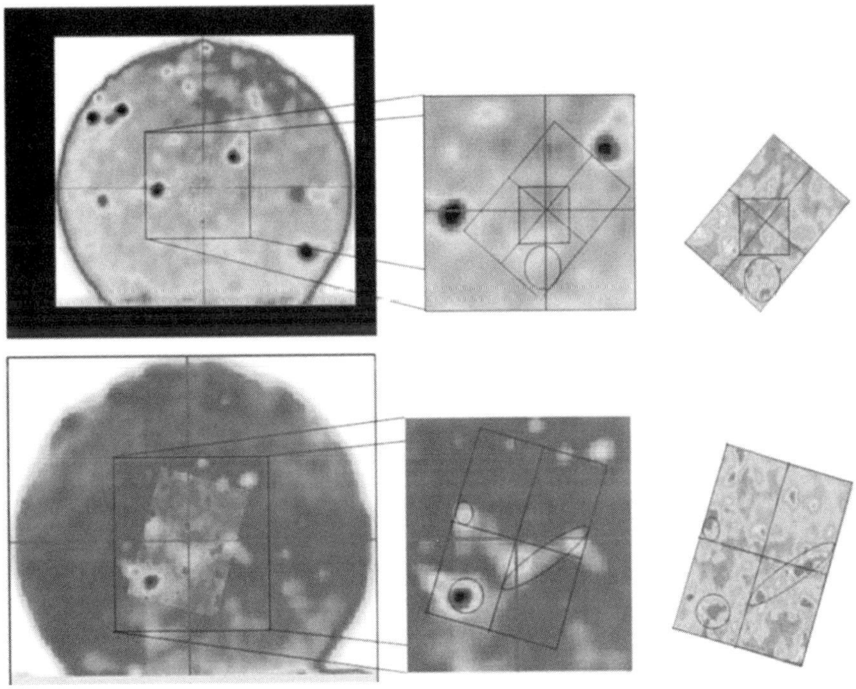

Figure 7.8 Images showing the results from FTIR spectroscopy compared with images from X-ray microtomography. The top set of images shows the results for the 100- 125-μm particle size of caffeine. The bottom set shows the data for the 125- 150-μm particle size. The FTIR data is shown on the right-hand side, the key area of the X-ray tomography is shown in the middle, and the FTIR is layered over the X-ray data for comparison on the left-hand side. Wray et al. [31]. Reproduced with permission of John Wiley and Sons. (*See color plate section for the color representation of this figure.*)

matrix [35] and differences in spatial distribution of curdlan (a bacterial polysaccharide) in hydrogels formed at different temperatures [36]. FTIR imaging has also been applied to study blends of poly(ε-caprolactone)/chitosan fibers to assess their homogeneity [37].

7.5.2 Polymer Processes

ATR–FTIR imaging is a highly versatile technology. The micro-ATR element may be temporarily removed from the microscope (depending on the manufacturer and the design of the microscope objective) to allow *in situ* measurements [38]. We have utilized this opportunity to study live cells with micro-ATR–FTIR imaging by growing live cells directly onto the micro-ATR element and imaging them periodically to study the chemical changes in living cells as a function of time [39]. A similar approach has been employed in a recent study of polymer degradation under UV irradiation in air as a function of exposure time by micro-ATR–FTIR imaging [40]. In this study, a polymer film was cast directly onto the ATR element, which was removed from the microscope objective. The polymer film on the ATR element was then subjected to degradation conditions. The molecular changes were monitored by placing the ATR element under the objective for the measurement. The flexibility of the removable ATR element has been clearly demonstrated. Apart from ATR mode, rapid FTIR imaging in reflection mode has also been used to study the effect of oxygen on the UV photopolymerization of a micropatterned hydrogel on a gold surface. The reaction was monitored *in situ*, with FTIR images measured while the sample was exposed to UV light [41]. The demonstrated rate of imaging was on the order of 10 s. Higher FTIR imaging rates can be achieved by optimizing the data collection. Up to eight FTIR images can be collected every second to capture transient images of a fast reaction. This rapid FTIR imaging method has also been used to investigate diffusion and reactions in microfluidic devices [42, 43], which can, in turn, be used to study fast reactions in polymer processes.

In addition, measurement of polymer processes (or degradation, in this case) can also be studied in transmission mode. Photooxidation of polypropylene has recently been studied by using FTIR imaging. Polypropylene film was exposed in UV radiation for up to 60 h and cross sections of the film were microtomed and analyzed using FTIR imaging at 4-h intervals. The results have caused the heterogeneous progression of oxidation in the polymer film to be reevaluated. When combined with principle component analysis, the results revealed that different oxidation products were produced at different irradiation times [44]. This work demonstrated that FTIR imaging can be a powerful tool in the study of the oxidation process in polymer films. Other work has shown that FTIR imaging can be combined with near-IR imaging and principal component analysis (PCA) to study spherulite growth in PHB and cellulose acetate butyrate (CAB) blends (see Fig. 7.10). The combined method has been demonstrated as a useful tool in obtaining time-resolved data to detect

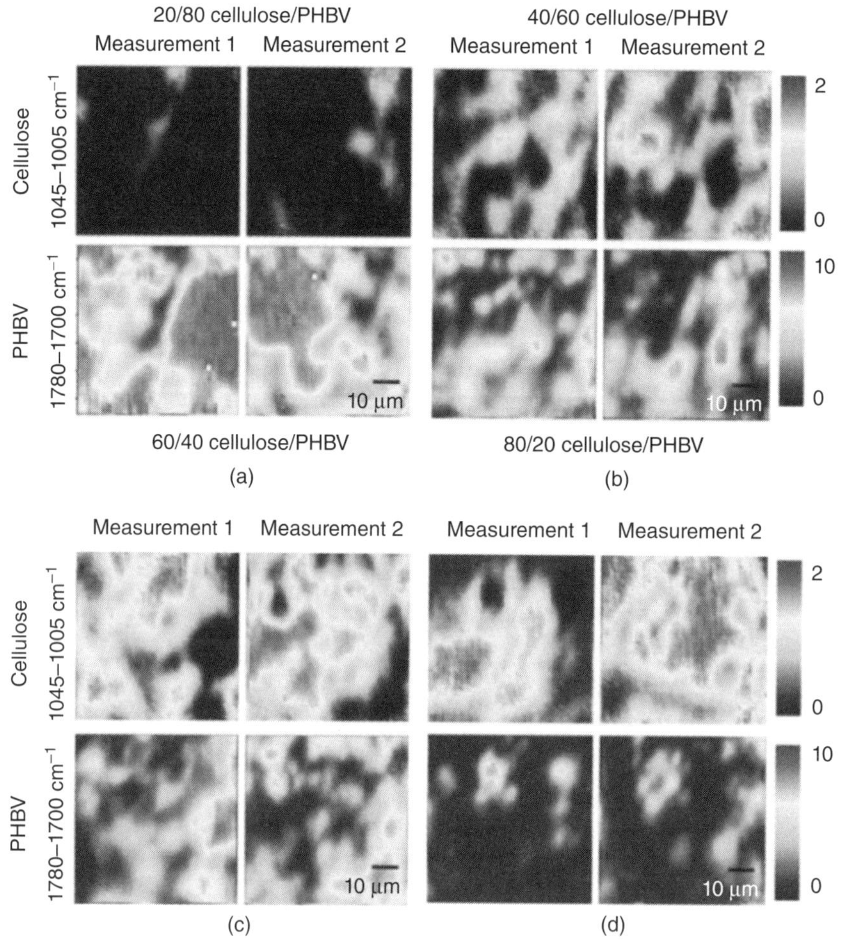

Figure 7.9 Micro-ATR–FTIR images of cellulose/PHBV blends at different compositions. Cellulose/PHBV: (a) 20/80, (b) 40/60, (c) 60/40, and (d) 80/20. The size of each image is ca. 63 μm × 63 μm. Hameed 2011 [32]. Reproduced with permission of Elsevier. (*See color plate section for the color representation of this figure.*)

changes in the molecular structures during the crystallization process [45].

Our earlier work has shown that phase separations of a polymer blend comprised PS and PVME under changing temperatures and pressures of CO_2 can be studied *in situ* using the ATR-FTIR imaging mode [19]. Vogel et al. have shown that the phase separation of a polymer blend of poly(3-hydroxybutyrate (PHB), poly(L-lactic acid) (PLA) and poly(ε-caprolactone) as a result of a change in blend composition can be studied with FTIR imaging in transmission mode in an *ex situ* manner [46]. In addition, with the use of a carefully designed temperature-controlled device, it is also possible to perform an *in situ* study of the polymer processes in transmission mode. Recently, the phase separation of a polymer blend comprising 50:50 wt% PHB and PLA under variable temperature has been studied *in situ* using both transmission FTIR and Raman mapping. FTIR images were used to detect the blend phase separation into PHB-rich and PLA-rich domains at lower temperatures, based on the first overtone of the ν(C=O) band ratio of PHB and PLA, and the process of melting of the polymer blend at elevated temperature (25–175 °C) (see Fig. 7.11). Raman mapping was used to resolve the smaller phase-separated domains [47]. Other recent applications of FTIR imaging on polymer processes also include the study of micropatterning process [48], and micropatterning with polymer brushes[49] demonstrating that FTIR imaging is a useful tool in analysis of polymeric systems.

7.5.3 Polarized FTIR Imaging for Orientation Studies

Apart from studying the spatial distribution of blends of different polymers, FTIR imaging can be a useful tool to distinguish between different polymorphic forms of a pure polymer and to study the orientation of the molecule in a crystalline structure. One of the early demonstrations combining polarized radiation with FTIR imaging was the study of the stretching of poly(vinylidene fluoride) (PVDF) film under different temperatures to observe the changes in molecular orientation as a result of the different treatments. The polarized FTIR imaging method has been clearly shown to be able to characterize the difference in the change of

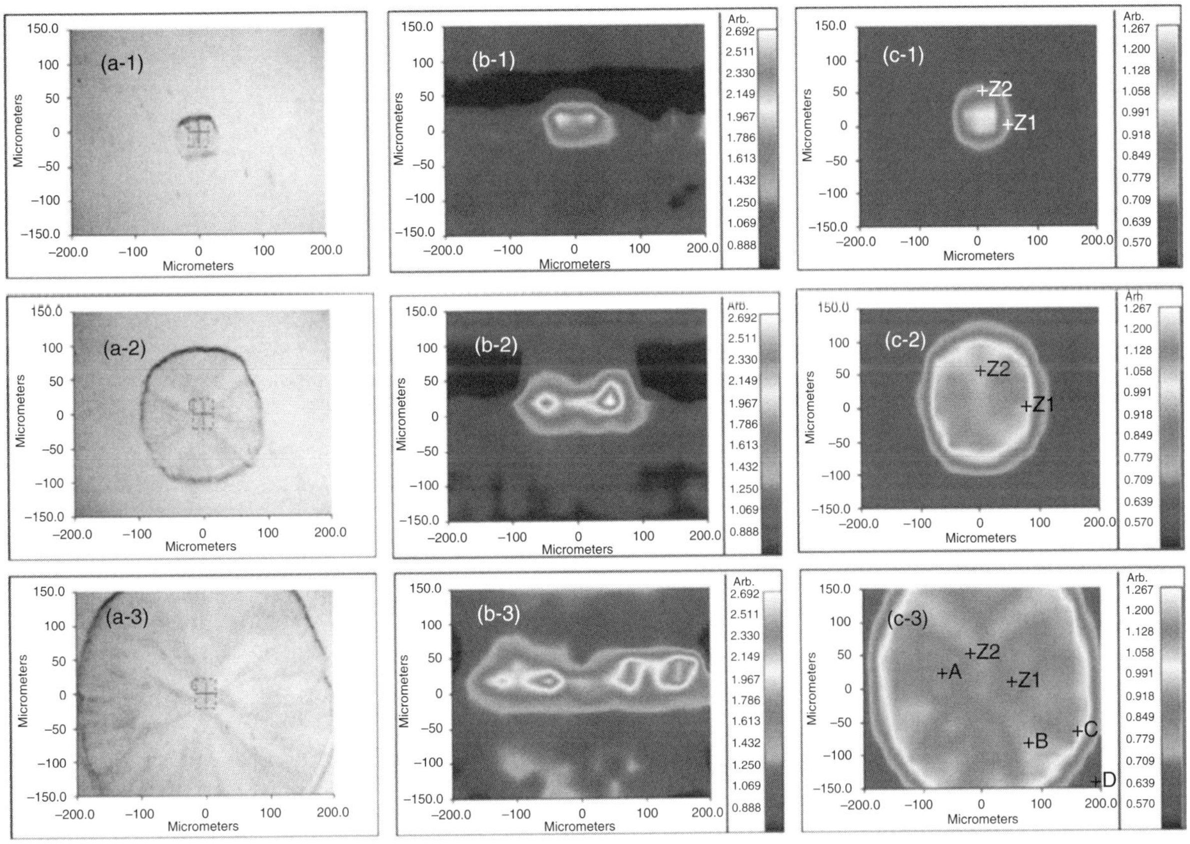

Figure 7.10 (a) Visual images, (b) NIR images, and (c) IR images of the PHB/CAB blend (80:20 wt%) during isothermal crystallization at 125°C at around 18 min (a-1, b-1, c-1), 36 min (a-2, b-2, c-2), and 66 min (a-3, b-3, c-3). Suttiwijitpukdee 2012 [45]. Reproduced with permission of the American Chemical Society. (*See color plate section for the color representation of this figure.*)

molecular conformation of the polymer film when it is stretched at high (no change) and low temperatures (a change from form II (α) to form I (β) is observed) [50].

In the study of spherulites formed in poly(ester urethane) multiblock copolymer [51], FTIR imaging was used to reveal valuable information about both the orientation of the polymer chain and the composition of the spherulites. This information was not available from the measurements obtained from atomic force microscopy or polarized light microscopy. A separate study has also demonstrated that linear polarized FTIR imaging is a powerful tool for the investigation of the crystalline and amorphous structures and chain orientation of spherulites of PHB and isotactic poly(propylene oxide) [52]. Recently, a novel multipolarization calculation method has been proposed and applied to obtain FTIR images showing band structure in poly(L-lactic acid) and PHB spherulites with the indication of local molecular chain orientation [53].

7.6 CONCLUSIONS

FTIR imaging is a highly versatile imaging approach providing opportunities for a wide range of polymeric studies. This chapter has shown that FTIR imaging provides label-free measurements that give highly chemically specific information. It can be used to obtain quantitative, spatially, and, in some cases, temporally resolved data. The number of different measurement modes available in FTIR imaging makes this imaging method highly adaptive to different applications whether the sample is a hard powder, a thin film, a soft solid or in a liquid form. FTIR imaging in micro-ATR mode further improves the spatial resolution to a few micrometers, allowing the observation of small features that were previously deemed to be too small to be studied by FTIR imaging. The combination of variable angle ATR with imaging also allows the study of thin films within a polymer laminate, where the depth resolution is not limited by diffraction. Thin-film layers of thicknesses of 240–400 nm were successfully studied by this method, which has demonstrated that it can potentially be applied to study the uniformity of thin-film thickness in coatings or laminates. FTIR imaging with the diamond ATR accessory provides an opportunity to study high-pressure and high-temperature processes *in situ*. This has been demonstrated by the *in situ* monitoring of the phase separation behavior of a polymer blend under high-pressure CO_2.

A number of applications in the study of polymer blends including pharmaceutical materials and biopolymers, *in situ*

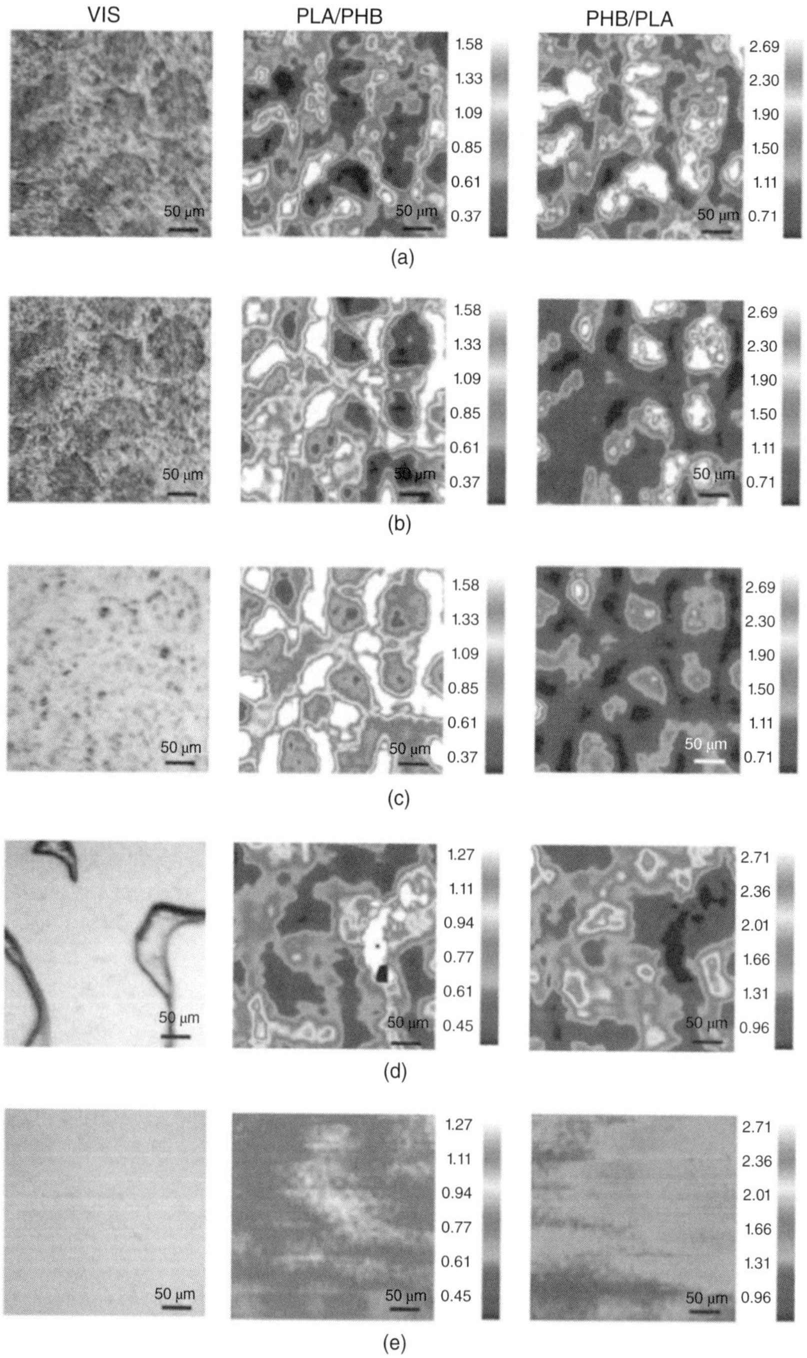

Figure 7.11 Visual images, PLA–PHB band-ratio, and PHB–PLA band-ratio FTIR images of the PHB–PLA (50:50 wt%)-blend film at 25 (a), 125 (b), 165 (c), 170 (d), and 175 °C (e). Vogel et al. [46]. Reproduced with permission of the American Chemical Society. (*See color plate section for the color representation of this figure.*)

monitoring of polymer processing such as UV degradation, polymerization, diffusion and crystallization, intermolecular interactions, and molecular orientation using polarized light have been realized. As the spatial resolution, sensitivity, and speed of FTIR imaging measurement is constantly being improved, new exciting experimental approaches and applications of FTIR imaging in polymeric systems are expected to be developed.

REFERENCES

1. Gupper A, Wilhelm P, Schmied M, Kazarian SG, Chan KLA, Reussner J. Combined application of imaging methods for the characterisation of a polymer blend. Appl Spectrosc 2002;56:1515–1523.
2. Wessel E, Vogel C, Kolomiets O, Hoffmann U, Siesler HW. FT-IR and NIR Spectroscopic Imaging: Principles, Practical

Aspects and Applications in Material and Pharmaceutical Sciences. In: Salzer R, Siesler HW, editors. *Infrared and Raman Spectroscopic Imaging*. Germany: Wiley-VCH; 2009, p 297–345.

3. Snively CM, Koenig JL. Studying anomalous diffusion in a liquid crystal/polymer system using fast FTIR imaging. J Appl Polym Sci B 1999;37:2261–2268.
4. Ribar T, Koenig JL, Bhargava R. FTIR imaging of polymer dissolution. 2 Solvent/nonsolvent mixtures. Macromolecules 2001;34:8340–8346.
5. Gupper A, Chan KLA, Kazarian SG. FT-IR imaging of solvent-induced crystallization in polymers. Macromolecules 2004;37:6498–6503.
6. Gupper A, Kazarian SG. Study of solvent diffusion and solvent-induced crystallization in syndiotactic polystyrene using FT-IR spectroscopy and imaging. Macromolecules 2005;38:2327–2332.
7. Venyaminov SY, Prendergast FG. Water (H_2O and D_2O) molar absorptivity in the 1000–4000 cm^{-1} range and quantitative infrared spectroscopy of aqueous solutions. Anal Biochem 1997;248:234–245.
8. Coutts-Lendon C, Koenig JL. Investigation of the aqueous dissolution of semicrystalline poly(ethylene oxide) using infrared chemical imaging: The effects of molecular weight and crystallinity. Appl Spectrosc 2005;59:976–985.
9. Wessel E, Vogel C, Siesler HW. Fourier transform infrared imaging spectroscopy of the diffusion process of D_2O into polyamide 11. Appl Spectrosc 2009;63:1–5.
10. Carr GL. Resolution limits for infrared microspectroscopy explored with synchrotron radiation. Rev Sci Instrum 2001;72:1613–1619.
11. Wetzel DL. A new approach to the problem of dispersivewindows in infrared microspectroscopy. Vib Spectrosc 2002;29:291–297.
12. Chan KLA, Kazarian SG. Correcting the effect of refraction and dispersion of light in FT-IR spectroscopic imaging in transmission through thick infrared windows. Anal Chem 2013;85:1029–1036.
13. Tobin MJ, Puskar L, Barber RL, Harvey EC, Heraud P, Wood BR, Bambery KR, Dillon CT, Munro KL. FTIR spectroscopy of single live cells in aqueous media by synchrotron IR microscopy using microfabricated sample holders. Vib Spectrosc 2010;53:34–38.
14. Nasse MJ, Ratti S, Giordano M, Hirschmugl CJ. Demountable liquid/flow cell for in vivo infrared microspectroscopy of biological specimens. Appl Spectrosc 2009;63:1181–1186.
15. Harrick NJ. *Internal Reflection Spectroscopy*. New York: Harrick Scientific Corporation; 1987.
16. Kazarian SG, Chan KLA, Tay FH. ATR-FT-IR Imaging for Pharmaceutical and Polymeric Materials: From Micro to Macro Approaches. In: Salzer R, Siesler HW, editors. *Infrared and Raman Spectroscopic Imaging*. Germany: Wiley-VCH; 2009, p 347–375.
17. Kazarian SG, Chan KLA. Chemical photography of drug release. Macromolecules 2003;36:9866–9872.
18. Fleming OS, Chan KLA, Kazarian SG. High-pressure CO_2-enhanced polymer interdiffusion and dissolution studied with in situ ATR-FTIR spectroscopic imaging. Polymer 2006;47:4649–4658.
19. Kazarian SG, Chan KLA. FTIR imaging of polymeric materials under high pressure carbon dioxide. Macromolecules 2004;37:579–584.
20. Chan KLA, Kazarian SG. Attenuated total reflection Fourier transform infrared imaging with variable angles of incidence: A three-dimensional profiling of heterogeneous materials. Appl Spectrosc 2007;61:48–54.
21. Frosch T, Chan KLA, Wong HC, Cabral JT, Kazarian SG. Nondestructive three-dimensional analysis of layered polymer structures with chemical imaging. Langmuir 2010;26: 19027–19032.
22. Wessel E, Heinsohn G, Schmidt-Lewer-Kuehne H, Wittern KP, Rapp C, Siesler HW. Observation of a penetration depth gradient in attenuated total reflection Fourier transform infrared spectroscopic imaging applications. Appl Spectrosc 2006;60:1488–1492.
23. Nasse MJ, Walsh MJ, Mattson EC, Reininger R, Kajdacsy-Balla A, Macias V, Bhargava R, Hirschmugl CJ. High-resolution Fourier-transform infrared chemical imaging with multiple synchrotron beams. Nat Methods 2011;8:413–416.
24. Chan KLA, Kazarian SG. New opportunities in micro- and macro-attenuated total reflection infrared spectroscopic imaging: Spatial resolution and sampling versatility. Appl Spectrosc 2003;57:381–389.
25. Chan KLA, Tay FH, Poulter G, Kazarian SG. Chemical imaging with variable angles of incidence using a diamond attenuated total reflection accessory. Appl Spectrosc 2008;62:1102–1107.
26. Chan KLA, Kazarian SG. ATR-FTIR spectroscopic imaging with expanded field of view to study formulations and dissolution. Lab Chip 2006;6:864–870.
27. Chan KLA, Kazarian SG. ATR-FTIR imaging of large areas using inverted prism crystals and combining imaging and mapping. Appl Spectrosc 2008;62:1095–1101.
28. Chan KLA, Gulati S, Edel JB, de Mello AJ, Kazarian SG. Chemical imaging of microfluidic flows using ATR-FTIR spectroscopy. Lab Chip 2009;9:2909–2913.
29. Kazarian SG, Chan KLA. Sampling approaches in Fourier transform infrared imaging applied to polymers. Progr Colloid Polym Sci 2006;132:1–6.
30. Kazarian SG, Chan KLA. Combined applications of ATR-IR imaging and confocal Raman microscopy to polymeric materials. In: Grundke K, Stamm M, Adler HJ, editors. *10th Dresden Polymer Meeting*. Berlin, Dresden, Germany: Springer-Verlag; 2005.
31. Wray P, Chan KLA, Kimber J, Kazarian SG. Compaction of pharmaceutical tablets with different polymer matrices studied by FTIR imaging and X-ray microtomography. J Pharm Sci 2008;97:4269–4277.
32. Hameed N, Guo QP, Tay FH, Kazarian SG. Blends of cellulose and poly(3-hydroxybutyrate-co-3-hydroxyvalerate) prepared from the ionic liquid 1-butyl-3-methylimidazolium chloride. Carbohydr Polym 2011;86:94–104.
33. Kapse SV, Gaikwad RV, Samad A, Devarajan PV. Self nanoprecipitating preconcentrate of tamoxifen citrate for enhanced bioavailability. Int J Pharm 2012;429:104–112.
34. Trey S, Jafarzadeh S, Johansson M. In situ Polymerization of Polyaniline in Wood Veneers. ACS Appl Mater Interfaces 2012;4:1760–1769.

35. Wang JH, Zhang BY, Xi M, Xu XY. Study on dispersion of liquid crystal droplets in PDCLC film by 2D/3D-FTIR spectrum. Mol Cryst Liq Cryst 2011;548:17–27.
36. Gagnon MA, Lafleur M. Comparison of the structure and the transport properties of low-set and high-set curdlan hydrogels. J Colloid Interface Sci 2011;357:419–427.
37. Malheiro VN, Caridade SG, Alves NM, Mano JF. New poly(epsilon-caprolactone)/chitosan blend fibers for tissue engineering applications. Acta Biomater 2010;6:418–428.
38. Kazarian SG, Chan KLA. ATR-FTIR spectroscopic imaging: recent advances and applications to biological systems. Analyst 2013;138:1940–1951.
39. Kuimova MK, Chan KLA, Kazarian SG. Chemical imaging of live cancer cells in the natural aqueous environment. Appl Spectrosc 2009;63:164–171.
40. Nagle DJ, George GA, Rintoul L, Fredericks PM. Use of micro-ATR/FTIR imaging to study heterogeneous polymer oxidation by direct solvent casting onto the ATR IRE. Vib Spectrosc 2010;53:24–27.
41. Biswal D, Hilt JZ. Analysis of oxygen inhibition in photopolymerizations of hydrogel micropatterns using FTIR imaging. Macromolecules 2009;42:973–979.
42. Chan KLA, Niu X, deMello AJ, Kazarian SG. Generation of chemical movies: FT-IR spectroscopic imaging of segmented flows. Anal Chem 2011;83:3606–3609.
43. Chan KLA, Kazarian SG. FT-IR spectroscopic imaging of reactions in multiphase flow in microfluidic channels. Anal Chem 2012;84:4052–4056.
44. Zhou YM, Li BB, Zhang PD. Fourier transform infrared (FT-IR) imaging coupled with principal component analysis (PCA) for the study of photooxidation of polypropylene. Appl Spectrosc 2012;66:566–573.
45. Suttiwijitpukdee N, Sato H, Unger M, Ozaki Y. Effects of hydrogen bond intermolecular interactions on the crystal spherulite of poly(3-hydroxybutyrate) and cellulose acetate butyrate blends: Studied by FT-IR and FT-NIR imaging spectroscopy. Macromolecules 2012;45:2738–2748.
46. Vogel C, Wessel E, Siesler HW. FT-IR imaging spectroscopy of phase separation in blends of poly(3-hydroxybutyrate) with poly(L-lactic acid) and poly(epsilon-caprolactone). Biomacromolecules 2008;9:523–527.
47. Unger M, Sato H, Ozaki Y, Fischer D, Siesler HW. Temperature-dependent Fourier transform infrared spectroscopy and Raman mapping spectroscopy of phase-separation in a poly(3-hydroxybutyrate)-poly(L-lactic acid) blend. Appl Spectrosc 2013;67:141–148.
48. Park Y, Felipe MJ, Advincula RC. Facile patterning of hybrid CdSe nanoparticle films by photoinduced surface defects. ACS Appl Mater Interfaces 2011;3:4363–4369.
49. Tria MCR, Advincula RC. Electropatterning of binary polymer brushes by surface-initiated RAFT and ATRP. Macromol Rapid Commun 2011;32:966–971.
50. Vogel C, Wessel E, Siesler HW. Fourier transform infrared spectroscopic imaging of anisotropic poly(vinylidene fluoride) films with polarized radiation. Appl Spectrosc 2008;62: 599–603.
51. Wang W, Jin Y, Yang XN, Su ZH. Chain orientation and distribution in ring-banded spherulites formed in poly(ester urethane) multiblock copolymer. J Polym Sci B 2010;48: 541–547.
52. Merten C, Kowalik T, Asshoff SJ, Hartwig A. FTIR imaging of poly(3-hydroxybutyrate) and isotactic poly(propylene oxide) spherulites. Macromol Chem Phys 2010;211: 1627–1631.
53. Hikima Y, Morikawa J, Hashimoto T. Wavenumber dependence of FT-IR image of molecular orientation in banded shperulites of poly(3-hydroxybutyrate) and poly(L-lactic acid). Macromolecules 2013;46:1582–1590.

8

NMR ANALYSIS OF MORPHOLOGY AND STRUCTURE OF POLYMERS

Takeshi Yamanobe and Hiroki Uehara
Department of Chemistry and Chemical Biology, Gunma University, Gunma 376-8515, Japan

8.1 INTRODUCTION

Investigation of the relationship between the structure and physical properties of polymers is one of the most important themes for polymer chemists and engineers. Polymer structures have been determined through many types of spectroscopy. Nuclear magnetic resonance (NMR) is now an indispensable tool for analyzing polymer structures [1–3].

When polymers are designed as a material, polymer morphology, which can be defined as the structure and the phase separation of polymers on a large scale, plays an important role in the physical properties of polymers. Polymer morphology involves the crystallinity, crystallite, and polymer macro-conformation, such as fringed micelles and their formation kinetics, and so on. The morphology/property relationship of polymers has been discussed for a long time, but not fully interpreted. Problems lie in the analytical difficulty caused by the complicated combination of the crystalline and amorphous phases, including their contents, arrangements, difference in physical properties, and the existence of an interface region between them. For example, each method for the determination of the crystallinity, such as thermal analysis, density, and X-ray measurements, focuses on its own view scale, namely, "where is the borderline between crystalline and amorphous" in terms of physical properties. This is one reason the crystallinity evaluations by different methods often disagree with each other. An advantage of NMR is attributed to its ability for simultaneous estimation of both phase contents and properties (relaxation times) on the same scale. On the other hand, the disadvantage of NMR exists in the difficulty of the spatial information, especially, between nano- and micrometer scales.

This disadvantage can be overcome by a combination of NMR and other techniques. Combination of NMR with X-ray diffraction and electron microscopy is able to analyze the structure from atomic level to macro scale. In this chapter, the morphology/property relationship, the dynamics of morphological transition, the kinetics of crystallization, and so on, analyzed by the combination of NMR and other tools are introduced.

8.2 BASIC CONCEPTS IN NMR

8.2.1 Principles of NMR

In the NMR measurements, samples are placed in a strong magnetic field and irradiated with radiofrequency radiation. The spin state of a nucleus splits corresponding to the spin number I, which is called Zeeman splitting. For the nuclei of $I = 1/2$, ^1H, ^{13}C, ^{15}N, and so on, which are popular in polymer analysis, the spin state splits into two energy states. The energy difference is expressed by

$$\Delta E = \gamma h H_0 / 2\pi \qquad (8.1)$$

where γ and H_0 are the magnetogyric ratio and the applied magnetic field, respectively. The transition between two energy states can be induced by the use of appropriate electromagnetic radiation. The frequency of the radiation, ν, is expressed as follows.

$$\nu = \gamma H_0 / 2\pi \qquad (8.2)$$

Polymer Morphology: Principles, Characterization, and Processing, First Edition. Edited by Qipeng Guo.

Because the magnitude of γ depends on the nucleus, the resonance frequency of v is characteristic of the nucleus, which enables the independent measurement of constituent elements such as ^{1}H and ^{13}C.

From NMR measurements, a lot of parameters can be obtained. Recent development of the hardware and the pulse sequences enables us to measure structural parameters such as the bond length, the dihedral angles, and so on, directly [3]. However, the measurements sometimes require isotopic labeling and long measuring time. NMR chemical shift and the relaxation times are suitable for the practical analysis of polymer morphology.

NMR chemical shift is the peak position and corresponds to the resonance frequency of the nuclei. Even for the same nucleus, the electronic environment is dependent on the chemical structure and the functional groups, which is reflected in the resonance frequency. In other words, the chemical shift is affected by the structural changes and is the measure for the conformation and the crystal structures [4].

There are two kinds of popular relaxation times, spin–lattice relaxation time T_1 and spin–spin relaxation time T_2. Dipolar interactions between nuclear spins in polymers predominantly govern relaxation times. According to the Bloembergen–Purcell–Pound (BPP) theory [5], T_1 and T_2 vary against the correlation time (τ_c) of the molecular motion, as indicated in Figure 8.1. T_1 has the minimum at $\tau_c = 1/\omega$, in which ω is the resonance angular frequency. If the T_1 minimum is observed, the absolute correlation time of the molecular motion can be obtained. On the other hand, T_2 decreases with an increment of τ_c. From these behaviors of the relaxation times, the dynamics of polymer chain can be obtained.

For the morphological discussions, NMR measurements must be carried out in the solid state. The cross-polarization and magic angle sample spinning (CPMAS) for ^{13}C and the solid echo pulse sequence for ^{1}H are practical methods. From CPMAS measurements, the spectrum similar to the solution NMR spectrum is observed. ^{13}C NMR chemical shifts are easily obtained from the CPMAS spectrum. Furthermore, the application of the sequence developed by Torchia supplies T_1 in the solid state. However, ^{13}C NMR measurements are time consuming because of the low natural abundance of ^{13}C nucleus. T_2 of ^{1}H can be observed easily by the solid echo sequence with ^{1}H pulse NMR. As the sensitivity of ^{1}H is high, the measurement time is short enough to obtain data with high signal-to-noise (*S/N*) ratio. ^{1}H T_2 is useful to discuss about the morphology.

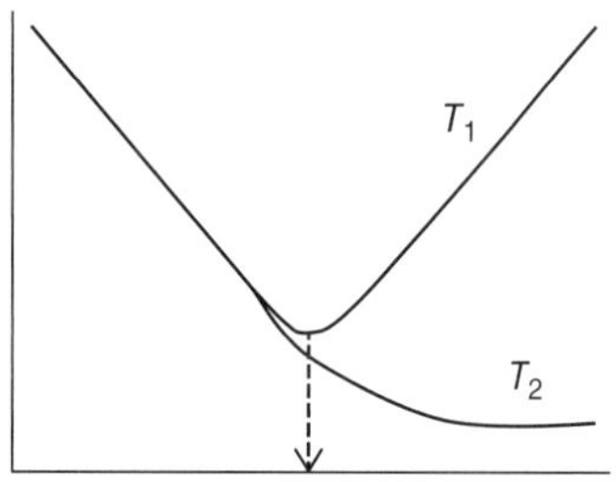

Figure 8.1 Correlation time dependence of T_1 and T_2.

8.2.2 Analysis of the Free Induction Decay (FID)

From ^{1}H pulse NMR, the free induction decay (FID) with a large number of data points can be obtained. The curve fitting for the observed FIDs gives the individual T_2 characteristics in the crystalline, amorphous, and interfacial phases [6–12]. Such a resolution into several components has been attempted on a broad-line spectrum of solid polyethylene (PE) [13–16], which reflect the sample morphologies.

Generally, in order to fit the observed FID, a series of exponential functions (Eq. 8.3) are used because the distribution of dipole interaction is expressed by Lorentzian function. This is true for the solution, melt, and amorphous phases of the polymers. Actually, a PE melt with a low molecular weight (MW) exhibits a single exponential curve [17–20]. On the other hand, Weibullian functions (Eq. 8.4) fit for the phase with partially restricted motion such as the interfacial phase [8, 21].

$$I_1(t) = A_1 \exp\left\{-\frac{k_1 t}{2}\right\} \tag{8.3}$$

$$I_2(t) = A_2 \exp\left\{-\frac{(k_2 t)^{d_2}}{2}\right\} \tag{8.4}$$

where t is the decay time and A is the component ratio.

Figure 8.2 shows typical ^{1}H-NMR FIDs for ultrahigh-molecular-weight polyethylene (UHMW-PE) (solution-grown crystallized (SGC), melt-grown crystallized (MGC), and the nascent powder (POW)) samples. Each FID has a beat (a drop and the subsequent recovery) at about 20–30 μs. The beat component is observed for the samples with high crystallinity [6–12, 22, 23]. The beat arises from the distribution of the dipole interaction with the shape of a bell or a trapezoid, which deviates from the Lorentzian function when the molecular motion is extremely restricted. It has been already recognized that this kind of FID can be characterized by multiplication of Gaussian and sine functions expressed for crystals of small molecules [24, 25].

For the polymeric materials, not only the dipole interactions between the nearest neighbors but also those between the proton pairs with longer distances contribute to the FID profile. Therefore, FID can be fitted by Equation 8.5 [26],

$$I_3(t) = A_3 \exp\left\{-\frac{(k_3 t)^{d_3}}{2}\right\} \frac{\sin b_3 t}{b_3 t} \tag{8.5}$$

where d_3 is the power coefficient for the Weibullian part. Here, d_3 of 2 can give the Gaussian/sine function. The actual FID was represented by the sum of these functions ascribed to different relaxation systems.

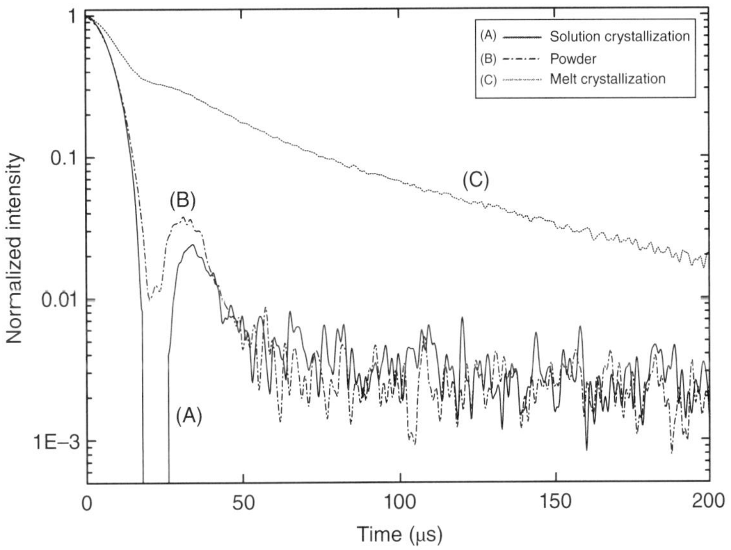

Figure 8.2 ^{1}H FIDs observed at RT for SGC (A), MGC (B), and POW (C) samples.

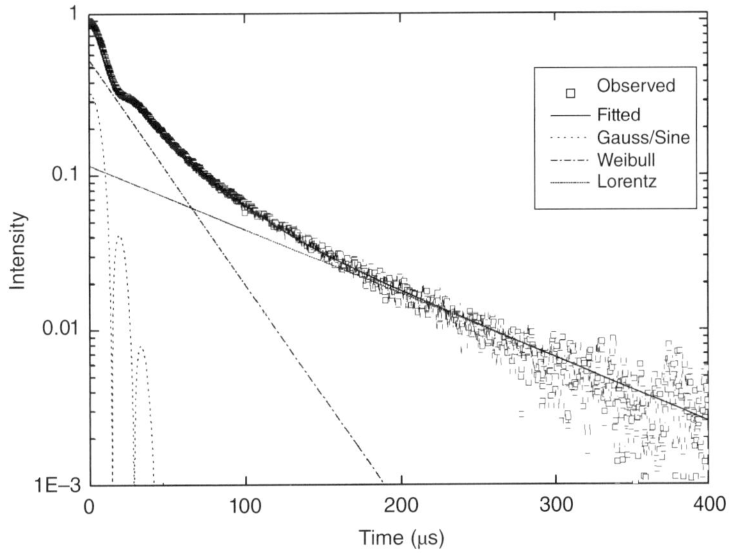

Figure 8.3 Observed FID and fitting with the Lorentzian, Weibullian, and Gauss/Sine functions.

Figure 8.3 illustrates the FID fitted with two amorphous and one crystalline relaxation components for an MGC. The fitting is carried out by the following procedure. First, the longest relaxation was fitted in the time region of 200–400 μs. After subtraction of the first component from FID, the residual plots were further fitted by another amorphous component. The resultant crystalline relaxation is well represented by Equation 8.5. The sum of these curves is in perfect agreement with the observed FID, as shown in Figure 8.3.

The molecular mobility is usually discussed by T_2 on the basis of BPP theory [5]. However, parameters, k_1–k_3 can no longer be directly compared. The mobility should be discussed by using the width of broad-line spectrum. By using these functions and the procedure, FID can be perfectly fitted without ambiguity. The obtained component ratio and integral width or T_2 can be used to discuss the morphology/property relationship directly.

In case of the FIDs with the low *S/N* ratio, the fitting described sometimes produces arbitrariness. To avoid this ambiguity, the explicit value determined from FID should be used. If the maximum value of FID is normalized to 1 and FID is composed of the sum of the exponentials, the area under FID can be obtained by integration as follows:

$$\int_0^\infty M(t)dt = \int_0^\infty \sum_i A_i e^{-\frac{t}{T_{2i}}} dt = \sum_i A_i T_{2i} = MP \quad (8.6)$$

Equation 8.6 is the sum of the amplitude multiplied by T_2 and the increment of the longer components increases the value of Equation 8.6. Here, this integration is defined as MP (the mobility parameter).

As the MP reflects T_2 value and the component ratio directly and the moment of NMR absorption line relates to the integrated value of FID theoretically [27, 28], MP can be a good measure for the whole mobility or softness of the material. Although it is impossible to integrate FID from 0 to infinite, integration of FID to a limited time gives the rough parameter for the mobility of the material as a whole.

8.3 MORPHOLOGY AND RELAXATION BEHAVIOR OF POLYETHYLENE

8.3.1 Morphology and Molecular Mobility

It is well known that the solid morphologies of semicrystalline polymers are composed of several structural levels of typical modifications; a parallel arrangement of folded molecular chains leading to lamellar crystals, and the actual sample is filled with a sandwiched structure of amorphous layers and such lamellae. Melt crystallization produces a spherulite structure having a distortion of these combined crystal/amorphous phases, but precipitation of separated lamellae is obtained for SGC. Figure 8.4 shows the sets of TEM micrographs for SGC, MGC, and POW samples. For SGC (see Fig. 8.4a), the regular stacking of lamellar crystals of ~10-nm thickness is clearly observed. These crystalline lamellae are located between dark layers (the amorphous region) and lie laterally within several micrometer lengths. For the MGC (see Fig. 8.4b), the random arrangement of curved lamellae consists of a spherulite structure. The lamellar thickness of 30 nm for the MGC is much larger than that for the SGC mats. The POW has quite a different morphology, compared to the SGC or MGC. The morphology of POW mainly consists of particles connected by fibrils, which is called a "cobweb" structure [29, 30]. The POW does not have any typical lamellar morphology, but has a domain structure where the crystalline domains distribute within the whole powder globule (Fig. 8.4c). The domain size has a wide range of several tens of nanometers.

FID profiles of these samples at ambient temperature are shown in Figure 8.2. The beat components are clearly observed in the 20–30 μs region on these FIDs, for higher crystalline SGC mats, and the POW. In addition, FIDs for these two samples decay faster than those for MGC, which indicate that the SGC and POW are more rigid than the MGC.

TABLE 8.1 Spin–Spin Relaxation Characteristics of Components Resolved from FID Observed at RT for UHMW-PE Samples[a]

Morphologies	Crystalline Integral Width	Amorphous Integral Width
SGC	66.2 kHz (91%)	41.2 kHz (9%)
MGC	70.4 kHz (61%)	17.4 kHz (28%) + 5.8 kHz (11%)
POW	65.8 kHz (85%)	38.3 kHz (15%)

[a]Value in parentheses represents component ratio for corresponding relaxation.

The relaxation characteristics are summarized in Table 8.1 [31]. For high-crystalline samples of SGC and POW, the two components are required to fit FID. Further, one component is needed for MGC. The crystalline component ratio is highest for the SGC. For the MGC, it is the lowest value of ~50%. The crystalline integral width for the MGC is the largest of these samples, suggesting constrained crystalline chain motions for the MGC that may be caused by the thick lamellar size. Regarding the amorphous relaxation, the integral widths are larger for the high-crystalline samples (SGC and POW), compared to the usual value of 5–10 kHz, which is general for the PE amorphous phase. This means that the amorphous chains in the SGC and POW are restricted. Such a poor molecular mobility of the amorphous phases for the POW has also been observed by ^{13}C-NMR measurement [32]. For the amorphous phase of the MGC, the usual relaxation of ~6 kHz and hindered motion similar to that for the other high-crystalline samples exist. The latter amorphous relaxation could be ascribed to the interfacial molecular motion. Because WAXD and electron microscopy cannot resolve these structural components, the combination with NMR results supplies enough information about the detailed morphology.

8.3.2 Lamellar Thickening by Annealing

Morphological differences affect the temperature dependence of FID. In Figure 8.5 is shown the temperature dependence of the FIDs for the SGC [31]. The beat component is gradually suppressed as the temperature increases, and it almost disappears around 110 °C.

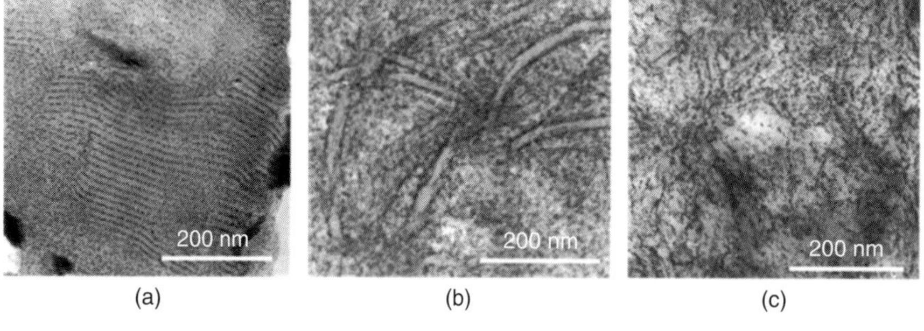

Figure 8.4 Electron micrographs of UHMW-PE prepared under different conditions: (a) SGC, (b) MGC, and (c) POW.

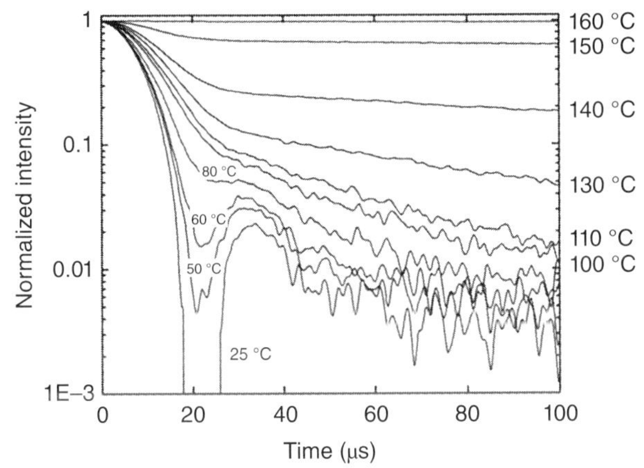

Figure 8.5 Changes in the FIDs during heating for SGC.

The integral width, except the smallest component of the MGC, is plotted in Figure 8.6 for the SGC, MGC, and POW [31]. In Figure 8.6, the larger integral width corresponds to the slower chain motion. For the SGC, the crystalline integral width is held at the same level below 60 °C, followed by a gradual decrease with increasing temperature. This shows that crystalline chain motion starts at a lower temperature, reflecting it is a less entangled lamellar morphology. In contrast, the crystalline chain mobility is accelerated above a higher temperature of 90 °C for the POW, due to a sudden release of the chain motion restricted below this critical temperature. For the poor crystalline MGC with the largest lamellar thickness, the largest integral widths are observed below 90 °C, compared to the other morphologies.

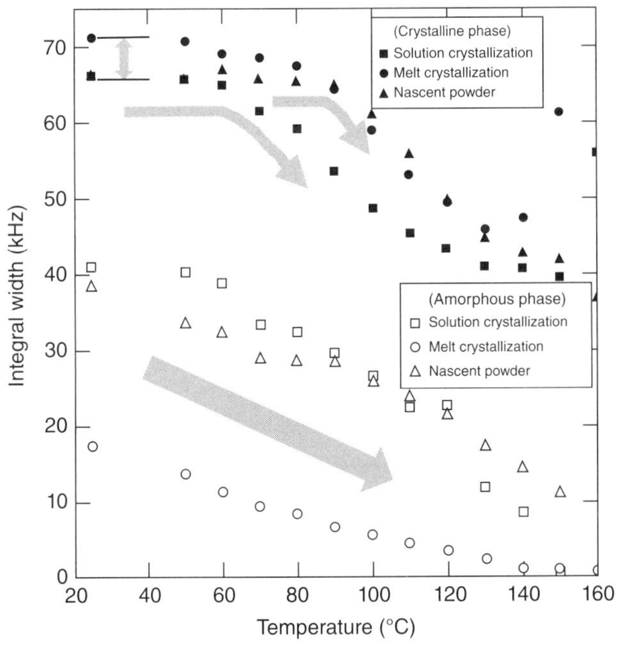

Figure 8.6 Temperature dependencies of integral widths during heating for SGC, MGC, and POW.

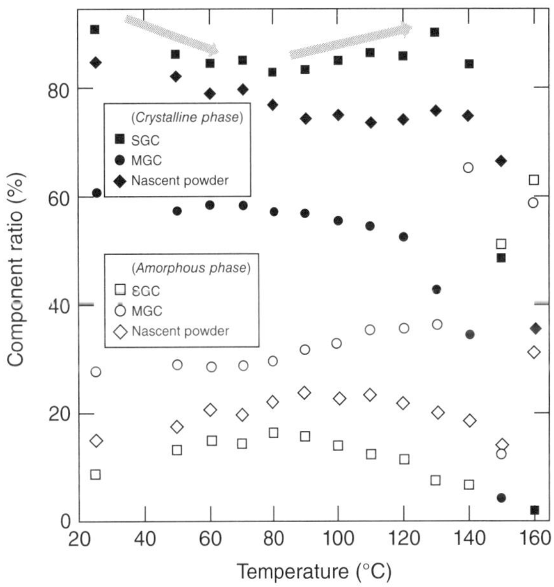

Figure 8.7 Temperature dependencies of component ratios during heating for SGC, MGC, and POW.

Regarding the amorphous relaxation, the integral width of all samples decreases with temperature and the amorphous molecular motion is gradually accelerated during heating, independent of the sample morphology. The amorphous integral widths of highly crystalline SGC and POW are twice as large as those of the MGC.

The component ratios of both crystalline and amorphous phases are plotted in Figure 8.7 [31]. For the MGC, the crystallinity gradually decreases and abruptly drops at about 120 °C, followed by reaching the 0% crystallinity at 150 °C, due to the complete sample melting. The crystallinity decreased at the early stage of heating (25–80 °C) for SGC and POW. Between 90 and 130 °C, the crystallinity increased for the SGC. This recovery of crystallinity for the SGC is attributed to the lamellar thickening. In DSC measurements, no exotherm caused by lamellar thickening has ever been observed, and this has been a long-standing question about the interpretation of the annealing process for semicrystalline polymers. Combination of synchronized evaluations of both chain mobility and component ratio on the same molecular scale, characterized by ^{1}H-NMR relaxation, shows the first and a clear evidence of lamellar thickening.

Changes in the amorphous relaxation behavior with rising temperature exhibited a monotonous reduction in the integral width independent of sample morphology, while a crystalline relaxation change is unique for each sample. Thus, the complete analysis of ^{1}H FID allows us to discuss the chain mobility of the amorphous and crystalline phases independently.

Based on the TEM micrographs and T_2 behaviors, the temperature dependence of SGC morphology can be divided into three regimes. Figure 8.8 shows the broad-line spectra converted from the crystalline decays of SGC, which exhibits the most typical characteristics of the annealing process [31].

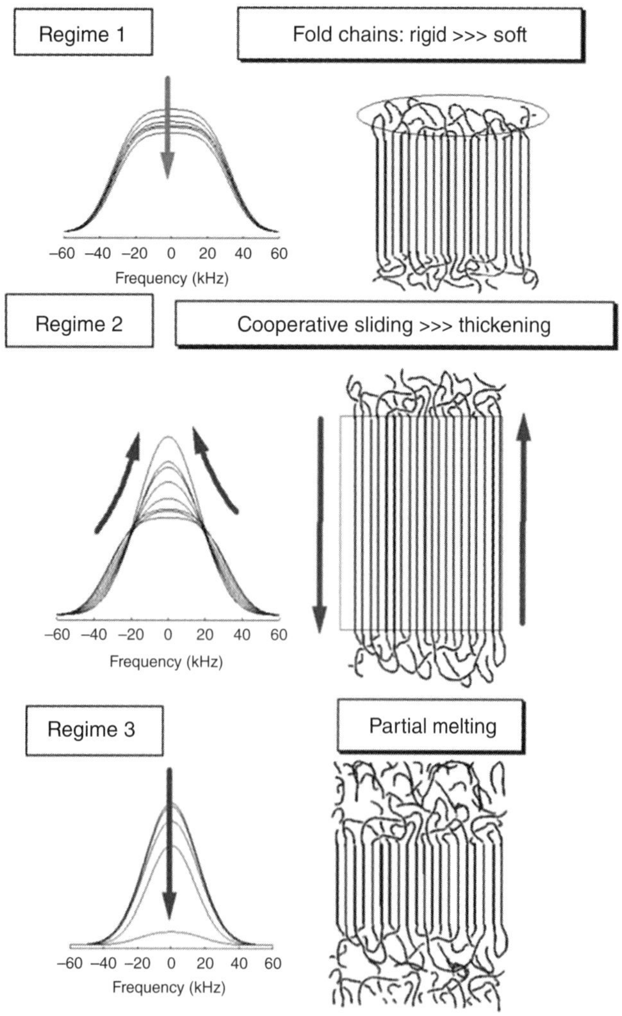

Figure 8.8 Schematic broad-line spectrum changes through three relaxation processes, divided into different temperature regions.

Regime 1: The interfacial chains, like amorphous chains, get the high mobility, which leads to the decrement of the component ratios (crystallinity) below 60 °C. However, the residual crystalline molecules inside the lamellae still maintain their hindered molecular motion; thus, the integral width of the spectrum maintains a constant level.

Regime 2: Above 60 °C, a remarkable crystallinity development occurs, which has not been detected by any other method. Lamellar growth, causing the taking in of the surrounding amorphous chains, is required for the crystallinity increase. Considering the regular stacking of single lamellae several micrometer wide, this erosion into the sandwiched amorphous layer chains progresses parallel to the lamellar normal. Therefore, cooperative molecular motion within the entire crystalline lamella from the surface to the interior takes place and leads to remarkable lamellar thickening.

Regime 3: Over 130 °C, rapid progress of partial melting causes a decreasing of the crystalline component. The residual crystalline region still retains its restricted molecular motion until the complete melting.

The MGC skipped Regime 1, due to the limited relaxation of its entangled molecules on the lamellar surfaces, rather than the adjacent reentry folding for the SGC.

Annealing of the POW passed through Regime 1; and above 90 °C, dynamic molecular motion starts, as defined by Regime 2. This critical temperature is slightly higher than that of SGC. This difference indicates the restricted crystalline chain motion for the domain-network structure crystallized during polymerization. In Regime 2, the crystallinity lay at a constant level for the POW. This shows that the lamellar thickening is limited for the POW.

8.3.3 Entanglement in the Amorphous Phase

Polymer entanglement has been an important concept for the physical properties of polymers. For example, UHMW-PE single-crystal mat can be super-drawn up to a draw ratio of 300 [33]. This high drawability is assumed in the existence of entanglement [34]. The entanglement was evaluated by many researchers [35, 36] and it was indicated that the molecular weight between entanglements in the amorphous region is identical to that in the theta state [37, 38].

Many ^{1}H NMR analyses have been carried out for the molten or solution state of PE to characterize the random coiled molecular chain [17–20, 23, 39–42]. The relaxation evaluation for the amorphous phase of PE in the solid state remains ambiguous because of the strong crystalline component that obscures the entanglement component in the amorphous phase. The perfect FID fitting and combination with TEM and SAXS was applied to evaluate the effects of prior concentration on the amorphous chain characteristics.

In SGC mats precipitated from dilute solutions (Fig. 8.4a), the lamellae are stacked and oriented parallel to the mat surface. The thicknesses for the amorphous and crystalline layers determined by TEM are 9 nm for the crystalline layer and 1.5 nm for the amorphous one independent of the solution concentration. The sum of the thicknesses of the crystalline and amorphous phases coincides with the long periods of SAXS profiles [43]. Although the morphologies seem to be independent of polymer concentration, the peak intensity of SAXS indicates that an increase in electron density in amorphous layers with decreasing prior polymer concentration [44].

Such a difference of amorphous characteristic was explained by NMR relaxation analysis. Figure 8.9 compares the FIDs for a series of SGC mats prepared from different prior polymer concentrations. The negative beating drop observed was folded back to the positive side at around 21 μs between two troughs. A backfolding gradually drops downward with increasing prior polymer concentration. Also, the position

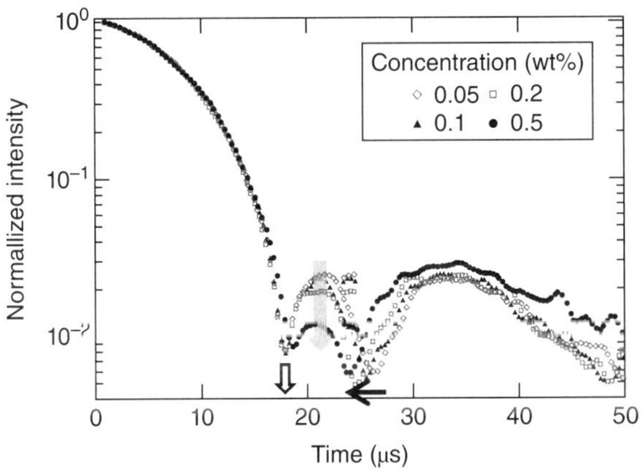

Figure 8.9 A comparison of ^{1}H-NMR FID profiles between different prior polymer concentrations.

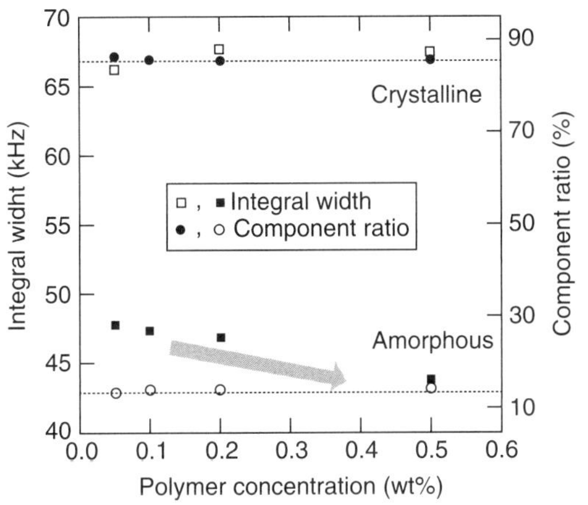

Figure 8.10 Prior polymer concentration dependence of the integral width and the component ratio for SGC.

of one of the troughs around 25 μs shifts to shorter time side with increasing prior polymer concentration. These two characteristics mean the increment of the beat component for the lower prior polymer concentration. In contrast, the position of another trough around 19 μs locates always the same point.

From the fitting of these FIDs, the integral width and component ratios for the relaxation in the crystalline and amorphous phases were plotted in Figure 8.10. The crystallinity lies around 86%, independent of prior polymer concentration, which is well coincident with the WAXD result. Although the integral width for the crystalline component is a constant value around 67 kHz, that for the amorphous decreased gradually with increasing prior polymer concentration. This trend of the amorphous relaxation exhibits the restricted molecular motion in the amorphous phase for a lower prior polymer concentration system.

The positions of troughs are a good measure of the state of the crystalline and amorphous phases. The unchanged position of the first trough around 19 μs in Figure 8.10, which is independent of prior polymer concentration, means the constant level of crystalline relaxation. The position of the second trough near 25 μs is significantly influenced by the slope of amorphous relaxation decay. Shorter relaxation time in the amorphous phase causes rapid decreasing FID, which rather emphasizes the crystalline beating at the lower prior polymer concentration.

8.4 MORPHOLOGY AND STRUCTURE OF THE NASCENT POWDERS

8.4.1 Etching by Fuming Nitric Acid

The POW has quite a different morphology from the SGC or MGC. For the morphology of the POW, the major structure consists of particles having a radius of ~3 μm, and several fibrils between these particles. The origin of these "cobweb" structures, composed of particles and fibrils, is ascribed to the internal expansion stress during polymerization [29, 30]. Much polymer chains synthesized inside the earlier powder globule causes extensive force, which stretches the outer polymer membrane in the polymerization process.

One of the analytical methods for the POW is the application of fuming nitric acid etching. The fuming nitric acid selectively removes the disordered structure. Application of the fuming nitric acid etching can eliminate various structures existing in the POW and reveal the structure of the POW. The application of the etching at an elevated temperature has been reported. Although the etching at elevated temperature accelerates the elimination of the structures, the treatment time is too short to classify the multiple amorphous phases [45, 46]. As the etching at room temperature (RT) requires a longer time to get a fully etched state [47], NMR analysis of a series of the etched samples can reveal the various states of the amorphous phases.

The initial M_w and M_n of POW gradually decrease with increasing etching period at RT [26]. Even for the POW etched for 18 months, certain amorphous components still survived from the TEM observation of the crystalline stem length. From SEM observations, unique "cobweb" structures, composed of globules and fibrils, are observed on the surfaces of the untreated POW [29, 30, 45, 47]. The paste-like morphologies cover the surface of this POW. The major powder structure consists of globules with radii of a few micrometers, independent of etching time. After etching for more than 4 months, the paste-like structure on the powder surface almost disappeared. The grain structures with radii of ~30 nm are recognized clearly by etching for 12 months. From the TEM micrographs of the POW, the crystalline components exhibit the distributed domain morphologies, which correspond well to SEM images.

The ^{1}H-NMR FIDs for a series of etched POW [26] show an increasing beat profile with the etching time, which means the sample crystallinity increases with treatment time. T_2 is plotted in Figure 8.11 as a function of etching time. For the POW etched for 18 months, the amorphous component disappeared. Longer etching completely removes this mobile amorphous

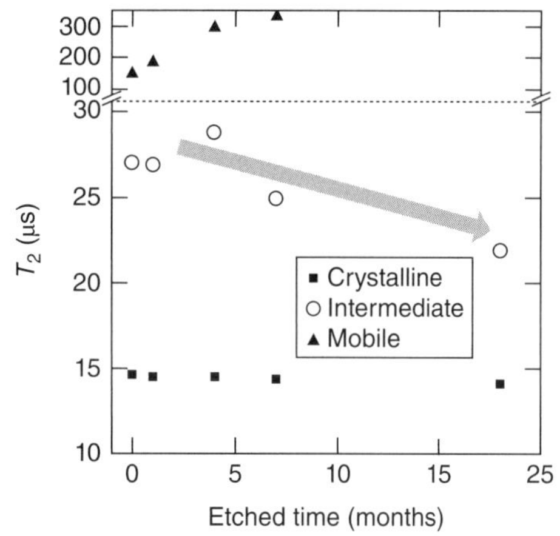

Figure 8.11 Etching time dependencies of T_2 for a series of etched POWs.

component. As the paste-like morphology on the powder surface disappeared in the SEM images, the mobile amorphous component corresponds to the paste-like morphology.

In Figure 8.11, the crystalline T_2 has a constant value around 14.5 µs, independent of etching time. The reason for the constant crystalline T_2 is that the mobility of the crystalline component is not affected by the etching or that the correlation time for the crystalline component is in the extremely slow region where T_2 is insensitive to the chain mobility. In both cases, the mobility of the crystalline component is slow enough to maintain the crystalline structure. This means that the crystalline region remains undamaged during mild etching. In contrast, T_2 of the intermediate amorphous decreases rapidly with etching time. Etching for 18 months finally gives the shortest T_2 of the intermediate amorphous near 22 µs. This phase simultaneously decreased from 17% for the POW to 7% for the 18-month-etched POW. With increasing etching time, the regions located between the crystalline grains were gradually removed. The decreasing T_2 value for this component suggests that the molecular motion of the surviving intermediate phase is restricted. Among the intermediate chains, the region far from the crystalline domains can be destroyed earlier, but more restricted chains that are connected with crystalline domains just at phase boundaries may survive even after prolonged etching.

8.4.2 Structural Change by Annealing

The higher ductility of polyethylene nascent powders suggested the coexistence of less entangled amorphous phases located between the crystalline and amorphous phases [26, 47]. This arises from the nonequilibrium crystallization during polymerization. Therefore, the polymerization temperature affects the structure and the morphology of the nascent powder.

Figure 8.12 presents the TEM images for a series of $T_{poly} = 70\,°C$ (polymerization temperature is 70 °C) powders annealed at an annealing temperature (T_a) of 80, 100, and 120 °C [46]. At $T_a = 80\,°C$ (Fig. 8.12a), the typical lamellar morphology of an MGC or SGC is not observed. Rather, the crystalline domains of 15 nm are distributed throughout the powder. At $T_a > 100\,°C$, the crystalline and amorphous phases are aggregated and the crystalline domain size increases (Fig. 8.12b). Above 120 °C, the lamellar structure appeared (Fig. 8.12c).

From ^{1}H-NMR analysis, the crystallinity decreases and exhibits a minimum value around 60 °C, then increases as the temperature increases. These decrement and subsequent increment in crystallinity coincide with those of the reflection intensities observed for the variable-temperature WAXD profile. Heating produces maximum crystallinity at about 120 °C followed by the abrupt drops in crystallinity, which is attributed to the sample melting. These temperatures of the maximum crystallinity coincide well with the temperatures at which the lamellae appear in the TEM observations. These results indicate that the lamellar formation originates from melt recrystallization during annealing.

In Figure 8.13, the annealing temperature dependences of the integral width for the crystalline, intermediate and amorphous phases for the POW of $T_{poly} = 20$ and 70 °C are shown [26]. The integral widths of the crystalline and amorphous phases show remarkable differences, depending on T_{poly}. For the crystalline phase, the integral width is always lower in the POW of $T_{poly} = 20\,°C$ below 90 °C. The larger crystallite size gives the higher integral width for the UHMW-PE prepared under different crystallization conditions [26]. Because the nascent powder contains a monoclinic form at polymerization temperature below 60 °C, the smaller crystallite size of the monoclinic crystals within the POW of $T_{poly} = 20\,°C$ is attributed to the lower integral width.

In contrast, the integral width of the amorphous phase is always larger for the POW of $T_{poly} = 20\,°C$ than for the POW of $T_{poly} = 70\,°C$ below 90 °C. This means that the molecular mobility of the POW of $T_{poly} = 20\,°C$ is lower, due to the more restricted progress of the structural formation at the lower T_{poly}. The amorphous integral width of the POW of $T_{poly} = 20\,°C$ begins to decrease at 40 °C. The aggregation of the crystalline domains is evident in the TEM observations above 40 °C for the POW of $T_{poly} = 20\,°C$. Similarly, the integral width for the amorphous component of the POW of $T_{poly} = 70\,°C$ starts to decrease above 90 °C, which is 20 °C higher than the polymerization temperature. This synchronization of the morphological and molecular mobility changes suggests that the spatial motion of the amorphous chains induces an increase in the size of the crystalline domains during annealing. Therefore, the constrained molecular motion of the as-polymerized amorphous chains surrounding the crystalline domains is released when the annealing temperature reached 20 °C higher than the polymerization temperature. In other words, the nascent powder was exposed at the temperature 20 °C higher than the polymerization temperature.

In Figure 8.14, from a combination of all the information obtained from TEM, WAXD, and NMR [26], the structural development model during polymerization is illustrated.

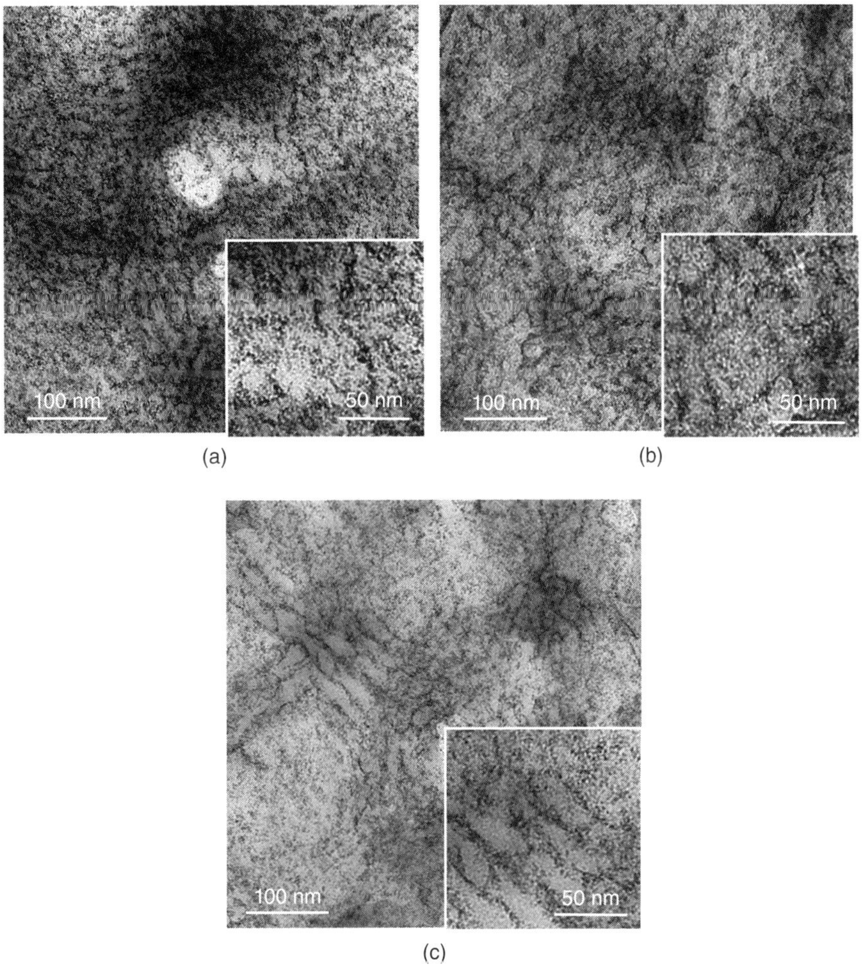

Figure 8.12 TEM micrographs of POW (T_{poly} = 70 °C) annealed at T_as (a) 80 °C, (b) 100 °C and (c) 120 °C. Enlarged images are attached at bottom right.

The results of the morphological changes during annealing demonstrate that the actual polymerization commenced when the temperature was 20 °C higher than the preparation T_{poly} for each of the nascent powders. Rapid crystallization occurs for the lower T_{poly} as soon as the shorter chains are produced at the active sites, and the resultant crystalline domains retain the constraints of the molecular chains. Therefore, the crystalline phase of the lower T_{poly} powder contains a larger amount of the monoclinic form within a smaller domain. The molecular mobility of the amorphous phase is also restricted in the nascent powder prepared at the lower T_{poly}. In contrast, the higher T_{poly} generates the stepwise processes of the initial chain growth and subsequent crystallization of the grown chains. Therefore, larger crystalline domains with the usual orthorhombic form are produced with a less-constrained amorphous phase.

8.4.3 Nascent Isotactic Polypropylene Powder

The morphology of the nascent iPP powder (iPP-POW) depends on the polymerization condition [48, 49]. iPP-POW polymerized at high temperature has a low melting temperature and the enthalpy of fusion [50]. From the WAXS study, it was concluded that iPP-POW polymerized at high temperature consists of a small number of large crystallites dispersed into a great number of smaller ones that are affected by the structural disorder [51]. These results indicate that iPP-POW polymerized at high temperature is primarily composed of a metastable crystallite. If it is metastable, the molecular mobility will be easily activated even at temperatures lower than the phase transition temperature such as the melting point. To clarify this point, CPMAS and ^{1}H pulse NMR was applied to iPP-POW by comparing with the isothermally crystallized iPP pellet (iPP-IC), the melt-quenched iPP pellet (iPP-MQ) and the melt-quenched iPP nascent powder (iPP-PMQ) [52]. iPP-POW has the lowest T_m and crystallinity (33.4%) of all samples (iPP-MQ: 43.8%, iPP-PMQ: 46.7%, iPP-IC: 63.3%). As the crystallinity of iPP-POW is lower than iPP-MQ and iPP-PMQ, this structure cannot be achieved by usual thermal treatment. As pointed out by Ferracini et al. [51], for iPP-POW, the disordered structure is unique and the crystallization does not appreciably proceed in polymerization process of this condition.

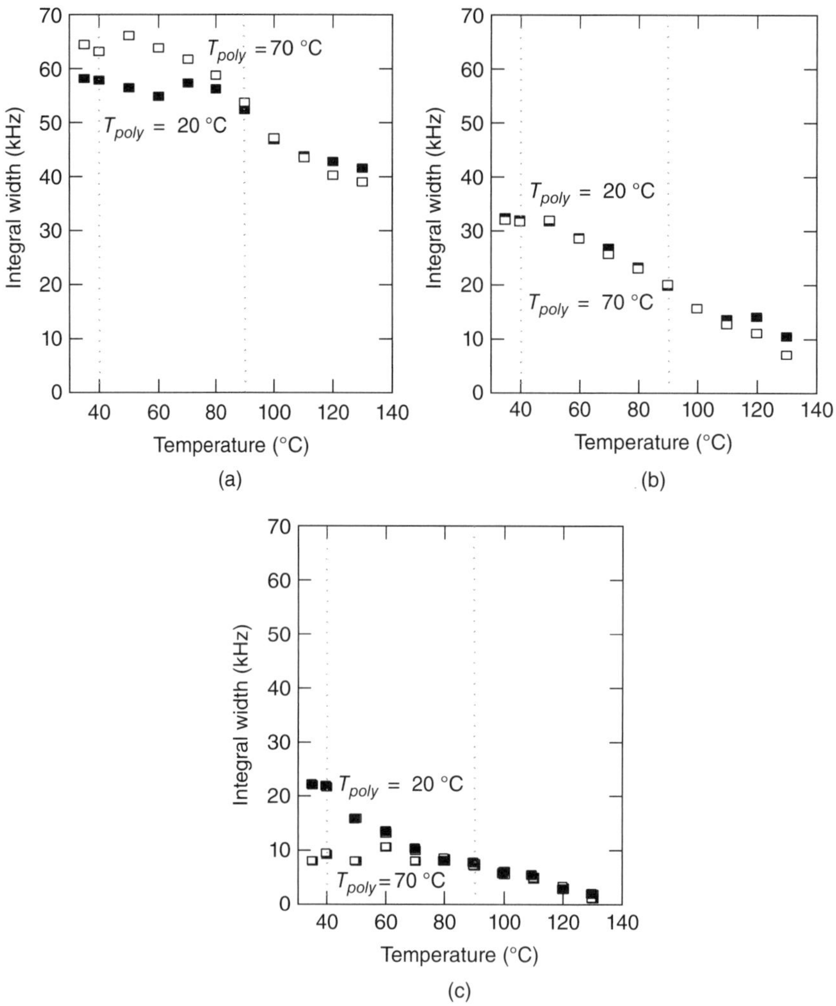

Figure 8.13 Annealing temperature dependence of the integral width for the POW of $T_{poly} = 20\,°C$ and $70\,°C$ (a) crystalline, (b) intermediate, and (c) amorphous components.

Because polypropylene has three chemically inequivalent carbons, the use of ^{13}C NMR supplies the detailed information [53–56]. As the CPMAS NMR spectrum at RT does not show the amorphous peaks of iPP, measurements at high temperature are required to discuss the crystalline and amorphous phases at the same time. Figure 8.15 shows CPMAS spectra of iPP-POW, iPP-PMQ, iPP-MQ, and iPP-IC at 100 °C [52]. At 100 °C, for all samples, the amorphous peaks are recognized at the lower field side of the crystalline peak for CH_2 and CH. It is worth noting that the amorphous peaks of iPP-POW are stronger than others. Because the crystallinity of iPP-POW is the lowest of all samples, the amount of the amorphous component should be highest for iPP-POW. However, the peak area of the amorphous component of iPP-POW is higher than that expected from its crystallinity. The peak area of ^{13}C CPMAS depends on not only the amount but also on the molecular mobility [57–59]. This suggests that the molecular mobility of iPP-POW is different from that of the others. The first possibility is that, in the iPP-POW, the molecular mobility of the amorphous component is higher than that in the others and the second one is that the molecular mobility of crystalline component is higher in the iPP-POW. In order to decide which possibility is suitable in this case, a discussion about the relaxation time is necessary.

In Figure 8.16 is shown the temperature dependence of T_2 for the amorphous and crystalline components. In Figure 8.16a, for all samples, T_2 for the amorphous component increases with temperature commonly for all samples. This corresponds to the similar enlargement of the amorphous peak for CH_2 and CH resonance in the VT-CPMAS spectra, which means that the molecular mobility of amorphous component is activated with similar degree for all samples.

In Figure 8.16b, the temperature dependence of T_2 of the crystalline component for iPP-POW is clearly different from that of the others. The T_2 of the crystalline component for iPP-POW increases remarkably with temperature. On the other hand, crystalline T_2s of the crystalline component for iPP-PMQ, iPP-MQ, and iPP-IC are almost constant and/or slightly increase against temperature. These results indicate that the mobility of the crystalline component of iPP-POW is

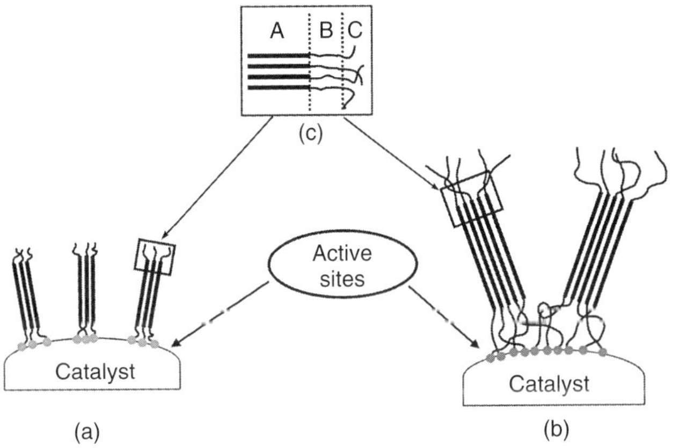

Figure 8.14 Structural development model during polymerization including chain growth and crystallization at $T_{poly} = 20\,°C$ (a) and 70 °C (b). (c) Indicates the intermediate region B sandwiched between crystalline A and amorphous C phases.

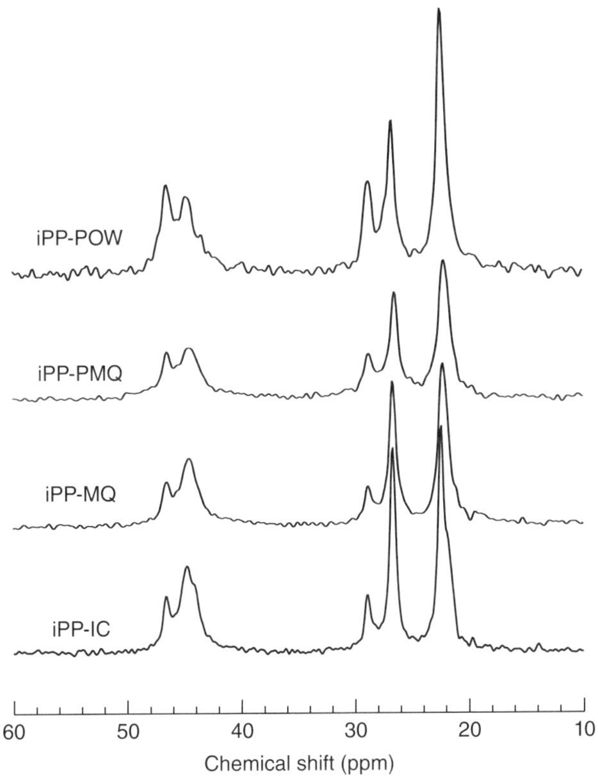

Figure 8.15 ^{13}C CPMAS NMR spectra of iPP-POW, iPP-PMQ, iPP-MQ, and iPP-IC at 100 °C.

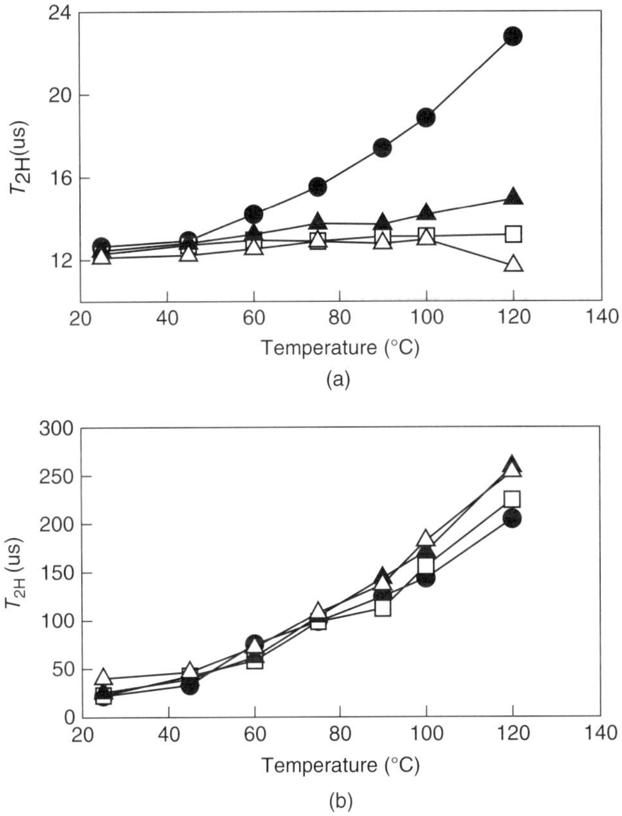

Figure 8.16 Temperature dependence of T_2 for the amorphous (a) and the crystalline (b) components for (•) iPP-POW, (▲) iPP-PMQ, (□) iPP-MQ, and (Δ) iPP-IC.

easily enhanced with temperature compared with the others. This result corresponds to the second possibility.

This difference is reflected in the morphologies of samples. Figure 8.17 shows TEM micrographs of all samples. A typical thick lamellar structure is observed for iPP-IC. For iPP-PMQ and iPP-MQ, thinner lamellae than iPP-IC and a cross-hatched structure are observed. On the other hand, iPP-POW has no lamellar structure, which means the crystalline component is too small to be observed by TEM micrograph. Therefore, it is suggested that the crystalline size is small or the unstable crystalline component contains irregularity. This indicates that there is remarkable difference in the crystalline size between iPP-POW and the others. Considering the result of pulse NMR measurement, iPP-POW has small crystals surrounded by the intermediate and amorphous components.

8.5 KINETICS OF DYNAMIC PROCESS OF POLYMERS

8.5.1 Melt Drawing of Polyethylene

UHMW-PE can be drawn even from the molten state due to its high melt viscosity [60–64]. Transient crystallization into a hexagonal form occurs during oriented crystallization into the orthorhombic form during melt drawing. This behavior is investigated by *in situ* WAXD using synchrotron radiation [65], which revealed that the structural transformation of the amorphous chains induces transient crystallization into the hexagonal form through disentanglement. Precise investigation of the amorphous phase is limited to a few spectroscopies such as infrared (IR) [66–68], Raman [69, 70], and NMR [17, 18, 71–73]. NMR spectroscopy has been applied in *in*

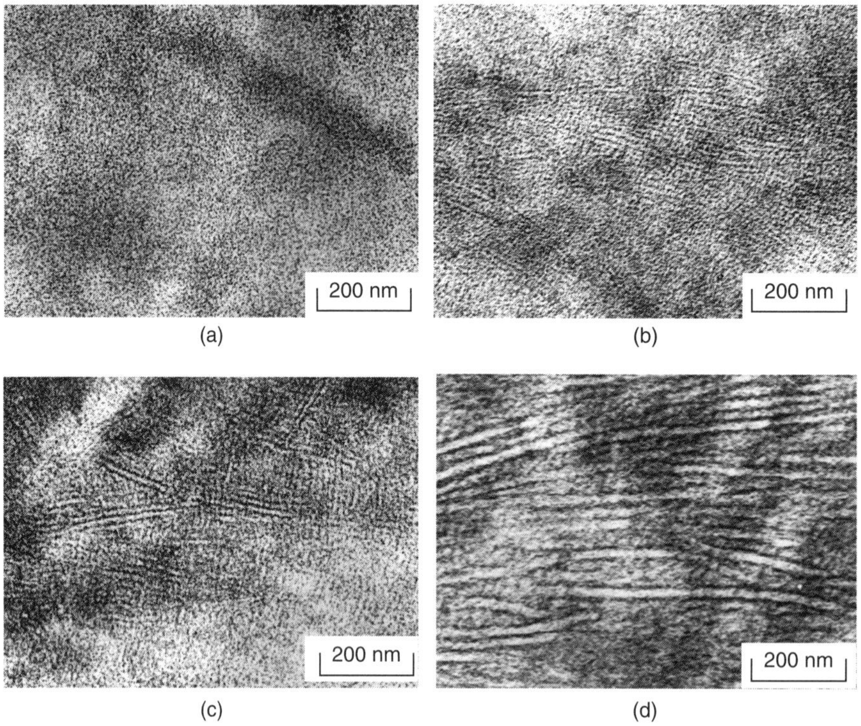

Figure 8.17 TEM micrographs of (a) iPP-POW, (b) iPP-PMQ, (c) iPP-MQ, and (d) iPP-IC.

situ measurements during deformation [74–78]. Gleason et al. investigated the effect of uniaxial deformation near the glass transition temperature on the chain mobility in the amorphous region of deuterated nylon 6 [75]. Rault et al. studied the stress-induced crystallization and melting of cross-linked rubber [78]. Quantitative evaluation of chain entanglement or disentanglement during melt drawing has been analyzed by the improved *in situ* NMR method [79].

An *in situ* NMR measurement for evaluating molecular mobility during melt drawing can be carried out by using uniaxial cross-head extension device. From CPMG measurements of the UHMW-PE films prepared by the metallocene and Ziegler catalysts, three amorphous components were obtained as rigid (shortest T_2), intermediate, and mobile (longest T_2) components. Figure 8.18 illustrates the changes in the component ratio of each component for the metallocene and Ziegler films during melt drawing [79]. A difference in component ratio between the metallocene and Ziegler films is observed for the intermediate and mobile amorphous components (Fig. 8.18a, b). Comparing the initial drawing stage, the mobile amorphous component is predominant for the metallocene film. The structural transformation from intermediate into mobile amorphous components occurs during melt drawing. In particular, this structural transformation is more significant for the Ziegler film. However, the rigid amorphous component hardly changes during melt drawing for both films. It is suggested that extensive disentanglement with chain slippage proceeds at the initial stage of melt drawing for a Ziegler film with broader molecular weight distribution [79–81]. The mobile amorphous component is attributed to the disentangled chains transformed from the prior intermediate amorphous component, because this component significantly increases at the same stage for the Ziegler film. In contrast, the intermediate amorphous component with less mobility than the mobile amorphous component can be construed as networked amorphous chains connected by entanglements and distributed homogeneously over the whole chain. The rigid amorphous component has much lower chain mobility, that is, shorter T_2. Thus, this component consists of entanglements tightly confining each other.

Models for the structural transformation of the amorphous chains are illustrated in Figure 8.19 based on the *in situ* NMR measurements. The rigid amorphous component with T_2 ~1 ms can be defined as amorphous chains with tight entanglements that transfer the drawing stress. The intermediate amorphous component with T_2 ~5 ms can be defined as amorphous chains with loose entanglements, and the mobile amorphous component with T_2 ~15 ms can be defined as disentangled amorphous chains. Significant structural transformation from intermediate into mobile amorphous components occurs during melt drawing for the Ziegler film. Consequently, the intermediate amorphous component for the Ziegler film has looser entanglements (Fig. 8.19a) than that for the metallocene film (Fig. 8.19b). Thus, the disentanglement takes places more smoothly during melt drawing for the Ziegler film (Fig. 8.19a). Melt drawing is also applied to the ultrahigh molecular weight polypropylene and concluded that disentanglement and oriented crystallization depends on the drawing temperature [82].

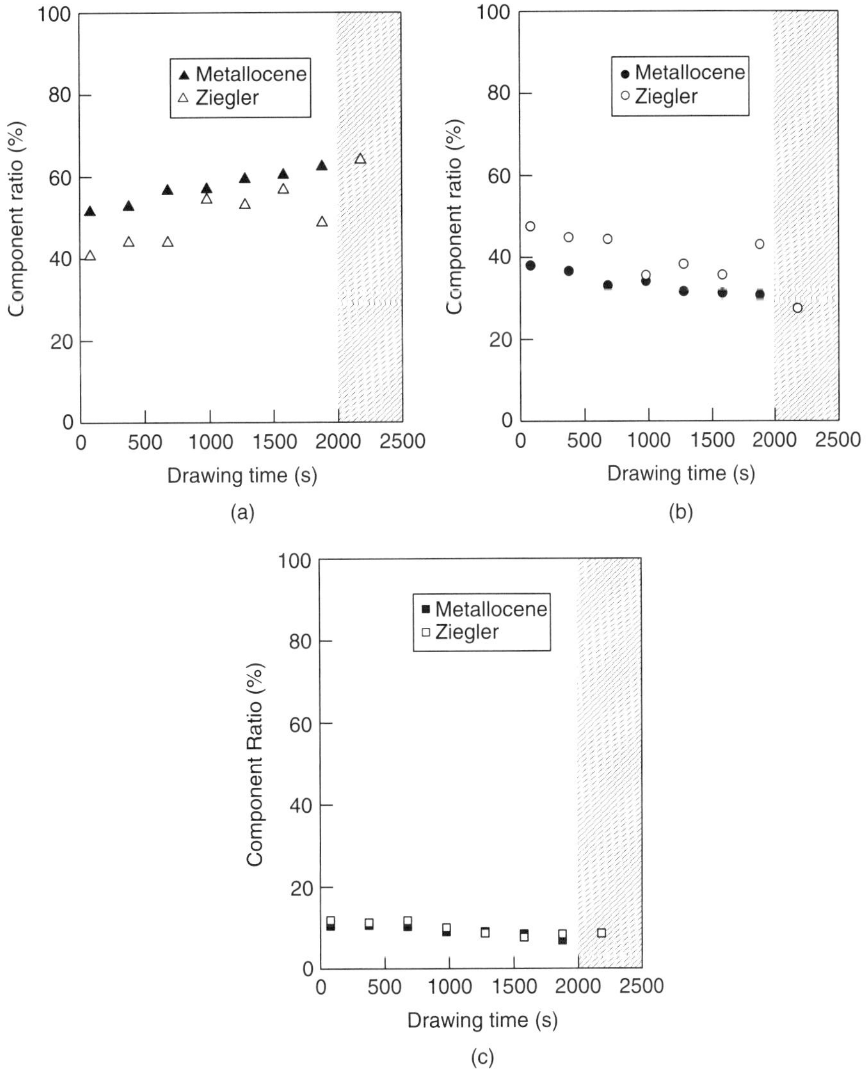

Figure 8.18 Drawing time dependence of the in-component ratio for metallocene (filled symbols) and Ziegler films (open symbols): (a) mobile, (b) intermediate, and (c) rigid amorphous components. The shaded area indicates the plateau stress region.

8.5.2 Crystallization Mechanism of Nylon 46

For the processing of polymeric materials, thermal behavior of structure and the crystallization mechanism are the important factors for the final physical properties. For nylon 46, the combination NMR and TEM was applied to investigate the crystallization mechanism [83].

In Figure 8.20 is shown the TEM micrographs of the SGC and the MGC for nylon 46 [83]. For SGC, the regular stacking of the lamellar crystals is clearly observed (Fig. 8.20a). The lamellar thickness is uniform compared to that of the MGC. For the MGC0m (where the number and alphabet after MGC means crystallization time), the stacking of the lamellar crystal is not clear and the random arrangement of the curved lamellae with a short length is observed (Fig. 8.20b). As the crystallization time increases, many long lamellar crystals are arranged parallel and the lamellar stacking grows (Fig. 8.20c). A closer look at the micrographs revealed the change in the lamellar thickness, which is the sum of the thickness of the crystalline and amorphous phases. The lamellar thickness is independent of the crystallization time for nylon 46. On the other hand, the thicknesses of the crystalline and amorphous phases increase and decrease, respectively [83], and become almost constant after crystallization for 6 h.

As the crystallization proceeds, the molecular mobility is also affected. Figure 8.21a and b shows the crystallization time dependences of T_1 for the crystalline and amorphous phases, respectively. In Figure 8.21a, the T_1s of MGC0m are less than 35 s. As the crystallization time increases, the T_1s of all the carbons rapidly increase and then become almost constant after 6 h. The T_1s of all the carbons for MGC24h become about four times greater than those for MGC0m. Because T_1 decreases with the activation in the molecular mobility based on the BPP theory for the solid polymers, the molecular mobility is restricted by the crystallization [5].

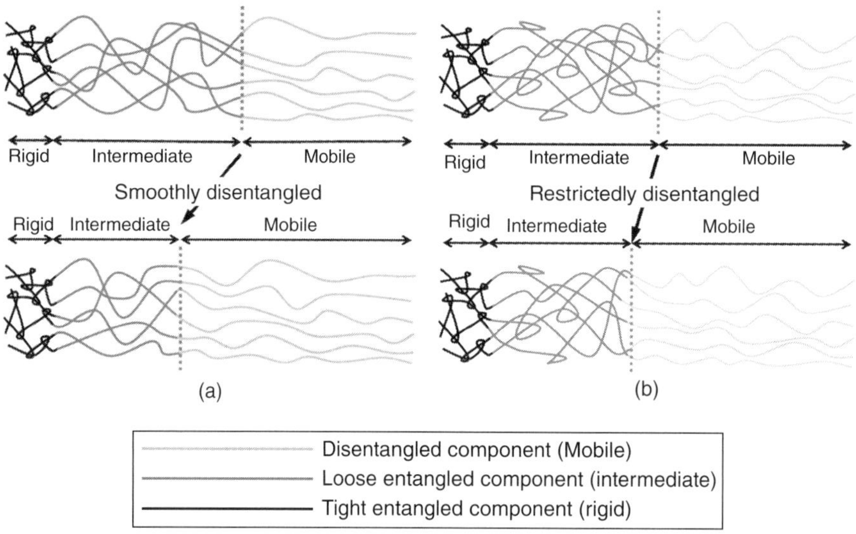

Figure 8.19 Models for structural transformation of amorphous chains for (a) Ziegler and (b) metallocene films. Top and bottom models represent the state before and after drawing, respectively. The vertical dotted lines indicate the boundary between intermediate and mobile amorphous components. Chain shape and length for each component in each model reflect T_2 and the component ratio estimated from *in situ* NMR measurement. There is a small disentangled region for metallocene film (b) with lower chain mobility because disentanglement of the intermediate amorphous component is "restricted." In contrast, the corresponding region is larger for the Ziegler film (a) with higher chain mobility due to "smooth" disentanglement of the intermediate amorphous component.

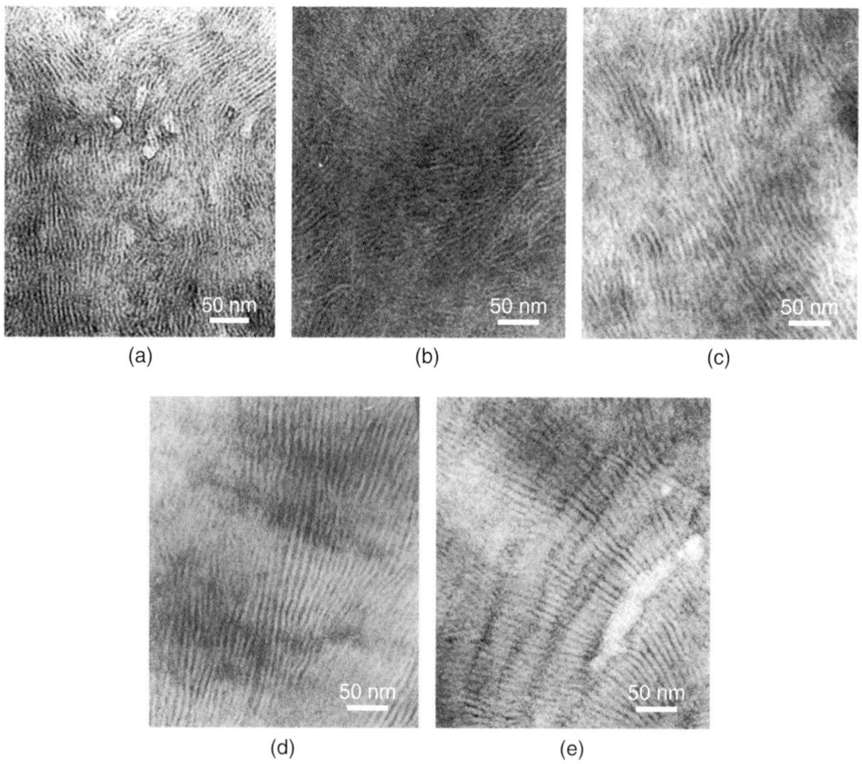

Figure 8.20 TEM micrographs of SGC and crystallized samples. (a) SGC, (b) MGC0m, (c) MGC1m, (d) MGC1h, and (e) MGC24h.

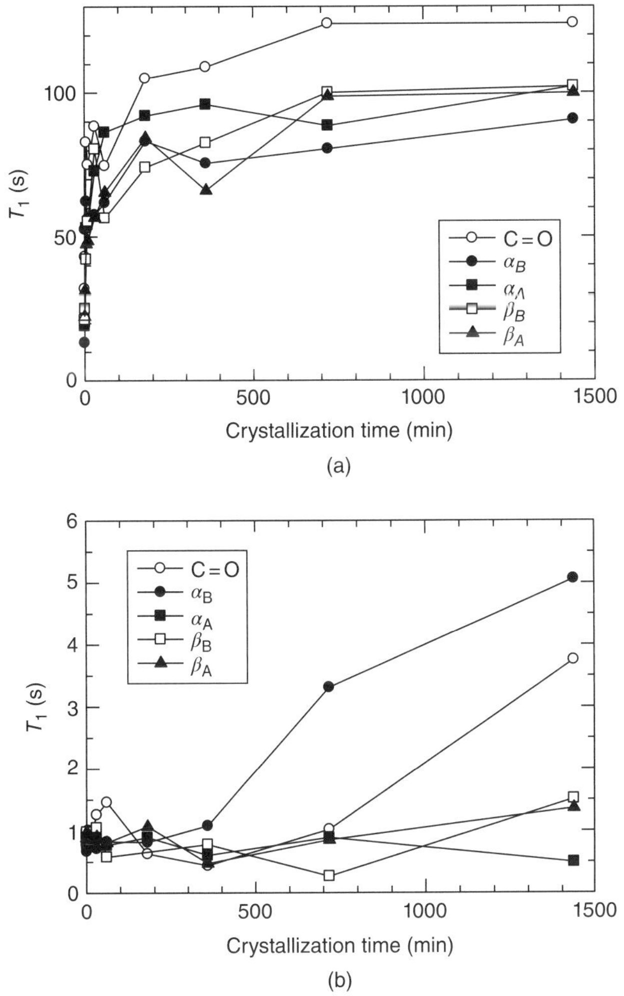

Figure 8.21 Crystallization time dependence of T_1 for nylon 46. (a) the crystalline phase, (b) the amorphous phase.

The crystallization time dependence of the T_1s is similar to the behavior of crystallinity. This means that the molecular motion of the crystalline phase is restricted as soon as the stable crystalline structure is formed. In other words, after the size of the crystallite reached a certain size, the molecular mobility of the crystalline phase is fixed. In Figure 8.21b, the crystallization time dependence of T_1 for the amorphous phase is shown. The T_1s for the amorphous phase are less than 5 s, which is much shorter than those for the crystalline phase, and means that the molecular mobility is high compared with that for the crystalline phase. The T_1s for all the carbons in the amorphous phase are about 1 s or less and do not change very much until 6 h. After crystallization for 6 h, the T_1s of α_B (CH_2 directly bonded to nitrogen) and CO increased. This means that the molecular motion of the amorphous phase is restricted after the T_1s for the crystalline phase became constant. The mobility of the amorphous phase is affected by the surrounding crystalline phase. From the chemical shift differences, the gauche conformation exists and the distribution of the structure is large in the amorphous phase, although the all trans-zigzag conformation is concluded in the crystalline phase.

Based on high-resolution solid-state NMR and TEM observations, the mechanism for the crystallization of nylon 46 is summarized in Figure 8.22 [83]. When the temperature of the melt sample decreased to the crystallization temperature, the hydrogen bond is formed with the nearby chains. At this stage, the size of the crystalline phase is small and the crystal structure is not energetically stable. The amorphous phase is mobile and the conformation of the amorphous phase contains the gauche conformation in the internal CH_2. As the crystallization proceeds, the crystal structure becomes stable and forms the hydrogen bond sheet structure. At the same time, the crystalline phase grows by the conformational change in the amorphous phase. As the sliding of the main chain does not occur, the thickening of the lamellae does not take place. The thickness of the crystalline phase increases with the decrease in the amorphous phase thickness. During the final stage of the crystallization, the mobility of the amorphous phase is restricted.

8.5.3 Degree of Curing of Novolac Resins

The physical properties of thermosetting resins such as novolac resin depend on the degree of curing because the thermosetting resins are amorphous and have the cross-linked network structure [84]. For instance, if the degree of curing is insufficient, the thermal stability decreases. Although the curing mechanism of the novolac resin was investigated by high-resolution solid-state NMR as a point of the reaction mechanisms [85–88], the estimation method of the degree of curing has not been reported. ^1H pulse NMR is suitable for this purpose from the practical point of view [31].

At ambient temperature, the solid novolac is the self-associated aggregates by hydrogen bonding network [89]. The curing of novolac starts by the destruction of the hydrogen bond network followed by the formation of HMTA (the cross-linking agent)-novolac adducts. At curing temperature, curing and hardening occur with the elimination of ammonia. This curing behavior is explained with the change in the molecular mobility analyzed by the fitting of FID.

Paying attention to only the degree of curing sometimes leads to ambiguity in understanding the resolution into several components. For the purpose, MP is the suitable parameter for measuring the degree of curing.

Figure 8.23 shows temperature dependences of MP for the orthorich novolac (o-novolac) and random novolac (r-novolac) with and without cross-linking agent HMTA. MP is calculated by the numerical integration of FID from 0 to 1000 μs. From this figure, between 30 and 60 °C, MP values for four samples are almost same. The difference of MPs for novolacs with and without HMTA comes into focus in the temperature between 70 and 100 °C. In this temperature range, the HMTA-novolac adduct is formed. MP for o- and r-novolac without the cross-linking agent abruptly increased with increasing temperature above 100 °C. On the other hand, MP for both novolacs with cross-linking agent gradually

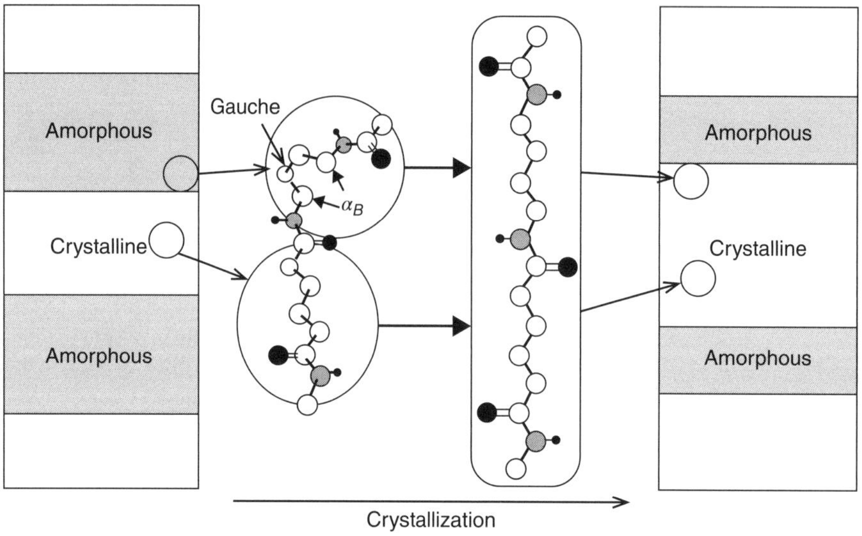

Figure 8.22 Crystallization mechanism of nylon 46.

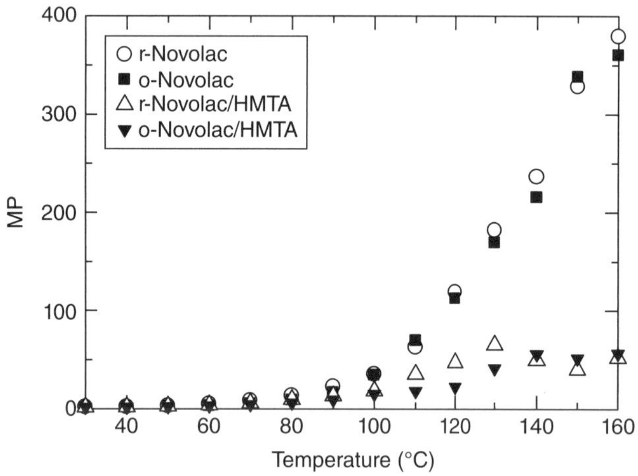

Figure 8.23 Temperature dependence of MP for o- and r-novolacs with and without HMTA.

increased with increasing temperature and reached a plateau above 140 °C. In the plateau region, novolacs finished the curing reaction. This result clearly indicates that the low MP values in the high-temperature region are accounted for by the high content of the rigid component. As MP can be calculated from FID directly, this parameter can be used for *in situ* detection of the degree of curing.

8.6 CONCLUSIONS

Practical application of NMR and other techniques such as microscopy and diffraction that supply information about the large-scale structure of the investigation into polymer morphology, structure, and physical properties are introduced. NMR has been thought to be an insensitive spectroscopy. Recent development of hardware improved the stability, sensitivity, and resolution of NMR, especially, the solid-state NMR. Based on the principle, the information about the morphology, the large-scale structure, is averaged as the bulk material in the NMR signal. However, the difference in the morphology or structure is accumulated in the signal. Careful and perfect fit or analysis of the NMR signal and the combination with other techniques that complement NMR disadvantages certainly offers the nature of the polymer.

REFERENCES

1. Ando I, Kobayashi M, Kanekiyo M, Kuroki S, Ando S, Matsukawa S, Kurosu H, Yasunaga H, Amiya S. NMR spectroscopy in polymer science. In: Tanaka T, editor. *Experimental Methods in Polymer Science*. Academic Press; 2000. p 261.
2. Ando I, Asakura T. *Solid State NMR of Polymers*. Elsevier; 1998.
3. Schmidt-Rohr K, Spiess HW. *Multidimensional Solid-State NMR and Polymers*. Academic Press; 1994.
4. Ando I, Webb GA. *Theory of NMR Parameters*. Academic Press; 1983.
5. Bloembergen N, Purcell EM, Pound RV. Relaxation Effects in Nuclear Magnetic Resonance Absorption. Phys Rev 1948;73:679.
6. Klüver W, Ruland W. NMR-studies of semicrystalline polymers using pulse techniques. Prog Colloid Polym Sci 1978;64:255.
7. Bergmann K. Determination of polymer crystallinity from proton solid-echo NMR measurements. Polym Bull 1981;5:355.
8. Bergmann K, Schmiedberger H, Unterforsthuber K. Meßverfahren zur Untersuchung der Protonen-Spin-Spin-Relaxation ein und zweiphasiger Polymerer in einem Zeitbereich von fünf Dekaden. Colloid Polym Sci 1984;262:283.
9. Dadayli D, Harris RK, Kenwright AM, Say BJ, Sunnetcioglu MM. Solid-state ^{1}H n.m.r. studies of polypropylene. Polymer 1994;35:4083.
10. Hansen EW, Kristiansen PE, Pedersen B. Crystallinity of Polyethylene Derived from Solid-State Proton NMR Free Induction Decay. J Phys Chem B 1998;102:5444.

11. Kristiansen PE, Hansen EW, Pedersen B. Phase Distribution during Isothermal Crystallization of Polyethylene Probed by Solid-State Proton NMR Free Induction Decay. J Phys Chem B 1999;103:3552.
12. Kristiansen PE, Hansen EW, Perdersen B. Phase distribution in polyethylene versus temperature probed by solid-state proton NMR free induction decay. Polymer 2000;41:311.
13. Loboda-Čačković VJ, Hosemann R, Wilke W, Kolloid ZZ. Kernresonanzuntersuchungen an verstrecktem und getempertem linearem Polyäthylen. Polymere 1969;235:1253.
14. Phaovibul O, Loboda-Čačković J, Čačković H, Hosemann R. Chain conformation in linear polyethylene and paraffins defined by four NMR-components. Makromol Chem 1974;175:2991.
15. Kitamaru R, Horii F, Hyon S–H. Proton magnetic resonance studies of the phase structure of bulk-crystallized linear polyethylene. J Polym Sci B 1977;15:821.
16. Bergmann K. Study of the molecular motions of polyethylene by line-shape analysis of broad-line proton NMR spectra. J Polym Sci B 1978;16:1611.
17. Folland R, Charlesby A. Entanglement effects on the NMR spin-spin relaxation in polyethylene melts. J Polym Sci C 1978;16:339.
18. Folland R, Charlesby A. Effect of previous history on the NMR relaxation of polyethylene. Eur Polym J 1979;15:953.
19. Koch H, Bachus R, Kimmich R. Molecular fluctuations in polyethylene melts. Dependence of the longitudinal and transverse proton relaxation on the chain length. Polymer 1980;21:1009.
20. Brereton MG, Ward IM, Boden N, Wright P. Nature of the proton NMR transverse relaxation function of polyethylene melts. 1. Monodispersed polyethylenes. Macromolecules 1991;24:2068.
21. Kenwright AM, Say BI. Solid-state proton NMR. In: Ibbett RN, editor. *NMR Spectroscopy of Polymers*. London: Chapman and Hall; 1993.
22. Look DC, Lowe IJ, Norhtby JA. Nuclear Magnetic Resonance Study of Molecular Motions in Solid Hydrogen Sulfide. J Chem Phys 1966;44:3441.
23. Cohen-Addan JP, Feio G, Peguy A. High-molecular-weight polyethylene: an NMR approach to partial crystallinity and swelling properties of thermoreversible gels. Polym Commun 1987;28:252.
24. Abragam A. *Principles of Nuclear Magnetism*. Oxford: Claredon; 1961.
25. Pake GE. Nuclear Resonance Absorption in Hydrated Crystals: Fine Structure of the Proton Line. J Chem Phys 1948;16:327.
26. Uehara H, Aoike T, Yamanobe T, Komoto T. Solid-State 1H NMR Relaxation Analysis of Ultrahigh Molecular Weight Polyethylene Reactor Powder. Macromolecules 2002;35:2640.
27. Dianoux AJ, Sykora S, Gutowsky HS. Moments of NMR Absorption Lines from the Free Induction Decay or Echo of Solids. J Chem Phys 1971;55:4768.
28. Parker GW. Calculation of moments of NMR absorption lines from free – induction decay curves. J Chem Phys 1973;58:3274.
29. Graff RJL, Kortleve G, Vonk CG. On the size of the primary particles in ziegler catalysts. J Polym Sci C 1970;8:735.
30. Nooijen GAH. On the importance of diffusion of cocatalyst molecules through heterogeneous ziegler/natta catalysts. Eur Polym J 1994;30:11.
31. Uehara H, Yamanobe T, Komoto T. Relationship between solid-state molecular motion and morphology for ultrahigh molecular weight polyethylene crystallized under different conditions. Macromolecules 2000;33:4861.
32. Morin FG, Delmas G, Gilson DFR. Variable-Temperature Solid-State ^{13}C NMR Studies of Nascent and Melt-Crystallized Polyethylene. Macromolecules 1995;28:3248.
33. Kanamoto T, Tsuruta A, Tanaka K, Takeda M, Porter RS. Super-drawing of ultrahigh molecular weight polyethylene. 1. Effect of techniques on drawing of single crystal mats. Macromolecules 1988;21:470.
34. Smith P, Lemstra PJ, Booij HC. Ultradrawing of high-molecular-weight polyethylene cast from solution. II. Influence of initial polymer concentration. J Polym Sci B 1981;19:877.
35. Porter RS, Johnson JF. Viscosity of polyethylenes: Dependence on molecular weight and temperature. J Appl Polym Sci 1960;3:194.
36. Porter RS, Johnson JF. The Entanglement Concept in Polymer Systems. Chem Rev 1966;66:1.
37. Higgins JS, Stein RS. Recent developments in polymer applications of small-angle neutron, X-ray and light scattering. J Appl Crystallogr 1978;11:346.
38. Hadziioannou G, Wang LH, Stein RS, Porter RS. Small-angle neutron scattering studies on amorphous polystyrene oriented by solid-state coextrusion. Macromolecules 1982;15:880.
39. Voigt G, Kimmich R. Chain fluctuations in the amorphous regions of polyethylene as indicated in proton relaxation spectroscopy. Polymer 1980;21:1001.
40. Kimmich R, Koch H. Three component model of molecular dynamics in polyethylene melts – a nuclear magnetic relaxation study. Colloid Polym Sci 1980;258:261.
41. Kamel I, Charlesby A. NMR spin–spin relaxation in solid and molten polyethylene structures . J Polym Sci B 1981;19:803.
42. Cohen-Addan JP, Dupeyre R. Strongly entangled polymer chains in a melt. Description of n.m.r. properties associated with a submolecule model. Polymer 1983;24:400.
43. Uehara H, Matsuda H, Aoike T, Yamanobe T, Komoto T. Lamellar characteristics controlled by prior polymer concentration for solution-crystallized ultra-high molecular weight polyethylene. Polymer 2001;42:5893.
44. Spells SJ, Hill MJ. Morphological changes on annealing polyethylene single crystals. Polymer 1991;32:2716.
45. Cook JTE, Klein PG, Ward IM, Brain AA, Farrar DF, Rose J. The morphology of nascent and moulded ultra-high molecular weight polyethylene. Insights from solid-state NMR, nitric acid etching, GPC and DSC. Polymer 2000;41:8615.
46. Weeks NE, Mori S, Porter RS. The morphology of ultradrawn polyethylene. I. Nitric acid etching plus gel permeation chromatography. J Polym Sci B 1975;13:2031.
47. Uehara H, Nakae M, Kanamoto T, Ohtsu O, Sano A, Matsuura K. Structural characterization of ultrahigh-molecular-weight polyethylene reactor powders based on fuming nitric acid etching. Polymer 1998;39:6127.
48. Wristers J. Nascent polypropylene morphology: Polymer fiber. J Polym Sci B 1973;11:1601.
49. Kakugo M, Sadatoshi H, Sakai J, Yokoyama M. Growth of polypropylene particles in heterogeneous Ziegler-Natta polymerization. Macromolecules 1989;22:3172.

50. Loos J, Arndt-Rosenau M, Weingarten U, Kaminsky W, Lemstra PJ. Melting behavior of nascent polyolefins synthesized at various polymerization conditions. Polym Bull 2002;48: 191.
51. Ferracini E, Ferrero A, Malta V, Martelli S, Vogel W. Advanced x-ray study on crystalline and amorphous components of as-polymerized isotactic polypropylene. Colloid Polym Sci 1991;269:1241.
52. Yamanobe T, Ueki K, Komoto T, Uehara H. Structure and molecular mobility of nascent isotactic polypropylene powders. Polym J 2012;44:757.
53. Bunn A, Cudby MEA, Harris RK, Packer KJ, Say BJ. Solid-state high-resolution ^{13}C n.m.r. spectra of polypropene. J Chem Soc Chem Commun 1981:15.
54. Bunn A, Cudby MEA, Harris RK, Packer KJ, Say BJ. 54. High resolution 13C n.m.r. spectra of solid isotactic polypropylene. Polymer 1982;23:694.
55. VanderHart DL, Alamo RG, Nyden MR, Kim M-H, Mandelkern L. Observation of Resonances Associated with Stereo and Regio Defects in the Crystalline Regions of Isotactic Polypropylene: Toward a Determination of Morphological Partitioning. Macromolecules 2000;33:6078.
56. Miyoshi T, Mamun A, Hu W. Molecular Ordering and Molecular Dynamics in Isotactic-Polypropylene Characterized by Solid State NMR. J Phys Chem B 2010;114:92.
57. Lyerla JR, Yannoni CS. High-Resolution Carbon-13 NMR of Polymers in the Solid State. IBM J Res Dev 1983;27:302.
58. Takegoshi K, Hikichi K. Effects of blending on local chain dynamics and glass transition: Polystyrene/poly(vinyl methyl ether) blends as studied by high–resolution solid–state 1 3C nuclear magnetic resonance spectroscopy. J Chem Phys 1991;94:3200.
59. Naito Y, Komoto T, Yamanobe T. Molecular motion of the main chain for a series of Poly(alkyl L-glutamate)s as studied by ^{2}H NMR. J Mol Struct 2002;602–603:437.
60. Bashir Z, Keller A. Melt drawing as a route to high performance polyethylene. Colloid Polym Sci 1989;267:116.
61. Uehara H, Nakae M, Kanamoto T, Zachariades AE, Porter RS. Melt Drawability of Ultrahigh Molecular Weight Polyethylene. Macromolecules 1999;32:2761.
62. Nakae M, Uehara H, Kanamoto T, Ohama T, Porter RS. Melt drawing of ultra-high molecular weight polyethylene: Comparison of Ziegler- and metallocene-catalyzed reactor powders. J Polym Sci B 1999;37:1921.
63. Nakae M, Uehara H, Kanamoto T, Zachariades AE, Porter RS. Structure Development upon Melt Drawing of Ultrahigh Molecular Weight Polyethylene: Effect of Prior Thermal History. Macromolecules 2000;33:2632.
64. Kakiage M, Sekiya M, Yamanobe T, Komoto T, Sasaki S, Murakami S, Uehara H. In situ SAXS analysis of extended-chain crystallization during melt-drawing of ultra-high molecular weight polyethylene. Polymer 2007;48:7385.
65. Uehara H, Kakiage M, Yamanobe T, Komoto T, Murakami S. Phase development mechanism during drawing from highly entangled polyethylene melts. Macromol Rapid Commun 2006;27:966.
66. Sasaki S, Tashiro K, Kobayashi M, Izumi Y, Kobayashi K. Microscopically viewed structural change of PE during the isothermal crystallization from the melt: II. Conformational ordering and lamellar formation mechanism derived from the coupled interpretation of time-resolved SAXS and FTIR data. Polymer 1999;40:7125.
67. Zhang J, Duan Y, Sato H, Tsuji H, Noda I, Yan S, Ozaki Y. Crystal Modifications and Thermal Behavior of Poly(l-lactic acid) Revealed by Infrared Spectroscopy. Macromolecules 2005;38:8012.
68. Watanabe S, Dybal J, Tashiro K, Ozaki Y. A near-infrared study of thermally induced structural changes in polyethylene crystal. Polymer 2006;47:2010.
69. Kurelec L, Rastogi S, Meier RJ, Lemstra PJ. Chain Mobility in Polymer Systems: On the Borderline between Solid and Melt. 3. Phase Transformations in Nascent Ultrahigh Molecular Weight Polyethylene Reactor Powder at Elevated Pressure As Revealed by in Situ Raman Spectroscopy. Macromolecules 2000;33:5593.
70. Brambilla L, Zerbi G. Local Order in Liquid n-Alkanes: Evidence from Raman Spectroscopic Study. Macromolecules 2005;38:3327.
71. Ito M, Kanamoto T, Tanaka K, Porter RS. Pulsed NMR studies of crystalline-state extrusion of high-density polyethylene. Macromolecules 1981;14:1779.
72. Tanaka H, Nishi T. Study of crystallization process of polymer from melt by a real–time pulsed NMR measurement. J Chem Phys 1986;85:6197.
73. Takenaka M, Yamanobe T, Komoto T, Ando I, Sato H, Sato K. Proton NMR relaxation and molecular motion of long-chain cyclic paraffins in the solid state. J Polym Sci B 1987;25:2165.
74. Loo LS, Cohen RE, Gleason KK. Deuterium Nuclear Magnetic Resonance of Phenol-d5 in Nylon 6 under Active Uniaxial Deformation. Macromolecules 1999; 32:4359.
75. Loo LS, Cohen RE, Gleason KK. Chain Mobility in the Amorphous Region of Nylon 6 Observed under Active Uniaxial Deformation. Science 2000;288:116.
76. Hedden RC, Tachibana H, Duncan TM, Cohen C. Effects of Molecular Structure on Segment Orientation in Poly(diethylsiloxane) Elastomers. 2. NMR Measurements from Uniaxially Stretched Samples. Macromolecules 2001;34:5540.
77. Kameda T, Asakura T. Structure and dynamics in the amorphous region of natural rubber observed under uniaxial deformation monitored with solid-state ^{13}C NMR. Polymer 2003;44:7539.
78. Rault J, Marchal J, Judeinstein P, Albouy PA. Stress-Induced Crystallization and Reinforcement in Filled Natural Rubbers: ^{2}H NMR Study. Macromolecules 2006;39:8356.
79. Kakiage M, Uehara H, Yamanobe T. Novel in situ NMR Measurement System for Evaluating Molecular Mobility during Drawing from Highly Entangled Polyethylene Melts. Macromol Rapid Commun 2008;29:1571.
80. Kakiage M, Yamanobe T, Komoto T, Murakami S, Uehara H. Effects of molecular characteristics and processing conditions on melt-drawing behavior of ultrahigh molecular weight polyethylene. J Polym Sci B 2006;44:2455.
81. Kakiage M, Yamanobe T, Komoto T, Murakami S, Uehara H. Transient crystallization during drawing from ultra-high molecular weight polyethylene melts having different entanglement characteristics. Polymer 2006;47:8053.

82. Yamanobe T, Kakiage M, Miyazaki N, Morita S, Uehara H. Development and Applications of in situ Pulse NMR measurement System for Drawing of Polymeric Materials. Jpn J Polym Sci Tech 2012;69:235.
83. Yamanobe T, Kurihara H, Uehara H, Komoto T. Structure and characterization of nylon 46. J Mol Struct 2006;829:80.
84. Knop A, Pilato LA. *Phenolic Resins*. New York: Springer-Verlag; 1985.
85. Fyfe CA, Rudin A, Tchir WJ. Application of High-Resolution ^{13}C NMR Spectroscopy Using Magic Angle Spinning Techniques to the Direct Investigation of Solid Cured Phenolic Resins. Macromolecules 1980;13:1320.
86. Bryson RL, Hatfield GR, Early TA, Palmer AR, MAciel GE. Carbon-13 NMR studies of solid phenolic resins using cross polarization and magic-angle spinning. Macromolecules 1983;16:1669.
87. Zhang X, Looney MG, Solomon DH, Whittaker AK. The chemistry of novolac resins: 3. ^{13}C and ^{15}N n.m.r. studies of curing with hexamethylenetetramine. Polymer 1997;38:5835.
88. Lim ASC, Solomon DH, Zhang X. Chemistry of novolac resins. X. Polymerization studies of HMTA and strategically synthesized model compounds. J Polym Sci A 1999;37:1347.
89. Katović Z, Štefanić M. Intermolecular hydrogen bonding in novolacs. Ind Eng Chem Process Des Dev 1985;24:179.

PART II

MORPHOLOGY, PROPERTIES, AND PROCESSING

9

SMALL-ANGLE X-RAY SCATTERING FOR MORPHOLOGICAL ANALYSIS OF SEMICRYSTALLINE POLYMERS

ANNE SEIDLITZ AND THOMAS THURN-ALBRECHT

Department of Physics, Martin-Luther University Halle-Wittenberg, Halle, Germany

9.1 INTRODUCTION

The fact that polymers crystallize only partially and exhibit a nanoscopic semicrystalline morphology largely determines their macroscopic properties. During crystallization from the melt, thin crystalline lamellae with thickness d_c grow, which are separated by amorphous regions with thickness d_a. The sum of both is the long period L with a typical size being of the order of a few 10 nm. In the lateral direction, the lamellar crystals can be much larger. Typically after crystallization from the quiescent melt, the lamellar stacks are part of larger spherulitic superstructures and are isotropically distributed within the sample. Figure 9.1 shows the lamellar structure in an exemplary atomic force microscopy (AFM) image of a sample of linear low-density polyethylene (LLDPE). The brighter parts correspond to the crystalline lamellae, while the dark parts belong to the softer amorphous phase. While microscopic techniques like transmission electron microscopy (TEM) and AFM give detailed qualitative insights about the semicrystalline morphology typically from a small part of the sample, small-angle X-ray scattering (SAXS) is able to give representative quantitative information about the structure, obtained from an average over macroscopic samples. Generally, features on a length scale of 1 to 100 nm are accessible by SAXS, which corresponds to the typical size for the structure parameters of semicrystalline polymers. SAXS is therefore the method of choice if detailed quantitative information about the semicrystalline morphology is desired. In this chapter, the basic principles of a SAXS experiment are discussed, beginning with a short description of the typical experimental setup, followed by a mathematical description of the relation between structural properties and scattered intensity. Different methods to analyze SAXS data obtained from semicrystalline polymers are introduced and examples of their application are given.

9.2 SMALL-ANGLE X-RAY SCATTERING

9.2.1 Typical Experimental Setup

This section provides a short overview of some general aspects of the experimental setup of a SAXS experiment, more detailed descriptions can, for example, be found in Refs [2–4]. The basic setup of a scattering experiment is shown schematically in Figure 9.2. The incident beam with wavelength λ and flux J_0 irradiates a sample. The scattered beam with flux J is detected as a function of the scattering angle 2θ at a certain distance from the sample. X-rays can be generated with a conventional sealed X-ray tube or a rotating anode, giving higher intensity. Today, another common source for X-rays is synchrotron radiation. All techniques produce a certain spectrum of wavelengths, while in a scattering experiment usually a monochromatic beam, although for SAXS not of a very narrow bandwidth, is required. Monochromatization is realized in the simplest case by a filter, nowadays usually either by a multilayer optics or a crystal monochromator. In a SAXS experiment the scattered intensity is detected at small angles, requiring a good collimation of the primary beam in order to separate the much weaker scattering signal from the primary beam. Pinholes, slits, or focusing optics are used for this purpose. For a typical wavelength in the range of 1 or

Polymer Morphology: Principles, Characterization, and Processing, First Edition. Edited by Qipeng Guo.

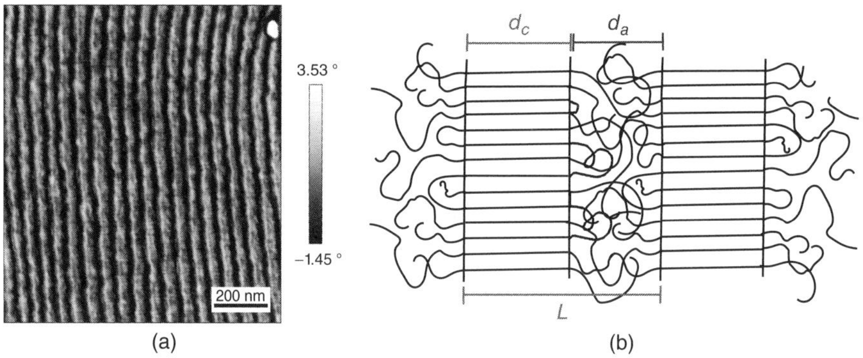

Figure 9.1 (a) AFM phase image of the lamellar structure of a sample of linear low-density polyethylene, Reprinted with permission from [1]. Copyright 1999 IOP Publishing Ltd. (b) Schematic of the lamellar structure of a semicrystalline polymer, consisting of crystalline lamellae with thickness d_c and amorphous regions with thickness d_a. The long period is $L = d_a + d_c$.

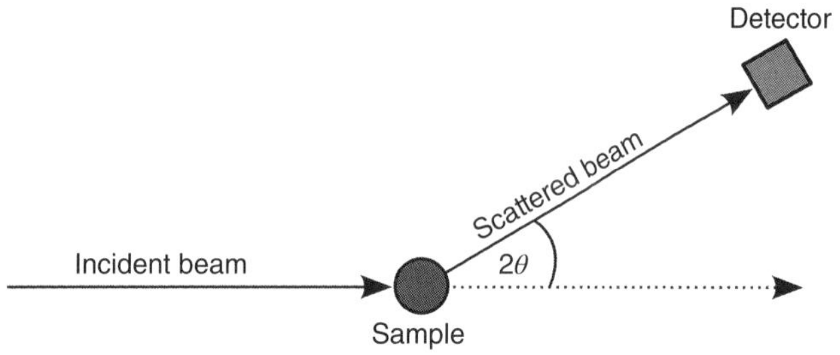

Figure 9.2 Basic setup of a scattering experiment. The incident beam (plane wave) irradiates the sample. The intensity of the scattered beam (spherical wave) is measured with a detector at an angle of 2θ.

1.5 Å, Bragg scattering angles 2θ in SAXS reach up to about 5°. The detection of the scattered intensity is usually realized with electronic detectors as, for example, proportional counters, scintillation counters, or semiconductor-based counters. Point detectors and position-sensitive devices are common.

9.2.2 Basic Formalism Describing the Relation between Real-Space Structure and Scattering Intensity in a SAXS Experiment

After a short description of the experimental setup in the previous section, we now focus on the basic mathematical description of the intensity measured in a SAXS experiment. X-rays are electromagnetic waves. Assume an incident plane wave with flux J_0, which is scattered by the electrons in the sample. The scattered spherical waves interfere with each other, resulting in an angle-dependent flux J of scattered radiation. The flux J_0 of the incident plane wave corresponds to energy transmitted per unit area per unit time and the flux J of the scattered radiation to energy transmitted per unit solid angle per unit time. Now the differential scattering cross section or scattered intensity[1] is defined as the ratio J (scattered beam)/J_0 (incident beam). It has the dimension of area per solid angle

$$I(\vec{s}) = \frac{d\sigma}{d\Omega} = \frac{J}{J_0} \tag{9.1}$$

Here, the scattering vector $\vec{s}$ is defined as the difference of the wave vector $\vec{\boldsymbol{k}}$ of the scattered beam and the wave vector $\vec{\boldsymbol{k}}_0$ of the incident beam.

$$\vec{s} = \frac{1}{2\pi}(\vec{\boldsymbol{k}} - \vec{\boldsymbol{k}}_0) \tag{9.2}$$

where elastic scattering is $|\vec{\boldsymbol{k}}_0| \approx |\vec{\boldsymbol{k}}|$ and the absolute value of the scattering vector is

$$|\vec{s}| = s = \frac{2}{\lambda}\sin(\theta) \tag{9.3}$$

Alternatively, $q = 2\pi s$ is also used. The advantage of s is that the equations for the Fourier transform used further below become symmetrical.

The flux J of scattered radiation is the square of the amplitude A of the scattered wave field

$$J(\vec{s}) = |A(\vec{s})|^2 = A(\vec{s}) \cdot A(\vec{s})^* \tag{9.4}$$

$A(\vec{s})$ can be calculated as the sum of the amplitudes of the scattered waves originating from N scattering centers in the sample

$$A(\vec{s}) = A_0 b_e \sum_{j=1}^{N} e^{-i2\pi\vec{s}\cdot\vec{r}} \tag{9.5}$$

Here, A_0 is the amplitude of the incident beam and $b_e = r_e\left(\frac{1+\cos^2 2\theta}{2}\right)^{\frac{1}{2}} \approx r_e$ is the scattering length of an electron. The classical radius of an electron is $r_e = 2.818 \cdot 10^{-15}$ m. The polarization factor (second term) is approximately 1 for small angles ($2\theta \leq 8°$). For small-angle scattering, generally a continuous electron density ρ is used to describe the structure of the sample and the sum in

[1] *Intensity* is generally used to denote a *flux J* in units of energy per unit area and per unit time. We use here a similar notation as [3] in order to distinguish clearly between the different quantities.

Equation 9.5 can be replaced by an integral

$$A(\vec{s}) = A_0 b_e \int_V \rho(\vec{r}) e^{-i2\pi\vec{s}\cdot\vec{r}} d\vec{r} \tag{9.6}$$

The integration is performed over the scattering volume V (the illuminated sample volume) and $\rho(\vec{r})$ represents the electron density. In mathematical terms, $A(\vec{s})$ is proportional to the three-dimensional Fourier transform of the electron density. In an experiment, a time-averaged intensity $I(\vec{s})$ is measured, which in an equilibrium system is equivalent to the ensemble average

$$I(\vec{s}) = \frac{J(\vec{s})}{J_0} = \frac{d\sigma}{d\Omega}(\vec{s}) = \left\langle \frac{|A(\vec{s})|^2}{|A_0|^2} \right\rangle$$
$$= \left\langle \left| b_e \int \rho(\vec{r}) e^{-i2\pi\vec{s}\cdot\vec{r}} d\vec{r} \right|^2 \right\rangle \tag{9.7}$$

With

$$g(\vec{r}) = \langle \rho(\vec{r}\,')\rho(\vec{r}\,' + \vec{r}) \rangle_{\vec{r}\,'} = \frac{1}{V} \int \rho(\vec{r}\,')\rho(\vec{r}\,' + \vec{r}) d\vec{r}\,' \tag{9.8}$$

follows [3]

$$I(\vec{s}) = b_e^2 V \int_V g(\vec{r}) e^{-i2\pi\vec{s}\cdot\vec{r}} d\vec{r} \tag{9.9}$$

That is to say the intensity $I(\vec{s})$ is proportional to the Fourier transform of the three-dimensional electron density correlation function $g(\vec{r})$. In a further step, $\rho(\vec{r})$ can be replaced by the deviation from the average electron density

$$\delta\rho(\vec{r}) = \rho(\vec{r}) - \rho \tag{9.10}$$

leading to

$$g(\vec{r}) = \langle \delta\rho(\vec{r}\,')\delta\rho(\vec{r}\,' + \vec{r}) \rangle_{\vec{r}\,'} + \langle \rho^2 \rangle \tag{9.11}$$

The constant $\langle \rho^2 \rangle$ gives only an unmeasurable contribution to forward scattering ($\vec{s} = 0$).

Finally, the absolute intensity $I_{abs}(\vec{s})$ is defined as the intensity normalized to the scattering of a single electron and to the irradiated volume

$$I_{abs}(\vec{s}) = \frac{1}{b_e^2 V} \frac{d\sigma}{d\Omega}(\vec{s}) = \int_V g(\vec{s}) e^{-i2\pi\vec{s}\cdot\vec{r}} d\vec{r} \tag{9.12}$$

It describes the scattering power of a material per volume in units of [e.u./nm^3].

9.2.3 Methods of Analysis Used for SAXS on Semicrystalline Polymers

In a good approximation, the morphology of a semicrystalline polymer can be described as a lamellar two-phase system consisting of crystalline and amorphous phases with different electron densities and with sharp boundaries between them. This structure gives rise to a peak in the scattering intensity at a scattering vector s_p from which the average long period $L = \frac{1}{s_p}$ can be determined. For this purpose, usually the Lorentz-corrected [5] data $Is^2(s)$ are used, although $Is^4(s)$ is also possible, as it will become clearer below. To gain more detailed structural information from the scattering curve, as, for example, the thickness of the crystallites and the amorphous layers, more advanced methods, all based on an extensive analysis of the correlation function or its second derivative, exist. The original methods were already introduced in 1970s and 1980s [6–8] of the past century. The corresponding formalism plus some more recent refinements and additions are described in the following sections. As it will become evident, the correlation function, and, respectively, its second derivative is especially well suited to analyze the scattering data from semicrystalline polymers, as their semicrystalline structure is periodic in character but limited to short-range order.

9.2.3.1 Correlation Function and Interface Distribution Function Typically, the lateral dimensions of the lamellar stacks in the sample are large compared to the interlamellar distance L; therefore, only the electron density distribution along the normal of the lamellar stacks, here denoted as z-direction, changes within the relevant length scale of a SAXS experiment (1–100 nm). Hence, $g(\vec{r})$ from Equation 9.11 reduces to the one-dimensional correlation function $K(z)$.

$$K(z) = \langle \delta\rho(z')\delta\rho(z' + z) \rangle_{z'} \tag{9.13}$$

Here, the unobservable contribution for $\vec{s} = 0$ is ignored. Figure 9.3 shows the structure of an ideal lamellar stack and the corresponding electron density along the z-direction. The crystalline regions have a higher electron density ρ_c than the average value ρ, whereas the amorphous parts have a lower electron density ρ_a. Inserting Equation 9.13 into Equation 9.12 yields

$$I_{abs}(\vec{s}) = \int_{x,y,z} K(z) e^{-i2\pi\vec{s}\cdot\vec{r}} dxdydz \tag{9.14}$$

After integration over x and y with $\delta(t) = \frac{1}{2\pi}\int_{-\infty}^{\infty} e^{-i\omega t} d\omega$, Equation 9.14 results in

$$I_{abs}(\vec{s}) = \delta(s_x)\delta(s_y) \int_{-\infty}^{\infty} K(z) e^{-i2\pi s_z z} dz \tag{9.15}$$

As the lamellar stacks are isotropically distributed inside the sample, the intensity in $\pm s_z$ direction is distributed over the surface of a sphere with radius s_z

$$I_{abs}(s) = \frac{2}{4\pi s_z^2} \int_{-\infty}^{\infty} K(z) e^{-i2\pi s_z z} dz \tag{9.16}$$

The reverse Fourier relation, using the nomenclature of Weaver [9], gives

$$K(z) = \int_{-\infty}^{\infty} 2\pi s^2 I_{abs}(s) e^{i2\pi s z} ds \tag{9.17}$$

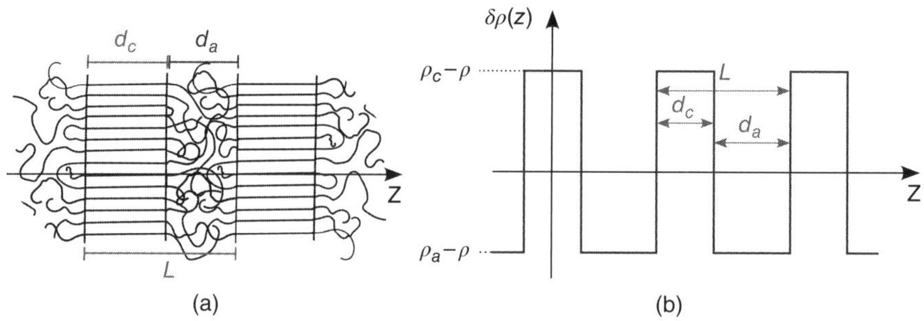

Figure 9.3 Ideal lamellar stack with z-direction along the lamellar normal (a) and electron density difference along the z-direction (b).

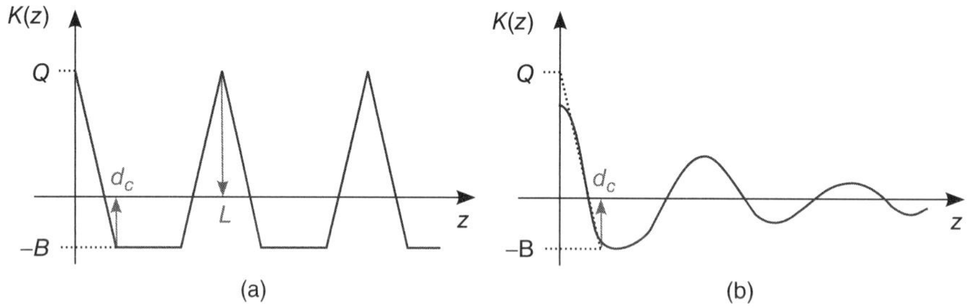

Figure 9.4 (a) Correlation function for an ideal lamellar stack. (b) Effect of deviations from ideality.

$K(z)$ and $I_{abs}(s)$ are both even functions ($f(x) = f(-x)$). Then the Fourier transform becomes a cosine transform

$$K(z) = 2\int_0^{\infty} 2\pi s^2 I_{abs}(s)\cos(2\pi sz)ds \qquad (9.18)$$

The one-dimensional correlation function $K(z)$ can be calculated directly from $I(s)$ [2,5,7,8]. Figure 9.4a shows $K(z)$ for an ideal lamellar stack. The "self-correlation triangle," centered at the origin, reflects the electron density correlation within a lamella. For a two-phase system, the maximum Q at $z = 0$ is

$$Q = K(0) = \delta\rho^2 = \Delta\rho^2\phi_a\phi_c \qquad (9.19)$$

It depends only on the volume fraction of the two phases and the electron density difference $\Delta\rho = \rho_c - \rho_a$, that is, Q does not depend on the detailed structure. Therefore, Q is often called the *invariant*. It can be calculated as

$$Q = \int I_{abs}(\vec{s})d\vec{s} = 4\pi\int_0^{\infty} s^2 I_{abs}(s)ds \qquad (9.20)$$

Of course, the integration here should be limited to the small-angle range, which for a two-phase system follows a power law $\sim s^{-4}$ at large s, as discussed below. From the maximum at the origin, $K(z)$ decreases linearly to a minimum value, namely, the baseline B, with the value

$$-B = -\phi_c^2\Delta\rho^2 = \frac{\phi_c}{1-\phi_c} \qquad (9.21)$$

If the crystallinity is smaller than 50% ($\phi_c \leq 0.5$), this minimum position corresponds to the crystalline thickness d_c. Otherwise, it corresponds to the amorphous thickness d_a and ϕ_c has to be replaced by $\phi_a = (1 - \phi_c)$, in accordance with Babinet's principle.

The slope of $K(z)$ at the origin is

$$\frac{dK}{dz} = -\frac{O_{ac}}{2}\Delta\rho^2 = \frac{\Delta\rho^2}{L} \qquad (9.22)$$

with O_{ac} as the specific inner surface per unit volume of the interfaces between crystalline and amorphous regions

$$O_{ac} = \frac{2}{L} = \frac{2\phi_c}{d_c} \qquad (9.23)$$

The second maximum in $K(z)$ is located at the long period L. The crystallinity and the square of the electron density then follow by

$$\phi_c = \frac{d_c}{L} = \frac{B}{B+Q} \qquad (9.24)$$

$$\Delta\rho^2 = \frac{Q}{\phi_c(1-\phi_c)} = \frac{(B+Q)^2}{B} \qquad (9.25)$$

In a real system, deviations from the ideal structure exist, for example, variations of the lamellar and crystalline thicknesses or diffuse interphases. Figure 9.4b illustrates this case. Around the origin, $K(z)$ is flattened. The first minimum does not necessarily reach the value of the baseline anymore. In addition, higher order maxima of $K(z)$ are damped due to the thickness variations of the crystalline and amorphous phase, leading to an overestimated long period [10]. These circumstances have to be taken into account when evaluating $K(z)$.

Another way of analyzing SAXS data of a lamellar two-phase system is the interface distribution function (IDF) established by Ruland [6]. For an ideal lamellar two-phase

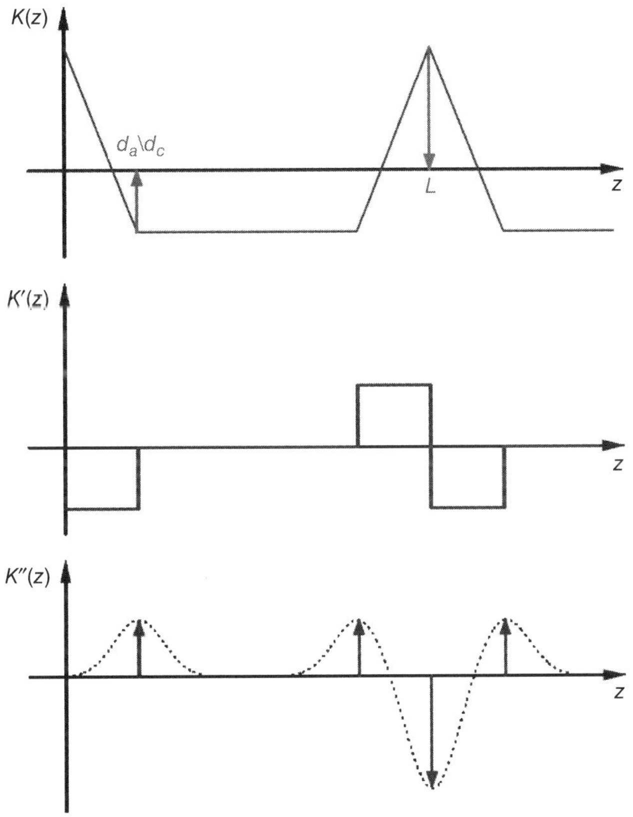

Figure 9.5 Correlation function $K(z)$ for an ideal lamellar stack and the first and second derivatives $K'(z)$ and $K''(z)$. Deviations from the ideal lamellar structure lead to a broadening (dotted line) of the δ-Peaks (arrows).

system, the second derivative of the one-dimensional correlation function $K''(z)$ has only contributions (δ-functions) at positions that correspond to the distance of any two interphases, with the first three giving the structure parameters d_a, d_c, and L. Figure 9.5 shows the correlation function and the first and second derivative $K'(z)$ and $K''(z)$. Deviations from an ideal lamellar structure, for example, varying amorphous and crystalline thicknesses, lead to a broadening of the δ-peaks. Thus, $K''(z)$ can be described as a sum of distribution functions, where $h_c(z)$ denotes the distribution of crystalline thicknesses and $h_a(z)$ the distribution of amorphous thicknesses. $h_{ac}(z)$ is the distribution of the long period, that is, one amorphous plus one crystalline thickness. Higher order distributions are defined analogously.

$$K''(z) = \frac{O_{ac}\Delta\rho^2}{2}(h_a(z) + h_c(z) - 2h_{ac}(z) + h_{aca}(z) + h_{acc}(z) - \dots) \quad (9.26)$$

Also, $K''(z)$ can be calculated directly from $I_{abs}(s)$ using the common expression for the Fourier transform of derivatives [9]

$$\mathcal{F}[K''(z)] = (i2\pi s)^2 \mathcal{F}[K(z)] \quad (9.27)$$

$$K''(z) = -8\pi^3 \int_{-\infty}^{\infty} s^4 I_{abs}(s) e^{i2\pi sz} ds \quad (9.28)$$

The triangular shape of $K(z)$ around the origin leads to a δ-function for $K''(z)$ at $z = 0$, which gives a constant contribution in reciprocal space for $s \to \infty$ [11], which has to be subtracted

$$\lim_{s\to\infty} I_{abs}(s)s^4 = \text{const} = P \quad (9.29)$$

The Porod parameter P is related to the specific inner surface O_{ac} and the electron density difference $\Delta\rho$

$$P = \frac{O_{ac}\Delta\rho^2}{8\pi^3} \quad (9.30)$$

A decrease of $I_{abs}(s) \propto s^{-4}$ is characteristic for a two-phase system with sharp boundaries [12]. After subtraction of the contribution at $z = 0$, Equation 9.28 yields

$$K''(z) = 8\pi^3 \int_{-\infty}^{-\infty} \left[\lim_{s\to\infty} I_{abs}(s)\, s^4 - s^4 I_{abs}(s)\right] e^{i2\pi sz} ds \quad (9.31)$$

The inverse Fourier transform is

$$\int_{-\infty}^{\infty} \frac{K''(z)}{8\pi^3} e^{-i2\pi sz} dz = \left[\lim_{s\to\infty} I_{abs}(s)\, s^4 - s^4 I_{abs}(s)\right] \quad (9.32)$$

With $K''(z)$ being an even function, the latter equation is equivalent to

$$2\int_{0}^{\infty} \frac{K''(z)}{8\pi^3} \cos(2\pi z) dz = \left[\lim_{s\to\infty} I_{abs}(s)\, s^4 - s^4 I_{abs}(s)\right] \quad (9.33)$$

With $s^4 I_{abs}(s)$ being an even function, one can rewrite Equation 9.31

$$K''(z) = 16\pi^3 \int_{0}^{\infty} \left[\lim_{s\to\infty} I_{abs}(s)\, s^4 - s^4 I_{abs}(s)\right] \cos(2\pi sz) ds \quad (9.34)$$

For an ideal system, $K''(z = 0)$ as defined in Equation 9.34 is zero, corresponding to the fact that there are no interfaces with zero distance, as they would occur, for example, at the edges of lamellar crystals. For laterally extended lamellae, this contribution is negligible also in a real lamellar system [13].

The scattering intensity $I_{abs}(s)$ is measured in absolute units as defined in Equation 9.12. In addition to scattering, absorption takes place in the sample, following the Lambert–Beer law [3] as it does for visible light. For SAXS, the transmitted and the scattered beams are attenuated by the same amount. The transmission measured for the primary beam can therefore be used to correct the scattering signal $I_{\exp}(s)$ for absorption effects. The absorption factor A is the ratio of the primary beam intensity with and without the sample.

$$A = \frac{J_{0(Sample)}}{J_{0(Background)}} \quad (9.35)$$

$$I_{\exp,abs}(s) = A^{-1} I_{\exp}(s) \quad (9.36)$$

Absorption also needs to be taken into account when subtracting the background $I_{\exp,BG}(s)$, caused, for example,

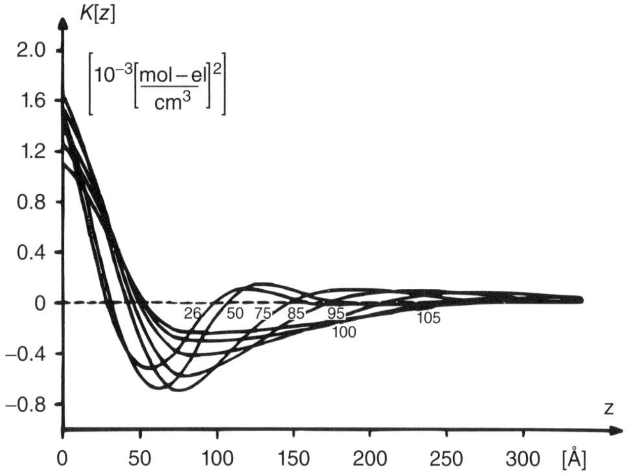

Figure 9.6 Correlation function $K(z)$ obtained for LDPE during cooling from the melt. Reprinted with permission from [14]. Copyright 1980.

by the empty sample holder. This is the last step by which the experimental data have to be corrected before calculating the correlation function or the IDF.

$$I_{abs}(s) = I_{\exp,abs}(s) - I_{\exp,BG}(s) \quad (9.37)$$

9.2.3.2 Examples for the Analysis of SAXS Data via the Correlation Function and the Interface Distribution Function In this section, examples for the analysis of SAXS data via the correlation function and the IDF selected from the literature are presented. Figure 9.6 shows several correlation functions $K(z)$ for a polyethylene sample (low-density polyethylene Lupolen 1800 S from BASF) measured during cooling at different temperatures. The sample was isothermally crystallized at $T_C = 100$ °C and then cooled to room temperature at a rate of 4 °C/min. Structure changes during the following heating and cooling cycle were investigated. The general evaluation of the "self-correlation triangle" is described in Section 9.2.3.1. Only for high temperatures (85 – 105 °C), the baseline B shows up and a direct evaluation of the crystallinity and average crystalline thickness d_c is possible. Below 85 °C the baseline is not visible. Hence, the SAXS experiment has to be supplemented by other data, for example, information about the crystallinity or the electron densities $\rho_{c/a}$ of the sample to determine the structure parameters, which makes the results obtained from $K(z)$ less conclusive. The second maximum in $K(z)$ is associated with the long period L. For high temperatures, the maximum is very broad and a decrease of L during cooling, that is, a shift of the maximum toward lower values, is clearly visible.

In the next example, the IDF is used to investigate the structure of a linear polyethylene sample with a molecular weight of 17 kg/mol [13]. The sample was measured at different temperatures during cooling, subsequent to an isothermal crystallization at $T_C = 120$ °C. Figure 9.7a shows the SAXS curves for 120 and 78 °C. The continuous lines represent the Porod scattering and a background due to density fluctuations. The data treatment to obtain $K''(z)$ from the scattering data is described in detail in Section 9.2.3.4. The corresponding IDFs for both temperatures are shown in Figure 9.7b. For 120 °C, the peaks partially overlap with each other, but the maximum positions are still clearly visible.

The first peak is attributed to the thickness of the amorphous layers; the second peak represents the thickness of the crystalline lamellae. The following negative peak corresponds to the long period. Upon cooling from 120 to 78 °C, the first (amorphous) contribution shifts to smaller values, which means the amorphous thickness decreases. The crystalline contribution is shifted toward higher values. The latter shift leads to an overlap with the negative contribution due to the

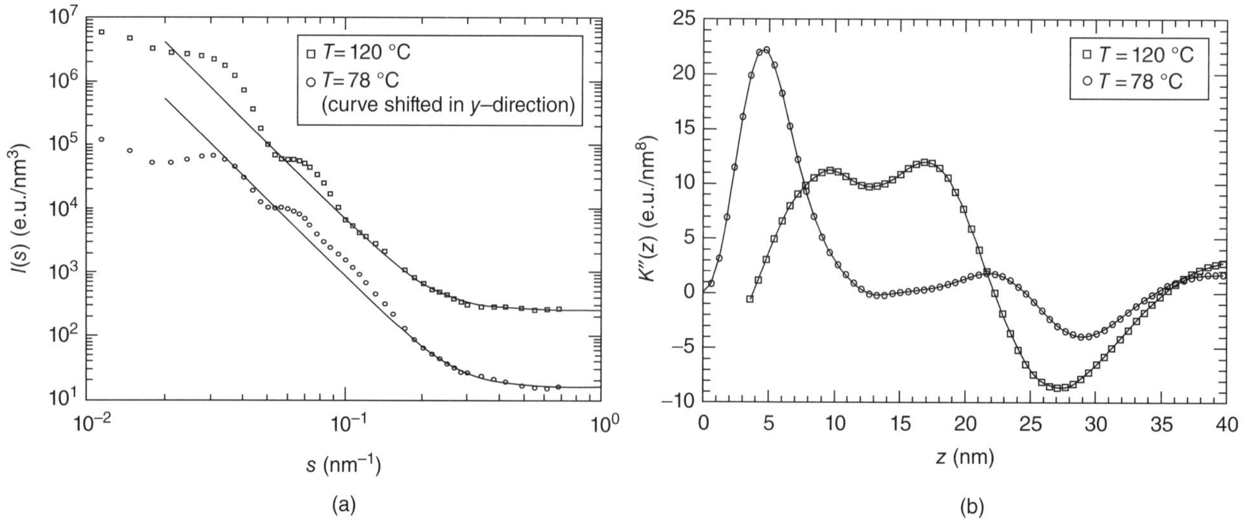

Figure 9.7 (a) Scattered absolute intensity for $T_C = 120$°C and $T = 78$°C for PE. The continuous lines represent the Porod fit. (b) Corresponding $K''(z)$ for both temperatures. Reprinted with permission from [13]. Copyright 1995 American Chemical Society.

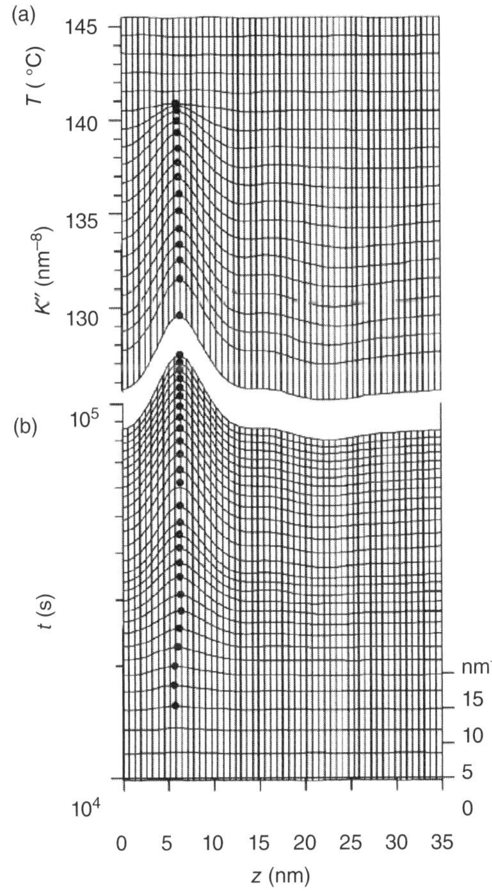

Figure 9.8 Evolution of $K''(z)$ during isothermal crystallization of *s*-poly(propylene-co-octene) at 125.5 °C (b) for different times t and during subsequent heating up to the melt (a). Reprinted with permission from [15]. Copyright 1999 IOP Publishing Ltd.

long period and thus to an apparent decrease in the amplitude of the crystalline component. In summary, with the analysis of the IDF, clear changes in the structure parameters (decrease of d_a and increase of d_c) during cooling can be observed. This example showed a $K''(z)$ with well-separated peaks, where the structure parameters can be derived directly from the peak positions. Often, this is not so straightforward, for example, when the peaks strongly overlap or have a very broad distribution. The second example, a time- and temperature-dependent SAXS experiment for a sample of *s*-poly(propylene-co-octene) (sPP-c-O) [15], will demonstrate these limitations. Figure 9.8 shows a series of IDF's $K''(z)$, measured during isothermal crystallization at $T_C = 125.5$ °C (Fig. 9.8b) and the subsequent melting during heating (Fig. 9.8a). The first peak corresponds to the crystalline thickness because the crystallinity of the sample is below 50%. As the peak position at about 7 nm does not change, the crystal thickness stays constant during crystallization and heating. The contributions of the amorphous layers (maximum around 16 nm) and the long spacing (minimum around 23 nm) are broadened and much weaker. A correct determination of the structure parameters d_a and L is therefore hardly possible.

9.2.3.3 Modeling the Interface Distribution Function for a 1D Lamellar Stack As demonstrated in the last section, the nonideality of a real semicrystalline polymer can lead to a broadening and overlapping of the peaks in $K''(z)$, which makes it difficult to extract the correct structure parameters simply from the peak positions. The one-dimensional paracrystalline stack has been suggested as an analytical model for the semicrystalline structure [2, 13, 16]. We here present a procedure that allows simulating and modeling the measured IDF based on this model. A simulated IDF $K''_{Sim}(z)$ is introduced

$$K''_{Sim}(z) = \frac{O_{ac}\Delta\rho^2}{2}(h_a(z) + h_c(z) - 2h_{ac}(z) + h_{aca}(z) + h_{acc}(z) - \dots) \quad (9.38)$$

for which $h_a(z)$ and $h_c(z)$ are taken to be normalized Gaussian functions, that is, the thicknesses $d_{a,c}$ have a Gaussian distribution with width $\sigma_{a,c}$

$$h_{a,c}(z) = \frac{1}{\sqrt{2\pi}\sigma_{a,c}} e^{-\frac{(z-d_{a,c})^2}{2\sigma_{a,c}^2}} \quad (9.39)$$

Higher order distributions $h_{aca\dots}(z)$ are assumed to be convolutions ($*$) of $h_a(z)$ and $h_c(z)$, here the order of indices is irrelevant. Every Gaussian function has two free parameters ($d_{a,c}$ and $\sigma_{a,c}$); that is, all together, the model function has four free parameters. The cosine transform of $K''(z)$ (Eq. 9.33) is approximated by the real part $\Re$ of the Fourier transform of $K''_{Sim}(z)$

$$\tilde{K}''_{Sim}(s) = \Re(\mathcal{F}[K''_{Sim}(z)]) \quad (9.40)$$

The calculation is simplified by the fact that the Fourier transform of a Gaussian function is again a Gaussian function

$$\mathcal{F}[h(z)] = \tilde{h}(s) = \int_{-\infty}^{\infty} h(z)e^{-i2\pi sz}dz = e^{-2\pi^2 s^2\sigma^2}e^{i2\pi sd} \quad (9.41)$$

and that the convolutions become products. The sum in Equation 9.38 can be represented as a geometric series and an analytical expression is obtained for $\tilde{K}''_{Sim}(s)$, which can be fitted to the experimental data (for details, see Appendix)

$$\begin{aligned}
&\tilde{K}''_{Sim}(s) \\
&= 16\pi^3 P \frac{1}{1 - 2e^{-2\pi^2 s^2(\sigma_a^2+\sigma_c^2)}cos(2\pi s(d_a + d_c)) + e^{-4\pi^2 s^2(\sigma_a^2+\sigma_c^2)}} \\
&\times \Big\{ e^{-2\pi^2 s^2\sigma_a^2}cos(2\pi s d_a) + e^{-2\pi^2 s^2\sigma_c^2}cos(2\pi s d_c) \\
&\quad -2e^{-2\pi^2 s^2(\sigma_a^2+\sigma_c^2)}cos(2\pi s(d_a + d_c)) \\
&\quad -e^{-4\pi^2 s^2\sigma_a^2}e^{-2\pi^2 s^2\sigma_c^2}cos(2\pi s d_c) \\
&\quad -e^{-4\pi^2 s^2\sigma_c^2}e^{-2\pi^2 s^2\sigma_a^2}cos(2\pi s d_a) + 2e^{-4\pi^2 s^2(\sigma_a^2+\sigma_c^2)} \Big\} \\
&= 16\pi^3 \left[\lim_{s\to\infty} I_{abs}(s)\, s^4 - s^4 I_{abs}(s)\right]
\end{aligned} \quad (9.42)$$

Finally, the calculation of $K''_{Sim}(z)$ is realized by a cosine transform of $\tilde{K}''_{Sim}(s)$.

For the Fourier transform of the experimental data, it is necessary to multiply the right side of Equation 9.42, that is, $16\pi^3\left[\lim_{s\to\infty} I_{abs}(s)\,s^4 - s^4 I_{abs}(s)\right]$, abbreviated $16\pi^3[P - I_{abs}s^4]$, with a window function

$$\omega(s) = e^{-4\pi\sigma^2 s^2} \tag{9.43}$$

to suppress contributions due to noise, which become very strong at high s due to the multiplication with s^4. Therefore, to compare the measurement with the simulation, both are multiplied with $\omega(s)$. This procedure is equivalent to a convolution of $K''(z)$ and $K''_{Sim}(z)$ with $\omega(z)$ in real space, resulting in a certain broadening of both. Typical values for σ are in the range between 1 and 2 nm. In addition, taking into account the finite resolution of the measuring instrument, the simulated $\tilde{K}''_{Sim}(s)$ has to be convoluted with a resolution function

$$A(s) = \frac{1}{\sqrt{2\pi}\sigma_A} e^{-\frac{s^2}{2\sigma_A^2}} \tag{9.44}$$

For the data shown below, a value of $\sigma_A = 0.00325\ \text{nm}^{-1}$ was assumed.

The final expression of the simulation function, in the following abbreviated with $\tilde{K}''_{Sim,final}(s)$, is then

$$\tilde{K}''_{Sim,final}(s) = A(s) * (\tilde{K}''_{Sim}(s)\omega(s)) \tag{9.45}$$

Figure 9.9 shows an exemplary simulated IDF (black straight line) and the first three Gaussian contributions h_a (light gray), h_c (dark gray) and $-2h_{ac}$ (black) as dotted lines.

While this approach allows to formulate a relatively simple analytical model for $K''_{Sim}(z)$, it is, strictly speaking, not correct, due to the fact that $K''(z \le 0) \ne 0$, which is an unphysical assumption, as in reality the smallest possible distance between two interfaces is zero. Furthermore, there is no simple analytical solution of the cosine transformation of a Gaussian function which extends into the range of negative z. A possible solution for this problem is to cut out the Gaussian function for $z \le 0$ by multiplying it with the Heaviside function [17], but an analytical solution is not possible. The approach taken here, namely, to take the real part $\Re$ of the Fourier transformation $\mathcal{F}$ as an approximation of the cosine transformation leads to an additional contribution for $K''_{Sim}(z)$ at small z. The light gray area in Figure 9.10 illustrates the additional contribution. One, therefore, has to make sure that this additional contribution is small. We suggest as a criterion that the integrated contribution of the Gaussian function $h_a^{\omega}(z)$ (convoluted with the window function)

$$h_a^{\omega}(z) = \mathcal{F}(\tilde{h}_a(s)\omega(s)) = \frac{1}{\sqrt{2\pi}\sqrt{(\sigma_a^2 + 2\sigma^2)}} e^{-\frac{(z-d_a)^2}{2\sigma_a^2+4\sigma^2}} \tag{9.46}$$

for negative z is smaller than 5% of the whole area underneath $h_a^{\omega}(z)$. In other words,

$$I_0 = \frac{1}{\sqrt{2\pi}\sqrt{(\sigma_a^2 + 2\sigma^2)}} \int_{-\infty}^{\infty} e^{-\frac{(z-d_a)^2}{2\sigma_a^2+4\sigma^2}} dz, \text{ and}$$

$$I = \frac{1}{\sqrt{2\pi}\sqrt{(\sigma_a^2 + 2\sigma^2)}} \int_{0}^{\infty} e^{-\frac{(z-d_a)^2}{2\sigma_a^2+4\sigma^2}} dz \tag{9.47}$$

must fulfill $I \ge 0.95 I_0$.

It can be shown that this is true if the relation $(\sigma_a^2 + 2\sigma^2) \le d_a^2/2.71$ is fulfilled.

9.2.3.4 Calculation and Modeling of the Interface Distribution Function In this section, the calculation of $K''(z)$ is demonstrated step by step on the basis of a real set of scattering data, followed by the modeling of the data by the simulated IDF $K''_{Sim,final}(s)$ in the reciprocal space. We analyze the scattering data of a sample of poly (ε-caprolactone) (PCL) with

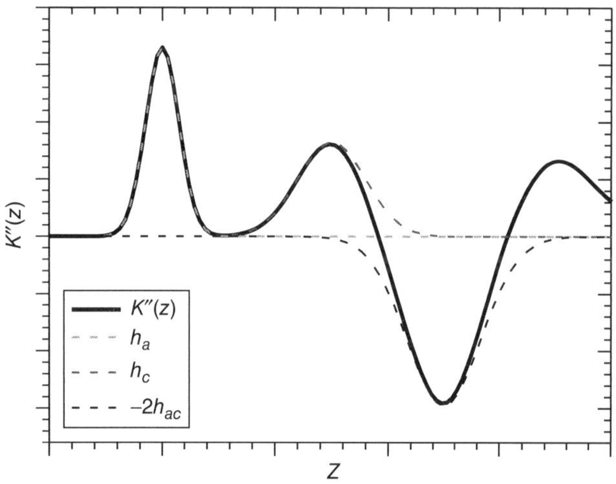

Figure 9.9 Simulated interface distribution function (black straight line). The Gaussian functions h_a (light gray) and h_c (dark gray) and their negative double convolution (black) $-2h_{ac}$ are indicated with dotted lines.

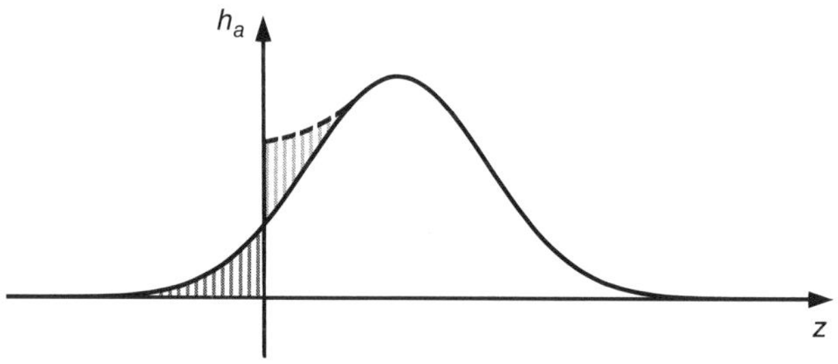

Figure 9.10 Gaussian function (black line) with nonzero values for $z \le 0$ (dark gray area). Calculating the real part of the Fourier transformation instead of a cosine transformation leads to an additional contribution (light gray area) for $h(z \ge 0)$.

$M_w = 60\,\text{kg/mol}$, isothermally crystallized at $T_C = 45\,°\text{C}$, abbreviated in the following with PCL60. The calculation of the IDF is realized by a discrete cosine transform (DCT) applied to the discrete data points from the experiment. The s-range, over which experimental data are measured, sets the resolution with which the IDF is determined. For a given highest scattering vector s_{max}, the IDF is calculated with a distance between the data points $\Delta z = \frac{1}{s_{max}}$. For a reliable analysis, Δz should be smaller than the scale of the smallest features to be determined, that is, d_a or d_c, whatever is smaller. Then the corresponding contribution in the IDF will be well-resolved. In reciprocal space, this requires that s_{max} is in the Porod range for which $I_{abs}(s) = Ps^{-4}$. If a smaller s_{max} is chosen, small-scale features in the IDF and also the correlation function are effectively averaged out with a resolution corresponding to $1/s_{max}$.

Figure 9.11 shows the scattering data for PCL60. The scattering intensity $I_{abs}(s)$ is measured in absolute units and describes the scattering power of the sample per volume in units of [e.u./nm^3]. For real data, the intensity contains, in addition, a constant contribution from thermal density fluctuations within the amorphous phase c_2 and a contribution proportional to s^2 from the amorphous halo. The latter term is often negligible. Thus, for large s

$$I_{abs}(s) \approx Ps^{-4} + c_1 s^2 + c_2 \tag{9.48}$$

To determine P, c_1 and c_2 , the expression in Equation 9.48 is fitted to the data in the range of high-scattering vectors s under the side condition $K''(0) = 0$ as explained above, with

$$K''(0) = 0 = 16\pi^3 P \int_0^{s_{max}} [P - (I_{abs}(s) - c_2)s^4 + c_1 s^6] \times e^{-4\pi^2\sigma^2 s^2} ds \tag{9.49}$$

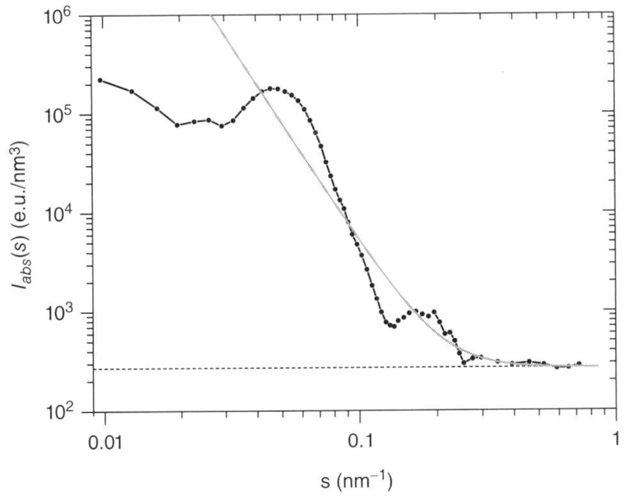

Figure 9.11 Scattered intensity (black dotted line) of PCL60 with Porod fit (light gray line). The continuous line gives the sum of the Porod scattering and a background (dashed line) due to density fluctuations.

As the scattered intensity at $s = 0$ is not measurable, an extrapolation is required. The window function $e^{-4\pi^2\sigma^2 s^2}$, on the other hand, avoids truncation effects at large s due to the finite interval $[0, s_{max}]$ and suppresses noise. As Figure 9.11 and Equation 9.48 show again, a large enough s-range is needed to determine c_2 correctly. If only data for a more limited range are available, a larger value of σ in the window function $\omega(s)$ can be used to make the function $16\pi^3[P - I_{abs}s^4]\omega(s)$ decrease to zero, but, as mentioned above, the resolution is limited for such data. The last step to obtain $K''(z)$ is to perform the DCT of $16\pi^3[P - I_{abs}s^4]\omega$ (cf. Eq. 9.34). Figure 9.12 shows $16\pi^3[P - I_{abs}s^4]\omega$ as the dotted black line and the resulting IDF. For the simulation, the value of the Porod constant, here $P = 0.507\,\text{e.u./nm}^7$, was taken from the corresponding fit and for the window function a width of $\sigma = 1.0$ nm was chosen, the same value as used during the calculation of the IDF from the scattering data. For the resolution function $A(s)$, a value of $\sigma_A = 0.00325\ \text{nm}^{-1}$ was taken. The four parameters d_a, d_c, σ_a, and σ_c are fitted. The resulting model function is shown in Figure 9.12a in light gray. The overall shape of the scattering data is described very well by the simulated curve. Only slight deviations, due to noise/experimental uncertainty containing no real structure information, are visible. In a final step, the IDF of the result of the simulation is calculated and shown along with the IDF from the experimental data in Figure 9.12b. The first three contributions h_a, h_c, and $-2h_{ac}$ are indicated as light gray, dark gray, and dashed black line. Here, the amorphous and crystalline peaks strongly overlap due to the crystallinity of approximately 50% (measured by DSC). This is an example where it is impossible to obtain the crystalline and amorphous thicknesses simply from the peak position of $K''(z)$. The separation into crystalline and amorphous contribution is only possible with the help of the simulation. To assign the two peaks to the amorphous and crystalline contribution, additional information (e.g., about crystallinity) is needed. In this case, the identification of the crystalline and amorphous contribution was done with a heating experiment by assuming partial melting. For semicrystalline polymers without crystalline α-process like PCL, d_a generally increases more strongly during heating than d_c [11]. Only small deviations between the simulated IDF and data can be recognized. Especially for $z > 20\,\text{nm}$, $K''(z)$, and $K''_{Sim,final}(z)$ deviate somewhat from each other. At this point, the model might become too simple. Nevertheless, the simple paracrystalline model can conveniently and well describe the important features of the IDF from the scattering experiment.

To illustrate the advantage of this evaluation procedure, the correlation function $K(z)$ derived from the same data set is shown in Figure 9.13. The information content of $K(z)$ is strongly limited due to the absence of the baseline and the crystallinity of approximately 50%. This makes it impossible to distinguish d_a from d_c. In addition, the information about the thickness distributions is hidden. Only the long period, which is about 1 nm larger than the value derived from the IDF, can be extracted.

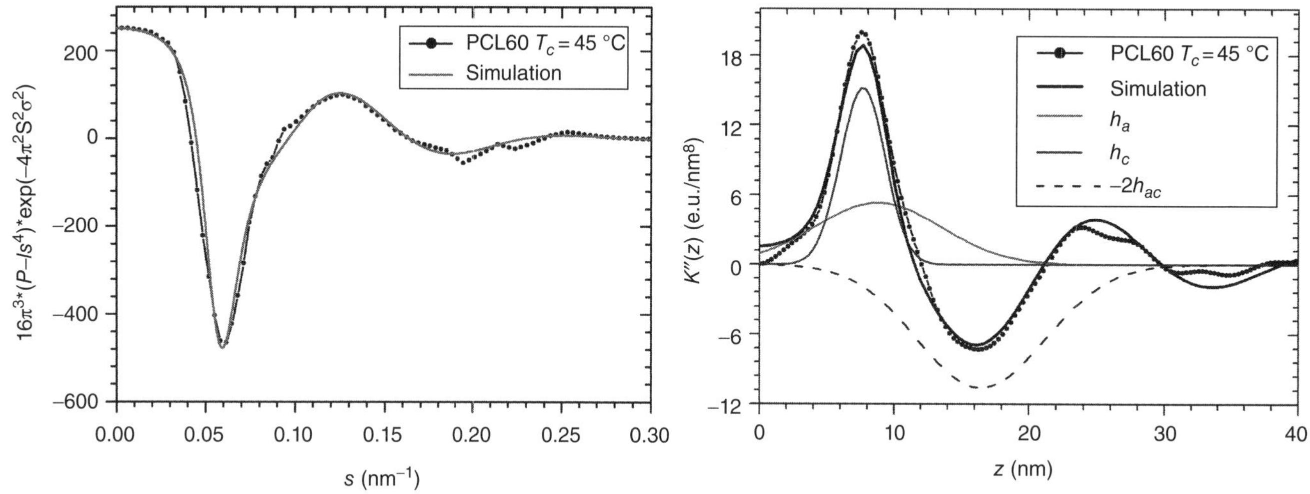

Figure 9.12 (a) Function $16\pi^3[P - Is^4]\omega$ as calculated from $I(s)$ (dotted lines) for PCL60. The continuous line is the result of the simulation, giving the result $d_a = 8.80\,\mathrm{nm}$, $\sigma_a = 4.20\,\mathrm{nm}$, $d_c = 7.65\,\mathrm{nm}$, $\sigma_c = 0.85\,\mathrm{nm}$. (b) Resulting interface distribution function $K''(z)$ after a discrete cosine transformation for PCL60 (dotted lines) and simulation (thick continuous lines). For the simulation the first three contributions h_a, h_c, and $-2h_{ac}$ are indicated additionally.

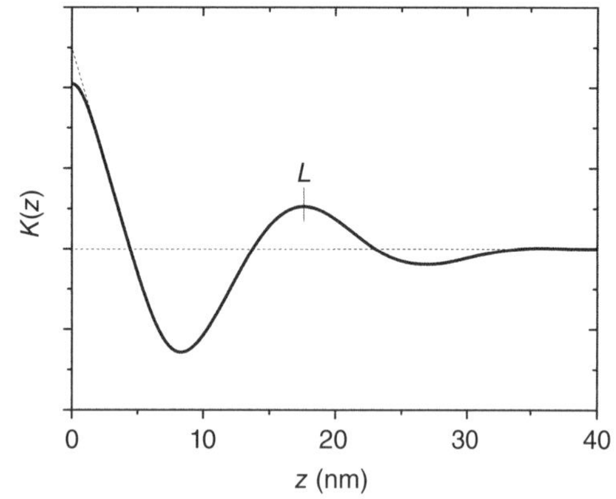

Figure 9.13 Correlation function for PCL60.

9.2.3.5 Alternative Approaches While the analysis of SAXS data of semicrystalline polymers based on the calculation of the correlation function or the IDF is well established, the parameters describing the semicrystalline morphology are usually simply read off from the correlation function $K(z)$ or the IDF $K''(z)$. A quantitative analysis based on modeling the scattering intensity or the IDF as described above is much less common. Some attempts along these lines exist, of course, and at the end of this contribution we give a short overview of such alternative approaches. In several cases, simulations of the IDF are used to illustrate the effect of different assumptions about the semicrystalline structure on the IDF, which is then compared qualitatively to experimental results obtained from scattering data [17–20]. A few authors perform a quantitative modeling of the data but with some simplifying assumptions, as, for example, truncating the series of contributions to the IDF after a few terms [21–23]. But in other cases also, significant modifications [24] or more complex models using other than simple Gaussian distributions were designed and used [17, 22]. In most cases though, these were only applied to a very limited number of data sets, so that all together it is still today not very common to analyze the complete information contained in the IDF including the information about the thickness distributions. To our experience, for that purpose, the simple model introduced above is in many cases sufficient.

9.3 CONCLUDING REMARKS

After introducing the basic principles of a SAXS experiment, we described the analysis of the morphology of semicrystalline polymers based on the direct calculation of the correlation function or the IDF from scattering data. A simple and analytically solvable model for $K''(z)$ was introduced and its application to real scattering data was demonstrated. Advantages of this analysis as well as its limits were discussed. The method provides detailed insight into the structure of the investigated sample, including information not only about the average thickness of the crystalline and amorphous phases but also about the width of the corresponding distributions. A prerequisite for the detailed data analysis is, in any case, the availability of scattering data over a large enough range in scattering vectors in order to collect the information about the full structure and to determine the Porod constant and the contribution from density fluctuations, which need to be subtracted beforehand. The fact that the IDF vanishes for $z = 0$ can be used as a side condition to perform these corrections in a much more consistent way. This last point is important as errors in these preparatory steps of the analysis can introduce significant changes in the IDF.

Appendix: Calculation of the Model Function $\tilde{K}''_{Sim}(s)$

The starting point to calculate the model function $\tilde{K}''_{Sim}(s)$ is Equation 9.38, with $h_{a/c}(z)$ being normalized Gaussian distributions as defined in Equation 9.39. Higher order distributions are assumed to be convolutions ($*$) of $h_a(z)$ and $h_c(z)$ and the order of indices is irrelevant

$$h_{\mu a\ \upsilon c}(z) = h_a(z)^{*\mu} * h_c(z)^{*\upsilon} \tag{9.50}$$

(e.g., $h_{aac} = h_a(z)^{*2} * h_c(z)^{*1} = h_a * h_a * h_c$).

Inserting Equation 9.50 into Equation 9.38 leads to

$$K''_{Sim}(z) = \frac{O_{ac}\Delta\rho^2}{2}\left\{\left[h_a(z) + h_c(z)\right] * \sum_{\mu=0}^{\infty}\left[h_a(z) * h_c(z)\right]^{*\mu} - 2\sum_{\mu=1}^{\infty}\left[h_a(z) * h_c(z)\right]^{*\mu}\right\} \tag{9.51}$$

By a Fourier transform of Equation 9.51 the convolutions become products

$$\mathcal{F}\left[K''_{Sim}(z)\right] = \frac{O_{ac}\Delta\rho^2}{2}\left\{\left[\tilde{h}_a(s) + \tilde{h}_c(s)\right]\sum_{\mu=0}^{\infty}\left[\tilde{h}_a(s)\tilde{h}_c(s)\right]^{\mu} - 2\sum_{\mu=1}^{\infty}\left[\tilde{h}_a(s)\tilde{h}_c(s)\right]^{\mu}\right\} \tag{9.52}$$

Rewriting the sums as geometric series

$$\begin{aligned}\sum_{\mu=0}^{\infty} a_0 x^k = \frac{a_0}{1-x} &\to \left[\tilde{h}_a(s) + \tilde{h}_c(s)\right]\sum_{\mu=0}^{\infty}\left[\tilde{h}_a(s)\tilde{h}_c(s)\right]^{\mu} = \frac{\tilde{h}_a(s) + \tilde{h}_c(s)}{1 - \tilde{h}_a(s)\tilde{h}_c(s)},\\ \sum_{\mu=1}^{\infty} a_0 x^k = \frac{a_0 x}{1-x} &\to -2\sum_{\mu=1}^{\infty}\left[\tilde{h}_a(s)\tilde{h}_c(s)\right]^{\mu} = \frac{-2\tilde{h}_a(s)\tilde{h}_c(s)}{1 - \tilde{h}_a(s)\tilde{h}_c(s)}\end{aligned} \tag{9.53}$$

results in

$$\mathcal{F}\left[K''_{Sim}(z)\right] = \frac{O_{ac}\Delta\rho^2}{2}\left\{\frac{\tilde{h}_a(s) + \tilde{h}_c(s) - 2\tilde{h}_a(s)\tilde{h}_c(s)}{1 - \tilde{h}_a(s)\tilde{h}_c(s)}\right\} \tag{9.54}$$

The cosine transform of $K''(z)$ (Eq. 9.33) is approximated by the real part $\Re$ of Equation 9.54, the Fourier transform of $K''_{Sim}(z)$

$$\left[\lim_{s\to\infty} I_{abs}(s)\,s^4 - s^4 I_{abs}(s)\right] = \frac{O_{ac}\Delta\rho^2}{8\pi^3}\Re\left\{\frac{\tilde{h}_a(s) + \tilde{h}_c(s) - 2\tilde{h}_a(s)\tilde{h}_c(s)}{1 - \tilde{h}_a(s)\tilde{h}_c(s)}\right\} \tag{9.55}$$

Expansion with the complex conjugated denominator yields

$$\begin{aligned}\left[\lim_{s\to\infty} I_{abs}(s)\,s^4 - s^4 I_{abs}(s)\right] &= \frac{P}{1 - 2\Re\left[\tilde{h}_a(s)\tilde{h}_c(s)\right] + \left|\tilde{h}_a(s)\tilde{h}_c(s)\right|^2}\\ &\times\Big\{\Re\left[\tilde{h}_a(s) + \tilde{h}_c(s)\right] - 2\Re\left[\tilde{h}_a(s)\tilde{h}_c(s)\right]\\ &\quad - \left|\tilde{h}_a(s)\right|^2\Re\left[\tilde{h}_c(s)\right] - \left|\tilde{h}_c(s)\right|^2\Re\left[\tilde{h}_a(s)\right]\\ &\quad + 2\left|\tilde{h}_a(s)\tilde{h}_c(s)\right|^2\Big\}\end{aligned} \tag{9.56}$$

The relation $\Re\{\mathcal{F}[h(z)]\} = \Re\{\tilde{h}(s)\} = e^{-2\pi^2 s^2 \sigma^2}\cos(2\pi s d)$ finally leads to Equation 9.42 for the simulation function $\tilde{K}''_{Sim}(s)$.

REFERENCES

1. Henze T, Schröter K, Thurn-Albrecht T. Investigation of the different stable states of the cantilever oscillation in an atomic force microscope. Nanotechnology 2012;24:245702.
2. Glatter O, Kratky O. *Small Angle X-Ray Scattering*. London: Academic Press; 1982.
3. Roe R-J. *Methods of X-Ray and Neutron Scattering in Polymer Science*. Oxford University Press; 2000.
4. Stribeck N. *X-Ray Scattering of Soft Matter*. Berlin Heidelberg: Springer; 2007.
5. Debye P, Buecher AM. Scattering by an inhomogeneous solid. J Appl Phys 1949;20:518–525.
6. Ruland W. The evaluation of the small-angle scattering of lamellar two-phase systems by means of interface distribution functions. Colloid Polym Sci 1977;255:417.
7. Vonk CG, Kortleve G. X-ray small-angle scattering of bulk polyethylene. Colloid Polym Sci 1967;220:19–24.
8. Strobl G, Schneider M. Direct evaluation of the electron density correlation function of partially crystalline polymers. J Polym Sci Polym Phys Ed 1980;18:1343–1359.
9. Butz T. *Fourier Transformation for Pedestrians*. Springer; 2006.
10. Santa Cruz C, Stribeck N, Zachmann HG, Baltá Calleja FJ. Novel aspects in the structure of poly(ethylene terephthalate) as revealed by means of small-angle X-ray scattering. Macromolecules 1991;24:5980–5990.

11. Strobl G. *The Physics of Polymers: Concepts for Understanding Their Structures and Behavior*. Berlin Heidelberg: Springer-Verlag; 2007.
12. Porod G. Die Röntgenkleinwinkelstreuung von dichtgepackten kolloiden Systemen. Colloid Polym Sci 1951;124(2):83–114.
13. Albrecht T, Strobl G. Temperature-dependent crystalline-amorphous structures in linear polyethylene: Surface melting and the thickness of the amorphous layers. Macromolecules 1995;28:5827–5833.
14. Strobl GR, Voigt-Martin IG. Model of partial crystallization and melting derived from small-angle X-ray scattering and electron microscopic studies on low-density polyethylene. J Polym Sci Polym Phys Ed 1980;18:1361–1381.
15. Heck B, Hugel T, Iijima M, Sadiku E, Strobl G. Steps in the transition of an entangled polymermelt to the partially crystalline state. New J Phys 1999;1:17.1–17.29.
16. Hosemann R, Bagchi SN. *Direct Analysis of Diffraction by Matter*. Amsterdam: North-Holland; 1962.
17. Ivanov D. Exploring the melting of a semirigid-chain polymer with temperature-resolved small-angle X-ray scattering. Eur Phys J E 2004;13:363–378.
18. Christ B. Analysis of small-angle X-ray scattering patterns. J Macromol Sci Phys 2007;39(4):493–518.
19. Stribeck N, Ruland W. Determination of the interface distribution function of lamellar two-phase systems. J Appl Cryst 1978;11:535–539.
20. Sun Y-S. Temperature-resolved SAXS studies of morphological changes in melt-crystallized poly(hexamethylene terephthalate) and its melting upon heating. Polymer 2006;47: 8032–8043.
21. Slusarczyk C. Structure development during isothermal crystallisation of high-density polyethylene: Synchrotron small-angle X-ray scattering study. Radiat Phys Chem 2013;93: 104–110.
22. Stribeck N. SAXS data analysis of a lamellar two-phase system. Layer statistics and compansion. Colloid Polym Sci 1993;271(11):1007–1023.
23. Stribeck N. Complete SAXS data analysis and synthesis of lamellar two–phase systems. Deduction of a simple model for the layer statistics. J Phys IV 1993;3(C8):507–510.
24. Miyamoto Y, Fukao K, Miyaji H. Small-angle X-ray scattering of isotactic polystyrene. Colloid Polym Sci 1995;273:66–75.

10

CRYSTALLINE MORPHOLOGY OF HOMOPOLYMERS AND BLOCK COPOLYMERS

Shuichi Nojima and Hironori Marubayashi

Department of Organic and Polymeric Materials, Graduate School of Science and Engineering, Tokyo Institute of Technology, Tokyo 152-8552, Japan

10.1 INTRODUCTION

It is well known through the long history of polymer science that the morphology formed in polymer materials is mainly controlled by several factors such as liquid–liquid phase separation (or macrophase separation), microphase separation, crystallization, liquid crystallization, gelation, and so on (Fig. 10.1). Theoretical and experimental studies on static and dynamic aspects of such morphology formation have made it possible to understand its detailed mechanism from the viewpoint of long-molecular motion based on scientific principles. The crystallization of homopolymers is one of the important research subjects in polymer science, and many new facts have been disclosed up to now for the crystallization behavior of long chains and resulting crystalline morphology [1–3]. Furthermore, it is possible to suppose crystallization under various situations described in Figure 10.1 (right), and actually many unique crystallization behaviors and crystalline morphologies have been found using more complicated systems. For example, the crystalline morphology of block copolymers, which is described in Section 10.3, is significantly different from that of homopolymers.

In this chapter, we first summarize the crystalline morphology formed in homopolymers, which is usually explained in many textbooks of polymer science. Next, we describe more complicated morphology, that is, the crystalline morphology of block copolymers. This subject is relatively new as compared with that of homopolymers, so that it is not fully understood at present.

10.2 CRYSTALLINE MORPHOLOGY OF HOMOPOLYMERS

Crystalline homopolymers form a series of characteristic morphologies in the system when they are quenched from a homogeneous melt. These morphologies are largely different in their dimension but intimately interrelated with each other in their formation process. We call these morphologies a *hierarchical structure* formed in crystalline homopolymers. This hierarchical structure consists of the crystal structure (with a repeating distance in the order of 10^{-1}–10^{0} nm), lamellar morphology (10^{1}–10^{2} nm), and spherulite structure ($>10^{3}$ nm), as schematically rendered in Figure 10.2. The dimension or repeating distance of these morphologies is widely different, so that we must employ suitable experimental methods to observe them. For example, we usually use wide-angle X-ray diffraction (WAXD) for observing crystal structures, small-angle X-ray scattering (SAXS), small-angle neutron scattering (SANS), or transmission electron microscope (TEM) for lamellar morphologies, and small-angle light scattering (SALS) or optical microscope (OM) for spherulite structures. In this section, we briefly describe the characteristics of the crystal structure, lamellar morphology, and spherulite structure from the viewpoint of interrelation among them, and also experimental methods to quantitatively analyze these morphologies.

10.2.1 Crystal Structure

In crystalline states, some parts of each polymer are regularly arranged with three-dimensional order to form *crystals*

Polymer Morphology: Principles, Characterization, and Processing, First Edition. Edited by Qipeng Guo.

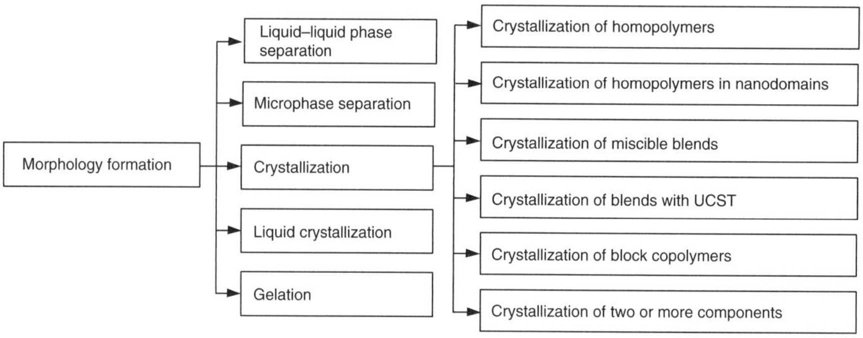

Figure 10.1 Major factors of morphology formation working in polymer materials and several kinds of crystallization under various situations (right).

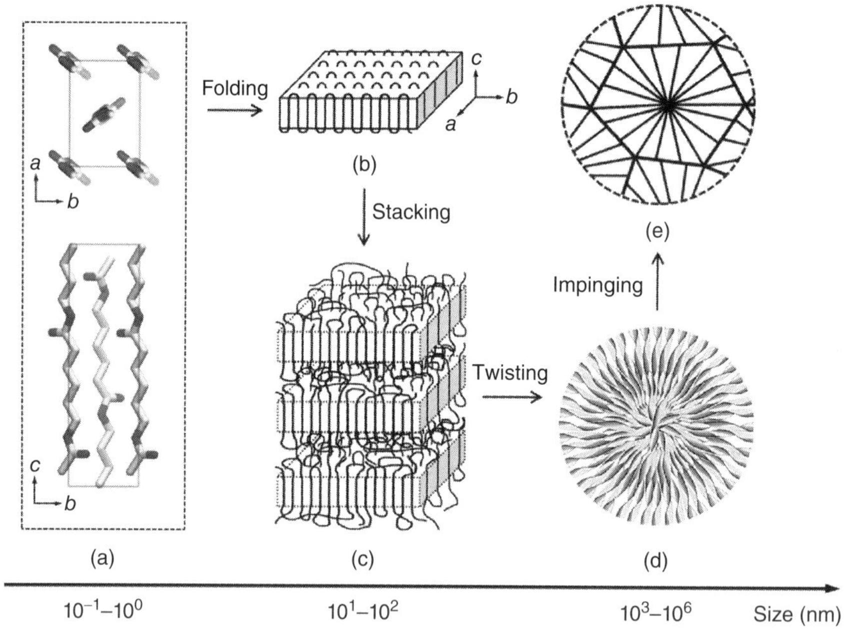

Figure 10.2 Schematic illustration showing a hierarchical structure formed in crystalline homopolymers when quenched from a homogeneous melt. (a) Crystal structure, (b) crystalline lamella, (c) lamellar morphology, (d) spherulite, and (e) spherulite structure.

(Fig. 10.2a). X-ray diffraction from such crystals occurs discontinuously in specific directions, as represented by the Bragg condition:

$$2d_{hkl} \sin\theta = n\lambda \quad (10.1)$$

where d_{hkl} is the lattice spacing of a given (*hkl*) plane, θ is the Bragg angle, *n* is the order of reflections, and λ is the X-ray wavelength. The diffraction intensity $I(\boldsymbol{s})$ is given by

$$I(\boldsymbol{s}) = |F(\boldsymbol{s})|^2 \frac{\sin^2(N_1\pi\boldsymbol{s}\cdot\boldsymbol{a})}{\sin^2(\pi\boldsymbol{s}\cdot\boldsymbol{a})}\cdot\frac{\sin^2(N_2\pi\boldsymbol{s}\cdot\boldsymbol{b})}{\sin^2(\pi\boldsymbol{s}\cdot\boldsymbol{b})}\cdot\frac{\sin^2(N_3\pi\boldsymbol{s}\cdot\boldsymbol{c})}{\sin^2(\pi\boldsymbol{s}\cdot\boldsymbol{c})} \quad (10.2)$$

$$F(\boldsymbol{s}) = \sum f(\boldsymbol{s})T(\boldsymbol{s})\exp\{2\pi i\,(\boldsymbol{s}\cdot\boldsymbol{r})\} \quad (10.3)$$

where $\boldsymbol{s}$ is the scattering vector ($|\boldsymbol{s}| = s = 2\sin\theta/\lambda$), N_1, N_2, and N_3 are numbers of unit cells existing in the *a*-, *b*-, and *c*-axis directions, respectively, $\boldsymbol{a}$ is the vector with a magnitude of *a*-axis length and a direction parallel to the *a*-axis, $\boldsymbol{b}$ and $\boldsymbol{c}$ are defined in the same manner, $F(\boldsymbol{s})$ is the structure factor, $f(\boldsymbol{s})$ is the atomic scattering factor, $T(\boldsymbol{s})$ is the Debye–Waller factor, and $\boldsymbol{r} = n_1\boldsymbol{a} + n_2\boldsymbol{b} + n_3\boldsymbol{c}$ (n_i; integer with $0 \le n_i \le N_i$, $i = 1, 2$, or 3). The diffraction condition is given by the Laue condition:

$$\boldsymbol{s}\cdot\boldsymbol{a} = h, \quad \boldsymbol{s}\cdot\boldsymbol{b} = k, \quad \boldsymbol{s}\cdot\boldsymbol{c} = l \quad (10.4)$$

where *h*, *k*, and *l* are integers (Miller indices). If the Laue condition is true, $I(\boldsymbol{s})$ is equal to $|F(\boldsymbol{s})|^2(N_1 N_2 N_3)^2$. Otherwise, $I(\boldsymbol{s})$ can be regarded as zero for relatively large values of N_1, N_2, and N_3. The Laue condition gives not only diffraction directions but also diffraction intensities.

The unit cell of a given crystal can be assigned to one of seven crystal systems shown in Table 10.1. If lattice points are only on vertexes of the unit cell, we call it the primitive lattice (*P*), whose total number is 7. If not, it is called the complex lattice (*C*: base-centered, *I*: body-centered, and *F*: face-centered lattices). The introduction of complex lattices makes it easy to analyze the crystal structure because of higher symmetry as compared with corresponding primitive lattices. As a result,

TABLE 10.1 Seven Crystal Systems and Corresponding Bravais Lattices

Crystal System	Lattice Constants	Bravais Lattice
Triclinic	$a \neq b \neq c,\ \alpha \neq \beta \neq \gamma \neq 90°$	P
Monoclinic	$a \neq b \neq c,\ \alpha = \gamma = 90° \neq \beta$	P, C
Orthorhombic	$a \neq b \neq c,\ \alpha = \beta = \gamma = 90°$	P, C, I, F
Tetragonal	$a = b \neq c,\ \alpha = \beta = \gamma = 90°$	P, I
Rhombohedral	$a = b = c,\ \alpha = \beta = \gamma \neq 90°$	P
Hexagonal	$a = b \neq c,\ \alpha = \beta = 90°,\ \gamma = 120°$	P
Cubic	$a = b = c,\ \alpha = \beta = \gamma = 90°$	P, I, F

we obtain 14 Bravais lattices composed of 7 primitive and 7 complex lattices. By considering 32 point groups, symmetrical elements with translation (screw axes and glide planes), and 14 Bravais lattices, it can be concluded that the total number of arrangement manners of points in a crystal (i.e., a space group) is 230. The details of space groups are found in International Tables for Crystallography [4]. A given crystal can be assigned to 1 of 230 space groups depending on its symmetry. The systematic absence of (*hkl*) reflections (i.e., extinction rule) helps us judge the space group (e.g., only even number (0*k*0) reflections are observed for $P12_11$). On the aforementioned X-ray crystallography, more detailed information will be found in several textbooks [2, 5–9].

The crystal structure of homopolymers (i.e., molecular conformation and packing manner) is governed by both intra- and intermolecular interactions. The molecular conformation is mainly determined by the intramolecular interaction. For example, it is known that polyethylene forms a planer zig-zag (all-trans) conformation [10], whereas isotactic polypropylene takes a helical conformation due to steric repulsion between side substituents [11]. Many different types of helical conformations have been reported depending on the tacticity, rigidity of backbones, and bulkiness of side chains. The packing manner is, on the other hand, mainly driven by the intermolecular interaction between neighboring chains such as van der Waals interaction and hydrogen bonding (e.g., poly(vinyl alcohol) [12], polyamides [13], or poly(ethylene terephthalate) [14]). The lists of unit cell parameters of many crystalline homopolymers including the molecular conformation and packing manner are found in Polymer Handbook [15].

WAXD provides information on crystal structures in a 10^{-1}–10^{0} nm scale. In general, X-ray diffraction from synthetic polymers is poor in both intensities and numbers, as compared with low-molecular-weight compounds and proteins, due to many kinds of disorders such as the molecular weight distribution, tacticity, branching, folding, entanglement, and so on. Even after annealing for a long time, perfect crystals cannot be obtained, that is, there are both crystalline and amorphous regions in crystallized polymers. The former is observed as diffraction peaks and the latter as a broad scattering (amorphous halo). The crystallinity can be evaluated from a ratio of total areas for the amorphous halo and diffraction peaks. The Ruland method simultaneously provides the crystallinity and the degree of crystal imperfections [2, 7, 9, 16]. It is generally difficult to determine polymer crystal structures (i.e., atomic coordinates) only from powder diffraction data of unoriented samples. Uniaxially oriented samples (film or fiber) are alternatively used to determine the crystal structure, which is known as an X-ray fiber diffraction method [8, 9, 17].

10.2.2 Lamellar Morphology

Crystalline homopolymers generally form a *lamellar morphology*, an alternating structure consisting of crystalline and amorphous layers (Fig. 10.2c). The crystalline layer is made up of perfect crystals described in Section 10.2.1 with a typical thickness of several nanometers (Fig. 10.2b). Constituent chains in the amorphous layer are similar in conformation to amorphous homopolymers except that one or both ends are tethered at interfaces between both layers. It is widely recognized that the entanglement of long chains in a homogeneous melt, which significantly affects the dynamics of polymer crystallization, is responsible for the formation of such lamellar morphology. As a result, the crystallinity (i.e., volume (or weight) fraction of crystalline layers) is significantly smaller than one, which can moderately be controlled by changing crystallization conditions applied to the system.

The details of the lamellar morphology, such as crystalline layer thickness l_c and amorphous layer thickness l_a, are quantitatively evaluated using SAXS [18]. For example, they are conveniently derived from the one-dimensional correlation function, that is, Fourier transform of SAXS curves, assuming an ideal lamellar morphology without any distribution for l_c and l_a [19]. More detailed information on the lamellar morphology can be obtained by fitting theoretical scattering curves (or theoretical one-dimensional correlation functions) calculated from some appropriate model to SAXS curves (or Fourier transform of SAXS curves) experimentally obtained. The Hosemann model in reciprocal space [20] and the Vonk model in real space [7, 21] are often employed for such purposes.

In the Hosemann model, which is based on a paracrystalline model, each crystalline region consists of N alternating stacks of crystalline and amorphous layers. The scattered intensity $I(s)$ calculated from this model as a function of s is given by

$$I(s) \propto \frac{1}{2\pi^2 s^2}\left[\mathrm{Re}\left\{N\frac{(1-f_c)(1-f_a)}{1-f_cf_a}\right\} + \mathrm{Re}\left\{f_a\left(\frac{1-f_c}{1-f_cf_a}\right)^2\left(1-f_c^Nf_a^N\right)\right\}\right] \tag{10.5}$$

where N represents the average number of repeating units of both layers, and f_c and f_a are the Fourier transform of thickness distribution functions $h_c(x)$ (with average thickness l_c and deviation β_c) and $h_a(x)$ (l_a and β_a) for the crystalline and amorphous layers, respectively, which are given by Gaussian functions:

$$h_c(x) = \frac{1}{\beta_c(2\pi)^{1/2}} \exp\left\{ -\frac{(x-l_c)^2}{2\beta_c^2} \right\} \tag{10.6}$$

$$h_a(x) = \frac{1}{\beta_a(2\pi)^{1/2}} \exp\left\{ -\frac{(x-l_a)^2}{2\beta_a^2} \right\} \tag{10.7}$$

The parameters included in this model can be evaluated by fitting Equation 10.5 to $I(s)$ experimentally obtained.

The Vonk model is characterized by an infinite number of alternating units of crystalline and amorphous layers. The one-dimensional correlation function $\gamma(x)$ derived from this model is expressed as

$$\gamma(x) = \frac{\chi_c}{1-\chi_c} \left\{ \frac{1}{\chi_c^2} \int_0^\infty (x_c - x)\, h_c(x_c)\, \mathrm{d}x_c + P_{cac} + P_{cacac} + \cdots - 1 \right\} \tag{10.8}$$

where χ_c is a linear crystallinity defined as $\chi_c = l_c/(l_c+l_a)$ and the terms $P_{ca\ldots c}$ indicate convolution products of the type $q_c(x)h_a(x)h_c(x)\ldots q_c(x)$, in which

$$q_c(x) = \int_x^\infty h_c(x_c)/\chi_c\, \mathrm{d}x_c \tag{10.9}$$

Four parameters, l_c, l_a, β_c, and β_a, can be determined by curve-fitting procedures using the long period L $(=l_c + l_a)$ determined from the first maximum of experimental correlation functions. However, the parameters obtained do not have an equal accuracy.

10.2.3 Spherulite Structure

The *spherulite* is a sphere-shaped superstructure (Fig. 10.2d), and usually observed when crystalline polymers are quenched or slowly cooled from a homogeneous melt into low temperatures. The typical dimension of one spherulite is comparable to the wavelength of light, so that the appearance of spherulite structures makes polymer films cloudy. The diameter of one spherulite is sometimes larger than 1 mm, and we can observe it directly without using any microscope. In general, many spherulites come out simultaneously in the system, and the growth of these spherulites stops completely when neighboring spherulites impinge on their interface (Fig. 10.2e). The spherulite has two characteristic patterns when it is observed using a polarized microscope (PM); extinction rings and Maltese crosses, as shown in Figure 10.3a. One spherulite consists of many twisted crystalline layers radiating from a spherulite center, as schematically illustrated in Figure 10.3b, although the origin of this twisting of crystalline layers is not fully understood. The refractive index ellipsoid of crystalline layers periodically takes an isotropic state during twisting, where banded rings appear under PM. These banded rings can be observed in crystalline homopolymers, binary homopolymer blends, and crystalline block copolymers [22–29], and it is known that molecular characteristics (e.g., molecular weight or molecular weight distribution) and crystallization conditions (e.g., crystallization temperature or impurities included in the system) critically affect the periodic distance of banded rings and their regularity. Maltese crosses are observed when the long axis of refractive index ellipsoids is parallel or perpendicular to the polarizer or analyzer in PM. Therefore, it appears at every 90°, which is the same direction to the polarizer or analyzer.

The details of spherulite structures are usually evaluated using SALS [30] as well as PM. The SALS result shows a four-leaf clover pattern, from which it is possible to obtain an average size of spherulites existing in the system. Therefore, SALS makes it possible to pursue the growth behavior of spherulites (i.e., average size of spherulites as a function of crystallization time) during isothermal crystallization. PM is employed to intuitively understand a difference in the characteristics of each spherulite structure formed under different crystallization conditions. It is also possible to evaluate the average growth rate of spherulites from a persevering observation using PM.

10.2.4 Crystalline Morphology of Homopolymers Confined in Isolated Nanodomains

It is revealed from recent experimental studies that the crystallization of homopolymers spatially confined in small isolated spaces (nanodomains), such as nanospheres or nanocylinders with a dimension of ~10 nm, is extremely different from that of bulk homopolymers with no spatial confinement. That is, existing nanodomains significantly affect the crystallization behavior of confined homopolymers to yield a unique crystalline morphology in the system. For example, it is intuitively supposed that the nanodomain will prevent the formation of morphologies larger than its dimension (e.g., lamellar morphology or spherulite structure). The nanodomain in which crystalline homopolymers are completely confined can be prepared using several methods; micelles in solution [31, 32], anodic aluminum oxides (AAO) [33–43], or microdomain structures provided by crystalline–amorphous diblock copolymers [44–47], where the dimension and shape of nanodomains critically affect the crystalline morphology such as crystallinity, crystalline layer thickness, crystal orientation, and so on. Here, we briefly describe the crystal orientation of homopolymers confined in nanocylinders.

The crystalline homopolymers spatially confined in nanocylinders can be prepared using the microphase separation of block copolymers, followed by the photocleavage of block junctions (Fig. 10.4), where *o*-nitrobenzyl (ONB) groups inserted between different blocks are conveniently used for the photocleavage reaction [48]. Nojima et al. [44–47] synthesized poly(ε-caprolactone)-*block*-polystyrene (PCL-*b*-PS) diblock copolymers with ONB between PCL and

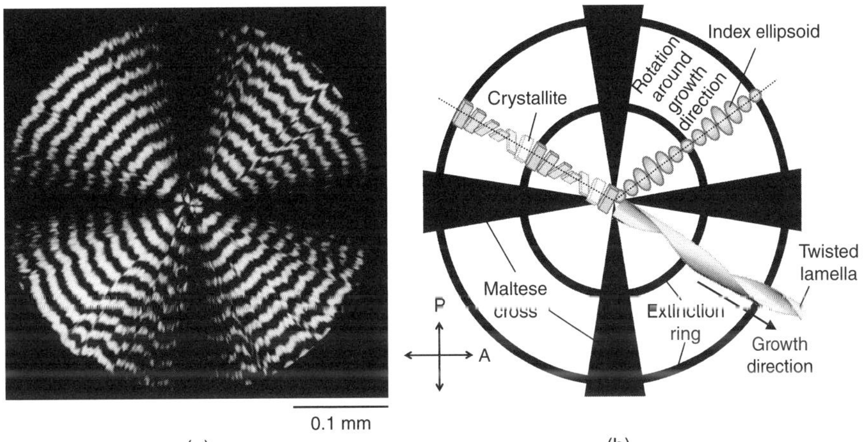

Figure 10.3 (a) A typical spherulite of poly(ϵ-caprolactone) (PCL) homopolymers in a PCL/poly(vinyl chloride) blend, and (b) a schematic illustration showing the inside of spherulites.

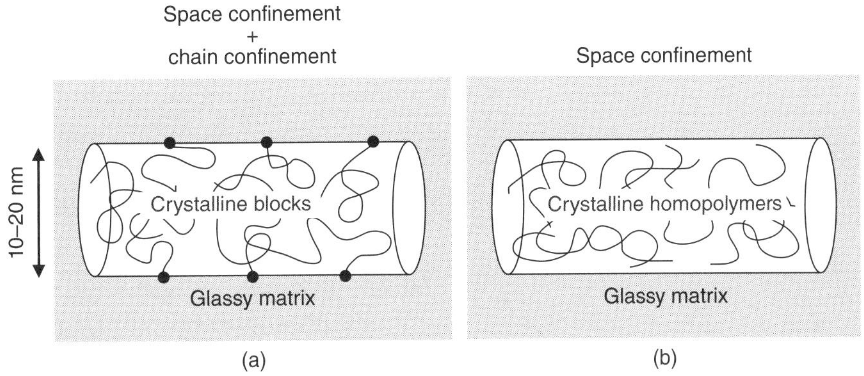

Figure 10.4 Schematic illustration showing (a) crystalline blocks and (b) crystalline homopolymers both confined in an identical nanocylinder. The crystalline homopolymer is obtained using the photocleavage reaction of ONB inserted between different blocks (closed circles in a).

PS blocks. After forming nanocylinders by the microphase separation of block copolymers (Fig. 10.4a), ONB was cleaved using UV irradiation to get PCL homopolymers confined in nanocylinders surrounded with PS matrices (Fig. 10.4b), where the vitrification of PS blocks prevented macrophase separation between PCL and PS homopolymers after the photocleavage. As a result, they obtained crystalline PCL blocks and PCL homopolymers both confined in identical nanocylinders with varying diameters. Therefore, it was possible to evaluate the difference in crystallization behavior and crystalline morphology between PCL blocks and PCL homopolymers, and also to clarify the characteristics of homopolymer crystallization spatially confined in nanocylinders.

The crystal orientation of PCL homopolymers spatially confined in nanocylinders showed that the *c*-axis of PCL crystals (stem direction of PCL chains) was always perpendicular to the cylinder axis but the crystal growth direction depended significantly on nanocylinder diameter *D* and crystallization temperature T_c; the *b*-axis (fastest growth axis of PCL crystals) was parallel to the cylinder axis in nanocylinders with smaller *D* ($\leq$13.0 nm), whereas the (110) plane of PCL crystals (fastest growth plane of PCL crystals) was normal in those with larger *D* ($\geq$14.9 nm). That is, the growth mode of PCL crystals was definitely different between smaller and larger nanocylinders to yield a moderate difference in the PCL crystallinity. In addition, the degree of crystal orientation increased remarkably with increasing T_c for PCL homopolymers (Fig. 10.5), whereas it improved slightly for PCL blocks, suggesting that the restricted mobility of PCL blocks with one chain end tethered at nanocylinder interfaces suppresses favorable crystal growth during isothermal crystallization at higher T_c. Effects of spatial confinement on the crystallization behavior and crystalline morphology are not fully understood at present, and therefore extensive studies are strongly anticipated.

10.2.5 Crystalline Morphology of Polymer Blends

The crystallization of homopolymers in binary blends is more complicated as compared with that of neat homopolymers. This is because the second homopolymer, usually amorphous homopolymer, may accelerate the crystallization by working as a diluent or decelerate it by disturbing the diffusion of

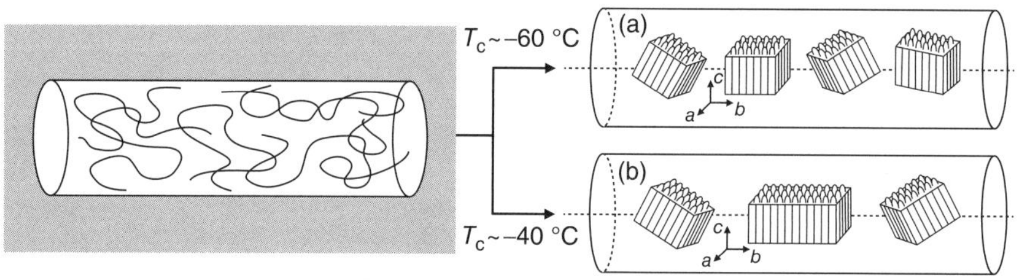

Figure 10.5 Schematic illustration showing the difference in crystal orientation between cylindrically confined PCL homopolymers crystallized at −60 °C (a) and −40 °C (b).

crystalline homopolymers during nucleation and/or growth processes. The crystalline morphology formed in miscible binary blends has extensively been studied up to now, which is summarized in several reviews [49, 50].

In miscible binary blends, amorphous homopolymers are completely accommodated within amorphous layers of the lamellar morphology formed after the crystallization of crystalline homopolymers. Stein et al. [51], for example, observed the lamellar morphology formed in a miscible blend of PCL and poly(vinyl chloride) (PVC) using SAXS as a function of composition. They found that PVC existed in amorphous layers of the lamellar morphology to yield a linear increase in the amorphous layer thickness with increasing PVC composition, whereas the crystalline layer thickness remained constant irrespective of composition. Wenig et al. [52] obtained similar results for a miscible blend of poly(2,6-dimethylphenylene oxide) (PPO) and isotactic polystyrene (iPS). However, a different result was reported for a miscible blend of iPS and atactic polystyrene [53], where the amorphous layer thickness was almost constant irrespective of composition. Stein et al. [51] explained this difference in terms of the diffusion rate of amorphous homopolymers and the crystal growth rate of crystalline homopolymers.

The lamellar morphology formed in miscible blends is usually observed using SAXS as described in Section 10.2.2, because its typical repeating distance is in the order of ~10 nm. In addition, a combination of SAXS and differential scanning calorimetry (DSC) results straightforwardly provides the crystalline layer thickness and amorphous layer thickness if we assume a perfect lamellar morphology with no transition zone between both layers.

More complicated situation arises when the binary blend has a liquid–liquid phase-separated region (such as an upper critical solution temperature (UCST)-type phase diagram) at temperatures where constituent polymers crystallize, as shown in Figure 10.6. In this case, we suppose an intimate interplay between crystallization and phase separation. That is, the phase separation may assist the crystallization of constituent homopolymers (case A in Fig. 10.6a) or the crystallization may promote further phase separation between components (case B). That is, existing phase-separated regions will significantly influence the crystallization. Consequently, the resulting morphology is unique; it consists of two domains, one is rich in crystalline homopolymers to form the

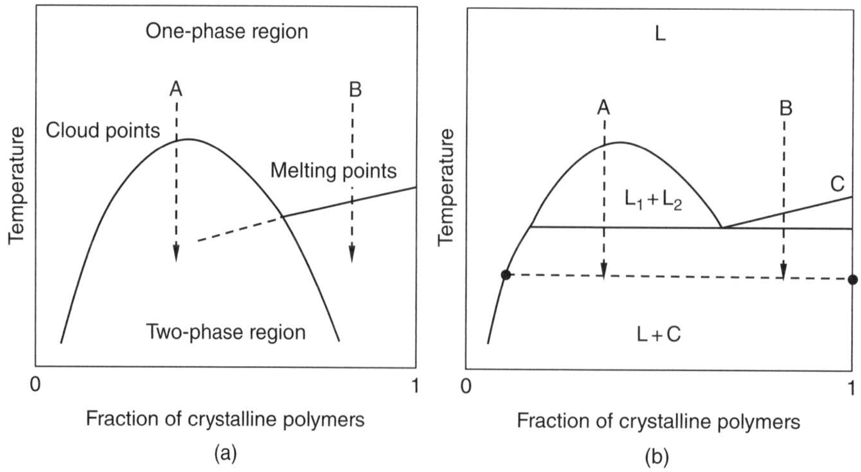

Figure 10.6 (a) Cloud points and melting points for a binary crystalline/amorphous homopolymer blend having an UCST-type phase-separated region. (b) Equilibrium phase diagram expected for this binary blend.

lamellar morphology inside and the other rich in amorphous homopolymers with no crystallization, although the phase diagram at equilibrium predicts a coexistence of solid (or perfect crystal) and liquid phases (Fig. 10.6b). Tanaka and Nishi [54, 55], for example, observed the local phase separation at the growth front of spherulites in a binary blend of PCL and PS homopolymers with an UCST-type phase diagram. They explained this phenomenon in terms of an intimate coupling of PCL crystallization and phase separation between PCL and PS. Some unique morphologies have been reported for several binary blends with an UCST-type phase diagram [56–61].

The miscible blends consisting of two crystalline homopolymers may form unique morphologies in the system, and several studies have been reported [62–67]. However, very few such blends are miscible, so that information on the crystalline morphology formed in crystalline/crystalline polymer blends is very limited.

10.3 CRYSTALLINE MORPHOLOGY OF BLOCK COPOLYMERS

It is widely recognized that amorphous–amorphous diblock copolymers form a variety of microdomain structures when the segregation strength between different blocks is moderately large. When one block is crystalline and the other is amorphous (i.e., crystalline–amorphous diblock copolymers), it is easily supposed that the morphology formation at low temperatures is driven by a close interplay between crystallization and microphase separation. That is, the crystallization will start from some microdomain structure already formed in the block copolymer, where the stability of this microdomain structure against crystallization is an important factor in controlling final crystalline morphologies (Fig. 10.7). That is, two different morphologies, that is, *lamellar morphology* and *crystalline microdomain structure*, are formed depending on static (or thermodynamic) and dynamic reasons listed in Figure 10.7. When the microdomain structure is not stable, it is completely replaced with the lamellar morphology by the crystallization (break-out crystallization, upper route in Fig. 10.7), whereas the crystalline microdomain structure is formed when microdomain structures in the melt are sufficiently stable against crystallization or it is substantially frozen due to dynamic reasons (e.g., glass transition of amorphous blocks) (confined crystallization, lower route in Fig. 10.7). The crystallization behavior and resulting morphology of many crystalline–amorphous diblock copolymers have extensively been studied during the past two decades, and several review articles are now available for this subject [68–77].

When both blocks are crystalline, the morphology formation is more complicated because two kinds of crystallization, as well as microphase separation, intervene during morphology formation. It is useful to summarize this morphology formation in terms of a difference in the crystallizable temperature T_c of both blocks (Fig. 10.8). When T_c of one block is considerably higher than that of the other block, two crystallizations occur almost independently (two-step crystallization). That is, the crystallization of higher T_c blocks

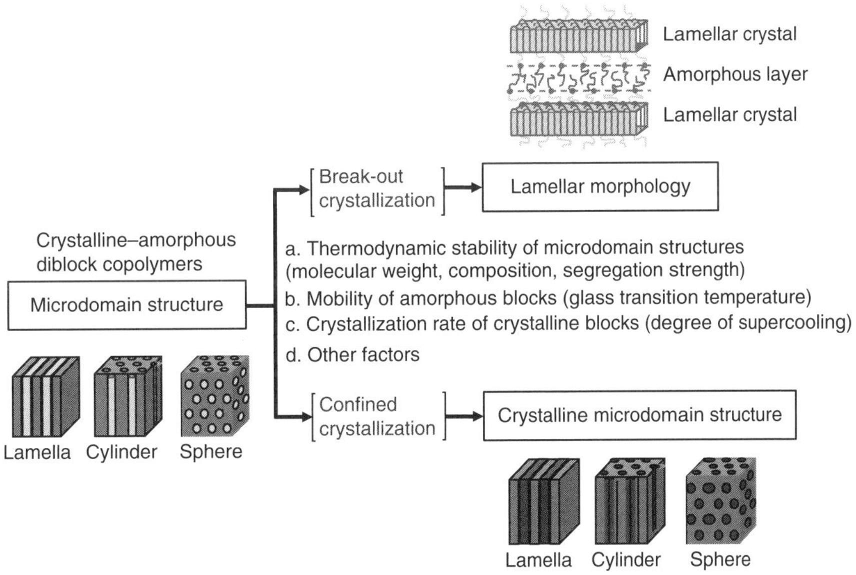

Figure 10.7 Schematic illustration showing the possible morphology formation in crystalline–amorphous diblock copolymers by the crystallization of constituent blocks. The upper route represents break-out crystallization, that is, the microdomain structure is completely replaced with the lamellar morphology, whereas the lower route shows confined crystallization, where the microdomain structure is preserved after crystallization. a–d indicate driving factors for the morphology formation.

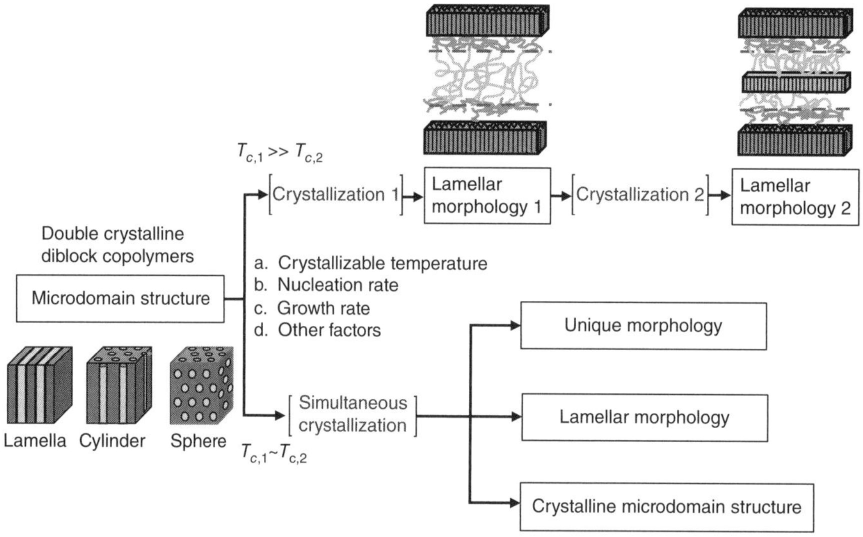

Figure 10.8 Schematic illustration showing the possible morphology formation in crystalline–crystalline diblock copolymers by two kinds of crystallization. The upper route represents two-step crystallization, whereas the lower route shows simultaneous crystallization, where an interactive crystallization is usually observed. a–d indicate driving factors for the morphology formation.

yields the lamellar morphology, and subsequently lower T_c blocks crystallize within this morphology (upper route in Fig. 10.8). When T_c values of both blocks are sufficiently close, on the other hand, we expect a simultaneous crystallization, that is, the crystallization processes of two blocks interact with each other to yield a unique morphology in the system (lower route in Fig. 10.8).

10.3.1 Crystalline Morphology of Weakly Segregated Block Copolymers

The thermodynamic stability of microdomain structures is usually expressed using a parameter χN (χ: the Flory–Huggins interaction parameter between different blocks, N: the overall degree of polymerization of block copolymers). When χN is not large, that is, the microdomain structure is not stable against subsequent crystallization, it is completely replaced with the lamellar morphology after crystallization (upper route in Fig. 10.7), which is an alternating structure consisting of crystalline and amorphous layers. This lamellar morphology is different from the usual lamellar morphology of crystalline homopolymers (Section 10.2.2) in that amorphous blocks are forced to lie in amorphous layers and the conformation of these blocks is moderately enlarged in a direction perpendicular to lamella surfaces.

If we suppose the conformation of crystalline and amorphous blocks in the lamellar morphology from a thermodynamic viewpoint, random coil conformations are most favorable for amorphous blocks and perfect crystals with no chain folding for crystalline blocks. However, these two conditions are not compatible in the lamellar morphology of block copolymers, because both blocks are linked with a covalent bond and hence enough space is not available for the amorphous block to take a random coil conformation on the crystalline layer with small chain-folding numbers (Fig. 10.9). As a result, the conformation of both blocks at equilibrium is driven by a critical balance between an entropic contribution of amorphous blocks and an energetic one of crystalline blocks; amorphous blocks are slightly enlarged in the direction perpendicular to lamella surfaces to reduce substantial areas on crystalline layers, whereas crystalline blocks take extra chain foldings to provide some space for the accommodation of amorphous blocks.

DiMarzio et al. [78] and Whitmore and Noolandi [79] theoretically predicted an equilibrium lamellar morphology formed in crystalline–amorphous diblock copolymers. The long period of the lamellar morphology L, that is, a sum of crystalline layer thickness and amorphous layer thickness, is expressed by a scaling form:

$$L \propto (M_c + M_a)M_a^{\beta} \quad (10.10)$$

where M_c and M_a are the molecular weight of crystalline and amorphous blocks, respectively. The value of β is predicted to be −1/3 by DiMarzio et al. [78] and −5/12 by Whitmore and Noolandi [79]. That is, L decreases with increasing M_a when the total molecular weight ($=M_c + M_a$) is constant, which is intuitively understood by considering that crystalline blocks take thinner crystalline layers to accommodate larger amorphous blocks to yield smaller L with increasing M_a. Many experimental results are reported on the crystallization

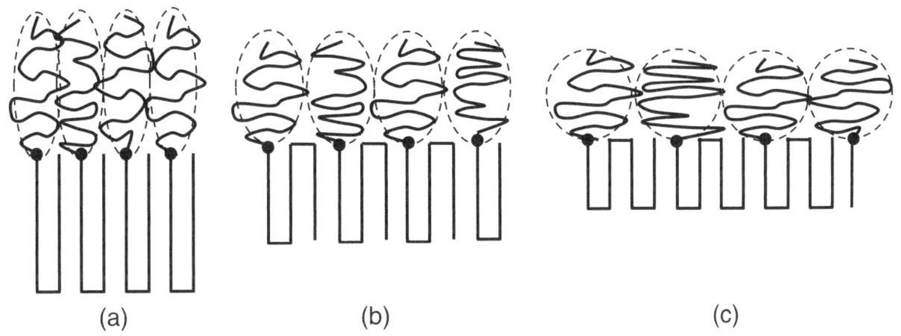

Figure 10.9 Illustration showing possible conformations of crystalline and amorphous blocks in the lamellar morphology of crystalline–amorphous diblock copolymers. (a) $n_f = 1$, (b) $n_f = 2$, and (c) $n_f = 3$, where n_f represents the chain-folding number of crystalline blocks.

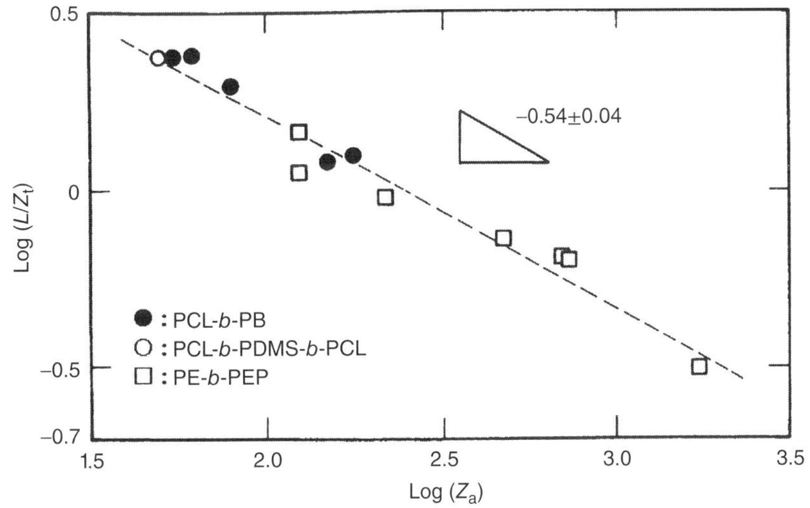

Figure 10.10 Plots of log (L/Z_t) $(Z_t = Z_c + Z_a)$ against log Z_a. Z_c ($\propto M_c$) and Z_a ($\propto M_a$) represent the segment number of crystalline and amorphous blocks, respectively. Closed circle: data for PCL-*b*-PB [85], open square: data for PE-*b*-PEP [84], open circle: data for PCL-*b*-PDMS-*b*-PCL [83]. Nojima et al. [85]. Reproduced with permission of The Society of Polymer Science, Japan.

behavior and the resulting morphology of weakly segregated crystalline–amorphous diblock copolymers [80–92], and the lamellar morphology at (quasi-)equilibrium is quantitatively examined using several diblock copolymers. It is confirmed by plotting log $\{L/(M_c + M_a)\}$ against log M_a that Equation 10.10 holds successfully for PCL-*block*-polybutadiene (PCL-*b*-PB), PCL-*block*-poly(dimethyl siloxane)-*block*-PCL (PCL-*b*-PDMS-*b*-PCL), and polyethylene-*block*-poly(ethylene-*alt*-propylene) (PE-*b*-PEP) copolymers (Fig. 10.10). The value of β thus evaluated is -0.54 ± 0.04, which is roughly consistent with the theoretical predictions (Eq. 10.10). It is also possible to estimate the chain-folding number of crystalline blocks n_f in the lamellar morphology. For example, Lee and Register [93] reported using a series of hydrogenated polynorbornene-*block*-hydrogenated poly(ethyl norbornene) copolymers (hPV-*b*-hPEN) that n_f increased discontinuously with increasing M_a of hPEN.

10.3.2 Crystalline Morphology of Block Copolymers with Glassy Amorphous Blocks

When the glass transition temperature of amorphous blocks T_g is significantly higher than the crystallizable temperature of crystalline blocks, amorphous blocks are vitrified and cannot move substantially during crystallization. Consequently, the existing microdomain structure is completely preserved through the crystallization process, and eventually crystalline blocks crystallize within this microdomain structure to yield a crystalline microdomain structure (lower route in Fig. 10.7). Amorphous domains in the crystalline microdomain structure are supposed to be hard, so that any deformation of crystalline domains is strictly prohibited. This fact intuitively explains unique experimental results such as a large supercooling necessary for the crystallization and extremely low crystallinity.

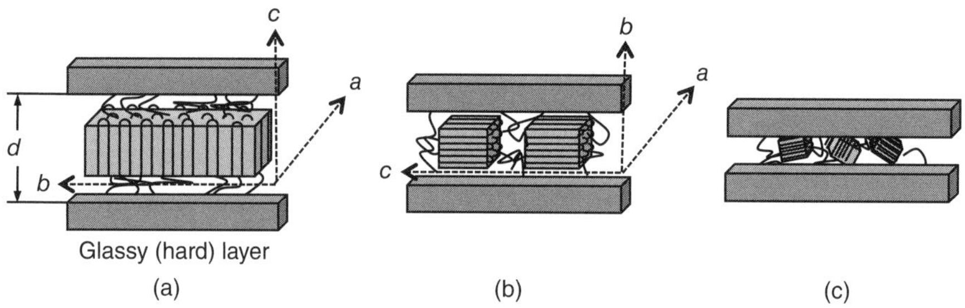

Figure 10.11 Schematic illustration showing the crystal orientation of crystalline blocks confined in hard nanolamellae with varying *d*. (a) $d \sim 11$ nm, (b) $d \sim 9$ nm, and (c) $d \sim 6$ nm.

Many experimental results are reported for crystalline microdomain structures formed in crystalline–amorphous diblock copolymers with high-T_g blocks [94–105], where the orientation of lamellar crystals spatially confined in nanocylinders or nanolamellae is interesting because the orientation depends significantly on confinement conditions (such as confinement shape or size) as well as crystallization conditions (crystallization temperature). Chung et al. [105], for example, investigated a relationship between the crystal orientation of PCL blocks and nanocylinder diameter D for cylindrically microphase-separated PCL-*block*-poly(4-vinylpyridine) (PCL-*b*-P4VP) block copolymers. They found that the crystal growth direction (*b*-axis) was parallel to the cylinder axis in nanocylinders with larger D, whereas the crystal orientation was random in those with smaller D. They explained this difference in terms of the dynamics of crystal growth within restricted nanocylinders.

The crystal orientation of crystalline blocks confined in nanolamellae is also reported [100, 103]. Sun et al. [103], for example, investigated the crystal orientation of PCL blocks confined in lamellar microdomains of PCL-*b*-P4VP as a function of crystalline layer (i.e., PCL layer) thickness d. When d was larger (~11 nm), the *c*-axis of PCL crystals was perpendicular to lamella surfaces (Fig. 10.11a), but parallel at smaller d (~9 nm) (Fig. 10.11b), and finally random at extremely small d (~6 nm) (Fig. 10.11c). This difference in crystal orientation is successfully explained by the dynamics of crystal growth within confined nanolamellae. In addition, the crystal orientation also depends significantly on the crystallization temperature T_c. Zhu et al. [100] reported the crystal orientation of poly(ethylene oxide) (PEO) blocks spatially confined in lamellar microdomains of PEO-*b*-PS as a function of T_c. They found that the *c*-axis of PEO crystals was perpendicular to lamella surfaces at higher T_c (similar to Fig. 10.11a), whereas it was parallel at lower T_c (Fig. 10.11b).

It is clear from many experimental results that crystalline blocks confined in isolated nanodomains surrounded by glassy (or hard) matrices show unique crystal orientations as compared with those by soft matrices (Section 10.3.3.). However, the physical properties of such block copolymers with crystalline microdomain structures have scarcely been investigated up to now.

10.3.3 Crystalline Morphology of Strongly Segregated Block Copolymers

When χN of crystalline–amorphous diblock copolymers is sufficiently large, the soft microdomain structure is stable against the subsequent crystallization. Therefore, this structure is preserved through the crystallization process, that is, constituent blocks crystallize within the soft microdomain structure, to yield a crystalline microdomain structure (lower route in Fig. 10.7). Amorphous domains in the crystalline microdomain structure are not hard in this case, so that crystalline domains can deform moderately during crystallization in order to get a larger crystallinity and/or favorable crystal orientation, which is critically different from the crystallization of block copolymers with high-T_g amorphous blocks, as described in Section 10.3.2.

Several experimental studies are reported on the crystalline morphology and crystallization behavior of crystalline blocks confined in soft nanospheres, nanocylinders, or nanolamellae [106–110]. Loo et al. [107] examined the crystallization behavior of PE blocks confined in a nanosphere of a PE-*block*-poly(styrene-ethylene-butene) copolymer using time-resolved SAXS, and found that the isothermal crystallization behavior showed a first-order kinetics, sharply contrasting with a sigmoidal change usually observed in crystalline homopolymers. Chen et al. [108] studied the crystalline morphology formed in a nanosphere provided by a binary blend of PEO-*b*-PB copolymers and PB homopolymers using SAXS and TEM, and showed a moderate deformation and volume contraction of PEO nanospheres upon crystallization to form ellipsoid objects with an aspect ratio of ca. 1.3. Ho et al. [110] investigated the crystal orientation of poly(L-lactide) (PLLA) blocks in PLLA-*b*-PS diblock copolymers using two-dimensional SAXS (2D-SAXS) and 2D-WAXD, where the glass transition temperature of PS blocks was close to the crystallizable temperature of PLLA blocks. Therefore, it was possible to investigate the crystal orientation of PLLA blocks confined in soft PS nanolamellae by setting appropriate crystallization temperatures T_c. They found that the *c*-axis of PLLA crystals was always perpendicular to lamella surfaces irrespective of T_c and dimensions of confinement spaces (i.e., PLLA layer thickness), but soft PS

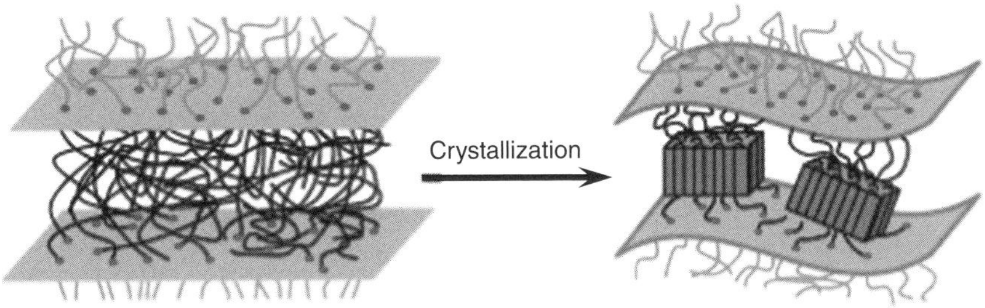

Figure 10.12 Schematic illustration showing undulated nanolamellae after the crystallization of constituent blocks

layers were undulated after the crystallization of PLLA blocks (Fig. 10.12). They speculated that this undulation might be ascribed to a change in crystallization mechanisms due to a change in the confinement nature (hard or soft). Actually, this change coincided with a transition from homogeneous nucleation (for hard confinement) to heterogeneous nucleation (for soft-confinement).

Model crystalline–amorphous diblock copolymers with a sufficiently large χN are not easy to synthesize. Therefore, experimental studies on the confined crystallization within soft nanodomains are very limited, and a relationship between the detailed crystalline morphology and the deformation of soft nanodomains is unclear.

10.3.4 Crystalline Morphology of Double Crystalline Block Copolymers

The crystalline morphology formed in crystalline–crystalline diblock copolymers is more complicated as compared with that in crystalline–amorphous diblock copolymers, because two kinds of crystallization start from some microdomain structure existing in the melt. It is useful to classify this crystallization into two cases in terms of the crystallizable temperature T_c of both blocks (Fig. 10.8): two-step crystallization when T_c of one block is significantly higher than that of the other, and simultaneous crystallization when both T_c values are sufficiently close.

The two-step crystallization has been examined using several crystalline–crystalline diblock copolymers [111–124]. Essential points of this crystallization are summarized in the upper route of Figure 10.8; first, the crystallization of higher T_c blocks forms the lamellar morphology (Fig. 10.2c) in which lower T_c blocks are completely accommodated, and subsequently they crystallize within this lamellar morphology. Therefore, the lamellar morphology is a kind of spatial confinement against the crystallization of lower T_c blocks. This confinement is expected to be intermediate between hard confinement by glassy layers (Section 10.3.2) and soft confinement by rubbery layers (Section 10.3.3), because confinement layers consist of hard crystals covered with soft (or uncrystallized) chains. Nojima et al. [111–117] investigated the crystalline morphology formed in PCL-*b*-PE copolymers by the two-step crystallization of PE and PCL blocks, where PE blocks always crystallized first by quenching from a microphase-separated melt to yield the lamellar morphology consisting of crystalline PE layers and amorphous PE+PCL layers (PE lamellar morphology), and subsequently PCL blocks crystallized. They found that PCL blocks crystallized completely in the PE lamellar morphology at lower T_c (<30 °C), whereas it crystallized by deforming the PE lamellar morphology at higher T_c (>40 °C). Furthermore, it was found that the orientation of PCL crystals confined in the PE lamellar morphology depended significantly on PCL layer thickness d and T_c (Fig. 10.13). This crystal orientation at each d and T_c should be intimately related to crystallization mechanisms of PCL blocks confined in the PE lamellar morphology. It is also reported that existing lamellar morphologies significantly affect the crystallization of lower T_c blocks [124].

The simultaneous crystallization has extensively been examined using PCL-*b*-PEO copolymers [125–133], where several characteristic features were disclosed. Shiomi et al. [128] and Sun et al. [130], for example, reported the formation of interesting spherulites consisting of PCL cores

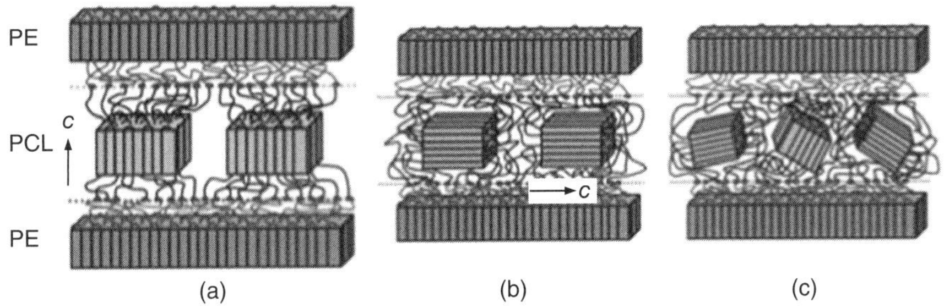

Figure 10.13 Schematic illustration showing the orientation of PCL crystals spatially confined in the PE lamellar morphology at different d and T_c. (a) 16.5 nm $\geq d \geq$ 10.7 nm (45 °C $\geq T_c \geq$ 0 °C), (b) $d \sim$ 8.8 nm (45 °C $\geq T_c \geq$ 25 °C), and (c) $d \sim$ 8.8 nm (20 °C $\geq T_c \geq$ 0 °C).

and PEO coronas, which was ascribed to a moderate difference in crystallization rates between PCL and PEO blocks. However, the electron density of amorphous PEO blocks is approximately equal to that of amorphous PCL blocks, so that SAXS data do not provide information on the phase state of molten PCL-*b*-PEO copolymers (i.e., miscible or microphase separated). Therefore, the effects of microdomain structures on the simultaneous crystallization behavior and the resulting crystalline morphology cannot be clarified only from SAXS measurements. The simultaneous crystallization is also investigated using other crystalline–crystalline diblock copolymers [124, 134–141], and it is found from these results that the crystallization behavior of one block is significantly influenced by the crystallization process of the other block (interactive crystallization) to yield a complicated crystalline morphology. However, the resulting crystalline morphology is not fully understood in terms of an intimate interplay between microphase separation and two kinds of crystallization.

10.4 CONCLUDING REMARKS

The crystallization of homopolymers yields a hierarchical structure in polymer materials, which substantially controls their physical properties. Therefore, the crystalline morphology of homopolymers has been one of the important research subjects in polymer science. In addition, the crystallization of homopolymers spatially confined in various nanodomains, such as micelles, AAO, or microdomain structures, may bring new information on crystallization mechanisms of homopolymers, because it will be possible to highlight a specific crystallization mechanism (e.g., nucleation or crystal growth) in the overall crystallization process consisting of several combined mechanisms. Furthermore, the crystallization in nanodomains has the possibility of providing new polymer materials, and their physical properties should be unique as compared with usual polymer materials. This is because the substantial control of nano-ordered structures formed in polymer materials will be possible by this crystallization, which is never achieved by the crystallization of neat homopolymers.

The basic research on the crystallization in more complicated systems started recently to find out unique morphologies formed in polymer systems. The crystallization of block copolymers is a striking example of such crystallization, which is intimately dependent on the molecular characteristics of crystalline block copolymers. For example, the crystallization of crystalline–amorphous diblock copolymers yields the lamellar morphology or crystalline microdomain structure depending on χN of block copolymers, T_g of amorphous blocks, crystallization conditions, and so on. These kinds of crystallization have the possibility of developing new crystalline polymer materials. Therefore, we strongly anticipate future advances in this research field.

REFERENCES

1. Flory PJ. *Principles of Polymer Chemistry*. USA: Cornell University Press; 1953.
2. Wunderlich B. *Macromolecular Physics Vol. 1 Crystal Structure, Morphology, Defects*. New York: Academic Press; 1973.
3. Mandelkern L. *Crystallization of Polymers Vol. 1 Equilibrium Concepts*. UK: Cambridge University Press; 2010.
4. Hahn T. *International Tables for Crystallography Vol. A: Space-Group Symmetry*. Dordrecht: Kluwer Academic Publishers; 2002.
5. Buerger MJ. *X-Ray Crystallography: An Introduction to the Investigation of Crystals by their Diffraction of Monochromatic X-Radiation*. New York: Wiley; 1942.
6. Klug HP, Alexander LE. *X-Ray Diffraction Procedures for Polycrystalline and Amorphous Materials*. New York: Wiley; 1974.
7. Baltá-Calleja FJ, Vonk CG. *X-Ray Scattering of Synthetic Polymers*. Amsterdam: Elsevier; 1989.
8. Hukins DWL. *X-Ray Diffraction by Disordered and Ordered Systems: Covering X-Ray Diffraction by Gases, Liquids, and Solids and Indicating How the Theory of Diffraction by these Different States of Matter Is Related and How It Can be Used to Solve Structural Problems*. Oxford: Pergamon Press; 1981.
9. Bodor G. *Structural Investigation of Polymers*. Chichester: Ellis Horwood; 1991.
10. Bunn CW. The crystal structure of long-chain normal paraffin hydrocarbons. The "shape" of the $>CH_2$ group. Trans Faraday Soc 1939;35:482–491.
11. Natta G, Corradini P. General considerations on the structure of crystalline polyhydrocarbons. II Nuovo Cimento 1960;15:9–39.
12. Bunn CW. Crystal structure of polyvinyl alcohol. Nature 1948;161:929–930.
13. Bunn CW, Garner EV. The crystal structures of two polyamides ('nylons'). Proc R Soc London A 1947;189:39–68.
14. de Daubeny RP, Bunn CW, Brown CJ. The crystal structure of polyethylene terephthalate. Proc R Soc London A 1954;226:531–542.
15. Brandrup J, Immergut EH. *Polymer Handbook*. 3rd ed. New York: Wiley; 1989.
16. Ruland W. X-ray determination of crystallinity and diffuse disorder scattering. Acta Crystallogr 1961;14:1180–1185.
17. Alfred DF, KennCorwin HG. *Fiber Diffraction Methods, ACS Symposium Series*. Vol. 141. Washington, D.C.: ACS Publications; 1980.
18. Roe RJ. *Methods of X-Ray and Neutron Scattering in Polymer Science*. New York: Oxford University Press; 2000.
19. Strobl GR, Schneider M. Direct evaluation of the electron density correlation function of partially crystalline polymers. J Polym Sci B 1980;18:1343–1359.
20. Hosemann R, Bagchi SN. *Direct Analysis of Diffraction by Matter*. Amsterdam: North-Holland; 1962.
21. Vonk CG, Kortleve G. X-ray small-angle scattering of bulk polyethylene. Colloid Polym Sci 1967;220:19–24.

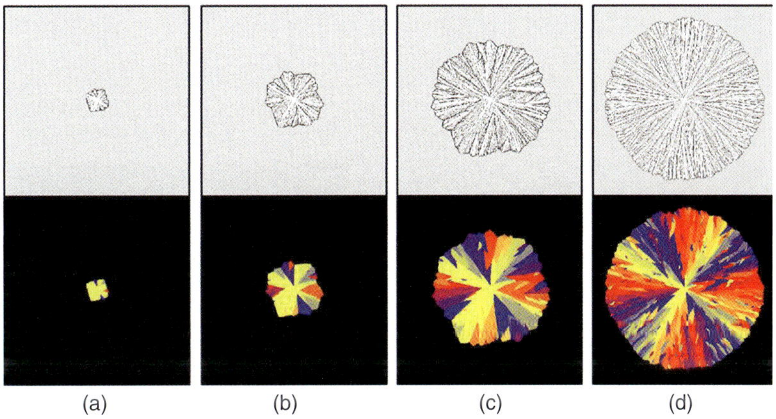

Figure 1.13 Phase field simulations of the growth of a spherulite. Upper row: composition maps. A grayscale map was used to increase the contrast. Lower row: orientation map. Schultz [56]. Reproduced with permission of American Physical Society.

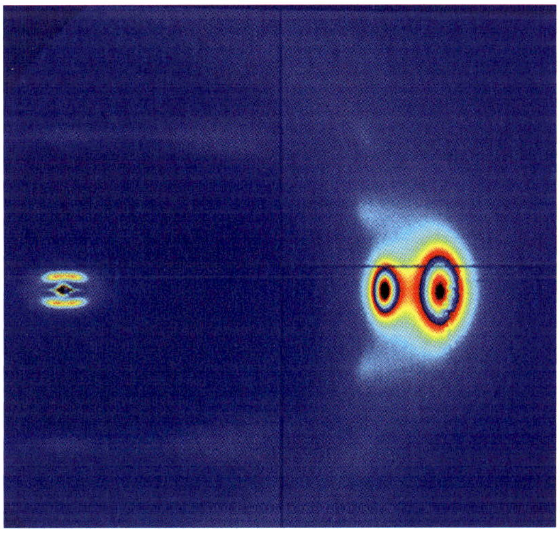

Figure 2.3 A combined SAXS and WAXS pattern from a drawn PA6 fiber. The small-angle pattern near the origin covers about 2.5° 2θ. The d-spacing of the SAXS reflection is ~100 Å. The d-spacings of the two wide-angle reflections along the equator are 4.4 and 3.8 Å [9].

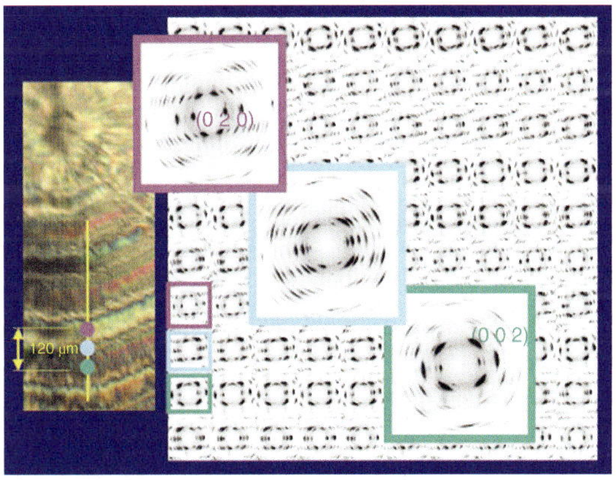

Figure 2.9 A series of wide-angle X-ray diffraction photographs from crystals located along the vertical line within a spherulite of poly(hydroxy butarate) shown in the left inset. The enlarged diffractograms are from three areas separated by 60 m as shown in the optical micrograph. Courtesy of C. Riekel, ESRF.

Polymer Morphology: Principles, Characterization, and Processing, First Edition. Edited by Qipeng Guo.

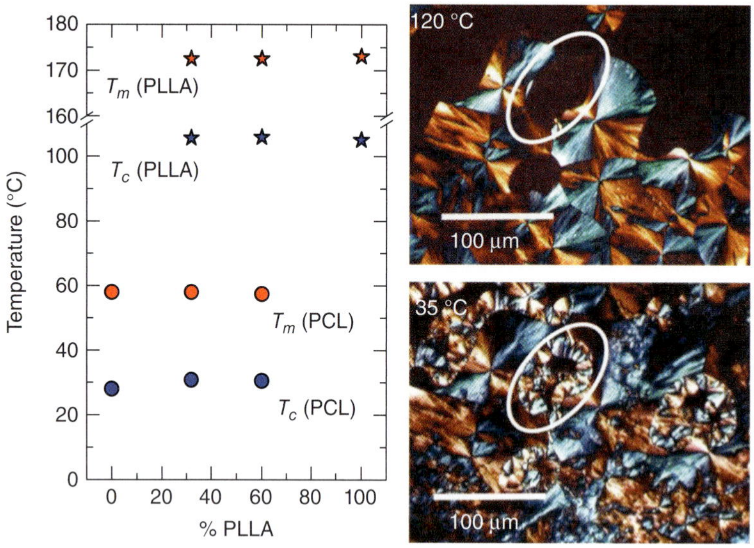

Figure 5.16 (a) Crystallization and melting temperatures (obtained from the data reported in Ref. [211]) for PLLA and PCL blends as a function of PLLA content. (b) Photomicrograph obtained by PLOM for the 32/68 PLLA/PCL blend. Castillo et al. [212]. Reproduced with permission of American Chemical Society.

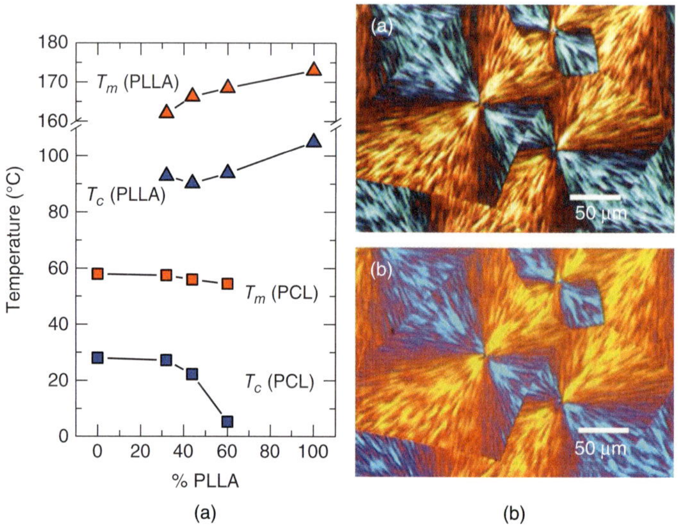

Figure 5.18 (A) Crystallization and melting temperatures (obtained from the data reported in Ref. [211]) for the PLLA and PCL block within all copolymers versus PLLA content. (B) Photomicrograph obtained by PLOM for the (a) $L_{32}{}^{7}C_{68}{}^{15}$ after 30 min at 122 °C. (b) $L_{32}{}^{7}C_{68}{}^{15}$ after 15 min at 42 °C. Castillo et al. [212]. Reproduced with permission of American Chemical Society.

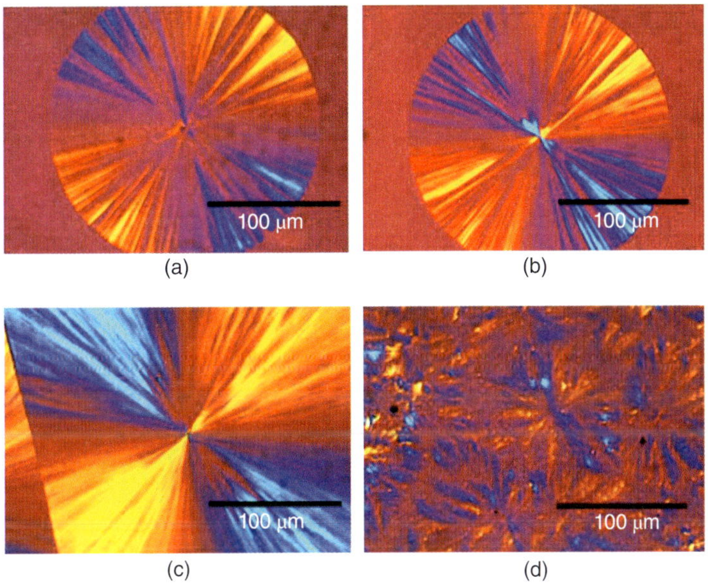

Figure 5.19 Polarized light optical micrographs during isothermal crystallization: (a) PLLA[24], after 8 °min at 140 °C. (b) $L_{81}C_{19}$[21], after 10 min at 140 °C. (c) $L_{60}C_{40}$[21], after 30 min at 140 °C. (d) $L_{10}C_{90}$[24] after 10 min at 100 °C. Castillo et al. [212]. Reproduced with permission of American Chemical Society.

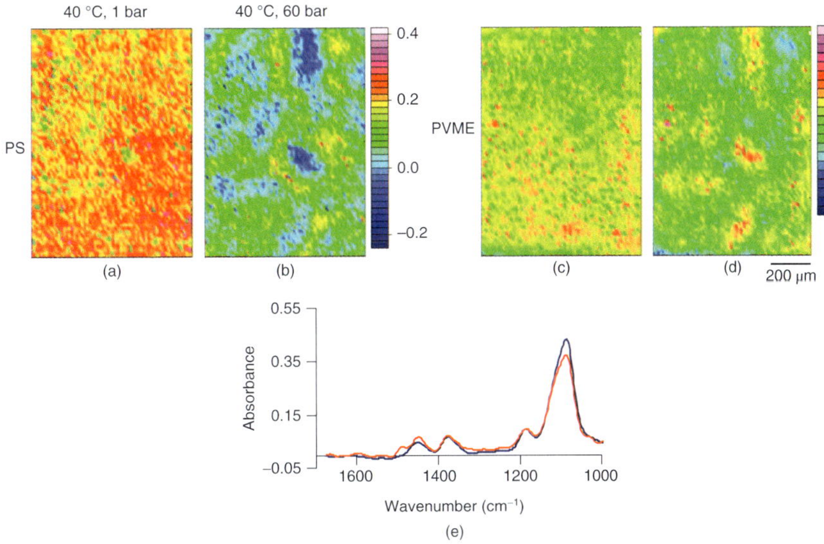

Figure 7.5 ATR–FTIR images of a PS-PVME blend before (a and c) and during exposure to 60 bar of CO_2 (b and d). The images (a and b) are based on the spectral band of PS, while images c and d are based on the spectral band of PVME. (e) Spectra extracted from PVME-rich (blue) and PS-rich (red) domains. Kazarian and Chan [19]. Reproduced with permission of the American Chemical Society.

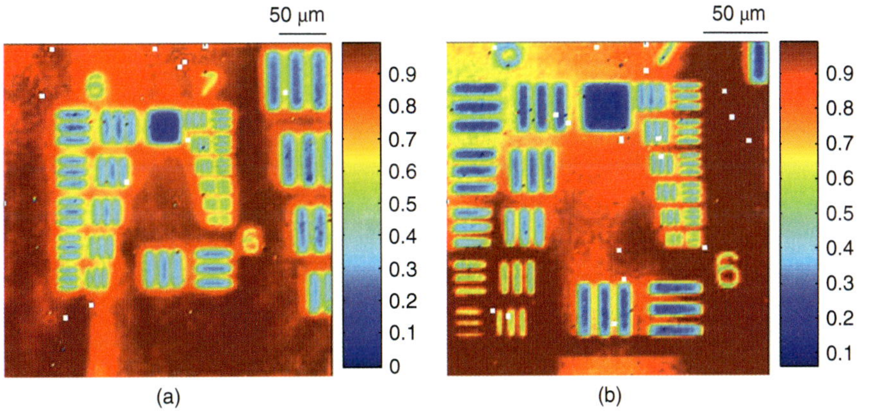

Figure 7.7 Transmission FTIR image of a USAF spatial resolution target (a) without lens (imaging size of 341 μm × 341 μm) and (b) with lens (imaging size of 240 μm × 240 μm). Chan and Kazarian [12]. Reproduced with permission of the American Chemical Society.

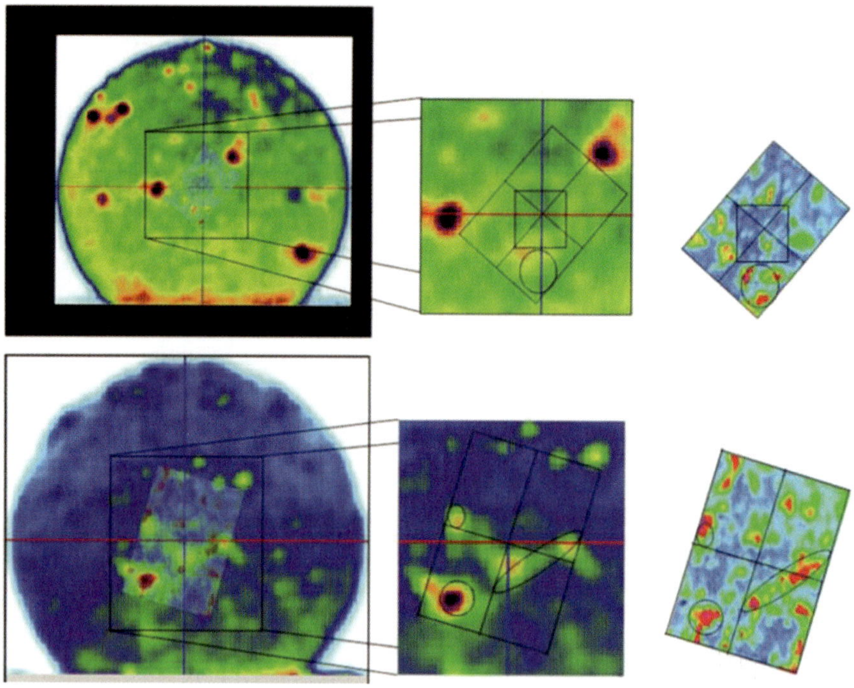

Figure 7.8 Images showing the results from FTIR spectroscopy compared with images from X-ray microtomography. The top set of images shows the results for the 100- 125-μm particle size of caffeine. The bottom set shows the data for the 125- 150-μm particle size. The FTIR data is shown on the right-hand side, the key area of the X-ray tomography is shown in the middle, and the FTIR is layered over the X-ray data for comparison on the left-hand side. Wray et al. [31]. Reproduced with permission of John Wiley and Sons.

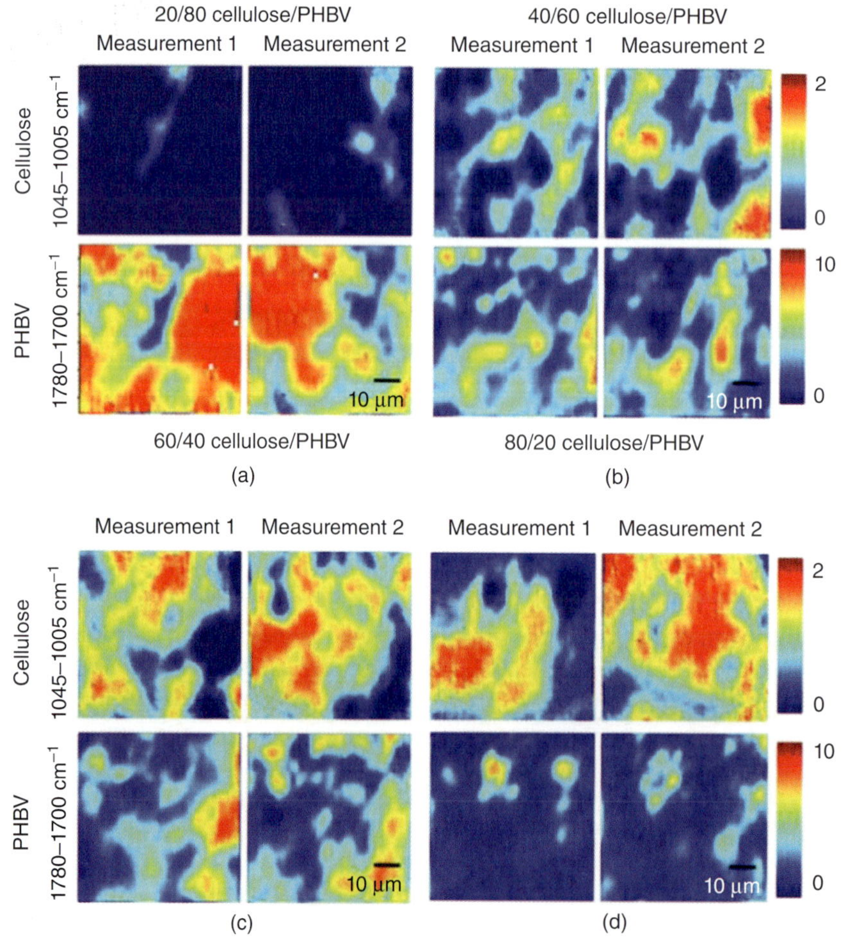

Figure 7.9 Micro-ATR–FTIR images of cellulose/PHBV blends at different compositions. Cellulose/PHBV: (a) 20/80, (b) 40/60, (c) 60/40, and (d) 80/20. The size of each image is ca. 63 μm × 63 μm. Hameed 2013 [32]. Reproduced with permission of Elsevier.

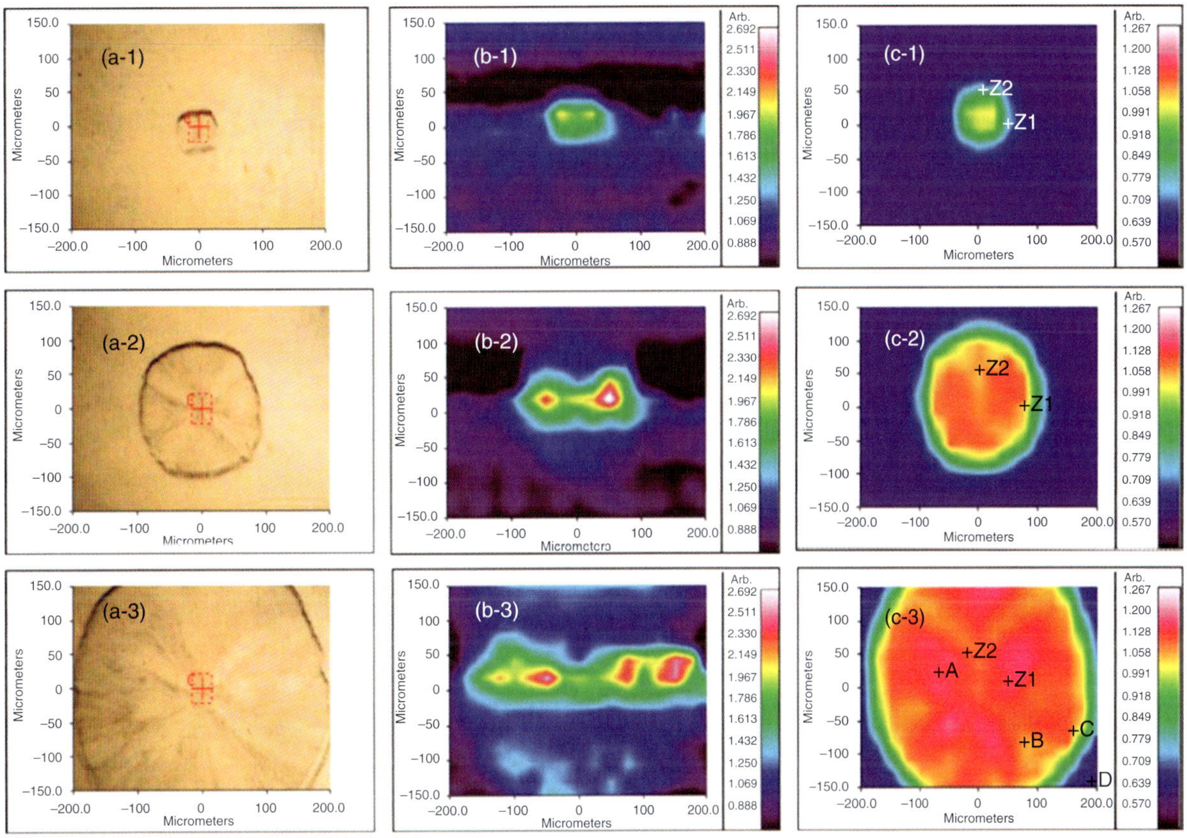

Figure 7.10 (a) Visual images, (b) NIR images, and (c) IR images of the PHB/CAB blend (80:20 wt%) during isothermal crystallization at 125°C at around 18 min (a-1, b-1, c-1), 36 min (a-2, b-2, c-2), and 66 min (a-3, b-3, c-3). Suttiwijitpukdee 2013 [45]. Reproduced with permission of the American Chemical Society.

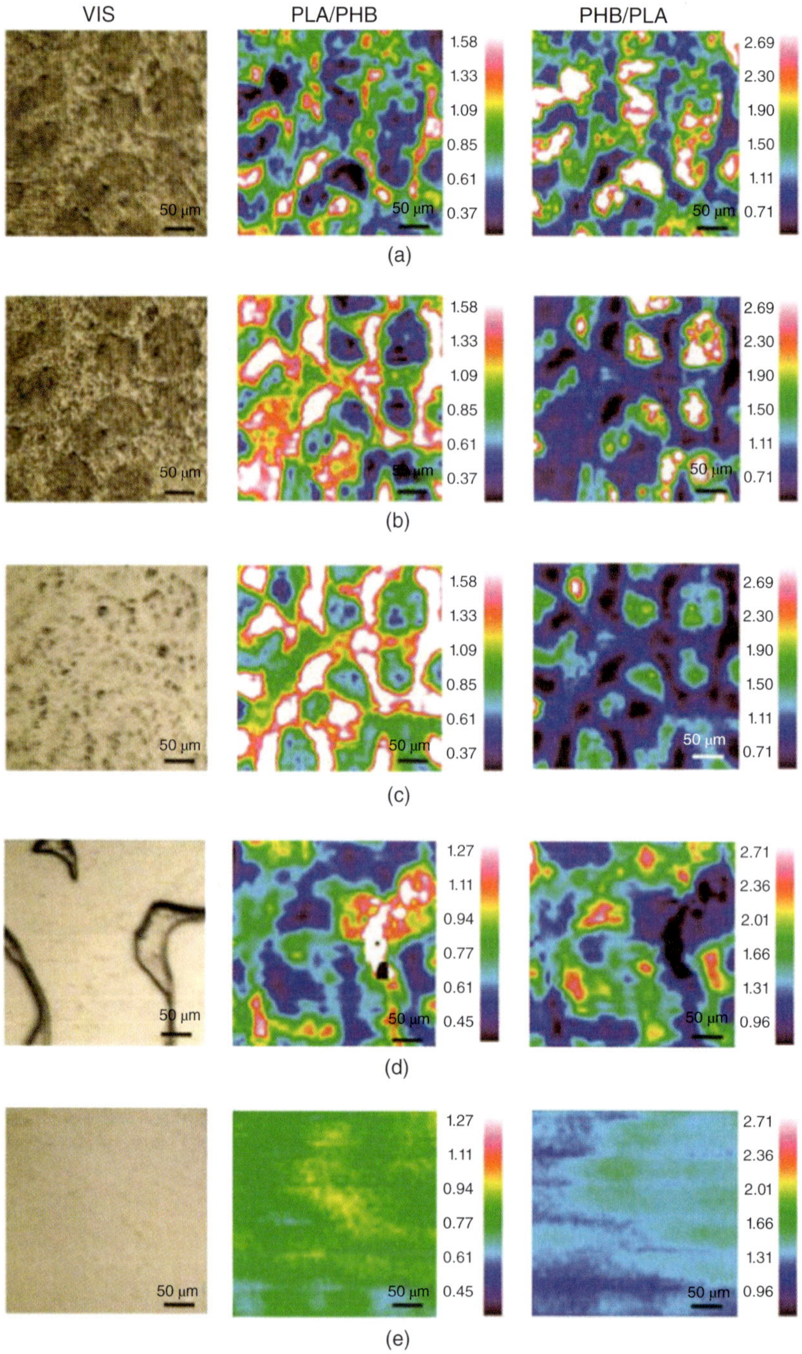

Figure 7.11 Visual images, PLA–PHB band-ratio, and PHB–PLA band-ratio FTIR images of the PHB–PLA (50:50 wt%)-blend film at 25 (a), 125 (b), 165 (c), 170 (d), and 175 °C (e). Vogel et al. [46]. Reproduced with permission of the American Chemical Society.

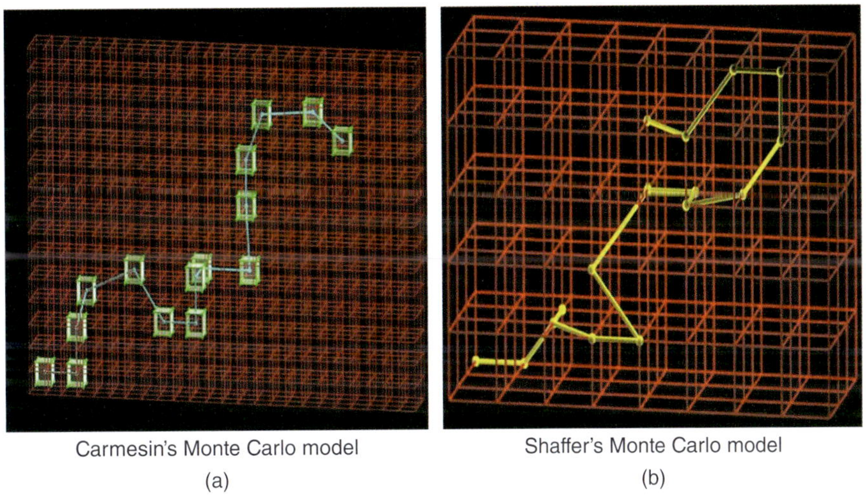

Figure 15.3 Two popular routines proposed by Carmesin and Kremer [13] (Reproduced with permission of American Chemical Society) and Shaffer [14] (Reproduced with permission of Elsevier) to generate the lattice model in MC simulation.

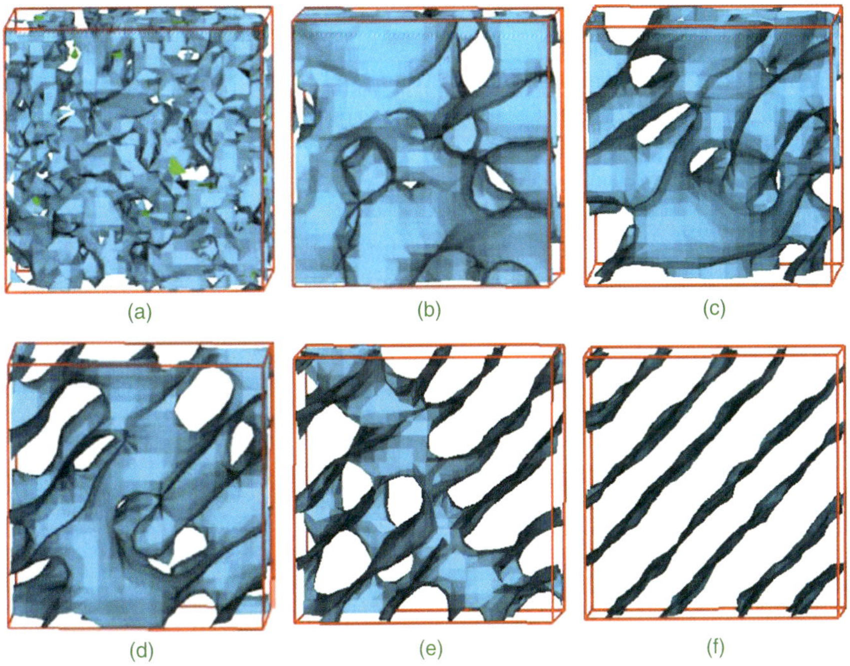

Figure 15.4 Sequential snapshots of the formation of lamellar morphology from the PS-*b*-PI diblock copolymer system at (a) $t = 0.006\ \mu s$, (b) $t = 0.12\ \mu s$, (c) $t = 0.17\ \mu s$, (d) $t = 0.29\ \mu s$, (e) $t = 0.58\ \mu s$, and (f) $t = 0.70\ \mu s$. Li et al. [33]. Reproduced with permission of Elsevier.

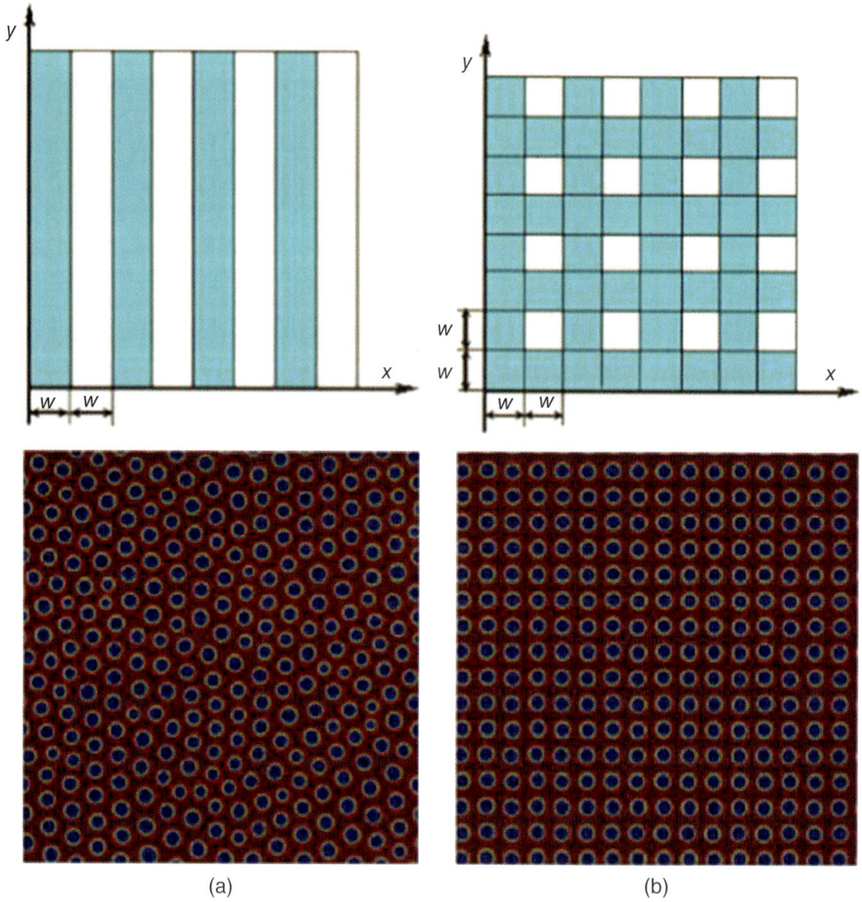

Figure 15.7 Self-assembled structures under (a) strip nanopatterned surface and (b) mosaic nanopatterned surface. Wu and Dzenis [43]. Reproduced with permission of Elsevier.

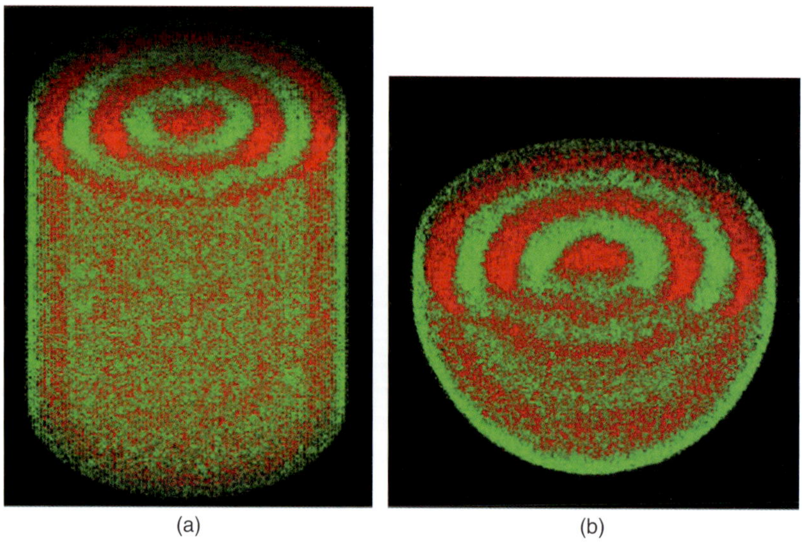

Figure 15.8 The lamellae structures formed by diblock copolymers in constrained spaces predicted by Monte Carlo simulation [48]. (a) Barrel, (b) spherical (cut view). He et al. [48]. Reproduced with permission of Elsevier.

22. Khambatta FB, Warner F, Russell T, Stein RS. Small-angle X-ray and light scattering studies of the morphology of blends of poly(ε-caprolactone) with poly(vinyl chloride). J Polym Sci B 1976;14:1391–1424.
23. Nojima S, Watanabe K, Zheng Z, Ashida T. Spherulite structure in compatible mixtures of poly(ε-caprolactone) and poly(vinyl chloride). Polym J 1988;20:823–826.
24. Nojima S, Wang D, Ashida T. Ringed spherulite in binary blends of poly(ε-caprolactone) and ε-caprolactone-butadiene diblock copolymer. Polym J 1991;23:1473–1482.
25. Tsuji H, Ikada Y. Blends of isotactic and atactic poly(lactide)s: 2 Molecular-weight effects of atactic component on crystallization and morphology of equimolar blends from the melt. Polymer 1996;37:595–602.
26. Ho RM, Ke KZ, Chen M. Crystal structure and banded spherulite of poly(trimethylene terephthalate). Macromolecules 2000;33:7529–7537.
27. Ma D, Zhang J, Wang M, Ma J, Luo X. The comparison of the ringed spherulite morphology of PCL blends with poly(vinyl chloride), poly(bisphenol A carbonate) and poly(hydroxyether of bisphenol A). Macromol Chem Phys 2001;202:961–966.
28. Xiao Q, Yan S, Rogausch KD, Petermann J, Huang Y. Ring-banded spherulites in poly(ε-caprolactone) blended with hydroxyethyl cellulose acetate as an indication for partial miscibility. J Appl Polym Sci 2001;80:1681–1686.
29. Woo EM, Wu PL. Effects of miscible diluent of poly(ether imide) on ring-banded morphology of poly(trimethylene terephthalate). Colloid Polym Sci 2006;284:357–365.
30. Meeten GH, Navard P. Small-angle scattering of polarized light. I. Comparison of theoretical predictions for isotropic and anisotropic spheres. J Polym Sci B 1989;27:2023–2035.
31. Yin L, Hillmyer MA. Disklike micelles in water from polyethylene-containing diblock copolymers. Macromolecules 2011;44:3021–3028.
32. Li Z, Liu R, Mai B, Wang W, Wu Q, Liang G, Gao H, Zhu F. Temperature-induced and crystallization-driven self-assembly of polyethylene-*b*-poly(ethylene oxide) in solution. Polymer 2013;54:1663–1670.
33. Steinhart M, Senz S, Wehrspohn RB, Gösele U, Wendorff JH. Curvature-directed crystallization of poly(vinylidene difluoride) in nanotube walls. Macromolecules 2003;36:3646–3651.
34. Steinhart M, Göring P, Dernaika H, Prabhukaran M, Gösele U, Hempel E, Thurn-Albrecht T. Coherent kinetic control over crystal orientation in macroscopic ensembles of polymer nanorods and nanotubes. Phys Rev Lett 2006;97:027801 (4 pages).
35. Wu H, Wang W, Yang H, Su Z. Crystallization and orientation of syndiotactic polystyrene in nanorods. Macromolecules 2007;40:4244–4249.
36. Shin K, Woo E, Jeong YG, Kim C, Huh J, Kim KW. Crystalline structures, melting, and crystallization of linear polyethylene in cylindrical nanopores. Macromolecules 2007;40:6617–6623.
37. Martin J, Mijangos C, Sanz A, Ezquerra TA, Nogales A. Segmental dynamics of semicrystalline poly(vinylidene fluoride) nanorods. Macromolecules 2009;42:5395–5401.
38. Lutkenhaus JL, McEnnis K, Serghei A, Russell TP. Confinement effects on crystallization and curie transitions of poly(vinylidene fluoride-*co*-trifluoroethylene). Macromolecules 2010;43:3844–3850.
39. Michell RM, Lorenzo AT, Müller AJ, Lin MC, Chen HL, Blaszczyk-Lezak I, Martin J, Mijangos C. The crystallization of confined polymers and block copolymers infiltrated within alumina nanotube templates. Macromolecules 2012;45: 1517–1528.
40. Li M, Wu H, Huang Y, Su Z. Effects of temperature and template surface on crystallization of syndiotactic polystyrene in cylindrical nanopores. Macromolecules 2012;45:5196–5200.
41. Wu H, Su Z, Takahara A. Isotactic polystyrene nanorods with gradient crystallite states. Soft Matter 2012;8:3180–3184.
42. Maiz J, Schäfer H, Rengarajan GT, Hartmann-Azanza B, Eickmeier H, Haase M, Mijangos C, Steinhart M. How gold nanoparticles influence crystallization of polyethylene in rigid cylindrical nanopores. Macromolecules 2013;46:403–412.
43. Noirez L, Stillings C, Bardeau JF, Steinhart M, Wendorff JH, Pepy G. What happens to polymer chains confined in rigid cylindrical inorganic (AAO) nanopores. Macromolecules 2013;46:4932–4936.
44. Nojima S, Ohguma Y, Namiki S, Ishizone T, Yamaguchi K. Crystallization of homopolymers confined in spherical or cylindrical nanodomains. Macromolecules 2008;41:1915–1918.
45. Nojima S, Ohguma Y, Kadena K, Ishizone T, Iwasaki Y, Yamaguchi K. Crystal orientation of poly(ε-caprolactone) homopolymers confined in cylindrical nanodomains. Macromolecules 2010;43:3916–3923.
46. Nakagawa S, Kadena K, Ishizone T, Nojima S, Shimizu T, Yamaguchi K, Nakahama S. Crystallization behavior and crystal orientation of poly(ε-caprolactone) homopolymers confined in nanocylinders: Effects of nanocylinder dimension. Macromolecules 2012;45:1892–1900.
47. Nakagawa S, Tanaka T, Ishizone T, Nojima S, Kakiuchi Y, Yamaguchi K, Nakahama S. Crystallization behavior of poly(ε-caprolactone) chains confined in nanocylinders: Effects of block chains tethered to nanocylinder interfaces. Macromolecules 2013;46:2199–2205.
48. Zhao H, Sterner ES, Coughlin EB, Theato P. *o*-Nitrobenzyl alcohol derivatives: Opportunities in polymer and materials science. Macromolecules 2012;45:1723–1736.
49. MacKnight WJ, Karasz FW, Fried JR. Solid state transition behavior of blends. In: Paul DR, Newman S, editors. *Polymer Blends*. Vol. 1. New York: Academic Press; 1978. p 185–242.
50. Martuscelli E. Morphology, crystallization phenomena and transitions in crystallizable polymer alloys. In: Martuscelli E, Palumbo R, Kryszewski M, editors. *Polymer Blends*. New York: Plenum Press; 1980. p 23–48.
51. Stein RS, Khambatta FB, Warner FP, Russell T, Escala A, Balizer E. X-ray and optical studies of the morphology of polymer blends. J Polym Sci Polym Symp 1978;63:313–328.
52. Wenig W, Karasz FE, MacKnight WJ. Structure and properties of the system: Poly(2,6-dimethylphenylene oxide) isotactic polystyrene small-angle X-ray studies. J Appl Phys 1975;46:4194–4198.

53. Warner FP, MacKnight WJ, Stein RS. A small-angle X-ray scattering study of blends of isotactic and atactic polystyrene. J Polym Sci B 1977;15:2113–2126.
54. Tanaka H, Nishi T. New types of phase separation behavior during the crystallization process in polymer blends with phase diagram. Phys Rev Lett 1985;55:1102–1105.
55. Tanaka H, Nishi T. Local phase separation at the growth front of a polymer spherulite during crystallization and nonlinear spherulitic growth in a polymer mixture with a phase diagram. Phys Rev A 1989;39:783–794.
56. Nojima S, Terashima Y, Ashida T. Small-angle X-ray scattering study of the morphology of blends of poly(ε-caprolactone) and polystyrene oligomer. Polymer 1986;27:1007–1013.
57. Nojima S, Satoh K, Ashida T. Morphology formation by combined effect of crystallization and phase separation in a binary blend of poly(ε-caprolactone) and polystyrene oligomer. Macromolecules 1991;24:942–947.
58. Nojima S, Kato K, Ono M, Ashida T. Time-resolved SAXS study of morphological change in a binary blend of poly(ε-caprolactone) and polystyrene oligomer. Macromolecules 1992;25:1922–1928.
59. Zhang X, Man X, Han CC, Yan D. Nucleation induced by phase separation in the interface of polyolefin blend. Polymer 2008;49:2368–2372.
60. Luo J, Liang Y, Yang J, Niu H, Dong JY, Han CC. Effects of liquid–liquid phase separation on crystallization kinetics and morphology of isotactic polypropylene/poly(ethylene-*co*-octene) in-reactor alloy. Polymer 2012;53:2465–2475.
61. Arai F, Takeshita H, Dobashi M, Takenaka K, Miya M, Shiomi T. Effects of liquid–liquid phase separation on crystallization of poly(ethylene glycol) in blends with isotactic poly(methyl methacrylate). Polymer 2012;53:851–856.
62. Lee JC, Tazawa H, Ikehara T, Nishi T. Crystallization kinetics and morphology in miscible blends of two crystalline polymers. Polym J 1998;30:780–789.
63. Chiu HJ, Chen HL, Lin JS. Crystallization induced microstructure of crystalline/crystalline poly(vinylidene fluoride)/poly(3-hydroxy butyrate) blends probed by small angle X-ray scattering. Polymer 2001;42:5749–5754.
64. Qiu Z, Ikehara T, Nishi T. Miscibility and crystallization in crystalline/crystalline blends of poly(butylene succinate)/poly(ethylene oxide). Polymer 2003;44:2799–2806.
65. Kaito A, Shimomura M, Akaba M, Nojima S. Lamellar structural changes in miscible crystalline polymer blends during melting and crystallization processes as studied by real-time small-angle X-ray scattering measurements. J Polym Sci B 2007;45:1959–1969.
66. Liu J, Jungnickel BJ. Crystallization kinetical and morphological peculiarities in binary crystalline/crystalline polymer blends. J Polym Sci B 2007;45:1917–1931.
67. Kaito A, Li Y, Shimomura M, Nojima S. Oriented lamellar structures in uniaxially drawn films of poly(vinylidene fluoride) and poly(3-hydroxybutyrate) blends studied by small-angle X-ray scattering measurements. J Polym Sci B 2009;47:381–392.
68. Hamley IW. *The Physics of Block Copolymers*. New York: Oxford University Press; 1998. p 278–330.
69. Loo YL, Register RA. Crystallization within block copolymer mesophases. In: Hamley IW, editor. *Developments in Block Copolymer Science and Technology*. England: John Wiley & Sons; 2004. p 213–243.
70. Müller AJ, Arnal ML, Balsamo V. Crystallization in block copolymers with more than one crystallizable block. In: Reiter G, Strobl GR, editors. *Progress in Understanding of Polymer Crystallization* Lecture Notes in Physics 714. Berlin: Springer-Verlag; 2007. p 229–259.
71. Müller AJ, Balsamo V, Arnal ML. Nucleation and crystallization in diblock and triblock copolymers. Adv Poly Sci 2005;190:1–63.
72. Nandan B, Hsu JY, Chen HL. Crystallization behavior of crystalline-amorphous diblock copolymers consisting of a rubbery amorphous block. Polym Rev 2006;46:143–172.
73. Castillo RV, Müller AJ. Crystallization and morphology of biodegradable or biostable single and double crystalline block copolymers. Prog Polym Sci 2009;34:516–560.
74. Lin MC, Nandan B, Chen HL. Mediating polymer crystal orientation using nanotemplates from block copolymer microdomains and anodic aluminum oxide nanochannels. Soft Matter 2012;8:7306–7322.
75. He WN, Xu JT. Crystallization assisted self-assembly of semicrystalline block copolymers. Prog Polym Sci 2012; 37:1350–1400.
76. Michell RM, Blaszczyk-Lezak I, Mijangos C, Müller AJ. Confinement effects on polymer crystallization: From droplets to almina nanopores. Polymer 2013;54:4059–4077.
77. Takeshita H, Shiomi T, Takenaka K, Arai F. Crystallization and higher-order structure of multicomponent polymeric systems. Polymer 2013;54:4776–4789.
78. DiMarzio EA, Guttman CM, Hoffman JD. Calculation of lamellar thickness in a diblock copolymer, one of whose components is crystalline. Macromolecules 1980;13:1194–1198.
79. Whitmore MD, Noolandi J. Theory of crystallizable block copolymer blends. Macromolecules 1988;21:1482–1496.
80. Douzinas KC, Cohen RE, Halasa AF. Evaluation of domain spacing scaling laws for semicrystalline diblock copolymers. Macromolecules 1991;24:4457–4459.
81. Unger R, Beyer D, Donth E. Phase behavior in poly(ethylene oxide-*b*-t-butyl methacrylate) block copolymers. Polymer 1991;32:3305–3312.
82. Vilgis T, Halperin A. Aggregation of coil-crystalline block copolymers: equilibrium crystallization. Macromolecules 1991;24:2090–2095.
83. Lovinger AJ, Han BJ, Padden FJ Jr, Mirau PA. Morphology and properties of polycaprolactone-poly(dimethyl siloxane)-polycaprolactone triblock copolymers. J Polym Sci B 1993;31:115–123.
84. Rangarajan P, Register RA, Fetters LJ. Morphology of semicrystalline block copolymrs of ethylene-(ethylene-*alt*-propylene). Macromolecules 1993;26:4640–4645.
85. Nojima S, Yamamoto S, Ashida T. Crystallization of block copolymers IV. Molecular weight dependence of the morphology formed in ε-caprolactone-butadiene diblock copolymers. Polym J 1995;27:673–682.

86. Nojima S, Kato K, Yamamoto S, Ashida T. Crystallization of block copolymers. 1. Small-angle X-ray scattering study of an ε-caprolactone-butadiene diblock copolymer. Macromolecules 1992;25:2237–2242.
87. Nojima S, Nakano H, Takahashi Y, Ashida T. Crystallization of block copolymers: 3 Crystallization behavior of an ε-caprolactone-butadiene diblock copolymer. Polymer 1994;35:3479–3486.
88. Rangarajan P, Register RA, Adamson DH, Fetters LJ, Bras W, Naylor S, Ryan AJ. Dynamics of structure formation in crystallizable block copolymers. Macromolecules 1995;28:1422–1428.
89. Ryan AJ, Hamley IW, Bras W, Bates FS. Structure development in semicrystalline diblock copolymers crystallizing from the ordered melt. Macromolecules 1995;28:3860–3868.
90. Ryan AJ, Fairclough JPA, Hamley IW, Mai SM, Booth C. Chain folding in crystallizable block copolymers. Macromolecules 1997;30:1723–1727.
91. Quiram DJ, Register RA, Marchand GR, Ryan AJ. Dynamics of structure formation and crystallization in asymmetric diblock copolymers. Macromolecules 1997;30:8338–8343.
92. Zhu L, Mimnaugh BR, Ge Q, Quirk RP, Cheng SZD, Thomas EL, Lotz B, Hsiao BS, Yeh F, Liu L. Hard and soft confinement effects on polymer crystallization in microphase separated cylinder-forming PEO-*b*-PS/PS blends. Polymer 2001;42:9121–9131.
93. Lee LBW, Register RA. Equilibrium control of crystal thickness and melting point through block copolymerization. Macromolecules 2004;37:7278–7284.
94. Hamley IW, Fairclough JPA, Terrill NJ, Ryan AJ, Lipic PM, Bates FS, Towns-Andrews E. Crystallization in oriented semicrystalline diblock copolymers. Macromolecules 1996;29:8835–8843.
95. Liu L, Jiang B, Zhou E. Study of semicrystalline-amorphous diblock copolymers: 1. Microphase separation, glass transition and crystallization of tetrahydrofuran-methyl methacrylate diblock copolymers. Polymer 1996;37:3937–3943.
96. Nojima S, Tanaka H, Rohadi A, Sasaki S. The effect of glass transition temperature on the crystallization of ε-caprolactone-styrene diblock copolymers. Polymer 1998;39: 1727–1734.
97. Quiram DJ, Register RA, Marchand GR, Adamson DH. Chain orientation in block copolymers exhibiting cylindrically confined crystallization. Macromolecules 1998;31:4891–4898.
98. Weimann PA, Hajduk DA, Chu C, Chaffin KA, Brodil JC, Bates FS. Crystallization of tethered polyethylene in confined geometries. J Polym Sci B 1999;37:2053–2068.
99. Shiomi T, Tsukada H, Takeshita H, Takenaka K, Tezuka Y. Crystallization of semicrystalline block copolymers containing a glass amorphous component. Polymer 2001;42:4997–5004.
100. Zhu L, Calhoun BH, Ge Q, Quirk RP, Cheng SZD, Thomas EL, Hsiao BS, Yeh F, Liu L, Lotz B. Initial-stage growth controlled crystal orientations in nanoconfined lamellae of a self-assembled crystalline-amorphous diblock copolymer. Macromolecules 2001;34:1244–1251.
101. Huang P, Guo Y, Quirk RP, Ruan J, Lotz B, Thomas EL, Hsiao BS, Avila-Orta CA, Sics I, Cheng SZD. Comparison of poly(ethylene oxide) crystal orientations and crystallization behaviors in nano-confined cylinders constructed by a poly(ethylene oxide)-*b*-polystyrene diblock copolymer and a blend of poly(ethylene oxide)-*b*-polystyrene and polystyrene. Polymer 2006;47:5457–5466.
102. Vasilev C, Reiter G, Pispas S, Hadjichristidis N. Crystallization of block copolymers in restricted cylindrical geometries. Polymer 2006;47:330–340.
103. Sun YS, Chung TM, Li YJ, Ho RM, Ko BT, Jeng US. Crystal orientation within lamellae-forming block copolymers of semicrystalline poly(4-vinylpyridine)-*b*-poly(ε-caprolactone). Macromolecules 2007;40:6778–6781.
104. Gitsas A, Floudas G, Butt HJ, Pakula T, Matyjaszewski K. Effects of nanoscale confinement and pressure on the dynamics of pODMA-*b*-p*t*BA-*b*-pODMA triblock copolymers. Macromolecules 2010;43:2453–2462.
105. Chung TM, Wang TC, Ho RM, Sun YS, Ko BT. Polymeric crystallization under nanoscale 2D spatial confinement. Macromolecules 2010;43:6237–6240.
106. Nojima S, Hashizume K, Rohadi A, Sasaki S. Crystallization of ε-caprolactone blocks within a crosslinked microdomain structure of poly(ε-caprolactone)-*block*-polybutadiene. Polymer 1997;38:2711–2718.
107. Loo YL, Register RA, Ryan AJ. Polymer crystallization in 25-nm spheres. Phys Rev Lett 2000;84:4120–4123.
108. Chen HL, Li HC, Huang YY, Chiu FC. Crystallization-induced deformation of spherical microdomains in block copolymer blends consisting of a soft amorphous phase. Macromolecules 2002;35:2417–2422.
109. Nojima S, Toei M, Hara S, Tanimoto S, Sasaki S. Size dependence of crystallization within spherical microdomain structures. Polymer 2002;43:4087–4090.
110. Ho RM, Lin FH, Tsai CC, Lin CC, Ko BT, Hsiao BS, Sics I. Crystallization-induced undulated morphology in polystyrene-*b*-poly(L-lactide) block copolymer. Macromolecules 2004;37:5985–5994.
111. Nojima S, Akutsu Y, Washino A, Tanimoto S. Morphology of melt-quenched poly(ε-caprolactone)-*block*-polyethylene copolymers. Polymer 2004;45:7317–7324.
112. Nojima S, Akutsu Y, Akaba M, Tanimoto S. Crystallization behavior of poly(ε-caprolactone) blocks starting from polyethylene lamellar morphology in poly(ε-caprolactone)-*block*-polyethylene copolymers. Polymer 2005;46:4060–4067.
113. Nojima S, Ito K, Ikeda H. Composition dependence of crystallized lamellar morphology formed in crystalline-crystalline diblock copolymers. Polymer 2007;48:3607–3611.
114. Ikeda H, Ohguma Y, Nojima S. Composition dependence of crystallization behavior observed in crystalline-crystalline diblock copolymers. Polym J 2008;40:241–248.
115. Sakurai T, Ohguma Y, Nojima S. Morphological evolution during isothermal crystallization observed in a crystalline-crystalline diblock copolymer. Polym J 2008;40:971–978.
116. Higa T, Nagakura H, Sakurai T, Nojima S. Crystal orientation of poly(ε-caprolactone) blocks confined in crystallized polyethylene lamellar morphology of poly(ε-caprolactone)-*block*-polyethylene copolymers. Polymer 2010;51:5576–5584.
117. Sakurai T, Nagakura H, Gondo S, Nojima S. Crystallization of poly(ε-caprolactone) blocks confined in crystallized lamellar morphology of poly(ε-caprolactone)-*block*-polyethylene copolymers: effects of polyethylene crystallinity and confinement size. Polym J 2013;45:436–443.

118. Sun L, Liu Y, Zhu L, Hsiao BS, Avila-Orta CA. Self-assembly and crystallization behavior of a double-crystalline polyethylene-*block*-poly(ethylene oxide) diblock copolymer. Polymer 2004;45:8181–8193.

119. Yang J, Zhao T, Liu L, Zhou Y, Li G, Zhou E, Chen X. Isothermal crystallization behavior of the poly(L-lactide) block in poly(L-lactide)-poly(ethylene glycol) diblock copolymers: Influence of the PEG block as a diluted solvent. Polym J 2006;38:1251–1257.

120. Hamley IW, Parras P, Castelletto V, Castillo RV, Müller AJ, Pollet E, Dubois P, Martin CM. Melt structure and its transformation by sequential crystallization of the two blocks within poly(L-lactide)-*block*-poly(ε-caprolactone) double crystalline diblock copolymers. Macromol Chem Phys 2006;207:941–953.

121. Yang J, Zhao T, Cui J, Liu L, Zhou Y, Li G, Zhou E, Chen X. Nonisothermal crystallization behavior of the poly(ethylene glycol) block in poly(L-lactide)-poly(ethylene glycol) diblock copolymers: Effect of the poly(L-lactide) block length. J Polym Sci B 2006;44:3215–3226.

122. Weiyu C, Tashiro K, Hanesaka M, Takeda S, Masunaga H, Sasaki S, Takata M. Relationship between morphological change and crystalline phase transitions of polyethylene-poly (ethylene oxide) diblock copolymers, revealed by the temperature dependent synchrotron WAXD/SAXS and infrared/Raman spectral measurements. J Phys Chem B 2009; 113:2338–2346.

123. Li S, Myers SB, Register RA. Solid-state structure and crystallization in double-crystalline diblock copolymers of linear polyethylene and hydrogenated polynorbornene. Macromolecules 2011;44:8835–8844.

124. Lin MC, Chen HL, Su WB, Su CJ, Jeng US, Tzeng FY, Wu JY, Tsai JC, Hashimoto T. Interactive crystallization kinetics in double-crystalline block copolymer. Macromolecules 2012;45:5114–5127.

125. Nojima S, Ono M, Ashida T. Crystallization of block copolymers II. Morphological study of poly(ethylene glycol)-poly(ε-caprolactone) block copolymers. Polym J 1992;24:1271–1280.

126. Gan Z, Jiang B, Zhang J. Poly(ε-caprolactone)/poly(ethylene oxide) diblock copolymer. 1. Isothermal crystallization and melting behavior. J Appl Polym Sci 1996; 59:961–967.

127. Bogdanov B, Vidts A, Schacht E, Berghmans H. Isothermal crystallization of poly(ε-caprolactone-ethylene glycol) block copolymers. Macromolecules 1999;32:726–731.

128. Shiomi T, Imai K, Takenaka K, Takeshita H, Hayashi H, Tezuka Y. Appearance of double spherulites like concentric circles for poly(ε-caprolactone)-*block*-poly(ethylene glycol)-*block*-poly(ε-caprolactone). Polymer 2001;42:3233–3239.

129. Ghoroghchian PP, Li G, Levine DH, Davis KP, Bates FS, Hammer DA, Therien MJ. Bioresorbable vesicles formed through spontaneous self-assembly of amphiphilic poly(ethylene oxide)-*block*-polycaprolactone. Macromolecules 2006;39:1673–1675.

130. Sun J, Chen X, He C, Jing X. Morphology and structure of single crystals of poly(ethylene glycol)-poly(ε-caprolactone) diblock copolymers. Macromolecules 2006;39:3717–3719.

131. Takeshita H, Fukumoto K, Ohnishi T, Ohkubo T, Miya M, Takenaka K, Shiomi T. Formation of lamellar structure by competition in crystallization of both components for crystalline-crystalline block copolymers. Polymer 2006;47:8210–8218.

132. Du ZX, Yang Y, Xu JT, Fan ZQ. Effect of molecular weight on spherulite growth rate of poly(ε-caprolactone) and poly(ε-caprolactone)-*b*-poly(ethylene glycol). J Appl Polym Sci 2007;104:2986–2991.

133. Li L, Meng F, Zhong Z, Byelov D, Jeu WH, Feijen J. Morphology of a highly asymmetric double crystallizable poly(ε-caprolactone-*b*-ethylene oxide) block copolymer. J Chem Phys 2007;126:024904(7 pages).

134. Albuerne J, Marquez L, Müller AJ, Raquez JM, Degee P, Dubois P, Castelletto V, Hamley IW. Nucleation and crystallization in double crystalline poly(*p*-dioxanone)-*b*-poly(ε-caprolactone) diblock copolymers. Macromolecules 2003;36: 1633–1644.

135. Müller AJ, Albuerne J, Marquez L, Raquez JM, Degee P, Dubois P, Hobbs J, Hamley IW. Self-nucleation and crystallization kinetics of double crystalline poly(*p*-dioxanone)-*b*-poly(ε-caprolactone) diblock copolymers. Faraday Discuss 2005;128:231–252.

136. Müller AJ, Castillo RV, Hillmyer M. Nucleation and crystallization of PLDA-*b*-PE and PLLA-*b*-PE diblock copolymers. Macromol Symp 2006;242:174–181.

137. Castillo RV, Müller AJ, Lin MC, Chen HL, Jeng US, Hillmyer MA. Confined crystallization and morphology of melt segregated PLLA-*b*-PE and PLDA-*b*-PE diblock copolymers. Macromolecules 2008;41:6154–6164.

138. Myers SB, Register RA. Crystalline-crystalline diblock copolymers of linear polyethylene and hydrogenated polynorbornene. Macromolecules 2008;41:6773–6779.

139. Nojima S, Fukagawa Y, Ikeda H. Interactive crystallization of a strongly segregated double crystalline block copolymer with close crystallizable temperatures. Macromolecules 2009;42:9515–9522.

140. Huang L, Kiyofuji G, Matsumoto J, Fukagawa Y, Gong C, Nojima S. Isothermal crystallization of poly(β-propiolactone) blocks starting from lamellar microdomain structures of double crystalline poly(β-propiolactone)-*block*-polyethylene copolymers. Polymer 2012;53:5856–5863.

141. Hijikawa R, Huang L, Kiyofuji G, Marubayashi H, Nojima S. Crystallization behavior of poly(β-propiolactone)-*block*-polyethylene copolymers with varying polyethylene crystallinities. Polymer 2014;55:6960–6966.

11

ISOTHERMAL CRYSTALLIZATION KINETICS OF POLYMERS

Alejandro J. Müller[1,2,3], Rose Mary Michell[3], and Arnaldo T. Lorenzo[4]

[1]*POLYMAT and Polymer Science and Technology Department, Faculty of Chemistry, University of the Basque Country UPV/EHU, Paseo Manuel de Lardizabal 3, 20018, Donostia-San Sebastián, Spain*
[2]*IKERBASQUE, Basque Foundation for Science, Bilbao, Spain*
[3]*Grupo de Polímeros USB, Departamento de Ciencia de los Materiales, Universidad Simón Bolívar, Apartado 89000, Caracas 1080-A, Venezuela*
[4]*The Dow Chemical Company, Performance Plastics R&D, Freeport, TX, USA*

11.1 INTRODUCTION

Polymer crystallization can be described as a phase transition process from the disorder isotropic melt to the order semicrystalline one. The disorder state is characterized by the randomly coiled chains, while the order state is complex because it is formed by crystalline chain-folded lamellae surrounded by the amorphous chains that constitute the fold surfaces and the interlamellar regions. The amorphous interfaces are formed by entanglements, end groups, bulky substituent groups, and chain defects, all of which cannot be included into the crystalline lattice. Polymers form metastable (thin) lamella separated by intervening amorphous layers and are nearly always semicrystalline.

The nanoscopic dimension of the crystalline lamella is its thickness, because the other two dimensions are usually much larger. The lamellar thickness depends on the supercooling at which it was grown (so it also depends on the crystallization temperature). The polymer crystallization process is a function of both thermodynamic and kinetic factors, but the latter ones usually dominate. The crystal grown at a given temperature is the crystal with the highest growth rate, not necessarily the structure with the lowest free energy.

The supercooling ΔT can be defined by Equation 11.1. It is always necessary to apply a finite supercooling in order to induce crystallization, and the crystallization rate is directly proportional to the supercooling.

$$\Delta T = T_m^0 - T_c \tag{11.1}$$

Polymer crystals are metastable (i.e., instead of forming extended chain-length infinite crystals, they form thin lamellae). As they are nonequilibrium structures, they may rearrange and modify their structure over time. Many variables can affect the metastability of polymer crystals, such as temperature, pressure, stress, or solvent vapors. This gives great importance to the history of preparation of a crystal.

Polymer crystallization is of great theoretical and practical significance and an extensive literature has been published that spans many decades. Some of the classic textbooks that have reviewed fundamental aspects of polymer crystallization are those by Wunderlich [1–3], Mandelkern [4, 5], Schultz [6], Gedde [7], and Hiemenz and Lodge [8]. More specialized books review recent literature; highlight controversies on polymer nucleation, crystallization theories, and simulation; and also present new experimental results on multiphasic materials [9–13].

In this chapter, we take a practical approach to briefly explain how to experimentally determine both spherulitic growth rates by polarized light optical Microscopy (PLOM) and overall isothermal crystallization kinetics by differential scanning calorimetry (DSC). We give examples on how to fit the data using both the Avrami theory and the Lauritzen and Hoffman theory. Both theories provide useful analytical equations that when properly handled represent valuable tools to understand crystallization kinetics and its relationship with morphology. They also have several shortcomings that are pointed out.

Polymer Morphology: Principles, Characterization, and Processing, First Edition. Edited by Qipeng Guo.

11.2 CRYSTALLIZATION PROCESS

The crystallization starts by a nucleation process and it continues by crystal growth [4, 10]. It is a matter of controversy if spinodal decomposition can occur during crystallization. The nucleation process is still under extensive investigation, because the advances in experimental techniques and new theoretical developments and simulations are facilitating the study of the early stages of crystallization. Many recent studies are dedicated to the melt organization prior to crystallization, and whether the melt can possess precursors or early metastable liquid crystalline-like structures that can later give origin to crystalline lamellae [10, 11, 14–19]. Computer simulations have also been extensively discussed, especially those by Muthukumar [11, 16, 20] and Hu et al. [13, 21, 22]. Strobl [10, 11, 23] has proposed a new nucleation and crystallization phenomenological model to try to explain controversial experimental results related to the early stages of chain organization.

Homogeneous nucleation is assumed to occur when a small group of aligned segments forms an embryo or cluster with appropriate size. This local chain alignment (embryo) is metastable. Embryos generate from random fluctuations and can melt again. Only those clusters that are larger than a critical size would act as nuclei. The way these nuclei are formed from the disorder state, that is, their relation to the density fluctuations, is still a matter of discussion.

On the other hand, heterogeneous nucleation is commonly originated by foreign nuclei, that is, any (low mass) particle with the correct size and surface, because it requires a much lower energy barrier than homogeneous nucleation. Commercial polymers typically nucleate in catalytic debris and any other impurities that are left over from their synthesis and/or their first processing before the material is commercialized.

Crystal growth is initially restricted to lamellae growing away from the nucleus, into the three-dimensional space. Each lamella grows linearly, creating the skeleton of the spherical semicrystalline entity. In order to fill the three-dimensional space, the lamellae must split and branch out to form a superstructure composed of lamellar stacks and intervening amorphous regions. This superstructural semicrystalline unit is called the spherulite. The spherulitic radial growth is stopped by the – also growing – neighboring spherulites. The primary crystallization occurs before spherulite impingement, when free growth of the superstructural units occurs. After impingement, a secondary crystallization process takes place where crystallization mainly occurs in the interspherulitic regions, although some intraspherulitic crystallization is also possible (within the inter-lamellar regions).

The crystallization of polymers has been studied both in solution and from the melt. Crystallization also occurs during polymerization and during cooling from the melt with imposed orientation (e.g., injection molding). In this chapter, we focus particularly on the isothermal melt crystallization under quiescent conditions.

The topotactic or crystallization in the solid state is one of the few methods that can produce single crystals that approach 100% crystallinity degree. The monomers should also be able to form a single crystal that can, during exposure to an ionizing radiation, polymerize and transform into a polymer single crystal [24]. The final properties of crystals formed by solid-state polymerization can be very interesting. For example, poly (sulfur nitride) crystals prepared by solid-state polymerization can form macroscopic single crystals that conduct electricity along the crystal axis (corresponding to the chain *c* direction) and can even have superconducting properties at sufficiently low temperatures [2, 4]).

The crystallization induced by orientation consists of stretching polymer chains to form fibrous crystals or fibers [25]. The formation of such fiber-like morphology is accompanied by the formation of a typical "shish–kebab" or bottlebrush morphology [2, 4–12, 25–27]. A relevant reference for crystallization under orientation during different polymer processing operations has been recently published [28].

The crystallization of polymeric chains under quiescent conditions has been observed and studied for a very large number of synthetic and natural polymers from both dilute solution (leading to the preparation of single crystals and single-crystal mats) and the melt (usually yielding superstructural 3D structures like spherulites, although 2D structures like hedrites and axialites are also possible).

The crystallization from dilute solutions has been used extensively to study the fundamental aspects of the structure and morphology at a molecular level [1–12, 29]. On the other hand, the crystallization from the melt is more complex, because diffusion and kinetic effects can often dominate, but it is closer to the solidification conditions applied during processing operations.

The nucleation, growth, and kinetics of development of the spherulitic superstructural aggregates are of fundamental and practical interest, and it is the main topic of this chapter.

11.3 CRYSTALLIZATION KINETICS

Spherulites constitute the most common morphological texture of polymers crystallized from the melt. The nucleation and crystallization kinetics controls the spherulitic texture, which in turn can greatly affect the mechanical properties of the polymer.

Generally, growth rates of polymers are evaluated by visually following the growth of spherulites developing at constant temperature (using PLOM). The radii of spherulites are usually found to be a linear function of time; that is, the growth rate remains constant during isothermal crystallization (see Figure 11.1). This procedure gives accurate growth rate values, but takes a long time and may not be useful for polymers with high nucleation densities in which spherulites are too small to be continuously monitored during crystallization through an optical microscope (e.g., high-density polyethylene) [2, 4, 6–8, 12, 25].

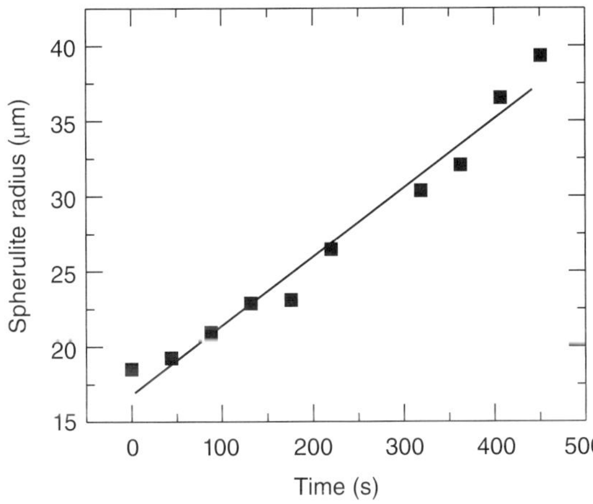

Figure 11.1 Spherulite radius versus isothermal crystallization time, at a fixed isothermal temperature (41 °C) for a PCL with a $M_n = 3.800$ g/mol.

The PLOM technique has several limitations: (i) the presence of additives with high refractive indices hinder the observation of polymer samples under light transmission microscopes [4] and (ii) if spherulites with different growth rates are present in various concentrations during transformation (as for polymorphs and polymer blends), the measure of growth rates may not be easy, especially for less abundant spherulite types.

In order to bypass such limitations, the overall crystallization kinetics may be determined by DSC. However, in this case, both primary nucleation and crystal growth will make a contribution to the overall isothermal crystallization rate [6, 7, 12]. Ideally, it would be better to determine both spherulitic growth rate and overall isothermal crystallization kinetics in separate experiments, if possible. In the literature, the most commonly reported [30] type of isothermal crystallization kinetic data is that measured by DSC, because it is the easier to obtain. The DSC experimental approach can be very useful and in some cases, the DSC data thus obtained can provide not only the overall crystallization rate but also the separation of the individual contributions of the primary nucleation and growth rate (more details to follow).

11.3.1 The Avrami Equation [31]

Crystallization theories have been developed for low-molecular-weight substances and later adapted to polymers. One of the first theories to study such phenomena was the "free growth" theory formulated by Göler and Sachs [5, 8, 32, 33]. They established that once a given nuclei or center is initiated, it grows unrestrained or without the influence of others that may have also been nucleated and could be growing within the same time scale. If N' is the steady-state nucleation rate per unit of untransformed mass (where all the material is in the liquid state or yet to be converted to the solid state), $w(t,\tau)$ is the mass of a given center at time t, that was initiated at time τ ($\tau \leq t$), then:

$$1 - \lambda(t) = \int_0^t w(t,\tau) \cdot N'(\tau) \cdot \lambda(\tau) \cdot dt \qquad (11.2)$$

where $\lambda(t)$ is the relative untransformed fraction at time t. The value of $\lambda(t)$ ranges from 1 to 0, indicating that all the material is initially fully amorphous (at $t = 0$, $\lambda = 1$) and then it will be progressively converted, until it achieves a saturated normalized degree of crystallinity. This corresponds to a relative untransformed fraction that varies from 0 (fully amorphous) to 1, or to complete conversion to the semicrystalline state.

Assuming that the polymer structure can be described by a two-phase model and considering the densities of the crystalline (ρ_c) and liquid phases (ρ_l), Equation 11.2 can be alternatively written as

$$1 - \lambda(t) = \frac{\rho_c}{\rho_l} \int_0^t v(t,\tau) \cdot N(\tau) \cdot \lambda(\tau) \cdot dt \qquad (11.3)$$

where $N(\tau)$ is the nucleation frequency per untransformed volume and $v(t, \tau)$ is the volume of a given center at time t, that was started at time τ ($\tau \leq t$). It is generally agreed [5, 34–39] that in order to solve this integral equation (Eq. 11.3), it is necessary to specify the details of the nucleation and growth rates. One simple solution is to consider a constant nucleation rate of nuclei that is characterized by free growth linearly in 1 to 3 dimensions.

The Avrami equation, also referred to as the Kolmogorov–Johnson–Mehl–Avrami equation [34–37, 40], can be considered to be one of the possible solutions of Equation 11.3, and in its simplest form it can be expressed as [31, 34–39, 41]

$$1 - V_c(t) = \exp(-kt^n) \qquad (11.4)$$

where V_c is the relative volumetric transformed fraction, n is the Avrami index, and k the overall crystallization rate constant (i.e., it includes contributions from both nucleation and growth). Equation 11.4 is the general Avrami relationship, taking into consideration a constant nucleation rate and constant linear growth. There are obviously many different possibilities that can be considered, resulting in a wide variety of expressions, considered to be derived Avrami equations. We only consider here the practical application of this simple equation for those cases where it can describe the experimental data obtained by DSC [31, 41].

The Avrami theory usually provides a good fit of the experimental data at least in the conversion range up to the end of the primary crystallization, that is, up to the impingement of spherulites at approximately 50% conversion to the solid semicrystalline state.

In Equation 11.4, the Avrami index can be considered as a first approximation to be composed of two terms [42]:

$$n = n_d + n_n \qquad (11.5)$$

where n_d represents the dimensionality of the growing crystals, which should correspond to 1, 2, or 3 for one-, two-, or three-dimensional entities. In the case of polymers, only 2 and 3 are commonly obtained as they represent axialites (two-dimensional lamellar aggregates) and spherulites (superstructural three-dimensional aggregates), respectively. The time dependence of the nucleation is represented by n_n. In principle, its limiting values should be either 0 or 1, where 0 corresponds to purely instantaneous nucleation and 1 to purely sporadic nucleation. The nucleation may not be completely sporadic or completely instantaneous, a fact that may lead to noninteger contributions to the Avrami index.

Based on the above, the limiting values of the Avrami index for free growth (without any diffusion problems) should be an integer number between 1 and 4. The summary of the theoretically expected Avrami index values is presented in Table 11.1. However, in practice, the values normally measured are rarely integer numbers, and even values lower than 1 have been reported [6, 31, 42–44]. Those discrepancies are understandable, because the Avrami equation was developed for general crystallization, and therefore there are some considerations and limitations related to polymer crystallization. In those cases where the growth of spherulites is not linear with time, the crystallization process may be governed by diffusion and n_n can have a value of 0.5, which indicates the Fickian dependence of growth with the square root of time [6, 31, 43, 44].

In the past few years, we have been employing the Avrami equation [31, 42, 45–76] to analyze DSC overall isothermal crystallization kinetics in many different types of polymeric materials and we have encountered many practical problems [31] that are sometimes apparent in published isothermal crystallization calorimetry data. Some of those problems are treated here as examples.

Isothermal crystallization of a model hydrogenated polybutadiene, HPB (M_n of 25 Kg/mol, polydispersity index of 1.01, ~10 wt% ethyl branches) was carried out on a Perkin-Elmer® DSC7 (the synthesis and experimental procedure for the preparation of the HPB has been described in Refs [77, 78]).

All the calculations needed to perform the Avrami fits and comparisons between the experimental data and the predictions of the theory were performed with an Origin® application software that was developed by our group [31]. This Origin® plugin was designed to analyze the *DSC* isotherms, establish the baseline, calculate integrals, perform the linear fit according to the Avrami equation, calculate fitting errors, and perform graphical comparisons between the experimental data and the predictions. This Avrami plugin is available for free distribution upon request [79]. All fits, except when indicated in the text, were performed for a relative volumetric conversion range encompassing 0.03 to 0.2, that is, from 3% to 20% relative conversion to the semicrystalline state.

Figure 11.2 presents the experimental data obtained by DSC of the crystallization of the HPB sample at 93 °C. The initial rise in the curve corresponds to the thermal stabilization of the equipment; the sample was quickly cooled at 60 °C/min from the melt (120 °C) down to the crystallization temperature and the isothermal run was immediately started. The induction time (that, in this case, includes the stabilization time) has been named t_0 and was subtracted from the time t (i.e., the time at which the experiment was started).

The Avrami equation can be adjusted to adapt better to experimental polymer crystallization data. Firstly, the relative volumetric fraction (V_c) used in the Avrami equation can be calculated as [7, 31]

$$V_c = \frac{W_c}{W_c + \frac{\rho_c}{\rho_a}(1 - W_c)} \tag{11.6}$$

where ρ_c and ρ_a are the fully crystalline and fully amorphous polymer densities, respectively (see Table 11.2). In this case, we have employed the following values that correspond to polyethylene: $\rho_c = 1.004$ and $\rho_a = 0.853$. W_c is the crystalline mass fraction that can be calculated as

$$W_c = \frac{\Delta H(t)}{\Delta H_{total}} \tag{11.7}$$

taking ΔH_{total} as the maximum enthalpy value reached at the end of the isothermal crystallization process, while $\Delta H(t)$ is the enthalpy variation as function of the time spent at a given crystallization temperature. Both quantities can be obtained from the integration of the experimental data presented in Figure 11.2.

Applying logarithmic properties to both sides of Equation 11.4, the following equation can be obtained:

$$\log[-\ln[1 - V_c(t)]] = \log(k) + n\log(t) \tag{11.8}$$

This equation is used to construct the so-called Avrami plot shown in Figure 11.3. The experimental data are

TABLE 11.1 Values of the Avrami Index *n* for the Different Cases of Nucleation and Possible Crystal Dimensionalities [5, 7]

Dimension	Geometry	Instantaneous Nucleation	Sporadic Nucleation	Sporadic Nucleation, Diffusion Controlled
1D	Line	1	2	1
2D	Circular	2	3	2
3D	Spherical	3	4	5/2
	Fibril	≤1	≤2	
	Circular lamellar	≤2	≤3	

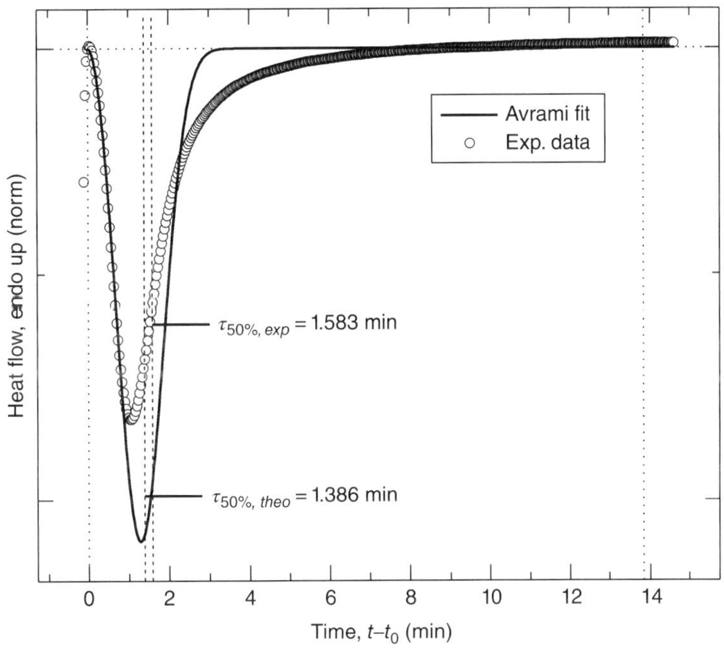

Figure 11.2 Isothermal DSC scan of HPB (open circles). The sample was previously melted at 120 °C for 3 min and then cooled at 60 °C/min to the isothermal crystallization temperature of 93 °C. The solid line was calculated after an Avrami fit performed to the data. The solid vertical lines indicate the values of the experimental half-crystallization times.

TABLE 11.2 Density Values, ρ_c and ρ_a, for Some Selected Common Polymers [30]

Polymer	ρ_c (g/cm^3)	ρ_a (g/cm^3)
Poly ε-caprolactone (PCL)	1.175	1.09
Poly(lactic acid) (PLA)	1.29	1.248
Poly(ethylene terephthalate) (PET)	1.455	1.333
Poly(ethylene oxide) (POE)	1.239	1.124
Polyamide 6 (PA6)	1.12–1.14	1.09
Polystyrene (isotactic) (iPS)	1.111	1.04–1.065
Polyethylene (PE)	1.004–0.9953	0.854
Polypropylene (isotactic) (iPP)	0.932–0.943	0.850–0.854

obtained from the integration of the DSC isothermal data of Figure 11.1. From the slope and intercept of the linear fit presented in Figure 11.3 plotted using Equation 11.8, the Avrami index "n" and the overall crystallization rate constant "k" can be obtained. A conversion range (V_c range) needs to be selected because the Avrami equation rarely describes the whole conversion process (V_c[0–100%]).

In general, the equation can describe the overall transformation process until the primary crystallization ends (V_c <40–50%). The software that we have developed allows choosing the conversion degree or V_c values for the fit.

As the calculation of n and k is performed by fitting the data in a double-logarithmic plot, the correlation coefficient of the fit has to be very high if a good fit is to be obtained. We propose, based in our experience on this type of curve fitting, that the R^2 value must be 0.9990 or larger in order to represent a very good fit of the data in a specific conversion range. Therefore, one can choose different conversion ranges in order to maximize the value of R^2. For the fitting presented in Figure 11.2, the best R^2 value (0.9995) was obtained with the relative conversion range of 3–20%, and this fit can be easily examined by

i) a comparison between the experimental crystallization exotherm obtained by DSC with that computed using the predicted values given by the Avrami equation with the fitted n and k values (as presented in Figure 11.2) and
ii) a comparison between the experimental and the predicted relative untransformed fraction (i.e., $1 - V_c$) as function of time (see Figure 11.4). Figures 11.2–11.4 show that the quality of the Avrami fit was excellent, until a V_c value of approximately 0.5% or 50% conversion. This is in agreement with the common assumption that primary crystallization usually ends at around 50% conversion.

In a previous work [31], practical guidelines were given for the Avrami equation fitting to DSC isothermal polymer crystallization kinetics. It was recommended that:

i) Controlled cooling rates from the melt to the isothermal crystallization temperature of 60 °C/min or higher be used, in order to prevent previous nonisothermal crystallization of the sample in the case of polyolefins or polymers that can rapidly crystallize. In fact, it is recommended that some preliminary tests be performed to check for possible crystallization during the cooling stage from the melt to T_c. This can be easily performed

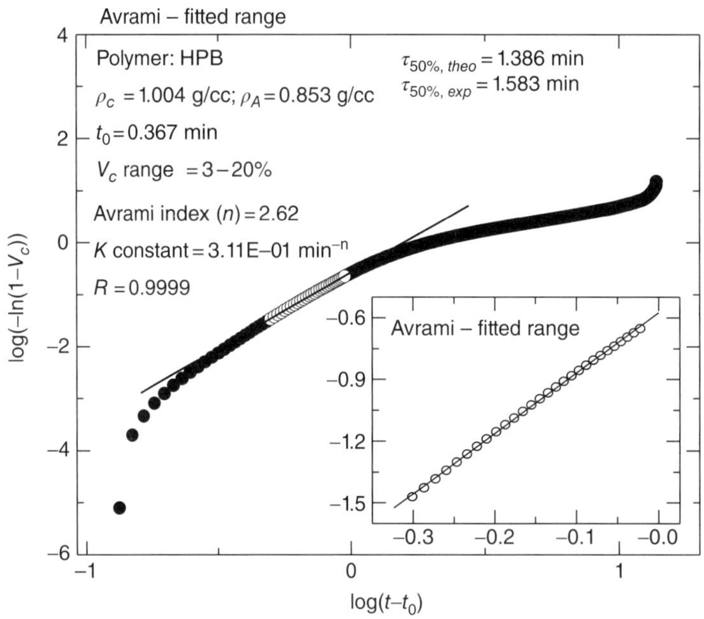

Figure 11.3 Avrami plot of the experimental data (circles) obtained from the isothermal crystallization of the HPB at 93 °C. The solid line represents the Avrami fit (Eq. 11.8) for a conversion range of 3–20% (open symbols. Insert: Avrami plot for the selected conversion range only.

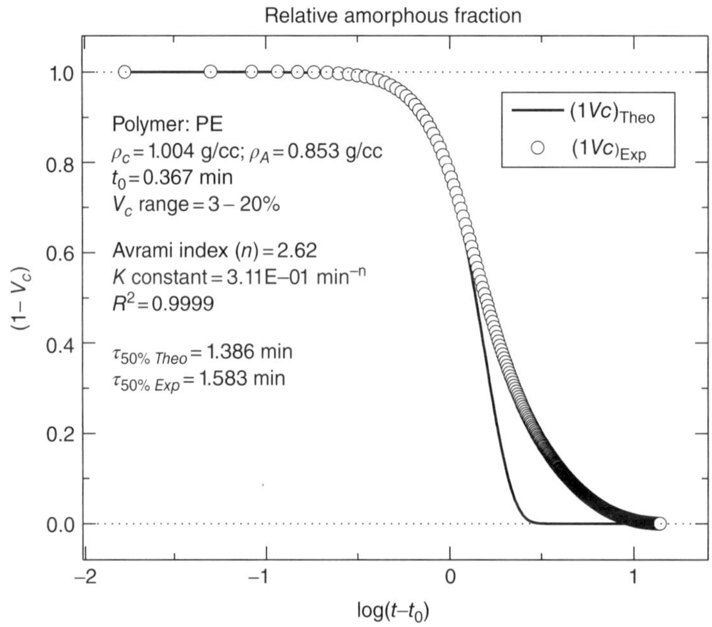

Figure 11.4 Untransformed relative volumetric fraction, $1-V_c$, as a function of time for the isothermal crystallization of HPB at 93 °C. Comparison between the experimental data and the Avrami fit employing a relative conversion range between 3% and 20%.

by cooling the sample from the melt at 60 °C/min to T_c and then, without any waiting period, immediately recording a heating scan from T_c to the melt. If any melting signal is present, then the sample has crystallized during cooling and the T_c temperature chosen cannot be employed. A higher T_c should then be tested. In this way, the minimum T_c value can be chosen.

ii) A complete DSC record of the isothermal crystallization exotherm must be obtained in order to guarantee the establishment of a good baseline and avoid errors associated with the integration of incomplete traces.

iii) The relative conversion range from the melt to the semicrystalline state must be chosen within the primary crystallization range of the sample under study and can be tuned according to the correlation coefficient obtained in the Avrami plots.

iv) The induction time, t_0, must also be determined and subtracted from the data.

The last two items (iii and iv) are crucial for a good fit of the Avrami equation to the experimental DSC data, and those items will be explained in more detail below.

11.3.1.1 The Role of the Induction Time, t_o The Avrami equation, in its simplest form (Eq. 11.4), gives a value of $1 - V_c$ equal to 1 when the crystallization time is 0; any value lower than 1 at any later time implies that crystallization has started. Therefore, in mathematical terms, the equation is only defined when crystallization starts. The equation cannot account for a time period where there is no crystallization. For simplicity, the initial time when the Avrami fit is applied will be denoted as the initial crystallization time or the induction time for the beginning of the crystallization "t_0." In other words, if we denote the absolute time t as the time at which the temperature reaches the desired crystallization temperature, a certain time may elapse before crystallization starts (this time is precisely t_0) and this time must be subtracted from the absolute time or, in other words, time count must start from time equal to t_0.

Therefore, a minor modification into the Avrami classical equation has to be introduced, in order to take into consideration the experimental induction time t_0. Equation 11.4 can be then rewritten as

$$1 - V_c(t - t_0) = \exp(-k(t - t_0)^n) \tag{11.9}$$

and after applying logarithmic properties:

$$\log[-\ln[1 - V_c(t - t_0)]] = \log(k) + n\log(t - t_0) \tag{11.10}$$

The incorporation of the induction time into the Avrami equation allows us to achieve reasonable results concerning the parameters obtained after the abovementioned linear fit. This can be observed if we compare the exotherms in Figure 11.5. Curves (a) and (c) presented in Figure 11.5 are quite similar Avrami parameters after the initial crystallization time is adjusted to t_0.

11.3.1.2 Effect of the Conversion Range Chosen to Fit the Experimental Data It is recommended that a reasonable conversion range that is within primary crystallization is chosen in order to fit the data. Table 11.3 shows the effect of varying the relative conversion to the semicrystalline state expressed by the values of V_c on the correlation coefficient of the Avrami plot fit (R^2) and also on the relative error with respect to the ideal case (which, in this case, was taken as the fitting with the largest R^2 value, that is, the fitting in the range 3–20%).

Evaluating the results presented in Table 11.3 two major conclusions can be drawn:

i) It is advisable to neglect the initial data points ($V_c < 3\%$) due to experimental errors during the first stages of the crystallization process regarding the stabilization of the equipment and the small quantity of heat evolved.
ii) The secondary crystallization processes produces nonlinearity in the Avrami plot, and the errors are clearly appreciated beyond 50% conversion in the case of the HPB employed here as an example.

It is also worth noting in Table 11.3 that the Avrami index experiences more variations than k upon changes in the relative conversion range employed for the fit. If the entire DSC isotherm is employed (0–100% relative conversion

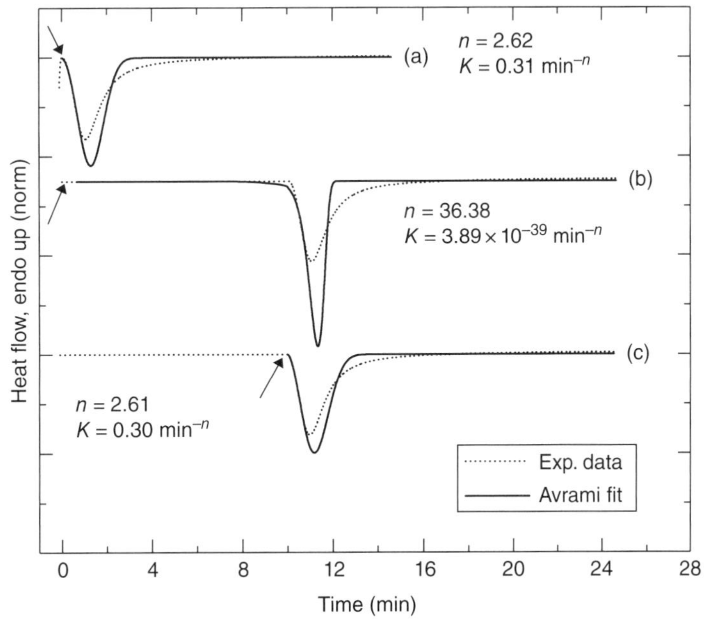

Figure 11.5 Effect of the induction time on the Avrami fits; see explanation in text. The arrows indicate the t_0 value employed in the Avrami fit.

TABLE 11.3 Comparison of the Avrami Parameters as a Function of the Relative Crystal Volumetric Fraction (V_c) Range

V_c Range	Avrami Index		k Constant		R^2
	n	% Error[a]	*k* (min^{-n})	% Error[a]	
3–20	*2.62*	*0*	*0.311*	*0*	*0.9999*
0–10	2.47	6	0.267	8	0.9983
0–20	2.50	5	0.284	9	0.9985
5–20	2.60	2	0.309	1	0.9991
5–35	2.49	5	0.296	5	0.9990
5–50	2.84	8	0.284	9	0.9954
0–100	1.59	39	0.19	36	0.9327

[a]Absolute relative errors were calculated on the basis of the control sample: 3–20% conversion range and are set in italics.

range), a relative error greater than 35% was detected for both *n* and *k*, these values represent the largest errors in Table 11.3.

11.3.2 Nucleation and Crystal Growth: Lauritzen–Hofmann Theory

According to classical crystallization theories, the temperature dependence of the linear spherulitic growth rate (*G*) can be expressed by two exponential factors: the molecular transport term and the nucleation term [2, 5, 7–10, 25]. It has been shown that these two terms have opposite crystallization temperature dependence behavior, thereby producing a maximum in growth rate (known as G_{MAX}). Many polymeric materials display a bell-shaped temperature dependence of the crystal growth rate, exhibiting a maximum growth rate at T_{cmax} [2, 4, 7–10, 25, 43, 80] (Fig. 11.6). From studies of the crystallization behavior of several polymeric materials, it has been demonstrated that the molecular transport term is of considerable importance at high supercoolings (left-hand side of the bell-shaped curve) and can be expressed by a WLF (Williams–Landel–Ferry)-type function. Therefore, the macromolecules diffusion will become very difficult as temperature approaches the glass transition temperature, T_g, and the growth rate decreases to zero at temperatures near or below T_g, as shown in Figure 11.6. On the other hand, the secondary nucleation (or crystal growth) term is dominant in the high-crystallization temperature range (right-hand side of the bell-shaped *G* curve of Fig. 11.6). Considering that in order to initiate the formation of a stable nuclei the melt has to be supercooled, it can be established that the greater the degree of supercooling the more favorable will be the thermodynamic conditions for nuclei generation.

Figure 11.6 also shows a schematic comparison between primary nucleation rate (*I*) and spherulitic growth rate (*G*), where it can be appreciated that both the primary nucleation (*I*) and the spherulitic growth (*G*) rates exhibit a bell-shaped curve with a maximum rate with crystallization temperature. The nucleation behavior is generally more complicated as compared to crystal growth. In fact, the spherulitic growth rate is mainly governed by the supercooling, while primary nucleation could be affected by the crystallization temperature, the density of heterogeneities present in the melt, and by the presence or absence of nucleating agents. Umemoto et al. [81] reported experimental data for the primary nucleation rate and growth rate in a wide range of temperature for poly(ethylene succinate) (PESU) that exemplifies the trend shown schematically in Figure 11.6, a trend that has been demonstrated many times in the literature [80].

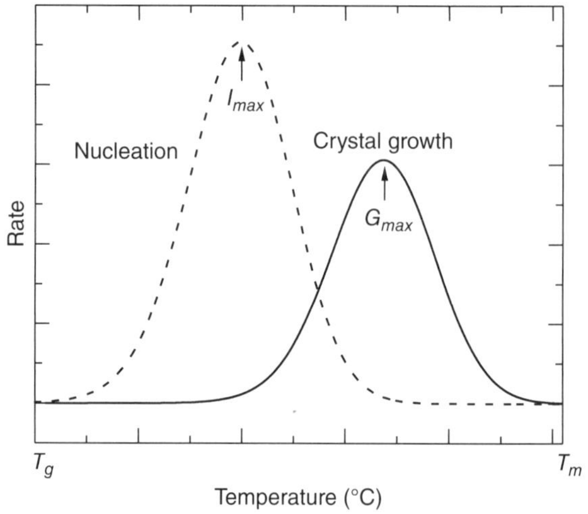

Figure 11.6 Schematic plots for the primary nucleation rate (*I*) and crystal growth rate (*G*) as a function of the isothermal crystallization (or nucleation) temperature. Adapted from Lorenzo and Müller [60].

If the crystallization kinetics is determined by DSC, both primary nucleation and crystal growth will make a contribution, resulting in a mathematical superposition of both bell-shaped curves shown in Figure 11.6. It is useful to be able to extract from DSC data not only the commonly obtained overall crystallization rate but also the separation of the individual contributions of the primary nucleation and the growth rate.

Isothermal crystallization kinetics data can be analyzed with a variety of models in order to quantify the energy barriers associated with the nucleation and crystal growth:

i) The Lauritzen and Hoffman (LH) model, which has the advantage of providing analytical expressions for the growth rate as a function of supercooling [2, 4, 7–10, 25].

ii) The Sadler and Gilmer (SG) theory, inspired by the thermal roughening phenomenon observed in low molecular weight molecules, where the removal of shorter stems is the controlling factor for growth kinetics in polymeric materials [7, 41, 82].

iii) Strobl has recently promoted a discussion in the polymer crystallization field by arguing that a mesomorphic precursor phase had to be formed before the more stable crystalline phase appears [10, 23, 83].

iv) Extensive computer simulations have also been presented, especially those of Muthukumar [11, 16, 20] and Hu et al. [13, 21, 22], among others [12].

Even though the model of nucleation and growth proposed by Lauritzen and Hoffman (LH model) has been under much criticism lately in view of recent morphological findings that lend some support to theories (ii) to (iv) [5, 7–10, 12, 16, 20, 23, 25, 83], it is still one of the few models that provides easy-to-use mathematical expressions that are capable of describing well the experimental data (even though the physical meaning of some of the fitting parameters could be questionable) and therefore it is widely employed.

When the isothermal crystallization is determined by spherulitic growth experiments, the energy barrier determined by applying the LH model refers exclusively to secondary nucleation or crystal growth. Instead, when the inverse of half-crystallization time ($1/\tau_{50\%}$) values obtained from DSC isothermal overall crystallization kinetic data is considered, both primary nucleation and crystal growth are considered. Therefore, the energetic parameters that we obtained after applying any classical kinetic crystallization theory to DSC data will include contributions from both processes.

The generalized LH model equation can be expressed as [5, 7–10, 25, 84, 85]

$$A(T) = A_o \exp\left(\frac{-U*}{R\left(T_c - T_\infty\right)}\right) \exp\left(\frac{-K_g^{\ A}}{T_c \Delta T f}\right) \quad (11.11)$$

where

$A(T)$ is a measure for crystal growth ($G(T)$, from PLOM), or a measure for the overall crystallization kinetics ($1/\tau_{50\%}$, from DSC).

A_o is a crystallization rate constant (G_o for crystal growth or τ_0^{-1} for the overall crystallization).

$U*$ is the activation energy for the transport of the chains to the growing front (1500 cal/mol is usually employed).

R is the gas constant.

T_c is the isothermal crystallization temperature.

T_∞ is the temperature at which chain mobility ceases, and it is usually taken as ($T_g - 30$) (K).

ΔT is the supercooling defined as ($T_m^{\ 0} - T_c$), where $T_m^{\ 0}$ is the equilibrium melting point.

f is a temperature correction term equal to: $2T_c/(T_c + T_m^{\ 0})$.

$K_g^{\ A}$ is the energy barrier for the crystallization process ($K_g^{\ G}$ for crystal growth, and $K_g^{\ \tau}$ for the overall nucleation and crystal growth kinetics).

Polymers usually exhibit spherulitic growth from the melt state and they often show a maximum growth rate at a given crystallization temperature, although in very flexible polymers the measurement of the left-hand side of the bell-shaped curve (the diffusion control side) is hindered by experimental conditions (i.e., at such high supercoolings, polymers like polyethylene crystallize during cooling to T_c almost regardless of the quench rate employed). Recently, chip calorimeters have provided unprecedented cooling rates that have allowed the determination of the full crystallization range overall kinetics in polymers like PCL [86] and PP [87].

Using a commercial sample of poly(1,4-dioxan-2-one) (PPDX[15], M_n of 15 Kg/mol and polydispersity of 1.2, more details in Refs [53, 88, 89]), which exhibited well-developed spherulites whose growth rate can be easily determined by PLOM [52, 90], Figure 11.8 prepared and shows the experimentally determined spherulitic growth rates (G) as a function of crystallization temperature (Fig. 11.8a). The data points clearly exhibit the classical bell-shaped curve discussed above [52, 90]. We used our "*LH Model fit*" Origin® plugin, that was designed to analyze crystallization kinetic data (G or $1/\tau$), performing the linear fit according to the LH theory model. This "*LH Model Fit*" is available for free distribution upon request [79].

Figure 11.8b shows the usual way of fitting the experimental data to the LH theory by plotting $\ln G + \frac{U*}{R(T_c - T_\infty)}$ versus $\frac{1}{T_c \cdot \Delta T \cdot f}$, a straight line should be obtained and $K_g^{\ G}$ can be determined from the slope of the plot. We have assumed that crystal growth occurs under Regime II [7, 52, 84, 85, 88–90] and the values of all parameters employed for the calculation are listed at the bottom of Table 11.4 [52]. Once we determined $K_g^{\ G}$, we calculated the theoretical prediction from Equation 11.11 and plotted the results over the experimental data points in Figure 11.7.

TABLE 11.4 Lauritzen–Hofmann K_g Values Obtained (by PLOM and DSC) for the PPDX[15]

	$(K^2) \times 10^{-4}$	
Method	K_g	R^2
PLOM	17.2	0.9982
DSC		
Neat	31.0	0.9987
Self-nucleated at 117 °C	17.0	0.9995
Nucleation contribution (%)	44.5	—
Crystal growth contribution (%)	55.5	—

Values employed for the LH fitting: $U* = 1500$ cal/mol, $T_m^{\ 0} = 127.1$ °C, $\Delta H_m^{\ 0} = 141.2$ J/g, $T_g = -10$ °C, $T_\infty = -30$ °C. Errors in K_g are estimated to be less than 3%.

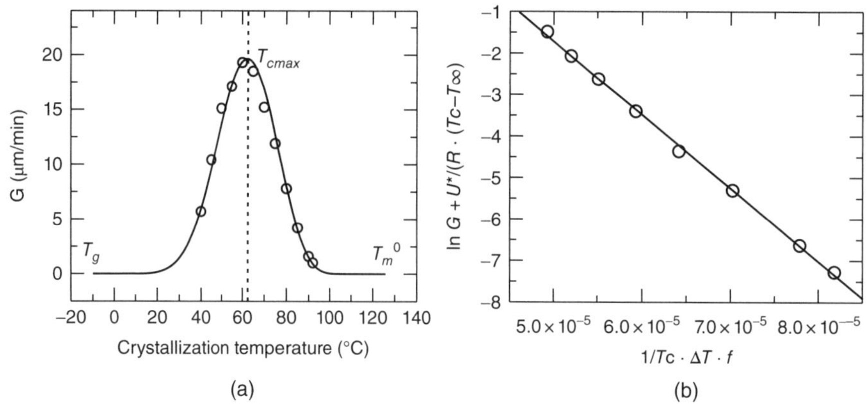

Figure 11.7 (a) Spherulitic growth rates for PPDX[15]. (b) Lauritzen–Hoffman kinetic theory plot for PPDX[15]. Solid lines represent the fit to the Lauritzen–Hoffman theory, according to Equation 11.11. Adapted from Lorenzo and Müller [60].

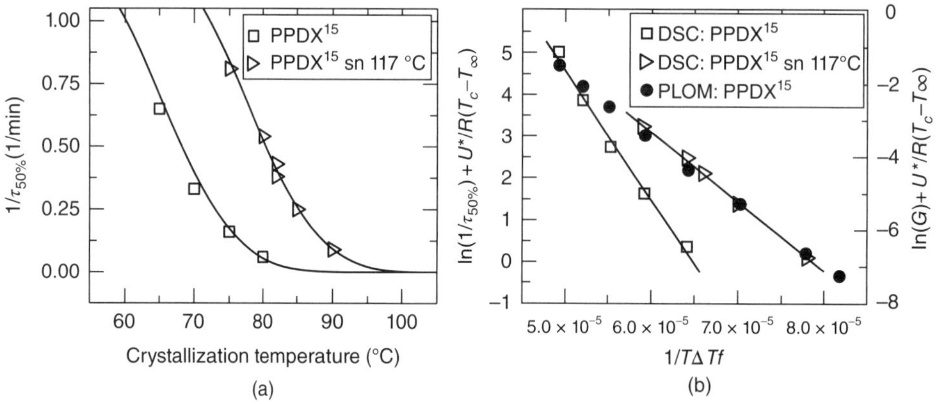

Figure 11.8 (a) Overall crystallization rate ($1/\tau_{50\%}$) for PPDX[15] and self-nucleated PPDX[15]. (b) Lauritzen–Hoffman kinetic theory plot for the overall crystallization of PPDX[15] and self-nucleated PPDX[15], and for the spherulitic growth of PPDX[15]. Solid lines are the fits to the Lauritzen–Hoffman theory, according to Equation 11.11. Adapted from Lorenzo and Müller [60].

In a similar fashion, DSC isothermal scans were recorded in order to study the crystallization kinetics of the PPDX[15] homopolymer after melting the samples for 3 min at 150 °C and quenching them (at 80 °C/min) to the desired crystallization temperature (T_c). After the crystallization was complete, the inverse of the half -crystallization time, $\tau_{50\%}$ (i.e., the time needed for 50% relative conversion to the crystalline state [31, 60]), was taken as a measure of the overall crystallization (nucleation and crystal growth) rate and its dependence on the crystallization temperature was analyzed.

Figure 11.8a shows how the LH theory was applied to fit overall crystallization kinetics data of $1/\tau_{50\%}$ versus T_c for the PPDX[15]. Here, we have also assumed that crystal growth occurs under Regime II and the values of all parameters employed for the LH theory calculation are listed as a footnote in Table 11.4.

DSC isothermal scans were also recorded in order to study the crystallization kinetics of PPDX after self-nucleating the sample at 117 °C. This self-nucleation temperature was determined by analyzing the self-nucleation domains and choosing an intermediate temperature within Domain II (more on the self-nucleation protocol can be found in Ref [52, 60, 91]).

The ideal self-nucleation temperature is probably the best possible choice, because the use of the ideal self-nucleation temperature will provide the polymer with an extremely high nucleation density (typically on the order of 10^{12} nuclei/cm^3), which can produce a full completion of the nucleation step before any isothermal crystallization is performed. This novel treatment could be used to study the relative contributions of nucleation and growth in semicrystalline homopolymers and semicrystalline components within diblock copolymers or polyblends.

The thermal protocol employed here for PPDX is summarized in Figure 11.9 and comprises the following steps:

(a) Melting at 150 °C for 3 min in order to erase any crystalline thermal history;
(b) Controlled cooling (10 °C/min) to 0 °C, and this step will create a standard crystalline history on the sample;
(c) Controlled heating (10 °C/min) to 117 °C (a T_s temperature within *Domain II* of the PPDX[15]) and then holding the sample at that temperature for 5 min;
(d) Quenching (at a controlled cooling rate of 80 °C/min) to an isothermal crystallization temperature (T_c), and

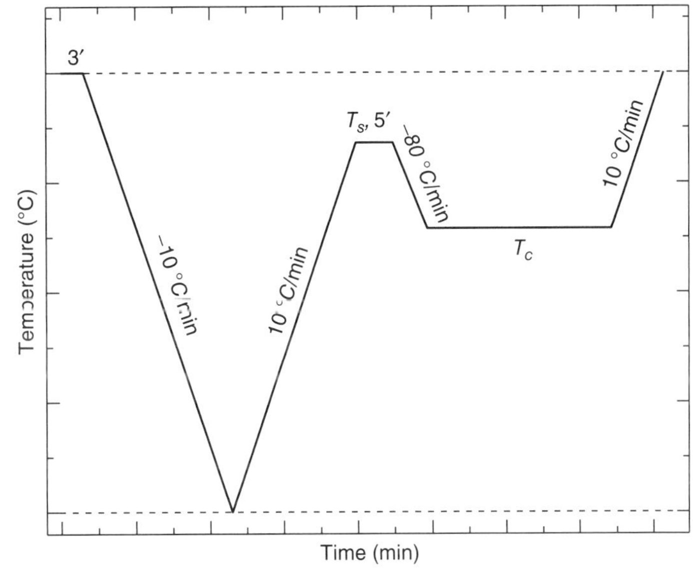

Figure 11.9 Typical thermal protocol employed to evaluate the isothermal crystallization after a *SN* thermal treatment was applied to the sample.

holding at that temperature for a period of time long enough to allow isothermal crystallization to take place while the DSC records the exothermic heat evolved in the process as a function of time;

(e) Finally, a controlled heating scan (10 °C/min) to 150 °C.

Even though the LH treatment was originally developed for describing crystal growth only, it has been employed to describe overall crystallization data because it is capable of fitting the data remarkably well [52, 60]. In these cases, such as when isothermal crystallization kinetics data obtained by DSC is employed, the energy barrier for crystallization reflected in the K_g value contains contributions from both primary nucleation and growth (K_g^{τ}). Figure 11.10 presents a schematic detailing this experimental protocol in a more visual way.

Figure 11.8a shows the overall crystallization kinetics of PPDX[15] obtained by DSC for neat and self-nucleated samples. When the sample is self-nucleated, the isothermal DSC data contain information of crystal growth only (assuming the self-nucleation process applied was 100% efficient in creating all necessary nuclei previously). In fact, the acceleration of the overall crystallization kinetics caused by the self-nucleation treatment is evident in Figure 11.8a, because the rates are higher for the self-nucleated sample as compared to neat PPDX at identical crystallization temperatures and also crystallization at lower supercoolings can be achieved in the self-nucleated samples.

Figure 11.8b shows the LH plots and it is evident that the slope has decreased in value substantially after self-nucleation, when the self-nucleated and the neat sample data are compared. Also, from Table 11.4 we can observe the excellent correlation found between the values of K_g obtained by the LH treatment from spherulitic growth rate data (PLOM) and those measured in self-nucleated samples by DSC (this can also be observed in Figure 11.8b, where the slopes of the LH plots for the self-nucleated PPDX DSC data and those taken by PLOM are nearly identical).

It can be concluded that the self-nucleation treatment was 100% efficient and therefore the data obtained by DSC after self-nucleation contain information on crystal growth only; hence, the identical values of K_g obtained. In samples that were not self-nucleated, the K_g values obtained by the LH treatment from DSC data were much larger than those obtained by PLOM; this can be explained by the contribution of both primary nucleation and growth.

The results presented in Table 11.4 are fully consistent with our hypothesis of possible separation of primary nucleation and growth contributions, as schematically represented in Figure 11.10. The K_g value obtained by the LH model for PPDX is reduced from 31.0 to 17.0 ($\times 10^4$ K^2) once the self-nucleation step is performed before the isothermal crystallization (see Table 11.4). This indicates that the energy barrier for primary nucleation is about 14.0×10^4 K^2; in other words, the energy barrier for nucleation of the PPDX[15] is about 45% of the total energy barrier for the overall crystallization process.

11.4 ISOTHERMAL CRYSTALLIZATION KINETICS – MORPHOLOGY RELATIONSHIP

11.4.1 Linear PS-*b*-PCL versus Miktoarm (PS_2)-*b*-(PCL_2) Block Copolymers

The $(PS)_2$-*b*-$(PCL)_2$ miktoarm star copolymers were synthesized in two steps, using high-vacuum techniques and a heterofunctional initiator derived from pentaerythritol with two free and two protected –OH groups. The linear PS-*b*-PCL block

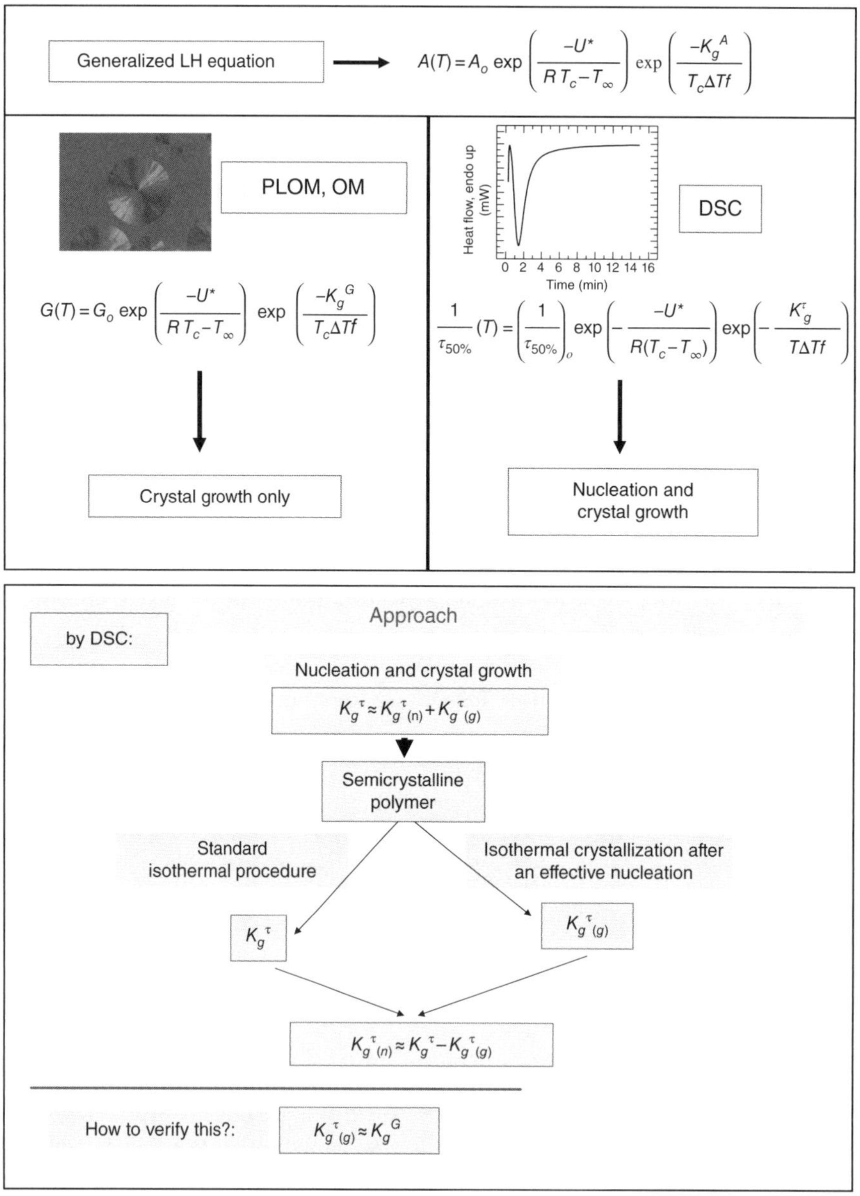

Figure 11.10 Schematic representation of the use of PLOM and DSC to extract information on both nucleation and growth kinetic parameters (see text).

copolymers were also prepared in two steps including the ROP of ε-caprolactone (CL) with a 1-decanol/Sn(Oct)2 initiating system, followed by the transformation of the end-OH group of PCL with 2-bromoisobutyryl bromide and polymerization of styrene with the newly formed Br-end groups. Full details are given in a previous paper (see Refs [65, 92]).

The molecular characteristics of the materials evaluated in this paper are provided in Table 11.5. In the notation: A_x-*b*-B_y^m here employed, the subscript numbers denote the mass fraction in weight percent and the superscripts give the number-averaged molecular weight M_n in Kg/mol of the entire copolymer.

Both linear and miktoarm star copolymers were found to undergo microphase segregation (see details in Refs [65, 92]). The $(PS)_2$-*b*-$(PCL)_2$ miktoarm star copolymers were found to display more compact morphologies for equivalent compositions than linear PS-*b*-PCL diblock copolymers. As a consequence, the crystallization of the PCL component always experiences much higher confinement in the miktoarm stars case than in the linear diblock copolymer case [65].

Figure 11.11 shows that the degree of supercooling needed for PCL crystallization increases with the content of PS in the copolymers. Furthermore, the PCL component within the miktoarm star block copolymers needs much larger supercoolings for crystallization than the PCL component within the analog linear diblock copolymers at similar PCL contents. This result confirms that miktoarm molecular architecture causes much larger confinement effects

TABLE 11.5 Molecular Characteristics of the (PCL_2)-*b*-(PS_2) Miktoarm Star and PCL-*b*-PS Linear Block Copolymers

Copolymer	M_w/M_n [a]	εCL [b] (wt%)	M_n [c] (Kg/mol)
PCL_{80}-*b*-PS_{20} [36]	1.70	79.8	36.1
PCL_{41}-*b*-PS_{59} [73]	1.35	41.2	72.8
PCL_{20}-*b*-PS_{80} [153]	1.46	19.6	152.8
$(PCL_2)_{72}$-*b*-$(PS_2)_{28}$ [34]	1.15	71.8	34.2[d]
$(PCL_2)_{39}$-*b*-$(PS_2)_{61}$ [62]	1.18	39.3	61.8
$(PCL_2)_{27}$-*b*-$(PS_2)_{73}$ [100]	1.41	27.1	100.1

[a]by SEC in THF at 40 °C.
[b]by ^{1}H-NMR spectroscopy in $CDCl_3$ at 25 °C.
[c]by MO in toluene at 37 °C.

at similar PCL contents when compared to linear analog diblock copolymers. Figure 11.11 also shows that in the PCL-containing copolymers (miktoarm and diblocks), where confinement is highest (cylinders and spheres morphologies), the overall crystallization rates are strongly affected, a clear evidence of the high degree of confinement upon the overall crystallization of the PCL component in comparison with neat PCL. In fact, in the case of cylinder- or sphere-forming materials, the overall crystallization had to be determined by the "isothermal step crystallization" (ISC) technique [48, 93], because their crystallization was so slow that conventional isothermal DSC cannot be employed. When the K_g^τ values obtained from the LH model fit were examined, the results indicated that as confinement increased, larger K_g^τ values were obtained (see Refs [65, 92] for more details).

The Avrami index trend as a function of crystallization temperature for the PCL-containing copolymers can be observed in Figure 11.12. The Avrami index increases markedly with T_c when the PCL content within the copolymers samples is nearly 40 wt% (also for the PCL homopolymer). Such a trend has been reported in the literature for PCL and many other polymers and it has been related to changes in the nucleation mechanism from instantaneous to sporadic nucleation when growth dimensionality is kept constant [5–8, 94]. When the PCL content is low (confined PCL copolymer samples), the Avrami index values are not so sensitive to changes in T_c, and they remain almost at a constant value.

When the PCL content within the copolymer samples is below 40 wt%, the Avrami index exhibits a decreasing trend as the PCL content within the samples decreases. There is a clear correlation between a decrease in the Avrami index and an increase in the confinement level. The observed shift in T_c range is an additional evidence of these topological restrictions, because the higher the confinement degree, the larger the supercooling (lower T_c range) needed for the crystallization of the PCL component.

Recently, a comparison between the miktoarm stars and the linear block copolymers was performed and three important effects were found [65]:

i) A decrease in Avrami index was found due to the morphological change at similar compositions when comparing linear versus miktoarm star block copolymers.
ii) If similar morphologies are considered, the Avrami index also decreased in the miktoarm star copolymers as compared to the linear ones.
iii) When the PCL component is highly confined (PCL content below 30 wt%), Avrami indexes between 1.5 and 1 were found. These results indicate that crystallization could start from surface or homogeneous nucleation because of the extreme supercooling needed for crystallization.

The change in the order of the crystallization kinetics to values on the order of 1 (i.e., Avrami index of 1) has been

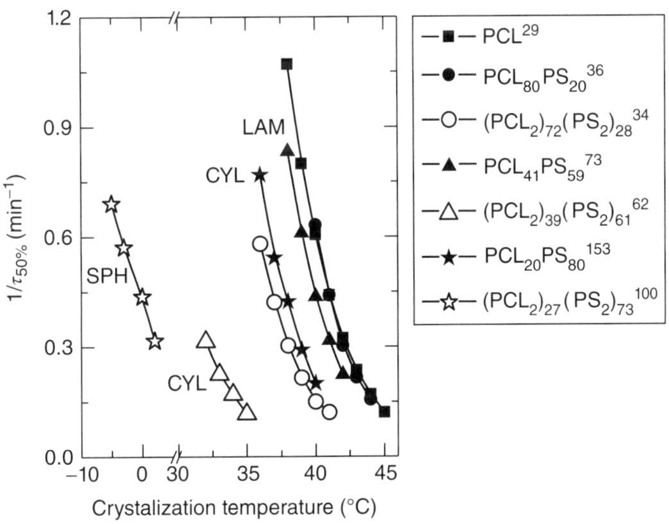

Figure 11.11 Overall crystallization rate as a function of isothermal crystallization temperature for the linear PCL-*b*-PS diblock and (PCL_2)-*b*-(PS_2) miktoarm star copolymers. Morphologies: LAM (lamellar), CYL (cylinders) and SPH (spheres). *Source:* Adapted from Lorenzo et al. [65].

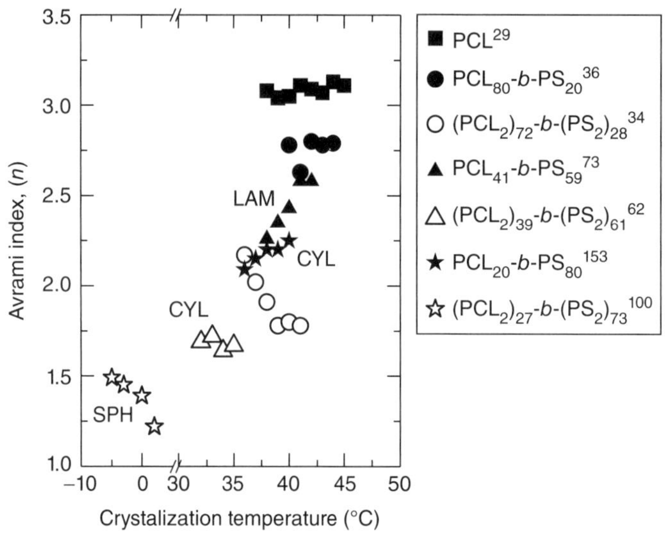

Figure 11.12 Avrami index values as a function of isothermal crystallization temperature for linear PCL-*b*-PS diblock copolymers and (PCL_2)-*b*-(PS_2) miktoarm star copolymers. *Morphologies: LAM (lamellar), CYL (cylinders), and SPH (spheres). Source:* Adapted from Lorenzo et al. [65].

successfully explained by the fact that confinement induces nucleation control within micro- or nanodomains. More details can be found in the following Refs [42, 61, 76, 95–97].

11.4.2 Crystallization Kinetics and Morphology of PLLA-*b*-PCL Diblock Copolymers

Poly(L-lactide)-*b*-poly(ε-caprolactone) (PLLA-*b*-PCL) diblock copolymers were synthesized by controlled/"living" sequential block copolymerization as initiated by aluminum trialkoxides in toluene solution. These procedures were reported in detail previously [50, 98]. Table 11.6 lists the molecular weight characterization data obtained by size exclusion chromatography (SEC) and by ^{1}H NMR. The diblock nomenclature we have used denotes the PLLA block as L and the PCL block as C, the subscripts indicate the approximate composition in wt% and the superscripts the approximate number average molecular weight of the entire block copolymer in kg/mol.

PLLA-*b*-PCL linear diblock copolymers were found to be either melt miscible or weakly segregated in the melt. This means that upon cooling the material will be able to form spherulites (see details in Refs [50, 51, 66, 95]).

TABLE 11.6 Molecular Characteristics of PLLA-*b*-PCL Block Copolymers and Homopolymers

Sample	PLLA/PCL Exp. Comp.[a]	$M_{n,exp}$ [b](Kg/mol) PLLA block	$M_{n,exp}$ [c](Kg/mol) PCL block	I^d
$PLLA^{24}$	100/0	23,900	—	1.1
$L_{93}C_{07}{}^{18}$	93/07	15,700[e]	1,700[e]	1.3
$L_{81}C_{19}{}^{21}$	81/19	16,700[e]	3,900[e]	1.3
$L_{60}C_{40}{}^{21}$	60/40	12,400	8,500	1.1
$L_{55}C_{45}{}^{18}$	55/45	9,500	8,100	1.4
$L_{44}C_{56}{}^{25}$	44/56	11,100	14,200	1.3
$L_{32}C_{68}{}^{22}$	32/68	6,900	14,900	1.4
$L_{10}C_{90}{}^{24}$	10/90	2,400	21,500	1.4
PCL^{29}	0/100	—	28,900	1.3

[a]Experimental composition as determined by ^{1}H NMR.
[b]Calculated M_n estimated by ^{1}H NMR for the PLLA knowing the M_n of the PCL block determined by SEC.
[c]Experimental M_n estimated by SEC for the PCL block.
[d]Polydispesrsity index of the final copolymer (determined by SEC).
[e]Theoretical M_n estimated considering monomers conversion (calculated by gravimetry).

11.4.2.1 Homo-PLLA and PLLA Block within the PLLA-*b*-PCL Diblock Copolymers

Crystallization kinetics of PLLA homopolymers have been analyzed in a wide temperature range by DSC and PLOM, from slightly above T_g to just below the apparent melting point [99–108]. Some peculiarities have been observed in the crystallization behavior of PLLA, because it exhibits a discontinuity in the crystallization kinetics at around 100–120 °C. The crystallization rate is very high at temperatures between 100 and 120 °C, showing a clear deviation from the usual bell-shaped curve of polymer crystal growth. This discontinuity had been correlated to a transition in regimes II–III growth of spherulites [104], to crystal polymorphism [18, 99, 109, 110], and to a sudden acceleration in spherulite growth rate not associated with morphological changes in the appearance of PLLA spherulite [106].

The results of fitting the Avrami equation to the experimental data reported in Ref [66] yielded values of n between 2.5 and 3.0, indicating instantaneous spherulitic growth for compositions of PLLA higher than 32 wt%. In the case of $L_{10}C_{90}{}^{24}$diblock copolymer, the values of n were lower and close to 2, a fact indicative of the crystallization into instantaneous two-dimensional aggregates (like axialites or 2-D lamellar aggregates).

The experimentally determined inverse half-crystallization time ($1/\tau_{50\%}$) is presented in Figure 11.13 as a function of the

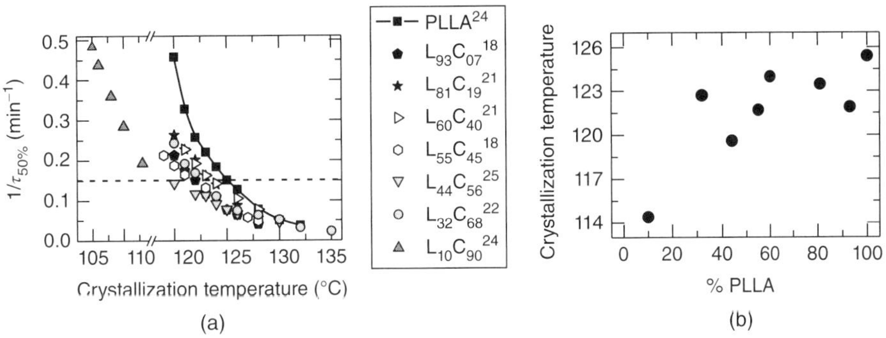

Figure 11.13 (a) Inverse of half-crystallization times ($1/\tau_{50\%}$) for the PLLA block within the block copolymers and (b) isothermal crystallization temperature (T_c) needed to obtain a value of $1/\tau_{50\%} = 0.15\,\text{min}^{-1}$ (dashed line in Fig. 11.13a) versus PLLA content. Adapted from Castillo et al. [66].

crystallization temperature. The conversion range employed was 3–30% and the correlation coefficients were always greater than 0.999 (see Ref [66]). It has to be noted that these results only deal with the crystallization kinetics at temperatures above 120 °C (PLLA block crystallization only; the PCL block is molten at such high temperatures).

The crystallization rates reported in Figure 11.13a reveal that a higher supercooling is needed to crystallize the PLLA block within $L_{10}C_{90}$[24] than homo-PLLA. The values of $1/\tau_{50\%}$ also indicate that the PLLA block within $L_{10}C_{90}$[24] crystallizes at much slower rates than homo-PLLA when similar crystallization temperatures are considered by extrapolation. To illustrate the composition dependence of the crystallization kinetics of the PLLA block within the PLLA-*b*-PCL diblock copolymers with PLLA contents higher than 10 wt%, we plotted the isothermal crystallization temperature ($T_{c(0.15)}$) needed to achieve an arbitrary value of $1/\tau_{50\%}$ (i.e., $0.15\,\text{min}^{-1}$) (Fig. 11.13b). Figure 11.13b shows that this isothermal crystallization temperature ($T_{c(0.15)}$) tends to decrease when the PLLA content decreases. Thus, a higher supercooling is needed to crystallize at the same crystallization rate the PLLA block than homo-PLLA; which is a result of the diluent effect of the molten PCL chains during PLLA block crystallization. However, at higher PLLA contents, $T_{c(0.15)}$ scarcely decreases with composition, but drops quickly at contents lower than 32 wt%. In summary, the effect of the molten PCL covalently bonded block was to slow down the crystallization rate of PLLA chains, slowly at first and then more rapidly as the content of PCL increased [66].

The crystallization kinetics of PLLA has been reported to be very much dependent on the molecular weight (MW). With increasing MW, the crystallization rate drops greatly [107]. The decrease in the crystallization kinetics shown in Figure 11.13 cannot be attributed to changes in MW (because the MW of the PLLA block is also decreasing) but to the presence of the molten PCL chains. Therefore, the decrease in crystallization rate for miscible systems can be related to a dilution effect that reduces the number of crystallizable segments on the front of the growing spherulite, and also to a decrease in supercooling due to the melting point depression. Similar effects have been reported for miscible block copolymers such as polyethylene-*b*-poly(ethylene-*alt*-propylene) where a stronger dilution effect induced by amorphous poly(ethylene-*alt*-propylene) chains cause the need for much higher supercoolings in order to crystallize the PE block chains but at compositions close to 50 wt% [77, 111].

11.4.2.2 Homo-PCL and PCL Block within the PLLA-b-PCL Diblock Copolymers The overall crystallization rate, expressed as $1/\tau_{50\%}$ for homo-PCL and for the PCL block within the PLLA-*b*-PCL diblock copolymers are shown in Figure 11.14. It is important to note that the isothermal crystallization of the PCL block was performed after crystallizing the PLLA block until saturation. At this condition, the PCL block chains are nucleated by the PLLA block crystals where the PCL peak crystallization temperature ($T_{c,PCL}$) shifted to higher values.

It is clear from Figure 11.14, that the overall crystallization rate of the PCL block decreases with PLLA content. In addition, larger supercoolings are needed to crystallize the PCL block within the diblock copolymer as compared to the parent homopolymer. This is another evidence of the restrictions that the PCL block is facing in order to crystallize, in spite of the nucleation effect produced by the PLLA block crystals. In the case of $L_{81}C_{19}$[21], the evolution of enthalpy with time was determined by ISC, following the procedure reported in Ref [48]. Our preceding works [50, 51] showed by SAXS experiments that during isothermal PCL crystallization within the diblock copolymer with PCL contents higher than 40 wt%, rearrangements of PLLA crystal stems occurred to accommodate the previously amorphous PCL chains, disturbing the initial PLLA lamellar structure. In fact, lateral contraction of the PLLA unit cell and local melting by rearrangement of the lamellar structure was reported for $L_{60}C_{40}$[21], $L_{44}C_{56}$[25], $L_{32}C_{68}$[22]. Although PCL crystallization effectively disturbed the previous PLLA lamellar structure (as demonstrated by SAXS experiments), it is not enough to avoid the confinement effect, as revealed by the DSC experiments performed. In addition, after isothermal crystallization of the PLLA block at 122 °C during SAXS experiments [50, 51], measured

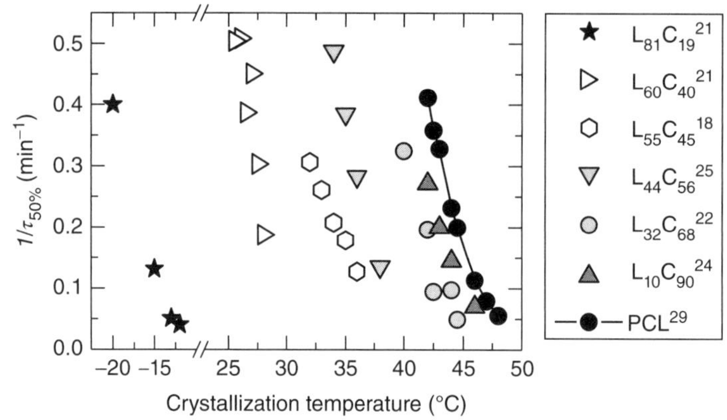

Figure 11.14 Inverse of half-crystallization times ($1/\tau_{50\%}$) for the PCL block within the block copolymers. Adapted from Castillo et al. [66].

domain spacings were $d = 335$, 309, and 267 Å for $L_{32}C_{68}^{22}$, $L_{44}C_{56}^{25}$, and $L_{60}C_{40}^{21}$, respectively. In consequence, it is expected that increasing PLLA composition increases the confinement during PCL crystallization.

Figure 11.15 shows how the Avrami index corresponding to the crystallization of the PCL block depends on PLLA content. The Avrami index is found to decrease with PLLA content. Such a decrease can be taken as a reduction in the dimensionality growth of the PCL superstructures as more PLLA previously crystallized is present in the sample. In fact, for the $L_{81}C_{19}^{21}$ sample, a first-order kinetics is revealed by Avrami indexes close to 1. This result confirms that the PCL block crystallizes by surface or homogeneous nucleation, that is, that crystal growth must be essentially instantaneous so the kinetics is entirely controlled by the nucleation process [51]. Values of Avrami index lower than 1 have been interpreted as due to a nucleation process that is in between sporadic and instantaneous [51]. These results were the first report (at the time) of surface or homogeneous nucleation of a crystallizable component within a miscible or weakly segregated diblock copolymer.

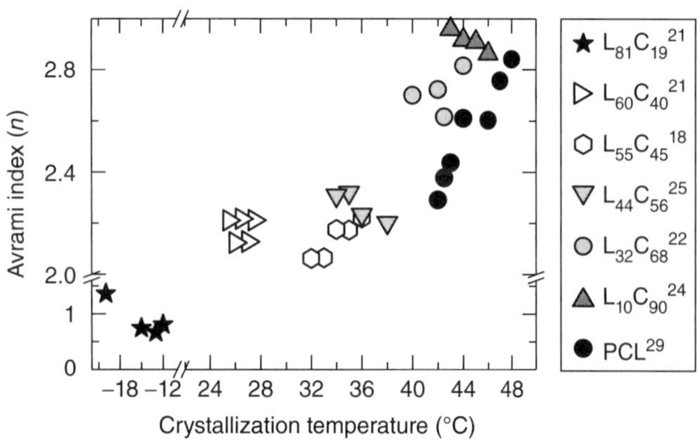

Figure 11.15 Avrami index values obtained by fitting isothermal crystallization data obtained by DSC for the PCL block within the block copolymers indicated. Adapted from Castillo et al. [66].

The morphology of the homo-PLLA and PLLA-*b*-PCL diblock copolymers was studied by PLOM [50, 51, 66, 112]. It was found that: (i) PLLA crystallized in spherulitic morphology regardless of composition; and (ii) the PLLA block templates the morphology in such a way that when the PCL block is able to crystallize at lower temperatures, it has to fit in between the PLLA radially grown lamellae and just a subtle change in the magnitude of birefringence occurs during PCL crystallization.

Figure 11.16a presents the spherulitic growth rates (G) for the homo-PLLA and PLLA blocks within three diblock copolymers that possess well-separated compositions, as a function of the isothermal crystallization temperature. Figure 11.16b reports the isothermal crystallization temperature ($T_{c(4)}$) needed for G to reach an arbitrary value of 4 μm/min. The spherulitic growth rates are fully dependent on both the composition of the block copolymers and on the crystallization temperature.

Initially, increasing the PCL content in the copolymers, the bell-shaped curves of Figure 11.16 are slightly shifted to lower temperatures, that is, the maximum value of G shifts to lower crystallization temperatures with PCL content. In addition, the $T_{c(4)}$ values also decrease slowly initially with PCL content. Finally, for PCL contents higher than 68 wt%, the spherulitic growth rate of the PLLA block is strongly depressed as compared to homo-PLLA. In general, G values decreased as the PLLA content decreases, in view of the diluent effect provoked by the PCL block molten chains, as previously explained.

11.4.3 Nucleation and Crystallization Kinetics of Double Crystalline Polyethylene/Polyamide (PE/PA) Blends

In the following interesting example, the isothermal crystallization kinetics of a blend composed of two crystallizable polymers is presented on the basis of the work of Córdova et al. [69].

PE/PA blends were prepared in a co-rotating twin-screw micro-compounder (see Table 11.7 for blends compositions), using two functionalized polyethylenes and two polyamide 6 homopolymers [69]:

i) PE-3 is a random copolymer of ethylene, acrylate, and maleic anhydride (MAH). It had a MAH weight fraction of 2.9 wt% and an acrylate weight fraction of 17 wt%.

ii) PE-1 is also a random copolymer of ethylene, acrylate, and MAH, but with a MAH weight fraction of 0.8 wt%, and 9 wt% of acrylate.

iii) A Polyamide 6 (PA6) synthesized by polycondensation, terminated at one end by the reactive functional group NH_2 and at the other end by a nonreactive CH_3 group. This PA6 had a M_n of 2500 g/mol and a polydispersity index of 2.

iv) A commercial PA6 (Domamid 24), terminated by 35 mol % of NH2 and 65 mol% of COOH, with a M_n = 15,000 g/mol and labeled as cPA6.

Figure 11.17 shows representative TEM images of the prepared polymer blends (MA to MC). The micrographs for the MA and MB polyblends exhibit a droplet-like morphology characteristic of immiscible polymer blends, where the minority component is dispersed in a matrix of the major component. The spherical microdomains, in both TEM micrographs, possess an average particle diameter range of 250–350 nm. Decreasing the cPA6 content from 30 wt% (MA) to 20 wt% (MB) did not change the morphology of the blends, as cPA6 spheres of a particular distribution of sizes is the predominant morphology.

In the case of the sample MC, a macroscopically homogeneous mesophase constituted by interconnected sheets of PA6 was observed with a characteristic thickness of about 12 nm. It has been reported [69, 113] that this peculiar morphology does not depend on mixing conditions. Also, the interconnected structure is stable and does not change with annealing above the melting point of the PA6 component, or after several processing operations such as injection molding, extrusion, or compression molding.

11.4.3.1 Isothermal Crystallization: Polyamide Phase

Figure 11.18 shows the inverse of the half-crystallization time as a function of crystallization temperature. The trend

TABLE 11.7 Composition for the Reactively Extruded Blends [69]

Sample Label	Polyamide (wt%)	Polyethylene (wt%)	$MAH:NH_2$ ratio
PA_6	100	—	—
cPA_6	100	—	—
PE-3	—	100	—
PE-1	—	100	—
MA	30% cPA_6	70% PE-1	4:1
MB	20% cPA_6	80% PE-1	7:1
MC	20% PA_6	80% PE-3	3:1

displayed by PA_6 and cPA_6 is expected, because the lower molecular weight PA_6 crystallizes much faster at equivalent supercoolings.

A larger supercooling is needed in order to crystallize the PA_6 phase within the MC blend as compared to neat PA_6. Accordingly, this means that if the samples could be compared at identical crystallization temperatures, the PA_6 phase within the blend would crystallize at a much lower rate than neat PA_6. This result corroborates the impact of the co-continuous morphology and the presence of graft copolymer on the crystallization kinetics of the PA_6 within the compatibilized blend. The PA_6 chains encountered topological restrictions for crystallization when they are within the submicron-percolated and compatibilized co-continuous morphology as compared to bulk neat PA_6 chains.

The impact of the blend morphology on the isothermal crystallization kinetics is even more striking for MA and MB. The data corresponding to cPA_6 phase in Figure 11.18 had to be plotted on a different y scale (hence the shadowing over these data) because their crystallization rates are much slower. A very large supercooling (on the order of 100 °C) is needed to crystallize the cPA_6 droplets as compared to bulk neat cPA_6. This is a direct consequence of dispersing cPA_6 into a large number of droplets that are virtually free of heterogeneities leading to surface or homogeneous nucleation.

The Avrami index (corresponding to the polyamide components) depends on the isothermal crystallization temperature

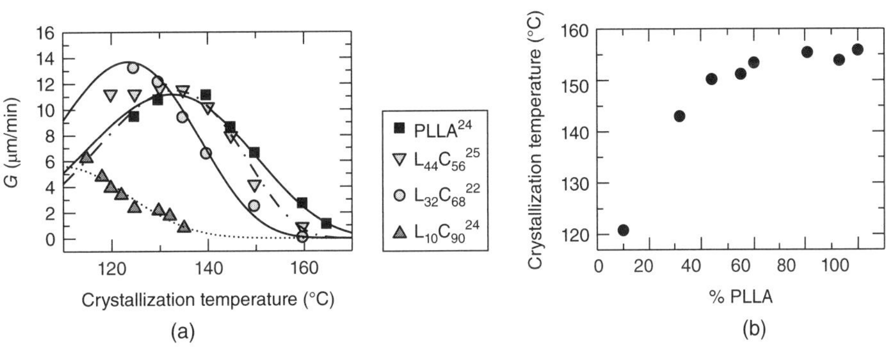

Figure 11.16 (a) Spherulitic growth rate (G) of the PLLA block as a function of the isothermal crystallization temperature (T_c) for the indicated samples and (b) isothermal crystallization temperature (T_c) needed to achieve a fixed value of G = 4 μm/min versus PLLA content. Adapted from Castillo et al. [66].

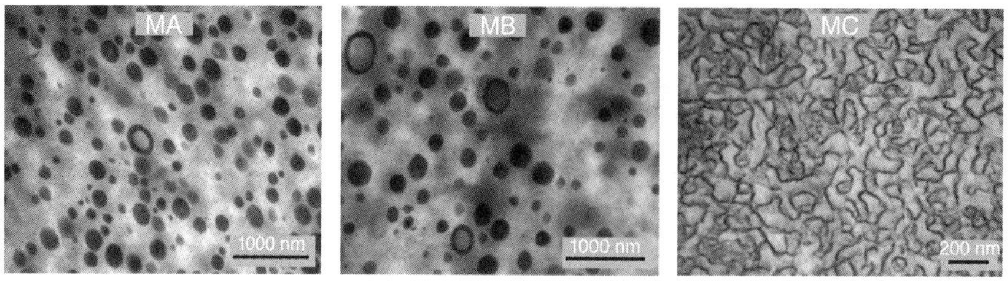

Figure 11.17 TEM images for PE/PA blends prepared by reactive extrusion; compositions are given in Table 11.7 (MA to MC). Adapted from Córdova et al. [[70], p. 16], Figure 1. Reproduced with permission of John Wiley and Sons.

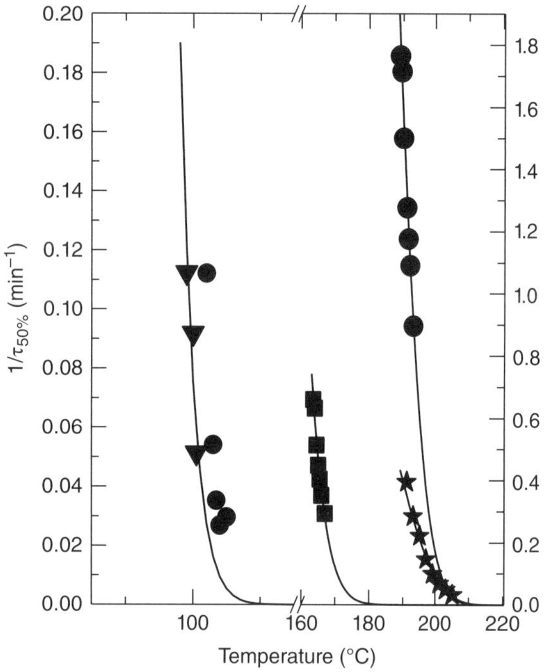

Figure 11.18 Variation of the inverse of the half-crystallization time ($1/\tau_{50\%}$) as a function of the crystallization temperature for the polyamide component within the studied blends and for the neat polyamide homopolymers as function of the crystallization temperature (T_c); see Table 11.7. Adapted from Córdova et al. [[70], p. 16], Figure 18. Reproduced with permission of John Wiley and Sons.

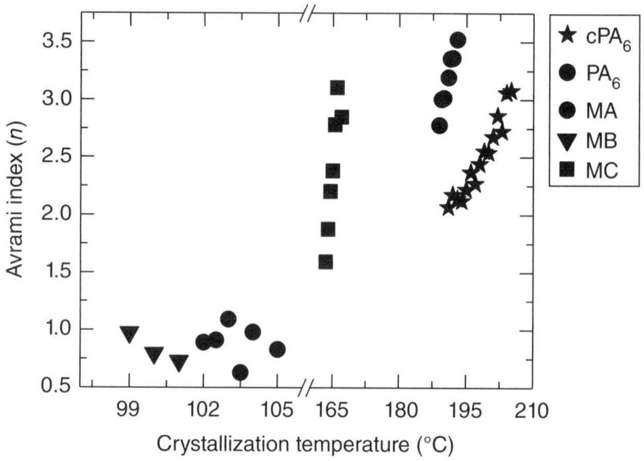

Figure 11.19 Avrami indexes for the PA component within the studied blends and for the neat PA homopolymers as a function of the crystallization temperature (T_c); see Table 11.7. Adapted from Córdova et al. [69].

for all the samples examined here (see Fig. 11.19). There is a clear correlation between the value of the Avrami index and the morphology. The highest values were obtained for polyamide homopolymers. Intermediate values were obtained when the PA_6 phase is a co-continuous phase within the MC blend. Finally, the Avrami index decreases to values close to 1 when the cPA_6 phase is dispersed in submicron droplets. Also, the higher the confinement degree the larger the supercooling needed to be applied in order to crystallize the polyamide phase. Similar results have been reported for block copolymers and nanocomposites, where the Avrami index values decrease as confinement increases upon changing composition from a matrix, to a lamellar morphology, cylinders, and then spheres [42, 60, 64, 65, 72, 76, 96, 97, 114, 115]. However, this example was the first time such a trend had been reported in polymer blends [69].

The Avrami index increases with T_c when the polyamide is found in a percolated morphology (MC) or in the bulk homopolymers (i.e., PA_6 and cPA_6), (see Figure 11.19). Such a trend has been related to changes in the nucleation rate from instantaneous to sporadic nucleation when growth dimensionality is kept constant [7, 31].

When the polyamide is confined into submicron droplets within the MA and MB blends, the Avrami index is not sensitive to changes in T_c (see Fig. 11.19); while its value is on the order of 1 or less than 1 [42, 96, 97]. This implies that crystallization starts from surface or homogeneous nucleation phenomena because of the extreme supercooling (as compared to other samples) needed for crystallization and because of the change in the order of the crystallization kinetics as indicated by the change in the Avrami index. When first-order kinetics are obtained using the Avrami fit (i.e., $n = 1$), the crystallization rate has to be proportional to the fraction of crystallizable material not yet crystallized, that is, the crystal growth must be essentially instantaneous within isolated microdomains (a likely situation in view of the extreme supercooling and the small sizes of the MDs). Therefore, the nucleation will be the rate-determining step in the crystallization process of isolated phases, like spheres [42, 96, 97]. This change from a sigmoidal crystallization kinetics (i.e., Avrami indexes in between 2 and 4) to a first-order one (i.e., $n = 1$) has also been observed in the

TABLE 11.8 Parameters Derived from the LH Model Fit to the Isothermal Crystallization Kinetics Data, for the Overall Isothermal Crystallization of the PA/PE Blends and Homopolymers

Sample	$K_g^\tau \times 10^{-4}$ (K^2)	R^2
PA_6[a]	16.7	0.996
cPA_6[a]	11.4	0.991
PE-1[b]	74.9	0.999
PE-3[b]	47.7	0.979
PE-1 (MA)[b]	61.3	0.934
PE-1 (MB)[b]	55.2	0.914
PE-3 (MC)[b]	115.6	0.991
cPA_6 (MA)[a]	61.6	0.837
cPA_6 (MB)[a]	—	—
PA_6 (MC)[a]	68.0	0.986

Constant used in the Lauritzen–Hoffman theory for PE and PA_6:
[a] $\rho_c = 1.004$ g/cm^3, $\rho_a = 0.853$ g/cm^3, a_0: 4.55 nm, b_0: 4.15 nm, T_g: −42.2 °C, T_∞: $T_g - 30$, T_m^0: 144.6 °C, $\Delta H_{100\%}$: 280 J/g.
[b] $\rho_c = 1.23$ g/cm^3, $\rho_a = 1.084$ g/cm^3, a_0:4.78 nm, b_0: 3.70 nm, T_g: 47 °C, T_∞: $T_g - 30$, T_m^0: 232 °C, $\Delta H_{100\%}$: 188.1 J/g.

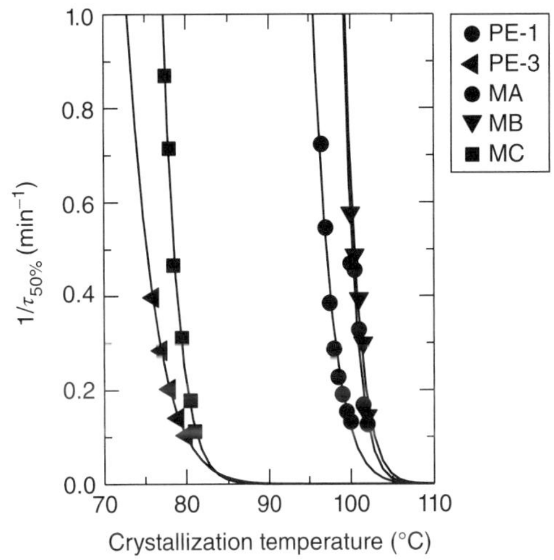

Figure 11.20 Variation of the inverse of the half-crystallization time ($1/\tau_{50\%}$) as a function of the crystallization temperature for the polyethylene component within the studied blends and for the neat polyethylene copolymers as a function of the crystallization temperature (T_c); see Table 11.7. Adapted from Córdova et al. [69].

crystallization of isolated phases within block copolymers and in polymers infiltrated within nanoporous alumina templates [42, 96, 97].

The solid lines shown in Figure 11.18 are fits of the data to the LH model. We have assumed that crystal growth occurs under Regime II and the values of all parameters employed for the calculation are listed in Table 11.8, the fits obtained were excellent in all cases.

Similar K_g^τ values were obtained for both PA_6 and cPA_6 (as expected for similar homopolymers; see Table 11.8). However, for the blends the values of K_g^τ are nearly 5–6 times larger. These results indicate that the energy barriers for the nucleation and growth processes (as the data was obtained by DSC, the energy barrier contains terms coming from nucleation and growth) are much larger in this case due to the confinement effects encountered by the polyamide phases within the blends.

11.4.3.2 *Isothermal Crystallization: Polyethylene Phase*

Figure 11.20 shows crystallization rate data as a function of the crystallization temperature for both polyethylene copolymers and for the polyethylene component within the blends. The linear low-density-type PE-1 crystallizes at much lower supercoolings than the very low-density-type PE-3. In the case of the blends, a nucleation effect caused by the previously crystallized polyamide phases was reported (the PA phases crystallize at higher temperatures than the PE phases; see Ref [69]). This nucleation effect accelerates the overall crystallization kinetics and therefore the polyethylene component in the blends crystallizes faster than the corresponding neat polyethylene material, as shown in Figure 11.20.

The K_g^τ values for the PE phases within these blends and their analog neat PE are shown in Table 11.8. It can be seen how the energy barrier for the overall crystallization of the polyethylene chains within MA and MB blends is lower (by 20–25%) than the value obtained for the PE-1 neat material. This lower energy barrier can be explained by the nucleation effect caused by the cPA6 droplets on the polyethylene matrix. In the case of the MC blend, the crystallization of the PE-3 component is restricted as compared to a bulk neat PE-3, and the energy barrier for crystallization is actually larger for the PE-3 phase within the blend than for the PE-3 neat material.

Figure 11.21 shows the Avrami index values obtained for the PE components as a function of T_c. PE-1 exhibits the typical behavior of linear low-density polyethylenes. It displays Avrami values that increase with T_c from roughly 2 to 3 (as already stated, this increase with temperature is expected because the nucleation becomes more sporadic as

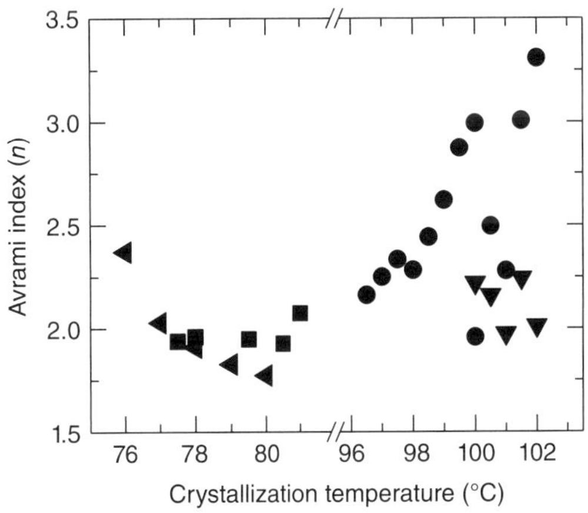

Figure 11.21 Avrami indexes for the polyethylene component within the studied blends and for the neat polyethylene copolymers as a function of the crystallization temperature (T_c). *Source:* Adapted from Córdova et al. [69].

T_c increases, a valid observation if the growth dimensionality remains constant) [69].

The PE-3 is a very low density material with short linear methylene sequences *t*. Very low density polyethylenes crystallize with ill-defined morphologies frequently exhibiting fringe-like lamellar aggregates [69, 116]. They usually display Avrami index values lower than or equal to 2, as shown by PE-3 in Figure 11.21. An inverse dependence of the Avrami index values with T_c could be due to dimensionality changes induced by temperature while the nucleation rate remains constant.

11.4.4 Crystallization Kinetics of Poly(ε-Caprolactone)/Carbon Nanotubes (PCL/CNTs) Blends

Our last example will be on nanocomposites [72]. For this study, PCL/MWNT nanocomposites were prepared by melt mixing a commercial PCL (M_n ~50 Kg/mol) with multiwalled carbon nanotubes (MWNTs or CNTs) at different compositions. Those nanocomposites contained the following quantities of MWNTs in wt%: 0.3; 0.5; 0.7; 1; and 3. The samples will be denoted PCL_xM_y. The *x* and *y* subscripts denote the PCL and MWNT content in wt%, respectively.

Figure 11.22a shows experimentally determined values for the inverse of the half-crystallization time as a function of the isothermal crystallization temperatures for PCL_xM_y nanocomposites. Three general observations can be made:

(a) The crystallization temperature range differs between neat PCL and the nanocomposites. This is the first indication that neat PCL experiences more difficulties in crystallizing than the nanocomposites, because it needs a larger degree of supercooling.

(b) The overall crystallization rate increases with MWNT loading for the systems explored. This is probably a consequence of the nucleation effect. For this material, in fact, a supernucleation effect was detected. Supernucleation of MWCNT occurs when the nanofiller is more efficient than the matrix's own nuclei in generating self-nucleation (details of this effect can be found in Refs [56, 64, 68, 72]). Supernucleation accelerates the primary nucleation rate and contributes to increase the overall crystallization kinetics.

(c) The temperature dependence of the overall crystallization rate is not the same for the two systems explored. In order to see this observation more clearly, Figure 11.22b represent the same data on a semi-log scale.

Figure 11.22 also shows that for MWNT contents of 0.7% or larger, there is a clear transition in crystallization rate with T_c, where the crystallization rate suddenly increases below a specific T_c temperature. This behavior can be interpreted as a change in the overall crystallization regime.

As can be seen, most of the experimental data presented in Figure 11.23 can be described by employing Regime II in the LH equation. This regime considers that the rate of secondary (or tertiary) nucleation is comparable to that of the rate of nuclei spreading across the growth front. Table 11.9 shows that K_g^{τ} is lower for the nanocomposites as compared to the value of K_g^{τ} for neat PCL. This result could be interpreted by considering that the reported supernucleation effect (see Ref [72] for full details on this) is contributing to decrease the energy barrier for overall crystallization (which includes both nucleation and growth).

Figure 11.23 also shows that for CNT content greater than 0.5%, the crystallization regime changed from Regimes II to III. This change is necessary to explain the sudden enhancement in the overall crystallization kinetics as the supercooling is increased (Fig. 11.22), where Regime III is characterized by a much faster nucleation than spreading rate [72].

Based on Figure 11.23 and Table 11.9, it would appear that beyond 0.5% CNT loading, the reported supernucleation effect is so large that a transition to Regime III occurs, because nucleation is much faster than spreading rate. The isothermal crystallization of neat PCL has been extensively studied in the past and only Regime II has been observed [60, 64, 65, 72, 114, 117–122], and this was the first time that such a regime transition was documented as a result of CNT addition.

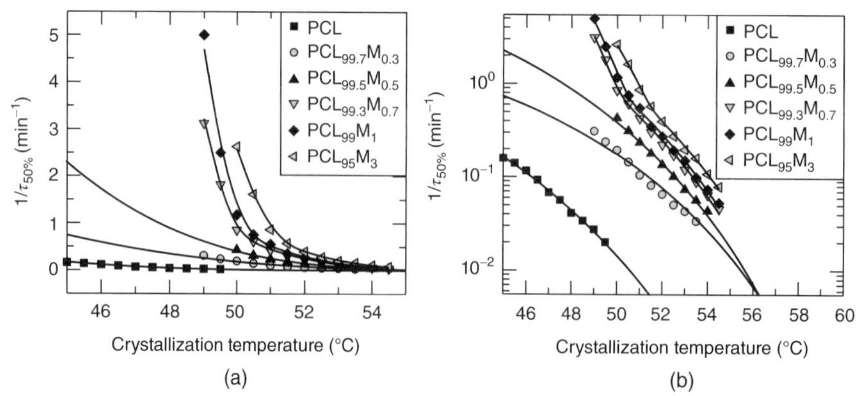

Figure 11.22 Variation of $1/\tau_{50\%}$ values as a function of the crystallization temperature for PCL_xM_y ((b) presents the same data as (a), but in a semi-log scale). Solid lines indicate the fittings to the LH model. *Source:* Adapted from Trujillo et al. [72].

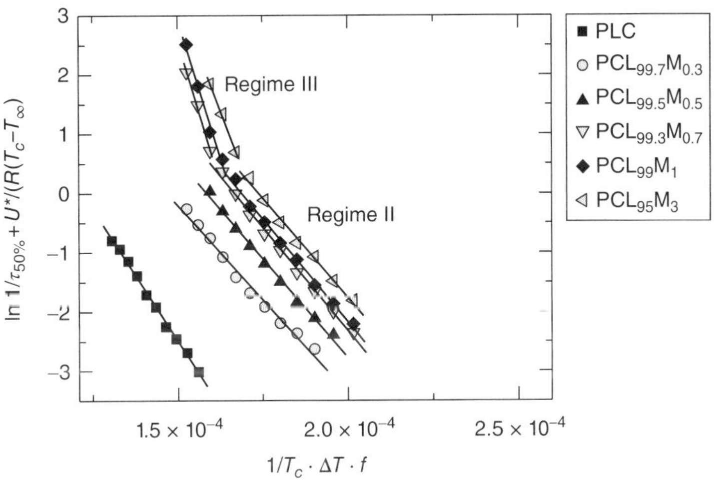

Figure 11.23 LH model plots for PCL_xM_y nanocomposites. Solid lines represent the fit to the LH equation. For selected samples, the change from Regimes II to III is obvious (see text). *Source:* Adapted from Trujillo et al. [72].

TABLE 11.9 Parameters of the LH Theory for PCL_xM_y nanocomposites.

Systems	PCL	$PCL_{99.7}\ M_{0.3}$	$PCL_{99.5}\ M_{0.5}$	$PCL_{99.3}\ M_{0.7}$		$PCL_{99}\ M_1$		$PCL_{97}\ M_3$	
Regimen	II	II	II	II	III	II	III	II	III
$K_g^{\tau} \times 10^{-4} \pm 0.2(K^2)$	9.4	6.4	6.7	7.2	19.3	7.2	18.6	6.7	13.7
R^2	0.995	0.986	0.995	0.993	0.993	0.993	0.985	0.993	0.993
$K_g^{\tau}{}_{III}/K_g^{\tau}{}_{II}$	—	—	—	2.6		2.7		2.0	

11.5 CONCLUSIONS

In this chapter, several guidelines were given on how to interpret spherulitic growth rates by PLOM and overall crystallization kinetics by DSC. In addition, examples of isothermal crystallization kinetics, and its relation with morphology, of different materials, ranging from neat polymers, to blends, block copolymers, and nanocomposites were provided. The fittings of the data were performed by simple plugins developed for the commercially available software Origin® that can be supplied to any interested users upon request. It was demonstrated that the Avrami and LH models can be used in order to correctly interpret the crystallization kinetics data with sound correlations to the specific morphology of each system examined. Experimental ways to separate the contributions of nucleation and growth were also highlighted.

ACKNOWLEDGMENTS

We acknowledge fruitful collaborations and discussions with professors María Luisa Arnal, Hsin-Lung Chen, Philippe Dubois, Nikos Hadjichristidis, and Ian Hamley.

REFERENCES

1. Wunderlich B. *Macromolecular Physics, Volume 1: Crystal Structure, Morphology and Defects*. New York: Academic Press; 1973.
2. Wunderlich B. *Macromolecular Physics Volume 2: Crystal nucleation, Growth, Annealing*. New York: Academic Press; 1976.
3. Wunderlich B. *Macromolecular Physics, Volume 3: Crystal Melting*. New York: Academic Press; 1980.
4. Mandelkern L. *Crystallization of Polymers, Volume 1: Equilibrium Concepts*. 2nd ed. Cambridge: Cambridge University Press; 2002.
5. Mandelkern L. *Crystallization of Polymers, Volume 2: Kinetics and Mechanisms*. 2nd ed. Cambridge: Cambridge University Press; 2004.
6. Schultz JM. *Polymer Crystallization: The Development of Crystalline Order in Thermoplastic Polymers*. New York: Oxford University Press and American Chemical Society; 2001.
7. Gedde UW. *Polymer Physics*. London: Chapman and Hall; 1995.
8. Hiemenz PC, Lodge TP. *Polymer Chemistry*. 2nd ed. Boca Raton: CRC Press; 2007.
9. Reiter G, Sommer JU, editors. *Lecture Notes in Physics: Polymer Crystallization: Observations, Concepts and Interpretations*. Berlin: Springer; 2003.
10. Strobl GR. *The Physics of Polymers: Concepts for Understanding Their Structures and Behavior*. 3rd ed. Berlin: Springer; 2007.
11. Reiter G, Strobl GR, editors. *Lecture Notes in Physics: Progress in Understanding of Polymer Crystallization*. Vol. 714. Berlin: Springer-Verlag; 2007.
12. Piorkowska E, Rutledge GC, editors. *Handbook of Polymer Crystallization*. New Jersey: Wiley; 2013.

13. Hu W. *Polymer Physics: A Molecular Approach*. Berlin: Springer; 2013.
14. Imai M, Kaji K, Kanaya T. Macromolecules 1994;27:7103.
15. Olmsted PD, Poon WCK, McLeish TCB, Terrill NJ, Ryan AJ. Phys Rev Lett 1998;81:373.
16. Muthukumar M, Welch P. Polymer 2000;41:8833.
17. Kaji K, Nishida K, Matsuba G, Kanaya T, Imai M. J Macromol Sci B 2003;42:709.
18. Zhang J, Duan Y, Sato H, Shen D, Yan S, Noda I, Ozaki Y. J Phys Chem B 2005;109:5586.
19. Gee RH, Lacevic N, Fried LE. Nat Mater 2006;5:39.
20. Muthukumar M. Eur Phys J E 2000;3:199.
21. Hu W, Frenkel D. Adv Polym Sci 2005;191:1.
22. Hu W, Cai T, Ma Y, Hobbs J, Farrance O, Reiter G. Faraday Discuss 2009;143:129.
23. Strobl GR. Eur Phys J E 2000;3:165.
24. Wegner G. Faraday Discuss Chem Soc 1979;68:494.
25. Sperling LH. *Introduction to Physical Polymer Science*. New Jersey: Wiley Interscience; 2006.
26. Pennings A. *Crystal Growth*. Oxford: Pergamon; 1967.
27. Keller A. Faraday Discuss Chem Soc 1979;68:145.
28. Janeschitz-Kriegl H. *Crystallization Modalities in Polymer Melt Processing: Fundamental Aspects of Structure Formation*. New York: Springer; 2010.
29. Geil PH. *Polymer Single Crystals*. New York: Interscience Publishers; 1963.
30. Brandrup J, Inmergut E, Grulke E. *Polymer Handbook*. New York: John Wiley and Sons; 1999.
31. Lorenzo AT, Arnal ML, Albuerne J, Müller AJ. Polym Test 2007;26:222.
32. Von Göler F, Sachs ZG. Physik 1932;77:281.
33. Ergoz E, Fatou JG, Mandelkern L. Macromolecules 1972;5:147.
34. Johnson WA, Mehl RF. Trans Am Inst Min Metall Eng 1939;135:416.
35. Avrami M. J Chem Phys 1939;7:1103.
36. Avrami M. J Chem Phys 1941;9:177.
37. Avrami M. J Chem Phys 1940;8:212.
38. Evans UR. Trans Faraday Soc 1945;41:365.
39. Mandelkern L, Quinn FA, Flory PI. J Appl Phys 1954;25:830.
40. Kolmogorov AN. Izv Akad Nayuk CCCR 1937;2:355.
41. Piorkowska E, Galeski A, Haudin JM. Prog Polym Sci 2006;31:549.
42. Müller AJ, Balsamo V, Arnal ML. Adv Polym Sci 2005;190:1.
43. Fatou JG. Crystallization kinetics. In: Bikales M, Menges O, editors. *Encyclopedia of Polymer Science and Engineering*. 2nd ed. Vol. Supplement. New York: Wiley-Interscience; 1986.
44. Fatou JG. Morphology and crystallization of polyolefins. In: Vasile C, Seymour RB, editors. *Handbook of Polyolefins*. New York: Marcel Dekker; 1993.
45. Balsamo V, Calzadilla N, Mora G, Müller AJ. J Polym Sci Polym Phys 2001;39:771.
46. Albuerne J, Márquez L, Müller AJ, Raquez JM, Degée P, Dubois P, Castelletto V, Hamley IW. Macromolecules 2003;36:1633.
47. Müller AJ, Arnal ML, Spinelli A, Cañizales E, Puig C, Wang H, Han CC. Macromol Chem Phys 2003;204:1497.
48. Balsamo V, Urdaneta N, Pérez L, Carrizales O, Abetz V, Müller AJ. Eur Polym J 2004;40:1033.
49. Müller AJ, Albuerne J, Esteves LM, Márquez L, Raquez JM, Degée P, Dubois P, Collings S, Hamley I. Macromol Symp 2004;215:369.
50. Hamley IW, Castelletto V, Castillo RV, Müller AJ, Martin CM, Pollet E, Dubois P. Macromolecules 2005;38:463.
51. Hamley IW, Parras P, Castelletto V, Castillo RV, Müller AJ, Pollet E, Dubois P, Martin CM. Macromol Chem Phys 2006;207:941.
52. Müller AJ, Albuerne J, Marquez L, Raquez JM, Degée P, Dubois P, Hobbs J, Hamley IW. Faraday Discuss 2005;128:231.
53. Sabino MA, Feijoo JL, Müller AJ. Polym Degrad Stab 2001;73:541.
54. Müller AJ, Castillo RV, Hillmeyer M. Macromol Symp 2006;242:174.
55. Müller AJ, Lorenzo AT, Castillo RV, Arnal ML, Boschetti-de-Fierro A, Abetz V. Macromol Symp 2006;1:154.
56. Trujillo M, Arnal ML, Müller AJ, Laredo E, Bredeau S, Bonduel D, Dubois P. Macromolecules 2007;40:6268.
57. Castillo RV, Arnal ML, Müller AJ, Hamley IW, Castelletto V, Schmalz H, Abetz V. Macromolecules 2008;41:879.
58. Boschetti-de-Fierro A, Lorenzo AT, Müller AJ, Schmalz H, Abetz V. Macromol Chem Phys 2008;209:476.
59. Trujillo M, Arnal ML, Müller AJ, Laredo E, Bredeau S, Bonduel D, Dubois P, Hamley IW, Castelleto V. Macromolecules 2008;41:2087.
60. Lorenzo AT, Müller AJ. J Polym Sci B Polym Phys 2008; 46:1478.
61. Castillo RV, Müller AJ. Prog Polym Sci 2009;34:519.
62. Vega JF, Martínez-Salazar J, Trujillo M, Arnal ML, Müller AJ, Bredeau S, Dubois P. Macromolecules 2009;42:4719.
63. Fernández CE, Bermudez M, Versteegen RM, Meijer EW, Müller AJ, Muñoz-Guerra S. J Polym Sci Pol Phys 2009;14:1368.
64. Priftis D, Sakellariou G, Hadjichristidis N, Penott EK, Lorenzo AT, Müller AJ. J Polym Sci A Polym Chem 2009;47:4379.
65. Lorenzo AT, Müller AJ, Lin MC, Chen HL, Jeng US, Priftis D, Pitsikalis M, Hadjichristidis N. Macromolecules 2009;42:8353.
66. Castillo RV, Müller AJ, Raquez JM, Dubois P. Macromolecules 2010;43:4149.
67. Michell RM, Müller AJ, Deshayes G, Dubois P. Eur Polym J 2010;46:1334.
68. Müller AJ, Arnal ML, Trujillo M, Lorenzo AT. Eur Polym J 2011;47:614.
69. Córdova ME, Lorenzo AT, Müller AJ, Gani L, Tence-Girault S, Leibler L. Macromol Chem Phys 2011;212:1335.
70. Michell RM, Müller AJ, Spasova M, Dubois P, Burattini S, Greenland BW, Hamley IW, Hermida-Merino D, Cheval N, Fahmi A. J Polym Sci Pol Phys 2011;49:1397.
71. Quero E, Müller AJ, Signori F, Coltelli MB, Bronco S. Macromol Chem Phys 2012;213:36.
72. Trujillo M, Arnal ML, Müller AJ, Mujica MA, Urbina de Navarro C, Ruelle B, Dubois P. Polymer 2012;53:832.

73. Michell RM, Müller AJ, Boschetti-de-Fierro A, Fierro D, Lison V, Raquez JM, Dubois P. Polymer 2012;53:5657.
74. Su HH, Chen HL, Díaz A, Casas MT, Hoskins JN, Grayson SM, Pérez RA, Puiggalí J, Müller AJ. Polymer 2013;54:846.
75. Córdova ME, Lorenzo AT, Müller AJ, Hoskins JN, Grayson SM. Macromolecules 2011;44:1742.
76. Michell RM, Blaszczy-Lezak I, Mijangos C, Müller AJ. Polymer 2013. DOI: 10.1016/ j.polymer.2013.05.029.
77. Lorenzo AT, Arnal ML, Müller AJ, Boschetti de Fierro A, Abetz V. Eur Polym J 2006;42:516.
78. Lorenzo AT, Arnal ML, Müller AJ, Boschetti A, Abetz V. Macromol Chem Phys 2006;207:39.
79. Lorenzo A. T., Avrami and LH plugin for origin 2013. Available at https://sites.google.com/a/usb.ve/ajmuller/downloads/plugins. Accessed 2015 Nov 4.
80. Okui N, Umemoto S, Kawano R, Mamun A. Temperature and molecular weight dependencies. In: Reiter G, Strobl GR, editors. *Progress in Understanding of Polymer Crystallization*. Berlin: Springer; 2007.
81. Umemoto S, Hayashi R, Kawano R, Kikutani T, Okui N. J Macromol Sci 2003;42:421.
82. Sadler DM, Gilmer GH. Phys Rev B 1988;38:5684.
83. Strobl G. A multiphase model describing polymer crystallization and melting. In: Reiter G, Strobl G, editors. *Progress in Understanding of Polymer Crystallization, Series: Lecture Notes in Physics*. Vol. 714. Heidelberg: Springer; 2007.
84. Lauritzen JI, Hoffman J. J Res Natl Bur Stand Sect A 1961;64:73.
85. Hoffman JD, Lauritzen JI. J Res Natl Bur Stand Sect A 1961;65:297.
86. Zhuravlev E, Schmelzer JW, Wunderlichb B, Schick C. Polymer 2011;52:1983.
87. Silvestre C, Cimmino S, Duraccio D, Schick C. Macromol Rapid Commun 2007;28:875.
88. Sabino MA, Ronca G, Müller AJ. J Mater Sci 2000;35:5071.
89. Sabino MA, Feijoo JL, Müller AJ. Macromol Chem Phys 2000;201:2687.
90. Sabino MA, Albuerne J, Müller AJ, Brisson J, Prud'homme RE. Biomacromolecules 2004;5:358.
91. Fillon B, Wittmann J, Lotz B, Thierry A. J Polym Sci B 1993;31:1383.
92. Lorenzo AT, Müller AJ, Priftis D, Pitsikalis M, Hadjichristidis N. J Polym Sci Polym Chem 2007;45:5387.
93. Galante MJ, Mandelkern L, Alamo RG, Lehtinen A, Paukkeri R. J Therm Anal 1996;47:913.
94. Strobl G. Prog Polym Sci 2006;31:398.
95. Müller AJ, Balsamo V, Arnal ML. Crystallization in block copolymers with more than one crystallizable block. In: Reiter G, Strobl G, editors. *Lecture Notes in Physics: Progress in Understanding of Polymer Crystallization*. Vol. 714. Berlin: Springer; 2007.
96. Loo YL, Register RA. Crystallization within block copolymer mesophases. In: Hamley IW, editor. *Developments in Block Copolymer Science and Technology*. New York: Wiley; 2004.
97. Müller AJ, Arnal ML, Lorenzo AT. Crystallization in nano-confined polymeric systems. In: Piorkowska E, Rutledge G, editors. *Handbook of Polymer Crystallization*. New York: Wiley; 2013.
98. Jacobs C, Dubois P, Jerome R, Teyssie P. Macromolecules 1991;24:3027.
99. Pan P, Kai W, Zhu B, Dong T, Inoue Y. Macromolecules 2007;40:6898.
100. Iannace S, Nicolais L. J Appl Polym Sci 1997;64:911.
101. Marega C, Marigo A, Di Noto V, Zannetti R. Makromol Chem 1992;193:1599.
102. Vasanthakumari R, Pennings AJ. Polymer 1983;24:175.
103. Miyata T, Masuko T. Polymer 1998;39:5515.
104. Di Lorenzo ML. Polymer 2001;42:9441.
105. Abe H, Kikkawa Y, Inoue Y, Doi Y. Biomacromolecules 2001;2:1007.
106. Di Lorenzo ML. Eur Polym J 2005;41:569.
107. He Y, Fan Z, Hu Y, Wu T, Wei J, Li S. Eur Polym J 2007;43:4431.
108. Pan P, Zhu B, Kai W, Dong T, Inoue Y. J Appl Polym Sci 2008;107:54.
109. Kawai T, Rahman N, Matsuba G, Nishida K, Kanaya T, Nakano M, Okamoto H, Kawada J, Usuki A, Honma N, Nakajima K, Matsuda M. Macromolecules 2007;40:9463.
110. Ohtani Y, Okumura K, Kawaguchi A. J MacromolSci B Phys 2003;42:875.
111. Müller AJ, Lorenzo AT, Arnal ML, Boschetti de Fierro A, Abetz V. Macromol Symp 2006;240:114.
112. Laredo E, Prutsky N, Bello A, Grimau M, Castillo RV, Müller AJ, Dubois P. Eur Phys J E 2007;23:295.
113. Pernot H, Baumert M, Court F, Leibler L. Nat Mater 2002;1:54.
114. Goulet L, Prud'homme RE. J Polym Sci B Polym Phys 1990;28:2329.
115. Chen HL, Wu JC, Lin TL, Lin JS. Macromolecules 2001;34:6936.
116. Mathot V. *Calorimetry and Thermal Analysis of Polymers*. New York: Hanser Publishers; 1994.
117. Guo Q, Groeninckx G. Polymer 2001;42:8647.
118. Kuo SW, Chan SC, Chang FC. J Polym Sci B Polym Phys 2004;42:117.
119. Nie K, Zheng SLF, Zhu Q. J Polym Sci B Polym Phys 2005;43:2594.
120. Liu Y, Yang X, Zhang W, Zheng S. Polymer 2006;47:6814.
121. Liu H, Huang Y, Yuan L, He P, Cai Z, Shen Y, Xu Y, Yu Y, Xiong H. Carbohydr Polym 2010;79:513.
122. Phillips PJ, Rensch GJ, Taylor KD. J Polym Sci B Polym Phys 1987;25:1725.

12

SURFACE-INDUCED POLYMER CRYSTALLIZATION

Xiaoli Sun and Shouke Yan

State Key Laboratory of Chemical Resource Engineering and Beijing Key Laboratory of Membrane Science and Technology, Beijing University of Chemical Technology, Beijing, China

12.1 INTRODUCTION

Polymeric materials offer many advantages for sophisticated technologies, which lead to a permanent place for them in many application fields. This has been generally correlated to their low density, easy fabrication (e.g., not necessary for special clean-room and/or high-temperature processes), low cost, and so on. Actually, the effective utilization of polymeric materials is related not only to the above-mentioned factors but also to a most important fact that the properties of the polymeric materials can fulfill the requests of specific applications. Polymer materials exhibit a great potential to meet the requirements from the market better than other materials because their structures at different scales control their physical and chemical properties. For example, the chemical structure at molecular scale determines the essential property and/or functionality of a polymeric material. A subtle manipulation of the chemical composition, functional group, and chain architecture may give rise to completely new polymeric materials with dramatically different properties or functionalities. Moreover, multiscale morphological structures of polymeric materials in condensed state also show pronounced influences on their mechanical and physical properties of the polymers [1–6]. As an example, the stiffness and strength of highly oriented polymeric materials can exceed those of their isotropic counterparts by orders of magnitude [1, 2], while an increase of more than a factor of 100 has been reported for the electrical conductivity of aligned conjugated macromolecules compared with their nonoriented counterparts [3–5]. For crystalline polymers, the crystal structure and orientation provide another most important key factor in regulating the property and/or functionality of a polymer. For this aspect, poly(vinylidene fluoride) (PVDF) provides an excellent example. While its α-form crystals constructed with helical chains can only be used as general thermoplastics, its β-form crystals composed of planar zigzag chains exhibit exceptional piezo- and pyroelectric properties [7]. Therefore, the study on the crystallization of semicrystalline polymers under various conditions is an everlasting research topic in the field of polymer physics.

The crystallization of polymers takes place generally in two steps, that is, the nucleation and crystal growth processes. The nucleation process can take place homogeneously or heterogeneously at the presence of heterogeneities. Surface-induced polymer crystallization displays a typical case of heterogeneous nucleation. It is also quite frequently encountered, as the polymers are frequently in contact with different types of substrates in many scientific and technical applications [8–11]. For example, layer-multiplying coextrusion is becoming a popular technology to produce polymer composites with improved properties. The design and fabrication of ultrathin polymer layers, which concern a lot of polymer/polymer interface, are of increasing importance because of the rapid development of nanoscience and nanotechnology. Moreover, in the fiber-reinforced polymer systems, induced crystallization of the polymer matrix at the fiber surface is an unavoidable phenomenon. Therefore, the study of surface-induced polymer crystallization is of particular interest from both practical and scientific points of view. It is now clear that the existence of a foreign surface can alter the crystallization kinetics, the resultant crystal structure, and crystalline morphology of a polymer. This provides an efficient way for fabricating a unique structure with improved property or/and special functionality of polymeric materials.

Polymer Morphology: Principles, Characterization, and Processing, First Edition. Edited by Qipeng Guo.

The purpose of this chapter is to provide a context as to how the foreign surface can affect the crystallization process of a polymer. It contains a brief introduction on the possible influences of foreign surfaces on the crystallization kinetics of polymers and the resultant unique crystal structure and morphology of polymers in contact with different surfaces. The advantages of surface-induced polymer crystallization to fabricate polymeric materials with improved property or enhanced efficiency is described with some elaborately selected examples.

12.2 INFLUENCE OF FOREIGN SURFACE ON THE CRYSTALLIZATION KINETICS OF POLYMERS

The crystallization of polymer thin films on a solid surface is most frequently encountered in many application fields, such as coating, electronics, and optoelectronics. It is well known that the existence of interaction between the substrate and the polymer can strongly affect the physical properties of the polymer, such as the glass transition temperature [12–19] and the molecular chain mobility [20–23], which in turn influence the crystallization kinetics of the polymer [24–30]. The influence of a foreign surface on the crystallization kinetics of polymers is multifarious, depending on the film thickness and the used polymer/substrate system, which determines the interaction between the substrate and the polymer melt. Generally speaking, due to the enhanced nucleation ability caused by the presence of a heterogeneous surface, the overall crystallization rate will be increased remarkably, especially for those with difficulty in homogeneous nucleation. For example, isotactic poly(methyl methacrylate) (PMMA) is one of the polymers with an extremely low crystallization rate [31–33]. The temperature of its maximum crystallization rate has been reported to be 120 °C, at which its bulk crystallization takes tens of days [34]. The crystallization of PMMA from the glassy state on highly oriented polyethylene (PE) surfaces is, however, much faster. It takes only tens of hours for a complete crystallization [35]. This is attributed to the enhanced nucleation of PMMA on the PE substrate surface. The enhanced nucleation abilities of polymers on a foreign surface are best revealed by the formation of transcrystalline layers in the fiber-reinforced polymer systems. Of course, the situation for thin or ultrathin polymer films on a substrate with strong interaction may be quite different in some cases. It has been reported that there is a substantial decrease in the lateral diffusion coefficient of polymer chains in film thinner than 150 nm [36]. This reduces the crystal growth velocity tremendously and therefore slows down the overall crystallization rate [24, 26, 28]. Actually, for some particular polymers in extreme cases, to cultivate crystallinity in thin or ultrathin polymer films becomes impossible. As an example, the crystallization of poly(di-*n*-hexylsilane) is completely prohibited in ultrathin films less than 15 nm in thickness [27]. Another example is the inhibition of the crystallization in ultrathin films of poly(3-hydroxybutyrate) (PHB) [37], which provides a way to produce *in vivo*-like amorphous PHB [38].

12.3 INFLUENCE OF FOREIGN SURFACE ON THE CRYSTAL STRUCTURE AND MORPHOLOGY OF POLYMERS

A brief description about the possible influences of foreign surfaces on the crystallization kinetics of polymeric materials is given in the last section. It may provide useful information for setting the processing conditions. However, the most important part of surface-induced polymer crystallization should be actually the influence of a foreign surface on the crystal structure and morphology of the polymers, which are the key factors in tailoring the property and even the functionality of the polymeric materials. Therefore, attention should be focused on the surface-induced special structure and morphology of polymers at different surfaces.

12.3.1 Crystallization of Thin Polymer Films on Amorphous Foreign Surface

It is well recognized that the crystal structure and molecular chain orientation depend strongly on the film thickness and the crystallization temperature [39–46]. Generally, thin polymer films of tens to hundreds of nanometers crystallized at relatively lower supercoolings encourage a specific orientation of the typical flat-on lamellar crystals, in which the molecular chains aligned perpendicular to the substrate (i.e., the lamellae lie "flat-on" against the substrate) [47–63]. Figure 12.1 shows the AFM height and amplitude images of regular flat-on single crystals of syndiotactic polypropylene (sPP) crystallized on amorphous carbon surface at 125 °C

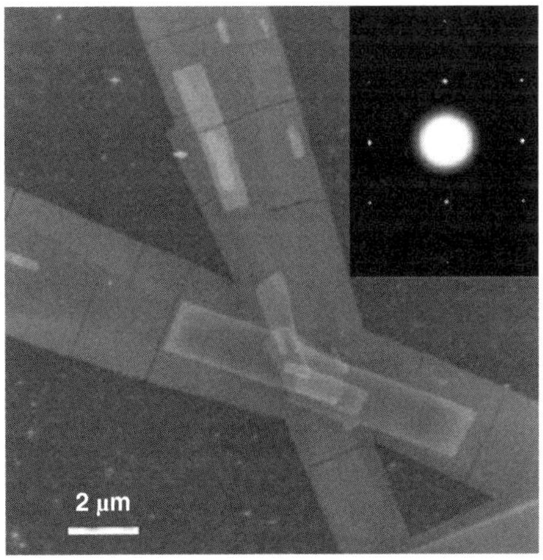

Figure 12.1 Atomic force microscopy (AFM) amplitude image of sPP crystallized on an amorphous carbon surface at 125 °C. The inset presents the electron diffraction pattern of the single crystal, which indicates an upright chain orientation. The single layer of the crystal is about 15 nm in thickness. The observed transverse microcracks, are associated with an approximate order-of-magnitude higher thermal expansion coefficient between the (020) lattice planes than between their (200) counterparts. Reproduced with permission from [64], copyright © 2011, American Chemical Society.

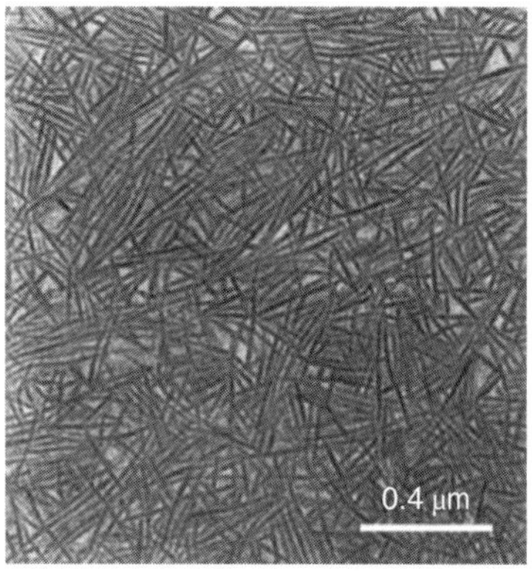

Figure 12.2 Transmission electron micrograph showing the edge-on lamellar structure of syndiotactic polypropylene crystallized on a carbon surface at 90 °C. The film thickness is about 50 nm. Reproduced with permission from [70], copyright © 2013, American Chemical Society.

[64]. The inserted electron diffraction pattern confirms the upright orientation of sPP molecular chains in the crystal. On the contrary, when crystallizing a polymer thin film at higher supercoolings, spherulites made up of radically arranged edge-on lamellae (chain axes oriented parallel to the substrate) are generally observed [65–69]. This can be seen in Figure 12.2, which shows the spherulitic structure of sPP composed of edge-on lamellae grown on amorphous carbon film at 90 °C [70].

It should be pointed out that the feature of the substrate surface is also important in determining the crystal orientation of the polymers. Recent dynamic Monte Carlo simulation indicates that thin polymer films on a slippery wall exhibit dominantly edge-on lamellar crystals even at high temperatures, while on a sticky wall they show mainly flat-on lamellar crystals [39]. The variation in crystallization behavior is related to the different interaction between the substrate and polymer film. Strong interaction existing between the substrate and the polymer melt at the interface, such as a hydrogen bond, plays a very important role in determining the crystallization habit of polymers. An example presented here is the crystallization of polymers on amorphous poly(vinyl phenol) (PVPh) sublayer, which exhibits high capability of forming hydrogen bonds with a series crystalline polymers [71, 72]. Taking the crystallization of thin PHB film on a 165-nm amorphous PVPh sublayer as an example [71], if the PHB films are thinner than 175 nm, an amorphous PHB film was always fabricated irrespective of crystallization temperature. When the film thickness is increased to 185 nm, the PHB can crystallize at temperatures above 53 °C. The resultant PHB crystals orient with their *b*-axes preferentially perpendicular to the substrate surface. This has been related to the effect of formation of hydrogen bond.

FTIR measurement was used to identify the formation of hydrogen bond. For the C=O groups of PHB, there are several bands related to its different states. The weak intramolecular hydrogen bonds between the C=O and C–H in the CH_3 side group of the PHB within its crystals, that is, $C = O \cdots H - C$ hydrogen bond, gives rise to the characteristic C=O stretching band at 1723 cm^{-1}, designated hereafter as *intra* C=O. The amorphous bands at 1750 and 1737 cm^{-1} assigned to the free C=O groups of PHB, designated hereafter as *free* C=O. The band at 1715 cm^{-1} is associated with the formation of hydrogen bond, designated as *inter* C=O. Therefore, the 1750/1737, 1723, and 1715 cm^{-1}bands are used to calculate the *Free*, *crystalline*, *and inter* C=O fractions through $f_i = \frac{A_i/\varepsilon_i}{\sum_i (A_i/\varepsilon_i)}$, where A_i and ε_i are the absorbance and absorption coefficient of the elemental spectrum. It can be seen that the hydrogen-bonded C=O groups, designated as f_{inter}, decreases almost linearly with the thickening of the PHB layer (see Fig. 12.3). At the same time, the fraction of non-hydrogen-bonded free C=O groups of PHB, designated as f_{free}, increases with the thickening of the PHB layer, indicating more and more PHB chain segments remain free from hydrogen bonding with the thickening PHB layer. Taking these into account, the crystallization of PHB in films thicker than 185 nm is associated to the PHB chains or chain segments free from hydrogen bonding.

One can realize from Figure 12.3 that the f_{inter} in the 175-nm-thick PHB film is much lower than that in the 30-nm-thick PHB film, indicating the existence of more PHB chain segments in "*free*" state in the thicker film. Both films on the PVPh layer are, however, all in amorphous state. This may be explained in terms of the high density of hydrogen-bonded C=O groups of PHB, leading to the formation of only short free PHB chain segments incapable of crystallization. Another possible explanation is that the PHB chains or chain segments free from hydrogen bonding in the 175-nm-thick film form only a very thin upper layer and the crystallization of this ultrathin PHB layer is inhibited as reported in Ref. [37]. In this case, the inhibition of PHB crystallization in a film of 175 nm thickness should be attributed to the effect of both thickness confinement and hydrogen-bonding interaction

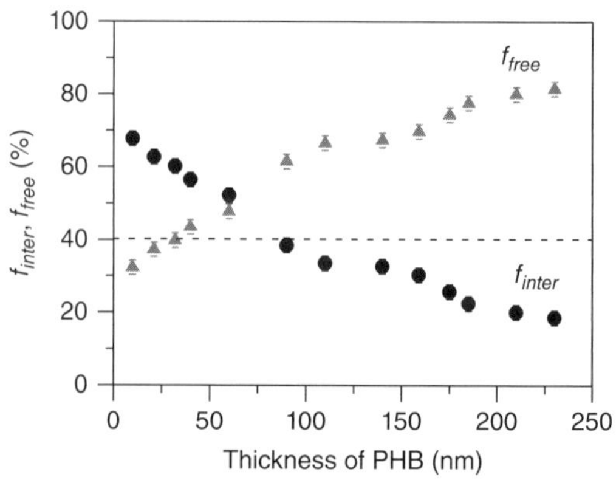

Figure 12.3 The PHB layer thickness dependence of the inter and free C=O fractions in the double-layered PHB/PVPh samples. Sun et al. [71]. Reproduced with permission of American Chemical Society.

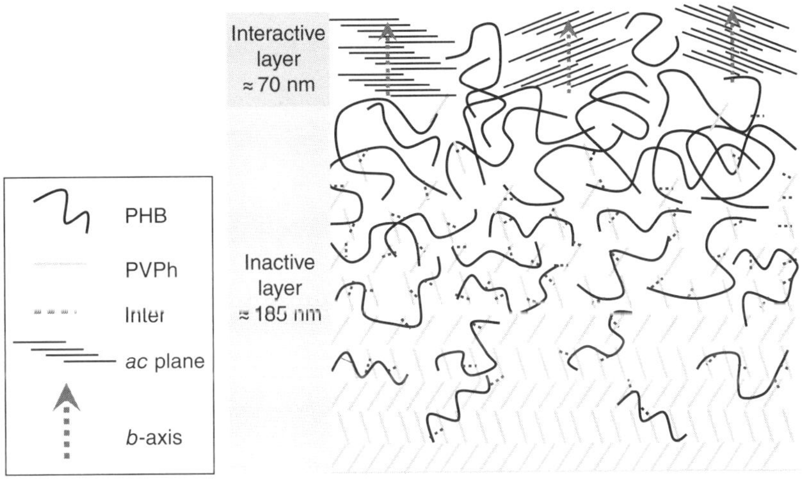

Figure 12.4 A sketch showing the existence of inactive and interactive layers in PHB films thicker than 185 nm. The PHB in the inactive layer remains always in stable amorphous state, while the PHB in the interactive layer can crystallize with the *b*-axes of the PHB crystals oriented perpendicular to substrate surface. Sun et al. [71]. Reproduced with permission of American Chemical Society.

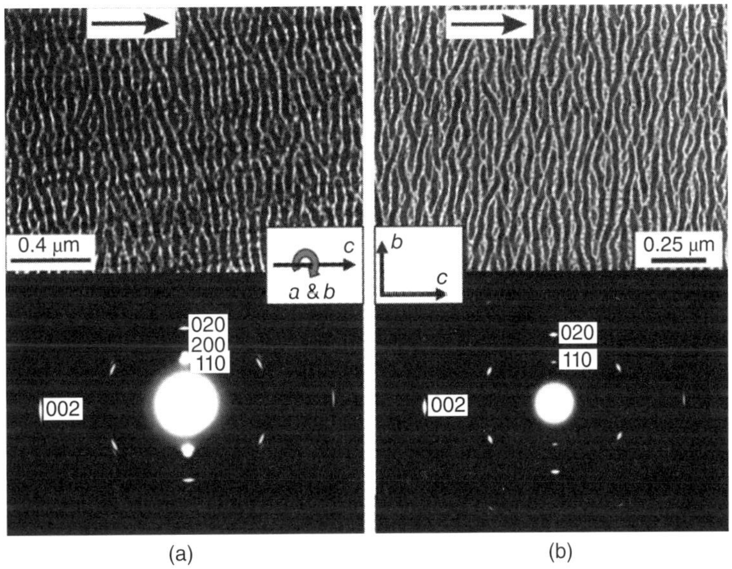

Figure 12.5 Electron micrographs and corresponding electron diffraction patterns of an oriented PE thin film vacuum evaporated with a thin carbon layer (a) before and (b) after heat treatment. The heat treatment was performed by heating the sample to 150 °C for 15 min and subsequent crystallization isothermally at 120 °C for 2 h. The arrows in the picture represent the chain directions of the PE crystals. Yan [76]. Reproduced with permission of American Chemical Society.

between the PVPh and PHB. From the above description, thicker PHB films (over 185 nm thick) on the PVPh surface can be divided into two layers: an inactive layer with high fraction of *inter* C=O remains in stable amorphous state and an interactive layer in which crystallization of PHB takes place depending on crystallization temperature with the *b*-axes of PHB crystals oriented perpendicular to the substrate, as sketched in Figure 12.4.

The effect of strong interfacial interaction on the crystallization behavior of polymers can be well illustrated by the melt recrystallization behavior of oriented polymer thin films with vacuum-evaporated amorphous carbon layers [73–79]. Presented in Figure 12.5 are the transmission electron micrographs and the corresponding electron diffraction patterns of carbon-coated PE-oriented thin films before and after subsequent melting followed by crystallization (melt recrystallization). The same chain orientation of pristine and recrystallized PE films can be easily identified. With careful inspection, even an improved orientation of their crystallized PE film can be recognized. As sketched in Figure 12.5 (the insets), while the pristine PE thin film exhibits only a uniaxial *c*-axis orientation with the *a*- and *b*-axes rotated randomly

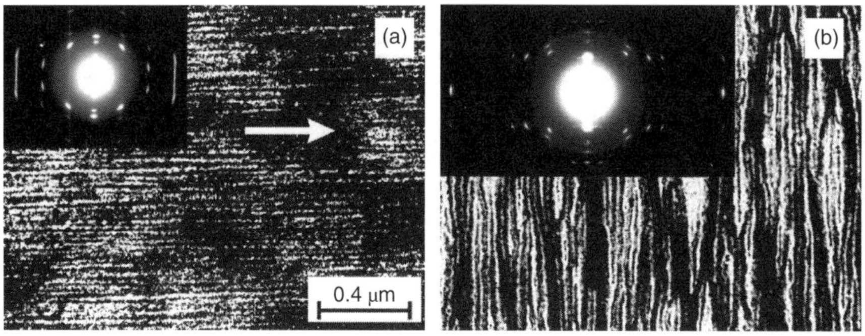

Figure 12.6 Electron micrographs and corresponding electron diffraction patterns (insets) of an oriented PB-1 thin film vacuum evaporated with a thin carbon layer (a) before and (b) after heat treatment. The heat treatment was performed by heating the sample to 150 °C for 15 min and then cooling to room temperature. The arrow in the picture represents the chain direction of the PB-1 crystals. Yan [76]. Reproduced with permission of American Chemical Society.

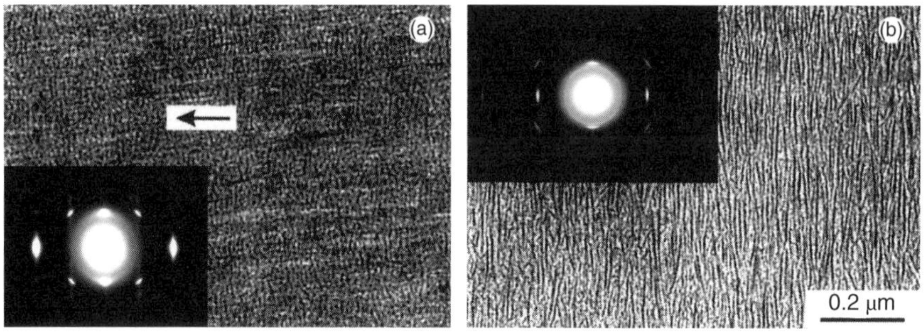

Figure 12.7 Electron micrographs and corresponding electron diffraction patterns (insets) of an oriented PVDF thin film vacuum evaporated with a thin carbon layer (a) before and (b) after heat treatment. The heat treatment was performed by heating the sample to 200 °C for 10 min and then cooling at a rate of 2 °C/min to room temperature. The arrow represents the chain direction of the PVDF crystals. Reproduced with permission from [75], copyright © 2011, American Chemical Society.

about the *c*-axis, the melt-recrystallized carbon-coated PE film contains PE crystals with both *b*- and *c*-axes in the film plane and the *c*-axis is still highly oriented in the same way as the pristine PE thin film. One may want to associate this simply to the occurrence of graphoepitaxy. This is actually not the case, because, besides the preserved chain-axis orientation, both the crystal structure and the morphology of some carbon-coated polymer films can be altered after the melt recrystallization. For example, polybutene-1 changes from highly oriented extended-chain crystals in hexagonal form into oriented folded-chain crystals intetragonal form (Fig. 12.6) [76], whereas the PVDF transforms from oriented α-crystals into oriented β-crystals by melt recrystallization, which normally promotes the β to α phase transition of PVDF (Fig. 12.7) [75]. Sophisticated experimental work demonstrates the existence of a strong fixing effect of the vacuum-evaporated carbon layer on the surface polymer chain segments via covalent bonding [79]. It is this strong fixing effect that has retarded the relaxation of the polymer chains at the spot and, therefore, preserves the original orientation of the polymer stems at high temperature, which in turn derives the recrystallization of the polymer chains in an oriented structure.

The aforementioned recrystallization behavior of polymer thin films with controlled crystal structure and orientation may be of great significance. Through this process, micropatterned polymer thin films with well-defined crystal structure and orientation can be simply prepared by selective carbon coating with the help of a mask and subsequent melt recrystallization of the preoriented polymer films [78]. The patterned polymer thin films exhibit exceptional anisotropic optical properties and may potentially find applications in the fields of nanometer- and/or micrometer-sized structures for optoelectronics and optical applications [80, 81].

In summary, the factors of film thickness, crystallization temperature, and interfacial interaction play a cooperative role in the crystallization and morphology of thin polymer films. Thinner films crystallized at high temperatures favors the formation of flat-on lamellae. The existence of strong interactions, such as hydrogen bonding, may suppress the crystallization of ultrathin polymer films or change the crystallization behavior of polymer thin films. If these factors exert conflicting effects, the final lamellar orientation will be determined by the competing ability of the factors.

12.3.2 Crystallization of Polymer Thin Films on Crystalline Foreign Surface with Special Crystallographic Interaction

The above describes only a general phenomenon of polymer thin-film crystallization on different amorphous surfaces at different temperatures. It does not concern any kind of crystallographic interaction between the substrate and polymer crystals. However, when a polymer crystallizes on a crystalline substrate, favorable crystallographic interaction between the substrate and overgrowth polymers in the contact planes may exist. The existence of the favorable crystallographic interaction can result in the crystallization of the overgrowth polymers in an unexpected manner and therefore leads to the formation of unique crystal structure and crystalline morphology of the crystallizing polymer, a phenomenon known as polymer epitaxy. The epitaxy is defined as the crystals of one phase (guest crystal) growing on the surface of a crystal of another phase (host crystal) in strictly defined crystallographic orientations [82]. For a better understanding of the occurrence of polymer epitaxy, the origin of polymer epitaxy will be first discussed.

12.3.2.1 Mechanism of Polymer Epitaxy The favorable crystallographic interaction generating epitaxial crystallization is commonly based on the similarity of crystal structures through some kind of matching in crystallographic geometry such as the coincidence of unit cell dimensions or molecular distances in the contact lattice planes. The amount of mismatching, sometimes also referred to as disregistry or discrepancy, is measured by the quantity $\Delta = 100 \times (d_o - d_s)/d_s$, expressed in the form of a percentage, where d_o and d_s are the lattice periodicities of overgrowth and substrate crystals, respectively. A 10–15% disregistry was considered as the upper limit for the occurrence of the surface-induced epitaxial growth [83]. Taking this into account, the polymer epitaxial crystallization phenomenon was first illustrated by the epitaxy of polymers on low-molecular weight organic or inorganic substrates because of the nearly endless possible variations in the unit cell parameters of these kinds compounds, which leads to the epitaxial growth of a series of polymers, such as polyolefins and aliphatic polyesters, as well as polyamides, on organic and inorganic substrates [84–93]. The epitaxial crystallization of polymers on organic and inorganic substrates is generally caused by two-dimensional lattice matchings [87]. For example, when crystallizing the PE on the single-crystal surface of anthracene, the best matching between the 0.494 nm interchain distance of PE in its (100) lattice plane and the 0.493 nm interfurrow distance of anthracene in its (001) lattice plane leads to an epitaxial crystallization of PE on the anthracene substrate with its (100) plane in contact with the (001) anthracene lattice plane. A second close matching is found between the intermolecular distances of the substrates in the <110> direction and twice the PE axis repeat, 0.490–0.520 nm for the substrate versus 0.508 nm for the PE.

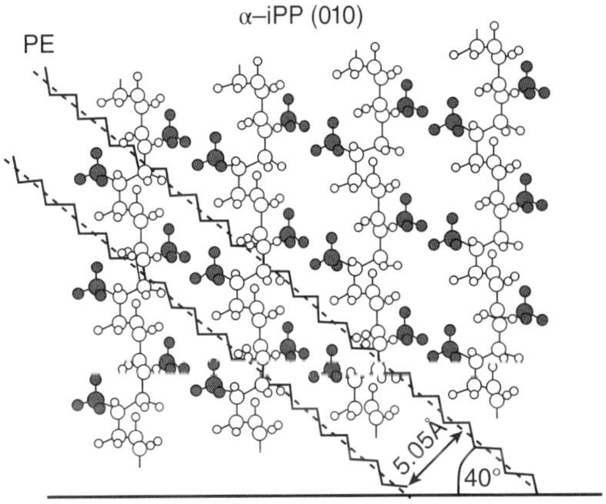

Figure 12.8 A schematic representation of PE/iPP epitaxial relationship. The contact planes are $(100)_{PE}$ and $(010)_{iPP}$. Li and Yan [64]. Reproduced with permission of American Chemical Society.

For polymer/polymer epitaxial systems, the matching is less pronounced compared to the crystallization of polymers on low-molecular weight organic or inorganic substrates. Nevertheless, one-dimensional lattice matching is always found for the epitaxial systems. Taking the PE and isotactic polypropylene (iPP) system as an example [94], as illustrated in Figure 12.8 [64, 70], the characteristic geometry in the (010) lattice plane of the monoclinic iPP crystals is the rows composed of out-sticking methyl groups along different directions. The densest population of methyl groups is found in the direction parallel to the [95] direction, with average distance between adjacent methyl groups being about 0.425 nm. The rows parallel to the $[\bar{1}01]$ direction come next with the methyl groups 0.505 nm apart from each other, followed by those in the *a*- and *c*-axis directions, respectively. Therefore, the distances between two adjacent methyl group rows in the [101] and $[\bar{1}01]$ directions are 0.505 and 0.425 nm, respectively. On the other hand, the distance between two adjacent PE molecule chains in the (100) plane is 0.494 nm. As a result, excellent one-dimensional matching between distances of methyl group rows along the [101] direction in the (010) iPP lattice plane and molecular chains of PE along the [010] direction in its (100) lattice plane is realized with a mismatch of only about 2%. This creates a special circumstance of PE and iPP molecular chains with favorable interaction, which results in the crystallization of PE on oriented iPP substrate in an unexpected manner with the molecular chains of both polymers 50° apart from each other. Consequently, thin PE films with fixed chain orientation and morphology can be produced.

It should be pointed out that two-dimensional matching can also be found occasionally in polymer epitaxial systems, even though very seldom. For example, when crystallizing polycaprolactone (PCL) on the oriented PE substrate, based on the almost identical orthorhombic unit cells of PCL with parameters $a = 0.747$, $b = 0.498$, and $c = 1.705$ nm and PE with parameters $a = 0.74$, $b = 0.494$, and $c = 0.253$ nm, perfect matching can be found between interplane distances for every (hk0) pairs. For example, the mismatching between interplane distances of the (010) lattice planes of both polymers is only 0.8%, while that between the (100) lattice planes is ca. 0.9%. Moreover, the 0.122 nm interatomic space along the chain axis of PCL matches that of PE, 0.123 nm, also very well [96]. The two-dimensional geometric matching results in the expitaxial crystallization of PCL on PE substrate with both polymer chains aligned parallel.

12.3.2.2 Structure Control through Interfacial Crystallographic Interaction From the definition of epitaxy, it can be recognized that the most important significance of epitaxy is the precise control of crystal orientation of the overgrowing polymer through using different kinds of substrates. For polymorphous polymers, regulation of crystal structure of the overgrowing polymer is also possible. In the following paragraphs, attention is focused on what kind of structures of semicrystalline polymers can be mastered by crystallization on a foreign surface for the systems with special crystallographic interactions. The advantages of thus produced polymer thin films with improved property or enhanced efficiency are also described with some elaborately selected examples.

Control of the chain orientation As the epitaxial crystallization is based on the crystallographic matches between the overgrowing and substrate crystals, the epitaxial crystallization of polymers can only be achieved with the side edge of the crystalline lamellae in contact with the substrate crystal. This means that the occurrence of polymer epitaxy should lead always to an alignment of the polymer chains in the film plane. The chain orientation in the film can, however, be different depending on the substrate. Therefore, the chain orientation of a deposited polymer can be easily controlled through surface-induced epitaxy on different substrates. As mentioned in Section 12.3.2.1, when crystallizing PCL on oriented PE substrate, based on a two-dimensional geometric matching between the interchain distances in every (hk0) lattice planes of PCL and PE as well as the inter atomic space along their chain axes, a parallel chain alignment is always observed whenever epitaxial crystallization takes place [96]. On the other hand, when crystallizing the PCL on oriented iPP substrate, excellent one-dimensional matching between distances of methyl group rows along the [101] direction in the (010) iPP lattice plane and molecular chains of PCL along the [010] direction in its (100) lattice plane is realized with a mismatch of about 1.4%. This results in the crystallization of PCL on oriented iPP substrate in a manner that the molecular chains of both polymers are ±50° apart from each other. Other typical examples are presented in Figure 12.9, which shows the different chain orientations of PE crystallized on highly oriented iPP and poly(tetrafluoroethylene) (PTFE) substrates, respectively [64]. The parallel alignment of the PE chains in the (100) lattice plane along the methyl group rows of iPP in the [101] direction of the (010) lattice plane creates a favorable circumstance with a chain row matching.

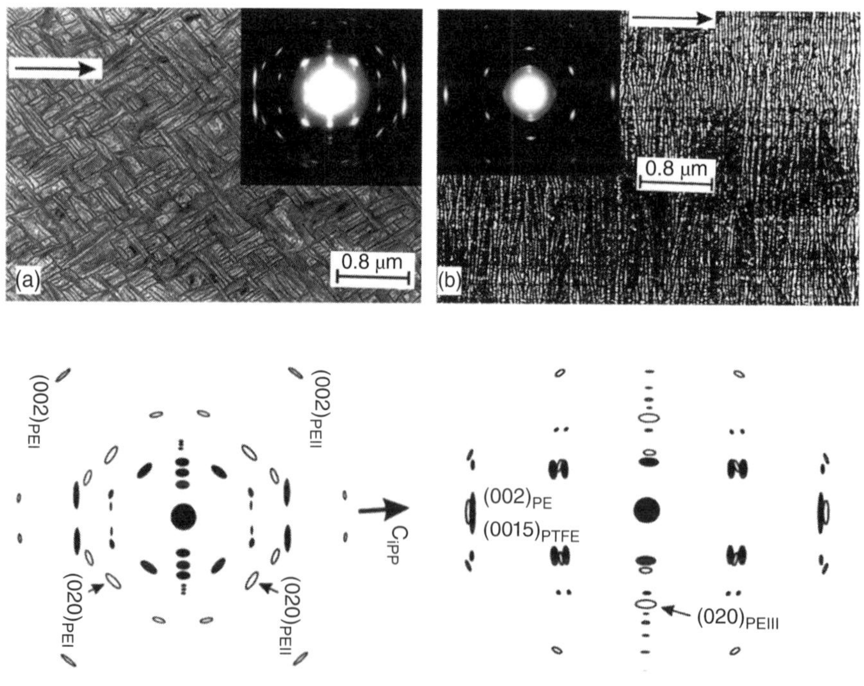

Figure 12.9 Bright-field (BF) electron micrographs and the corresponding electron diffraction patterns (insets of the BF images) of (a) PE/iPP and (b) PE/PTFE double-layered samples, which were heat treated at 150 °C for 10 min and subsequently cooled to room temperature. The arrows show the chain directions of the corresponding substrate crystals. The sketches in the bottom part of each BF image show the corresponding diffraction patterns with the main reflections indexed. Li and Yan [64]. Reproduced with permission of American Chemical Society.

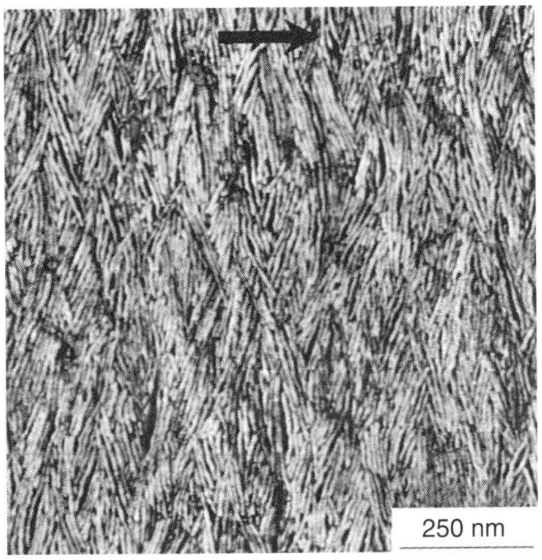

Figure 12.10 An AFM phase image showing the fine structural feature of PEA crystallized from the solution on a highly oriented melt-drawn PE substrate. The molecular chain direction of PE is indicated by an arrow. Yan et al. [97]. Reproduced with permission of Springer.

This results in the orientation of PE on the iPP substrate with the $(100)_{PE}$ and $(010)_{iPP}$ planes in contact and both molecular chains ±50°apart from each other, producing a cross-hatched lamellar structure of PE (Fig. 12.9a). On the other hand, matching between the interchain distances of the PE/PTFE system causes a parallel chain alignment of the PE with PTFE, inducing a parallel aligned lamellar structure of PE (Fig. 12.9b).

In order to avoid misunderstanding, we emphasize here that the parallel aligned lamellar structure is not the sole morphology of parallel chain polymer epitaxy. As an example, when crystallizing the poly(ethylene adipate) (PEA) on uniaxially oriented PE substrate [97], a cross-hatched edge-on lamellar structure of PEA with its lamellae ca. ±66.5° apart from the chain direction of the PE substrate film is observed (see Fig. 12.10), indicating the occurrence of epitaxial crystallization of PEA on oriented PE substrate. Known from the experience of PE/iPP epitaxy, where the ±50° intersect of both polymer chain directions leads to the formation of a cross-hatched lamellar structure of PE (see Fig. 12.9a), one may conclude that the PEA and PE molecular chains are not parallel aligned. It is not true, as judged from the electron diffraction results. As presented in Figure 12.11a, the appearance of well-defined reflection spots of both PEA and PE demonstrates the high degree of orientation of both PEA and PE crystals, indicating the preferred orientation of PEA on oriented PE substrate. All of the reflection spots contributed by PEA crystals can be accounted for by a monoclinic unit cell with parameters $a = 0.547$, $b = 0.724$, $c = 1.155$ nm, and $\beta = 113.5°$ [95, 98, 99]. A sketch corresponding to the electron diffraction pattern of Figure 12.11a with some reflection spots being indexed is shown in Figure 12.11c. The angle between the [00*l*] diffraction directions for PE and PEA crystals are measured to be ±23.5°. Actually, the diffraction pattern contributed by the PEA crystals is somewhat similar to its fiber pattern, as presented in Figure 12.11b, simulated by using Cerius^2 program according to the same monoclinic unit cell. From Figure 12.11a, one can see that the strongest (110) and (020) diffractions in the fiber pattern (see Fig. 12.11b) are not observed for the epitaxially crystallized PEA on PE substrate, indicating that the (110) and (020) lattice planes of the epitaxially grown PEA crystals are out of Bragg diffraction condition. This suggests that the *ac* plane of the PEA is in contact with the PE substrate. In this case, as depicted in Figure 12.12, a parallel chain alignment of both PEA and PE results in the location of (00*l*) PEA diffractions in the direction exactly ±23.5° apart from the chain direction of both PE and PEA. According to the alignment shown in Figure 12.12, the $(200)_{PEA}$ diffraction spots of the PEA are expected to locate along the [hk0] direction of PE. It can, however, not be identified in the electron diffraction pattern. This seems hard to be understood. But considering the almost identical lattice spaces of $(200)_{PEA}$ and $(020)_{PE}$ (0.25 vs 0.247 nm), overlapping of the $(200)_{PEA}$ diffraction with the $(020)_{PE}$ diffraction is highly possible. Taking all these into account, it is concluded that the epitaxial crystallization of PEA on oriented PE substrate takes place with the $(010)_{PEA}$ lattice plane in contact with the PE substrate and a parallel chain alignment of both polymers.

The cross-hatched lamellar arrangement of PEA grown on the PE substrate with a parallel chain orientation can be understood in the following way. The cross-hatched lamellar

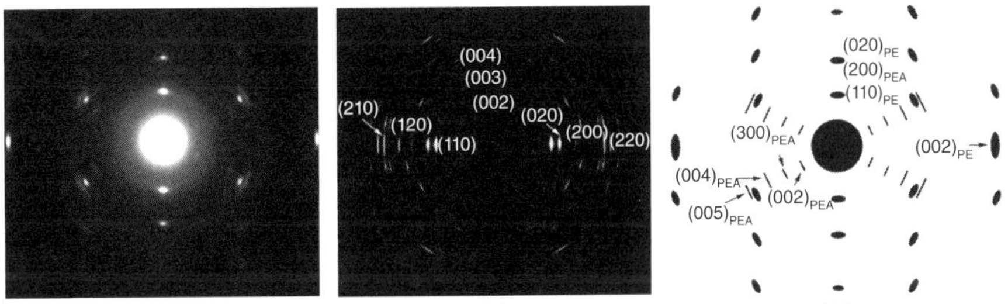

(a) (b) (c)

Figure 12.11 (a) An electron diffraction pattern of a PEA/PE double layer, which has been heat treated at 85 °C for 5 min and then cooled to 27 °C for isothermal crystallization; (b) a simulated diffraction pattern of the PEA with fiber orientation; and (c) a sketch of the electron diffraction pattern shown in Figure 11a. The solid ellipses represent the reflection spots of PE, while the short lines represent the electron diffractions of PEA. Yan et al. [97]. Reproduced with permission of Springer.

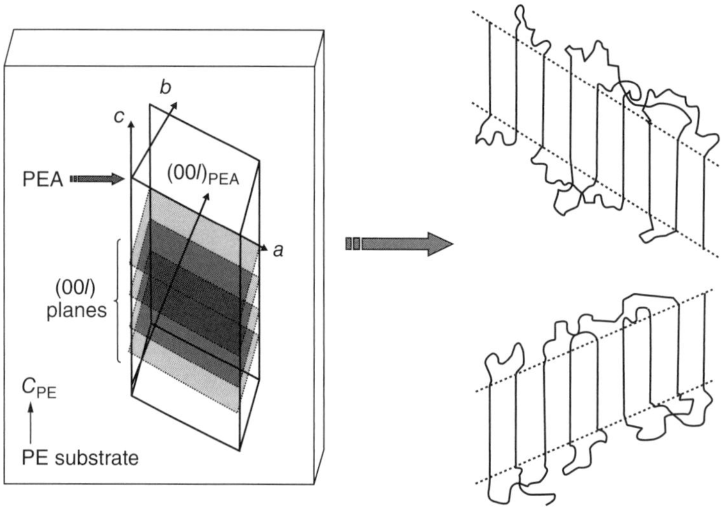

Figure 12.12 A sketch showing the crystal orientation of PEA with respect to the PE substrate and the chain folding way of the PEA lamellar crystals. Yan et al. [97]. Reproduced with permission of Springer.

structure shown in the AFM image indicates that the PEA lamellae are ca. ±66.5° apart from the chain direction of the PE substrate crystals. It causes a ±23.5° inclination of the PEA chains with respect to its lamellar normal direction. This is reasonable when the lamellae with {00*l*} fold surfaces are considered. As illustrated in the right part of Figure 12.12, a chain folding in the {00*l*} surface leads exactly to the formation of PEA lamellae 66.5° apart from the chain direction of the PE substrate crystals. This cross-hatched structure should take the response from a remarkable property improvement of the PE/PEA system reported in Ref. [100], as also found for the PE/iPP system [101, 102] which is discussed later.

Another aspect that should be addressed here is the parallel-aligned lamellar structure originating from a nonparallel chain polymer epitaxy. As shown in Figure 12.13, cold crystallization of poly(L-lactic acid) (PLLA) on a uniaxially oriented PE substrate results in an epitaxial growth of PLLA on the oriented PE surface with molecular chains in film plane and highly oriented perpendicular to the chain direction of PE (see Figure 12.13b) [103]. This provides a first example of perpendicular chain orientation of polymer epitaxy and produces also a parallel-aligned lamellar structure with the PLLA lamellae oriented in the chain direction of PE substrate crystals (see Figure 12.13c and d.)

From the above discussion, it is clear that there are two kinds of mutual chain orientations of the overgrowing polymers with respect to the oriented substrates for polymer epitaxies, that is, the parallel chain alignment of the epitaxial pairs and the intersect chain alignment of polymers at fixed angles. Tables 12.1 and 12.2 listed the known parallel and nonparallel epitaxial polymer systems reported in the literature to date.

It should be pointed out that the substrate and deposit polymers for the epitaxial polymer pairs are interactive. In other words, the substrate and deposited polymers of different epitaxial systems listed in Tables 12.1 and 12.2 can be mutually exchanged. Taking the PE/iPP epitaxial system as an example

TABLE 12.1 Epitaxial Crystallization Polymer Systems with Parallel Chain Alignment

Deposit Polymers	Substrate Polymers	References
Wax, Paraffin	PE	[81, 105]
	Poly(ethylene sebacate)	
	Polyoxymethylene	
PE	PTFE	[105–107]
	Polyoxymethylene	
PCL	PE	[106–108]
	PTFE	
	Polyoxymethylene	
PBA	PE	[109, 110]
PEA	PE	[97]

TABLE 12.2 Epitaxial Crystallization Polymer Systems with Non-Parallel Chain Alignment

Deposit Polymers	Substrate Polymers	Chain Cross Angle	References
PE	iPP	50°	[111]
iPP	Nylon-6, 6.6, 11, 12, ...	50°	[111]
Polybutadiene	iPP	50°	[108]
poly-ε-caprolactone	iPP	50°	[112]
Polyoctenamer	iPP	50°	[113]
PE	sPP	37°	[114]
sPP	PE	37°	[115]
sPP	Nylon-12	37°	[116]
iPP	PTFE	57°	[117]
Poly(L-lactide acid)	PE	90°	[104]
PBA	iPP	50°	[118]

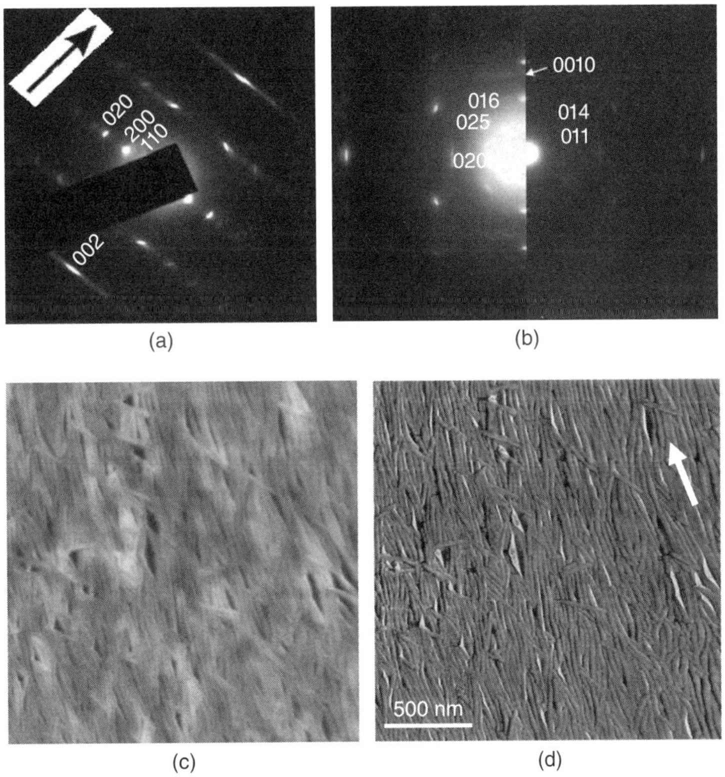

Figure 12.13 Electron diffraction patterns of a PLLA/PE double-layered thin film (a) before and (b) after the thermal treatment. The PE substrate film was prepared by the melt-draw technique. Reproduced with permission from [103], copyright © 2013, American Chemical Society. (c) and (d) The AFM images of PLLA epitaxially grown on PE substrate. An et al. [104]. Reproduced with permission of Springer.

again, the crystallization of iPP on uniaxially oriented PE substrate results in the same mutual chain orientation as PE crystallized on uniaxially oriented iPP substrate with both polymer chains ±50°apart from each other [119]. It leads, however, to the formation of a cross-hatched lamellar structure of iPP rather than PE, as presented in the bottom part of Figure 12.14. The epitaxial crystallization of iPP on an oriented PE substrate from the melt is realized by vacuum evaporating an ultrathin carbon layer on the other side of the PE substrate film, which helps maintain the chain orientation of the PE thin film after melt recrystallization, as described in Section 12.3.1 [76, 119]. In this case, when both the iPP and the oriented PE substrate are molten, a fast cooling of the carbon-decorated PE/iPP double layers will lead to the epitaxial crystallization of iPP on the early crystallized PE-oriented substrate based on the fact that the PE crystallizes always earlier than iPP when the cooling rate is faster than 20 °C/min.

Taking the interactivity of substrate and deposit polymers into account, we see from Table 12.2 that a single polymer substrate can lead to different orientations of different polymers. Figure 12.14a shows the electron micrograph and its corresponding electron diffraction pattern of sPP epitaxially grown on a highly oriented PE substrate. A cross-hatched lamellar structure of sPP with its molecular chains aligned in the directions ±37°apart from the chain direction of PE is obtained [115, 120]. The different chain orientation of sPP on the PE substrate compared with iPP on the PE substrate is based on a different favorable crystallographic interaction between sPP and PE [114].

Control of the crystal modification It is well known that most of polymers exhibit pronounced polymorphisms [72, 75, 121–126]. The polymers with different crystal form can show different property or even functionality. Therefore, the control of crystal modification of polymers is frequently encountered for improving the performance or realizing special function of polymeric materials. To this end, a special nucleating agent is most frequently used, for example, the β-crystallization of iPP [127–129]. Furthermore, special crystallization procedures, such as existence of thermal gradient [130], orientation-induced crystallization encountered in most shaping processes of polymers [131–135], and so on, show also great influence on the polymorphic behavior of polymers. For example, it was found that crystallization of sheared iPP melt encourages the formation of β-iPP crystals [136, 137]. Also, a kinetically favored α to β growth transition can occur during crystallization of iPP in temperature gradient [130, 138].

It should be pointed out here that the epitaxy caused by special crystallographic interaction provides another efficient way to control the crystal modification of semicrystalline polymers together with molecular chain orientation [83, 88, 139–142]. As examples, while the crystal structure of isotactic poly(1-butene) (iPB-1) is successfully controlled through using substrates with different crystal structures

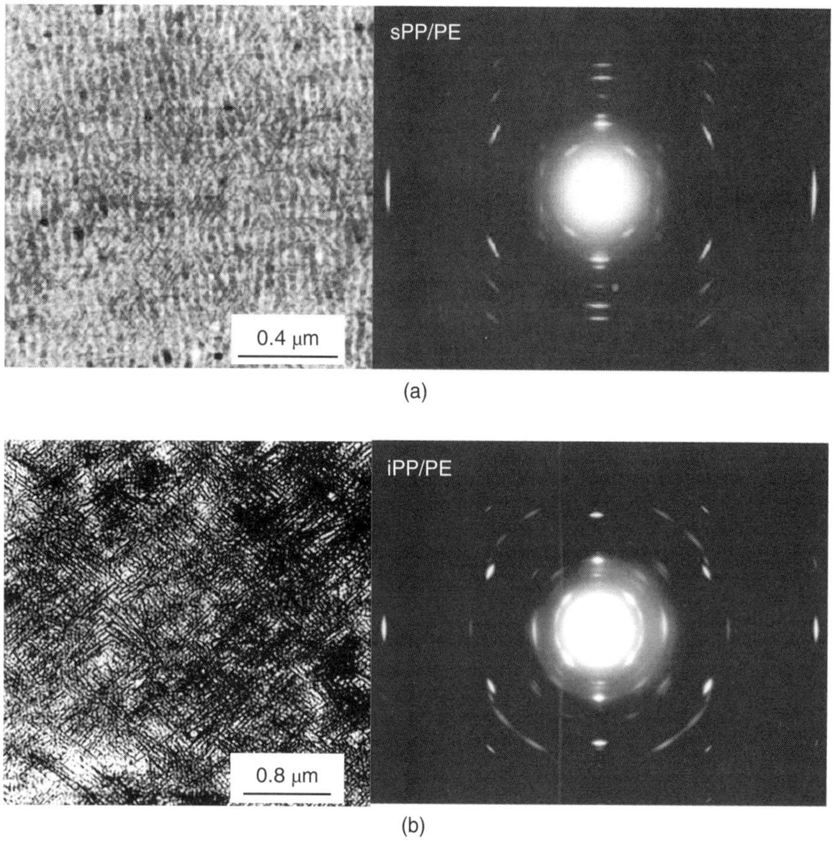

Figure 12.14 Bright-field (BF) electron micrographs and the corresponding electron diffraction patterns of sPP (a) and iPP (b) epitaxially crystallized on a highly oriented PE substrate. The molecular chain directions of the PE substrates are horizontal. The corresponding electron diffraction patterns are shown in (b). Wittmann and Lotz [87].

during melt crystallization of iPB-1 by Lotz et al. [140, 141], Lovinger [142] has skillfully controlled the crystallization of PVDF from melt in its piezoelectric and pyroelectric β-form at atmospheric pressure through epitaxy based on specific crystallographic interaction. Figure 12.15 shows the controlled crystallization of PBA, an aliphatic biodegradable polyester, through its epitaxy on an oriented PE substrate. The PBA exhibits two modifications designated as α(monoclinic unit cell with dimensions $a=0.67$, $b=0.80$, $c=1.42$ nm, and $\beta=45.5°$) and β(orthorhombic unit cell with dimensions $a=0.506$, $b=0.735$, and $c=1.467$ nm) [143–145]. Studies on the crystallization of PBA from melt indicate that, depending on crystallization temperature, PBA can grow in different polymorphic crystals [146–149]. It was reported that PBA chains pack in α-form by melt crystallization at temperature above 32 °C, while the β-form crystal is formed at temperature below 27 °C. A mixture of α and β phases can be obtained when crystallizing the PBA at temperatures intermediate of these two temperatures. Because the crystal structure affects the degradation rate of PBA, for example, the α-PBA degraded relatively faster than its β-counterpart [150], a successful domination of its crystalline structure is of great importance. From Figure 12.15a, it can be seen that the PBA film grown on PE substrate consists of highly oriented edge-on lamellae aligned perpendicularly to the chain direction of the PE substrate [109]. This implies that the existence of favorable crystallographic interaction between PBA and PE leads a parallel chain alignment for them. The corresponding electron diffraction pattern of the PBA/PE double-layered films confirms the aforementioned mutual chain orientation. The appearance of sharp and well-defined reflection spots of both PBA and PE on the electron diffraction pattern (see Fig. 12.15b) reveals the high orientation of both PBA and PE layers. The alignment of [001] of PBA along [002] of the PE reflects a parallel chain orientation between PBA and PE chains, just like many other epitaxial systems with PE listed in Table 12.1. Moreover, all of the PBA reflection spots in Figure 12.15b are accounted for by its orthorhombic unit cell in β-form, indicating the occurrence of epitaxial crystallization of PBA on the PE substrate in its β-form. What should be emphasized here is that the crystal structure of PBA crystallized on the PE substrate is not temperature dependant [109]. This is associated with the similar interchain distances of PBA along a- and b-axes with that of PE along b- and a-axes, respectively, which results in perfect matches between many (hk0) lattice planes of β-PBA and PE crystals. It is, however, unclear whether the same orthorhombic unit cells of both β-PBA and PE plays an important role in governing the epitaxial growth of PBA. The crystallization of PEA on the oriented PE substrate has clarified this issue.

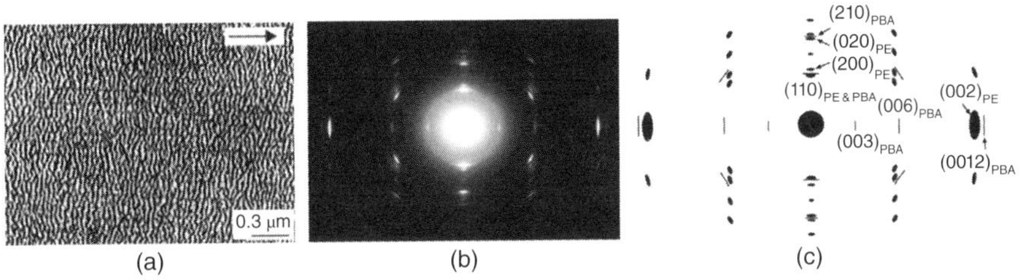

Figure 12.15 (a) Transmission electron micrograph, (b) the corresponding electron diffraction pattern, and (c) a sketch illustrating the mutual chain orientation of PBA crystallized from the melt onto a highly oriented PE substrate. Sun et al. [109]. Reproduced with permission of American Chemical Society.

It was reported that the PEA exhibits two different crystalline forms [95, 98, 99, 151, 152]. One of them is characterized by the aforementioned monoclinic unit cell, while the other one is reported to be the type of an orthorhombic unit cell with the dimensions of $a = 0.512$, $b = 0.841$, and $c = 1.228$ nm [151, 152]. For the PEA/PE system, one can find excellent matching of both kinds of PEA with the PE substrate. For example, the interchain distance of PEA in its orthorhombic form along the a-axis matches very well with the interchain distance of PE along the b-axis. The mismatching is 3.5%. On the other hand, even excellent matching is found between the interchain distances of PEA in its monoclinic form along the [100] direction and PE along its b-axis direction. The discrepancy is only 1.2%. The better matching results in the epitaxial growth of PEA on the PE substrate in its monoclinic form. This may imply that the crystallographic matching is more important for polymer epitaxy than the molecular chain packing way in the unit cell. The epitaxial crystallization of PBA in its β-form on oriented iPP substrate regardless of crystallization temperature further confirms that similar unit cell geometry is not important in polymer epitaxy [118]. The chain orientation of the produced β-PBA crystals on the iPP substrate is different from the case of parallel chain alignment on the PE substrate. In PBA/iPP system, a cross-hatched lamellar structure of PBA with its molecular chains ±50° apart from the chain direction of iPP crystals is obtained, as in the case of PE-iPP epitaxy.

It should be pointed out that the simultaneous control of crystal structure and orientation through surface-induced epitaxial crystallization under specific crystallographic interaction exhibits an advantage compared to the procedure of using nucleation agents. For example, even though the β-form iPP can be achieved by bulk crystallization with nucleation agents, oriented β-form iPP cannot be achieved by mechanical means due to its $\beta\alpha$ transition on stretching. This can, however, be easily achieved through crystallizing the iPP from melt on the γ-quinacridone and dicyclohexylterephthalamide, which produces a biaxially oriented metastable β-iPP [153].

Control of the spatial arrangement of backbone chain plane
It is clear from the examples given in the last section that the matching in crystallographic unit cell dimensions or molecular distances in the contact lattice planes determines the crystal orientation of the overgrowing polymers with respect to the orientation of the substrate crystals. For polymorphic polymers, epitaxial crystallization can govern their growth in a certain crystalline modification having excellent matching with the substrate crystals. In some circumstances, excellent matching may be realized by adjusting the contact lattice plane of the overgrowing polymer. This can be utilized to control the spatial arrangement of the planar molecules. The expitaxial crystallization of PE on an oriented PP substrate helps understanding this aspect.

As illustrated in Figure 12.9a, the favorable interaction between the PE and iPP crystals is realized in the (100) lattice plane of the PE crystal. Therefore, when crystallizing the PE on a highly oriented iPP substrate, the PE crystal in contact with the iPP substrate is its (100) lattice plane, which was confirmed by the electron diffraction inserted in Figure 12.9a through the absence of (200) diffractions [154–156]. However, the favorable interaction between the PE and sPP is realized in the (110) lattice plane of the PE crystal. As a result, when

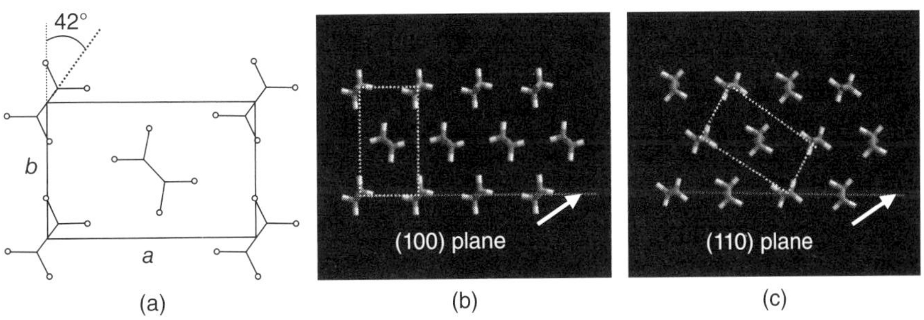

Figure 12.16 Sketches illustrating the (a) unit cell structure, (b) the backbone chain plane orientations of the (100) lattice plane orientation, and (c) the (110) lattice plane orientation of PE crystals. Wittmann and Lotz [87].

crystallizing the PE on a highly oriented sPP substrate, the lattice plane of PE crystal in contact with the sPP substrate is its (110) lattice plane [120]. According to the crystal structure of PE, as illustrated in Figure 12.16a, two chains are involved in one unit cell. The setting angle, defined as the angle between the trans-planar C–C chain skeleton and (0k0) lattice planes, is about ±42° [157, 158]. Taking this structure feature into account, the alignment of PE molecular chains on the iPP substrate leads to the trans-planar C–C chain skeleton of PE about ±42° away from the iPP film surface (see Fig. 12.16b), while the trans-planar C–C chain skeleton of PE on the sPP substrate (see Fig. 12.16c) is either about 8.3° or about 81.7° away from the sPP film plane. Therefore, combining with the different mutual chain orientation described in the previous section, the arrangement of PE molecular chains including chain axis orientation and spatial trans-planar C–C chain skeleton alignment has been well controlled. This also works for conjugated rigid chain polymers. For example, the regioregular poly(3-hexylthiophene) (P3HT) can adopt different chain alignment depending on crystallization conditions. It was well reported that different properties can be achieved with various molecular chain arrangement. This stimulates a lot of study on the crystallization behavior of P3HT under different conditions. The results of previous studies indicate that uniaxially oriented P3HT thin films can be fabricated through the friction-transfer technique or the surface-induced epitaxial way. For the friction-transferred P3HT thin films, they are generally composed of edge-on crystalline lamellae

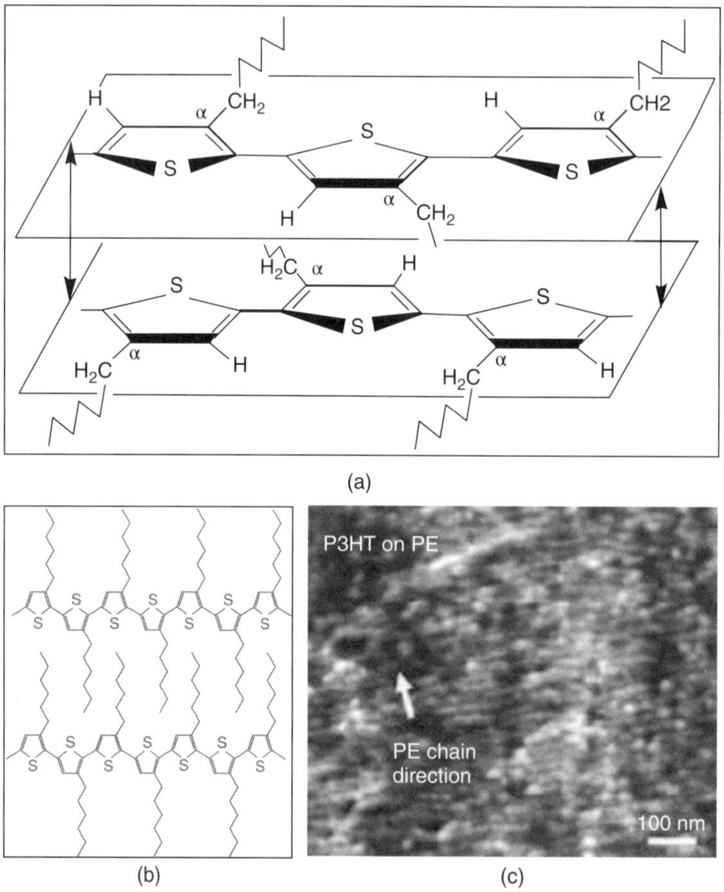

Figure 12.17 Sketches illustrate the (a) face-on and (b) side-on backbone chain plane orientations of the regioregular P3HT. (c) AFM image of the epitaxially crystallized P3HT on PE substrate with side-on molecular orientation. Zhou et al. [161]. Reproduced with permission of John Wiley and Sons.

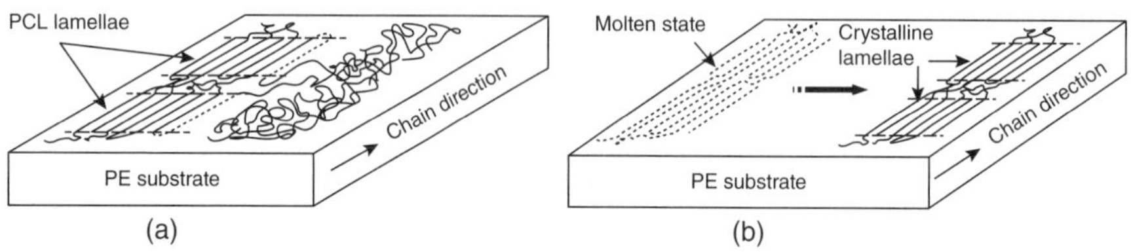

Figure 12.18 Sketches showing the possible chain ordering process of PCL on an oriented PE substrate. Yan et al. [166]. Reproduced with permission of American Chemical Society.

with molecular chains aligned in the film plane and along the sliding direction [159, 160]. Moreover, it is confirmed that the molecular backbone planes in the friction-transferred thin films are parallel to the film surface, that is, a face-on orientation as shown in Figure 12.17a.

On the other hand, the epitaxial crystallization of P3HT on different substrates produces also edge-on lamellae with molecular chain arranged in film plane and oriented in a special manner depending on the used substrate, such as the organic 1,3,5-trichlorobenze single-crystal substrate and the highly oriented polymer thin films [161–165]. It was found that crystallizing the P3HT on an oriented PE substrate results in the formation of parallel aligned edge-on lamellar structure (see Fig. 12.17c), in which the P3HT molecular chain axis is aligned in film plane and along the chain direction of PE [161]. Meanwhile, it is confirmed that the (100) lattice plane of P3HT is in contact with the PE substrate. This demonstrates that the epitaxy of P3HT on the oriented PE thin film provides a side-on orientation of P3HT, as revealed in Figure 12.17b.

In summary, when crystallizing a polymer on a crystalline substrate, some favorable crystallographic interactions based on the similarity of crystal structures, such as the matching in crystallographic geometries, often exist between the deposit and substrate polymer crystals. This leads to the occurrence of polymer epitaxy, which strictly governs the crystal orientation of the overgrowing polymer. It can, therefore, be utilized to control the structure of semicrystalline polymers in several important aspects, including molecular chain orientation, crystal structure of the polymers with polymorphisms, and the spatial arrangement of the planar backbone molecular chains.

Molecular chain ordering process of the overgrowing polymer
In the last section, possible structure regulation of the overgrowing polymer through epitaxy has been described on the basis of mutual orientation relationship of the involved polymer pairs. The molecular dynamics before or during the expitaxial crystallization of a polymer on an oriented substrate was, however, not concerned. For example, the epitaxial crystallization of PE on highly oriented iPP substrate through a chain-row matching results in the alignment of PE with molecular chains ±50°apart from the chain direction of iPP. But how the PE chains are well aligned from the random coiled melt on the iPP substrate during the cooling process is not clear at all. To disclose the expitaxial crystallization process of polymers on a molecular scale, the molecular dynamics during the epitaxial ordering process of a crystalline polymer on a highly oriented polymer substrate in its molten state has been studied. Taking the epitaxial crystallization of PCL on oriented PE as an example, it was demonstrated that PCL chains aligned parallel to the chain direction of the PE oriented film after epitaxial crystallization. As sketched in Figure 12.18, there may be two possible ways to realize the parallel alignment of PCL on PE substrate [166]. One possibility is that the random coiled PCL chain segments aligned parallel to the PE chains while registered on the crystal growth front during crystallization to fulfill the lattice matching (see the dotted lines in Fig. 12.18a). The other possibility is that the PCL chain segments in the supercooled molten state possess already a roughly parallel alignment with respect to PE chains (see left part of Fig. 12.18b with dotted lines representing the polymer chains in the molten state), which results in a parallel chain epitaxy during the cooling process, as indicated in the right part of Figure 12.18b.

It should be pointed out that the process illustrated in Figure 12.18a is difficult to be monitored. However, the preordering process, if it exists, can be easily identified by polarized infrared (IR) spectroscopy as it is very sensitive both to chain conformation and local molecular environment of a polymer [167]. To find out whether there is a preordering process of the PCL chain in the supercooled melt on an oriented PE substrate, one need only to select a temperature at which the crystallization of PCL is suppressed. Fourier transform infrared (FTIR) spectroscopy in polarization mode can then be used to check the orientation status of the PCL chain segments during the annealing process.

Figure 12.19 presents the time- and temperature-dependent X-ray diffraction profiles of PCL on an oriented PE substrate annealed at 59 °C, which is about 4 °C higher than its nominal melting point (ca. 55 °C), for 22 h and subsequently cooled to room temperature. It is clear that the X-ray profiles do not display any diffraction peaks during the isothermal annealing process (Fig. 12.19a). This indicates that the used PE-oriented thin film (ca. 50 nm) is too thin to provide a diffraction signal under the present experimental condition and no crystallization of PCL takes place by keeping its melt on the PE substrate at 59 °C for more than 20 h. During the cooling process, it is found that the crystallization of PCL on the PE substrate starts by cooling the samples down to about 49 °C. Therefore, the molecular dynamics of PCL on the highly oriented PE substrate at 59 °C can be followed by IR spectroscopy.

For IR spectroscopic characterization, band assignments given in the literatures are summarized in Table 12.3 [168–170]. It was reported that by melting the PCL sample, the 710 cm^{-1} band completely disappeared, while the 731 cm^{-1} peak was reduced to about one-third its original height. This indicates that these two bands are primarily the crystalline peaks. Taking into account that the PCL has an almost identical crystal structure as the PE crystal, especially for the a and b unit cell dimensions [171, 172], these bands are associated with the factor group splitting due to the intermolecular interaction of the CH_2 sequences packed in an orthorhombic unit cell as in the case of PE crystals. A bigger setting angle between the planar *trans* C–C chain skeleton and the (0*k*0) lattice plane of PCL, ca. 62° as indicated in Figure 12.20b, compared with that of PE (about 42° as indicated in Fig. 12.20a), makes the CH_2 groups in the PCL unit cell closer than in the PE unit cell [171, 172]. Therefore, a wider splitting of these bands has been seen in PCL crystals than in PE crystals. On the basis of the above analysis, the 731 and 710 cm^{-1} bands can be correlated to the crystalline-sensitive bands originating from the three-dimensional ordered structure of PCL. The other bands at 1295, 1245, and 1192 $cm^{-1}$are associated with the

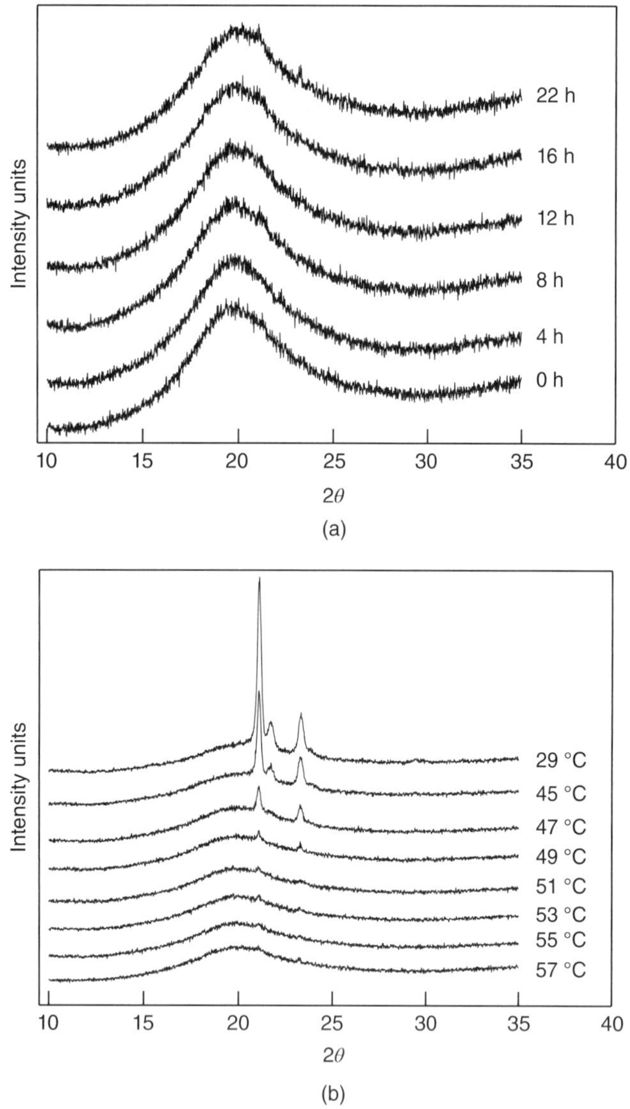

Figure 12.19 WAXD intensity profile of PCL melt on oriented PE substrate at 59 °C (a) for different time and (b) during the subsequent cooling process . The sample was first heated to 85 °C for 10 min to erase the possible thermal history and then cooled direct to 59 °C for 22 h. Yan et al. [166]. Reproduced with permission of American Chemical Society.

TABLE 12.3 Characteristic Infrared Bands of PCL in the Crystalline Phase

Wave number	Assignments	Direction	Abbreviation
2944	Asymmetric CH_2 stretching	⊥ to the *CC* axis	$\nu_{as}(CH_2)$
2865	Symmetric CH_2 stretching	⊥ to the *CC* axis	$\nu_s(CH_2)$
1727	Carbonyl stretching	Intersect the *CC* axis	ν(CdO)
1295	C–O and C–C stretching	// to the *CC* axis	ν(C–OC–C)
1245	Asymmetric COC stretching	// to the *CC* axis	ν_{as}(COC)
1192	OC–O stretching	I to the *CC* axis	î(OC–O)
731	CH_2 rocking in-phase	^ to the *CC* axis	çin(CH_2)
710	CH_2 rocking out-of-phase	^ to the *CC* axis	çout(CH_2)

ν = stretching vibration; *s* = symmetric; *as* = asymmetric.

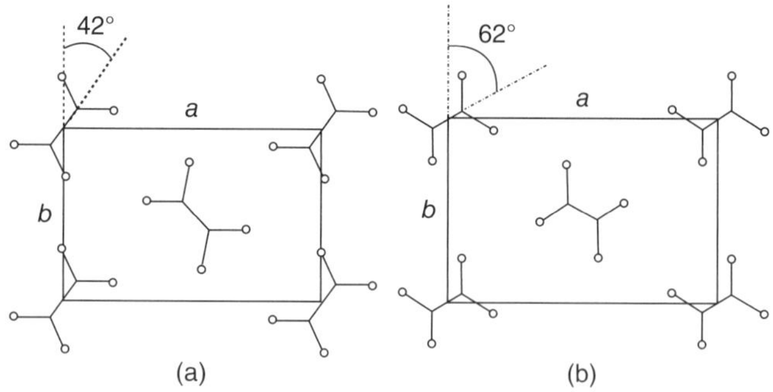

Figure 12.20 Sketches showing the unit cells of (a) PE and (b) PCL in the *ab* projection. Yan et al. [166]. Reproduced with permission of American Chemical Society.

specific chain conformation in the crystalline phase. These bands are used for the structure analysis of PCL on the PE substrate at 59 °C. Here, attention is paid to the chain or chain segment orientation status of the PCL. To this end, polarized IR spectroscopy measurements were conducted during the annealing process of PCL melt on a PE substrate at 59 °C.

Figure 12.21 shows the time-dependent FTIR spectra of PCL melt on an oriented PE substrate at 59 °C with polarized IR beam aligned parallel or perpendicular to the PE chain direction. It is evident that when the polarized IR beam parallels to the chain direction of PE (see Figure 12.21a), the absorbance of conformational bands at 1295, 1245, and 1192 cm^{-1} increases significantly with time. On the contrary, as shown in Figure 12.21b, the IR bands at those positions decrease a little with time if the polarized beam is perpendicular to the PE chain backbone. Considering that the vibrations of all these bands are parallel to the chain backbone of PCL, the obtained results indicate that the PCL chains are aligned gradually along the oriented PE chains. This means that the PCL chains in the molten state have already possessed a relative orientation with respect to the PE substrate before crystallization, that is, the occurrence of soft epitaxy of PCL on the highly oriented PE substrate.

To get more detailed information about the orientation process of the PCL chains, the orientation function *f* of PCL was calculated using different bands according to the follow equation [173]:

$$f = \frac{R-1}{R+2} \times \frac{2}{3\cos^2\alpha - 1}$$

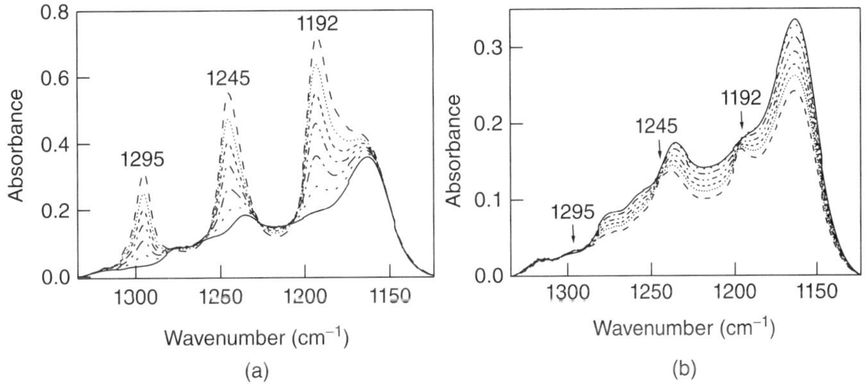

Figure 12.21 Time-dependent polarized FTIR spectra of PCL melts kept on an oriented PE substrate at 59 °C with polarized beam aligned (a) parallel and (b) perpendicular to the PE chain direction. Yan et al. [166]. Reproduced with permission of American Chemical Society.

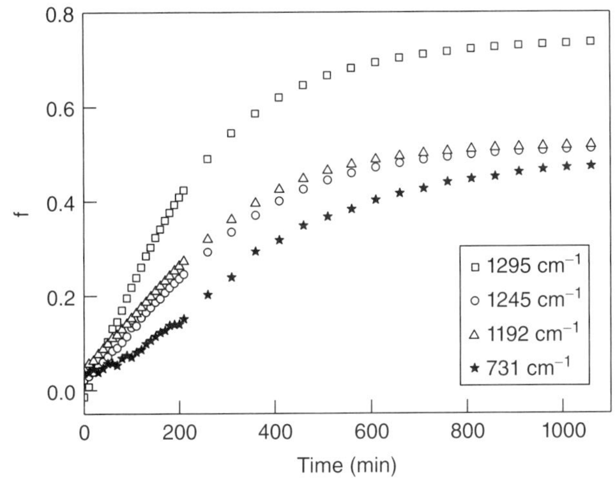

Figure 12.22 Time dependence of orientation function change of PCL annealed on the oriented PE substrate at 59 °C. Yan et al. [166]. Reproduced with permission of American Chemical Society.

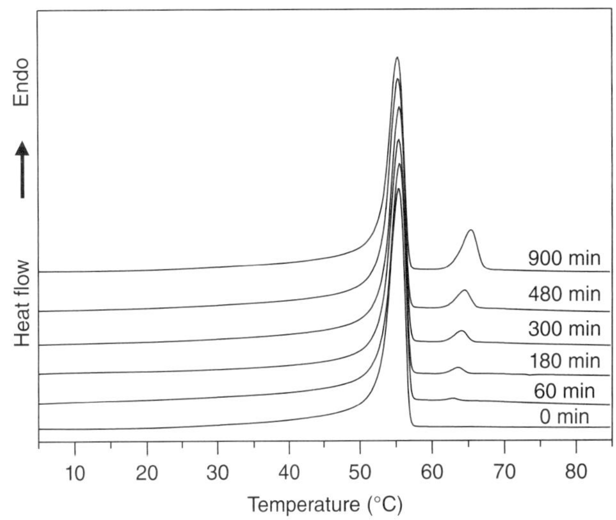

Figure 12.23 DSC heating scans of the PCL/PE samples cooled from melt to and kept at 60 °C for different times as indicated on the right side and then cooled at 10 °C/min to room temperature before heating. The heating rate is 10 °C/min. Chang et al. [174]. Reproduced with permission of American Chemical Society.

where α is the angle between the chain axis and the transition moment associated with the IR band used for the measurement; R is the measurable IR dichroic ratio defined as $R = A_{\|}/A_{\perp}$, with the $A_{\|}$ and $A_{\perp}$ reflecting the absorbance when the IR beam polarized parallel and perpendicular to the chain direction, respectively. The variation in f corresponding to different bands of PCL with time is presented in Figure 12.22. From Figure 12.22, it can be seen that the PCL chains align quickly along the PE chain direction in the first 400 min, and then reach an equilibrium state gradually. After 20 h annealing of PCL on PE substrate at 59 °C, the value of orientation function f of PCL reaches approximately 0.5. It is much lower than that of the oriented PE substrate, which is around 0.9. This may indicate a poorer orientation of PCL chains with respect to the PE chains in the crystalline state and is reasonable because the PCL on PE at 59 °C is in amorphous state as supported by the WAXD result. Another reason may be associated with the limited layer thickness of PCL affected by the PE substrate. In this case, the PCL out of this layer will remain still in the random coil state and the nonoriented PCL layer will unambiguously reduce the overall orientation factor f. To check how far off the PCL melt from the interface can be affected by the PE substrate, differential scanning calorimetry (DSC) study of the PCL/PE layered samples annealed at 60 °C for different periods of time has been performed.

Figure 12.23 shows the DSC heating scans of the PCL/PE samples, which have been heat treated at 85 °C for 15 min, annealed at 60 °C for different times, and then cooled to room temperature before heating scan [174]. From Figure 12.23, it is clear that all the samples exhibit a lower temperature endothermic peak at about 55 °C. The appearance of only this low temperature endothermic peak of nonannealed sample tells us that this peak is contributed by the melting of crystals formed during the cooling of the sample. For the samples kept at 60 °C for more than 3 h, extra melting peaks

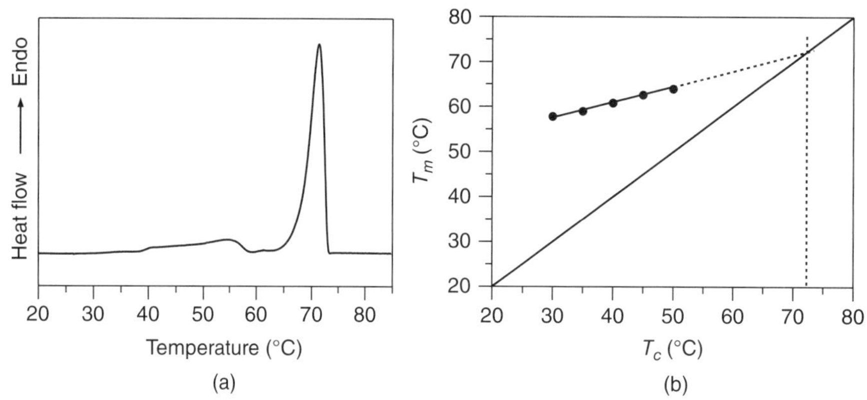

Figure 12.24 DSC heating scans of the PCL/PE samples cooled from melt to and kept at 60 °C for 14 days and then cooled at 10 °C/min to room temperature before heating. The heating rate is 10 °C/min. Chang et al. [174]. Reproduced with permission of American Chemical Society.

can be clearly seen at relatively higher temperatures. These high-temperature peaks become larger and shift to a higher temperature with increasing time at 60 °C. This reflects the influence of the isothermal process on the crystallization of PCL on the PE substrate through the formation of ordered molecular structures at a temperature above the melting point. The enhancement of the high-temperature melting peak with time implies an increase in the ordered PCL chain structure. It is highly suggested that the ordering of PCL chains does not only occur at the interface contacting with the PE substrate; rather, it propagates also from the interface into the bulk away from the contact surface. In other words, the PCL layer thickness affected by the PE substrate at 60 °C increases with increasing annealing time. It is found that the PE can affect a PCL layer of micrometers through annealing at 60 °C for sufficient time. As shown in Figure 12.24a, a PCL layer of 5 μm in thickness annealed at 60 °C for 14 days shows almost only one high-temperature melting peak, indicating almost that the whole PCL layer is arranged in the same way as the interfacial layer. As for its high-temperature shift, it should be related to the increase in stem length of the ordered chain PCL stems by keeping at 60 °C, as the melting point of a polymer is proportional to its lamellar thickness [175]. Tracz et al. have confirmed the formation of highly ordered, unusually thick PE lamellae in contact with atomically flat solid surfaces [176–178]. The melting point of the sample annealed at 60 °C for 14 days reaches ca. 71 °C. It is quite close to the equilibrium melting temperature of the used PCL, which is about 72 °C (see Figure 12.24b). This may suggest the formation of extent chain shish crystals of PCL.

According to the above discussion, the ordering process of PCL from melt on the PE substrate at temperatures higher than its bulk melting point can be schematically presented in Figure 12.25. At the initial stage, the PCL is in a random coil molten state (Fig. 12.25a). With the passage of time, the PCL starts to organize into an ordered structure following the

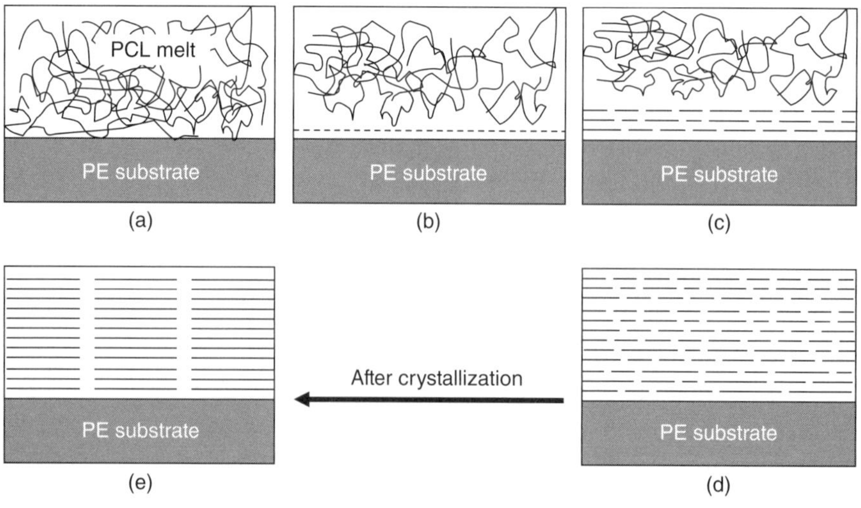

Figure 12.25 (a–d) Sketches describing the chain ordering process of PCL on highly oriented PE substrate during the isothermal process at 60 °C and (e) its final epitaxial crystalline structure after cooling to room temperature (e). Chang et al. [174]. Reproduced with permission of American Chemical Society.

orientation of the substrate crystals. The organization process may firstly take place in a monolayer contacted directly with the PE substrate (Fig. 12.25b). With increasing time, the ordered PCL chain sequence length increases. At the same time, the ordered layer propagates from the interface into the PCL bulk (Fig. 12.25c). The increase in the ordered PCL chain sequence length results in the shift of the melting temperature toward high temperature, while the increase in the layer thickness contributes to the increment of the melting peak. For a sufficient time, all of the PCL chains in a 5-μm thick melt can be organized into an ordered structure with quite long segment length or even possibly form roughly extend chains (Fig. 12.25d). It is these ordered molecular chains that correspond to the formation of a highly oriented structure with unusually thick PCL lamellae.

The abovementioned chain segment preordering of overgrowing polymers has been also found in the cold epitaxial crystallization process of PLLA on an oriented PE substrate [103]. As presented in Figure 12.13a, the electron diffraction pattern of the as-prepared PLLA/PE double layers shows only the electron diffraction spots of oriented PE substrate, indicating the PLLA in the as-prepared PLLA/PE double layers is in amorphous state. After isothermally annealing at 130 °C for 24 h, cold crystallization of PLLA takes place with its molecular chains aligned perpendicular to the direction of PE as judged from the diffraction pattern presented in Figure 12.13b. The perpendicular chain orientation of PLLA epitaxially crystallized on the PE substrate is also confirmed by the FTIR analysis. For FTIR study, the PLLA/PE double-layered samples were prepared by spin coating the PLLA solution on the highly oriented PE film at 2000 rpm for 30 s in air at room temperature. The thickness of the PLLA layer was regulated by using 1 and 0.25 wt% chloroform solutions, respectively. The thicknesses of the thus prepared PLLA layers were estimated to be about 100 nm with a 1 wt% chloroform solution and less than 30 nm with 0.25 wt% chloroform solution.

As summarized in Table 12.4, it was reported that the C=O stretching vibration region of 1860–1660 cm^{-1} and the region of 1500–1000 cm^{-1} corresponding to the CH_3, C–H bending, and C–O–C stretching vibrations are sensitive to the structural changes during crystallization of PLLA. The spectral change of the C–H stretching vibration in the high wavenumber region of 3000–2800 cm^{-1} is relatively small after crystallization. However, a combination of these bands with the bands at 2918 and 2848 cm^{-1} of PE corresponding to the total asymmetric and symmetric CH_2 stretching vibrations helps reveal the chain orientation of the PLLA with respect to the oriented PE substrate directly. Figure 12.26a shows the polarized FTIR spectra of a PLLA/PE double-layered film annealed at 130 °C for 24 h in the wavenumber range from 2800 to 3000 cm^{-1} with the electron vector perpendicular and parallel to the PE chain direction. As expected, the absorption intensities of the FTIR band at 2918 cm^{-1} characteristic of PE with perpendicular polarization is higher than that with parallel polarization. However, the absorption intensities of the FTIR band at 2997 cm^{-1} corresponding to the PLLA in its crystalline state with perpendicular polarization is lower than that with parallel polarization. Considering that the FTIR band at 2997 cm^{-1} represents the asymmetric CH_3 stretching vibration with also a perpendicular transition moment related to the main chain of PLLA, the different absorption intensity change of the 2997 cm^{-1} band confirms a perpendicular alignment of the PLLA chains with respect to PE chain direction. A similar result was also observed in the polarized FTIR spectra with the stretching vibration of C=O groups, as presented in Figure 12.26b.

TABLE 12.4 Band Assignments of Amorphous and Semicrystalline PLLA

Wavenumbers (cm^{-1})		Polarization	Assignments
Amorphous	Semi-crystalline		
2995	2997	⊥	$\nu_{as}(CH_3)$
2945	2946	Unclear	$\nu_s(CH_3)$
1757	1759	⊥	$\nu(C{=}O)$
1454	1458	‖	$\delta_{as}(CH_3)$
1384	1386	⊥	$\delta_s(CH_3)$
1364	1368	‖	δ(CH), CH wagging (blending)
	1360	⊥	
1268	1268	None	$\nu(CH) + \nu(COC)$
1212	1215	⊥	$\nu_{as}(COC) + \gamma_{as}(CH3)$
1183	1187	‖	$\gamma_s(CH3) + \nu_s(COC)$
1133	1133	None	$\gamma_s(CH_3)$
1090	1090	‖	$\nu_s(COC)$
1044	1044	None	$\nu(C{-}CH_3)$

δ = deformation vibration; γ = in-plane rocking vibration; ν = stretching vibration; s = symmetric; as = asymmetric.

The aim of this paragraph is to explain when the alignment of PLLA chains in the perpendicular direction of the PE chain starts. In other words, to find out whether there are already oriented chains of PLLA in the amorphous phase or not. This will provide different crystallization scenarios about the epitaxial cold crystallization of PLLA on the PE substrate. If some PLLA chains oriented perpendicular to the chain direction of PE in the as-prepared sample due to intermolecular interaction, these oriented PLLA chains will act as "prenuclei." The transformation of these "prenuclei" into stable crystal nuclei in the subsequent cold crystallization process initiates the growth of fold chain lamellae with high orientation. Otherwise, the alignment of the PLLA chains in the random coil amorphous state should start in the nucleation process during cold crystallization. For this purpose, the orientation status of PLLA in the as-prepared sample, that is, in amorphous phase, has been followed by polarized FTIR.

Figure 12.27 presents the polarized FTIR spectra of spin-coated PLLA thin films with and without PE substrate in the 3050–2820 cm^{-1} region. The appearance of 2995 and 2945 cm^{-1} bands instead of the 2997 and 2946 cm^{-1} bands indicates that the PLLA thin films with and without PE substrate are all in the amorphous state. From Figure 12.27a,

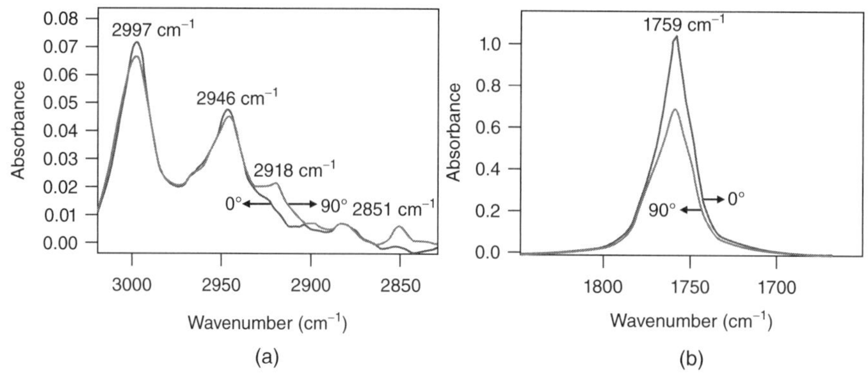

Figure 12.26 Polarized FTIR spectra of the PLLA/PE double-layered sample in the 3050–2800 cm^{-1} and 1850–1650 cm^{-1} regions with the electron vector perpendicular (denoted as 90°) and parallel (denoted as 0°) to the PE drawing direction, respectively. The PLLA/PE double-layered sample has been annealed at 130 °C for 24 h. Tu et al. [103]. Reproduced with permission of American Chemical Society.

the identical position and height of these bands with parallel and perpendicular polarizations demonstrate the lack of chain orientation in the pure PLLA spin-coated thin film. Figure 12.27b presents the polarized FTIR spectra of PLLA spin-coated on oriented PE film from a 1 wt% PLLA/chloroform solution. It can be seen that the PE substrate film exhibits a high orientation as judged from the different intensities of the 2918 cm^{-1} band under parallel and perpendicular polarizations. With close inspection, different intensities of the 2995 and 2945 cm^{-1} bands with parallel and perpendicular polarizations are also be identified, which are more clearly observed in the enlarged inset of Figure 12.27b. This indicates that a small amount of PLLA chains in the amorphous state are already oriented in the direction perpendicular to the PE chain. Considering that the chain orientation of PLLA in the amorphous state on the PE substrate reflects the intermolecular interaction between them, these oriented PLLA chains should exist only in a very thin layer directly in contact with the PE substrate. Therefore, the difference in band intensity of the spin-coated film from a 1 wt% PLLA/chloroform solution should be limited because the thick nonoriented PLLA layer covered on the thin oriented layer shows stronger absorption. To magnify the band intensity difference of the oriented PLLA chains, a thinner PLLA film was spin coated from a 0.25 wt% PLLA/chloroform solution on the surface of the oriented PE substrate. As shown in Figure 12.27c, now the total absorption is really reduced to ca. 1/4 of the sample prepared with 1 wt% PLLA/chloroform solution. However, the intensity differences of the 2995 and 2945 cm^{-1} bands with parallel and perpendicular polarizations are quite evident. This leads to the conclusion that there is a very thin oriented PLLA layer in contact with the PE substrate. It is these oriented PLLA chains in amorphous state that trigger the oriented cold crystallization of PLLA on the PE substrate, that is, the cold epitaxial crystallization.

The analysis of the polarized FTIR spectra of PLLA/PE as-prepared samples in the stretching vibration region of C=O groups provides a similar result. As shown in Figure 12.28, the absorption band that appeared at 1757 cm^{-1} instead of 1759 cm^{-1} demonstrates that the PLLA in the as-prepared samples are in amorphous state regardless of whether they are on the KBr plate or on the oriented PE thin films. The total overlapping of the FTIR bands with parallel and perpendicular polarizations in Figure 12.28a indicates the random orientation of the as-prepared pure PLLA sample. The existence of a very thin oriented PLLA layer spin coated on the PE substrate is reflected by a tiny intensity difference of the bands with parallel and perpendicular polarizations (see Figure 12.28b), which has been magnified by a reduction of the spin-coated PLLA layer thickness as shown in Figure 12.28c.

12.3.2.3 Possible Impact of the Controlled Structure on the Property of Polymer Films As described in the introduction, the properties of a semicrystalline polymeric material depend strongly on its crystal structure as well as the morphology in the condensed state. Therefore, the controlled structure through surface-induced crystallization will show great impact on the thin-film properties. First of all, the different crystal orientation will influence the property of the layered composites. For the nonparallel alignment of the crystalline lamellae, the cross-hatched lamellar structure was tested to improve the mechanical properties of the layered materials significantly [101, 179]. This rests on the fact that the mechanically soft amorphous inter-lamellar regions of one phase are bridged by the crystalline lamellae of another phase, as schematically depicted in Figure 12.29. It was also found that the adhesion between the polymer sheets with epitaxial crystallization can be enhanced remarkably. For example, by dipping the iPP sheets in a PE solution before thermal bonding, the adhesion between the laminate was enhanced enormously [180, 181].

The parallel alignment of the crystals can be employed to fabricate functional polymeric materials with unique ordered structures. For example, one of the efforts devoted to organic field-effect transistors (OFETs) based on soluble conjugated polymers is to improve the key performance parameters of OFETs, such as carrier mobility, on/off ratio,

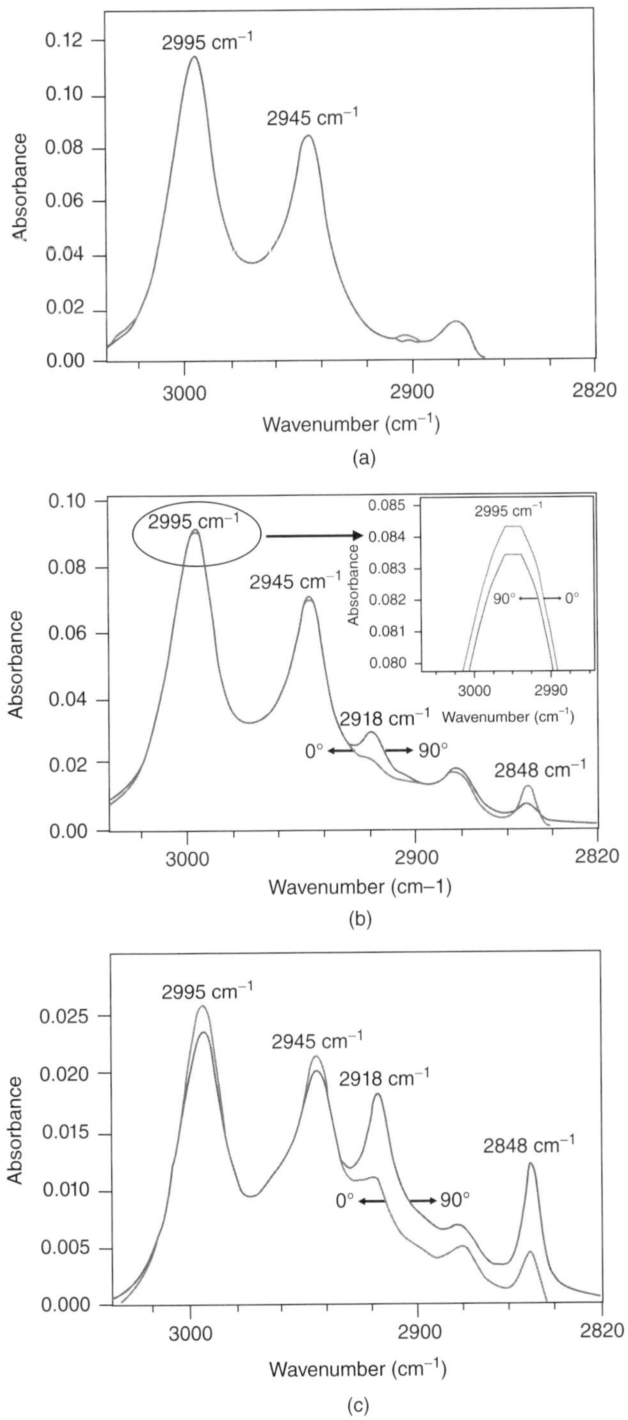

Figure 12.27 Polarized FTIR spectra of the as-prepared (a) PLLA, (b) and (c) PLLA/PE double-layered samples in the 3050–2820 cm^{-1} region with the electron vector perpendicular (denoted as 90°) and parallel (denoted as 0°) to the PE chain direction (b and c), respectively. The concentrations of the solution used for spin coating the PLLA thin films were 1 wt% for (a and b) and 0.25 wt% for (c). An enlarged part of the 2995 cm^{-1} band is inserted in (b). Tu et al. [103]. Reproduced with permission of American Chemical Society.

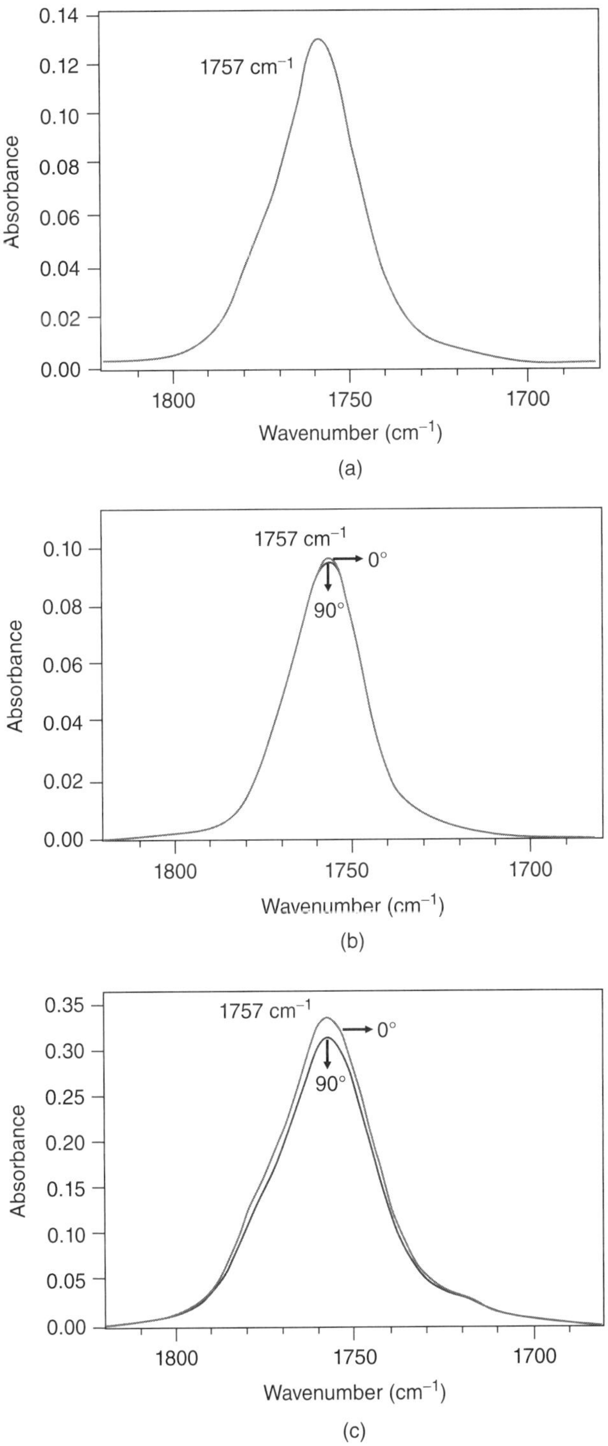

Figure 12.28 Polarized FTIR spectra of the as-prepared (a) PLLA, (b) and (c) PLLA/PE double-layered samples in the 1820–1680 cm^{-1} region with the electron vector perpendicular (denoted as 90°) and parallel (denoted as 0°) to the PE chain direction (b and c), respectively. The concentrations of the solution used for spin coating the PLLA thin films are 1 wt% for (a and b) and 0.25 wt% for (c). Tu et al. [103]. Reproduced with permission of American Chemical Society.

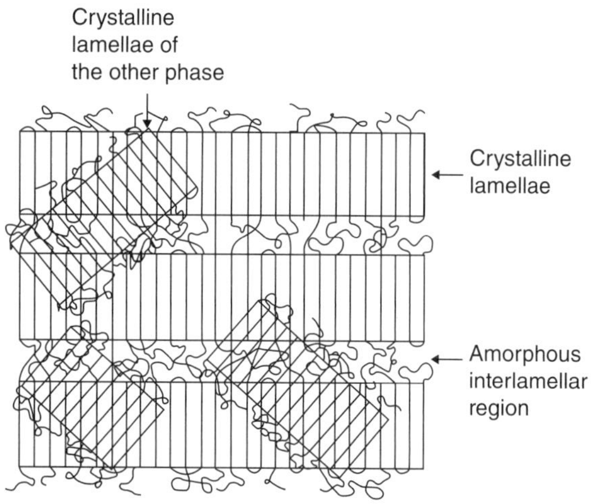

Figure 12.29 Schematic representation of epitaxially oriented cross-hatched lamellar structure, in which the mechanically soft amorphous inter-lamellar regions of one phase are bridged by the crystalline lamellae of the other phase. Lotz and Wittmann [94]. Reproduced with permission of American Chemical Society.

and threshold voltage. Excellent properties of OFETs were mostly achieved by individual nanowires with single-crystal structure [182, 183]. The performance of OFETs fabricated using thin films with random orientation is normally very poor when compared with that using single-crystal nanowires [182]. Considering that the preparation of polymer single crystals with sufficient size is hardly possible, from a practical standpoint, it is of great significance for preparing large-area films with controlled structural anisotropy, so as to fabricate high-performance electrical micro devices. Surface-induced epitaxial crystallization has successfully been used to fabricate thin films of the semiconducting materials with unique crystal orientation [162–165, 184].

Figure 12.30 Optical micrograph and the corresponding electron diffraction pattern (inset of the optical micrograph) show the epitaxial morphology of PTH on highly oriented PE substrate. The arrow indicates the molecular chain direction of the PE substrate crystals. Jiang et al. [185]. Reproduced with permission of American Chemical Society.

Figure 12.30 shows an optical micrograph of a semiconducting material with unique organized single-crystal nanowires, which is fabricated through epitaxy on a highly oriented PE thin film [185]. The electron diffraction of the obtained film shows exactly the same pattern as the single-crystal nanowires superimposed on the diffraction pattern of the oriented PE substrate. Through epitaxial crystallization on highly oriented PTFE, a significant improvement on the switch on/off ratio of the photoswitchers based on well-aligned poly(para-phenylene ethynylene)s derivative molecules with thioacetate end groups (TA-PPE) was demonstrated [186]. As illustrated in Figure 12.31, the switch on/off ratio increases from 8–12 for the nonoriented film up to 330–400 when the molecules are aligned along the carrier transport direction of the devices. Taking this into account, great efforts have already been made in the epitaxial growth of organic semiconductor molecules on highly oriented substrates [187–190]. For example, Brinkmann and coworkers have recently done excellent work in fabricating ordered structure of conjugated polymers, such as poly(3-alkylthiophene)s, through the epitaxial way on oriented aromatic crystal surfaces, friction-transferred polymer substrates and directional epitaxial crystallization [162–165, 184, 188, 191].

Second, the control of spatial arrangement of the planar backbone chains is also of particular interest. For example, the highly crystalline rigid-rod conjugated TA-PPE (see Fig. 12.32) possesses a planar backbone chain. For such molecules, except for the chain-axis orientation, the backbone chain plane exists also in different spatial arrangements. Figure 12.33 presents two extreme arrangements of the planar TA-PPE molecules with the backbone chain planes perpendicular (side-on) and parallel (face-on) to the thin-film plane. Property investigations show that the switch on/off ratio of the photoswitchers with TA-PPE molecules aligned along the carrier transport direction of the devices is much higher than that for the devices with randomly oriented molecules [186]. Moreover, charge-carrier mobility of the OFETs based on TA-PPE nanowires with the backbone molecular chains aligned parallel to the long axis of the nanowire and the planar molecules standing on the substrate, that is, a side-on orientation as shown in Figure 12.33a, is confirmed to be 3–4 orders of magnitudes higher than that of thin-film transistors made by the same polymer with random orientation. In other words, the control of molecular spatial arrangement is an important factor for a maximal utilization of polymeric materials. Considering that the favorable crystallographic interaction is generally related to a certain lattice plane of the deposit polymer, the surface-induced epitaxy based on special crystallographic interaction can be utilized to control the spatial arrangement of the planar molecules, as illustrated by the expitaxial crystallization of PE on polypropylenes (iPP and sPP) and the side-on orientation of P3HT epitaxially grown on the oriented PE substrate. This provides an efficient way for fabricating devices with significantly improved performance.

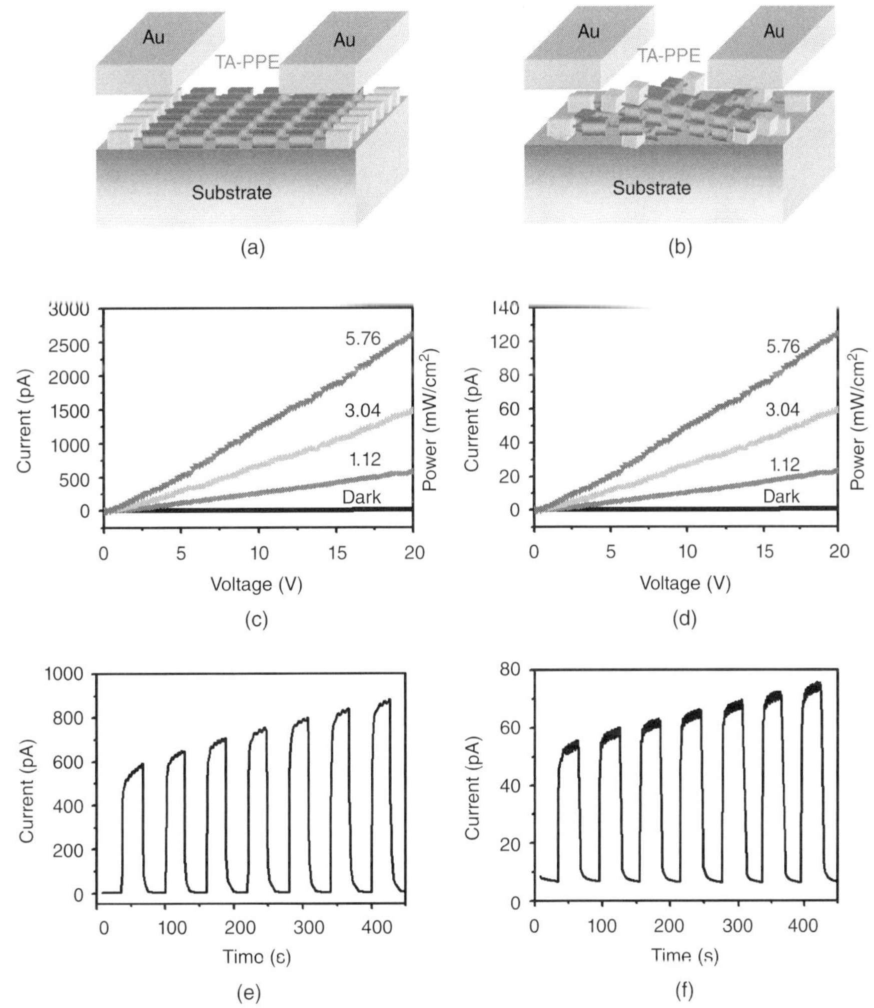

Figure 12.31 (a) Photoswitchers of aligned TA-PPE films, (b) photoswitchers of TA-PPE films without alignment, (c, e) photoresponse and photoswitch behavior of TA-PPE films with alignment, and (d, f) photoresponse and photoswitch behavior of TA-PPE films without alignment. Dong et al. [186]. Reproduced with permission of American Chemical Society.

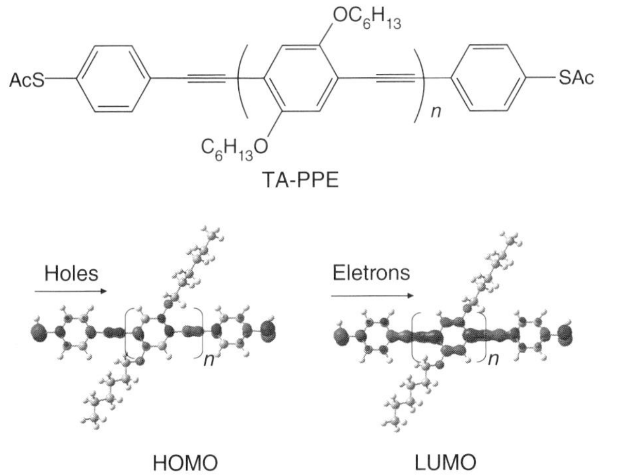

Figure 12.32 Molecular structure of TA-PPE. Dong et al. [186]. Reproduced with permission of American Chemical Society.

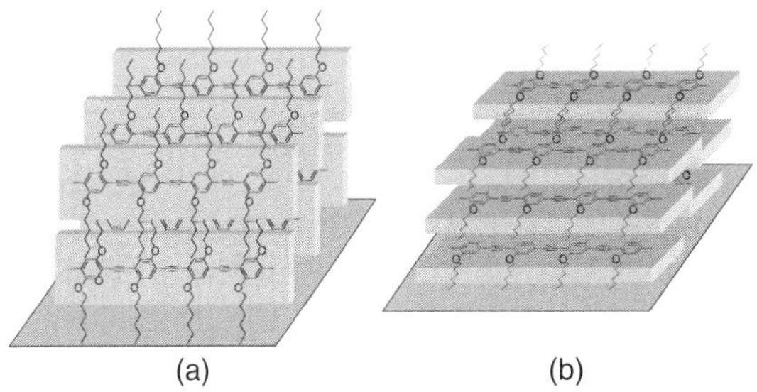

Figure 12.33 Sketches illustrating the (a) side-on and (b) face-on backbone chain arrangement of the TA-PPE molecules. Reproduced with permission from [70], copyright © 2013, American Chemical Society.

12.4 BULK CRYSTALLIZATION OF POLYMERS IN CONTACT WITH A FOREIGN SURFACE

In the last section, the surface-induced crystallization behavior of polymer thin films was described. Another frequently encountered phenomenon is the crystallization of polymer bulk in contact with a foreign surface. A typical example is the fiber-induced crystallization of polymers in fiber-reinforced thermoplastic systems [135, 192–200]. As illustrated in Figure 12.34, the well-known transcrystalline structure often emerges after crystallization of the polymer in fiber-reinforced materials. In such cases, an oriented polymer layer close to the fiber surface is produced, with the polymer chains generally parallel to the fiber axis. The formation of this specific structure is a result of a high density of active nuclei at the fiber surface, which hinder the lateral development of spherulites and therefore the crystal grows unidirectionally normal to the fiber axis. From Figure 12.34, it is clear that the transcrystallization is a nucleation-controlled process and that the enhanced nucleation at the fiber surface is the key factor for the formation of transcrystalline morphology. Even though extensive work has been devoted to this field since the 1950s, the exact mechanism of the enhanced nucleation is still not quite clear.

The situation for polymer transcrystallization around fibers is somewhat more complicated than the aforementioned flat surface-induced polymer thin-film crystallization. There are a number of factors that can control the occurrence of polymer transcrystallization. First of all, the intrinsic surface characteristics of the fibers play a very important role in generating polymer transcrystallization. For example, high-modulus carbon fibers are able to induce the transcrystallization of iPP [192], while their high-strength counterparts hardly exhibit an effect on the crystallization of iPP. It was demonstrated that high-energy surfaces can increase the density of nuclei at the surface and result in the formation of transcrystals in the interfacial region [193, 194]. Also, the chemical composition of the fiber surface is important. Strong interactions between the fiber and polymer molecules were found to be in favor of transcrystallization. Second, the presence of interfacial residual stresses or a temperature gradient at the interface also encourages the formation of transcrystallization layer [195, 196]. Third, flow field, which is unavoidable during many processing procedures, will increase the degree of polymer chain orientation near the fiber, and is therefore another important factor affecting the nucleation of polymer at the fiber surface. It was well demonstrated that transcrystallization of polymers at fiber surfaces will always take place under shearing conditions, regardless of the fiber used. This has been best revealed by the crystallization of iPP in PP/fiber systems with the fibers being pulled along their long axes, a process developed by Varga et al. [128, 197–200]. For example, it has been demonstrated that Kevlar 49 fiber exhibits no nucleation ability toward iPP in a quiescent environment (see Fig. 12.35a), resulting in the formation of iPP spherulites in its monoclinic α-form. However, oriented iPP column structures were always observed in the vicinity of the Kevlar 49 fiber whenever it was pulled along its axis (see Fig. 12.35b,c), even though the induced column structures may be different, depending on the fiber pulling rate or/and pulling time. As illustrated in Figure 12.35b and c, when the Kevlar 49 fiber was pulled at a rate of 30 μm/s for 20 s, an oriented α-iPP column structure was obtained (see Fig. 12.35b). On the contrary, if the Kevlar 49 fiber was pulled at a rate of 30 μm/s for 60 s, an oriented β-iPP column structure was observed (see Fig. 12.35c). This has been unambiguously confirmed by the selective melting of the β-iPP crystals at 158 °C, as presented in Figure 12.35d [135].

Moreover, if the fiber is in the crystalline state, as in the case of surface-induced polymer epitaxy, lattice matching between the polymer and fiber crystals provides another favorable situation for transcrystallization. For example, the existence of excellent matching of nylons and iPP with high-modulus graphitic carbon fibers has led to the transcrystallization of nylons and iPP around the carbon fibers [135]. Taking this into account, the case in which both fiber and matrix are of

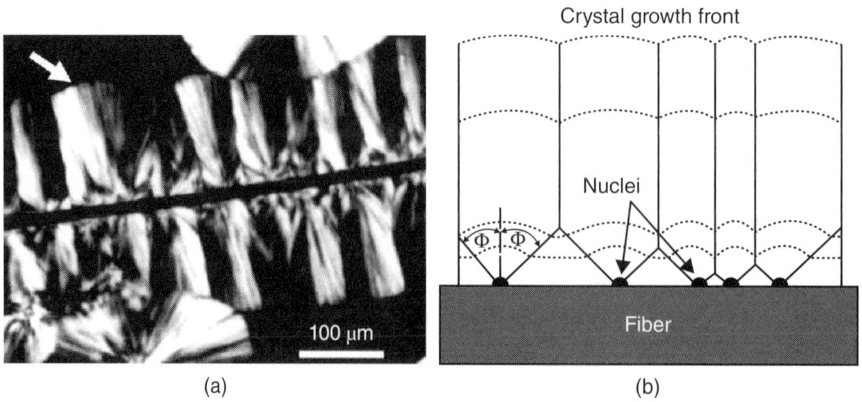

Figure 12.34 Optical micrograph showing (a) the growing transcrystalline zone (indicated by a white arrow) of iPP at the carbon fiber (center black part) surface, and (b) a sketch illustrating its nucleation and crystal growth processes. Lotz and Wittmann [94]. Reproduced with permission of American Chemical Society.

Figure 12.35 Optical micrographs showing the interfacial morphologies of iPP/fiber (Kevlar 49) composites. The samples were heat treated at 210 °C for 5 min and then quickly cooled to 137 °C for isothermal crystallization after fiber pulling. The fibers were pulled at a rate of 30 μm/s for (a) 0, (b) 20, and (c) 60 s. (d) Displays the selective melting of (c) at 158 °C. Sun et al. [135]. Reproduced with permission of American Chemical Society.

the same polymer (single-polymer matrix/fiber composites) is most conducive to surface-induced transcrystallization due to the identical chemical composition and perfect lattice matching [132–134, 193, 201–203]. As an example, Figure 12.36 presents the morphologies of iPP crystallized around iPP fibers [204]. Comparing Figure 12.36a with Figure 12.34a, one can easily find the different nucleation densities of iPP formed on carbon and iPP fiber surfaces. While individual nucleation sites can be recognized at the carbon fiber surface with careful inspection, the nucleation and crystal growth of iPP at the iPP fiber surface starts simultaneously from every place on the fiber surface. As a result, the oriented transcrystallization layer of the former case is caused by the restricted crystal growth in the direction parallel to the fiber axis, whereas that of the latter case originates from the induced crystallization of each crystalline lamella by the presented fiber. The high-magnification scanning electron micrograph has confirmed the above conclusion [133]. As shown in Figure 12.36b, one can clearly see the growth of individual iPP lamellae from the iPP fiber surface, producing parallel aligned lamellae with their long axes perpendicular to the fiber axis. In this case, the crystal structure of the induced transcrystalline layers is generally the same as the solid fiber due to the epitaxial mechanism. The transcrystals of iPP shown in Figure 12.36 are attested indeed to be in the α-form. This is reasonable, because the iPP fibers can exist only in the α-form due to the $\beta\alpha$ transition on stretching.

For single-polymer composites, if the fiber is partially surface molten, the situation is quite different. The oriented lamellar structures around the iPP fiber are determined by the molten state and orientation degree of fibers. For this aspect, iPP single-polymer fiber/matrix composites were produced by a procedure sketched in Figure 12.37 [132]. The iPP matrix thin film was first heated to 200 °C for 10 min to erase possible effects of thermal history on the subsequent crystallization of the sample and then moved to a preheated hot plate, where the iPP matrix was kept in the molten or supercooled molten state at the moment of fiber introduction. As the iPP molten or supercooled molten thin layer reached equilibrium at the desired temperature, homogeneous iPP fibers tightly fixed on a metal frame (see upper part of Fig. 12.37) were introduced into the iPP matrix. After introduction of the fibers, the prepared fiber/matrix single iPP composites were subsequently moved quickly to another hot plate set at a desired temperature for isothermal crystallization. In the subsequent paragraphs, the recent results are reviewed and their implications for the orientation-induced β-iPP crystallization mechanism are discussed.

The crystallization rate of the β-iPP plays a very important role. When the iPP fiber melts at temperatures slightly over

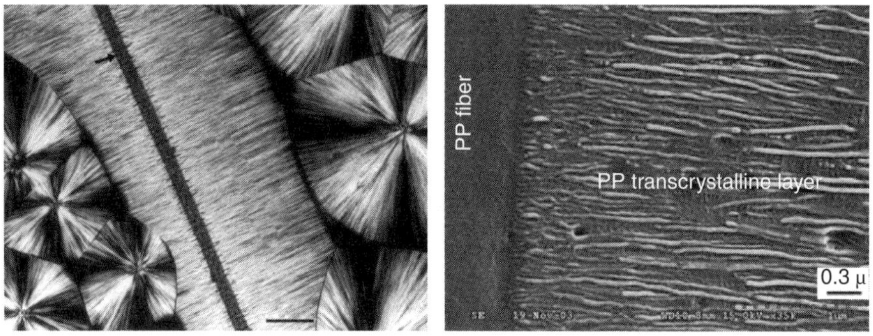

Figure 12.36 (a) Optical micrograph showing the growing transcrystalline layer of iPP on an iPP fiber surface and (b) a magnified scanning electron micrograph showing the interfacial structure on a lamellar scale. The samples were prepared by introducing the iPP fiber into the supercooled iPP matrix at 138 °C and then isothermally crystallizing for 6 h. Li et al. [132]. Reproduced with permission of American Chemical Society.

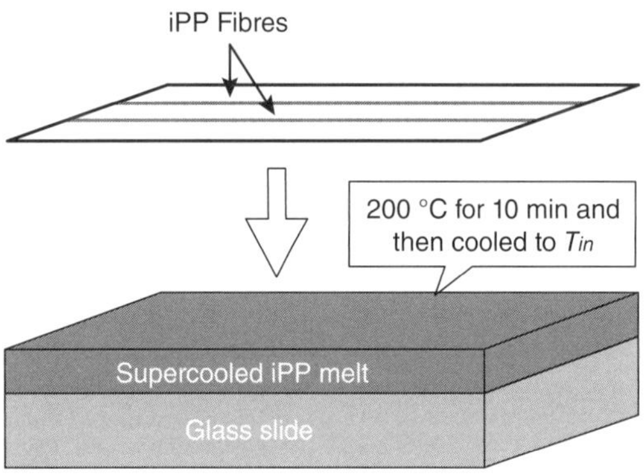

Figure 12.37 A sketch shows the sample preparation procedure of single iPP fiber/matrix composites. Li et al. [132]. Reproduced with permission of American Chemical Society.

its nominal melting point for a short time, mainly β-iPP cylindrical structures are observed after crystallization in the temperature window of 100–138 °C (see Fig. 12.38a–c), while α-iPP cylindrical structures are generated at crystallization temperatures above 140 °C (see Fig. 12.38d) [205]. The interfacial morphology created by crystallization at 139 °C under otherwise unchanged parameters is particularly helpful for elucidating the effect of crystal growth rate on the β-iPP crystallization. As presented in Figure 12.39, now both α-iPP and fan-shaped β-iPP crystals are observed in the transcrystalline layers of iPP. Comparing the growth fronts of the α- and β-iPP crystals, the β-iPP crystals possess an evident higher growth rate than its α-counterpart. The formation of β-iPP transcrystalline structure is, however, suppressed. This implies that a higher crystal growth rate of β-iPP may be necessary, but not sufficient, for promoting β-iPP crystallization. The nucleation of β-iPP crystals is a prerequisite for β-iPP crystallization.

The nucleation of β-iPP crystals can be controlled by adjusting the fiber introduction temperature. For this purpose, the crystallization temperatures in all experiments were set in the temperature range for a higher crystal growth rate of β-iPP. Optical microscopy observations show that if the iPP fibers were introduced into the supercooled iPP matrix at temperatures far below the melting point of iPP fiber ($T_m \approx 170$ °C), for example, below 160 °C, the iPP fiber is kept in the solid state without any melting. In this case, it illustrates simply the crystallization of iPP matrix induced by its homogeneous fiber. The α-transcrystallization layer surrounding its single-polymer fibers is formed [206].

With increasing fiber introduction temperature, fan-shaped β-iPP domains sporadically inlaid in the α-iPP column layers have been occasionally observed. When the fibers were put into the matrix at temperatures close to its nominal melting point, for example, 168 °C, substantive β-iPP crystals can be created. Figure 12.40 shows the representative optical micrographs with β-iPP crystals distributed unevenly along the fibers. Around the top part of the left fiber (see Fig. 12.40a), the iPP grows mainly in its crystalline β phase, whereas the fiber on the right side of the figure is predominantly surrounded by the α-iPP crystals. In the middle, different supermolecular structures of iPP matrix are generated on different sides of the same iPP fiber, that is, the α-iPP and β-iPP on the right and left sides of the same iPP fiber, respectively. The different crystallization manners may be associated with either the different local thermal condition of the sample or the different local nature of the fiber. Anyway, this is quite helpful for exploring the different growth mechanisms of the α-iPP and β-iPP crystals. With careful comparison, the different responses of the iPP fibers in different areas during thermal treatment can be found. The fibers in the areas where β-iPP crystals are generated seem to be molten (see upper left corner of Figure 12.40a). The fiber located at the middle of the figure becomes thinner and thinner from its bottom to top ends, suggesting a partial melting of the iPP fibers. This phenomenon is more clearly seen from the magnified optical micrograph, as illustrated in Figure 12.40b. At the right side of the bottom end of the iPP fiber, the existence of a well-defined boundary line between the iPP fiber and the

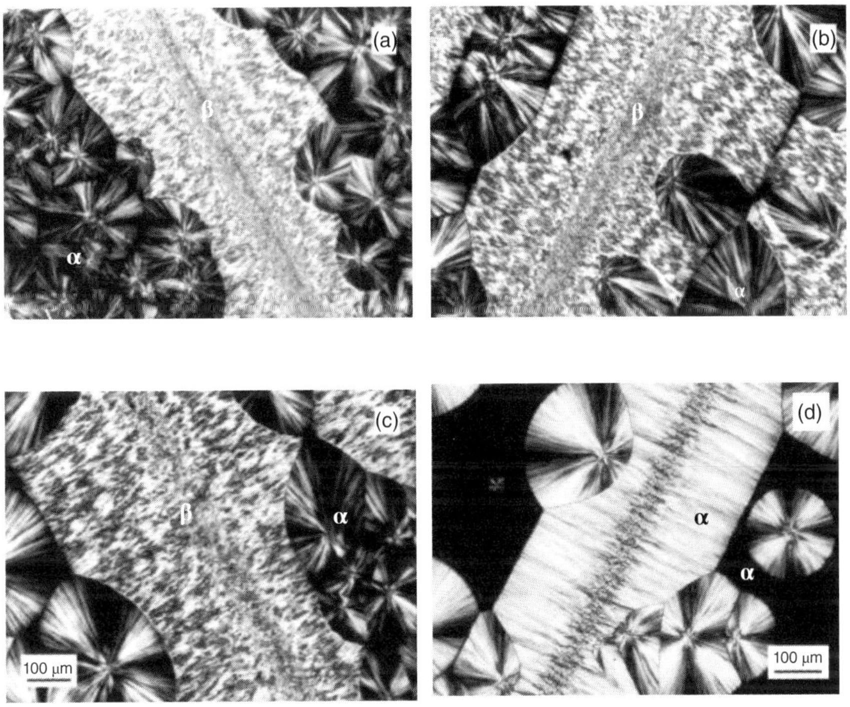

Figure 12.38 Optical micrographs of iPP single-polymer composites with α and β iPP crystals indicated. The fiber introduction temperature was 173 °C and the isothermal crystallization temperatures were (a) 126 °C, (b) 130 °C, (c) 133 °C, and (d) 141 °C. Li et al. [205]. Reproduced with permission of Elsevier.

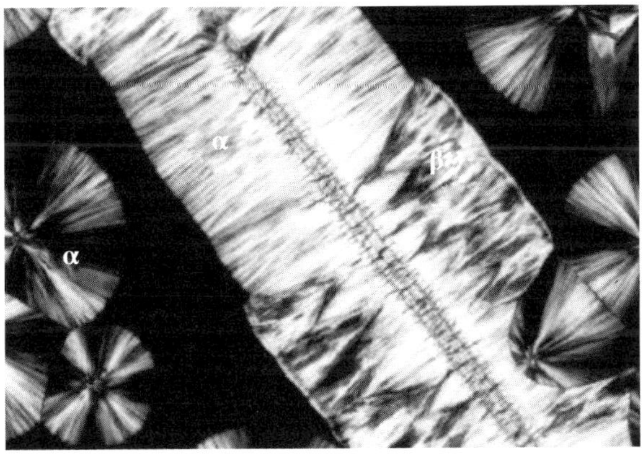

Figure 12.39 Optical micrographs of an iPP single-polymer composite with α and β iPP crystals indicated. The fiber introduction temperature was 173 °C and the isothermal crystallization temperature was 139 °C. Li et al. [205]. Reproduced with permission of Elsevier.

induced crystalline α-iPP (as indicated by an "α") indicates that the iPP fiber remains intact during the thermal treatment. On the contrary, the top part of the iPP fiber surrounded by β-iPP crystals can hardly be distinguished from the iPP matrix, indicating the inosculation of the iPP fiber with its homogeneous molten matrix through melting or at least surface melting. This implies that melting or at least partial melting of the iPP fiber is in favor of the formation of β-iPP in the fiber/matrix interfacial layer. Therefore, as expected, interfacial layers composed of purely β-iPP crystals were obtained when the fibers were introduced into the matrix melt at their nominal melting point (see Figure 12.41).

Scanning electron microscopy (SEM) observations provide further evidence for the occurrence of fiber melting and illustrate the growth process of β-iPP crystalline lamellae [133]. As shown in Figure 12.42b, the fiber consists of edge-on lamellae with most of them well arranged in the direction perpendicular to the fiber axis. A few of the iPP lamellae aligned more or less in the axial direction of the fiber, as indicated by a small white arrow, indicates the occurrence of the unique lamellar branching of α-iPP. All these should result from the melting and recrystallization of the iPP fiber during sample preparation. Figure 12.42c and d provides detailed structural information about β-iPP crystal growth at the initial stage. The growth of β-iPP crystals starts from several single lamellas embedded in the rich α-iPP lamellar region, as indicated by the white arrows. At the early stage, the β-iPP lamellae are loosely packed with some α-iPP inclusion, as indicated by the ellipses. These α- and β- iPP lamellae propagate for a few micrometers without interference. Subsequently, the β-iPP lamellae start to branch and splay out leading to the formation of fan-shaped structures, which stop the growth of α-iPP crystals, and finally pure β-iPP crystalline lamellae are observed with nonperiodic cracks, as indicated by white arrows (Fig. 12.42e). Further propagation of the β-iPP crystals

Figure 12.40 Optical micrographs of an iPP single-polymer composite with α and β iPP crystals indicated. The fiber introduction temperature was 168 °C and the isothermal crystallization temperature was 138 °C. Li et al. [132]. Reproduced with permission of American Chemical Society.

Figure 12.41 Optical micrographs showing the morphologies of an iPP fiber/matrix single-polymer composite. The fiber introduction temperature was 173 °C. The crystallization temperature was 138 °C. Li et al. [132]. Reproduced with permission of American Chemical Society.

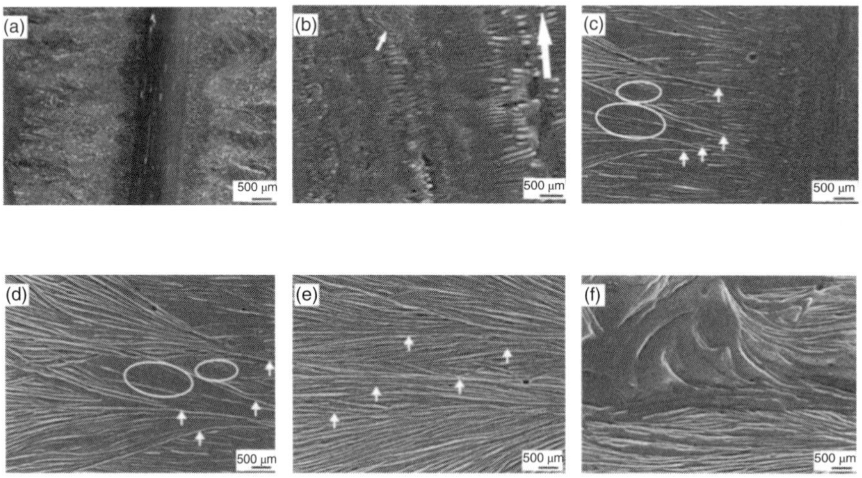

Figure 12.42 SEM micrographs of an iPP fiber/matrix single-polymer composite prepared by introducing the iPP fiber into its supercooled homogeneity matrix at 168 °C and isothermally crystallized at 135 °C for 1 h. (a) An overall view at low magnification. (b) Surface lamellar structure of the iPP fiber with its axis being indicated by a big white arrow; a small white arrow labels the cross-hatched lamellar structure. (c) Interfacial structure with the β-iPP lamellae indicated by white arrows. (d) Growth of β-iPP at early stage. (e) Continuous growth of edge-on β-iPP lamellae with the nonperiodic cracks being indicated by white arrows. (f) Twisting of the β-iPP lamellae. Li et al. [133]. Reproduced with permission of American Chemical Society.

results in the occurrence of the typical lamellae twisting (see Figure 12.42f).

From the above discussion, the chain orientation status in molten iPP fiber is expected to play an important role in producing β-iPP crystals. The results obtained from iPP fiber/matrix single-polymer composites with iPP fibers of varying molecular weight indeed show different induced crystallization behavior owing to the different relaxation of molecular chains [134]. Figure 12.43 shows three representative optical micrographs of iPP single-polymer composites with the same matrix material and crystallization temperature but different iPP fibers in molecular weight. For a high-molecular-weight fiber system (see Figure 12.43a), pure β-iPP crystalline interfacial layers can be produced in a wide temperature range near its nominal melting point. On the contrary, in the sample with a low-molecular-weight fiber (see Figure 12.43c), only α-iPP crystalline interfacial layers have been observed at all chosen temperatures. Interfacial morphologies with coexistence of both α- and β-iPP crystals have been observed for the systems with the iPP fibers having moderate molecular weights (see Figure 12.43b). These discrepant interfacial morphologies should originate from the different relaxation behavior of the used fibers, which depends strongly on the molecular weight [207, 208]. The following equivalent experiments have confirmed the validity of the above conclusion. Figure 12.44 shows the optical micrographs of the iPP matrix/fiber (same as that used in Fig. 12.43a) samples prepared by introducing the fiber into the matrix at 175 °C. After holding at 175 °C for different times, the samples were subsequently cooled to 135 °C for isothermal crystallization. It can be seen that a 5-min maintenance of the sample at 175 °C after fiber introduction (Fig. 12.44a) induces still the crystallization of iPP in its β-form. Holding the sample at 175 °C for 10 min, not only the total nucleation ability but also the ability in generating β-iPP crystals have decreased obviously (see Figure 12.44b). For a 15-min placement of the sample at 175 °C (see Figure 12.44c), the nuclei formed along the molten fiber become sparse. The individual nucleus along the iPP fiber is now recognizable. Moreover, the appearance of fully α iPP crystals indicates undoubtedly that the more relaxed molten iPP fiber has lost its ability in triggering its β-crystallization. These experimental results indicate that the relaxation extent of the originally highly oriented iPP chains in the molten iPP fiber is an important parameter for controlling the formation of β-iPP transcrystalline layers. This leads to the conclusion that the formation of β-crystals is associated with the oriented or stretched macromolecular chains survived during incomplete melting of the highly oriented iPP fiber. From this point of view, the melting and recrystallization process of the iPP fiber should be involved to get a better understanding of the orientation-induced β-iPP crystallization.

The effect of molecular mass of iPP matrix on the crystallization of β-iPP has also been checked by introducing iPP fibers with $M_w = 1.94 \times 10^5$ g/mol and draw ratio = 4 into the matrixes with different molecular weight [203]. As shown in Figure 12.45a, it is found that β-cylindrites formed around the fiber in the matrix with $M_w = 1.94 \times 10^5$ g/mol. On the other hand, the interfacial morphology for the matrix with $M_w = 4.46 \times 10^5$ g/mol consists of mainly α-crystals

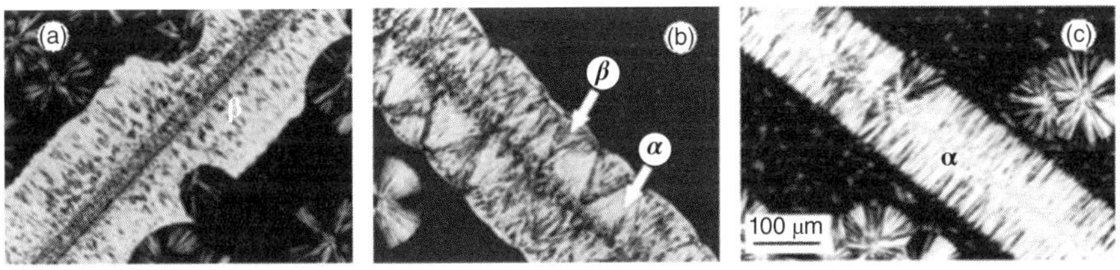

Figure 12.43 Optical micrographs of iPP fiber/matrix single-polymer composites. The iPP fibers with molecular weight (a) 1.94×10^5, (b) 1.85×10^5, and (c) 1.15×10^5 were introduced into the same iPP melts at (a) 178 °C, (b) 168 °C, and (c) 165 °C. The samples were isothermally crystallized at 135 °C. Sun et al. [134]. Reproduced with permission of American Chemical Society.

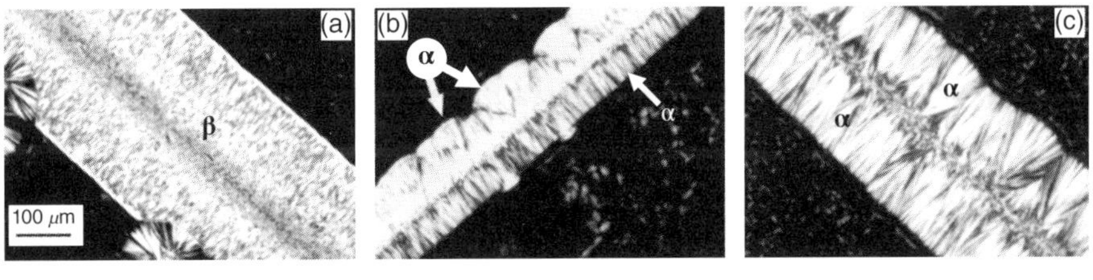

Figure 12.44 Optical micrographs of iPP fiber/matrix single-polymer composites. The iPP fibers with molecular weight of 1.94×10^5 were introduced into iPP melts at 175 °C, then kept at that temperature for (a) 5, (b) 10, and (c) 15 min and finally cooled to 135 °C for isothermal crystallization. Sun et al. [134]. Reproduced with permission of American Chemical Society.

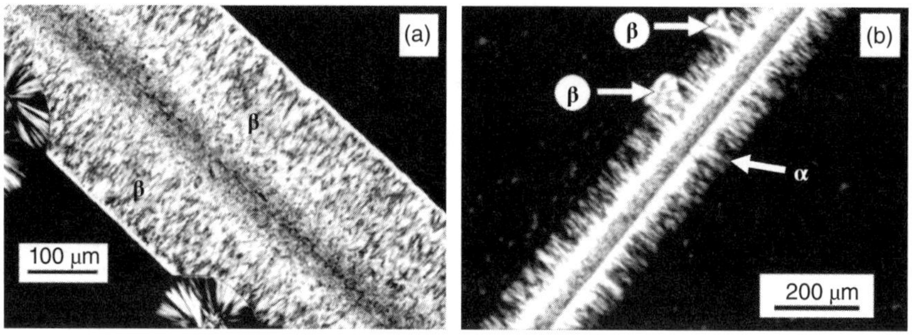

Figure 12.45 Optical micrographs of the iPP fiber/matrix composites prepared by introducing the same iPP fiber into molten iPP matrix with molecular weight (a) $M_w = 1.94 \times 10^5$ and (b) $M_w = 4.46 \times 10^5$ at 180 °C and subsequently isothermally crystallized at 138 °C for 2 h. Sun et al. [203]. Reproduced with permission of American Chemical Society.

(Fig. 12.45b). There are a few fan-shaped β-iPP regions (indicated by the arrows). It has to be noted that both iPP matrixes can crystallize in β-form at the existence of the β-nucleating agent. This indicates that the iPP chains in the matrix participate in the nucleation process of β-iPP. A phenomenological hypothesis for the nucleation of the β-phase is that the necessary condition for the formation of β-nuclei requires the participation of iPP chains in the matrix with the oriented iPP chains at the partially relaxed fiber surface. In this case, some extent of iPP chain relaxation at the fiber surface is necessary for the penetration of the matrix chains. The chain orientation of iPP in the molten fiber should also be maintained above a critical level for creating β-nuclei. Lacking such relaxation and interdiffusion, the solid fiber surface will act as a substrate for homoepitaxy of the α-phase. Considering that the diffusivity is inversely proportional to the square of the molecular mass [209, 210], the diffusivity of matrix with $M_w = 1.94 \times 10^5$ is estimated ca. five times greater than that of the matrix with $M_w = 4.46 \times 10^5$. Therefore, the required chain interpenetration and relaxation for β-nucleation could be easily realized in the lower molecular weight matrix.

It is well known that the melting behavior of macromolecules is different from that of the low-molecular-weight compounds. Due to the long chain character, the macromolecular chains should experience a "recoiling" or "relaxation" process during melting. The recoiling process may take quite a long time depending on the melting temperature and the molecular weight. Extensive studies on the relaxation and crystallization of sheared polymer melt indicated that metastable oriented structures in sheared polymer melt might resist relaxation upon cessation of shearing and could act as "prenuclei" from which fold chain lamellae grow, with the chain axis highly oriented along the shear direction [211–214]. These "prenuclei" can remain active only for a certain period of time after cessation of shear flow at the temperatures above its melting point. Azzurri and Alfonso reported that the lifetime of the nucleation precursors is very sensitive to both relaxation temperature and molecular weight [207, 208]. The melting and recrystallization processes of highly oriented iPP fiber are actually similar to the sheared iPP melt. Considering that the fiber introduction was performed in quite a short time and at temperatures close to or slightly above its nominal melting point, a complete recoiling of the molten iPP fiber could hardly be attained before recooling of the samples. Varga [197] found that some local order of the molecular chains previously included in a crystal lattice could be preserved as "prenuclei" or sometimes referred to as "nucleation precursors" through studying the effect of melting history on the recrystallization of iPP by optical microscopy. Taking this into account, one

Figure 12.46 Polarized optical micrographs of an iPP fiber/matrix composite crystallized isothermally at 116 °C for 30 min. The temperature of fiber introduction was 173 °C. (a) As-prepared sample and (b) after melting of the β-iPP crystals at 158 °C. Sun et al. [134]. Reproduced with permission of American Chemical Society.

may argue that the partially or incompletely molten iPP fibers exist actually in the form of amorphous domains with oriented or stretched macromolecular chain segments as in the sheared melt. These oriented molecular segments or chains, in turn, serve as nucleation precursors and initiate the β-iPP crystallization during the cooling process just like the case of shear-induced crystallization. Selective melting of the β-crystals displays, however, a scenario different from the shear-induced iPP crystallization by fiber pulling.

Figure 12.46 shows the optical micrographs of an iPP single-polymer composite before and after melting of the β-iPP crystals. In the interface layer, some randomly dispersed leaf-shaped α-iPP inclusions are clearly visible. Comparing Figure 12.46a and b, one may notice that a transition of α-iPP spherulites from mixed type (α_m) into negative radial type (α_{II}) has been realized by annealing at 158 °C for a short time [138]. From Figure 12.46b, the similar diameter and relatively smooth surface of the recrystallized iPP fiber compared with the used original one implies that the growth of the β-iPP transcrystals starts directly from the fiber surface. This is different from what was observed in heterogeneous fiber/matrix composites produced by fiber pulling, where α-iPP layers with zigzag edges on both sides of the heterogeneity fiber were observed after melting of the β-iPP cylindrites. Moreover, it was occasionally observed that the iPP fibers could be broken off by selective melting at temperatures above the melting point of β-iPP but below that of α-iPP (see the circled part in Figure 12.46b). Because the used iPP fibers are originally in their α-form, the breakage of the iPP fiber unambiguously implies that some local parts of the original α-iPP fiber have transformed into β-form through melting and recrystallization. From this point, one may conclude that the melting status, or, in other words, the degree of local chain orientation of the molten iPP fiber plays a very important role in generating transcrystalline β-iPP crystals.

It should be noted that due to the amazing characteristics of nanofibers and nanotubes, such as extraordinary mechanical and electrical properties coupled with very large surface area to volume ratio, polymer-based composites with nanofibers and nanotubes have also attracted great attention in recent years [215–220]. Taking the high aspect ratio into account, nanofiber- or nanotube-induced polymer crystallization could be totally different from traditional fiber-induced polymer crystallization. Therefore, the crystallization behavior of polymers around nanoscale fibers or tubes is another aspect deserving comment. In this field, the majority of work has been concerned with carbon nanotubes (CNTs), as reviewed by Thostenson et al. [220], while iPP was the most frequently used polymer matrix, owing to its wide applications. As an example, Figure 12.47a shows a micrograph of an iPP/CNT ultrathin composite film [221]. One can clearly see that while a common cross-hatched lamellar structure is observed in the intact area (see upper left corner of Fig. 12.47a), the iPP grown at the CNT surface forms transcrystals, indicating the high efficiency of CNT in nucleating iPP. It should be pointed out that unlike the carbon-fiber-induced iPP crystallization, where edge-on iPP lamellae propagated outward from the fiber surface with the molecular chain parallel to the long axis of the fiber were obtained, the iPP in the CNT-induced transcrystalline zone grows in flat-on form with molecular chains oriented perpendicular to the long axis of the CNT (see Fig. 12.47b) [222]. Moreover, it has once been reported that the CNT could induce the growth of the less-preferred β-form of iPP, even though it has not been confirmed by others [204].

It has been confirmed that CNT exhibits high nucleation efficiency toward a number of polymers, promoting the formation of transcrystals of the semicrystalline polymers, even though with different orientation depending on the polymer used, as reviewed in Ref. [223]. In this field,

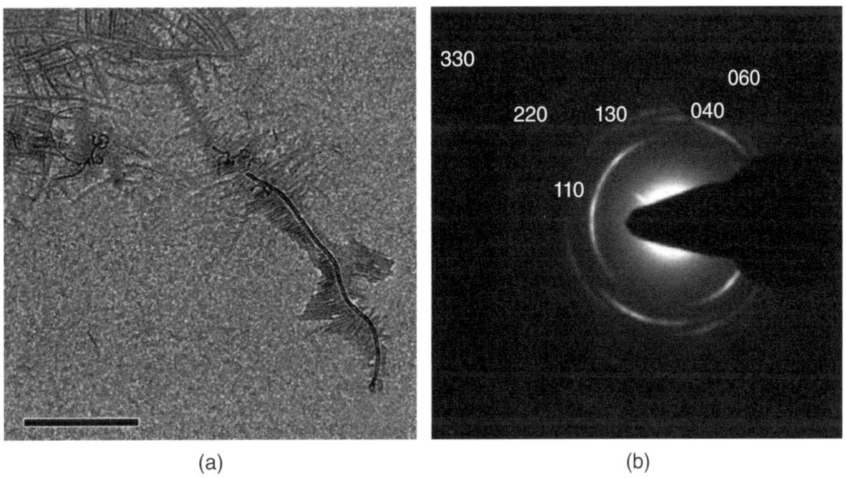

Figure 12.47 (a) Bright-field transmission electron micrograph of an iPP/CNT composite thin film nonisothermally crystallized from the melt during cooling at 5 °C/min; scale bar: 500 nm. Miltner et al. [221]. Reproduced courtesy of American Chemical Society. (b) Representative select-area electron diffraction pattern of the transcrystalline-grown iPP crystals around the CNTs in an ultrathin film. Lu et al. [222]. Reproduced with permission of American Chemical Society.

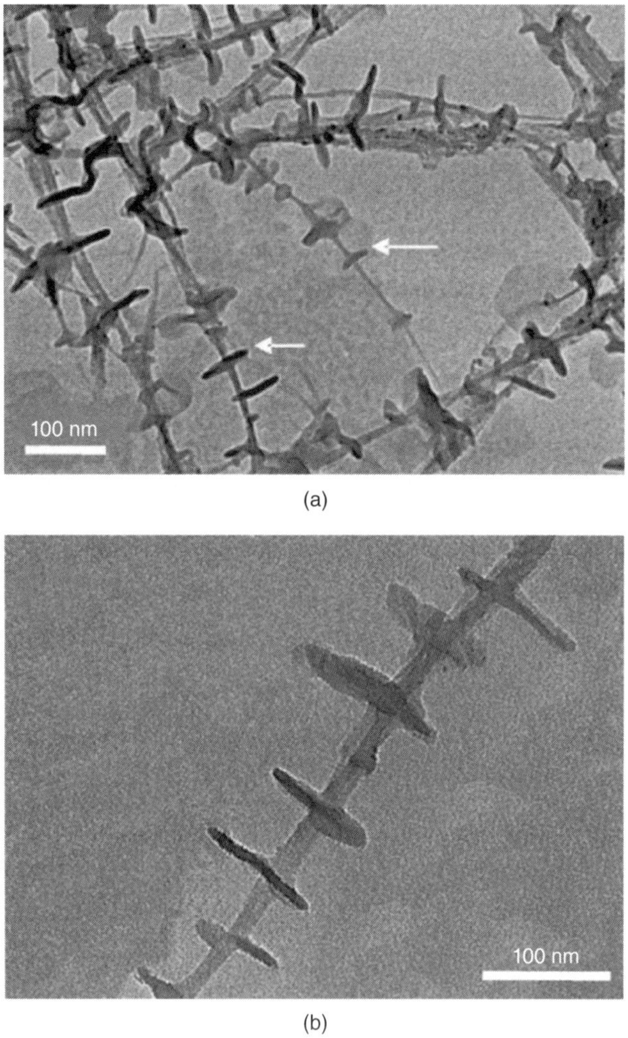

Figure 12.48 Bright-field transmission electron micrographs of SWNTs periodically patterned with PE lamellar crystals produced by crystallization of PE on SWNTs at 104 °C in p-xylene for 0.5 h. The PE and SWNT concentrations are (a) 0.01 and (b) 0.002 wt %, respectively. SWNT bundles can be seen in NHSK as shown in (b). Li et al. [225]. Reproduced with permission of American Chemical Society.

Li et al. [224–226] have done very good work on the crystallization of polymers from solution at the surface of CNTs. With controlled crystallization conditions, disklike edge-on lamellar crystals surrounding the CNT have been successfully produced. Figure 12.48 shows two representative transmission electron micrographs of the PE crystallized from solution on the single-walled CNTs [225]. Similar structures are also obtained when crystallizing polymers in supercritical CO_2 on the CNTs [227–231]. In those structures, lamellar single crystals of polymers were formed and periodically spaced along entire CNTs with an orthogonal orientation between polymer lamellar surface and the long axis of CNTs. The periodicity of the polymer lamellar single crystals can be well controlled through regulating the crystallization conditions of the polymers, such as solution concentration, crystallization temperature, solvent, and so on.

The mechanism of CNT-induced polymer crystallization rests most likely on the epitaxial growth of the polymer on the CNT surface, a situation like the traditional graphitic-carbon-fiber-induced polymer crystallization. On the other hand, taking the small diameter of CNTs into account, it has also been suggested that the CNT itself serves as a macromolecule, leading to the preferred alignment of polymer chains along its long axis [225]. Whatever the mechanism is, the morphologies presented in Figure 12.48 closely resemble the shish–kebab structure of polymers formed in elongated or sheared melts (see Fig. 12.49) [232–234]. Actually, the formation of shish–kebab structures is also a kind of surface-induced polymer epitaxy, in which a stack of folded chain lamellae form around a prealigned long chain central part. This has been well explained through simulation work as performed by Hu et al. [235].

12.5 SUMMARY

The straightforward dependence of properties on the crystal structure and morphology of semicrystalline polymers provides an opportunity to tailor the properties of these

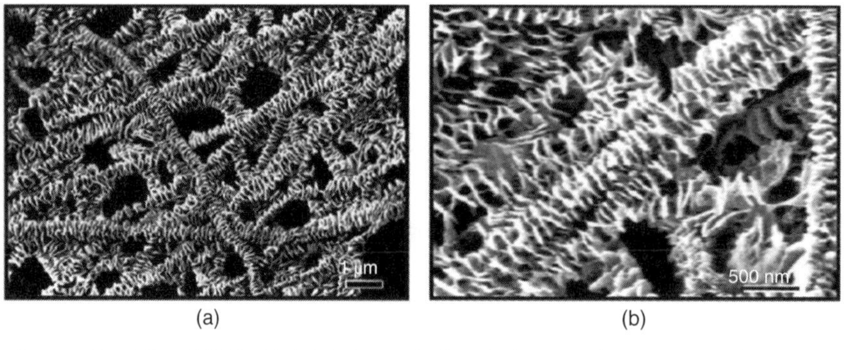

Figure 12.49 Electron micrographs of UHMWPE shish–kebabs by field emission scanning electron microscopy (FE-SEM). Somani [232]. Reproduced with permission of Elsevier.

materials through crystal engineering. Therefore, a full understanding of the crystallization habits of polymers under different environments and conditions is essential for purposeful control of their structures and morphologies. This has stimulated a mass of research work on polymer crystallization since the 1920s. Considering that polymeric materials are directly in contact with some kinds of solid surfaces in a variety of applications (such as microelectronic devices and composites), surface-induced crystallization of polymers has attracted considerable attention during the past decades. It has been shown that solid surfaces most generally accelerate the crystallization of the polymers. Moreover, they can, in some cases, also promote the crystallization of polymers in unexpected manners and lead to the formation of unique crystal structure and morphology, for example, the polymer epitaxy and transcrystallization as summarized here. It should be pointed out that, to date, most of the work in this field focuses mainly on the mechanisms of the unusual crystallization behavior and the resultant special structure and morphology. Through sophisticated studies, an understanding of some of these phenomena on a molecular level has been reached, for example, the molecular epitaxy of polymers. On the other hand, although the positive effects of the resultant specific structures and morphologies on the mechanical and physical properties have clearly been demonstrated (e.g., the improved photoresponse characteristics and the improved mechanical properties of fiber-reinforced polymer systems), the use of these techniques to design and fabricate the polymeric materials with properties matching specific product requirement has not been well developed. Therefore, further challenges in this field are how to utilize the induced unusual crystallization behavior of polymers for preparing high-performance polymeric materials. With an in-depth understanding of surface-induced polymer crystallization, novel technological pathways for preparing polymeric materials for advanced applications can be expected.

REFERENCES

1. Ciferri A, Ward IM. *Ultra-High Modulus Polymers*. London: Applied Science; 1979.
2. Smith P, Lemstra P. Ultra-high-strength polyethylene filaments by solution spinning/drawing. J Mater Sci 1980;15:505–514.
3. Oh SY, Akagi K, Shirakawa H. Synthesis and properties of polyacetylene films prepared by soluble vanadium-based catalysts. Synth Met 1989;32:245–252.
4. Gagnon DR, Karasz FE, Thomas EL, Lenz RW. Electrochemical studies of fused-pyrrole systems. Synth Met 1987;20:245–258.
5. Andreatta A, Tokito S, Smith P, Heeger A. High performance fibers of conducting polymers. J Mol Cryst Liq Cryst 1990;189:169–182.
6. Lovinger AJ. Ferroelectric polymers. Science 1983;220: 1115–1121.
7. Wada Y. *Electronic Properties of Polymers*. New York: Wiley & Sons; 1982.
8. Klein J. The interdiffusion of polymers. Science 1990;250: 640–646.
9. Green PF, Palmstrøm CJ, Mayer JW, Kramer EJ. Marker displacement measurements of polymer-polymer interdiffusion. Macromolecules 1985;18:501–507.
10. Brown HR, Char K, Deline VR, Green PF. Effects of a diblock copolymer on adhesion between immiscible polymers. 1. Polystyrene (PS)-PMMA copolymer between PS and PMMA. Macromolecules 1993;26:4155–4163.
11. Liu FP, Gardner DJ, Wolcot MP. A model for the description of polymer surface dynamic behavior 1. Contact angle vs polymer surface properties. Langmuir 1995;11:2674–2681.
12. Keddie JL, Jones RAL, Cory RA. Size-dependent depression of the glass transition temperature in polymer films. Europhys Letters 1994;27:59–64.
13. Frank CW, Rao V, Despotopoulou MM, Rease RFW, Hinsberg WD, Miller RD, Rabolt JF. Structure in thin and ultrathin spin-cast polymer films. Science 1996;273:912–915.
14. Forrest JA, Dalnoki-Veress K, Dutcher JR. Interface and chain confinement effects on the glass transition temperature of thin polymer films. Phys Rev E 1997;56:5705–5716.
15. DeMaggio GB, Frieze WE, Gidley DW, Zhu M, Hristov HA, Yee AF. Interface and surface effects on the glass transition in thin polystyrene films. Phys Rev Lett 1997;78:1524–1527.
16. Forrest JA, Mattson J. Reductions of the glass transition temperature in thin polymer films, Probing the length scale of cooperative dynamics. Phys Rev E 2000;61:R53.
17. Fryer DS, Nealey PF, de Pablo JJ. Thermal probe measurements of the glass transition temperature for ultrathin polymer films as a function of thickness. Macromolecules 2000;33:6439–6447.
18. Torres JA, Nealey PF, de Pablo JJ. Molecular simulation of ultrathin polymeric films near the glass transition. Phys Rev Lett 2000;85:3221–3224.
19. Kim JH, Jang J, Zin WC. Thickness dependence of the glass transition temperature in thin polymer films. Langmuir 2001;17:2703–2710.
20. Zheng X, Sauer BB, van Alsten JG, Schwarz SA, Rafailovich MH, Sokolov J, Rubinstein M. Reptation dynamics of a polymer melt near an attractive solid interface. Phys Rev Lett 1995;74:407–410.
21. Zheng X, Rafailovich MH, Sokolov J, Strzhemechny Y, Schwarz SA, Sauer BB, Rubinstein M. Long-range effects on polymer diffusion induced by a bounding interface. Phys Rev Lett 1997;79:241–244.
22. Russell TP, Kumar SK. The one that got away. Nature 1997;386:771–772.
23. Lin EK, Kolb R, Satija SK, Wu WL. Reduced polymer mobility near the polymer/solid interface as measured by neutron reflectivity. Macromolecules 1999;32:3753–3757.
24. Reiter G, Sommer J-U. Crystallization of adsorbed polymer monolayers. Phys Rev Lett 1998;80:3771–3774.
25. Despotopoulou MM, Frank CW, Miller RD, Rabolt JF. Kinetics of chain organization in ultrathin poly (di-n-hexylsilane) films. Macromolecules 1996;29:5797–5804.
26. Schönherr H, Frank CW. Ultrathin films of poly(ethylene oxides) on oxidized silicon. 1. Spectroscopic characterization of film structure and crystallization kinetics. Macromolecules 2003;36:1188–1198.

27. Despotopoulou MM, Frank CW, Miller RD, Rabolt JF. Role of the restricted geometry on the morphology of ultrathin poly (di-n-hexylsilane) films. Macromolecules 1995;28:6687–6688.
28. Schönherr H, Frank CW. Ultrathin films of poly (ethylene oxides) on oxidized silicon. 2. In situ study of crystallization and melting by hot stage AFM. Macromolecules 2003;36:1199–1208.
29. Schönherr H, Waymouth RM, Frank CW. Nucleation and crystallization of low-crystallinity polypropylene followed in situ by hot stage atomic force microscopy. Macromolecules 2003;36:2412–2418.
30. Schönherr H, Wiyatno W, Pople J, Frank CW, Fuller GG, Gast AP, Waymouth RM. Morphology of thermoplastic elastomers, elastomeric polypropylene. Macromolecules 2002;35:2654–2666.
31. Brinkhuis RHR, Schouten AJ. Thin-film behavior of poly(methyl methacrylates). 3. Epitaxial crystallization in thin films of isotactic poly(methyl methacrylate) using crystalline Langmuir-Blodgett layers. Macromolecules 1992;25:2717–2724.
32. De Boer A, Alberda van Ekenstein GOR, Challa G. Crystallization of isotactic poly(methyl methacrylate) from the melt. Polymer 1975;16:930–932.
33. Zhao D, Li L, Che B, Cao Q, Lu Y, Xue Q. Metastable isotactic poly(methyl methacrylate) prepared by freeze-extracting solutions in poly(ethylene glycol). Macromolecules 2004;37: 4744–4747.
34. Schneider B, Štokr J, Spěváček J, Baldrian J. Infrared spectra and crystallinity of isotactic poly (methyl methacrylate) and of its deuterated analogues. Macromol Chem Phys 1987;188:2705–2711.
35. Liu J, Wang J, Li H, Shen D, Zhang J, Ozaki Y, Yan S. Epitaxial crystallization of isotactic poly (methyl methacrylate) on highly oriented polyethylene. J Phys Chem B 2006;110:738–742.
36. Frank B, Gast AP, Russell TP, Brown HR, Hawker C. Polymer mobility in thin films. Macromolecules 1996;29:6531–6534.
37. Capitán MJ, Rueda DR, Ezquerra TA. Inhibition of the crystallization in nanofilms of poly (3-hydroxybutyrate). Macromolecules 2004;37:5653–5659.
38. Calvert P. Tender morsels for microbes. Nature 1992;360: 535–535.
39. Ma Y, Hu W, Reiter G. Lamellar crystal orientations biased by crystallization kinetics in polymer thin films. Macromolecules 2006;39:5159–5164.
40. Reiter G, Sommer JU. Polymer crystallization in quasi-two dimensions. I. Experimental results. J Chem Phys 2000;112: 4376–4379.
41. Sommer JU, Reiter G. Polymer crystallization in quasi-two dimensions. II. Kinetic models and computer simulations. J Chem Phys 2000;112:4384–4393.
42. Vonau F, Aubel D, Bouteiller L, Reiter G, Simon L. Cooperative rearrangements leading to long range order in monolayers of supramolecular polymers. Phys Rev Lett 2007;99:086103.
43. Ma Y, Zha L, Hu W, Reiter G, Han CC. Crystal nucleation enhanced at the diffuse interface of immiscible polymer blends. Phys Rev E 2008;77:061801.
44. Grozeva N, Botizb I, Reiter G. Morphological instabilities of polymer crystals. Eur Phys J E 2008;27:63–71.
45. Sommer JU, Reiter G. Morphogenesis of lamellar polymer crystals. Europhys Lett 2001;56:755–761.
46. Xu J, Ma Y, Hu W, Rehahn M, Reiter G. Cloning polymer single crystals through self-seeding. Nat Mater 2009;8:348–353.
47. Keller A, O'Connor A. Large periods in polyethylene, the origin of low-angle X-ray scattering. Nature 1957;180:1289–1290.
48. Bassett DC, Frank FC, Keller A. Evidence for distinct sectors in polymer single crystals. Nature 1959;184:810–811.
49. Keller A. Polymer single crystals. Polymer 1962;3:393–421.
50. Lovinger AJ, Keith HD. Electron diffraction investigation of a high-temperature form of poly (vinylidene fluoride). Macromolecules 1979;12:919–924.
51. Kovacs AJ, Straupe C. Isothermal growth, thickening and melting of poly(ethylene oxide) single crystals in the bulk. Part 4—Dependence of pathological crystal habits on temperature and thermal history. Faraday Discuss 1979;68:225–238.
52. Keith HD, Padden FJ, Lotz B, Wittmann JC. Asymmetries of habit in polyethylene crystals grown from the melt. Macromolecules 1989;22:2230–2238.
53. Bu Z, Yoon Y, Ho R-M, Zhou W, Jangchud I, Eby RK, Cheng SZD, Hsieh ET, Johnson TW, Geerts RG, Palackal SJ, Hawley GR, Welch MB. Crystallization, melting, and morphology of syndiotactic polypropylene fractions. 3. Lamellar single crystals and chain folding. Macromolecules 1996;29:6575–6581.
54. Sakai Y, Imai M, Kaji K, Tsuji M. Growth shape observed in two-dimensional poly (ethylene terephthalate) spherulites. Macromolecules 1996;29:8830–8834.
55. Sutton SJ, Izumi K, Miyaji H, Miyamoto Y, Miyatashi S. The morphology of isotactic polystyrene crystals grown in thin films: the effect of substrate material. J Mater Sci 1997;32:5621–5627.
56. Abe H, Kikkawa Y, Iwata T, Aoki H, Akehata T, Doi Y. Microscopic visualization on crystalline morphologies of thin films for poly [3-hydroxybutyric acid] and its copolymer. Polymer 2000;41:867–874.
57. Taguchi K, Miyaji H, Izumi K, Hoshimo A, Miyamoto Y, Kokawa R. Growth shape of isotactic polystyrene crystals in thin films. Polymer 2001;42:7443–7447.
58. Zhang B, Yang D, De Rosa C, Yan S, Petermann J. Single crystal structure of form I syndiotactic poly(butene-1). Macromolecules 2001;34:5221–5223.
59. Zhang B, Yang D, De Rosa C, Yan S. TEM studies on single crystal structure of syndiotactic poly(propene-co-butene-1)s. Macromolecules 2002;35:4646–4652.
60. Hu Z-J, Huang H-Y, Zhang F-J, Du B-Y, He TB. Thickness-dependent molecular chain and lamellar crystal orientation in ultrathin poly(di-n-hexylsilane) films. Langmuir 2004;20:3271–3277.
61. Duan Y, Jiang Y, Jiang S, Li L, Yan S, Schultz JM. Depletion-induced nonbirefringent banding in thin isotactic polystyrene thin films. Macromolecules 2004;37:9283–9286.
62. Mareau VH, Prud'homme RE. In-situ hot stage atomic force microscopy study of poly(ε-caprolactone) crystal growth in ultrathin films. Macromolecules 2005;38:398–408.
63. Jiang S, Li H, De Rosa C, Auriemma F, Yan S. Structure and morphology of syndiotactic poly(propene-co-1-butene)s with 1-butene as a rich component. Macromolecules 2010;43:1449–1454.

64. Li H, Yan S. Surface-induced polymer crystallization and the resultant structures and morphologies. Macromolecules 2011;44:417–428.
65. Theodorou DN. Variable-density model of polymer melt surfaces, structure and surface tension. Macromolecules 1989;22:4578–4589.
66. Mansfield KF, Theodorou DN. Atomistic simulation of a glassy polymer surface. Macromolecules 1990;23:4430–4445.
67. Mansfield KF, Theodorou DN. Molecular dynamics simulation of a glassy polymer surface. Macromolecules 1991;24:4295–4309.
68. Harmandaris VA, Kaoulas KC, Mavrantzas VG. Molecular dynamics simulation of a polymer melt/solid interface, local dynamics and chain mobility in a thin film of polyethylene melt adsorbed on graphite. Macromolecules 2005;38:5796–5809.
69. Liu T, Yan S, Bonnet M, Lieberwirth I, Rogausch KD, Petermann J. DSC and TEM investigations on multiple melting phenomena in isotactic polystyrene. J Mater Sci 2000;35:5047–5055.
70. Zhou H, Yan S. Can we control through interfacial crystallographic interaction. Macromol Chem Phys 2013;214:639–653.
71. Sun X, Chen Z, Wang F, Yan S, Takahashi I. Influence of poly(vinylphenol) sublayer on the crystallization behavior of poly (3-hydroxybutyrate) thin films. Macromolecules 2013;46:1573–1581.
72. Sun X, Pi F, Zhang J, Takahashi I, Ozaki Y, Wang F, Yan S. Study on the phase transition behavior of poly (butylene adipate) in its blends with poly (vinyl phenol). J Phys Chem B 2011;115:1950–1957.
73. Lü K, Yang D. Surface fixing effect of vacuum evaporated particles on polyethylene single crystals. Macromol Rapid Commun 2005;26:1159–1162.
74. Taniguchi N, Kawaguchi A. Melt crystallization of constrained polyethylene chains in uniaxially drawn film and solution-grown lamellae. Macromolecules 2005;38:4761–4768.
75. Wang J, Li H, Liu J, Duan Y, Jiang S, Yan S. On the $\alpha \rightarrow \beta$ transition of carbon-coated highly oriented PVDF ultrathin film induced by melt recrystallization. J Am Chem Soc 2003;125:1496–1497.
76. Yan S. Origin of oriented recrystallization of carbon-coated preoriented ultrathin polymer films. Macromolecules 2003;36:339–345.
77. Liu J, Li H, Duan Y, Jiang S, Miao Z, Wang J, Wang D, Yan S. Morphologies of iPP induced by its partially carbon-coated homogeneity fibers. Polymer 2003;44:5423–5428.
78. Yan S, Lieberwirth I, Katzenberg F, Petermann J. Microstructured ultrathin HDPE films prepared by selective oriented recrystallization. J Macromol Sci B 2003;42:641–652.
79. Chang H, Guo Q, Shen D, Li L, Qiu Z, Wang F, Yan S. Study on the oriented recrystallization of carbon-coated polyethylene oriented ultrathin films. J Phys Chem B 2010;114: 13104–13109.
80. Tumala RR, Pamaszewski PJ. *Microelectronics Packaging Handbook*. New York: Van Nostrand Reinhold; 1989.
81. Grebel H, Iskandar B, Pien P, Sheppard K. In situ monitoring of a laser-induced etched grating on InP, thin-film cell configuration. Appl Phys Lett 1990;57:2959–2961.
82. Bonev I. On the terminology of the phenomena of mutual crystal orientation. Acta Crystallogr A 1972;28:508–512.
83. Richards RB. Oriented overgrowth on cold-drawn polymers. J Polym Sci A 1951;6:397–402.
84. Mathieu C, Stocker W, Thiery A, Wittmann JC, Lotz B. Epitaxy of isotactic poly (1-butene), new substrates, impact and attempt at recognition of helix orientation in form I′ by AFM. Polymer 2001;42:7033–7047.
85. Wittmann JC, John Manley RS. Polymer–monomer binary mixtures. I. Eutectic and epitaxial crystallization in poly (ε-caprolactone)–trioxane mixtures. J Polym Sci B 1977;15:1089–1100.
86. Wittmann JC, John Manley RS. Epitaxial crystallization of polyesters on aromatic hydrocarbon substrates. J Polym Sci B 1978;16:1891–1895.
87. Wittmann JC, Lotz B. Epitaxial crystallization of polyethylene on organic substrates, A reappraisal of the mode of action of selected nucleating agents. J Polym Sci B 1981;19:1837–1851.
88. Wittmann JC, Lotz B. Epitaxial crystallization of aliphatic polyesters on trioxane and various aromatic hydrocarbons. J Polym Sci B 1981;19:1853–1864.
89. Wittmann JC, Hodge AM, Lotz B. Epitaxial crystallization of polymers onto benzoic acid, polyethylene and paraffins, aliphatic polyesters, and polyamides. J Polym Sci B 1983;21:2495–2509.
90. Wittmann JC, Lotz B. Epitaxial crystallization of monoclinic and orthorhombic polyethylene phases. Polymer 1989;30:27–34.
91. Haubruge HG, Daussin R, Jonas AM, Legras R, Wittmann JC, Lotz B. Epitaxial nucleation of poly (ethylene terephthalate) by talc, Structure at the lattice and lamellar scales. Macromolecules 2003;36:4452–4456.
92. Plank H, Resel R, Sitterb H, Andreev A, Sariciftci NS, Hlawacek G, Teichert C, Thierry A, Lotz B. Molecular alignments in sexiphenyl thin films epitaxially grown on muscovite. Thin Solid Films 2003;443:108–114.
93. Mathieu C, Thierry A, Wittmann JC, Lotz B. Specificity and versatility of nucleating agents toward isotactic polypropylene crystal phases. J Polym Sci B 2002;40:2504–2515.
94. Lotz B, Wittmann JC. Structural relationships in blends of isotactic polypropylene and polymers with aliphatic sequences. J Polym Sci B 1986;24:1559–1575.
95. Turner-Jones A, Bunn CW. The crystal structure of polyethylene adipate and polyethylene suberate. Acta Crystallogr 1962;15:105–113.
96. Takahashi T, Teraoka F, Tsujimoto IJ. Epitaxial crystallization of crystalline polymers on the surface of drawn polytetrafluoroethylene. J Macromol Sci B 1976;B12:303–305.
97. Yan C, Guo L, Chang H, Yan S. Induced crystallization behavior of poly(ethylene adipate) by highly oriented polyethylene. Chin J Polym Sci 2013;31:1173–1182.
98. Woo EM, Wu PL, Wu MC, Yan KC. Thermal behavior of ring-band versus Maltese-cross spherulites, case of monomorphic poly(ethylene adipate). Macromol Chem Phys 2006;207:2232–2243.
99. Hobbs SY, Billmeyer FW Jr. Crystal unit-cell dimensions and densities of linear aliphatic polyesters. J Polym Sci B 1969;7:1119–1121.

100. Im SS, Kim TJ, Han S, Moon TJ, Bae YC. Physical properties of poly (ethylene adipate)/low-density polyethylene blends. J Appl Polym Sci 1997;65:1745–1750.
101. Gross B, Petermann J. Synergisms of mechanical properties in blends of semi-crystalline polymers. J Mater Sci 1984;19:105–112.
102. Petermann J, Broza G, Rieck U, Kawaguchi A. Epitaxial interfaces in semi-crystalline polymers and their applications. J Mater Sci 1987;22:1477–1481.
103. Tu C, Jiang S, Li H, Yan S. Origin of epitaxial cold crystallization of poly(l-lactic acid) on highly oriented polyethylene substrate. Macromolecules 2013;46:5215–5222.
104. An YK, Jiang SD, Yan SK, Sun JR, Chen XS. Crystallization behavior of polylactide on highly oriented polyethylene thin films. Chin J Polym Sci 2011;29:513–519.
105. Horio M. The theory of crimp of textile fibers. Mem Fac Eng, Kyoto Univ 1964;26:222.
106. Fenwick D, Smith P, Wittmann JC. Epitaxial and graphoepitaxial growth of materials on highly orientated PTFE substrates. J Mater Sci 1996;31:128–131.
107. Takahashi T, Ogata N. Epitaxial crystallization of polyoxymethylene in the course of polymerization in aqueous solution. Polym Lett 1971;9:895–899.
108. Takahashi T, Inamura M, Tsujimoto I. Epitaxial growth of polymer crystals on uniaxially drawn polymers. J Polym Sci B 1970;8:651–657.
109. Sun Y, Li H, Huang Y, Chen E, Zhao L, Gan Z, Yan S. Epitaxial crystallization of poly (butylene adipate) on highly oriented polyethylene thin film. Macromolecules 2005;38:2739–2743.
110. Liu Q, Sun X, Li H, Yan S. Orientation-induced crystallization of isotactic polypropylene. Polymer 2013;54:4404–4421.
111. Wittmann JC, Lotz B. Epitaxial crystallization of polymers on organic and polymeric substrates. Prog Polym Sci 1990;15:909–948.
112. Tao X, Yan S, Yang D. Chin Chem Lett 1993;4:1093.
113. Xu Y, Asano T, Kawaguchi A, Rieck U, Petermann J. Epitaxial crystallization of polyoctenamer on uniaxially oriented polypropylene films. J Mater Sci Lett 1989;8:675–677.
114. Schumacher M, Lovinger AJ, Agarwal P, Wittmann JC, Lotz B. Heteroepitaxy of syndiotactic polypropylene with polyethylene and homoepitaxy. Macromolecules 1994;27:6956–6962.
115. Petermann J, Xu Y, Loos J, Yang D. Epitaxial crystallization of syndiotactic polypropylene on uniaxially oriented polyethylene. Polym Commun 1992;33:1096–1098.
116. Yan S, Petermann J, Yang D. Epitaxial crystallization in the sPP/nylon-12 semicrystalline polymer system. Polymer 1996;37:2681–2685.
117. Yan S, Katzenberg F, Petermann J, Yang D, Shen Y, Straupe C, Wittmann JC, Lotz B. A novel epitaxy of isotactic polypropylene (α phase) on PTFE and organic substrates. Polymer 2000;41:2613–2625.
118. Sun Y, Li H, Huang Y, Chen E, Gan Z, Yan S. Epitaxial crystallization of poly (butylene adipate) on highly oriented isotactic polypropylene thin film. Polymer 2006;47:2455–2459.
119. Yan S, Katzenberg F, Petermann J. Epitaxial and graphoepitaxial growth of isotactic polypropylene (iPP) from the melt on highly oriented high density polyethylene (HDPE) substrates. J Polym Sci B 1999;3:1893–1898.
120. Yan S, Bonnet M, Petermann J. A comparison of the crystallization behavior of HDPE on oriented iPP and sPP substrates. Polymer 2000;41:1139–1145.
121. Sun X, Li H, Lieberwirth I, Yan S. α and β interfacial structures of the iPP/PET matrix/fiber systems. Macromolecules 2007;40:8244–8249.
122. Orts WJ, Marchessault RH, Bluhm TL, Hamer GK. Observation of strain-induced β form in poly (β-hydroxyalkanoates). Macromolecules 1990;23:5368–5370.
123. Zhang J, Duan Y, Sato H, Tsuji H, Noda I, Yan S, Ozaki Y. Crystal modifications and thermal behavior of poly(L-lactic acid) revealed by infrared spectroscopy. Macromolecules 2005;38:8012–8021.
124. Hoogsteen W, Postema AR, Pennings AJ, ten Brinke G, Zugenmaier P. Crystal structure, conformation and morphology of solution-spun poly(L-lactide) fibers. Macromolecules 1990;23:634–642.
125. Sasaki S, Asakura T. Helix distortion and crystal structure of the α-form of poly(L-lactide). Macromolecules 2003;36:8385–8390.
126. Wang T, Li H, Yan S. Effect of poly(butylene succinate) on the morphology evolution of poly(vinylidene fluoride) in their blends. Chin J Polym Sci 2012;30:269–277.
127. Varga J. β-Modification of polypropylene and its two-component systems. J Therm Anal 1989;35:1891–1912.
128. Varga J. Melting memory effect of the β-modification of polypropylene. J Therm Anal 1986;31:165–172.
129. Shi G, Zhang X, Qiu Z. Crystallization kinetics of β-phase poly (propylene). Makromol Chem 1992;193:583–591.
130. Lovinger AJ, Chua JO, Gryte CC. Studies on the α and β forms of isotactic polypropylene by crystallization in a temperature gradient. J Polym Sci B 1977;15:641–656.
131. Grubb DT, Odell JA, Keller A. Characterization of superheatable flow-induced polymer crystals by optical microscopy. J Mater Sci 1975;10:1510–1518.
132. Li H, Jiang S, Wang J, Wang D, Yan S. Optical microscopic study on the morphologies of isotactic polypropylene induced by its homogeneity fibers. Macromolecules 2003;36:2802–2807.
133. Li H, Zhang X, Kuang X, Wang D, Li L, Yan S. A scanning electron microscopy study on the morphologies of isotactic polypropylene induced by its own fibers. Macromolecules 2004;37:2847–2853.
134. Sun X, Li H, Zhang X, Wang J, Wang D, Yan S. Effect of fiber molecular weight on the interfacial morphology of iPP fiber/matrix single polymer composites. Macromolecules 2006;39:1087–1092.
135. Sun X, Li H, Wang J, Yan S. Shear-induced interfacial structure of isotactic polypropylene (iPP) in iPP/fiber composites. Macromolecules 2006;39:8720–8726.
136. Crissman JM. Mechanical relaxation in polypropylene as a function of polymorphism and degree of lamella orientation. J Polym Sci 1969;7:389–404.
137. Fujiwara Y. Das Doppelschmelzverhalten der a-Phase des isotaktischen Polypropylens. Colloid Polym Sci 1975;253:273–282.
138. Varga J, Ehrenstein GW. Formation of β-modification of isotactic polypropylene in its late stage of crystallization. Polymer 1996;37:5959–5963.

139. Zhang J, Yang D, Thierry A, Wittmann JC, Lotz B. Isochiral form II of syndiotactic polypropylene produced by epitaxial crystallization. Macromolecules 2001;34:6261–6267.
140. Kopp S, Wittmann JC, Lotz B. Epitaxial crystallization and crystalline polymorphism of poly (1-butene), forms III and II. Polymer 1994;35:908–915.
141. Kopp S, Wittmann JC, Lotz B. Epitaxial crystallization and crystalline polymorphism of poly (1-butene), form I. Polymer 1994;35:916–924.
142. Lovinger AJ. Crystallization of the β phase of poly(vinylidene fluoride) from the melt. Polymer 1981;22:412–413.
143. Fuller CS, Erickson CL. An X-ray study of some linear polyesters. J Am Chem Soc 1937;59:344–351.
144. Fuller CS, Erickson CL. X-ray investigation of the decamethylene series of polyesters. J Am Chem Soc 1939;61:2575–2580.
145. Fuller CS, Erickson CL. J Phys Chem 1939;43:323–334.
146. Gan Z, Kuwabara K, Abe H, Iwata T, Doi Y. Metastability and transformation of polymorphic crystals in biodegradable poly (butylene adipate). Biomacromolecules 2004;5:371–378.
147. Pouget E, Almontassir A, Casas MT, Puiggali J. On the crystalline structures of poly(tetramethylene adipate). Macromolecules 2003;36:698–705.
148. Gan Z, Abe H, Doi Y. Temperature-induced polymorphic crystals of poly(butyleneadipate). Macromol Chem Phys 2002;203:2369–2374.
149. Woo EM, Wu MC. Thermal and X-ray analysis on polymorphic crystals in poly(butylene adipate). J Polym Sci B 2005;43:1662–1672.
150. Gan Z, Kuwabara K, Abe H, Iwata T, Doi Y. The role of polymorphic crystal structure and morphology in enzymatic degradation of melt-crystallized poly (butylene adipate) films. Polym Degrad Stab 2005;87:191–199.
151. Yang J, Pan P, Dong T, Inoue Y. Crystallization kinetics and crystalline structure of biodegradable poly (ethylene adipate). Polymer 2010;51:807–815.
152. Satoh M. Soc Polym Sci, Japan 2008;57:460.
153. Stocker W, Schumacher M, Graff S, Thierry A, Wittmann JC, Lotz B. Epitaxial crystallization and AFM investigation of a frustrated polymer structure, isotactic poly (propylene), β phase. Macromolecules 1998;31:807–814.
154. Yan S, Spath T, Petermann J. Morphological studies of PE crystallized between iPP and PTFE oriented films. Polymer 2000;41:4863–4868.
155. Yan S, Yang D, Petermann J. Controlling factors for the occurrence of heteroepitaxy of polyethylene on highly oriented isotactic polypropylene. Polymer 1998;39:4569–4578.
156. Yan S, Petermann J. Nucleation and overgrowth of PE on PTFE/iPP interfaces. J Polym Sci B 2000;38:80–83.
157. Abbate S, Gussoni M, Zerbi G. Infrared and Raman intensities of polyethylene and perdeuteropolyethylene: Factor group splittings. J Chem Phys 1979;70:3577–3585.
158. Yan C, Zhang J, Yan S. Structure characterization of melt drawn polyethylene ultrathin films. Chin Sci Bull 2006;51:2844–2850.
159. Nagamatsu S, Takashima W, Kaneto K, Yoshida Y, Tanigaki N, Yase K. Backbone arrangement in "Friction-Transferred" regioregular poly(3-alkylthiophene)s. Macromolecules 2003;36:5252–5257.
160. Hartmann L, Tremel K, Uttiya S, Crossland E, Ludwigs S, Kayunkid N, Vergnat C, Brinkmann M. 2D versus 3D crystalline order in thin films of regioregular poly (3-hexylthiophene) oriented by mechanical rubbing and epitaxy. Adv Funct Mater 2011;21:4047–4057.
161. Zhou H, Jiang S, Yan S. Epitaxial crystallization of poly (3-hexylthiophene) on a highly oriented polyethylene thin film from solution. J Phys Chem B 2011;115:13449–13454.
162. Brinkmann M, Wittmann JC. Orientation of regioregular poly (3-hexylthiophene) by directional solidification, a simple method to reveal the semicrystalline structure of a conjugated polymer. Adv Mater 2006;18:860–863.
163. Brinkmann M, Rannou P. Effect of molecular weight on the structure and morphology of oriented thin films of regioregular poly(3-hexylthiophene) grown by directional epitaxial. Adv Funct Mater 2007;17:101–108.
164. Kayunkid N, Uttiya S, Brinkmann M. Structural model of regioregular poly (3-hexylthiophene) obtained by electron diffraction analysis. Macromolecules 2010;43:4961–4967.
165. Brinkmann M, Contal C, Kayunkid N, Djuric T, Resel R. Highly oriented and nanotextured films of regioregular poly (3-hexylthiophene) grown by epitaxy on the nanostructured surface of an aromatic substrate. Macromolecules 2010;43:7604–7610.
166. Yan C, Li H, Zhang J, Ozaki Y, Shen D, Yan D, Shi AC, Yan S. Surface-induced anisotropic chain ordering of polycaprolactone on oriented polyethylene substrate, epitaxy and soft epitaxy. Macromolecules 2006;39:8041–8048.
167. Esposito LD, Koening JL. Application of Fourier transform infrared spectroscopy to the study of semicrystalline polymers: Poly(ethylene terephthalate). J Polym Sci B 1976;12:1731–1741.
168. Coleman MM, Zarian J. Fourier-transform infrared studies of polymer blends. II. Poly (ε-caprolactone)–poly (vinyl chloride) system. J Polym Sci B 1979;17:837–850.
169. Hubble D, Cooper S. Segmental orientation in blends of poly (ε-caprolactone) with poly (vinyl chloride) and nitrocellulose. J Polym Sci B 1977;15:1143–1149.
170. Elzein T, Nasser-Eddine M, Delaite C, Bistac S, Dumas P. FTIR study of polycaprolactone chain organization at interfaces. J Colloid Interface Sci 2004;273:381–387.
171. Bittiger H, Marchessault RH, Niegisch WD. Crystal structure of poly-ε-caprolactone. Acta Crystallogr 1970;B23:1923–1927.
172. Chatani Y, Okita Y, Tadokoro H, Yamoshita Y. Structural studies of polyesters. III. Crystal structure of poly-ε-caprolactone. Polym J 1970;1:555–562.
173. Fraser RDB. The structure of the ω-form of poly-β-benzyl-l-aspartate. J Chem Phys 1953;21:1511–1518.
174. Chang H, Zhang J, Li L, Wang Z, Yang C, Takahashi I, Ozaki Y, Yan S. A study on the epitaxial ordering process of the polycaprolactone on the highly oriented polyethylene substrate. Macromolecules 2010;43:362–366.
175. Wunderlich B. *Macromolecular Physics*. New York: Academic; 1973.
176. Tracz A, Jeszka JK, Kucinska I, Chapel JP, Boiteux G, Kryszewski M. Influence of the crystallization conditions on the morphology of the contact layer of polyethylene crystallized on graphite atomic force microscopy studies. J Appl Polym Sci 2002;86:1329–1336.

177. Tracz A, Kucinska I, Jeszka JK. Formation of highly ordered, unusually broad polyethylene lamellae in contact with atomically flat solid surfaces. Macromolecules 2003;36:10130–10132.
178. Takenaka Y, Miyaji H, Hoshino A, Tracz A, Jeszka JK, Kucinska I. Surface adsorption-induced conformational ordering and crystallization of polyethylene oxide. Macromolecules 2004;37:9667–9669.
179. Jaballah A, Rieck U, Petermann J. Properties of layered structured PE/PP and PB-1/PP films having ultra-thin layers. J Mater Sci 1990;25:3105–3110.
180. Lee IH, Schultz JM. Adhesion in laminates of highly oriented polypropylene sheets. Polymer 1986;27:1219–1227.
181. Lee IH, Schultz JM. Bonding of highly oriented polypropylene sheets by epitaxial crystallization of polyethylene. J Mater Sci 1988;23:4237–4243.
182. Sun Y, Tan L, Jiang S, Qian H, Wang Z, Yan D, Di C, Wang Y, Wu W, Yu G, Yan S, Wang C, Hu W, Liu Y, Zhu D. High-performance transistor based on individual single-crystalline micrometer wire of perylo[1,12-*b*,*c*,*d*]thiophene. J Am Chem Soc 2007;129:1882–1883.
183. Dong H, Jiang S, Jiang L, Liu Y, Li H, Hu W, Wang E, Yan S, Wei Z, Xu W, Gong X. Nanowire crystals of a rigid rod conjugated polymer. J Am Chem Soc 2009;131:17315–17320.
184. Brinkmann M, Rannou P. Molecular weight dependence of chain packing and semicrystalline structure in oriented films of regioregular poly(3-hexylthiophene) revealed by high-resolution transmission electron microscopy. Macromolecules 2009;42:1125–1130.
185. Jiang S, Qian H, Liu W, Wang C, Wang Z, Yan S, Zhu D. Vapor phase epitaxy of perylo[1,12-b,c,d]thiophene on highly oriented polyethylene thin films. Macromolecules 2009;42:9321–9324.
186. Dong H, Li H, Wang E, Wei Z, Xu W, Hu W, Yan S. Ordering rigid rod conjugated polymer molecules for high performance photoswitchers. Langmuir 2008;24:13241–13244.
187. Guo D, Sakamoto K, Miki K, Ikada S, Saiki K. Alignment-induced epitaxial transition in organic-organic heteroepitaxy. Phys Rev Lett 2008;101:236103–236106.
188. Brinkmann M, Pratontep S, Chaumont C, Wittmann JC. Oriented and nanostructured polycarbonate substrates for the orientation of conjugated molecular materials and gold nanoparticles. Macromolecules 2007;40:9420–9426.
189. Prelipceanu M, Tudose OG, Prelipceanu OT, Schrader S, Grystsenko K. Study of oriented growth of oligofluorene–thiophene films onto aligned vacuum-deposited polytetrafluoroethylene layers. Mater Sci Semicond Process 2007;10:24–35.
190. Kang SJ, Noh YY, Baeg KJ, Ghim J, Park JH, Kim JS, Park JH, Cho K. Effect of rubbed polyimide layer on the field-effect mobility in pentacene thin-film transistors. Appl Phys Lett 2008;92:052107.
191. Brinkmann M, Chandezon F, Julien-Raban C, Pansu RP. Epitaxial growth of highly oriented fibers of semiconducting polymers with a shish-kebab-like superstructure. Adv Funct Mater 2009;19:2759–2766.
192. Thomason JL, Van Rooyen AA. Transcrystallized interphase in thermoplastics composites. J Mater Sci 1992;27:889–896.
193. Ishida H, Bussi P. Surface induced crystallization in ultrahigh-modulus polyethylene fiber-reinforced polyethylene composites. Macromolecules 1991;24:3569–3577.
194. Cho K, Kim D, Yoon S. Effect of substrate surface energy on transcrystalline growth and its effect on interfacial adhesion of semicrystalline polymers. Macromolecules 2003;36:7652–7660.
195. Thomason JL, van Rooyen AA. The transcrystallized interphase in thermoplastic composites. In: Ishida H, editor. *Controlled Interphases in Composite Materials*. New York: Elsevier Science; 1990. p 423.
196. Hsiao BS, Chen EJ. Transcrystalline interphase in advanced polymer composites. In: Ishida H, editor. *Controlled Interphases in Composite Materials*. New York: Elsevier Science; 1990. p 613.
197. Varga J. Review of supermolecular structure of isotactic polypropylene. J Mater Sci 1992;27:2557–2579.
198. Varga J, Fujiwara Y, Ille A. $\beta\alpha$-Bifurcation of growths during the spherulitic crystallization of polypropylene. Period Polytech, Chem Eng 1990;34:255–271.
199. Varga J. Crystallisation, melting and supermolecular structure of isotactic polypropylene. In: Karger-Kocsis J, editor. *Poly(propylene), Structure, Blends and Composites*. Vol. 1. London: Chapman & Hall; 1995. p 56.
200. Varga J, Karger-Kocsis J. Rules of supermolecular structure formation in sheared isotactic polypropylene melts. J Polym Sci B 1996;34:657–670.
201. Hobbs SY. Row nucleation of isotactic polypropylene on graphite fibres. Nature 1971;234:12–13.
202. Li H, Sun X, Yan S, Schultz JM. Initial stage of iPP β to α growth transition induced by stepwise crystallization. Macromolecules 2008;41:5062–5064.
203. Sun X, Li H, Zhang X, Wang D, Schultz JM, Yan S. Effect of matrix molecular mass on the crystallization of β-form isotactic polypropylene around an oriented polypropylene fiber. Macromolecules 2010;43:561–564.
204. Grady BP, Pompeo F, Shambaugh RL, Resasco DE. Nucleation of polypropylene crystallization by single-walled carbon nanotubes. J Phys Chem B 2002;106:5852–5858.
205. Li H, Zhang X, Duan Y, Wang D, Li L, Yan S. Influence of crystallization temperature on the morphologies of isotactic polypropylene single-polymer composite. Polymer 2004;45:8059–8065.
206. Loos J, Schimanski T, Hoffman J, Peijs T, Lemstra PJ. Morphological investigations of polypropylene single-fibre reinforced polypropylene model composites. Polymer 2001;42:3827–3834.
207. Alfonso GC, Scardigli P. Melt memory effects in polymer crystallization. Macromol Symp 1997;118:323–328.
208. Azzurri F, Alfonso GC. Lifetime of shear-induced crystal nucleation precursors. Macromolecules 2005;38:1723–1728.
209. de Gennes PG. *Scaling Concepts in Polymer Physics*. Ithaca, NY: Cornell University Press; 1979.
210. Klein J. Evidence for reptation in an entangled polymer melt. Nature 1978;271:143–145.
211. Somani RH, Hsiao BS, Nogales A, Srinivas S, Tsou AH, Sics I, Balta-Calleja FJ, Ezquerra TA. Structure development during shear flow-induced crystallization of i-PP: *In situ* small-angle X-ray scattering study. Macromolecules 2000;33:9385–9394.
212. Somani RH, Hsiao BS, Nogales A, Fruitwala H, Srinivas S, Tsou AH. Structure development during shear flow induced crystallization of i-PP: In situ wide-angle X-ray diffraction study. Macromolecules 2001;34:5902–5909.

213. Somani RH, Yang L, Sics I, Hsiao BS, Pogodina NV, Winter HH, Agarwal P, Fruitwala H, Tsou A. Orientation-induced crystallization in isotactic polypropylene melt by shear deformation. Macromol Symp 2002;185:105–117.
214. Nogales A, Hsiao BS, Somani RH, Srinivas S, Tsou AH, Balta-Calleja FJ, Ezquerra TA. Shear-induced crystallization of isotactic polypropylene with different molecular weight distributions: in situ small- and wide-angle X-ray scattering studies. Polymer 2001;42:5247–5256.
215. Andrews R, Jacques D, Rao AM, Rantell T, Derbyshire F, Chen Y, Chen J, Haddon RC. Nanotube composite carbon fibers. Appl Phys Lett 1999;75:1329–1331.
216. Li B, Li CY. Alternating patterns on single-walled carbon nanotubes. J Am Chem Soc 2007;129:13.
217. Lourie O, Cox DM, Wagner HD. Buckling and collapse of embedded carbon nanotubes. Phys Rev Lett 1998;81:1638–1641.
218. Regev O, ElKati PNB, Loos J, Koning CE. Preparation of conductive nanotube–polymer composites using latex technology. Adv Mater 2004;16:248–251.
219. Koganemaru A, Bin YZ, Agari Y, Matsuo M. Composites of polyacrylonitrileand multiwalled carbon nanotubes prepared by gelation/crystallization from solution. Adv Funct Mater 2004;14:842–850.
220. Thostenson ET, Ren ZF, Chou TW. Advances in the science and technology of carbonnanotubes and their composites. Compos Sci Technol 2001;61:1899–1912.
221. Miltner HE, Grossiord N, Lu K, Loos J, Koning CE, Van Mele B. Isotactic polypropylene/carbon nanotube composites prepared by latex technology. Thermal analysis of carbon nanotube-induced nucleation. Macromolecules 2008;41:5753–5762.
222. Lu K, Grossiord N, Koning CE, Miltner HE, Van Mele B, Loos J. Carbon nanotube/isotactic polypropylene composites prepared by latex technology: Morphology analysis of CNT-induced nucleation. Macromolecules 2008;4:8081–8085.
223. Li L, Li B, Hood MA, Li CY. Carbon nanotube induced polymer crystallization: The formation of nanohybrid shish–kebabs. Polymer 2009;50:953–965.
224. Li CY, Li LY, Cai WW, Kodjie SL, Tenneti KK. Polymer crystallization-driven, periodic patterning on carbon nanotubes. Adv Mater 2005;17:1198–1202.
225. Li CY, Li LY, Ni CY. Polymer crystallization-driven, periodic patterning on carbon nanotubes. J Am Chem Soc 2006;128:1692–1699.
226. Li CY, Li LY, Ni CY, Rong LX, Hsiao B. Structure and crystallization behavior of Nylon 66/multi-walled carbon nanotube nanocomposites at low carbon nanotube contents. Polymer 2007;48:3452–3460.
227. Yue J, Xu Q, Zhang ZW, Chen ZM. Periodic patterning on carbon nanotubes: Supercritical CO_2-induced polyethylene epitaxy. Macromolecules 2007;40:8821–8826.
228. Zhang ZW, Xu Q, Chen ZM, Yue J. Nanohybrid shish-kebabs: Supercritical CO_2-induced PE epitaxy on carbon nanotubes. Macromolecules 2008;41:2868–2873.
229. Zhang F, Zhang H, Zhang ZW, Chen ZM, Xu Q. Modification of carbon nanotubes: Water-soluble polymers nanocrystal wrapping to periodic patterning with assistance of supercritical CO_2. Macromolecules 2008;41:4519–4523.
230. Zheng X, Xu Q. Comparison study of morphology and crystallization behavior of polyethylene and poly (ethylene oxide) on single-walled carbon nanotubes. J Phys Chem B 2010;114:9435–9444.
231. He L, Zheng X, Xu Q. Modification of carbon nanotubes using poly (vinylidene fluoride) with assistance of supercritical carbon dioxide: The impact of solvent. J Phys Chem B 2010;114:5257–5262.
232. Somani RH, Yang L, Zhu L, Hsiao BS. Flow-induced shish-kebab precursor structures in entangled polymer melts. Polymer 2005;46:8587–8623.
233. Yang L, Somani RH, Sics I, Hsiao BS, Kolb R, Fruitwala H, Ong C. Shear-induced crystallization precursor studies in model polyethylene blends by in-situ rheo-SAXS and rheo-WAXD. Macromolecules 2004;37:4845–4859.
234. Hsiao BS, Yang L, Somani RH, Avila-Orta CA, Zhu L. Unexpected shish-kebab structure in a sheared polyethylene melt. Phys Rev Lett 2005;94:117802–117806.
235. Hu W, Frenkel D, Mathot VBF. Simulation of shish-kebab crystallite induced by a single prealigned macromolecule. Macromolecules 2002;35:7172–7174.

13

THERMODYNAMICS AND KINETICS OF POLYMER CRYSTALLIZATION

WENBING HU AND LIYUN ZHA

Department of Polymer Science and Engineering, School of Chemistry and Chemical Engineering, Nanjing University, Nanjing, China

13.1 INTRODUCTION

Vast natural and synthetic polymers are semicrystalline, and such a structural character dominates their physical performance, such as mechanical, optical, and thermal properties. Therefore, it is necessary to understand the process of polymer crystallization. Still, polymer crystallization appears as complicated. It can be influenced by many factors, such as chemical structures of polymer chains, compositions, temperatures, thermal history, spatial confinements, and pressures. In order to gain a better understanding about these factors in polymer crystallization, theoretical and simulation approaches are often concerted with experimental approaches.

This review chapter makes a brief survey focusing mainly on the fundamental thermodynamics and kinetic aspects of polymer crystallization. Thermodynamics address why, or under which circumstance, polymers perform crystallization, or, in the opposite direction, polymer crystals start to melt. Kinetics address how fast polymer crystallization can be initiated, be developed, or be further improved. On thermodynamics, we introduce the thermodynamic concepts such as the melting points, the metastable states, phase diagrams, mesophase formation, as well as the statistical thermodynamic theory and those factors governing melting points. On kinetics, we introduce the classical nucleation theory and the related kinetic equations governing crystal nucleation, crystal growth, and crystal annealing.

13.2 THERMODYNAMICS OF POLYMER CRYSTALLIZATION

Semicrystalline states of polymers mean polymers partially in both the amorphous and the crystalline states. Polymer crystallization and melting are typically first-order phase transitions between the amorphous phase and the crystalline phase. When these two phases are in thermodynamic equilibrium, two phase transitions are reversible under a certain temperature. This temperature is referred to as the equilibrium melting point $T_m{}^0$ of polymer crystals. In the bulk system of polymers, the free energy change of melting becomes zero at its equilibrium melting point, as given by

$$\Delta F_m = \Delta Q_m - T_m^0 \Delta S_m = 0 \tag{13.1}$$

The equilibrium melting point can accordingly be calculated by

$$T_m^0 = \Delta Q_m / \Delta S_m \tag{13.2}$$

The crystalline states of polymer materials contain a low heat of fusion owing to weak van der Waals packing interactions, as well as a high entropy of fusion owing to large conformational entropy change. Therefore, according to Equation 13.2, their melting points are commonly as low as 100 °C, not like the conventional metals that require 1000 °C for melting. Synthetic polymers are easily reshaped at the temperatures not so far away from the room temperature, and hence are endowed with the name "plastics."

In the amorphous state of homogeneous polymer solutions or melt, polymer chains are fully disordered, as described by the random-coil model. The random-coil model was first proposed by Kuhn [1] as well as by Guth and Mark [2] to predict the entropic elasticity of polymer chains, and then was used to describe the amorphous state of polymers by Flory [3]. Polymer crystallization commonly chooses a pathway favoring its kinetics, which results in metastable semicrystalline states [4]. It takes a long time for people to figure out the structural

Polymer Morphology: Principles, Characterization, and Processing, First Edition. Edited by Qipeng Guo.

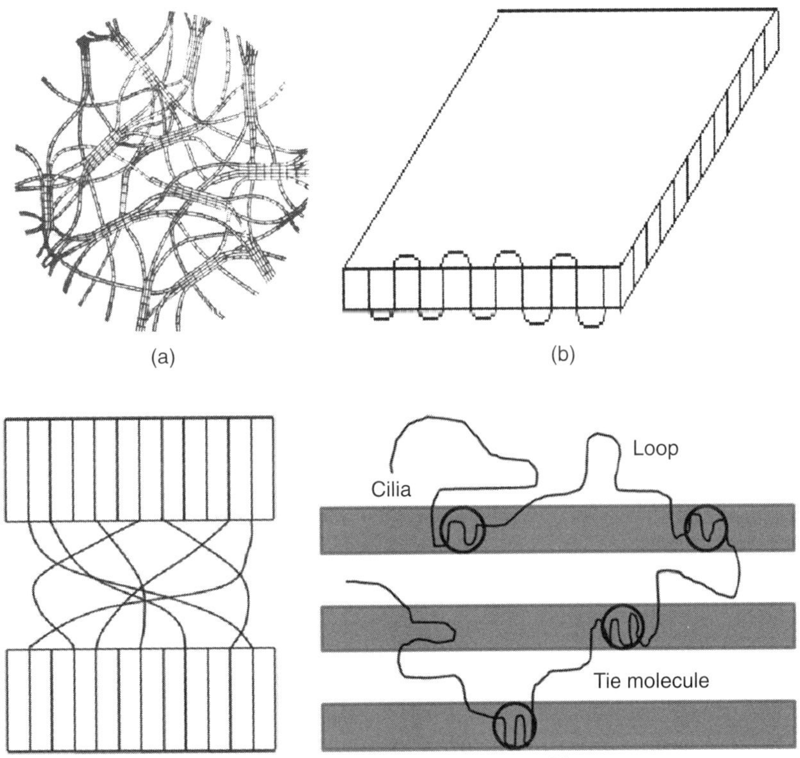

Figure 13.1 Illustration of the models of (a) fringed micelle [5], (b) adjacent chain folding [6], (c) switchboard model [7], and (d) the variable cluster model [8].

features of semicrystalline polymers, with two dominant models, which are, fringed-micelle and chain-folding models, as illustrated in Figure 13.1. In 1930, Hermann et al. set up the fringed-micelle model [5, 9] to describe the high elasticity of low-density polyethylene products. But later on, this model could not explain the spherulite morphologies of polymer crystals often observed under optical microscopy. In 1957, Keller set up *the adjacent chain-folding model* [6] on the basis of the facts that polymer chains in lamellar single crystal grown from dilute solutions are aligned normal to the lamellar surface, and the lamellar thickness was only in the scale of one-tenth of chain lengths. This model was then confirmed by the small-angle neutron scattering experiments [10], as well as by the integer folding of short chains in lamellar crystals grown in the melt [11–14]. For lamellar crystals grown in the long-chain melt, Flory proposed the "switchboard" model [7], as illustrated in Figure 13.1 as well, which was then developed into the interzonal switchboard model as discussed by Mandelkern [15]. As both adjacent chain-folding and switchboard models had their own interpretations on the same neutron-scattering results, there was a hot debate at the 68th Faraday Discussion meeting. It is now well accepted that either model describes a certain aspect of structural features on metastable lamellar crystals grown in the melt. The variable cluster model combining both features of local chain folding and global random coil was discussed by Hoffman [8], as also illustrated in Figure 13.1. The loops, cilia, and tie molecules are restricted at the fold-end surfaces of lamellar crystals, which constitute the *rigid amorphous* phase near the crystalline region, as the third part besides the mobile amorphous phase and the crystalline phase [16].

The lamellar crystals intend to perform thickening into a more thermodynamically stable state upon annealing. Before they are able to reach the most stable state, their melting points are dominated by the limited lamellar thickness, as described by the well-known *Gibbs–Thomson equation*,

$$T_m = T_m^0 - \frac{2\sigma_e T_m^0}{l\Delta h} \tag{13.3}$$

where σ_e is the free energy density of the fold-end surface, l is the lamellar thickness, and Δh is the heat of fusion. A wide distribution of lamellar thickness in polymer crystals results in a broad range of melting temperature, shown as a wide melting peak in the heating curve of differential scanning calorimetry (DSC). We commonly take the peak temperature as the experimentally observed melting point of polymer crystals. Hoffman and Week proposed that lamellar crystals thicken into a metastable state with the thickness several times larger than the minimum thickness at T_c, and derived the equilibrium melting point as the crossover point of the extrapolated T_m versus T_c curve at $T_m = T_c$ [17]. Although the fixed increasing times of lamellar thicknesses were a big assumption, this method has been pragmatically applied to derive the equilibrium melting points of polymers.

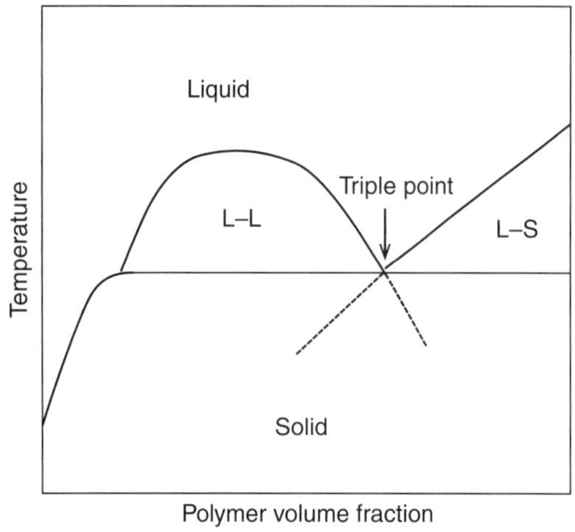

Figure 13.2 Schematic illustration of a typical phase diagram combining both L–L binodal and L–S coexistence in polymer solutions.

In polymer solutions, liquid–liquid (L–L) demixing is another common phase transition besides crystallization. The thermodynamic boundary conditions for both of them behave as the functions of polymer concentrations and temperatures, demonstrated as *phase diagrams*. The schematic L–L binodal and liquid–solid (L–S) coexistence curves in polymer solutions and their interception are shown in Figure 13.2. The illustrated L–L binodal contains an upper critical solution temperature. Some other solutions also contain binodals with a lower critical solution temperature. When the L–S curve intersects with the L–L curve in the overlapping temperature windows, both curves are terminated at the intersection point, which is referred to as the monotectic triple point.

In between the amorphous phase and the crystalline phase, there sometimes exists an intermediate-phase of anisotropic group (called mesogen group) carried either on the backbone or on the branches of polymer chains, which is referred to as *liquid crystal* (LC) *mesophase*. The mesogen groups become orientational-ordered under suitable thermodynamic conditions. The transition temperature from LC mesophase to melt is often named as the clearing point or the isotropization point T_i. When L–S phase transition is accompanied by LC transition, the above phase diagram of polymer solution becomes more complicated, as illustrated in Figure 13.3.

According to the phase diagram schematically shown in Figure 13.3, two basic types of liquid crystals formed along different dimensions can be categorized. One is the lyotropic liquid crystal prepared by changing the concentration in solutions. The other is the thermotropic liquid crystal prepared by changing the temperature in the concentrated solutions or in the bulk phase. According to Onsager's interpretation [19], the lyotropic liquid crystal results from an entropy-driven phase transition due to anisotropic excluded-volume interactions of rodlike mesogens. When the concentration of rodlike molecules becomes high enough, the space for anisotropic particles to move freely appears limited, resulting in a significant entropy loss. In this case, if part of the rods aligned in parallel with each other in a domain of higher concentration, they sacrifice part of their space for more free movement of other rods and thus increase the total entropy. This entropy-driven phase transition makes the ordered system stable. Also, according to Maier and Saupe's theory [20], the thermotropic liquid crystal is a result of orientation-dependent dispersion interactions between rodlike mesogens. With the decrease in temperature, spontaneous ordering lowers the attractive potential energy related to the orientational ordering. The phase diagram of LC polymers in solutions can be understood better by a combination of the two theoretical approaches above [21–28].

The phase diagrams of solutions can be calculated by the statistical thermodynamics of lattice models [29]. The lattice model of polymer solutions was first considered by Chang [30] and Meyer [31]. The well-known Flory–Huggins equation was then successfully derived by Flory [32, 33] and Huggins [34] to deal with the solutions of flexible polymers by using a mean-field approximation.

Following Flory's treatment for semiflexible polymer solutions [35], the total partition function of polymer solutions can be described as

$$Z = \left(\frac{n}{n_1}\right)^{n_1} \left(\frac{n}{n_2}\right)^{n_2} \left(\frac{q}{2}\right)^{n_2} e^{-(r-1)n_2} z_c^{(r-2)n_2} z_m^{rn_2} z_p^{(r-1)n_2} \tag{13.4}$$

where n_1 is the number of solvent molecules, n_2 is the number of polymer chains, each chain composed of r monomers, and $n = n_1 + n_2 r$ is the total lattice sites occupied by solvent and polymer chains, and q is the coordination number of the lattice space. Here, the conformational partition function z_c is defined as

$$z_c = 1 + (q-2)\exp\left(-\frac{E_c}{kT}\right) \tag{13.5}$$

where E_c is the energy penalty for the $q-2$ noncollinear consecutive bonds along the polymer chain with a reference to the collinear connection, k is Boltzmann constant, and T is system temperature. The partition function related with the mixing enthalpy of solvent and monomers is

$$z_m = \exp\left[-(q-2)\frac{n_1}{n} \times \frac{B}{kT}\right] \tag{13.6}$$

where B is the energy change for a pair of solvent and monomer before and after mixing as defined by

$$B = E_{12} - \frac{E_{11} + E_{12}}{2} \tag{13.7}$$

As polymer chains in the crystals are commonly featured with parallel packing, the driving force for crystallization can be reflected as the parallel packing of bonds in the lattice model [36], as the energy penalty for nonparallel packing of two bonds E_p. The partition function related to nonparallel

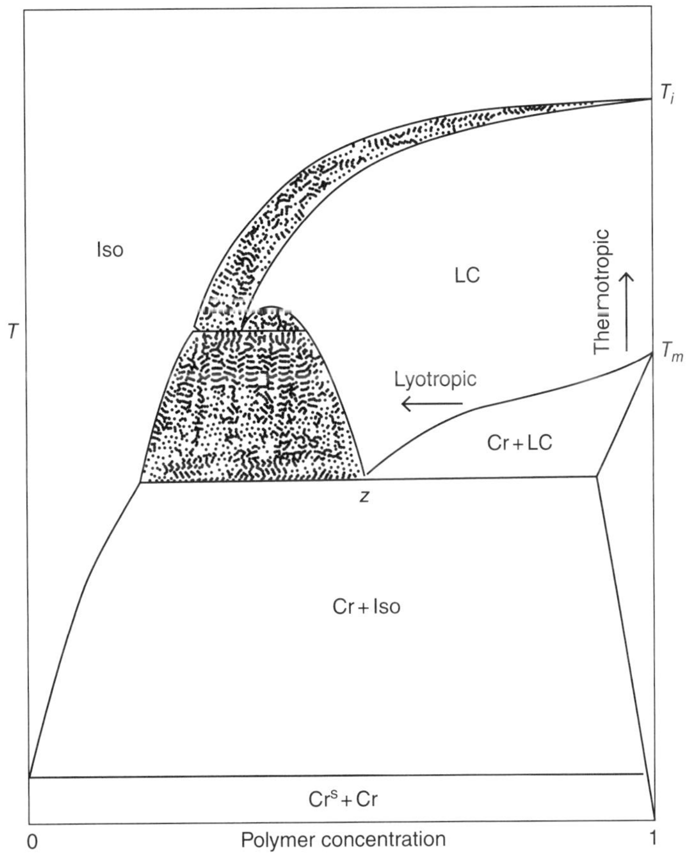

Figure 13.3 Typical phase diagram of polymer solutions combining isotropic solutions (Iso), crystalline phase (Cr), and the intermediate liquid crystal phase (LC). Two arrows show the dimensions for thermotropic liquid crystal and lyotropic liquid crystal, respectively. Keller [18]. Reproduced with permission of ©IUPAC.

packing of neighboring bonds around a chain bond is given by [37]

$$z_p = \exp\left\{-\frac{q-2}{2}\left[1-\frac{2(r-1)n_2}{qn}\right]\frac{E_p}{kT}\right\} \qquad (13.8)$$

The corresponding free energy density is calculated according to Boltzmann's relationship, as given by

$$\begin{aligned}\frac{f(\phi)}{k_BT} &= (1-\phi)\ \ln(1-\phi)+\frac{\phi}{r}\ln\phi \\ &+\phi\left(-\frac{\ln(qr/2)}{r}-(1-2/r)\ln z_c+(1-1/r)\right. \\ &\left.+(q-2)\frac{B}{k_BT}+(1-1/r)\frac{q-2}{2}\frac{E_p}{k_BT}\right) \\ &-\phi^2\left((q-2)\frac{B}{k_BT}+(1-1/r)^2\frac{q-2}{q}\frac{E_p}{k_BT}\right) \qquad (13.9)\end{aligned}$$

Equation 13.9 can be used to calculate various thermodynamic properties of polymers, in particular, the equilibrium melting point and phase diagrams. It can be used to calculate the mixing free energy and the binodal L–L curve, as well as the coexistence L–S curve in polymer solutions, as introduced below. Furthermore, Monte Carlo simulations of lattice polymer can be realized by following a micro-relaxation model [38]. Its micro-relaxation is accepted by the Metropolis importance sampling algorithm [39]. Monte Carlo simulations of lattice polymers hold the parallel parameters to the theoretical phase diagrams, and appear as a powerful tool in understanding the practical process of polymer crystallization [37].

Taking the fully ordered extended chains as the ground state, and considering bulk polymers with infinitely long polymer chains, $r \to \infty$, $n_1 = 0$, $n = rn_2$, one can get the free energy of the amorphous state from Equation 13.9, and let it to equal that of the ground crystalline state (zero). The equilibrium melting point of bulk polymers is approximately derived as

$$T_m \approx \frac{E_c + \frac{(q-2)^2}{2q}E_p}{k\ \ln(q-2)-k} \qquad (13.10)$$

We can clearly see that a larger E_c (a higher rigidity of polymer chains) favors a higher equilibrium melting point. Melting point of polymers with extremely high rigidity may be even higher than its thermal degradation temperature, making their melt phase unreachable. We can also see that a larger E_p (a stronger binding of polymer chains) favors a higher equilibrium melting point. Small side groups, polar side groups, and hydrogen bonding will raise the melting points.

The chain ends can be regarded as the crystalline defects in the infinitely large crystals, inducing a depression of melting points with the decrease of chain lengths. Flory and Vrij divided the free energy change of melting into three parts [40], that is, the free energy change for the melting of infinitely long polymers, the free energy change by introducing chain-end defects in the crystals, and the conformational entropy change when the infinitely long polymer chain is cut into the segments with limited chain length r. So the total free energy change of melting can be described as

$$\Delta F_m = r\Delta F_u + \Delta F_e - kT_m \ \ln r = 0 \tag{13.11}$$

ΔF_u is the free energy change of fusion, which, according to Equation 13.10, can be roughly estimated by

$$\begin{aligned}\Delta F_u &= \Delta h_u - T_m \Delta s_u \\ &= E_c + \frac{(q-2)^2}{2q}E_p - k_B T_m[\ln(q-2) - 1]\end{aligned} \tag{13.12}$$

ΔF_e is the extra free energy change due to the existence of chain ends, which can be estimated from the equilibrium condition $f=0$ in Equation 13.9 by setting the chain length $r=2$ in the melt phase.

$$\Delta F_e = \frac{(q-2)(q-1)}{2q}E_p - k_B T_m(\ln q - 1) - 2\Delta f_u \tag{13.13}$$

The term $\ln(r)$ represents a change in conformational entropy upon cutting. Inserting Equations 13.12 and 13.13 into Equation 13.11, the equilibrium melting point predicted by the Flory–Vrij theory can be calculated.

One can also directly calculate the equilibrium melting point from Equation 13.8. Under the equilibrium melting point, the chemical potentials of solutions equal the crystals (zero), so the melting point can be obtained from the equation

$$\begin{aligned}&(1-r)\frac{n_2 r}{n} + \ln\frac{qn}{2n_2} + (r-2) \\ &\times \ln\left[1 + (q-2)\exp\left(-\frac{E_c}{k_B T_m}\right)\right] \\ &= \frac{(r-1)(q-2)}{2}\left[1 - \frac{2(r-1)n_2(n+n_1)}{qn^2}\right] \\ &\times \frac{E_p}{k_B T_m} + \frac{rn_1^2}{n^2}\frac{(q-2)B}{k_B T_m}\end{aligned} \tag{13.14}$$

Figure 13.4 compares the melting temperatures of bulk polymers with different chain lengths derived separately by the lattice theory, the Flory–Vrij equation, and Monte Carlo simulations. The dimension of the temperature unit is reduced as $E_c/(k_B T_m)$. Their agreements are satisfying, although they hold various assumptions.

From Equation 13.9, the theoretical curves of L–S coexistence and L–L binodal can be separately calculated, provided by the absence of each counterpart. The equilibrium melting point (L–S coexistence curve) can be estimated from the solution chemical potential equal to zero,

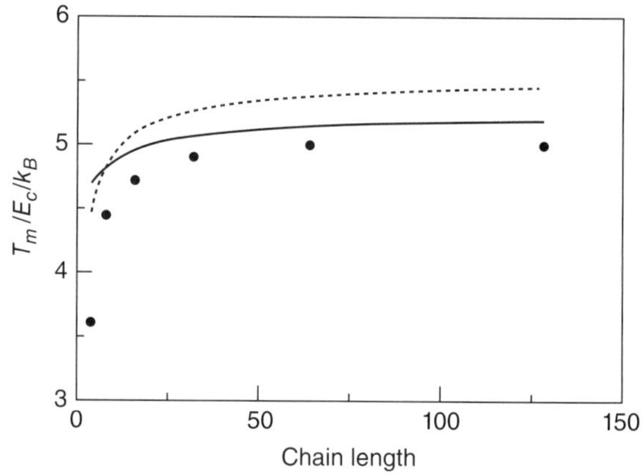

Figure 13.4 Melting temperature ($T_m/E_c/k_B$) of bulk polymers with different chain lengths for three theoretical approaches sharing the same sets of molecular parameters. The solid line is calculated from Equation 13.14, and the dashed line is resulted from the Flory–Vrij equation. Circles are the results of Monte Carlo simulations of polymer solutions with a 0.9375 concentration in a 32^3 cubic lattice for short chains and a 64^3 cubic lattice for long chains [37].

$$\mu^s - \mu^0 = \frac{\partial f}{kT\partial n_2} \approx 0 \tag{13.15}$$

The L–L binodal curve can be calculated by the chemical potential equivalence of components between the dense phase and the dilute phase after phase separation, on the basis of the mixing free energy from the bulk amorphous state to the solution state of polymers, as given by

$$\begin{aligned}\frac{\Delta F_{mix}}{kT} &= \frac{F_{solution} - F_{bulk}}{kT} = -(\ln Z - \ln Z_{n_1=0}) \\ &= n_1 \ln\phi_1 + n_2 \ln\phi_2 + n_1\phi_2 \\ &\times \left[(q-2)\frac{B}{kT} + \left(1-\frac{2}{q}\right)\left(1-\frac{1}{r}\right)^2 \frac{E_P}{kT}\right]\end{aligned} \tag{13.16}$$

Figure 13.5 shows the parallel results of liquid–liquid demixing curves and liquid–solid coexistence curves in polymer solutions with different energy parameter sets, obtained from the lattice mean-field theory and Monte Carlo simulations. Theoretical calculations and Monte Carlo simulations agree well with each other, validating the mean-field assumption in theoretical approaches.

In Figure 13.5, we can see that the L–S curve is mainly determined by E_p/E_c, and the L–L curve is mainly controlled by B/E_c but is also slightly affected by E_p/E_c. By changing the values of E_p/E_c and B/E_c, the interplay between crystallization and liquid–liquid demixing can be studied in a combination of the lattice theory and parallel molecular simulations [42–44], which has been introduced in the book on Polymer Physics [45].

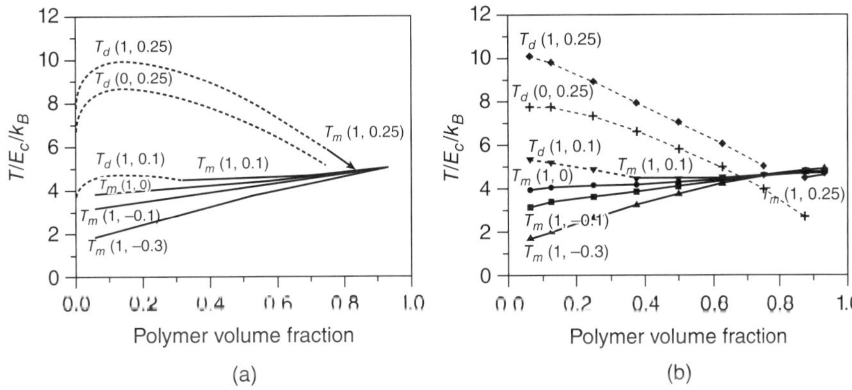

Figure 13.5 Liquid–liquid demixing curves (dashed lines) and liquid–solid curves (solid lines) of polymer solutions with different energy parameter combinations denoted by $T(E_p/E_c, B/E_c)$, separately obtained from (a) mean-field theory and (b) Monte Carlo simulations of parallel lattice models [41].

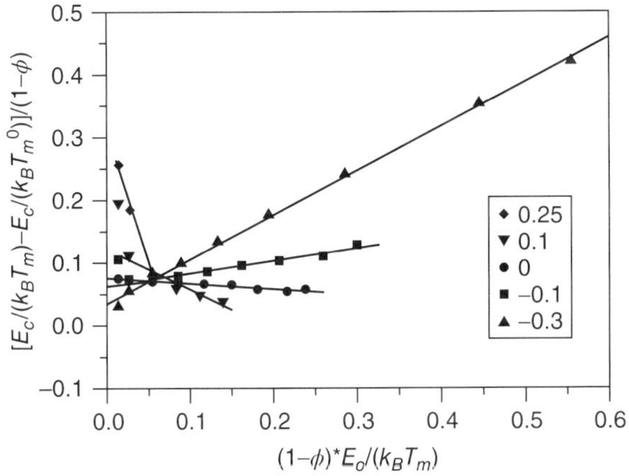

Figure 13.6 Rescaled data from Figure 13.5b according to Equation 13.17. The bulk equilibrium melting temperature $(E_c/k_BT_m^0)$ is chosen to be approximately 0.2. Lines are linear regressions of symbols at the same values of B/E_c as labeled [41].

An empirical equation for the liquid-solid boundary curve has been derived by Flory, as given by

$$\frac{1}{T_m} - \frac{1}{T_m^0} = \frac{R}{\Delta H_u}(\phi_1 - \chi\phi_1^2) \tag{13.17}$$

This equation fits well with experiments [46], and be also consistent with Monte Carlo simulations [41]. Figure 13.6 shows the linear relationship according to Equation 13.17, with the data points adopted from Figure 13.5b.

Real polymers often contain various kinds of sequence irregularities, such as different chemical compositions, different geometrical connections, or different stereo optical isomers of monomers. Because of the spatial mismatch between irregular and regular sequences for the compact packing in crystalline order, the melting points will be lowered. If the regular sequences are referred to monomer A and the irregular sequences are referred to comonomer B, Flory [47] treated this AB random copolymer as an ideal solution and derived

$$\frac{1}{T_m} - \frac{1}{T_m^0} = \frac{R}{\Delta H_u} \ln X_A \tag{13.18}$$

Flory assumed all the comonomers staying only in the amorphous phase. By considering comonomers coexist with monomers in the crystallites, Sanchez and Eby [48] gave another expression of melting point for copolymers, shown as

$$T_m = T_m^0\left(1 - \frac{\Delta H_B}{\Delta H_u}X_B\right) \tag{13.19}$$

where ΔH_B is the heat of fusion for each comonomer as a defect in the crystalline phase, and X_B is the mole fraction of comonomers.

13.3 CRYSTAL NUCLEATION

Polymer crystallization is conventionally initiated by crystal nucleation. According to the Gibbs classical nucleation theory, when thermal fluctuations generate the ordered domains with the sizes large enough from a homogeneous phase, the trend to increase the surface free energy can be overcome by the trend to decrease the body free energy, and larger domains intend to be more stable [49]. Assuming nucleus as a sphere with a radius r, the free energy change of nucleation can be estimated as

$$\Delta G = -\Delta g \times \frac{4}{3}\pi r^3 + \sigma \times 4\pi r^2 \tag{13.20}$$

Δg is the melting free energy of unit volume and σ is the specific surface free energy. A schematic plot for the free energy change with the increasing radius of nucleus is shown in Figure 13.7.

Δg in Equation 13.20 is calculated as

$$\Delta g = \Delta h - T_c\Delta s \approx \Delta h - T_c\frac{\Delta h}{T_m} = \Delta h\frac{T_m - T_c}{T_m} \propto \Delta T \tag{13.21}$$

In practice, to get over the nucleation barrier via thermal fluctuations, the initiation of primary crystal nucleation

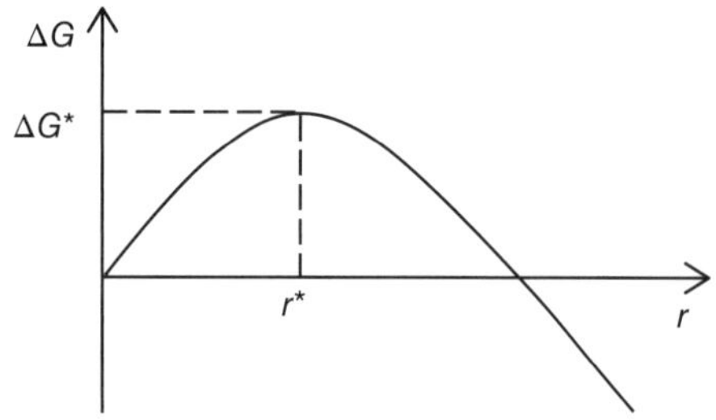

Figure 13.7 Schematic curve of the free energy change as the radius of nucleus formed in the amorphous phase.

requires a crystallization temperature T_c much lower than $T_m{}^0$. Thus, the parameter of *supercooling* for crystallization is defined as

$$\Delta T = T_m^0 - T_c \tag{13.22}$$

Three basic types of nucleation are considered at various dimensions: primary nucleation, secondary nucleation, and tertiary nucleation. Primary nucleation is a three-dimensional nucleus newly formed by thermal fluctuations, with six extra nucleus surfaces if the nucleus is considered to be cubic. Secondary nucleation is two-dimensional nucleation on the advancing surface of nucleus, with four extra surfaces produced. Secondary nucleation is much easier than primary nucleation as its free energy barrier is lower, making crystallization kinetics like a self-acceleration process. Tertiary nucleation is one-dimensional nucleation at the step edge of the spreading layer on the advancing smooth surface of nucleus, with only two extra surfaces produced. Tertiary nucleation is so fast that it can rarely be observed. The schematic pictures of different types of nucleation are shown in Figure 13.8.

Primary nucleation is the most observable phenomenon for the initiation of polymer crystallization, which can be categorized further into homogeneous nucleation and heterogeneous nucleation. In homogeneous nucleation, polymer nuclei can be treated as a cylindrical bunch of stems due to the anisotropic molecular structure, as depicted in Figure 13.9. Thus, its free energy change during nucleation is

$$\Delta G = -\pi r^2 l \Delta g + 2\pi r l \sigma + 2\pi r^2 \sigma_e \tag{13.23}$$

Here, r and l are the radius and the length of the cylinder, respectively; σ is the specific free energy on the lateral surface, and σ_e is the specific free energy on the end surface.

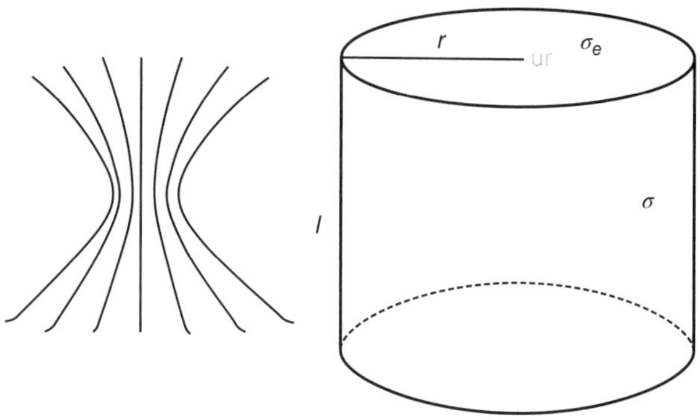

Figure 13.9 Schematic picture of nucleus treated as cylindrical bunch of stems.

By taking the minimum of ΔG with respect to r and l, the critical free energy barrier for nucleation is derived as

$$\Delta G^* = \frac{8\pi\sigma^2\sigma_e}{\Delta g^2} \propto \Delta T^{-2} \tag{13.24}$$

The critical sizes are separately calculated as

$$r^* = \frac{2\sigma}{\Delta g} \propto \Delta T^{-1} \tag{13.25}$$

$$l^* = \frac{4\sigma_e}{\Delta g} \propto \Delta T^{-1} \tag{13.26}$$

So the length-to-radius ratio of the critical nucleus is

$$\frac{l^*}{r^*} = \frac{2\sigma_e}{\sigma} \tag{13.27}$$

For polymers, homogeneous nucleation can be realized through two typical styles. One is the so-called intramolecular nucleation featured with adjacent chain-folding, which can be called as chain-folding nucleation. The other is the so-called intermolecular nucleation composed of parallel stacking among neighboring chains, which can be called as fringed-micelle nucleation. Schematic illustration of these two models can be found in Figure 13.10.

Let us first consider the intramolecular nucleation. The specific free energy on the lateral surfaces of PE crystals was estimated to be 11.8 erg/cm^2, and its specific free

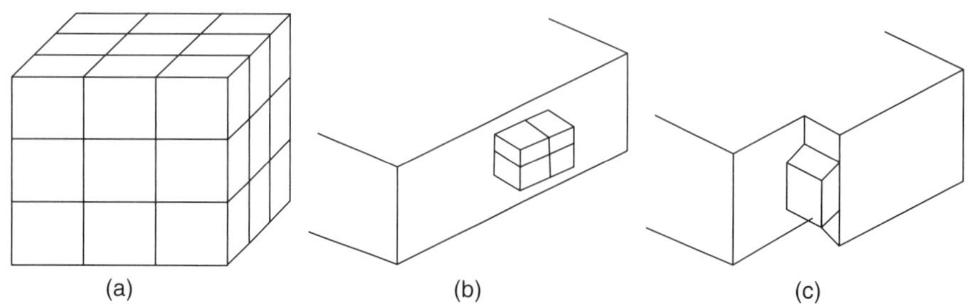

Figure 13.8 Schematic illustration of (a) primary nucleation, (b) secondary nucleation, and (c) tertiary nucleation.

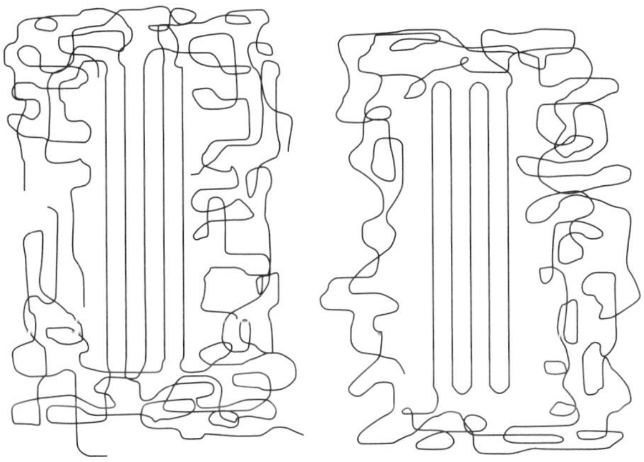

Figure 13.10 Schematic illustration of (a) intermolecular nucleation and (b) intramolecular nucleation. Adapted from Wunderlich [50].

energy on the fold-end surface was about 90 erg/cm^2 [51]. By Equation (13.27), the optimized aspect ratio of critical nucleation is 15.3. Now look at the intermolecular nucleation. The end surface has an extra free energy of about 245 erg/cm^2 as estimated by Zachmann [50, 52, 53], due to the entropy loss of disordered chains. So the aspect ratio of critical nucleus in intermolecular nucleation is as high as 56.8, and it appears very difficult to produce such a fibril nucleus through thermal fluctuations on primary nucleation. Moreover, by Equation 13.22, the higher end-surface free energy results in a higher nucleation barrier than the chain-folding model. So from both kinetic aspects of the critical nucleation barrier and of thermal fluctuations, the intramolecular nucleation is preferred in the process of primary nucleation. This selection decides the preference of chain folding during polymer crystallization, if we consider that the crystal growth is dominated by secondary nucleation favoring intramolecular nucleation as well.

Heterogeneous nucleation is primary crystal nucleation on the foreign surfaces of other materials such as catalysts, dusts, and content walls. Because less extra surface free energy is required, heterogeneous nucleation is much easier than homogeneous nucleation. There is also another type of heterogeneous nucleation called self-nucleation investigated first by Blundell et al. [54]. The foreign surfaces for self-nucleation are provided by crystals of the same species that survived during thermal history. As there is no extra surface free energy change during self-nucleation, it is also called athermal nucleation. This type of nucleation is an important source of memory effects for polymer crystallization.

The nucleation rate is dominated by two factors. One is the critical free energy barrier of nucleation. Its exponential dependence was first proposed by Volmer and Weber [55]. Another is the diffusion energy barrier for molecules crossing over the liquid–solid interfaces. Its exponential dependence was first proposed by Becker and Döring [56]. The quantitative expression of the prefactor I_0 in the kinetic equation of the nucleation rate is given by Turnbull and Fisher [57] as

$$I = I_0 \exp\left(-\frac{\Delta E + \Delta G^*}{kT}\right) \qquad (13.28)$$

where ΔE is the activation barrier for short-range diffusion over the liquid–solid boundary, and I_0 is the prefactor. The critical free energy barrier is proportional to the inverse square supercooling of primary nucleation. So when temperature is high with a high free energy barrier, the nucleation rate is small. However, when temperature is low with a high activation barrier for polymer diffusion, the nucleation rate is again small. Thus, the temperature-dependence curve of the nucleation rate is somewhat like a bell shape between the glass transition temperature and the melting temperature, as illustrated in Figure 13.11.

At high temperatures, the kinetic studies of polymer crystallization are mostly focused on the nucleation. In this region, heterogeneous nucleation takes place and the resulted mechanical properties of the semicrystalline polymers are usually hard and brittle. However, as a result of the high density of small crystallites in the region of low temperatures, the semicrystalline polymers become soft and tough.

Intramolecular nucleation is preferred in both primary and secondary polymer crystal nucleation, which results in lamellar shapes during crystal growth. The typical intramolecular nucleation model has been established by considering crystal nucleation of a single-chain system [58]. Taking the extended single chain in a crystal composed of extended parallel polymer chains as the ground state, and assuming the number of melting bonds n, the free energy change of the chain according to the classical Gibbs nucleation theory is

$$\Delta F = \Delta f n + \sigma(N - n)^{2/3} \qquad (13.29)$$

where $\Delta f n$ is the bulk free energy change and $\sigma(N-n)^{2/3}$ is the surface free energy change. The free energy change of one bond during melting is

$$\Delta f = \frac{q-2}{2} E_p - kT \ \ln(q-1) \qquad (13.30)$$

The first term on the right-hand side means $q-2$ parallel bonds around the bond in the ground state, and the denominator "2" is the symmetrical factor. The number of total conformation of a chain with n melting bonds is $(q-1)^n$, so the second term is the average conformational entropy change of each bond during melting (Fig. 13.12).

When the system is in equilibrium, the free energy in the disordered state is equal to that in the ordered state, then

$$\Delta f_e = \sigma N^{-1/3} \qquad (13.31)$$

And the equilibrium free energy barrier for primary nucleation is

$$\Delta F_e = \frac{4\sigma^3}{27\,\Delta f_e^2} \qquad (13.32)$$

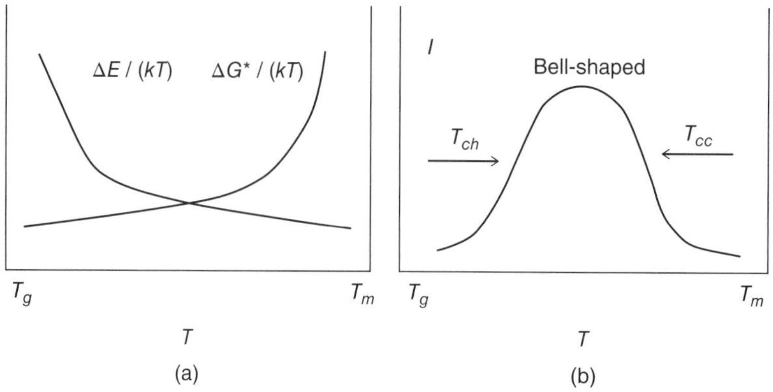

Figure 13.11 Schematic illustration of the temperature dependence (a) of the critical free energy barrier and the activation barrier for diffusion, and (b) of the bell-shaped curve of the nucleation rates.

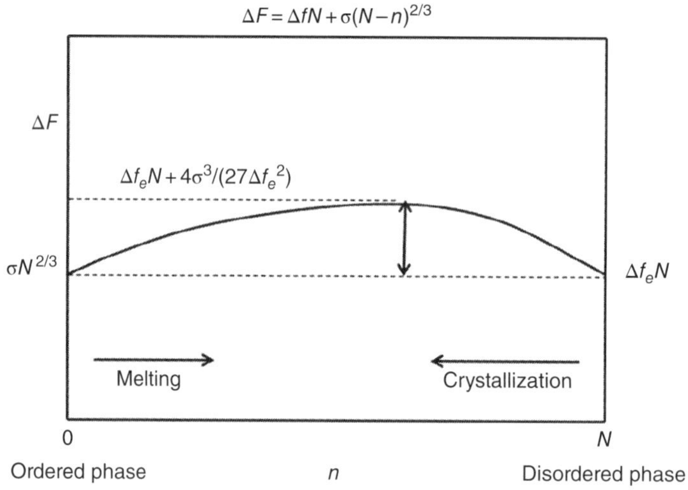

Figure 13.12 Schematic free energy change of single chain as a function of melting bonds under thermal equilibrium state. Hu et al. [58]. Reproduced with permission of American Chemical Society.

The free energy barriers for crystallization and melting of a single chain are separately calculated as follows,

$$\Delta F_c = \frac{4\sigma^3}{27\,\Delta f^2} \tag{13.33}$$

$$\Delta F_m = \frac{4\sigma^3}{27\,\Delta f^2} + \Delta f N - \sigma N^{2/3} \tag{13.34}$$

Single chains with different lengths will crystallize at the same temperature on cooling, but will melt at different temperatures on heating, with higher melting temperatures for higher chain lengths. In bulk polymers, the nucleation rates appear as chain-length dependent, which can be attributed to the prefactor in the kinetic equation of nucleation.

In practice, intermolecular nucleation could coexist with intramolecular nucleation. Intermolecular nucleation is often observed with short chains, rigid chains, or when chains are stretched. Recently, upon stretching network polymers, the transition from intramolecular nucleation to intermolecular nucleation was observed in dynamic Monte Carlo simulations [59]. By analyzing the probability of adjacent chain folding of those newly formed crystallites with a size between 50 and 200 parallel packed bonds at each step of stretching, an obvious reduction of chain folding was observed in its evolution curve under each temperature as shown in Figure 13.13a. The corresponding critical strain was considered to be the transition point under which intramolecular nucleation is the favorite and above which intermolecular nucleation becomes the dominant. Comparing the critical strain with the onset strain of crystallization during stretching under each temperature as shown in Figure 13.13b, one can see that at low temperatures (≤4.0) intramolecular nucleation dominates the initiation of polymer crystallization with a strain smaller than the critical value, yielding a high amount of chain folding; when temperature becomes higher, intermolecular nucleation dominates the initiation of crystallization, yielding a low amount of chain folding.

A preordered structure in polymer melt before nucleation was proposed by Imai and coworkers, and be attributed to spinodal decomposition during orientational fluctuations at low temperatures for cold crystallization of PET [60, 61]. The main evidence came from the observation of small-angle

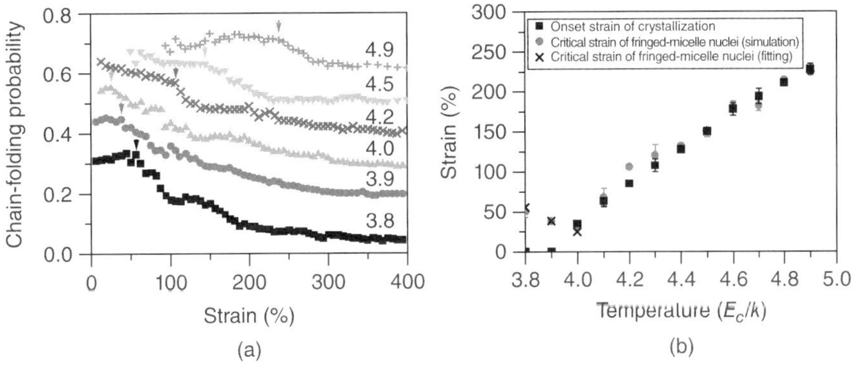

Figure 13.13 (a) Strain–evolution curves of chain-folding probability of small crystalline clusters containing 50 to 200 parallel packed bonds under different temperatures. The lines are vertically shifted by 0, 0.15, 0.25, 0.35, 0.45, and 0.55. (b) Comparison between the onset strains of crystallization and the critical strain for fringed-micelle nuclei under different temperatures. Nie et al. [59]. Reproduced with permission of Elsevier.

X-ray scattering (SAXS) signal before wide-angle X-ray diffraction (WAXD) during isothermal crystallization at low temperatures [62]. However, the X-ray evidence for the pre-ordered structure could be attributed to improper experiment treatment or limited instrument sensitivity. Howard Wang thought that a signal like that of spinodal decomposition could be expected as a result of improper overreduction in the empty correction [63]. Zhigang Wang et al. attributed the phenomenon to a limited instrument sensitivity of WAXD for the small number of crystallites in the early stage of crystallization [64]. Indeed, after improving the sensitivity of WAXD by four magnitudes, the difference between SAXS and WAXD disappears at low temperatures as observed by Heeley et al. [65]. The remaining difference at high temperatures can be associated with heterogeneous nucleation rather than homogeneous nucleation. Sirota [66, 67] observed a mesophase of chain cluster at the early stage of nucleation of long alkane chains and there is almost no supercooling for nucleation, although the mesophase disappears as soon as chain folding starts in the crystallization of long-enough chains [68]. Olmsted and coworkers [69] thought spinodal decomposition was a result of coupling between orientational-order fluctuations and density fluctuations at low temperatures. The spinodal decomposition will enhance crystal nucleation at a certain supercooling. In molecular dynamics simulations, Gee et al. [70] observed that the crystallization behaviors of PVDF under 600K and PE under 450K in a time scale of nanoseconds appear as spinodal decomposition. Milner calculated the free energy change of PE crystal nucleation through a rotated mesophase and it is lower than the surface free energy of orthogonal crystalline phase. He thought this could be the free energy barrier for crystal nucleation [71].

13.4 CRYSTAL GROWTH

Crystal growth starts after primary crystal nucleation, which can be controlled by either long-range diffusion or interface process. If crystal grows under a large supercooling in dilute polymer solutions or ultrathin films, the growth rate is mainly controlled by a long-distance diffusion from the far-away bulk solution to the crystal growth front. The diffusion-controlled mechanism means that the crystal size is not linearly dependent on the growth time [72]. In the commonly practical cases, the linear crystal growth rate is independent of time, and crystal growth is controlled by the process at the advancing surface of the crystal, referred to be the interface-controlled mechanism. In experiments, for example, by small-angle laser scattering (SALS) on small crystallites, or by polarized light microscope (PLM) on large crystallites, it is found that the growth rate of lamellar crystals is independent of time. This behavior implies the surface-controlled mechanism for crystal growth. The interface-controlled mechanism can be further separated into three categories, which are, secondary nucleation growth, screw dislocation growth, and surface roughing growth. The secondary nucleation prevails in the description of the kinetics of lamellar polymer crystal growth [73–76].

Thickening at the growth front is also observed in many experiments [77–79]. So the crystal growth process can be treated as two steps at the wedge-shaped growth front, secondary nucleation occurs first, followed by instant thickening until thickness becomes larger than the minimum thickness for further growth of lamellar crystals. Secondary nucleation dominates the temperature dependence of the growth rate, and thickening provides the driving force for crystal growth. The linear crystal growth rate of lamella by side-surface advancing can be treated as the competition result between advancing rate and melting rate of the growth front. Taking secondary nucleation as the rate-determining steps for both crystal growth and melting, they need to overcome the same nucleation energy barriers from mutually opposite directions. It is the difference of the two energy barriers ΔG that determines the linear growth rate, as derived by

$$v = v_{growth} - v_{melting} = v_{growth}\left(1 - \frac{v_{melting}}{v_{growth}}\right) = v_{growth}\left(1 - \exp\left(-\frac{\Delta G}{kT_c}\right)\right) \quad (13.35)$$

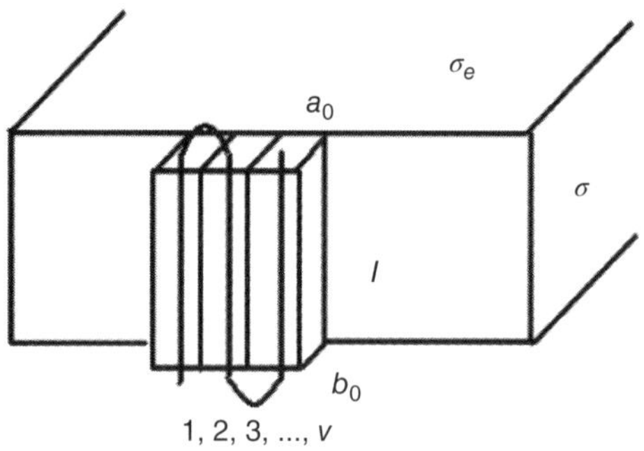

Figure 13.14 Schematic picture of folded stems at the growth front considered in the LH theory.

Assuming ΔG is very small, and replacing the exponential term by the first two terms of Maclaurin expansion ($\exp(x) = 1 + x + x^2/2 + x^3/6 + \cdots + x^i/i! + \cdots$), we can get the linear growth rate as

$$v \approx v_{growth}\frac{\Delta G}{kT_c} = v_{growth}(l - l_{\min})\frac{b^2\Delta g}{kT_c} \tag{13.36}$$

In Equation 13.36, the term $l - l_{min}$ determines the net free energy for crystal growth, b is the average distance between stems inside the crystal and Δg is the free energy change for melting of unit volume. Thus, we can treat v_{growth} in Equation 13.36 as the free energy barrier for crystal growth and the rest part as the driving force for crystal growth [80]. Under low temperatures, $l > l_{min}$ and the crystal will grow, while under high temperatures, $l < l_{min}$ and the crystal will melt. There are different models in the detailed description of the free energy barrier for secondary nucleation, as introduced below.

Lauritzen–Hoffman (LH) theory is the most widely used theoretical model in the explanation of the growth kinetics of lamellar crystals [81–83]. The following picture schematically shows the basic assumptions of folded stems at the growth front made in the LH theory (Fig. 13.14).

The LH theory holds four basic assumptions as listed below.

1. The growth front of polymer crystal is smooth. Secondary nucleation begins with a first stem deposited at the growth front, and follows with lateral spreading until reaching the lateral edges of the front substrate.
2. The chain-folded length l is constant during crystal growth. The width, thickness, and number of stems is a_0, b_0, and v.
3. The number of stems N_v is in a steady-state distribution.
4. Each stem should go through an activation state before entering the crystal lattice and the fraction of stems successfully entering the crystal lattice is ϕ.

Thus, the average chain-folding length of all the crystals could be derived as

$$\langle l\rangle = \frac{2\sigma_e}{\Delta g} + \frac{kT}{2b_0\sigma} \times \frac{2 + (1 - 2\phi)a_0\Delta g/(2\sigma)}{[1 - \phi a_0\Delta g/(2\sigma)][1 + (1 - \phi)a_0\Delta g/(2\sigma)]} \tag{13.37}$$

where $2\sigma_e/\Delta g$ is the minimum length for steady growth. Equation 13.37 is used to predict the average thickness of lamellar crystals grown at different temperatures.

Besides an explanation of chain-folding lengths, the phenomenon of regime transitions can also be explained on the basis of the LH theory, although crystal thickening during crystal growth was omitted in the LH theory [84]. If crystal growth is controlled by secondary nucleation, the temperature dependence of crystal growth rates is dominated by

$$G = G_0 \exp\left(-\frac{U}{T - T_0}\right)\exp\left(-\frac{K_g}{T\Delta T}\right) \tag{13.38}$$

The first exponential term is attributed to short-range diffusion across the interface and the second term is attributed to secondary nucleation with the nucleation barrier proportional to the inverse supercooling. When temperature is high enough, crystal growth rate becomes mainly controlled by secondary nucleation. In this case, the curve of $\lg G + U/(T - T_0)$ versus $-1/T/\Delta T$ can be divided into three linear regimes with the lowering of temperature, and the ratios of K_g among three regimes are

$$K_g(\mathrm{I}) : K_g(\mathrm{II}) : K_g(\mathrm{III}) = 2 : 1 : 2 \tag{13.39}$$

The above ratio is the so-called regime-transition phenomenon, evidenced by many experimental observations as well as molecular simulations.

At very high temperatures, the fraction of high molecular weights will crystallize first. In order to explain the molecular weight effect in crystallization of PE and other polymers, Wunderlich and Mehta put forward the concept of molecular nucleation [85–88], as illustrated in Figure 13.15, and proposed that each molecule entered the crystal with an additional nucleation barrier. Chain-folded secondary nucleus formed by molecules with enough lengths can only be stabilized over the critical size, while short-chain nucleus will be melted again. The concept of molecular nucleation can be regarded as a patch on the LH model.

The intramolecular nucleation model was then developed by supposing that all the secondary nucleation is mainly controlled by nucleation inside single-chain systems [58, 90] and the basic crystal lamella is resulted due to the preference of chain folding in this unique style of secondary nucleation. Assuming secondary nucleation of a single chain at the two-dimensional locally smooth growth front, the free energy change based on the classical nucleation theory is

$$\Delta F = \Delta f n + \sigma(N - n)^{1/2} \tag{13.40}$$

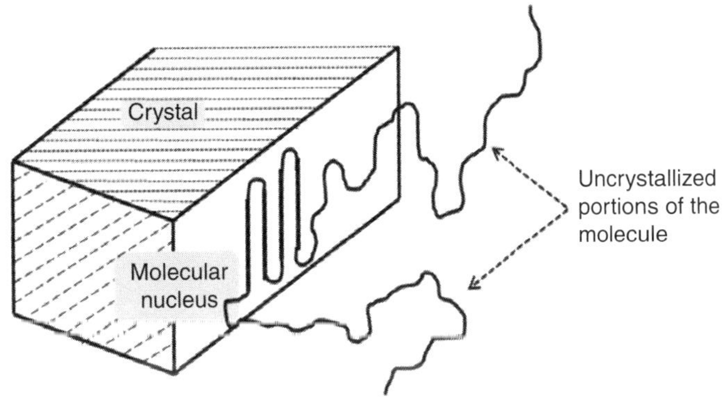

Figure 13.15 Schematic picture of molecular nucleation. Adapted from Wunderlich [89].

The free energy barrier for crystal nucleation is

$$\Delta F_c = \frac{\sigma^2}{4\Delta f} \tag{13.41}$$

The free energy barrier for critical intramolecular nucleation under a certain temperature is independent of chain length, but not for the opposite direction for melting. Therefore, a critical molecular length exists for the equilibrium intramolecular melting at the growth front under each crystallization temperature. For a polymer sample featured with a high polydispersity of molecular weights, only the fractions of chain lengths larger than the critical length can be stable and thus enter the crystal during secondary nucleation. This is the reason why molecular segregation occurs upon lamellar crystal growth. Molecular segregation can be observed only in polymer crystallization at very high temperatures. Besides molecular segregation [91], many other phenomena unique to polymer crystal growth also favor the intramolecular nucleation model, such as cocrystallization of long and short chains [92] as well as the interpretation of regime transitions [93]. Under a low temperature, the critical chain length for secondary nucleation may be much smaller than the chain length, and several events of intramolecular nucleation happen along the same chains [94]. If they occur in the same lamellar crystal, loops are formed; and if they occur in different lamellar crystals, tie molecules are formed. The intramolecular nucleation model allows a statistical treatment on the semicrystalline texture.

Under pressures higher than 3 kbar, Wunderlich and Arakawa observed the extended-chain PE crystals with the crystallinity almost 100% [95,96]. The wedge-shaped growth front and a thickening growth mechanism under high pressures were proposed [50]. Molecular nucleation or secondary nucleation first took place at the growth front, and later on developed into extended-chain crystals (ECCs) by fast thickening. By observing the growth process of PE folded-chain crystal (FCC) and ECC, Hikosaka developed the growth mechanism with chain-sliding diffusion for thickening, on the basis of the LH theory [97,98]. In his equation of nucleation rate, the free energy barrier for short-range diffusion across the interface was also considered besides the free energy barrier of critical nucleus. The growth appears as two-dimensional, which holds both lateral and longitudinal growth of the chain stems. The two-dimensional nucleation growth mechanism can be used to explain the dependence of lamella thickness on supercooling near the triple point of high pressures [99].

The (200) growth front of PE single crystals becomes curvature when crystallization temperatures are very high, which could not be explained by the LH theory based on secondary nucleation growth on smooth surface. Also, a pair of concave (110) surfaces was observed in the twin single crystal, which means the free energy barrier from side surfaces may not be the main problem for the advancing of the growth front [100]. Based on these observations, Sadler and Gilmer (SG) proposed the row model of continuous growth along the direction perpendicular to the growth front [101], as illustrated in Figure 13.16.

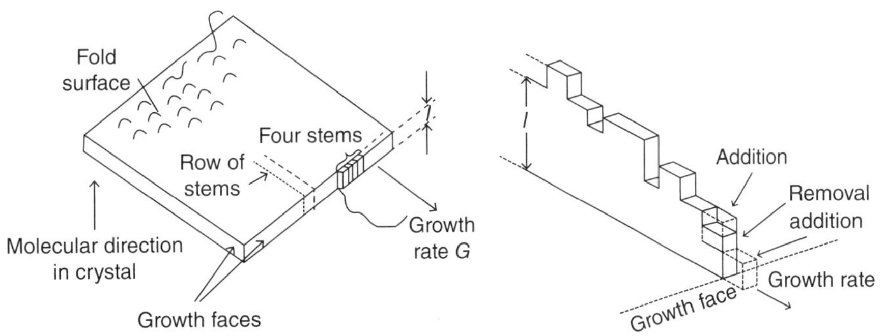

Figure 13.16 Schematic pictures of the row model.Sadler and Gilmer [101]. Reprinted with permission of American Physical Society.

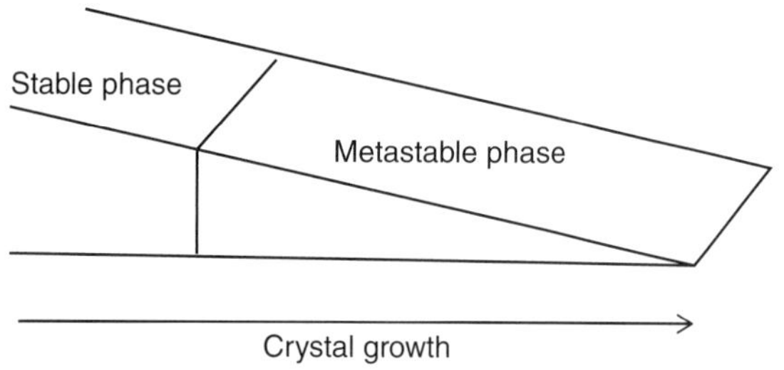

Figure 13.17 Schematic picture of wedge-shaped growth front of PE lamella.

In both LH and SG models, the growth front was supposed to directly reach the critical thickness in the growth process, and the thickening after growth was neglected. Keller and coworkers [18, 102] proposed a wedge-shaped growth-front model considering an obvious thickening in the crystal growth process of PE, as illustrated in Figure 13.17.

In the wedge-shaped growth model, the melt may first grow into a mesophase (hexagonal phase) in the thinnest region of the growth front. The mesophase is stable because of the large specific surface energy of small crystallites, which is referred as the finite-size effect. Then the thin lamella thickens into the stable orthogonal phase, which decides the lamellar thickness. There is a triple point Q, as illustrated in Figure 13.18. If temperature is above the triple point temperature, the melt will grow into the orthogonal phase directly. If temperature is below the triple point, there will be mesophase.

Keller's model was later on expanded by Strobl [103–106] to general polymers, in order to explain the experimental observations of his group, as illustrated in Figure 13.19. There is a linear relationship between the crystallization temperatures and the inverse lamella thicknesses, which is quite in accordance with the Gibbs–Thomson equation. There is also a linear relationship between the melting temperatures and the inverse lamella thicknesses. Crossover of these two linear curves is considered to be the triple point of mesophase transition.

Allegra [107–110] proposed statistical thermodynamic theory for the mesophase of small crystallites or crystal cluster in the metastable disordered phase before crystallization. The cluster first grows into a stable size and then joins into the crystal, as illustrated in Figure 13.20. The thickness of the lamella is decided by the cluster size. Zhang and Muthukumar [111] performed simulations of clusters to form single crystals grown in dilute solutions, consistent with the experimental observations. Muthukumar and coworkers [112, 113] then gave a thermodynamic explanation to the lamella thickness in detail. He thought the finite lamella thickness was the result of the largest thermodynamic stability of small crystallites. The crystalline chain will find the folded length related to the minimum free energy of the whole system, though there is a free energy barrier for the thickening of integer folding.

13.5 CRYSTAL ANNEALING

Crystal growth of polymers at the growth front often brings a lot of defects besides the limited folding length. Annealing relaxes the inner stress and removes defects to make the crystal more stable. Crystal thickening of lamellar crystals happens as well for more stable crystals. If the annealing temperature is very high, even higher than the melting point of metastable crystals, crystals may melt and recrystallize into a more stable state unless the temperature is close to the equilibrium melting point of infinitely large crystals. Therefore, there are two mechanisms of lamellar thickening: one is solid-chain sliding-diffusion mechanism and the other is melting-recrystallizaiton mechanism.

Peterlin [114] proposed an activation energy barrier for sliding diffusion of folded chains in monolayer crystals, to explain why the folded length of polymer chains increases linearly with the logarithm of time. The sliding-diffusion mechanism was then developed into a more general theory by Sanchez and his collaborators [115, 116]. Dreyfus and Keller proposed the fold-dislocation thickening model, in which the lamellar thickness can be doubly increased, as schematically demonstrated in Figure 13.21.

The phenomenon that lamella thickness increases with the logarithmic time has been observed in many experiments

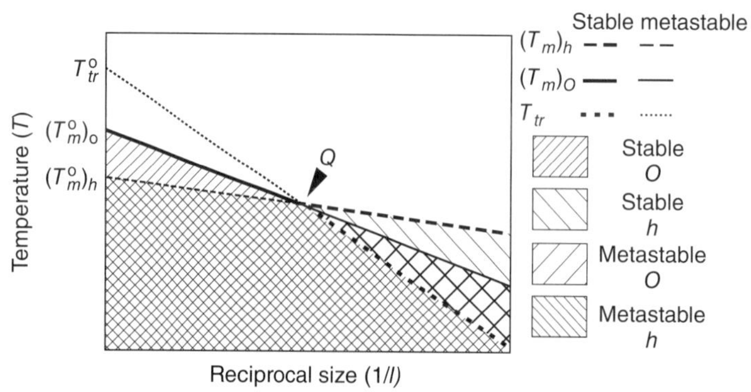

Figure 13.18 Schematic picture showing temperature against reciprocal size of polyethylene. Subscripts refer to phases (o, orthorhombic; h, hexagonal) or phase transition (tr, transition from hexagonal phase to orthorhombic phase) of polyethylene. Keller et al. [102]. Reproduced with permission of The Royal Society.

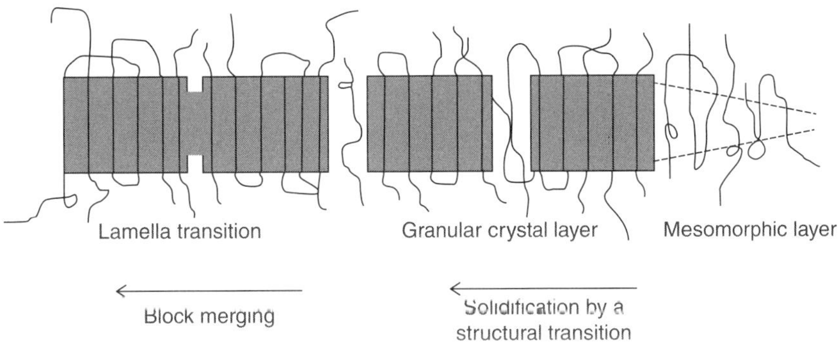

Figure 13.19 Schematic picture of mesophase at the growth front in lamella. Strobl [106]. Reproduced with permission of Springer.

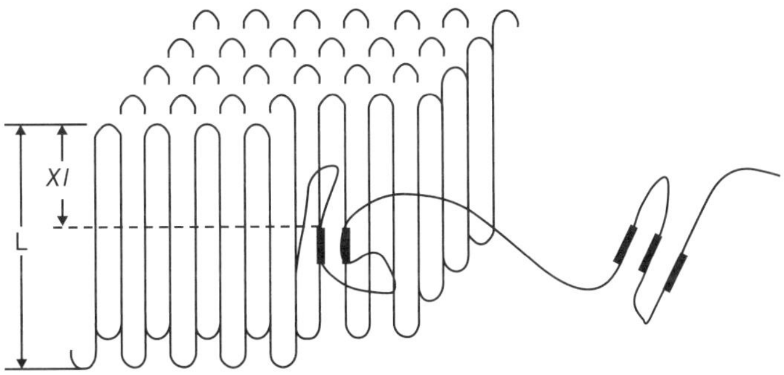

Figure 13.20 Schematic picture of equilibrium-sized cluster growth model. Allegra and Meille [110]. Reproduced with permission of AIP Publishing LLC.

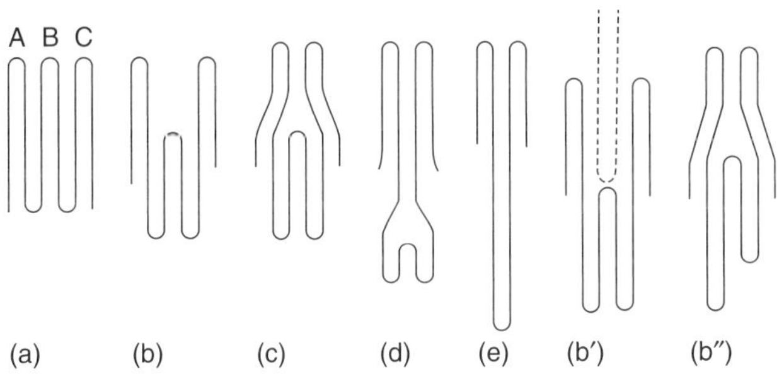

Figure 13.21 Schematic illustration of the fold-dislocation thickening model [117]. (a)–(e) shows the thickening process of folded chain in a lamellar crystal. The space generated in this process should be filled by other stems (b′) or be discharged by merging of stems (b″).

[77, 118]. The logarithmic-time dependence of crystal thickness can be deduced by assuming a frictional barrier (ΔE_s) for chain-sliding diffusion proportional to the lamella thickness (l); thus, the thickening rate of monolayer lamellar crystal under a certain temperature is

$$\frac{dl}{dt} \propto be^{-al/k_bT} \tag{13.42}$$

Equation 13.42 can be solved by

$$l = c \ \ln t + d \tag{13.43}$$

In the above equations, a, b, c, and d are the proportional coefficients. This continuous thickening of mobile high-molecular-weight polymers was observed as well by Monte Carlo simulations [119].

The melting-recrystallization mechanism was first reviewed by Fisher [120] and then introduced by Wunderlich [50] in his famous book. This mechanism was confirmed by several experimental phenomena, such as decrease-then-increase of crystallinity in the annealing process [121]. Also, when PE single crystal is annealed at a temperature higher than 130 °C, an extra peak occurs in its DSC melting curve and the peak area decreases with longer annealing time. This observation means the thinner lamellae melted and meanwhile the thicker lamellae generated.

13.6 SUMMARY

We summarized current thermodynamics and kinetics understanding of polymer crystallization. There are still many controversial arguments, in particular, on the kinetics of polymer crystallization. We first introduced basic thermodynamic concepts, including the melting point, the phase diagrams of polymer solutions, the metastable crystal structures, and the mesophase. Those molecular factors governing the melting points were discussed. We then introduced crystal nucleation and crystal growth. The classical nucleation theory as well as recently developed intramolecular nucleation model can be regarded as an updated version of secondary-nucleation models beyond the LH model. Some other models based on nonnucleation ideas were also introduced. We also introduced crystal annealing and its kinetics for crystal thickening via chain-sliding diffusion. We hope that our survey can stimulate further fundamental research on polymer crystallization, to gain a better understanding of the complicated crystal morphologies.

REFERENCES

1. Kuhn W. Über die gestalt fadenförmiger moleküle in lösungen. Kolloid Z Z Polym 1934;68:2–15.
2. Guth E, Mark H. Zur innermolekularen statistik, insbesondere bei Ketten-molekiilen I. Monatsh Chem 1934;65:93–121.
3. Flory PJ. *Principles of Polymer Chemistry*. Ithaca, NY: Cornell University Press; 1953.
4. Cheng SZ. *Phase Transitions in Polymers: The Role of Metastable States*. Oxford: Elsevier Science Ltd; 2008.
5. Hermann K, Gerngross O, Abitz W. Zur rontgenographischen strukturforschung des gelatinemicells. Z Phys Chem B 1930;10:371–394.
6. Keller A. A note on single crystals in polymers: Evidence for a folded chain configuration. Philos Mag 1957;2:1171–1175.
7. Flory PJ. On the morphology of the crystalline state in polymers. J Am Chem Soc 1962;84:2857–2867.
8. Hoffman JD. Regime III crystallization in melt-crystallized polymers: The variable cluster model of chain folding. Polymer 1983;24:3–26.
9. Hermann K, Gerngross O. Die elastizitat des kautschuks. Kautschuk 1932;8:181.
10. Spells SJ, Sadler DM, Keller A. Chain trajectory in solution grown polyethylene crystals: correlation between infra-red spectroscopy and small-angle neutron scattering. Polymer 1980;21:1121–1128.
11. Arlif J, Spegt P, Skoulios A. Etude de la cristallisation des polymères I. Structure lamellaire de polyoxyéthylènes de faible masse moléculaire. Makromol Chem 1966;99:160–174.
12. Arlie PJ, Spegt P, Skoulios A. Etude de la cristallisation des polymères. II. Structure lamellaire et repliement des chaines du polyoxyéthylène. Makromol Chem 1967;104:212–229.
13. Ungar G, Stejny J, Keller A, Bidd I, Whiting M. The crystallization of ultralong normal paraffins: The onset of chain folding. Science 1985;229:386–389.
14. Organ SJ, Keller A. The onset of chain folding in ultralong n-alkanes: An electron microscopic study of solution-grown crystals. J Polym Sci B 1987;25:2409–2430.
15. Mandelkern L. *Crystallization of Polymers*. New York: McGraw-Hill; 1964.
16. Wunderlich B. Reversible crystallization and the rigid–amorphous phase in semicrystalline macromolecules. Prog Polym Sci 2003;28:383–450.
17. Hoffman JD, Weeks JJ. X-ray study of isothermal thickening of lamellae in bulk polyethylene at the crystallization temperature. J Chem Phys 1965;42:4301–4302.
18. Keller A. Morphology of polymers. Pure Appl Chem 1992;64:193–204.
19. Onsager L. The effects of shape on the interaction of colloidal particles. Ann N Y Acad Sci 1949;51:627–659.
20. Maier W, Saupe A. A simple molecular statistical theory of the nematic crystalline-liquid phase. Z Naturf 1959;14:882–889.
21. Jähnigb F. Molecular theory of lipid membrane order. J Chem Phys 1979;70:3279–3290.
22. Ronca G, Yoon D. Theory of nematic systems of semiflexible polymers. I. High molecular weight limit. J Chem Phys 1982;76:3295–3299.
23. Ronca G, Yoon D. Theory of nematic systems of semiflexible polymers. II. Chains of finite length in the bulk. J Chem Phys 1984;80:925–929.
24. Bosch T, Maissa AP, Sixou P. Molecular model for nematic polymers in liquid crystal solvents. J Chem Phys 1983;79:3462–3466.
25. Bosch T, Maissa A, Sixou P. A Landau-de Gennes theory of nematic polymers. J Phys Lett 1983;44:105–111.
26. Khokhlov A, Semenov A. On the theory of liquid-crystalline ordering of polymer chains with limited flexibility. J Stat Phys 1985;38:161–182.
27. Gupta AM, Edwards SF. Mean-field theory of phase transitions in liquid-crystalline polymers. J Chem Phys 1993;98:1588–1596.
28. Lekkerkerker H, Vroege G. Lyotropic colloidal and macromolecular liquid crystals. Philos Trans R Soc London A 1993;344:419–440.
29. Prigogine I. *The Molecular Theory of Solution*. Amsterdam: North-Holland Publishing Co.; 1957.
30. Chang TS. Statistical theory of absorption of double molecules. Proc R Soc London A 1939;169:512–531.
31. Meyer K. The compound entropies of systems with long-chain compounds and their statistical explanation. Z Phys Chem Abt B 1939;44:383–391.
32. Flory PJ. Thermodynamics of high polymer solutions. J Chem Phys 1941;9:660.
33. Flory PJ. Thermodynamics of high polymer solutions. J Chem Phys 1942;10:51–61.
34. Huggins ML. Thermodynamic properties of solutions of long-chain compounds. Ann N Y Acad Sci 1942;43:1–32.
35. Flory P-J. Statistical thermodynamics of semi-flexible chain molecules. Proc R Soc London Ser A 1956;234:60–73.
36. Hu W. Chain folding in polymer melt crystallization studied by dynamic Monte Carlo simulations. J Chem Phys 2001;115:4395–4401.
37. Hu W, Frenkel D. Polymer crystallization driven by anisotropic interactions. Adv Polym Sci 2005;191:1–35.
38. Hu W. Structural transformation in the collapse transition of the single flexible homopolymer model. J Chem Phys 1998;109:3686–3690.
39. Metropolis N, Rosenbluth AW, Rosenbluth MN, Teller AH, Teller E. Equation of state calculations by fast computing machines. J Chem Phys 1953;21:1087–1092.
40. Flory P, Vrij A. Melting points of linear-chain homologs. The normal paraffin hydrocarbons. J Am Chem Soc 1963;85:3548–3553.
41. Hu W, Frenkel D, Mathot VB. Lattice-model study of the thermodynamic interplay of polymer crystallization and liquid–liquid demixing. J Chem Phys 2003;118: 10343–10348.

42. Hu W, Frenkel D. Effect of metastable liquid–liquid demixing on the morphology of nucleated polymer crystals. Macromolecules 2004;37:4336–4338.
43. Ma Y, Hu W, Wang H. Polymer immiscibility enhanced by thermal fluctuations toward crystalline order. Phys Rev E 2007;76:031801.
44. Ma Y, Zha L, Hu W, Reiter G, Han CC. Crystal nucleation enhanced at the diffuse interface of immiscible polymer blends. Phys Rev E 2008;77:061801.
45. Hu W. *Polymer Physics: A Molecular Approach*. Wien: Springer-Verlag; 2012.
46. Prasad A, Mandelkern L. Equilibrium dissolution temperature of low molecular weight polyethylene fractions in dilute solution. Macromolecules 1989;22:914–920.
47. Flory PJ. Theory of crystallization in copolymers. Trans Faraday Soc 1955;51:848–857.
48. Sanchez I, Eby R. Thermodynamics and crystallization of random copolymers. Macromolecules 1975;8:638–641.
49. Kelton K. Crystal nucleation in liquids and glasses. Solid State Phys 1991;45:75–177.
50. Wunderlich B. *Macromolecular Physics. Vol 2. Crystal Nucleation*. New York: Academic Press; 1976.
51. Hoffman JD, Miller RL. Kinetic of crystallization from the melt and chain folding in polyethylene fractions revisited: Theory and experiment. Polymer 1997;38:3151–3212.
52. Zachmann H. Der einfluss der konfigurationsentropie auf das kristallisations und schmelzverhalten von hochpolymeren stoffen. Kolloid Z Z Polym 1967;216:180–191.
53. Zachmann H. Statistische thermodynamik des kristallisierens und schmelzens von hochpolymeren stoffen. Colloid Polym Sci 1969;231:504–534.
54. Blundell D, Keller A, Kovacs A. A new self-nucleation phenomenon and its application to the growing of polymer crystals from solution. J Polym Sci B 1966;4:481–486.
55. Volmer M, Weber A. Nucleus formation in supersaturated systems. Z Phys Chem 1926;119:277–301.
56. Becker R, Döring W. Kinetische behandlung der keimbildung in übersättigten dämpfen. Ann Phys 1935;416: 719–752.
57. Turnbull D, Fisher JC. Rate of nucleation in condensed systems. J Chem Phys 1949;17:71–73.
58. Hu W, Frenkel D, Mathot VB. Intramolecular nucleation model for polymer crystallization. Macromolecules 2003;36:8178–8183.
59. Nie Y, Gao H, Ma Y, Hu Z, Reiter G, Hu W. Competition of crystal nucleation to fabricate the oriented semi-crystalline polymers. Polymer 2013;54:3402–3407.
60. Imai M, Kaji K, Kanaya T. Orientation fluctuations of poly (ethylene terephthalate) during the induction period of crystallization. Phys Rev Lett 1993;71:4162–4165.
61. Imai M, Kaji K, Kanaya T. Structural formation of poly (ethylene terephthalate) during the induction period of crystallization. 3. Evolution of density fluctuations to lamellar crystal. Macromolecules 1994;27:7103–7108.
62. Terrill NJ, Fairclough PA, Towns-Andrews E, Komanschek BU, Young RJ, Ryan AJ. Density fluctuations: The nucleation event in isotactic polypropylene crystallization. Polymer 1998;39:2381–2385.
63. Wang H. SANS study of the early stages of crystallization in polyethylene solutions. Polymer 2006;47:4897–4900.
64. Wang Z-G, Hsiao BS, Sirota EB, Agarwal P, Srinivas S. Probing the early stages of melt crystallization in polypropylene by simultaneous small-and wide-angle X-ray scattering and laser light scattering. Macromolecules 2000;33:978–989.
65. Heeley E, Maidens A, Olmsted P, Bras W, Dolbnya I, Fairclough J, Terrill N, Ryan A. Early stages of crystallization in isotactic polypropylene. Macromolecules 2003;36:3656–3665.
66. Sirota E, Herhold A. Transient phase-induced nucleation. Science 1999;283:529–532.
67. Kraack H, Sirota E, Deutsch M. Measurements of homogeneous nucleation in normal-alkanes. J Chem Phys 2000;112:6873–6885.
68. Sirota E. Supercooling and transient phase induced nucleation in n-alkane solutions. J Chem Phys 2000;112:492–500.
69. Olmsted PD, Poon WC, McLeish T, Terrill N, Ryan A. Spinodal-assisted crystallization in polymer melts. Phys Rev Lett 1998;81:373–376.
70. Gee RH, Lacevic N, Fried LE. Atomistic simulations of spinodal phase separation preceding polymer crystallization. Nat Mater 2005;5:39–43.
71. Milner ST. Polymer crystal-melt interfaces and nucleation in polyethylene. Soft Matter 2011;7:2909–2917.
72. Holland V, Lindenmeyer P. Morphology and crystal growth rate of polyethylene crystalline complexes. J Polym Sci 1962;57:589–608.
73. Flory P, McIntyre A. Mechanism of crystallization in polymers. J Polym Sci 1955;18:592–594.
74. Burnett BB, McDevit W. Kinetics of spherulite growth in high polymers. J Appl Phys 1957;28:1101–1105.
75. Wunderlich B, Cormier C. Seeding of supercooled polyethylene with extended chain crystals. J Phys Chem 1966;70:1844–1849.
76. Wunderlich B, Melillo L, Cormier C, Davidson T, Snyder G. Surface melting and crystallization of polyethylene. J Macromol Sci B 1967;1:485–516.
77. Wunderlich B, Mielillo L. Morphology and growth of extended chain crystals of polyethylene. Die Makromol Chem 1968;118:250–264.
78. Abo el Maaty MI, Bassett MD. Evidence for isothermal lamellar thickening at and behind the growth front as polyethylene crystallizes from the melt. Polymer 2005;46:8682–8688.
79. Mullin N, Hobbs JK. Direct imaging of polyethylene films at single-chain resolution with torsional tapping atomic force microscopy. Phys Rev Lett 2011;107:197801.
80. Ren Y, Ma A, Li J, Jiang X, Ma Y, Toda A, Hu W. Melting of polymer single crystals studied by dynamic Monte Carlo simulations. Euro Phys J E 2010;33:189–202.
81. Lauritzen JI, Hoffman JD. Theory of formation of polymer crystals with folded chains in dilute solution. J Res Natl Bur Stand A 1960;64:73–102.
82. Hoffman JD, Lauritzen J. Crystallization of bulk polymers with chain folding: theory of growth of lamellar spherulites. J Res Natl Bur Stand A 1961;65:297–336.
83. Hoffman JD, Davis GT, Lauritzen JI Jr. The Rate of Crystallization of Linear Polymers with Chain Folding. In: Hannay NB, editor. *Treatise on Solid State Chemistry*. Vol. 3. New York: Springer; 1976.

84. Armistead K, Goldbeck-Wood G. Polymer crystallization theories. Adv Polym Sci 1992;100:219–312.
85. Wunderlich B, Mehta A. Macromolecular nucleation. J Polym Sci 1974;12:255–263.
86. Mehta A, Wunderlich B. A study of molecular fractionation during the crystallization of polymers. Colloid Polym Sci 1975;253:193–205.
87. Wunderlich B. Molecular nucleation and segregation. Faraday Discuss Chem Soc 1979;68:239–243.
88. Cheng SZ, Wunderlich B. Molecular segregation and nucleation of poly (ethylene oxide) crystallized from the melt. I. Calorimetric study. J Polym Sci B 1986;24:577–594.
89. Wunderlich B. *Thermal Analysis of Polymeric Materials*. Berlin: Springer; 2005.
90. Hu W. Intramolecular crystal nucleation. In: Reiter G, Strobl GR, editors. *Lecture Notes Physics: Progress in Understanding of Polymer Crystallization*. Springer-Verlag; 2007. p 47–63.
91. Hu W. Molecular segregation in polymer melt crystallization: Simulation evidence and unified-scheme interpretation. Macromolecules 2005;38:8712–8718.
92. Cai T, Ma Y, Yin P, Hu W. Understanding the growth rates of polymer co-crystallization in the binary mixtures of different chain lengths. J Phys Chem B 2008;112:7370–7376.
93. Hu W, Cai T. Regime transitions of polymer crystal growth rates: Molecular simulations and interpretation beyond Lauritzen–Hoffman model. Macromolecules 2008;41:2049–2061.
94. Hu W, Frenkel D, Mathot VB. Sectorization of a lamellar polymer crystal studied by dynamic Monte Carlo simulations. Macromolecules 2003;36:549–552.
95. Wunderlich B, Arakawa T. Polyethylene crystallized from the melt under elevated pressure. J Polym Sci A 1964;2: 3697–3706.
96. Geil PH, Anderson FR, Wunderlich B, Arakawa T. Morphology of polyethylene crystallized from the melt under pressure. J Polym Sci A 1964;2:3707–3720.
97. Hikosaka M. Unified theory of nucleation of folded-chain crystals and extended-chain crystals of linear-chain polymers. Polymer 1987;28:1257–1264.
98. Hikosaka M. Unified theory of nucleation of folded-chain crystals (FCCs) and extended-chain crystals (ECCs) of linear-chain polymers: 2. Origin of FCC and ECC. Polymer 1990;31:458–468.
99. Hikosaka M, Okada H, Toda A, Rastogi S, Keller A. Dependence of the lamellar thickness of an extended-chain single crystal of polyethylene on the degree of supercooling and the pressure. J Chem Soc Faraday Trans 1995;91:2573–2579.
100. Sadler DM, Barber M, Lark G, Hill MJ. Twin morphology: 2. Measurements of the enhancement in growth due to re-entrant corners. Polymer 1986;27:25–33.
101. Sadler DM, Gilmer GH. Selection of lamellar thickness in polymer crystal growth: A rate-theory model. Phys Rev B 1988;38:5684–5693.
102. Keller A, Hikosaka M, Rastogi S, Toda A, Barham P, Goldbeck-Wood G, Windle A, Thomas E, Bassett D. The size factor in phase transitions: Its role in polymer crystal formation and wider implications [and discussion]. Philos Trans R Soc London Ser A 1994;348:3–17.
103. Strobl G. From the melt via mesomorphic and granular crystalline layers to lamellar crystallites: A major route followed in polymer crystallization? Eur Phys J E 2000;3:165–183.
104. Strobl G. A thermodynamic multiphase scheme treating polymer crystallization and melting. Eur Phys J E 2005;18:295–309.
105. Strobl G. Crystallization and melting of bulk polymers: New observations, conclusions and a thermodynamic scheme. Prog Polym Sci 2006;31:398–442.
106. Strobl G. Colloquium: Laws controlling crystallization and melting in bulk polymers. Rev Mod Phys 2009;81:1287–1300.
107. Allegra G. Chain folding and polymer crystallization: A statistical–mechanical approach. J Chem Phys 1977;66:5453–5463.
108. Allegra G. Polymer crystallization: The bundle model. Ferroelectrics 1980;30:195–211.
109. Allegra G, Meille SV. The bundle theory for polymer crystallisation. Phys Chem Chem Phys 1999;1:5179–5188.
110. Allegra G, Meille SV. Pre-crystalline, high-entropy aggregates: A role in polymer crystallization? Adv Polym Sci 2005;191:87–135.
111. Zhang J, Muthukumar M. Monte Carlo simulations of single crystals from polymer solutions. J Chem Phys 2007;126:234904–234921.
112. Welch P, Muthukumar M. Molecular mechanisms of polymer crystallization from solution. Phys Rev Lett 2001;87:218302.
113. Muthukumar M. Modeling polymer crystallization. Adv Polym Sci 2005;191:241–274.
114. Peterlin A. Thickening of polymer single crystals during annealing. J Polym Sci B 1963;1:279–284.
115. Sanchez I, Colson J, Eby R. Theory and observations of polymer crystal thickening. J Appl Phys 1973;44:4332–4339.
116. Sanchez I, Peterlin A, Eby R, McCrackin F. Theory of polymer crystal thickening during annealing. J Appl Phys 1974;45:4216–4219.
117. Dreyfus P, Keller A. A simple chain refolding scheme for the annealing behavior of polymer crystals. J Polym Sci B 1970;8:253–258.
118. Fischer E, Schmidt G. Long periods in drawn polyethylene. Angew Chem Int Ed Engl 1962;1:488–499.
119. Wang M, Gao H, Zha L, Chen E-Q, Hu W. Systematic kinetic analysis on monolayer lamellar crystal thickening via chain-sliding diffusion of polymers. Macromolecules 2012;46:164–171.
120. Fischer E. Zusammenhänge zwischen der Kolloidstruktur kristalliner Hochpolymerer und ihrem Schmelz-und Rekristallisationsverhalten. Kolloid Z Z Polym 1969;231:458–503.
121. Matsuoka S. The effect of pressure and temperature on the specific volume of polyethylene. J Polym Sci 1962;57:569–588.

14

SELF-ASSEMBLY AND MORPHOLOGY IN BLOCK COPOLYMER SYSTEMS WITH SPECIFIC INTERACTIONS

Anbazhagan Palanisamy and Qipeng Guo
Institute for Frontier Materials, Deakin University, Geelong, Australia

14.1 INTRODUCTION

Since Parkes patented the first polymer blend in 1846, numerous synthetic polymer blends with novel properties were commercialized. On the other hand, macroscopic phase separation between incompatible homopolymers due to low entropy of mixing emerged as a major hurdle in further development. Later on, block copolymers were used as macromolecular surfactants in compatibilizing homopolymers, which led to creation of materials with improved properties [1]. Further, nanostructured polymers were produced using block copolymers in higher concentration by exploiting their microphase separation of incompatible blocks. The bulk properties like thermal and mechanical properties of polymers hinge on the structural parameters and supramolecular organization of constituent block copolymers [2]. Hence it is necessary to control the self-organization, morphology, and phase behavior of block copolymers to obtain the desired property.

Self-assembly and morphology of block copolymers depend on their architecture and composition [3]. Several equilibrium phases like lamellae, gyroid, hexagonal-packed cylinders, and body-centered cubic phases were observed in melts. In thin films, microphase separation resulted in formation of lamellae, stripes, and circular domains. Various types of micellar structures and arrangements were seen in dilute solutions [4]. These phase behaviors were dictated by Flory–Huggins interaction parameter (χ), copolymer degree of polymerization (N), and composition (f) in melts and thin films. In addition to these parameters, amphiphilicity was the most important property of block copolymers enabling them to self-assemble into various structures in dilute solutions [3]. In all these instances, morphology change was achieved through synthesizing new monomers and polymers. Development and commercialization of new polymers require higher capital in terms of research and development and scale-up procedures. Polymer blending offers an alternate way of developing new materials with novel properties through mixing two or more homopolymers or block copolymers. The miscibility between corresponding blocks in a multicomponent block copolymer system is important to ensure production of miscible blends. Component miscibility in multicomponent systems was brought via adding a block-selective second component. For example, incorporation of parent homopolymer A into AB diblock copolymer can result in a binary blend. Here, the final property of blend relies on the interplay between macrophase and microphase separation of the blend components. Morphology and phase behavior of block copolymer multicomponent systems containing single block-selective polymers has been reviewed in depth by Spontak et al. [5].

In addition to conventional blending through noninteracting components, block copolymer systems with specific interactions have received attention in recent years [6]. Noncovalent interactions such as ionic bonding interaction, hydrogen bonding interaction, and metal complexation were exploited to increase compatibility between blend components and modulate morphology in bulk, as well as in solution. The traditional covalent bond building chemistry was replaced with noncovalent interactions to exploit its dynamic nature in building macromolecules [7]. Control of supramolecular ordering in block copolymer blends was envisaged through judicious selection of functional polymers that can interact via noncovalent bonds. Self-assembly and morphology of bock copolymer blends interacting via electrostatic interaction,

Polymer Morphology: Principles, Characterization, and Processing, First Edition. Edited by Qipeng Guo.

hydrogen bonding interaction and metal–ligand coordination bonds are discussed in this chapter.

Selective miscibility among blend components is essential for tuning the morphology that dictates end property of blends. Hence, miscibility is brought through choosing block copolymers that contain complementary functional groups that can interact via noncovalent interactions. These blending strategies are applied both in solid and solution state to develop functional materials with tuneable properties [8, 9].

14.2 BLOCK COPOLYMER SYSTEMS WITH HYDROGEN BONDING INTERACTION IN SOLID STATE

14.2.1 Diblock Copolymer/Homopolymer Systems

Self-assembly tuning coupled with improving material property was the key issue addressed in polymer blending studies via hydrogen bonding interaction. Although the blends of block copolymer with parent homopolymers provide tailored nanostructures, block copolymer/homopolymer blends through specific interactions between the components show more complex morphologies [10–13]. Zhao et al. [14] first studied the blends of polystyrene-*b*-poly(4-hydroxystyrene) (PS-*b*-PHOST) diblock copolymer with hydrogen bond accepting homopolymers poly(ethylene oxide) (PEO), poly(4-vinyl pyridine) (P4VP), and poly(*n*-butyl acrylate) (PnBA) and identified composition-dependant microphase separation. They also reported that homopolymer molecular weight effect is not important if there is sufficient interaction with the block copolymer for morphology development. In conventional polymer blending with no specific interaction within components, miscibility depends on the molecular weight of the homopolymer and corresponding block(s) of the block copolymer [15]. While in hydrogen-bonded polymer blends, miscibility is mediated through the hydrogen-bonding strength between the interacting functional groups. Painter–Coleman [16] association model (PCAM) was used to determine the interassociation equilibrium constant (K_A) of hydrogen-bonded systems. A typical interaction scheme can be depicted as follows,

$$\mathrm{A + B} \underset{}{\overset{K_A}{\rightleftharpoons}} \mathrm{AB} \quad \mathrm{A + A} \underset{}{\overset{K_B}{\rightleftharpoons}} \mathrm{AA}$$

where A and B denote the components that interact through hydrogen bonds. K_A and K_B are equilibrium constants of self-associating and interassociating components, respectively. An interplay between K_A, K_B and solvent medium decides the complex formation via hydrogen bonding [17]. Block copolymers with poly(2-vinyl pyridine) (P2VP) and P4VP blocks containing basic nitrogen that is capable of forming strong hydrogen bonds with carboxylic acid and phenolics have been investigated in recent years to discover new morphologies and properties [18–22]. Chen et al. [23] studied the effect of hydrogen-bonding strength on phase behavior of various block copolymer/homopolymer blends. Homopolymers P4VP, poly(methyl methacrylate) (PMMA), and poly(4-vinylphenol) (PVPh) show different hydrogen-bonding strength with PVPh block of the block copolymer PVPh-*b*-PS. The ratio of interassociation equilibrium constant (K_A) over self-association equilibrium constant (K_B) in hydrogen-bonded groups was used as a guide to estimate the hydrogen-bonding strength and respective phase behavior of blends. Order–order phase transition was observed in PVPh-*b*-PS/P4VP system, as observed in pure block copolymers, while order–disorder phase transition was witnessed in systems involving PMMA and PVPh homopolymers due to weak hydrogen-bonding interaction. The schematic representation is shown in Figure 14.1.

As evidence of strong hydrogen bonds $K_A/K_B \approx 9$, higher miscibility between PVPh and P4VP was observed. In contrast, $K_A/K_B \approx 0.6$ observed for blends involving PVPh and PMMA blocks showing weak interaction. Polymer blend miscibility without macrophase separation over a wide composition range was evidenced through strong hydrogen bonding between blend components. Dobrosielska et al. [24], for instance, have demonstrated that they could induce morphological transition in binary blends of PS-*b*-P2VP and PHOST homopolymer even at higher volume fractions of homopolymer. Macrophase separation was not observed in blends with homopolymer/P2VP ratio 4.1, while volume fraction of styrene was $\phi_s = 0.28$. Figure 14.2 shows various morphologies of PS-*b*-P2VP/PHOST blends. They also reported that highly miscible blends can be produced with homopolymers having molecular weight higher than the corresponding block of the block copolymer that forms hydrogen bonding. Nanophase-separated structures were reported for PS-*b*-P2VP/PHOST blend systems having extremely high homopolymer/P2VP ratio 28, with $\phi_s = 0.11$ [17].

Higher miscibility among block copolymer/homopolymer blends through hydrogen bonding makes the blend behave as a neat block copolymer that exhibits microphase separation into a variety of ordered morphologies [25]. Strong hydrogen bonding between PEO and poly(acrylic acid) (PAA) was exploited to tune lattice spacing and orientation of microdomains in thin films of PS-*b*-PEO/PAA system. PEO forms cylindrical microdomains in PS matrix, whereas homopolymer PAA is localized in the center of PEO microdomains. Morphology transition from cylindrical structures to lamellar morphology was evidenced from atomic force microscope (AFM) analysis shown in Figure 14.3 with an increase in the PAA content [26].

Long-range-ordered microphase separation was observed in block copolymer/homopolymer blends only if there was strong repulsion within block copolymer blocks and selective attraction with homopolymer. For example, in poly(ε-caprolactone-*b*-P4VP (PCL-*b*-P4VP)/PVPh blend, PVPh has attraction for both PCL and P4VP blocks, hence long-range ordering was absent. But after incorporating PS block, PCL-*b*-P4VP/PVPh-*b*-PS blends showed sharp and multiple peaks in SAXS analysis, indicating long-range

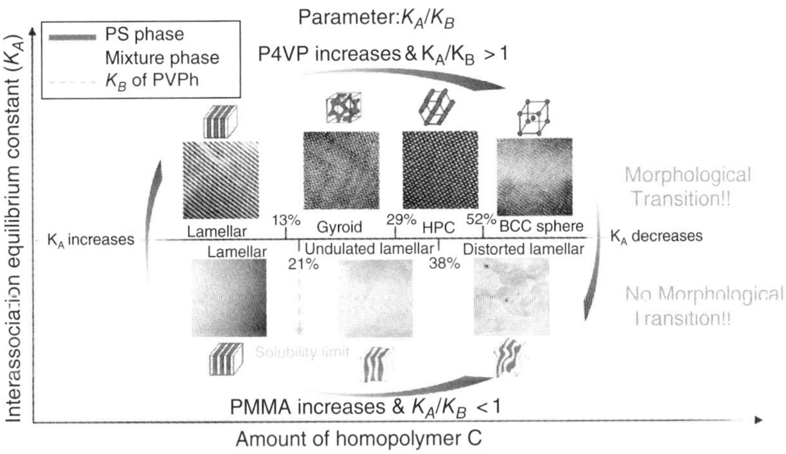

Figure 14.1 Schematic representation with TEM images of PVPh-*b*-PS/P4VP blend system showing phase transition from lamellar, gyroid, and hexagonally packed cylinder to body-centered cubic structures with increase in P4VP volume fractions. Undulated and distorted lamellar structures observed in PVPh-*b*-PS/PMMA blends due to weak hydrogen bonding between PMMA and PVPh. Chen et al. [23]. Reproduced with permission of American Chemical Society.

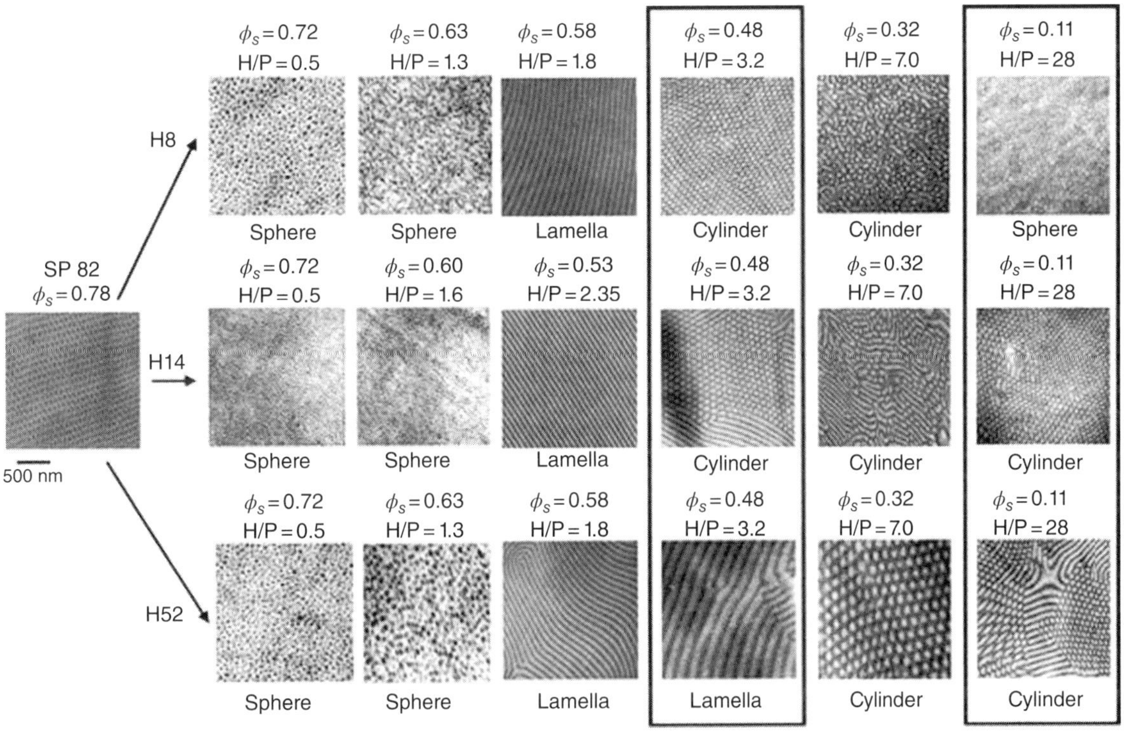

Figure 14.2 TEM images of PS-*b*-P2VP (SP)/poly(4-hydroxystyrene) (H) blends with different block copolymer/homopolymer ratios. Homopolymer H with three different molecular weights 8k, 14k, and 52k (H8, H14, H52) were blended with block copolymer SP (82k). Microphase-separated morphologies such as spheres, lamellae, and cylinders were observed as the amount of homopolymer increased. Dobrosielska et al. [24]. Reproduced with permission of American Chemical Society.

microphase separation [27]. Block copolymers that have attractive interaction within their bocks leads to homogeneous phase in bulk. Microphase separation was induced in these types of block copolymers by adding homopolymers having selective interaction with the block copolymer. But microphase separation was observed only in a narrow window of homopolymer weight fraction. For instance, blends of PEO-*b*-poly(N-vinylcarbazole) (PEO-*b*-PVK) and PAA show microphase separation only between 40 and 70 wt% of PAA. With sufficient homopolymer concentration, morphology transformation was observed, although long-range ordering was absent [28].

In contrast to selective interaction of homopolymer "C" with one of the blocks A or B of the immiscible block copolymer "A-*b*-B" discussed earlier, homopolymers simultaneously interacting with both blocks A and B of the

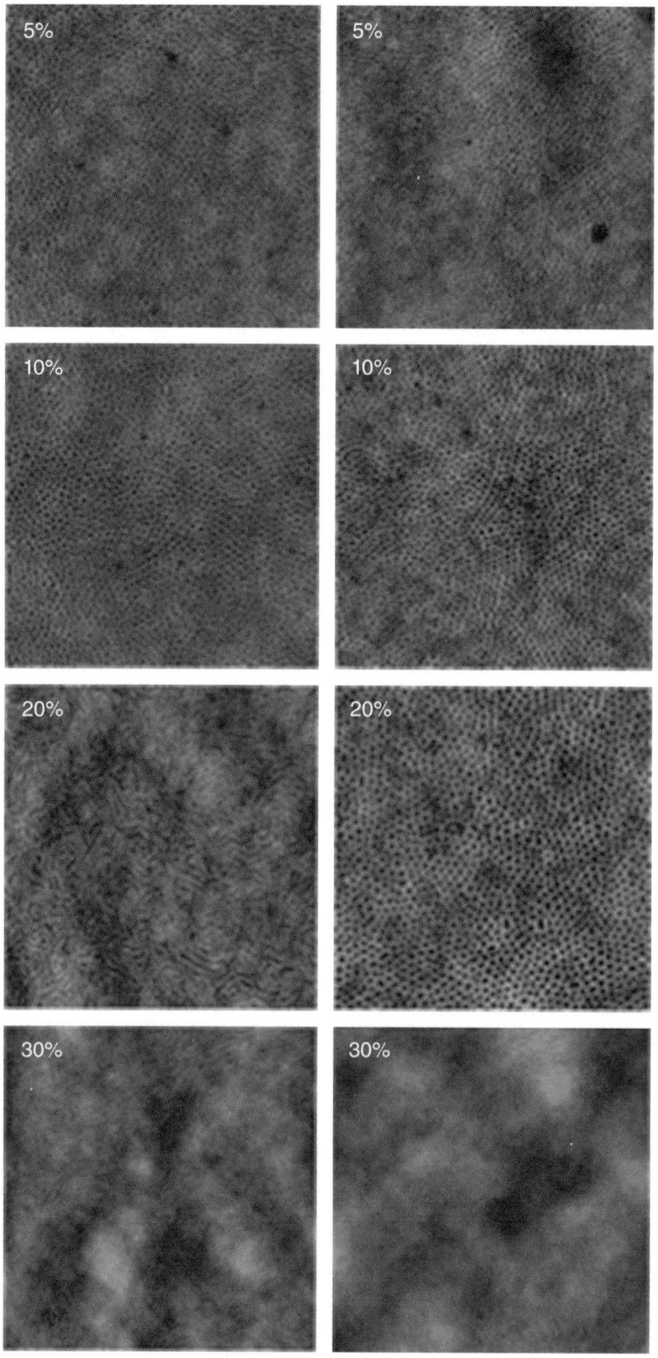

Figure 14.3 Atomic force microscopic (AFM) image showing morphology transition in thin films of mixtures of PS_{19}-*b*-$PEO_{6.4}$ with different weight fractions of $PAA_{5.3}$ (left column) and PAA_{20} (right column). Lefèvre et al. [26]. Reproduced with permission of American Chemical Society.

block copolymer tend to phase separate into a variety of morphologies due to the imbalance in intermolecular interactions. For instance, Hameed et al. [29] studied microphase separation induced by competitive hydrogen bonding in A-*b*-B/C diblock copolymer/homopolymer complexes, where A-*b*-B diblock copolymer is immiscible and homopolymer C can interact with both A and B blocks unequally through hydrogen bonding. Morphology transformation was studied in PVPh/P2VP-*b*-PMMA complexes as a function of different composition, and phase relationship was predicted, as shown in Scheme 14.1. The P2VP-*b*-PMMA bock copolymer shows spherical micelles, wormlike micelles, and hierarchical nanostructures at 10–20 wt%, 30–60 wt%, and 80 wt% of PVPh concentrations, as shown in Figure 14.4. At higher concentrations of PVPh, near-homogeneous morphology was observed due to PVPh hydrogen bonding with both

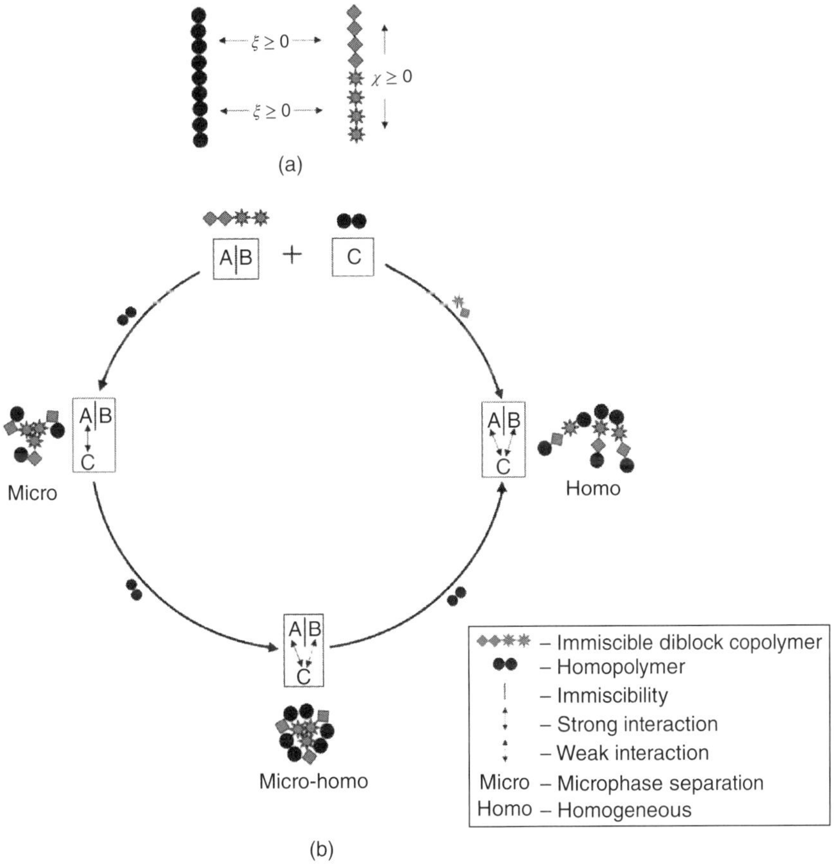

Scheme 14.1 Schematic representation of A-*b*-B/C diblock copolymer/homopolymer system showing microphase separation and homogeneous phase formation mediated by competitive hydrogen bonding, where "χ" represents repulsive interaction and "ξ" represents attractive interaction. Hameed et al. [29]. Reproduced with permission of AIP Publishing.

PMMA and P2VP blocks. The morphologies were further investigated using small-angle X-ray scattering (SAXS) technique. Well-defined scattering peaks were seen in SAXS profiles as an evidence of microphase-separated morphology in PVPh/P2VP-*b*-PMMA complexes. SAXS profiles are included in Figure 14.4. The average distance between the neighboring microdomains was found to increase in the order 23.2, 27.4, 31, and 35.9 for 20, 40, 60, and 80 wt%. From the SAXS profile, it was concluded that ordered nanostructures exist only below 40 wt% PVPh content.

In the same vein, nanostructure formation and morphology transformation depending on relative strength of hydrogen bonding interaction between each block of block copolymer PEO-*b*-PCL with homopolymer PVPh was investigated by Salim et al. [30]. The plain block copolymer shows an ordered cubic structure with spherical PEO domains arranged in cubic lattices. After addition of 20, 40, and 60 wt% of PVPh homopolymer, the morphology of block copolymer changed to hexagonal cylinder, disordered bicontinuous, and near-homogeneous phases. Schematic representation of phase separation along with transmission electron microscopy (TEM) images are shown in Figure 14.5. Furthermore, microphase separation through competitive interactions in A-*b*-B/C blend systems, PVPh-*b*-PMMA/poly(vinylpyrrolidone) (PVP)[31], PMMA-*b*-PVP/PVPh [32] and PCL-*b*-P4VP/PVPh [33] were also reported. All these studies show morphology transformation as a function of varying homopolymer concentration. For higher homopolymer concentrations, absence of sharp and multiple peaks in SAXS was observed, indicating the absence of long-range ordering due to multiple interactions in the blends. This analogy applies to all block copolymer/homopolymer systems involving competitive hydrogen bonding. It is further evidenced via formation of homogeneous phases at higher homopolymer concentrations.

Hameed and Guo [34] studied competitive hydrogen bonding in PCL-*b*-P2VP (A-*b*-B diblock copolymer) with poly(hydroxyether of bisphenol A) (C homopolymer), where both bocks are miscible with poly(hydroxyether of bisphenol A) (phenoxy). The value of χ_{AB} is positive, that is, A and B blocks are immiscible and χ_{AC} and χ_{BC} are negative. But χ_{BC} is more negative than χ_{AC}, because the P2VP block forms strong hydrogen bonding with the phenoxy bock than the PCL block. This disparity in hydrogen bonding leads to formation

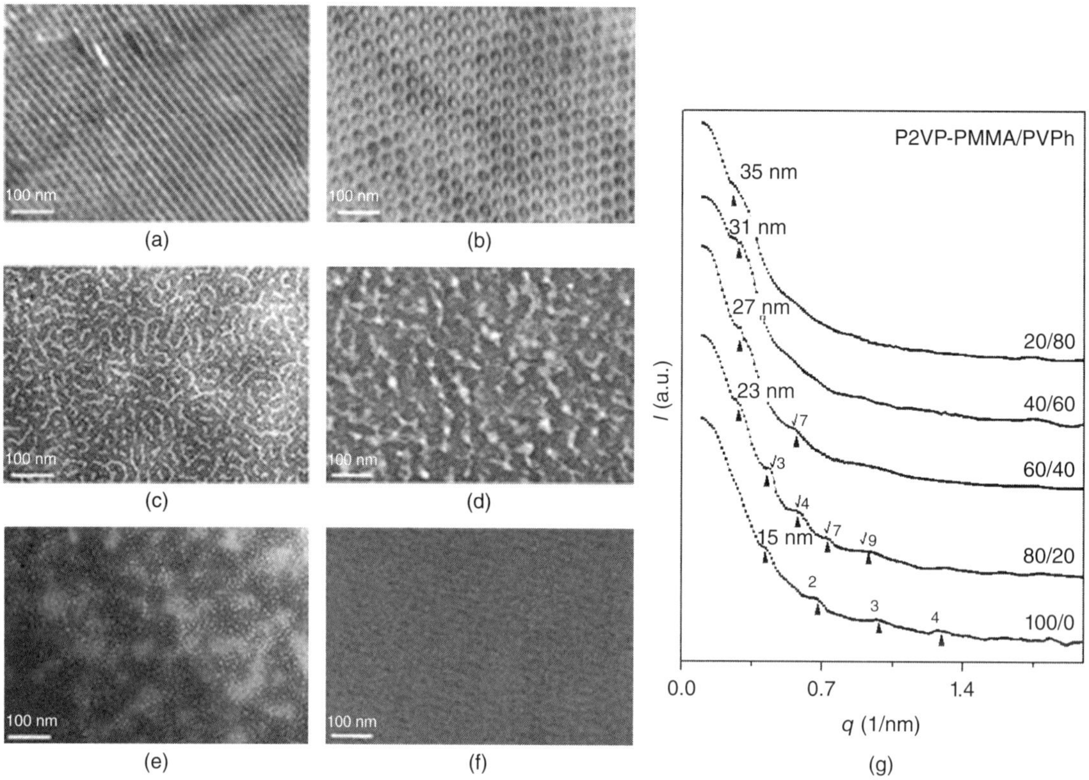

Figure 14.4 TEM images at different weight ratios (a) 100/0, (b) 80/20, (c) 60/40, (e) 20/80, and (f) 10/90 of P2VP-*b*-PMMA/PVPh complexes. (g) SAXS profiles of P2VP-*b*-PMMA/PVPh complexes at different weight ratios. Hameed et al. [29]. Reproduced with permission of AIP Publishing.

of a variety of composition-dependant self-assembly. Competitive hydrogen bonding and self-assembly in P2VP-*b*-PMMA/phenoxy blends was investigated by Salim et al. [20]. Imbalance in the intermolecular interaction and repulsion between both blocks of block copolymer lead to composition-dependant morphology transformation. With increasing phenoxy concentration in block copolymer system, spherical micelles observed initially were transformed into wormlike micelles and finally to a homogeneous morphology. Blending of block copolymer polyisoprene-*b*-poly(2-vinyl pyridine) (PI-*b*-P2VP) with novolac resin was reported by Kosonen et al. [35]. Hydrogen-bonding interaction between pyridine groups of P2VP and hydroxyl groups of novolac resulted in a miscible phase. As the weight fraction of PI decreased from 0.4, 0.3, to 0.05, morphology transformation was observed from lamellar, cylindrical, to spherical structures, respectively. These morphologies were preserved even after blends cured at higher temperatures. A number of similar studies pertaining to nanostructuration of synthetic resins using block copolymers have been extensively discussed elsewhere [36–38].

14.2.2 Diblock/Triblock Copolymer Systems

Blends involving multiple blocks produce a variety of morphology depending on the compatibility between the blocks. Blending A-*b*-B block copolymer with A-*b*-C block copolymer, where B and C have favorable interaction will produce a biphasic system. Compared to synthesizing a pristine diblock copolymer to achieve a tailored nanostructure, morphology can be controlled easily by simple mixing of A-*b*-B and A-*b*-C block copolymers. Highly asymmetric lamellar patterns were produced in this study from mixing PS-*b*-P2VP with PS-*b*-PHOST, where PHOST and P2VP formed single-phase lamellae of width 7 nm and PS lamellae of 28 nm [39, 40]. A triphasic system can be produced through mixing A-*b*-B and C-*b*-D block copolymers, where B and D has favorable interaction. For instance, Asari et al. [41] studied blends of PI-*b*-P2VP (IP) and PS-*b*-PHOST (SH) that self-assembled into complex three-phase microphase-separated structures. An IP/SH system with symmetric diblocks and another IP/SH system with asymmetric diblocks were mixed at 50/50 weight ratio, where hydrogen-bonded P and H blocks formed lamellar in the former, as shown in Figure 14.6, and isolated cylinders in the latter, as shown in Figure 14.7. Construction of highly ordered complex structures were demonstrated by simply combining two block copolymers that can interact through selective hydrogen bonding.

As mentioned in the previous section, strong hydrogen bonding within a small fraction of blending components can avoid macrophase separation. In the same vein, Dobrosielska et al. have proved, nanophase separation can be achieved even though the interacting components between two block copolymers are less than a mere 4%. Hierarchical three-phase

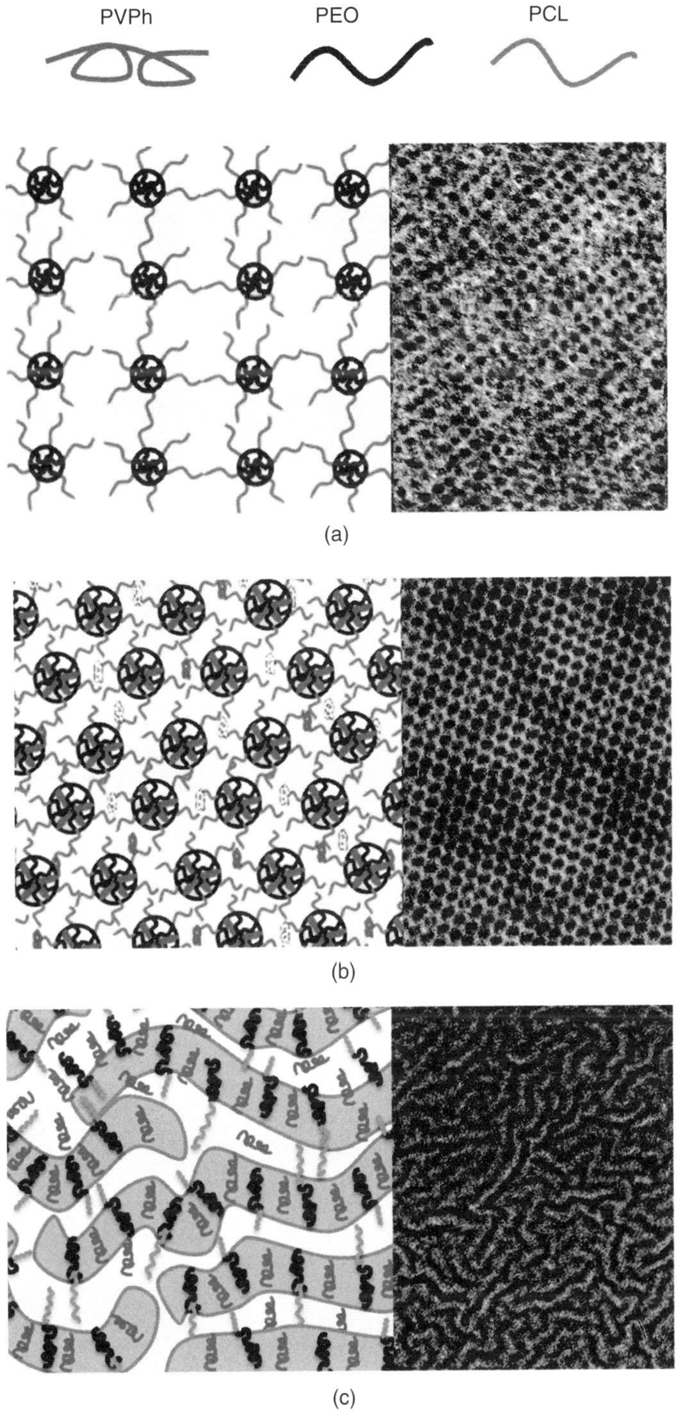

Figure 14.5 The schematic representation and TEM images of PEO-*b*-PCL/PVPh blends showing different morphologies in (a) pristine block copolymer, (b) 20 wt%, and (c) 40 wt% of PVPh concentration. Salim et al. [30]. Reproduced with permission of American Chemical Society.

lamellar structures were observed in blends of two asymmetrical block copolymers PS-*b*-PHOST and PI-*b*-P2VP having short (0.04 mol fraction) PHOST and P2VP blocks, respectively [42].

Asari et al. have also studied blends of same-block copolymer PS-*b*-PHOST with a triblock copolymer P2VP-*b*-PI-*b*-P2VP. New self-assembled mesoscopic Archimedean tiling patterns were identified as a result of strong hydrogen bonding between P2VP and PHOST blocks. These highly ordered hierarchical 2D nanostructures were regulated by the bridged conformation of the PI block in the triblock copolymer [43]. Furthermore, partially modified polystyrene-*b*-poly(1,2-butadiene)-*b*-poly(*tert*-butyl methacrylate) (SBT) triblock copolymer blended with

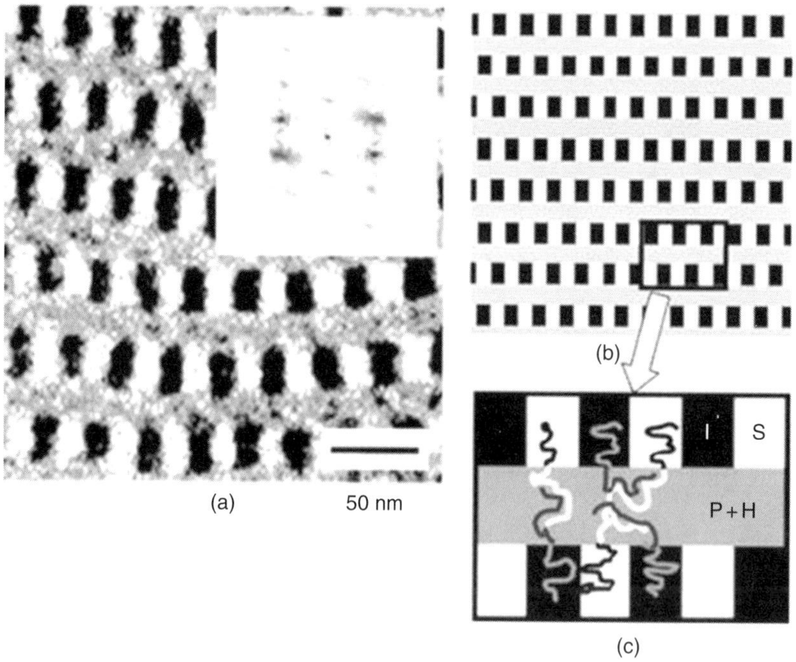

Figure 14.6 (a) TEM image of IP/SH 50/50 blend with symmetric diblock copolymer; inset shows the fast Fourier transform pattern for the whole TEM image. (b) Schematic illustration of microdomain arrangement for corresponding TEM image. (c) Enlarged domain assembly and possible molecular arrangements in this blend. Asari et al. [41]. Reproduced with permission of American Chemical Society.

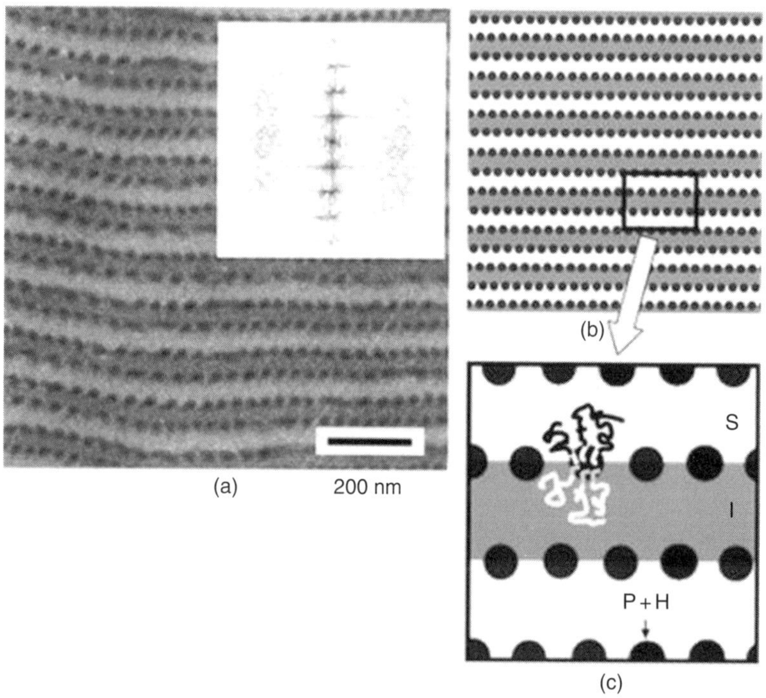

Figure 14.7 (a) TEM image of IP/SH 50/50 blend with asymmetric diblock copolymer; inset shows the fast Fourier transform pattern for whole TEM image. (b) Schematic illustration of microdomain arrangement for corresponding TEM image. (c) Enlarged domain assembly and possible molecular arrangements in this blend. Asari et al. [41]. Reproduced with permission of American Chemical Society.

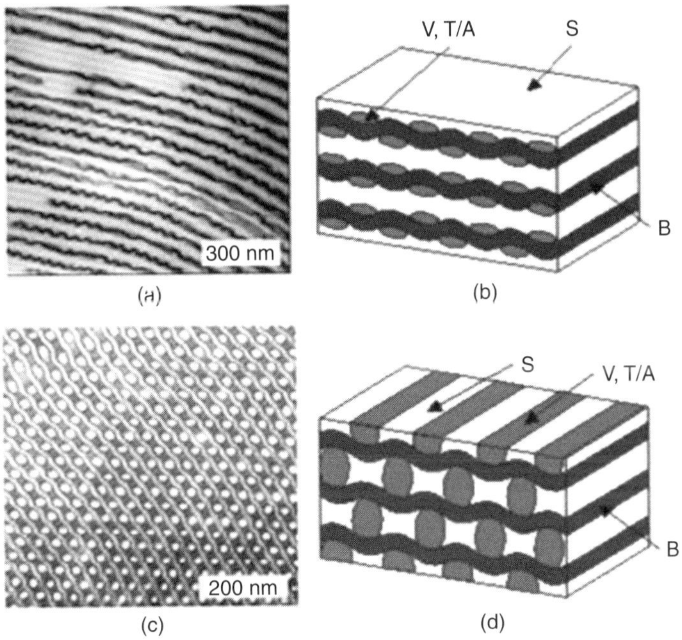

Figure 14.8 (a) and (b) TEM image and schematic representation of SB(T88/A12)/$S_{45}V_{55}$ blends showing curved lamellae within lamellar structure. (c) and (d) TEM image and schematic representation of SB(T82/A18)/$S_{45}V_{55}$ blends showing hexagonally packed cylinders within curved lamellar structure. Jiang et al. [44]. Reproduced with permission of American Chemical Society.

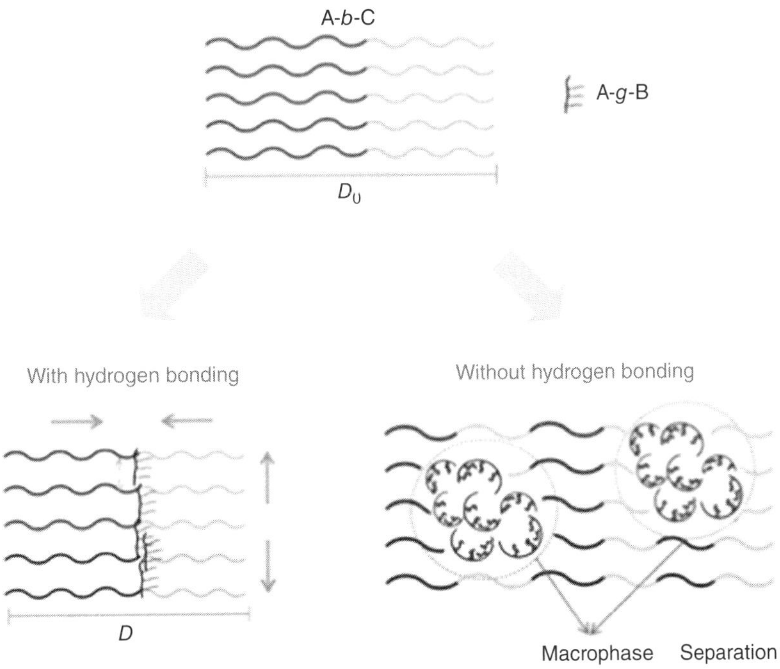

Figure 14.9 Schematic representation of microphase and macrophase separation in block copolymer/graft copolymer blends due to hydrogen bonding. Chen and Lo [3]. Reproduced with permission of the Royal Society of Chemistry.

PS-*b*-P2VP (SV) produced novel morphologies depending on the degree of saponification of poly(*tert*-butyl methacrylate) block, as reported by Jiang et al. [44]. Poly(methacrylic acid) (A) obtained as the result of saponification acts as proton donor to form hydrogen bonding with P2VP block of the diblock copolymer. Figure 14.8 shows various morphologies obtained from SB(T/A)/SV complex system.

The importance of hydrogen bonding in compatibilization of polymer blends was shown in blends of poly(styrene-*graft*-acrylic acid) (PS-*g*-PAA) with other polystyrene-based block copolymers [3]. Microphase separation accompanied by morphology tuning was observed in the blends where hydrogen bonding existed between blend components. As shown in Figure 14.9, graft copolymer can be positioned at the domain interface, which helps in tuning the morphology, whereas absence of hydrogen bonding between the graft copolymer and the block copolymer resulted in macrophase separation.

14.3 BLOCK COPOLYMER SYSTEMS WITH HYDROGEN-BONDING INTERACTION IN SOLUTION

14.3.1 Single-Component Block Copolymer Systems

Asymmetric diblock copolymers with two blocks interacting through interchain or intrachain hydrogen bonds can lead to micellar self-assembly. For instance, Štěpánek et al. [45] demonstrated spherical nanoparticles of PHOS-*b*-PEO in aqueous solution and water/THF mixtures, as shown in Figure 14.10. The interchain hydrogen bonding between hydroxyl groups of PHOS and oxygen atoms of PEO led to formation of spherical nanoparticles. Furthermore, these nanoparticles were electrostatically stabilized by a fraction of ionized PHOS units on the surface. Similarly, in block copolymers containing polyacid block, micelles were stabilized by dissociated acid components on the surface. pH-sensitive micelles were prepared through interchain hydrogen bonding in poly(methacrylic acid-*b*-PEO) (PMAA-*b*-PEO) block copolymer in aqueous system. In these complexes, hydrogen bonding was stabilized till a certain degree of dissociation of the acid component, above which no complexation was observed [46]. Holappa et al. [47] investigated the self-complexation mechanism in a similar PEO-*b*-PMAA system as discussed previously. As complexes are formed via interchain hydrogen bonding between PMAA and PEO blocks, the different degrees of ionization of carboxylic acid groups determine the nature of the complexes. At pH lower than 4.5, spherical micelles were observed due to strong intermolecular hydrogen bonding. The PMMA/PEO complex forms the micellar core surrounded by free PMMA chains to form corona. At pH between 4.5 and 5.5, contracted free polymer chains were observed due to intramolecular hydrogen bonding. Due to complete ionization of PMMA above pH 5.5, completely dissociated polymer chains were observed.

Micellar self-assembly of a single-component ABC triblock copolymer system was reported by Cai et al. [48] in aqueous system. The triblock copolymer poly(ethylene oxide-*b*-diethylamino ethyl methacrylate-*b*-succinyloxyethyl methacrylate) (PEO-*b*-PDEA-*b*-PSEMA) forms three different micellar aggregates in aqueous solution as a function of pH. The driving forces for the formation of micelles were hydrogen bonding, interpolyelectrolyte complexation, and hydrophobic interactions. At low pH, hydrogen bonding between PEO/PSEMA forms a complex core with cationic PDEA corona. At intermediate pH, partial ionization of PSEMA and PDEA leads to formation of a complex core via electrostatic interaction with PEO corona. At relatively higher pH, PDEA forms an insoluble core with anionic EO/PSEMA corona. In single-component micellar systems, morphology transformation can be achieved via changing pH, solvent, ionic strength, and temperature. Kyeremateng et al. [49] investigated self-assembly of a triblock copolymer incorporating poly(glycerol monomethacrylate) (PGMA), poly(propylene oxide) (PPO), and perfluoroalkyl (PF) blocks. Large cigar-shaped supramolecular aggregates were observed due to combination of hydrogen bonding between PGMA blocks and the hydrophobic effect.

Low-molecular-weight cross-linkers were also used in single-component systems to construct micelles with a cross-linked core surrounded by soluble corona. This approach was used to construct PS-*b*-P4VP micelles with a P4VP core cross-linked using bisphenol A. Strong hydrogen bonding between pyridine and phenolic groups led to formation of micelles in a nonselective solvent [19]. In another study, Chen et al. [18] used octyl gallate (OG) to modulate self-assembly of PS-*b*-P4VP block copolymer in different solvents. A series of morphological transition from spheres, pearl-necklace-like rods, wormlike rods, vesicles, to core–shell–corona aggregates were observed for molar ratios of OG to 4VP ranging from

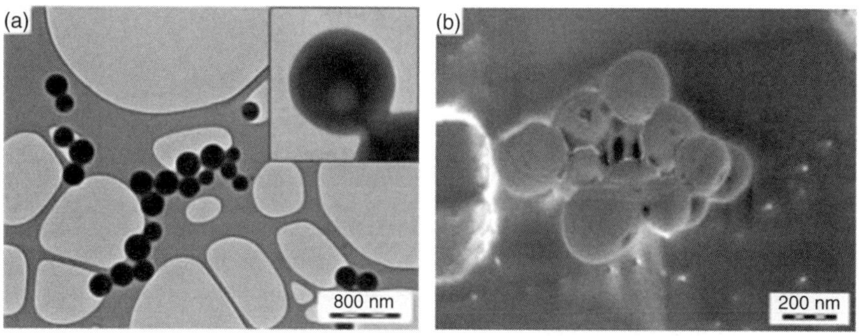

Figure 14.10 (a) and (b) TEM and Cryo-FESEM (cryogenic field-emission scanning electron microscopy) images of PHOS-*b*-PEO nanoparticles, respectively. The inset in (a) shows a detailed view of the nanoparticle. Štěpánek et al. [45]. Reproduced with permission of American Chemical Society.

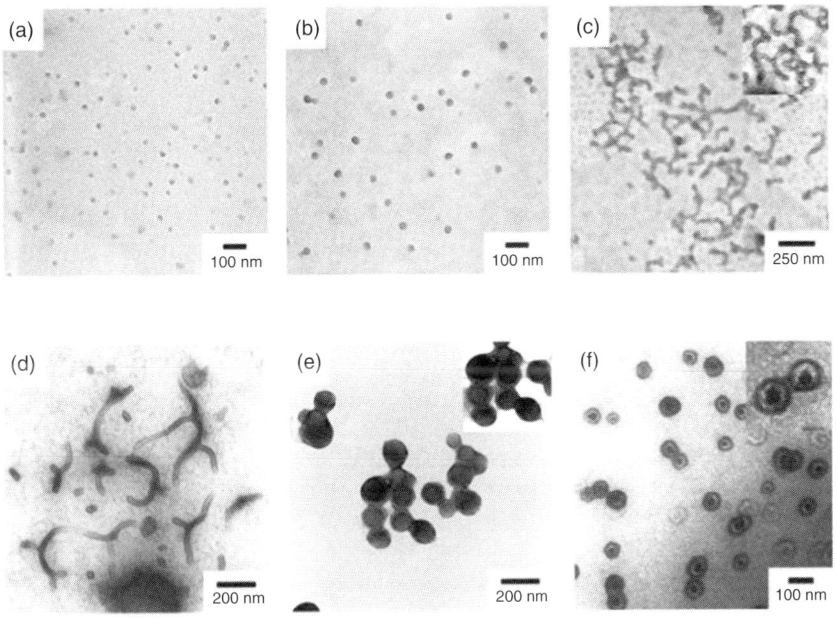

Figure 14.11 TEM images showing various morphologies of (a) pure PS-*b*-P4VP block copolymer and mixtures at (b) 1/50, (c) 1/20, (d) 1/10, (e) 1/5, and (f) 1 molar ratios of OG to 4VP in THF solvent. Chen et al. [18]. Reproduced with permission of American Chemical Society.

1/50, 1/20, 1/10, 1/5, to 1, respectively. Morphological transitions are shown in Figure 14.11.

14.3.2 Diblock Copolymer/Homopolymer Systems

Usually, micellization of block copolymers occurs in block selective solvents, where an insoluble block forms the core and a soluble block forms the corona. But, interpolymer complexation through noncovalent secondary interactions between complementary bonding sites of AB block copolymers and C homopolymers can also lead to formation of micellar aggregates in nonselective solvents. Secondary interactions should be established between C and B blocks, leading to aggregates stabilized by still-soluble A block of the block copolymer. These secondary interactions alter the solubility and conformation of polymer components to form various self-assembled nanostructures. The solvent used must be a nonsolvent of the C + B complex and nonselective for copolymer and homopolymer. Generally, complexation studies were performed in aprotic solvents to avoid competitive hydrogen bonding of the proton acceptor block with solvent. The choice of solvent plays an important role in complex formation. For instance, PVPh/poly(*N*,*N*-dimethyl acrylamide) (PDMA) can form interpolymer complexes in dioxane, but no complexation was observed in DMF solvent. This is because dioxane is a weak hydrogen bond acceptor compared to DMF. Competition between solvent molecules and PDMA to form hydrogen bonding with hydroxyl groups of PVPh determines complex formation [50]. Yao et al. [51] investigated complexation of AB-type block copolymers with small molecular acids to form core–shell micelles. Later on, homopolymers were used to trigger micellization of block copolymers through hydrogen-bonding interactions. Spherical core–shell micelles of PS-*b*-PAA with homopolymer P4VP was reported by Zhang et al. [52]. P4VP unimers were adsorbed into self-assembled PS-*b*-PAA micelles and penetrated inside the micellar shell. This adsorption led to folding of PAA chains due to strong hydrogen bonding between PAA and P4VP and led to shrinkage of micelles. A similar phenomenon was reported by Hameed and Guo [25] in the PCL-*b*-P2VP/PAA system. As the content of PAA increased (up to 40 wt%), hydrodynamic size of micelles decreased initially due to hydrogen bonding between P2VP and PAA chains. Later on, above 50 wt%, PAA content micelle size increased due to adsorption of PAA chains on the surface of the micelles. In another study, PS-*b*-P4VP/PAA system was reported to form core–shell micelles due to complexation of P4VP and PAA blocks. Morphology transition from cylindrical rodlike micelles to spherical micelles was observed as the result of complex formation. Hydrodynamic size of micelles were found to increase as PAA concentration increased in the system due to selective swelling of the micellar core [53]. These authors have also studied complexation of PVPh and PCL-*b*-P2VP in THF [54]. Both the blocks of block copolymer had favorable interaction with PVPh homopolymer in this system. But P2VP block had significantly stronger interaction with PVPh than PCL and PVPh. This disparity in competitive interaction led to formation of a variety of nanostructures depending on the concentration of PVPh. Spherical micelles were obtained at 10 wt% PVPh, wormlike micelles were observed at 30–60 wt% PVPh concentration, and a homogeneous phase was observed above 80 wt% of PVPh.

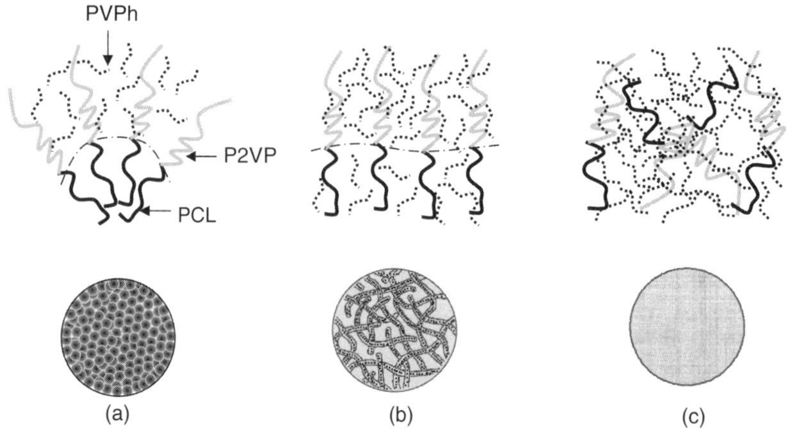

Scheme 14.2 Schematic representation of PVPh/PCL-*b*-P2VP complexes showing various self-assembled nanostructures (a) spherical micelles at 10 wt% PVPh concentration (b) wormlike micelles at 30–60 wt% PVPh, and (c) homogeneous phase at 80 wt% PVPh concentration. Hameed et al. [54]. Reproduced with permission of American Chemical Society.

At higher concentration of PVPh, it acts as a nonselective solvent for both the blocks of block copolymer, and hence no phase separation was observed. The schematic representation in Scheme 14.2 shows the morphology transformation in PVPh/PCL-*b*-P2VP complexes.

Zhang et al. [55] studied comicellization of PEO-*b*-PAA and P4VP in ethanol medium. Because PAA is a polyacid and P4VP is a polybase, there exists a strong hydrogen-bonding interaction between PAA/P4VP than PAA/PEO blocks, which led to the formation of micellar complexes. The insoluble PAA/P4VP complex forms a core, while the ethanol-soluble PEO forms the corona. This system follows the same pattern of morphology transformation as PS-*b*-P4VP/PAA system [53], that is, an increase in hydrodynamic diameter of micelles was observed with an increase in homopolymer content due to swelling of micellar core. Liu et al. [56] prepared stable micelles made of insoluble PMMA/poly(styrene-co-[*p*-(2,2,2-trifluro-1-hydroxy-1-trifluromethyl) ethyl-α-methylstyrene]) (PSOH) core and soluble PS corona. These hydrogen-bonded complexes prepared from PS-*b*-PMMA/PSOH system was stable as long as the content of PSOH was greater than 8 mol%. Increase in micellar size and molar mass with increase in initial polymer concentration indicates that complexation is a diffusion-controlled process. Apart from simple micellar self-assembly, more complex morphologies have been reported form block copolymer/homopolymer systems in the literature. For instance, Gao et al. [57] reported formation of unilamellar vesicles (ULVs) and multivesicular vesicles (MVVs) in PAA/poly(ethylene oxide-*b*-butadiene) (PEO-*b*-PB) system in a mixed solvent of tetrahydrofuran (THF) and *n*-dodecane. Figure 14.12 shows TEM images of ULVs and MVVs. Hydrogen-bonding complexation between PAA and PEO triggers self-assembly into various nanostructures depending on the ratio of PAA to PEO.

In addition to vesicles, formation of hollow spheres were reported by Kuang et al. [58] based on the interaction between rigid homopolymer poly(amic acid) ester (PAE) and PS-*b*-P2VP block copolymer in a common solvent chloroform or THF. Due to hydrogen bonding between PAE

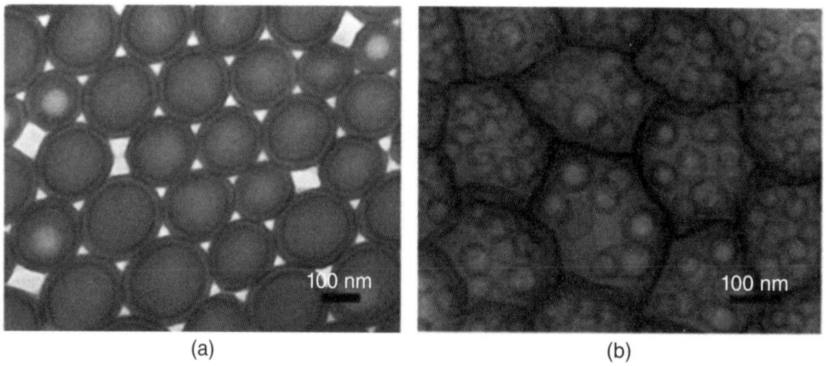

Figure 14.12 TEM images of PAA/PEO-*b*-PB system showing (a) unilamellar vesicles at PAA/PEO-*b*-PB weight ratio $W_A/W_{EB} = 2.0$ (b) multivesicular vesicles at $W_A/W_{EB} = 10.0$. Gao et al. [57]. Reproduced with permission of American Chemical Society.

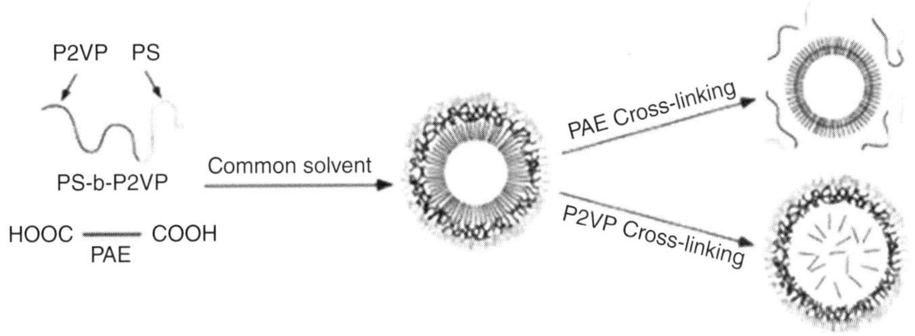

Scheme 14.3 Schematic representation of formation of hollow spheres through complexation of PS-*b*-P2VP block copolymer and PAE and structure stabilization via cross-linking. Kuang et al. [58]. Reproduced with permission of American Chemical Society.

and P2VP, a unique self-assembly behavior was observed. As a common trend, the hydrodynamic radius of hollow spheres decreased with increasing the content of the proton-donating homopolymer. In addition, nanocages with entrapped PAE were obtained by cross-linking PS-*b*-P2VP block copolymer. Also, polyimide-based hollow spheres with a central cavity were produced by cross-linking PAE, as shown in Scheme 14.3. The same authors reported formation of hollow spheres in complex aggregates of PS-*b*-P4VP/polyimide system [59]. Solvent cast films of PS-*b*-P4VP/polyimide complex aggregates self-assembled into microporous films are shown in Figure 14.13.

Interpolymer-complexation-induced morphology evolution in selective solvents was reported by Salim et al. [60]. Morphology transition from spheres to vesicles in PS-*b*-PEO/PAA complexes was observed as a function of different molar ratios of PAA to PEO in aqueous solution. The complexation and morphology transformation were driven by hydrogen bonding between the PAA homopolymer and PEO block of the block copolymer. In order to decrease the dissociation of carboxylic acid of PAA and in turn increase the capability of hydrogen bonding, complexation was carried out at lower pH ($pH < 4.8$) values. The charge ratio between carboxylic groups of PAA and ether oxygen of PEO is an important parameter that controls micelle/vesicle formation. At lower content of PAA (acrylic acid to ethylene oxide ratio, $[AA]/[EO] = 0.2$), spherical micelles were observed. Further, increasing the acrylic acid to ethylene oxide ratio [AA]/[EO] to 1.0, 4.0, and 8.0, vesicles, compound vesicles, and large compound vesicles were observed. Figure 14.14 shows morphology transition in PS-*b*-PEO/PAA complex system.

As alluded earlier, polyelectrolyte complexes formed through hydrogen bonding are sensitive to pH, ionic strength, temperature, and nature of solvent used. Morphology transformation was achieved as a function of changing pH. Lefevre et al. [61, 62] investigated the self-assembly of PS-*b*-P4VP block copolymer and PAA homopolymer complexes in organic solvents. Complexation occurs via hydrogen bonding between PAA and P4VP blocks. The insoluble complex aggregates of PAA/P4VP form a core with soluble PS chains as corona in DMF. Because hydrogen bonding is sensitive to the pH of the medium, reorganization of these micelles were observed in acidic and basic medium. Reorganization was induced by breaking of hydrogen bonds between PAA and P4VP through protonation of P4VP or ionization of carboxylic acid groups of PAA, leading to formation of inverted micelles with PS core and P4VP corona in aqueous medium, as shown in Scheme 14.4. A pH-controlled reversible micellar system was developed by Lee et al. [63] through hydrogen bonding between poly(ε-caprolactone-*b*-methacrylic acid) (PCL-*b*-PMAA) and PEO. Below pH 3.9, protonation of PMAA helps complex formation through hydrogen bonding between PEO and PMAA. The individual micelles observed at pH 7.4 were transformed into long-range interconnected micelles due to hydrogen bonding between PEO and PMAA at pH 3.9. Matejicek et al. [64] demonstrated complexation of core–shell micellar system in mixed selective solvents. Interconnected block copolymer micelles of PS-*b*-PMAA were prepared through hydrogen bonding between PMAA shell of the micelles and poly(2-vinylpyridine) (P2VP) homopolymer. Staikos at al. [65] studied hydrogen-bonded interpolymer complexation between homopolymer PAA and poly(*N*,*N*-dimethyl acrylamide) (PDMAM) side chains of graft copolymer poly(acrylic acid-*b*-2-acrylamido-2-methyl-1-propane sulfonic acid)-graft-poly(*N*,*N*-dimethyl acrylamide) (PAA-*b*-AMPSA)-*g*-PDMAM) in aqueous solution. At lower pH ($pH < 3.75$), negatively charged colloidal particles were observed with hydrogen-bonded PAA/PDMAM core surrounded by hydrophilic PAA-*b*-AMPSA chains.

14.3.3 Diblock/Diblock Copolymer Systems

Compared to diblock/homopolymer complexes, more complex morphologies are expected in diblock/diblock complexes due to multiple factors involved in controlling self-assembly of these systems. The two possible systems in diblock/diblock mixture may be one containing a homogeneous corona by mixing AB and AC type copolymers (forms $B + C$ hydrogen-bonded complex core) and another having a heterogeneous corona by mixing AB and CD type copolymers (forms $A + C$ hydrogen-bonded complex core). Depending on the nature of interaction between the blocks and their

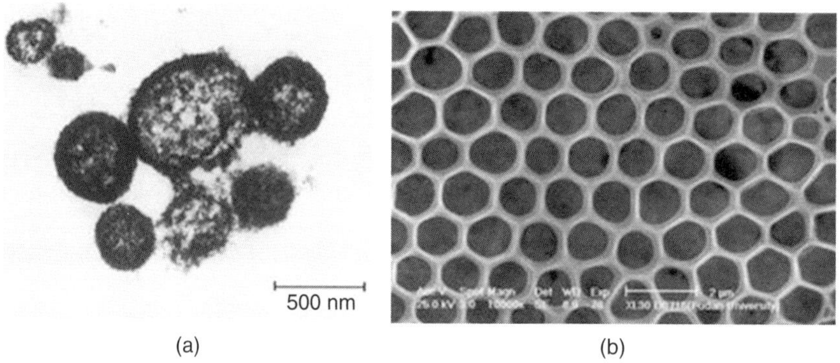

Figure 14.13 TEM image of hollow spheres of PS-*b*-P4VP/PI complex aggregates (a) and SEM image of microporous structures observed in solvent cast film of hollow sphere solution (b). Duan et al. [59]. Reproduced with permission of American Chemical Society.

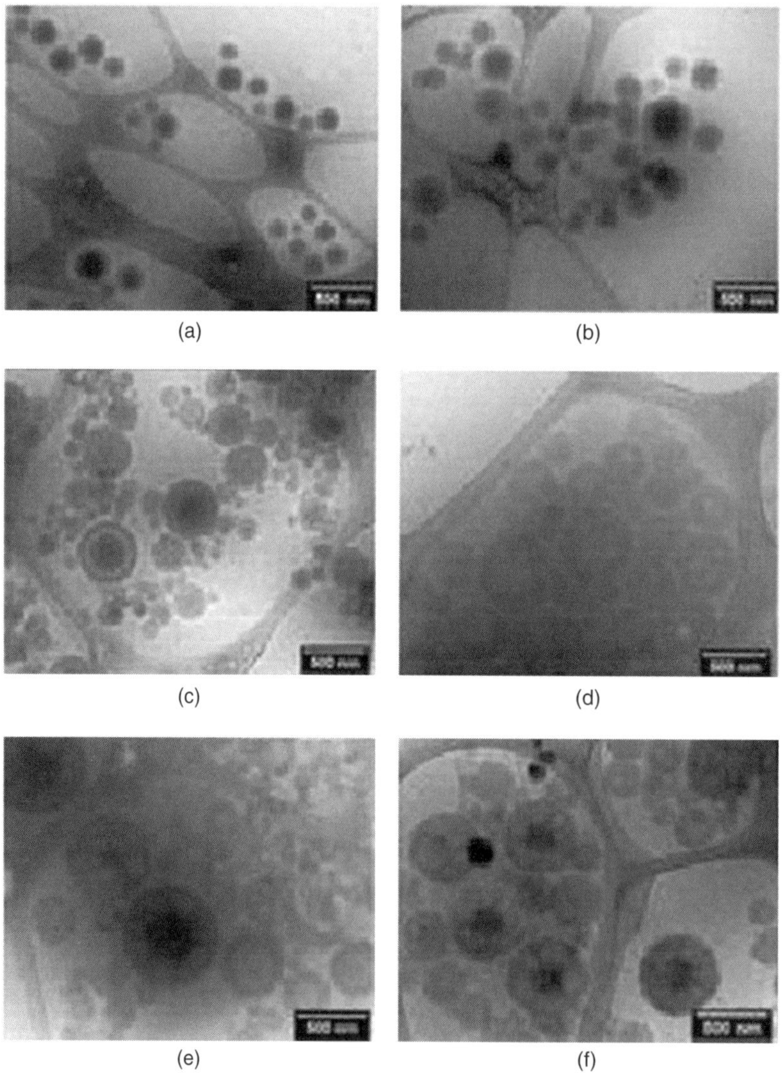

Figure 14.14 Cryo-TEM images of (a) plain PS-*b*-PEO block copolymer showing spherical micelles and PS-*b*-PEO/PAA complex with various [AA]/[EO] ratios of (b) 0.2 shows polydisperse spherical micelles, (c) 0.6 shows both spherical micelles and vesicles, (d) 1.0 shows vesicles, (e) 4.0 shows compound vesicles, and (f) 8.0 shows large compound vesicles. Salim et al. [60]. Reproduced with permission of John Wiley and Sons.

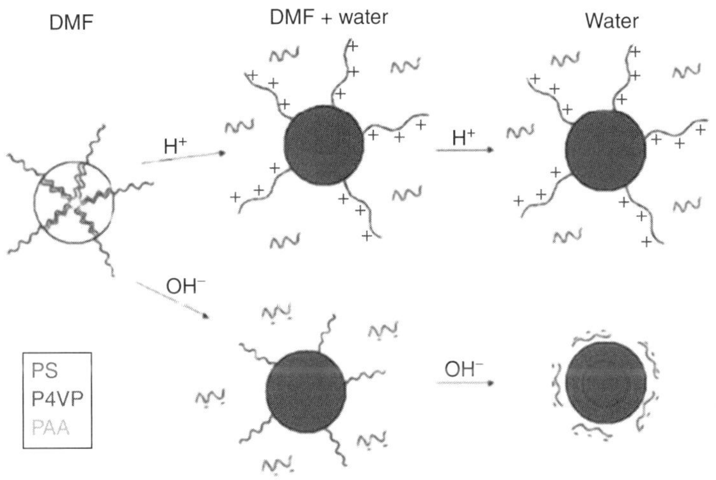

Scheme 14.4 Schematic illustration of micellar reorganization due to addition of acidic or basic water . Lefèvre et al. [61]. Reproduced with permission of American Chemical Society.

solubility, core/shell micelles [62, 63, 66], core/shell/corona micelles [57, 67], compartmentalized core [68, 69] or corona [63, 66] micelles, vesicles [70], and multilamellar vesicles (MLVs) [71] were reported. Furthermore, reverse micelles were also reported if the hydrogen-bonded complex forms the corona and the insoluble block forms the core.

Mixed micellar clusters were reported from a mixture of PS-*b*-PEO and PS-*b*-PAA block copolymers in toluene. Hydrogen-bonding complexation between PEO and PAA blocks in core and repulsion between coronal PS blocks as a function of molar ratio of PEO to PAA manipulate the structural evolution [72]. The same system was studied by Salim and Guo [71] and they identified a wealth of morphologies in selective solvent. In contrast to the previous study, hydrogen-bonded PAA/PEO blocks form the corona and water insoluble PS the core. A variety of bilayer aggregates, such as vesicles, MLVs, thick-walled vesicles (TWVs), interconnected compound vesicles (ICCVs), and irregular aggregates were identified at various molar ratios of PEO to PAA as shown in Figure 14.15.

Spherical micelles formed from complexation of PS-*b*-PAA and PMMA-*b*-PEO showed time-dependant evolution into hyperbranched structures. The interplay between segregation in PMMA/PS corona and complexation of PEO/PAA core as a function of molar ratio of PEO to PAA determines the structural evolution [72]. Gao et al. [73] studied hydrogen-bonding complexation in PS-*b*-P4VP/PS-*b*-PAA system. Complex aggregates were formed due to hydrogen bonding between P4VP and PAA blocks. Upon mixing both block copolymers with equal mole ratio of P4VP and PAA, spherical micelles and soluble and insoluble complexes coexisted in solution. After spin coating the polymer mixture, fibrillar aggregates coexisting with spherical aggregates were observed. Fibrillar aggregates were absent in solutions studied at different concentration and different evaporation rates. This study led to the conclusion that solubility of complexes and kinetic factor determine the morphology of the aggregates. Complex micelles with tunable channels were prepared by Li et al. [74] using two diblock copolymers, namely, poly(*tert*-butyl acrylate)-*b*-poly(*N*-isopropyl acrylamide) (PtBA-*b*-PNIPAM) and poly(*tert*-butyl acrylate)-*b*-poly(4-vinylpyridine) (PtBA-*b*-P4VP). The hydrophobic PtBA blocks of the two polymers associated to form core, while water-soluble P4VP and PNIPAM blocks act as shell at low pH. By varying pH or temperature, the size and permeability of the channels have been regulated, which is shown in Scheme 14.5. Wormlike aggregates were observed in another PNIPAM-based block copolymer PNIPAM-*b*-P4VP on mixing with PEO-*b*-PAA. Hydrogen-bonded PAA/P4VP forms a complex core and a PEO/PNIPAM mixed shell in ethanol. The length of wormlike aggregates was adjusted by altering the weight ratio of PNIPAM-*b*-P4VP to PEO-*b*-PAA. The length changed from

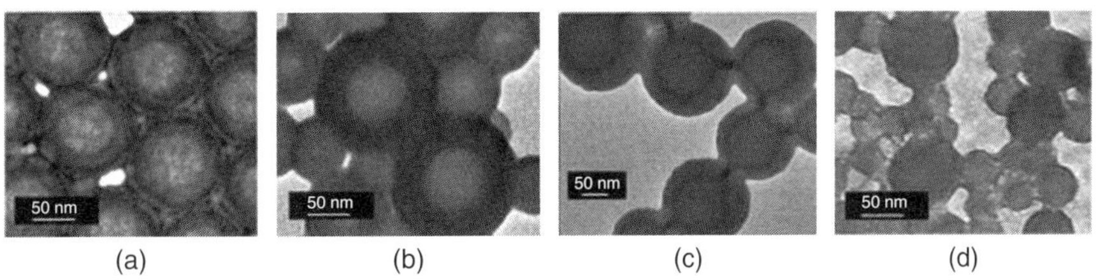

Figure 14.15 TEM images of PS-*b*-PAA/PS-*b*-PEO complex in water showing (a) MLVs, (b) TWVs, (c) ICCVs, and (d) irregular aggregates at [EO]/[AA] = 1, 2, 6, and 8. Salim and Guo [71]. Reproduced with permission of American Chemical Society.

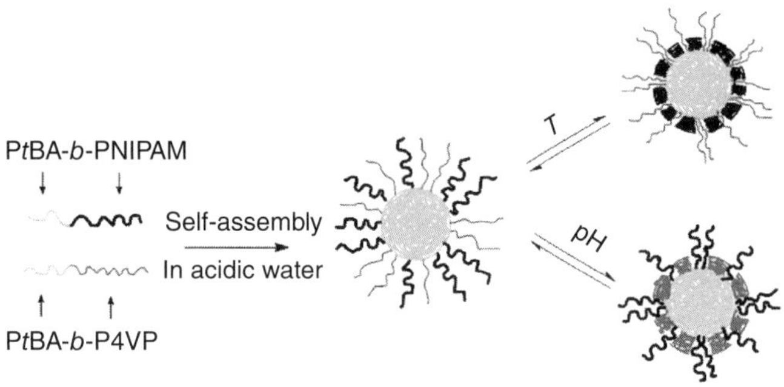

Scheme 14.5 Schematic representation of self-assembled complex micelles of (PtBA-*b*-PNIPAM) and (PtBA-*b*-P4VP). Li et al. [74]. Reproduced with permission of John Wiley and Sons.

100 nm to several microns for 20–150% weight ratio change, but the diameter remained constant at about 15 nm [75].

In block copolymer or block copolymer/homopolymer complex systems, core/shell micelles with unicompartmental corona have been observed so far. Systems involving complexes of two block copolymers have an opportunity to form compartmentalized corona due to difference in hydrogen-bonding strength between two coronal blocks. Kuo et al. [76] reported patched spherical structures from the intermolecular complexes of PS-*b*-PVPh and PMMA-*b*-P4VP block copolymers. The hydrogen-bonding strength between PVPh and P4VP in DMF solvent was weak, compared to those in THF, and led to patched spherical structures. Strong intermolecular hydrogen bonding in THF resulted in vesicle formation, which is shown in Figure 14.16. The difference in hydrogen-bonding strength in different solvent has been exploited to modulate the morphology. The same group reported preparation of well-defined micelles through complexation of block copolymer and a random block copolymer in a single nonselective solvent. The block copolymer PMMA-*b*-P4VP mixed with random copolymer PS-r-PVPh at various ratios of P4VP to PVPh and morphology transition was studied in two different solvents THF and DMF. It was concluded that the strength of hydrogen bonding between P4VP/PVPh and PVPh/PMMA complexes in different solvents at various block copolymer/random block copolymer compositions affect the self-assembly behavior [77].

Core–shell–corona or onion-like micelles were reported by Zhang et al. [78]. Spherical core–shell micelles were first prepared from the amphiphilic block copolymer PS-*b*-PAA in ethanol. Further addition of PEO-*b*-P4VP block copolymer

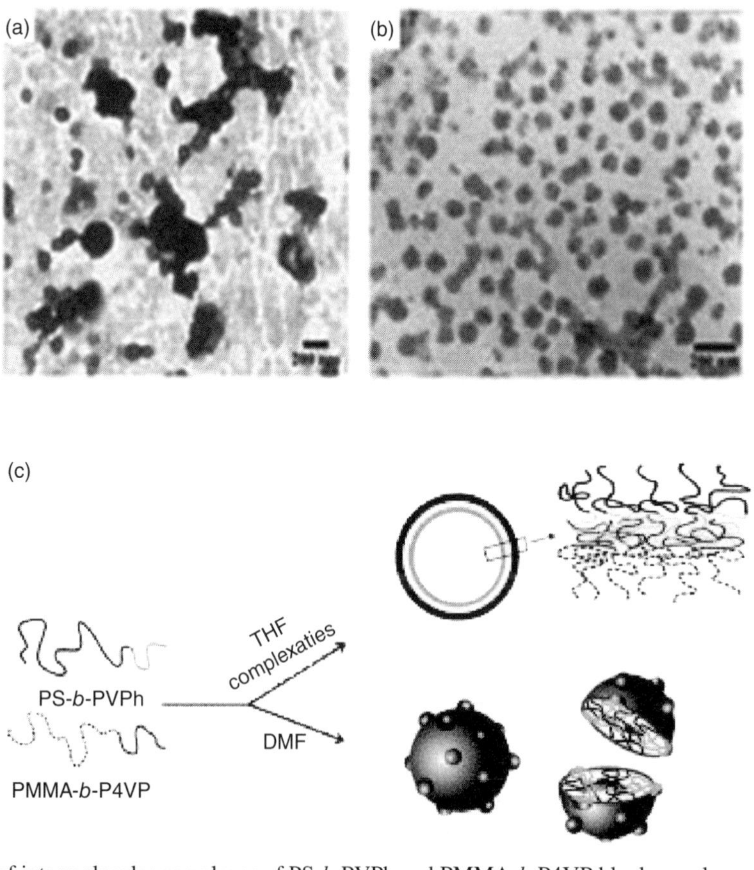

Figure 14.16 TEM images of intermolecular complexes of PS-*b*-PVPh and PMMA-*b*-P4VP block copolymers showing (a) vesicular structures in THF and (b) patched spherical structures in DMF. Schematic illustration of vesicle and patched sphere formation in different solvents is shown in (c). Kuo et al. [76]. Reproduced with permission of John Wiley and Sons.

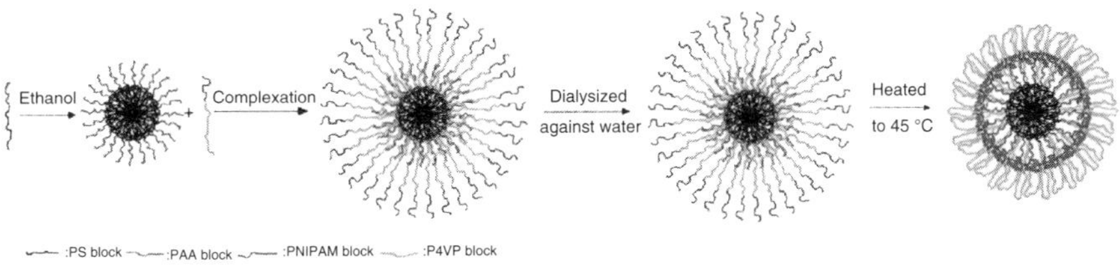

Scheme 14.6 Schematic illustration of micelle formation, complexation, and temperature response in PS-*b*-PAA/P4VP-*b*-PNIPAM complex system. Xiong et al. [79]. Reproduced with permission of Elsevier.

into the micellar solution resulted in penetration of P4VP blocks into previously formed PAA shell due to hydrogen bonding between P4VP and PAA. A three-layered spherical core–shell–corona complex composed of PS core, PAA/P4VP shell, and PEO corona was formed. These types of micelles belong to inverse micelles, where insoluble block forms core with interpolymer complex shell. A similar study by Xiong et al. [79] led to synthesis of a temperature-sensitive multilayered micelle made up of PS-*b*-PAA/P4VP-*b*-PNIPAM complexes in acidic aqueous solution. The insoluble PS block forms the core, P4VP/PAA complex forms the shell, and PNIPAM forms the corona. As P4VP block length was much greater than PAA block, at higher temperatures, excessive P4VP blocks extend into acidic water after PNIPAM collapsed onto PAA/P4VP shell to form corona reversed micelles, which is illustrated in Scheme 14.6.

14.3.4 Triblock Copolymer Systems

More complex morphologies were observed in complexes involving a triblock copolymer and a homopolymer or diblock copolymer. For instance, flowerlike morphologies were observed from self-assembled complexes of PEO-*b*-P2VP-*b*-PEO triblock copolymer and PAA homopolymer at low pH in the presence of 1 M citric acid [22]. Huang et al. [80] followed the morphology evolution and solvent-induced crystallization process of spherical micelles formed from PS-*b*-PAA/PS-*b*-P2VP-*b*-PEO blends. The micellar core was made of hydrogen-bonded P2VP and PAA blocks in DMF solvent. Different morphologies such as rupture of film, formation of cylindrical aggregates due to PEO chain folding, and square lamellae due to nucleation of PEO blocks at the corners of quasi-square lamellae were observed when the films were treated in DMF vapor for different times. The unique distribution of crystallisable blocks and phase separation between PEO and PS blocks in corona play an important role in crystallization process. As discussed in the previous section, patched spherical structures have also been reported, as shown in Figure 14.17 in complexes of PS-*b*-P2VP-*b*-PEO triblock copolymer with a tapered triblock copolymer P(PEO-MA)-*b*-PAA-*b*-P(PEO-MA) in DMF. Nanophase separation between corona-forming PS and PEO blocks resulted in patched spheres on the surface of PAA/P2VP core [66].

14.4 BLOCK COPOLYMER SYSTEMS WITH IONIC INTERACTION

Water-soluble interpolyelectrolyte complexes (IPECs) can be prepared by complexation of amphiphilic block copolymers, which comprise a hydrophobic block and an ionic or hydrophilic nonionic block [81, 82]. Double hydrophilic block copolymers having an ionic block and a nonionic one can be prepared for IPECs even in 1:1 charge-to-charge ratio of the polymeric components in aqueous solution [81]. IPECs normally have a core/corona structure and are often referred to as polyion complex micelles, complex coacervate micelles, or block ionomer complexes [83, 84].

14.4.1 Diblock Copolymer/Homopolymer Systems

Similar to core–shell structures reported in hydrogen-bonded block copolymer/homopolymer systems, core–shell micelle formation through electrostatic interaction were also reported. These complexes were generally formed at equimolar ratio of polycations and polyanions. But non-stoichiometric mixtures can also be employed to increase the stability

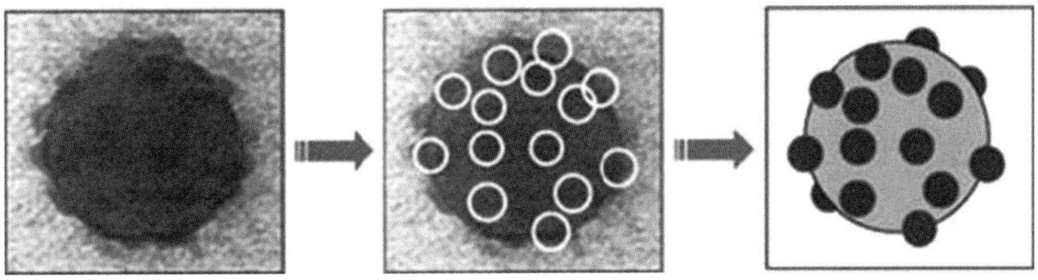

Figure 14.17 TEM images showing collapsed PS domains on the corona of PS-*b*-P2VP-*b*-PEO/P(PEO-MA)-*b*-PAA-*b*-P(PEO-MA) micelles. Gohy et al. [66]. Reproduced with permission of John Wiley and Sons.

of the system. Block ionomer complexes formed between poly(*N*-ethyl-4-vinylpyridinium bromide) (PEVP) homopolymer and poly(ethylene oxide)-*b*-poly(sodium methacrylate) (PEO-*b*-PMANa) block copolymer were studied by Kabanov et al. [84] Complexation between poly(methacrylate) anion (PMA^-) and poly(*N*-ethyl-4-vinylpyridinium) cation ($PEVP^+$) leads to formation of core–shell micelles with a complex core surrounded by water-soluble PEO corona. Stability of these IPECs was found to depend on pH of the medium and salt concentration. Complexation can be reversed by addition of electrolytes that screen electrostatic interaction. Different types of core–shell micelles were observed as a function of pH and molar ratios of anionic and cationic species [85, 86]. It is proposed that the size of as-formed core–shell micelles through electrostatic interaction can be manipulated by varying the amount of homopolyelectrolyte available for complexation and ionic concentration. Hofs et al. [87] prepared complex coacervate core microemulsion from poly(acrylic acid-*b*-acrylamide) (PAA-*b*-PAAm) and poly(*N*,*N*-dimethyl aminoethyl methacrylate) (PDMAEMA). The hydrodynamic radius of micelles was found to increase in phosphate buffer solution, because of the weakening electrostatic interaction between polyanions and polycations. But a general phenomenon of decrease in hydrodynamic size of micelles as a function of increasing homopolymer concentration was observed in both hydrogen-bonded and electrostatically formed complexes. Complexes of double hydrophilic block copolymer (sulfamate-carboxylate) isoprene-*b*-ethylene oxide with quaternized poly(2-vinylpyridine) were studied in aqueous solution. Spherical and ellipsoid shaped core–shell micellar structures were observed depending on the added salt concentration and mixing ratio between components [88]. Stuart et al. [89] studied complexation of poly([dimethyl amino]ethyl methacrylate)-*b*-poly(glyceryl methacrylate) (PDMAEMA-*b*-PGMA) with PAA. Vesicle-like objects, as shown in Figure 14.18, filled with micelles were observed, which totally disappear at higher salt concentrations.

As introduced previously, onion-like or core–shell–corona micelles were also reported in complexes formed via electrostatic interaction. Usually, homopolymers are added to micellar solution to form IPECs. For example, micellization of polyisobutylene-block-poly(methacrylic acid) (PIB-*b*-PMAA) block copolymer with quaternized poly(4-vinylpyridine) (P4EVP) forms a hydrophobic PIB core surrounded by a shell of PMAA/P4EVP complex and an ionic corona of PMAA not involved in complexation. These micelles were "dynamic" in nature, that is, dissociation of P4EVP chains observed at different pH or electrolyte concentration [90, 91]. Poly(sodium styrene sulfonate) (PSSNa) was added to micellar solutions of PS-*b*-P2VP, where PSSNa/P2VP complex forms micellar corona at low pH values [92]. The former system with a soft PIB core and later with a rigid PS core produced dynamic micelles responding to change in pH and ionic strength regardless of the T_g of core-forming blocks.

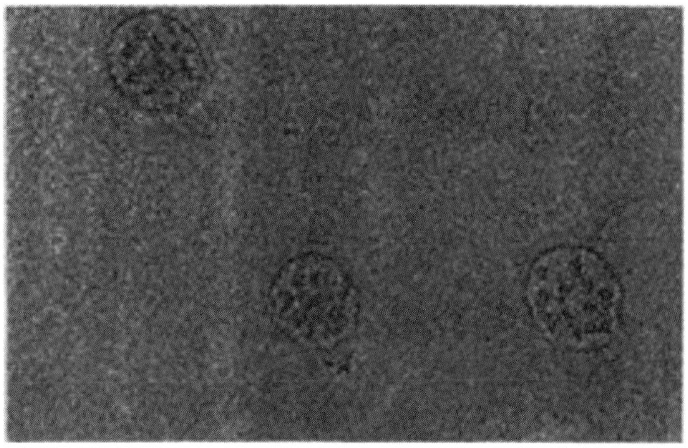

Figure 14.18 Cryo-TEM micrograph of IPEC micelles formed from PDMAEMA-*b*-PGMA/PAA system. Stuart et al. [89]. Reproduced with permission of American Chemical Society.

In addition to polymer systems with flexible coillike polymers discussed above, polymer complexes involving rigid rodlike polymers such as polyaniline was reported recently. The chain rigidity in conjugated polymers may provide the extra degree of freedom to control the self-assembly. Large compound vesicles were observed to form polyaniline/amphiphilic block copolymer complexes in aqueous solution. The morphology transformation was controlled via changing the composition or polyaniline content [93]. Similarly, multimicellar vesicles were observed from PMMA-*b*-PAA/polyaniline complexes [94]. The electrostatic interaction between polyaniline and polyacrylic groups of block copolymer was responsible for morphology evolution. Interestingly, the morphology transformation was observed in low concentrations of polyaniline. Rigid polymers with inherent properties such as hydrophobicity, chain rigidity, π–π stacking, and strong inter/intrachain electrostatic interactions may be responsible for the unique self-assembly behavior. Figure 14.19 shows the schematic representation of morphology evolution in amphiphilic block copolymer/polyaniline complexes.

14.4.2 Diblock/Triblock Copolymer Systems

Janus-type micelles were reported via electrostatic interaction between two double-hydrophilic block copolymers poly(acrylic acid)-*b*-poly(acryl amide) (PAA-*b*-PAAm), and poly(2-methylvinylpyridinium iodide)-*b*-poly(ethylene oxide) (P2MVP-*b*-PEO). The aggregate forms disc-shaped micelle with PAA/P2MVP complex coacervate core and microphase-separated asymmetric corona, with two distinct domains formed by PEO and PAAm. Figure 14.20 shows a TEM image of Janus-type micelles [95]. Temperature-sensitive complex coacervate micelles were prepared using two oppositely charged block copolymers poly(*N*-methyl-2-vinyl pyridinium iodide)-*b*-PEO (P2MVP-*b*-PEO) and PAA-*b*-PNIPAAm by Voets et al. [96] Initially, micelles with P2MVP/PAA complex core and randomly mixed PEO/PNIPAM corona were observed. At temperatures greater than 60 °C, onion-like micelles with PNIPAM core, P2MVP/PAA complex shell, and PEO

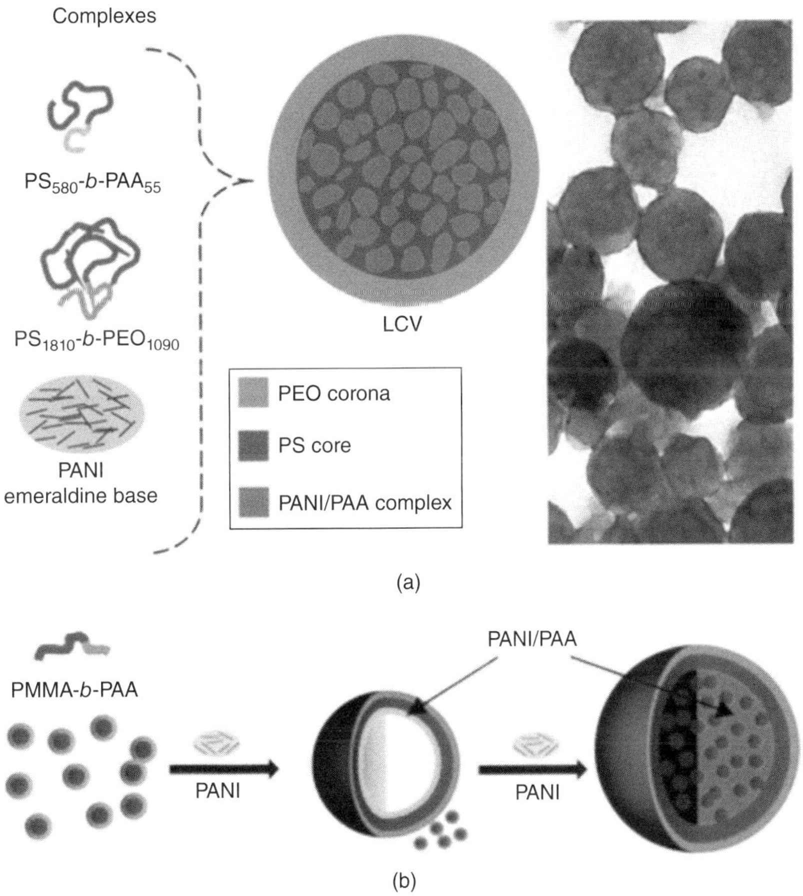

Figure 14.19 Schematic illustrations of morphology evolution in (a) PS-*b*-PAA/PS-*b*-PEO/PANI and (b) PMMA-*b*-PAA/PANI complex aggregates in aqueous solution. [(a) Reprinted with permission from Palanisamy, A. and Q. Guo, *J Phys Chem B* (2014), 118, 12796-12803. Copyright (2014) American Chemical Society.] Palanisamy and Guo [94]. Reproduced with permission of The Royal Society of Chemistry.

corona were observed. A shell cross-linked core–shell–corona micelle was prepared by exploiting the electrostatic interaction between poly(2-acrylamido-2-methylpropanesulfonate) (PAMPS) and poly([3-acrylamidopropyl]trimethylammonium chloride) (PAMPTMA) of two different block copolymers PNIPAM-*b*-PAMPS and PEO-*b*-PAMPTMA. The hydrophobic PNIPAM block collapses into the core with IPEC complex shell and the hydrophilic PEO chains form the corona. The temperature-sensitive PNIPAM core provides the advantage of changing hydrodynamic size of these micelles by varying temperature [97].

Triblock copolymers containing oppositely charged blocks within the block copolymer can form intramolecular IPECs (im-IPECs). A polyampholytic polybutadiene-*b*-poly(1-methyl-2-vinylpyridinium iodide)-*b*-PMAA (PB-*b*-P2VPq-*b*-PMAA) triblock copolymer containing a neutral PB block, a cationic P2VPq block, and an anionic PMAA block was found to self-assemble into micelles with "raspberry" morphology [98]. Multicompartmental micelles with PB core, P2VPq/PMAA im-IPEC shell, and excess PMAA-forming corona were observed. Thinner micellar corona were observed in systems with shorter PMAA blocks, because of complete complexation of PMAA with P2VPq. Morphologies observed

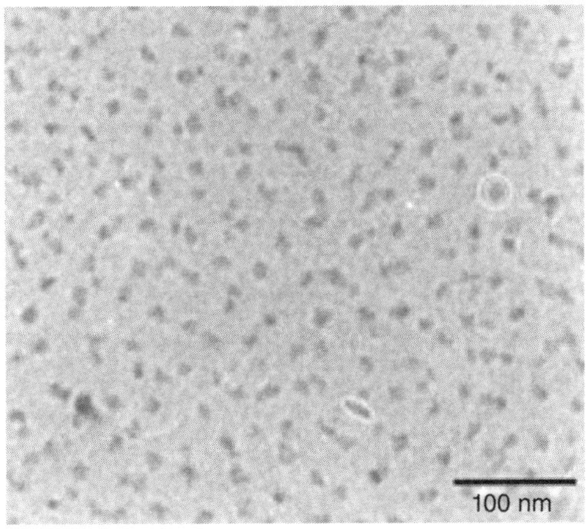

Figure 14.20 Cryo-TEM image showing nonspherical morphology of Janus-type micelles formed from PAA-*b*-PAAm/P2MVP-*b*-PEO complex. Voets et al. [95]. Reproduced with permission of John Wiley and Sons.

at different block length were shown in Figure 14.21. In another study, PB-*b*-PMAA-*b*-poly(2-(dimethylamino)ethyl methacrylate) (PB-*b*-PMAA-*b*-PDMAEMA) triblock copolymer with cationic PDMAEMA as a terminal block and anionic

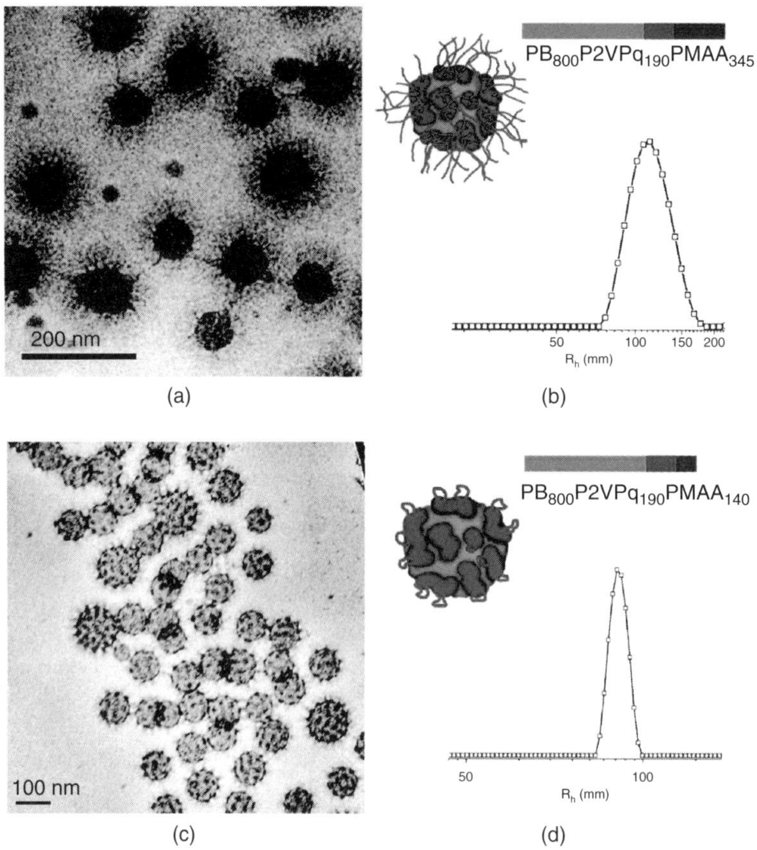

Figure 14.21 (a) and (c) Cryo-TEM images of triblock copolymers PB_{800}-*b*-$P2VPq_{190}$-*b*-$PMAA_{345}$, PB_{800}-*b*-$P2VPq_{190}$-*b*-PMAA140 and their schematic block length depiction with proposed solution structure were shown in (b) and (d), respectively . Schacher et al. [98]. Reproduced with permission of American Chemical Society.

PMAA as a middle block was used. A similar morphology was observed, as reported in the previous triblock copolymer system. Both the systems shown change in aggregation number with varying pH and ionic strength [99].

14.5 BLOCK COPOLYMER BLENDS VIA METAL–LIGAND COORDINATION BONDS

Self-assembly of block copolymers manipulated by adding another homopolymer or block copolymer that selectively interacts via secondary interactions such as hydrogen bonding or electrostatic interaction has been discussed in previous sections. Selective interaction of metal salts via metal–ligand coordination with one of the blocks of block copolymer was found to been an effective method in producing organic–inorganic hybrid materials. The block copolymers show morphology transition with increase in metal salt concentration. For example, Noro et al. [100] studied morphology transformation in PS-*b*-P4VP/$FeCl_3$ blends as a function of increasing $FeCl_3$ metal salt concentration. Block copolymer morphology transition from spherical, cylindrical, and alternating lamellar structures was observed as the weight ratio of $FeCl_3$ to pyridine was increased from 0.4, 0.6, and 0.9, which is shown in Figure 14.22. The metal salt plays role similar to that of a homopolymer in morphology development. Even though $FeCl_3$ is an inorganic compound, it is mixed in P4VP phase up to 0.9 molar equivalent of $FeCl_3$ to pyridine units. In the same vein, morphology transformations were observed in P4VP-*b*-PCL/gold [21], P2VP-*b*-PI/palladium [101], and PS-*b*-P4VP/cadmium [102] hybrids.

14.6 CONCLUDING REMARKS

Spontaneous self-assembly of block copolymers has provided an opportunity to create a variety of nanostructures both in bulk and in solution. Even though a number of synthetic techniques yield tailored macromolecules that can self-assemble into specific morphology, block copolymer blending is comparatively a simpler method. A step ahead, blending block copolymers through exploiting secondary interactions between dissimilar blending components provided flexibility in tuning morphology of self-assembled structures. Therefore, it is necessary to understand the effect of secondary interactions such as hydrogen bonding, electrostatic interaction, and metal–ligand coordination bonds on the morphology evolution of block copolymer blends. In this chapter, self-assembly and morphology of block copolymer blends both in solution and bulk were considered for discussion. This includes morphology

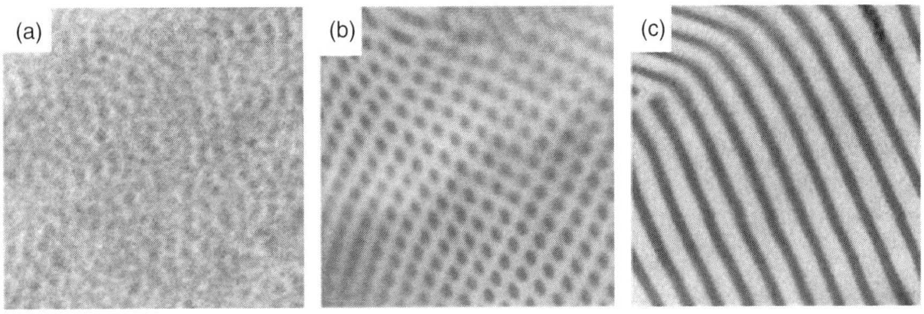

Figure 14.22 TEM images of PS-*b*-P4VP/$FeCl_3$ blends at weight ratio (a) 0.4, (b) 0.6, and (c) 0.9 of $FeCl_3$. Noro et al. [100]. Reproduced with permission of American Chemical Society.

evolution in blends of diblock and triblock copolymers with small molecules, homopolymers, and diblock copolymers. Composition-dependant morphology evolution of bock copolymer blends was discussed. The dynamic nature of secondary interactions plays a significant role in developing smart and responsive nanostructures [103]. Developing environment-sensitive micelles for drug delivery application is one of the active areas of research, which utilizes secondary interactions that respond to change in pH, ionic strength, temperature, and so on. Identifying new morphologies and properties in block copolymer blend systems by utilizing these noncovalent interactions are active areas of research. Complete understanding of these noncovalent interactions may help polymer scientist develop materials with tailored properties.

REFERENCES

1. Ruzette A-V, Leibler L. Block copolymers in tomorrow's plastics. Nat Mater 2005;4(1):19–31.
2. Zhang Y et al. Structure and properties of poly(butyl acrylate-block-sulfone-block-butyl acrylate) triblock copolymers prepared by ATRP. Macromol Chem Phys 2005;206(1):33–42.
3. Chen YT, Lo CT. Self-assembled structures in block copolymer/graft copolymer blends with hydrogen bonding interaction. Soft Matter 2013;9(6):1756–1760.
4. Hamley IW. *The Physics of Block Copolymers*. Oxford University Press; 1998.
5. Spontak RJ, Patel NP. In: Hamley IW, editor. *Developments in Block Copolymer Science and Technology*. New York: Wiley; 2004.
6. Eckelt A, Eckelt J, Wolf BA. Interpolymer complexes and polymer compatibility. Macromol Rapid Commun 2012;33(22): 1933–1937.
7. Pollino JM, Weck M. Non-covalent side-chain polymers: design principles, functionalization strategies, and perspectives. Chem Soc Rev 2005;34(3):193–207.
8. Aida T, Meijer EW, Stupp SI. Functional supramolecular polymers. Science 2012;335(6070):813–817.
9. Nunes SP et al. Switchable pH-responsive polymeric membranes prepared via block copolymer micelle assembly. ACS Nano 2011;5(5):3516–3522.
10. Bendejacq D et al. Well-ordered microdomain structures in polydisperse poly(styrene)-poly(acrylic acid) diblock copolymers from controlled radical polymerization. Macromolecules 2002;35(17):6645–6649.
11. Han YK, Pearce EM, Kwei TK. Poly(styrene-b-vinylphenyl dimethylsilanol) and its blends with homopolymers. Macromolecules 2000;33(4):1321–1329.
12. Holoubek J et al. Self-assembled structures in blends of block copolymer A-block-B with homopolymer A: SAXS and SANS study. Macromol Chem Phys 2006;207(20):1834–1841.
13. Jinnai H et al. 3D nanometer-scale study of coexisting bicontinuous morphologies in a block copolymer/homopolymer blend. Macromol Rapid Commun 2006;27(17):1424–1429.
14. Zhao JQ, Pearce EM, Kwei TK. Binary and ternary blends of polystyrene-block-poly(p-hydroxystyrene). Macromolecules 1997;30(23):7119–7126.
15. Wang C, Wang TM, Wang QH. Solvent annealing assisted self-assembly of hydrogen-bonded interpolymer complexes of AB block copolymer/C homopolymer in thin film. Polymer 2010;51(21):4836–4842.
16. Lee JY, Painter PC, Coleman MM. Hydrogen bonding in polymer blends. 4. Blends involving polymers containing methacrylic acid and vinylpyridine groups. Macromolecules 1988;21(4):954–960.
17. Dobrosielska K et al. Nanophase-separated structures of AB block copolymer/C homopolymer blends with complementary hydrogen-bonding interactions. Macromolecules 2008;41(20):7695–7698.
18. Chen S-C, Kuo S-W, Chang F-C. On modulating the self-assembly behaviors of poly(styrene-b-4-vinylpyridine)/octyl gallate blends in solution state via hydrogen bonding from different common solvents. Langmuir 2011;27(16): 10197–10205.
19. de Luzuriaga AR et al. Design and stabilization of block copolymer micelles via phenol–pyridine hydrogen-bonding interactions. Polymer 2010;51(6):1355–1362.
20. Salim NV, Hameed N, Guo Q. Competitive hydrogen bonding and self-assembly in poly(2-vinyl pyridine)-block-poly(methyl methacrylate)/poly(hydroxyether of bisphenol A) blends. J Polym Sci B 2009;47(19):1894–1905.
21. Lin T, Ho R-M, Ho J-C. Phase behavior in self-assembly of inorganic/poly(4-vinylpyridine)-b-poly(ε-caprolactone) hybrid. Macromolecules 2009;42(3):742–751.
22. Karanikolas A et al. Stimuli-responsive poly(ethylene oxide)-b-poly(2-vinylpyridine)-b-poly(ethylene oxide) triblock

copolymers and complexation with poly(acrylic acid) at low pH. Eur Phys J E 2008;27(3):335–343.

23. Chen SC et al. On modulating the phase behavior of block copolymer/homopolymer blends via hydrogen bonding. Macromolecules 2010;43(2):1083–1092.
24. Dobrosielska K et al. Effect of homopolymer molecular weight on nanophase-separated structures of AB block copolymer/C homopolymer blends with hydrogen-bonding interactions. Macromolecules 2009;42(18):7098–7102.
25. Hameed N, Guo Q. Nanostructure and hydrogen bonding in interpolyelectrolyte complexes of poly(epsilon-caprolactone)-block-poly(2-vinyl pyridine) and poly(acrylic acid). Polymer 2008;49(24):5268–5275.
26. Lefèvre N et al. Self-assembly in thin films of mixtures of block copolymers and homopolymers interacting by hydrogen bonds. Macromolecules 2010;43(18):7734–7743.
27. Chen W-C, Kuo S-W, Chang F-C. Self-assembly of an A–B diblock copolymer blended with a C homopolymer and a C–D diblock copolymer through hydrogen bonding interaction. Polymer 2010;51(18):4176–4184.
28. Hu N et al. Interchain hydrogen bonding interaction induced phase behaviors of poly(ethylene oxide)-b-poly(N-vinylcarbazole)/poly(acrylic acid) blend. Macromolecules 2012; 45(13):5546–5555.
29. Hameed N, Salim NV, Guo Q. Microphase separation through competitive hydrogen bonding in self-assembled A-b-B/C diblock copolymer/homopolymer complexes. J Chem Phys 2009;131(21):214905–214905.
30. Salim NV, Hanley T, Guo Q. Microphase separation through competitive hydrogen bonding in double crystalline diblock copolymer/homopolymer blends. Macromolecules 2010;43(18):7695–7704.
31. Chen WC et al. Self-assembly through competitive interactions of miscible diblock copolymer/homopolymer blends: Poly(vinylphenol-b-methyl methacrylate)/poly(vinyl pyrrolidone) blend. Macromolecules 2008;41(4):1401–1410.
32. Lee HF et al. Hydrogen-bonding interactions mediate the phase behavior of an A-B/C block copolymer/homopolymer blend comprising poly(methyl methacrylate-b-vinylpyrrolidone) and poly(vinylphenol). Macromolecules 2006;39(16):5458–5465.
33. Chen WC et al. Self-assembly structures through competitive interactions of crystalline-amorphous diblock copolymer/homopolymer blends: Poly(epsilon-caprolactone-b-4-vinyl pyridine)/poly(vinyl phenol). Macromolecules 2009;42(10): 3580–3590.
34. Hameed N, Guo Q. Selective hydrogen bonding and hierarchical nanostructures in poly(hydroxyether of bisphenol A)/poly(ϵ-caprolactone)-block-poly(2-vinyl pyridine) blends. Polymer 2008;49(4):922–933.
35. Kosonen H et al. Self-organized cross-linked phenolic thermosets: Thermal and dynamic mechanical properties of novolac/block copolymer blends. Polymer 2001;42(23): 9481–9486.
36. Wu S et al. Toughening epoxy thermosets with block ionomer complexes: A nanostructure–mechanical property correlation. Macromolecules 2012;45(9):3829–3840.
37. Wu S et al. A new route to nanostructured thermosets with block ionomer complexes. Soft Matter 2012;8(3):688–698.
38. Guo Q. *Thermosets: Structure, Properties and Applications.* Woodhead Pub Limited; 2012.
39. Han SH et al. Highly asymmetric lamellar nanopatterns via block copolymer blends capable of hydrogen bonding. ACS Nano 2012;6(9):7966–7972.
40. Han SH et al. Phase behavior of binary blends of block copolymers having hydrogen bonding. Macromolecules 2011;44(12):4970–4976.
41. Asari T et al. Three-phase hierarchical structures from AB/CD diblock copolymer blends with complemental hydrogen bonding interaction. Macromolecules 2005;38(21):8811–8815.
42. Dobrosielska K, Takano A, Matsushita Y. Creation of hierarchical nanophase-separated structures via supramacromolecular self-assembly from two asymmetric block copolymers with short interacting sequences giving hydrogen bonding interaction. Macromolecules 2009;43(2):1101–1107.
43. Asari T et al. Archimedean tiling structures from ABA/CD block copolymer blends having intermolecular association with hydrogen bonding. Macromolecules 2006;39(6):2232–2237.
44. Jiang SM, Gopfert A, Abetz V. Novel morphologies of block copolymer blends via hydrogen bonding. Macromolecules 2003;36(16):6171–6177.
45. Štěpánek M et al. Association of poly(4-hydroxystyrene)-block-poly(ethylene oxide) in aqueous solutions: block copolymer nanoparticles with intermixed blocks. Langmuir 2011;28(1):307–313.
46. Koňák Č, Sedlák M. pH-sensitive micelles formed by interchain hydrogen bonding of poly(methacrylic acid)-block-poly(ethylene oxide) copolymers. Macromol Chem Phys 2007;208(17):1893–1899.
47. Holappa S et al. Self-complexation of poly(ethylene oxide)-block-poly(methacrylic acid) studied by fluorescence spectroscopy. Macromolecules 2004;37(18):7008–7018.
48. Cai Y, Armes SP. A zwitterionic ABC triblock copolymer that forms a “Trinity” of micellar aggregates in aqueous solution. Macromolecules 2004;37(19):7116–7122.
49. Kyeremateng SO et al. Synthesis and self-organization of poly(propylene oxide)-based amphiphilic and triphilic block copolymers. Macromolecules 2011;44(3):583–593.
50. Wang LF, Pearce EM, Kwei TK. Glass transitions in hydrogen-bonded polymer complexes. J Polym Sci B 1991;29(5):619–626.
51. Yao X, Chen D, Jiang M. Micellization of PS-b-P4VP/formic acid in chloroform without or with the premixing of the copolymer with decanoic acid. Macromolecules 2004;37(11): 4211–4217.
52. Zhang WQ et al. Adsorption of poly(4-vinyl pyridine) unimers into polystyrene-block-poly(acrylic acid) micelles in ethanol due to hydrogen bonding. Macromolecules 2004;37(8):2924–2929.
53. Hameed N, Guo Q. Self-assembled complexes of poly(acrylic acid) and poly(styrene)-block-poly(4-vinyl pyridine). J Polym Sci B 2009;47(12):1192–1202.
54. Hameed N, Liu J, Guo Q. Self-assembled complexes of poly(4-vinylphenol) and poly(epsilon-caprolactone)-block-poly(2-vinylpyridine) via competitive hydrogen bonding. Macromolecules 2008;41(20):7596–7605.

55. Zhang W et al. Comicellization of poly(ethylene glycol)-block-poly(acrylic acid) and poly(4-vinylpyridine) in ethanol. Macromolecules 2005;38(3):899–903.
56. Liu S et al. Interpolymer hydrogen-bonding complexation induced micellization from polystyrene-b-poly(methyl methacrylate) and PS(OH) in toluene. Langmuir 2000;16(8): 3712–3717.
57. Gao W-P et al. Controlling vesicle formation via interpolymer hydrogen-bonding complexation between poly(ethylene oxide)-block-polybutadiene and poly(acrylic acid) in solution. Macromolecules 2006;39(14):4894–4898.
58. Kuang M et al. Structural factors of rigid-coil polymer pairs influencing their self-assembly in common solvent. J Phys Chem B 2004;108(41):16023–16029.
59. Duan H et al. Self-assembly of rigid and coil polymers into hollow spheres in their common solvent. J Phys Chem B 2004;108(2):550–555.
60. Salim NV et al. A simple and effective approach to vesicles and large compound vesicles via complexation of amphiphilic block copolymer with polyelectrolyte in water. Macromol Rapid Commun 2012;33(5):401–406.
61. Lefevre N, Fustin C-A, Gohy J-F. Reorganization of hydrogen-bonded block copolymer complexes. Langmuir 2007;23(8): 4618–4622.
62. Lefevre N et al. Self-assembly of block copolymer complexes in organic solvents. Polymer 2007;48(8):2306–2311.
63. Lee SC et al. pH-induced reversible complexation of poly(ethylene glycol) and poly(epsilon-caprolactone)-b-poly (methacrylic acid) copolymer micelles. Macromolecules 2005; 38(22):9291–9297.
64. Matejicek P et al. Interpolymer complexes based on the core/shell micelles. Interaction of polystyrene-block-poly (methacrylic acid) micelles with linear poly(2-vinylpyridine) in 1,4-dioxane water mixtures and in aqueous media. J Phys Chem B 2007;111(29):8394–8401.
65. Sotiropoulou M, Oberdisse J, Staikos G. Soluble hydrogen-bonding interpolymer complexes in water: A small-angle neutron scattering study. Macromolecules 2006;39(8):3065–3070.
66. Gohy JF et al. Segregation of coronal chains in micelles formed by supramolecular interactions. Macromol Rapid Commun 2004;25(17):1536–1539.
67. Zhang WQ et al. Formation of core-shell-corona micellar complexes through adsorption of double hydrophilic diblock copolymers into core-shell micelles. Macromol Rapid Commun 2005;26(16):1341–1345.
68. Kubowicz S et al. Multicompartment micelles formed by self-assembly of linear ABC triblock copolymers in aqueous medium. Angew Chem Int Ed 2005;44(33):5262–5265.
69. Li ZB et al. Multicompartment micelles from ABC miktoarm stars in water. Science 2004;306(5693):98–101.
70. Talingting MR et al. Onion-type micelles from polystyrene-block-poly(2-vinylpyridine) and poly(2-vinylpyridine)-block-poly(ethylene oxide). Macromolecules 1999;32(5):1593–1601.
71. Salim NV, Guo Q. Multiple vesicular morphologies in AB/AC diblock copolymer complexes through hydrogen bonding interactions. J Phys Chem B 2011;115(31):9528–9536.
72. Xie D et al. Structural evolution of mixed micelles due to interchain complexation and segregation investigated by laser light scattering. J Phys Chem B 2007;111(4):778–781.
73. Gao J et al. Fabrication of fibril like aggregates by self-assembly of block copolymer mixtures via interpolymer hydrogen bonding. Polymer 2008;49(9):2354–2361.
74. Li G et al. Formation of complex micelles with double-responsive channels from self-assembly of two diblock copolymers. Angew Chem Int Ed 2006;45(30):4959–4962.
75. Xiong Da et al. Composite worm-like aggregates formed from a pair of block-copolymers containing hydrogen-bonding donor and acceptor. Macromol Rapid Commun 2007;28(2):194–199.
76. Kuo S-W et al. Supramolecular micellization of diblock copolymer mixtures mediated by hydrogen bonding for the observation of separated coil and chain aggregation in common solvents. Macromol Rapid Commun 2008;29(3):229–233.
77. Hsu C-H et al. Self-assembly behavior of A-B diblock and C-D random copolymer mixtures in the solution state through mediated hydrogen bonding. Langmuir 2008;24(15):7727–7734.
78. Zhang WQ et al. Core-shell-corona micellar complexes between poly(ethylene glycol)-block-poly(4-vinyl pyridine) and polystyrene-block-poly(acrylic acid). Macromol Chem Phys 2005;206(23):2354–2361.
79. Xiong Da et al. Temperature-responsive multilayered micelles formed from the complexation of PNIPAM-b-P4VP block-copolymer and PS-b-PAA core-shell micelles. Polymer 2008;49(10):2548–2552.
80. Huang W et al. Solvent-induced crystallization of spherical micelles formed by copolymer blends. Macromolecules 2006;39(23):8075–8082.
81. Lefèvre N, Fustin C-A, Gohy J-F. Polymeric micelles induced by interpolymer complexation. Macromol Rapid Commun 2009;30(22):1871–1888.
82. Kuo S-W. Hydrogen bond-mediated self-assembly and supramolecular structures of diblock copolymer mixtures. Polym Int 2009;58(5):455–464.
83. Harada A, Kataoka K. Formation of polyion complex micelles in an aqueous milieu from a pair of oppositely-charged block copolymers with poly(ethylene glycol) segments. Macromolecules 1995;28(15):5294–5299.
84. Kabanov AV et al. Soluble stoichiometric complexes from poly(N-ethyl-4-vinylpyridinium) cations and poly(ethylene oxide)-block-polymethacrylate anions. Macromolecules 1996; 29(21):6797–6802.
85. Gohy JF et al. Water-soluble complexes formed by sodium poly(4-styrenesulfonate) and a poly(2-vinylpyridinium)-block-poly(ethyleneoxide) copolymer. Macromolecules 2000;33(25): 9298–9305.
86. Weaver JVM, Armes SP, Liu SY. A "Holy Trinity" of micellar aggregates in aqueous solution at ambient temperature: Unprecedented self-assembly behavior from a binary mixture of a neutral-cationic diblock copolymer and an anionic polyelectrolyte. Macromolecules 2003;36(26):9994–9998.
87. Hofs B et al. Complex coacervate core micro-emulsions. Soft Matter 2008;4(7):1473–1482.
88. Pispas S. Complexes of polyelectrolyte-neutral double hydrophilic block copolymers with oppositely charged surfactant and polyelectrolyte. J Phys Chem B 2007;111(29): 8351–8359.
89. Cohen Stuart MA, Besseling NAM, Fokkink RG. Formation of micelles with complex coacervate cores. Langmuir 1998;14(24):6846–6849.

90. Pergushov DV et al. Novel water-soluble micellar interpolyelectrolyte complexes. J Phys Chem B 2003;107(32): 8093–8096.
91. Pergushov DV et al. Micelles of polyisobutylene-block-poly(methacrylic acid) diblock copolymers and their water-soluble interpolyelectrolyte complexes formed with quaternized poly(4-vinylpyridine). Polymer 2004;45(2):367–378.
92. Talingting MR et al. Observation of massive overcompensation in the complexation of sodium poly(styrenesulfonate) with cationic polymer micelles. Macromolecules 2000;33(26): 9612–9619.
93. Palanisamy A, Guo Q. Large compound vesicles from amphiphilic block copolymer/rigid-rod conjugated polymer complexes. J Phys Chem B 2014;118(44):12796–12803.
94. Palanisamy A, Guo Q. Self-assembled multimicellar vesicles via complexation of a rigid conjugated polymer with an amphiphilic block copolymer. RSC Adv 2014;4(97): 54752–54759.
95. Voets IK et al. Double-faced micelles from water-soluble polymers. Angew Chem Int Ed 2006;45(40):6673–6676.
96. Voets IK et al. Temperature responsive complex coacervate core micelles With a PEO and PNIPAAm corona. J Phys Chem B 2008;112(35):10833–10840.
97. Masci G, De Santis S, Cametti C. Dielectric properties of micellar aggregates due to the self-assembly of thermoresponsive diblock copolymers. J Phys Chem B 2011;115(10):2196–2204.
98. Schacher F, Walther A, Müller AHE. Dynamic multicompartment-core micelles in aqueous media. Langmuir 2009;25(18):10962–10969.
99. Betthausen E et al. Dual stimuli-responsive multicompartment micelles from triblock terpolymers with tunable hydrophilicity. Soft Matter 2011;7(19):8880–8891.
100. Noro A et al. Preparation and morphology control of block copolymer/metal salt hybrids via solvent-casting by using a solvent with coordination ability. Macromolecules 2010;43(12): 5358–5364.
101. Tsutsumi K et al. Selective incorporation of palladium nanoparticles into microphase-separated domains of poly(2-vinylpyridine)-block-polyisoprene. Langmuir 1999;15(16): 5200–5203.
102. Lee DH et al. Swelling and shrinkage of lamellar domain of conformationally restricted block copolymers by metal chloride. Macromolecules 2006;39(6):2027–2030.
103. Pietsch C et al. Thermo-induced self-assembly of responsive poly(DMAEMA-b-DEGMA) block copolymers into multi- and unilamellar vesicles. Macromolecules 2012;45(23):9292–9302.

15

DYNAMICS SIMULATIONS OF MICROPHASE SEPARATION IN BLOCK COPOLYMERS

Xuehao He[1], Xuejin Li[2], Peng Chen[3], and Haojun Liang[4]

[1] *Department of Chemistry, School of Science, Tianjin University, Tianjin, China*
[2] *Division of Applied Mathematics, Brown University, Providence, RI, USA*
[3] *Department of Materials, School of Chemistry and Chemical Engineering, Anhui University, Hefei, China*
[4] *Department of Polymer Science and Engineering, University of Science & Technology of China, Hefei, China*

15.1 INTRODUCTION

Block copolymers are a class of macromolecules in which each polymer chain consists of two or more chemically distinct blocks and the blocks are covalently linked together. According to the connecting ways and numbers of blocks, they are classified as linear diblock or multiblock copolymer, star block copolymer, branching block, circle block copolymer, and so on (Fig. 15.1). Due to the repulsive interactions among the different blocks, the different polymeric blocks are usually incompatible and phase separation occurs when decreasing temperature or increasing macromolecular weight. Critically, the phase separation on the larger scale cannot proceed due to the covalent bond connections of different blocks and such a class of copolymer has an ability to form rich microphase structures such as spherical, cylinder, lamellar, and gyroid structures by phase separation at the scale of 10–100 nm. The patterns of ordered microphases strongly depend on the molecular parameters of block copolymers, such as component, interaction energies between distinct blocks, and the architectures of block copolymers. Using different block copolymers, variously ordered microstructures, including body-cubic (BCC) spheres, hexagonally ordered cylinders (HEX), lamellae (LAM), knitting pattern (KP), and the core–shell bicontinuous gyroid phase structure, have been discovered in experiments [1–5].

Systematic study of the segregation behaviors of block copolymers is necessary for an in-depth understanding of the formation of new phase structures and the transition mechanism of various phase structures, besides being valuable in guiding the potential application in the fields of advanced materials such as nanochip and photocrystal manufacture. It is critical to develop the effective technologies to control the shapes and sizes of microstructures. The rich phase structures of block copolymers also attracted attention toward theories [6–8] and computer simulations [9–11]. Compared with experiments, computer simulations have become powerful tools for investigating various phenomena in complex systems, such as block copolymers. They captured the essential features characterizing microphase ordering of block copolymers at different levels.

In the field of macromolecular simulation, the coarse-grained (CG) method, based on a viewpoint that the phase behaviors of various polymers depend on a few parameters and not sensitively on the details of molecular structures, is often used. The general property of phase separation only depends on the chain length and Flory interaction in the mean field. For example, the transition point of phase separation for the symmetric diblock copolymer is $\chi N \simeq 10.5$ [12]. A specific polymer chain can be represented using different CG models. For example, a polystyrene-*b*-polyethyleneoxide (PS-*b*-PEO) chain can be represented using lattice model, off-lattice model, or even string model (Fig. 15.2).

The CG method not only provides general models for studying a class of block copolymers but also conducts efficient algorithms for simulation. In this chapter, we overview the theoretical and computational approaches toward the simulations of dynamics of microphase separation of block copolymers with the focus on the recent contributions applying Monte Carlo (MC), dissipation particle dynamics

Polymer Morphology: Principles, Characterization, and Processing, First Edition. Edited by Qipeng Guo.

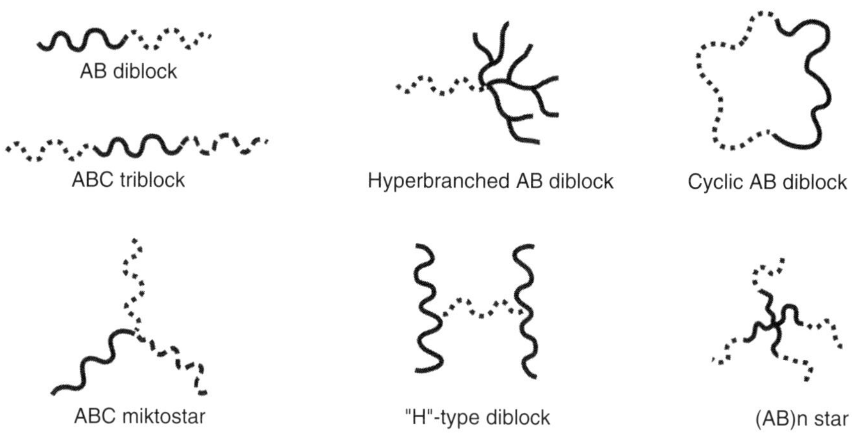

Figure 15.1 Block copolymer architectures.

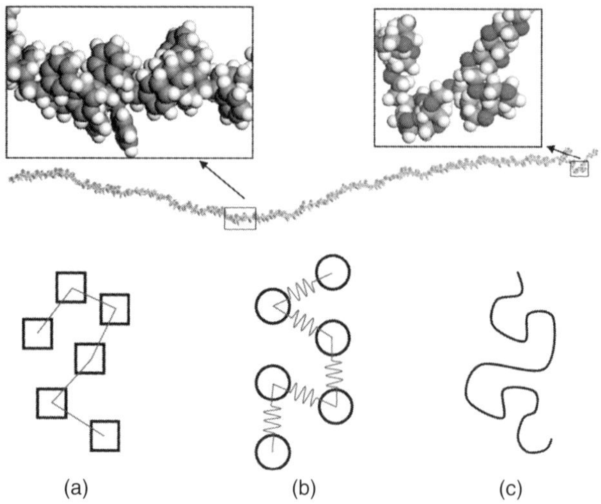

Figure 15.2 Three coarse-grained (CG) models to represent a PS-*b*-PEO polymer chain. (a) Lattice model; (b) bead–spring model; (c) string model.

(DPD), and dynamic polymeric self-consistent field (SCF) methods.

15.2 POLYMER MODEL AND SIMULATION ALGORITHM

15.2.1 Monte Carlo Method

The MC method has been extensively applied for studying the microphase separation of block copolymers. Basically, there are lattice model and off-lattice model in MC algorithms for simulating the microphase separation of block copolymers. A lattice model means that the movement of CG bead and energy calculation in MC simulation is formulated strictly on an underlying lattice. A squall potential is usually involved in the energy calculation. In contrast, the off-lattice model employs a kind of bead–spring potential to describe the interaction between neighboring CG beads, and the movement of a CG bead is decided by the interaction potential. Generally, at molecular level, the off-lattice MC simulation considers more detail than lattice MC simulation does, such as bond angle, bond length, and chain flexibility. However, the researchers pay more attention to the morphology than to the molecular details in the field of microphase separation of block copolymers. Furthermore, the lattice MC simulation runs much faster than the off-lattice MC simulation does. Therefore, many MC simulations employed the lattice model to explore the microphase separation of block copolymers.

There are two popular routines to realize the lattice model in MC simulation of polymer microphase separation (see Figure 15.3). One was first proposed by Carmesin and Kremer [13] and after that it was improved by Deutsch and Binder [15]. In this routine, a polymer chain is modeled as a set of bead units and every bead occupies a cubic cell (eight sites) in lattice space. In this polymer model, 108 bond vector and 5 bond lengths should be considered. It makes the MC simulation very fine-grained. The other one was proposed by Shaffer, in which a bead unit only occupies one site in lattice space [14]. Both of these routines have been used to reproduce experimental observations and to make reasonable predictions [16].

In recent decades, researchers have developed various novel MC algorithms to explore the phase separation of polymer. Detcheverry et al. proposed a new MC sampling formalism, in which the interaction between beads is not the conventional pairwise one but a "grid-mediated" potential [17, 18]. The uncrossability of the conventional bond is broken up in this formalism, while the physically realistic dynamics may be well recovered by restricting the trial moves in MC formalism to physically realistic local moves, according to the authors. This sampling formalism has been successfully applied to study the block copolymers and nanocomposites. Using "soft" repulsive potentials, Wang proposed fast off-lattice and lattice MC algorithms [19, 20]. The basic idea of these algorithms is to allow CG beads overlapping instead of hard-core repulsion in the conventional off-lattice MC algorithm and the self-avoiding walk in the conventional lattice one. By this disposition, the chain relaxation and

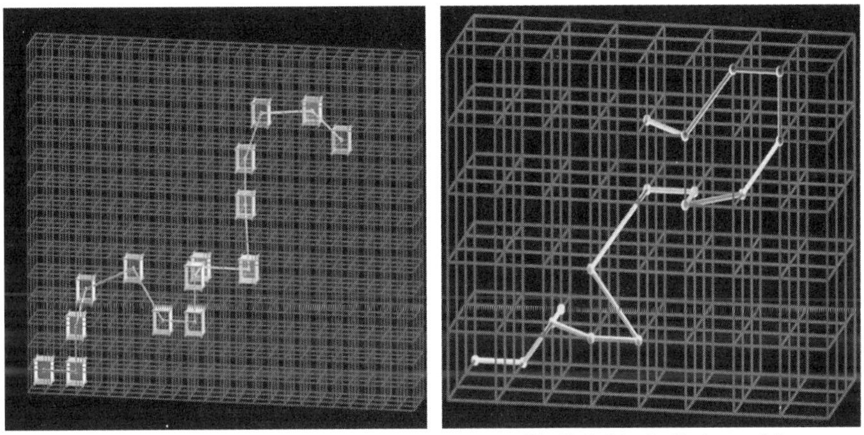

Figure 15.3 Two popular routines proposed by Carmesin and Kremer [13] (Reproduced with permission of American Chemical Society) and Shaffer [14] (Reproduced with permission of Elsevier) to generate the lattice model in MC simulation. (*See color plate section for the color representation of this figure.*)

sampling of polymer chain configuration in simulation are greatly fastened and it is particularly suitable for the study of equilibrium properties of polymer microphase separation [21].

15.2.2 Dissipative Particle Dynamics Method

DPD is a mesoscopic simulation method, introduced in 1992 by Hoogerbrugge and Koelman [22]. It is a simple but intrinsically promising simulation method that allows the study of the phase behaviors of block copolymers. In a DPD simulation, a particle represents the center of mass in a cluster of atoms, and the position and momentum of the particle are updated in a continuous phase but spaced at discrete time steps. Particles i and j at positions r_i and r_j interact with each other via pairwise conservative, dissipative, and random forces, which are given by

$$F_{ij}^{C} = a_{ij}\omega(r_{ij})n_{ij} \tag{15.1}$$

$$F_{ij}^{D} = -\gamma\omega^2(r_{ij})(n_{ij}\cdot v_{ij})n_{ij} \tag{15.2}$$

$$F_{ij}^{R} = \sigma\omega(r_{ij})\varsigma_{ij}\Delta t^{-1/2}n_{ij} \tag{15.3}$$

where $r_{ij} = r_i - r_j$, $r_{ij} = |r_{ij}|$, $n_{ij} = r_{ij}/r_{ij}$, and $v_{ij} = v_i - v_j$. The parameters a_{ij}, γ, and σ define the strength of conservative, dissipative, and random forces. In addition, ς_{ij} is a random number with zero mean and unit variance.

$$\langle\varsigma_{ij}(t)\rangle = 0 \tag{15.4}$$

$$\langle\varsigma_{ij}(t)\varsigma_{kl}(\tau)\rangle = (\delta_{ik}\delta_{jl} + \delta_{il}\delta_{jk})\delta(t-\tau) \tag{15.5}$$

The weight function $\omega(r_{ij})$ is given by

$$\omega(r_{ij}) = \begin{cases} 1 - r_{ij}/r_c & r_{ij} \le r_c \\ 0 & r_{ij} > r_c \end{cases} \tag{15.6}$$

where r_c is the cutoff radius, which gives the extent of interaction range. In the DPD method, the issipative force and the random force act as heat sink and source, respectively. The combined effect of the two forces acts as a thermostat. It conserves momentum and provides the correct description of hydrodynamics to the system. Also, a common choice of the soft repulsion for the conservative force permits us to use larger integration time steps than are usually allowed by the molecular dynamics (MD) simulation technique; thus, DPD is a simple but intrinsically promising simulation method that correctly represents hydrodynamic interactions.

To model the block copolymers, the total force can also have an elastic contribution, which is derived from the harmonic force used to connect two consecutive particles in the chains of polymer [10]. This contribution is expressed as

$$F_{ij}^{S} = k_s\left(\frac{r_s - r_{ij}}{r_s}\right)n_{ij} \tag{15.7}$$

where k_s and r_s are the spring constant and equilibrium bond length between two consecutive particles, respectively. To control the chain flexibility, an extra bond-bending force between consecutive bonds is added.

$$F^{\theta} = -\nabla v_{bend} \tag{15.8}$$

$$v_{bend} = \frac{1}{2}k_\theta(\theta - \theta_0)^2 \tag{15.9}$$

where k_θ and θ_0 are the bending constant and the equilibrium angle between two consecutive bonds, respectively.

The dynamics of the DPD particles are followed by solving Newton's equation of motion, with the forces above using a modified version of velocity-Verlet algorithm:

$$\begin{aligned}
r_i(t+\delta t) &= r_i(t) + v_i\delta t + 0.5f_i(t)\delta t^2 \\
\widetilde{v}_i(t+\delta t) &= \widetilde{v}_i(t) + \lambda f_i(t)\delta t \\
f_i(t+\delta t) &= f_i(r_i(t+\delta t), \widetilde{v}_i(t+\delta t)) \\
v_i(t+\delta t) &= v_i(t) + 0.5(f_i(t) + f_i(t+\delta t))\delta t
\end{aligned} \quad (15.10)$$

where f_i is the total force on a particle i and $\widetilde{v}_i$ denotes a predicted velocity that is corrected in the last step of the algorithm.

15.2.3 Polymeric Self-Consistent Field Theory

Instead of the MC and MD methods using explicit particles, another method, that is, polymeric self-consistent field theory (SCFT) proposed by Edwards, is often used to study the phase separation of block copolymers. In SCFT, a polymer chain is treated as a Gaussian string, which is exposed to a set of effective chemical potentials (ω). The chemical potentials are used instead of the actual interactions between different components. Importantly, the relation between the external potentials and the concentration field (ϕ) is bijective.

In the melting system of diblock AB copolymer, the free energy of such a system per volume (in units of k_BT) equals

$$\begin{aligned}
\frac{F}{v} = -\frac{1}{N}\ln\left(\frac{Q_P}{v}\right) + \frac{1}{v}\int \Big[\chi_{AB}\phi_A\phi_B - \omega_A\phi_A \\
-\omega_B\phi_B + \frac{\kappa_H}{2}(\phi_A+\phi_B-1)^2\Big]
\end{aligned} \quad (15.11)$$

where N is the length of a copolymer chain, χ_{ij} is the Flory–Huggins parameter between blocks i and j, and κ_H is the coefficient of compression energy. Q_P is the partition function of copolymer chain ($Q_P = \int drg(r,s)$). $g(r,s)$ is the distribution probability of a section in a string chain with a free chain end and the "connected end" located at r, and s is the contour length variable increasing continuously from 0 to 1 from one end of a block component to the other. Similarly, a second distribution function $g'(r,s)$, starting from another chain end, can be defined. The function $g(r,s)$ and $g'(r,s)$ satisfies a diffusion equation (Eq. 15.12) with the initial condition $g(r,0) = 1$ and $g'(r,0) = 1$.

$$\left.\begin{aligned}
\frac{\partial g_i(r,s)}{\partial s} &= \nabla^2 g_i(r,s) - N\omega_i g_i(r,s) \\
\frac{\partial g_i'(r,s)}{\partial s} &= \nabla^2 g_i'(r,s) - N\omega_i g_i'(r,s)
\end{aligned}\right\} (I = A, B) \quad (15.12)$$

The density of each component is calculated by the following equation.

$$\begin{aligned}
\phi_A(r) &= \frac{v}{Q_P}\int_0^{f_A} dsg(r,s)g'(r,1-s) \\
\phi_B(r) &= \frac{v}{Q_P}\int_{f_A}^{1} dsg(r,s)g'(r,1-s)
\end{aligned} \quad (15.13)$$

Here, f_A is the volume fraction of A block in diblock copolymer. To study the dynamics of phase separation, the polymeric external potential dynamics (EPD) method can be employed, which was proposed by Maurits and Fraaije [23] in dynamic density functional theory (DDFT) method (bead–string model). In EPD, the monomer concentration is a conserved quantity, and the polymer dynamics is inherently of Rouse type. The external dynamical equation in terms of the potential field ω_i is expressed as

$$\frac{\partial \omega_i(r)}{\partial t} = -D\nabla^2(\mu_i + \eta_i) \quad (15.14)$$

Here $\mu_i = \delta F/\delta\phi_i$ and D is a constant diffusion coefficient of copolymer. η_i is a Gaussian distributed random noise corresponding to the thermal fluctuations in the experiment and fulfills:

$$\begin{aligned}
\langle \eta_i(r,t)\rangle &= 0 \\
\langle \eta_i(r,t)\eta_j(r',t')\rangle &= \beta\delta_{ij}\delta(t-t')\delta(r-r')
\end{aligned} \quad (15.15)$$

The functional derivative $\delta F/\delta\phi_i$ can be calculated from Equation 15.11, which is given as

$$\begin{aligned}
\frac{\delta F}{\delta \phi_A} &= \chi_{AB}\phi_B + \kappa_H - \omega_A \\
\frac{\delta F}{\delta \phi_B} &= \chi_{AB}\phi_A + \kappa_H - \omega_B
\end{aligned} \quad (15.16)$$

Such a framework can be applied straightforward to the other copolymers with complex structures such as multiblock copolymer or star-block copolymer. The critical point is to rebuild the partition functions of the polymer chain with different chain structures. For miktoarm star polymer, the expression of partition calculation $g_i(r,s)$ and $g_i'(r,s)$ is the same as in Equation 15.12. But the contour length variable s in Equations 15.12 and 15.13 is from 0 to f_i (f_i is the average volume fraction of the ith component in the system; i represents A, B, and C; and $f_A + f_B + f_C = 1$). Meanwhile, the initial condition is $g_i(r,0) = 1$ and $g_A'(r,0) = g_B(r,f_B)g_C(r,f_C)$, $g_B'(r,0) = g_A(r,f_A)g_C(r,f_C)$, and $g_C'(r,0) = g_A(r,f_A)g_B(r,f_B)$. For the solution system, the partition function of a solvent molecule is $Q_S = \int dr\,\exp(-\omega_S)$. As a result, the free energy of ABC-miktoarm star terpolymer solution should equal:

$$\begin{aligned}
\frac{F}{v} &= -f_S\ln\left(\frac{Q_S}{v}\right) - \frac{f_P}{N}\ln\left(\frac{Q_P}{v}\right) \\
&+ \frac{1}{v}\int [\chi_{AB}\phi_A\phi_B + \chi_{AC}\phi_A\phi_C + \chi_{BC}\phi_B\phi_C + \chi_{AS}\phi_A\phi_S \\
&+ \chi_{BS}\phi_B\phi_S + \chi_{CS}\phi_C\phi_S - \omega_A\phi_A - \omega_B\phi_B - \omega_C\phi_C \\
&-\omega_S\phi_S + \frac{\kappa_H}{2}(\phi_A+\phi_B+\phi_C+\phi_S-1)^2\Big]
\end{aligned} \quad (15.17)$$

Here, f_P and f_S are the volume fractions of ABC-miktoarm star terpolymer and solvent in the system, respectively, and

$f_P + f_S = 1$. According to the chemical potential field, the density of each component is calculated as follows.

$$\phi_A(r) = \frac{vf_P}{Q_P}\int_0^{f_A} ds g_A(r,s) g_A^{'}(r, f_A - s)$$
$$\phi_B(r) = \frac{vf_P}{Q_P}\int_0^{f_B} ds g_B(r,s) g_B^{'}(r, f_B - s)$$
$$\phi_C(r) = \frac{vf_P}{Q_P}\int_0^{f_C} ds g_C(r,s) g_C^{'}(r, f_C - s)$$
$$\phi_S(r) = \frac{vf_S \exp(-\omega_S(r))}{Q_S} \tag{15.18}$$

Similarly, the functional derivative $\delta F/\delta\phi_i$ can be calculated. When the initial state of the system is assumed to be disordered or homogeneous, the initial external potential fields can be defined as $\omega_i^{ini} = \sum_{j \neq i} \chi_{ij} f_j^0$, $(f_A^0 = f_P f_A, f_B^0 = f_P f_B, f_C^0 = f_P f_C, f_S^0 = f_S)$. It is worthy of note that other methods hybrid with the Lattice Boltzmann model or multifluid model in the polymeric SCFT framework have also been developed to explore the hydrodynamic effects [24, 25].

15.3 DYNAMICS OF SELF-ASSEMBLY OF BLOCK COPOLYMERS

15.3.1 Phase Separation of Linear Block Copolymers

Linear block copolymers consist of a single main chain. Diblock copolymers are the simplest case of linear block copolymers in which only two distinct chemical blocks are involved (an A-block and a B-block). Typically, the polymers or blocks that make up block copolymers are immiscible and ideal for phase separation. Compared to a complete separation of the binary mixtures into a single A-rich domain and a single B-rich domain, the existing chemical bonds between two blocks in diblock copolymers make complete separation impossible. As a consequence, microphase separation occurs, resulting in complex morphologies, such as spherical, cylindrical, and lamellar microstructures. Groot and Madden studied the microphase separation of linear diblock copolymer melts by DPD simulation method [10], and the order in which the phases appear is in line with expectation based on experiments and on the SCFT approach. The phase behaviors of diblock copolymers by the off-lattice MC simulation have been studied by Besold et al. [26]. Ginzburg et al. [27] developed a new CG model for diblock copolymers to investigate the influence of nanoscale particles on the phase separation and the morphologies of symmetric diblock copolymer films. In a recent series of papers, Addison and coworkers [28] proposed a model of systematic CG representation of block copolymer by modeling the chain as two soft blobs tethered by an entropic spring. Pierleoni et al. [29] studied the self-assembly of diblock copolymer through a two-step CG strategy to build blocks of supermolecular structures. In addition, few studies have looked into the phase structures of specific diblock copolymers [30–32]. The morphologies and mechanical properties of poly(styrene-*b*-isoprene) (PS-*b*-PI) diblock copolymers have recently been studied by Soto-Figueroa et al. [31, 32]. Li et al. [33] developed a CG force field and simulate the phase separation of PS-*b*-PI diblock copolymers. Starting from a symmetric diblock copolymer consisting of S and I blocks of equal length, they observed the LAM, gyroid, HEX, and BCC morphologies by changing the fraction of the two blocks.

A direct observation of the dynamic processes for the formation of these complex microstructures is important for developing skills to control the sizes and shapes of these microstructures. The DPD simulation provides a good choice to understand these processes. In Ref. [33], Li et al. study the dynamic formation process of these morphologies. A typical dynamic formation process of lamellar microstructure is provided in Figure 15.4. It reveals that the block copolymers initially rapidly aggregate into localized spherical and cylindrical microstructures (Fig. 15.4a). Then these microstructures come into close contact and form gyroid morphology (Fig. 15.4b). Next, an irregular hexagonal cylindrical morphology (Fig. 15.4c) appears. This morphology subsequently evolves first, into a perforated hexagonal lamellar morphology (Fig. 15.4d), and second, into an irregular lamellar morphology (Fig. 15.4e) in which some microdomains between two alternating layers are connected. Finally, the connections between two alternating layers are broken and the lamellar microstructure appears (see Fig. 15.4f). The process of HEX morphology also initiates the formation of spherical and cylindrical microstructures, and then the gyroid morphology; however, irregular and aligned cylindrical microstructures then formed. Finally, the sideward connections are broken and the hexagonal packed cylindrical morphology appears. And the BCC morphology can be directly evolved from the spherical and cylindrical microstructures after a much larger time scale.

15.3.2 Self-Assembly of Star Block Copolymers in Melt

When the copolymer is star shaped such as the ABC miktoarm star copolymer in which there are three different blocks (A–C blocks) with a common joint junction, the richer phase structure can be found [34, 35]. When the interaction and length are the same, a symmetrical honeycomb structure can be found (Fig. 15.5a, only phases of blocks A and B are shown) and each block self-assembles into cylinder phase. The junction points are inhomogeneously distributed over the intermaterial dividing surface around cylinders formed by the respective blocks A-C. The region I shown in Figure 15.5a is the interface of different domains, which contains the mixtures of blocks A-C and star junctions.

It indicates the existence of localization behavior of junction points in the ABC miktoarm star terpolymer system. By keeping each block length unchangeable ($N_A = N_B = N_C = 10$. Here, N_i is the segment number of the ith arm), but improving

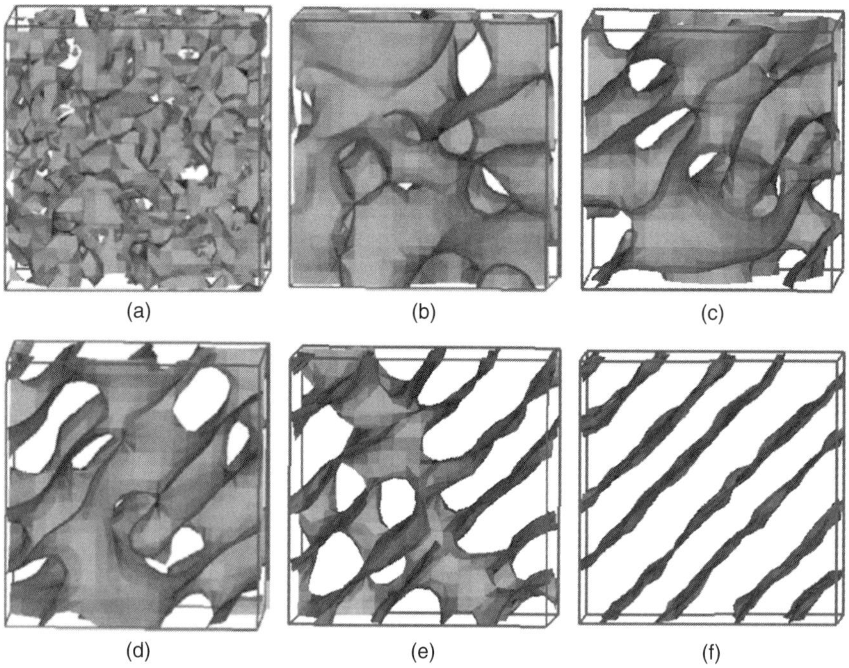

Figure 15.4 Sequential snapshots of the formation of lamellar morphology from the PS-*b*-PI diblock copolymer system at (a) $t = 0.006\,\mu s$, (b) $t = 0.12\,\mu s$, (c) $t = 0.17\,\mu s$, (d) $t = 0.29\,\mu s$, (e) $t = 0.58\,\mu s$, and (f) $t = 0.70\,\mu s$. Li et al. [33]. Reproduced with permission of Elsevier. *(See color plate section for the color representation of this figure.)*

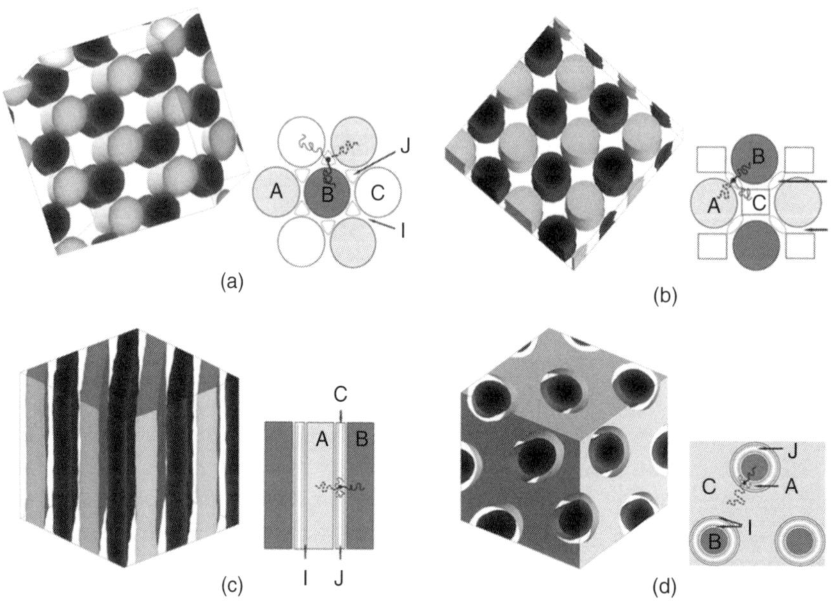

Figure 15.5 Ordered microdomain structures of 3-miktoarm star terpolymers in three-dimensional space. A and B ($\rho_A > 0.5$, $\rho_B > 0.5$) are shown in gray and black, respectively. (a) Honeycomb structure; (b) tetragonally packed cylinders; (c) parallel plane structure; (d) cylinder-shell structure. Schematic of section perpendicular to the cylinders, J is the rich junctions region and I the region containing the mixtures of blocks A–C and junctions. He et al. [35]. Reproduced with permission of Elsevier.

the compatibility between components A, C and B, C, namely, and $\chi_{AB} = 1.2$, $\chi_{AC} = \chi_{BC} = 0.667$ (χ_{ij} is the Flory–Huggins parameter between the *i*th and *j*th kind of segments), the microstructures shown in Figure 15.5b consisting of cylinders of block A and B domains and the rectangular-prism-shaped C domains are formed. In the case of $N_A = N_B = N_C = 10$, and $\chi_{AB} = 2.0$, $\chi_{AC} = \chi_{BC} = 0.4$, that is, $\chi_{AB} \gg \chi_{AC} = \chi_{BC}$, the compatibilities between blocks A, C and B, C are further improved. The microstructures with a set of parallel planes consisting of blocks A, B, and C, respectively (shown in Fig. 15.5c), are formed. The phase behavior of the 3-miktoarm star terpolymer in this case should be similar to the case of ACB linear triblock copolymers. The microstructures of the ABC 3-miktoarm star terpolymer depend on both its molecular architecture and interaction parameters between distinct blocks. If three blocks are strongly segregated, the junction points would have a tendency to be restricted on the lines of mutual intersection of different domains, that is, the realization of the localization of junction points. With the high improvement of compatibility between one block, for example, block C with two other blocks, the junction points are absorbed into the domains of block C. When $N_A = 10$, $N_B = 12$, $N_C = 23$, $\chi_{AB} = 0.42$, $\chi_{AC} = 0.14$, $\chi_{BC} = 0.63$, in which blocks A-C correspond to the PS, PI, and PMMA, respectively, a three-dimensional periodic microstructures of inner block B cylinders with a surrounding block A shell in matrix of block C are formed (see Fig. 15.5d). The simulation results are in agreement with those of Thomas et al. [36, 37]. In this system, due to the better compatibility between block A and two other blocks, junction points are reasonably expected to be concentrated in the center of domain A (Fig. 15.5d), that is, set on the intermaterial dividing surface between domains B and C. It indicates that microstructures and distributions of junction points depend not only on the architecture of the macromolecules' chain but also on the interaction parameters between distinct blocks.

15.3.3 Self-Assembly of Block Copolymers in Constrained Systems

The microphase separation of block copolymers in a constrained situation has attracted interest both from experiments and theories. There are several important characters to make this system favorable in fabricating nanoscale features. Firstly, the constrained situation helps confine the dimension of microphase separation into microscale or nanoscale, so the materials formed in such a system spontaneously possess microstructures. Secondly, the constrained situation can suppress the fluctuation of phase separation and help create an ordered pattern at large scale. Thirdly, the surface of the constrained situation can be deliberately modified to direct the microphase separation of block copolymers with attention to make nonbulk structures.

Geometrically, the constrained situation can be a one-dimensional, two-dimensional, or three-dimensional. Most previous simulation studies concentrated on the self-assembly in a one-dimensional constrained situation, that is, a polymer film being confined between two planar surfaces. A pioneering theoretic work by Matsen evaluated the possible structures formed in a confined symmetric diblock copolymer film [38]. He predicted the parallel, perpendicular, and mixed lamellar structures in confined films. These structures are observed in the following experiments. Huinink et al. studied asymmetric block copolymers confined in a thin film using a DDFT [39]. They predicted a variety of nanostructures, such as parallel cylinders, perpendicular cylinders, parallel lamellae, and parallel perforated lamellae (Fig. 15.6). The effects of system parameters (film thickness and surface preference) on the microphase separation are investigated in their study. For cylinder-forming diblock copolymer self-assembly between two surfaces, Wang et al. drew a phase diagram of microphase-separated structures as a function of film thickness and surface properties using MC simulation [40]. The cylinders, a hybrid of cylinders and lamellae, and perforated lamellae structures, and so on are listed in the phase diagram. Using a simulated annealing MC algorithm, Yin et al. studied the structural evolution of symmetric diblock copolymer confined between two homogeneous surfaces [41]. The film of a gyroid-forming diblock copolymer was also studied using annealing MC simulation [42].

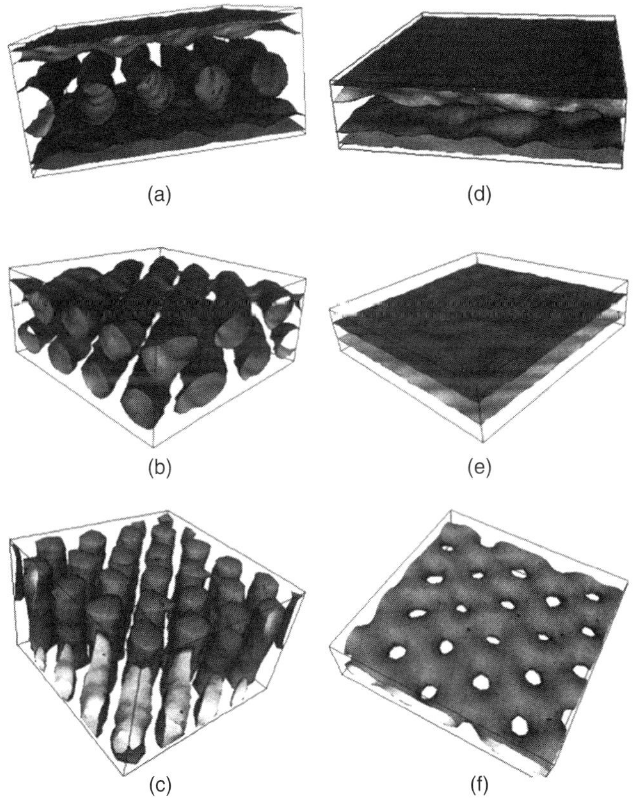

Figure 15.6 Self-assembled structures of asymmetric diblock copolymers confined in a thin film. (a) and (b) Parallel cylinder structures, (c) perpendicular cylinder structures, (d) and (e) lamellar structures, and (f) perforated lamellae structure.Huinink et al. [39]. Reproduced with permission of Elsevier.

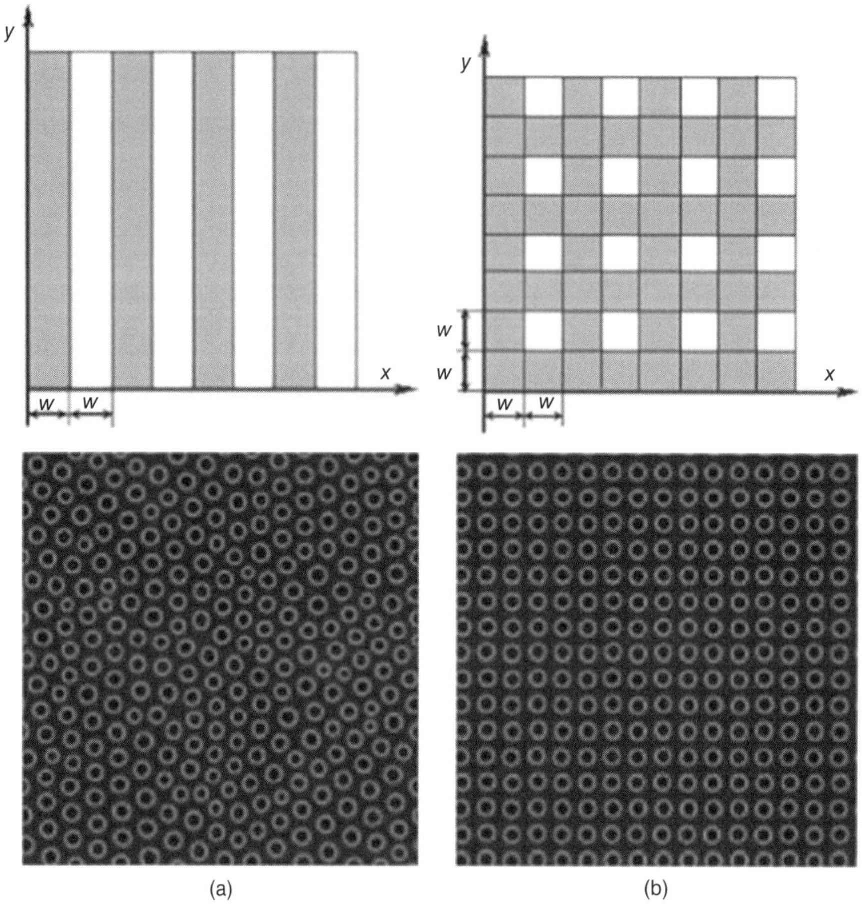

Figure 15.7 Self-assembled structures under (a) strip nanopatterned surface and (b) mosaic nanopatterned surface. Wu and Dzenis [43]. Reproduced with permission of Elsevier. (*See color plate section for the color representation of this figure.*)

Recently, researchers paid more attention to the guided self-assembly of block copolymer thin films on a patterned surface. The patterned surface means the surface of a constrained situation is chemically or physically modified to form a pattern with specific property and size. A series of exquisite structures are found in the microphase separation of block copolymer under the patterned surface. In the theoretic work of Wu and Dzenis [43], they designed two kinds of patterned surface to direct the block copolymer self-assembly (Fig. 15.7). The self-assembled structures are found strongly influenced by the commensurability of polymer bulk period and pattern period. With mismatched patterns on two surfaces, both MC simulation [44] and SCFT researching [45] predicted the titled lamellae and perforated lamellae structures for symmetric diblock copolymers. Petrus et al. carried out a detailed investigation on the microphase separation of symmetric and asymmetric diblock copolymers confined between two planar surfaces using DPD simulation [46, 47]. It is found that various nonbulk nanostructures can be fabricated by the nanopatterns on the surfaces.

The two-dimensional constraint situation means that the polymer is confined in a nanopore. He et al. first explored the microphase separation of a symmetric diblock copolymer in a circular nanopore using MC simulation [48] (Fig. 15.8). The rolled-up lamellae structure predicted by the simulation was testified by the experiments and other simulation research. The self-assembly of cylinder-forming block copolymers in circular nanopores received great interest because of the helix and circle structures rising in these circular constraint situations. Yu et al. predicted the helix structure by diblock copolymers under cylindrical confinement using MC simulation. They showed the evolution of nanostructures with the pore diameter increasing [49, 50]. Li et al. drew a phase diagram for the asymmetric diblock copolymers in a circular nanopore using SCFT [51]. Besides the circular nanopore, the theoretic researchers also considered the nanopore with noncircular geometries, such as triangular or square ones [52, 53]. Detcheverry et al. studied the directed assembly of cylinder-forming diblock copolymer in a trench [54]. They theoretically fabricated defect-free arrays with a high degree of uniformity across the trench using MC simulation. When block copolymers being constrained between two plate substrates containing pillars with different shapes, richer phase structures can be found. The work can be found in the concerned references [55–59].

The three-dimensional constraint situation is relatively less studied compared with the one-dimensional and

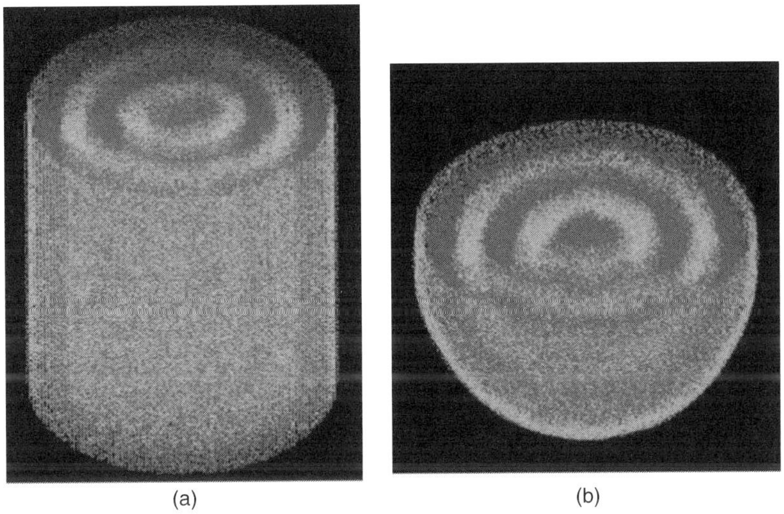

Figure 15.8 The lamellae structures formed by diblock copolymers in constrained spaces predicted by Monte Carlo simulation [48]. (a) Barrel, (b) spherical (cut view). He et al. [48]. Reproduced with permission of Elsevier. (*See color plate section for the color representation of this figure.*)

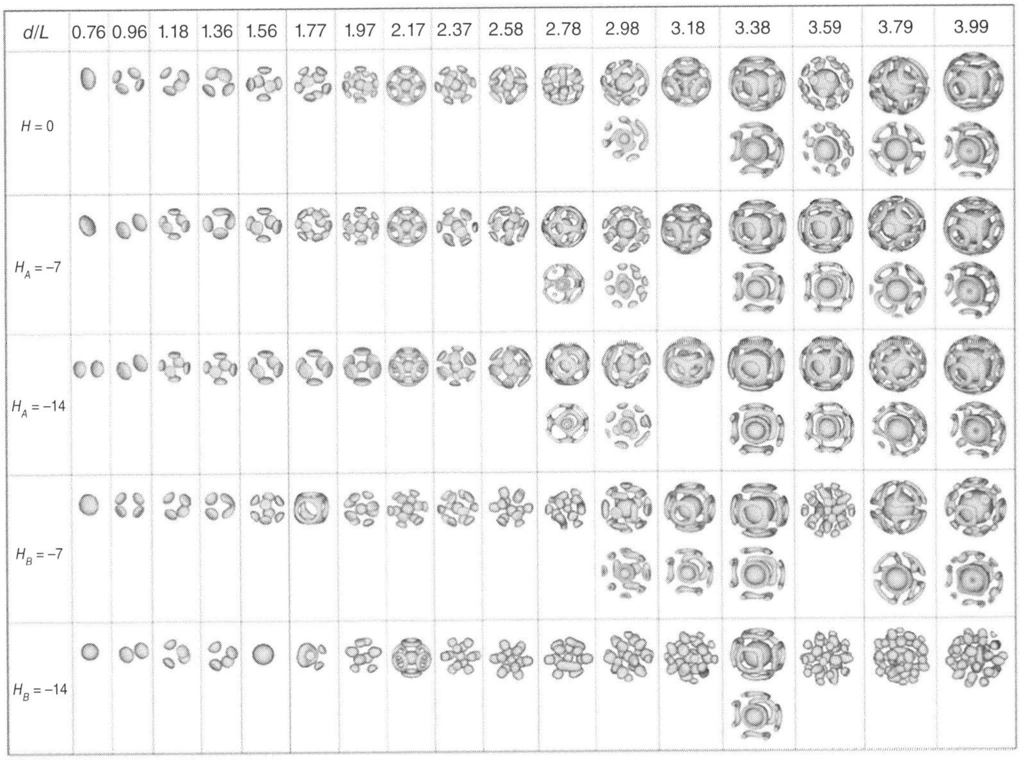

Figure 15.9 The microphase-separated structures for cylinder-forming diblock copolymers confined in a spherical space. Chen et al. [61]. Reproduced with permission of American Chemical Society.

two-dimensional ones. In He's MC simulation (Fig. 15.8), he first predicted the core–multishell structures for symmetric diblock copolymers confined in a spherical constraint situation [48]. Besides the core–multishell structures, Yu et al. observed a series of novel structures under this three-dimensional spherical confinement using annealing MC simulation [60]. Chen et al. studied the microphase separation of cylinder-forming asymmetric diblock copolymers confined in a spherical space using SCFT (Fig. 15.9) [61]. Some nanostructures predicted by these theoretic researching has been observed by experiments.

By combining the experimental and theoretical researches, people make clear the reason under experimental observation, and correspondingly the experimental result can testify the theoretical prediction. There is more and more cooperation between experimental researchers and theoretic researchers. Just as Tsarkova and Sevink et al. once observed, the evolution of cylinder structure varies from vertical orientation into parallel orientation in a block copolymer film using an *in situ* scanning force microscopy technology and SCFT simulation [62]. A good agreement in the nanostructure and dynamic process is found in the conjunction study. Ludwigs et al. studied the formation of a perforated lamella structure by triblock terpolymers both in experiment and simulation [63]. Nealey and de Pablo have cooperated for a long time in the field of polymer self-assembly under constraint situation to combine the experiment with simulation. For example, the directed assembly of lamellae-forming block copolymers on chemically patterned substrates was studied by a combination of theoretical and experimental approaches [64–66]. The simulation helps explain the experimental observation and explore much larger parameter spaces than those that are hardly realized by present experiment technology.

15.3.4 Micellization of Amphiphilic Block Copolymer in Solution

For a selective solvent, amphiphilic block copolymers with hydrophilic and hydrophobic blocks can self-assemble into a wide variety of morphologies, such as spherical or cylindrical micelles, bilayers, vesicles, and nanotubes [67–71]. In the past decade, the self-assembly behaviors and aggregate structures of amphiphilic block copolymers have attracted considerable attention in experimental studies. The micellization of amphiphilic molecules is determined not only by the amphiphilic nature of the molecules but also by selective solvents. Besides these extensive experimental investigations, simulation and theoretical studies on the self-assembly of amphiphilic block copolymers in selective solvents have also been performed by a few groups [72–78]. Studies on linear diblock copolymers have made much progress as the topology is relatively simple. Linear triblock copolymers and nonlinear block copolymers with unique architectures have attracted less attention, especially in terms of computer simulation. The existing few simulation studies on the self-assembly of linear triblock copolymers in selective solvents are mostly based on MC [78, 79], DPD [77], SCFT [75, 80, 81] and DDFT [82]. For instance, the kinetics of vesicle formation of amphiphilic triblock copolymers from an initially homogeneous state was theoretically and experimentally investigated by Han et al. [78]. It has been recognized that the formation of a threefold "Y-like" junction can be inferred from the interconnection of a cylinder with an end-cap or fusion of three enlarged end-caps [72, 73]. Some possible pathways for the dynamic process of vesicle formation have been described in the literature: bending of membranes [83–85], fusion of micelles with the rearrangement of amphiphilic molecules [86–88], and fusion of vesicles [86, 89]. For the formation of toroidal micelles, theoretical studies and experimental observations have suggested that this process can proceed via either a conventional micelle coalescence pathway [77, 90] or a growth pathway [90–92].

Despite recent progress on the different pathways for amphiphilic block copolymer micelles in dilute solutions [79, 88, 92, 93], the route by which amphiphilic molecules in a selective solution self-assemble into various micelles, as well as the dynamic properties of amphiphilic block copolymers in solution, is still not fully understood. With the purpose of developing skills to control these microstructures in sizes and shapes, He et al. presented an extensive simulation study on the spontaneous formation of complex micelles from amphiphilic triblock copolymers in a dilute solvent [94]. To illustrate the effects of the selectivity of a solvent for amphiphilic molecules on microstructure formation, they systematically varied the repulsive interactions between the solvent and the hydrophilic/hydrophobic particles. Figure 15.10 presents the corresponding diagram of morphologies. A plethora of complex morphologies, including spheres, cylinders, "Y-like" junctions, toroids, and vesicles, is revealed by the DPD simulations.

It is generally believed that three factors, core-chain stretching entropy, the interfacial energy, and repulsion among intercoronal chains, determine the equilibrium size and shape of micelles. Changing the selectivity of the solvent disturbs the force balance governing the size and shape of micelles, which can lead to the morphology transformation from one to another. The combined contribution from all the interactions balanced the contact area between hydrophilic and hydrophobic blocks and defined the local interfacial curvature. The dominating effect of improving the selectivity of the solvent by increasing the values of a_{BS} and/or decreasing the values of a_{AS} is the shrinkage of hydrophobic blocks and expansion of hydrophilic corona, which results in micelle structures with increased interfacial curvature. Therefore, varying the solvent content provides an effective way to control morphologies in our amphiphilic system.

Recently, more attention has been given to rod–coil amphiphilic block copolymers composed of rigid-rod blocks bonded to flexible-coil blocks. Besides the chemical composition, the existence of an anisotropic orientation of rigid-rod blocks, the effect of chain topology on conformational entropy, and molecular packing geometries also show profound influence on the self-assembly of the block copolymers. As such, the self-assembly of rod–coil block copolymers is fundamentally different from that of classical coil–coil amphiphilic block copolymers. Based on the functionality of the rigid-rod blocks, there is growing interest in amphiphilic block copolymers with rod–coil structures, and a number of intriguing phases, including wavy lamellae, zigzags, arrowheads, straight lamellae, perforated lamellae, hexagonal strips, honeycombs, and hollow spherical micelles, have been observed through self-assembly from them [95–98]. A number of theoretical models have been developed to understand the self-assembly behavior of rod–coil amphiphilic block copolymers [99–104]. These theoretical studies help researchers better understand both ordered structures and ordering transition.

In addition to these theoretical studies, particle-based mesoscopic simulation studies on the self-assembly behavior of rod–coil block copolymers have also been performed

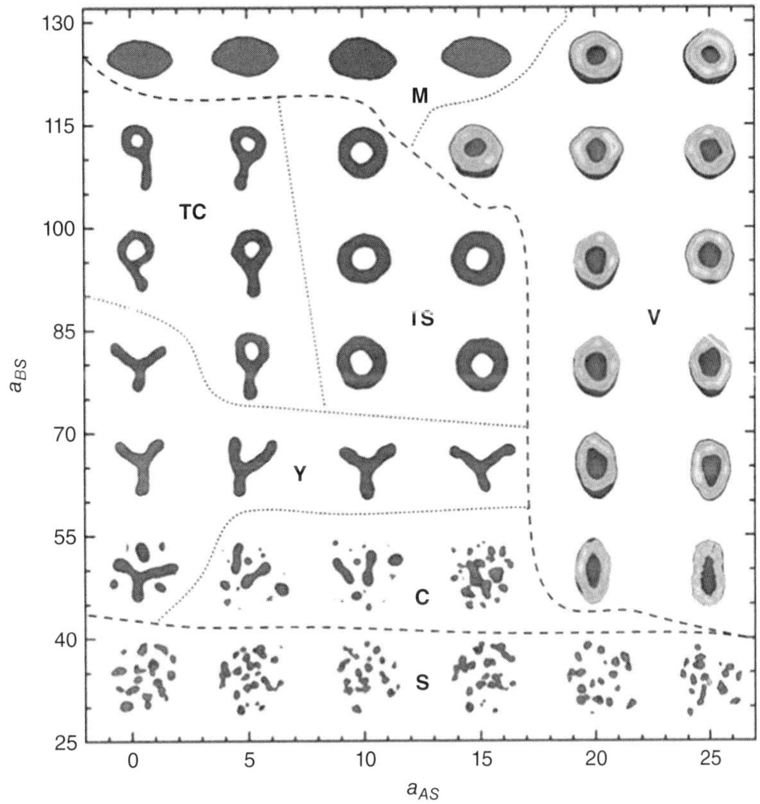

Figure 15.10 Morphological diagram of amphiphilic triblock copolymers in terms of DPD repulsive parameters a_{AS} and a_{BS}. These microstructures included spheres (S), cylinders (C), "Y-like" junctions (Y), toroids (T), vesicles (V), and disklike aggregates (M). Toroids have the form of either a single loop (TS) or one loop with one end-cap (TC). The dashed line divides the assembled morphologies into three regimes: spheres, cylinders, and bilayers. The dotted lines serve as guides. He et al. [94]. Reproduced with permission of Elsevier.

by a few groups using various computer techniques, such as Brownian dynamics [105, 106], DPD [107], and MC simulations [108, 109]. However, most of them focus on the liquid-crystalline behavior of rod–coil block copolymer systems, and papers on the computer simulation of solution-state self-assembly of rod–coil block copolymers are seldom reported. Glotzer and coworkers developed a generic model of tethered nano building blocks to simulate amphiphilic block copolymers and surfactants microphase separation and liquid-crystal phase ordering using the Brownian dynamics method [110–112]. They predicted a variety of fascinating microstructures, including sheets, wires, tubes, honeycombs, helical scrolls, and alternating cylinders, from the self-assembly of rod–coil amphiphilic block copolymers in solutions. Song et al. applied lattice MC simulation to study the phase behavior of symmetric ABA-type triblock copolymers [113]. He et al. studied the self-assembled microstructures of coil–rod–coil ABA-type amphiphilic triblock copolymers depending on the selectivity of solvents [114]. They focus on how rigid-rod hydrophobic blocks affect aggregations of the microstructures and the dynamic properties of amphiphilic block copolymers in selective solvents. By changing the selectivity of solvents, defined in terms of the repulsive interactions between the solvent and the hydrophilic/hydrophobic particles, they found that the aggregation morphology changes from bundle-like micelles to spherical and cylindrical micelles to elongated micelles and then to ringlike toroidal micelles, revealing that the selectivity of solvents is a key factor that determines aggregation morphology. The potential mechanism behind the formation of these complex micelles may be helpful in explaining how these aggregates are formed and in understanding the self-assembly behavior and dynamics of amphiphilic block copolymers.

The formation of the vesicle or ringlike micellar structures of amphiphilic block copolymer is controlled by kinetic trap from the competition of the interactions between hydrophilic or hydrophobic part and the solvent. These metastable states provide new opportunities in designing soft self-assembly materials with enhanced functionalities in applications, such as the complex micro-release system, a template for nanodevice fabrication. In the past decades, a number of experimental works focused on the diversities of the morphologies of micelles using various amphiphilic compounds. Besides spherical, rodlike (or wormlike micelles) and lamellar micelles, special topological micelles, such as unilamellar or multilamellar vesicle, complex vesicle, onion vesicle, genus vesicle, and toroidal micelles with single and multirings, and cage micelles, have also been discovered [83, 93, 115–119]. The process of spontaneous formation of topological micelles is interesting.

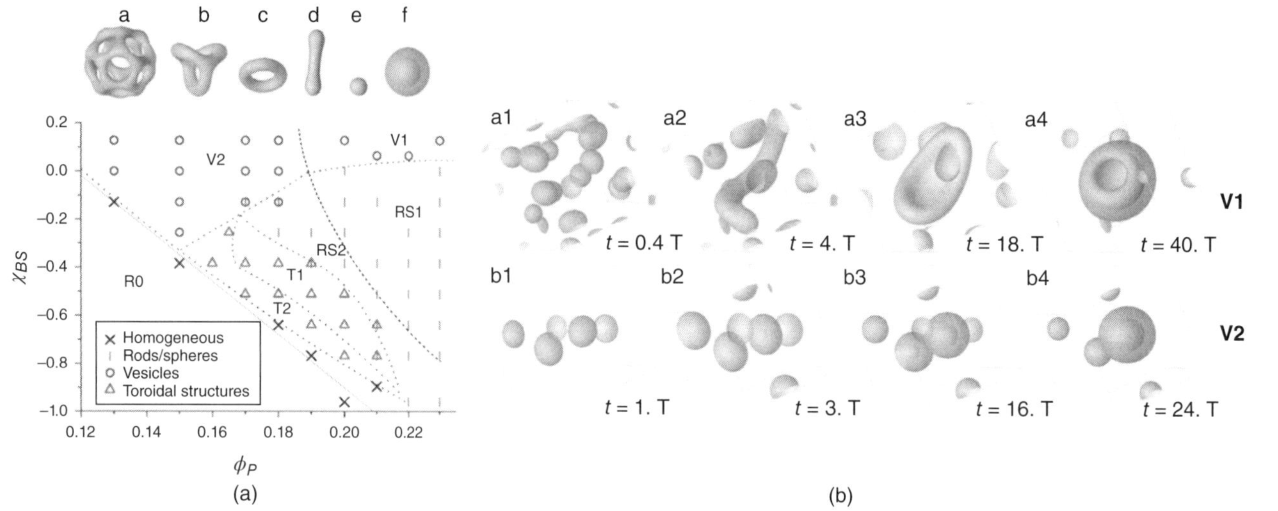

Figure 15.11 Phase diagram of (a) various micelles and (b) pathways of vesicle formation. He and Schmid [92]. Reproduced with permission of American Physical Society.

Using EPDs in 3D, the pathways of vesicle formation of amphiphilic diblock copolymer in solution from a homogeneous initial state was revealed (Fig. 15.11) [74, 87, 92, 120]. Two distinct pathways of spontaneous vesicle formation were simultaneously discovered and various toroidal micelles were found when changing initial concentration and molecular interaction. The topological micelles are assembled under the mechanisms and together controlled by amphiphilic molecular structures and segregation kinetics. Amphiphlic diblock copolymers aggregate from homogeneous distribution in solution corresponding to the quenching processes in reality by various methods, such as changing temperature and adding nonsolvents or salts. Figure 15.11a shows the phase diagram of various structures of micelles obtained from the initial homogeneous state depending on the interaction χ_{BS} and concentration ϕ_P. The random phase (region R0) represents no micellar formation and is located at the left side of dotted line. When the quenched point locates in region V1 far from the spinodal curve, it is found that the vesicle is formed through the conventional path, that is, coalescence and curvature, which is controlled by minimization rim energy. The amphiphilic copolymers quickly aggregate into many small micelles from homogeneous distribution, then small micelles coalescence to form the rod; after that, rod micelles transform into bilayer, and finally the bilayer curves to form the vesicle (Fig. 15.11b).

The ways of decreasing rim energy varied at different stages: in the initial stage, it is finished by the rim-to-rim coalescence of small spherical or disk micelles that increase the length of rod micelles or the area of the individual bilayer; later, the further decrease of system energy is realized by the curvature and closure of the individual bilayer through a continuous shrink path till the vesicle is fully formed. When the quenched points locate into the region V2 near the spinodal curve, the vesicle comes forth through another path controlled by the nucleation and growth-swelling mechanism. For instance, at $\phi_p = 0.15$, $\chi_{BS} = 0.128$ in Figure 15.11, after an incubation time, several stable nuclei occur and then they grow larger and larger, and transform into a semivesicle; finally, the semivesicle grows and swells into a vesicle. A small semivesicle as an intermediate state comes from relatively fast aggregation compared with slow separation of solvent-philic and solvent-phobic parts so as to make the size of aggregation body grow larger than the radius of stable sphere micelles and promote amphiphilic molecule flip-flop. The solvent-philic part of the inner micelles may further absorb solvent into the semivesicle to form the vesicle. Then the small vesicle grows larger and larger till exhausting the remaining copolymer in solution. Meanwhile, ringlike micelles also show similar nucleation growth pathways. When the amphiphilic block copolymer is star copolymer, for example, ABC miktoarm star copolymer (C is hydrophilic and A and B are hydrophobic), more complex multicompartment micelles such as the hamburger micelle, toroidal micelle, raspberry micelle, worm micelle, and laterally structured vesicle can be formed and discovered from experiments [121–123]. Simulation shows that they formed also through a pathway similar to that of coalescence and growth during aggregation (Fig. 15.12).

15.4 OUTLOOK

The exploration of block copolymers still has a long way to go, although intensive work has been carried out in the past. Till now, block copolymers have been widely applied in the fields of material template manufacture and biology drug delivery. The complex phase structure of block copolymers and their dynamical response to the external signal and fields propose a new challenge to soft matter science. It also provides the opportunity for the fabrication of intelligence materials and multifunctional device. It is known that the formation of the

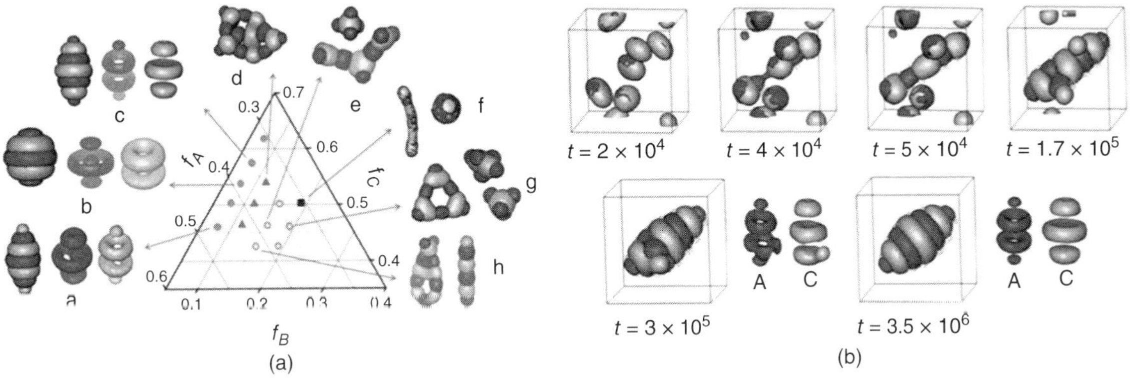

Figure 15.12 Phase diagram of miktoarm ABC star copolymer in solution S with (a) the interaction parameters $\chi_{AS} = \chi_{CS} = 1.548$ and $\chi_{BS} = -0.172$ and (b) the formation pathway of the laterally structured vesicle. The solvophilic B block is shown with small particles. Wang et al. [124]. Reproduced with permission of The Royal Society of Chemistry.

defect in phase structures is not always negative to us from the viewpoint of practical application. Novel and rich structures can be designed when the defect can be controlled or tailored in its position and amount. It is no doubt that the phase diagrams of block copolymers in various systems are important to researchers. However, it is difficult to achieve only from experiments when the objective system concerns many parameters and the scanning space is very large. Although the simulation and theory have the capability to describe and predict the phase diagram of a general system using CG models, the quantitative prediction of the properties of a realistic polymer system such as transition temperature is not satisfying. Developing a multiscale model to describe and predict the phase structure quantitatively becomes urgent. A new simulation framework using the CG force field with good transfer in various systems is highly expected.

REFERENCES

1. Darling SB. Directing the self-assembly of block copolymers. Prog Polym Sci 2007;32:1152.
2. Breiner U, Krappe U, Thomas EL, Stadler R. Structural characterization of the "knitting pattern" in polystyrene-block-poly(ethylene-co-butylene)-block-poly(methylmethacrylate) triblock copolymers. Macromolecules 1998;31:135.
3. Chen H, Chakrabarti A. Morphology of thin block copolymer films on chemically patterned substrates. J Chem Phys 1998;108:6897.
4. Bates FS, Fredrickson GH. Block copolymer thermodynamics - theory and experiment. Annu Rev Phys Chem 1990;41:525.
5. Fredrickson GH, Bates FS. Dynamics of block copolymers: theory and experiment. Annu Rev Mater Sci 1996;26:501.
6. Matsen MW, Bates FS. Unifying weak- and strong-segregation block copolymer theories. Macromolecules 1996;29:1091.
7. Matsen MW, Bates FS. Conformationally asymmetric block copolymers. J Polym Sci, Part B: Polym Phys 1997;35:945.
8. Matsen MW. Effect of architecture on the phase behavior of AB-type block copolymer melts. Macromolecules 2012;45:2161.
9. Bedrov D, Ayyagari C, Smith GD. Multiscale modeling of poly(ethylene oxide)-poly(propylene oxide)-poly(ethylene oxide) triblock copolymer micelles in aqueous solution. J Chem Theory Compt 2006;2:598.
10. Groot RD, Madden TJ. Dynamic simulation of diblock copolymer microphase separation. J Chem Phys 1998;108:8713.
11. Horsch MA, Zhang ZL, Iacovella CR, Glotzer SC. Hydrodynamics and microphase ordering in block copolymers: Are hydrodynamics required for ordered phases with periodicity in more than one dimension? J Chem Phys 2004;121:11455.
12. Leibler L. Theory of microphase separation in block copolymers. Macromolecules 1980;13:1602.
13. Carmesin I, Kremer K. The bond fluctuation method: a new effective algorithm for the dynamics of polymers in all spatial dimensions. Macromolecules 1988;21:2819.
14. Shaffer JS. Effects of chain topology on polymer dynamics: bulk melts. J Chem Phys 1994;101:4205.
15. Deutsch HP, Binder K. Interdiffusion and self-diffusion in polymer mixtures: a Monte Carlo study. J Chem Phys 1991;94:2294.
16. Subramanian G, Shanbhag S. On the relationship between two popular lattice models for polymer melts. J Chem Phys 2008;129:144904.
17. Detcheverry FA, Kang H, Daoulas KC, Muller M, Nealey PF, de Pablo JJ. Monte Carlo simulations of a coarse grain model for block copolymers and nanocomposites. Macromolecules 2008;41:4989.
18. Detcheverry FA, Pike DQ, Nagpal U, Nealey PF, de Pablo JJ. Theoretically informed coarse grain simulations of block copolymer melts: method and applications. Soft Matter 2009;5:4858.
19. Wang Q, Yin Y. Fast off-lattice Monte Carlo simulations with "soft" repulsive potentials. J Chem Phys 2009;130:104903.
20. Wang Q. Studying soft matter with "soft" potentials: fast lattice Monte Carlo simulations and corresponding lattice self-consistent field calculations. Soft Matter 2009;5:4564.
21. Zhang P, Li B, Wang Q. Quantitative study of fluctuation effects by fast lattice Monte Carlo simulations. 2. Homopolymer brushes in an implicit, good solvent. Macromolecules 2011;44:7837.

22. Hoogerbrugge PJ, Koelman JMVA. Simulating microscopic hydrodynamic phenomena with dissipative particle dynamics. Europhys Lett 1992;19:155.
23. Maurits NM, Fraaije JGEM. Mesoscopic dynamics of copolymer melts: From density dynamics to external potential dynamics using nonlocal kinetic coupling. J Chem Phys 1997;107:5879.
24. Hall DM, Lookman T, Fredrickson GH, Banerjee S. Hydrodynamic self-consistent field theory for inhomogeneous polymer melts. Phys Rev Lett 2006;97:114501.
25. Zhang LS, Sevink A, Schmid F. Hybrid lattice Boltzmann/dynamic self-consistent field simulations of microphase separation and vesicle formation in block copolymer systems. Macromolecules 2011;44:9434.
26. Besold G, Hassager O, Mouritsen OG. Monte Carlo simulation of diblock copolymer microphases by means of a fast off-lattice model. Comput Phys Commun 1999;121:542.
27. Ginzburg VV, Gibbons C, Qiu F, Peng GW, Balazs AC. Modeling the dynamic behavior of diblock copolymer/particle composites. Macromolecules 2000;33:6140.
28. Addison CI, Hansen JP, Krakoviack V, Louis AA. Coarse-graining diblock copolymer solutions: a macromolecular version of the Widom-Rowlinson model. Mol Phys 2005;103:3045.
29. Pierleoni C, Addison C, Hansen JP, Krakoviack V. Multiscale coarse graining of diblock copolymer self-assembly: From monomers to ordered micelles. Phys Rev Lett 2006;96:128302.
30. Ortiz V, Nielsen SO, Discher DE, Klein ML, Lipowsky R, Shillcock J. Dissipative particle dynamics simulations of polymersomes. J Phys Chem B 2005;109:17708.
31. Soto-Figueroa C, Vicente L, Martinez-Magadan JM, Rodriguez-Hidalgo MD. Self-organization process of ordered structures in linear and star poly(styrene)-poly(isoprene) block copolymers: Gaussian models and mesoscopic parameters of polymeric systems. J Phys Chem B 2007;111:11756.
32. Soto-Figueroa C, Rodriguez-Hidalgo MDR, Martinez-Magadan JM, Vicente L. Dissipative particle dynamics study of order-order phase transition of BCC, HPC, OBDD, and LAM structures of the poly(styrene)-poly(isoprene) diblock copolymer. Macromolecules 2008;41:3297.
33. Li XJ, Guo JY, Liu Y, Liang HJ. Microphase separation of diblock copolymer poly(styrene-b-isoprene): A dissipative particle dynamics simulation study. J Chem Phys 2009;130:074908.
34. He XH, Huang L, Liang HJ, Pan CY. Self-assembly of star block copolymers by dynamic density functional theory. J Chem Phys 2002;116:10508.
35. He XH, Huang L, Liang HJ, Pan CY. Localizations of junction points of ABC 3-miktoarm star terpolymers. J Chem Phys 2003;118:9861.
36. Sioula S, Hadjichristidis N, Thomas EL. Novel 2-dimensionally periodic non-constant mean curvature morphologies of 3-miktoarm star terpolymers of styrene, isoprene, and methyl methacrylate. Macromolecules 1998;31:5272.
37. Sioula S, Hadjichristidis N, Thomas EL. Direct evidence for confinement of junctions to lines in an 3 miktoarm star terpolymer microdomain structure. Macromolecules 1998;31:8429.
38. Matsen MW. Thin films of block copolymer. J Chem Phys 1997;106:7781.
39. Huinink HP, Brokken-Zijp JCM, van Dijk MA, Sevink GJA. Asymmetric block copolymers confined in a thin film. J Chem Phys 2000;112:2452
40. Wang Q, Nealey PF, de Pablo JJ. Monte Carlo simulations of asymmetric diblock copolymer thin films confined between two homogeneous surfaces. Macromolecules 2001;34:3458.
41. Yin Y, Sun P, Chen T, Li B, Jin Q, Ding D, Shi A-C. Simulated annealing study of diblock copolymer thin films confined between two homogeneous surfaces. ChemPhysChem 2004;5:540.
42. Yin Y, Sun P, Jiang R, Li B, Chen T, Jin Q, Ding D, Shi A-C. Simulated annealing study of asymmetric diblock copolymer thin films. J Chem Phys 2006;124:184708.
43. Wu X-F, Dzenis YA. Guided self-assembly of diblock copolymer thin films on chemically patterned substrates J Chem Phys 2006;125:174707.
44. Wang Q. Morphology of symmetric diblock copolymers confined between two stripe-patterned surfaces – tilted lamellae and more. Macromol Theory Simul 2005;14:96.
45. Ye X, Edwards BJ, Khomami B. Elucidating the formation of block copolymer nanostructures on patterned surfaces: A self-consistent field theory study. Macromolecules 2010;43:9594.
46. Petrus P, Lisal M, Brennan JK. Self-assembly of symmetric diblock copolymers in planar slits with and without nanopatterns: Insight from dissipative particle dynamics simulations. Langmuir 2010;26:3695.
47. Petrus P, Lisal M, Brennan JK. Self-assembly of lamellar-and cylinder-forming diblock copolymers in planar slits: insight from dissipative particle dynamics simulations. Langmuir 2010;26:14680.
48. He XH, Song M, Liang HJ, Pan CY. Self-assembly of the symmetric diblock copolymer in a confined state: Monte Carlo simulation. J Chem Phys 2001;114:10510.
49. Yu B, Jin Q, Ding D, Li B, Shi A-C. Confinement-induced morphologies of cylinder-forming asymmetric diblock copolymers. Macromolecules 2008;41:4042.
50. Yu B, Sun PC, Chen TH, Jin QH, Ding DT, Li BH, Shi A-C. Confinement-induced novel morphologies of block copolymers. Phys Rev Lett 2006;96:138306.
51. Li W, Wickham RA. Self-assembled morphologies of a diblock copolymer melt confined in a cylindrical nanopore. Macromolecules 2006;39:8492.
52. Hur SM, Garcia-Cervera CJ, Kramer EJ, Fredrickson GH. SCFT simulations of thin film blends of block copolymer and homopolymer laterally confined in a square well. Macromolecules 2009;42:5861.
53. Yu B, Sun PC, Chen TH, Jin QH, Ding DT, Li BH, Shi A-C. Self-assembled morphologies of diblock copolymers confined in nanochannels: Effects of confinement geometry J Chem Phys 2007;126:204903.
54. Detcheverry FA, Nealey PF, de Pablo JJ. Directed assembly of a cylinder-forming diblock copolymer: Topographic and chemical patterns. Macromolecules 2010;43:6495.
55. Zou Z, He X, Wang L. Phase segregation of a symmetric diblock copolymer in constrained space with a square-pillar array. J Chem Phys 2012;136.

56. Yang G, Tang P, Yang Y, Cabral JT. Self-Assembly of AB Diblock Copolymers under Confinement into Topographically Patterned Surfaces. J Phys Chem B 2009;113:14052.
57. Maniadis P, Tsimpanogiannis IN, Kober EM, Lookman T. Phase segregation of diblock copolymers in nanopore geometries. Europhys Lett 2008;81.
58. Man X, Andelman D, Orland H. Block copolymer at nano-patterned surfaces. Macromolecules 2010;43:7261.
59. Li S, Chen P, Wang X, Zhang L, Liang H. Surface-induced morphologies of lamella-forming diblock copolymers confined in nanorod arrays. J Chem Phys 2009;130:014902.
60. Yu B, Li B, Jin Q, Ding D, Shi A-C. Self-assembly of symmetric diblock copolymers confined in spherical nanopores. Macromolecules 2007;40:9133.
61. Chen P, Liang H, Shi A-C. Microstructures of a cylinder-forming diblock copolymer under spherical confinement. Macromolecules 2008;41:8938.
62. Horvat A, Knoll A, Krausch G, Tsarkova L, Lyakhova KS, Sevink GJA, Zvelindovsky AV, Magerle R. Time evolution of surface relief structures in thin block copolymer films. Macromolecules 2007;40:6930.
63. Ludwigs S, Boker A, Voronov A, Rehse N, Magerle R, Krausch G. Self-assembly of functional nanostructures from ABC triblock copolymers. Nature Materials 2003;2:744.
64. Detcheverry FA, Liu G, Nealey PF, de Pablo JJ. Interpolation in the directed assembly of block copolymers on nanopatterned substrates: simulation and experiments. Macromolecules 2010;43:3446.
65. Ramírez-Hernández A, Liu G, Nealey PF, de Pablo JJ. Symmetric diblock copolymers confined by two nanopatterned surfaces. Macromolecules 2012;45:2588.
66. Liu G, Ramirez-Hernandez A, Yoshida H, Nygård K, Satapathy DK, Bunk O, de Pablo JJ, Nealey PF. Morphology of lamellae-forming block copolymer films between two orthogonal chemically nanopatterned striped surfaces. Phys Rev Lett 2012;108:065502.
67. Alexandridis P, Lindman B. *Amphiphilic block copolymers: self-assembly and applications*. New York: Elsevier; 2000.
68. Won YY, Davis HT, Bates FS. Giant wormlike rubber micelles. Science 1999;283:960.
69. Discher BM, Won YY, Ege DS, Lee JCM, Bates FS, Discher DE, Hammer DA. Polymersomes: Tough vesicles made from diblock copolymers. Science 1999;284:1143.
70. Discher DE, Eisenberg A. Polymer vesicles. Science 2002;297:967.
71. Cui HG, Chen ZY, Zhong S, Wooley KL, Pochan DJ. Block copolymer assembly via kinetic control. Science 2007;317:647.
72. Tlusty T, Safran SA. Defect-induced phase separation in dipolar fluids. Science 2000;290:1328.
73. Tlusty T, Safran SA, Strey R. Topology, phase instabilities, and wetting of microemulsion networks. Phys Rev Lett 2000;84:1244.
74. He XH, Liang HJ, Huang L, Pan CY. Complex microstructures of Amphiphilic diblock copolymer in dilute solution. J Phys Chem B 2004;108:1731.
75. Zhu JT, Jiang Y, Liang HJ, Jiang W. Self-assembly of ABA amphiphilic triblock copolymers into vesicles in dilute solution. J Phys Chem B 2005;109:8619.
76. Pam LS, Spell LL, Kindt JT. Simulation and theory of flexible equilibrium polymers under poor solvent conditions. J Chem Phys 2007;126:134906.
77. Li XJ, Deng MG, Liu Y, Liang HJ. Dissipative particle dynamics simulations of toroidal structure formations of amphiphilic triblock copolymers. J Phys Chem B 2008;112:14762.
78. Han YY, Yu HZ, Du HB, Jiang W. Effect of selective solvent addition rate on the pathways for spontaneous vesicle formation of ABA amphiphilic triblock copolymers. J Am Chem Soc 2010;132:1144.
79. Kong WX, Li BH, Jin QH, Ding DT, Shi AC. Complex micelles from self-assembly of ABA triblock copolymers in B-selective solvents. Langmuir 2010;26:4226.
80. Tang P, Qiu F, Zhang HD, Yang YL. Morphology and phase diagram of complex block copolymers: ABC linear triblock copolymers. Phys Rev E 2004;69:031803.
81. Li XA, Tang P, Qiu F, Zhang HD, Yang YL. Aggregates in solution of binary mixtures of amphiphilic diblock copolymers with different chain length. J Phys Chem B 2006;110:2024.
82. Xia JF, Sun MZ, Qiu F, Zhang HD, Yang YL. Microphase ordering mechanisms in linear ABC triblock copolymers. A dynamic density functional study. Macromolecules 2005;38:9324.
83. Egelhaaf SU, Schurtenberger P. Micelle-to-vesicle transition: A time-resolved structural study. Phys Rev Lett 1999;82:2804.
84. Yamamoto S, Maruyama Y, Hyodo S. Dissipative particle dynamics study of spontaneous vesicle formation of amphiphilic molecules. J Chem Phys 2002;116:5842.
85. Weiss TM, Narayanan T, Gradzielski M. Dynamics of spontaneous vesicle formation in fluorocarbon and hydrocarbon surfactant mixtures. Langmuir 2008;24:3759.
86. Bernardes AT. Computer simulations of spontaneous vesicle formation. Langmuir 1996;12:5763.
87. He XH, Schmid F. Dynamics of spontaneous vesicle formation in dilute solutions of amphiphilic diblock copolymers. Macromolecules 2006;39:2654.
88. Wolf C, Bressel K, Drechsler M, Gradzielski M. Comparison of vesicle formation in zwitanionic and catanionic mixtures of hydrocarbon and fluorocarbon surfactants: phase behavior and structural progression. Langmuir 2009;25:11358.
89. Lipowsky R. The Conformation of Membranes. Nature 1991;349:475.
90. Yu HZ, Jiang W. Effect of shear flow on the formation of ring-shaped ABA amphiphilic triblock copolymer micelles. Macromolecules 2009;42:3399.
91. Jiang Y, Zhu JT, Jiang W, Liang HJ. Cornucopian cylindrical aggregate morphologies from self-assembly of amphiphilic triblock copolymer in selective media. J Phys Chem B 2005;109:21549.
92. He XH, Schmid F. Spontaneous formation of complex micelles from a homogeneous solution. Phys Rev Lett 2008;100:137802.
93. Schmolzer S, Grabner D, Gradzielski M, Narayanan T. Millisecond-range time-resolved small-angle x-ray scattering studies of micellar transformations. Phys Rev Lett 2002;88:258301.
94. He PT, Li XJ, Kou DZ, Deng MG, Liang HJ. Complex micelles from the self-assembly of amphiphilic triblock copolymers in selective solvents. J Chem Phys 2010;132:204905.

95. Chen JT, Thomas EL, Ober CK, Mao GP. Self-assembled smectic phases in rod-coil block copolymers. Science 1996;273:343.
96. Jenekhe SA, Chen XL. Self-assembled aggregates of rod-coil block copolymers and their solubilization and encapsulation of fullerenes. Science 1998;279:1903.
97. Jenekhe SA, Chen XL. Self-assembly of ordered microporous materials from rod-coil block copolymers. Science 1999;283:372.
98. Mao M, Turner SR. Aggregation of rod-coil block copolymers containing rigid polyampholyte blocks in aqueous solution. J Am Chem Soc 2007;129:3832.
99. Williams DRM, Fredrickson GH. Cylindrical micelles in rigid-flexible diblock copolymers. Macromolecules 1992;25:3561.
100. Reenders M, ten Brinke G. Compositional and orientational ordering in rod-coil diblock copolymer melts. Macromolecules 2002;35:3266.
101. Chen JZ, Zhang CX, Sun ZY, Zheng YS, An LJ. A novel self-consistent-field lattice model for block copolymers. J Chem Phys 2006;124:104907.
102. Chen JZ, Zhang CX, Sun ZY, An LJ, Tong Z. Study of self-assembly of symmetric coil-rod-coil ABA-type triblock copolymers by self-consistent field lattice method. J Chem Phys 2007;127:024105.
103. Chen JZ, Sun ZY, Zhang CX, An LJ, Tong Z. Self-assembly of rod-coil-rod ABA-type triblock copolymers. J Chem Phys 2008;128:074904.
104. Xia YD, Chen JZ, Sun ZY, Shi TF, An LJ, Jia YX. Self-assembly of T-shaped rod-coil block copolymer melts. J Chem Phys 2009;131:144905.
105. Kim KH, Huh J, Jo WH. Wirelike micelle formed by a T-shaped graft copolymer with a rigid backbone. Macromolecules 2004;37:676.
106. Iacovella CR, Glotzer SC. Complex crystal structures formed by the self-assembly of ditethered nanospheres. Nano Lett 2009;9:1206.
107. AlSunaidi A, den Otter WK, Clarke JHR. Microphase separation and liquid-crystalline ordering of rod-coil copolymers. J Chem Phys 2009;130:124910.
108. Diplock R, Sullivan DE, Jaffer KM, Opps SB. Nematic-isotropic phase transition in diblock fused-sphere chain fluids. Phys Rev E 2004;69:026701.
109. Movahed HB, Hidalgo RC, Sullivan DE. Phase transitions of semiflexible hard-sphere chain liquids. Phys Rev E 2006;73:032701.
110. Horsch MA, Zhang ZL, Glotzer SC. Self-assembly of polymer-tethered nanorods. Phys Rev Lett 2005;95:056105.
111. Horsch MA, Zhang Z, Glotzer SC. Self-assembly of laterally-tethered nanorods. Nano Lett 2006;6:2406.
112. Nguyen TD, Glotzer SC. Switchable helical structures formed by the hierarchical self-assembly of laterally tethered nanorods. Small 2009;5:2092.
113. Song JH, Shi TF, Li YQ, Chen JZ, An LJ. Rigidity effect on phase behavior of symmetric ABA triblock copolymers: a Monte Carlo simulation. J Chem Phys 2008;129:054906.
114. He PT, Li XJ, Deng MG, Chen T, Liang HJ. Complex micelles from the self-assembly of coil-rod-coil amphiphilic triblock copolymers in selective solvents. Soft Matter 2010;6:1539.
115. Yu K, Bartels C, Eisenberg A. Trapping of intermediate structures of the morphological transition of vesicles to inverted hexagonally packed rods in dilute solutions of PS-b-PEO. Langmuir 1999;15:7157.
116. Shen HW, Eisenberg A. Control of architecture in block-copolymer vesicles. Angew Chem Int Ed 2000;39:3310.
117. Pochan DJ, Chen ZY, Cui HG, Hales K, Qi K, Wooley KL. Toroidal triblock copolymer assemblies. Science 2004;306:94.
118. Weiss TM, Narayanan T, Wolf C, Gradzielski M, Panine P, Finet S, Helsby WI. Dynamics of the self-assembly of unilamellar vesicles. Phys Rev Lett 2005;94:038303.
119. He YY, Li ZB, Simone P, Lodge TP. Self-assembly of block copolymer micelles in an ionic liquid. J Am Chem Soc 2006;128:2745.
120. He XH, Schmid F. Using prenucleation to control complex copolymeric vesicle formation in solution. Macromolecules 2006;39:8908.
121. Li ZB, Hillmyer MA, Lodge TP. Morphologies of multicompartment micelles formed by ABC miktoarm star terpolymers. Langmuir 2006;22:9409.
122. Li ZB, Hillmyer MA, Lodge TP. Laterally nanostructured vesicles, polygonal bilayer sheets, and segmented wormlike micelles. Nano Lett 2006;6:1245.
123. Li ZB, Kesselman E, Talmon Y, Hillmyer MA, Lodge TP. Multicompartment micelles from ABC miktoarm stars in water. Science 2004;306:98.
124. Wang L, Xu R, Wang ZL, He XH. Kinetics of multicompartment micelles formation by self-assembly of ABC miktoarm star terpolymer in dilute solution. Soft Matter 2012;8:11462.

16

MORPHOLOGY CONTROL OF POLYMER THIN FILMS

JIANGANG LIU[1], XINHONG YU[1], LONGJIAN XUE[2], AND YANCHUN HAN[1]

[1]*State Key Laboratory of Polymer Physics and Chemistry, Changchun Institute of Applied Chemistry, Chinese Academy of Sciences, Changchun, China*
[2]*Dynamic Biointerfaces, Max Planck Institute for Polymer Research, Mainz, Germany*

Polymer thin films have become an important part of our everyday life and have been extensively explored from both the fundamental and applied points of view. Polymer materials have the ability of controllable morphology and porosity ratio, adherence to the flexible substrate, and ease of large-area processing. It makes the polymer films satisfy the requirements of broad range of applications. These films are increasingly used in technological applications like protective coatings; antireflection coatings; lubricants; decorative paints; adhesives; biomembranes; sensors; as active layers in devices such as organic light-emitting diodes (OLEDs), organic field-effect transistors (OFETs), organic solar cells (OSCs), and organic laser devices, and numerous other fields [1–7].

A large number of studies revealed that various physical properties of thin polymer films strongly deviate from those in the three-dimensional bulk state when the film thickness is less than the unperturbed dimension of the polymer chain [8–15]. Polymer chains in the bulk typically take on a so-called random coil conformation, where the physical orientation of each repeat unit along the chain's backbone is essentially uncorrelated with the orientations of the chain's other repeat units. In thin films, the confinement of molecules between two interfaces (solid substrate and free surface) reduces the degree of freedom of the polymer chains such that the random coil conformation will be modified due to the spatial confinement and a transition from the 3D to the 2D case will take place with decreasing film thicknesses [16, 17]. The conformation of single polymer chain and the interactions between the polymer chain and the interface are crucial factors to determine the characteristic properties of the thin film. Therefore, polymers in thin-film state have attracted great attention and have been extensively studied by many researchers [18–21].

The stability and the wetting behavior of thin polymer films are of technological and scientific importance in a broad range of applications [22, 23]. Understanding and controlling the factors leading to the dewetting of thin films is of critical importance in obtaining continuous, uniform, stable, and defect-free coatings. Meanwhile, in many applications, such as paints, adhesives, and OSCs, the films are composed of multiple polymeric systems. Hence, it is also important to investigate the structure and stability of a thin polymer film containing other component or additives [24]. In general, most of binary polymer blends are immiscible and will demix (phase separation). The degree of the phase separation in blends greatly affects the domain morphology, which has a strong effect on the mechanical and electrical properties of the film [25]. Many studies have been performed to elucidate the fundamental mechanism and dynamics of dewetting and phase separation in efforts to generate functional structures. In this chapter, a brief introduction to the fundamental theory and the influencing factors on dewetting in thin polymer film and phase separation in thin polymer blend film are introduced in Sections 16.1 and 16.2, respectively. The control of morphology in polymer thin film and its application in OSC are discussed in Section 16.3.

16.1 WETTING

Different phenomena happen when some liquid is dropped onto a solid surface. In case the amount of liquid is large, the liquid can fully cover the solid surface even if the liquid does not like the solid surface. And the thick liquid film remains stable due to the effect of gravity. When the thickness of the

Polymer Morphology: Principles, Characterization, and Processing, First Edition. Edited by Qipeng Guo.

liquid film gets thinner, two competing effects will determine the spreading of the liquid on the solid surface. For a solid surface, it is energetically favorable to be wetted by a liquid with lower surface energy. However, the spreading of the liquid increases its surface area and the contacting areas between the liquid and the solid surface, which means an increase in interfacial energy and surface energy of the liquid. When the surface energy of the solid surface dominates, the liquid completely wets the solid surface; otherwise, the solid surface can only get partially wetted, defining a contact angle θ of the liquid. The partial wetting is described by Young's equation, which shows a balance of macroscopic forces:

$$\cos\theta = \frac{(\gamma_{sg} - \gamma_{sl})}{\gamma_{lg}} \tag{16.1}$$

where γ_{sg}, γ_{lg}, and γ_{sl} are interfacial energies of solid/gas, liquid/gas, and solid/liquid, respectively. For $\theta = 0$, the liquid completely spreads on the solid surface. $0 < \theta < \pi$ indicates a partial wetting, which happens in most cases. The nonwetting case happens with $\theta = \pi$.

In practice, liquids (e.g., polymers discussed here) are normally forced to spread onto solid surfaces, including onto nonwettable surfaces, using techniques like spin coating. The resulted film is in a nonequilibrium state. If the thickness of the polymer film is smaller than the capillary length of the liquid, λ_{cap}, which is defined by $\lambda_{cap} = (\gamma_{lg}/\rho g)^{1/2}$ with ρ the density of the liquid, molecular forces dominate the gravitational forces that the liquid film is unstable [26–34]. As we know, the state of droplets with equilibrium contact angle θ_e represents the state with lowest system energy, which the system is approaching to. How will the liquid film decay into droplets and at what speed will this process take place at a given temperature? Depending on the thickness, the answers to the above questions will be different. Moreover, how will the physical and chemical heterogeneities affect the stability (instability) of a liquid film as they always present on a real solid surface? In the following, we try to give a brief answer to these questions based on polymer thin films.

16.1.1 Dewetting Mechanisms

It is important to clarify the distinction among stable, metastable, and unstable films using an easily detectable parameter like the thickness h of a thin polymer film, defined by the distance separating the interfaces of substrate/polymer and the air/polymer. The stability of a film depends on the effective interface potential $\varphi(h)$ as a function of h, defined as the excess free energy per unit area that is necessary to bring two interfaces (solid–liquid and liquid–gas interface) from infinity to the distance h.

As shown in Figure 16.1, $h \to \infty$ induces $\varphi(h) \to 0$, indicating the stability of a film with infinite film thickness. For the stable films (as shown in curve (1) of Fig. 16.1), the free energy is always positive and the global minimum lies at the infinite thickness. If the second derivative of φ with respect to the film

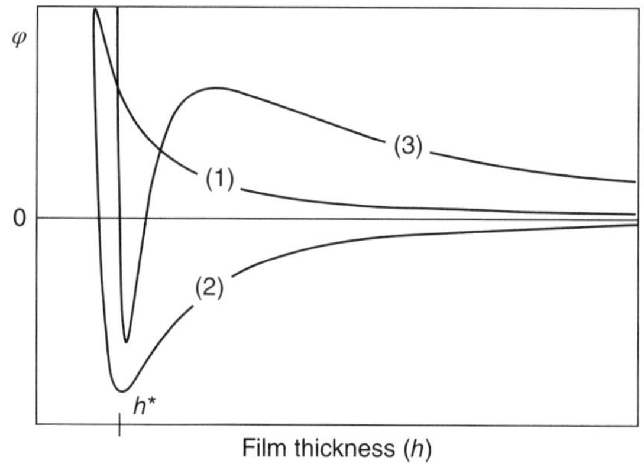

Figure 16.1 Effective interface potential ($\varphi(h)$) as a function of the film thickness (h) for stable (1), unstable (2), and metastable (3) films. Seemann [35]. Reproduced with permission of American Physical Society.

thickness is negative, $\varphi''(h_0) < 0$, where h_0 is the initial thickness of a homogeneous film, the system is referred to as unstable (curve (2) of Figure 16.1). According to Brochard-Wyart and Daillant [27] the surface fluctuation of a film with thickness h, can be described by the equation

$$\zeta(x,t) = h + u\exp(ipx)\exp\left(\frac{-t}{\tau}\right) \tag{16.2}$$

with amplitude u, wave vector q, time t, relaxation time τ, and a coordinate parallel to the surface x. The perturbations of the surface induce a pressure gradient caused by Laplace pressure P_L and disjoining pressure Π, which leads to a Poiseuille flow in the film

$$P_L = \gamma_{lg}\left(\frac{d^2\zeta}{dx^2}\right) \tag{16.3}$$

and for nonretarded van der Waals forces

$$\Pi = \frac{-A_{eff}}{(6\pi\zeta^3)} \tag{16.4}$$

where A_{eff} is the effective Hamaker constant. The sign of A_{eff} indicates whether molecules of the film are more attractive by molecules or by the substrate. A negative sign of A_{eff} indicates the tendency of the film to get thinner, suggesting the film is unstable.

Considering mass conservation, surface fluctuation has a critical wave vector $\boldsymbol{q}_c$, $\boldsymbol{q}_c = 3a/h^2$, where a is a molecular length defined by

$$a^2 = \frac{|A_{eff}|}{(6\pi\gamma_{lg})} \tag{16.5}$$

For $\boldsymbol{q} < \boldsymbol{q}_c$, the fluctuations of the film surface are amplified exponentially with a fastest growing wave vector $\boldsymbol{q}_m$

$$\boldsymbol{q}_m = \left(\frac{3}{2}\right)^{1/2}\left(\frac{a}{h^2}\right) \tag{16.6}$$

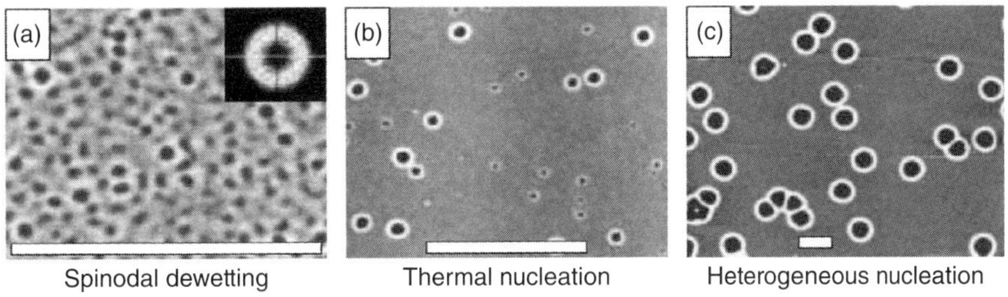

Figure 16.2 (a–c) AFM images of dewetting PS (2k) films. The scale bars indicate 5 μm, the z-scale ranges from 0 (black) to 20 nm (white): (a) spinodal dewetting of 3.9 nm PS on type C wafer. The inset shows a Fourier transform of the image. (b) Thermal nucleation dewetting of 4.1 nm PS on type B wafer. (c) Heterogeneous nucleation dewetting of 6.6 nm PS on a type B wafer. Seemann [35]. Reproduced with permission of American Physical Society.

As the amplified fluctuation touches the substrate, a hole forms in the film. This rupture mechanism is called "spinodal dewetting" [26, 27, 36].

In curve (3) of Figure 16.1, a region exists for very small film thickness where $\varphi''(h_0) < 0$, the film is unstable, whereas for thicker films ($\varphi''(h_0) > 0$), the film is metastable. In the metastable case, the system has to overcome a potential barrier to reach the state of lowest energy at $h = h^*$. Some kind of nuclei in the form of dust particles or other heterogeneities are required to lower $\varphi(h)$ and can therefore induce rupture of film, termed as the mechanism of "heterogeneous nucleation" [37, 38]. Numerical simulations have also indicated that there may be a region where both mechanisms are involved [39, 40]. For the unstable system, there is a specific wavelength (λ_s), whose amplitude increases rapidly, possessing a characteristic dewetting pattern of the liquid film, as shown in Figure 16.2a. The wavelength λ_s and $\varphi''(h_0)$ are related via Equation (16.7) [26]:

$$\lambda_s = \left[\frac{-8\pi\gamma_{lg}}{\varphi''(h)}\right]^{1/2} \tag{16.7}$$

Close to the sign reversal of $\varphi''(h_0) > 0$, no heterogeneity that thermal activation is sufficient to overcome the potential barrier for the nucleation of holes is required. It is called "thermal nucleation" or "homogeneous nucleation" representing a third dewetting mechanism [35]. The characteristic feature of homogeneous nucleation is the continuous breakup of holes (Fig. 16.2b), whereas heterogeneous nucleation causes holes to emerge only within a sharp time window (Fig. 16.2c). Such hole opening is most likely to happen at the locations having the smallest thickness, which can be caused by the thickness fluctuation of film thickness or the elevated region on the substrate. In practice, characters from different mechanisms could show up in a single system, making the mechanism much more complex. For example, a stable film with $\varphi''(h_0) > 0$ may become unstable locally on an elevated heterogeneity if the local thickness declines below the critical thickness. In some composite polymer film [41], continuously opened holes in a metastable film resulted in a pattern showing the wavelength of spinodal dewetting. On a complex substrate combined with wettable and less wettable (or unwettable) regions, holes generate at these unfavored locations and the wettability contrast drives the liquid flow from the less wettable areas to the more wettable areas, lowering the total energy.

In addition, thin films can also rupture and dewet due to the release of residual stress accumulated during film preparation [42, 43]. In a typical case of film preparation from solution, the solvent evaporates quickly allowing only a short time for the rearrangement of polymer chains and the formation of entanglements. Residual stresses are therefore introduced into the film by the nonequilibrium state and the partially entangled conformations of chains. The release of residual stresses was found to play a prominent role in controlling nucleation and growth of dewetting holes in thin polymer films. By aging the film at a temperature lower than glass transition temperature (T_g), the number of generated holes decreased exponentially with aging time. The exponential decay of number of holes allows for the possibility that no holes will be generated in the film after a sufficiently long period of aging, assuming no other sources of rupture. It was even found that residual stresses can be significantly larger than capillary forces, indicated by a highly asymmetric rim around holes [44]. Upon releasing, the progressive decreasing contribution of residual stresses to the driving force causes a rapid decay of the dewetting velocity [42].

16.1.2 Dewetting Dynamics

The driving force F_d, which is the negative spreading coefficient S,

$$\begin{aligned} F_d &= -S = \gamma_{lg} + \gamma_{sl} - \gamma_{sg} \\ &= \frac{1}{2}\gamma_{lg}\theta_e^2 \end{aligned} \tag{16.8}$$

causes the growth of opened holes, uncovering the solid substrate (dry areas). The accumulation of dewetted polymers builds up rims, circle protrusions, around dry areas. Molecular and interfacial interactions define the profile of rims, showing either a symmetrical profile with a trough at the "wet" side [45, 46] or an asymmetrical profile monotonically decaying into the undisturbed film.

A stronger molecular interaction of liquid and solid molecules, that is, an ideal (Newtonian) liquid on a solid

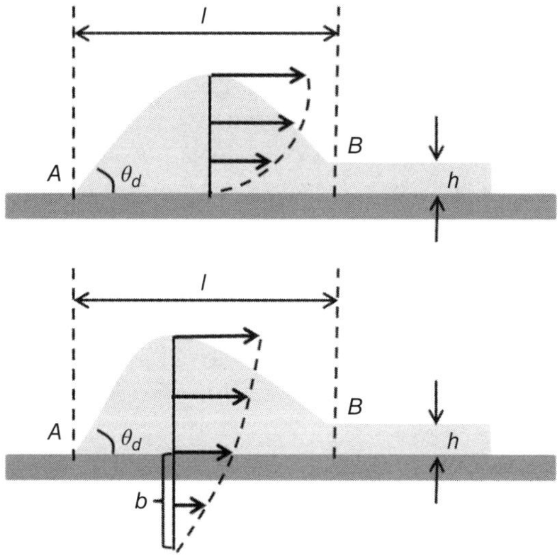

Figure 16.3 The two possible mechanisms of friction and the growth law of the holes.

surface, causes viscous friction inside the liquid film dissipating energy and no slippage at the liquid–solid interface, that a symmetrical profile of the rim forms. It means a balance between F_d and viscous friction force F_v. In the lubrication approximation, in a film with thickness h much larger than the extrapolation length b, which is defined as the distance from the interface at which the velocity extrapolates to zero, F_v of a moving rim at velocity V with its contour $\xi(x)$:

$$F_v = \int_{x_A}^{x_B} \frac{\eta V}{\xi/3 + b} dx \tag{16.9}$$

where η is viscosity of the liquid film and two contact lines A and B define the rim width, l (Fig. 16.3). The balance between F_d and F_v then gives:

$$\int_{x_A}^{x_B} \frac{\eta V}{\xi/3 + b} dx = \frac{1}{2}\gamma\theta_e^2 \tag{16.10}$$

The profile near liquid wedges, where the dissipation dominates, can be estimated approximately by the dynamic contact angle θ_d via $\xi = \theta_d x$. It then leads to

$$\frac{6\eta V}{\theta_d} |\log(\theta_d x + 3b)|_{x=0}^{x=l/2} = \frac{1}{2}\gamma\theta_e^2 \tag{16.11}$$

Taking into account that $\theta_d = \theta_e^{1/2}$, the moving velocity of the rim is given by [47–50]:

$$V = \frac{1}{12\sqrt{2}\kappa} \frac{\gamma}{\eta} \theta_e^3 \tag{16.12}$$

The numerical dissipation factor $\kappa = \ln(\theta_e l/b)$ is expected to be of order 10 [27]. Considering the substrate is ideal flat, the radius of hole, R, grows linearly with time t, $R \sim t$.

A material like polymer with large molecular weight, however, does not flow like an ideal liquid. The interfacial interactions between liquid molecules and the solid surface cause slippage of liquid on the smooth, passive solid surfaces [51–53]. Quite a lot of parameters like the nature of the liquid molecules [54, 55], substrate roughness [56–59], nanoscale air bubbles in the liquid film [60, 61], and the amount of water depletion on the surface [62] have influences on the slippage. A plug flow occurs in the thin film such that Equation (16.10) can then be simplified as

$$\frac{\eta V l}{b} = \frac{1}{2}\gamma\theta_e^2 \tag{16.13}$$

The velocity decreases with time because the friction is now proportional to the width l of the rim. As the materials accumulated from dewetted areas continuously add to the volume of the rim, width l of the rim can be calculated from the conservation equation, leading to

$$l \propto \left(\frac{eR}{\theta_e}\right)^{1/2} \tag{16.14}$$

Substituting Equation (16.14) into Equation (16.13) leads to [47]

$$R \sim t^{2/3} \tag{16.15}$$

In reality, however, both viscous flow and slippage are expected to dissipate energy in thin polymer films. It is found that slippage plays a prominent role in a small rim (either for a tiny hole or for a rim in a very thin film), whereas viscous dissipation dominates the hole growth for a large rim. Thus, a hole would undergo the shifting from a slippage ($R \sim t^{2/3}$) to a nonslip mode ($R \sim t$) [49, 50].

In devices like thin-film transistors (TFTs), multilayered structures are often introduced so that the stability/instability of a multilayered polymer film is of even more practical importance. However, the dewetting of a thin polymer film on top of another polymer layer is much more complicated than the common liquid–solid case, as both the polymer–polymer interface and the free film surface are deformable. Taking an immiscible bilayer as an example, dewetting dynamics of the top layer depends mainly on the relative viscosities of the two liquids, the thicknesses of respective liquid layers, and the surface and interfacial tensions involved [63, 64]. For a very viscous sub-layer ($\eta_{sub} > \eta_{up}/\theta_e$, where η_{sub} and η_{up} are viscosity of sub- and upper layer, respectively), the sublayer behaves solid-like. The dewetting velocity of the upper layer on such a solid-like sublayer depends mainly on the surface tension and viscosity of the upper layer [65, 66]:

$$V = \frac{1}{12\kappa\sqrt{2}} \frac{\gamma_{up}}{\eta_{up}} \theta_e^3 \tag{16.16}$$

As the polymeric sublayer is also prepared from solution, the short-range roughness of the sublayer, which would be different depending on the preparation conditions, was found to increase the effective contact area, accelerating the dewetting [67].

The sublayer behaves liquid-like if $\eta_{sub} < \eta_{up}/\theta_e$ that the upper layer extends considerably into the sublayer beneath the rim [68], which leads to a hydrodynamic resistance against the growth of holes in the upper layer. The deformation in the sublayer levels off by increasing the molecular weight of the sublayer (within the scope of liquid-like behavior), which reduces the flow resistance. Therefore, the dewetting velocity of the upper layer is more dependent on the molecular weight of the sublayer than on that of the upper layer:

$$V = \left(\frac{\gamma^2 h_{sub}^2 \theta_e}{\eta_{sub}^2 h_{up}} \right)^{1/3} t^{-1/3} \tag{16.17}$$

where $\gamma^{-1} = \gamma_{up}^{-1} + \gamma_{sub}^{-1}$ and h_{sub} and h_{up} are the thickness of sub- and upper layer, respectively. Moreover, the dewetting velocity has a dependence on the thicknesses of both layers that the dewetting velocity increases with a thinner upper layer or a thicker sublayer [66, 69], different from that on a solid-like sublayer where the dewetting velocity is even independent on the thickness of the upper layer.

16.1.3 Rim Instability

In a slipping case, the increasing volume of the rim and, therefore, the friction forces slows down the dewetting velocity. However, the dewetting always prefers a constant dewetting velocity, which results in a fluctuation in the rim [70]. Disjoining pressure (van der Waals forces) drives materials from thinner to thicker parts of the rim and tends to enlarge the fluctuation in the rim. Meanwhile, Laplace pressure would flatten the fluctuation. The competition between disjoining pressure and Laplace pressure then determines the stability of the rim.

When disjoining pressure dominates, materials accumulate to the thicker parts of the rim, causing a sharp increase in the resist force locally. The local resist force is so high that the three-phase line at the thicker parts of the rim can hardly move. The thinner parts keep the same dewetting velocity by transferring materials continuously to the pinned sites, leaving behind the formation of a strip perpendicular to the moving front of the hole. The formed strips are called fingers [71]. The distance between the fingers, λ_f, is a constant after their maturation:

$$\lambda_f = \left(\frac{M_e \gamma_{lg}}{|S|} \right)^{1/3} \frac{h_0^{7/6}}{M^{1/2}} \tag{16.18}$$

where M and M_e are the molecular weight of polymer and molecular weight between entanglements, respectively. Eventually these fingers decay into droplets by Rayleigh instability [72], and the mean diameter (D) of droplets grows linearly with the initial film thickness h_0, $D \sim h_0$ [73].

16.1.4 Factors Affecting the Stability of Polymer Thin Films

Understanding and controlling the factors causing the instability of thin films is of critical importance in obtaining uniform, continuous, defect-free, and stable coatings. From the point of view of thermodynamics and kinetics, all the parties involved in the system influence the stability of the film, including the nature of polymer chains (molecular weight and interactions with polymer chains and substrate), the film (thickness, homogeneity, the solvent used, the way it is prepared, thermal history), the substrate (roughness, surface energy, homogeneity) and the environment (temperature, solvent vapor). For more details about the stabilization of polymer thin films on a substrate, we draw your attention to a recent review [74].

16.1.4.1 Polymer Chains As the component of a film, the polymer used surely is vitally important for the stability of the film. Different types of polymers (homopolymer or copolymer, linear or branched, amorphous or semicrystalline, etc.) are used to form films according to the designed function of the film such that the structures of the formed thin films differ quite a lot. In order to understand the behavior of dewetting, the simplest polymers (amorphous linear homopolymers, like polystyrene and poly(methyl methacrylate)) are widely used as the model systems. As a simplified system, molecular weight plays a very important role in determining the stability of the resulted film. When the molecular weight is above the critical entanglement molecular weight, the entanglement of molecular chains induces a higher viscosity of the film. According to Equation (16.12), dewetting velocity is reduced. Moreover, if the polymer chains have strong affinity to the substrate, anchored chains forming loops at the surface of substrate interlock the passing-through chains, which provides an additional brake on the liquid flow in the region of greatest thinning. This additional dissipative effect due to the entanglement coupling may lead to an "effective" viscosity in the thinned region and thereby prevents the formation of holes.

Introducing special interactions, such as H-bond and dipolar and electrostatic interactions, greatly changes the interactions among polymer chains and with the substrate, which therefore changes the stability of the film or the dynamics of dewetting. For example, the ionic-end-functionalized polymer chains, together with the presence of metal counterions, form physical cross-links in the film arresting the hole opening in the film and greatly slowing down the dewetting velocity. The same effect could also be achieved by chemical cross-linking; however, it may cause the loss of desired properties of the polymer, especially when the thin polymer film serves as a functional layer in the device.

16.1.4.2 Thin Film As already discussed in the previous section, the thickness of a thin film has a critical influence on the stability of the film and also the mechanism of dewetting. We will not discuss more about it here; but one thing should be mentioned here that the glass transition temperature may be increased or decreased when the thickness of thin film is below a certain limit, depending on the interaction between the polymer and the substrate. For example, T_g of polystyrene (PS) was found to decrease linearly with thickness $h \leq 70$ nm, with a reduction of 70 K for a film with $h = 29$ nm [75]. T_g of thin PMMA film on silicon with native oxide layer, however, was found to be slightly higher than the bulk value [76]. The temperature of T_g is of great importance to the dewetting under thermal stimulation because the polymer films are usually heated to a temperature above T_g. It allows polymer chains to gain enough mobility, facilitating the observation of dewetting in a proper time window. Moreover, the morphologies of dewetted film with thickness less than the radius of gyration of the polymer may change greatly due to the confinement effect.

Fabrication by solution processing is the main strategic advantage organic electronics can offer over inorganic electronics as it allows for the possibility of large-area, low-cost flexible electronics and manufacture by roll-to-roll processing [77]. Microscopically, different morphologies of thin films form because of the different evaporation rates and the wettabilities of solvents on substrate. Moreover, the configurations of polymers in the film are determined by the combination of solubility, boiling point of the solvent, and the way of preparing the film. Fast evaporation and short thermal annealing create much more stresses inside the film because the fast solidification of polymer chains leaves the chains fixed in nonequilibrium states. The location with concentrated stresses can be considered as a defect in the film, where the rupture of the film could be initiated by the release of the stresses.

Another kind of defect in films is impurities, like small particles from air and air bubbles originated from the evaporation of solvent. They can lower the potential barrier, causing the rupture of the thin film by heterogeneous nucleation. On the other hand, however, the addition of nanoparticles or nanotubes is found to have the ability to enhance the stability of polymer thin films. The mechanism of stabilization was hypothesized to be the result of the change of wettability by modifying the surface of the substrate with segregated additives and the contact line pinning arose from the increased interfacial roughness. It means the stabilization effect of nanoadditives has both thermodynamic (change of surface energy of substrate) and kinetic (contact line pinning) contributions. The dispersion state of the additive, which is related to its surface chemistry, size, and concentration, plays a key role in the stabilization effect. In order to inhibit and eventually eliminate the dewetting, additives with smaller sizes and segregating to the substrate interface seem to be more efficient. And the amount of additives should be sufficient to form at least one monolayer at the film–substrate interface and not too much to form obvious aggregates in the film.

16.1.4.3 Substrate The properties of interface between polymer thin film and substrate have a great impact on the spreading and wetting behavior of the polymer thin film. First, it depends on the spreading coefficient S of the substrate, which is related to the chemical composition of the substrate. As self-assembly techniques have been developed quite well, it is convenient to modify the surface with different chemical groups and thus the sign of S and its value. However, it is not necessary to consider that a negative S must cause a dewetting of the film as the film may be dynamically stable. Also, dewetting of polymer thin film could happen on the substrate with a positive S. For example, PMMA is thermal stable on silicon wafer with a thin layer of native silicon oxide. However, when the film is exposed to a good solvent of PMMA, like acetone, dewetting happens [78]. A new interface establishes between the anchored and movable PMMA chains; and entropy effects caused the autophobic dewetting on the newly established interface. Anyway, a strong affinity of polymer to the substrate is helpful to stabilize the film. Moreover, there should be enough entanglements between the anchored and the free chains. Second, the physical roughness of the substrate is also a critical factor. A big roughness, comparable to or larger than the thickness of the film, reduces the local thickness facilitating the rupture of the film. However, both the effects of accelerating and inhibition of dewetting dynamics caused by the short-range roughness have been reported.

Furthermore, the chemical- or physical-patterned substrate-induced dewetting has been proved to be a convenient and effective way to pattern surface on a micro/nanometer scale without a lithographic process. About this topic, we draw your attention to a review article [34].

16.1.4.4 Environment As we know, the dewetting of polymer thin film means the polymer chains gain enough mobility under thermal or solvent vapor stimulations. Thus, keeping the film at a temperature well below the T_g of polymer and avoiding the exposure of the film to solvent vapor can keep the film in a continuous, homogeneous state within a quite long period, though the film maybe not be thermal stable. Luckily, most polymers have T_gs above room temperature, that the polymer thin films could stay homogeneous. In some applications of polymer thin films, however, exposing to a high temperature cannot be avoided. Then the physical or chemical cross-linking can be used to partially or permanently reduce the mobility of polymer chains, even under elevated temperatures.

16.2 THIN FILM OF POLYMER BLEND

Assuming a polymer thin film is prepared from a single-solute solution, three interaction pairs, namely, polymer–solvent, polymer–substrate, and solvent–substrate, determine the spreading and the stability of the film. By adding another polymer into the solution, six interaction pairs contribute to the final structure of the resulted film, which is much more complex. On the other hand, in order to get better performance or multifunctions, multicomponents (here, we focus on polymer mixtures) in different structures are required in more and more devices. Meanwhile, the sizes of these devices are getting smaller and thinner, aiming at energy saving and portability. Therefore, it requires a much better understanding of the phase behavior of polymer blends in the state of thin films. As most blends of polymers with high molecular weight are intrinsically immiscible, phase separation occurs under appropriate conditions because of the vanishing entropy of mixing. Phase separation of polymer blends offers the possibility of obtaining various morphologies in thin polymer films, like bicontinuous structure, islands, and holes by controlling system parameters such as composition, molecular weight and architecture, film thickness, solvent and exterior environment including substrate, pressure, temperature, and external fields. In this section, we briefly introduce the fundamental theory of phase separation of polymer blends and the phase behavior in thin-film state. In the following section, the relationship between a phase-separated structure and the performance of devices, taking OSC as the example, is discussed in detail and we try to give a clue about how to manipulate phase separation in order to get a better performance.

16.2.1 Fundamentals of Polymer Blends

The change of free energy per segment ΔG_m, which is estimated through Gaussian polymer chains on an incompressible ($\varphi_A + \varphi_B = 1$) lattice [79, 80],

$$\frac{\Delta G_m}{k_B T} = \frac{\varphi_A}{N_A} \ln \varphi_A + \frac{1-\varphi_A}{N_B} \ln(1-\varphi_A) + \varphi_A(1-\varphi_A)\chi \tag{16.19}$$

determines the miscibility of two dissimilar linear homopolymers (polymer A and B). In the equation, k_B is the Boltzmann constant, T is the temperature, φ_A (φ_B) is the volume fraction of polymer A (B), N_A (N_B) is the polymerization degree of polymer A (B), and χ is the Flory–Huggins segment–segment interaction parameter. In general, a positive ΔG_m causes the demixing (phase separation) of the mixture; otherwise, mixing occurs. The first two terms in the right-hand side account for the combinatorial entropy of mixing ΔS_m. ΔS_m naturally increases ($\Delta S_m > 0$) as mixing increases the system's randomness, thereby decreasing the free energy of mixing. However, ΔS_m could not be large because of the long chain of polymers and the entanglement of polymer chains; and ΔS_m decreases with increasing polymerization degree, N. The third term represents the enthalpy of mixing, ΔH_m, and can either increase or decrease ΔG_m depending on the sign of χ, which is approximated by the following equation [81]:

$$\chi = \frac{1}{k_B T}\left[\varepsilon_{AB} - \frac{1}{2}\left(\varepsilon_{AA} + \varepsilon_{BB}\right)\right] \tag{16.20}$$

where ε_{ij} is the contact energy between i and j segments (components). Negative values of χ occur for certain types of specific interactions between A and B, such as hydrogen bonding, that is, A–B segment–segment pairs on average have a lower system energy than the sum of A–A and B–B pairs. A positive value of χ represents the increase in net system energy upon forming A–B contact pairs. For example, the mixing of nonpolar polymers, such as polystyrene (PS) and polyisoprene (PI), governed solely by dispersive interactions (van der Waals interactions) is thermodynamically unfavorable, which means $\chi \geq 0$ assuming no volume change or preferential segment orientation happens during mixing [82]. In practice, anisotropic monomer structures may lead to nonrandom segment packing, which must be absorbed in χ as an excess entropy of mixing given by

$$\chi = \Im T^{-1} + \varsigma \tag{16.21}$$

where $\Im$ and ς represent experimentally determined enthalpy and excess entropy coefficients for a particular composition. In homopolymer mixtures, $\Im$ and ς are affected by φ, N, T and molecular architecture of the component. The phase behavior can then be predicted with Equation (16.19) based on the standard criteria for equilibrium, and stability evaluated at constant temperature and pressure by a phase diagram (take a symmetric case, $N_A = N_B$, as the example, as shown in Figure 16.4) [83]. The solid and dash curves represent the solutions of the usual equilibrium and stability equations, respectively:

$$\text{Equilibrium}: \frac{\partial \Delta G_m(\varphi_A')}{\partial \varphi_A} = \frac{\partial \Delta G_m(\varphi_A'')}{\partial \varphi_A} \tag{16.22}$$

$$\text{Stability}: \frac{\partial^2 \Delta G_m}{\partial \varphi_A^2} = 0 \tag{16.23}$$

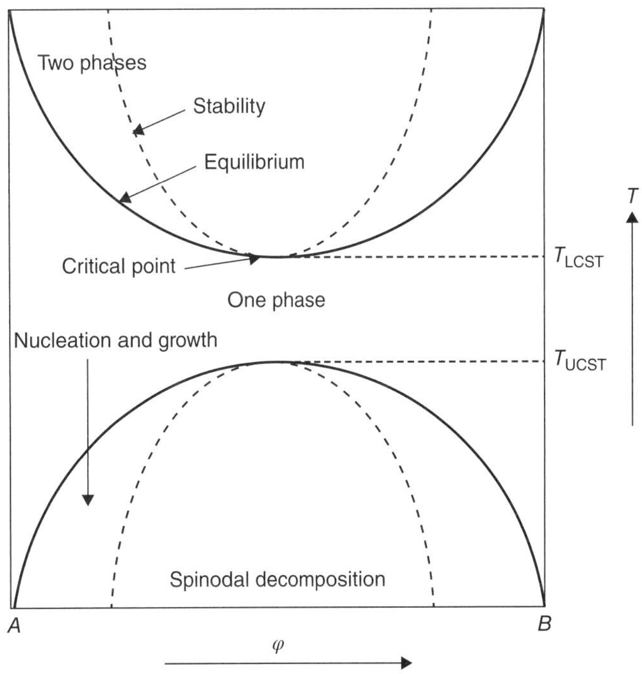

Figure 16.4 Schematic phase diagram for a symmetric ($N_A = N_B$) binary mixture of linear homopolymers showing the LCST (TLCST) and UCST (TUCST).

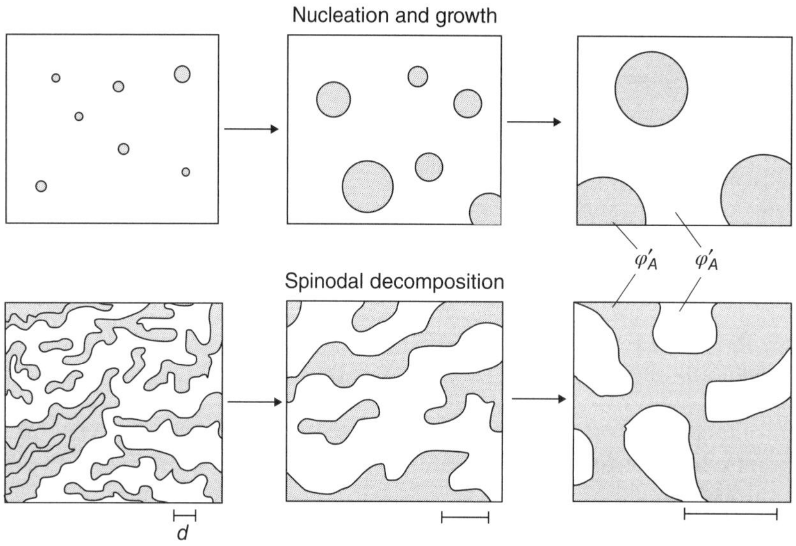

Figure 16.5 Time evolution of structure in phase-separating binary homopolymer mixtures. Nucleation and growth result when a homogeneous mixture is thrust into the metastable region of the phase diagram. Spinodal decomposition occurs when a mixture is placed in a thermodynamically unstable state.

where φ'_A and φ''_A are the two volume fractions of component A. The region between equilibriums in the phase diagram represents miscible of the blend. If $\Im$ is negative and ς is positive, then a lower critical solution temperature (LCST) may result depending on the magnitude of N and ς. Phase separation happens in the blend by increasing T. Positive $\Im$ and negative ς always increase χ and result in an upper critical solution temperature (UCST). However, phase separation will not be triggered by decreasing the temperature because polymer chains can hardly move at low temperatures.

There have two mechanisms for phase separation: (i) spinodal decomposition and (ii) nucleation and growth. In the thermodynamically unstable state (the region enveloped by the dash line in phase diagram), mixtures phase separate spontaneously, resulting in a disordered, bicontinuous two-phase structure. This behavior is known as spinodal decomposition (Fig. 16.5). The quench depth $\chi_s - \chi$, where χ_s corresponds to the stability limit (dashed curve in Figure 16.4), controls the initial size d_0 of the spinodal structure. Deeper quench produces better structures. Almost immediately after the initiation of bicontinuous pattern, interfacial tension favors a reduction in surface area by increasing d. However, the equilibrium morphology can hardly be reached because the viscosities of the polymer melts are extremely large. The mixture locating in the region between the equilibrium and stability in phase diagram is metastable. Phase separation of a metastable mixture is governed by the mechanism of nucleation and growth. The nucleation of a new phase in homogeneous mixtures is triggered only when a free energy barrier is overcome, which proceeds by diffusion of material from the supersaturated continuum. Once the composition of the supernatant reaches equilibrium, the domains of the new phase increase their size by coalescence or Ostwald ripening; the latter refers to the growth of large domains through the absorption of smaller ones. The further growth of the domain size may be extremely slow because of the low diffusivity and enormous viscosity of polymers, and may result in wide distribution of particle sizes.

16.2.2 Phase Separation in Thin Polymer Films

It has been demonstrated that the phase separation of polymer blend in thin-film state is quite different from that of polymer blends in bulk, and is much complicated by the presence of the interfaces of substrate/film and film/air. A miscible blend of PS/PVME in bulk was even found to undergo phase separation when the blend is in a state of thin film with its thickness less than twice the radius of gyration of an unperturbed chain, $2Rg$ [84, 85]. Moreover, the temperature for phase separation [84], the Flory–Huggins parameter χ [86], and breakup mechanism [87] could also be different from that in bulk.

The phase-separated structure in thin polymer film is sensitive to a variety of parameters including substrate (polymer substrate interactions) [88, 89], film thickness [90, 91], Flory–Huggins parameter χ [92–94], molecular weight of polymer [95, 96], component ratio [97], surface tension of polymer [97], the relevant solvent parameter (polymer solubility [91, 98] or evaporation speed [99]), addition of additives [100], and so on. Some general mechanism could be applied to different systems: (i) the component with lower surface energy prefers to migrate to the free surface [101], while the component possessing higher affinity with the substrate would segregate to the substrate interface [102]. Such preferential segregation is called vertical phase separation, creating layered structures [103] or gradient distribution [87] in the thickness direction. While a neutral substrate may cause a lateral phase separation, in which the phase boundaries are perpendicular to the surface. Furthermore, a substrate with ordered pattern will guide the component with higher affinity to the corresponding area copying the pattern on the substrate, thus creating an ordered, instead of random, pattern of phase separation [83]. (ii) A better compatibility between compositions results in uniform morphologies with smaller size scale [87, 104]. By changing the values x and y in the PBr_xS/PBr_yS system, the degree of compatibility of the mixture can be very well adjusted [105]. It was found that the mixture with better compatibility preferred to form pore structures with smaller size scale at the beginning. (iii) The change in molecular weight also affects the compatibility that a smaller molecular weight increases the miscibility of polymer blend resulting in a decreased size of microdomains [96, 106]. Meanwhile, the difference in surface energy between chain end (γ_e) and main chain (γ_m) groups for a low-molecular-weight polymer has a dramatic influence on the phase-separated structure. If $\gamma_e < \gamma_m$, the chain end groups are preferentially segregated at the surface [107, 108]. This effect becomes more remarkable with the decrease of M_n, due to an increase in the number density of chain end groups. Furthermore, the molecular weight affects the movement of polymer chains and therefore the morphology of phase separation.

As a thin polymer film is usually prepared from a solution using techniques like spin coating, the solvent plays a critical role in determining the phase-separated morphology. In the thin film forming process, phase separation occurs for immiscible blends with the loss of solvent by evaporation. The different solubilities of polymers in the casting solvent causes demixing of the polymers as the solvent evaporates. The rapid increase in the viscosity of the film captures the phase-separated structure in place, apparently in a nonequilibrium state. A component with higher solubility enriches at the film surface and results in a vertical phase separation [91, 109]. This effect may then result in a vertical phase separation with the higher surface-tension component on top of the film [91, 105]. As for selective solvent, the phase domain with better solubility is still swollen when the phase domains of the component with less solubility is solidified. Further evaporation of the solvent collapses the swollen phase to a level below that of the vitrified phase and results in relief structures. For the same kind of solvent, the solvent with lower vapor pressure will take a longer time to reach polymer solidification; therefore, the phase separation process is closer to the thermodynamic equilibrium state [110].

16.3 THE INTRODUCTION OF POLYMER BLEND FILM IN SOLAR CELLS

Polymer blend film usually not only possesses excellent properties of each single composition but also presents some extra functions that are related to the micro- and nanostructures of the polymer blend film. Therefore, researching on different structures, discussing the relationship between the morphology and property of blend film consciously is of great importance in polymer science. The phase separation of all-conjugated polymer blend system is one of the important fields in polymer science. From the basic research perspective, it is well worth illustrating the effect of rigid polymer chain on the phase-separation mechanism of polymer film; in addition, the competitive relationship between crystallization and phase separation also needs further debate. From the applied perspective, conjugated polymer is an important constituent to fabricate TFT, OSCs, and so on. It is easy to achieve some new photoelectric property and to improve the efficiency and stability of photoelectric devices effectively through blending conjugated polymers [111, 112]. As a result, relations between phase separation and photovoltaic properties of conjugated polymer blend system are not only the requisites for development of discipline but also have a definite application prospect.

In the following, we try to illustrate the strategies for adjusting phase-separated structures using OSCs as the example, which have received much attention because they have potential advantages over other solar energy technologies [113], such as mechanical flexibility, light weight, and low cost. As well known, the morphology of active layer based on organic materials has a profound influence on the device performance. Organic semiconductor thin films show high-absorption coefficients exceeding $10^5\,cm^{-1}$ and high charge-carrier mobility [114, 115]. The electronic band gap of organic semiconductors also could be engineered by chemical synthesis [116]. Recently, the power conversion efficiency (PCE) for all polymer blend systems has exceeded 4.0%; this is the highest reported in literatures [117]. While the PCE could reach to as high as 5.20% reported by Polyera Corporation when it adopted ActivInk PV2400 as p-type material and ActivInk NV 2400 as n-type material.

The nanomicrostructure of active layer in all-polymer thin-film solar cells plays a crucial role in determining the PCE. There are four physical processes in the power conversion of all-polymer solar cells, as shown in Figure 16.6: photon absorption, exciton diffusion, dissociation of charge-transfer states, charge-carrier transport, and collection. These processes are controlled by the order of bulk heterojunction, the domain size and purity, the structure of phase separation and the diffused structure at the interface. For example, the domain size should be less than 20 nm so that the exciton can diffuse to the interface. The important premise for enhancement of charge-transfer state dissociation and charge-carrier transport is to promote the domain purity and molecular order within domains. To ensure the charge-carrier transport, improve the charge-collection efficiency and decrease charge recombination at the electrode interface, the active layer should be bicontinuous for donor and acceptor, which in turn enhances internal quantum efficiency and short-circuit current and decreased device series resistor. The performance of organic photovoltaic devices can be greatly improved by optimizing the nanomicrostructure of the active layer. Thermal annealing and cosolvents/additives are effective methods to change morphology and promote device performance of polymer solar cells. For example, in poly(3-hexylthiophene)

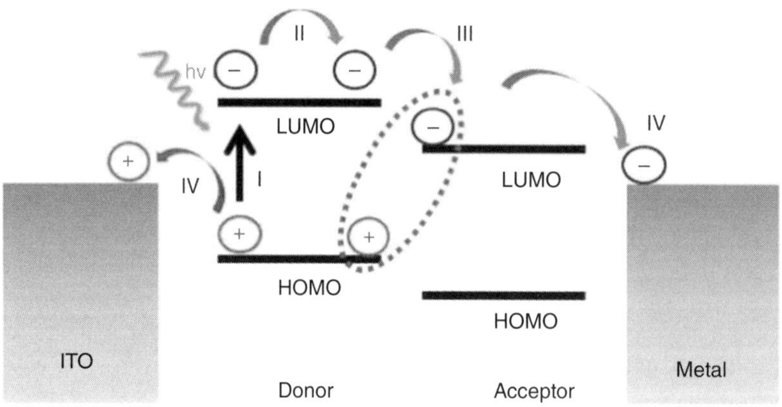

Figure 16.6 Four physical processes in the power conversion of all-polymer solar cells: photon absorption, exciton diffusion, dissociation of charge-transfer states, and charge-carrier transport and collection.

(P3HT): poly[9,9-dioctylfluorene-2,7-diyl-alt-*N*,*N*′-dihexyl-1,4,5,8-naphthalene diimide-2,6-diyl](PF-NDI) blend system, 1,8-octanedithiol (DIO) was employed to promote the self-assembly of P3HT and to decrease the domain sizes. The enhancement of photon absorption efficiency and exciton diffusion efficiency led to the improvements of device short-circuit current from 2.02 to 3.63 mA/cm^2 and device performance from 0.76 to 1.63% [118]. Friend et al. chose 4-bromoanisole (BrAni) as solvent additive, which prefers to dissolve donor polymer rather than acceptor polymer, to promote the polymer self-assembly and construct an interpenetrating structure, which would facilitate charge-carrier transport and decrease charge-carrier recombination. The short-circuit current improved from 0.53 to 4.32 mA/cm^2 and the fill factor also improved from 21.71 to 33.74 [119]. Applying thermal annealing at the temperature above both glass transition temperatures of P3HT: poly[2,7-(9,9-didodecylfluorene)-alt-5,5-(4′,7′-bis(2-thienyl)-2′,1′,3-benzothiadiazole)] (PF12TBT), the increased polymer molecular movement led to the diffusion of molecules deviating from different phases, promoting the phase purification. The increased domain purity led to improved charge-transfer state dissociation and decreased charge-carrier nongeminate recombination and therefore the increase of short-circuit current from 0.8 to 2.0 mA/cm^2 [120].

In the following, we show in detail the strategies to improve the performance of all-polymer photovoltaic cells, which is directly related to the nanostructures of phase separation in active layer, in four aspects: (i) Establish interpenetrating network structure by controlling phase separation, (ii) control the domain size and purify the domains, (iii) adjust the diffused structure at the interface between donor and acceptor, (iv) and construct the relationship between film morphology and device performance.

16.3.1 Establish Interpenetrating Network Structure by Controlling Phase Separation

The bicontinuous pathways formed by donor and acceptor materials in OSCs will ensure the separation of charge-transfer (CT) state and transportation of free charges, while the formation of island-type phase-separated structure will induce the increase of nongeminate charge-recombination rate and space-limited charge density. Therefore, establishing a donor/acceptor interpenetrating network that facilitates effective charge transport is the necessary prerequisite to fabricate high-efficiency photovoltaic cells.

The researches on the model of polymer blends suggest two types of phase-separation mechanisms: the nucleation and growth type and the spinodal decomposition type. The nucleation and growth type of phase separation takes place in the metastable system producing island-like structures, while the spinodal decomposition undergoes within the unstable system, signified by a bicontinuous morphology. The basic phase diagram for polymer/polymer blends system indicates that the achievement of the unstable region by adjusting blend ratio ψ and treatment temperature T is the prerequisite for the bicontinuous morphology of phase separation. In the case of amorphous/amorphous polymer blends model, for example, poly(styrene)(PS)/poly(vinylmethylether)(PVME) blend system, with 20% PS weight ratio corresponding to the critical composition under critical temperature, was annealed under vacuum at critical temperature for 17 min and then moderately quenched in the two-phase region of the phase diagram (Fig. 16.7. The post-treatment finally induced a bicontinuous spinodal pattern. When the weight ratio of PS was less than 20%, an island-like phase-separated structure formed [121]. In the case of crystalline/amorphous polymer blends, there is competition between the crystallization and phase-separation process. Therefore, the phase-separation behavior for such a model is more complex. In order to study the relationship between the crystallization and phase-separation process, Shiomi et al. effectively controlled the crystallization of polyethylene glycol (PEG) before or after liquid–liquid phase-separation process for PEG/polymethyl methacrylate (PMMA) blends by single quenching or double quenching [122]. The crystallization of PEG would occur simultaneously with liquid–liquid phase separation (LLPS) when the sample was directly quenched to crystallization temperature

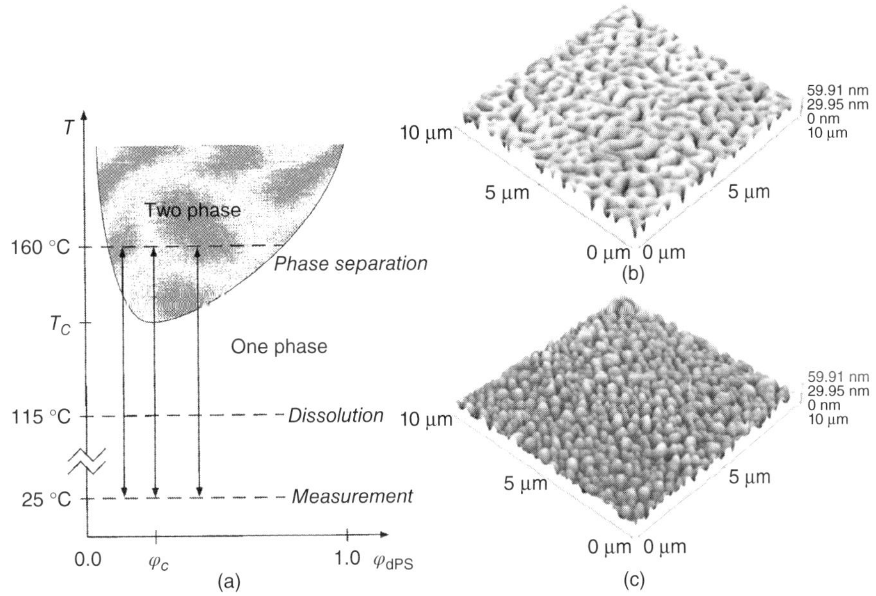

Figure 16.7 (a) Schematic phase diagram of deuterated PS and PVME. The top horizontal lines indicate the two-phase annealing temperature, the middle horizontal line indicates the reverse temperature jump, while the bottom horizontal line indicates the temperature where AFM measurements were performed. (b) AFM image of PS/PVME film at near-critical composition ($\Phi_{PS} = 0.20$) after annealing into the two-phase region of the phase diagram for 17 min at 160 °C. (c) AFM image of PS/PVME film at near-critical composition ($\Phi_{PS} = 0.10$) after annealing into the two-phase region of the phase diagram for 17 min at 160 °C. Erni [121]. Reproduced with permission of John Wiley and Sons.

(single quench). The heterogeneous nucleation was induced by the concentration fluctuation and the crystallization rate increased apparently, thus facilitating the formation of polymer blend film with high crystallinity. The crystallization of PEG would occur after LLPS when isothermal crystallization was carried out via isothermal step for LLPS (double quench). In this case, the restricted size of the phase-separated domain depresses the crystallization rate, thus forming a polymer blend film with low crystallinity (Fig. 16.8).

In the case of crystalline/crystalline blends, the driving force of phase separation is usually lower than that of the crystallization process. Therefore, the crystallization process is always prone to induce phase separation. The study focused on the crystalline polymer/fullerene blend system that has been intensively investigated. Considering that fullerene is a small molecule, ΔG_m for this blend system is quite low, thus tending to form a uniform mixing structure. Only when the crystallization process occurs in either component, could a distinct phase-separated structure form. Taking poly(3-octylthiophene) (P3HT)/[6,6]-phenyl C_{61}-butyric acid methyl ester (PCBM) as an example, with volatilization of solvent, the crystallization of P3HT occurred before the aggregation of PCBM; thus, the phase-separation process was induced by crystallization rather than spinodal decomposition, as shown in Figure 16.9 [123]. Because both components could crystallize, the relationship between the crystallization and phase-separation process is interdependent and interactional.

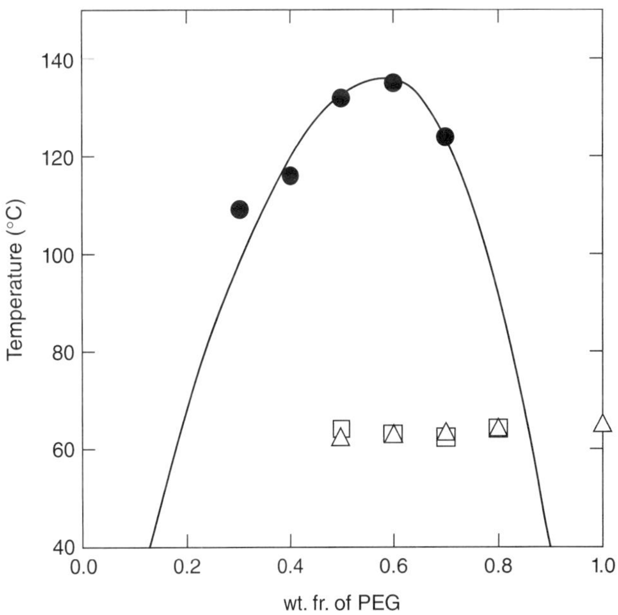

Figure 16.8 Phase diagram of i-PMMA/PEG blends. The filled circles are experimental cloud points obtained by light scattering measurements. The open triangles and squares are equilibrium melting temperatures T_m for single quench and double quench, respectively, obtained by extrapolation of the Hoffman–Weeks plots. Arai [122]. Reproduced with permission of Elsevier.

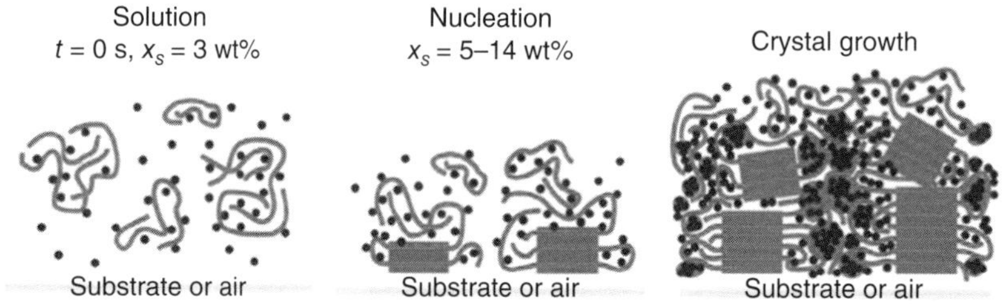

Figure 16.9 Scheme shows the stages of molecular arrangement during solvent evaporation with nucleation at the substrate or air interface. Lines represent P3HT chains, squares are crystalline P3HT domains, and dots are PCBM molecules. Schmidt-Hansberg [123]. Reproduced with permission of American Chemical Society.

On the one hand, the crystallization of P3HT could induce phase separation. Because PCBM molecules are dispersed in P3HT amorphous domain, part of the PCBM molecules will spread out when P3HT polymer chains turn from random to ordered arrangement. On the other hand, P3HT crystalline network will restrict the growth of PCBM aggregates and suppress the formation of large-sized PCBM aggregates, thus inhibiting the extent of phase separation [124, 125].

In the crystalline/crystalline polymer blends, the interaction between polymer chains depends on molecular structure; thus, the phase-separation behavior is much more complex. Generally, the energy barrier for phase separation is higher than that of crystallization. Therefore, the crystallization process always occurs first [126]. With the growth of crystals, the concentration fluctuation will become even larger at the growth interface of crystals. Meanwhile, the driving force of phase separation will become large, thus initiating the phase separation [127]. In the conjugated polymer blend system, the molecular rigidness results in entanglement among polymer chains that restrict the migration ability of the polymer. Therefore, the self-assembly of the conjugated polymer is quite weak. In this case, the thermodynamic drive force, rather than the crystallization process, is the dominant factor in the phase-separation process. When the solubility parameter is similar between polymers, phase separation for this polymer blend system will not occur. For example, spin-coated film of P3HT/poly((9,9-dioctyfluorene)-2,7-diyl-alt-[4,7-bis(3-hexylthien-5-yl)-2,1,3-benzothiadiazole]-2′,2′′-diyl) (F8TBT) blends is well mixed. On the contrary, when the solubility parameters are quite different, P3HT/poly[2,7-(9,9-didodecylfluorene)-alt-5,5-(4′,7′-bis(2-thienyl)-2′,1′,3-benzothiadiazole)] (PF12TBT) blends, for example, the interaction between polymers will induce local concentration fluctuation, resulting in spinodal decomposition. The molecule diffusion between phase domains (uphill diffusion) induces occurrence of the nucleation process. In other words, the phase separation occurs first and then induces the crystallization process.

Up to now, the study about phase-separation theory is mainly focused on amorphous/amorphous polymer blend and crystalline/amorphous polymer blend models. There is still lack of systemically theoretical guidance and support for crystalline/crystalline polymer blend models. Therefore, the establishment of the phase-separation theory for all-conjugated-polymer systems is of great importance for the balance of competition between crystallization and phase separation, construction of bicontinuous structure, and regulation of film crystallinity and domain purity.

16.3.2 Control the Domain Size and Purify of the Domains

Domain size plays a crucial role in determining exciton-dissociation efficiency and charge-collection efficiency. With domain size decreasing, the interfacial area between the donor and acceptor increases, which will effectively facilitate exciton dissociation. However, small domain size will suppress the charge-collection efficiency resulting from the increased opportunity of geminate recombination in the transport process. Consequently, domain size should be proper so as to facilitate exciton dissociation and charge collection at the same time [128]. It is believed that a domain size of 10–20 nm could effectively promote the PCE of photovoltaic solar cells. In an all-polymer system, the interaction between polymers can hardly be controlled. Due to the large molecular weight, polymers exhibit high compatibility when the interaction between repeating units of each polymer is small; the phase separation can hardly appear, as a result of the difficulty in chain movement. On the contrary, polymers exhibit poor compatibility when the interaction between repeated units is so large that the tendency of macro-sized phase separation can appear. As a result, it is important to control the degree of phase separation by adjusting the dynamics and thermodynamic of film solidification. At present, the way to adjust the domain size of the all-polymer system is to control the dynamics of film solidification. By choosing a solvent with proper boiling point, the time for polymer movement could be controlled, allowing to gain different nonequilibrium states and thus the domain size. Mori et al. employed solvents with various boiling points to prepare the films of PF12TBT and P3HT and the domain size increased with the increasing boiling point, as shown in Figure 16.10. The films cast from O-dichlorobenzene and chlorobenzene exhibited large domain sizes and their PCEs were merely 1.0%. The film cast from chloroform exhibited nano-sized phase-separated domains that facilitate exciton dissociation and charge collection and its PCE could reach to

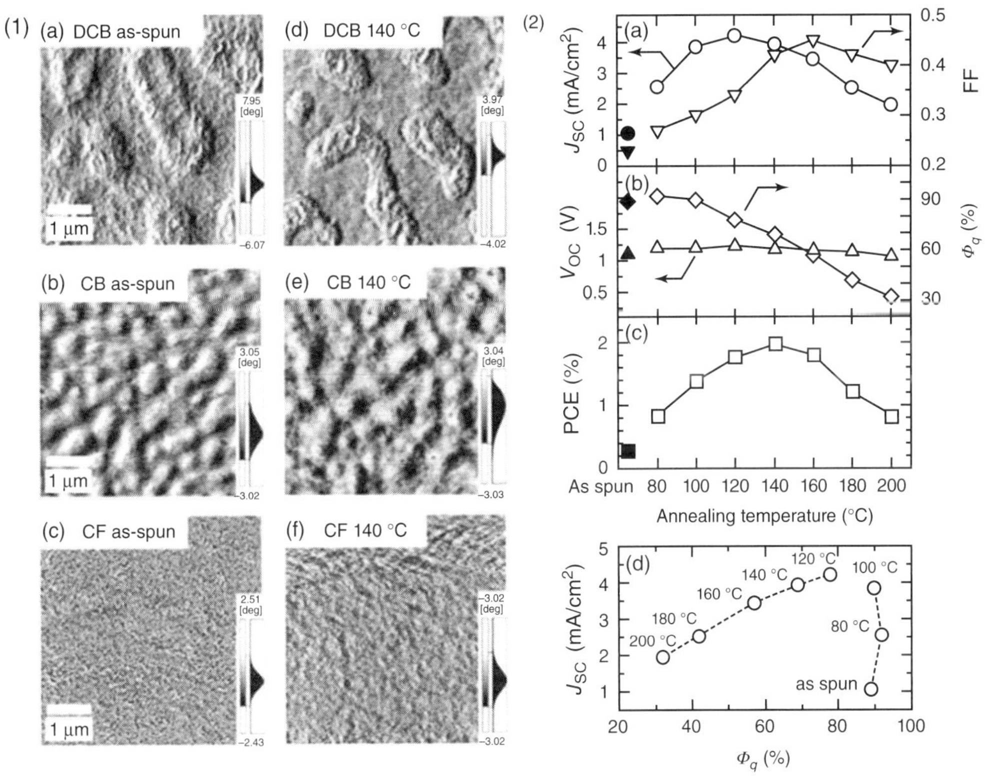

Figure 16.10 (1) Tapping-mode AFM phase images of P3HT/PF12TBT blend films spin coated from (a and d) DCB, (b and e) CB, and (c and f) CF solutions on glass substrates: (a–c) as-spun films; (d–f) annealed films at 140°C for 10 min. The scale bar corresponds to a length of 1 μm. (2) Dependence of the device parameters and Φ_q on the annealing temperature: (a) J_{SC} (open circles) and FF (open inverted triangles), (b) V_{OC} (open triangles) and Φ_q (open diamonds), and (c) PCE (open squares). The closed symbols represent the device parameters and Φ_q before thermal annealing. (d) Plots of J_{SC} against their respective Φ_q. These parameters were measured for P3HT/PF12TBT solar cells fabricated by spin coating from CF. Mori [129]. Reproduced with permission of American Chemical Society.

2.0% [129]. Using a cosolvent is another effective method to control film solidification. Zhou et al. found that the device efficiency could be improved by incorporation of chloroform into toluene. When chloroform was introduced, the movement time of polymer molecules decreased and the interpenetrating morphology with small domain size was formed, which led to the improvement of PCE from 1.85 to 2.23% [130].

Phase-separated hierarchical structures commonly exist in the solution-processing polymer blend system. Due to the chain entanglement among polymer molecules in polymer blend, the ability of molecular diffusion is limited. In the process of phase separation, partial molecules exist in a different phase, leading to the formation of phase-separated hierarchical structures. In addition, the self-organization of polymer molecules are inhibited resulting from the appearance of different polymer molecules, which lead to the decreased ability of crystalline-induced phase separation and enhance the possibility of hierarchical phase separation. The hierarchical phase separation will be detrimental for charge-carrier transport because of the increased possibility of geminate recombination in the process of charge-carrier transport. At present, the purification of domains can be achieved by promoting diffusion of polymer molecules deviating from a different phase. In PF12TB and P3HT blend system, the aggregation of donor and acceptor decreases the phase purification. Applying thermal annealing at a temperature above the glass transition temperatures of both polymers, the increased polymer molecular movement leads to the diffusion of molecules deviating from a different phase and promotes the phase purification. The fluorescent-quenching efficiency decreases from 83 to 55% and the short-circuit current increases from 0.8 to 2.0 mA/cm^2. Besides, enhancement of polymer crystallinity can promote the phase purification. Lam et al. prepared P3HT nanofibers by incorporating F8BT into the aged P3HT/*p*-xylene solution and found the crystallinity of P3HT nanofibers was greatly increased. It decreases the diffusion of F8BT molecules, promotes the phase purification and leads to enhanced the short-circuit current from 0.029 to 0.291 mA/cm^2. The PCE could reach to 0.055% [131].

At present, it is rare to control domain purification by adjusting film dynamics and thermodynamic state. The researches on domain purification mainly focus on process methods, such as emulsion processing and nanoimprinting [132, 133]. However, those methods are fussy and complicated and therefore inadaptable for commercialization of solar cells. As a result, it is crucial to promote the domain purification by controlling the crystallization and phase separation of polymers in all-polymer solar cells.

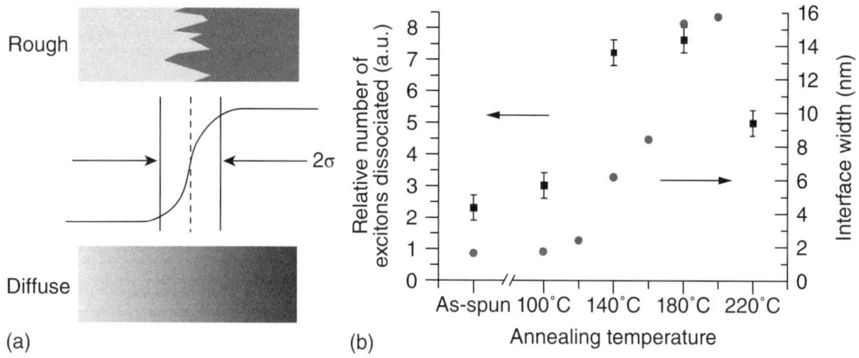

Figure 16.11 (a) Schemes of the rough and diffuse interface. (b) Changes in the relative number of excitons dissociated at the PFB/F8BT interface with annealing (squares); changes in the interface width of the PFB/F8BT bilayer with annealing (circles). Yan [137]. Reproduced with permission of John Wiley and Sons.

16.3.3 Adjust the Diffused Structure at the Interface Between Donor and Acceptor

In an all-polymer system, the excitons generate in the donor phase after the absorption of photons in active layer and then transform to the charge-transfer state, which will dissociate into free charge or recombine geminately. Those two processes compete with each other. A promoted efficiency of OSCs requires the inhibition of geminated recombination and the transformation of more excitons into free charges [134, 135].

Based on simulation, Lyons et al. studied the influence of interfacial diffusion on the efficiency of charge collection. Due to the high molecular disorder in the high-diffused interface, charge-carrier transport decreased and the possibility of geminated recombination increased by 50% compared to that of low-diffused interface [136]. However, the interface between donor and acceptor tends to form a diffused structure resulting from the interentanglement and low crystallinity of polymer molecules. The diffused structure, which increases drawbacks in the films, could decrease the charge-carrier transfer and increase the possibility of geminated recombination. At present, the main methods to adjust interface include promoting crystallinity of donor or acceptor and nonequilibrium processing. The increased crystallinity can inhibit the diffusion of different polymer molecules into crystalline domain and, in turn, restrain the diffused interface. From the theoretical analysis, McNeill holds that the interfacial diffusion can be prevented by using crystalline or semicrystalline polymers because of the high ability of self-organization [137]. Liu et al. employed 4-bromoanisole, which had different solubilities for P3HT and acceptors, as the solvent additive to promote the crystallinity of P3HT. With the increasing ratio of additive, crystallinity of P3HT had a gradual enhancement, which led to a decreased interfacial diffusion. For the blends of P3HT/F8TBT and P3HT/PCDTBT, their PCEs improved from 0.1 to 0.33% and from 1.23 to 1.33%, respectively. In the nonequilibrium process, the movement of polymer molecules can hardly reach the equilibrium state and the diffused interface can be inhibited [119]. McNeill et al. prepared the apparent interface by a two-step method, in which F8BT and PFB were made into bilayer structures. Thermal annealing enhanced the molecular diffusion and therefore expanded the interfacial width, which improved the number of exciton dissociation by 200%. However, the efficiency of charge collection decreased from 36.7 to 24.9% [137]. As a result, it is crucial to inhibit the diffusion at the interface between donor and acceptor in order to enhance the device efficiency, as shown in Figure 16.11.

16.3.4 Construct the Relationship Between Film Morphology and Device Performance

There is a close relationship between the morphology of active layers and performance (physical processes) of conjugated-polymer-based photovoltaic cells, which is quite important for the further performance improvement. The morphological characteristics include ordered packing of donor/acceptor materials, domain size, phase-separation structure and interfacial diffusion structure, while physical processes include exciton diffusion, charge-transfer state separation, lifetime of charge carrier, and carrier mobility.

In the polymer blend film, the level of ordered packing determines the efficiency of photon absorption and the diffusion length of exciton. When the crystallinity of polymer is comparatively strong, the diffusion length of excitons and the absorption efficiency of photons increase, facilitating the improvement of short-circuit current. Besides, the level of ordered packing has a great impact on charge transport. For example, the amorphous part in donor phase will induce the formation of hole-traps that restrict charge-transport process. Because the π–π stacking of donor molecules determines the distribution of HOMO energy level to some extent, the decreased π–π stacking will broaden the distribution of HOMO energy levels, thus increasing the rate of deep HOMO energy level distribution (the deeper the HOMO level, the easier for hole to be trapped) [138]. The domain size and phase-separated structure determine the efficiencies of exciton diffusion and charge transport. Phase separation with small domain size will facilitate the diffusion of exciton toward the interface, the formation of charge-transfer state, and therefore the separation into free charges. If the domain size is much too small, the recombination rate of non-geminate charges in the carrier transport process will increase, thus restricting carrier transport. On the other hand, large domain size will benefit the carrier transport process. However, the limited interfacial area is adverse to exciton diffusion and separation. The phase separation of blended film usually includes bicontinuous structure and island-like structure. The bicontinuous pathways formed by donor/acceptor materials establish a direct transportation path for electron/hole to their corresponding collection electrode, while the island-type phase separation increases the non-geminate charge recombination rate and then increases space-limited charge density. What is more, domain purity also has a great influence on carrier transport. The impurity domain will increase the charge-carrier recombination rate. In systems of all-conjugated polymer blends, when excitons diffuse to the donor/acceptor interface, electron transfer from acceptor materials to donor materials occurs and the charge-transfer state is formed. A part of the charge-transfer states separates into free carriers, while another part recombines into ground states or triplet states. In order to improve device performance, much more free charges should be formed and the recombination of charge-transfer states must be suppressed. Finally, the diffusion structure existing in the donor/acceptor interface leads to the molecular-scale mixing. The diffusion structure impedes the formation of pure ordered aggregates. Therefore, the ionization potential will not migrate toward the lower position (the electron affinity will not migrate toward the higher position). Analyzed by thermodynamics, the diffusion structure will not supply much more driving force for the separation of the charge-transfer state.

16.4 SUMMARY AND OUTLOOK

It has been demonstrated in many areas that material structures play a critical role in determining corresponding functions. Therefore, the control of shape, size, distribution, homogeneity, and stability of the micro- and nanostructures is important for both basic research and applications. For example, in OFET, the functional layer needs to be continuous and stable. Also, a proper phase-separated structure of conjugated polymer blend is the core issue for higher performance of OSCs. To further improve the device performance, however, carefully designed new materials are needed. For example, the development of donor polymers in OSCs with enhanced optical absorption in the red or near- infrared part of the solar spectrum is one of the main targets for enhancing cell efficiency. In addition, the structure of the polymer molecule is also very important as it affects the assembled structures, which in turn have a strong influence on the device performance, such as carrier mobility and photon absorption. In order to gain the designed function or to enhance the device performance, the careful design should start from the smallest unit, the molecule, which affects the assembled nano- and microstructures and therefore the functions.

REFERENCES

1. Cowie JMG. *Polymers: Chemistry and Physics of Modern Materials*. New York: Chapman and Hall; 1991.
2. Singh J, Agrawal KK. Polymeric materials for contact lenses. J Macromol Sci: Rev Macromol Chem Phys 1992;32:521–534.
3. Helgesen M, Søndergaard R, Krebs FC. Advanced materials and processes for polymer solar cell devices. J Mater Chem 2010;20:36–60.
4. Lu G, Usta H, Risko C, Wang L, Facchetti A, Ratner MA, Marks TJ. Synthesis, characterization, and transistor response of semiconducting silole polymers with substantial hole mobility and air stability. Experiment and theory. J Am Chem Soc 2008;130:7670–7685.
5. Armstrong NR, Wang W, Alloway DM, Placencia D, Ratcliff E, Brumbach M. Organic/organic heterojunctions: organic light emitting diodes and organic photovoltaic devices. Macromol Rapid Commun 2009;30:717–731.
6. Li X, Yu XH, Han YC. Polymer thin films for antireflection coatings. J Mater Chem C 2013;1:2266–2285.
7. Krebs FC. Fabrication and processing of polymer solar cells: A review of printing and coating techniques. Sol Energy Mater Sol Cells 2009;93:394–412.
8. Keddie JL, Jones RAL, Cory RA. Size-dependent depression of the glass transition temperature in polymer films. Europhys Lett 1994;27:59.
9. Tanaka K, Taura A, Ge SR, Takahara A, Kajiyama T. Molecular weight dependence of surface dynamic viscoelastic properties for the monodisperse polystyrene film. Macromolecules 1996;29:3040–3042.
10. Kajiyama TK, Tanaka A. Surface molecular motion of the monodisperse polystyrene films. Macromolecules 1997;30:280–285.
11. Tsui OKC, Zhang HF. Effects of chain ends and chain entanglement on the glass transition temperature of polymer thin films. Macromolecules 2001;34:9139–9142.
12. Grohens Y, Hamon L, Reiter G, Soldera A, Holl Y. Some relevant parameters affecting the glass transition of supported ultra-thin polymer films. Eur Phys J E 2002;8:217–224.
13. Tanaka K, Tsuchimura Y, Akabori K, Ito F, Nagamura T. Time- and space-resolved fluorescence study on interfacial mobility of polymers. Appl Phys Lett 2006;89:061916.
14. Si L, Massa MV, Dalnoki-Veress K, Brown HR, Jones RAL. Shear deformation in thin free-standing polymer films as a probe of entanglement in confined systems. Phys Rev Lett 2005;94:127801.
15. Aoki H, Morita S, Sekine R, Ito S. Conformation of single poly(methyl methacrylate) chains in an ultra-thin film studied by scanning near-field optical microscopy. Polym J 2008;40:274–280.
16. Binder K. Phase transitions in reduced geometry. Annu Rev Phys Chem 1992;43:33–59.
17. Kraus J, MÄuller-Buschbaum P, Kuhlmann T, Schubert DW, Stamm M. Confinement effects on the chain conformation in thin polymer films. Europhys Lett 2000;49(2):210–216.
18. Baschnagel J, Varnik F. Computer simulations of supercooled polymer melts in the bulk and in confined geometry. J Phys Condens Matter 2005;17:R851–R953.
19. Alcoutlabi M, McKenna GB. Effects of confinement on material behaviour at the nanometre size scale. J Phys Condens Matter 2005;17:R461.
20. Forrest JA, Dalnoki-Veress K. The glass transition in thin polymer films (PDF). Adv Colloid Interface Sci 2001;94:167–195.
21. Roth CB, Dutcher JR. *Soft Condensed Matters: Structure and Dynamics*. New York: Dekker; 2004.
22. Kargupta K, Konnur R, Sharma A. Instability and pattern formation in thin liquid films on chemically heterogeneous substrates. Langmuir 2000;16:10243–10253.

23. Russell TP. X-ray and neutron reflectivity for the investigation of polymers. Mater Sci Rep 1990;5:171–271.
24. Akpalu YA, Karim A, Satija SK, Balsara NP. Suppression of lateral phase separation in thin polyolefin blend films. Macromolecules 2001;34:1720–1729.
25. Krausch G. Surface induced self assembly in thin polymer films. Mater Sci Eng, R 1995;14:1–94.
26. Vrij A. Possible mechanism for the spontaneous rupture of thin, free liquid films. Discuss Faraday Soc 1966;42:23–33.
27. Brochard-Wyart F, Daillant J. Drying of solids wetted by thin liquid films. Can J Phys 1990;68:1084–1088.
28. Reiter G. Dewetting of thin polymer films. Phys Rev Lett 1992;68:75–78.
29. Reiter G. Unstable thin polymer films: rapture and dewetting process. Langmuir 1993;9:1344–1351.
30. Reiter G. Dewetting as a probe of polymer mobility in thin films. Macromolecules 1993;27:3046–3052.
31. Herminghaus S, Jacobs K, Mecke K, Bischof J, Fery A, Ibn-Elhaj M, Schlagowski S. Spinodal dewetting in liquid crystal and liquid metal films. Science 1998;282:916–919.
32. Xie R, Karim A, Douglas JF, Han CC, Weiss RA. Spinodal dewetting of thin polymer films. Phys Rev Lett 1998;81:1251–1254.
33. Meredith JC, Smith AP, Karim A, Amis EJ. Combinatorial materials science for polymer thin film dewetting. Macromolecules 2000;33:9747–9756.
34. Xue LJ, Han YC. Pattern formation by dewetting of polymer thin film. Prog Polym Sci 2011;36:269–293.
35. Seemann R, Herminghaus S, Jacobs K. Dewetting patterns and molecular forces: a reconciliation. Phys Rev Lett 2001;86:5534–5537.
36. Ruckenstein E, Jain RK. Spontaneous rupture of thin liquid films. J Chem Soc, Faraday Trans 2 1974;70:132–147.
37. Mitlin VS. Dewetting of solid surface: analogy with spinodal decomposition. J Colloid Interface Sci 1993;156: 491–497.
38. Mitlin VS. On dewetting conditions. Colloids Surf, A 1994;89:97–101.
39. Sharma A, Khanna R. Pattern formation in unstable thin liquid films under the influence of antagonistic short- and long-range forces. J Chem Phys 1999;110:4929–4936.
40. Reiter G, Sharma A, Khanna R, Casoli A, David M. The strength of long-range forces across thin liquid films. J Colloid Interface Sci 1999;214:126–128.
41. Xue LJ, Cheng ZY, Fu J, Han YC. Dewetting behavior of polystyrene film filled with (C6H5C2H4NH3)(2)PbI4. J Chem Phys 2008;129:054905.
42. Reiter G, Hamieh M, Damman P, Sclavons S, Gabriele S, Vilmin T, Raphaël E. Residual stresses in thin polymer films cause rupture and dominate early stages of dewetting. Nat Mater 2005;4:754–758.
43. Vix ABE, Müller-Buschbaum P, Stocker W, Stamm M, Rabe JP. Crossover between dewetting and stabilization of ultrathin liquid crystalline polymer films. Langmuir 2000;16:10456–10462.
44. Akhrass SA, Reiter G, Hou SY, Yang MH, Chang YL, Chang FC, Wang CF, Yang ACM. Viscoelastic thin polymer films under transient residual stresses: two-stage dewetting on soft substrates. Phys Rev Lett 2008;100:178301/1-4.
45. Reiter G. Dewetting of highly elastic thin polymer films. Phys Rev Lett 2001;87:186101/1-4.
46. Damman P, Baudelet N, Reiter G. Dewetting near the glass transition: transition from a capillary force dominated to a dissipation dominated regime. Phys Rev Lett 2003;91:216101/1-4.
47. de Gennes PG. Wetting: statics and dynamics. Rev Mod Phys 1985;57:827–863.
48. Redon C, Brochard-Wyart F, Rondelez F. Dynamics of dewetting. Phys Rev Lett 1991;66:715–718.
49. Brochard F, de Gennes PG, Hervert H, Redon C. Wetting and slippage of polymer melts on semi-ideal surfaces. Langmuir 1994;10:1566–1572.
50. Redon C, Brzoka JB, Brochard-Wyart F. Dewetting and slippage of microscopic polymer films. Macromolecules 1994;27:468–471.
51. Barrat JL, Bocquet L. Large slip effect at a nonwetting fluid-solid interface. Phys Rev Lett 1999;82:4671–4674.
52. Pit R, Hervet H, Léger L. Direct experimental evidence of slip in hexadecane: Solid interfaces. Phys Rev Lett 2000; 85:980–983.
53. de Gennes PG. Viscometric flows of entangled polymers. C R Acad Sci, Ser B 1979;228:219–227.
54. Schmatko T, Hervet H, Léger L. Friction and slip at simple fluid-solid interfaces: The roles of the molecular shape and the solid–liquid interaction. Phys Rev Lett 2005;94:244501/1-4.
55. Cho JHJ, Law BM, Rieutord F. Dipole-dependent slip of Newtonian liquids at smooth solid hydrophobic surfaces. Phys Rev Lett 2004;92:166102/1-4.
56. Zhu Y, Granick S. Limits of the hydrodynamic no-slip boundary condition. Phys Rev Lett 2002;88:106102/1-4.
57. Kunert C, Harting J. Roughness induced boundary slip in microchannel flows. Phys Rev Lett 2007;99:176001/1-4.
58. Cottin-Bizonne C, Barrat JL, Bocquet L, Charlaix E. Low-friction flows of liquid at nanopatterned interfaces. Nat Mater 2003;2:237–240.
59. Xu L, Sharma A, Joo SW. Substrate heterogeneity induced instability and slip in polymer thin films: Dewetting on silanized surfaces with variable grafting density. Macromolecules 2010;43:7759–7762.
60. Tretheway DC, Meinhart CD. A generating mechanism for apparent fluid slip in hydrophobic microchannels. Phys Fluids 2004;16:1509–1515.
61. Hendy SC, Lund NJ. Effective slip lengths for flows over surfaces with nanobubbles: The effects of finite slip. J Phys Condens Matter 2009;21:144202/1-4.
62. de Gennes PG. On fluid/wall slippage. Langmuir 2002;18: 3413–3414.
63. Segalman RA, Green PF. Dynamics of rims and the onset of spinodal dewetting at liquid/liquid interfaces. Macromolecules 1999;32:801–807.
64. Harris M, Appel G, Ade H. Surface morphology of annealed polystyrene and poly(methyl methacrylate) thin film blends and bilayers. Macromolecules 2003;36:3307–3314.
65. Xue LJ, Hu BH, Han YC. Effect of interfacial roughness on dewetting behavior of polystyrene/poly(methyl methacrylate) bilayer film. J Chem Phys 2008;129:214902/1-6.
66. Brochard-Wyart F, Martin P, Redon C. Liquid/liquid dewetting. Langmuir 1993;9:3682–3690.
67. Limary R, Green PF. Dynamics of droplets on the surface of a structured fluid film: late-stage coarsening. Langmuir 2003;19:2419–2424.
68. Lambooy P, Phelan KC, Haugg O, Krausch G. Dewetting at the liquid–liquid interface. Phys Rev Lett 1996;76:1110–1113.

69. Wang C, Krausch G, Geoghegan M. Dewetting at a polymer–polymer interface: Film thickness dependence. Langmuir 2001;17:6269–6274.
70. Reiter G, Sharma A. Auto-optimization of dewetting rates by rim instabilities in slipping polymer films. Phys Rev Lett 2001;87:166103/1-4.
71. Brochard-Wyart F, Redon C. Dynamics of liquid rim instabilities. Langmuir 1992;8:2324–2329.
72. Besancon BM, Green PF. Moving fronts in entangled polymeric films. Phys Rev E 2004;70:051808/1-8.
73. Rayleigh L. On the instability of jets. Proc London Math Soc 1878;10:4–13.
74. Xue LJ, Han YC. Inhibition of dewetting of thin polymer films. Prog Mater Sci 2012;57:947–979.
75. Forrest JA, Dalnoki-Veress K, Stevens JR, Dutcher JR. Effect of free surfaces on the glass transition temperature of thin polymer films. Phys Rev Lett 1996;77:2002–2005.
76. Keddie JL, Jones RAL, Cory RA. Interface and surface effects on the glass-transition temperature in thin polymer films. Faraday Discuss 1994;98:219–230.
77. Newby C, Lee J-K, Ober CK. The solvent problem: Redissolution of macromolecules in solution-processed organic electronics. Macromol Res 2013;21(3):248–256.
78. Xue LJ, Han YC. Autophobic dewetting of a poly(methyl methacrylate) thin film on a silicon wafer treated in good solvent vapor. Langmuir 2009;25:5135–5140.
79. Flory PJ. Thermodynamics of high polymer solutions. J Chem Phys 1942;10:51–61.
80. Huggins M. Theory of solutions of high polymers. J Am Chem Soc 1942;64:1712–1719.
81. Flory PJ. *Principles of Polymer Chemistry*. Ithaca, NY: Cornell University Press; 1953.
82. Bastes FS. Polymer–polymer phase behavior. Science 1991;251:898–905.
83. Xue LJ, Zhang JL, Han YC. Phase separation induced ordered patterns in thin polymer blend films. Prog Polym Sci 2012;37:564–594.
84. Tanaka K, Yoon J-S, Takahara A, Kajiyama T. Ultrathinning-induced surface phase separation of polystyrene/poly(vinyl methyl ether) blend film. Macromolecules 1995;28:934–938.
85. Li X, Wang Z, Cui L, Xing RB, Han YC, An L. Phase separation of PS/PVME blend films induced by capillary force. Surf Sci 2004;571:12–20.
86. Neserov A, Horichko V, Lipatov Y. Phase-separation of poly(vinyl acetate)-poly(methyl methacrylate) mixtures in thin-films. Macromol Chem Rapid Commun 1991;12: 571–574.
87. Li L, Sosnowski S, Chaffey CE, Balke ST, Winnik MA. Surface morphology of a polymer blend examined by laser confocal fluorescence microscopy. Langmuir 1994;10:2495–2497.
88. Affrossman S, O'Neill SA, Stamm M. Topography and surface composition of thin films of blends of polystyrene with brominated polystyrenes: effects of varying the degree of bromination and annealing. Macromolecules 1998;31:6280–6288.
89. Slep D, Asselta J, Rafailovich MH, Sokolov J, Winesett DA, Smith AP, Strzhemechny Y, Schwarz SA, Sauer BB. Phase separation of polystyrene and bromo-polystyrene mixtures in equilibrium structures in thin films. Langmuir 1998;14: 4860–4864.
90. Reich S, Cohen Y. Phase separation of polymer blends in thin films. J Poly Sci Poly Phys Ed 1981;19:1255–1267.
91. Tanaka K, Takahara A, Kajiya T. Film thickness dependence of the surface structure of immiscible polystyrene/poly(methyl methacrylate) blends. Macromolecules 1996;29:3232–3239.
92. Gutmann JS, Müller-Buschbaum P, Stamm M. Complex pattern formation by phase separation of polymer blends in thin films. Faraday Discuss Chem Soc 1999;112:285–297.
93. Raczkowska J, Bernasik A, Budkowski A, Sajewicz K, Penc B, Lekki J, Lekka M, Rysz J, Kowalski K, Czuba P. Structures formed in spin-cast films of polystyrene blends with poly(butyl methacrylate) isomers. Macromolecules 2004; 37:7308–7315.
94. Genzer J, Composto RJ. Effect of molecular weight on the interfacial excess, tension, and width in a homopolymer/binary polymer blend system. Macromolecules 1998;31:870–878.
95. Hariharan A, Kumar SK, Russell TP. Surface segregation in binary polymer mixtures: A lattice model. Macromolecules 1991;24:4909–4917.
96. Li X, Han Y, An L. Surface morphology control of immiscible polymer-blend thin films. Polymer 2003;44:8155–8165.
97. Muller-Buschbaum P, Gutmann JS, Stamm M. Influence of blend composition on phase separation and dewetting of thin polymer blend films. Macromolecules 2000;33:4886–4895.
98. Hopkinson I, Myatt M. Phase separation in ternary polymer solutions induced by solvent loss. Macromolecules 2002;35:5153–5160.
99. Müller-Buschbaum P, Gutmann JS, Wolkenhauer M, Kraus J, Stamm M, Smilgies D, Petry W. Solvent-induced surface morphology of thin polymer films. Macromolecules 2001;34:1369–1375.
100. Tanaka H, Lovinger AJ, Davis DD. Pattern evolution caused by dynamic coupling between wetting and phase separation in binary liquid mixture containing glass particles. Phys Rev Lett 1994;72:2581–2584.
101. Kajiyama T, Tanaka K, Takahara A. Depth dependence of the surface glass transition temperature of a poly(styrene-block-methyl methacrylate) diblock copolymer film on the basis of temperature-dependent X-ray photoelectron spectroscopy. Macromolecules 1995;28:3482–3484.
102. Krausch G, Dai C-A, Kramer EJ, Marko JF, Bates FS. Interference in spinodal waves in thin polymer films. Macromolecules 1993;26:5566–5571.
103. Ogawa H, Kanaya T, Nshida K, Matsuba G. Composition fluctuations before dewetting in polystyrene/poly(vinyl methyl ether) blend thin films. Polymer 2008;49:2553–2559.
104. Gutmann JS, Mulluer-Buschbaum P, Schubert DW, Stribeck N, Stamm M. Influence of the blend compatibility on the morphology of thin polymer blend films. J Macromol Sci Part B: Phys 1999;38:563–576.
105. Ton-That C, Shard AG, Teare DOH, Bradley RH. XPS and AFM surface studies of solvent-cast PS/PMMA blends. Polymer 2001;42:1121–1129.
106. Sferrazza M, Xiao C, Bucknall DG, Jones RAL. Interface width of low-molecular-weight immiscible polymers. J Phys Condens Matter 2001;13:10269–10277.
107. Affrossman S, Bertrand P, Hartshorne M, Kiff T, Leonard D, Pethrick RA, Richards RW. Surface segregation in blends of polystyrene and perfluorohexane double end capped polystyrene studied by static SIMS, ISS, and XPS. Macromolecules 1996;29:5432–5437.

108. Kawaguchi D, Tanaka K, Torikai N, Takahara A, Kajiyama T. Surface and interfacial segregation in blends of polystyrene with functional end groups and deuterated polystyrene. Langmuir 2007;23:7269–7275.
109. Walheim S, Böltau M, Mlynek J, Krausch G, Steiner U. Structure formation via polymer demixing in spin-cast films. Macromolecules 1997;30:4995–5003.
110. Cui L, Ding Y, Li X, Wang Z, Han Y. Solvent and polymer concentration effects on the surface morphology evolution of immiscible polystyrene/poly(methyl methacrylate) blends. Thin Solid Films 2006;515:2038–2048.
111. McNeill CR. Morphology of all-polymer solar cells. Energy Environ Sci 2012;5(2):5653–5667.
112. Duan C, Huang F, Cao Y. Recent development of push-pull conjugated polymers for bulk-heterojunction photovoltaics: Rational design and fine tailoring of molecular structures. J Mater Chem 2012;22(21):10416–10434.
113. Xue JG. Perspectives on organic photovoltaics. Poly Rev 2010;50(4):411–419.
114. Hoppe H, Sariciftci NS. Organic solar cells: An overview. J Mater Res 2004;19(7):1924–1945.
115. Gunes S, Neugebauer H, Sariciftci NS. Conjugated polymer-based organic solar cells. Chem Rev 2007;107(4):1324–1338.
116. Beaupre S, Breton AC, Dumas J, Leclerc M. Multicolored electrochromic cells based on poly(2,7-carbazole) derivatives for adaptive camouflage. Chem Mater 2009;21(8):1504–1513.
117. Mori D, Benten H, Okada I, Ohkita H, Ito S. Low-bandgap donor/acceptor polymer blend solar cells with efficiency exceeding 4%. Adv Energy Mater 2014;4(3):1301006.
118. Zhou E, Cong J, Zhao M, Zhang L, Hashimoto K, Tajima K. Synthesis and application of poly(fluorene-alt-naphthalene diimide) as an n-type polymer for all-polymer solar cells. Chem Commun 2012;48(43):5283–5285.
119. Liu X, Huettner S, Rong Z, Sommer M, Friend RH. Solvent additive control of morphology and crystallization in semiconducting polymer blends. Adv Mater 2012;24(5):669–674.
120. Mori D, Benten H, Ohkita H, Ito S, Miyake K. Polymer/polymer blend solar cells improved by using high-molecular-weight fluorene-based copolymer as electron acceptor. ACS Appl Mater Interfaces 2012;4(7):3325–3329.
121. Ermi BD, Karim A, Douglas JF. Formation and dissolution of phase-separated structures in ultrathin blend films. J Polym Sci B 1998;36:191–200.
122. Arai F, Takeshita H, Dobashi M, Takenaka K, Miya M, Shiomi T. Effects of liquid–liquid phase separation on crystallization of poly(ethylene glycol) in blends with isotactic poly(methyl methacrylate). Polymer 2012;53(3):851–856.
123. Schmidt-Hansberg B, Sanyal M, Klein MFG, Pfaff M, Schnabel N, Jaiser S, Vorobiev A, Mueller E, Colsmann A, Scharfer P, Gerthsen D, Lemmer U, Barrena E, Schabel W. Moving through the phase diagram: morphology formation in solution cast polymer-fullerene blend films for organic solar cells. ACS Nano 2011;5(11):8579–8590.
124. Lin C, Pan W-C, Tsai F-Y. Optimization of the active-layer morphology with a non-halogenic solvent for bulk-heterojunction polymer solar cells. Synth Met 2010;160(23–24):2643–2647.
125. Oh JY, Lee TI, Myoung J-M, Jeong U, Baik HK. Coating on a cold substrate largely enhances power conversion efficiency of the bulk heterojunction solar cell. Macromol Rapid Commun 2011;32(14):1066–1071.
126. Tsuburaya M. Crystallization of polycarbonate induced by spinodal decomposition in polymer blends. Polymer 2004;45(3):1027–1032.
127. Shi W, Han CC. Dynamic competition between crystallization and phase separation at the growth interface of a PMMA/PEO blend. Macromolecules 2012;45(1):336–346.
128. Liu F, Gu Y, Jung JW, Jo WH, Russell TP. On the morphology of polymer-based photovoltaics. J Polym Sci B 2012;50(15):1018–1044.
129. Mori D, Benten H, Kosaka J, Ohkita H, Ito S, Miyake K. Polymer/polymer blend solar cells with 2.0% efficiency developed by thermal purification of nanoscale-phase-separated morphology. ACS Appl Mater Interfaces 2011;3(8):2924–2927.
130. Zhou E, Cong J, Wei Q, Tajima K, Yang C, Hashimoto K. All-polymer solar cells from perylene diimide based copolymers: Material design and phase separation control. Angew Chem Int Ed 2011;50(12):2799–2803.
131. Salim T, Sun S, Wong LH, Xi L, Foo YL, Lam YM. The role of poly(3-hexylthiophene) nanofibers in an all-polymer blend with a polyfluorene copolymer for solar cell applications. J Phys Chem C 2010;114(20):9459–9468.
132. Kietzke T, Neher D, Kumke M, Montenegro R, Landfester K, Scherf U. A nanoparticle approach to control the phase separation in polyfluorene photovoltaic devices. Macromolecules 2004;37(13):4882–4890.
133. He X, Gao F, Tu G, Hasko D, Hüttner S, Steiner U, Greenham NC, Friend RH, Huck WTS. Formation of nanopatterned polymer blends in photovoltaic devices. Nano Lett 2010;10(4):1302–1307.
134. Tong M, Coates NE, Moses D, Heeger AJ, Beaupre S, Leclerc M. Charge carrier photogeneration and decay dynamics in the poly(2,7-carbazole) copolymer PCDTBT and in bulk heterojunction composites with PC(70)BM. Phys Rev B 2010;81(12):125210.
135. Etzold F, Howard IA, Mauer R, Meister M, Kim T-D, Lee K-S, Baek NS, Laquai F. Ultrafast exciton dissociation followed by nongeminate charge recombination in PCDTBT:PCBM photovoltaic blends. J Am Chem Soc 2011;133(24):9469–9479.
136. Lyons BP, Clarke N, Groves C. The relative importance of domain size, domain purity and domain interfaces to the performance of bulk-heterojunction organic photovoltaics. Energy Environ Sci 2012;5(6):7657–7663.
137. Yan H, Swaraj S, Wang C, Hwang I, Greenham NC, Groves C, Ade H, McNeill CR. Influence of annealing and interfacial roughness on the performance of bilayer donor/acceptor polymer photovoltaic devices. Adv Funct Mater 2010;20(24):4329–4337.
138. Beiley ZM, Hoke ET, Noriega R, Dacuna J, Burkhard GF, Bartelt JA, Salleo A, Toney MF, McGehee MD. Morphology-dependent trap formation in high performance polymer bulk heterojunction solar cells. Adv Energy Mater 2011;1(5):954–962.

17

POLYMER SURFACE TOPOGRAPHY AND NANOMECHANICAL MAPPING

HAO LIU[1], SO FUJINAMI[2], DONG WANG[3], KEN NAKAJIMA[4], AND TOSHIO NISHI[5]

[1]*School of Materials Science and Engineering, Zhengzhou University, Zhengzhou, China*
[2]*RIKEN SPring-8 Center, RIKEN, Hyogo, Japan*
[3]*State Key Laboratory of Organic-Inorganic Composites, College of Materials Science and Engineering, Beijing University of Chemical Technology, Beijing, China*
[4]*Department of Organic and Polymeric Materials, Graduate School of Science and Engineering, Tokyo Institute of Technology, Tokyo, Japan*
[5]*International Division, Tokyo Institute of Technology, Tokyo, Japan*

17.1 INTRODUCTION

The invention of scanning tunneling microscopy (STM) about three decades ago [1, 2] opened a new age in which various related scanning probe microscopy (SPM) methods were developed and some of those were widely accepted as key tools for nanotechnology. Although the original STM limited its application to conductive materials, atomic force microscopy (AFM), which appeared soon after the invention of STM [3], proved applicable to various types of materials, including polymeric materials.

AFM has a cantilever as its most basic component, with a very sharp probe at its free end in order to interact with sample surfaces. The interaction force induces the deflection of the cantilever, which is the most important signal of concern. The original AFM by the developers used STM to detect the cantilever deflection, while an optical beam method is mostly used nowadays due to easier handling.

Applications of AFM to polymeric materials started mainly with the aim of studying the surface structures with superior lateral resolution [4–7]. For these purposes, AFM retains an advantage over other microscopic techniques because its environmental limitation is very little; it can work in air, water, and some other solutions. What AFM users need to keep in mind is that the obtained image may be affected by interactions such as surface force and elastic repulsion that work between the sample and probe. For example, contact force depresses soft surface, which may lessen the quality of topographic image [8, 9].

However, the fact that AFM is sensitive to these weak forces can be turned into an advantage. One of the most common ways to measure the interactions is a force–distance curve measurement, in which the scanner starts to move vertically to make contact with the probe and surface and then reverse to make them separated. During the process, the cantilever deflection is recorded as a function of vertical scanner movement. One can obtain the relation between normal load P and indentation depth δ using simple relations (the detailed procedure is discussed later). Applying an appropriate contact mechanics theory, one can estimate the mechanical properties of the sample surfaces.

17.2 CONTACT MECHANICS

Contact mechanics plays a key role in understanding the phenomena occurring around the contact between the AFM probe and sample surfaces. It can give answers about, for example, how much contact area and indentation depth are induced by the given load and how much stress is applied around the apex of the probe. Especially, the theory is necessary to obtain the mechanical properties from experimental data such as force–distance curves.

Polymer Morphology: Principles, Characterization, and Processing, First Edition. Edited by Qipeng Guo.

This section introduces several theories for the aim of applying them to contact in AFM measurement. The following discussion assumes contact between two spheres, in which the reduced radius R is defined as

$$R \equiv \left(\frac{1}{R_1} + \frac{1}{R_2}\right)^{-1}$$

and the reduced modulus E^* as

$$E^* \equiv \left(\frac{1-\nu_1^2}{E_1} + \frac{1-\nu_2^2}{E_2}\right)^{-1}$$

where R_1 and R_2 are curvature radii, E_1 and E_2 are Young's moduli, and ν_1 and ν_2 are Poisson's ratios associated with each body. Because the material of AFM probes, which is typically made of silicon or silicon nitride, is much stiffer than most polymeric materials, the reduced modulus is often approximated to $E^* = E/(1-\nu^2)$, where E and ν are represented by Young's modulus and Poisson's ratio of the sample, respectively. Also, for practical reasons, the sample surface is often assumed to be flat; then R is simply represented by the curvature radius of the cantilever probe.

17.2.1 Hertzian Theory (Repulsion between Elastic Bodies)

Hertz established his famous theory about the contact between two elastic spheres in 1882 [10]. The theory assumes normal stress p_1 has an elliptical distribution on the radial coordinate r in the contact area, that is,

$$p_1(r) = \frac{3P_1}{2\pi a^2}\left(1 - \frac{r^2}{a^2}\right)^{1/2} \tag{17.1}$$

where P_1 is the applied normal load and a is the radius of the contact circle, respectively. The Hertzian theory relates the contact radius a to the normal load P_1 by the equation

$$P_1 = \frac{4E^*a^3}{3R} \tag{17.2}$$

Equation 17.1 implies that the maximum contact pressure, $p(r=0)$, is 150% of the "averaged pressure," $P_1/\pi a^2$. It also relates indentation length δ_1 to the contact radius a by

$$\delta_1 = \frac{a^2}{R} = \left(\frac{9P_1^2}{16RE^{*2}}\right)^{1/3} \tag{17.3}$$

The profile of this nonadhesive contact is shown in Figure 17.1a.

17.2.2 Bradley Model (Interaction between Rigid Bodies)

While adhesion between spheres was not considered in the Hertzian contact model, understanding of interactive forces between solids, and particularly colloids, showed a big progress in the early twentieth century. By integrating attractive interactions that follow the power law, Bradley found the total force between two rigid spheres separated at an equilibrium distance z_0 is given by $P_{pull\text{-}off} = -2\pi wR$, where w is the work of adhesion [11]. While Bradley only took z_0 to be the atomic equilibrium separation, that is, a constant value, one can obtain

$$P(z) = 2\pi wR\left[-\frac{4}{3}\left(\frac{z}{z_0}\right)^{-2} + \frac{1}{3}\left(\frac{z}{z_0}\right)^{-8}\right] \tag{17.4}$$

by introducing the Lennard-Jones potential to an equation in the article, where z is the atomic separation between two spheres. Equation 17.4 is often referred to as the Bradley equation, although it does not appear in the original article.

17.2.3 Johnson–Kendall–Roberts (JKR) Model

By the 1960s, several experimental results that are contradictory to the Hertz theory had been reported, especially at low loads [12]. These observations strongly suggested the intervention of attractive surface forces in the elastic contact of two bodies.

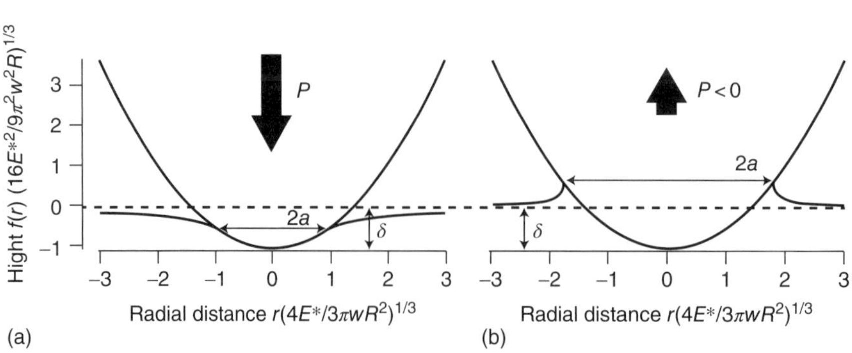

Figure 17.1 Profile of elastic plane deformed by rigid indenter. (a) Hertzian/DMT model, $\overline{a} = 1$, $\overline{\delta}_1 = 1$, $\overline{P}_1 = 1$ in Hertzian (or $\overline{P} = -1$ in the DMT-M). (b) The JKR model, $\overline{a} = 1.78$, $\overline{\delta} = 1$, $\overline{P} = -0.16$. See Equation 17.14 for nondimensionalized parameters.

Incorporating the effect of adhesion in Hertzian contact, Johnson et al. formulated a theory of adhesive contact using a balance between the stored elastic energy and the loss in surface energy, which is now renowned as the JKR theory [12]. In this model, adhesion works only inside the contact area and the net traction $p(r)$ acting on the contact area is given by the sum of Hertzian repulsion p_1 and adhesive traction p_a, that is,

$$p(r) = p_1(r) + p_a(r) \tag{17.5}$$

where p_1 is given by Equation 17.1 and p_a is given by

$$p_a(r) = -\left(\frac{2wE^*}{\pi a}\right)^{1/2}\left(1 - \frac{r^2}{a^2}\right)^{-1/2} \tag{17.6}$$

The approach distance between the two spheres is given by

$$\delta = \frac{a^2}{R} - \sqrt{\frac{2\pi a w}{E^*}} \tag{17.7}$$

The contribution of the surface energy modifies the area of contact a from Hertzian value, which is given by

$$a^3 = \frac{3R}{4E^*}\left[P + 3\pi wR + \sqrt{6\pi wRP + (3\pi wR)^2}\right] \tag{17.8}$$

These equations give the pull-off force by the form

$$P_{pull\text{-}off} = -\frac{3}{2}\pi wR \tag{17.9}$$

Figure 17.1b shows the profile of the deformed elastic plane predicted by the JKR model. The profile is depicted under the condition of the same indentation δ as in the Hertzian model in Figure 17.1a. Owing to the existence of adhesion, the JKR model requires more contact region and less loading force than the Hertz model.

17.2.4 Derjaguin–Muller–Toporov (DMT) Model

Derjaguin et al. proposed a different model for adhesive contact in 1975 [13]. The model assumed, in addition to Hertzian repulsion, the existence of molecular attraction forces that would not be able to change the contact profile appreciably. The net load P is given by $P = P_1 + P_a$, where P_1 is Hertzian repulsion given by Equation 17.2 and P_a is the molecular attraction (< 0). Thus, the profile predicted by the DMT model is the same as in Figure 17.1a, the Hertzian profile, but a smaller net load is required. Owing to the assumption, the deformation length δ has the identical form with Equation 17.3, thus P and δ can be directly related by the form

$$P = \frac{3}{4}E^* R^{1/2}\delta^{3/2} + P_a \tag{17.10}$$

The attractive interaction P_a depends on the profile near the contact perimeter and is typically represented as a function of the contact radius a, that is, $P_a(a)$. At the point contact ($a \to 0$), elastic displacement on the surface profile vanishes; hence, P_a becomes the same as Bradley's value for the pull off,

$$P_{pull\text{-}off} = -2\pi wR \tag{17.11}$$

Derjaguin et al. originally calculated the attractive interaction P_a by a "thermodynamic" approach, which yielded the result that as a increased, P_a steadily decreased from the Bradley value to $-\pi wR$. In 1983, Derjaguin and his colleagues [14] demonstrated a "force" approach, which deduced the opposite functional dependency, that is, as a increased P_a steadily increased, and noted that the "force method should be preferred". The dependency and its defects of these approaches were discussed by Pashley and Greenwood [15, 16]. Read them for further details.

Although P_a has functional dependency on a, near the Bradley limit ($a \to 0$) it may be approximated to the constant ($P_a \approx P_{pull\text{-}off} = -2\pi wR$). In such a case, the net load P can be simply expressed as

$$\begin{aligned} P &= \frac{4E^* a^3}{3R} - 2\pi wR \\ &= \frac{4}{3}E^* R^{1/2}\delta^{3/2} - 2\pi wR \end{aligned} \tag{17.12}$$

This simplified form is often referred to as the DMT equation, although it is, to be exact, different from the original model. As discussed by Greenwood [16], Maugis showed this form as the limit of the DMT model at the Bradley limit but labeled simply (and somewhat misleadingly) "DMT" [17]. Greenwood relabeled it the "DMT-M" model to make a clear distinction.

17.2.5 The JKR–DMT transition and Maugis–Dugdale (MD) Model

The emergence of these two theories, which are apparently contradictory to each other, induced a long dispute. A turning point of the argument was an indication by Tabor in 1977 [18]; he noted that the JKR model becomes invalid and the forces outside the contact zone must be considered if the value,

$$\mu \equiv \left(\frac{Rw^2}{E^{*2}z_0{}^3}\right)^{1/3} \tag{17.13}$$

is close to a unit or less. μ is referred to as the Tabor parameter. In 1980, Muller showed that the Tabor parameter indeed governs the transition of pull-off force from the value of the Bradley/DMT ($-2\pi wR$) to that of the JKR ($-1.5\pi wR$) using numerical calculation [19]. For $\mu \ll 1$ (hard solid of small radius and low surface energy), the DMT theory would be valid and for $\mu \gg 1$ (soft material with large radius and high surface energy) the JKR theory.

Another remarkable contribution to Tabor's idea is an analytical solution obtained by Maugis [17]. The key assumption in his model was that the traction form was represented in terms of the Dugdale approximation, in which the attractive force is a constant σ_0 for separation h satisfying $z_0 \le (z_0 + h) \le (z_0 + h_0)$. These parameters relate to each other by $\sigma_0 = w/h_0$. The model assumed that Hertzian repulsion acts only inside the intimate contact whose radius is a and the forms of load P_1 and stress p_1 follow Hertzian law, Equations 17.1 and 17.2; in the vicinity of the contact perimeter, the surfaces are slightly separated but still a constant attraction σ_0 works. The "outer radius" c is defined to satisfy that the atomic separation h reaches $h + h_0$ at $r = c$, and when $r > c$ one can ignore the contribution of traction.

Maugis introduced nondimensionalized parameters defined below:

$$\bar{a} = a\left(\frac{4E^*}{3\pi wR^2}\right)^{1/3}; \bar{c} = c\left(\frac{4E^*}{3\pi wR^2}\right)^{1/3};$$
$$\bar{\delta} = \delta\left(\frac{16E^{*2}}{9\pi^2 w^2 R}\right)^{1/3}; \bar{P} = \frac{P}{\pi wR} \quad (17.14)$$

He also introduced the transition parameter λ as

$$\lambda \equiv \sigma_0\left(\frac{9R}{2\pi wE^{*2}}\right)^{1/3} \quad (17.15)$$

As Maugis defined σ_0 to be equal to the stress of the Lennard-Jones potential, which gives $h_0 = 0.97z_0$, λ is related to μ by $\lambda = 1.16\mu$.

In the model, both the net contact force P and indentation depth δ are represented by the sum of two terms, one is contributed to by the Hertzian repulsion (subscript "1") and the other is attractive interaction (subscript "a"). They are respectively given by

$$\bar{P} = \bar{P}_1 + \bar{P}_a = \bar{a}^3 - \lambda\bar{a}^2\left(\sqrt{m^2-1} + m^2 \operatorname{arcsec} m\right) \quad (17.16)$$

and

$$\bar{\delta} = \bar{\delta}_1 + \delta_a = \bar{a}^2 - \frac{4}{3}\lambda\bar{a}\sqrt{m^2-1} \quad (17.17)$$

where $m \equiv c/a$ is given by

$$\frac{\lambda\bar{a}^2}{2}\left[\left(m^2-2\right)\operatorname{arcsec} m + \sqrt{m^2-1}\right] + \frac{4\lambda^2\bar{a}}{3}\left[\sqrt{m^2-1}\operatorname{arcsec} m - m + 1\right] = 1 \quad (17.18)$$

Figure 17.2 shows the force ($\bar{P}$) –indentation ($\bar{\delta}$) curves obtained by the MD theory for various λ, together with those given by the Hertz, DMT, and JKR models. One can observe the transition of pull-off force $\bar{P}_{pull\text{-}off}$ from −2 to −1.5, as λ increases. The MD curve for $\lambda = 2$ is traced in close proximity of the JKR curve. For smaller λ, the MD curves appear to be similar to the DMT curve near the pull-off point. However, as the indentation depth δ increases, the curve tends to be close to the JKR curve; it is because larger intimate contact area relatively diminishes the contribution of the long-range traction.

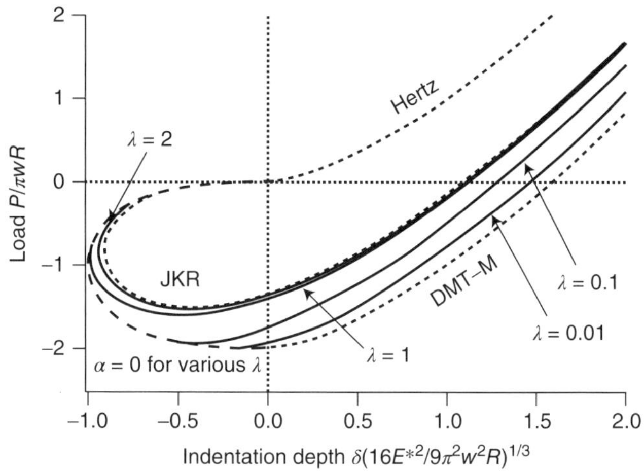

Figure 17.2 Relation between force $\bar{P}$ and indentation $\bar{\delta}$ for various λ (solid lines), together with those given by the Hertz, DMT-M, and JKR models (broken lines).

17.2.6 Adhesion Map

Based on the theories above, Johnson and Greenwood drew a map of elastic contact between spheres, shown in Figure 17.3, which guides to find the most appropriate theory for a given condition [20]. One coordinate of the map is the elasticity parameter λ (μ) and the second coordinate is normalized load $\bar{P}$, which approximates to the ratio of the load to the adhesive pull-off force.

The boundaries of respective zones are provided by the following rules:

- The adhesive contribution can be neglected at a sufficiently high load. In his map, the Hertz zone is defined to be $|P_a/P| < 0.05$, which can be calculated by Equations 17.2, 17.16, and 17.18.

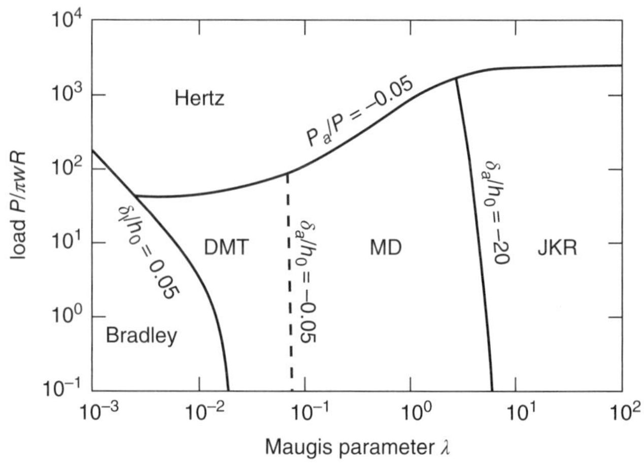

Figure 17.3 Adhesion map. Johnson [20]. Reproduced with permission of Elsevier.

- By analogy with Tabor's consideration about necking [18], the ratio of the elastic deformation of the surfaces caused by the adhesive traction, δ_a, to its effective range, h_0, may govern the JKR–DMT transition. The JKR zone in the map satisfies $|\delta_a/h_0| > 20$, while the DMT zone $|\delta_a/h_0| < 0.05$. One can draw the boundaries by Equations 17.17 and 17.18 and $\overline{h}_0 = 2/\pi\lambda$.
- When the material is stiff and applied load can deform less than atomic separation distance, the macroscopic Hertzian elastic theory may be invalid; in this case, the Lennard-Jones-type interaction should be considered. In the map, the boundary is depicted by the condition that $|\delta_1/h_0| < 0.05$.

17.3 APPLICATION OF CONTACT MECHANICS TO EXPERIMENTAL DATA

17.3.1 Consideration of Contact Models

In literature, plenty of studies proved that in most of the cases the JKR contact theory is appropriate to describe the contact of polymers and polymer composites [21–24], whereas the DMT contact theory is also successfully applied in several rigid polymers [25–27].

The adhesion map shown in the previous section gives a theoretical guideline about this problem. The z_0 for polymers is usually in the range of $0.1 - 0.2$ nm [28], and we choose $z_0 = 0.2$ nm here for simple calculation. For the Lennard-Jones potential, the constant value for the attractive force $\sigma_0 = 1.03w/z_0$, and thus the transition parameter λ can be estimated by Equation 17.15. In the case of isoprene (IR) rubber, an example of soft polymers where $E^* = 2$ MPa, $w = 0.15$ N/m, $R = 15$ nm, the Maugis parameter $\lambda = 250 \gg 1$. The usage of the JKR model is very convincing for soft materials. The situation changes when dealing with hard polymers. Consider the case of polystyrene (PS) as an example. $E^* = 3$ GPa, $w = 0.07$ N/m, $R = 15$ nm, for PS and thus the Maugis parameters $\lambda = 1.2$, which is in the MD zone between JKR and DMT in the adhesion map. However, the MD model cannot be solved without knowing the value of λ. Accordingly, either the JKR or DMT model is chosen somewhat arbitrarily in most practical use, but it may induce error in obtained values.

It should be also reminded that plastic deformation is not considered in any models mentioned above. For a punch that includes both the elastic and plastic deformation, the Oliver–Pharr model is widely known [29]. However, it does not take adhesion into account, and therefore it is often inadequate to analyze AFM experimental data. In many cases, it is preferable to operate at possibly light load to avoid damage on both the probe and sample surfaces.

One can make a rough estimate of the loading pressure given by the AFM cantilever by Equation 17.1 or 17.5. The Hertzian model predicts that, by a punch with a 40-nm-diameter spherical probe under 10 nN load, a plastic material whose reduced modulus is 2.0 GPa will be indented by 0.89 nm, with the contact radius of 4.2 nm and maximum pressure of 270 MPa (the JKR model gives less pressure owing to the adhesive contribution), which is higher than the yield strength of 178 MPa reported in PS films measured by AFM [30]. It implies that the pressure working at the apex of the probe may damage the probe or sample. In such a case, wear may happen both on the sample and probe. On the other hand, by a punch with the same 40-nm-diameter probe under much lower applied force of 1 nN, a rubbery material whose reduced modulus is 2 MPa will be deformed by 19 nm, with the contact radius of 20 nm and maximum pressure of 1.2 MPa. In the condition, the contact radius is close to the curvature radius of the probe, which may lessen the spacial resolution.

17.3.2 Force–Distance Curve Conversion

A force–distance curve measurement is a mode that has been widely used to obtain the force response of the sample surface [31]. This mode records the cantilever deflection d as a function of vertical scanner displacement Z. Figure 17.4a and c shows the schematic relations of d and Z on the assumption of the DMT-M and JKR contact, respectively. (In the DMT model, attraction does not affect the surface profile, and therefore the relation in the noncontact state may be given by the Bradley equation, Eq. 17.4 [14],) When the cantilever approaches the sample from far away, it starts to sense attractive force and in the case of a sufficiently soft cantilever gradually bends to the sample. When the potential of the interaction reaches over that of the cantilever spring (or at the point contact ($\delta = 0$) in the JKR model, because the model ignores the long-range interaction), the sudden jump into the sample surface occurs (point A to A'). At that moment, the probe bends down and its tip is often indented by the attractive force. After passing through the point where the surface attraction and elastic repulsion are counterbalanced (point B), it comes into the repulsive region. The loading process is ended when the scanner displacement Z or cantilever deflection d has achieved a user-specified value (point C). The unloading process reduces the repulsion and goes through the counterbalanced point (point D) into the attractive region. Soon after showing the peak in attractive force (point E), the probe is rapidly detached from the surface (point F to F'). The force at point E is often referred to as the pull-off force.

To convert the force–distance curve data to a relation that can be applied to the contact mechanics model, the following procedure is typically used. One can obtain the load P by Hooke's law,

$$P = kd \tag{17.19}$$

where d and k are the deflection distance and spring constant of the cantilever, respectively. Although the spring constant can be nominally given from the geometry of the cantilever and the elasticity of the materials, it may be often inaccurate, due to the inaccuracy of nanofabrication or the effect of backside coating, which is intended to improve the laser reflectivity. Several

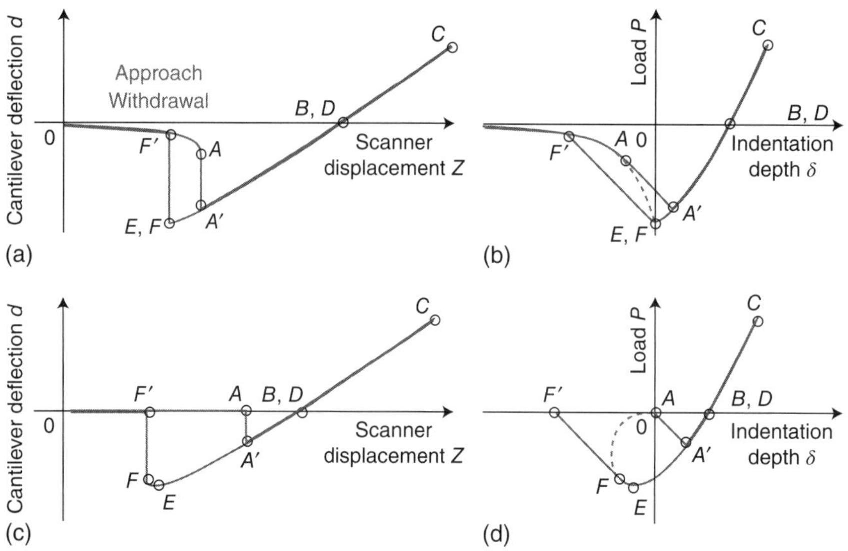

Figure 17.4 Displacement–deflection curves and load–indentation curves theoretically produced by the the DMT-M model (a), (b) and the JKR model (c) (d).

methods are known to calibrate the spring constant [32–34]. As the cantilever detects the force by deflecting itself, to obtain the indentation distance δ one needs to subtract this contribution from the piezoscanner displacement Z, that is,

$$\delta = Z - Z_0 - d \tag{17.20}$$

where Z_0 is the scanner displacement at which the apex of the probe is on the level of the unperturbed sample surface. One can obtain the relation between load P and indentation δ of Figure 17.4b and d from Figure 17.4a and c respectively, using Equations 17.19 and 17.20.

17.3.3 Analysis of Load–Indentation Curves

Applying the appropriate contact mechanics to $P - \delta$ relation, one can obtain the mechanical properties of elastic surface. The adhesion map indicates that even very stiff polymeric materials lie far away from the DMT zone, while most other types of polymeric materials, from rubber and gel to materials of intermediate stiffness, reside in the JKR zone. One of the reasons why sometimes the DMT model is still used regardless of this fact may be due to mathematical handiness, as expressed in Equation 2.4. The AFM cannot directly observe the contact area a; hence, the form of the JKR model, which does not have an explicit solution for $P(\delta)$, is evaded.

Another problem in AFM measurements is the difficulty to find the unperturbed surface level, Z_0. One can theoretically deduce Z_0 from the force–distance curve; in the DMT model, it is the value of Z at point E, while in the JKR model it is point A in Figure 17.4. However, in practice, the attractive interaction is often not able to be attributed to just one of the two, adhesion inside intimate contact zone and long-range attraction at the perimeter of the contact. Therefore, to determine δ in absolute terms is difficult. An approach based on the DMT-M model that can avoid the problem was described by Sahin [35]. Here the so-called two-point method proposed by Sun et al. is briefly introduced, which uses the JKR model and also avoids those two difficulties mentioned above [36].

Using Equation 17.9, one can obtain the work of adhesion from the pull-off force P_E,

$$w = -\frac{2P_E}{3\pi R} \tag{17.21}$$

To obtain the elasticity, this method requires the values at two points; one is point D, where the force is equilibrated, and the other is point E, the point at the pull-off force. The JKR model gives

$$\delta_E = -\frac{1}{3}\left(\frac{9P_E^2}{16E^{*2}R}\right)^{1/3} \tag{17.22}$$

and

$$\delta_D = -\frac{1}{3}\left(\frac{9P_E^2}{E^{*2}R}\right)^{1/3} \tag{17.23}$$

hence the following expression is obtained:

$$E^* = -\frac{3}{4}\left(\frac{1+\sqrt[3]{16}}{3}\right)^{3/2}\frac{P_E}{\sqrt{R(\delta_D - \delta_E)^3}} \tag{17.24}$$

As this formula uses the difference of indentation depth δ, one can obtain the reduced modulus E^* without ambiguous determination of Z_0.

17.3.4 Nanomechanical Mapping

A single force–distance curve measurement gives the mechanical information at the contact area, typically of order of nanometers. However, polymeric materials often have lateral distribution in mechanical properties. For these materials, one can operate sequential force–distance curve measurement on different locations in a user-specified area, which is referred to as force volume (FV) or force mapping [8, 21, 37–39]. In the measurement, each curve is usually controlled to have the same maximum load, P_C, and the scanner displacement at the load, Z_C, is mapped to the location of the curve, which composes the topographic image under the load of P_C. Therefore, the topographic image in this mode is affected by the exerted load, like contact mode or intermittent contact mode images. However, the degree of deformation in this "nanomechanical mapping" mode can be calculated. Therefore, one can reconstruct the surface topography of undeformed condition [8, 21].

Meanwhile, from a set of force–distance curves one can produce the lateral distribution of the surface elasticity and adhesion energy, using appropriate contact mechanics.

17.4 APPLICATION EXAMPLES

Based on the procedures described in the previous sections, one can obtain nanomechanical maps of a wide variety of polymeric and biological materials, including carbon black (CB)-reinforced natural rubber (NR) [40], carbon nanotube (CNT)-reinforced NR [41, 42], reactive polymer blend [43], block copolymers [9, 21, 44, 45], deformed plastics [46, 47], human hair [48, 49], honeycomb-patterned polymer films [50–52], CNT-reinforced hydrogel [53], and diffusion front of polymer [54, 55]. The detailed descriptions are also found in other literatures [56–59]. Hereafter, several example studies are reviewed.

17.4.1 Effect of Processing Conditions on Morphology and Mechanical Properties of Block Copolymers

Thermoplastic elastomers (TPEs), which combine the elastic response like rubber vulcanizates with the processability of thermoplastics, are becoming one of the industrially important polymeric materials [60–62]. The morphology control by processing conditions is a key issue to improve mechanical properties of TPEs. Therefore, the effects of processing conditions on morphology and microscopic mechanical properties of poly(styrene-*b*-ethylene-*co*-butylene-*b*-styrene) (SEBS) triblock copolymer, one of the most widely used TPEs, is investigated by nanomechanical mapping [45].

The weight-average molecular weight, M_w of SEBS sample is 70,000 and the molecular weight distribution, M_w/M_n is about 1.1. The weight fraction of polystyrene (PS) for the triblock copolymer is 29%. SEBS samples are prepared by two methods. One is solution casting; an SEBS film with thickness of about 10 μm is prepared by casting of 4 mg/mL toluene solution on a clean glass slide. The as-prepared film is then subjected to common drying procedure to remove residual solvent. The other is melt compounding using a micro-compounder that consists of a vertical barrel with two conical screws. The processing is performed at a rotational speed of 400 rpm at 230 °C for 5 min under nitrogen protection. The maximum and minimum shear rates calculated at the rotation speed are 4250 s^{-1} and 1863 s^{-1}, respectively. The samples are then cut by ultramicrotome at −120 °C to obtain flat surfaces for AFM characterization. The microtomed surface of an as-received unprocessed sample (pellet form) is also examined for comparison.

The triblock copolymer SEBS consisting of hard and soft blocks that usually exhibits a phase-separated morphology has been widely studied by AFM techniques [63]. Figure 17.5a and e shows the AFM FV-mode original topographic images of SEBS samples processed under different conditions (data on the unprocessed sample are not shown because they are similar to those of the melt-compounded sample). All the samples show the characteristic phase-separated morphologies consisting of bright and dark nanoscale domains as expected. However, the film prepared by solution casting shows well-ordered structures having an intermediate between the cylinder and lamella, while these ordered structures cannot be seen in the case of processed samples. The morphological difference between them is due to the degree of microphase separation. Molecular mobility is limited for melt-processed samples, while sufficient mobility is achieved in the solution casting due to slow solvent evaporation. Radial-averaged power spectral density (PSD), W_r of the images shown in Figure 17.6 confirms a highly ordered structure formed in the cast film, while the results of the melt-processed sample indicates the absence of long-range order. The specific value of periodicity distance of the film sample determined by Gaussian fit to PSD solid curve is 31.3 ± 3.7 nm, with measured periodicity of SEBS samples by Motomatsu et al. [64] and Bhowmick et al. [65] , respectively. As discussed in the previous section, the topographic image obtained by FV mode is distorted by exerted load. In order to compensate it and to obtain a true topographic image, a sample deformation image as in Figure 17.5b is added to the original topographic image in Figure 17.5a. The result is shown in Figure 17.5c. By comparing these two images, it is obvious that true height difference between hard and soft domains (2.12 nm in root-mean-square roughness) is smaller than the apparent one (3.47 nm).

Young's modulus maps obtained using the above-mentioned procedure based on JKR analysis (Section 3.4) on SEBS samples processed under different conditions are presented in Figure 17.5d and f. Typical force–deformation curves both on hard and soft domains are given in Figure 17.7a and b, respectively. The solid lines superimposed on unloading curves are JKR theoretical curves, which are in good agreement with experimental data on both domains. The Young's modulus maps with characteristic phase-separated morphologies consisting of high and low Young's modulus regions are exhibited, resembling the topographic images. Therefore, such a Young's modulus map enables us to directly link the morphology and mechanical properties of the materials. In the Young's modulus maps, brighter areas with higher Young's

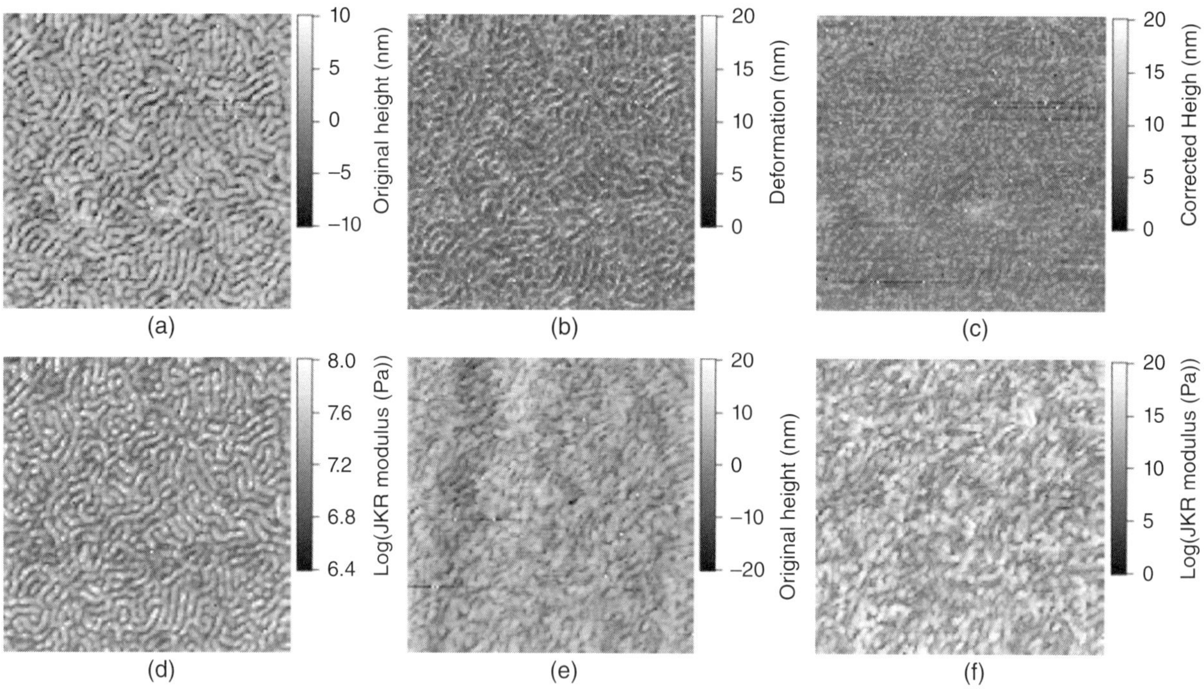

Figure 17.5 Nanomechanical mapping results in FV-mode on SEBS samples. (a) Original topographic, (b) sample deformation, (c) true topographic, and (d) JKR modulus (log-scale) images on cast sample. (e) Original topographic and (f) JKR modulus (log-scale) images on melt-compounded sample. The scan size of each image is 1.0 μm.

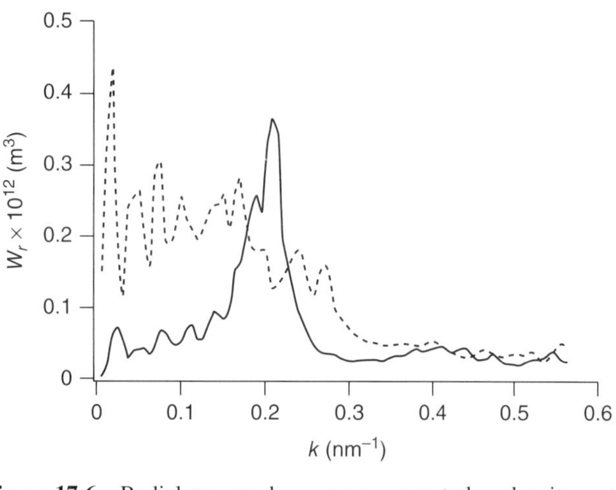

Figure 17.6 Radial-averaged power spectral density of Figure 17.5a (solid line) and Figure 17.5e (dashed line). The wave number corresponding to the peak of solid line is 0.209 nm^{-1}.

modulus are considered to be the hard PS blocks, while darker areas with lower Young's modulus are considered to be the soft poly(ethylene-*co*-butylene) (PEB) blocks. In a film sample, Young's modulus is calculated as 37.8 ± 6.1 MPa for PS blocks and 8.1 ± 2.5 MPa for PEB blocks. Using the same evaluation method, we also investigated the Young's modulus of the pure PS and PEB films. The measured modulus value of glassy PS and rubbery PEB are 2.02 ± 0.59 GPa and 10.1 ± 0.5 MPa, respectively. Therefore, the observed modulus on PEB block agrees with the value of the pure sample, while PS block's shows a dramatic decrease. This decrease is due to the microstructure effect that the soft PEB blocks surround and support the PS blocks underneath [64, 66].

Figure 17.8 shows the Young's modulus distribution for cast film (filled circle), melt compound (open circle), and unprocessed pellet (filled triangle) samples. The topographic feature of unprocessed sample does not differ so much from that of the melt compound sample as mentioned previously, although the difference in Young's modulus distribution is evident: a clear decrease in the ratio of high Young's modulus regions is observed. The results indicate that the high shear rate leads to the degradation of SEBS. This conclusion is further demonstrated by GPC results [45]. As the increase of rotational speed from 0 to 400 rpm, the molecular weight, M_n and M_w decrease from 67,100 to 26,000 and 70,000 to 43,000, respectively. The M_n below 10,000 also increases from 0% to 5.8%. Then, both the decreased ratio of high Young's modulus regions in the Young's modulus maps and decreased molecular weight indicate that the mechanical property of the processed SEBS sample might deteriorate with the increase of rotational speed. However, tensile testing reveals that the initial tensile modulus (Young's modulus) for each sample is not so different to each other (data not shown). At the same time, tensile strength at break dramatically decreases for the melt-compounded sample. Thus, it might be possible to correlate the tensile strength at break to the high Young's modulus regions and initial tensile modulus to the peak Young's modulus. Actually as shown in Figure 17.8, the distribution becomes narrower as the rotational speed increases, while its peak values keep almost constant at

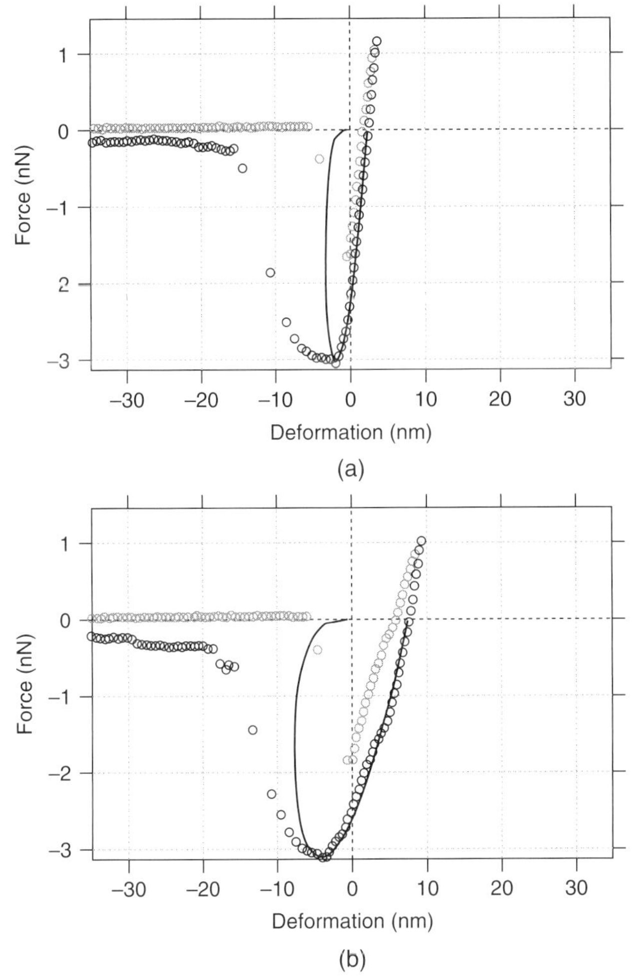

Figure 17.7 Typical force–deformation curves on (a) hard domain and (b) soft domain of SEBS cast film sample.

about 13.0 MPa. In conclusion, Young's modulus of SEBS is independent of the rotational speed until 400 rpm, at a processing temperature of 230 °C for 5 min. The unchanged Young's modulus peak values can be explained as an intrinsic property that depends on intermolecular bonding of a material and is related to molecular weight to some extent.

The results obtained here with AFM nanomechanical mapping indicate that this technique is very valuable in evaluating the mechanical properties of polymer materials at a micro or nanoscale and in visualizing the distribution of mechanical properties, which are not available by any other conventional characterizing methods.

17.4.2 Measuring the Deformation of Both Ductile and Fragile Polymers

It is of great interest to investigate the evolution of submicron-scale local mechanical properties of polymers under deformation for the better understanding of deformation mechanisms, while AFM nanomechanics might be a good solution for this purpose. Here, submicron-scale mechanical

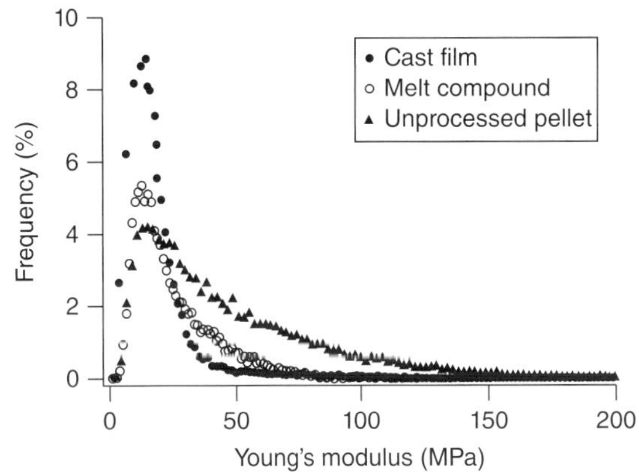

Figure 17.8 Young's modulus distribution of cast film (filled circle), melt compound (open circle), and unprocessed pellet (filled triangle) samples.

studies on uniaxial drawing of poly(ε-caprolactone) (PCL) and poly(lactic acid) (PLA) at room temperature is introduced. Figure 17.9 shows the typical stress–strain curve of PCL and PLA. At room temperature, PCL shows ductile nature with high elongation at break, while PLA is brittle and breaks just after yielding. AFM nanomechanical mapping is applied to investigate the deformation process of the two

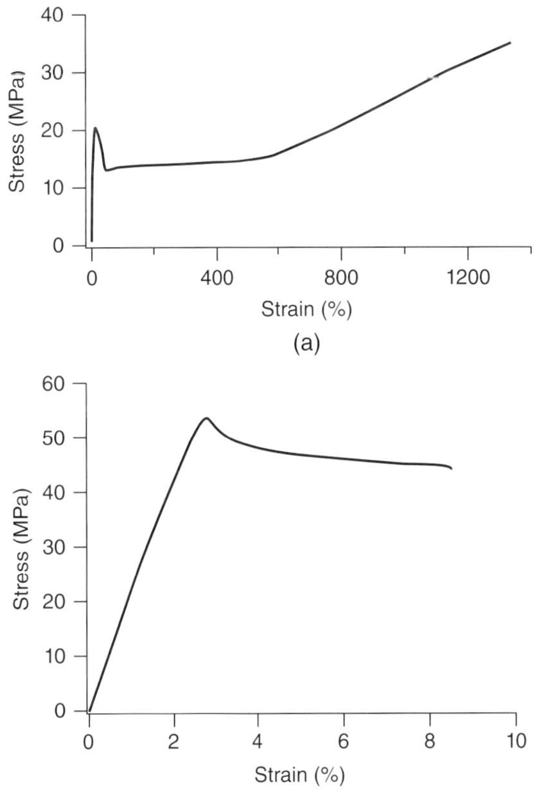

Figure 17.9 Typical stress–strain curve of (a) PCL and (b) PLA measured at room temperature.

polymers [46, 47]. It is preferable that the force–distance curves are collected on a flat surface. Thus, neat and deformed PCL and PLA are cut at −80 °C using cryo-microtome. The sections of sample face on the machine direction are cut to avoid possible relaxation. Uniaxial tensile tests are also performed for comparison with nanomechanical properties.

Figure 17.10 shows the topographic and nanomechanical Young's modulus maps of neat and deformed PCL together with Young's modulus distribution. JKR theory is used to analyse the unloading force–deformation curves. Neat PCL presents a relative uniform morphology and modulus distribution, where a sharp single peak is shown in the modulus distribution in Figure 17.10e. The Gaussian fitting to the distribution curve gives 140 ± 26 MPa, which is almost the same with 124 ± 14 MPa calculated from the uniaxial tensile test, which assures the accuracy of AFM nanomechanical mapping together with the fact that macroscopic initial modulus is well described by the peak Young's modulus for the neat sample.

Contrary to the uniform distribution of the neat sample, the topography of the lateral section of necking part shows obvious orientation along the tensile direction; similar morphology was reported in deformed polypropylene [67, 68] and polyamide [69] matrix. The necking process is a heterogeneous deformation and results in the formation of fibrillar structures. The cryo-microtome process is supposed to make the cross sections in perfect plane. However, oriented height fluctuation can also be seen due to the uneven fibrillar structure. Moreover, a slight relaxation of the oriented nanofibrils might also contribute to the height difference. The shape of the nanofibrils can be distinguished clearly with a width of about 15–40 nm, very similar to 20–40 nm reported by Koike et al. [68]. More detailed information of the necking part can be seen from the Young's modulus map shown in Figure 17.10d. The surface topology is filled with nanofibrils, which are oriented along the tensile direction. The Young's modulus distribution shows a much broader peak in the case of deformed PCL as shown in Figure 17.10f. The nanofibrils have higher modulus and make the modulus distribution peak widened to the higher direction. As a result, the necking parts have a Young's modulus higher than the neat PCL matrix. The Young's modulus peak obtained from

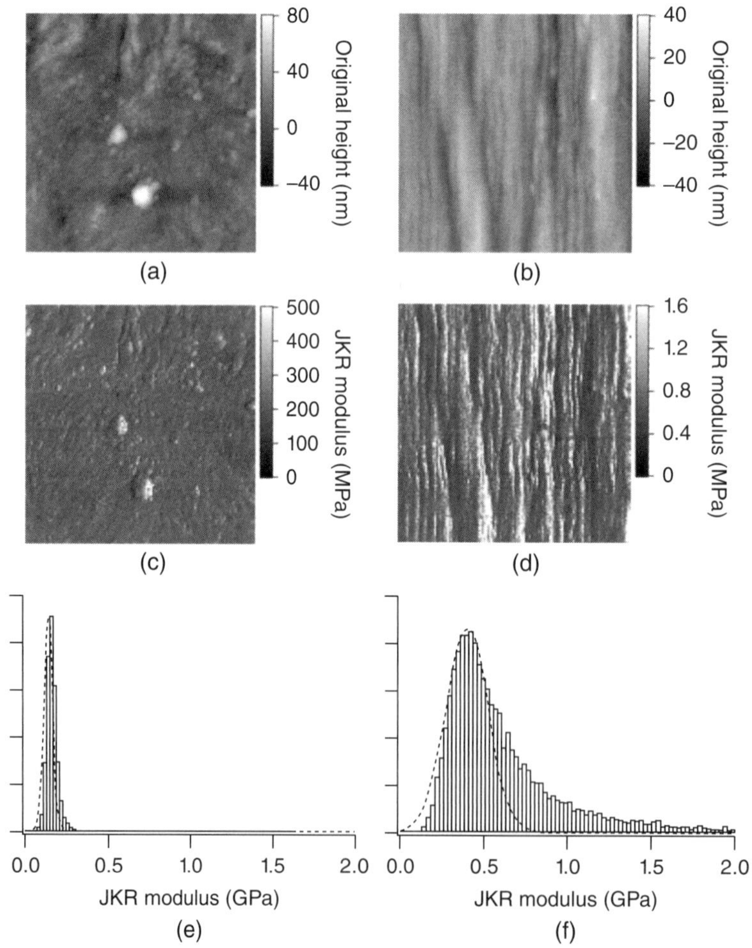

Figure 17.10 (a) and (b) Topographic and (c) and (d) nanomechanical Young's modulus maps of both neat (left) and deformed (right) PCL. The scan size is 5.0 μm for neat PCL and 3.0 μm for deformed PCL. (e) and (f) Young's modulus distribution of both samples corresponding to (c) and (d), respectively.

Gaussian fitting on the deformed PCL is 397 ± 126 MPa, whereas the value calculated form the uniaxial tensile test of the necking part is 727 ± 160 MPa. This discrepancy is due to wide distribution of Young's modulus. Actually, the statistical average of Young's modulus obtained from Figure 17.10d is 903 ± 121 MPa, more closer to the macroscopic value. Thus, AFM nanomechanical mapping provides a good method to quantitatively evaluate the mesoscopic mechanical properties of materials and to compare with macroscopic properties. In this specific case, it reveals the highly heterogeneous nature of deformed plastics. In other words, the deformation process strongly enhances an intrinsic heterogeneous nature of plastics.

The story is different in the case of PLA, where fractures happen just after the yielding. When PLA is under uniaxial deformation, the delocalization of polymer chains is induced by weakening the structure perpendicular to the loading direction, which gives rise to small crazes. The crazes tend to emerge in the adjacent positions and leave large spaces without any crazes, which can be ascribed to the dynamical heterogeneities of amorphous polymers [70]. One can monitor the microscopic structural/mechanical evolution simultaneously by applying AFM nanomechanical mapping on deformed PLA as in the case of PCL. The results are shown in Figure 17.11. As shown in Figures 17.11b and 17.12, microscopic Young's modulus of the neat PLA sample had a sharp single peak distribution with peak value of 3.2 ± 0.8 GPa. When PLA is deformed below its glass-transition temperature, catastrophic localization in terms of craze formation perpendicular to the loading direction occurs. Figure 17.11c and (d) shows the data collected at the location of an initial craze. Similar to flexible polymers, fibrillar structure appears in the origin of cracks. The fibrillar structures in PLA only occur in the middle part of stress concentration region. Young's modulus of the fibrillate part is 2.7 ± 1.1 GPa, which is slightly lower than neat PLA. This observation is supported by the tail in the lower modulus part of Young's modulus distribution in Figure 17.12. The other parts in Figure 17.11d

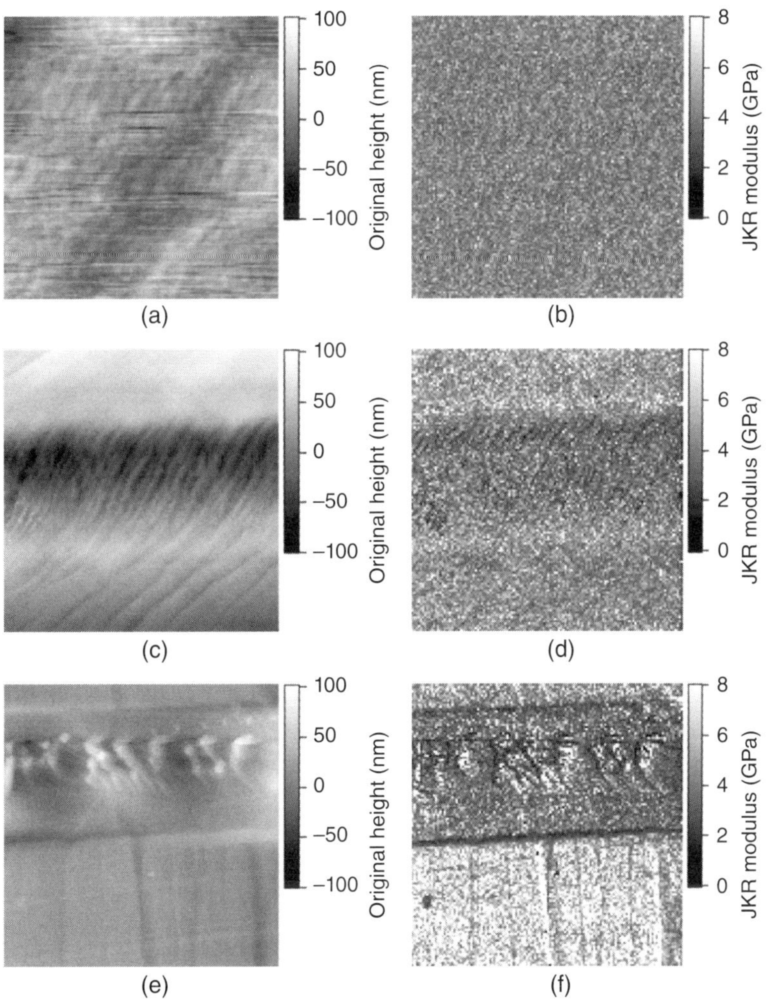

Figure 17.11 Topographic (left) and nanomechanical Young's modulus (right) maps of (a) and (b) neat PLA, (c) and (d) initial craze of PLA and (e) and (f) mature craze of PLA. The scan size is 1.0 μm for neat PLA and 5.0 μm for craze parts of PCL.

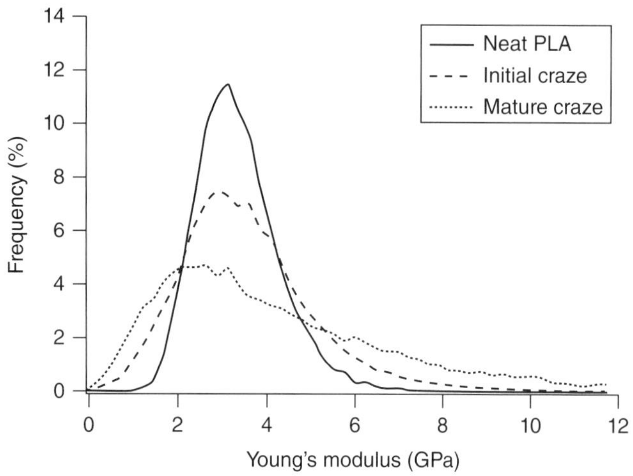

Figure 17.12 Young's modulus distribution of neat PLA (solid line), initial craze (dashed line), and mature craze (dotted line).

have peak Young's modulus of 3.5 ± 1.1 GPa, which is a value almost similar to that in the neat sample, indicating that these regions experience elastic recovery after the external load is released. A similar recovery is observed in the portion far from craze areas for pre-yield, post-yield, and even the fractured samples [47].

Similarly, but much more interestingly, the reliable nanomechanical properties of mature craze are characterized and shown in Figure 17.11e and f. The AFM nanomechanical mapping results show that the modulus in the edge of mature crazes keeps on increasing to 5.2 ± 2.4 GPa as shown in the lower half of the Young's modulus map (also in the long tail in higher modulus part of modulus distribution in Fig. 17.12). The strain hardening occurs, which could be ascribed to the stress concentration and possible orientation. While in the upper half part the matrix remains fibrillar structures with a reduced Young's modulus of 2.3 ± 1.2 GPa, the possible explanation of this phenomenon is that chain scission/slippage must occur during crazing to permit the generation of the fibrillar structure, and modulus is lowered as a consequence. Furthermore, the applied local stress may lead to an increase in segmental mobility, giving rise to the decreases in viscosity and moduli. The standard derivation of the whole part is increased accordingly, and the whole area in Figure 17.11c has the statistical-average Young's modulus of 4.5 ± 2.6 GPa, much broader distribution than that of the neat PLA. It is consistent with the finding of Rottler et al. [71] that the tension is mostly carried by the covalent backbone bonds in the craze and that the exponential distribution of tension is found in the craze zone. The overlapped modulus distribution of neat PLA, initial craze, and mature craze is compared in Figure 17.12 to describe the local modulus change introduced by craze as a summary. The deformed PLA shows a highly heterogeneous nature of mechanical property distribution. With the initial and growth of crazes, the modulus distribution is much widened in both increase and decrease directions.

17.4.3 Nanorheological AFM on Rubbers

Polymeric materials exhibit viscoelastic phenomena, which must be taken into account in designing the materials' applications. For example, rubber in a tire receives stimuli over a wide frequency and temperature range from the road surface. In the case of bulk samples, the frequency and temperature can be converted mutually based on the time-temperature superposition (TTS) principle [72]. However, TTS is a kind of empirical rule and, consequently, an actual measurement method with a wide frequency and temperature range is necessary to precisely predict the properties of practical products. Various AFM-based conventional methods have been proposed to measure viscoelasticity such as lateral force microscopy (LFM) [73–75], force modulation (FM) [76–78], and contact resonance (CR) [79–81]. Even tapping mode can report energy-dissipative phenomena [44, 82–84] and further offers loss tangent mapping [85, 86].

Herein is introduced as one of the examples is a modified FV-based AFM method, hereafter called as nanorheological AFM, to measure the viscoelastic properties of polymeric materials, especially rubbers, over a wide frequency range, which was originally developed by the authors [87]. To encompass a wide-frequency range, a tiny piezoelectric actuator, measuring $3\,mm \times 3\,mm \times 2\,mm$, is fixed to a metallic sample holder located on the AFM scanner. The actuator is driven by the built-in oscillator of the lock-in amplifier, which is electrically isolated from all the AFM circuits. The lock-in amplifier is used to measure the amplitude and phase shift of the cantilever-deflection oscillatory signal to the drive signal.

The method combines the above-mentioned oscillation detection and FV operation, the sequential force–distance curve measurements introduced in Section 3.4. When the cantilever deflection reaches a certain preset value (1.5 nm in this experiment) during each force–distance curve measurement, the synchronization signal from AFM controller tells the piezoelectric actuator to oscillate, and the cantilever is maintained in contact with the sample surface while the frequency sweep is completed in 30 (7.2) s for a single-point (mapping) measurement using surface delay control during which the AFM scanner does not move. The drive signal is regulated such that the frequency responses of the amplitude and phase shift on mica, which is regarded as a sufficiently hard reference sample, become flat. Appropriate time constants of the lock-in amplifier are selected according to the drive frequencies.

Figure 17.13 shows the relationships between the oscillation of the piezoelectric actuator and that of the cantilever on mica (reference) and polymeric sample. A_p and $\omega_p = 2\pi f_p$ are the amplitude and angular frequency of the actuator oscillation, respectively. Because the oscillation of the actuator cannot be experimentally measured, the amplitude A_r and phase shift ϕ_r of the cantilever oscillation are monitored on mica beforehand. A_s and ϕ_s are the measured amplitude and phase shift of the cantilever oscillation on the sample, respectively.

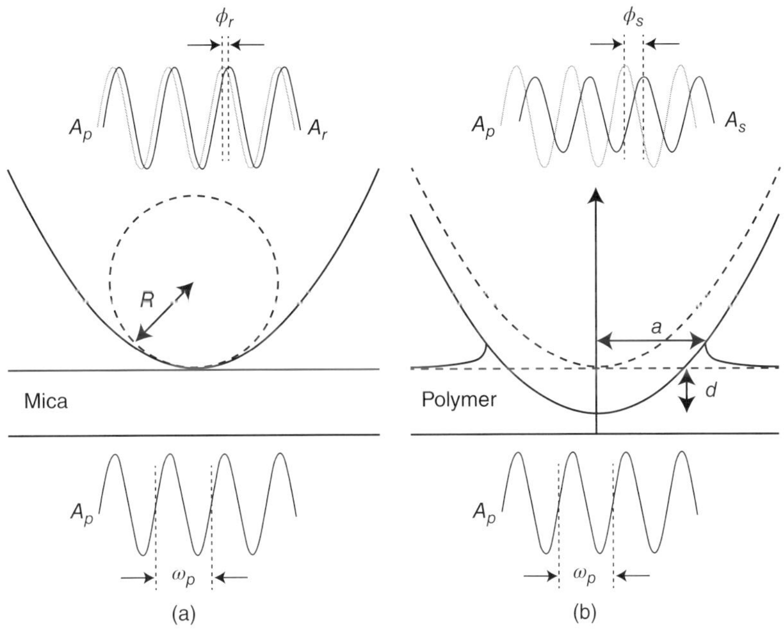

Figure 17.13 Schematic of a nanorheological measurement on hard standard sample (a) and soft polymeric sample (b). Probe tip radius, R, sample deformation, d, and contact radius, a, are defined as in this figure.

Then, sample deformation, Δ is defined as

$$\Delta = A_r \cos(\omega_p t + \phi_r) - A_s \cos(\omega_p t + \phi_s) \equiv A \cos(\omega_p t + \phi) \tag{17.25}$$

where A and ϕ are the amplitude and phase shift of the sample deformation oscillation, respectively. By solving Equation 4.3, A and ϕ can be expressed as

$$A = \sqrt{{A_r}^2 + {A_s}^2 - 2A_r A_s \cos(\phi_s - \phi_r)} \tag{17.26}$$

$$\tan\phi = \frac{A_r \sin\phi_r - A_s \sin\phi_s}{A_r \cos\phi_r - A_s \cos\phi_s} \tag{17.27}$$

Note that all quantities in Equations 17.26 and 17.27 are experimentally acquired. As a viscoelastic property, the dynamic stiffness, S' and S'' are defined as

$$S' = k\frac{A_s}{A}\cos(\phi_s - \phi) \tag{17.28}$$

$$S'' = k\frac{A_s}{A}\sin(\phi_s - \phi) \tag{17.29}$$

where k is the spring constant of the cantilever. Moreover, the loss tangent ($\tan\delta$), which is defined as the ratio between loss and storage stiffness, can be expressed as

$$\tan\delta \equiv \frac{S''}{S'} = \tan(\phi_s - \phi) \tag{17.30}$$

Like in the case of static measurement, elastic moduli can be determined from stiffness if the contact radius, a, is known. Ideally, the quantity should be directly measured or analyzed *in situ* under a dynamic experiment. However, nanometer-scale contact area under AFM probe cannot be visualized at the current state of the art. Furthermore, the contact mechanics for viscoelastic material is not sufficiently mature. Therefore, the contact radius, a, is estimated using JKR elastic theory assuming that the viscoelastic effect on the contact area can be regarded as perturbation. Assuming JKR contact (Equations 17.7 and 17.8), the relationship between the dynamic stiffness and the dynamic modulus is expressed as

$$E' = \frac{(1-\nu^2)S'}{2a_t}\frac{1-(1/6)(a_i/a_t)^{3/2}}{1-(a_i/a_t)^{3/2}} \tag{17.31}$$

$$E'' = \frac{(1-\nu^2)S''}{2a_t}\frac{1-(1/6)(a_i/a_t)^{3/2}}{1-(a_i/a_t)^{3/2}} \tag{17.32}$$

where ν is Poisson's ratio, which is 0.5 for typical rubber materials, and a_i is the contact radius at the point D in Figure 17.4, where the force is equilibrated. The oscillation measurement is executed at $a = a_t$ (usually repulsive-force region).

Vulcanized styrene-butadiene rubber (SBR), IR and SBR/IR 7/3 blend samples are subjected to nanorheological AFM as model samples. Figure 17.14 shows E', E'', and $\tan\delta$ measured by this method. The measurement temperature is fixed at 25 °C. From the result of SBR homopolymer in Figure 17.14a, E' increases from 3.0 to 410 MPa and E'' increases from 1.0 to 248 MPa as the frequency is increased from 1.0 Hz to 20 kHz. The frequency dependence of $\tan\delta$ has a peak at 300 Hz. These results indicate that SBR encountered the glass-transition phenomena at the measurement temperature. On the other hand, from the result of IR homopolymer

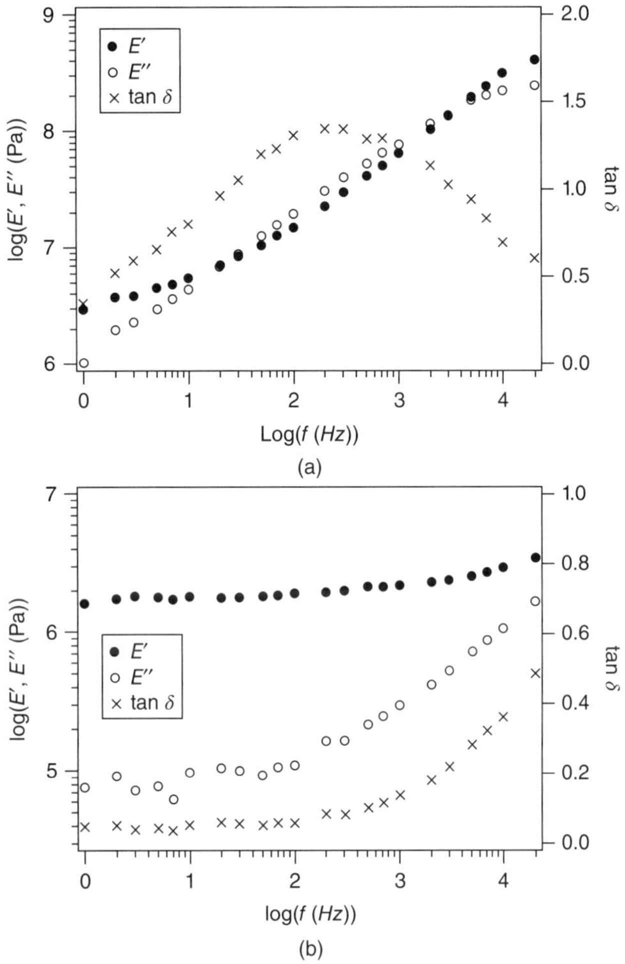

Figure 17.14 Frequency dependence of E', E'', and tan δ measured by AFM nanorheological method described in the main text. (a) Data at a point of SBR homopolymer and (b) IR homopolymer.

in Figure 17.14b, E' only increases from 1.6 to 3.4 MPa, while E'' increases from 0.07 to 1.6 MPa as the frequency is increased from 1.0 Hz to 20 kHz. tan δ does not have a peak in this frequency range. These results indicate that IR is in the rubbery plateau region at 25 °C.

Figure 17.15 shows master curves of E' and tan δ obtained for the bulk SBR sample by conventional dynamic mechanical analysis (DMA), with the frequency range of 0.05 to 50 Hz over a temperature range of −65 to +45 °C. One returns to the TTS principle, which can be expressed by the Williams–Landel–Ferry (WLF) equation [72] to build the master curves,

$$\log a_T = -\frac{C_1(T - T_r)}{C_2 + (T - T_r)} \tag{17.33}$$

where a_T is a shift factor, T is measurement temperature. T_r is the reference temperature, which is related to the glass-transition temperature, T_g, as $T_r = T_g + 50\,°\mathrm{C}$, in order to use universal constants ($C_1 = 8.86$ and $C_2 = 101.6$). T_g of SBR is measured to be −5 °C by differential scanning calorimetry (DSC). The use of universal constants means that a kind of empirical shifting is not performed this time and also that thermorheological simplicity for SBR is assumed. By this way, one can use the same shift factor for nanorheological AFM data, which is only taken at a fixed temperature (25 °C). The data is also superimposed in Figure 17.15, the raw data of which is the same data with Figure 17.14a. E' and tan δ of SBR obtained by nanorheological AFM corresponded well to the master curves obtained by DMA. The peaks of tan δ are perfectly in good agreement to each other. The small deviation of the value of tan δ is mainly attributed to the deviation in E''. Although further elaboration is necessary, it is speculated that the reason for this deviation mainly comes from the assumption of JKR contact. As the viscoelastic nature dominates at the glass-transition region, the force–deformation curve at such a situation always shows a less negligent deviation from the JKR theoretical curve [80, 88, 89].

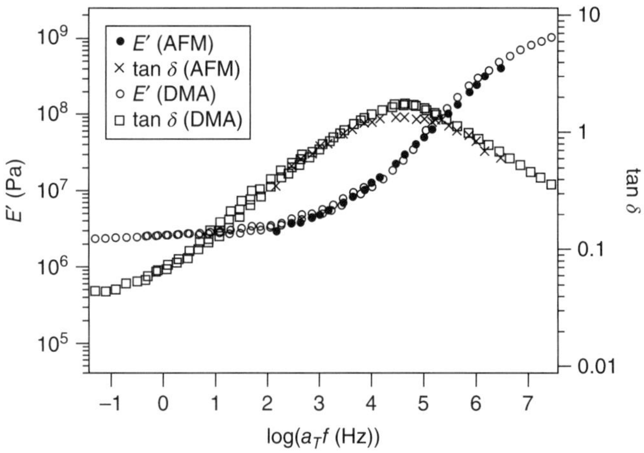

Figure 17.15 Comparison between bulk master curve (E' and tan δ) with the data by nanorheological AFM.

Figure 17.16 shows the nanorheological AFM loss tangent maps of an SBR/IR blend sample at eight representative frequencies. The nanorheological AFM data is taken as a 64 pixel × 64 pixel image for 2.0 μm × 2.0 μm square region, where each pixel contained the data measured at 13 different oscillation frequencies (0.01, 0.05, 0.1, 0.3, 0.5, 0.7, 1.0, 2.0, 3.0, 5.0, 7.0, 10, and 20 kHz). Although approximately 8 h are required to capture one image, the figures in Figure 17.16 are obtained at the same time and at the same location. The AFM images revealed a sea-island-like structure for the sample because SBR and IR are immiscible. On the basis of the blend ratio, the sea should be SBR and the islands should be IR. In the loss tangent image acquired at 1.0 kHz, the smallest IR domain had a diameter of 90 nm. Therefore, this nanorheology AFM has at least a lateral resolution of 90 nm. The contrast of the tan δ images is enhanced at 300 Hz (similar to Figure 17.16d taken at 500 Hz), whereas the sea and islands are almost indistinguishable at 20 kHz. E', E'', and tan δ values have an almost perfect coincidence with the

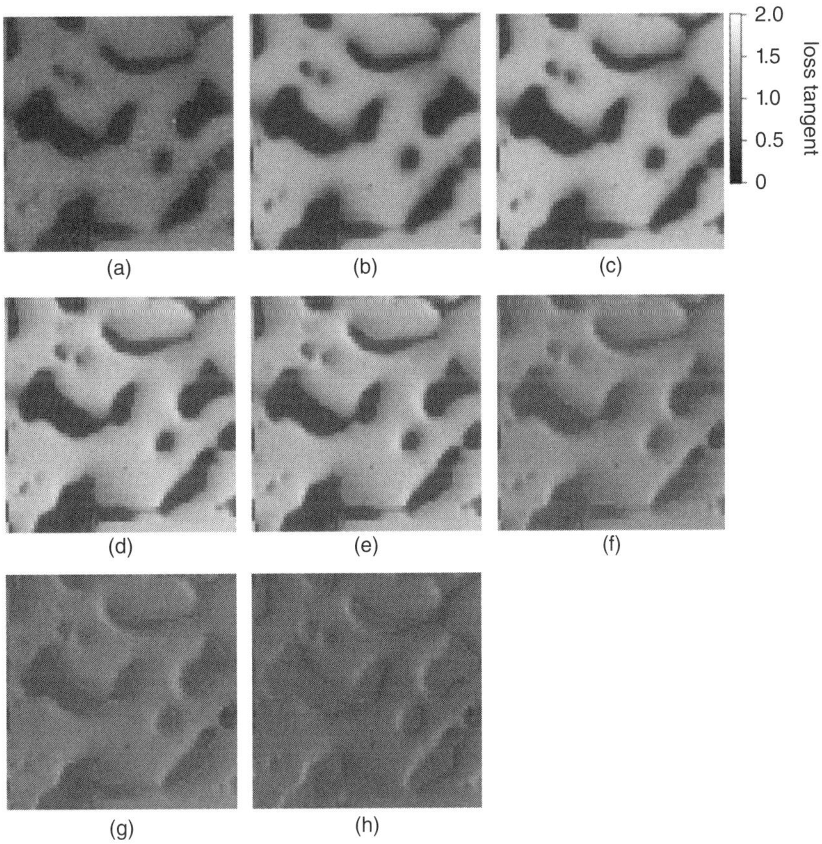

Figure 17.16 Nanorheological AFM loss tangent maps of a SBR/IR blend sample. (a), (b), (c), (d), (e), (f), (g), and (h) correspond 10 Hz, 50 Hz, 100 Hz, 500 Hz, 1.0 kHz, 5.0 kHz, 10 kHz, and 20 kHz, respectively. The scan size is 2.0 μm.

data on homopolymers in Figure 17.14, indicating that at first the blend sample used in this study is really immiscible and that furthermore this measurement can distinguish the viscoelastic nature of different samples at the level of nanometer resolution. This statement would be particularly important to investigate more complex systems where the blend is not totally immiscible. Nanorheological AFM may play a very important role in investigating such samples. Quite recently, the comparison between nanorhelogical AFM and tapping-mode AFM loss tangent imaging has been reported [90], which can be a further guideline for the readers who are interested in viscoelastic imaging by AFM.

17.5 CONCLUSION

In this chapter, the theoretical background and several applications of nanomechanical mapping using the atomic force microscope are introduced. It enables us to image mechanical properties of materials at nanometer scale. The surface mechanical properties including Young's modulus, adhesion, viscoelasticity, and so on can be investigated during probe–sample interactions by employing nanomechanical mapping. The applications of nanomechanical mapping on polymers can help us investigate microscopic structures and mechanical properties evolutions simultaneously due to processing, deformation, degradation, and so on. The authors anticipate that the nanomechanical observation provided here will open a new way to study surface properties and explore the microstructure–properties relationship in a large range of polymeric and biological materials.

REFERENCES

1. Binnig G, Rohrer H, Gerber C, Weibel E. Surface studies by scanning tunneling microscopy. Phys Rev Lett 1982;49:57–61.
2. Binnig G, Rohrer H, Gerber C, Weibel E. 7 × 7 reconstruction on Si(111) resolved in real space. Phys Rev Lett 1983;50:120–123.
3. Binnig G, Quate CF, Gerber C. Atomic force microscope. Phys Rev Lett 1986;56:930–933.
4. Zhong Q, Inniss D, Kjoller K, Elings VB. Fractured polymer/silica fiber surface studied by tapping mode atomic force microscopy. Surf Sci Lett 1993;290:L688–L692.
5. Kim DY, Tripathy SK. Lian Li, and J. Kumar. Laserâ€ induced holographic surface relief gratings on nonlinear optical polymer films. Appl Phys Lett 1995;66:1166–1168.
6. Rochon P, Batalla E, Natansohn A. Optically induced surface gratings on azoaromatic polymer films. Appl Phys Lett 1995;66:136–138.
7. Rockford L, Liu Y, Mansky P, Russell TP, Yoon M, Mochrie SGJ. Polymers on nanoperiodic, heterogeneous surfaces. Phys Rev Lett 1999;82:2602–2605.

8. Nukaga H, Fujinami S, Watabe H, Nakajima K, Nishi T. Nanorheological analysis of polymer surfaces by atomic force microscopy. Jpn J Appl Phys 2005;44:5425–5429.
9. Wang D, Fujinami S, Nakajima K, Nishi T. True surface topography and nanomechanical mapping measurements on block copolymers with atomic force microscopy. Macromolecules 2010;43:3169–3172.
10. Hertz H. Ueber die berührung fester elastischer körper. J Reine Angew Math 1882;92:156–171.
11. Bradley RS. LXXIX. The cohesive force between solid surfaces and the surface energy of solids. Philos Mag 1932;13:853–862.
12. Johnson KL, Kendall K, Roberts AD. Surface energy and the contact of elastic solids. Proc R Soc London A 1971;324:301–313.
13. Derjaguin BV, Muller VM, Toporov YP. Effect of contact deformations on the adhesion of particles. J Colloid Interface Sci 1975;53:314–326.
14. Muller VM, Derjaguin BV, Toporov YP. On two methods of calculation of the force of sticking of an elastic sphere to a rigid plane. Colloids Surf 1983;7:251–259.
15. Pashley MD. Further consideration of the dmt model for elastic contact. Colloids Surf 1984;12:69–77.
16. Greenwood JA. Adhesion of elastic spheres. Proc R Soc London A 1997;453:1277–1297.
17. Maugis D. Adhesion of spheres: The JKR–DMT transition using a dugdale model. J Colloid Interface Sci 1992;150:243–269.
18. Tabor D. Surface forces and surface interactions. J Colloid Interface Sci 1977;58:2–13.
19. Muller VM, Yushchenko VS, Derjaguin BV. On the influence of molecular forces on the deformation of an elastic sphere and its sticking to a rigid plane. J Colloid Interface Sci 1980;77:91–101.
20. Johnson KL, Greenwood JA. An adhesion map for the contact of elastic spheres. J Colloid Interface Sci 1997;192:326–333.
21. Wang D, Fujinami S, Liu H, Nakajima K, Nishi T. Investigation of true surface morphology and nanomechanical properties of poly(styrene-b-ethylene-co-butylene-b-styrene) using nanomechanical mapping: Effects of composition. Macromolecules 2010;43:9049–9055.
22. Oláh A, Hillborg H, Vancso GJ. Hydrophobic recovery of UV/ozone treated poly(dimethylsiloxane): Adhesion studies by contact mechanics and mechanism of surface modification. Appl Surf Sci 2005;239:410–423.
23. Nagai S, Fujinami S, Nakajima K, Nishi T. Nanorheological investigation of polymeric surfaces by atomic force microscopy. Compos Interfaces 2009;16:13–25.
24. Dokukin ME, Sokolov I. Quantitative mapping of the elastic modulus of soft materials with HarmoniX and PeakForce QNM AFM modes. Langmuir 2012;28:16060–16071.
25. DelRio FW, Jaye C, Fischer DA, Cook RF. Elastic and adhesive properties of alkanethiol self-assembled monolayers on gold. Appl Phys Lett 2009;94:131909 (3pp).
26. Schön P, Bagdi K, Molnár K, Markus P, Pukánszky B, Vancso GJ. Quantitative mapping of elastic moduli at the nanoscale in phase separated polyurethanes by AFM. Eur Polym J 2011;47:692–698.
27. Wu S, Guo Q, Zhang T, Mai Y-W. Phase behavior and nanomechanical mapping of block ionomer complexes. Soft Matter 2013;9:2662–2672.
28. Israelachvili JN. *Intermolecular and Surface Forces*, chapter 13.13, third edition. London: Academic Press; 2011. p 275.
29. Oliver WC, Pharr GM. An improved technique for determining hardness and elastic modulus using load and displacement sensing indentation experiments. J Mater Res 1992;7:1564–1583.
30. Riggleman RA, Lee H-N, Ediger MD, de Pablo JJ. Heterogeneous dynamics during deformation of a polymer glass. Soft Matter 2010;6:287–291.
31. Burnham NA, Colton RJ. Measuring the nanomechanical properties and surface forces of materials using an atomic force microscope. J Vac Sci Technol A 1989;7:2906–2913.
32. Hutter JL, Bechhoefer J. Calibration of atomic-force microscope tips. Rev Sci Instrum 1993;64:1868–1873.
33. Sader JE, Chon JWM, Mulvaney P. Calibration of rectangular atomic force microscope cantilevers. Rev Sci Instrum 1999;70:3967–3969.
34. Cleveland JP, Manne S, Bocek D, Hansma PK. A nondestructive method for determining the spring constant of cantilevers for scanning force microscopy. Rev Sci Instrum 1993;64:403–405.
35. Sahin O, Erina N. High-resolution and large dynamic range nanomechanical mapping in tapping-mode atomic force microscopy. Nanotechnology 2008;19:445717 (9pp).
36. Sun Y, Akhremitchev B, Walker GC. Using the adhesive interaction between atomic force microsscopy tips and polymer surfaces to measure the elastic modulus of compliant samples. Langmuir 2004;20:5837–5845.
37. Haga H, Sasaki S, Kawabata K, Ito E, Ushiki T, Sambongi T. Elasticity mapping of living fibroblasts by AFM and immunofluorescence observation of the cytoskeleton. Ultramicroscopy 2000;82:253–258.
38. Lemieux M, Usov D, Minko S, Stamm M, Shulha H, Tsukruk VV. Reorganization of binary polymer brushes: Reversible switching of surface microstructures and nanomechanical properties. Macromolecules 2003;36:7244–7255.
39. Jiao Y, Schäffer TE. Accurate height and volume measurements on soft samples with the atomic force microscope. Langmuir 2004;20:10038–10045.
40. Nishi T, Nukaga H, Fujinami S, Nakajima K. Nanomechanical mapping of carbon black reinforced natural rubber by atomic force microscopy. Chin J Polym Sci 2007;25:35–41.
41. Wang D, Fujinami S, Nakajima K, Niihara K, Inukai S, Ueki H, Magario A, Noguchi T, Endo M, Nishi T. Production of a cellular structure in carbon nanotube/natural rubber composites revealed by nanomechanical mapping. Carbon 2010;48:3708–3714.
42. Wang D, Fujinami S, Nakajima K, Inukai S, Ueki H, Magario A, Noguchi T, Endo M, Nishi T. Visualization of nanomechanical mapping on polymer nanocomposites by AFM force measurement. Polymer 2010;51:2455–2459.
43. Wang D, Fujinami S, Liu H, Nakajima K, Nishi T. Investigation of reactive polymer–polymer interface using nanomechanical mapping. Macromolecules 2010;43:5521–5523.
44. Wang D, Liang X-B, Liu Y-H, Fujinami S, Nishi T, Nakajima K. Characterization of surface viscoelasticity and energy dissipation in a polymer film by atomic force microscopy. Macromolecules 2011;44:8693–8697.
45. Wang D, Nakajima K, Fujinami S, Shibasaki Y, Wang J-Q, Nishi T. Characterization of morphology and mechanical properties of block copolymers using atomic force microscopy: Effects of processing conditions. Polymer 2012;53:1960–1965.

46. Liu H, Fujinami S, Wang D, Nakajima K, Nishi T. Nanomechanical mapping on the deformed poly(ε-caprolactone). Macromolecules 2011;44:1779–1782.
47. Liu H, Chen N, Fujinami S, Louzguine-Luzgin D, Nakajima K, Nishi T. Quantitative nanomechanical investigation on deformation of poly(lactic acid). Macromolecules 2012;45:8770–8779.
48. Kitano H, Yamamoto A, Niwa M, Fujinami S, Nakajima K, Nishi T. Young's modulus mapping on hair cross-section by atomic force microscopy. Compos Interfaces 2009;16:1–12.
49. Nawa T, Kawaguchi A, Kitano H, Yamamoto T, Fujinami S, Asao N, Nakajima K. Alkaline peroxide treatment induces acquired unruly hair by apparently affecting distinct macrofibrils. J Cosmet Sci 2013;64:261–271.
50. Yabu H, Saito Y, Nakamichi Y, Hirai Y, Fujinami S, Nakajima K, Nishi T, Shimomura M. Self-assembled porous templates allow pattern transfer to poly(dimethyl siloxane) sheets through surface wrinkling. Polym J 2012;44:573–578.
51. Kawano T, Nakamichi Y, Fujinami S, Nakajima K, Yabu H, Shimomura M. Mechanical regulation of cellular adhesion onto honeycomb-patterned porous scaffolds by altering the elasticity of material surfaces. Biomacromolecules 2013;14:1208–1213.
52. Fujie T, Ahadian S, Liu H, Chang H, Ostrovidov S, Wu H, Bae H, Nakajima K, Kaji H, Khademhosseini A. Engineered nanomembranes for directing cellular organization toward flexible biodevices. Nano Lett 2013;13:3185–3192.
53. Ramón-Azcón J, Ahadian S, Estili M. X Liang, S. Ostrovidov, H. Kaji, H. Shiku, M. Ramalingam, Ken Nakajima, Y. Sakka, A. Khademhosseini, and T. Matsue. Dielectrophoretically aligned carbon nanotubes to control electrical and mechanical properties of hydrogels to fabricate contractile muscle myofibers. Adv Mater 2013;25:4028–4034.
54. Wang D, Russell TP, Nishi T, Nakajima K. Atomic force microscopy nanomechanics visualizes molecular diffusion and microstructure at an interface. ACS Macro Lett 2013;2:757–760.
55. Wang D, Liang X-B, Russell TP, Nakajima K. Visualization and quantification of the chemical and physical properties at a diffusion-induced interface using AFM nanomechanical mapping. Macromolecules 2014;47:3761–3765.
56. Bhowmick AK, editor. *Current Topics in Elastomers Research*. New York: CRC Press; 2008. p 579.
57. Utracki LA, Jamieson AM, editors. *Polymer Physics: From Suspensions to Nanocomposites and Beyond*. New Jersey: Wiley; 2010. p 129.
58. Mittal V, editor. *Characterization Techniques for Polymer Nanocomposite*. Weinheim: Wiley; 2013. p 185.
59. Nakajima K, Ito M, Wang D, Liu H, Nguyen HK, Liang X-B, Kumagai A, Fujinami S. Nano-palpation AFM and its quantitative mechanical property mappingmechanical. Microscopy 2014;63:193–208.
60. Bates FS, Fredrickson GH. Block copolymer thermodynamics: Theory and experiment. Annu Rev Phys Chem 1990;41:525–557.
61. Adhikari R, Michler GH. Influence of molecular architecture on morphology and micromechanical behavior of styrene/butadiene block copolymer systems. Prog Polym Sci 2004;29:949–986.
62. Mamodia M, Panday A, Gido SP, Lesser AJ. Effect of microdomain structure and process conditions on the mechanical behavior of cylindrical block copolymer systems. Macromolecules 2007;40:7320–7328.
63. McLean RS, Sauer BB. Tapping-mode AFM studies using phase detection for resolution of nanophases in segmented polyurethanes and other block copolymers. Macromolecules 1997;30:8314–8317.
64. Motomatsu M, Mizutani W, Tokumoto H. Microphase domains of poly (styrene-block-ethylene/butylene-block-styrene) triblock copolymers studied by atomic force microscopy. Polymer 1997;38:1779–1785.
65. Ganguly A, De Sarkar M, Bhowmick AK. Morphological mapping and analysis of poly [styrene-b-(ethylene-co-butylene)-b-styrene] and its clay nanocomposites by atomic force microscopy. J Polym Sci B 2007;45:52–66.
66. Zhou T, Zhang A, Zhao C, Liang H, Wu Z, Xia J. Molecular chain movements and transitions of SEBS above room temperature studied by moving-window two-dimensional correlation infrared spectroscopy. Macromolecules 2007;40:9009–9017.
67. Magonov SN, Sheiko SS, Deblieck RAC, Moller M. Atomic-force microscopy of gel-drawn ultrahigh-molecular-weight polyethylene. Macromolecules 1993;26:1380–1386.
68. Koike Y, Cakmak M. The influence of molten fraction on the uniaxial deformation behavior of polypropylene: Real time mechanoâ€ optical and atomic force microscopy observations. J Polym Sci B 2006;44:925–941.
69. Ferreiro V, Pennec Y, Séguéla R, Coulon G. Shear banding in polyamide 6 films as revealed by atomic force microscopy. Polymer 2000;41:1561–1569.
70. Park J-W, Ediger MD, Green MM. Chiral studies in amorphous solids: The effect of the polymeric glassy state on the racemization kinetics of bridged paddled binaphthyls. J Am Chem Soc 2001;123:49–56.
71. Rottler J, Robbins MO. Shear yielding of amorphous glassy solids: Effect of temperature and strain rate. Phys Rev E 2003; 68:011507 (10pp).
72. Williams ML, Landel RF, Ferry JD. The temperature dependence of relaxation mechanisms in amorphous polymers and other glass-forming liquids. J Am Chem Soc 1955;77:3701–3707.
73. Overney RM, Meyer E, Frommer J, Brodbeck D, Lüthi R, Howald L, Güntherrodt H-J, Fujihira M, Takano H, Gotoh Y. Friction measurements on phase-separated thin films with a modified atomic force microscope. Nature 1992;359:133–135.
74. Overney RM, Meyer E, Frommer J, Güentherodt H-J, Fujihira M, Takano H, Gotoh Y. Force microscopy study of friction and elastic compliance of phase-separated organic thin films. Langmuir 1994;10:1281–1286.
75. Kajiyama T, Tanaka K, Takahara A. Surface molecular motion of the monodisperse polystyrene films. Macromolecules 1997;30:280–285.
76. Maivald P, Butt HJ, Gould SAC, Prater CB, Drake B, Elings VB, Gurley JA, Hansma PK. Using force modulation to image surface elasticities with the atomic force microscope. Nanotechnology 1991;2:103–106.
77. Radmacher M, Tillmann RW, Gaub HE. Imaging viscoelasticity by force modulation with the atomic force microscope. Biophys J 1993;64:735–742.
78. Galuska AA, Poulter RR, McElrath KO. Force modulation afm of elastomer blends: Morphology, fillers and cross-linking. Surf Interface Anal 1997;25:418–429.

79. Yuya PA, Hurley DC, Turner JA. Contact-resonance atomic force microscopy for viscoelasticity. J Appl Phys 2008;104:074916 (7pp).
80. Killgore JP, Yablon DG, Tsou AH, Gannepalli A, Yuya PA, Turner JA, Proksch R, Hurley DC. Viscoelastic property mapping with contact resonance force microscopy. Langmuir 2011;27:13983–13987.
81. Yablon DG, Gannepalli A, Proksch R, Killgore J, Hurley DC, Grabowski J, Tsou AH. Quantitative viscoelastic mapping of polyolefin blends with contact resonance atomic force microscopy. Macromolecules 2012;45:4363–4370.
82. Cleveland JP, Anczykowski B, Schmid AE, Elings VB. Energy dissipation in tapping-mode atomic force microscopy. Appl Phys Lett 1998;72:2613–2615.
83. Garcia R, Gómez CJ, Martinez NF, Patil S, Dietz C, Magerle R. Identification of nanoscale dissipation processes by dynamic atomic force microscopy. Phys Rev Lett 2006;97:016103 (4pp).
84. Wang D, Liu Y-H, Nishi T, Nakajima K. Length scale of mechanical heterogeneity in a glassy polymer determined by atomic force microscopy. Appl Phys Lett 2012;100:251905 (4pp).
85. Proksch R, Yablon DG. Loss tangent imaging: Theory and simulations of repulsive-mode tapping atomic force microscopy. Appl Phys Lett 2012;100:073106 (3pp).
86. Yablon DG, Grabowski J, Chakraborty I. Measuring the loss tangent of polymer materials with atomic force microscopy based methods. Meas Sci Technol 2014;25:055402 (7pp).
87. Igarashi T, Fujinami S, Nishi T, Asao N, Nakajima K. Nanorheological mapping of rubbers by atomic force microscopy. Macromolecules 2013;46:1916–1922.
88. Young TJ, Monclus MA, Burnett TL, Broughton WR, Ogin SL, Smith PA. The use of the PeakForce™ quantitative nanomechanical mapping AFM-based method for high-resolution Young's modulus measurement of polymers. Meas Sci Technol 2011;22:125703 (6pp).
89. Johnson KL. *Contact mechanics*, chapter 4. Cambridge: Cambridge University Press; 1985. p 84.
90. Nguyen HK, Ito M, Fujinami S, Nakajima K. Viscoelasticity of inhomogeneous polymers characterized by loss tangent measurements using atomic force microscopy. Macromolecules 2014;47:7971–7977.

18

POLYMER MORPHOLOGY AND DEFORMATION BEHAVIOR

Masanori Hara
Department of Chemical and Biochemical Engineering, Rutgers University, Piscataway, NJ, USA

18.1 INTRODUCTION

The study of the mechanical deformation of polymers is important for practical applications as seen in standard stress–strain experiments. More importantly, deformation mode is influenced by morphology of polymers; and, by changing the morphology, deformation behavior and mechanical properties of polymers can be modified. This chapter deals with deformation behavior of polymers and its relationship with *morphology*. This is the only chapter in this book dealing with deformation of polymers, although many chapters deal with morphology and its characterization. As the deformation behavior of all polymers cannot be covered in this chapter, materials are restricted to the following: amorphous polymers, semicrystalline polymers, and block copolymers (relevant to the deformation). Also, literature is chosen for the work that deals with mechanical properties and their relationship with deformation behavior and morphology. Simple deformation, coupled with various morphological characterization techniques, is important, but such work is not included in this chapter.

Deformation behavior of amorphous polymers has been studied for years, and explained in terms of structure of polymers. Especially, great progress has been made by using the concept of "*network (strand) density*" to understand the deformation behavior (such as crazing and shear deformation) [1, 2]. When the concept of a critical thickness of the polymer layer, below which a sample behaves in a ductile manner even for normally brittle polymers like polystyrene (PS), has been added to the network density concept, more comprehensive understanding has become possible [3–8]; and it provides a great opportunity for developing ductility of otherwise brittle polymeric materials. This subject is explained in detail in Section 18.2.

Deformation behavior of semicrystalline polymers is more complex [9], because the morphology of such polymers are inherently complex and thus influences the deformation behavior [10]. For example, the degree of crystallinity changes the amount of amorphous region, which tends to deform easily. Furthermore, higher order structure, such as spherulites (e.g., size), can influence the deformation mode. The control of morphology is not straightforward for semicrystalline polymers, unlike for amorphous polymers, making the deformation behavior of semicrystalline polymers more difficult to understand. However, a general trend has been noted and summarized by various researchers [10, 11]. Also, recent studies explain large-strain deformation of semicrystalline polymers in terms of *entanglement density* of the amorphous phase [12]. This approach is consistent with the successful approach used to understand the deformation behavior of amorphous polymers in terms of strand density. This is explained in detail in Section 18.3.

Finally, as an example showing significant influence of morphology on deformation behavior, block copolymers are chosen. Because of the microphase separation occurring between different blocks, block copolymers form various structures (*morphologies*), which in turn influence their deformation behavior. This is described in Section 18.4, in which only a small number of examples have been chosen to demonstrate the morphology–deformation relationships, because a vast amount of literature and work cannot be covered. Morphology and characterization of block copolymers are also described in several chapters in this book.

Polymer Morphology: Principles, Characterization, and Processing, First Edition. Edited by Qipeng Guo.

18.2 DEFORMATION BEHAVIOR OF AMORPHOUS POLYMERS

18.2.1 Deformation Behavior of Thin Films

Although polymers are finding increasing use in engineering applications because of their many advantages, such as low density, low cost, ease of fabrication, and unique properties that are not found in other materials, their use is limited by the tendency of many of these polymers to fail in a brittle fashion [10]. An important consideration for the use of polymers in engineering applications is an understanding of the way they respond to mechanical loading, including small- and large-strain properties, deformation mechanisms, and ultimate failure modes.

The deformation behavior of amorphous polymers has been studied extensively, partly because the structure is rather simple as compared with semicrystalline polymers; thus, the relationship between structure and properties can be established with relative ease. It is well known that two major micromechanisms are involved in the deformation and subsequent fracture of glassy polymers [1, 2, 13] (see Figs. 18.1 and 18.2). These are *crazing* and *shear yielding*, and both involve localized plastic deformation and some energy is dissipated during the deformation. In a craze, polymer chains are stretched along the stress direction and form plastically deformed fibrils that are interspersed with voids. Thus, crazing is a cavitational process, leading to an increase in volume. S*hear yielding* occurs essentially at constant volume (no voids) and leads to a permanent change in shape, where polymer chains are stretched due to shear component of the stress. Shear yielding can take place either in highly localized shear bands or diffuse shear deformation zones [4]. Brittle polymers, such as PS, usually deform by crazing, while tougher polymers, such as polycarbonate (PC), usually deform by shear yielding. Although crazes are load bearing, as they consist of many small-oriented fibrils, with diameter in the range of 5–30 nm, cracks can be formed within crazes by breakdown of the fibril structure, leading to decreased fracture resistance [14]. Therefore, one way to make polymers more fracture resistant is to control crack nucleation in crazes by suppressing craze formation in favor of shear deformation [1].

There has been considerable progress over recent years in the study of molecular mechanisms involved in crazing and shear yielding [1, 2]. When sample preparation and test conditions (temperature, stress state, strain, strain rate, and thermal history) are chosen to be the same, the effect of molecular variables can be determined. It is now well established that (network) *strand density* plays an important role in determining the deformation mechanism and in affecting the craze–shear deformation transition. Polymers are

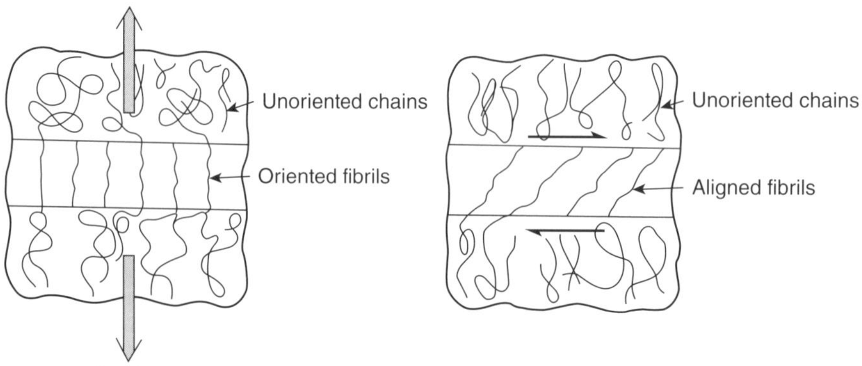

Figure 18.1 Deformation modes in amorphous polymers: crazing (normal yielding) and shear yielding. Hertzberg [13]. Reproduced with permission of John Wiley and Sons.

Figure 18.2 A craze formed under tension and shear bands formed under compression for PS. Kinloch and Young [10]. Reproduced with permission of Elsevier.

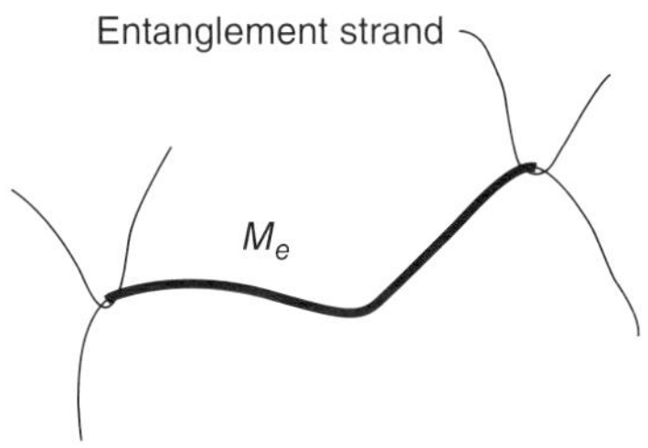

Figure 18.3 Entanglement strand in amorphous polymers. Ma et al. [15]. Reproduced with permission of American Chemical Society

considered as a network made of strands and entanglements (see Fig. 18.3); and entangled network strand density, ν_e, can be calculated as

$$\nu_e = \rho \frac{N_A}{M_e} \tag{18.1}$$

where ρ is the polymer density, M_e is the entanglement molecular weight, and N_A is Avogadro's number. Thus, the network strand density of linear polymers, which is simply the (physical) entanglement density calculated from the entanglement molecular weight, is considered to be a material constant, a characteristic value for any specific polymer.

The value of ν_e can determine whether the polymer deforms by crazing or shear yielding. The larger the ν_e value, the larger the crazing stress, but shear yield stress is little changed with ν_e; thus, by increasing the ν_e, yield stress becomes smaller than the crazing stress, and this makes shear deformation easier to occur. Thus, polymers with low ν_e, like PS and poly(tert-butylstyrene)(PTBS) deform by crazing, whereas polymers with higher ν_e, like PC and poly(phenylene oxide)(PPO), deform by shear yielding. In polymers of intermediate values of ν_e, such as poly(styrene-*co*-acrylonitrile)(SAN), both crazing and shear deformation are simultaneously observed in thin films [16]. Because crazing is a cavitational process and can be a precursor of crack formation, but shear yielding is not, PC and PPO are more fracture resistant than PS and PTBS.

The (*theoretical*) maximum extension ratio of a strand can be determined as (see Fig. 18.4)

$$\lambda_{max} = \frac{l_e}{d} \tag{18.2}$$

where l_e is the chain contour length between entanglements (i.e., maximum length of a stretched strand) calculated as

$$l_e = l_0 \frac{M_e}{M_0} \tag{18.3}$$

where l_0 is the average projected length of a stiff unit along the chain and M_0 is its molecular weight. d is the mesh size, that is, the root-mean-square end-to-end distance of the chain, determined by

$$d = k(M_e)^{1/2} \tag{18.4}$$

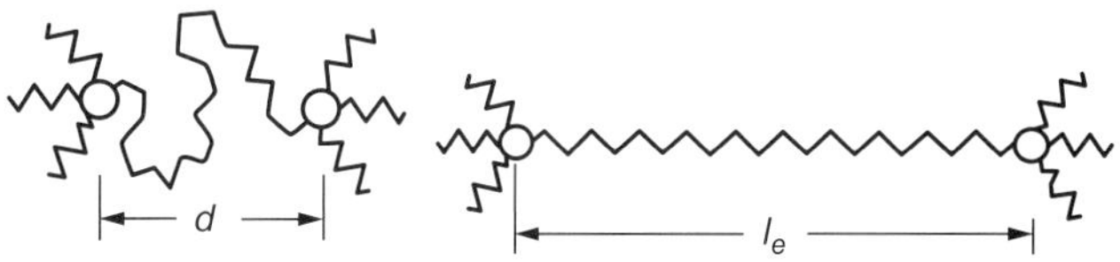

Figure 18.4 Polymer chain (strand) before and after extension. Donald and Kramer [17]. Reproduced with permission of Elsevier.

where k is a constant determined by small-angle neutron scattering (glassy solid) or light scattering experiment (dilute solution) in a Θ state. It has been observed that the *measured* craze fibril extension ratio, λ, follows the relation, $\lambda \sim 0.8\lambda_{max}$, for glassy polymers; and the *measured* extension ratio of microscopic shear deformation zones is $\lambda \sim 0.6\lambda_{max}$. Thus, large-scale extension of polymer chains, which is expected from theoretical calculation, occurs within the localized regions, although overall extension of the sample is very small in brittle polymers.

The idea described above for glassy amorphous homopolymers can be extended to include miscible amorphous polymer blends, such as PS/PPO. Furthermore, a low degree of covalent cross-links can be considered as equivalent to entanglements for controlling the deformation mode. The strand density of cross-linked polymers is defined as the sum of the entanglement density and the covalent cross-link density [18] as

$$\nu = \nu_e + \nu_x \tag{18.5}$$

where ν_x is the cross-linked-strand density. Figure 18.5 shows a systematic comparison of such polymers as a function of ν. As the figure indicates [2, 18], deformation mechanism of amorphous glassy polymers changes from crazing only, to crazing plus shear yielding, and to shear yielding only, as the strand density of the polymer network is increased. The transitions from crazing to crazing plus shear and from crazing plus shear to shear only take place at almost identical values of ν for all polymers studied. The craze–shear transition may be predicted by a craze growth model [2]. Craze suppression, by either increasing ν_x in the cross-linked polymer or ν_e in the uncross-linked ones, is due to the extra energy required to break more main-chain bonds to form these craze fibril surfaces [18].

We can further extend the concept of *strand density* to include the *ionic cross-linking effects* observed for ionomers [19, 20]. Ionomers, usually containing ionic groups of up to 10–15 mol% along backbone chains, form ionic aggregates. These aggregates can work as ionic cross-links [21–25] (see Fig. 18.6). Therefore, for linear ionomers, ionic cross-linking effect should be considered in addition to the entanglements of the base polymers. Like a covalent cross-linked polymer, ionic cross-link density can be calculated as

$$\nu_i = \rho \frac{N_A}{M_i} = f\rho \frac{N_A}{M_0} \tag{18.6}$$

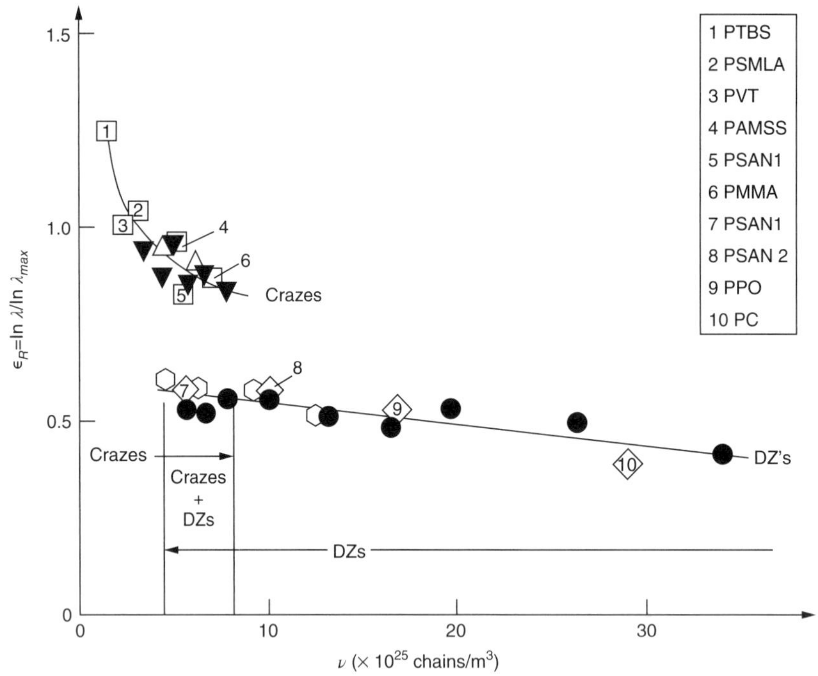

Figure 18.5 True strain ratio in craze and shear deformation zones as a function of network strand density. The open squares and open diamonds represent homopolymers and copolymers, open triangles and hexagons uncross-linked blends of PS and PPO, and filled triangles and circles cross-linked PS. Henkee and Kramer [18]. Reproduced with permission of John Wiley and Sons.

where M_i is the average molecular weight between ionic groups and f is the ion content (fraction). The total effective strand density can be written as

$$\nu_{eff} = \nu_e + k\nu_i \tag{18.7}$$

where ν_{eff} is the *effective* strand density and k is a coefficient reflecting the effectiveness of ionic cross-links as compared with covalent cross-links or physical entanglements (at temperatures far below T_g). As the strength of ionic cross-links is smaller than that of covalent cross-links, the range of k is $0 < k < 1$. The k value of monovalent (e.g., Na) salt ionomers is low, while the k value of divalent (e.g., Ca) salt ionomers of the same ion content is higher, leading to higher ν_{eff} values. As ν_i increases with f, as seen from Equation 18.6, ν_{eff} value increases with increasing ion content f, and this increase is expected to produce changes in deformation modes of ionomers.

While crazing is the only observed deformation mechanism of PS, introduction of ionic groups (and ionic cross-links) into PS can alter the deformation mechanisms induced in strained thin films and lead to the development of smaller, more stable crazes; and, at ion contents of about 6 mol% or above, shear deformation develops in addition to crazing [26, 27]. A similar trend has been observed for other ionomers, such as poly(methyl methacrylate) (PMMA) ionomers [15]. Although covalent cross-linking can also alter the deformation mechanism from crazing to shear yielding [18], as described above, this is accompanied by a limitation in processability. In contrast, ionomers, having thermolabile ionic cross-linking,

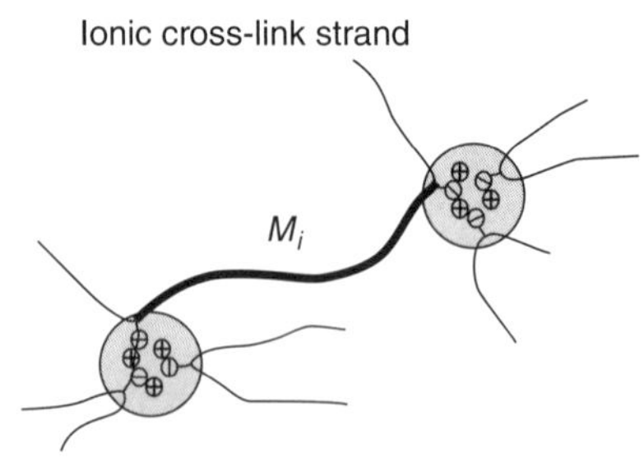

Figure 18.6 Ionic cross-link strand formed in ionomer. Ma et al. [15]. Reproduced with permission of American Chemical Society.

maintain good processability, and therefore can be used as thermoplastic polymers with enhanced mechanical properties. As the structure of ionic aggregates and the strength of ionic bonds can be altered through changes in the *ion content* or the *nature of the counterion* [19], there are more varieties in ionically cross-linked polymers than covalently cross-linked polymers.

18.2.2 Deformation Behavior of Bulk Polymers

Although the results described above have been observed for thin films under tension and large-strain plastic deformation is usually observed only for localized regions – crazes or shear deformation zones – Meijer has demonstrated that *macroscopic plastic deformation* can be indeed observed for these amorphous polymers under tension, as far as the thickness of the deformed polymer is less than a critical value [3–8]. Thin

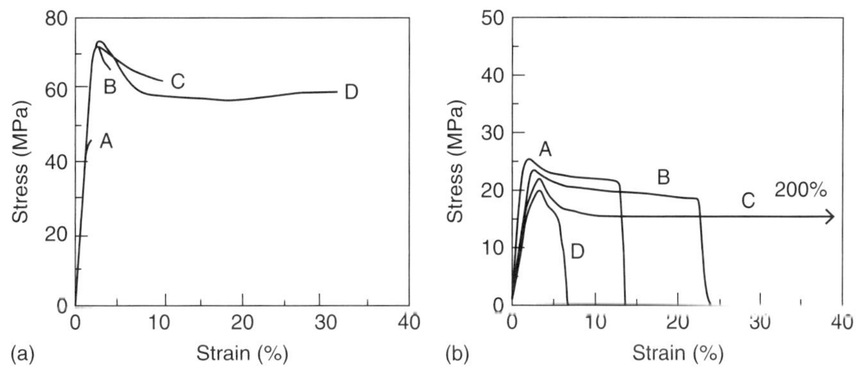

Figure 18.7 Stress–strain behavior of the PS layer in (a) PS/PE multilayers (PS layer thickness: A, 35 μm; B, 4.5 μm; C, 1.5 μm; D, 0.8 μm) and (b) PS containing nonadhering rubber particles (rubber content: A, 40 wt%; B, 50 wt%; C, 60 wt%). Significant ductility is developed for PS layers with reduced thickness. van der Sanden et al. [7]. Reproduced with permission of Elsevier.

layers were formed either by (i) coextrusion of an amorphous polymer (i.e., a polymer of interest) and a poor-bonding polymer – for example, PS/polyethylene (PE) (Fig. 18.7a) or (ii) addition of nonadhering rubber particles into the amorphous matrix polymer, where the average thickness of the polymer layer between rubber particles can be decreased by increasing the rubber content (Fig. 18.7b). Large extension (strain at break), for example, over 200% for PS! (Fig. 18.7b), can be observed when the thickness of the film is less than the critical value (0.05 μm for PS). Figure 18.8 clearly shows significant increase in ductility (strain at break) as the thickness is reduced to the critical thickness. The behavior is explained in terms of network strand density developed for localized deformation. Like localized deformation discussed by Kramer, the idea can be applied to amorphous glassy polymers, polymer blends, and cross-linked polymers. For example, polymer blends made of PS and PPO show the strain at break of 75% for PS/PPO (40/60) blend when the blend layer thickness is less than 0.3 μm (see Fig. 18.9). The same trend is observed for cross-linked thermosetting polymers. From the volume measurements, it is shown that the deformation mode changes from crazing for thicker layers to shear deformation for thinner layers (below the critical thickness). Thus, large extension can be achieved not only for a localized (inhomogeneous) deformation but also for macroscopic (homogeneous) deformation. In fact, observed macroscopic extension ratio, λ_{macr}, of shear deformation is related to the theoretical maximum extension ratio, λ_{max} (defined by Eq. 18.2) by $\lambda_{macr} = 0.6\ \lambda_{max}$ (see Fig. 18.10). This relation is the same as that obtained for localized shear deformation zones [18]. The large extension of polymer chains observed in a localized zone is now observed macroscopically for a whole sample with the same extension ratio.

By using the energy criteria for a brittle-to-ductile transition of amorphous polymers, Meijer derived the quantitative relation between the critical thickness and ν, showing that the larger the ν value, the larger the critical thickness. Thus, polymers that tend to craze due to low ν value (e.g., PS) need thinner layers to develop large macroscopic extension than polymers that tend to shear due to high ν value (e.g., PC).

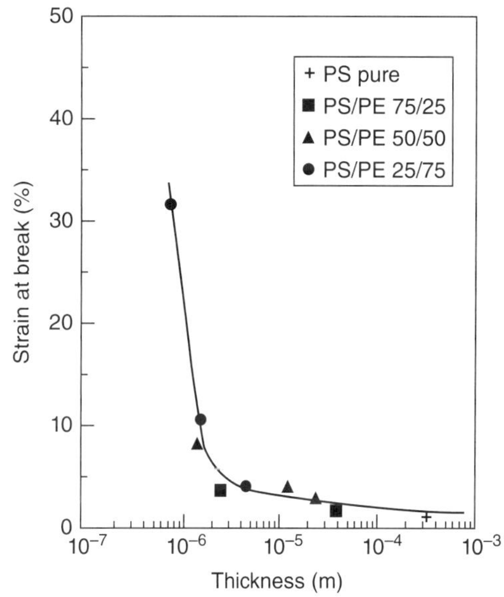

Figure 18.8 Thickness dependence of strain at break for the PS layer in PS/PE multilayers. Below the critical value of thickness, fracture strains increases significantly. van der Sanden et al. [7]. Reproduced with permission of Elsevier.

Many interesting results can be found in their treatments and readers are encouraged to read their original papers for more details [3–8]. The effect of reduced layer thickness and the resulting ductile behavior is later identified for block copolymers, too, in which thin PS layers in lamellar morphology deform plastically when the thickness of the PS layer is smaller than the critical value (named as "thin-layer yielding") [28](see Section 18.4).

18.3 DEFORMATION BEHAVIOR OF SEMICRYSTALLINE POLYMERS

Crystalline polymers show various morphologies, ranging from single crystals (100% crystalline) to semicrystalline polymers (made of a crystalline phase and an amorphous

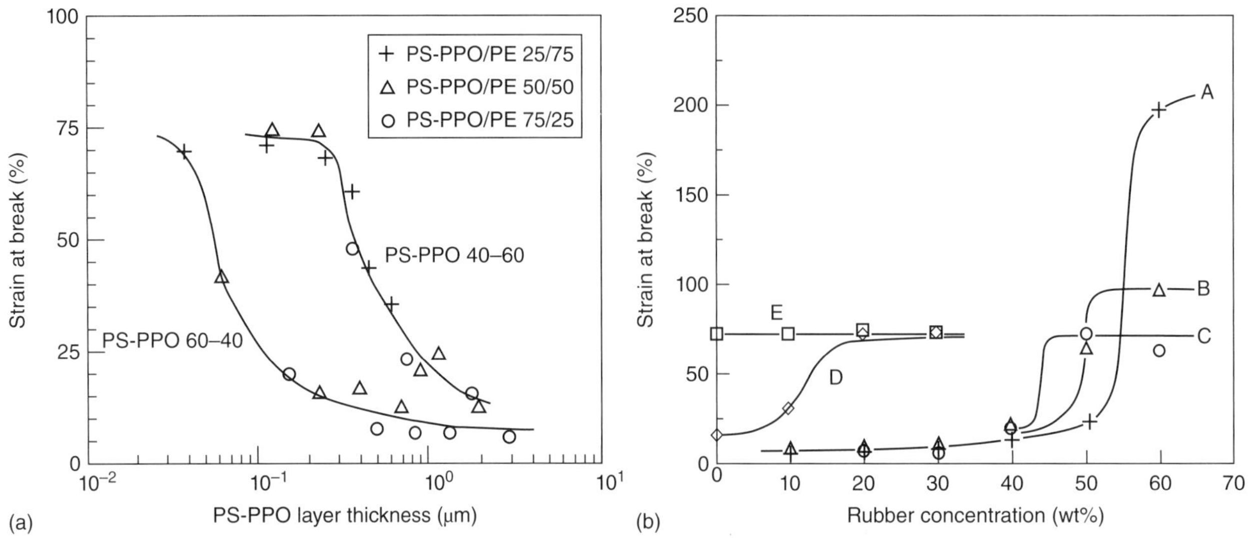

Figure 18.9 Significant enhancement in strain at break is seen for layers that are thinner than the critical value for PS/PPO blends: (a) Strain at break versus layer thickness for PS–PPO blend/PE multilayers; (b) strain at break versus rubber content for PS–PPO blends that contain nonadhering rubber particles (PS content in PS–PPO: A, 100 wt%, B, 80 wt%; C, 60 wt%; D, 40 wt%; E, 20 wt%). PPE (poly(dimethyl phenylene ether)) is used for PPO in the original paper. van der Sanden et al. [8]. Reproduced with permission of Elsevier.

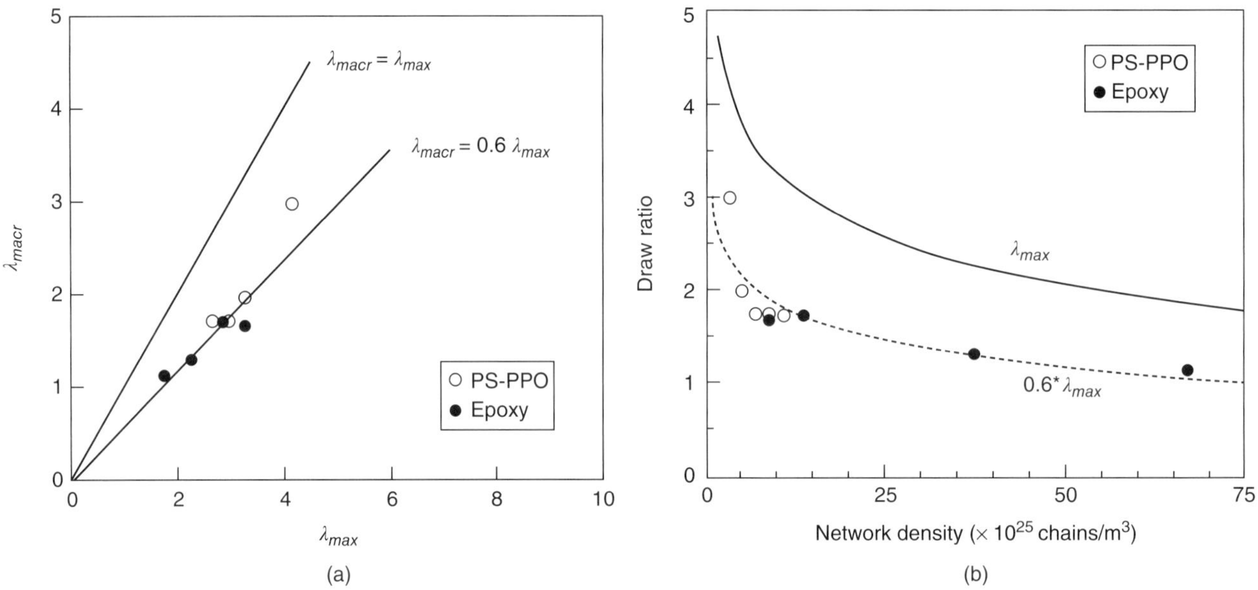

Figure 18.10 (a) The relation $\lambda_{macr} = 0.6\lambda_{max}$ is established for PS–PPO blends and cross-linked epoxies (both contain nonadhering rubber particles to reduce the polymer layer thickness); (b) draw ratio versus network strand density. van der Sanden and Meijer [5]. Reproduced with permission of Elsevier.

phase), where the amorphous phase is either rubbery or glassy (depending on whether the temperature is above T_g or below T_g). Also, semicrystalline polymers can be either in an unoriented form or an oriented (fiber) form, with contrasting behavior. Semicrystalline polymers are often ductile, unlike many amorphous polymers that are brittle, and deformation is nonlinear showing large-scale yielding [10].

Typical stress–strain curves for crystalline polymers are shown in Figure 18.11 [10]. Polymer single crystals, for example, polydiacetylene, develop very high modulus and strength with small deformation. Semicrystalline polymer fibers show high modulus due to a high degree of molecular orientation, with little plastic deformation. Unoriented semicrystalline polymers behave differently, according to the temperature. At a temperature below T_g, the amorphous phase does not flow, showing higher yield stress with lower strains. At a temperature above T_g, the amorphous phase is ductile, yield, and flow [10]. In this section, single crystals and oriented fibers are excluded from consideration.

As described above, semicrystalline polymers are more complex, including the two-phase structure, with the crystalline phase having various degrees of crystallinity and

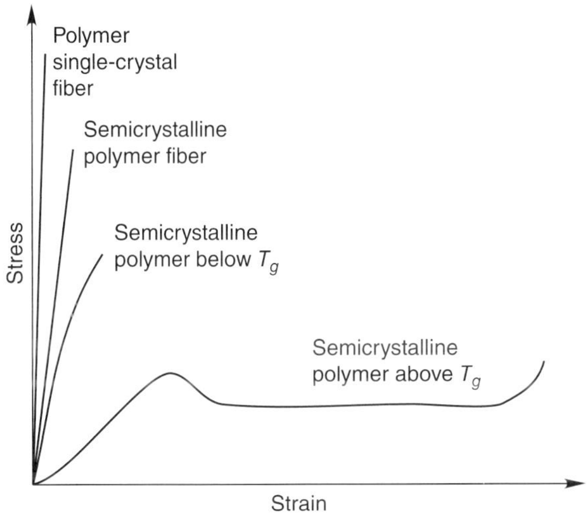

Figure 18.11 Schematic stress–strain curves in different morphological forms of crystalline polymers. Kinloch and Young [10]. Reproduced with permission of Elsevier.

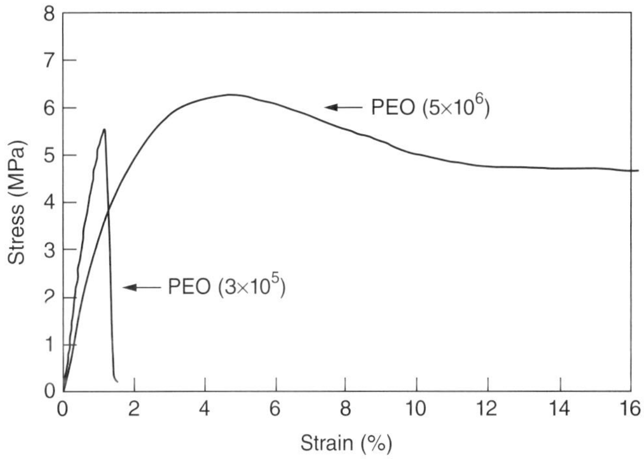

Figure 18.12 Stress–strain curves for low MW PEO showing small ductility and high MW PEO showing fracture strain of well over 100%. Tsou et al. [32]. Reproduced with permission of John Wiley and Sons.

various sizes of spherulites (or lamellar thickness). More recently, an explanation of deformation behavior has been given (or reemphasized) in terms of an *entanglement network* existing in the amorphous phase of the semicrystalline polymer. The idea has been successfully used for ultrahigh-molecular-weight polymers [29–31]. Meijer et al. used the same concept to explain strain hardening in semicrystalline polymers. By considering the huge success in explaining the deformation behavior of amorphous polymers in terms of network density, as explained in Section 18.2, such an explanation seems attractive, not to mention the simplicity and the systematic approach achieved in such an approach.

18.3.1 Deformation of Unoriented Semicrystalline Polymers

At room temperature (i.e., above T_g for many semicrystalline polymers), semicrystalline polymers are tough and show large plastic deformation before fracture. A typical morphology (at the highest level) found in semicrystalline polymers is spherulite, which influences the deformation behavior of polymers.

The effect of spherulite on the deformation behavior of semicrystalline polymers can be seen for a semicrystalline polymer, poly(ethylene oxide)(PEO), with different molecular weights. One is small (300,000) and the other is large (5,000,000) (see Fig. 18.12). Large ductility is seen for high-molecular-weight PEO, while low values of the ductility and energy to fracture is seen for low-molecular-weight PEO. One possible cause is the presence of large spherulites in the low-molecular-weight PEO. In semicrystalline polymers, a high degree of extensibility is generally associated with small spherulites [10, 33]. An optical view of the spherulitic structure in the 300,000 PEO shows that the average spherulite size is large, 300–350 mm [32]; and the boundaries between the spherulites contain many imperfections. Such regions provide relatively easy paths for crack propagation [10]. The SEM image shows a crack that has propagated along the weak spherulite boundary regions. The detrimental effect of large spherulites and weak boundaries, as noted here for low-molecular-weight PEO, is consistent with findings that, in another crystalline polymer, isotactic polypropylene (PP), there is a transition from ductile to brittle behavior [33] as the spherulite size is increased.

Another reason for the low values of strain to fracture for the 300,000 PEO is a reduced number of *tie molecules* between adjacent spherulites and between the crystalline lamellae that are located in the radial arms of the spherulites. Tie molecules increase in number with increasing molecular weight and have a significant effect on resistance to fracture [10, 34, 35]. Hence, for high-molecular-weight grades of PEO, not only is the average spherulite smaller but there are many more stress-carrying tie molecules. These interspherulitic and intraspherulitic links provide the semicrystalline polymer with a much greater resistance to crack development and fracture. As a result, the high-molecular-weight PEO develops large values of strain before fracture and have high values of fracture energy or toughness. Although the word "tie molecules" is often used to explain the mechanical properties of semicrystalline polymers, it is of interest to consider these as "effective entanglements," [11], because the concept has been successfully used to explain the behavior of amorphous polymers.

As seen above, the problem in dealing with semicrystalline polymers is the difficulty in controlling the morphology (varying one when others are fixed). One promising approach has been reported more recently, as explained next.

18.3.2 Strain Hardening and Network Density

Although understanding of the deformation behavior of amorphous polymers is well established in terms of molecular structure, as described in Section 18.2, the deformation behavior of semicrystalline polymers, in particular, large-strain behavior,

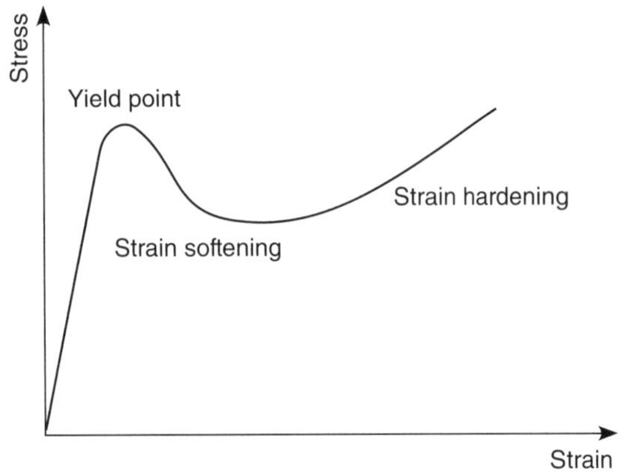

Figure 18.13 A stress–strain behavior of semicrystalline polymers under compression.

such as strain softening and strain hardening (see Fig. 18.13), is still not well understood. Although many studies are focused on the crystalline structures, such as degree of crystallinity and spherulite size, the role of amorphous phase and its chain structure are less emphasized.

More recently, Meijer and coworkers have presented systematic experimental results and explanations about large-deformation behavior by focusing on the amorphous phase of semicrystalline polymers [12]. They reviewed the various reported results, stating that most plausible explanation for semicrystalline polymers is the concept of the rubber-elasticity theory, that is, strain hardening being proportional to *entanglement (strand) density*. Similar to amorphous polymers, only the entangled polymer network, which depends on processing conditions and chain length, is the origin of the strain-hardening behavior.

The relation between the microstructure of semicrystalline polymers and their intrinsic deformation behavior is studied [12]. For this purpose, well-characterized poly(ethylene terephthalate)(PET), PE, and PP samples are used. Compression tests are used to study large-deformation behavior, because some samples may deform in a brittle fashion under the normal stress–strain testing condition. Two processing methods are used for making samples: melt crystallization and cold crystallization (i.e., quenching followed by room-temperature annealing, because T_g is lower than the room temperature). Melt crystallization provides very slow cooling, causing disentanglement during lamellar folding, whereas cold crystallization leads to constant *entanglement density* (same as that in amorphous melt). It has been demonstrated that the *entangled polymer network* is the origin of the strain hardening of semicrystalline polymers, as explained below.

First, it is observed that the strain hardening is same for amorphous PET and cold crystallized PET, although the crystalline phase is developed by cold crystallization. For example, the slope at large strain is used to define the strain-hardening modulus; and the lines are parallel and thus strain-hardening modulus values are the same for PET with 0%, 22%, and 29% crystallinity (Fig. 18.14a). Also, true stress–strain curves of annealed PE show that strain hardening is not affected by the degree of crystallinity.

Second, by contrast, melt crystallization leads to lower values of strain-hardening modulus. This observation proves that entanglements affect the strain-hardening behavior, because melt crystallization causes disentanglement during crystallization from melt (i.e., lamellar folding). Also, a lower molecular weight grade polymer shows lower strain hardening (see Fig. 18.14b for PE). The dependency on molecular weight also indicates that the *entangled polymer network* is the origin of strain hardening in semicrystalline polymers. The longer chains of a higher molecular weight grade result in a larger number of entanglements, which in turn increase the stress contribution of the network at large deformations. The similarity in the results observed on all polymers tested (PET, PE, and PP) supports the conclusion that the crystalline phase does not contribute to strain hardening, which is mainly controlled by the chain *entanglements*.

The importance of entanglements in the deformation and fracture of semicrystalline polymers is widely acknowledged, although by implicitly using such a concept as tie molecules [36]. The importance of entanglements in large-strain behavior is also well known for processing of ultrahigh-molecular-weight polymers [29–31]. Smith and coworkers were able to draw solution-crystallized samples (e.g., PE) up to very high strains; in contrast, melt-crystallized samples showed a much larger strain hardening and subsequent fracture of the sample. This difference in large-strain behavior is explained in terms of the difference in *entanglement density*. For the melt-crystallized samples, the entangled network causes fracture of chains. For the solution-crystallized samples, a large degree of chain disentanglement occurs, leading to polymer coil stretching [12]. Thus, the concept of entanglements is very important in understanding deformation behavior of semicrystalline polymers. Trapped chain entanglements in the amorphous phase and physical cross-links by crystallites have been used to explain the large-strain behavior of semicrystalline polymers [12].

18.4 DEFORMATION BEHAVIOR OF BLOCK COPOLYMERS

Block copolymers are made up of more than two different polymers connected by covalent bonds. Due to thermodynamic repulsion between the component polymers, microphase separation occurs and various types of structures (spheres, cylinders, lamellae, and gyroids) are formed [37–39] (see Fig. 18.15), which have been studied extensively and are described in several chapters in this book. Here, deformation of block copolymers has been described in terms of microstructure/morphology of block copolymers. As there are many reports concerning simple deformation, here the emphasis is placed on the mechanical deformation (mechanical properties) and their relationships with microstructures [41].

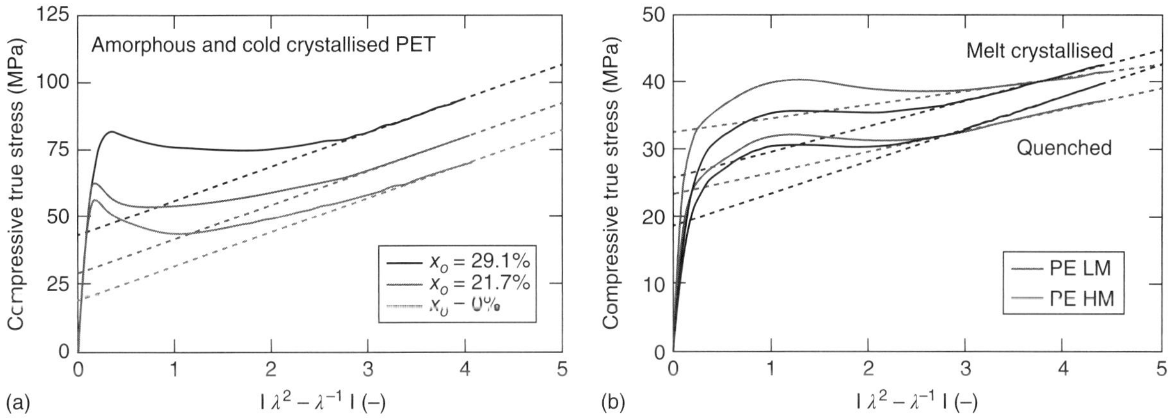

Figure 18.14 Effects on the strain-hardening behavior of semicrystalline polymers: (a) effect of the crystallinity for PET; (b) effect of the molecular weight and processing condition for PE (the slope of the dashed line is defined as strain-hardening modulus). Schrauwen et al. [12]. Reproduced with permission of American Chemical Society.

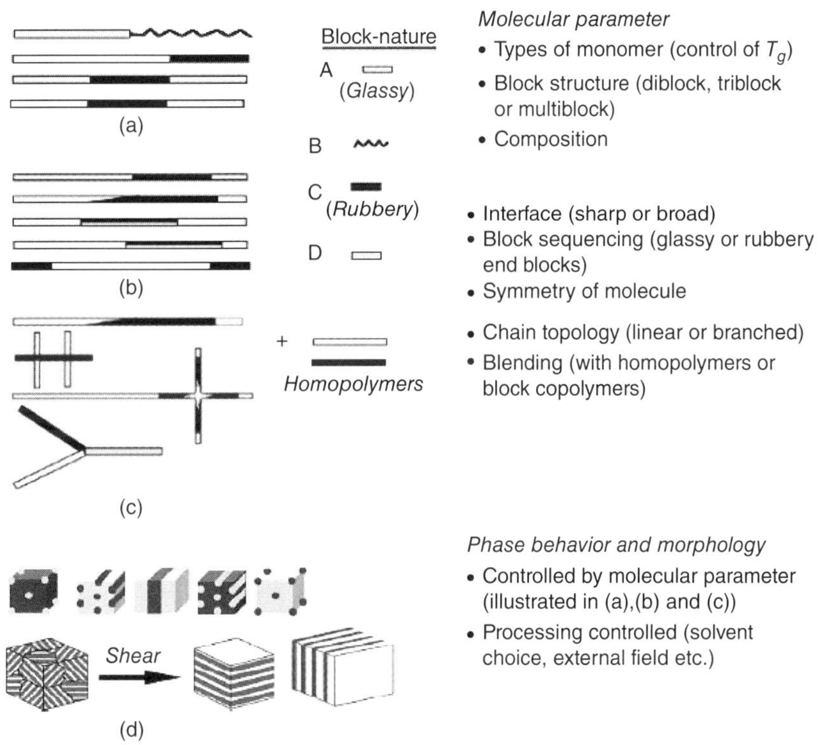

Figure 18.15 Molecular and morphological factors affecting the deformation behavior of block copolymers. Adhikari and Michler [40]. Reproduced with permission of Elsevier.

The main factors influencing the deformation behavior may be molecular weight, composition, *morphology*, chain architecture, orientation of the microdomains, and thermal history of the block copolymer. Two examples are chosen for demonstrating the effect of morphology on the deformation behavior of block copolymers.

18.4.1 Block Copolymers Based on S and B

Block copolymers, S–B and S–B–S, are chosen as a first example, where S is styrene and B is butadiene. The T_g values are 100 °C for PS and −100 °C for PB; thus, the former is glassy and the latter is rubbery at room temperature. The S–B (glassy/rubbery) diblock copolymer shows brittle behavior due to localized craze formation [42, 43]. Because of the lack of chain mobility, a hard PS phase works as "physical cross-links" for a rubbery PB phase. In the S–B block copolymers, physical cross-links are formed only at one PB chain end; thus, insufficient cross-links make the material weak against deformation.

By contrast, S–B–S triblock copolymers show different behaviors according to the content of the glassy phase. When

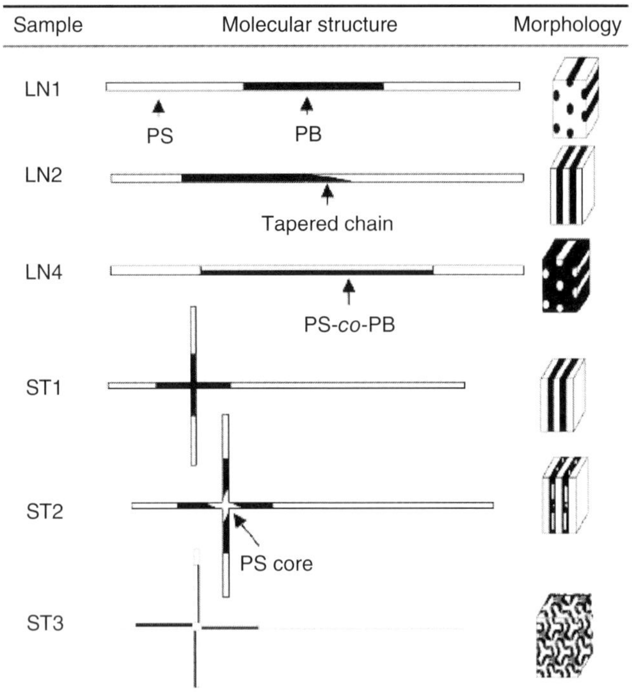

Figure 18.16 Molecular architecture of the block copolymers; the oblique line between the blocks represents a tapered transition. The volume fraction of PS is ~0.7. Adhikari [40]. Reproduced with permission of Elsevier.

a PS phase is a dispersed phase (at low PS content), rubbery deformation is seen. Strain mainly arises from the soft matrix PB phase. Rubbery chains orient along the stress direction and the glassy phase rotates its long axis along the stress direction. When the PS content is increased to be comparable to the PB content, a lamellar structure is formed. This alternating structure leads to the formation of necking under tension, leading to large plastic deformation of the glassy PS phase. It is expected that the rubbery block possesses sufficient *physical* "*cross-links*" on both sides of the rubbery molecule (due to the glassy nature of outer blocks) in the S–B–S triblock copolymers. When the PS content is further increased, reversal of the structure occurs (a PB phase in the PS matrix phase). The block copolymers deform in a brittle fashion due to crazing in the PS phase.

Michler also studied the effect of morphology on the deformation and mechanical behavior of S–B based block copolymers with different *molecular architectures*. By fixing molecular weight (ca. 100,000) and composition (ca. 70 vol% of PS), but changing the molecular architecture (S–B–S triblock, S–(B/S)–S triblock, star block, and star/tapered block), they observed morphology changes from cylinders, to lamellae, and to gyroids (see Fig. 18.16). These different morphologies lead to different deformation and mechanical behavior. Figure 17a and b shows the stress–strain behavior of linear block copolymers: LN1 (symmetric S–B–S), LN2 (asymmetric S–B–S with a tapered midblock), and LN4 (symmetric S–(B/S)–S block with a random B–S midblock). LN1 forms PB cylinders in the PS matrix, and shows yielding, strain softening, and fracture at rather small strain (20%). Deformation mode observed by TEM is a craze-like deformation zone due to cavitation of the matrix PS phase. LN2 forms lamellar morphology, and shows yielding at lower stress than LN1, with little strain softening, followed by extensive strain hardening (strain at break of ca. 400%). TEM observation indicates that alternating layers of PS and PB domains become thinner after large plastic deformation. The large ductility arises from "thin layer yielding" [44], as also described in Section 18.2 and noted by other researchers [5, 7, 8, 45]. Finally, LN4 shows behavior typically seen for elastomers. Tapping mode scanning force microscopy shows destruction of the randomly oriented PS phase and fragmentation of the glassy PS cylinders at high deformation (strain at break is ca. 550%). Figure 18.17c compares the stress–strain behavior of star block copolymers: both star block copolymers (ST2

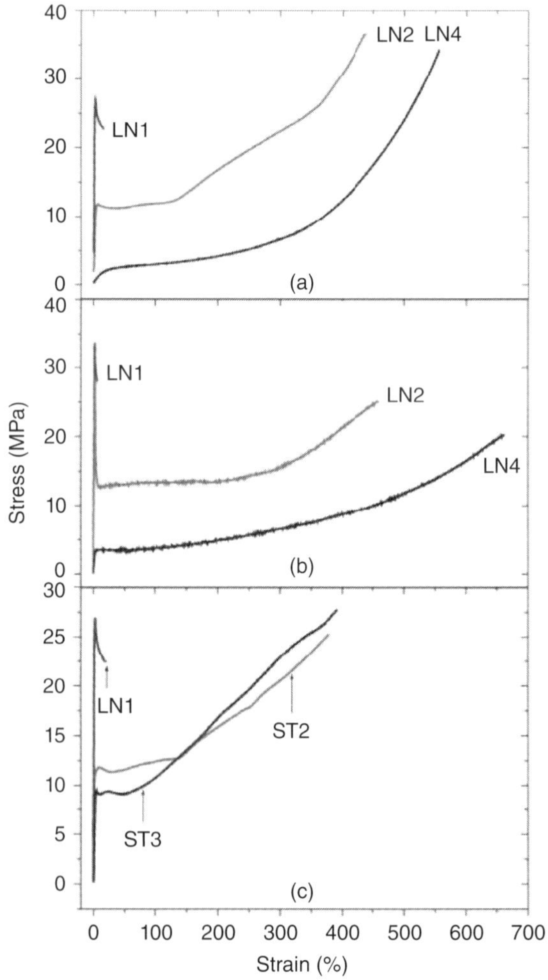

Figure 18.17 Stress–strain curves of block copolymers shown in Figure 18.16. The behavior of linear symmetric block copolymer LN1 is chosen as a reference: (a) solution cast films; (b) injection-molded bars; (c) solution cast films. Adhikari [40]. Reproduced with permission of Elsevier.

with lamellar morphology and ST3 with co-continuous structure) show ductile behavior characterized by large plastic deformation, explained in terms of thin-layer yielding of PS layers whose thickness is below the critical value [28]. These results clearly indicate the important role of *morphology* in determining the deformation behavior of block copolymers.

18.4.2 Block Copolymers Based on E and C (CHE)

The importance of *chain architecture* is studied by fixing the molecular weight (ca. 110,000) and composition (PE fraction of ca. 30 wt%) of block copolymers made of poly(vinylcyclohexane) (PCHE) and PE. Both triblock (C–E–C) and pentablock (C–E–C–E–C) copolymers show cylindrical morphology of the PE phase [46]. As explained in Section 18.2.1, PCHE is a glassy polymer and shows brittle deformation due to crazing, which is expected from high-entanglement molecular weight of 49,000 (or low-entanglement density). On the other hand, the amorphous phase of semicrystalline PE is expected to deform by shear deformation due to very low entanglement molecular weight of 1200 (or high-entanglement density).

The triblock copolymer C–E–C shows quite brittle behavior due to craze formation and craze breakdown. However, when the pentablock copolymer C–E–C–E–C with similar molecular weight and composition is deformed, significant improvement is noted: no fracture is noted up to the limit of the experiment using a cooper grid method, for example, 23% strain. TEM observation shows that deformation is primarily due to shear deformation, and often crazes and shear deformation zones coexist where crazes are blunted by shear, as often seen for polymers with a relatively high entanglement density (e.g., SAN [16]). Kramer proposed the following mechanism to explain the observation. The existence of the midblock C chain is critical, in that the midblock chains bridge between highly entangled PE domains (cylinders) and increase the *effective entanglement density*, thereby enhancing shear deformation in favor of the crazing and enhance fracture resistance. This leads to a brittle-to-ductile transition.

Also, experiments were conducted for triblock copolymers (C–E–C) to apply different sample history to show the effect of cylinder orientation [47]. In quenched (spun-cast) samples, PE cylinders are randomly oriented. Upon annealing the samples, more cylinders are arranged perpendicular to the film surface. The experimental results indicate that by applying thermal treatments, films change from being ductile to brittle. Also, by applying shear flow field, cylinders are aligned along shear direction; thus, film samples with cylinders either parallel or perpendicular to the film surface can be made. The former samples show ductile behavior, but the latter samples show brittle behavior. The orientation of cylinders clearly changes the deformation behavior of C–E–C triblock copolymers.

Finally, combining the abovementioned studies with additional experiments, a more complete picture has emerged as to the effect of various factors for deformation behavior of C–E-based block copolymers (see Fig. 18.18) [48]. When the second phase is *lamellar*, *chain architecture* plays a major role. Diblock copolymers show very small strain (brittle): all the films fail at 2% and deforms predominantly by crazing. In contrast, triblock and pentablock copolymers show ductility and toughness (no failure up to 27%, an upper limit of a copper grid method) and the deformation mode is mainly shear deformation, as explained above.

When the second phase is *cylindrical* (for C–E–C–E–C pentablocks), *microdomain orientation* becomes important. For example, when PE cylinders are randomly oriented (made by spun cast), samples are ductile and show no fracture up to 27%. When annealed and cylinders are more aligned to normal to film direction, more films failed at 27%; and when cylinders are near completely aligned normal to film surface upon further annealing, all the films fail at 27% strain. This clearly indicates the effect of thermal history (physical aging), which changes the orientation of cylinders from random to normal to the film surface. It is also shown that highly oriented C–E–C–E–C is tougher than the similarly oriented C–E–C. Bridging C midblock in the pentablock architecture is primarily responsible for the differences in toughness between the C–E–C and C–E–C–E–C blocks (Fig. 18.18).

Here again, the important idea for understanding the deformation behavior and mechanical properties of block copolymers is *entanglement (strand) density*. The entanglement density is increased by using block copolymers in which soft entangled domains (E) are anchored with hard domains (C), which work as physical cross-links for the PE phase, leading to an increase in *effective entanglement density*.

Again, the importance of *entanglement density* is noted for understanding the deformation of block copolymers: this is applicable to all polymeric materials from amorphous polymers, to semicrystalline polymers, and to block copolymers. It is expected that many of the deformation behavior will be explained systematically by using the concept of entanglement density. This needs more systematic investigation both experimentally and theoretically.

18.5 CONCLUSIONS AND OUTLOOK

Deformation behavior of amorphous polymers is well understood by using the concept of strand network density. Whether crazing or shear deformation is observed depends on the total network density, and the extension ratio of each deformation zone is proportional to the theoretical maximum extension ratio. Macroscopic shear deformation, leading to ductile behavior under tension, can be achieved by using thin layers whose thickness is below the critical value, even for normally brittle polymers like PS. The observed extension ratio of macroscopic shear zones is the same as that observed for localized shear zones.

Deformation of semicrystalline polymers is more difficult to understand due to complexity of structures made of an amorphous phase and a crystalline phase; yet, a simple explanation

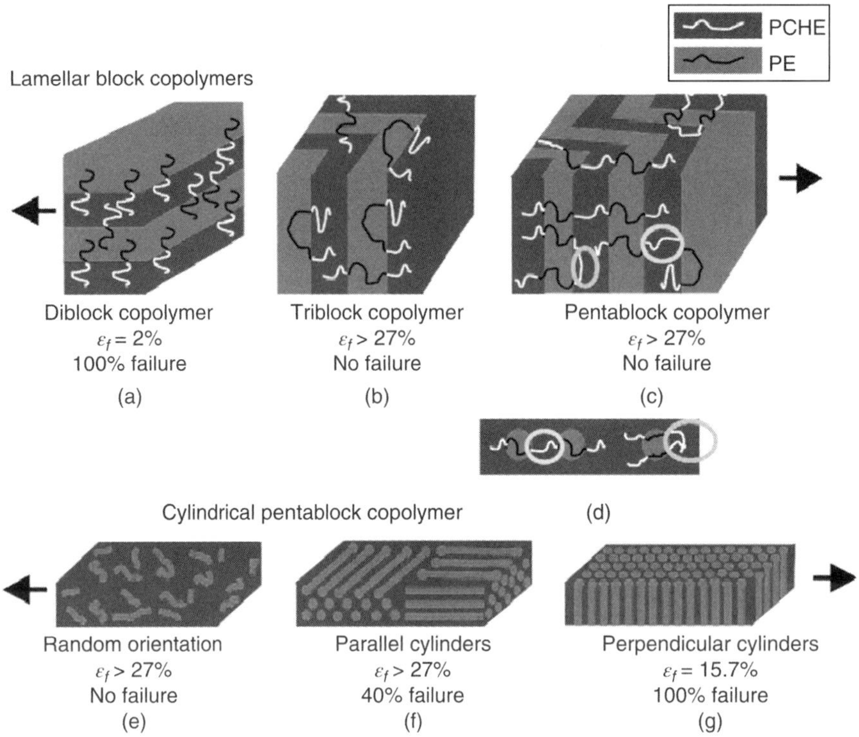

Figure 18.18 Block copolymers made of polyethylene (PE) and poly(vinylcyclohexane) (PCHE) with lamellar and cylindrical morphologies. Khanna et al. [48]. Reproduced with permission of American Chemical Society.

is given in terms of the strand density of amorphous phase to explain the behavior after yielding (strain softening and strain hardening). Entanglement effects between crystalline phases, typically expressed as tie molecules, are used to explain some deformation behavior of semicrystalline polymers.

Block copolymers provide various types of morphologies that influence deformation behavior. The concept of effective entanglement density due to the existence of a hard phase (and physical "cross-links") is used to explain the deformation behavior of block copolymers.

Although there are some differences of nuance, entanglement (network) density plays a major role in determining deformation modes in polymers, ranging all the way from amorphous polymers to semicrystalline polymers, including block copolymers. Further research will provide more evidence for this line of reasoning; also, some previous results may be reinterpreted in terms of the entanglement density.

REFERENCES

1. Kramer EJ. Microscopic and molecular fundamentals of crazing. Adv Polym Sci 1983;52/53:1–68.
2. Kramer EJ, Berger LL. Fundamental processes of craze growth and fracture. Adv Polym Sci 1990;91/92:1–68.
3. van der Sanden MCM, Buijs LGC, de Bie FO, Meijer HEH. Deformation and toughness of polymeric systems: 5. A critical examination of multilayered structures. Polymer 1994;35(13):2783–2792.
4. van der Sanden MCM, de Kok JMM, Meijer HEH. Deformation and toughness of polymeric systems: 7. Influence of dispersed rubbery phase. Polymer 1994;35(14):2995–3004.
5. van der Sanden MCM, Meijer HEH. Deformation and toughness of polymeric systems: 3. Influence of crosslink density. Polymer 1993;34(24):5063–5072.
6. van der Sanden MCM, Meijer HEH. Deformation and toughness of polymeric systems: 4. Influence of strain rate and temperature. Polymer 1994;35(13):2774–2782.
7. van der Sanden MCM, Meijer HEH, Lemstra PJ. Deformation and toughness of polymeric systems: 1. The concept of a critical thickness. Polymer 1993;34(10):2148–2154.
8. van der Sanden MCM, Meijer HEH, Tervoort TA. Deformation and toughness of polymeric systems: 2. Influence of entanglement density. Polymer 1993;34(14):2961–2970.
9. Bassett DC. *Principles of Polymer Morphology*. Cambridge: Cambridge University Press; 1981.
10. Kinloch AJ, Young RJ. *Fracture Behavior of Polymers*. New York: Applied Science; 1983.
11. Michler GH, Baltá-Calleja FJ. *Mechanical Properties of Polymers Based on Nanostructure and Morphology*. New York: Taylor & Francis; 2005.
12. Schrauwen BAG, Janssen RPM, Govaert LE, Meijer HEH. Intrinsic deformation behavior of semicrystalline polymers. Macromolecules 2004;37(16):6069–6078.
13. Hertzberg RW. *Deformation and Fracture Mechanics of Engineering Materials*. New York: John Wiley & Sons; 1983.
14. Kambour RP. A review of crazing and fracture in thermoplastics. J Polymer Sci: Macromol Rev 1973;7:1–154.

15. Ma X, Sauer JA, Hara M. Poly(methyl methacrylate) ionomers. 2. Deformation modes under simple tension. Macromolecules 1995;28(16):5526–5534.
16. Donald AM, Kramer EJ. Deformation zones and entanglements in glassy polymers. Polymer 1982;23(8):1183–1188.
17. Donald AM, Kramer EJ. Effect of molecular entanglements on craze microstructure in glassy polymers. J Polym Sci Polym Phys Ed 1982;20(5):899–909.
18. Henkee CS, Kramer EJ. Crazing and shear deformation in crosslinked polystyrene. J Polym Sci Part B Polym Phys 1984;22(4):721–737.
19. Hara M, Sauer JA. Mechanical-properties of ionomers. J Macromol Sci Rev Macromol Chem Phys 1994;C34(3):325–373.
20. Sauer JA, Hara M. Effect of molecular variables on crazing and fatigue of polymers. Adv Polym Sci 1990;91/92:69–118.
21. Holliday L. *Ionic Polymers*. London: Applied Science Publishers; 1975.
22. Eisenberg A, Kim J-S. *Introduction to Ionomers*. New York: John Wiley; 1998.
23. Eisenberg A, King M. *Ion-Containing Polymers*. New York: Academic Press; 1977.
24. MacKnight WJ, Earnest TR. The structure and properties of ionomers. J Polym Sci Macromol Rev 1981;16(1):41–122.
25. Schlick S. *Ionomers: Characterization, Theory, and Applications*. Boca Raton: CRC Press; 1996.
26. Hara M, Jar P. Deformation and fracture of ionomers under simple tension. 1. Sulfonated polystyrene film from THF solution. Macromolecules 1988;21:3187.
27. Ma X, Sauer JA, Hara M. Poly(methyl methacrylate) based ionomers. 1. Dynamic mechanical properties and morphology. Macromolecules 1995;28(11):3953–3962.
28. Adhikari R, Michler GH, Lebek W, Goerlitz S, Weidisch R, Knoll K. Morphology and micromechanical deformation behavior of SB-block copolymers. II. Influence of molecular architecture of asymmetric star block copolymers. J Appl Polym Sci 2002;85(4):701–713.
29. Smith P, Lemstra PJ, Booij HC. Ultradrawing of high-molecular-weight polyethylene cast from solution. II. Influence of initial polymer concentration. J Polym Sci Polym Phys Ed 1981;19(5):877–888.
30. Smith P, Lemstra PJ. Ultra-drawing of high molecular weight polyethylene cast from solution. Colloid Polym Sci 1980;258(7):891–894.
31. Termonia Y, Smith P. Kinetic model for tensile deformation of polymers. 2. Effect of entanglement spacing. Macromolecules 1988;21(7):2184–2189.
32. Tsou L, Sauer JA, Hara M. Molecular composites of poly(*p*-phenylene terephthalamide) anion and poly(ethylene oxide): Mechanical properties. J Polym Sci Part B Polym Phys 2000;38(10):1377–1385.
33. Kausch HH. *Polymer Fracture*. New York: Springer-Verlag; 1978.
34. Samuels RJ. *Structural Polymer Properties*. New York: Wiley; 1974.
35. Lustiger A, Markham RL. Importance of tie molecules in preventing polyethylene fracture under long-term loading conditions. Polymer 1983;24(12):1647–1654.
36. Plummer C. Microdeformation and fracture in semicrystalline polymers. In: Michler GH, Baltá-Calleja FJ, editors. *Mechanical Properties of Polymers Based on Nanostructure and Morphology*. Boca Raton: CRC Press; 2005. p 230–259.
37. Hamley IW. *The Physics of Block Copolymers*. Oxford: Oxford University Press; 1998.
38. Hadjichristidis N, Pispas S, Floudas G. *Block Copolymers, Synthetic Strategies, Physical Properties, and Applications*. New York: Wiley-Interscience; 2003.
39. Abetz V. *Block Copolymers*. New York: Springer; 2005.
40. Adhikari R, Michler GH. Influence of molecular architecture on morphology and micromechanical behavior of styrene/butadiene block copolymer systems. Prog Polym Sci 2004;29(9):949–986.
41. Argon AS, Cohen RE. Crazing and toughness of block copolymers and blends. Adv Polym Sci 1990;91/92:301–351.
42. Schwier CE, Argon AS, Cohen RE. Crazing in polystyrene–polybutadiene diblock copolymers containing cylindrical polybutadiene domains. Polymer 1985;26(13):1985–1993.
43. Adhikari R, Michler GH, Huy TA, Ivan'kova E, Godehardt R, Lebek W, et al. Correlation between molecular architecture, morphology, and deformation behaviour of styrene/butadiene block copolymers. Macromol Chem Phys 2003;204(3):488–499.
44. Michler GH, Adhikari R, Lebek W, Goerlitz S, Weidisch R, Knoll K. Morphology and micromechanical deformation behavior of styrene/butadiene-block copolymers. I. Toughening mechanisms in asymmetric star block copolymers. J Appl Polym Sci 2002;85(4):683–700.
45. Kerns J, Hsieh A, Hiltner A, Baer E. Mechanical behavior of polymer microlayers. Macromol Symp 1999;147(1):15–25.
46. Ryu CY, Ruokolainen J, Fredrickson GH, Kramer EJ, Hahn SF. Chain architecture effects on deformation and fracture of block copolymers with unentangled matrices. Macromolecules 2002;35(6):2157–2166.
47. Ruokolainen J, Fredrickson GH, Kramer EJ, Ryu CY, Hahn SF, Magonov SN. Effect of thermal history and microdomain orientation on deformation and fracture properties of poly(cyclohexylethylene)–polyethylene triblock copolymers containing cylindrical PE domains. Macromolecules 2002;35(25):9391–9402.
48. Khanna V, Ruokolainen J, Kramer EJ, Hahn SF. Deformation and fracture of lamellar and cylindrical block copolymers with unentangled glassy matrices: Effect of chain architecture and microdomain orientation. Macromolecules 2006;39(13):4480–4492.

19

MORPHOLOGY DEVELOPMENT IN IMMISCIBLE POLYMER BLENDS

Ruth Cardinaels[1,2] and Paula Moldenaers[1]

[1]*Department of Chemical Engineering, KU Leuven, Leuven, Belgium*
[2]*Department of Mechanical Engineering, TU Eindhoven, Eindhoven, The Netherlands*

19.1 INTRODUCTION

Since the discovery of the first synthetic polymer Bakelite [1], around a century ago, a diverse range of polymeric materials with versatile properties has been synthesized and is widely being used in industrial and household applications. However, even before the possibility to synthesize new polymer materials was realized, polymer blending was already recognized as a method to generate materials with improved properties [2]. By blending different polymers, materials can be obtained with properties that are a synergistic combination of those of the components. This allows to meet the increasingly stringent material requirements that are brought about by rising consumer demands and advancing technologies. Several characteristics such as mechanical, electrical, thermal and flow properties should often be tailored simultaneously, which is an extremely complicated goal to achieve with single polymers. Hence, in 2012, the global engineering plastics market of around 10 million tons, consisted for around 10% of polymer alloys and blends [3]. In addition to the development of new materials, polymer blending also becomes more and more important with respect to polymer recycling. The large amounts and variety of plastics that are used nowadays lead to huge mixed plastic waste streams that require high investments for separation and hence often end up in low value waste management processes such as landfill or energy recovery [4, 5]. Hence, recycling these mixed waste streams into polymer blend based products, would allow to increase the plastics recycling rate on the one hand and to create added value on the other hand. Development of a fundamental understanding of the effects of blending on the properties of polymers is thus essential to optimally utilize this promising class of materials.

As the high molecular weight of polymers results in a limited entropy gain upon mixing, most polymers are immiscible [6]. Nevertheless, a limited set of miscible polymer pairs exists, and several of them have found commercial applications [1]. In such systems, the level of molecular mixing is sufficient to yield macroscopic properties that are consistent with a single-phase material [6]. Nevertheless, concentration fluctuations and chain connectivity effects can lead to the presence of local microheterogeneities. The segmental dynamics in miscible blends is thus rather complex and not yet fully understood at present [7]. For many polymer pairs, miscibility is only obtained under certain conditions of concentration, temperature and pressure, in which case the polymers are termed partially miscible. For such systems, temperature-composition or pressure-composition diagrams can be drawn, that consist of three distinct miscibility regions namely a stable, a metastable and an unstable region. Figure 19.1a presents an example of a temperature-composition diagram for a partially miscible blend. When changing temperature, pressure or concentration as such that the blend undergoes a transition from the stable or one-phasic region to the unstable or metastable regions, phase separation will occur respectively spontaneously or driven by concentration fluctuations [6, 8]. This results in a two-phasic polymer blend system. Examples of two-phasic structures can be seen in Figure 19.1b–d whereas Figure 19.1e shows a miscible blend with a one-phasic structure. The thermodynamics and kinetics of phase separation in polymer blends is a rich and intriguing phenomenon that is still under

Polymer Morphology: Principles, Characterization, and Processing, First Edition. Edited by Qipeng Guo.

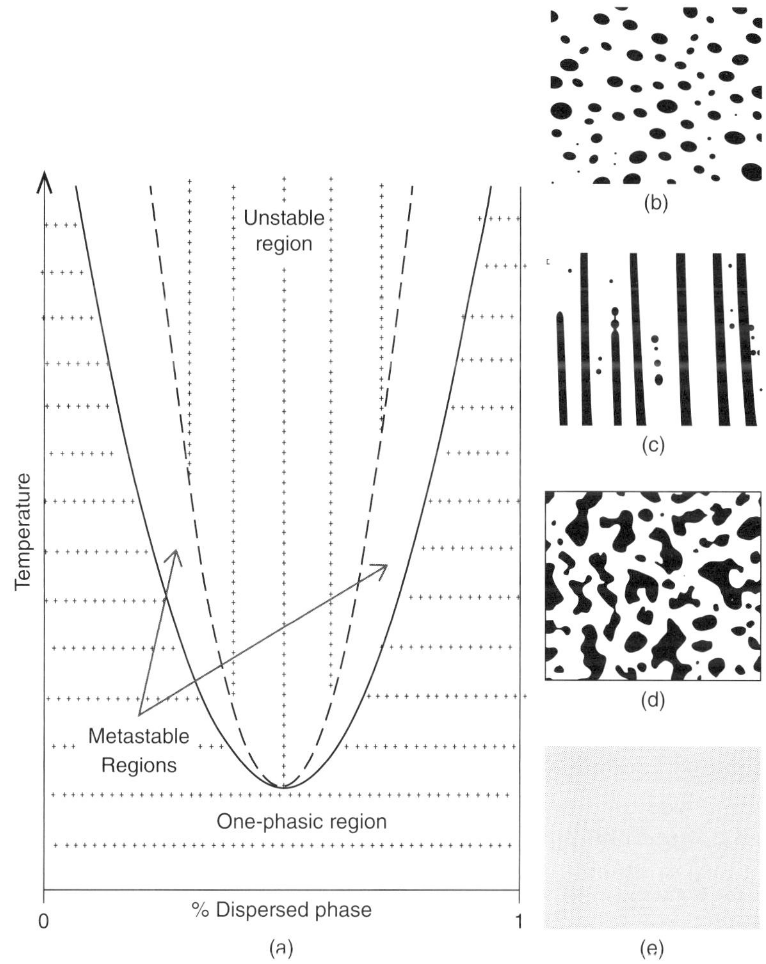

Figure 19.1 Schematic representations of (a) temperature-composition diagram for a partially miscible blend with a lower critical solution temperature and most common blend morphologies including (b) droplet–matrix structure, (c) fibrillar morphology, (d) cocontinuous morphology, and (e) miscible state.

intense research. However, this topic is beyond the scope of the present chapter, in which the focus will be on immiscible polymer blends.

Blending immiscible polymers results in multi-phasic materials with a certain phase morphology [8]. Although blending two polymers is the most common approach, also ternary blends have received a considerable amount of interest. The different morphologies occurring in ternary blends have recently been reviewed by Shokoohi and Arefazar [9]. In the present chapter, only binary immiscible polymer blends will be considered. Such systems consist of either dispersed domains in a continuous phase or of two cocontinuous phases, as shown in respectively Figure 19.1b, c and d. Polymer blends are generally solid at room temperature but they constitute a very viscous emulsion during their processing in the melt. This allows for morphology development during blending and further processing, while the blends are subjected to flow. After processing, cooling or cross-linking leads to a solidification of the material and hence a fixation of the generated microstructure. Once the relations between morphology development and flow conditions are derived, it is thus possible to tailor blend morphology by applying the appropriate processing conditions. It is well established that the properties of immiscible blends not only depend on the characteristics of the blend components and their concentration, but are also highly dependent on the phase morphology. Tailoring of the blend morphology thus allows to optimize product properties.

In the present chapter, the morphology development of immiscible binary polymer blends is discussed. First, morphology development in droplet–matrix structures is described. Subsequently, the dynamics of fibrillar structures is reviewed and finally cocontinuous structures are briefly discussed. Although the main aspects of polymer blending are well established and polymer blends are already widely used in commercial products, recent novel insights in the areas of miniaturization and particle stabilization have opened new research topics in the area of polymer blending. In the last part of this chapter, these recent advances in polymer blend systems are briefly discussed.

19.2 MORPHOLOGY DEVELOPMENT IN BULK FLOW

Depending on blend concentration, properties of the components, and flow history experienced by the blend, different morphology types can occur [8]. At low concentrations of dispersed phase, droplet–matrix structures (Fig. 19.1b) are the predominant morphology type. When the concentration of dispersed phase is increased and/or more intense flow conditions are applied, fibrillar or laminar morphologies (Fig. 19.1c) can form. Finally, in a certain range of concentrations, both phases can become continuous, leading to a cocontinuous structure (Fig. 19.1d). These microstructures can evolve dynamically during and after flow, with the specific processes and their kinetics depending on the morphology type. In this section, the main aspects of the dynamics of droplets, fibrils and cocontinuous structures during flow and in quiescent conditions will be described. In order to study these dynamic processes, both direct imaging techniques and indirect characterization techniques can be used, several of which are described in Chapters 2–9. However, most of the described techniques apply the post-mortem principle whereas the study of dynamic processes requires in situ time-resolved observations. These are commonly performed with rheological and rheo-optical techniques. In the latter techniques, controlled flow conditions are applied and material structure is probed by means of microscopy, light scattering, light transmission, and so on [10].

During processing, the material generally experiences a complex time-dependent flow field. For example, the flow in an extruder consists of a combination of shear and extension due to the changing cross-section of the interior of the device. This complexity of the flow field in processing equipment hampers the development of fundamental insights and limits the generalization of the results obtained from blending studies. Hence, morphology development in model type flows is often studied to map out the different aspects of blend behavior in flow and the effects of the relevant parameters. In the present chapter, the main focus will be on morphology development in shear flow, which is the main flow component in many processing, mixing and other operations that involve rotating or moving parts. A limited description of works dealing with extensional flow is also included. Using numerical simulations or flow visualization, the flow kinematics in processing equipment can readily be obtained. Hence, scaling relations and models for morphology development in different simple flows can be combined to describe the dynamics in complex flow fields [11–14]. This section deals with flows for which the presence of the walls does not affect morphology development, which is further referred to as bulk conditions. More detailed descriptions of specific aspects of morphology development in immiscible polymer blends can be found in literature reviews dealing with respectively droplet dynamics [13, 15–21], cocontinuous blends [22], compatibilized blends [15, 23–26], morphology development during processing [13, 27] and rheology–morphology relationships [11, 28]. Finally, it has to be noted that in case of blends of molten polymers aspects such as melting order of the components can also affect the generated morphology [8]. However, the discussion in the present chapter focuses on the hydrodynamic aspects of morphology development.

19.2.1 Droplet–Matrix Structures

Figure 19.2a depicts a droplet in shear flow. In bulk conditions, the gap spacing H is much larger than the droplet diameter $2R$. In blends with a droplet–matrix morphology, morphology development occurs through deformation, retraction, breakup and coalescence of droplets. These dynamic processes are schematically depicted in Figure 19.2b–d. The relation between flow conditions and droplet dynamics for blends with rheologically simple Newtonian fluids is well-known and models and numerical simulations are widely available [16–18]. As these relations allow to capture the major characteristics of droplet dynamics in polymer blends and to predict the main effects of flow intensity and component viscosity

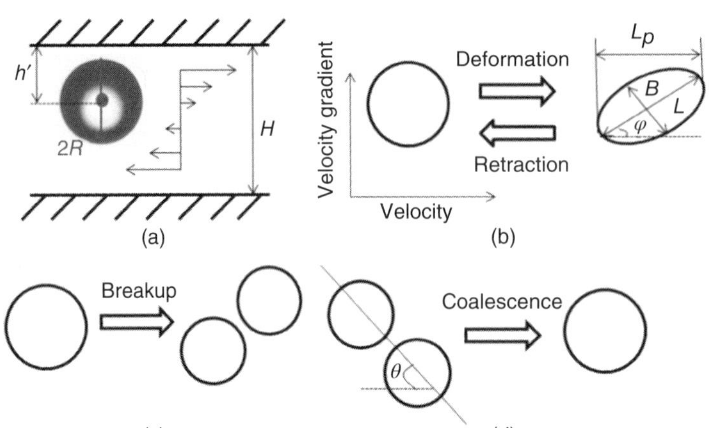

Figure 19.2 Schematic representations of (a) droplet in shear flow, (b) droplet deformation and retraction, (c) droplet breakup, and (d) droplet coalescence.

on blend morphology, they will be discussed first. However, in commercial polymer blends, additional complexities generally arise due to the non-Newtonian rheology of the blend components, interactions between neighboring droplets in nondilute blends and the presence of compatibilizers at the blend interface. Obviously, from a research point of view, such complex systems cannot be tackled in their total complexity at once. Therefore, researchers have studied certain model type problems that focus on a particular degree of complexity at a time. The effects of these complexities will also be briefly addressed here.

19.2.1.1 Droplet Deformation in Blends with Newtonian Components As most polymers are highly viscous and have closely-matched densities, buoyancy, inertia and Brownian motion can generally be neglected. In that case, the dynamics of isolated Newtonian droplets in a Newtonian matrix is determined by two independent dimensionless groups: the capillary number Ca and the viscosity ratio p [29]. The capillary number represents the ratio of the deforming hydrodynamic stresses over the restoring interfacial stresses:

$$Ca = \frac{\eta_m \cdot \dot{\gamma} \cdot R}{\Gamma} \tag{19.1}$$

with η_m the matrix viscosity, $\dot{\gamma}$ the shear rate, R the droplet radius and Γ the interfacial tension. For extensional or complex flows, similar expressions are available in which the shear rate is replaced by the relevant parameter that describes the flow intensity [13]. The viscosity ratio p is defined as:

$$p = \frac{\eta_d}{\eta_m} \tag{19.2}$$

with η_d and η_m the droplet and matrix viscosity respectively. For a wide range of Ca-numbers, the droplet shape can be approximated by an ellipsoid and the droplet deformation is then generally quantified by means of the deformation parameter D [17]:

$$D = \frac{L - B}{L + B} \tag{19.3}$$

with L and B the long and short axis of the droplet in the velocity–velocity gradient plane, as indicated in Figure 19.2b. When the droplet deformation remains limited, the mass and momentum conservation equations in both fluids, together with the appropriate boundary conditions at the droplet interface can be solved analytically, leading to analytical models for the droplet deformation [17]. Taylor was the first to obtain an expression for the droplet deformation parameter in bulk shear flow [29]:

$$D_{Taylor} = Ca\frac{16 + 19p}{16 + 16p} \tag{19.4}$$

It can be seen that the droplet deformation obtained from the Taylor model increases linearly with the Ca-number. The orientation angle φ of the droplet with respect to the velocity direction (Fig. 19.2b) is predicted to be 45° [29]. The equation obtained by Taylor provides the droplet deformation up to the first order in Ca. Based on this initial theory, several authors theoretically explored the droplet deformation problem, including the time-dependent and higher-order behavior, leading to more accurate expressions for the droplet deformation parameter and orientation angle [16, 17].

When the droplet deformation further increases, it is not possible anymore to find analytical solutions for the complex fluid mechanics problem of a deforming droplet in flow. For those cases, models that employ a shape tensor **S** are often used to describe the droplet dynamics. In a first approach, an ellipsoidal droplet shape is assumed and the tensor **S** is set to evolve with time according to the following equation [20]:

$$\dot{\mathbf{S}} = f(\mathbf{S}, \mathbf{D}, \mathbf{\Omega}) \tag{19.5}$$

with **D** the deformation rate tensor, **Ω** the vorticity tensor and f a function that represents the competition between flow and relaxation. This methodology has led to the development of several phenomenological models [20], with the Maffettone–Minale model the most widely used one [30, 31]:

$$\begin{aligned}\frac{d\mathbf{S}^*}{dt} &- Ca(\mathbf{\Omega}^* \cdot \mathbf{S}^* - \mathbf{S}^* \cdot \Omega^*)\\ &= -f_1[\mathbf{S}^* - g(\mathbf{S}^*)\mathbf{I}] + Ca\ f_2(\mathbf{D}^* \cdot \mathbf{S}^* + \mathbf{S}^* \cdot \mathbf{D}^*)\end{aligned} \tag{19.6}$$

In this evolution equation for the shape tensor, **S***, **D*,** and **Ω*** are the dimensionless versions of the shape tensor, rate of deformation tensor and vorticity tensor respectively. The eigenvalues of **S** represent the square semi-axes of the ellipsoid. The factors f_1 and f_2 depend on the viscosity ratio and Ca:

$$f_1 = \frac{40(p + 1)}{(2p + 3)(19p + 16)} \tag{19.7}$$

$$f_2 = \frac{5}{2p + 3} \tag{19.8}$$

These factors were chosen as such that the small deformation model of Taylor is recovered for small deformations. The left-hand side of Equation 19.6 represents the change of **S** with time and its rotation with the overall flow field. The first term on the right-hand side represents the restoring action of the interfacial tension, where $g(\mathbf{S})$ is introduced to preserve the droplet volume. The second term on the right-hand side captures the deformation of the droplet with the flow. This equation allows to accurately predict the droplet deformation in flows with an arbitrary but uniform velocity gradient. In case of shear flow, the following expression is obtained for the steady state deformation parameter [30]:

$$D_{MM} = \frac{\sqrt{f_1^2 + Ca^2} - \sqrt{f_1^2 + Ca^2 + f_2^2 Ca^2}}{f_2 Ca} \tag{19.9}$$

The evolution of the deformation parameter as a function of the Ca-number is presented in Figure 19.3a for droplet–matrix

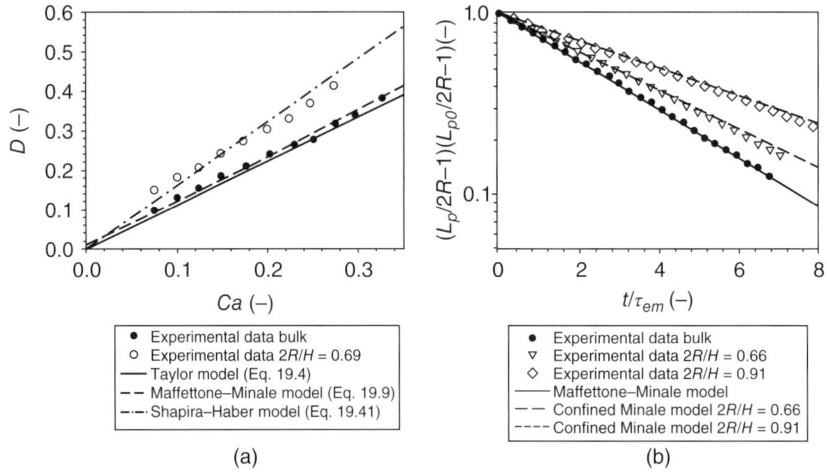

Figure 19.3 (a) Steady state droplet deformation parameter versus *Ca* number for Newtonian droplets in a Newtonian matrix with $p=1.0$: experimental data (symbols) and model predictions (lines) (Adapted from Vananroye et al. [32]), (b) droplet retraction after cessation of shear flow at $Ca=0.2$ for Newtonian droplets in a Newtonian matrix with $p=2.2$: experimental data (symbols) and model predictions (lines). Adapted from Vananroye et al. [33]. Reproduced with permission of American Institute of Physics.

systems with a viscosity ratio p of 1. From the results in bulk flow it can be seen that the experimental data are well described by both the Taylor model and the Maffettone–Minale model. For viscosity ratios that are either below or above 1, the Maffettone–Minale model provides a more substantially improved agreement with experimental data as compared to the Taylor model [30].

A second class of models directly relates flow to blend structure without the assumption of an ellipsoidal droplet shape. This description was initiated by Doi and Ohta for an equiviscous blend with equal compositions of both components [34]. Coupling this method with a constraint of constant volume of the inclusions, leads again to equations for microstructural dynamics in blends with a droplet–matrix morphology [35]. An alternative way to develop these microstructural theories is the use of nonequilibrium thermodynamics. This way, Grmela et al. showed that the phenomenological Maffettone–Minale model can be retrieved for a specific choice of the free energy [36]. An in-depth review of the different available models for droplet dynamics can be found in the work of Minale [20].

In addition to the steady state droplet deformation, the different models for droplet dynamics also provide the transient behavior of droplets during startup or after cessation of flow. An aspect that is of particular interest in the processing of polymer blends is the droplet retraction after cessation of flow. This retraction toward the spherical shape is driven by interfacial tension and the ratio of the time scale of retraction to that of solidification of the blend determines the remaining anisotropy after the morphology has been frozen in. For ellipsoidal droplets, the retraction occurs exponential in time, with the following time constant [37]:

$$\tau_{ellipsoid} = \frac{R\eta_m}{\Gamma}\frac{(19p+16)(2p+3)}{40(p+1)} \qquad (19.10)$$

The retraction time thus increases with droplet radius, viscosity of the matrix fluid and viscosity ratio. An example of such an exponential retraction together with the predictions of the Maffettone–Minale model is provided in Figure 19.3b for the projected droplet axis L_p (Fig. 19.2b). When the initial droplet shape is nonellipsoidal, shape retraction proceeds slower as compared to that of ellipsoidal droplets and can no longer be described by means of an exponential function [16, 28, 38].

19.2.1.2 Droplet Breakup in Blends with Newtonian Components As shown in Equations 19.4 and 19.9, the steady state droplet deformation increases with *Ca*. However, when a certain critical value of *Ca* is exceeded, the droplet will deform irreversibly under flow until breakup eventually occurs. For Newtonian–Newtonian systems in bulk flow, the critical values only depend on the viscosity ratio and the flow type. In shear flow, the following relation between the critical *Ca*-number Ca^b_{crit} for breakup and the viscosity ratio has been established [39, 40]:

$$\log(Ca^b_{crit}) = -0.506 - 0.0994\log(p) + 0.124\log^2(p) - \frac{0.115}{\log(p) - 0.611} \qquad (19.11)$$

From this equation, it follows that in pure shear flow, droplets with a viscosity ratio above 4 cannot be broken (infinite value of Ca^b_{crit}). Beyond $p=4$ the droplet continuously rotates in the shear plane or oscillates around the flow direction. This tumbling and wobbling reduces the stretch exerted on the droplet and inhibits breakup [41]. In extensional or complex flows, on the other hand, no such upper limit for the viscosity ratio exists. Hence, the generation of a fine droplet–matrix dispersion in typical blending and mixing equipment involving complex flows is possible, even when the viscosity ratio exceeds 4 [42]. Based on the value of the critical Ca-number, a limiting droplet size for droplet breakup R^b_{crit} can be determined:

$$R^b_{crit} = \frac{\Gamma Ca^b_{crit}}{\eta_m \dot{\gamma}} \tag{19.12}$$

All droplets with a size that exceeds this critical value, will break up during flow. However, breakup is not instantaneous and a breakup time t^b_{crit} is required during which the droplet stretches irreversibly and eventually develops a neck leading to the formation of two daughter droplets and some small satellite droplets in between. This breakup time for near-critical breakup scales as [43]:

$$t^b_{crit} \approx \frac{1}{\dot{\gamma}}\left(\frac{Ca}{Ca^b_{crit}} - 1\right)^{(-1/2)} \tag{19.13}$$

Hence, only when both the critical Ca-number and the breakup time are exceeded, droplet breakup will occur.

19.2.1.3 Droplet Coalescence in Blends with Newtonian Components Whereas droplet breakup generally prevails at high flow intensities, low flow intensities promote droplet coalescence [44]. Coalescence occurs when two or more droplets collide and merge together. A considerable amount of research has been devoted to the interaction and coalescence of isolated droplet pairs in different flow types [21, 44]. These studies provide fundamental insight in the coalescence process by mapping out the outcome of collision events as a function of the governing dimensionless parameters. Since coalescence is per definition more complex as compared to droplet deformation and breakup due to on the one hand the presence of at least two droplets and on the other hand the contribution of nonhydrodynamic forces, additional dimensionless groups come into play. These include parameters that characterize the relative droplet positions with respect to each other, for example the dimensionless distances between the droplet centers in the velocity, velocity gradient, and vorticity directions as well as the size ratio of the droplets and a dimensionless Hamaker constant [21]. Depending on whether the viscous or interfacial stress is used for the nondimensionalization, different dimensionless Hamaker parameters can be obtained, with the most simple one being [21]:

$$A' = \frac{A}{6\pi\Gamma R^2} \tag{19.14}$$

with A the effective Hamaker constant of the system, characterizing the Van der Waals interaction in the presence of a matrix phase. Numerical simulations of droplet coalescence are rather complex due to the large range of length scales that is involved. Nevertheless, several scaling relations derived from analytical and experimental work and sometimes even quantitative aspects of the coalescence process could be confirmed by numerical simulations [21].

At large inter-droplet distances, each droplet follows a streamline of the external flow field. In shear flow, each of the streamlines corresponds to a different velocity, thereby enabling approach of droplets on different streamlines. The collision frequency per unit volume of monodisperse spheres in shear flow was first derived by Smoluchowski [45]:

$$C(t) = \frac{24\phi^2\dot{\gamma}}{\pi^2(2R)^3} \tag{19.15}$$

with ϕ the volume fraction of spheres. The coalescence frequency of interacting droplets can then be obtained as the product of the collision frequency given by Equation 19.15 and the coalescence probability. The coalescence probability provides the fraction of colliding droplets that coalesces [46]. To determine the coalescence probability, the coalescence process is generally depicted as a process consisting of three stages namely approach, drainage and rupture [47]. In her derivation, Smoluchowski [45] assumed that the streamlines are not affected by the presence of neighboring droplets and thus approaching droplets in different shear planes separated by a distance less than the sum of the droplet radii will collide. However, when the droplets come close together, hydrodynamic interactions will cause them to deviate from their original path, which lowers the collision frequency as compared to the estimation of Equation 19.15. By using the creeping flow equations, trajectory analysis could be used to accurately determine the approach trajectories of droplets in shear flow [48]. This way, coalescence probabilities, taking into account the reduced number of collisions due to hydrodynamic interactions, were mapped out as function of viscosity ratio and size ratio of the droplet pair [48]. Once a droplet pair is formed, the doublet behaves as a solid ellipsoid [49] and tumbles in the shear flow according to Jeffrey's equations [50]. For two droplets in the same velocity–velocity gradient plane the evolution of the orientation angle θ of the droplet pair is given by [50]:

$$\tan(\theta - 90°) = 2\tan\left(\frac{2\dot{\gamma}t}{5}\right) \tag{19.16}$$

with θ the angle between the flow direction and the line connecting the droplet centers, as defined in Figure 19.2d. From this equation, it can be seen that the interaction time of a droplet pair in shear flow is inversely proportional to the shear rate [47]. During the interaction, the external flow exerts a hydrodynamic force on the droplet pair. The hydrodynamic force along the line of centers of the droplets is given by [51]:

$$F = 4.34\frac{2/3 + p}{1 + p}\eta_m\dot{\gamma}R^2\sin(2(90° - \theta)) \tag{19.17}$$

This force initially pushes the droplets together, and has a maximal value when $\theta = 45°$. For values of θ above 90° the hydrodynamic force becomes negative, thus pulling the droplets apart. Due to the compressive force, the matrix film in between the droplet interfaces is drained during droplet interaction, which corresponds to the second stage of the coalescence process. In addition to the hydrodynamic force, the kinetics of film drainage is affected by the shape and mobility of the interface. Although the interface in general develops a complex dimpled shape [21], either a spherical undeformed droplet shape or a parallel disk with a radius that may change during droplet interaction, is often assumed in the development of drainage models [47]. The mobility of the interface can range from fully mobile over partially mobile to immobile and determines the relative contributions to the drainage of the uniform plug flow portion due to slip at the interfaces and the parabolic flow portion due to pressure driven drainage flow [47]. Different combinations of the geometrical boundary conditions have resulted in a variety of drainage models. An elaborate discussion of the different models can be found in the work of Chesters [47]. Finally, coalescence occurs if the matrix film in between the droplet interfaces becomes sufficiently thin to rupture as a result of the Van der Waals forces. By equating the Van der Waals attraction between two spheres with the viscous force on a sphere in shear flow, Chesters estimated the critical film thickness as [47]:

$$h_c = \left(\frac{AR}{8\pi\Gamma}\right)^{1/3} \tag{19.18}$$

with A the effective Hamaker constant. Vrij found an alternative expression by assuming film rupture due to spontaneous thickness fluctuations [52].

Although the processes of droplet interaction and film drainage are intimately linked, for modeling purposes, the coalescence problem is often decoupled into an external flow and an internal flow problem [21, 47]. In that case, the external flow is assumed to only determine the collision frequency, interaction time and hydrodynamic force. Film drainage is modeled separately with the effects of the global droplet deformation and the fluid flow due to the external flow field assumed to be negligible, which is not always the case [53–55]. A drainage model that has proven to be useful for polymer blends is that for the drainage of parallel sided fluid films with a constant radius and partially mobile interfaces squeezed together at a constant contact force [47]. A comparison of the drainage time obtained with this drainage model to the interaction time provides the coalescence probability. Hence, a model for the evolution of the droplet size with time could be derived [56]:

$$\frac{dR}{dt} = 0.525\phi\dot{\gamma}R \cdot \exp\left[-\left(\frac{R}{R^c_{crit}}\right)^{5/2}\right] \tag{19.19}$$

in which ϕ is the volume fraction of the dispersed phase and R^c_{crit} is the steady state droplet size determined by coalescence at the applied shear rate, which corresponds to [56]:

$$R^c_{crit} = \left(\frac{4}{\sqrt{3}}\frac{h_c}{p}\right)^{2/5}\left(\frac{\Gamma}{\eta_m\dot{\gamma}}\right)^{3/5} \tag{19.20}$$

This model does not take into account hydrodynamic interactions, but nevertheless proved to be valuable in predicting the droplet size evolution in nondilute polymer blends [56]. More accurate expressions for the coalescence probability in nondilute emulsions and blends can be obtained by combining trajectory analysis with drainage models [55, 57]. Finally, it should be noted that also in quiescent conditions, coarsening of droplet–matrix morphologies can occur. Since the large viscosity and small density difference of polymers leads to limited driving forces for droplet approach, coalescence is in general a rare event during annealing of polymer blends. On the other hand, Ostwald ripening, which occurs via diffusion of droplet material throughout the matrix from the small toward the large droplets can occur and leads to coarsening with the following kinetics [8]:

$$R^3 = R_0^3 + bt \tag{19.21}$$

in which R_0 is the initial droplet radius and b is a factor that depends, among others, on the diffusion coefficient of the molecules of the dispersed phase in the matrix material.

Based on the critical droplet sizes for breakup and coalescence in Equations 19.12 and 19.20, the droplet size in polymer blends as a function of flow intensity (shear rate) can be mapped out [28], as shown in Figure 19.4. The critical droplet sizes for droplet breakup and coalescence become equal at a certain critical shear rate. For shear rates larger than this critical value, the critical droplet size for breakup is smaller than the critical droplet size for coalescence and the final droplet size is determined by a dynamic equilibrium between breakup and coalescence. However, below the critical shear rate, the critical droplet size for breakup is larger than the critical droplet size for coalescence, which results in a range of droplet sizes for which neither breakup nor coalescence will occur. This phenomenon is called morphological hysteresis and changing the flow conditions within this region

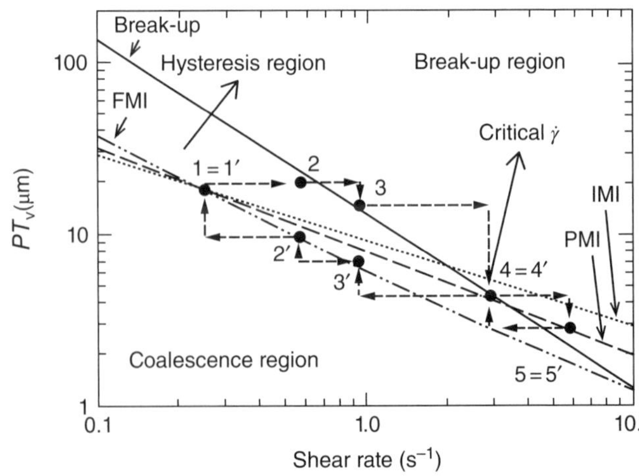

Figure 19.4 Morphology diagram; arrows indicate the order in which the shear rates have been applied. Adapted from Minale et al. [58]. Reproduced with permission of American Chemical Society.

will not alter the blend morphology [28, 58]. This hysteresis phenomenon is illustrated in Figure 19.4.

19.2.1.4 Effects of Component Viscoelasticity Contrary to the Newtonian–Newtonian systems that are dominantly used in fundamental research, polymer blends for industrial and household applications mostly contain components with a complex rheology. When viscoelastic components are present, a multitude of rheological parameters such as shear thinning, normal stress differences, elongational viscosity and relaxation times come into play [59]. The number and type of parameters that are relevant to describe the additional effects on the droplet dynamics depend on the complexity of the viscoelastic nature of the blend components [60]. However, the aspect that has most elaborately been investigated in this respect is droplet dynamics in systems with one viscoelastic component for which the viscoelasticity can be described with a viscosity and normal stress coefficients that are independent of shear rate [16, 61]. When the viscoelastic component obeys the second-order fluids model, as is the case for all materials in sufficiently slow and slowly varying flows [59], Greco found four additional parameters to be of influence [60]. The first two parameters are a Deborah number De for the matrix and droplet fluid [60]:

$$De = \frac{\Psi_1 \cdot \Gamma}{2R \cdot \eta^2} \tag{19.22}$$

in which Ψ_1 is the first normal stress coefficient. This Deborah number represents the ratio of the characteristic relaxation time of the fluid over the so-called emulsion time of the droplet, which characterizes the droplet deformation time scale and corresponds to the first fracture in Equation 19.10. Viscoelastic effects come into play when De has values on the order of 1 or higher [61]. The second two parameters that play a role are the ratios Ψ of the second to the first normal stress coefficient for each fluid [60]:

$$\Psi = \frac{\Psi_2}{\Psi_1} \tag{19.23}$$

in which Ψ_1 and Ψ_2 are the first and second normal stress coefficients respectively. Viscoelastic effects on the steady state droplet deformation only come into play at the second order in Ca [60]. For small droplet deformations, Greco obtained the steady state droplet deformation up to the second order in Ca for blends in which the components are second-order fluids [60]. Interestingly, though the predicted L and B of the droplets were dependent on the additional dimensionless parameters of Equations 19.22 and 19.23, the steady state droplet deformation parameter of the Taylor model (Eq. 19.4) was retrieved. To obtain predictions of droplet deformation in systems with viscoelastic components at larger values of Ca and during transient flow conditions, several of the phenomenological models for droplet dynamics in Newtonian–Newtonian systems, represented by Equation 19.5, were extended to include effects of component viscoelasticity, as reviewed by Minale [20]. Each of the models provides optimal results for a certain range of dimensionless parameters, but unfortunately none of them allows to describe experimental data over a wide range of parameter values [62].

Experimentally, constant viscosities and normal stress coefficients can be obtained by using so-called Boger fluids, which are dilute solutions of high molecular weight polymers in a viscous solvent [63]. Hence, an extensive amount of experimental research on the effects of component viscoelasticity on droplet dynamics has been performed with polymer blends containing a Boger fluid as one of the blend components [16, 61]. For such systems, it has been shown that matrix viscoelasticity leads to an increased orientation of sheared droplets with respect to the flow direction, a reduced droplet deformation and increased values of the critical Ca-number for breakup in shear flow [61, 64–67]. In addition, the droplet dynamics after startup of flow and after cessation of flow is slowed down and can show complex features [61, 64, 68–70]. For example, the retraction of a droplet toward the spherical shape no longer shows an exponential kinetics, as in Equation 19.10, but is severely slowed down at the longer time scales [64, 70]. Since the viscoelastic stresses in the matrix that reduce the droplet deformation with respect to that in a Newtonian matrix need some time to develop, the droplet deformation after startup of shear flow can show substantial overshoots [64, 69]. Numerical simulations have shown that the effects of matrix viscoelasticity on the droplet deformation result from a subtle interplay between flow modifications, viscoelastic stretch at the droplet poles and tensile stresses at the droplet equator [71]. The occurrence of these viscoelastic stresses is shown in Figure 19.5a. In order to describe such effects quantitatively, complex rheological constitutive equations should be used to describe the component rheology [65, 70]. Hence, the developed analytical theories and phenomenological models often do not provide quantitative predictions of the observed complex droplet dynamics [64, 69]. When considering droplet coalescence, the presence of viscoelastic stresses and modifications of the flow field around the droplet pair can affect the different aspects of the coalescence process including the interaction dynamics and the film drainage [72–74]. Hence, the outcome of a coalescence event in a viscoelastic matrix will be determined by a complex interplay of different factors and predictions are thus not straightforward. In shear flow, droplet viscoelasticity has far less impact as compared to viscoelasticity of the matrix phase [61, 65], due to the fact that the flow inside a sheared droplet is largely rotational, thereby limiting the development of viscoelastic stresses [65]. This is illustrated in Figure 19.5b showing the streamlines in and around a droplet in shear flow. Nevertheless, droplet viscoelasticity can stabilize droplets against breakup in shear flow, similar to matrix viscoelasticity [75–77].

Finally, an important aspect for polymer processing is the fact that the uniqueness of Stokes flow is lost when dealing with viscoelastic liquids. As a consequence, the breakup of droplets in systems with a viscoelastic matrix has been found to be largely dependent upon the shear flow history, with more

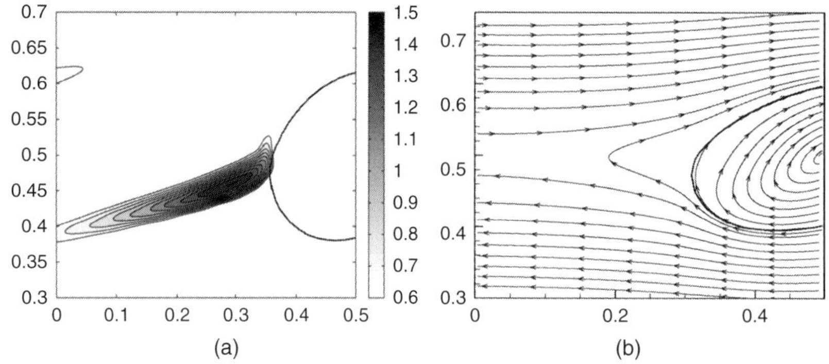

Figure 19.5 (a) Contours of the trace of the extra stress tensor representing the viscoelastic stresses at $Ca = 0.154$, $De_m = 6$ and $t/\tau_{em} = 30$, (b) streamlines in and around a droplet in shear flow at $Ca = 0.35$, $De_m = 4$ and $t/\tau_{em} = 30$. Verhulst et al. [65]. Reproduced with permission of Elsevier.

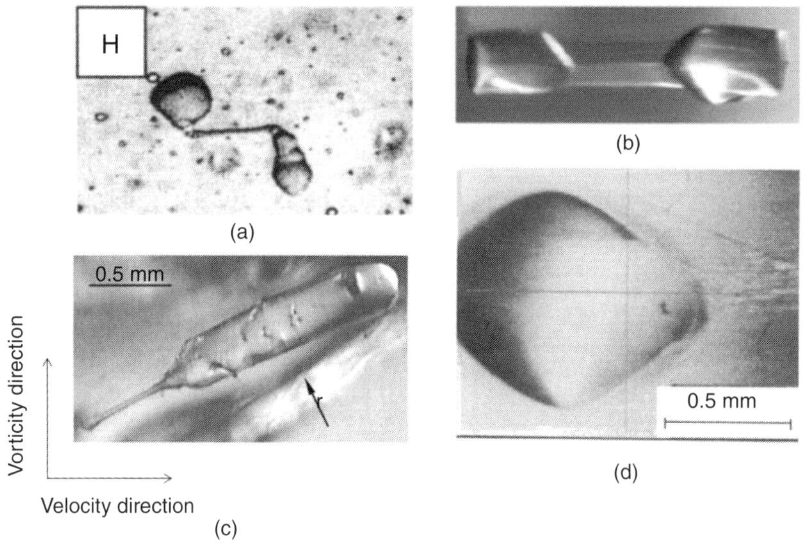

Figure 19.6 Microscopy images of droplet breakup in systems with viscoelastic components; (a) breakup along the vorticity direction for viscoelastic droplet in Newtonian matrix with $p = 0.64$. Migler [79]. Reproduced with permission of American Institute of Physics. (b) Parallel sheet breakup for viscoelastic droplet in viscoelastic matrix with $p = 8.6$. Lin et al. [82]. Reproduced with permission of John Wiley and Sons. (c) Sheet breakup via tip formation for viscoelastic droplet in viscoelastic matrix at $p = 5.7$. Lin and Sundararaj [83]. Reproduced with permission of Elsevier. (d) Droplet erosion for viscoelastic droplet in viscoelastic matrix at $p = 8.8$. Lin et al. [81]. Reproduced with permission of John Wiley and Sons.

gradual increases in shear rate leading to larger values of the critical *Ca*-number [70]. Since the flow history in industrial polymer processing operations is in general very complex, it becomes non-trivial to model and predict droplet breakup during processing in systems with viscoelastic components. Interestingly, when one or both of the blend components is viscoelastic, the droplets can stretch and eventually break up along the vorticity direction of the flow [76, 78–80]. Contrary to fully Newtonian systems, this breakup mechanism can cause breakup in shear flow at viscosity ratios above 4 [78]. For blends of molten polymers with large viscosity ratios several additional breakup mechanisms were discovered, showing that breakup of droplets at viscosity ratios above 4 is generally possible in sheared polymer blends with viscoelastic components [78, 81–83]. These breakup mechanisms are depicted in Figure 19.6.

In addition to investigations on elasticity, also the dynamics of droplets consisting of shear thinning or yield stress fluids has received some attention. Both for shear and extensional flow, it has been found that shear thinning of the droplet fluid reduces the deformation as compared to that of a Newtonian droplet with the same viscosity at the applied shear rate [84, 85]. Desse et al. [86] showed that the dependence of the critical *Ca*-number on viscosity ratio for a starch suspension droplet with a yield stress deviates substantially from that of Newtonian droplets. In conclusion, the precise relations between the rheological constitutive parameters of droplet and matrix fluid and the droplet dynamics for materials with a complex rheology are far from fully revealed.

19.2.1.5 Effects of Compatibilization Due to the inherent thermodynamic incompatibility of immiscible polymers [6],

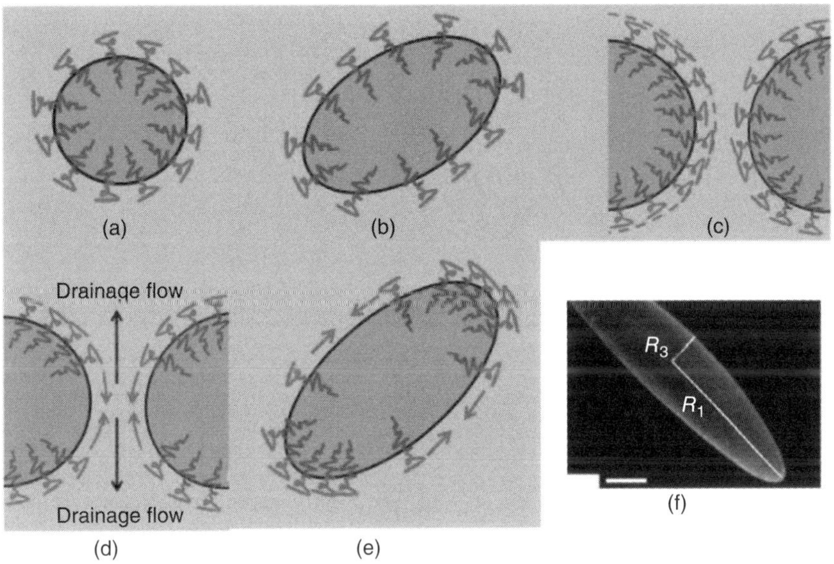

Figure 19.7 Overview of the mechanisms by which copolymers can affect droplet dynamics: (a) block copolymer at the interface lowering the interfacial tension, (b) dilution due to droplet deformation, (c) steric hindrance, (d) Marangoni stresses (light gray arrows) during film drainage, (e) Marangoni stresses (light gray arrows) on deformed droplets, (f) visualization of concentration gradients of a fluorescently labeled PS-*b*-PMMA block copolymer along the droplet interface for a PMMA droplet in a PS matrix. Jeon and Macosko [87]. Reproduced with permission of Elsevier.

they have the tendency to demix and to form macroscopically separated phases. Classically, physically added or in situ reactively formed copolymers that preferentially locate at the blend interface are used for compatibilization, as shown in Figure 19.7a [24–26]. The goal of compatibilization is threefold: refinement and stabilization of the blend morphology and improvement of the interfacial adhesion [25]. The presence of compatibilizer at the interface of a polymer blend has several implications on the properties of the interface [11, 23]. Compatibilizers will primarily reduce the interfacial tension [23–26]. This effect can be taken into account in models and simulations by using the equilibrium interfacial tension, which is the interfacial tension of a spherical droplet uniformly covered with compatibilizer, rather than the interfacial tension of the neat system, in Equations 19.1–19.20. From Equations 19.12 and 19.20 it is clear that this will reduce the critical droplet sizes for breakup as well as coalescence, thus causing morphology refinement. However, if the flow intensity is sufficiently high, leading to large droplet deformations, dilution of the compatibilizer at the interface can occur [23], as shown in Figure 19.7b. This renders the compatibilization effect dependent on the flow conditions. In addition, block copolymers can be swept toward the sides of the droplet, as shown in Figure 19.7f, which results in gradients in block copolymer concentration and hence in interfacial tension [11, 23]. Whether or not this convection process will occur, can be assessed by means of the dimensionless Peclet number *Pe*, which represents the ratio of convection relative to diffusion at the interface [88]:

$$Pe = \frac{\dot{\gamma} R^2}{D_s} \tag{19.24}$$

in which D_s is the surface diffusivity of the compatibilizer. Gradients in interfacial tension result in tangential stresses along the droplet interface, also called Marangoni stresses, that immobilize the interface [11, 23]. The relative importance of this interfacial immobilization can be expressed by means of the Marangoni number *Ma* which provides the ratio of the Marangoni stress to the viscous stress [89]:

$$Ma = \frac{(c_{eq}/R)|d\Gamma/dc|_{c_{eq}}}{\eta_m \dot{\gamma}} \tag{19.25}$$

in which c_{eq} is the compatibilizer concentration present at the interface of a spherical, uniformly covered droplet whereas c represents the local compatibilizer concentration at the interface. Since Marangoni stresses counteract both film drainage during coalescence as well as droplet extension and necking during breakup [11, 23, 90], as depicted in Figure 19.7d and e, it complicates predictions of the effects of compatibilization on droplet size. It is worthwhile mentioning that for blends with a low viscosity ratio, the increased concentration of compatibilizer at the droplet tips can lead to tip streaming, a process in which very small daughter droplets are ejected from the droplet tips at much lower values of *Ca* than the critical value for breakup of Newtonian droplets in a Newtonian matrix [11, 23]. A third effect of the presence of compatibilizer at the blend interface is the occurrence of steric hindrance, which causes coalescence suppression due to the presence of long compatibilizer chains sticking out of the interface and inhibiting close approach of the droplets [11, 23]. This effect is schematically depicted in Figure 19.7c. Finally, the presence of compatibilizer at the interface can endow the interface with complex interfacial rheological properties including shear as well as dilatational viscosity and elasticity [11, 23, 91]. Although the presence of an increased elasticity at long relaxation times, originating from the interface, has clearly been established in compatibilized polymer blends [23, 92], the detailed interfacial rheological properties of compatibilizer-covered droplets and their impact on droplet dynamics is still under intense research for the moment.

The effects of compatibilization are clearly governed by a complex interplay of several phenomena, the extent of which depends on the concentration, architecture, molecular weight and chemical nature of the compatibilizer in combination with the blend system [24, 26]. This renders a priori predictions of morphological parameters such as droplet size in compatibilized blends rather complex. Nevertheless, it is well established that the effects of compatibilization on the droplet size in polymer blends can be described by means of an emulsification curve [24], which shows a rather steep decrease of the droplet size at low concentrations of compatibilizer and then levels off toward a plateau value at a certain critical compatibilizer concentration. The typical saturation concentrations of copolymer range from 5 to 15 wt% of the dispersed phase, but depend on the system of interest [26]. In addition, it is clear that compatibilization hinders droplet coalescence and quiescent coarsening of droplet–matrix structures, thus stabilizing the blend morphology [8, 23]. Finally, it should be noted that although the aim is to have the compatibilizer located at the blend interface, both kinetic and thermodynamic factors play a role in the localization process which may lead to part of the compatibilizer residing in one or both bulk phases rather than at the interface [1]. This aspect mainly plays a role when compatibilization proceeds through the addition of premade copolymers. In case of block copolymers, micelles in the bulk phases are formed once the critical micelle concentration in the bulk is exceeded [1]. Obviously, this will reduce the efficiency of the added compatibilizer.

19.2.1.6 Effects of Hydrodynamic Interactions A last aspect that plays a role in droplet dynamics in realistic systems is the presence of neighboring droplets, which cause hydrodynamic interactions. The presence of other droplets around the droplet under observation will influence the behavior of the latter on a local and a global scale [93]. On a global scale, a surrounding medium filled with droplets has a higher overall viscosity as compared to the matrix viscosity. On a local scale, neighboring droplets cause fluctuations in the local velocity field and can even directly collide with the droplet under investigation. The first effect leads to an increased droplet deformation, increased droplet orientation with respect to the flow direction and lower values of the critical *Ca*-number for breakup [93–96]. The second effect causes fluctuations in the droplet deformation, asymmetric droplet shapes and sporadic breakup of droplets at conditions far below the critical ones [93, 94]. Choi and Schowalter [97] developed an analytical model to describe the effects of concentration on the droplet shape and orientation with respect to the flow direction in case of small droplet deformations. The deformation parameter and orientation angle in shear flow based on this model are given by [97]:

$$D = \frac{16 + 19p}{(16 + 16p)\sqrt{1 + Z^2}} Ca \left[1 + \varphi \left(\frac{5\,(2 + 5p)}{4(p + 1)} \right) \right] \tag{19.26}$$

and

$$\varphi = \frac{\pi}{4} - \frac{1}{2} \arctan Z \tag{19.27}$$

with

$$Z = \frac{(16 + 19p)(3 + 2p)}{40(1 + p)} \mathrm{Ca} \left(1 + \varphi \frac{5\,(16 + 19p)}{4(p + 1)(3 + 2p)} \right) \tag{19.28}$$

For more deformed droplets, the so-called mean-field approach has been proposed, in which the effect of concentration on the steady state deformation and critical conditions is taken into account by replacing the matrix viscosity in the dimensionless numbers by the emulsion viscosity [13, 94–96]. Numerical simulations suggest that, provided the concentration is not too high, the droplet deformation, orientation and the critical *Ca*-number for breakup in blends with different concentrations can approximately be superposed onto a master curve with this approach [95]. Experimentally, this approach has only been validated for rather dilute emulsions or with big marker droplets in a very finely dispersed emulsion [94, 96]. Figure 19.8 confirms the validity of the mean field approach for the determination of the critical *Ca*-number for droplet breakup in concentrated blends. In addition to the steady state droplet deformation and critical conditions for breakup, also the dynamics of droplets are affected by the presence of neighboring droplets. For example, retraction of the droplet shape after cessation of shear flow is known to slow down due to the presence of neighboring droplets [35, 97]. A model, taking into account hydrodynamic interactions on the dynamics of droplets at small deformations, has been developed by Palierne [98] and provides the following expression for the droplet retraction time in concentrated blends:

$$\tau_{Palierne} = \frac{R\eta_m}{4\Gamma} \frac{(19p + 16)(2p + 3 - 2\phi(p - 1))}{10(p + 1) - 2\phi(5p + 2)} \tag{19.29}$$

Finally, also droplet coalescence can be affected by the concentration of the droplet–matrix system. The effects of concentration on droplet coalescence are generally investigated by subjecting nondilute blends to a specific shear protocol, consisting of the application of a large shear rate used to generate a fine droplet morphology by droplet breakup followed by a substantial reduction of the shear rate, after which coalescence is the dominant mechanism [28]. In situ time-resolved observations of the droplet size during shear can be performed [57, 99], or the shear can be periodically interrupted to facilitate the determination of the droplet size by, for example, microscopy, light scattering or rheology (small amplitude oscillatory shear) [56, 100, 101]. This protocol is schematically depicted in Figure 19.9a. Some

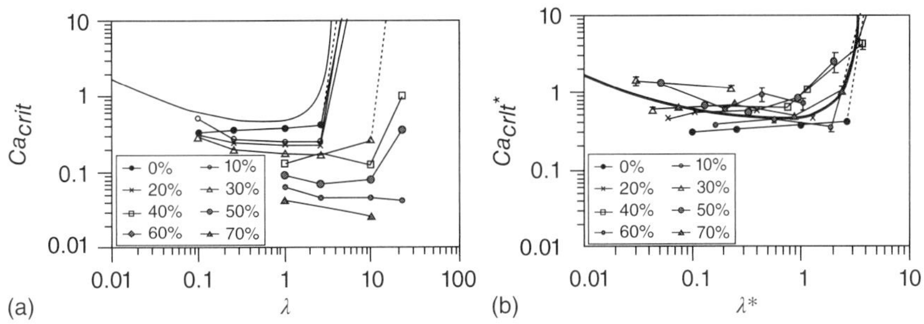

Figure 19.8 Effect of hydrodynamic interactions on the critical *Ca*-number for breakup, full line represents Equation 19.11 and lines with symbols are experimental data; (a) Ca_{crit} based on matrix viscosity, (b) Ca_{crit} and *p* obtained by replacing matrix viscosity by emulsion viscosity. Jansen et al. [96]. Reproduced with permission of American Institute of Physics.

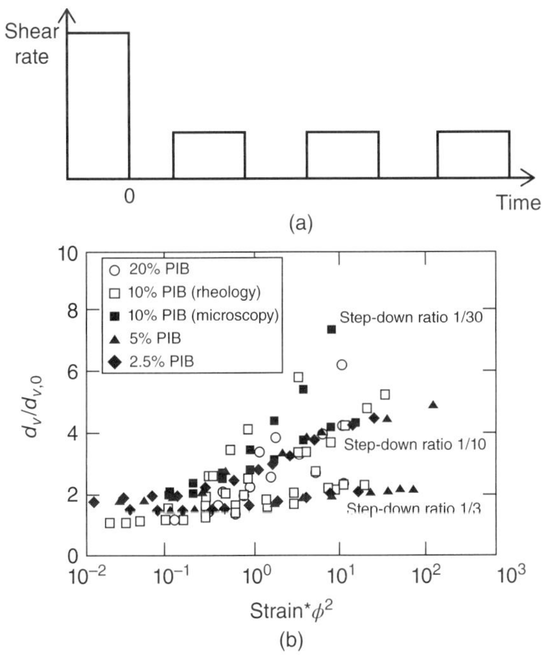

Figure 19.9 Coalescence kinetics after stepdown in shear rate: (a) schematic of the experimental protocol, (b) droplet diameter versus strain scaled with ϕ^2. Vinckier et al. [56]. Reproduced with permission of John Wiley and Sons.

authors found that the linear relation between coalescence rate and blend volume fraction, as predicted by Equation 19.19, remains valid up to rather large concentrations of dispersed phase [57, 100], indicating that coalescence is dominated by binary droplet collisions [100]. Since it can be expected that interacting droplets in a concentrated blend can separate less easily due to the presence of the neighboring droplets, leading to longer contact times, an increased coalescence efficiency can occur in concentrated blends [102]. This was confirmed by experimental results, showing an increase of the droplet size with increasing blend concentration [56, 99–102]. In addition, it was found that the coalescence rate scales with ϕ^2 rather than ϕ for a wide range of systems and concentrations [56, 99, 101]. An example of this scaling of the coalescence kinetics is shown in Figure 19.9b. Minale et al. [102] suggested that the increased coalescence efficiency can be taken into account in Equation 19.20 by using h_{crit} as a fitting parameter. These authors also showed that an increase of the blend concentration leads to a shift of the critical shear rate to lower values thus shifting the region of morphology hysteresis to lower values of the shear rate [102].

19.2.2 Fibrillar Structures

Whereas the dispersed phase is predominantly present as droplets in a continuous matrix when the flow intensity is rather mild, more intense shear or extension results in the formation of elongated fibrils. For *Ca*-numbers that largely exceed the critical value, the droplet dynamics is completely governed by the hydrodynamic forces [103]. If the viscosity ratio is sufficiently close to 1, the dispersed domains will follow the principal stretch of the matrix and their shape can be described by means of the affine deformation equations. Elemans et al. found that the ratio Ca/Ca_{crit} should exceed 2 in order to obtain affine deformation for a system with $p = 0.1$ [104]. For highly extended domains, the aspect ratio *L*/*B* is generally used to quantify the droplet deformation rather than the deformation parameter *D*. In case of affine deformation, the following expression provides the evolution of the aspect ratio *L*/*B* in shear flow [103]:

$$\frac{L}{B} = 1 + \frac{\gamma^2}{2} + \frac{\gamma}{2}\sqrt{\gamma^2 + 4} \tag{19.30}$$

with γ the strain. The third axis *W* on the other hand remains unaltered. This will result in a flattened pancake-like shape. After some time, when the curvature becomes large enough, the interfacial tension favors the formation of a circular cross-section, resulting in pseudo-affine deformation, for which the aspect ratio evolves according to [104]:

$$\frac{L}{B} = \left(1 + \frac{\gamma^2}{2} + \frac{\gamma}{2}\sqrt{\gamma^2 + 4}\right)^{(3/4)} \tag{19.31}$$

while *B* and *W* remain equal. In both cases, the fibril orientation with respect to the flow direction corresponds to [103]:

$$\varphi = \arctan\left(\frac{1}{2}\gamma + \frac{1}{2}\sqrt{4 + \gamma^2}\right)^{-1} \tag{19.32}$$

For blends with a viscosity ratio that differs from unity, the continuity of the stress at the interface causes deviations from affine deformation, with viscosity ratios below 1 leading to increased deformations and viscosity ratios above 1 causing the reverse effect [105]. In addition, for systems with low viscosity ratios, droplet widening can occur at sufficiently large *Ca*-numbers [106]. Contrary to vorticity stretching in systems with viscoelastic components, as illustrated in Figure 19.6a, this phenomenon is always transitory in Newtonian–Newtonian systems [106]. Due to droplet widening, sheets can be generated, an example of which is shown in Figure 19.10a [106]. By sufficiently rapid cooling, these transitory sheets can be frozen in, resulting in a lamellar blend morphology in the end product.

Whereas continuous application of supercritical flow conditions will inevitably lead to droplet breakup, cessation of the flow before breakup can generate different outcomes that can be classified into three main categories. When the dimensionless droplet length at the moment of flow cessation is below a critical value, the droplet will retract to the spherical shape. As mentioned before, for highly extended droplets such retraction can no longer be characterized by an exponential kinetics [38]. As compared to the critical droplet lengths for breakup during flow, quite large deformations are needed to generate breakup after cessation of flow, especially for systems with a very large or a very small viscosity ratio [109]. Hence, plotting the critical droplet length for breakup after flow cessation versus viscosity ratio results in a V-shaped curve with a minimum at viscosity ratios between 0.1 and 1 [109]. Contrary to droplet breakup in bulk shear flow, this curve does not show an asymptote at a viscosity ratio of 4, which means that if sufficiently long fibrils are generated due to the mixed flow conditions in processing equipment, breakup can subsequently occur in zones of less intense flow, even when the viscosity ratio is above 4. For droplet lengths slightly above the critical value, breakup occurs by endpinching, a process in which the ends of the droplet bulb up and pinch off [18, 109]. Finally, if very long fibrils are formed, growth of capillary waves becomes the dominant breakup mechanism, generating a string of small daughter droplets [18, 109].

Capillary waves originate from tiny disturbances along the surface of the fibril. Rayleigh showed as early as 1879 that such disturbances are unstable and will hence grow if their wavelength is larger than the circumference of the fibril [110]. This is caused by the fact that in that case growth of the disturbances leads to a reduction in surface area. Tomotika further extended the work of Rayleigh to the case of a liquid thread immersed in another liquid [111]. To determine the kinetics of fibril breakup, the distortions at the fibril surface were assumed to be sinusoidal (an assumption valid for any arbitrary axisymmetric distortion as such a distortion can be decomposed into a sum of sine waves) with an amplitude α, as indicated in Figure 19.10b. Then, the growth of the instabilities can be described by [111, 112]:

$$\alpha = \alpha_0 \exp(qt) \tag{19.33}$$

with α_0 the initial amplitude of the disturbance and q the disturbance growth rate. If the instabilities originate from thermal fluctuations, the initial disturbance amplitude is given by [113]:

$$\alpha_0 = \sqrt{\frac{21 k_B T}{8\pi^{3/2}\Gamma}} \tag{19.34}$$

with k_B the Boltzmann constant, T the absolute temperature and Γ the interfacial tension. Obviously, interactions with other droplets or fibrils or the presence of flow fluctuations might cause larger initial disturbances [112, 114]. The growth rate of the disturbances depends on the disturbance wavelength λ and

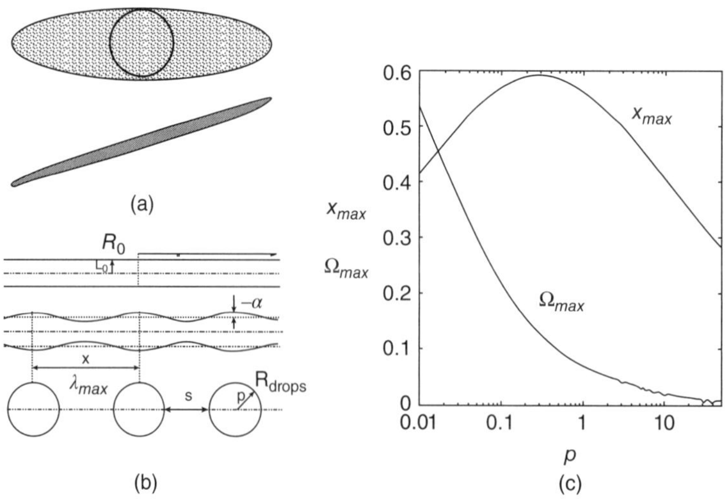

Figure 19.10 (a) Lamellar structure formed due to droplet widening. Cristini et al. [106]. Reproduced with permission of American Chemical Society. (b) Schematic representation of the breakup of a fibril by Rayleigh instabilities. Adapted from Mewis et al. [107]. Reproduced with permission of Elsevier. (c) Dimensionless wave number x_{max} and its corresponding dimensionless growth rate Ω_{max} as a function of the viscosity ratio p. Knops et al. [108]. Reproduced with permission of John Wiley and Sons.

the fibril diameter as well as the system properties [111]:

$$q = \frac{\Gamma}{2\eta_m R_0}\Omega(\lambda, p) \quad (19.35)$$

with R_0 the diameter of the undisturbed fibril as indicated in Figure 19.10b. The maximum of Ω versus λ provides the fastest growing or dominant wavelength λ_{max} which is the one that will eventually cause fibril breakup and determine the size of the daughter droplets, as indicated in Figure 19.10b. This final droplet size can be calculated from volume conservation, resulting in [103, 108]:

$$R_{drops} = R_0 \left(\frac{3\pi}{2x_{max}} \right)^{1/3} \quad (19.36)$$

With x_{max} the wavenumber corresponding to the wavelength λ_{max}. Figure 19.10c plots the values of x_{max} and the corresponding Ω as a function of viscosity ratio, as obtained by Tomotika from a linear (small amplitude) analysis. Based on this figure, it can be concluded that breakup occurs faster for systems with a lower viscosity ratio. In addition, Figure 19.10c and Equation 19.36 show that the smallest droplets are formed for viscosity ratios between 0.1 and 1. In addition to the primary daughter droplets, nonlinear effects at the end of the breakup process, which are not included in the Tomotika theory, result in the additional formation of small satellite droplets. The number and relative size of these satellite droplets is strongly dependent on the viscosity ratio [115]. Using Equation 19.33 and the fact that breakup occurs when the amplitude of the instability equals the average thread radius, the following expression is obtained for the breakup time of long fibrils in quiescent conditions [103, 104]:

$$t_{fibril}^{b,quiescent} = \frac{1}{q} \ln \left(\frac{\sqrt{\frac{2}{3}} R_0}{\alpha_0} \right) \quad (19.37)$$

Similar to the breakup of droplets, both the critical breakup time and the critical breakup length should be exceeded in order for fibril breakup to occur.

The above description and equations are valid for thread breakup under quiescent conditions. Obviously, if no flow cessation occurs after fibril formation, fibril extension continues while simultaneously disturbances on the fibril interface can grow and lead to breakup. It has been shown that the presence of flow postpones breakup of the fibril due to the fact that interfacial disturbances are damped [116]. In addition, the stretch of the fibrils results in a continuous shift of the wavelength with the fastest growth rate whereas nonaxisymmetric flows such as shear flow induce asymmetry in the disturbances [117]. Since fibrils are stretched further before breakup occurs, breakup during flow results in smaller daughter droplets as compared to fibril breakup after flow cessation [114]. In addition, for this transient breakup mechanism the finest dispersion can be generated for systems with a high viscosity ratio and at high flow intensities [114]. Khakhar and Ottino [117] developed a theory describing the breakup of fibrils in flow, which can be used to determine the breakup time and size of the daughter droplets for different flow conditions and viscosity ratios [114]. However, this model is not very practical to obtain breakup times as it is limited to the actual breakup stage, which is only initiated after fibril formation. Van Puyvelde et al. on the other hand derived a scaling relation for the breakup time that includes both the deformation and breakup stages [118]:

$$t_{fibril}^{b,flow} \dot{\gamma} \sim Ca^{2/3} \quad (19.38)$$

This model allows a more straightforward prediction of fibril breakup times during processing, for which the moment of initiation of interface disturbances can in general not be determined.

Similar to droplet–matrix systems, also for fibrils the effects of component viscoelasticity, compatibilization and concentration have been investigated to a certain extent. Lee et al. [119] show, both with a linear stability analysis and experimental work, that the growth rate of the instabilities on a long fibril is increased due to elasticity of either the matrix or the fibril fluid. In addition, the wavelength of the instabilities can significantly differ from that for systems with Newtonian components, with the effect of viscoelasticity being dependent on the viscosity ratio of the system [119]. In addition, at the end of the breakup process, nonlinear effects can emerge and significantly alter the breakup process. For example, in the case of viscoelastic fibrils, the growth of the instabilities can slow down and beads-on-a-string structures can be formed due to the presence of a substantial elongational viscosity [112, 119]. Nevertheless, it is worthwhile to note that in case of molten polymers, the deformation rates might become very slow due to the high viscosity of the materials. Hence, viscoelasticity effects on fibril breakup in processing of molten polymer blends can often be neglected [112]. Also compatibilizer has been shown to affect the growth rate of the disturbances and their wavelength [120]. Similar to droplet–matrix systems, several factors including a reduction of the surface tension and the presence of Marangoni stresses due to gradients in interfacial tension play a role [120]. From a practical point of view, an important consequence of the presence of compatibilizer is the fact that the generated daughter droplets are in general less uniform in size [121]. When multiple viscous fibrils are present, they can affect each other, leading to a delay of the development of large-amplitude distortions [122]. In addition, breakup will occur with the disturbances on the different threads being either in-phase or out-of-phase [108]. In conclusion, by tailoring the flow conditions, component properties and cooling speed, a whole range of fibrillar, lamellar and ordered droplet–matrix structures can be generated from extended fibrils.

19.2.3 Cocontinuous Structures

As described in the previous sections, blending two immiscible polymers generally results in a material that contains dispersed domains of one component in the other. It is obvious that the polymer with the highest volume fraction will preferentially form the continuous phase. Thus, when increasing the volume fraction of dispersed phase, the phases will reverse at a certain composition, which is termed the phase inversion composition [22]. However, also the viscosity ratio has an impact on the phase continuity. A lower viscosity increases the tendency of a phase to become continuous as this will minimize energy dissipation during flow [22]. Hence, when mapping out the phase continuity in a concentration–viscosity ratio diagram, the phase inversion concentration is situated around 50% dispersed phase for blends containing equiviscous components but shifts to higher or lower concentrations for other viscosity ratios [22], as shown in Figure 19.11. In a concentration region around the phase inversion composition, both phases can be continuous, resulting in a cocontinuous structure. In a fully cocontinuous structure, each phase forms a three-dimensional spatially continuous network throughout the material [22]. In some cases, only part of each component forms a continuous structure whereas the remaining part is present as dispersed domains [22, 123]. These partially cocontinuous structures can be characterized by their degree of cocontinuity, which represents the fraction of material that belongs to a continuous network [22]. Several models have been developed to predict the phase inversion concentration and hence provide an estimate of the blend compositions resulting in cocontinuous structures. Using the simple arguments leading to Figure 19.11, a straightforward relationship providing the phase inversion composition from the blend viscosity ratio has been derived [124]:

$$\phi_{1,PI}\eta_2 = \phi_{2,PI}\eta_1 \tag{19.39}$$

with $\phi_{1,PI}$ and $\phi_{2,PI}$ the volume fraction of components 1 and 2, respectively, at the phase inversion composition. Several empirical modifications to Equation 19.39 have been proposed, often providing better predictions of the phase inversion composition for a specific blend system. In addition, physical arguments based on maximum packing fraction, percolation theory and fibril stability have led to a range of theory-based models for the phase inversion composition. A comprehensive overview of these models can be found in the works of Pötschke and Paul [22] and Harrats et al. [8]. Despite the quantitative differences between these models, the phase inversion concentration versus viscosity ratio relation remains qualitatively the same as that presented in Figure 19.11 [22]. Finally, it should be noted that not only the viscosity ratio, but also other component properties can affect the phase inversion concentration and the composition range in which cocontinuous structures are obtained. The latter range has been shown to broaden for systems with a lower interfacial tension [125]. Although some models that include the effects of component elasticity on the phase inversion composition have already been proposed [126], the impact of component rheology on phase inversion and cocontinuity is far from fully understood [22].

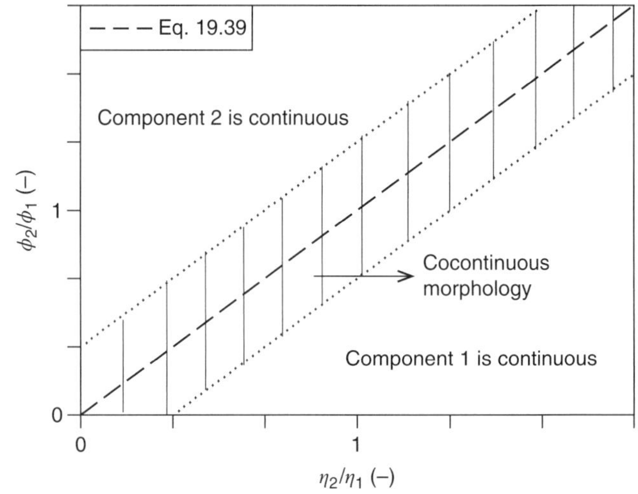

Figure 19.11 Volume fraction – viscosity ratio diagram indicating the occurrence of cocontinuous phase morphologies.

Polymer blending is often aimed at obtaining a cocontinuous structure since in such structures both components contribute more equally to the properties as compared to blends consisting of dispersed domains in a continuous matrix [22]. With respect to mechanical properties, it is important to note that cocontinuous structures lead to a more effective stress transfer between the different phases which is beneficial for, for example, mechanical modulus and impact strength [22]. Tailoring the flow conditions to generate cocontinuous structures requires fundamental knowledge of the development of cocontinuous structures in flow. During the initial stages of mixing, polymer pellets melt and deform into thin sheets or ribbons which subsequently undergo breakup resulting in extended irregular structures that may further break up into fibrils and/or dispersed droplets in a continuous matrix [127]. The mechanism of formation of cocontinuous structures is not yet fully understood but it has been established that cocontinuous structures are formed when either the irregular network formed in the initial stages of mixing is stable during processing or coalescence events reconnect dispersed fibrils and droplets into a network structure [22]. Clearly, in many systems, cocontinuous structures are an intermediate state in the early mixing stages whereas droplet–matrix morphologies are obtained after processing. Also when phase inversion occurs during processing, a cocontinuous morphology will be formed as an intermediate structure [22]. Phase inversion can occur when the minor phase melts first or when the viscosity ratio changes substantially during processing due to either reactions or changes in processing conditions (flow type and intensity or temperature). If mixing or processing is stopped while the intermediate cocontinuous structure is present, the morphology can be solidified by, for example, quenching, resulting in materials with a cocontinuous morphology.

Although solidification provides a possibility to maintain the cocontinuous morphology even though it is not stable, this approach is not practically feasible when additional mixing, annealing or processing steps are required. For such cases, the stability of the cocontinuous morphology is essential. There are two processes that contribute to the destabilization of cocontinuous morphologies namely breakup and coarsening. Since cocontinuous structures consist of a three-dimensional network of interconnected fibril-like structures, the mechanisms and predictive equations provided in Section 19.2.2 on fibrillar structures can be applied to predict stability of cocontinuous morphologies. Based on Equation 19.35 it can thus be concluded that the stability of cocontinuous structures against breakup is enhanced by decreasing interfacial tension and increasing matrix viscosity, viscosity ratio, and fibril diameter. The latter aspect is the reason for the decrease after annealing of the composition range over which cocontinuous structures are obtained since low contents of the dispersed phase generally result in cocontinuous structures with thinner threads [22]. Since flow is known to delay fibril breakup [116], it can also stabilize cocontinuous morphologies. An alternative process that can alter the characteristics of a cocontinuous morphology, eventually leading to macroscopic phase separation, is coarsening. Similar to breakup, this process is driven by interfacial tension and slowed down by an increase in viscosity. Based on these arguments, the following relation has been derived for the coarsening rate of cocontinuous structures [128]:

$$\frac{dr}{dt} = \frac{a\Gamma}{\eta_{blend}} \tag{19.40}$$

with r the average thickness of the network ligaments, Γ the interfacial tension, a a dimensionless parameter and η_{blend} the blend viscosity. Doi and Ohta [34] developed a model that describes the morphology and rheology of blends consisting of equiviscous components blended in equal amounts. This model is able to describe the dynamics of the interface both in the presence and in the absence of flow by means of the evolution of the interfacial area and the interface anisotropy. In case of quiescent coarsening, the scaling relation provided by Equation 19.40 is retrieved from this model [129]. Based on Equations 19.35 and 19.40 it can be concluded that lowering the interfacial tension stabilizes cocontinuous morphologies. Hence, addition of compatibilizer typically leads to more stable cocontinuous structures [8, 130]. Similar to droplet–matrix structures, as discussed in Section 19.2.1.5, other effects such as Marangoni stresses can also contribute to this stabilization [22]. However, due to the compatibilization action, the formed cocontinuous structures are in general also finer with thinner ligament sizes, which generally leads to a narrowing of the composition range over which cocontinuous structures are formed [8, 123].

19.3 RECENT ADVANCES IN POLYMER BLENDS

19.3.1 Immiscible Blends in Confined Flow

The past decades have witnessed a rapid development of micro- and nanotechnology that is being used in a wide variety of applications. The corresponding growing demand for microcomponents in electronical, biomedical and mechanical applications as well as the synthesis of high-performance materials in increasingly small quantities, has resulted in the widespread use of micro-scale polymer mixing and polymer processing equipment [131–133]. When multiphasic materials such as polymer blends are processed in these devices, the dimensions of the channels might be of the same order of magnitude as the size of the dispersed phase. Hence, deviations from bulk behavior can be expected, causing the need for incorporation of wall effects, or so-called confinement effects, in the design and modeling of miniaturized polymer processing equipment.

The research on confined multiphase liquid–liquid systems can roughly be divided into two main categories namely multiphase microfluidics and morphology development in confined shear flow. In the present work, a brief description of the main achievements of the second type of studies will be provided. It should be noted that morphology development in confined flow is an active research area at present and the effects of confinement on the dynamics of multiphase systems are far from fully understood. Detailed descriptions of the present state of the art can be found in several reviews [61, 134–136].

The deformation and orientation of sheared droplets in confinement have been studied by experimental, analytical and numerical methods [134]. Consensus exists that geometrical confinement increases the droplet deformation and its orientation with respect to the flow direction, mainly for systems with a large viscosity ratio [137]. Shapira and Haber determined an analytical solution for the droplet deformation in confined flow, up to the first order in Ca, which corresponds to the Taylor deformation parameter multiplied by an additional factor to take into account confinement [138]:

$$D_{SH} = D_{Taylor} \cdot \left[1 + C_S \frac{1 + 2.5p}{8(1+p)} \left(\frac{2R}{H}\right)^3\right] \tag{19.41}$$

with C_s a parameter that depends on the relative position of the droplet between both walls (h'/H with h' the distance from the droplet center to the closest wall and H the gap spacing as indicated in Figure 19.2a) and $2R/H$ the confinement ratio. Equation 19.41 shows that due to confinement two additional dimensionless parameters come into play. It was experimentally verified that the Shapira–Haber model provides good predictions of the droplet deformation parameter in confined shear flow [32, 137]. An example can be found in Figure 19.3a. However, the model only predicts an increase of the droplet deformation whereas the droplet shape and orientation angle are expected to remain unaltered. Starting from the Maffettone–Minale model, given by Equation 19.6, Minale [139] developed a phenomenological model for the dynamics of droplets in generic confined flows. In this confined Minale model [139], Equation 19.6 for **S** was taken over, but altered expressions for f_1 and f_2 as a function of λ, $2R/H$ and h'/H were derived. This adapted model provides good predictions of the droplet deformation up to large values of the confinement ratio and *Ca*-number [139]. Contrary to the models, numerical simulations and experimental studies showed that in confinement, the droplet shape can transform from ellipsoidal to sigmoidal [140]. This is caused by the large recirculation zones at the front and rear of confined droplets, large shear rates at the droplet tips and increased pressure drop in the wedge between the droplet and the walls [140]. In addition to altering the steady state droplet deformation, confinement can also have substantial effects on the droplet dynamics during startup of shear flow and after cessation of flow. More in particular, hydrodynamic interactions between the deforming droplet and the walls can lead to periodic tumbling and extension–retraction cycles after startup of shear flow, resulting in a droplet deformation that goes through several maxima and minima before reaching a steady state value [137, 140]. This effect can increase the time needed to reach a steady state droplet deformation with an order of magnitude [137, 140]. For confinement ratios above 0.5, also the retraction of the droplet shape after cessation of shear flow becomes substantially slower as compared to that in bulk conditions [134, 141]. This is shown in Figure 19.3b, which also illustrates the validity of the confined Minale model for predicting the dynamics of confined droplets.

Similar to bulk conditions, the droplet size in confined polymer blends is determined by a combination of droplet breakup and coalescence. The effects of geometrical confinement on droplet breakup are largely affected by the viscosity ratio of the system: for viscosity ratios below 1, droplet breakup is inhibited by confinement whereas for viscosity ratios above 1, it is enhanced by confinement [142]. Even droplets in droplet–matrix pairs with a viscosity ratio above 4, that are unbreakable in bulk shear flow, can be broken in confined shear flow. In addition, at confinement ratios close to 1, a transition has been observed from binary to ternary breakup [137]. Rather than generating two main daughter droplets with some small central fragments, confined breakup can provide three or more equal-sized droplets. This phenomenon provides a means of generating very monodisperse droplet–matrix systems [137]. In fibrillar systems geometrical confinement slows down thread breakup and causes a transition to nonaxisymmetric fibril shapes at large confinement ratios [143, 144]. However, the critical droplet elongation above which breakup occurs remains unaltered and overall stability of long quiescent threads is not obtained, as long as the thread diameter remains smaller than the gap width [143, 145, 146]. Contrary to droplet breakup, droplet coalescence is promoted by geometrical confinement, irrespective of the viscosity ratio [147, 148]. Hence, the overall effect of geometrical confinement on the steady state droplet size in polymer blends depends on the viscosity ratio and on the relative contributions of droplet breakup and coalescence.

In addition to the extensive amount of work on the dynamics of confined Newtonian droplets in a Newtonian matrix [134], also the combined effects of geometrical confinement and either compatibilization by means of block copolymers or viscoelasticity of one of the blend phases has received attention. Cardinaels et al. have shown that the effects of matrix viscoelasticity on droplet dynamics are much more pronounced in confined conditions as compared to bulk conditions [149, 150]. This was attributed to the substantial increase of the elongation rates and viscoelastic stresses in and around the droplet due to geometrical confinement [150]. For example, even for cases in which matrix viscoelasticity has limited effects on the bulk critical *Ca*-number, it can reduce the critical *Ca*-number, critical droplet length and breakup length in confined conditions by more than a factor 2 [149]. Whereas morphology development of polymer blends in bulk shear flow can generally be well-described by means of models for Newtonian–Newtonian systems, incorporation of effects of matrix viscoelasticity is thus expected to be essential in the case of confined blends. Viscoelasticity of the droplet phase and the presence of block copolymers at the droplet interface on the other hand have less pronounced effects on the dynamics of confined droplets [149, 151, 152].

The altered breakup and coalescence behavior in confined shear flow as compared to bulk shear flow, in combination with changed hydrodynamic interactions amongst droplets and between the droplets and the walls, leads to the formation of a range of interesting structures in nondilute blends in confined shear flow. Migler observed transitions from a droplet–matrix morphology, via ordered pearl necklace structures and strings to ribbons when the size of the droplets became comparable to the gap between the shearing surfaces [145]. In addition, confinement can lead to the formation of layered morphologies [153], squashed droplets [154] and vorticity bands [155]. The latter are alternating regions with high and low concentrations of droplets. Images of the different peculiar structures that can be exhibited by polymer blends in confined flow are provided in Figure 19.12. These structures have potential for the generation of polymeric materials for applications such as fibers, scaffolds or polymer sheets with anisotropic properties [145].

19.3.2 Blend Compatibilization by Nanoparticles

Addition of nanoparticles to polymers, resulting in polymer nanocomposites, has been used for decades to improve the mechanical, electrical and thermal properties of polymeric systems [156–162]. Recently, it has been realized that nanoparticles can also refine and stabilize phase morphology in polymer blends, similar to the case of low-viscous Pickering emulsions. This discovery has triggered a renewed research interest in the morphology and properties of particle-filled polymer blends [163]. A wide variety of nanoparticles, with differences in particle size, shape and surface properties are being used with carbon black, silica, clay and more recently carbon nanotubes and graphene sheets the most popular ones [163–166]. Nanoparticles have two main advantages as compared to copolymers. First, nanoparticles can combine the functionalities of a filler and a compatibilizer, thus enhancing certain blend properties such as electrical or mechanical characteristics in addition to refining and stabilizing the blend

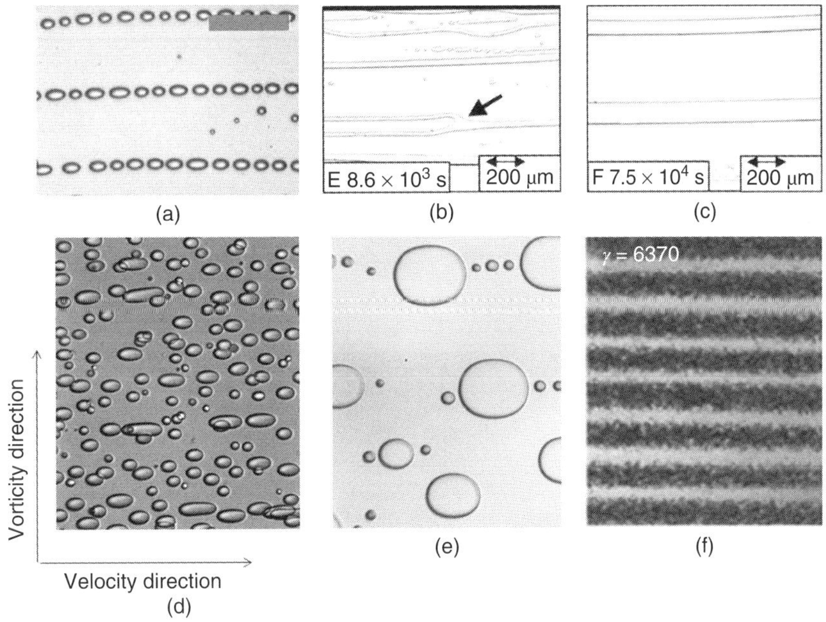

Figure 19.12 Typical structures obtained in confined polymer blends in shear flow; (a) pearl necklaces. Pathak et al. [153]. Reproduced with permission of Elsevier. (b) Strings. Migler et al. [145]. Reproduced with permission of American Physical Society. (c) Ribbons. Migler et al. [145]. Reproduced with permission of American Physical Society. (d) Layered morphology. Courtesy P. De Bruyn. (e) Squashed droplets. Courtesy P. De Bruyn. (f) Vorticity bands. Caserta and Guido [155]. Reproduced with permission of American Chemical Society.

morphology [159]. Second, contrary to copolymers, they do not have to be synthesized exclusively for a certain blend system but have a more general applicability.

Despite the large potential of nanoparticles in the development of multifunctional polymer blend materials, exploiting their full potential for different commercial polymer blend systems remains challenging. Mainly the dispersion and localization of nanoparticles in a polymer blend are essential for the blend morphology and properties [163]. Due to the large surface to volume ratio of nanoparticles, they have a tendency to aggregate due to Van der Waals and other attractive forces such as electrostatic interactions [156, 158, 161]. Hence, processing protocols have to be adapted in order to improve particle dispersion. As melt compounding is the most versatile and industrially valuable strategy for polymer blend production, most efforts have been devoted to optimizing extrusion conditions in order to enhance particle dispersion. Nevertheless, various other techniques such as solvent mixing, possibly combined with high shear mixing or ultrasonication, and in situ polymerization have proven to be of interest for the preparation of polymer nanocomposites [156, 157, 160, 161]. In addition to dispersion, also the localization of the nanoparticles in the blend is crucial [163]. It is known that morphology refinement and stabilization can be achieved with a minimum amount of particles when they are localized at the blend interface [163, 167]. On the other hand, in order to develop electrically conductive polymer blends, selective localization of nanoparticles in the continuous blend phase, leading to so-called double percolation is often aimed at [163, 164, 166, 168–170]. The equilibrium location of nanoparticles, as derived from the thermodynamics principle of minimization of interfacial energy, can be determined by means of the wetting parameter ω [163]:

$$\omega = \frac{\Gamma_{s1} - \Gamma_{s2}}{\Gamma_{12}} \tag{19.42}$$

in which Γ_{s1}, Γ_{s2}, and Γ_{12} are the interfacial tensions between respectively the particles and polymer 1, the particles and polymer 2 and between both polymers. When the wetting parameter has a value between −1 and 1, the particles will be preferentially localized at the blend interface. It should be noted here that in many cases, the nanoparticle location is determined by kinetic rather than thermodynamic effects [163]. When particles are initially not localized in the preferred phase, they tend to migrate [171, 172]. However, the time available for particle migration during processing might be too short to reach the equilibrium location. In addition, particles might be trapped in a nonpreferred phase due to specific irreversible interactions with one of the blend components [173–175]. Although kinetic effects complicate predictions of nanoparticle location after processing, they also provide a means of tailoring particle location by controlling mixing conditions such as mixing time, mixing speed and the use of master batches in which particles are predispersed in one of the phases [171, 176, 177]. Finally, it is important to keep in mind that in general only a small amount of nanoparticles is needed to fully saturate the interface [178]. Any excess amount of particles is driven to one of the blend phases [178–180].

Similar to copolymers, the effects of nanoparticles on blend morphology are governed by a complex interplay of

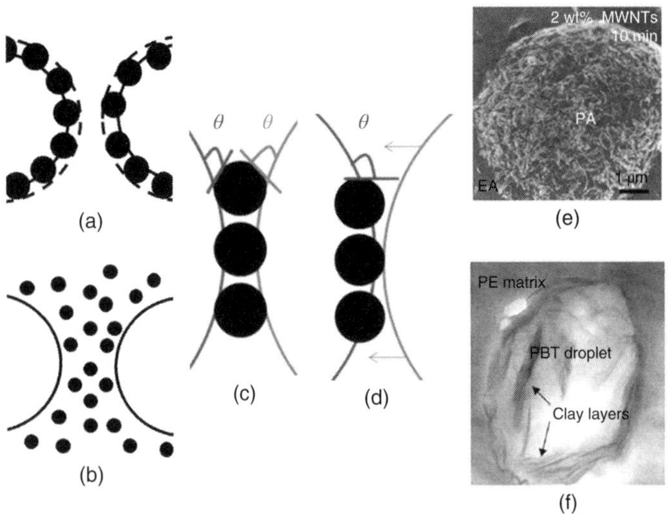

Figure 19.13 Different mechanisms by which particles can affect droplet dynamics; (a) steric hindrance from particles at the interface, (b) steric hindrance from particles in the bulk phase, (c) particle-induced droplet bridging, (d) bridging–dewetting, (e) multiwall carbon nanotubes at the droplet interface. Baudouin et al. [181]. Reproduced with permission of Elsevier. (f) Clay platelets at the droplet interface. Hong et al. [179]. Reproduced with permission of Elsevier.

multiple mechanisms. When localized at the blend interface, particles can suppress droplet coalescence due to steric hindrance caused by particles protruding into the matrix phase [163, 167, 179, 180], as shown in Figure 19.13a. However, protruding particles can evoke other phenomena, depending on whether the preferred phase is the continuous or the dispersed phase [167, 182]. When the particles are preferentially wetted by the continuous phase, droplet clusters can be formed due to particle-induced bridging in which a particle bridges two droplets while the location of the liquid–solid contact line, which is governed by the contact angle, prevents coalescence [169, 183, 184]. This is illustrated in Figure 19.13c. On the other hand, when the particle is preferentially wetted by the dispersed phase, the bridging–dewetting phenomenon can occur, in which the simultaneous wetting of the particle by two droplets leads to rapid coalescence [182]. The bridging–dewetting phenomenon is depicted in Figure 19.13d. Due to the presence of particles at the blend interface, part of the liquid–liquid interface is replaced by solid–liquid interface. For such systems, the concept of an effective interfacial tension has been introduced [163, 185]. This effective interfacial tension is defined as the interfacial energy per unit area over an area that is much larger than the particle size. Several authors have found that this effective interfacial tension reduces when particles are located at the blend interface [178, 180, 186, 187]. Hence, based on Equations 19.12 and 19.20 morphology refinement can be expected. However, whereas the interfacial tension reduction is often considered to be the main mechanism in copolymer compatibilization, it has been suggested that effects of nanoparticles on blend morphology might be dominated by the interfacial rheology of particle-covered interfaces [188]. Particles at interfaces can self-assemble into various structures thereby generating a strongly elastic and thus less mobile interface [189, 190]. This will substantially hinder film drainage during coalescence [188] but may also affect droplet breakup, an aspect that has received much less attention up to now.

Another substantial difference between nanoparticles and copolymers is their effect on the bulk phases. On the one hand, copolymers are less likely to become kinetically trapped in one of the bulk phases and on the other hand, their effect on the component rheology is much less pronounced. However, very small concentrations of nanoparticles can cause a substantial increase in the viscosity and elasticity of the blend components [160–162]. In addition, nanoparticles can form a percolated particle network which results in a solid-like behavior [160–162]. This obviously affects the different dynamic processes that are described in Section 19.2. Besides affecting morphology development by changes in rheology, which can to a large extent be taken into account in modeling by using adapted values for the viscosity and normal stresses in the equations that are provided in Section 19.2, particles in the matrix phase also provide steric effects, which are less easy to quantify [163, 191]. Figure 19.13b illustrates how steric hindrance due to particles in the matrix phase can affect film drainage. Finally, it should be noted that both in the case of particles located at the blend interface or in the droplet phase, geometry effects can play a role limiting the minimum size of the dispersed domain to the order of magnitude of the particle size [179, 192]. As a consequence, particle-covered droplets can take on a nonspherical equilibrium shape since the interface has to accommodate the particles, especially in the case of platelike fillers such as organoclay [179]. Figure 19.13e and f provides images of droplets covered with nanoparticles, illustrating the nonspherical droplet shape in case of nanoclay.

Based on the previous discussion about the various, often interfering, mechanisms by which nanoparticles can affect

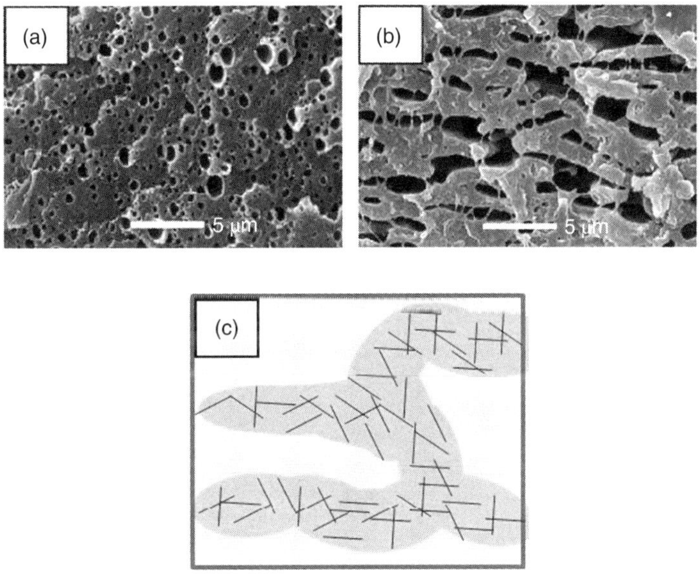

Figure 19.14 Transition from droplet–matrix structure to cocontinuous structure; (a) Neat 60/40 PPS/PA66 blend. Zou et al. [193]. Reproduced with permission of Elsevier. (b) 60/40 PPS/PA66 blend with 0,3 phr multiwall carbon nanotubes. Zou et al. [193]. Reproduced with permission of Elsevier. (c) Schematic representation of percolated particle network in a cocontinuous structure.

morphology development in polymer blends, it is clear that the effects of nanoparticles on blend morphology can often not be predicted a priori. In order to be able to use the equations presented in Section 19.2 to predict blend morphology, knowledge of the particle location and subsequently the component and interfacial rheology in the presence of particles is required. However, additional effects such as steric hindrance, bridging, and geometrical constrains should not be overlooked. Despite the complexity of particle-filled polymer blend systems, morphology refinement and stabilization, both in case of droplet–matrix structures and in case of cocontinuous structures has been achieved with a wide range of particle types, sizes and shapes and for various particle locations [163]. Nevertheless, also an increase of the droplet size after particle addition has been observed [163, 179, 182]. In addition to morphology refinement, the presence of nanoparticles can cause transitions from the droplet–matrix structure to a cocontinuous morphology [193–195]. This is exemplified in Figure 19.14a and b for a blend containing carbon nanotubes. The transition can originate from changes in the viscosity ratio due the presence of the particles, which alters the phase inversion concentration as shown in Equation 19.39 and Figure 19.11. But also the presence of a percolated particle network, as depicted in Figure 19.14c, can play a role [194]. Such transitions are, in addition to the particle properties, very sensitive to the processing conditions. For example, Hong et al. [172] show that addition of organoclay only leads to cocontinuous structures at relatively high shear rates whereas Wu et al. [196] illustrate the opposite trend. Though less common as compared to transitions from the droplet–matrix to a cocontinuous morphology, also the reverse transition has been observed [178, 193].

19.4 CONCLUSIONS

Polymer blending is a valuable strategy to meet the increasingly stringent material requirements that are brought about by the rising consumer demands and advancing technologies of the last decades. In addition, it provides an opportunity to create added value from mixed polymer waste streams. Since most polymers are immiscible, multiphasic materials are generally formed after blending. The properties of immiscible blends are, in addition to the concentration and component properties, determined by the phase morphology. The latter is formed during mixing and further processing, and is thus dependent on the processing conditions. Hence, a fundamental understanding of the relations between on the one hand blend composition, component properties and processing conditions and on the other hand blend morphology is essential to tailor the properties of immiscible polymer blends. In the present chapter, morphology development of immiscible polymer blends is reviewed for the three most prevalent blend morphologies namely droplet–matrix, fibrillar and cocontinuous structures, both in flow and under quiescent conditions. The focus is on morphology development in model flow conditions, and more in particular shear flow as this is the main flow component in many processing, mixing and other operations that involve rotating or moving parts. For the case of dilute or semi-dilute polymer blends consisting of components with limited elasticity and in absence of interfacial modifiers, an in-depth mechanistic understanding of the dynamics of droplets and fibrils including deformation, retraction, breakup and coalescence is available. In addition, combining the governing equations for each of these processes results in a predictive framework that allows modeling of morphology development in immiscible blends. However, when the blend concentration is close to the phase inversion concentration, a cocontinuous phase morphology is frequently displayed. Contrary to the availability of scientifically based design equations for the morphology of dilute and semi-dilute polymer blends, quantification of the conditions for the formation and stability of cocontinuous structures is less straightforward. Finally, despite the fact that polymer blending is industrially used for quite a long time to develop polymeric materials with properties that are a synergistic combination of those of the components, there still is a knowledge gap between the industrial processing of polymer blends and the scientifically derived design equations. First of all, the effects of component viscoelasticity, compatibilization and hydrodynamic interactions on morphology development are only known up to a certain extent and need to be further explored. Second, recent developments in the areas of micro- and nanotechnology have triggered the interest in on the one hand morphology development in confined flow conditions and on the other hand particle-filled blends. Both show potential for the development of multifunctional materials with specific morphologies. However, in order to utilize their full potential, additional research into the interplay between the various mechanisms that play a role is essential.

ACKNOWLEDGMENTS

Ruth Cardinaels is indebted to the Research Foundation Flanders (FWO) for a postdoctoral fellowship at KU Leuven.

REFERENCES

1. Utracki LA, editor. Vol. 1. *Polymer Blends Handbook*. Kluwer Academic Publishers; 2002.
2. Utracki LA. History of commercial polymer alloys and blends (from a perspective of the patent literature). Polym Eng Sci 1995;35(1):2–17.
3. Schlechter M. *Engineering Resins, Polymer Alloys and Blends: Global Markets*. BCC Research; 2013.
4. Fainleib A, Grigoryeva O, editors. *Recent Developments in Polymer Recycling*. Transworld Research Network; 2011.
5. Anyadike N. Mixed plastics creates recycling challenge in Europe. Eur Plast News 2013, http://www.plasticsnews.com/article/20130417/NEWS/130419923/mixed-plastics-creates-recycling-challenge-in-europe.
6. Lipatov YS, Nesterov AE. *Thermodynamics of Polymer Blends*. Basel: Technomic Publishing AG; 1997.
7. Colmenero J, Arbe A. Segmental dynamics in miscible polymer blends: Recent results and open questions. Soft Matter 2007;3(12):1474–1485.
8. Harrats C, Thomas S, Groeninckx G, editors. *Micro- and Nanostructured Multiphase Polymer Blend Systems*. Boca Raton: Taylor & Francis Group; 2006.
9. Shokoohi S, Arefazar A. A review on ternary immiscible polymer blends: Morphology and effective parameters. Polym Adv Technol 2009;20(5):433–447.
10. Füller G. *Optical Rheometry of Complex Fluids*. New York: Oxford University Press; 1995.
11. Van Puyvelde P, Moldenaers P. Rheology and morphology development in immiscible polymer blends. Rheol Rev 2005:101–145.
12. DeRoussel P, Khakhar DV, Ottino JM. Mixing of viscous immiscible liquids. Part 1: Computational models for strong-weak and continuous flow systems. Chem Eng Sci 2001;56(19):5511–5529.
13. Windhab EJ et al. Emulsion processing – From single-drop deformation to design of complex processes and products. Chem Eng Sci 2005;60(8–9):2101–2113.
14. Domingues N, Gaspar-Cunha A, Covas JA. Estimation of the morphology development of immiscible liquid–liquid systems during single screw extrusion. Polym Eng Sci 2010;50(11):2194–2204.
15. Fischer P, Erni P. Emulsion drops in external flow fields – The role of liquid interfaces. Curr Opin Colloid Interface Sci 2007;12(4–5):196–205.
16. Guido S, Greco F. Dynamics of a liquid drop in a flowing immiscible liquid. Rheol Rev 2004:99–142.
17. Rallison JM. The deformation of small viscous drops and bubbles in shear flows. Annu Rev Fluid Mech 1984;16:45–66.
18. Stone HA. Dynamics of drop deformation and breakup in viscous fluids. Annu Rev Fluid Mech 1994;26:65–102.
19. Briscoe BJ, Lawrence CJ, Mietus WGP. A review of immiscible fluid mixing. Adv Colloid Interface Sci 1999;81(1):1–17.
20. Minale M. Models for the deformation of a single ellipsoidal drop: A review. Rheol Acta 2010;49(8):789–806.
21. Janssen PJA, Anderson PD. Modeling film drainage and coalescence of drops in a viscous fluid. Macromol Mater Eng 2011;296(3–4):238–248.
22. Pötschke P, Paul DR. Formation of co-continuous structures in melt-mixed immiscible polymer blends. J Macromol Sci C 2003;43(1):87–141.
23. Van Puyvelde P, Velankar S, Moldenaers P. Rheology and morphology of compatibilized polymer blends. Curr Opin Colloid Interface Sci 2001;6(5–6):457–463.
24. Di Lorenzo ML, Frigione M. Compatibilization criteria and procedures for binary blends: A review. J Polym Eng 1997;17(6):429–459.
25. Utracki LA. Compatibilization of polymer blends. Can J Chem Eng 2002;80(6):1008–1016.
26. Koning C et al. Strategies for compatibilization of polymer blends. Prog Polym Sci 1998;23(4):707–757.
27. Utracki LA, Shi ZH. Development of polymer blend morphology during compounding in a twin-screw extruder: Part I. Droplet dispersion and coalescence – A review. Polym Eng Sci 1992;32(24):1824–1833.
28. Tucker CL, Moldenaers P. Microstructural evolution in polymer blends. Annu Rev Fluid Mech 2002;34:177–210.
29. Taylor GI. The formation of emulsions in definable fields of flow. Proc R Soc London A 1934;146:501–523.

30. Maffettone PL, Minale M. Equation of change for ellipsoidal drops in viscous flow. J Non-Newtonian Fluid Mech 1998;78(2–3):227–241.
31. Maffettone PL, Minale M. Equation of change for ellipsoidal drops in viscous flows (vol. 78, p. 227, 1998). J Non-Newtonian Fluid Mech 1999;84(1):105–106.
32. Vananroye A, Van Puyvelde P, Moldenaers P. Effect of confinement on the steady-state behavior of single droplets during shear flow. J Rheol 2007;51(1):139–153.
33. Vananroye A et al. Effect of confinement and viscosity ratio on the dynamics of single droplets during transient shear flow. J Rheol 2008;52(6):1459–1475.
34. Doi M, Ohta T. Dynamics and rheology of complex interfaces. J Chem Phys 1991;95(2):1242–1248.
35. Almusallam AS, Larson RG, Solomon MJ. A constitutive model for the prediction of ellipsoidal droplet shapes and stresses in immiscible blends. J Rheol 2000;44(5):1055–1083.
36. Grmela M, Bousmina M, Palierne JF. On the rheology of immiscible blends. Rheol Acta 2001;40(6):560–569.
37. Luciani A, Champagne MF, Utracki LA. Interfacial tension coefficient from the retraction of ellipsoidal drops. J Polym Sci Part B 1997;35(9):1393–1403.
38. Yamane H et al. Observation of deformation and recovery of poly(isobutylene) droplet in a poly(isobutylene)/poly(dimethyl siloxane) blend after application of step shear strain. J Rheol 1998;42(3):567–580.
39. Grace HP. Dispersion phenomena in high-viscosity immiscible fluid systems and applications of static mixers as dispersion devices in such systems. Chem Eng Commun 1982;14(3–6):225–277.
40. de Bruijn RA. *Deformation and Breakup of Droplets in Simple Shear Flows*. Eindhoven University of Technology; 1989.
41. Torza S, Mason SG, Cox RG. Particle motions in sheared suspensions XXVII. Transient and steady deformation and burst of liquid drops. J Colloid Interface Sci 1972;38(2):395–411.
42. Bentley BJ, Leal LG. An experimental investigation of drop deformation and breakup in steady, two-dimensional linear flows. J Fluid Mech 1986;167:241–283.
43. Cristini V et al. Drop breakup and fragment size distribution in shear flow. J Rheol 2003;47(5):1283–1298.
44. Leal LG. Flow induced coalescence of drops in a viscous fluid. Phys Fluids 2004;16(6):1833–1851.
45. Smoluchowski MV. Versuch einer mathematischen theorie der koagulationskinetik kolloider losungen. Z Phys Chem 1917;92:129–168.
46. Liao YX, Lucas D. A literature review on mechanisms and models for the coalescence process of fluid particles. Chem Eng Sci 2010;65(10):2851–2864.
47. Chesters AK. The modeling of coalescence processes in fluid liquid dispersions – A review of current understanding. Chem Eng Res Des 1991;69(4):259–270.
48. Wang H, Zinchenko AZ, Davis RH. The collision rate of small drops in linear flow-fields. J Fluid Mech 1994;265:161–188.
49. Bartok W, Mason SG. Particle motions in sheared suspensions VIII. Singlets and doublets of fluid spheres. J Colloid Sci 1959;14:13–26.
50. Jeffery GB. The motion of ellipsoidal particles immersed in a viscous fluid. Proc R Soc London A 1922;102:161–179.
51. Jaeger PT et al. Coalescence in emulsions containing inviscid drops with high interfacial mobility. Colloids Surf A 1994;85(2–3):255–264.
52. Vrij A. Possible mechanism for the spontaneous rupture of thin, free liquid films. Discuss Faraday Soc 1966;42:23–33.
53. Cristini V, Blawzdziewicz J, Loewenberg M. An adaptive mesh algorithm for evolving surfaces: Simulations of drop breakup and coalescence. J Comput Phys 2001;168(2):445–463.
54. Baldessari F, Leal LG. Effect of overall drop deformation on flow-induced coalescence at low capillary numbers. Phys Fluids 2006;18(1):1–20.
55. Rother MA, Davis RH. The effect of slight deformation on droplet coalescence in linear flows. Phys Fluids 2001;13(5):1178–1190.
56. Vinckier I et al. Droplet size evolution during coalescence in semiconcentrated model blends. AIChE J 1998;44(4):951–958.
57. Burkhart BE et al. Droplet growth by coalescence in binary fluid mixtures. Phys Rev Lett 2001;87(9):098304.
58. Minale M, Moldenaers P, Mewis J. Effect of shear history on the morphology of immiscible polymer blends. Macromolecules 1997;30(18):5470–5475.
59. Macosko CW. *Rheology Principles, Measurements and Applications*. New York: Wiley-VCH; 1994.
60. Greco F. Drop deformation for non-Newtonian fluids in slow flows. J Non-Newtonian Fluid Mech 2002;107(1–3):111–131.
61. Guido S. Shear-induced droplet deformation: Effects of confined geometry and viscoelasticity. Curr Opin Colloid Interface Sci 2011;16(1):61–70.
62. Cardinaels R, Verhulst K, Moldenaers P. Influence of confinement on the steady state behavior of single droplets in shear flow for immiscible blends with one viscoelastic component. J Rheol 2009;53(6):1403–1424.
63. James DF. Boger fluids. Annu Rev Fluid Mech 2009;41:129–142.
64. Verhulst K, Moldenaers P, Minale M. Drop shape dynamics of a Newtonian drop in a non-Newtonian matrix during transient and steady shear flow. J Rheol 2007;51(2):261–273.
65. Verhulst K et al. Influence of viscoelasticity on drop deformation and orientation in shear flow: Part 1. Stationary states. J Non-Newtonian Fluid Mech 2009;156(1–2):29–43.
66. Guido S, Simeone M, Greco F. Deformation of a Newtonian drop in a viscoelastic matrix under steady shear flow – Experimental validation of slow flow theory. J Non-Newtonian Fluid Mech 2003;114(1):65–82.
67. Sibillo V, Simeone M, Guido S. Break-up of a Newtonian drop in a viscoelastic matrix under simple shear flow. Rheol Acta 2004;43(5):449–456.
68. Sibillo V et al. Start-up and retraction dynamics of a Newtonian drop in a viscoelastic matrix under simple shear flow. J Non-Newtonian Fluid Mech 2006;134(1–3):27–32.
69. Maffettone PL et al. Analysis of start-up dynamics of a single drop through an ellipsoidal drop model for non-Newtonian fluids. J Non-Newtonian Fluid Mech 2005;126(2–3):145–151.
70. Verhulst K et al. Influence of viscoelasticity on drop deformation and orientation in shear flow. Part 2: Dynamics. J Non-Newtonian Fluid Mech 2009;156(1–2):44–57.
71. Yue PT et al. Viscoelastic effects on drop deformation in steady shear. J Fluid Mech 2005;540:427–437.

72. Yu W, Zhou CX. Coalescence of droplets in viscoelastic matrix with diffuse interface under simple shear flow. J Polym Sci Part B 2007;45(14):1856–1869.
73. Fortelny I, Juza J. Modeling of the influence of matrix elasticity on coalescence probability of colliding droplets in shear flow. J Rheol 2012;56(6):1393–1411.
74. Yue PT et al. Diffuse-interface simulations of drop coalescence and retraction in viscoelastic fluids. J Non-Newtonian Fluid Mech 2005;129(3):163–176.
75. Lerdwijitjarud W, Sirivat A, Larson RG. Influence of dispersed-phase elasticity on steady-state deformation and breakup of droplets in simple shearing flow of immiscible polymer blends. J Rheol 2004;48(4):843–862.
76. Li HP, Sundararaj U. Does drop size affect the mechanism of viscoelastic drop breakup? Phys Fluids 2008;20(5):1–6.
77. Mighri F, Carreau PJ, Ajji A. Influence of elastic properties on drop deformation and breakup in shear flow. J Rheol 1998;42(6):1477–1490.
78. Mighri F, Huneault MA. In situ visualization of drop deformation, erosion, and breakup in high viscosity ratio polymeric systems under high shearing stress conditions. J Appl Polym Sci 2006;100(4):2582–2591.
79. Migler KB. Droplet vorticity alignment in model polymer blends. J Rheol 2000;44(2):277–290.
80. Cherdhirankorn T et al. Dynamics of vorticity stretching and breakup of isolated viscoelastic droplets in an immiscible viscoelastic matrix. Rheol Acta 2004;43(3):246–256.
81. Lin B et al. Erosion and breakup of polymer drops under simple shear in high viscosity ratio systems. Polym Eng Sci 2003;43(4):891–904.
82. Lin B et al. Parallel breakup of polymer drops under simple shear. Macromol Rapid Commun 2003;24(13):783–788.
83. Lin B, Sundararaj U. Sheet formation during drop deformation and breakup in polyethylene/polycarbonate systems sheared between parallel plates. Polymer 2004;45(22):7605–7613.
84. Boufarguine M et al. Droplet deformation of a strongly shear thinning dense suspension of polymeric micelles. Rheol Acta 2010;49(6):647–655.
85. Favelukis M, Lavrenteva OM, Nir A. Deformation and breakup of a non-Newtonian slender drop in an extensional flow. J Non-Newtonian Fluid Mech 2005;125(1):49–59.
86. Desse M et al. Experimental study of the break-up of starch suspension droplets in step-up shear flow. J Rheol 2009;53(4):943–955.
87. Jeon HK, Macosko CW. Visualization of block copolymer distribution on a sheared drop. Polymer 2003;44(18):5381–5386.
88. Van Puyvelde P et al. Effect of Marangoni stresses on the deformation and coalescence in compatibilized immiscible polymer blends. Polym Eng Sci 2002;42(10):1956–1964.
89. Blawzdziewicz J, Cristini V, Loewenberg M. Stokes flow in the presence of a planar interface covered with incompressible surfactant. Phys Fluids 1999;11(2):251–258.
90. Bazhekov IB, Anderson PD, Meijer HEH. Numerical investigation of the effect of insoluble surfactants on drop deformation and breakup in simple shear flow. J Colloid Interface Sci 2006;298(1):369–394.
91. Miller R, Ligieri L, editors. *Interfacial Rheology. Progress in Colloid and Interface Science*. Leiden: Koninklijke Brill NV; 2009.
92. Riemann RE, Cantow HJ, Friedrich C. Interpretation of a new interface-governed relaxation process in compatibilized polymer blends. Macromolecules 1997;30(18):5476–5484.
93. Kaur S, Leal LG. Drop deformation and break-up in concentrated suspensions. J Rheol 2010;54(5):981–1008.
94. Caserta S et al. Drop deformation in sheared polymer blends. J Rheol 2007;51(4):761–774.
95. Loewenberg M. Numerical simulation of concentrated emulsion flows. J Fluids Eng 1998;120(4):824–832.
96. Jansen KMB, Agterof WGM, Mellema J. Droplet breakup in concentrated emulsions. J Rheol 2001;45(1):227–236.
97. Choi SJ, Schowalter WR. Rheological properties of non-dilute suspensions of deformable particles. Phys Fluids 1975;18(4):420–427.
98. Palierne JF. Linear rheology of viscoelastic emulsions with interfacial tension. Rheol Acta 1990;29(3):204–214.
99. Caserta S, Simeone M, Guido S. A parameter investigation of shear-induced coalescence in semidilute PIB–PDMS polymer blends: Effects of shear rate, shear stress volume fraction, and viscosity. Rheol Acta 2006;45(4):505–512.
100. Lyu SP, Bates FS, Macosko CW. Coalescence in polymer blends during shearing. AIChE J 2000;46(2):229–238.
101. Li YY et al. Morphology development in polypropylene/polystyrene blends during coalescence under shear. J Appl Polym Sci 2007;104(1):666–671.
102. Minale M, Mewis J, Moldenaers P. Study of the morphological hysteresis in immiscible polymer blends. AIChE J 1998;44(4):943–950.
103. Janssen JMH. *Dynamics of Liquid–Liquid Mixing*. Eindhoven: Technische Universiteit Eindhoven; 1993. p 117.
104. Elemans PHM et al. Transient phenomena in dispersive mixing. Chem Eng Sci 1993;48(2):267–276.
105. Comas-Cardona S, Tucker CL. Measurements of droplet deformation in simple shear flow with zero interfacial tension. J Rheol 2001;45(1):259–273.
106. Cristini V et al. A numerical and experimental investigation of lamellar blend morphologies. Ind Eng Chem Res 2002;41(25):6305–6311.
107. Mewis J et al. Small-angle light scattering study of droplet break-up in emulsions and polymer blends. Chem Eng Sci 1998;53(12):2231–2239.
108. Knops YMM et al. Simultaneous breakup of multiple viscous threads surrounded by viscous liquid. AIChE J 2001;47(8):1740–1745.
109. Stone HA, Bentley BJ, Leal LG. An experimental study of transient effects in the breakup of viscous drops. J Fluid Mech 1986;173:131–158.
110. Rayleigh L. On the capillary phenomena of jets. Proc R Soc London 1879;29:71–97.
111. Tomotika S. On the instability of a cylindrical thread of a viscous liquid surrounded by another viscous fluid. Proc R Soc London A 1935;150:322–337.
112. Elmendorp JJ. A study on polymer blending microrheology. Polym Eng Sci 1986;26(6):418–426.

113. Kuhn W. Spontane aufteilung von Flüssigkeitszylindern in kleine Kugeln. Kolloid Z 1953;132:84–99.
114. Janssen JMH, Meijer HEH. Droplet breakup mechanisms – Stepwise equilibrium versus transient dispersion. J Rheol 1993;37(4):597–608.
115. Tjahjadi M, Stone HA, Ottino JM. Satellite and subsatellite formation in capillary breakup. J Fluid Mech 1992;243: 297–317.
116. Mikami T, Cox RG, Mason SG. Breakup of extending liquid threads. Int J Multiphase Flow 1975;2(2):113–138.
117. Khakhar DV, Ottino JM. Breakup of liquid threads in linear flows. Int J Multiphase Flow 1987;13(1):71–86.
118. Van Puyvelde P et al. Breakup of filaments in blends during simple shear flow. J Rheol 2000;44(6):1401–1415.
119. Lee WK, Yu KL, Flumerfelt RW. Instability of stationary and uniformly moving cylindrical fluid bodies – II. Viscoelastic threads and experimental observations. Int J Multiphase Flow 1981;7(4):385–400.
120. Hansen S, Peters GWM, Meijer HEH. The effect of surfactant on the stability of a fluid filament embedded in a viscous fluid. J Fluid Mech 1999;382:331–349.
121. Milliken WJ, Stone HA, Leal LG. The effect of surfactant on the transient motion of Newtonian drops. Phys Fluids A 1993;5(1):69–79.
122. Elemans PHM, van Wunnik JM, van Dam RA. Development of morphology in blends of immiscible polymers. AIChE J 1997;43(6):1649–1651.
123. Lyngaae-Jorgensen J et al. Flow induced deformation of dual-phase continuity in polymer blends and alloys. Part I. Polym Eng Sci 1999;39(6):1060–1071.
124. Paul DR, Barlow JW. Polymer blends (or alloys). J Macromol Sci C 1980;18(1):109–168.
125. Willemse RC et al. Co-continuous morphologies in polymer blends: A new model. Polymer 1998;39(24):5879–5887.
126. Bourry D, Favis BD. Cocontinuity and phase inversion in HDPE/PS blends: Influence of interfacial modification and elasticity. J Polym Sci Part B 1998;36(11):1889–1899.
127. Scott CE, Macosko CW. Morphology development during the initial-stages of polymer–polymer blending. Polymer 1995;36(3):461–470.
128. Veenstra H, van Dam J, de Boer AP. On the coarsening of co-continuous morphologies in polymer blends: Effect of interfacial tension, viscosity and physical cross-links. Polymer 2000;41(8):3037–3045.
129. Lopez-Barron CR, Macosko CW. Rheological and morphological study of cocontinuous polymer blends during coarsening. J Rheol 2012;56(6):1315–1334.
130. Mekhilef N, Favis BD, Carreau PJ. Morphological stability, interfacial tension, and dual-phase continuity in polystyrene–polyethylene blends. J Polym Sci Part B 1997;35(2):293–308.
131. Yang C, Yin XH, Cheng GM. Microinjection molding of microsystem components: New aspects in improving performance. J Micromech Microeng 2013;23(9):1–21.
132. Heckele M, Schomburg WK. Review on micro molding of thermoplastic polymers. J Micromech Microeng 2004;14(3):R1–R14.
133. Son Y. Development of a novel microcompounder for polymer blends and nanocomposite. J Appl Polym Sci 2009;112(2):609–619.
134. Van Puyvelde P et al. Review on morphology development of immiscible blends in confined shear flow. Polymer 2008;49(25):5363–5372.
135. Guido S, Preziosi V. Droplet deformation under confined Poiseuille flow. Adv Colloid Interface Sci 2010;161(1–2):89–101.
136. Zhao CX, Middelberg APJ. Two-phase microfluidic flows. Chem Eng Sci 2011;66(7):1394–1411.
137. Sibillo V et al. Drop deformation in microconfined shear flow. Phys Rev Lett 2006;97(5):1–4.
138. Shapira M, Haber S. Low Reynolds-number motion of a droplet in shear flow including wall effects. Int J Multiphase Flow 1990;16(2):305–321.
139. Minale M. A phenomenological model for wall effects on the deformation of an ellipsoidal drop in viscous flow. Rheol Acta 2008;47(5–6):667–675.
140. Vananroye A et al. Microconfined equiviscous droplet deformation: Comparison of experimental and numerical results. Phys Fluids 2008;20(1):1–10.
141. Cardinaels R, Moldenaers P. Droplet relaxation in blends with one viscoelastic component: Bulk and confined conditions. Rheol Acta 2010;49(9):941–951.
142. Vananroye A, Van Puyvelde P, Moldenaers P. Effect of confinement on droplet breakup in sheared emulsions. Langmuir 2006;22(9):3972–3974.
143. Janssen PJA, Meijer HEH, Anderson PD. Stability and breakup of confined threads. Phys Fluids 2012;24(1):1–18.
144. Son Y et al. Suppression of capillary instability of a polymeric thread via parallel plate confinement. Macromolecules 2003;36(15):5825–5833.
145. Migler KB. String formation in sheared polymer blends: Coalescence, breakup, and finite size effects. Phys Rev Lett 2001;86(6):1023–1026.
146. Cardinaels R, Moldenaers P. Relaxation of fibrils in blends with one viscoelastic component: From bulk to confined conditions. J Polym Sci Part B 2010;48(12):1372–1379.
147. Chen DJ, Cardinaels R, Moldenaers P. Effect of confinement on droplet coalescence in shear flow. Langmuir 2009;25(22):12885–12893.
148. De Bruyn P, Cardinaels R, Moldenaers P. The effect of geometrical confinement on coalescence efficiency of droplet pairs in shear flow. J Colloid Interface Sci 2013;409:183–192.
149. Cardinaels R, Moldenaers P. Critical conditions and breakup of non-squashed microconfined droplets: Effects of fluid viscoelasticity. Microfluid Nanofluid 2011;10(6):1153–1163.
150. Cardinaels R et al. An experimental and numerical investigation of the dynamics of microconfined droplets in systems with one viscoelastic phase. J Non-Newtonian Fluid Mech 2011;166(1–2):52–62.
151. Cardinaels R et al. Breakup criteria for confined droplets: Effects of compatibilization and component viscoelasticity. Macromol Mater Eng 2011;296(3–4):214–222.
152. Vananroye A, Van Puyvelde P, Moldenaers P. Deformation and orientation of single droplets during shear flow: Combined effects of confinement and compatibilization. Rheol Acta 2011;50(3):231–242.

153. Pathak JA et al. Layered droplet microstructures in sheared emulsions: Finite-size effects. J Colloid Interface Sci 2002;255(2):391–402.
154. Pathak JA, Migler KB. Droplet-string deformation and stability during microconfined shear flow. Langmuir 2003;19(21):8667–8674.
155. Caserta S, Guido S. Vorticity banding in biphasic polymer blends. Langmuir 2012;28(47):16254–16262.
156. Pavlidou S, Papaspyrides CD. A review on polymer-layered silicate nanocomposites. Prog Polym Sci 2008;33(12):1119–1198.
157. Potts JR et al. Graphene-based polymer nanocomposites. Polymer 2011;52(1):5–25.
158. Bose S, Khare RA, Moldenaers P. Assessing the strengths and weaknesses of various types of pre-treatments of carbon nanotubes on the properties of polymer/carbon nanotubes composites: A critical review. Polymer 2010;51(5):975–993.
159. Paul DR, Robeson LM. Polymer nanotechnology: Nanocomposites. Polymer 2008;49(15):3187–3204.
160. Moniruzzaman M, Winey KI. Polymer nanocomposites containing carbon nanotubes. Macromolecules 2006;39(16): 5194–5205.
161. Kim H, Abdala AA, Macosko CW. Graphene/polymer nanocomposites. Macromolecules 2010;43(16):6515–6530.
162. Cassagnau P. Melt rheology of organoclay and fumed silica nanocomposites. Polymer 2008;49(9):2183–2196.
163. Fenouillot F, Cassagnau P, Majeste JC. Uneven distribution of nanoparticles in immiscible fluids: Morphology development in polymer blends. Polymer 2009;50(6):1333–1350.
164. Huang JC. Carbon black filled conducting polymers and polymer blends. Adv Polym Technol 2002;21(4):299–313.
165. Cao YW et al. Compatibilization of immiscible polymer blends using graphene oxide sheets. ACS Nano 2011;5(7):5920–5927.
166. Mao C, Zhu YT, Jiang W. Design of electrical conductive composites: Tuning the morphology to improve the electrical properties of graphene filled immiscible polymer blends. ACS Appl Mater Interfaces 2012;4(10):5281–5286.
167. Vermant J et al. Coalescence suppression in model immiscible polymer blends by nano-sized colloidal particles. Rheol Acta 2004;43(5):529–538.
168. Sumita M et al. Double percolation effect on the electrical conductivity of conductive particles filled polymer blends. Colloid Poly Sci 1992;270(2):134–139.
169. Bose S et al. Phase separation as a tool to control dispersion of multiwall carbon nanotubes in polymeric blends. ACS Appl Mater Interfaces 2010;2(3):800–807.
170. Gubbels F et al. Selective localization of carbon-black in immiscible polymer blends – A useful tool to design electrical conductive composites. Macromolecules 1994;27(7):1972–1974.
171. Elias L et al. Migration of nanosilica particles in polymer blends. J Polym Sci Part B 2008;46(18):1976–1983.
172. Hong JS et al. Shear-induced migration of nanoclay during morphology evolution of PBT/PS blend. J Appl Polym Sci 2008;108(1):565–575.
173. Baudouin AC, Devaux J, Bailly C. Localization of carbon nanotubes at the interface in blends of polyamide and ethylene-acrylate copolymer. Polymer 2010;51(6):1341–1354.
174. Fang ZP, Xu YZ, Tong LF. Effect of clay on the morphology of binary blends of polyamide 6 with high density polyethylene and HDPE-graft-acrylic acid. Polym Eng Sci 2007;47(5):551–559.
175. Chow WS et al. Compatibilizing effect of maleated polypropylene on the mechanical properties and morphology of injection molded polyamide 6/polypropylene/organoclay nanocomposites. Polymer 2003;44(24):7427–7440.
176. Gubbels F et al. Kinetic and thermodynamic control of the selective localization of carbon black at the interface of immiscible polymer blends. Chem Mater 1998;10(5):1227–1235.
177. Vo LT, Giannelis EP. Compatibilizing poly(vinylidene fluoride)/nylon-6 blends with nanoclay. Macromolecules 2007;40(23):8271–8276.
178. Ray SS et al. Role of organically modified layered silicate as an active interfacial modifier in immiscible polystyrene/polypropylene blends. Polymer 2004;45(25): 8403–8413.
179. Hong JS et al. The role of organically modified layered silicate in the breakup and coalescence of droplets in PBT/PE blends. Polymer 2006;47(11):3967–3975.
180. Huitric J et al. Rheological, morphological and structural properties of PE/PA/nanoclay ternary blends: Effect of clay weight fraction. J Rheol 2009;53(5):1101–1119.
181. Baudouin AC et al. Polymer blend emulsion stabilization using carbon nanotubes interfacial confinement. Polymer 2011;52(1):149–156.
182. Thareja P, Moritz K, Velankar SS. Interfacially active particles in droplet/matrix blends of model immiscible homopolymers: Particles can increase or decrease drop size. Rheol Acta 2010;49(3):285–298.
183. Thareja P, Velankar S. Particle-induced bridging in immiscible polymer blends. Rheol Acta 2007;46(3):405–412.
184. Nagarkar SP, Velankar SS. Morphology and rheology of ternary fluid–fluid–solid systems. Soft Matter 2012;8(32):8464–8477.
185. Levine S, Bowen BD. Capillary interactions of spherical particles adsorbed on the surface of an oil-water droplet stabilized by the particles: 3. Effective interfacial tension. Colloids Surf A 1993;70(1):33–45.
186. Elias L et al. Immiscible polymer blends stabilized with nano-silica particles: Rheology and effective interfacial tension. Polymer 2008;49(20):4378–4385.
187. Hong JS et al. Interfacial tension reduction in PBT/PE/clay nanocomposite. Rheol Acta 2007;46(4):469–478.
188. Vandebril S, Vermant J, Moldenaers P. Efficiently suppressing coalescence in polymer blends using nanoparticles: Role of interfacial rheology. Soft Matter 2010;6(14):3353–3362.
189. Boker A et al. Self-assembly of nanoparticles at interfaces. Soft Matter 2007;3(10):1231–1248.
190. Fuller GG, Vermant J. Complex fluid–fluid interfaces: Rheology and structure. Annu Rev Chem Biomol Eng 2012;3:519–543.
191. Khatua BB et al. Effect of organoclay platelets on morphology of nylon-6 and poly(ethylene-ran-propylene) rubber blends. Macromolecules 2004;37(7):2454–2459.
192. Si M et al. Compatibilizing bulk polymer blends by using organoclays. Macromolecules 2006;39(14):4793–4801.

193. Zou H et al. A change of phase morphology in poly(*p*-phenylene sulfide)/polyamide 66 blends induced by adding multi-walled carbon nanotubes. Polymer 2006;47(22):7821–7826.

194. Li YJ, Shimizu H. Novel morphologies of poly(phenylene oxide) (PPO)/polyamide 6 (PA6) blend nanocomposites. Polymer 2004;45(22):7381–7388.

195. Filippone G, Romeo G, Acierno D. Role of interface rheology in altering the onset of co-continuity in nanoparticle-filled polymer blends. Macromol Mater Eng 2011;296(7):658–665.

196. Wu DF et al. Morphology evolution of nanocomposites based on poly(phenylene sulfide)/poly(butylene terephthalate) blend. J Polym Sci Part B 2008;46(12):1265–1279.

20

PROCESSING, STRUCTURE, AND MORPHOLOGY IN POLYMER NANOCOMPOSITES

DURACCIO DONATELLA[1,2], CLARA SILVESTRE[1], SOSSIO CIMMINO[1], ANTONELLA MARRA[1], AND MARILENA PEZZUTO[1]

[1] *Istituto per i Polimeri, Compositi e Biomateriali, Consiglio Nazionale delle Ricerche, Pozzuoli, Italy*
[2] *Istituto per le Macchine Agricole e Movimento Terra (IMAMOTER) – Consiglio Nazionale delle Ricerche, Torino, Italy*

The term "polymer nanocomposite" (PNC) has evolved, since the first reports in the early 1990s to refer to a multicomponent system, where the major constituent is a polymer or blend thereof and the minor constituent exhibits a length scale below 100 nm [1–3]. The minor constituent is usually an inorganic filler, called nanofiller, nanoload, or, improperly, nanoparticle. The most commonly used are layered silicates (clays), carbon nanotubes (CNTs), and metals and various metal oxides (silica, titania, zirconia, zinc oxide, etc.).

PNCs have generated a significant amount of industrial and academic attention for the past 25 years. No matter the measure (articles, patents, or research and development funding), worldwide efforts in PNCs have been growing exponentially. For example, the total number of hits for "polymer" and "nanocomposite" on SciFinder (Chemical Abstract Service (CAS) of the American Chemical Society) from 1988 to 2005 is >9400, where the yearly number has approximately doubled every 2 years since 1992 [4].

The possible applications of PNCs are huge and several PNC products have been already marketed: medical products (bandages, heart valves, etc.); food packaging material; electronic components; scratch-free paint; articles sporting goods; and wrinkle- and stain-resistant fabrics. Analysts have estimated that PNCs' market is expected to reach USD 5.91 billion by 2018 [5]. Major revenues are forecast from large commercial opportunities, such as automobile, coatings, and packaging, where lower cost and higher performance resins would improve durability and design flexibility while lowering unit price. In this scenario, North America was the largest market for PNCs in 2011, accounting for approximately 40% of global demand. Europe accounted for a large portion of PNC sales. Asia Pacific and the rest of the worlds (RoW) are expected to be strong future markets for nanocomposites with growing industrialization and infrastructure. Asia Pacific is expected to show the highest growth in demand and the market for nanocomposites is forecast to grow at a CAGR (compounded annual growth rate) of 18.8% between 2012 and 2018 [5].

Given the extensive variety of nanoparticles now commercially accessible, the potential combinations of polymers and nanoparticles, and thus the tailorability of the property suite, is essentially endless. The diversity in scientific investigation, technology advancement, processing innovations, and product development is staggering. A significant number of excellent review papers (e.g., clays, [6–16], CNTs [12, 13, 17–20], and spherical particles [21, 22]) and books [23–28] that chronicle and summarize the status of various nanoparticle–polymer combinations and the broad scientific and technological challenges still to be overcome are available. The most important issue for a complete exploitation of nanotechnology is to obtain a homogeneous dispersion of the particles in the matrix. This chapter is a critical review of the most significant method of PNC preparation (processability) and the most representative structure/morphology of polymer nanocomposites in relation to the dimensionality of the nanofiller.

20.1 OVERVIEW

PNCs are shown to have remarkable property enhancements relative to conventionally scaled composites. The transition from microparticles to nanoparticles leads to a number of

Polymer Morphology: Principles, Characterization, and Processing, First Edition. Edited by Qipeng Guo.

changes in physical properties. These changes are associated with the following three concepts:

1. As shown in Figure 20.1, the number of filler particles at a given volume fraction (3% in the figures) rapidly increases as the size of particles decreases. As a consequence, compared to conventional polymer composites with conventional fillers, which normally require loadings of 20% or more, the distances between nanoparticles are drastically reduced to the nanometer range, even when relatively low concentrations of nanofillers are used.
2. The sizes of the nanoparticles are comparable to the radius of gyration of the macromolecules in the polymer matrix, so that the morphological development of polymer matrix can be affected by the dispersed nanoparticles.
3. The nanoparticles provide ultrahigh specific surfaces (high surface area to volume ratio). As the surface area of a particle increases (Fig. 20.2), the interface becomes large and the amount of modified polymer interface relative to the total volume is significantly increased. This means that the portion of its constituent atoms at or near the surface increases exponentially, creating more sites for bonding, catalysis, or reaction with surrounding materials, resulting in improved properties such as increased strength or chemical and/or heat resistance.

Nanoparticles can be classified as a function of their geometry:

1. One dimension less than 100 nm (layered silicates: thickness about 1 nm);
2. Two dimensions less than 100 nm (nanotubes: diameter less than 100 nm);
3. Three dimensions less than 100 nm (spherical nanoparticles such as metal oxide, metal, etc.).

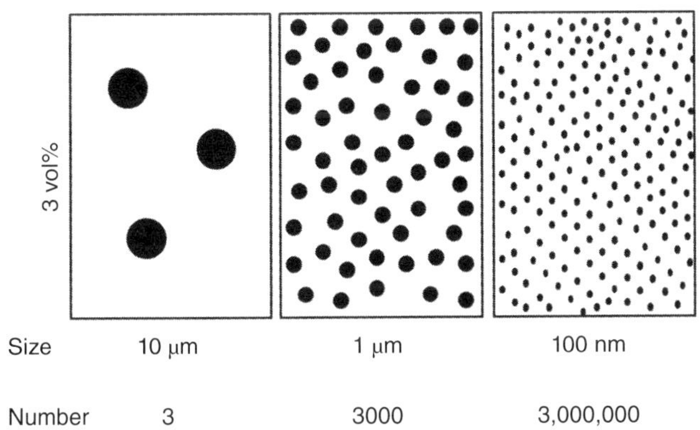

Figure 20.1 Illustration of how the number of particles (supposedly spherical) increases as the particle size decreases at a constant volume content.

The development of successful PNCs with a good balance of properties demands a high degree of coupling between processing and morphological and micromechanical control. That is, the balance of properties must be optimized. To this end, at least the essential morphological requirement should be met in the rational design of a nanocomposite: the uniform dispersion of nanofillers to avoid large agglomerates and high stress concentrations, and good or optimum interfacial bonding between them and the polymer matrix to achieve effective load transfer across the nanofiller and the matrix interface [29]. Unfortunately, it has been extensively reported that nanofillers are often dispersed in the polymer matrix in the form of agglomerates. This strong tendency to agglomerate significantly reduces their ability to bond with the polymer matrix because it reduces the contact area and decreases the effective aspect ratio (the aspect ratio of a nanoparticle is defined as the ratio between its sizes in different dimensions) of the reinforcement. Moreover, under external load the stress will be readily concentrated around such agglomerates, which, in turn, generally leads to the premature failure of the system and a worsening of the properties of the polymer. Good dispersion and distribution (Fig. 20.3) of nanoparticles in the polymer matrix are key factors for improved materials. The modification of the nanoparticles surface and/or the polymer, and the use of interfacial agents are the most used methodologies to achieve the best morphology [29].

20.2 NANOPARTICLES WITH ONE DIMENSION LESS THAN 100 NM (LAYERED SILICATES)

Layered silicates used in the synthesis of PNC are natural or synthetic minerals, consisting of very thin layers that are usually bound together with counterions. Their basic building blocks are tetrahedral sheets in which silicon is surrounded by four oxygen atoms, and octahedral sheets in which a metal like aluminum is surrounded by eight oxygen atoms. Therefore, in 1:1 layered structures (e.g., in kaolinite), a tetrahedral sheet is fused with an octahedral sheet, whereby the oxygen atoms are shared [30]. On the other hand, the crystal lattice of 2:1 layered silicates (or 2:1 phyllosilicates), consists of two-dimensional layers where a central octahedral sheet of alumina is fused to two external silica tetrahedra by the tip, so that the oxygen ions of the octahedral sheet also belong to the tetrahedral sheets, as shown in Figure 20.4.

The layer thickness is around 1 nm and the lateral dimensions may vary from 300 Å to several microns, and even larger, depending on the particulate silicate, the source of the clay, and the method of preparation. The aspect ratio of these layers (ratio length/thickness) is particularly high, with values greater than 1000 [7, 31–33].

Because, in their pristine state, layered silicates are only miscible with hydrophilic polymers, such as poly(ethylene oxide) (PEO) and poly(vinyl alcohol) (PVA), in order to render them miscible with other polymers, one must exchange the alkali counterions with a cationic-organic surfactant.

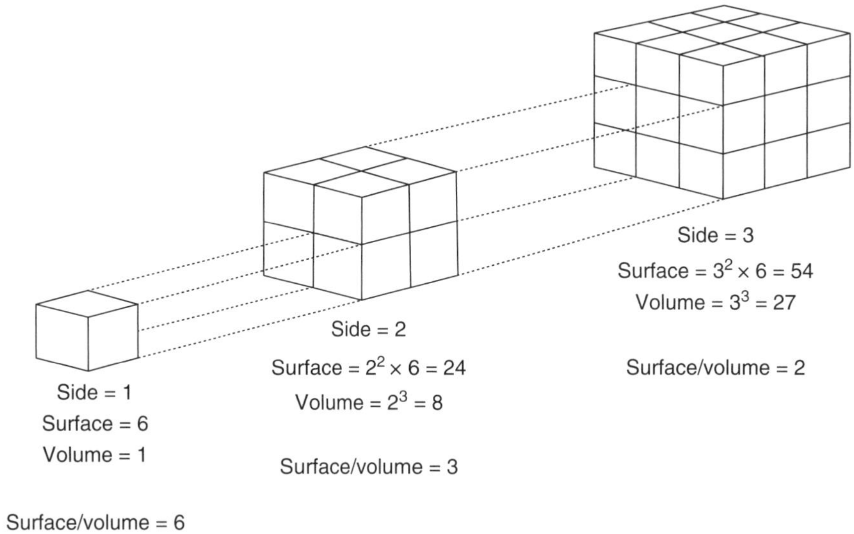

Figure 20.2 The dimensions, surface areas, and volumes of three different-sized cubes, which schematically represent particles, are compared here to show how surface area/volume ratio increases as the dimensions decrease.

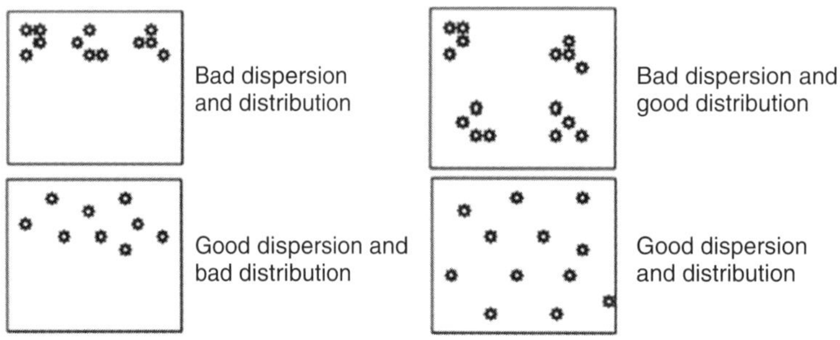

Figure 20.3 Schema of dispersion and distribution of nanoparticles in a polymer matrix, where dispersion and distribution are the level of nanoparticle conglomeration and homogeneity in the matrix, respectively.

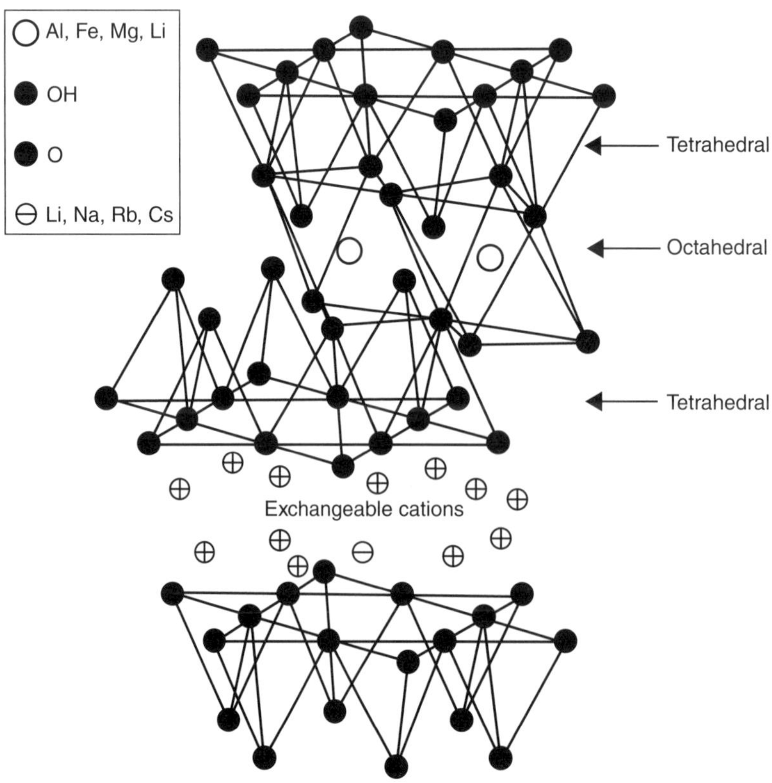

Figure 20.4 The structure of 2:1 layer silicates. Ray and Okamoto [8]. Reproduced with permission of Elsevier.

Alkylammonium ions are mostly used, although other salts can be used, such as sulfonium and phosphonium [7, 34, 35]. This can be readily achieved through ion-exchange reactions that render the clay organophilic [36]. In order to obtain the exchange of the salt ions with the cations in the galleries, water swelling of the silicate is needed. For this reason, alkali cations are preferred in the galleries because 2-valent and higher valent cations to prevent swelling by water. Indeed, the hydrate formation of monovalent intergallery cations is the driving force for water swelling.

Natural clays may contain divalent cations such as calcium and require exchange procedures with sodium prior to further treatment with ion salts [37]. The alkali cations, as they are not structural, can be easily replaced by other positively charged atoms or molecules, and thus are called exchangeable cations [38]. The organic cations lower the surface energy of the silicate surface and improve wetting with the polymer matrix [6, 39]. Moreover, the long organic chains of such surfactants, with positively charged ends, are tethered to the surface of the negatively charged silicate layers, resulting in an increase of the gallery height [37, 40], as shown in Figure 20.5. Conclusively, the surface modification both increases the basal spacing of clays and serves as a compatibilizer between the hydrophilic clay and the hydrophobic polymer [41].

Depending on the nature of components (polymer matrix, clay filler, and organic surfactant) and processing conditions, clay particles (organically modified or not) can be present in three configurations when incorporated in the polymer matrix (Fig. 20.6).

If the polymer is unable to intercalate into the galleries, a phase-separated composite is formed, whose properties are similar to that of traditional microcomposites; the poor interaction between the organic and the inorganic component results in relatively poor mechanical performance. Unseparated layers, after introduction in a polymer matrix, are often referred to as tactiodes [42]. On the other hand, an intercalated nanocomposite is obtained when extended polymer macromolecules diffuse between unchanged clay sheets, leading to a well-ordered multilayer structure of alternating polymeric and inorganic layers with a repeating distance of a few nanometers between them. The properties of this type of nanocomposites typically resemble those of ceramic materials. The most significant changes in physical properties are observed in exfoliated hybrids, where clay layers are separated and uniformly dispersed, thus maximizing the polymer–clay interactions [6, 43].

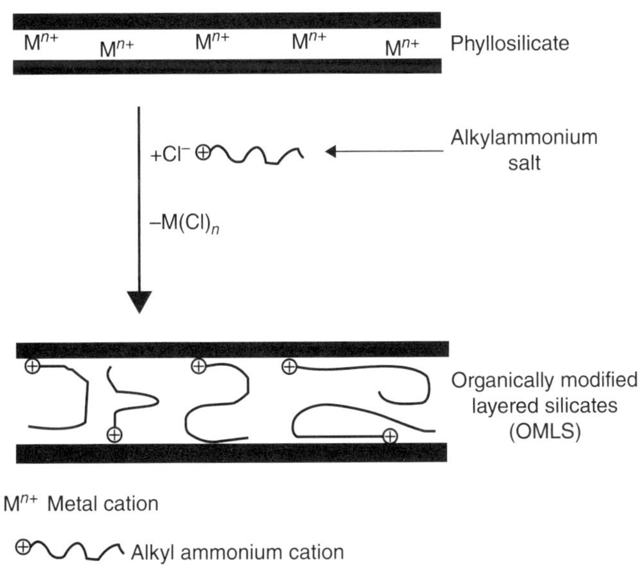

Figure 20.5 Schematic demonstration of clay organic modification. Zanetti et al. [37]. Reproduced with permission of John Wiley and Sons.

Both transmission electron microscopy (TEM) and wide-angle X-ray diffraction (WAXD) are essential tools [8] for evaluating PNC structures. With WAXD measurements, by monitoring the position, shape, and intensity of the basal reflections from the distributed layered silicates, the nanocomposite structure may be identified. In Figure 20.7 are reported TEM and WAXD for the different type of PNCs that can be generated. In the figure, the term flocculated nanocomposites indicates intercalated nanocomposites with hydroxylated edge–edge interaction between the silicate layers.

It is important to observe that the different structures can be generated simultaneously: an example [44] is reported in Figure 20.8.

20.3 NANOPARTICLES WITH TWO DIMENSIONS LESS THAN 100 NM (CARBON NANOTUBES)

CNTs were first reported by Iijima [45] in 1991, and the first PNCs using CNTs as a filler were reported in 1994 by Ajayan et al. [46] CNTs possess high flexibility [47], low mass density [48], and large aspect ratio (length/diameter of ~ 300–1000). CNT have a unique combination of mechanical, electrical, and thermal properties. Some nanotubes are stronger than steel, lighter than aluminum, and more conductive than copper. For example, theoretical and experimental results on individual single-walled carbon nanotubes (SWNTs) show extremely high tensile modulus [49] (640 GPa to 1 TPa) and tensile strength [50] (150–180 GPa). CNTs are long cylinders of covalently bonded carbon atoms. The ends of the cylinders may or may not be capped by hemifullerenes. There are two basic types of CNTs: single-walled carbon nanotubes (SWNTs) and multiwalled carbon nanotubes (MWNTs).

An SWNT can be considered as a single graphene sheet (graphene is a monolayer of sp^2-bonded carbon atoms) rolled into a seamless cylinder along an (m,n) lattice vector in the graphene plane (Fig. 20.9). The carbon atoms in the cylinder have partial sp^3 character that increases as the radius of curvature of the cylinder decreases. The (m,n) indices determine the diameter and chirality, which are key parameters of a nanotube. Depending on the chirality (the chiral angle between hexagons and the tube axis), SWNTs can be either

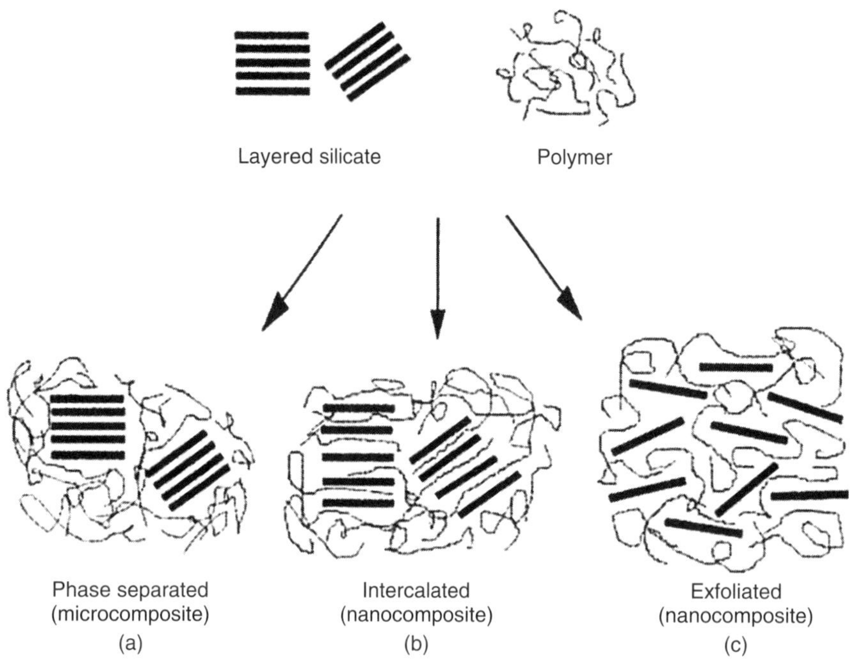

Figure 20.6 Possible polymer/layered silicate structures. Dubois [7]. Reproduced with permission of Elsevier.

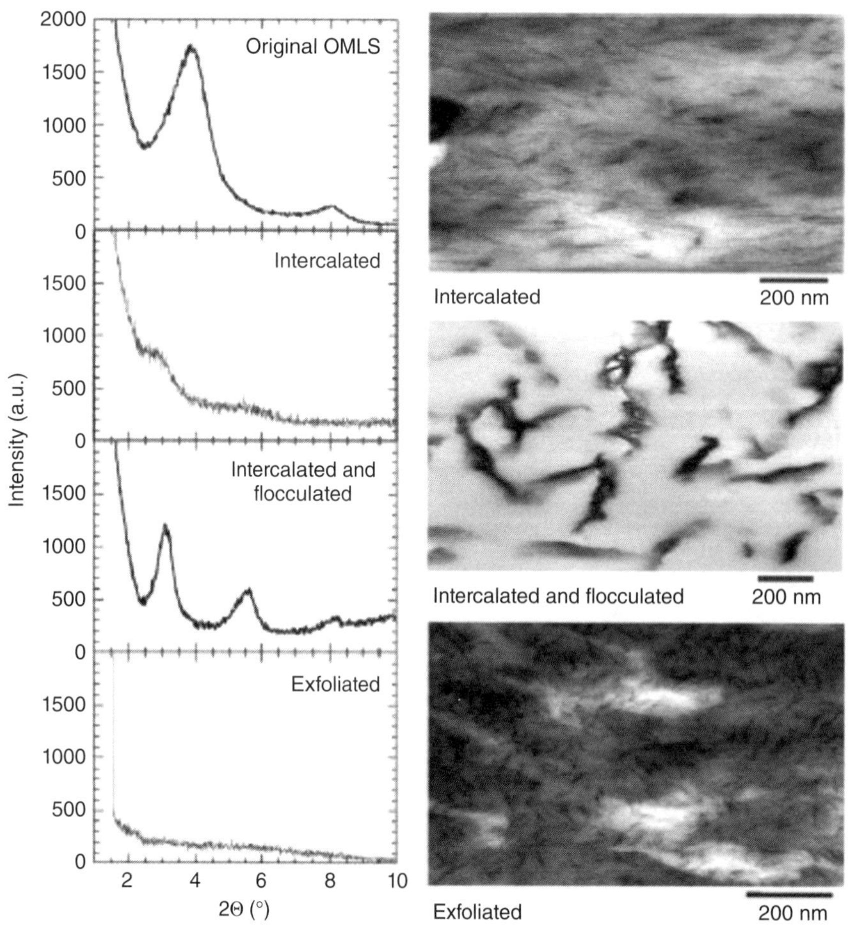

Figure 20.7 WAXD patterns and TEM images of three different PNC structures. Ray and Okamoto [8]. Reproduced with permission of Elsevier.

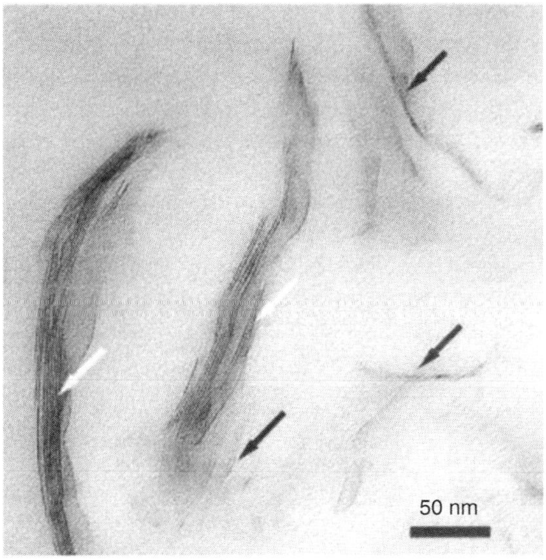

Figure 20.8 TEM image of an epoxy/clay nanocomposite that presents simultaneously an intercalated (white arrows) and exfoliated (black arrows) structure. Miyagawa et al. [44]. Reproduced with permission of John Wiley and Sons.

metals or semiconductors, with band gaps that are relatively large (~0.5 eV for typical diameter of 1.5 nm) or small (~10 meV), even if they have nearly identical diameters [51]. For same-chirality semiconducting nanotubes, the band gap is inversely proportional to the diameter. Thus, there are infinite possibilities in the type of carbon tube "molecules," and each nanotube could exhibit distinct physical properties.

MWNTs consist of nested graphene cylinders coaxially arranged around a central hollow core with interlayer separations of ~0.34 nm, indicative of the interplane spacing of graphite [52]. A special case of MWNT is double-walled nanotubes (DWNTs) that consist of two concentric graphene cylinders. DWNTs are expected to exhibit higher flexural modulus than SWNTs, due to the two walls and higher toughness than regular MWNTs because of their smaller size [53].

Presently, MWNTs and SWNTs are mainly produced by three techniques, each having their own nuances: arc discharge, laser ablation, and chemical vapor decomposition (CVD). A number of reviews [52, 54] are available on these production techniques. Arc discharge and laser ablation methods involve the condensation of hot gaseous carbon atoms generated from the evaporation of solid carbon. In CVD, a gaseous carbon source (hydrocarbon, CO) is decomposed catalytically, and the nanotubes are deposited on a substrate or they grow from a substrate. Compared with arc and laser methods, CVD might offer more control over the length and structure of the produced nanotubes. As mentioned earlier, at present, all known preparations of CNTs give mixtures of nanotube chiralities, diameters, and lengths along with different amount and type of impurities. These parameters vary significantly both within a sample and between samples from different batches and laboratories. Thus, it is very

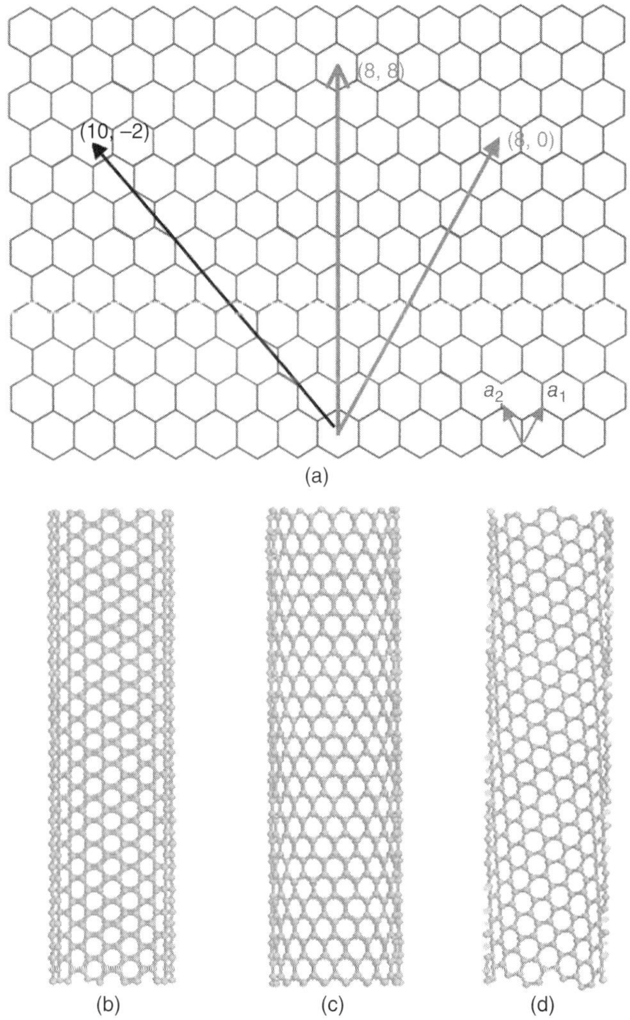

Figure 20.9 (a) Schematic honeycomb structure of a graphene sheet. Single-walled carbon nanotubes can be formed by folding the sheet along lattice vectors. The two basis vectors $\boldsymbol{a}_1$ and $\boldsymbol{a}_2$ are shown. Folding of the (8,8), (8,0), and (10,−2) vectors lead to (b) armchair, (c) zigzag, and (d) chiral tubes. Dai [51]. Reproduced with permission of American Chemical Society.

difficult to conduct reproducible control experiments with these inconsistent nanofillers and virtually impossible to compare results between different researchers.

Another great challenge in nanotube/polymer composites is the efficient translation of nanotube properties, both into the polymer matrix and between nanotubes.

This is why CNTs are functionalized with organic molecules before/during their mixing with a polymer matrix. Functionalization of CNTs can serve to not only improve their solubility and, therefore, dispersion but also to create attractive van der Waals interactions between the polymer matrix and the nanotube addend [55]. Several research groups have reported successful functionalization reactions for SWNT and MWNT CNTs [55–75].

These reactions may roughly be divided into two categories: a direct attachment of functional groups to the graphitic surface and the use of the nanotube-bound carboxylic acids. The two categories of functionalization reactions may have different effects on the structures and intrinsic properties of CNTs.

In the first category, Margrave, Smalley, and coworkers reported the fluorination of SWNTs [57, 58]. In various alcohol solvents, these functionalized SWNTs were solvated as individual tubes, making it possible to carry out further solution chemistry. Pekker and coworkers reported the hydrogenation of CNTs via the Birch reduction in ammonia [70]. Other examples include the derivatization of small-diameter (ca. 0.7 nm) SWNTs reported by Tour and coworkers [68, 69]; the interactions with anilines reported by Wilson and coworkers [67]; the reactions with nitrenes, carbenes, and radicals reported by Hirsch and coworkers [74]; and the 1,3-dipolar addition reported by Prato and others [75].

In the second category of functionalization reactions, the nanotube-bound carboxylic acids come from intrinsic or induced defects. The latter refer to the creation of terminal carbons in the shortening of nanotubes, which upon oxidation are converted to carboxylic acids [56, 76, 77]. Haddon and coworkers first reported the use of the acid groups for attaching long alkyl chains to SWNTs via amide linkages [56] or carboxylate-ammonium salt ionic interactions [59]. Sun and coworkers showed that the esterification of the carboxylic acids can also be applied to functionalize and solubilize nanotubes of any length [60, 65, 66]. An advantage with the ester linkages is that they can be facilely defunctionalized via acid- or base-catalyzed hydrolysis, allowing the recovery of CNTs from the soluble samples [66]. The nanotube-bound carboxylic acids are the sites to attach a variety of functional groups for the solubilization of both shortened and full-length CNTs.

20.4 NANOPARTICLES WITH THREE DIMENSIONS LESS THAN 100 NM (METAL, METAL OXIDE)

Isodimensional nanoparticles used for preparation of PNCs are mainly oxide and metal oxide. The most common are silver (Ag), gold (Au), zinc oxide (ZnO), silica (SiO_2), titanium dioxide (TiO_2), and alumina (Al_2O_3). A comprehensive coverage of all the different kinds of isodimensional nanoparticles would be beyond the scope of this review. Instead, this paragraph focuses, principally, on the most commonly used isodimensional nanoparticles, that is, silver, zinc oxide, titanium dioxide, and silica.

Ag nanoparticles are clusters of Ag atoms that range in diameter from 1 to 100 nm and are attracting interest because their most innovative characteristics are associated with their increased antimicrobial properties [22, 23, 78, 276–278]. Owing to these antimicrobial properties, Ag nanoparticles are currently the most commonly used nanoparticles in consumer and industrial applications. The use of Ag as an antimicrobial agent, however, is not a new concept. Ag pots and coins were used in ancient times to keep water sterile and, in particular, Ag-nitrate ($AgNO_3$) has been used since the seventeenth century as an essential multipurpose medicinal product [79, 80]. Examples of recent consumer applications using Ag as an antimicrobial agent consist of food supplements, materials for food packaging, coatings on medical devices, water disinfectants, air filters, electronic appliances, odor-resistant textile fabrics, and cosmetic products such as deodorants [81].

In general, there are two strategies to produce metal nanoparticles. One is the so-called bottom-up method, where nanostructures are obtained through assembly of smaller, basic units into larger structures. The other is the top-down method, which starts from bulk material, whose size is decreased during the process. Common top-down techniques include photolithography and electron beam lithography [82, 83].

The most extensively studied method for synthesizing Ag nanoparticles is the controlled reduction of Ag salts [84, 85]. This reduction is often accomplished using a chemical reducing agent [86–88] but also via electrochemical [89] or photochemical reduction [90, 91]. This wet-chemical method, which involves Ag salts reduced (in one or more steps) to metal Ag, can be steered by many different parameters, such as the choice of the reducing agent, the concentrations of reagents, temperature, pH, mixing rate, and reaction time [90]. In order to tailor the particle properties, many studies have focused on the generation of different shapes, such as rods, wires, cubes, thin plates, or bipyramids [92, 93]. In addition, substantial effort has been devoted to the construction of core–shell structures, such as Ag@Au (silver core particles with a gold shell) or Ag@SiO_2 nanoparticles [94, 95]. In general, strong reducing agents result in the formation of smaller NPs, whereas weaker reducing agents usually lead to larger Ag particles [90]. Another method makes use of a Tollens process to produce Ag nanoparticles in a one-step method [90]. This reaction has been long used in the electroless deposition of Ag to generate reflective mirrors on solid supports and combines reducing sugars and Ag ammoniacal solutions [86]. The obtained particle size distributions are generally narrow. The Tollens method is an example of a green synthesis of Ag nanoparticles [92].

Among metal oxides, nano-ZnO, nano-TiO_2, and nano-SiO_2 have been objects of an intense research. Nano-ZnO and nano-TiO_2 display antibacterial activity and act as UV blocking agents. Certain forms of TiO_2 are also photocatalytic, and have been incorporated in products such as glazing and cement to impart self-cleaning properties. Interest also extends to the electronic properties of TiO_2 and it is a key component of dye-sensitized solar cells. Interest in nanoscale ZnO extends to its electronic properties, which are being exploited for use in such products as transparent electrodes, LEDs, lasers, and piezoelectric materials. Nano-SiO_2 possess ordered porous structures with extremely high surface areas ($>1000\,m^2/g$), well-defined and tunable pore sizes (1.5–10 nm), and can be easily functionalized by many approaches. Recently, several groups have taken them as suitable reservoirs of drugs and biomolecules for biomedical applications [96].

Many different methods have been reported for the production of metal oxide nanoparticles. The most common processes have been developed to synthesize oxide are chemical solution decomposition (CSD) [97], spray pyrolysis [98], CVD [99–101], two-step wet chemical method [102], sol–gel [103], ultrasonic irradiation [104, 105], and ethanol thermal and hydrothermal method [106]. Literature revealed that sol–gel is the most commonly used method for the preparation of metal oxide nanoparticles. It facilitates the synthesis of nanometer-sized crystallized metal oxide powder of high purity at a relatively low temperature [103].

The sol–gel chemistry is based on the hydrolysis and condensation reactions of metal alkoxides, M(OR): where M is a metal ion and R an alkyl group (R = Me, Et). In the case of titanium, hydrolysis leads to the formation of Ti–OH groups:

$$\equiv \text{Ti–OR} + H_2O \rightarrow \equiv \text{Ti–OH} + \text{R–OH} \quad (20.1)$$

The condensation reactions lead to the formation of oxo bridges, Ti–O–Ti represented by the following reactions:

Oxalation or water elimination

$$2 \equiv \text{Ti–OH} \rightarrow \equiv \text{Ti–O–Ti} + H_2O \quad (20.2)$$

And alcoxalation or alcohol elimination

$$\equiv \text{Ti–OH} + \equiv \text{Ti–OR} \rightarrow \equiv \text{Ti–O–Ti} + \text{R–OH} \quad (20.3)$$

Finally, the overall generalized reaction can be written as

$$Ti(OR)_4 + 2H_2O \rightarrow TiO_2 + 4ROH \quad (20.4)$$

Many factors influence the kinetics of the hydrolysis and condensation reactions in the sol–gel process, which include the water/alkoxide ratio, temperature, the nature of solvent, and so forth [107].

The functionalization of these particles is always required for preparing a PNC; in fact nanoparticles have a strong tendency to agglomerate and, consequently, the so-called nanoparticle-filled polymers sometimes contain a number of loosened clusters of particles (Fig. 20.10a) and the final material exhibits properties even worse than conventional particle/polymer systems.

Noble metal nanoparticles were mainly modified by thiols [109–115], disulfides [116], amines [111–113, 117, 118], nitriles, carboxylic acids, and phosphines [113, 118, 119]. The use of organosulfur compounds for modifying noble metal nanoparticles is one of the more developed methods, because organosulfur groups strongly coordinate to various metals, such as Ag, Cu [120], Pt, Hg, Fe, or Au. Sulfur possesses a huge affinity for metal surfaces, and organosulfur compounds thus will adsorb spontaneously [121].

Various organic compounds are potential metal oxide modifiers, among them, as for metal nanoparticles, thiols, carboxylic acids, and amines [122–124]. However, amines and thiols are relatively rarely used; an example is thiourea to modify SnO_2 surfaces [122]. The main compounds used for modifying metal oxide nanoparticles are phosphonates or silanes. Among silanes, the most used include alkoxysilane, ≡Si–OR (R = alkyl), hydrogenosilane, ≡Si–H, or chlorosilane, ≡Si–Cl, reagents. As a matter of fact, there is a broad variety of commercially available silanes, and others are readily synthesized or modified. One of the main advantages of silanes is that they can bear numerous functionalities, for example, amino, cyano, carboxylic acid, epoxy groups, and so on [123, 125–132]. The postsynthesis grafting of silyl groups on a metal oxide surface is quite easy. Moreover, silanes can also be used for the one-step method, where the silane is introduced at the same time as the metal oxide precursors [110]. This method is applicable to numerous metal oxides such as SiO_2, Al_2O_3, TiO_2, SnO_2, ZrO_2, V_2O_5, and so on.

An irradiation grafting method was applied for the modification of nanosilica in the preparation of polypropylene

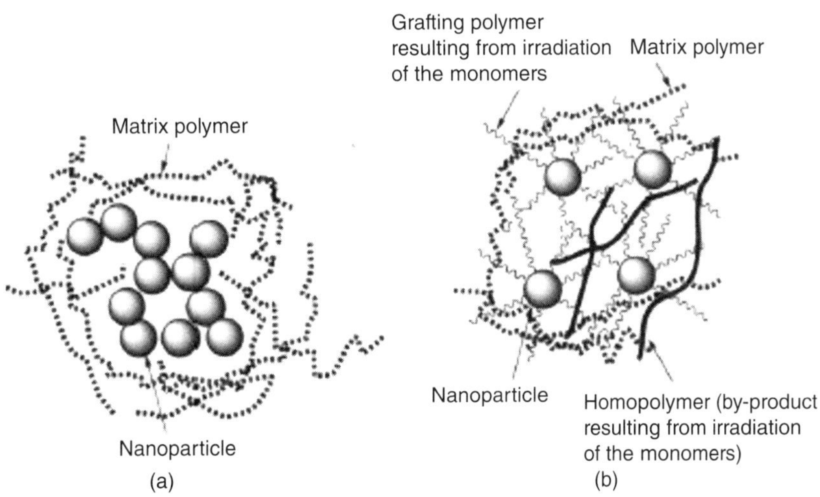

Figure 20.10 Schematic drawings of (a) agglomerated nanoparticles dispersed in a polymer matrix and (b) the possible structure of grafted nanoparticles dispersed in a polymer matrix. Rong et al. [108]. Reproduced with permission of Elsevier.

(PP)/nanosilica composites have been reported by Zhang and coworkers [108, 133]. Through irradiation grafting polymerization, nanoparticle agglomerates turned into a nanocomposite microstructure (comprising the nanoparticles and the grafted, homopolymerized secondary polymer; Fig. 20.10b), which in turn built up a strong interfacial interaction with the surrounding, primary polymeric matrix during the subsequent mixing procedure.

20.5 PREPARATIVE METHODS

The properties of a polymer nanocomposites and their improvement with respect to the traditional composites depend on the structure and morphology developed during the preparation step. In turn, structure and morphology depend on preparation conditions.

At present, the most used methods to prepare polymer nanocomposites are (i) solution processing; (ii) *in situ* polymerization; (iii) melt processing; and (iv) *in situ* sol–gel technology.

20.5.1 Solution Processing

In this approach, a dispersion of nanoparticles in a suitable solvent and polymer are mixed in solution. The PNC is formed by precipitation or by evaporation of the solvent, usually under vacuum.

When layered silicates are used, the weak forces that stack the layers together can be easily dispersed in an adequate solvent. After the clay/organoclay has swollen in the solvent, the polymer is added to the solution and intercalates between the clay layers (in this case, the method is called *intercalation of polymer from solution*).

Even though this technique has been mostly used with water-soluble polymers, such as PEO, polyvinyl ether (PVE), polyvinylpyrrolidone (PVP), and poly(acrylic acid) (PAA) [134–141], intercalation from nonaqueous solutions has also been reported [142–145]. For example, high-density polyethylene (HDPE)-based nanocomposites have been prepared by dissolving HDPE in a mixture of xylene and benzonitrile with dispersed organomodified layered silicates (OMLSs). The nanocomposite was then recovered by precipitation from tetrahydrofuran (THF) [143]. Polystyrene (PS)/OMLS-exfoliated nanocomposites have also been prepared by the solution intercalation technique, by mixing pure PS and organophilic clay with adsorbed cetyl pyridium chloride [146]. Similarly, several studies have focused on the preparation of polylactide (PLA)-layered silicate nanocomposites using intercalation from solution.

The first attempts by Ogata et al. [142] involved dissolving the polymer in hot chloroform in the presence of OMLS. However, TEM and WAXD analyses revealed that only microcomposites were formed and that an intercalated morphology was not achieved. In a later study, Krikorian and Pochan [147] prepared PLA nanocomposites using dichloromethane as the polymer solvent and as the OMLS dispersion medium. The authors obtained intercalated or exfoliated nanocomposites, depending on the type of OMLS used. That is, exfoliated nanocomposites were formed when diols were present in the organic modifier of the clay, due to the favorable enthalpic interaction between these diols and the C=O bonds in the PLA backbone. Chang et al. [148] reported the preparation of PLA-based nanocomposites with different kinds of OMLS via solution intercalation using *N*,*N*-dimethylacetamide (DMA).

In the case of polymeric materials that are infusible and insoluble even in organic solvents, the only possible route to produce nanocomposites with this method is to use polymeric precursors that can be intercalated in the layered silicate and then thermally or chemically converted to the desired polymer [149]. It is important to note that, in using this method, intercalation only occurs for certain polymer/clay/solvent systems, meaning that for a given polymer one has to find the right clay, organic modifier, and solvents [150].

When other nanoparticles (CNTs, metal oxide, etc.) are used, their dispersion in a solvent by simple stirring is very difficult. A high-power ultrasonication process is more effective in forming a dispersion of particles. Ultrasonic wave and mechanically stirring play important roles in the formation of the composites with a uniform particle size. The chemical effects of ultrasound are associated with the rapid (microsecond timescale), violent collapse of cavitation bubbles created as the ultrasonic waves pass through a liquid medium [151]. Sonochemical theory and the corresponding studies suggested that ultrasonic cavitation can generate a high local temperature of 5000 K and a local pressure of 500 atm [152], which is a very rigorous environment.

For example, Li et al. [153] used a simple solution–precipitation technique to improve the dispersion of CNTs in a polycarbonate solution by sonication at a frequency of 20 kHz for 10 min. They showed that the CNTs were uniformly dispersed in polycarbonate matrix on its consolidation. Safadi et al. [154] dispersed MWNTs in PS using ultrasonication and dismembrator at 300 W for 30 min. Uniform dispersions of CNTs in PS were achieved by using sonication. Recently, Cho and coworkers successfully prepared polyurethane (PU)/MWNT composites with better dispersion of CNTs up to 20 wt% in PU [155].

Solution processing has been used for preparation of many different polymer/CNTs such as PU/CNT [156], PS/CNT [157–159], epoxy/CNT [158, 160, 161], poly(vinyl alcohol) (PVA)/CNT [162], polyacrylonitrile (PAN)/CNT [163], and polyethylene (PE)/CNT [164].

By this method also, polymer/silica nanocomposites have been fabricated [165, 166]. Merkel et al. discovered that the addition of nanometer-sized fumed silica particles to certain high-free volume, glassy polymers could systematically increase gas permeability. Such high-permeability polymers included poly(4-methyl-2-pentyne) (PMP), poly[1-(trimethylsilyl)-1-propyne] (PTMSP), and poly(2,2-bis(trifluoromethyl)-4,5-difluoro-1,3-dioxole-*co*-tetrafluoroethylene) [107].

van Zyl et al. [167] reported the preparation of PA/silica nanocomposites via solution processing. Nylon 6 was first dissolved in formic acid, the pH was controlled at ca. 2, and the silica sol with particle sizes of 10–30 nm was added to the nylon solution and stirred gently at room temperature. The solution was then casted, and the solvent was evaporated. The nanocomposite was examined with TEM, which revealed that the silica particles were well dispersed and nonaggregated.

20.5.2 In situ Polymerization

In this method, the nanoparticles are dispersed in monomer followed by polymerization. This method is useful for the preparation of composites with polymers that cannot be processed by solution or melt mixing, for example, insoluble and thermally unstable polymers [168, 169].

In situ polymerization was the first method used to synthesize polymer–clay nanocomposites based on polyamide (PA) 6. In this technique, the modified layered silicate is swollen by a liquid monomer or a monomer solution. The monomer migrates into the galleries of the layered silicate, so that the polymerization reaction can occur between the intercalated sheets. The reaction can be initiated either by heat or radiation, by the diffusion of a suitable initiator or by an organic initiator or catalyst fixed through cationic exchange inside the interlayer before the swelling step by the monomer. Polymerization produces long-chain polymers within the clay galleries. Under conditions in which intra- and extra-gallery polymerization rates are properly balanced, the clay layers are delaminated and the resulting material possesses a disordered structure [33, 170]. The method has been applied for the preparation of nanocomposites based on thermoplastic polymers other than polyamides, including polymethylmethacrylate (PMMA) [171, 172], (PS) [173], polybenzoxale (PBO) [174], polyolefins (PP and PE), [175 178], and polyethylene terephthalate (PET) [179].

In situ polymerization has been widely used for the preparation of PMMA-CNT composites [180–183]. Conducting polymers are attached to CNT surfaces by *in situ* polymerization to improve the processability and electrical, magnetic, and optical properties of CNTs [184–187]. PU/MWNT [188, 189] and PA/MWNT [190] composites were also synthesized by this method.

An interesting example is proposed by Sahoo et al. [191], where polypyrrole (PPy)-coated MWNTs were synthesized by *in situ* polymerization of pyrrole on MWNT-COOH. The possible interaction of hydrogen bonding between PPy chain and MWCT-CCOH is reported in Figure 20.11(1). They investigated the effect of the monomer concentration on the coating and properties of the resulting complex nanotubes. By changing the pyrrole/MWNT ratio, the layer thickness of PPy could be easily controlled in MWNT-PPy complex nanotubes, as shown in Figure 20.11(2).

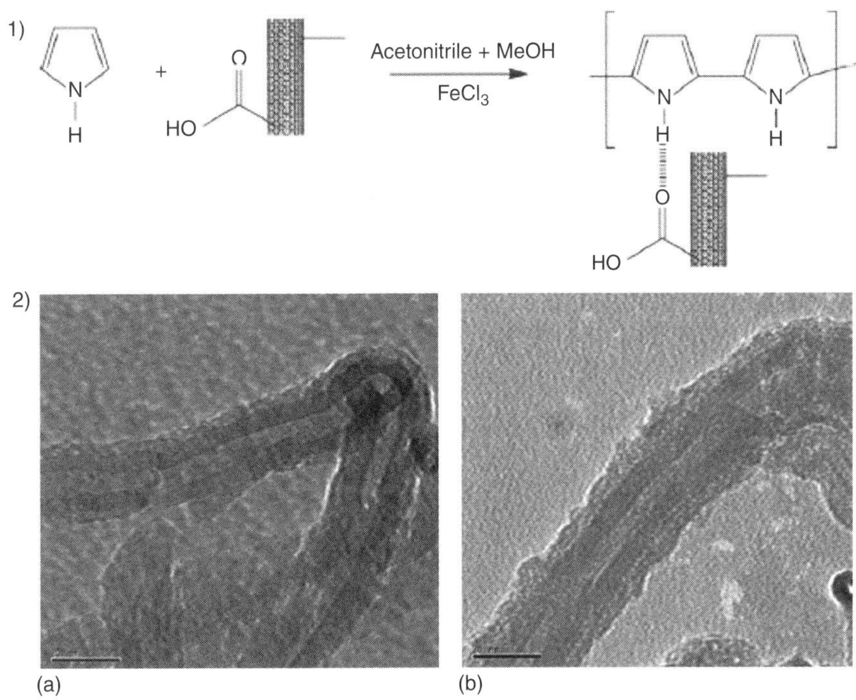

Figure 20.11 (1) Possible interaction of hydrogen bonding existed between for PPy chain and MWNT-COOH. (2) TEM micrographs of PPy-coated MWNTs: by changing the monomer concentration ((a) MWNT:PPy = 1:2; (b) MWNT:PPy = 1:5), the layer thickness of polymerized PPy changed, resulting in different electrical properties. Sahoo et al. [191]. Reproduced with permission of Elsevier.

20.5.3 Melt Processing

Melt processing is a common and simple method, particularly useful for thermoplastic polymers. This technique consists of blending the nanoparticles with the polymer matrix in the molten state. The nanoparticles are mechanically dispersed into a polymer matrix using a high-temperature and high-shear-force mixer or compounder [192].

When technologically important polymer/layer silicate nanocomposites are to be prepared, both *in situ* polymerization and intercalation from solution are limited because neither a suitable monomer nor a compatible polymer–silicate solvent system is always available. Moreover, they are not always compatible with current polymer processing techniques. These disadvantages drive the researchers to the direct melt intercalation method, which is the most versatile and environmentally benign among all the methods of preparing polymer–clay nanocomposites [6]. The advantages of forming nanocomposites by melt processing are quite appealing, rendering this technique a promising new approach that would greatly expand the commercial opportunities for nanocomposite technology [193–196]. If technically possible, melt compounding would be significantly more economical and simpler than *in situ* polymerization. It minimizes capital costs because of its compatibility with existing processes. That is, melt processing allows nanocomposites to be formulated directly using ordinary compounding devices such as extruders or mixers, without the necessary involvement of resin production. Therefore, it shifts nanocomposite production downstream, giving end-use manufacturers many degrees of freedom with regard to final product specifications (e.g., selection of polymer grade, choice of organoclay, level of reinforcement, etc.). At the same time, melt processing is environmentally sound because no solvents are required [193, 195]; and it enhances the specificity for the intercalation of polymer, by eliminating the competing host–solvent and polymer–solvent interactions [2]. Thus, the majority of thermoplastic polymers, including PA [32, 195, 197, 198], PET [199] (and recycled PET [200]), EVA [201, 202], thermoplastic PU [203], polyolefins [204, 205], PLA [206–208], polycaprolactone (PCL) [209, 210], and so on, have been used to prepare polymer/clay nanocomposites by this method.

Melt processing has been successfully applied for the preparation of different polymer–CNT composites such as PP/CNT [211–213], PE/CNT [212–216], polycarbonate (PC)/CNT [217–219], PMMA/CNT [220–223], polyoxymethylene/CNT [224], polyimide/CNT [225], PA6/CNT [226, 227], and so on. The shear forces from the mixers/extruders should help break nanoparticle aggregates or prevent their formation. Unfortunately, the dispersion of CNTs in a polymer matrix is quite poor compared to the dispersion that may be achieved through solution mixing. In addition, the CNTs must be lower due to the high viscosities of the composites at higher loading of CNTs.

Melt processing has been used also for polymer and polymer blend/silica nanocomposites. In particular, extensive studies are reported for PP [228], PP-based copolymer [229], PE [230], PE-based copolymer [231, 232], PS [233–235], PMMA [234, 235], PC [234], PC-based copolymer [236], polyethylene naphthalate (PEN) [237], perfluoropolymer [238], PET [239–241], PA [242], polyvinyl acetate (PVAc) [243], copolyetherester [244], styrene-butadiene rubber [245, 246], ethylene vinyl acetate (EVA) [247], PET/PS [248], PLLA [249], and many others.

20.5.4 In situ Sol–Gel Technology

An extensive literature survey reveals that almost all polymer–ceramic nanocomposites were prepared through mechanical blending of nanosized particles, polymer, and salt in a compatible solvent. Aggregation of these particles stemming from their high surface energy is a perennial problem that undermines the efficacy of the ceramic fillers. A simple and effective method to overcome such a problem is the sol–gel process, wherein the nanosized ceramic fillers are precipitated *in situ* in the polymer matrix through a series of hydrolysis and condensation reactions of suitable precursors. Thus, the ceramic fillers are uniformly distributed in the polymer and exhibit excellent properties.

These organic–inorganic hybrids are generally classified into two broad families: in type I nanocomposites, the interactions between the hosting matrix and the entrapped species are weak and based on hydrogen bonds and van der Waals forces; whereas in type II nanocomposites, the organic and inorganic components are connected through stronger chemical bonds.

The successful implementation of this versatile process in the synthesis of siloxane-derived composite polymer (i.e., polymer/silica nanocomposites) electrolytes has appeared in several recent reports [250–252]. The sol–gel method has already been reported for the preparation of nanocomposite-based silicone rubber [253, 254], polyisobutylene [255], and several diene rubbers [256–261]. Nevertheless, this method is suitable usually for amorphous polymers (rubbers and elastomers) as it is based on the polymer swelling by the inorganic precursor solution. Regarding polyolefin matrix, Sun et al. [262] reported the preparation of polypropylene/silica nanocomposites via a two-step method: (i) diffusion of the tetraethoxysilane (TEOS) through the polypropylene (PP) matrix using supercritical carbon dioxide as a carrier and swelling agent, (ii) hydrolysis/condensation reactions of the precursor molecules confined in polymer network. Jain et al. [263, 264] developed a new process, combining the sol–gel method with the solid-state modification consisting of grafting vinyl triethoxysilane (VTEOS). Following this, they prepared PP/silica nanocomposites, varying the degree of covalent interaction between silica nanofiller and matrix. Dou et al. [265] reported a preparation of PP/silica nanocomposite by *in situ* sol–gel process using hyperbranched polyethoxysiloxane.

Several works have been published on the efficiency of the sol–gel process in the production of high-quality titanium dioxide from titanium alkoxide precursor [266–270].

20.6 STRUCTURE AND MORPHOLOGY OF POLYMER NANOCOMPOSITES

Good dispersion and distribution (and in the case of clay nanoparticles an exfoliated structure) are the key factors for improving properties in a nanocomposites.

The different structure and morphologies that can be generated during the preparation of a nanocomposite depend on many parameters. The most important are method of preparations, nature of the particle functionalization, particle concentrations. A comprehensive coverage of all the different effect obtained by changing the parameters would be not possible and beyond the scope of this review. Anyway some examples in this paragraph are reported.

In Figure 20.12, different dispersions of CNTs in epoxy matrix, fabricated using different techniques are shown by TEM.

The techniques used for preparation of nanocomposites (sonication in water bath, shear mixing, probe sonicator,

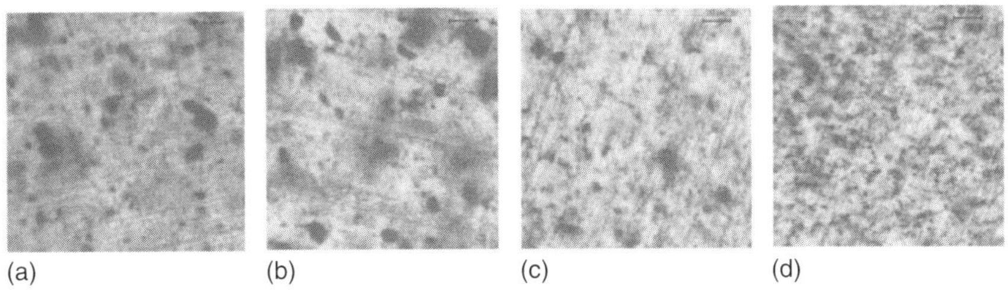

Figure 20.12 Dispersion of CNTs in nanocomposites fabricated using different dispersion techniques (a: sonication in water bath; b: shear mixing; c: probe sonicator; d: calendering; scale bar = 20 μm). Peng-Cheng et al. [271]. Reproduced with permission of Elsevier.

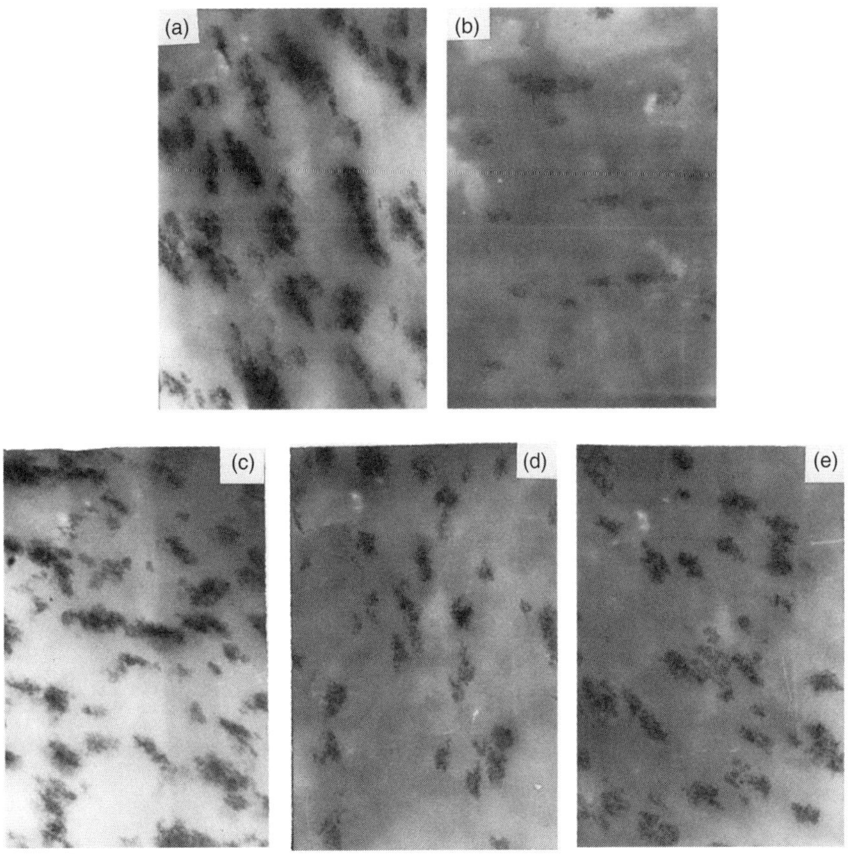

Figure 20.13 TEM micrographs of PP (MI = 6.7 g/10 min)-based nanocomposites filled with (a) SiO_2 as-received (content of SiO_2 1.96 vol%), (b) SiO_2-*g*-PS (content of SiO_2 1.96 vol%), (c) SiO_2-*g*-PS (content of SiO_2 6.38 vol%), (d) SiO_2-*g*-PMMA (content of SiO_2 1.96 vol%), and (e) SiO_2-*g*-PMMA (content of SiO_2 6.38 vol%). Magnification 2×10^4. Rong et al. [108]. Reproduced with permission of Elsevier.

and calendering) influences the dispersion of CNTs and, to a large extent, the mechanical properties of CNT/polymer nanocomposites. The best CNT dispersion was achieved by employing calendaring machine (Fig. 20.12d), and the nanocomposites fabricated using this technique exhibited the best flexural properties among the four techniques used [271, 272].

In the work of Rong et al. [108], various polymers were grafted on the surface of nanoscale silica filler particles through the simultaneous irradiation polymerization technique. In this way, the modified nanoparticles can be more effectively utilized in thermoplastics (such as PP) than conventional particulate fillers, when using the same direct compounding technology.

Morphology of PP/SiO_2 is reported in Figure 20.13, where different polymers (PS and PMMA) are grafter on the silica surface by irradiation treatment. Two different concentrations of SiO_2 are considered (1.96 and 6.38 vol%). From the figure it is clear that higher filler loading is detrimental to silica uniform dispersion in the polymer matrix. In fact, the size of the dispersed phases increases with increasing filler content in the nanocomposites. It should be noted that the dispersed phases in the nanocomposites illustrated in Figure 20.13b–e are clearly smaller than the untreated SiO_2.

As a result of these phenomena, reinforcing and toughening effects of the nanoparticles on polymeric materials could be fully brought into play. In fact, the tensile properties, such as strength, modulus, and elongation at break of PP were improved simultaneously when the modified nanoparticles were incorporated.

In the work of Bahloul [270], *in situ* generation of TiO_2 inorganic particles in a semicrystalline polymer (PP) is reported. Titanium n-butoxide was mixed with PP under molten conditions in a corotating twin screw extruder. At the die exit of the extruder, the extent of the hydrolysis was higher than 70% and fine particles dispersed in PP were observed by TEM analysis. Figure 20.14a and b shows a fine dispersion of few nanometers of TiO_2 particles, whereas Figure 20.14c, which presents the morphology of nanocomposites with higher percentage of silica particles, indicates larger scale structure (aggregates) at the nanometric scale.

When melt-mixing process is considered for preparation of nanocomposites, the effects of operating and processing conditions on the morphology of nanocomposites, and therefore

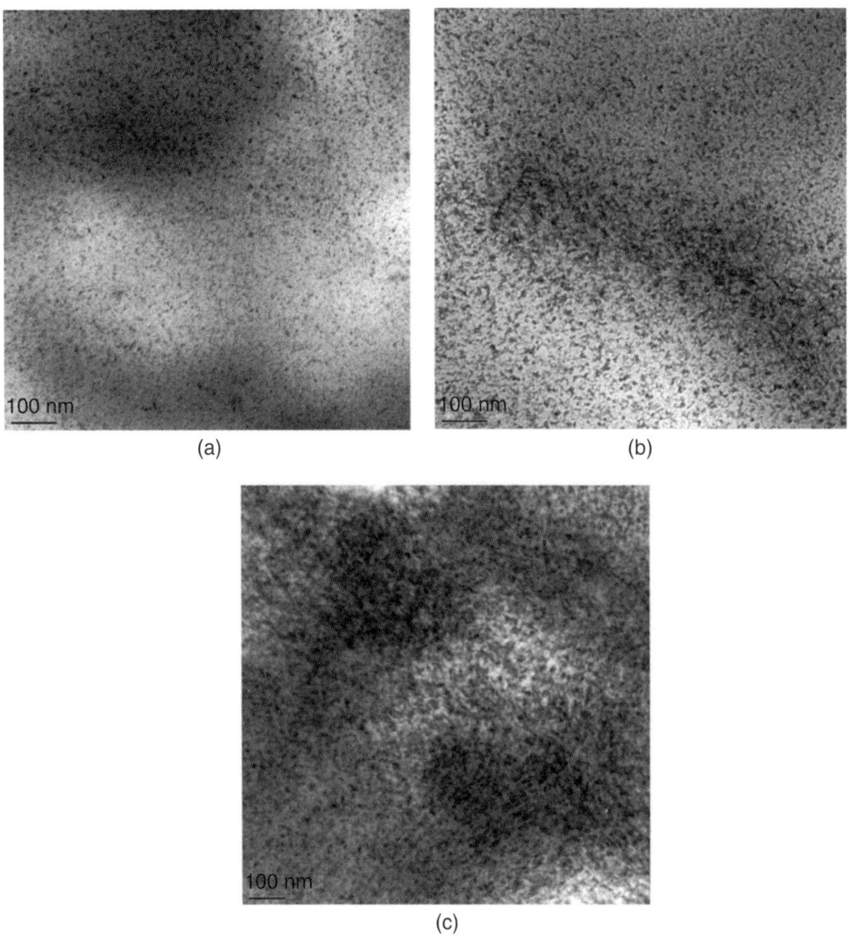

Figure 20.14 TEM of nanocomposites containing different amounts of nano-TiO_2: (a) PP-2.9%TiO_2, (b) PP-6%TiO_2, and (c) PP-9.3%TiO_2. Bahloul et al. [270]. Reproduced with permission of John Wiley and Sons.

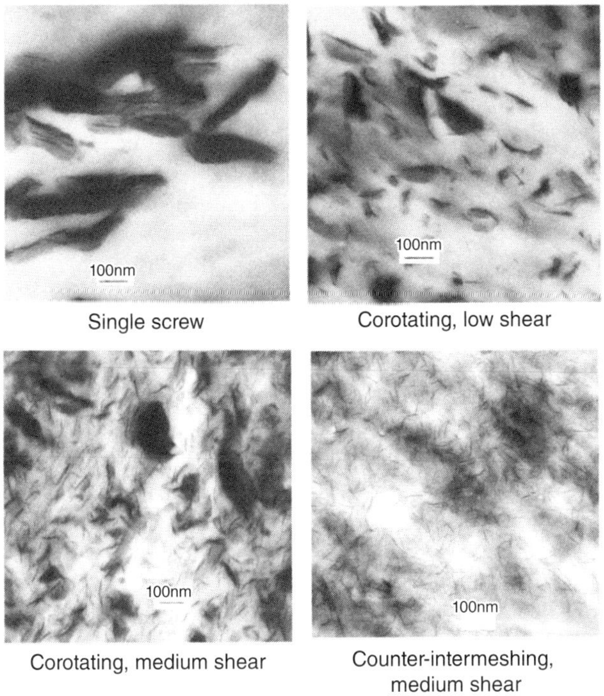

Figure 20.15 Transmission electron micrographs for selected nanocomposites prepared from clay and PA6 using different extruders and screw configurations. Dennis et al. [273]. Reproduced with permission of Elsevier.

on their properties, has to be considered. Dennis et al. [273] have studied the effect of the processing conditions on the formation of nanocomposites from PA6 and organoclay by using several combinations of extruders and screw configurations. In particular, they have used a single-screw extruder and two twin-screw extruders: corotating twin-screw extruder and counterrotating twin-screw extruder. For each screw extruder type, the shear intensity of mixing was changed by changing the screw configurations. Excessive shear intensity or backmixing apparently causes poorer delamination and dispersion and the best results are obtained by using a counterrotating twin screw extruder (Fig. 20.15).

In another work, Lertwimolnun and Vergnes [274] have prepared PP/organoclay nanocomposites by using a twin-screw extruder and have found that the state of intercalation, interpreted by interlayer spacing, is globally unaffected by processing parameters. A complete intercalated structure is obtained, even at high feed rate, in a very short period of time (~21 s). In parallel, the level of exfoliation is estimated by the determination of a melt yield stress. It increases when the feed rate (Q) decreases and the screw speed (N) increases. The effect of these two parameters can be expressed through the ratio Q/N. Figure 20.16 shows that the values of the melt yield stress, and thereby the degree of exfoliation, decrease regularly as a function of Q/N. The best exfoliation is thus obtained at high screw speed and low feed rate. Direct morphological

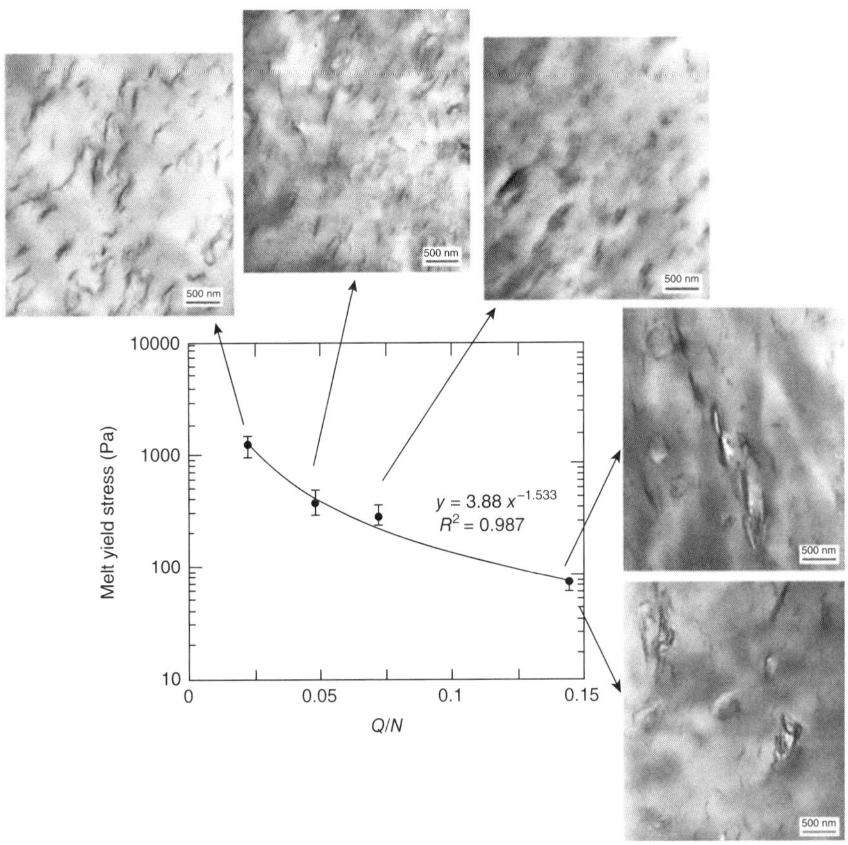

Figure 20.16 Evolution of melt yield stress as a function of Q/N for the nanocomposites prepared with different operating conditions. Lertwimolnun and Vergnes [275]. Reproduced with permission of Elsevier.

observations by TEM confirm the results found by SAOS measurements (rheological measurements in small-amplitude oscillatory shear) [275]. The better dispersion is observed at low *Q*/*N* ratio. The difference of morphology is evident between the highest and the lowest *Q*/*N* ratios.

20.7 CONCLUDING REMARKS

There are several approaches for developing high-performance PNCs. The critical challenge is the development of methods to improve the dispersion of nanoparticles in a polymer matrix because their enhanced dispersion in polymer matrices greatly improves the mechanical, electrical, and optical properties of nanocomposites. Despite various methods, such as melt processing, solution processing, *in situ* polymerization, sol–gel technology and chemical functionalization, there are still opportunities and challenges to be found in order to improve dispersion and modify interfacial properties. In particular, it is needed to identify stable procedures for homogeneous dispersion of nanoparticles, by directing the structure and properties of the polymer matrix during crystallization/solidification taking into consideration the materials' characteristics, the nanoparticles' shape, as well as the conditions of the processing.

REFERENCES

1. Fukushima Y, Okada A, Kawasumi M, Kurauchi T, Kamigaito O. Swelling behavior of montmorillonite by poly-6-amide. Clay Miner 1998;23:27–34.
2. Vaia RA, Ishii H, Giannelis EP. Synthesis and properties of two-dimensional nanostructures by direct intercalation of polymer melts in layered silicates. Chem Mater 1993;5:1694–1696.
3. Lan T, Pinnavaia TJ. Clay-reinforced epoxy nanocomposites. Chem Mater 1994;6:2216–2219.
4. Winey K, Vaia R, editors. *Polymer Nanocomposites*. Pittsburgh, PA: MRS Bulletin; Materials Research Society; 2007.
5. Nanocomposites (Carbon Nanotubes, Polymer Metal Fiber, Nanofibers, Graphene, Nanoplatelet and Others) Market for Automotive, Aviation, Electronics, Energy, Construction, Healthcare, Plastics, Military, Consumer Goods and Other Applications - Global Industry Analysis, Size, Share, Growth, Trends and Forecast, 2012–2018 Transparency Market design Albany, New York (PRWEB) August 09, 2013.
6. Giannelis EP. Polymer layered silicate nanocomposites. Adv Mater 1996;8:29–35.
7. Dubois MAP. Polymer–layered silicate nanocomposites: preparation, properties and uses of a new class of materials. Mater Sci Eng R 2000;R28:1–65.
8. Ray SS, Okamoto M. Polymer/layered silicate nanocomposites: A review from preparation to processing. Prog Polym Sci 2003;2:1539–1641.
9. Okamoto M. In: Nalwa HS, editor. *Encyclopedia of Nanoscience and Nanotechnology*. Volume 8. Stevenson Ranch, CA: American Scientific; 2004. p 1–52.
10. Thostenson ET, Li C, Chou TW. Nanocomposites in context. Compos Sci Technol 2005;65:491–516.
11. Drummy LF, Koerner H, Farmer BL, Vaia RA. *Advanced Morphology Characterization of Clay-Based Polymer Nanocomposites*, CMS Workshop Lecture Series, Vol. 14. Clay Minerals Society; 2006.
12. Tjong SC. Structural and mechanical properties of polymer nanocomposites. Mater Sci Eng R 2006;53:73–197.
13. Hussain F, Hojjati M, Okamoto M, Gorga RE. Review article: Polymer-matrix nanocomposites, processing, manufacturing, and application: An overview. J Compos Mater 2006;40(17): 1511–1575.
14. Pavlidou S, Papaspyrides CD. A review on polymer–layered silicate nanocomposites. Prog Polym Sci 2008;33:1119–1198.
15. Kiliaris P, Papaspyrides CD. Polymer/layered silicate (clay) nanocomposites: An overview of flame retardancy. Prog Polym Sci 2010;35:902–958.
16. Bitinis N, Hernandez M, Verdejo R, Kenny JM, Lopez-Manchado MA. Recent advances in clay/polymer nanocomposites. Adv Mater 2011;23:5229–5236.
17. Xie XL, Maia YW, Zhou XP. Dispersion and alignment of carbon nanotubes in polymer matrix: A review. Mater Sci Eng R 2005;49:89–112.
18. Moniruzzaman M, Winey KI. Polymer nanocomposites containing carbon nanotubes. Macromolecules 2006;39:5194–5205.
19. Sahooa NG, Ranab S, Chob JW, Li L, Chana SH. Polymer nanocomposites based on functionalized carbon nanotubes. Prog Polym Sci 2010;35:837–867.
20. Spitalskya Z, Tasisb D, Papagelisb K, Galiotis C. Carbon nanotube–polymer composites: Chemistry, processing, mechanical and electrical properties. Prog Polym Sci 2010; 35:357–401.
21. Domènech B, Bastos-Arrieta J, Alonso A, Muñoz M, Muraviev DN, Macanás J. Bifunctional polymer-metal nanocomposite ion exchange materials. In: Kilislioğlu A, editor. *Ion Exchange Technologies*. InTech 2012, chapter 3. p 35–72.
22. Silvestre C, Duraccio D, Cimmino S. Food packaging based on polymer nanomaterials. Prog Polym Sci 2011;36:1766–1782.
23. Silvestre C, Cimmino S, editors. *Ecosustainable Polymer Nanomaterials for Food Packaging*. Boca Raton, FL: CRC Press Taylor and Francis Group; 2013.
24. Pinnavaia TJ, Beall GWJ. *Polymer-Clay Nanocomposites*. New York: John Wiley & Sons; 2001.
25. Krishnamoorti R, Vaia RA, editors. *Polymer Nanocomposites: Synthesis, Characterization, and Modeling*; ACS Symposium Series. Washington, DC: American Chemical Society; 2001.
26. Ray SS, Bousmina M. *Polymer Nanocomposites and Their Applications*. Stevenson Ranch, CA: American Scientific; 2006.
27. Mai YW, Yu ZZ, editors. *Polymer Nanocomposites*. Boca Raton, FL: CRC; 2006.
28. Wilkie CA, Morgan AB, editors. *Fire Retardancy of Polymeric Materials*. 2nd ed. Baco Raton, FL, USA: CRC Press; 2009. ISBN 9781420083996.
29. Ruckenstein E, Li ZF. Surface modification and functionalization through the self-assembled monolayer and graft polymerization. Adv Colloid Interface Sci 2005;113:43–63.

30. Miranda-Trevino JC, Coles CA. Kaolinite properties, structure and influence of metal retention on pH. Appl Clay Sci 2003;23:133–139.
31. Beyer G. Nanocomposites: A new class of flame retardants for polymers. Plast Addit Compd 2002;4:22–27.
32. McNally T, Murphy WR, Lew CY, Turner RJ, Brennan GP. Polyamide12-layered silicate nanocomposites by melt compounding. Polymer 2003;44:2761–2772.
33. Solomon MJ, Almusallam AS, Seefeldt KF, Somwangthanaroj A, Varadan P. Rheology of polypropylene/clay hybrid materials. Macromolecules 2001;34:1864–1872.
34. Manias E, Touny A, Wu L, Strawhecker K, Lu B, Chung TC. Polypropylene/ montmorillonite nanocomposites. Review of the synthetic routes and materials properties. Chem Mater 2001;13:3516–3523.
35. Hedley CB, Yuan G, Theng BKG. Thermal analysis of montmorillonites modified with quaternary phosphonium and ammonium surfactants. Appl Clay Sci 2007;35:180–188.
36. Kornmann X, Lindberg H, Berglund LA. Synthesis of epoxy–clay nanocomposites: influence of the nature of the clay on structure. Polymer 2001;42:1303–1310.
37. Zanetti M, Lomakin S, Camino G. Polymer layered silicate nanocomposites. Macromol Mater Eng 2000;279:1–9.
38. Xie W, Gao Z, Liu K, Pan WP, Vaia R, Hunter D, et al. Thermal characterization of organically modified montmorillonite. Thermochim Acta 2001;367(368):339–350.
39. Kornmann X, Lindberg H, Berglund LA. Synthesis of epoxy–clay nanocomposites: influence of the nature of the clay on structure. Polymer 2001;42:1303–1310.
40. Kim CM, Lee DH, Hoffmann B, Kressler J, Stoppelmann G. Influence of nanofillers on the deformation process in layered silicate/polyamide12 nanocomposites. Polymer 2001;42:1095–1100.
41. Zerda AS, Lesser AJ. Intercalated clay nanocomposites: morphology, mechanics and fracture behavior. J Polym Sci Polym Phys 2001;39:1137–1146.
42. Lan T, Kaviratna PD, Pinnavia TJ. Mechanism of clay tactoid exfoliation in epoxy-clay nanocomposites. Chem Mater 1995;7:2144–2150.
43. Pandey JK, Reddy KR, Kumar AP, Singh RP. An overview of the degradability of polymer nanocomposites. Polym Degrad Stab 2005;88:234–250.
44. Miyagawa H, Drzal LT, Miyagawa H, Carsello JA. Intercalation and exfoliation of clay nanoplatelets in epoxy-based nanocomposites: TEM and XRD observations. Polym Eng Sci 2006;46:453–463.
45. Iijima S. Helical microtubules of graphitic carbon. Nature 1991;354:56–58.
46. Ajayan PM, Stephan O, Colliex C, Trauth D. Aligned carbon nanotube arrays formed by cutting a polymer resin—nanotube composite. Science 1994;265:1212–1214.
47. Cooper CA, Young RJ, Halsall M. Investigation into the deformation of carbon nanotubes and their composites through the use of Raman spectroscopy. Composites A 2001;32:401–411.
48. Gao G, Cagin T, Goddard WA III. Energetics, structure, mechanical and vibrational properties of single-walled carbon nanotubes. Nanotechnology 1998;9:184–191.
49. Uchida T, Kumar S. Single wall carbon nanotube dispersion and exfoliation in polymers. J Appl Polym Sci 2005;98: 985–989.
50. De Heer Walt A. Nanotubes and the pursuit of applications. MRS Bull 2004;29:281–285.
51. Dai H. Carbon nanotubes: Synthesis, integration, and properties. Acc Chem Res 2002;35:1035–1044.
52. Awasthi K, Srivastava A, Srivastava ON. Synthesis of carbon nanotubes. J Nanosci Nanotechnol 2005;5:1616–1636.
53. Monthioux M. Filling single-wall carbon nanotubes. Carbon 2002;40:1809–1823.
54. Thostenson ET, Zhifeng R, Chou TW. Advances in the science and technology of carbon nanotubes and their composites: A review. Compos Sci Technol 2001;61:1899–1912.
55. Dyke CA, Tour JM. Overcoming the insolubility of carbon nanotubes through high degree of sidewall functionalization. Chem Eur J 2004;10:812–817.
56. Haddon RC. Solution properties of single-walled carbon nanotubes. Science 1998;282:95–98.
57. Michelson ET, Huffman CB, Rinzler AG, Smalley RE, Hauge RH, Margrave JL. Fluorination of single-wall carbon nanotubes. Chem Phys Lett 1998;296:188–194.
58. Michelson ET, Chiang IW, Zimmerman JL, Boul PJ, Lozano J, Liu J, Smalley RE, Hauge RH, Margrave JL. Solvation of fluorinated single-wall carbon nanotubes in alcohol solvents. J Phys Chem B 1999;103:4318–4322.
59. Chen J, Rao AM, Lyuksyutov S, Itkis ME, Hamon MA, Hu H, Cohn RW, Eklund PC, Colbert DT, Smalley RE, Haddon RC. Dissolution of full-length single-walled carbon nanotubes. J Phys Chem B 2001;105:2525–2528.
60. Riggs JE, Guo Z, Carroll DL, Sun YP. Strong luminescence of solubilized carbon nanotubes. J Am Chem Soc 2000;122:5879–5880.
61. Riggs JE, Walker DB, Carroll DL, Sun Y-P. Optical limiting properties of suspended and solubilized carbon nanotubes. J Phys Chem B 2000;104:7071–7076.
62. Jin Z, Sun X, Xu G, Goh SH, Ji W. Nonlinear optical properties of some polymer/multiwalled carbon nanotube composites. Chem Phys Lett 2000;318:505–510.
63. Niyogi S, Hu H, Hamon MA, Bhowmik P, Zhao B, Rozenzhak SM, Chen J, Itkis ME, Meier MS, Haddon RC. Chromatographic purification of soluble single-walled carbon nanotubes (s-SWNTs). J Am Chem Soc 2001;123:733–734.
64. Zhao B, Hu H, Niyogi S, Itkis ME, Hamon MA, Bhowmik P, Meier MS, Haddon RC. Chromatographic purification and properties of soluble single-walled carbon nanotubes. J Am Chem Soc 2001;123:11673–11677.
65. Sun YP, Huang W, Lin Y, Fu K, Kitaygorodskiy A, Riddle LA, Yu YJ, Carroll DL. Soluble dendron-functionalized carbon nanotubes: Preparation, characterization, and properties. Chem Mater 2001;13:2864–2869.
66. Fu K, Huang W, Lin Y, Riddle LA, Carroll DL, Sun YP. Defunctionalization of functionalized carbon nanotubes. Nano Lett 2001;1:439–441.
67. Sun Y, Wilson SR, Schuster DI. High dissolution and strong light emission of carbon nanotubes in aromatic amine solvents. J Am Chem Soc 2001;123:5348–5349.

68. Bahr JL, Yang J, Kosynkin DV, Bronikowski MJ, Smalley RE, Tour JM. Functionalization of carbon nanotubes by electrochemical reduction of aryl diazonium salts: A bucky paper electrode. J Am Chem Soc 2001;123:6536–6542.
69. Bahr JL, Tour JM. Highly functionalized carbon nanotubes using in situ generated diazonium compounds. Chem Mater 2001;13:3823–3824.
70. Pekker S, Salvetat JP, Jakab E, Bonard JM, Forro L. Hydrogenation of carbon nanotubes and graphite in liquid ammonia. J Phys Chem B 2001;105:7938–7943.
71. Czerw R, Guo Z, Ajayan PM, Sun YP, Carroll DL. Organization of polymers onto carbon nanotubes: A route to nanoscale assembly. Nano Lett 2001;1:423–427.
72. Lin Y, Rao AM, Sadanadan B, Kenik EA, Sun YP. Functionalizing multiple-walled carbon nanotubes with aminopolymers. J Phys Chem B 2002;106:1294–1298.
73. Huang W, Lin Y, Taylor S, Gaillard J, Rao AM, Sun YP. Sonication-assisted functionalization and solubilization of carbon nanotubes. Nano Lett 2002;2:231–234.
74. Holzinger M, Vostrowsky O, Hirsch A, Hennrich F, Kappes M, Weiss R, Jellen F. Sidewall functionalization of carbon nanotubes. Angew Chem Int Ed 2001;40:4002–4005.
75. Georgakilas V, Kordatos K, Prato M, Guldi DM, Holzinger M, Hirsch A. Organic functionalization Organic functionalization of carbon nanotubes. J Am Chem Soc 2002;120:760–761.
76. Liu J, Rinzler AG, Dai H, Hafner JH, Bradley RK, Boul PJ, Lu A, Iverson T, Shelimov K, Huffman CB, Rodriguez-Macias F, Shon YS, Lee TR, Colbert DT, Smalley RE. Fullerene pipes. Science 1998;280:1253–1256.
77. Hamon MA, Hu H, Bhowmik P, Niyogi S, Zhao B, Itkis ME, Haddon RC. End-group and defect analysis of soluble single-walled carbon nanotubes. Chem Phys Lett 2001;347:8–12.
78. Behra R, Sigg L, Clift MJD, Herzog F, Minghetti M, Johnston B, Petri-Fink A, Rothen-Rutishauser B. Bioavailability of silver nanoparticles and ions: from a chemical and biochemical perspective. J R Soc Interface 2013:10.
79. Klasen H. Historical review of the use of silver in the treatment of burns. I. Early uses. Burns 2000;26:117–130.
80. Maillard JY, Hartemann P. Silver as an antimicrobial: Facts and gaps in knowledge. Crit Rev Microbiol 2012;39:373–383.
81. Rai M, Yadav A, Gade A. Silver nanoparticles as a new generation of antimicrobials. Biotechnol Adv 2009;27:76–83.
82. Eustis S, Krylova G, Smirnova N, Eremenko A, Tabor C, Huang W, El-Sayed MA. Using silica films and powders modified with benzophenone to photoreduce silver nanoparticles. J Photochem Photobiol A 2006;181:385–393.
83. Van Hoonacker A, Englebienne P. Revisiting silver nanoparticle chemical synthesis and stability by optical spectroscopy. Curr Nanosci 2006;2:359–371.
84. Pastoriza-Santos I, Liz-Marzan LM. N,N-dimethylformamide as a reaction medium for metal nanoparticle synthesis. Adv Funct Mater 2009;19:679–688.
85. Shirtcliffe N, Nickel U, Schneider S. Reproducible preparation of silver sols with small particle size using borohydride reduction: for use as nuclei for preparation of larger particles. J Colloid Interface Sci 1999;211:122–129.
86. Nickel U, Castell AZ, Poppl K, Schneider S. A silver colloid produced by reduction with hydrazine as support for highly sensitive surface-enhanced Raman. Langmuir 2000;16:9087–9091.
87. Fievet F, Francoise FV, Lagier JP, Dumont B, Figlarz M. Controlled nucleation and growth of micrometre-size copper particles prepared by the polyol process. J Mater Chem 1993;3:627–632.
88. Leopold N, Lendl B. A new method for fast preparation of highly surface-enhanced Raman scattering (SERS) active silver colloids at room temperature by reduction of silver nitrate with hydroxylamine hydrochloride. J Phys Chem B 2003;107:5723–5727.
89. Zhang Y, Peng H, Huang W, Zhou Y, Zhang X, Yan D. Hyperbranched poly(amidoamine) as the stabilizer and reductant to prepare colloid silver nanoparticles in situ and their antibacterial activity. J Phys Chem C 2008;112:2330–2336.
90. Yin Y, Li ZY, Zhong Z, Gates B, Xia Y, Venkateswaran S. Synthesis and characterization of stable aqueous dispersions of silver nanoparticles through the Tollens process. J Mater Chem 2002;12:522–527.
91. Liu YC, Lin LH. New pathway for the synthesis of ultrafine silver nanoparticles from bulk silver substrates in aqueous solutions by sonoelectrochemical methods. Electrochem Commun 2004;6:1163–1168.
92. Raveendran P, Fu J, Wallen SL. Completely 'green' synthesis and stabilization of metal nanoparticles. J Am Chem Soc 2003;125:940–941.
93. Cobley CM, Rycenga M, Zhou F, Li ZY, Xia Y. Etching and growth: An intertwined pathway to silver nanocrystals with exotic shapes. Angew Chem Int Ed 2009;48:4824–4827.
94. Sun Y, Xia Y. Shape-controlled synthesis of gold and silver nanoparticles. Science 2002;298:2176–2179.
95. Sharma VK, Yngard RA, Lin Y. Silver nanoparticles: Green synthesis and their antimicrobial activities. Adv Colloid Interface Sci 2009;145:83–96.
96. Vallet-Regi M, Balas F, Arcos D. Mesoporous materials for drug delivery. Angew Chem Int Ed 2007;46:7548–7558.
97. Ghorai TK, Dhak D, Biswas SK, Dalai S, Pramanik P. Photocatalytic oxidation of organic dyes by nano-sized metal molybdate incorporated titanium dioxide ($M_xMo_xTi_{1-x}O_6$) (M = Ni, Cu, Zn) photocatalysts. J Mol Catal A 2007;273:224–229.
98. Rossignol C, Verelst M, Dexpert-Ghys J, Rul S. Synthesis of undoped ZnO nanoparticles by spray pyrolysis. Adv Sci Technol 2006;45:237–241.
99. Babelon P, Dequidt AS, Mostefa-Sba H, Bourgeois S, Sibillot P, Sacilotti M. SEM and XPS studies of titanium dioxide thin films grown by MOCVD. Thin Solid Films 1998;322:63–67.
100. Kim BH, Lee JY, Choa YH, Higuchi M, Mizutani N. Preparation of TiO_2 thin film by liquid sprayed mist CVD method. Mater Sci Eng B 2004;107:289–294.
101. Okuyama K, Shimada M, Fujimoto T, Maekawa T, Nakaso K, Seto T. Effects of preparation conditions on the characteristics of titanium dioxide particles produced by a CVD method. J Aerosol Sci 1998;29(1):S907–S908.
102. Gu DE, Yang BC, Hu YD. V and N co-doped nanocrystal anatase TiO_2 photocatalysts with enhanced photocatalytic activity under visible light irradiation. Catal Commun 2008;9:1472–1476.

103. Zhang X, Liu Q. Preparation and characterization of titania photocatalyst co-doped with boron, nickel, and cerium. Mater Lett 2008;62:2589–2592.

104. Peng F, Cai L, Huang L, Yu H, Wang H. Preparation of nitrogen-doped titanium dioxide with visible-light photocatalytic activity using a facile hydrothermal method. J Phys Chem Solids 2008;69(7):1657–1664.

105. Peng F, Cai L, Huang L, Yu H, Wang H. Synthesis and characterization of substitutional and interstitial nitrogen-doped titanium dioxides with visible light photocatalytic activity. J Solid State Chem 2008;181:130–136.

106. Feng X, Wang Q, Wang G, Qiu F. Preparation of Nano-TiO_2 by ethanol-thermal method and its catalytic performance for synthesis of dibutyl carbonate by transesterification. Chin J Catal 2006;27(3):195–196.

107. Zou H, Wu S, Shen J. Polymer/silica nanocomposites: Preparation, characterization, properties, and applications. Chem Rev 2008;108:3893–3957.

108. Rong MZ, Zhang MQ, Zheng YX, Zeng HM, Walter R, Friedrich K. Structure–property relationships of irradiation grafted nano-inorganic particle filled polypropylene composites. Polymer 2001;42:167–183.

109. Neouze MA, Schubert U. Surface modification and functionalization of metal and metal oxide nanoparticles by organic ligands. Monatsh Chem 2008;139:183–195.

110. Kickelbick G, Schubert U. Organic functionalization of metal oxide nanoparticles. In: Baraton MI, editor. *Synthesis, Functionalization and Surface Treatment of Nanoparticles*. Stevenson Ranch; 2003. p 91.

111. Doty RC, Tshikhudo TR, Brust M, Fernig DG. Extremely stable water-soluble Ag nanoparticles. Chem Mater 2005; 17:4630–4637.

112. Schulz-Dobrick M, Sarathy KV, Jansen M. Surfactant-free synthesis and functionalization of gold nanoparticles. J Am Chem Soc 2005;127:12816–12817.

113. Warner MG, Hutchison JE. Synthesis and assembly of functionalized gold nanoparticles. In: Baraton MI, editor. *Synthesis, Functionalization and Surface Treatment of Nanoparticles*. Stevenson Ranch; 2003. p 67.

114. Roux S, Garcia B, Bridot JL, Salome M, Marquette C, Lemelle L, Gillet P, Blum L, Perriat P, Tillement O. Synthesis, characterization of dihydrolipoic acid capped gold nanoparticles, and functionalization by the electroluminescent luminol. Langmuir 2005;21:2526–2536.

115. Zayats M, Katz E, Baron R, Willner I. Reconstitution of apo-glucose dehydrogenase on pyrroloquinoline quinone-functionalized Au nanoparticles yields an electrically contacted biocatalyst. J Am Chem Soc 2005;127:12400–12406.

116. Ulman A. Formation and structure of self-assembled monolayers. Chem Rev 1996;96:1533–1534.

117. Fan H, Chen Z, Brinker CJ, Clawson J, Alam T. Synthesis of organo-silane functionalized nanocrystal micelles and their self-assembly. J Am Chem Soc 2005;127:13746–13747.

118. Ramirez E, Jansat S, Philippot K, Lecante P, Gomez M, Masdeu-Bulto AM, Chaudret B. Influence of organic ligands on the stabilization of palladium nanoparticles. J Organomet Chem 2004;689:4601–4610.

119. Woehrle GH, Hutchison JE. Thiol-functionalized undecagold clusters by ligand exchange: Synthesis, mechanism, and properties. Inorg Chem 2005;44:6149–6158.

120. Dong TY, Wu HH, Lin MC. Superlattice of octanethiol-protected copper nanoparticles. Langmuir 2006;22:6754–6756.

121. Gooding JJ, Mearns F, Yang W, Liu J. Self-assembled monolayers into the 21st century: Recent advances and applications. Electroanalysis 2003;15:81–96.

122. Liu F, Quan B, Liu Z, Chen L. Surface characterization study on SnO_2 powder modified by thiourea. Mater Chem Phys 2005;93:301–304.

123. Elbhiri Z, Chevalier Y, Chovelon JM, Jaffrezic-Renault N. Grafting of phosphonate groups on the silica surface for the elaboration of ion-sensitive field-effect transistors. Talanta 2000;52:495–507.

124. Hua F, Swihart MT, Ruckenstein E. Efficient surface grafting of luminescent silicon quantum dots by photoinitiated hydrosilylation. Langmuir 2005;21:6054–6062.

125. Rothon RN. *Particulate-Filled Polymer Composites*. Harlow, U.K.: Longman; 1995. p 123.

126. del Campo A, Sen T, Lellouche JP, Bruce IJ. Multifunctional magnetite and silica–magnetite nanoparticles: Synthesis, surface activation and applications in life sciences. J Magn Magn Mater 2005;293:33–40.

127. Kikukawa T, Kuraoka K, Kawabe K, Yamashita M, Fukumi K, Hirao K, Yazawa T. Stabilities and pore size effect of proton-conducting organic–inorganic hybrid membranes prepared through surface modification of porous glasses. J Membr Sci 2005;259:161–166.

128. Li JJ, Xu XY, Jiang Z, Hao ZP, Hu C. Nanoporous silica-supported nanometric palladium: Synthesis, characterization, and catalytic deep oxidation of benzene. Environ Sci Tech 2005;39:1319–1323.

129. Pere E, Cardy H, Latour V, Lacombe S. Low-temperature reaction of trialkoxysilanes on silica gel: A mild and controlled method for modifying silica surfaces. J Colloid Interface Sci 2005;281:410–416.

130. Reculusa S, Poncet-Legrand C, Perro A, Duguet E, Bourgeat-Lami E, Mingotaud C, Ravaine S. Hybrid dissymmetrical colloidal particles. Chem Mater 2005;17:3338–3344.

131. Wang YP, Pei XW, Yuan K. Reverse ATRP grafting from silica surface to prepare well-defined organic/inorganic hybrid nanocomposite. Mater Lett 2004;59:520–523.

132. Wu CG, Tzeng LF, Kuo YT, Shu CH. Enhancement of the photocatalytic activity of TiO_2 film via surface modification of the substrate. Appl Catal A 2002;226:199–211.

133. Ruan WH, Mai YL, Wang XH, Rong MZ, Zhang MQ. Effects of processing conditions on properties of nano-SiO_2/polypropylene composites fabricated by pre-drawing technique. Compos Sci Technol 2007;67:2747–2756.

134. Lagaly G. Introduction: from clay mineral-polymer interactions to clay mineral-polymer nanocomposites. Appl Clay Sci 1999;15:1–9.

135. Greenland DJ. Adsorption of poly(vinyl alcohols) by montmorillonite. J Colloid Sci 1963;18:647–664.

136. Ogata N, Kawakage S, Ogihara T. Poly(vinyl alcohol)/clay and poly(ethylene oxide)/clay blend prepared using water as solvent. J Appl Polym Sci 1997;66:573–581.

137. Parfitt RL, Greenland DJ. Adsorption of poly(ethylene glycols) on montmorillonites. Clay Miner 1970;8:305–315.
138. Ruiz-Hitzky E, Aranda P, Casal B, Galvan JC. Nanocomposite materials with controlled ion mobility. Adv Mater 1995;7:180–184.
139. Levy R, Francis CW. Interlayer adsorption of polyvinylpyrrolidone on montmorillonite. J Colloid Interface Sci 1975;50:442–450.
140. Billingham J, Breen C, Yarwood J. Adsorption of polyamine, polyacrylic acid and polyethylene glycol on montmorillonite: An in situ study using ATR-FTIR. Vib Spectrosc 1997;14:19–34.
141. Zhao X, Urano K, Ogasawara S. Adsorption of poly(ethylene vinyl alcohol) from aqueous solution on montmorillonite clays. Colloid Polym Sci 1989;267:899–906.
142. Ogata N, Jimenez G, Kawai H, Ogihara T. Structure and thermal/ mechanical properties of poly(l-lactide)–clay blend. J Polym Sci Polym Phys 1997;35:389–396.
143. Jeon HG, Jung HT, Lee SW, Hudson SD. Morphology of polymer silicate nanocomposites. High density polyethylene and a nitrile. Polym Bull 1998;41:107–113.
144. Jimenez G, Ogata N, Kawai H, Ogihara T. Structure and thermal/mechanical properties of poly(γ-caprolactone)–clay blend. J Appl Polym Sci 1997;64:2211–2220.
145. Srivastava SK, Pramanik M, Acharya H. Ethylene/vinyl acetate copolymer/clay nanocomposites. J Polym Sci Polym Phys 2006;44:471–480.
146. Tseng CR, Wu JY, Lee HY, Chang FC. Preparation and crystallization behavior of syndiotactic polystyrene–clay nanocomposites. Polymer 2001;42:10063–10070.
147. Krikorian V, Pochan D. Poly(l-lactide acid)/layered silicate nanocomposite: Fabrication, characterization, and properties. Chem Mater 2003;15:4317–4324.
148. Chang JH, Uk-An Y, Sur GS. Poly(lactic acid) nanocomposites with various organoclays. I. thermomechanical properties, morphology, and gas permeability. J Polym Sci Polym Phys 2003;41:94–103.
149. Yano K, Usuki A, Okada A, Kurauchi T, Kamigaito O. Synthesis and properties of polyimide/clay hybrid. J Polym Sci Polym Chem 1993;31:2493–2498.
150. Ray SS, Bousima M. Biodegradable polymers and their layered silicate nanocomposites: In greening the 21st century materials world. Prog Mater Sci 2005;50:962–1079.
151. Xia H, Wang Q, Qiu G. Polymer-encapsulated carbon nanotubes prepared through ultrasonically initiated in situ emulsion polymerization. Chem Mater 2003;15:3879–3886.
152. Suslick KS. Sonochemistry. Science 1990;247:1439–1445.
153. Li C, Pang XJ, Qu MZ, Zhang QT, Wang B, Zhang BL, et al. Fabrication and characterization of polycarbonate/carbon nanotubes composites. Composites A 2005;37:1485–1489.
154. Safadi B, Andrews R, Grulke EA. Multiwalled carbon nanotubes polymer composites: Synthesis and characterization of thin films. J Appl Polym Sci 2002;84:2660–2669.
155. Sahoo NG, Jung YC, Yoo HJ, Cho JW. Effect of functionalized carbon nanotubes on molecular interaction and properties of polyurethane composites. Macromol Chem Phys 2006;207:1773–1780.
156. Chen W, Tao X. Self-organizing alignment of carbon nanotubes in thermoplastic polyurethane. Macromol Rapid Commun 2005;26:1763–1767.
157. Qian D, Dickey EC, Andrews R, Rantell T. Load transfer and deformation mechanisms in carbon nanotube-polystyrene composites. Appl Phys Lett 2000;76:2868–2870.
158. Wong M, Paramsothy M, Xu XJ, Ren Y, Li S, Liao K. Physical interaction at carbon nanotube-polymer interface. Polymer 2003;44:7757–7764.
159. Kota AK, Cipriano BH, Duesterberg MK, Gershon AL, Powell D, Raghavan SR, et al. Electrical and reheological percolation in polystyre/MWCNT nanocomposites. Macromolecules 2007;40:7400–7406.
160. Sandler J, Shaffer MSP, Prasse T, Bauhofer W, Schulte K, Windle AH. Development of a dispersion process for carbon nanotubes in an epoxy matrix and the resulting electrical properties. Polymer 1999;40:5967–5971.
161. Song YS, Youn JR. Influence of dispersion states of carbon nanotubes on physical properties of epoxy nanocomposites. Carbon 2005;43:1378–1385.
162. Shaffer MSP, Windle AH. Fabrication and characterization of carbon nanotube/poly(vinyl alcohol) composites. Adv Mater 1999;11:937–941.
163. Chae HG, Sreekumar TV, Uchida T, Kumar S. A comparison of reinforcement efficiency of various types of carbon nanotubes in polyacrylonitrile fiber. Polymer 2005;46:10925–10935.
164. Bin Y, Kitanaka M, Zhu D, Matsuo M. Development of highly oriented polyethylene filled with aligned carbon nanotubes by gelation/crystallization from solutions. Macromolecules 2003;36:6213–6219.
165. Oberdisse J, Hine P, Pyckhout-Hintzen W. Structure of interacting aggregates of silica nanoparticles in a polymer matrix: small-angle scattering and reverse Monte Carlo simulations. Soft Matter 2007;3:476–485.
166. Zhang Q, Archer LA. Optical polarimetry and mechanical rheometry of poly(ethylene oxide) – silica dispersions. Macromolecules 2004;37:1928.
167. van Zyl WE, García M, Schrauwen BAG, Kooi BJ, De Hosson JM, Verweij H. Hybrid polyamide/silica nanocomposites: Synthesis and mechanical testing. Macromol Mater Eng 2002;287:106–110.
168. Hu N, Zhou H, Dang G, Rao X, Chen C, Zhang W. Efficient dispersion of multi-walled carbon nanotubes by in situ polymerization. Polym Int 2007;56:655–659.
169. So HH, Cho JW, Sahoo NG. Effect of carbon nanotubes on mechanical and electrical properties of polyimide/carbon nanotubes nanocomposites. Eur Polym J 2007;43:3750–3756.
170. Beyer G. Nanocomposites: A new class of flame retardants for polymers. Plast Addit Compd 2002;4(10):22–27.
171. Okamoto M, Morita S, Taguchi H, Kim YH, Kotaka T, Tateyama H. Synthesis and structure of smectic clay/poly (methyl methacrylate) and clay polystyrene nanocomposites via in situ intercalative polymerization. Polymer 2000;41:3887–3890.
172. Okamoto M, Morita S, Kotaka T. Dispersed structure and ionic conductivity of smectic clay/polymer nanocomposites. Polymer 2001;42:2685–2688.

173. Akelah A, Moet M. Polymer–claynanocomposites: Free radical grafting of polystyrene on to organophilic montmorillonite interlayers. J Mater Sci 1996;31:3589–3596.

174. Hsu SLC, Chang KC. Synthesis and properties of polybenzoxazole–clay nanocomposites. Polymer 2002;43: 4097–4101.

175. Sun T, Garces JM. High-performance polypropylene–clay nanocomposites by in-situ polymerization with metallocene/clay catalysts. Adv Mater 2002;14:128–130.

176. Bergman JS, Chen H, Giannelis EP, Thomas MG, Coates GW. Synthesis and characterization of polyolefin–silicate nanocomposites: A catalyst intercalation and in situ polymerization approach. J Chem Soc Chem Commun 1999;21:2179–2180.

177. Tudor J, Willington L, O'Hare D, Royan B. Intercalation of catalytically active metal complexes in phyllosilicates and their application as propene polymerization catalyst. Chem Commun 1996;17:2031–2032.

178. Jin YH, Park HJ, Im SS, Kwak SY, Kwak S. Polyethylene/clay nanocomposite by in situ exfoliation of montmorillonite during Ziegler–Natta polymerization of ethylene. Macromol Rapid Commun 2002;23:135–140.

179. Ke YC, Long CF, Qi ZN. Crystallization, properties, and crystal and nanoscale morphology of PET–clay nanocomposites. J Appl Polym Sci 1999;71:1139–1146.

180. Jia Z, Wang Z, Xu C, Liang J, Wei B, Wu D, et al. Study on poly(methyl methacrylate): Carbon nanotube composites. Mater Sci Eng A 1999;271:395–400.

181. Putz KW, Mitchell CA, Krishnamoorti R, Green PF. Elastic modulus of single-walled carbon nanotube/poly(methyl methacrylate) nanocomposites. J Polym Sci B 2004;42:2286–2293.

182. Sung JH, Kim HS, Jin HJ, Choi HJ, Chin IJ. Nanofibrous membranes prepared by multiwalled carbon nanotube/poly(methyl methacrylate) composites. Macromolecules 2004;37:9899–9902.

183. Hwang GL, Shieh YT, Hwang KC. Efficient load transfer to polymer grafted multiwalled carbon nanotubes in polymer composites. Adv Funct Mater 2004;14:487–491.

184. Karim MR, Lee CJ, Park YT, Lee MS. SWNTs coated by conducting polyaniline: Synthesis and modified properties. Synth Met 2005;151:131–135.

185. Long Y, Chen Z, Zhang X, Zhang J, Liu Z. Electrical properties of multiwalled carbon nanotube/polypyrrole nanocables: Percolation dominated conductivity. J Phys D Appl Phys 2004;37:1965–1969.

186. An KH, Jeong SY, Hwang HR, Lee YH. Enhanced sensitivity of a gar sensor incorporating single-walled carbon nanotube-polypyrrole nanocomposites. Adv Mater 2004;16:1005–1009.

187. Fan J, Wan M, Zhu D, Chang B, Pan Z, Xie S. Synthesis, characterizations, and physical properties of carbon nanotubes coated by conducting polypyrrole. J Appl Polym Sci 1999;74:2605–2610.

188. Yoo HJ, Jung YC, Sahoo NG, Cho JW. Electroactive shape memory polyurethane nanocomposites from in-situ polymerization with carbon nanotubes. J Macromol Sci B 2006;45:441–451.

189. Gao J, Itkis ME, Yu A, Bekyarova E, Zhao B, Haddon RC. Continuous spinning of a single-walled carbon nanotube-nylon composite fiber. J Am Chem Soc 2005;127:3847–3854.

190. Jung YC, Sahoo NG, Cho JW. Polymeric nanocomposites of polyurethane block copolymers and functionalized multi-walled carbon nanotubes as crosslinkers. Macromol Rapid Commun 2006;27:126–131.

191. Sahoo NG, Jung YC, So HH, Cho JW. Polypyrrole coated carbon nanotubes: synthesis, characterization, and enhanced electrical properties. Synth Met 2007;157:374–379.

192. Andrews R, Jacques D, Minot M, Rantell T. Fabrication of carbon multiwall nanotube/polymer composites by shear mixing. Macromol Mater Eng 2002;287:395–403.

193. Cho JW, Paul DR. Nylon 6 nanocomposites by melt compounding. Polymer 2001;42:1083–1094.

194. Vaia RA, Giannelis EP. Polymer melt intercalation in organically modified layered silicates: model predictions and experiment. Macromolecules 1997;30:8000–8009.

195. Fornes TD, Yoon PJ, Keskkula H, Paul DR. Nylon 6 nanocomposites: the effect of matrix molecular weight. Polymer 2011;42:9929–9940.

196. Huang JC, Zhu ZK, Yin J, Qian XF, Sun YY. Poly(etherimide)/montmorillonite nanocomposites prepared by melt intercalation: morphology, solvent resistance properties and thermal properties. Polymer 2001;42:873–877.

197. Fornes TD, Yoon PJ, Hunter DL, Keskkula H, Paul DR. Effect of organoclay structure on nylon 6 nanocomposite morphology and properties. Polymer 2002;43:5915–5933.

198. Phang IY, Liu T, Mohamed A, Pramoda KP, Chen L, Shen L, et al. Morphology, thermal and mechanical properties of nylon 12/organoclay nanocomposites prepared by melt compounding. Polym Int 2005;54:456–464.

199. Davis CH, Mathias LJ, Gilman JW, Schiraldi DA, Shields JR, Trulove P, et al. Effects of melt-processing conditions on the quality of poly(ethylene terephthalate) montmorillonite clay nanocomposites. J Polym Sci Polym Phys 2002;40:2661–2666.

200. Pegoretti A, Kolarik J, Peronic C, Migliaresi C. Recycled poly(ethylene terephthalate) layered silicate nanocomposites: morphology and tensile mechanical properties. Polymer 2004;45:2751–2759.

201. Srivastava SK, Pramanik M, Acharya H. Ethylene/vinyl acetate copolymer/clay nanocomposites. J Polym Sci Polym Phys 2006;44:471–480.

202. Tang Y, Hu Y, Wang J, Zong R, Gui Z, Chen Z, et al. Influence of organophilic clay and preparation methods on EVA/montmorillonite nanocomposites. J Appl Polym Sci 2004;91:2416–2421.

203. Finnigan B, Martin D, Halley P, Truss R, Campell K. Morphology and properties of thermoplastic polyurethane nanocomposites incorporating hydrophilic layered silicates. Polymer 2004;45:2249–2260.

204. Shin SYA, Simon LC, Soares JBP, Scholz G. Polyethylene–clay hybrid nanocomposites: in situ polymerization using bifunctional organic modifiers. Polymer 2003;44:5317–5321.

205. Gopakumar TG, Lee JA, Kontopoulou M, Parent JS. Influence of clay exfoliation on the physical properties of montmorillonite/polyethylene composites. Polymer 2002;43:5483–5491.

206. Thellen C, Orroth C, Froio D, Ziegler D, Lucciarini J, Farrell R, et al. Influence of montmorillonite layered silicate on plasticized poly(llactide) blown films. Polymer 2005;46:11716–11727.
207. Ray SS, Maiti P, Okamoto M, Yamada K, Ueda K. New polylactide/layered silicate nanocomposites. 1. Preparation, characterization and properties. Macromolecules 2002;35:3104–3110.
208. Pluta M. Morphology and properties of polylactide modified by thermal treatment, filling with layered silicates and plasticization. Polymer 2004;45:8239–8251.
209. Di Y, Iannace S, Maio ED, Nicolais L. Nanocomposites by melt intercalation based on polycaprolactone and organoclay. J Polym Sci Polym Phys 2003;41:670–678.
210. Gorrasi G, Tortora M, Vittoria V, Pollet E, Lepoittevin B, Alexandre M, et al. Vapor barrier properties of polycaprolactone montmorillonite nanocomposites: effect of clay dispersion. Polymer 2003;44:2271–2279.
211. Lopez Manchado MA, Valentini L, Biagiotti J, Kenny JM. Thermal and mechanical properties of single-walled carbon nanotubes polypropylene composites prepared by melt processing. Carbon 2005;43:1499–1505.
212. Valentini L, Biagiotti J, Kenny JM, Santucci S. Morphological characterization of single-wall carbon nanotube/polypropylene composites. Compos Sci Technol 2003;63:1149–1153.
213. Zhang H, Zhang Z. Impact behaviour of polypropylene filled with multi-walled carbon nanotubes. Eur Polym J 2007;43:3197–3207.
214. Tang W, Santare MH, Advani SG. Melt processing and mechanical property characterization of multi-walled carbon nanotube/high density polyethylene (MWNT/HDPE) composite films. Carbon 2003;41:2779–2785.
215. Bocchini S, Frache A, Camino G, Claes M. Polyethylene thermal oxidative stabilisation in carbon nanotubes based nanocomposites. Eur Polym J 2007;43:3222–3235.
216. McNally T, Potschkeb P, Halley P, Murphy M, Martin D, Bell SEJ, et al. Polyethylene multiwalled carbon nanotube composites. Polymer 2005;46:8222–8232.
217. Potschke P, Fornes TD, Paul DR. Rheological behavior of multiwalled carbon nanotube/polycarbonate composites. Polymer 2002;43:3247–3255.
218. Potschke P, Bhattacharyya AR, Janke A. Morphology and electrical resistivity of melt mixed blends of polyethylene and carbon nanotube filled polycarbonate. Polymer 2003; 44:8061–8069.
219. Potschke P, Bhattacharyya AR, Janke A. Melt mixing of polycarbonate with multiwalled carbon nanotubes: Microscopic studies on the state of dispersion. Eur Polym J 2004; 40:137–148.
220. Jin Z, Pramoda KP, Xu G, Goh SH. Dynamic mechanical behavior of melt-processed multi-walled carbon nanotube/poly(methyl methacrylate) composites. Chem Phys Lett 2001;337:43–47.
221. Zeng J, Saltysiak B, Johnson WS, Schiraldi DA, Kumar S. Processing and properties of poly(methyl methacrylate)/carbon nano fiber composites. Composites B 2004;35:173–178.
222. Haggenmueller R, Gommans HH, Rinzler AG, Fischer JE, Winey KI. Aligned single-wall carbon nanotubes in composites by melt processing methods. Chem Phys Lett 2000;330:219–225.
223. Gorga RE, Cohen RE. Toughness enhancements in poly(methyl methacrylate) by addition of oriented multiwall carbon nanotubes. J Polym Sci B 2004;42:2690–2702.
224. Zeng Y, Ying Z, Du J, Cheng HM. Effects of carbon nanotubes on processing stability of polyoxymethylene in melt-mixing process. J Phys Chem C 2007;111:13945–13950.
225. Siochi EJ, Working DC, Park C, Lillehei PT, Rouse JH, Topping CC, et al. Melt processing of SWNT-polyimide nanocomposite fibres. Composites B 2004;35:439–446.
226. Zhang WD, Shen L, Phang IY, Liu T. Carbon nanotubes reinforced nylon-6 composite prepared by simple melt-compounding. Macromolecules 2004;37:256–259.
227. Meincke O, Kaempfer D, Weickmann H, Friedrich C, Vathauer M, Warth H. Mechanical properties and electrical conductivity of carbon-nanotube filled polyamide-6 and its blends with acrylonitrile/ butadiene/styrene. Polymer 2004;45: 739–748.
228. Wu CL, Zhang MQ, Rong MZ, Friedrich K. Silica nanoparticles filled polypropylene: effects of particle surface treatment, matrix ductility and particle species on mechanical performance of the composite. Compos Sci Technol 2005;65:635–645.
229. Reddy CS, Das CK, Narkis M. Propylene-ethylene copolymer nanocomposites: Epoxy resin grafted nanosilica as a reinforcing filler. Polym Compos 2005;26:806–812.
230. Gao XW, Hu GJ, Qian ZZ, Ding YF, Zhang SM, Wang DJ, Yang MS. Immobilization of antioxidant on nanosilica and the antioxidative behavior in low density polyethylene. Polymer 2007;48:7309–7315.
231. Lee JA, Kontopoulou M, Parent JS. Synthesis and characterization of polyethylene-based ionomer nanocomposites. Polymer 2005;46:5040–5049.
232. Tanahashi M, Hirose M, Lee JC, Takeda K. Organic/inorganic nanocomposites prepared by mechanical smashing of agglomerated silica ultrafine particles in molten thermoplastic resin. Polym Adv Technol 2006;17:981–990.
233. Kontou E, Anthoulis G. The effect of silica nanoparticles on the thermomechanical properties of polystyrene. J Appl Polym Sci 2007;105:1723–1731.
234. Yang F, Nelson GL. Polymer/silica nanocomposites prepared via extrusion. Polym Adv Technol 2006;17:320–326.
235. Katsikis N, Zahradnik F, Helmschrott A, Munstedt H, Vital A. Thermal stability of poly(methyl methacrylate)/silica nano- and microcomposites as investigated by dynamic-mechanical experiments. Polym Degrad Stab 2007;92:1966–1976.
236. Nodera A, Kanai T. Flame retardancy of polycarbonate–polydimethylsiloxane block copolymer/silica nanocomposites. J Appl Polym Sci 2006;101:3862–3868.
237. Ahn SH, Kim SH, Lee SG. Surface modified silica nanoparticle–reinforced poly (ethylene 2,6-naphthalate). J Appl Polym Sci 2004;94:812–818.
238. Tanahashi M, Hirose M, Watanabe Y, Lee JC, Takeda K. Silica/perfluoropolymer nanocomposites fabricated by direct melt-compounding: a novel method without surface modification on nano-silica. J Nanosci Nanotechnol 2007;7: 2433–2442.
239. Chung SC, Hahm WG, Im SS, Oh SG. Poly(ethylene terephthalate)(PET) nanocomposites filled with fumed silicas by melt compounding. Macromol Res 2002;10:221–229.

240. Bikiaris D, Karavelidis V, Karayannidis G. A new approach to prepare poly(ethylene terephthalate)/silica nanocomposites with increased molecular weight and fully adjustable branching or crosslinking by SSP. Macromol Rapid Commun 2006;27:1199–1205.

241. Todorov LV, Viana JC. Characterization of PET nanocomposites produced by different melt-based production methods. J Appl Polym Sci 2007;106:1659–1669.

242. Zhang H, Zhang Z, Yang JL, Friedrich K. Temperature dependence of crack initiation fracture toughness of various nanoparticles filled polyamide 66. Polymer 2006;47:679–689.

243. Bogoslovov RB, Roland CM, Ellis AR, Randall AM, Robertson CG. Effect of silica nanoparticles on the local segmental dynamics in poly(vinyl acetate). Macromolecules 2008;41:1289–1296.

244. Aso O, Eguiazabal JI, Nazabal J. The influence of surface modification on the structure and properties of a nanosilica filled thermoplastic elastomer. Compos Sci Technol 2007;67:2854–2863.

245. Arrighi V, McEwen IJ, Qian H, Prieto MBS. The glass transition and interfacial layer in styrene-butadiene rubber containing silica nanofiller. Polymer 2003;44:6259–6266.

246. Gauthier C, Reynaud E, Vassoille R, Ladauce-Stelandre L. Analysis of the non-linear viscoelastic behaviour of silica filled styrene butadiene rubber. Polymer 2004;45:2761–2771.

247. Cassagnau P, Melis F. Non-linear viscoelastic behaviour and modulus recovery in silica filled polymers. Polymer 2003;44:6607–6615.

248. Wu TB, Ke YC. Melting, crystallization and optical behaviors of poly (ethylene terephthalate)-silica/polystyrene nanocomposite films. Thin Solid Films 2007;515:5220–5226.

249. Yan SF, Yin JB, Yang Y, Dai ZZ, Ma J, Chen XS. Surface-grafted silica linked with l-lactic acid oligomer: A novel nanofiller to improve the performance of biodegradable poly(l-lactide). Polymer 2007;48:1688–1694.

250. Kweon JO, Noh ST. Thermal, thermomechanical, and electrochemical characterization of the organic–inorganic hybrids poly(ethylene oxide) (PEO)–silica and PEO–silica–$LiClO_4$. J Appl Polym Sci 2001;81:2471–2479.

251. Mello NC, Bonagamba TJ, Panepucci H, Dahmouche K, Judeinstein P, Aegerter MA. NMR study of ion-conducting organic – inorganic nanocomposites poly(ethylene glycol) – silica – $LiClO_4$. Macromolecules 2000;33:1280–1288.

252. de Souza PH, Bianchi RF, Dahmouche K, Judeinstein P, Faria RM, Bonagamba TJ. Solid-state NMR, ionic conductivity, and thermal studies of lithium-doped siloxane – poly(propylene glycol) organic – inorganic nanocomposites. Chem Mater 2001;13:3685–3692.

253. Mark JE. *Science of Ceramic Chemical Processing*. New York: John Wiley & Sons; 1985. ; Chapter 17.

254. Erman B. *Structures and Properties of Rubberlike Networks*. New York: Oxford University Press; 1997. p 370.

255. Mark JE, Pan J. Reinforcement of polydimethylsiloxane networks by in-situ precipitation of silica: A new method for preparation of filled elastomers. Makromol Chem Rapid Commun 1982;3:681–685.

256. Kohjiya S, Ikeda Y. Reinforcement of general-purpose grade rubbers by silica generated in situ. Rubber Chem Technol 2000;73:534–550.

257. Hashim AS, Ikeda Y, Kohjya S. Moisture cure and in-situ silica reinforcement of epoxidized natural rubber. Polym Int 1995;38:111–117.

258. Ikeda Y, Tanaka A, Kohjya S. Effect of catalyst on in situ silica reinforcement of styrene–butadiene rubber vulcanizate by the sol–gel reaction of tetraethoxysilane. J Mater Chem 1997;7:455–458.

259. Ikeda Y, Tanaka A, Kohjya S. Reinforcement of styrene–butadiene rubber vulcanizate by in situ silica prepared by the sol–gel reaction of tetraethoxysilane. J Mater Chem 1997;7:1497–1503.

260. Ikeda Y, Kohjya S. In situ formed silica particles in rubber vulcanizated by the sol-gel method. Polymer 1997;38: 4417–4423.

261. Hashim AS, Azahari B, Ikeda Y, Kohjya S. The effect of Bis(3-triethoxysilylpropyl) tetrasulfide on silica reinforcement of styrene-butadiene rubber. Rubber Chem Technol 1998;71:289–299.

262. Sun D, Zhang R, Liu Z, Huang Y, Wang Y, He J, Han B, Yang G. Polypropylene/silica nanocomposites prepared by in-situ sol−gel reaction with the aid of CO_2. Macromolecules 2005;38:5617–5624.

263. Jain S, Goosens H, Picchioni F, Magusin P, Mezari B, Van Duin M. Synthetic aspects and characterization of polypropylene–silica nanocomposites prepared via solid-state modification and sol–gel reactions. Polymer 2005;46:6666–6681.

264. Jain S, Goossens JGP, Van Duin M. Synthesis, characterization and properties of (vinyl triethoxy silane-grafted PP)/silica nanocomposites. Macromol Symp 2006;233:225–234.

265. Dou Q, Zhu X, Peter K, Demco DE, Moller M, Melian C. Preparation of polypropylene/silica composites by in situ sol–gel processing using hyperbranched polyethoxysiloxane. J Sol-Gel Sci Technol 2008;48:51–60.

266. Yu K, Zhao J, Guo Y, Ding X, Yanhua Liu HB, Wang Z. Sol–gel synthesis and hydrothermal processing of anatase nanocrystals from titanium n-butoxide. Mater Lett 2005;59:2515–2518.

267. Barlier V, Bounor-Legare V, Boiteux G, Davenas J, Slazak A, Rybak A, Jung J. Photogeneration and photovoltaic effect in poly(N-vinylcarbazole): TiO_2 bulk heterojunction elaborated by hydrolysis-condensation reactions of TiO_2 precursors. Synth Met 2009;159:508–512.

268. Barlier V, Bounor-Legare V, Boiteux G, Leonard D, Davenas J. TiO_2: polymer bulk-heterojunction thin films made from a miscible new carbazole based TiO_2 precursor with poly(Nvinylcarbazole) for enhanced charge transfer properties. Mater Chem Phys 2009;115:429–433.

269. Mazzocchetti L, Scandola M, Pollicino A. Study of the organic-inorganic phase interactions in polyester-titania hybrids. Polymer 2008;49:5215–5224.

270. Bahloul W, Bounorlegare V, David L, Cassagnau P. Morphology and viscoelasticity of PP/TiO_2 anocomposites prepared by in situ sol–gel method. J Polym Sci B 2010;48:1213–1222.

271. Peng-Cheng M, Naveed A, Siddiqui GM, Jang-Kyo K. Dispersion and functionalization of carbon nanotubes for polymer-based nanocomposites: A review. Composites A 2010;41:1345–1367.

272. Ma PC, Siddiqui NA, Cao SG, Tang BZ, Kim JK. Effect of dispersion techniques and functionalization of CNTs on the properties of CNT/epoxy composites. In: *Proceedings of the 2nd*

international conference on advanced materials and structures; 2008 October; Nanjing, China; 2008.

273. Dennis HR, Hunter DL, Chang D, Kim S, White JL, Cho JW, Paul DR. Effect of melt processing conditions on the extent of exfoliation organoclay-based nanocomposites. Polymer 2001;42:9513–9522.
274. Lertwimolnun W, Vergnes B. Effect of processing conditions on the formation of polypropylene/organoclay nanocomposites in a twin screw extruder. Polym Eng Sci 2006;46:314–323.
275. Lertwimolnun W, Vergnes B. Influence of compatibilizer and processing conditions on the dispersion of nanoclay in a polymer matrix. Polymer 2005;46:3462–3471.
276. Panacek A et al. Silver colloid nanoparticles: synthesis, characterization, and their antibacterial activity. J Phys Chem B 2006;110:248–253.
277. Nair LS, Laurencin CT. Silver nanoparticles: Synthesis and therapeutic applications. J Biomed Nanotechnol 2007;3:301–316.
278. Sun Y, Xia Y. Shape-controlled synthesis of gold and silver nanoparticles. Science 2002;298:2176–2179.

21

MORPHOLOGY AND GAS BARRIER PROPERTIES OF POLYMER NANOCOMPOSITES

ABBAS GHANBARI[1], MARIE-CLAUDE HEUZEY[1], PIERRE J. CARREAU[1], AND MINH-TAN TON-THAT[2]

[1]*Chemical Engineering Department, Ecole Polytechnique de Montreal, Montreal, Canada*
[2]*Automotive and Surface Transport, National Research Council Canada, Montreal, Canada*

21.1 INTRODUCTION

Polymers filled with layered silicates, in comparison with unfilled polymers or common microcomposites, have shown improvements in several properties. The incorporation of nanolayers of high aspect ratio at very low volume fractions can improve the tensile modulus, strength, gas barrier properties, heat resistance, and decrease flammability of polymer nanocomposites. Thus, one should expect a reduction in thickness of commercial packaging containers and improvement in barrier properties by incorporation of nanoplatelets [1–3].

Generally, at least one critical dimension of dispersed particles in nanocomposites must be in the nanometer range (<100 nm). Typical nanomaterials, which are currently under investigation, are classified by their geometries. Silica nanoparticles and carbon black are examples of nanoparticles, while carbon nanotubes and nanofibers are classified as fibrous nanomaterials. Silicate layers with platelike structure belong to the family of layered nanomaterials [4].

21.2 STRUCTURE OF LAYERED SILICATES

Silicate nanolayers, which are widely used in polymer clay nanocomposites (PCNs), are montmorillonites from the smectite family, that is, 2:1 phyllosilicates composed of octahedral and tetrahedral crystalline sheets. The octahedral sheet consists of hydroxyl groups as well as oxygen, aluminum, iron, and magnesium atoms. On the other hand, the tetrahedral sheet comprises a central silicon atom and four oxygen atoms or hydroxyl groups [5]. The structure of 2:1 phyllosilicates is shown in Figure 21.1.

The lateral dimensions of these layers change over a wide range, from 30 nm to a few microns, while the thickness of the layer is around 1 nm. Interlayer or gallery spacing is the gap between the stacked layers. The ability of clay minerals to hold cations is called the cation exchange capacity (CEC), which is generally expressed as meq/100 g clay. Substitution with lower valence elements in the tetrahedral sheet (e.g., Si^{4+} is replaced by Al^{3+}) or in the octahedral sheet (e.g., replacement of Al^{3+} by Mg^{2+} or Fe^{2+}) leads to the creation of a net negative surface charge. The replacement of ions of relatively similar sizes is called isomorphous substitution. The negative charge can be counterbalanced by alkali and alkaline earth cations such as Na^+ or Ca^{2+} situated in the gallery [1].

Montmorillonite is highly polar and its surface hydrophilic. Therefore, delamination of montmorillonite in commodity and engineering polymeric matrices, which are hydrophobic, is problematic. In order to enhance the compatibility with polymers, it is essential to make the hydrophilic silicate surface more organophilic. Via this conversion, intercalation of most polymeric chains between silicate nanolayers becomes possible. Cationic surfactants such as alkylammonium and alkylphosphonium can penetrate into the interlayer space of silicate nanolayers via ion exchange reactions. The produced organosilicates have a lower surface energy, higher interlayer spacing, and more ability for polymer wetting in comparison with unmodified (pristine) montmorillonite (Na-MMT). In some cases, these organic modifiers or surfactants possess functional groups interacting with the polymer matrix [1, 3]. Figure 21.2 displays a schematic representation of a cation exchange reaction between pristine montmorillonite and a cationic surfactant.

Polymer Morphology: Principles, Characterization, and Processing, First Edition. Edited by Qipeng Guo.

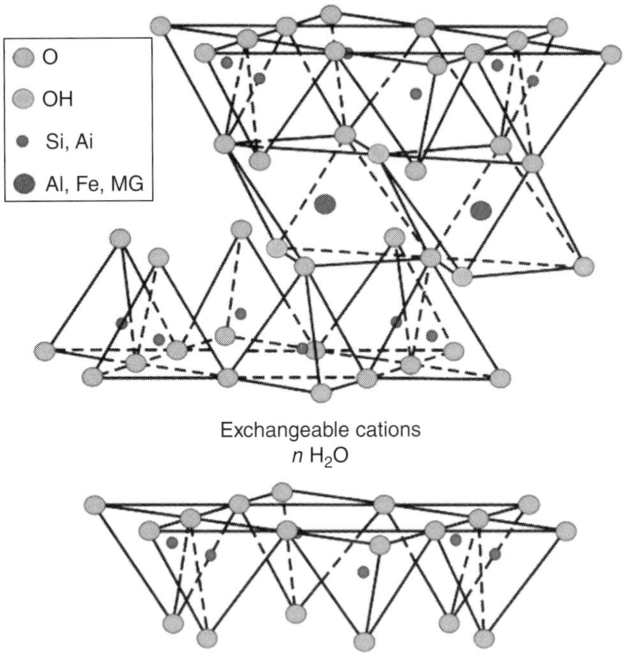

Figure 21.1 Structure of 2:1 phyllosilicates. Adapted from Grim [6]; Boczkowska [7].

21.3 MORPHOLOGIES OF POLYMER-LAYERED SILICATE COMPOSITES

Typically, and depending on the interfacial interaction strength between the polymer matrix and the nanolayers, three main types of polymer-layered silicate nanocomposite morphologies may be observed. In immiscible or phase-separated systems, polymer chains are not able to penetrate into the silicate galleries. Therefore, the basal *d*-spacing of the nanolayers remains unchanged and the properties of the resulting composites cannot be better than conventional microcomposites. Intercalated nanocomposites are obtained when polymer chains diffuse in the interlayer gallery between the silicate nanolayers. In the intercalated morphology, the stacking order of the nanolayers is preserved; however, due to the penetration of polymer chains, the interlayer *d*-spacing expands. Delaminated or exfoliated nanocomposites refer to the morphology in which silicate nanolayers are dispersed randomly and individually in the polymer host matrix. In this case, the distance between separated layers is typically 10 nm or higher, depending on clay loading. Remarkable improvements are expected to be obtained via exfoliation of silicate nanolayers within polymer matrices [2]. Figure 21.3 illustrates a schematic representation and the corresponding TEM images of the three common morphologies of polymer filled with layered silicates. A combination of these morphologies may also be observed in a single sample.

Generally, the final morphology of polymer-layered silicate nanocomposites is determined by the interplay of entropy and enthalpy changes. The entropy change includes an entropy loss due to the confinement of polymer chains within the interlayer spacing of nanolayers, and an entropy gain because of gallery enhancement and increased conformational freedom of the organic modifier chains. If the entropic penalty of polymer confinement is equal to or larger than the entropy gain, enthalpy will determine if intercalation is thermodynamically favorable. The enthalpy of mixing depends on the interactions between the polymer chains, the silicate nanolayer surface, and the organic modifier chains [1, 8]. From a mean-field statistical lattice model, it was reported that the entropy loss due to the polymer confinement is almost equal to the entropy gain; hence, a very small entropy net change is involved in the thermodynamic of nanocomposite formation [1, 9, 10]. Therefore, polymer intercalation into silicate nanolayers depends mostly on favorable enthalpic contributions.

21.4 NANOCOMPOSITE PREPARATION METHODS

Solution blending, *in situ* polymerization, and melt blending are three processing techniques used for preparing polymer-layered silicate nanocomposites. In solution blending, a polymeric solution is prepared and layered silicates are swollen in a cosolvent (water, toluene, chloroform, etc.). By mixing the polymeric and silicate nanolayers solutions,

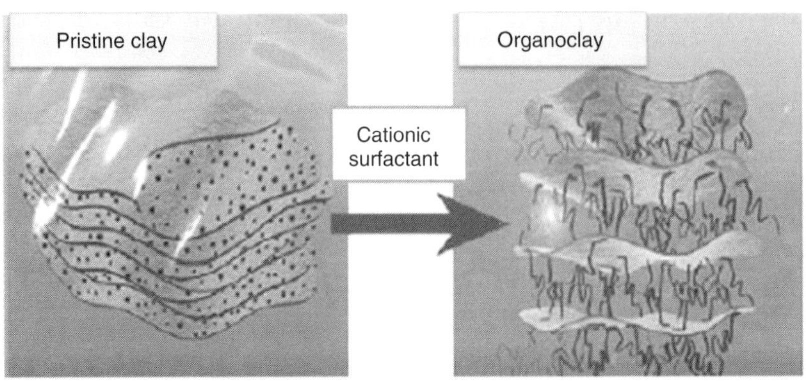

Figure 21.2 Schematic representation of organic modification of silicate nanolayers. Pavlidou and Papaspyrides [2]. Reproduced with permission of Elsevier.

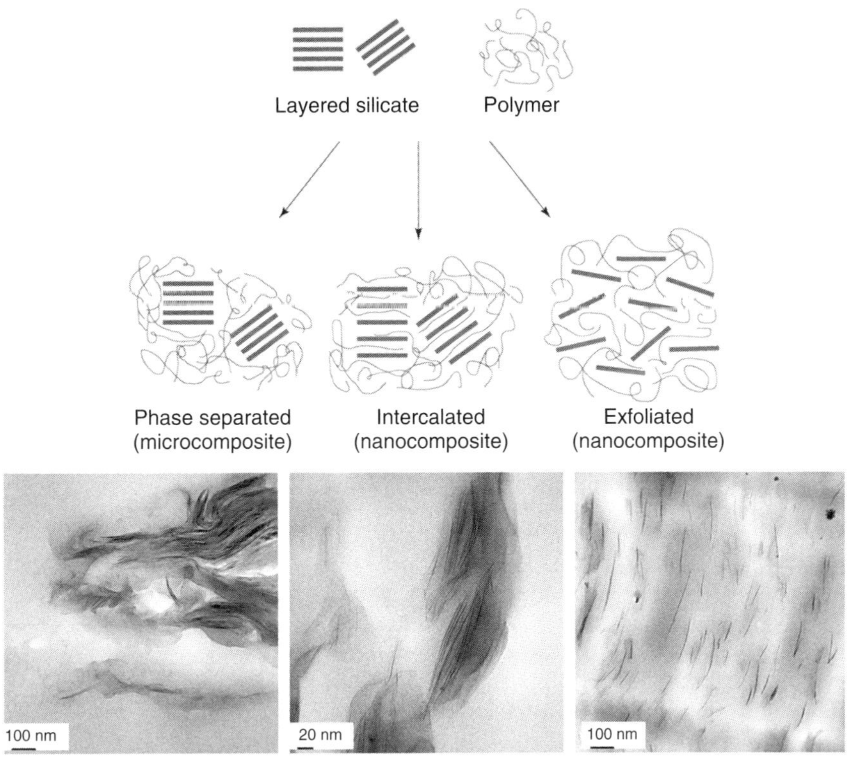

Figure 21.3 Scheme of different types of composite structures arising from the mixing of layered silicates and polymers, with corresponding TEM images. Adapted from Hussain et al. [4].

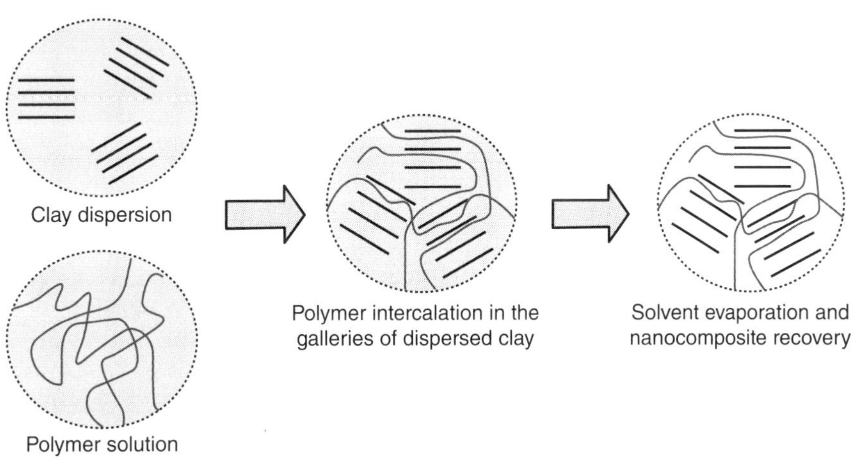

Figure 21.4 Schematic representation of nanocomposites obtained by solution blending. Pavlidou and Papaspyrides [2]. Reproduced with permission of Elsevier.

macromolecules can intercalate and displace the solvent within the gap of the silicate layers. Upon removal of the solvent, polymer-layered silicate nanocomposites can be achieved if the polymer and silicate are compatible [1]. A schematic representation of this method is shown in Figure 21.4.

Pramanik et al. [11] prepared ethylene vinyl acetate (EVA) nanocomposites containing 2, 4, 6, and 8 wt% dodecyl ammonium ion–intercalated montmorillonite by solution blending. EVA was dissolved in toluene and the organoclay was dispersed in *N,N*-dimethyl acetamide. The absence of any diffraction peak in X-ray diffraction (XRD) for nanocomposites containing up to 6 wt% clay was attributed to the homogeneous exfoliation and random dispersion of silicate layers within the polymer matrix. However, it is well established that the absence of XRD peaks, which can be due to clay dilution, is not sufficient to conclude about an exfoliated structure. It is also worth noting that EVA nanocomposites containing 8 wt% organoclay exhibited the basal reflection peak of the pristine organoclay. The authors reported a 36% improvement in tensile strength for the nanocomposites containing 2 wt% organoclay.

Another technique to prepare polymer-layered silicate nanocomposites is *in situ* polymerization. This method has attracted much academic and industrial interest after a Toyota research group synthesized polyamide-6 clay nanocomposites, with remarkable improvements in mechanical and thermal properties, via *in situ* polymerization. In this technique, a liquid monomer or a monomer solution is used to swell the silicate nanolayers. Upon starting the polymerization, macromolecule formation can occur between the silicate layers and even bond with the surfactant attached on the clay [1, 12]. Lee and coworkers [13] prepared PET nanocomposites containing trimethyl octadecyl ammonium ion–intercalated montmorillonite by ring-opening polymerization of ethylene terephthalate cyclic oligomers (ETCs). Figure 21.5 depicts the schematic of the *in situ* polymerization of a PET nanocomposite. The intercalation of the cyclic oligomer of low molecular weight and viscosity enhanced the gallery spacing of the organoclay. Subsequent polymerization resulted in the coexistence of exfoliated and intercalated clay particles rather than perfect exfoliation. Besides, this method led to the formation of a PET with a very low intrinsic viscosity (IV = 0.55 dL/g).

Melt intercalation involves direct mixing of layered silicates with a molten polymer. It is desirable that the molten polymer chains diffuse into the interlayer spacing of the nanolayers and lead to their delamination. Shear and high temperature can be used to facilitate the diffusion of polymer molecules into the clay galleries. The melt intercalation method has several advantages over the other techniques. First of all, this is an environmental friendly method because there is no need for using organic solvents. Secondly, common industrial processing techniques such as extrusion and injection molding can be used in this approach. It is a cost-effective method and can be employed for the preparation of nanocomposites of a wide range of polymers, from nonpolar to strongly polar ones. In addition, the high shear forces encountered in polymer processing equipment facilitate the dispersion of the nanoparticles in the molten polymer matrix. Therefore, this method is greener and commercially more viable in comparison with the two other approaches presented. Most results presented in this chapter were obtained using melt intercalation.

21.5 CHALLENGES OF THERMAL DEGRADATION IN MELT INTERCALATION

In spite of the several advantages of the melt-mixing approach, melt intercalation can lead to the thermal degradation of commercial organoclays. As melt processing and polymerization of PET are performed at high temperatures (250–300 °C), the organomodifiers of commercial organoclays could degrade below 200 °C [14]. It seems that the thermal stability of organosilicates have a critical role in the resulting nanocomposite morphology, because upon decomposition of the silicate organic modifier, the interface between the polymer matrix and the silicate nanolayers may change. Moreover, the decomposition of the organic modifier, which leads to the collapse of the silicate nanolayers and a decreased interlayer spacing, may impede the intercalation of polymer chains. Color formation, enhanced degradation of the polymer matrix, plasticization effects, and deterioration of general physical properties may result from the thermal degradation of organoclays during melt processing. Therefore, the thermal stability of the organic modifiers used for the modification of the nanosilicates should be improved to avoid an anticipated collapse and agglomeration of the organoclays. Beside thermal stability, compatibility of organosilicates with the polymer host matrix is another critical parameter that controls the level of dispersion and delamination of the silicate nanolayers and the final morphology of the nanocomposites [15, 16].

Fornes et al. [17] prepared nylon-6 nanocomposites containing various quaternary alkylammonium organoclays using a twin-screw extruder. Molecular weight reduction, determined via intrinsic viscosity measurements of unfiltered solutions, and color formation were reported for all nanocomposites, due to reactions between the matrix polymer chains and the organoclays. Further, matrix degradation was reported for the nanocomposites with higher platelet delamination, which was attributed to more exposure of the organoclays to the matrix.

Shah and Paul [18] prepared low-density polyethylene nanocomposites containing 5 wt% trimethyl hydrogenated-tallow ammonium montmorillonite, $M_3(HT)_1$, and dimethyl bis(hydrogenated-tallow) ammonium montmorillonite, $M_2(HT)_2$, using a twin-screw extruder. Melt compounding was performed at 150, 165, 180, 200, and 240 °C to study the effect of processing temperature on the degradation of the quaternary ammonium surfactants of the organoclays. Changes in the position of XRD peaks were assigned to platelet collapse, induced by the mass reduction of the surfactants within the clay galleries. Figure 21.6 illustrates the XRD patterns of polyethylene nanocomposites containing $M_3(HT)_1$ and $M_2(HT)_2$ organoclays. The authors did not present TEM images for direct visualization of the state of the dispersion of the organoclays, but XRD patterns do not show evidence of intercalation or exfoliation.

Polycarbonate nanocomposites were prepared by melt compounding and the effect of organoclay structure on color formation was examined. Darker colored nanocomposites were obtained for samples containing the organoclay with double bonds in the hydrocarbon tail, compared to those with saturated organic modifiers. The presence of both hydroxy-ethyl groups and tallow tails led to the most important color changes [19].

Scaffaro and coworkers [20] prepared PET nanocomposite using a twin-screw extruder. The intrinsic viscosity of the neat PET and extruded PET was reported to be 0.83 and 0.82 dL/g, respectively, while PET nanocomposites containing 10 wt% C30B exhibited an intrinsic viscosity of 0.63 dL/g (see the discussion on intrinsic viscosity measurements of nanocomposites in the next paragraph). This remarkable reduction in

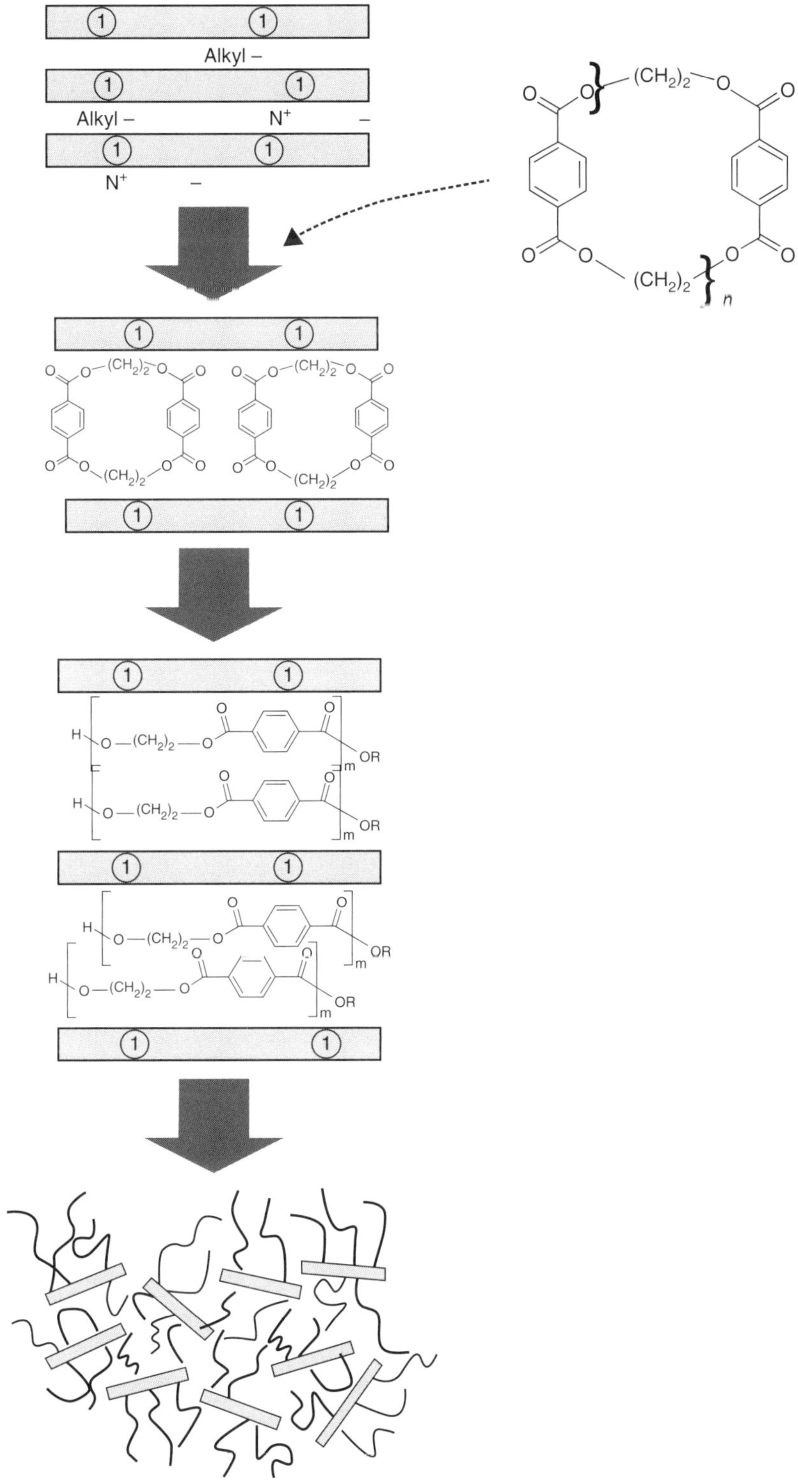

Figure 21.5 Ring-opening polymerization of ethylene terephthalate cyclic oligomers in the interlayer spacing of the organoclay. Lee et al. [13]. Reproduced with permission of Elsevier.

the intrinsic viscosity was attributed to the depolymerization of PET induced by the degradation of the C30B organic modifier. Hwang et al. [21] prepared PET nanocomposites containing Cloisite® 10A (C10A) by *in situ* polymerization at 285 °C. By incorporating 5 wt% C10A, the intrinsic viscosity of the neat PET was reduced from 0.76 to 0.70 dL/g. In another study, PET nanocomposites containing organomodified synthetic mica were prepared using a twin-screw extruder at 270 °C. The molecular weight of the extruded PET, characterized by gel permeation chromatography (GPC), was reported to decrease from 90,100 to 78,500 g/mol due to the incorporation of 5 wt% organoclay [22].

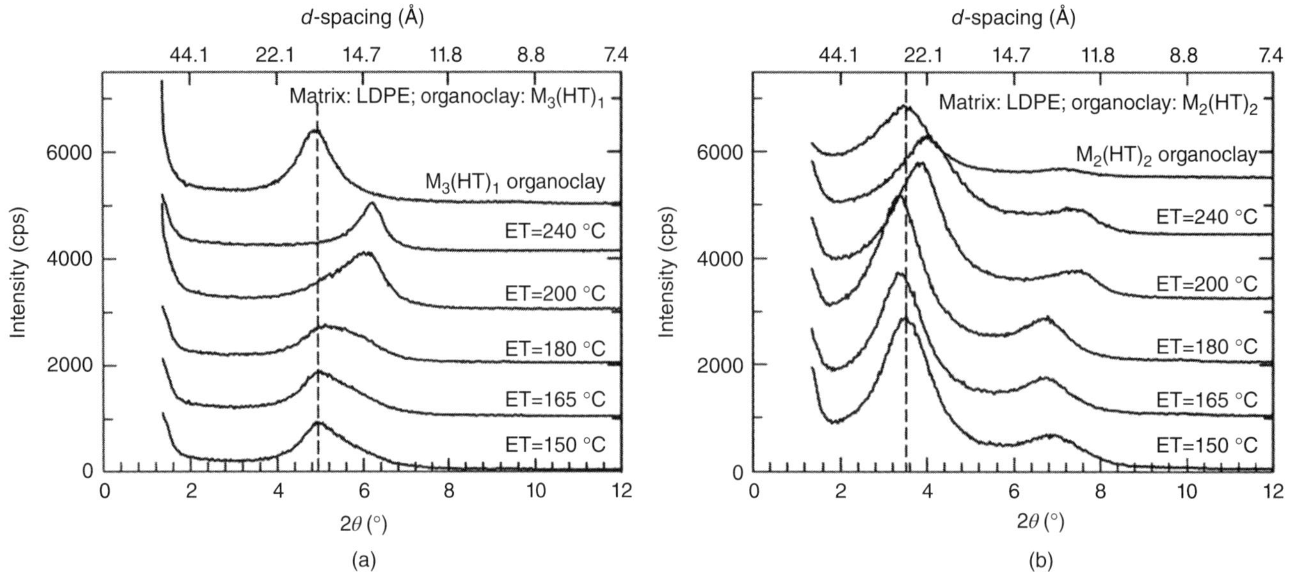

Figure 21.6 XRD patterns of low-density polyethylene nanocomposites processed at several extrusion temperatures and containing (a) $M_3(HT)_1$ and (b) $M_2(HT)_2$ organoclays. Shah and Paul [18]. Reproduced with permission of Elsevier.

Measuring the intrinsic viscosity of clay-containing nanocomposites requires the elimination of the clay particles from the dilute solutions, because the presence of the particles influences the flow time. We prepared PET nanocomposites containing 2, 4, 6, and 8 wt% C30B [23]. While the thermal degradation of the matrix for the nanocomposite containing 2 wt% C30B was already assessed by the rheological results and color formation, a larger intrinsic viscosity (0.74 dL/g) was obtained for this sample in comparison to the extruded neat PET (0.72 dL/g). Besides, nanocomposites containing 6 and 8 wt% C30B exhibited the same intrinsic viscosity (0.47 dL/g). This illustrates the difficulty in measuring the intrinsic viscosity of nanocomposites, and that caution should be exercised in interpreting results from the literature (the same concern applies to GPC measurements).

An approach to reduce the matrix degradation in the presence of organomodifiers is the purification of the organoclays by removing unbounded halogen impurities. Davis et al. [24] treated Na-MMT and synthetic mica (Na-SM) with dimethyl dioctadecyl ammonium bromide (DMDODA-Br). After the ion exchange reaction, both types of organoclays contained ionicaly bounded DMDODA, unbounded DMDODA-Br, and NaBr by-product. Both NaBr and DMDODA-Br decreased the thermal stability of the produced organosilicates. Therefore, it was considered necessary to remove these two sources of residual bromide. Davis et al. [24] found that the thermal stability of the organosilicates depended on the employed solvent for removing the bromide impurities. Hot water, ethanol, and tetrahydrofuran (THF) were used for washing the organoclays. While hot water washing had no effect on the thermal stability of the organoclays, a 40 °C enhancement in temperature stability at 5% mass fraction loss, $T_{5\%}$, was obtained for organoclays that experienced both ethanol and THF extraction. Ngo et al. [25] reported that the onset temperature of degradation of imidazolium cations was drastically reduced in the presence of a halide (by about 100 °C). Cui et al. [26] have shown that washing organoclays to remove the excess unbounded organic modifier salt improved the thermal stability of the organoclays. Stoeffler et al. [27] purified Cloisite® 20A (C20A) by successive washing with a water/ethanol (1:1) mixture at 70 °C to eliminate the residual chloride anions present in the organoclay. However, they only reported a small improvement (4 °C) in $T_{5\%}$ for the purified C20A.

Quaternary ammonium organic modifiers are less costly than thermally stable organic modifiers such as imidazoliums [24]. Besides, organoclays based on quaternary ammonium surfactants are widely commercialized, which makes purification of the alkyl ammonium–modified organoclays more conceivable. However, purification procedures are time consuming and require the use of solvents. Besides, this approach increases the final cost of the commercialized alkyl ammonium-modified organoclays. Modification of silicate layers with more thermally stable cationic surfactants is another approach to reduce matrix degradation. Kim et al. [14] modified a Na-MMT with a thermally stable organic modifier based on dimethyl imidazolium bromide (DMIBr). PET nanocomposites containing 3 wt% of this organoclay were prepared by a melt-compounding approach at 280 °C. PET nanocomposites containing C15A and C30B were also prepared for comparison. Thermogravimetric analysis (TGA) measurements were carried out for the organoclays. The onset of decomposition temperature of the imidazolium-based organoclay, C15A, and C30B was reported to be 255, 196, and 174 °C. TEM images of PET nanocomposites containing the imidazolium-based organoclay are presented in Figure 21.7 for different magnifications. Although imidazolium-based

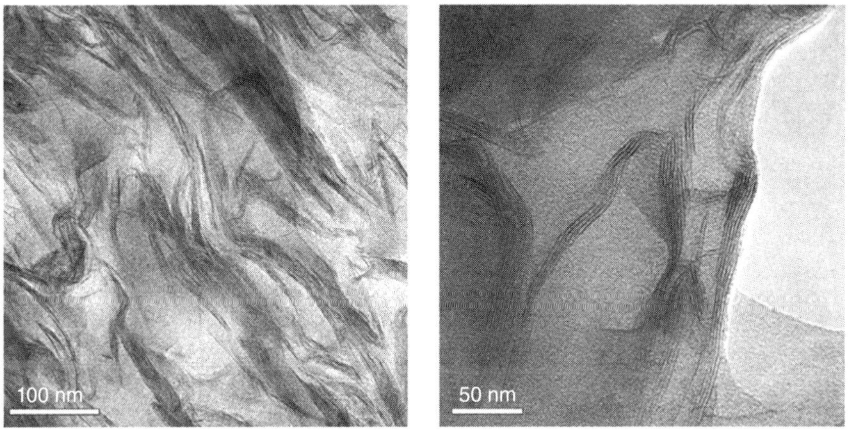

Figure 21.7 TEM images of PET nanocomposites containing imidazolium-based organoclay at two magnifications. Kim et al. [14]. Reproduced with permission of Springer.

organoclay exhibited a better thermal stability over C30B and C15A, the TEM images show an intercalated morphology containing tactoids of several silicate layers rather than an exfoliated structure.

Ghasemi et al. [28] prepared PET nanocomposites containing C30B, and thermally stable organoclays modified with imidazolium, pyridinium, and phosphonium surfactants, using a twin-screw extruder. TGA analysis confirmed that the $T_{5\%}$ of the more thermally stable organoclays was significantly higher than that of C30B. However, based on XRD results the highest gallery opening (i.e., PET intercalation) was obtained for C30B. Image analysis based on TEM images also confirmed a higher degree of intercalation for the nanocomposites containing C30B. Stoeffler et al. [29] prepared PET nanocomposites containing alkyl phosphonium, alkyl pyridinium, and dialkyl imidazolium-modified montmorillonite using an internal mixer at 280 °C. Based on TEM images, intercalated morphologies with tactoids composed of around 4–10 layers were reported for the nanocomposites. The presence of microaggregates, observed by optical microscopy, was also reported for all the nanocomposites. PET nanocomposites containing alkyl phosphonium and dialkyl imidazolium–modified montmorillonite exhibited a yellowish/brownish color, and those containing pyridinium-modified montmorillonite presented a dark brown color in comparison to the neat PET matrix. Na-Hect (hectorite) was also modified with thermally stable hexadecyl quinolinium bromide (Q16). PET nanocomposites were prepared with the organomodified and pristine clays at a loading of 3 wt% using an internal mixer at 280 °C [30]. Figure 21.8 shows XRD patterns of the pristine and modified hectorite, as well as PET nanocomposites containing the organoclay. According to the XRD results, the modification of Na-Hect with the quinolinium surfactant increased the d-spacing from 1.21 to 1.73 nm, and after melt blending with PET the gallery spacing reached 3.15 nm.

Low- and high-magnification TEM images of the obtained nanocomposites are presented in Figure 21.9. Based on the XRD and TEM results, a mixed morphology including tactoids, and intercalated and exfoliated silicate layer structures was reported [30].

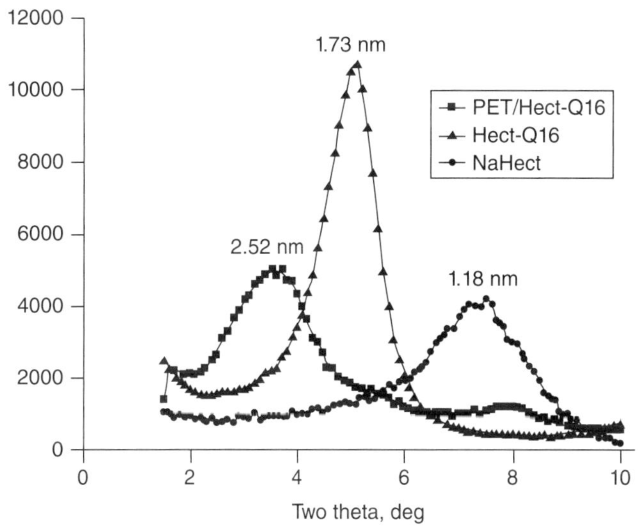

Figure 21.8 XRD patterns of Na-Hect, quinolinium-modified hectorite (Hect-Q16), and PET nanocomposite containing Hect-Q16. Costache et al. [30]. Reproduced with permission of John Wiley and Sons.

In another study, PET nanocomposites containing imidazolium-modified montmorillonite were synthesized by *in situ* polymerization. XRD and TEM results revealed the coexistence of large aggregates and intercalated silicate layers for the synthesized nanocomposites [31].

21.6 METHODS FOR IMPROVING GAS BARRIER PROPERTIES OF POLYMERS

Food and beverage producers require distributing their products all over the world. This increases the time interval between production and consumption of the products. The

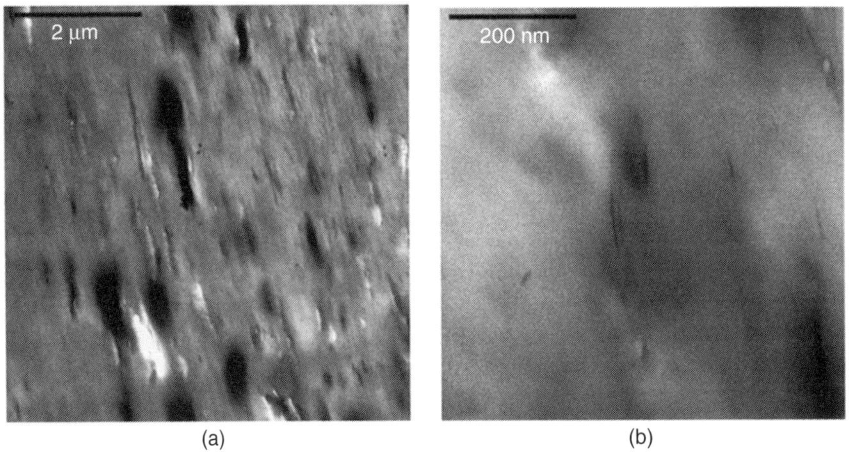

Figure 21.9 (a) Low- and (b) high-magnification TEM images of PET nanocomposites containing Hect-Q16. Costache et al. [30]. Reproduced with permission of John Wiley and Sons.

shelf life of many foodstuffs and beverages is strongly dependent on their packaging resistance to oxygen, carbon dioxide, and water vapor. Hence, various efforts have been devoted to the improvement of barrier properties of packaging materials. A thin layer of material with high intrinsic barrier properties such as aluminum and silicon oxide can be coated on polymer substrates to reduce gas permeability. The level of adherence between the plastic substrate and the coating controls the durability of the coated article. The barrier performance is significantly reduced if the interior coating delaminates or the exterior coating scratches off [32, 33].

The permeability of semicrystalline polymers strongly depends on the extent of crystallinity. Annealing (heat setting) of semicrystalline polymers results in crystal perfection and the enhancement of the degree of crystallinity by the reduction of defects [34]. Crystalline domains are impermeable and a higher degree of crystallinity reduces the volume of polymer available for gas penetration. Besides, impermeable crystals create a tortuous path and force the gas or liquid molecules to follow longer diffusion pathways within the amorphous phase [35, 36]. Therefore, crystalline domains affect both the solubility and diffusivity of gases in the polymer matrix [37]. The influence of annealing on the degree of crystallinity and oxygen permeability of PET was studied by Perkins [38]. A 33% improvement in barrier properties was reported when crystallinity of the matrix increased from about 25 to 55% [38]. Ghasemi et al. [39] showed that after 20 min of annealing at 150 °C, the degree of crystallinity and oxygen barrier properties of PET increased by seven times and 40%, respectively. However, it was reported that after annealing the sample became completely opaque and white. Another drawback of annealing is brittleness of the product.

Blending with high-barrier polymers is another approach to increase barrier properties of polymers like PET or polyolefins [40–42]. Yeo et al. [42] prepared a blend of polypropylene (PP)/ethylene-vinyl alcohol copolymer (EVOH) (85/15) using a single-screw extruder. Oxygen permeability of biaxially stretched PP/EVOH films decreased by about 10 times compared to the neat PP. A 19% improvement in oxygen barrier properties of biaxially stretched PET films blended with 5 wt% poly(*m*-xylene adipamide) MXD6 was also reported [41]. The main problem of this approach is the frequent immiscibility of the components in the blends, which deteriorates properties of the final product such as visual appearance and mechanical resistance. Besides, biaxial stretching is essential to deform the droplets of the high-barrier dispersed phase into parallel and extended layers to create a tortuous pathway and restrain the diffusion of gas molecules, which increases the processing costs.

Another approach to reduce gas permeability is employing intrinsically high-barrier polymers in a multilayer conformation. High-barrier polymers alone are not used as single-layer packaging materials because of low physical properties and cost issues [43]. Packaging containers with multilayer structures (three to nine layers) typically comprise a main constituent polymer (e.g., PET or high-density polyethylene (HDPE)) and high-barrier polymer (e.g., EVOH). In contrast to the previous method (i.e., blending with high-barrier polymers), the multilayer approach does not involve immiscibility and related visual appearance issues. However, adherence between layers can be a limitation, and an adhesive polymer is generally required to tie or bond the barrier layer to the main constituent polymer. Higher production and material costs is another disadvantage of this approach. Moreover, it is more economical and simpler to recycle a single-layer plastic component than a multilayer packaging material [33, 44, 45].

Introducing impermeable lamellar fillers with high aspect ratio in the polymer matrix is another method to enhance barrier properties [46–51]. A tortuous path is created in the polymer matrix due to the presence of the impermeable silicate layers. The impermeable obstacles force the gas permeant to travel a longer path to penetrate through the film and, consequently, the permeability decreases. Permeability is the product of diffusion and solubility. The kinetic aspect of the transport is described by the diffusion coefficient, D, and the thermodynamic aspect of the transport is reflected by the solubility coefficient, S, which implies the affinity between the polymer matrix and the gas molecules. The following equation shows the relationship between the permeability coefficient, P, and the solubility and diffusion coefficients:

$$P = DS \quad (21.1)$$

By ignoring the effect of the filler on the local characteristics of the host polymer matrix, the gas solubility in the nanocomposite is given by the following equation:

$$S_{PCN} = (1 - \phi)S_P \quad (21.2)$$

where ϕ is the clay volume fraction and S_{PCN} and S_P are the solubility coefficients of the PCN and the neat polymer, respectively. The presence of silicate layers not only decreases the available area for diffusion of the solutes but, due to the tortuosity, also reduces the diffusion rate in the nanocomposites according to the following equation:

$$D_{PCN} = \frac{D_P}{\tau} \quad (21.3)$$

where D_{PCN} and D_P are the permeability coefficients of the PCN and the neat polymer, respectively [52]. The tortuosity, τ, must be known in order to predict the permeability of the nanocomposite. There are several models describing tortuosity of filled systems [53–58]. For example, the following model was proposed by Bharadwaj [53]:

$$\tau = 1 + \frac{\alpha\phi}{2}\cos^2\theta \quad (21.4)$$

where α is the silicate layers' aspect ratio and θ is the angle between the direction of the penetrating flow and the normal of the layers, schematically illustrated in Figure 21.10.

The average aspect ratio of clay particles can be determined by the method described by Ghasemi et al. [39] using TEM images. Figure 21.11 shows the evaluation of the aspect ratio of silicate layers by image analysis. An exfoliated particle is shown in Figure 21.11a. For overlapped tactoids, the average thickness and overall length are used (Fig. 21.11b), while an end-to-end vector is considered for the length of curved tactoids (Fig. 21.11c).

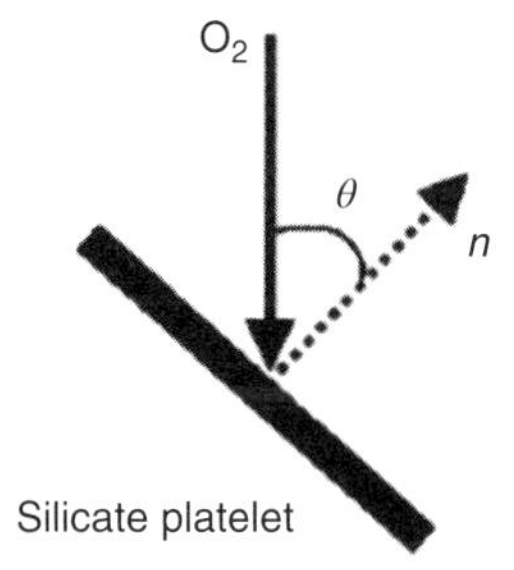

Figure 21.10 Orientation of a silicate layer with respect to oxygen flow. Russo et al. [59]. Reproduced with permission of American Chemical Society.

21.7 POLYAMIDE NANOCOMPOSITES

Russo et al. [59] prepared polyamide nanocomposites containing C30B by using nylon-6 and two copolyamides of nylon-6 of low and high molecular weights (denoted LADS and HADS, respectively). Percentage increments of O_2 barrier properties in nylon-6, LADS, and HADS are reported in Figure 21.12. All the three matrices exhibited improvements of O_2 barrier properties with increasing silicate content.

The highest improvement in the barrier properties obtained for the copolyamide HADS was attributed to the better dispersion of the organoclay particles within this matrix. Figure 21.13 shows TEM micrographs of HADS and nylon-6 nanocomposites containing 6 wt% C30B [59].

Figure 21.14 reports the effect of Nanomer® I.24TL loading on the oxygen transmission rate of polyamide 6. Around 80% improvement in oxygen barrier properties is obtained for nanocomposites containing 8 wt% organoclay [60].

Meng et al. [61] prepared polyamide 12 nanocomposites containing 5 wt% organoclay, modified with octadecylammonium chloride, using an internal mixer. 6, 31, and 32% reductions in hydrogen permeation were reported for the nanocomposites melt blended for 5, 20, and 40 min. It suggests that the dispersion of the organoclay improved with mixing time.

Swain et al. [62] prepared polyamide 6 nanocomposites containing C30B, C15A, and Cloisite® 93A (C93A) using a single-screw extruder. Based on XRD and TEM results, it was concluded that C30B was exfoliated within the matrix, while C15A and C93A were only intercalated. Figure 21.15 reports the oxygen permeability of the neat polyamide 6 and polyamide 6 nanocomposites containing 2.5, 5, and 10 wt% C30B. It is obvious that gas molecules have to pass through a more tortuous path when more silicate layers are incorporated in the matrix.

21.8 POLYOLEFIN NANOCOMPOSITES

PP nanocomposite films containing 3, 5, and 7 wt% C15A and 10 wt% ethylene vinyl acetate copolymer (EVA), as a compatibilizer, were prepared by melt mixing [63]. XRD and TEM results indicated an intercalated morphology. By increasing clay loading, the Young modulus and tensile strength increased continuously while the strain at break reduced. For example, 78 and 25% improvements in the Young modulus and tensile strength were obtained for the nanocomposites containing 7 wt% C15, respectively, while the strain at break reduced around 28%. Figure 21.16 presents the relative oxygen permeability of PP/EVA nanocomposites as a function of C15A concentration. A significant improvement in oxygen barrier properties was obtained even at low clay concentration.

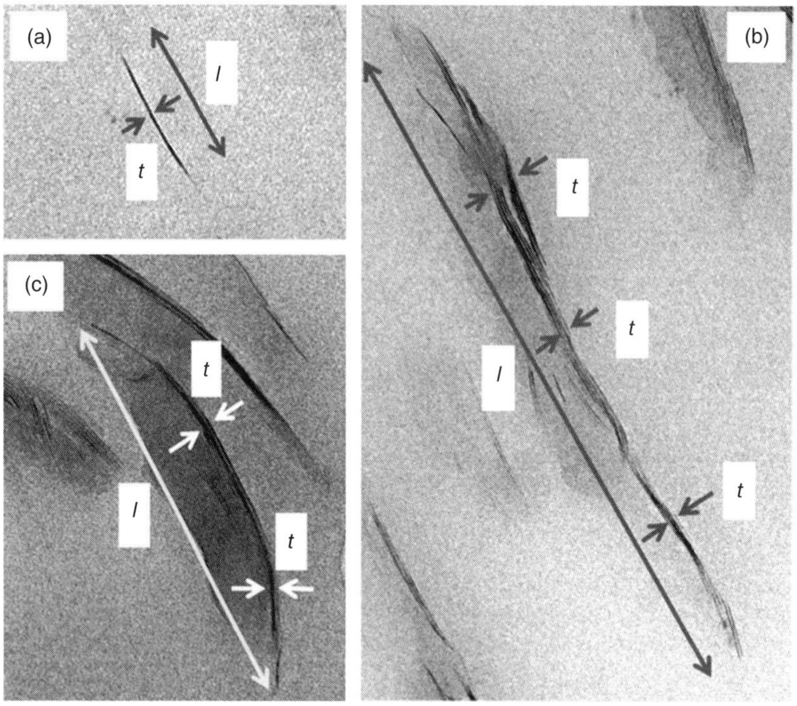

Figure 21.11 Measurement of the aspect ratio of silicate layers by image analysis. Ghasemi et al. [39]. Reproduced with permission of John Wiley and Sons.

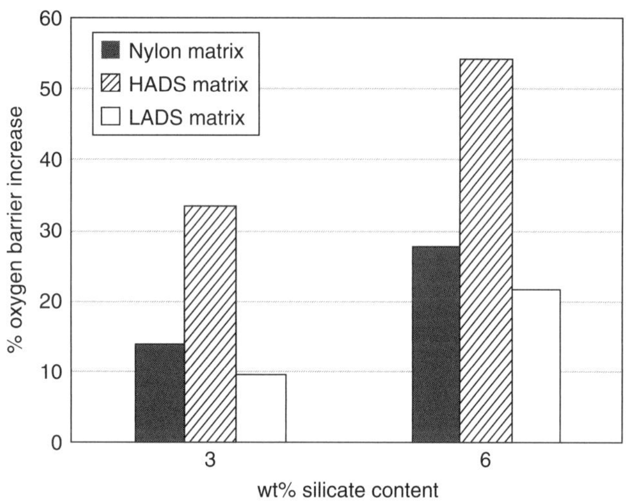

Figure 21.12 O_2 barrier increases for three nylon-based systems analyzed at different silicate contents. Russo et al. [59]. Reproduced with permission of American Chemical Society.

LDPE nanocomposite films were prepared by melt mixing, using a twin-screw extruder attached to a blown-film set, and polyethylene-grafted maleic anhydride (PEMA) was used as compatibilizer [64]. Di (hydrogenated tallow alkyl) dimethyl ammonium chloride was used as surfactant for clay modification and the ratio of PEMA to organoclay was kept constant for all nanocomposites at 3:1. Both tensile modulus and tensile strength of the nanocomposite blown-film organoclay increased in comparison to LDPE films. Figure 21.17 shows oxygen permeability results of LDPE and LDPE nanocomposites. Increasing the clay loading up to 7 wt% gradually decreases the oxygen permeability by 24% [64].

21.9 PET NANOCOMPOSITES

Ghasemi et al. [39] prepared PET nanocomposite films containing 3 wt% Cloisite® 30B (C30B) by cast extrusion using a twin-screw extruder. A partially exfoliated/intercalated morphology was reported for the nanocomposite films. In comparison to neat PET films, the oxygen barrier properties and Young modulus of the nanocomposite films exhibited 23% and 20% improvement, respectively. However, the presence of the silicate nanolayers increased the brittleness of the samples and resulted in more hazy films. In another study, they investigated the effect of processing conditions on the properties of PET nanocomposite films containing C30B [65]. Compared with the neat PET, a 27% reduction in oxygen permeability and a 30% improvement of the tensile modulus was reported for the nanocomposites containing 3 wt% C30B, processed in a severe screw profile in terms of mixing and high screw rotational speed, 250 rpm [65]. In another study, PET nanocomposites containing Cloisite® 15A (C15A) were prepared using a twin-screw extruder. A 16% reduction in oxygen permeability was reported for the nanocomposites containing 3 wt% C15A [66].

Biaxial stretching was reported to improve exfoliation of silicate layers within a PET matrix [67, 68]. Rajeev et al. [67] employed a twin-screw extruder to produce PET

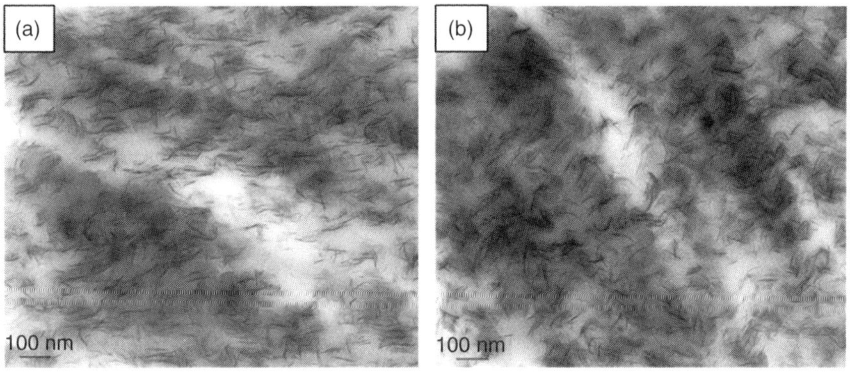

Figure 21.13 TEM images of (a) HADS and (b) nylon-6 matrices containing 6 wt% C30B. Russo et al. [59]. Reproduced with permission of American Chemical Society.

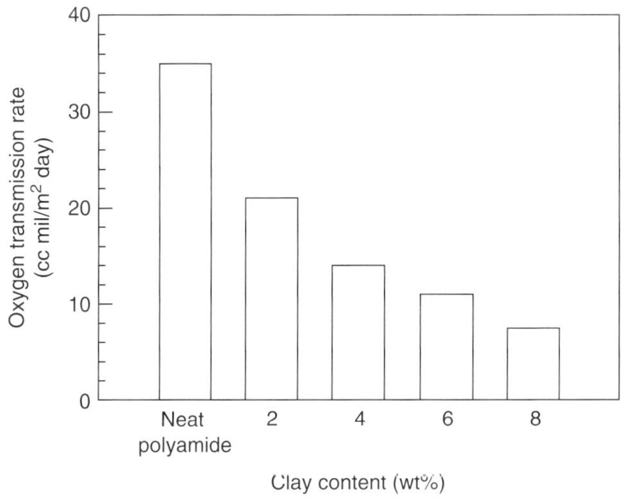

Figure 21.14 Oxygen transmission rate of polyamide 6 and polyamide 6 nanocomposites at various clay (Nanomer® I.24TL) loadings. Adapted from Lan et al. [60].

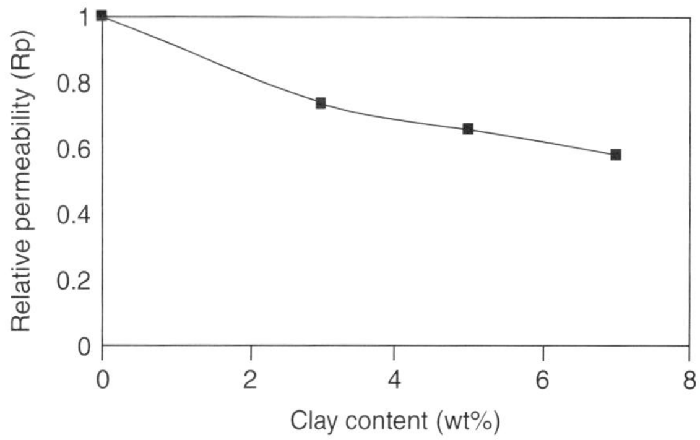

Figure 21.16 Relative oxygen permeability for the PP/EVA nanocomposites as a function of clay (C15A) loading. Shafiee et al. [63]. Reproduced with permission of Taylor and Francis.

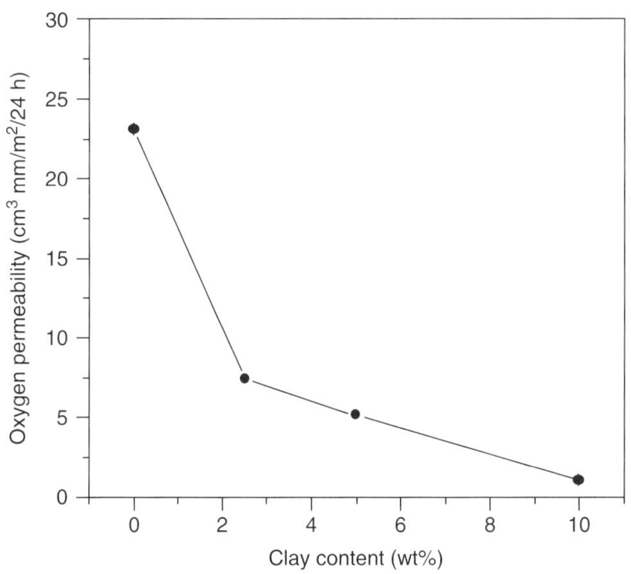

Figure 21.15 Oxygen permeability of polyamide 6 nanocomposites as a function of C30B concentration. Swain and Isayev [62]. Reproduced with permission of John Wiley and Sons.

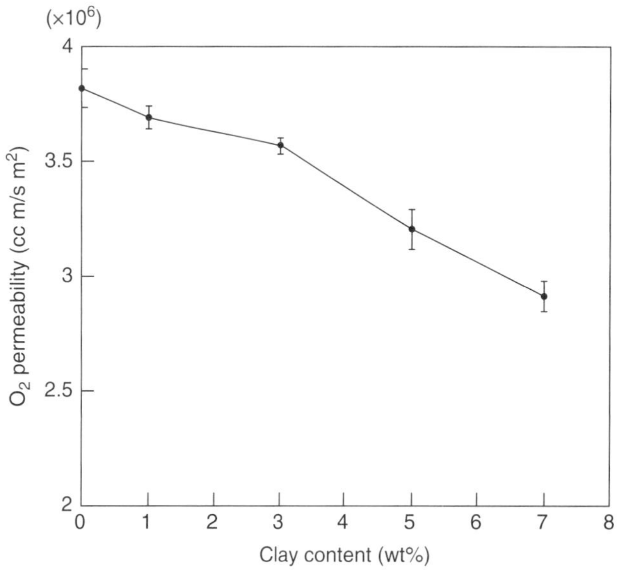

Figure 21.17 Oxygen permeability of LDPE and LDPE nanocomposites as a function of clay (C15A) concentration. Arunvisut et al. [64]. Reproduced with permission of John Wiley and Sons.

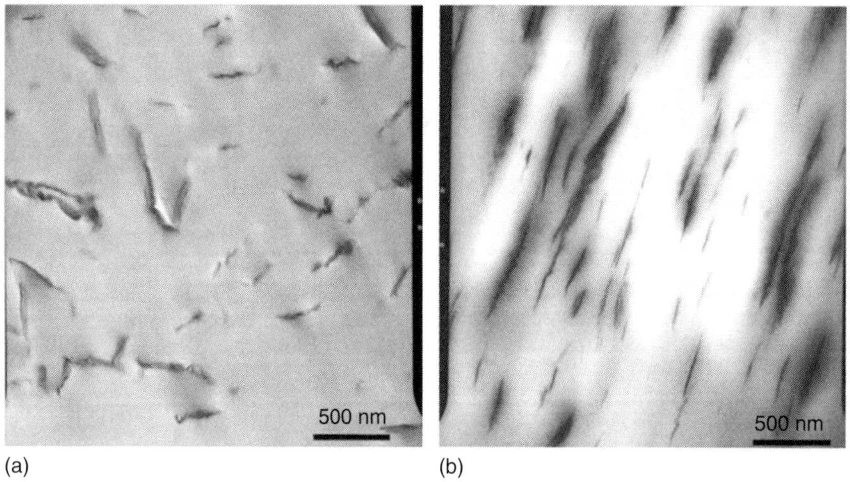

Figure 21.18 TEM images of PET-based nanocomposites containing 2 wt% Somasif® MAE (a) before and (b) after stretching. Rajeev et al. [67]. Reproduced with permission of Elsevier.

nanocomposites containing Somasif® MAE, which is an alkylammonium-modified synthetic clay. Platelets in tactoids slipped past each other due to biaxial stretching. As a result, the tactoid length and the frequency of single-layer and double-layer particles were increased. Figure 21.18 exhibits the effect of equi-biaxial stretching, with a stretch ratio equal to 3 in both directions, on the morphology of nanocomposites containing 2 wt% Somasif® MAE. A 22% reduction in oxygen permeability was reported for the nanocomposites, in comparison to the neat PET [67].

The Young modulus of samples with various clay loadings and stretch ratios is shown in Figure 21.19. Improvements of 10, 20, and 30% in the Young modulus were reported for the unstretched samples containing 1, 2, and 5 wt% Somasif® MAE. By stretching the samples, better mechanical properties were obtained, which was attributed to reduced agglomeration and a better tactoid alignment [68].

Efforts have also been devoted to improve the compatibility between organoclays and polyesters. For examples, pentaerythrytol and maleic anhydride were employed to improve the compatibility between silicate nanolayers and PET chains; however, an exfoliated morphology was not achieved [69]. Chisholm et al. [70] investigated the effect of sodium sulfonate functionalization of poly(butylene terephthalate) (PBT) on the properties of PBT nanocomposites containing 5 wt% Na-MMT, and an alkylammonium-modified montmorillonite. TEM images of PBT and PBT-ionomer samples containing the organoclay and pristine Na-MMT are shown in Figure 21.20. PBT and PBT-ionomer composites containing Na-MMT exhibited almost the same morphology, while the presence of sulfonated groups in the PBT-ionomer/organoclay nanocomposite led to a significant change in the morphology as compared to the reference PBT/organoclay nanocomposite. The higher exfoliation level obtained for the PBT-ionomer/organoclay nanocomposite was attributed to favorable interactions between the positively charged edges of montmorillonite and the negatively charged ionomer.

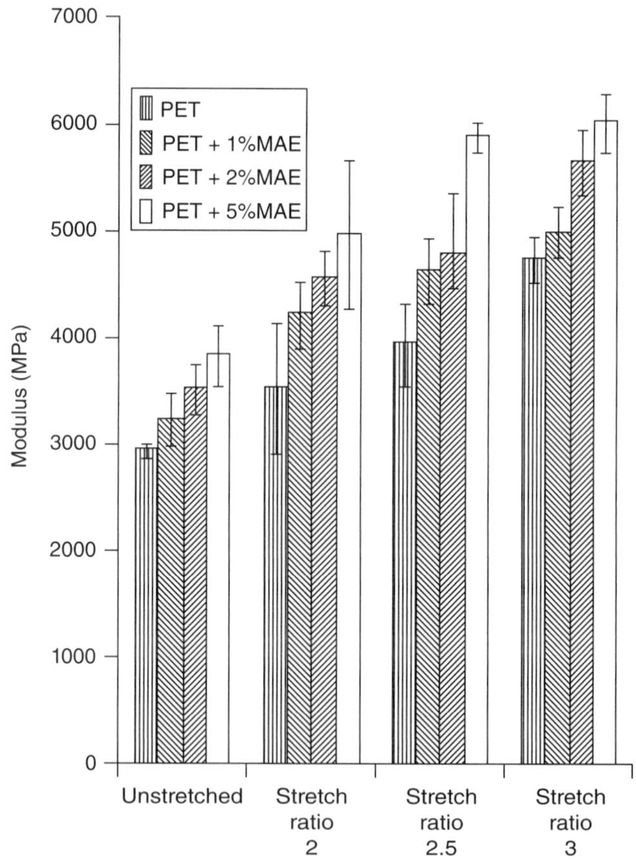

Figure 21.19 Effect of biaxial stretch ratio and clay (Somasif® MAE) concentration on the tensile modulus of PET and PET-based nanocomposites. Soon et al. [68]. Reproduced with permission of John Wiley and Sons.

Various blends of PET and a more polar polymer such as a PET-ionomer [71, 72] and polyamide 6 [73] were prepared to investigate the clay dispersion in the PET matrix. In all cases, immiscible blends were obtained. Figure 21.21 illustrates the

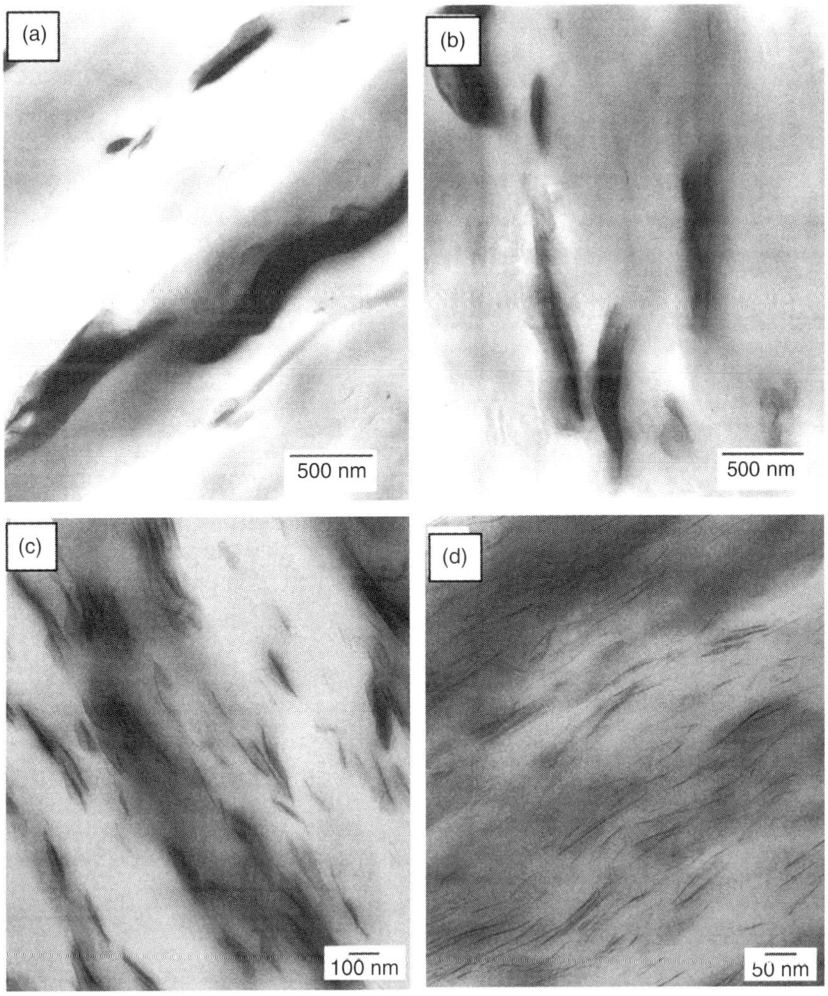

Figure 21.20 (a) TEM images of PBT/Na-MMT, (b) PBT-ionomer/Na-MMT, (c) PBT/organoclay, and (d) PBT-ionomer/organoclay. Chisholm et al. [70]. Reproduced with permission of American Chemical Society.

phase-separated morphologies (i.e., co-continuous and droplet matrix structures) for PET/PET-ionomer blends containing 50, 30, and 20 wt% PET-ionomer. The difference between the polarity of PET and that of the dispersed phases results in the preferential localization of the clay particles within the PET-ionomer or polyamide 6 domains, due to the higher affinity of the silicate layers for the more polar phase. As shown in Figure 21.22, the nanoclay is preferentially localized into the polyamide 6 and the PET-ionomer domains. The lack of a good distribution and the absence of clay particles outside of the more polar domains prevent remarkable enhancements of the barrier properties.

Ghanbari et al. [74] employed a multifunctional epoxy-based chain extender, Joncryl® ADR-4368F (Joncryl), to control the effect of thermal degradation of PET/organoclay nanocomposites and improve clay exfoliation using a master-batch approach. Based on Fourier transform infrared (FTIR0 spectroscopy and rheological results, the epoxy groups of Joncryl were shown to link the functional terminal groups of degraded PET chains and to rebuild the molecular weight and viscoelastic properties. XRD patterns of C30B and Nanomer® I.28E (N28E) and their corresponding nanocomposites containing 4 wt% organoclays with and without Joncryl (1 wt%) are shown in Figure 21.23. The interlayer spacing of both organoclays increased due to the intercalation of PET chains. The introduction of Joncryl in nanocomposites containing C30B and N28E also reduced the intensity of the XRD peaks significantly. The reduction in the peak intensity in clay-containing polymers is associated with a higher level of exfoliation, as no peak is expected for exfoliated structures, and consequently smaller domains of periodicity (smaller tactoids or single layers).

Figure 21.24 illustrates the effect of Joncryl on the morphology of PET nanocomposites containing N28E. For the nanocomposites containing the chain extender, we observe at low magnification a higher clay distribution density (number of particles per unit area) for the nanocomposite prepared with 1 wt% Joncryl (Fig. 21.24c compared to a), while at high magnification a significant reduction in tactoid sizes for the Joncryl-based nanocomposite is seen (Fig. 21.24d compared to b). The higher level of exfoliation in the nanocomposites containing Joncryl may be explained by the larger viscosity

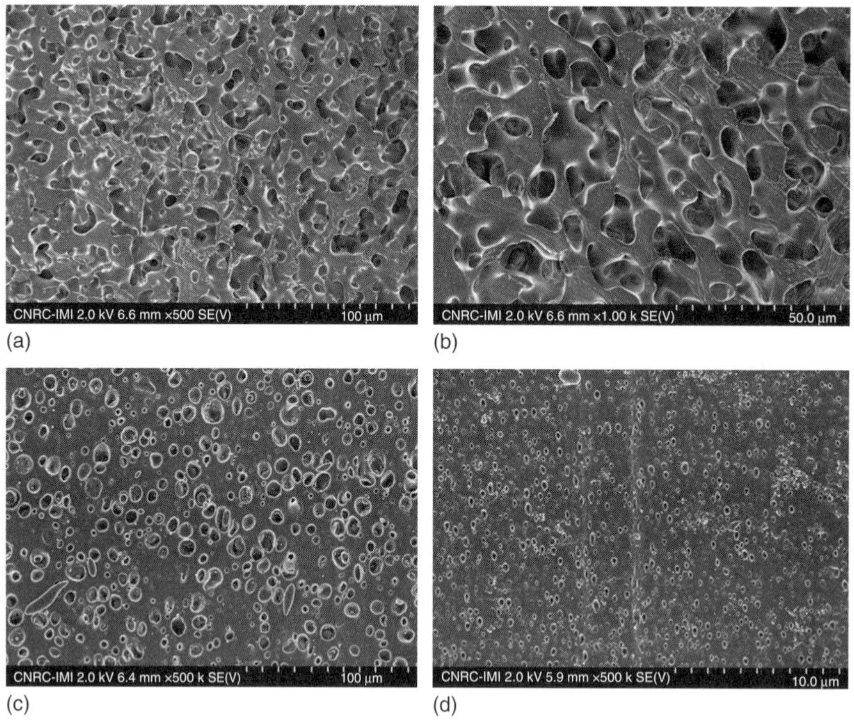

Figure 21.21 SEM micrographs of PET/PET-ionomer blends initially containing (a, b) 50 wt% PET-ionomer at different magnifications, (c) 30 wt% PET-ionomer, and (d) 20 wt% PET-ionomer. All micrographs were taken after extraction of the PET-ionomer phase. Ghanbari et al. [71]. Reproduced with permission of John Wiley and Sons.

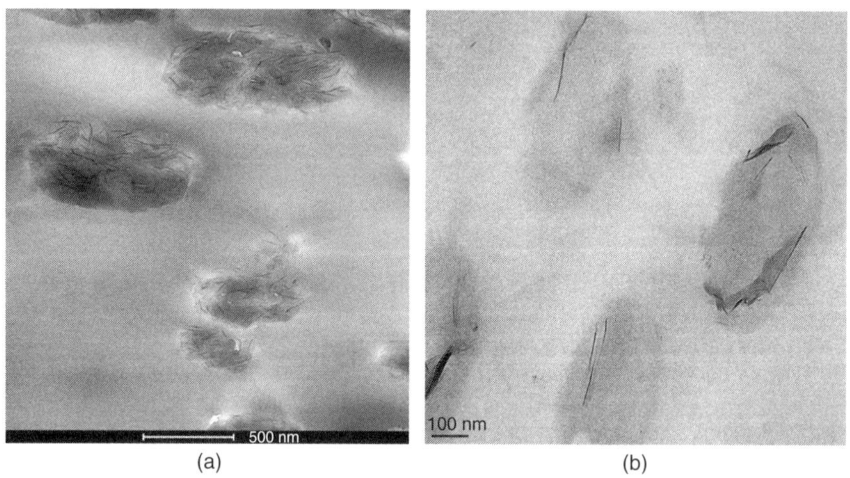

Figure 21.22 TEM images of PET nanocomposites containing (a) 20 wt% polyamide 6, Goitisolo et al. [73]. Reproduced with permission of Elsevier. and (b) 20 wt% PET-ionomer, Ghanbari et al. [71]. Reproduced with permission of John Wiley and Sons.

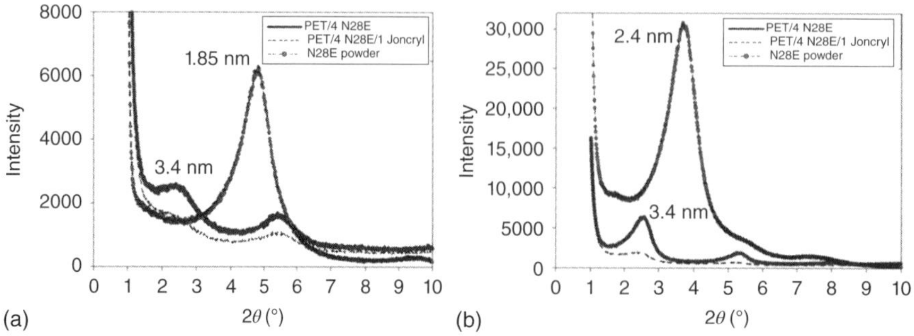

Figure 21.23 Effect of Joncryl: XRD patterns of the nanoclays and their PET-based nanocomposites: (a) C30B and its corresponding nanocomposites and (b) N28E and its corresponding nanocomposites Ghanbari et al. [74]. Reproduced with permission of Elsevier.

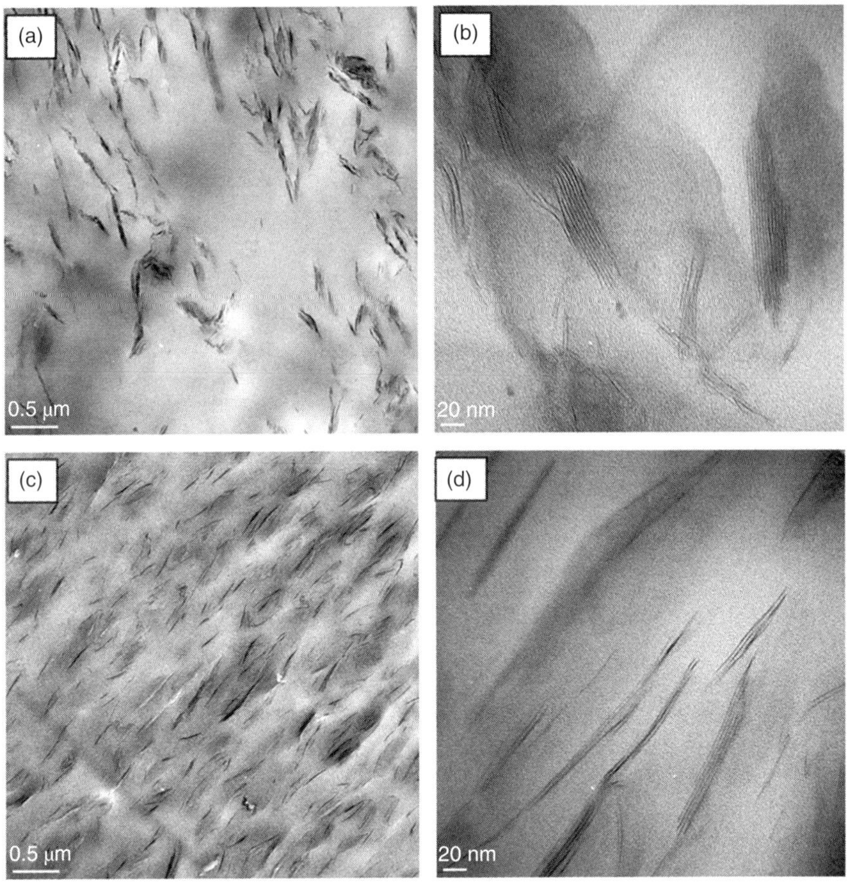

Figure 21.24 TEM images of (a, b) PET/4 N28E and (c, d) PET/4 N28E/1 Joncryl at various magnifications. Ghanbari et al. [74]. Reproduced with permission of Elsevier.

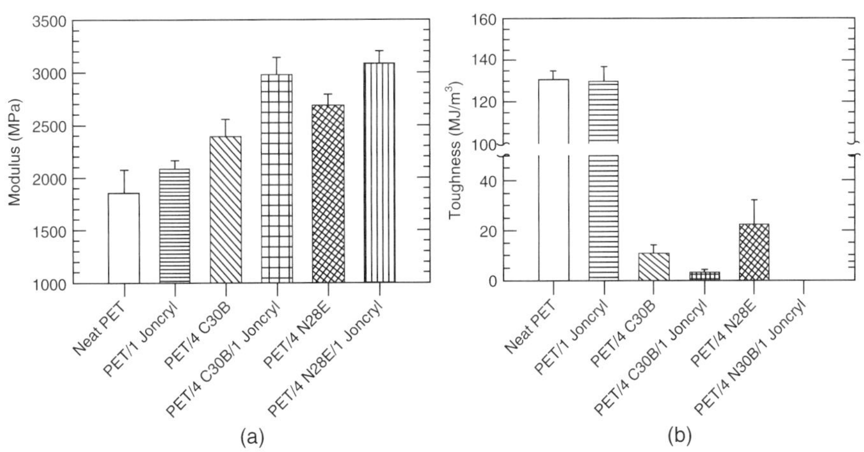

Figure 21.25 (a) Tensile modulus and (b) toughness data of the neat PET and PET-based nanocomposites with and without Joncryl. Ghanbari et al. [74]. Reproduced with permission of Elsevier.

of the matrix, which generates larger shear stresses that break down large clay tactoids into smaller ones.

Crystal content of the produced films also increased in the presence of the organoclays and higher degree of crystallinity was obtained for the samples containing the chain extender [74]. Mechanical properties in the machine direction of the neat PET and PET-based nanocomposites, with and without Joncryl, are reported in Figure 21.25. Clay-containing films were more brittle and exhibited larger tensile modulus. The larger Young modulus for the nanocomposites containing Joncryl might be attributed to larger molecular weight and increased crystallinity of the matrix, as well as a better dispersion of the silicate nanolayers in the matrix, as confirmed by XRD and TEM results [74].

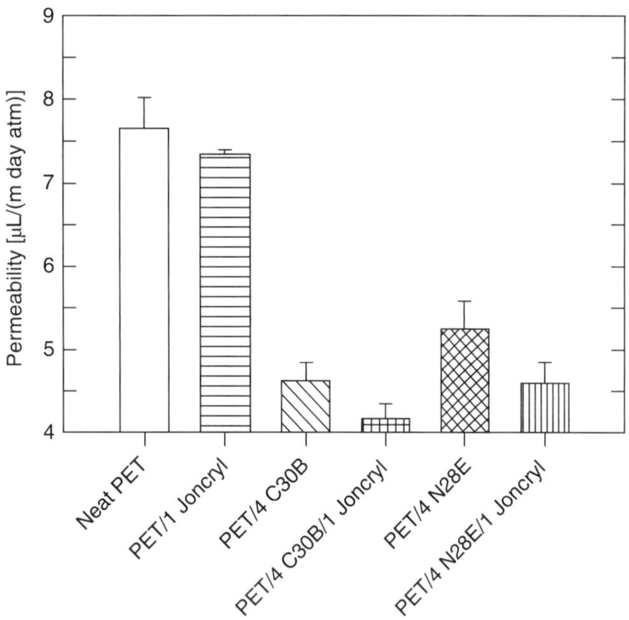

Figure 21.26 Oxygen permeability of the neat PET and PET-based nanocomposites with and without Joncryl. Ghanbari et al. [74]. Reproduced with permission of Elsevier.

The oxygen permeability of the neat PET and PET-based nanocomposite films, with and without Joncryl, is reported in Figure 21.26. The presence of impermeable silicate layers and larger crystallinity resulted in higher gas barrier properties for the nanocomposite films. The oxygen permeability of the nanocomposite films containing 4 wt% C30B and 1 wt% Joncryl decreased by 46% in comparison to the neat PET films. Better barrier properties of the nanocomposite films containing the chain extender were attributed to the better clay dispersion and higher crystallinity [74].

XRD, SEM, and TEM are widely used for morphological characterization of nanocomposites. Rheology has also been used extensively in complement to these techniques in several studies as it is very sensitive to the morphology of nanocomposites [23, 75–80]. The summary of the most significant results from these studies is the transition from liquid-like to solid-like viscoelastic behavior for nanocomposites, even at low-volume fractions of silicate layers, as well as a strong shear-thinning behavior. The solid-like behavior has been attributed to the formation of a percolated network of clay particles that occurs at relatively low clay loading, due to the anisotropy of the particles, which prevents their free rotation and the dissipation of stress.

By employing rheometry, detailed information about molecular-level changes in the structure (e.g., oxidation, degradation, cross-linking, etc.) can be obtained. In addition, the interactions between the nanoparticles and the matrix can also be detected and quantified. Another advantage of rheological methods is that they probe the bulk of nanocomposite materials on a significant volume of sample tested, which increases the reliability of the data. Finally, measurements are performed in the molten state, which can provide valuable information about the processability of the nanocomposites [1, 23, 80].

Ghanbari et al. [23] prepared PET nanocomposites containing C30B, C15A, Cloisite® 25A (C25A), N28E, and Somasif® ME100 (SM100) by melt compounding using a twin-screw extruder. The effects of the screw geometry, clay concentration, and surfactant chemistry of the organoclays on the morphology and rheology of the PET nanocomposites were investigated using XRD, SEM, TEM, and rheometry. The Maron–Pierce equation was used to determine the apparent viscosity of the degraded matrix in the nanocomposites. To estimate an apparent matrix molecular weight in order to quantify the degradation induced by the presence of the organoclay, the matrix apparent viscosity, $\eta^*_{M(100\text{rad/s})}$, was evaluated at 100 rad/s where the contribution of the matrix should be dominant. The apparent molecular weight of the PET matrix was found to decrease from 65 kg/mol for the neat PET to around 30 kg/mol for a PET nanocomposite containing 8 wt% C30B. The reduced complex viscosity and reduced storage modulus of the neat PET and the nanocomposites containing 2–8 wt% C30B as functions of angular frequency are presented in Figure 21.27. The

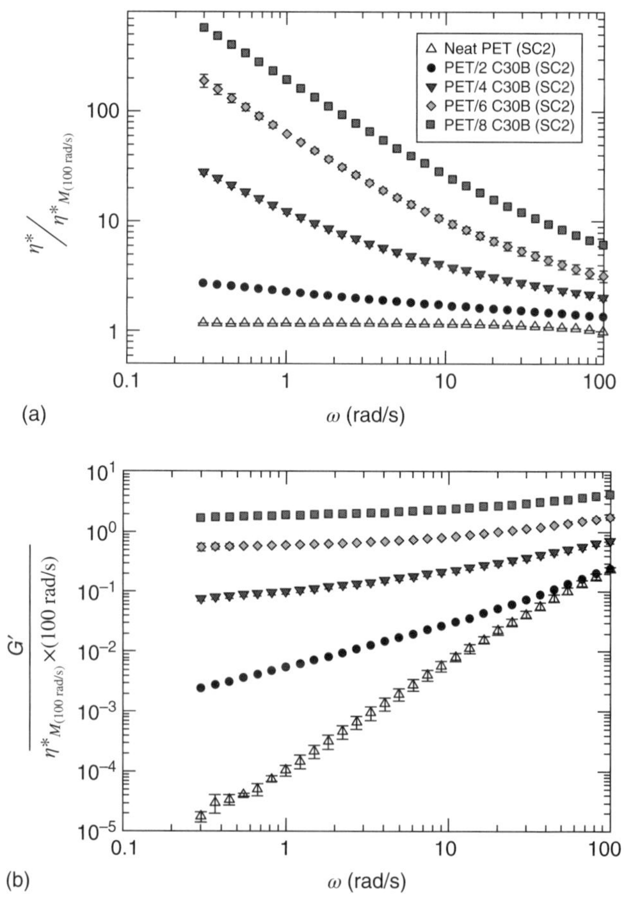

Figure 21.27 (a) Reduced viscosity and (b) reduced storage modulus of the neat PET and PET/C30B nanocomposites as functions of clay loading and frequency. Ghanbari et al. [23, figure 7]. Reproduced with permission of Springer.

reduced viscosity increased with clay content and exhibits a very strong solid-like behavior at high clay loadings, with a slope on the log–log plot approaching −1. The reduced storage modulus also increased with clay concentration and exhibited a low frequency plateau at high clay loadings. This pseudo-solid-like behavior was attributed to the formation of a space-filling interconnected network of clay particles. This percolated three-dimensional network acts as a weak solid and causes significant enhancement of the storage modulus, as well as the emergence of a low frequency plateau.

Larger apparent yield stress is correlated to stronger particle–particle and/or polymer–particle interactions. For the nanocomposites containing 4, 6, and 8 wt% C30B apparent yield stress was reported to be 25.7, 91.8, and 166 Pa, respectively. This value is 41.8 Pa for the nanocomposites containing 6 wt% N28E, indicating a better dispersion of C30B particles compared to N28E in the PET matrix. XRD, SEM, and TEM observations as well as image analysis confirmed the rheological results. By calculating the solubility parameter based on the Fedors group contribution method, the authors showed that C30B is more miscible with PET, as the solubility parameter of PET is closer to that of C30B organic modifier compared to N28E organic modifier [23].

21.10 POLYLACTIDE NANOCOMPOSITES

Najafi et al. [81] prepared PLA nanocomposites containing 1 wt% Joncryl and 2 wt% C30B via various compounding strategies using a twin-screw extruder. In the first strategy (S1), PLA, Joncryl, and C30B were extruded simultaneously. In the second strategy (S2), the nanocomposites were prepared using a master-batch approach. First, PLA and 4 wt% C30B were extruded to prepare a master batch. Then, the master batch was diluted in the second extrusion pass by adding Joncryl and additional PLA. In the third strategy (S3), PLA and Joncryl were blended in the first extrusion pass, while clay was added to the blend in the second extrusion pass. Morphological characterization revealed that a better dispersion of clay particles was obtained for the nanocomposites prepared by the second strategy (S2) (a master batch of PLA/C30B). The viscoelastic properties of PLA and the nanocomposites prepared by different compounding strategies are presented in Figure 21.28.

The complex viscosity and storage modulus of the PLA show a very impressive enhancement in the presence of the chain extender. However, due to the degradation of the matrix in the presence of the organoclay, the effect of Joncryl on the viscoelastic properties of the nanocomposites is not as spectacular. Among all the nanocomposites, the highest values of the viscoelastic properties at low frequencies were obtained for the Joncryl-based nanocomposite prepared by the second compounding strategy (S2), which indicates better clay dispersion. Figure 21.29 presents the oxygen gas permeability for PLA and PLA-based nanocomposites with and without Joncryl. All nanocomposites show higher barrier properties in comparison to the neat PLA. The lowest oxygen permeability was obtained for the nanocomposite produced by the second strategy due to further delamination and better distribution of clay particles [81].

Figure 21.30 reports the effects of C30B and Joncryl loadings on the Young modulus and maximum tensile strength of PLA/Joncryl nanocomposite films [82].

The Young modulus of the neat PLA increases from 3.0 to over 3.8 GPa by incorporating 6 wt% C30B and 0.3 wt% Joncryl. The effect of clay on tensile strength is less pronounced compared to the Young modulus. The presence of clay agglomerates and severe thermal degradation of the PLA matrix in the presence of the organoclay during melt mixing were mentioned as possible reasons for the tensile strength reduction of the sample containing 6 wt% C30B. Due to recovery of molecular weight, the tensile strength increases

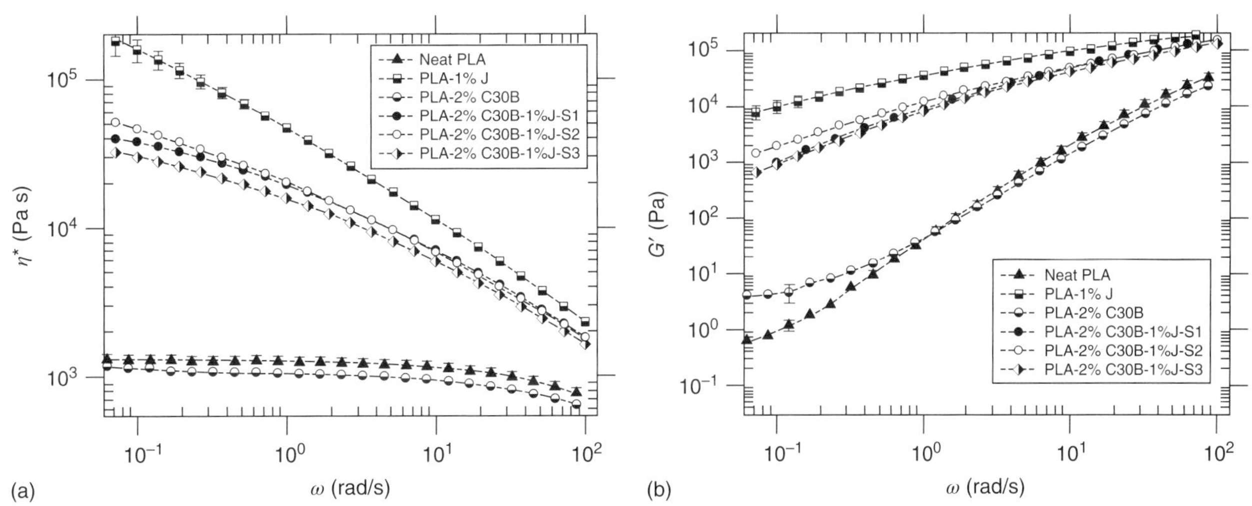

Figure 21.28 (a) Complex viscosity and (b) storage modulus of PLA and PLA-based nanocomposites prepared via various compounding strategies as functions of angular frequency. Najafi et al. [81]. Reproduced with permission of Elsevier.

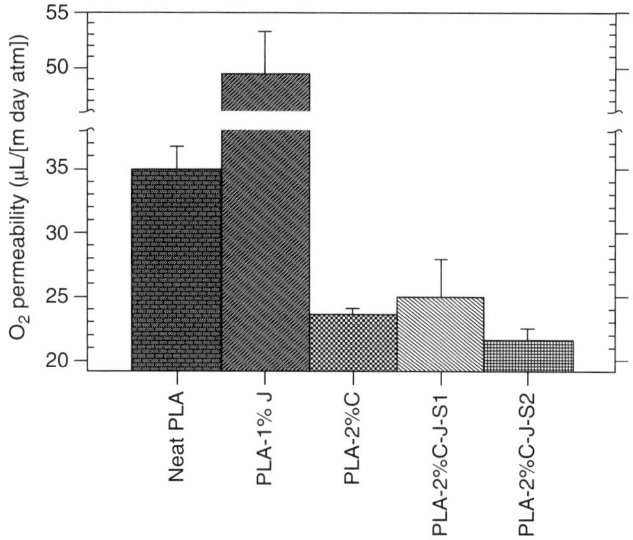

Figure 21.29 Oxygen permeability of PLA and PLA-based nanocomposites with and without Joncryl. Najafi et al. [81]. Reproduced with permission of Elsevier.

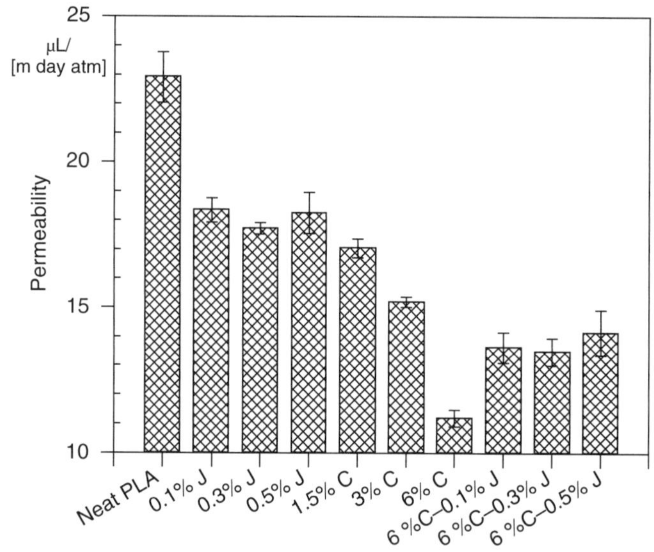

Figure 21.31 Oxygen permeability of PLA and PLA-based nanocomposites containing various loadings of Joncryl and C30B. Meng et al. [82]. Reproduced with permission of Carl Hanser Verlag.

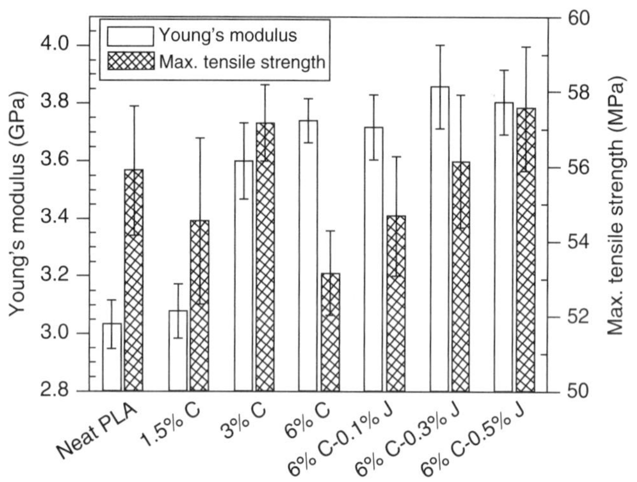

Figure 21.30 Effects of C30B and Joncryl loadings on the tensile properties of PLA/Joncryl nanocomposites. Meng et al. [82]. Reproduced with permission of Carl Hanser Verlag.

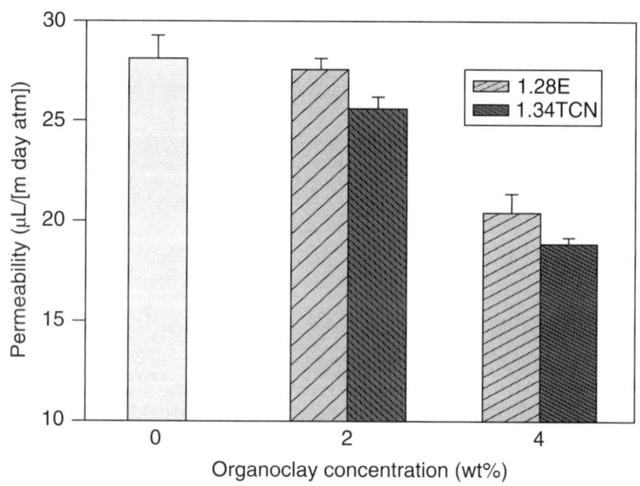

Figure 21.32 Oxygen permeability of neat PLA and PLA nanocomposites. Yourdkhani 2013 [83], pp. 47–53. Reproduced with permission of Elsevier.

monotonically with the chain extender concentration. The effects of clay and Joncryl on the oxygen barrier properties of PLA are reported in Figure 21.31. The oxygen permeability decreases from 22.9 to 11.2 μL/[m day atm] by incorporating 6 wt% C30B. The permeability was increased by introduction of Joncryl in the PLA nanocomposites containing 6 wt% C30B, which was attributed to the long-chain branched and local cross-linked structures of PLA matrix induced by the chain extender [82].

Yourdkhani et al. [83] prepared PLA nanocomposites containing 2 and 4 wt% N28E and Nanomer® I.34TCN (I.34TCN) using a twin-screw extruder. Oxygen permeability results for the neat PLA and PLA nanocomposites are shown in Figure 21.32. A larger reduction in oxygen permeation was obtained for the nanocomposites containing I.34TCN, which was attributed to the better dispersion of the clay particles within the matrix.

21.11 CONCLUSIONS AND PERSPECTIVES

Polymer-layered silicate nanocomposites are prepared at very low filler content (≤5 wt%), which results in far lighter products in comparison to conventional composites. Besides, the ease of preparation through simple processing such as melt extrusion and injection molding with low production costs makes nanocomposites good candidates for a wide range of applications, from automotive to food packaging. In most cases, it is essential to modify pristine silicate layers with surfactants to make them compatible with hydrophobic polymers, and hence increasing the initial gallery spacing of the silicate layers to facilitate polymer chain intercalation. The final morphology (i.e., the state of clay dispersion and distribution within the polymeric matrix) depends on many factors such as the type of polymer, layered silicate and organic modifier, the preparation technique, and processing conditions. In general, improvements in various properties are reported for PCNs, particularly those with exfoliated structures. For example, it has been observed by many authors that impermeable nanoplatelets improve the barrier properties of polymers due to lengthening of the diffusion path of the permeating gas molecules. In terms of mechanical properties, although the Young modulus is increased by the incorporation of silicate layers, contradictory results are reported concerning their strength, elongation at break, and toughness. Besides, improvements in some properties may be obtained at the expense of optical properties (e.g., haze and clarity), especially if thermal degradation takes place.

One of the major problems encountered in preparing PET and PLA organoclay nanocomposites is the severe degradation of the polymer matrix in the presence of the organoclay. Considerable success has been recently obtained by using chain extenders and, hence, recovering the molecular weight of the matrix and improving the properties of the nanocomposites. Nevertheless, more efforts should be devoted to the search for the most appropriate chain extender depending on the application.

As the developments of the nanocomposite technology area are still in their infancy, the performances are not yet meeting the expectations and significant improvements need to be achieved. Therefore, significant research efforts should be devoted to better understand the chemistry of filler modification, the thermodynamics of filler dispersion, and the complex structure–property relationships involved in the preparation of various nanocomposites. Further studies should be devoted to the preparation of more stable, but compatible organoclays. Synthetic clays of higher aspect ratios compared to natural clays, if properly compatibilized with hydrophobic polymers, are good candidates to create longer tortuous pathways and, consequently, could yield better barrier properties.

REFERENCES

1. Bhattacharya S, Gupta R, Kamal M. *Polymeric nanocomposites: theory and practice*. Hanser Gardner Publications; 2007.
2. Pavlidou S, Papaspyrides CD. A review on polymer-layered silicate nanocomposites. Prog Polym Sci 2008;33:1119–1198.
3. Ray SS, Okamoto M. Polymer/layered silicate nanocomposites: a review from preparation to processing. Prog Polym Sci 2003;28:1539–1641.
4. Hussain F, Hojjati M, Okamoto M, Gorga RE. Review article: Polymer-matrix nanocomposites, processing, manufacturing, and application: An overview. J Compos Mater 2006;40:1511–1575.
5. Mittal V. *In-situ Synthesis of Polymer Nanocomposites*. Wiley; 2011.
6. Grim RE. *Applied Clay Mineralogy*. McGraw-Hill Book Company Inc.; 1962.
7. Boczkowska A. *Advanced Elastomers - Technology*. InTech: Properties and Applications; 2012.
8. Giannelis EP. Polymer layered silicate nanocomposites. Adv Mater 1996;8:29–35.
9. Vaia RA, Giannelis EP. Lattice model of polymer melt intercalation in organically-modified layered silicates. Macromolecules 1997;30:7990–7999.
10. Giannelis EP. Polymer-layered silicate nanocomposites: Synthesis, properties and applications. Appl Organ Chem 1998;12:675–680.
11. Pramanik M, Srivastava SK, Samantaray BK, Bhowmick AK. EVA/Clay nanocomposite by solution blending: Effect of aluminosilicate layers on mechanical and thermal properties. Macromol Res 2003;11:260–266.
12. Esmaeili B, Dubois C, Carreau PJ, Heuzey MC. In situ polymerization of PET in the presence of pristine and organo-modified clays. Int Polym Process 2013;28:331–340.
13. Lee SS, Ma YT, Rhee HW, Kim J. Exfoliation of layered silicate facilitated by ring-opening reaction of cyclic oligomers in PET-clay nanocomposites. Polymer 2005;46:2201–2210.
14. Kim KH, Huh J, Jo WH. Synthesis of thermally stable organosilicate for exfoliated poly(ethylene terephthalate) nanocomposite with superior tensile properties. Macromol Res 2007;15:178–184.
15. Mittal V. *Thermally Stable and Flame Retardant Polymer Nanocomposites*. Cambridge University Press; 2011.
16. Utracki LA. *Clay-Containing Polymeric Nanocomposites*. Smithers Rapra Technology; 2004.
17. Fornes TD, Yoon PJ, Paul DR. Polymer matrix degradation and color formation in melt processed nylon 6/clay nanocomposites. Polymer 2003;44:7545–7556.
18. Shah RK, Paul DR. Organoclay degradation in melt processed polyethylene nanocomposites. Polymer 2006;47:4075–4084.
19. Yoon PJ, Hunter DL, Paul DR. Polycarbonate nanocomposites: Part 2. Degradation and color formation. Polymer 2003;44: 5341–5354.
20. Scaffaro R, Botta L, Ceraulo M, La Mantia FP. Effect of kind and content of organo-modified clay on properties of PET nanocomposites. J Appl Polym Sci 2011;122:384–392.
21. Hwang SY, Lee WD, Lim JS, Park KH, Im SS. Dispersibility of clay and crystallization kinetics for in situ polymerized PET/pristine and modified montmorillonite nanocomposites. J Polym Sci B 2008;46:1022–1035.
22. Soon KH, Harkin-Jones E, Rajeev RS, Menary G, McNally T, Martin PJ, Armstrong C. Characterisation of melt-processed poly(ethylene terephthalate)/synthetic mica nanocomposite sheet and its biaxial deformation behaviour. Polym Int 2009;58:1134–1141.

23. Ghanbari A, Heuzey MC, Carreau PJ, Ton-That MT. Morphological and rheological properties of PET/clay nanocomposites. Rheol Acta 2013;52:59–74.
24. Davis RD, Galman JW, Sutto TW, Callahan JH, Trulove PC, De Long H. Improved thermal stability of organically modified layered silicates. Clays Clay Miner 2004;52:171–179.
25. Ngo HL, LeCompte K, Hargens L, McEwen AB. Thermal properties of imidazolium ionic liquids. Thermochim Acta 2000;357:97–102.
26. Cui L, Dirnitri MKB, Christopher WBB, Hunter DL, Yoon PJ, Paul DR. Effect of organoclay purity and degradation on nanocomposite performance, Part 1: Surfactant degradation. Polymer 2008;49:3751–3761.
27. Stoeffler K, Lafleur PG, DenaUlt J. Effect of intercalating agents on clay dispersion and thermal properties in polyethylene/montmorillonite nanocomposites. Polym Eng Sci 2008;48:1449–1466.
28. Ghasemi H, Carreau PJ, Kamal MR, Uribe-Calderon J. Preparation and characterization of PET/clay nanocomposites by melt compounding. Polym Eng Sci 2011;51:1178–1187.
29. Stoeffler K, Lafleur PG, Denault J. Thermal decomposition of various alkyl onium organoclays: Effect on polyethylene terephthalate nanocomposites' properties. Polym Degrad Stab 2008;93:1332–1350.
30. Costache MC, Heidecker MJ, Manias E, Wilkie CA. Preparation and characterization of poly(ethylene terephthalate)/clay nanocomposites by melt blending using thermally stable surfactants. Polym Adv Technol 2006;17:764–771.
31. Monemian SA, Goodarzi V, Zahedi P, Angaji MT. PET/imidazolium-based OMMT nanocomposites via in situ polymerization: Morphological, thermal, and nonisothermal crystallization studies. Adv Polym Technol 2007;26:247–257.
32. Bishop C. *Roll-to-Roll Vacuum Deposition of Barrier Coatings*. Wiley-Scrivener; 2011.
33. Maul P. Barrier enhancement using additives. Fillers, Pigments and Additives for Plastics in Packaging Applications Pira International Conference. Brussels, Belgium, 2005.
34. Alger M. *Polymer Science Dictionary*. Springer; 1996.
35. Manas C, Roy SK. *Plastics Technology Handbook*. CRC Press; 2006.
36. Siracusa V. Food packaging permeability behaviour: A report. Int J Polym Sci 2012;2012:1.
37. Stanislav ES, Goldman AY. *Mass Transport & Reactive Barriers in Packaging: Theory, Applications, & Design*. DEStech Publications; 2007.
38. Perkins W. Effect of molecular weight and annealing temperature on the oxygen barrier properties of oriented PET film. Polym Bull 1988;19:397.
39. Ghasemi H, Carreau PJ, Kamal MR, Tabatabaei SH. Properties of PET/clay nanocomposite films. Polym Eng Sci 2012;52:420–430.
40. Donadi S, Modesti M, Lorenzetti A, Besco S. PET/PA nanocomposite blends with improved gas barrier properties: Effect of processing conditions. J Appl Polym Sci 2011;122:3290–3297.
41. Ozen I, Bozoklu G, Dalgicdir C, Yucel O, Unsal E, Cakmak M, Menceloglu YZ. Improvement in gas permeability of biaxially stretched PET films blended with high barrier polymers: The role of chemistry and processing conditions. Euro Polym J 2009;46:226–237.
42. Yeo JH, Lee CH, Park CS, Lee KJ, Nam JD, Kim SW. Rheological, morphological, mechanical, and barrier properties of PP/EVOH blends. Adv Polym Technol 2001;20:191–201.
43. Hannay F. *Rigid Plastics Packaging - Materials, Processes and Applications*. Smithers Rapra Technology; 2002.
44. Duncan TV. Applications of nanotechnology in food packaging and food safety: Barrier materials, antimicrobials and sensors. J Colloid Interface Sci 2011;363:1–24.
45. Hopewell J, Dvorak R, Kosior E. Plastics recycling: Challenges and opportunities. Philos Trans R Soc B 2009;364:2115–2126.
46. Alix S, Follain N, Tenn N, Alexandre B, Bourbigot S, Soulestin J, Marais S. Effect of highly exfoliated and oriented organoclays on the barrier properties of polyamide 6 based nanocomposites. J Phys Chem C 2012;116:4937–4947.
47. Dadfar SMA, Alemzadeh I, Dadfar SMR, Vosoughi M. Studies on the oxygen barrier and mechanical properties of low density polyethylene/organoclay nanocomposite films in the presence of ethylene vinyl acetate copolymer as a new type of compatibilizer. Mater Des 2011;32:1806–1813.
48. Beall GW, Powell CE. *Fundamentals of Polymer-Clay Nanocomposites*. Cambridge University Press; 2011.
49. Hong SI, Rhim JW. Preparation and properties of melt-intercalated linear low density polyethylene/clay nanocomposite films prepared by blow extrusion. Lwt-Food Sci Technol 2012;48:43–51.
50. Okamoto M. *Polymer/Layered Silicate Nanocomposites*. Smithers Rapra Press; 2003.
51. Svagan AJ, Akesson A, Cardenas M, Bulut S, Knudsen JC, Risbo J, Plackett D. Transparent films based on PLA and montmorillonite with tunable oxygen barrier properties. Biomacromolecules 2012;13:397–405.
52. Picard E, Vermogen A, Gerard JF, Espuche E. Barrier properties of nylon 6-montmorillonite nanocomposite membranes prepared by melt blending: Influence of the clay content and dispersion state - consequences on modelling. J Membr Sci 2007;292:133–144.
53. Bharadwaj RK. Modeling the barrier properties of polymer-layered silicate nanocomposites. Macromolecules 2001;34:9189–9192.
54. Cussler EL, Hughes SE, Ward Iii WJ, Aris R. Barrier membranes. J Membr Sci 1988;38:161–174.
55. Fredrickson GH, Bicerano J. Barrier properties of oriented disk composites. J Chem Phys 1999;110:2181–2188.
56. Gusev AA, Lusti HR. Rational design of nanocomposites for barrier applications. Adv Mater 2001;13:1641–1643.
57. Lape NK, Nuxoll EE, Cussler EL. Polydisperse flakes in barrier films. J Membr Sci 2004;236:29–37.
58. Nielsen LE. Models for the permeability of filled polymer systems. J Macromol Sci 1967;A1:929–942.
59. Russo GM, Simon GP, Incarnato L. Correlation between rheological, mechanical, and barrier properties in new copolyamide-based nanocomposite films. Macromolecules 2006;39:3855–3864.
60. Lan T, Cho J, Liang Y, Qian J, and Maul P. Applications of Nanomer in nanocomposites: from concept to reality. Proceedings of the nanocomposites. Chicago, IL, USA, 2001.

61. Meng X, Wang Z, Zhao Z, Du X, Bi W, Tang T. Morphology evolutions of organically modified montmorillonite/polyamide 12 nanocomposites. Polymer 2007;48:2508–2519.
62. Swain SK, Isayev AI. PA6/clay nanocomposites by continuous sonication process. J Appl Polym Sci 2009;114:2378–2387.
63. Shafiee M, Ramazani SAA, Danaei M. Investigation of the gas barrier properties of PP/clay nanocomposite films with EVA as a compatibiliser prepared by the melt intercalation method. Polym-Plast Technol Eng 2010;49:991–995.
64. Arunvisut S, Phummanee S, Somwangthanaroj A. Effect of clay on mechanical and gas barrier properties of blown film LDPE/clay nanocomposites. J Appl Polym Sci 2007;106:2210–2217.
65. Ghasemi H, Carreau PJ, Kamal MR, Chapleau N. Effect of Processing Conditions on Properties of PET/Clay Nanocomposite Films. Int Polym Process 2011;26:219–228.
66. Frounchi M, Dourbash A. Oxygen barrier properties of poly(ethylene terephthalate) nanocomposite films. Macromol Mater Eng 2009;294:68–74.
67. Rajeev RS, Harkin-Jones E, Soon K, McNally T, Menary G, Armstrong CG, Martin PJ. Studies on the effect of equi-biaxial stretching on the exfoliation of nanoclays in polyethylene terephthalate. Eur Polym J 2009;45:332–340.
68. Soon K, Harkin-Jones E, Rajeev RS, Menary G, Martin PJ, Armstrong CG. Morphology, barrier, and mechanical properties of biaxially deformed poly(ethylene terephthalate)-mica nanocomposites. Polym Eng Sci 2012;52:532–548.
69. Sanchez-Solis A, Garcia-Rejon A, Manero O. Production of nanocomposites of PET-montmorillonite clay by an extrusion process. Macromol Symp 2003;192:281–292.
70. Chisholm BJ, Moore RB, Barber G, Khouri F, Hempstead A, Larsen M, Olson E, Kelley J, Balch G, Caraher J. Nanocomposites derived from sulfonated poly(butylene terephthalate). Macromolecules 2002;35:5508–5516.
71. Ghanbari A, Heuzey MC, Carreau PJ, Ton-That MT. Morphology and properties of polymer/organoclay nanocomposites based on poly(ethylene terephthalate) and sulfopolyester blends. Polym Int 2013;62:439–448.
72. Xu XF, Ghanbari A, Leelapornpisit W, Heuzey MC, Carreau P. Effect of ionomer on barrier and mechanical properties of PET/organoclay nanocomposites prepared by melt compounding. Int Poly Process 2011;26:444–455.
73. Goitisolo I, Eguiazabal JI, Nazabal J. Stiffening of poly(ethylene terephthalate) by means of polyamide 6 nanocomposite fibers produced during processing. Compos Sci Technol 2010;70:873–878.
74. Ghanbari A, Heuzey MC, Carreau PJ, Ton-That MT. A novel approach to control thermal degradation of PET/organoclay nanocomposites and improve clay exfoliation. Polymer 2013;54:1361–1369.
75. Aubry T, Razafinimaro T, Mederic P. Rheological investigation of the melt state elastic and yield properties of a polyamide-12 layered silicate nanocomposite. J Rheol 2005;49:425–440.
76. Eslami H, Grmela M, Bousmina M. Linear and nonlinear rheology of polymer/layered silicate nanocomposites. J Rheol 2010;54:539–562.
77. Nazockdast E, Nazockdast H, Goharpey F. Linear and nonlinear melt-state viscoelastic properties of polypropylene/organoclay nanocomposites. Polym Eng Sci 2008;48:1240–1249.
78. Ray SS, Bousmina M. Poly(butylene sucinate-co-adipate)/montmorillonite nanocomposites: effect of organic modifier miscibility on structure, properties, and viscoelasticity. Polymer 2005;46:12430–12439.
79. Solomon MJ, Almusallam AS, Seefeldt KF, Somwangthanaroj A, Varadan P. Rheology of polypropylene/clay hybrid materials. Macromolecules 2001;34:1864–1872.
80. Vermant J, Ceccia S, Dolgovskij MK, Maffettone PL, Macosko CW. Quantifying dispersion of layered nanocomposites via melt rheology. J Rheol 2007;51:429–450.
81. Najafi N, Heuzey MC, Carreau PJ. Polylactide (PLA)-clay nanocomposites prepared by melt compounding in the presence of a chain extender. Compos Sci Technol 2012;72:608–615.
82. Meng QK, Heuzey MC, Carreau PJ. Effects of a multifunctional polymeric chain extender on the properties of polylactide and polylactide/clay nanocomposites. Int Polym Process 2012;27:505–516.
83. Yourdkhani M, Mousavand T, Chapleau N, Hubert P. Thermal, oxygen barrier and mechanical properties of polylactide-organoclay nanocomposites. Compos Sci Technol 2013;82:47–53.

22

FEATURES ON THE DEVELOPMENT AND STABILITY OF PHASE MORPHOLOGY IN COMPLEX MULTICOMPONENT POLYMERIC SYSTEMS: MAIN FOCUS ON PROCESSING ASPECTS

CHAREF HARRATS[1], MARIA-BEATRICE COLTELLI[2], AND GABRIEL GROENINCKX[3]

[1]*Laboratoire de Chimie Appliquée (LAC), DGRSDT, Institut des Sciences et Technologie, Ctr Univ Belhadj Bouchaib, Ain Temouchent, Algeria*
[2]*Department of Civil and Industrial Engineering, University of Pisa, Pisa, Italy*
[3]*Department of Chemistry, Division of Molecular and Nanomaterials, Laboratory of Macromolecular Structure Chemistry, Katholieke Universiteit Leuven, Heverlee, Belgium*

22.1 INTRODUCTION

Polymer blending is an economical strategy to design new materials from existing polymers as old as the polymer industry itself. After the great progress recorded in monomer synthesis and new polymer or copolymer design, the innovation was slowed down by the cost criteria. Designing new random, block, or graft copolymers is more expensive than developing novel polymer mixtures. Indeed, driven by tough competition, the development of new materials via blending of existing homopolymers and copolymers was initiated during the 1970s. The major advantages of blending can be summarized as follows:

- Opportunity to develop new properties or improve the existing ones to meet specific customer needs,
- Significant material cost reduction without substantial loss in properties,
- Improvement of the processability of some interesting polymers that are difficult to shape as neat,
- Ability of recycling two, three, or more polymers of different nature.

Among the crucial sides of blends development, one finds intensive research activity focusing mainly on the phase morphology generated during blending. A huge volume of literature, which we have avoided listing in this section, deals exclusively with the topic. This is a clear indication that rules are hard to standardize because many processing and formulation parameters are involved in blending.

Note that the development of phase morphology in polymer blends continues to be an important up-to-date research topic in many laboratories in universities or industries. The underlying challenge is to reduce the material development cost by keeping acceptable performances. Sometimes new approaches are explored to reach that objective.

Research developed for industry is quite different from that carried out for academic purposes of understanding the relationship between various parameters in blending. For example, as long as compatibilization of uncompatible polymer blends is concerned, the industry prefers the generation of the copolymer in a one-shot process during the blending of all the compounds using continuous machines as twin screw extruders (TSEs). The motivation is an active search for a substantial cost reduction. In contrast, academic laboratories adopt more controlled processing stages, where researchers worry about the molecular structure of the compatibilizer to be used. Thus, pre-forming copolymers and then adding them as "physical entities" to the blend partners is the method frequently used to develop and investigate polymer blends.

Polymer Morphology: Principles, Characterization, and Processing, First Edition. Edited by Qipeng Guo.

In this chapter, we choose to highlight the reports where attempts are made to identify acceptable rules on the effect of processing parameters on the phase morphology obtained. The effects of the composition of the blend, the processing shear rate, the processing mixing time, and the surface segregation on the generated phase morphology are discussed. In addition, the chemistry involved in the blending operation in terms of compatibilizer formation, catalysis, polymers bearing reactive groups, functionalized polymers, addition of reactive species to the blends are all discussed in relation to the morphology developed.

The literature reports on the case of blends composed of three components is also summarized and commented on. The stability of the phase morphology, which is critical for industrial application, is sufficiently debated. Furthermore, recent development on the new strategy of compatibilization using organoclays as interfacial agents is reviewed and the reported results highlighted.

22.2 PHASE MORPHOLOGY DEVELOPMENT IN POLYMER BLENDS

Because of thermodynamic criteria, the majority of existing homopolymers form immiscible mixtures constituted of two or more phases. Some couples form complex mixtures exhibiting partial and conditional miscibility. They display lower critical solution temperature (LCST) or upper critical solution temperature (UCST). Indeed, depending on temperature and composition (i.e., position in the phase diagram), the blend can be miscible or immiscible. In monophase blends, the final macroscopic properties are most often intermediate between those of the corresponding parent homopolymers. In contrast, additive properties of immiscible blends can only be obtained if a good level of interfacial adhesion, and appropriate particle size, shape, and distribution of the dispersed phase are reached via adequate compatibilization routes.

For a judicious control of the macroscopic properties of polymer blends, phase morphology constitutes a key parameter for many specific applications. The blending process of immiscible polymers in the melt state results in a heterogeneous morphology that is characterized by the shape, the size, and the distribution of the component phases. Depending on the composition, the homopolymer characteristics and the processing conditions used to mix them, two main types of morphologies are obtained, a dispersed type (a particle can be of any shape: rod, platelet, flacks, disc, sphere, etc.) or a co-continuous one.

22.2.1 Droplet-in-Matrix (Dispersed) Phase Morphology

A droplet-in-matrix phase morphology developed in immiscible polymer blends depends on the viscoelastic properties and composition of the two components of the blend in the melt state. The rheological formalism used for the non-Newtonian phases as polymer melts follows, with adjustment of the viscous properties, the analysis of the mechanisms of a Newtonian drop in a Newtonian liquid [1].

For readers who desire to have more details on the fundamentals of the process and the mechanism of dispersed phase establishments, we believe that the extensive works of Taylor [2, 3], Tomotika [4], Rayleigh [5], Van Oene [6], Elmondorp [7], Elemans [8], and Grace [9] on this subject provide a more complete investigation.

22.2.2 Co-continuous Phase Morphology

Contrary to the droplet-in-matrix, the mechanism and the control of the co-continuous phase morphology, where the two phases are continuous and interconnected throughout the whole volume of the blend, is still not well elucidated. The complexity arises mainly from the ambiguous effect of the viscoelastic characteristics of the components, their composition in the blends, and the magnitude of their interfacial tension. Several empirical expressions have been proposed so far to predict either the phase inversion or the conditions for which co-continuous morphology is generated.

In reference [10] are summarized the most important empirical models reported in literature to predict the phase inversion; that is, the critical composition at which the phases A and B of a binary blend could become co-continuous in A/B uncompatible blend. Sometimes phase inversion covers a window of compositions (with onset and offset limits) rather than a transient point. When a compatibilizer is added or formed *in situ* by suitable chemical reaction, the situation becomes much more complex and hardly obeying any model as the copolymer changes the interfacial tension and affects the viscoelastic behavior of the blend.

Note that co-continuous phase morphology can also be generated from systems exhibiting temperature- and composition-dependent miscibility and UCST and LCST phase diagrams. In the former, when rapidly decreasing the temperature from above the binodal (homogeneous region) through the metastable into the spinodal region, an initially homogeneous blend evolves through spinodal decomposition to a co-continuous and finely interlocked two-phase state. The system must be frozen in before reaching the last stages of the phase separation, where coalescence may alter the co-continuous character of the morphology. The same result is expected in the LCST diagram when the temperature is rapidly increased from below the binodal through the metastable into the spinodal region.

The rate of thermal treatment is thus a crucial parameter in controlling the morphology. A slow variation of temperature through the metastable region (enclosed by the binodal and the spinodal) favors a nucleation and growth process of phase separation. This mechanism results in dispersed-type morphology.

The kinetics of phase separation through spinodal decomposition is well documented. The most important ones are summarized elsewhere [11]. Indeed, in the limit of metastability of a homogeneous system, infinitesimal composition fluctuations start to grow and initiate a diffusion process that controls the phase separation.

In addition to the determination of the co-continuity diagram, that is, the determination of the composition window at which the blend exhibits phase co-continuity, the structural stability of the generated morphology upon post-annealing a formed blend at high temperatures under quiescent conditions or under shear during melting has been an important research topic during the past decade.

22.2.3 Phase Morphology in Ternary Blends

We decided to rank apart the phase morphology in ternary immiscible blends because it can be droplet-in-matrix, co-continuous, or a mixture of both and, in many situations, an encapsulated droplet-in-matrix structure.

The phase morphology of immiscible ternary polymer blends was the object of a review of Shokohooi et al. [12]. According to the generalized Harkins equation, in a ternary A/B/C blend, the spreading coefficient, λ_{CB}, is defined as the parameter showing the tendency of component C to encapsulate component B in a matrix of component A and is related to the interfacial tension of the components in the following manner:

$$\lambda_{CB} = \gamma_{BA} - \gamma_{CA} - \gamma_{BC} \quad (22.1)$$

where γ represents the interfacial tension for various polymer pairs and subindexes A, B, and C refer to each component.

In this ternary system, the tendency of B to encapsulate C can be considered by writing the spreading coefficient $\lambda_{BC} = \gamma_{CA} - \gamma_{BA} - \gamma_{BC}$. If λ_{CB} is positive and λ_{BC} is negative, the encapsulation of B with a C layer prevails (Fig. 22.1a). On the other hand, if λ_{BC} is positive and λ_{CB} is negative, the encapsulation of C with a B layer takes place (Fig. 22.1b). If both are negative, they will be separately dispersed as droplets in matrix A (Fig. 22.1c).

For a full understanding of the general behavior of the ternary system, the interfacial tension between matrix A and the C and the B phases should be considered, too. In particular, if the two interfacial tensions are similar, the morphology of Figure 22.1c can be predicted. If the interfacial tension between A and C is lower than the interfacial tension between A and B but the λ_{CB} and λ_{BC} are both negative, a stacked morphology, where the C droplet is adherent to the B droplet, is generated.

Recently, Rastin et al. [13] published a very interesting study on the reliability of the existing theoretical models predicting the phase morphology in ternary polymer blends by melt compounding a model blend comprising HDPE, PA6, and EVOH components. The spreading coefficient (SC), the relative interfacial energy (RIE), and the dynamic interfacial energy (DIE) conceptual models were employed to anticipate the type of phase morphology developed in the assigned ternary systems. Due to some assumptions involved, the original DIE model was unable to predict the type of morphology when HDPE is the matrix. The authors extended the DIE model by revising the dynamic interfacial tension concept. Also, a dimensionless parameter was simply defined to estimate the core–shell size in terms of viscosity and elasticity ratios over the whole composition range in the composite droplet. A more precise monitoring of particle size alteration, following the qualitative analysis of scanning electron microscopy (SEM) micrographs after treatment with suitable solvents, provided a clear insight into the development of microstructure in HDPE/PA6/EVOH ternary blends.

Either SC or RIE models were successful in anticipating transition from encapsulated to individually dispersed structures. It was also postulated that the ratio of average viscosity or elasticity of the core–shell particle to that of the matrix simply correlates with the size of the composite droplets. For ratios greater than unity, a direct relationship between mentioned parameters exists, whereas an inverse trend was observed when elasticity or viscosity ratio takes values less than unity.

A ternary polymer blend consisting of high-density polyethylene (HDPE), polystyrene (PS), and poly(methyl methacrylate) (PMMA) was selected as a model system, demonstrating complete wetting and four subcategories of morphologies were identified including: (i) matrix/core/shell dispersed phase; (ii) tri-continuous; (iii) matrix/two separate dispersed phases, and (iv) bi-continuous/dispersed phase morphologies [14]. Combination of electron microscopy, selective solvent extraction, and a technique based on focused

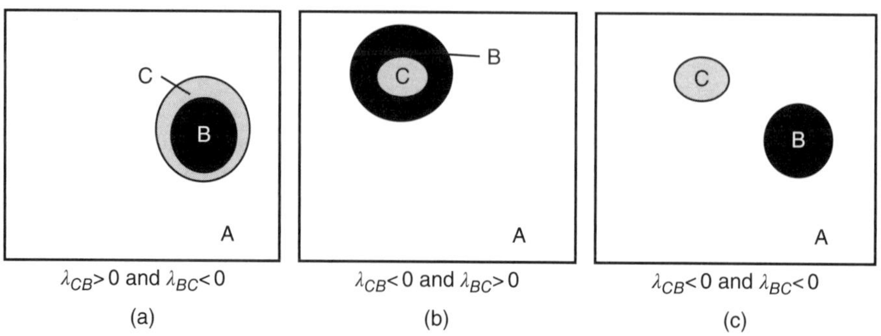

Figure 22.1 Some of the most common morphologies observed in ternary blends having A as matrix. (a) $\lambda_{CB} > 0$ and $\lambda_{BC} < 0$; (b) $\lambda_{CB} < 0$ and $\lambda_{BC} > 0$; (c) $\lambda_{CB} < 0$ and $\lambda_{BC} < 0$.

ion beam irradiation and atomic force microscopy were used to identify and quantify the various phases. Interesting triangular compositional diagrams were used to distinguish the various morphological regions (Fig. 22.2).

According to the authors, all ternary blends with complete wetting would result in similar morphological diagrams. The type of phase morphology was dependent on the composition of the phases and the spreading coefficients between pairs of them. The selected viscosities of the phases used do not affect the developed phase morphologies. HDPE/PS/PMMA of a fixed composition but varying PMMA viscosities (H-PMMA grade was 200 times more viscous than L-PMMA) led to the same morphological structures, where the PS always separates the HDPE and the PMMA phases (HDPE and PS form a co-continuous structure and PMMA droplets were dispersed preferentially in the PS phase (Fig. 22.3).

The recent studies of Ravati and Favis [15, 16] on ternary blends of biodegradable polymers showed that selecting a ternary blend from the four biodegradable polyesters (polybutylene terephthalate: PBT, polycaprolactone: PCL, polybutylene adipate-co-terephthalate: PBAT, and polylactide: PLA), lead to three very different and stable morphological wetting situations: partial wetting in PLA/(polybutylene succinate: PBS)/PCL blends in which the droplets of the minor phase (PLA or PBS) are located along the PCL/PBS or PCL/PLA interface; complete wetting in PBS/PLA/PBAT blends in which a tri-continuous phase morphology was observed; and combined partial-complete wetting in PBS/PBAT/PCL blends in which PBS(or PBAT) is located both at the interface as droplets and in the PBAT (or PBS) phase. This plurality of structures is possible because the interfacial tensions between the phases are very low and the spreading

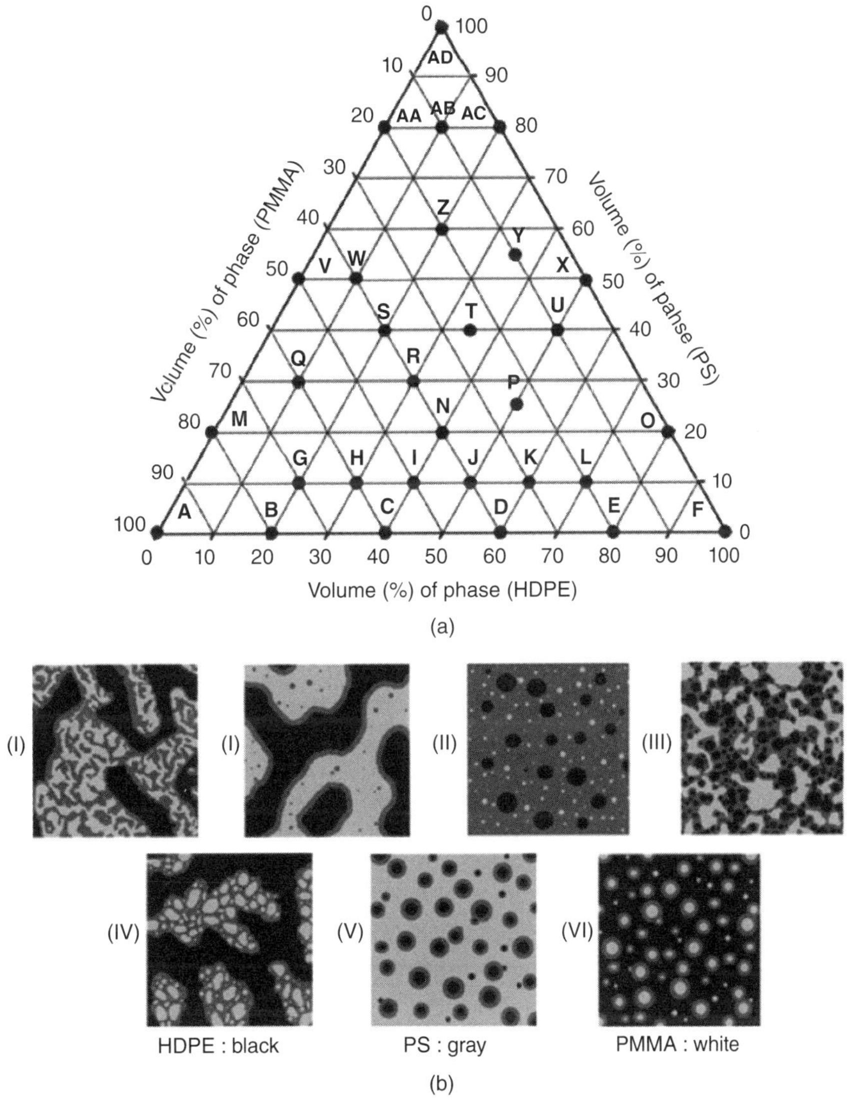

Figure 22.2 (a) Triangular concentration diagram showing the composition of the various ternary HDPE/PS/PMMA blends examined in this study, (b) various morphological states for ternary HDPE/PS/PMMA, and (c) triangular concentration diagram showing the various regions of the morphological states for ternary HDPE/PS/PMMA. Ravati and Favis [14]. Reproduced with permission of Elsevier.

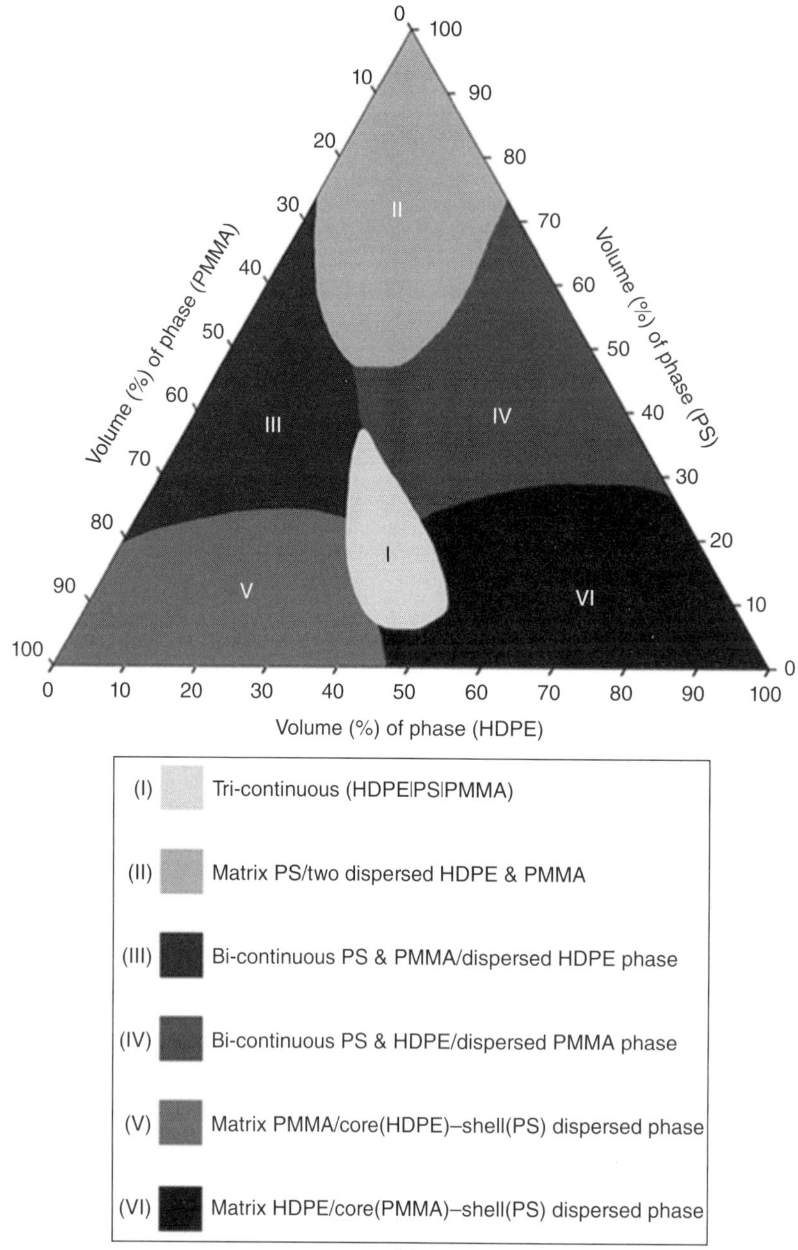

(c)

Figure 22.2 (*continued*)

coefficients are close to zero. Simply replacing one component with another can change the sign of the spreading coefficient and results in different wetting behavior. The observed phase morphologies were in agreement with Harkin's theory.

Wilkinson et al. [17] studied the development of phase morphology using a model of ternary blend comprising 70 wt% of polypropylene (PP) and 30 wt% of a dispersed phase containing a mixture of 15 wt% of polyamide 6 and 15 wt% of varying ratios (unR:R) of unreactive poly[styrene-*b*-(ethylene-co-butylene)-*b*-styrene] (SEBS) triblock copolymer and a reactive maleic anhydride-grafted SEBS-*g*-MA. In the absence of reactive SEBS-*g*-MA, that is, unR:R = 100:0, the phase morphology consists of individual and separated SEBS and PA6 droplets dispersed in PP matrix. However, when progressively replacing parts of SEBS with reactive SEBS-*g*-MA, it was possible to produce acorn-type composite droplets (unR:R = 75:25), then individual encapsulated core–shell structures (unR:R = 50:50) and finally to large agglomerates of core–shell structures at unR:R = (25:75 and 0:100), that is, in the presence of only reactive SEBS-*g*-MA copolymer. The formation *in situ* of SEBS-*g*-PA6 resulting from the reaction of PA6 and SEBS-*g*-MA was responsible for the reduction in the interfacial tension between SEBS and PA6 phases, leading to variation in the spreading coefficient which results in various extent of wetting of PA6 by the SEBS phase.

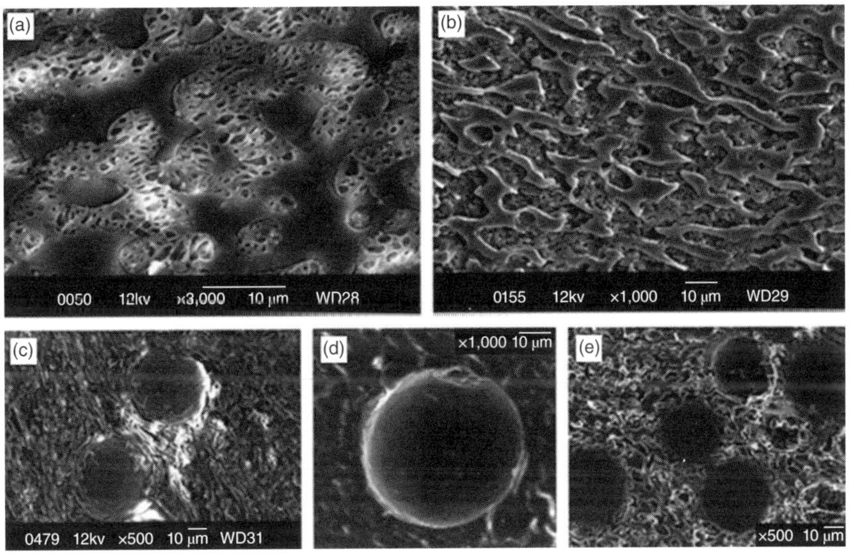

Figure 22.3 SEM micrograph of morphology of (a) 35/40/25 HDPE/PS/L-PMMA after extraction of L-PMMA by acetic acid, (b) 35/40/25 HDPE/PS/L-PMMA after extraction of PS by cyclohexane, (c) 35/40/25 HDPE/PS/H-PMMA (cryofracture), (d) 35/40/25 HDPE/PS/H-PMMA after extraction of PMMA by acetic acid, and (e) 35/40/25 HDPE/PS/H-PMMA after extraction of PS by cyclohexane. Ravati and Favis [14]. Reproduced with permission of Elsevier.

22.3 MELT PROCESSING OF POLYMER BLENDS

22.3.1 Morphology Buildup during Processing

Scientists and engineers dealing with blend processing and compounding were always curious about how an immiscible blend is formed inside mixing machines as extruders and when the phase morphology reaches the equilibrium state (even a dynamic one), although not stable. Of course, this question is legitimus and scientifically interesting as it can help solve or situate judiciously the mechanism of the melt blending of immiscible polymers. Many indirect experiments were performed to gain insight on the mechanism of phase morphology formation in binary immiscible blends. The mixing time effect in a Brabender mixer on PP/PC model immiscible blend has revealed that the most particle deformation takes place within 2 min of mixing. After that, the size reduction was less significant even if the blending operation was allowed to continue for 20 min. [18]

By compounding PS/PE immiscible model blend in a set of industrial mixers, Plochocki et al. [19] attributed the dispersion mechanism of morphology development to the abrasion of the softening polymer pellets against the walls of the mixing tools. But those qualitative observations, although interesting to depict the initial stages of deformation mechanism, were insufficient to give a clear picture on the final phase morphology development and setup.

For the same objective and logics of investigation, Scott and Macoskow [20], by using a batch mixer, could follow the evolution of the phase morphology of an A/B blend from the pellets state to the sub-micron scale. Similarly, Sundararaj et al. explored further the understanding of the morphology development [21–23] on various types of mixers, including a TSE. Binary blend models of PA/PS and PP/PS have been melt processed for that objective. Interestingly they showed that:

- The dispersed phase stretches into sheets and then subsequently breaks up further into smaller domains; either directly or via intermediate formation of filaments.
- Most of the significant morphology setup occurred within 1/(L/D) of the TSE of the first point of melting and within the first minute of mixing in a batch mixer.
- For single screw extruder, sheet breakup may occur over a longer section before droplets appear.
- Once the domains reach the micron size scale, they may be stretched into filaments and are then broken up into droplets.
- The sheet stretching takes place at the high-shear zones, whereas the filaments breakup occurs at low-shear zones via Rayleigh instabilities.

Many reports, either theoretical or experimental, have been published in literature accounting for the understanding and modeling of the phase morphology developments. Sundararaj has summarized elsewhere [24] the most important ones in a concise analysis and review chapter.

Bourry and Favis [25] proposed a model of blending experiment where three distinct situations were tested:

- 1: HDPE and PS were fed together from the hopper of a TSE in the classical procedure of melt-blending pellets of two dissimilar polymers.
- 2: HDPE was fed from the TSE hopper, but PS was fed in the melt state using a side single screw extruder (connected midway of the TSE).

– 3: PS melt is fed further downstream so that its residence time in the TSE is limited to 25 s.

The study revealed that at the three compositions investigated, that is, 10, 30, and 70% PS, the final blend morphology (particle diameter and its distribution) does not depend on the initial stages of the mixing procedure, but is built up mainly when both components are in the melt state. Furthermore, when the kinetics are considered, the phase morphology is developed very fast. For instance, the report showed that a residence time as short as 25 s in the extruder was sufficient to form the final blend morphology.

Two interesting investigations were carried out by Canto et al. [26] and Canto [27] on phase coarsening of dispersed particles of blends of polybutylene terephthalate (PBT)/acrylonitrile butadiene styrene terpolymer (ABS): 60/40 and blends of PP containing 20 wt% of ethylene vinyl acetate (EVA) copolymer, respectively. The co-continuous phase morphology developed in 60PBT/40ABS blends was found to coarsen along the extruder compared to (methyl-methacrylate)-(glycidyl-methacrylate) (MMA-GMA) reactively compatibilized blends, which showed moderate phase coarsening. The authors reported that through their flow inside the die at the exit of the extruder where extra shear exists, the two blends exhibit a refining (less coarse and finely structured) of the co-continuous phase morphology.

In the PP/EVA blend system, the effect of mixing time (converted as residence time) on phase coarsening has revealed a substantial linear particle size increase as a function of the mean residence time. As illustrated in Figure 22.4, the phase morphology is set up very fast (within a minute) and starts evolving by a more dominant coalescence process as the residence time is increased. The particle size is doubled after 2 min 30 s of residence time in the extruder. The pronounced coarsening effect is ascribed to a decrease in the viscosity of the dispersed phase; a low viscosity favors flow of particles, their subsequent collision, and merging.

Huang et al. [28] described a technique of sampling along a single screw extruder to spontaneously monitor the phase morphology developed in the blending process of a model blend of polyamide dispersed in polypropylene matrix. By selecting two blend combinations having viscosity ratios differing by a factor of 10 (0.1 and 10) because of the different PA6 molecular weight, it was shown that for the lowest viscosity ratio the phase morphology develops during the softening stage, whereas, for the other blend it continues to develop slowly along a larger part of the extruder. As illustrated in Figures 22.5 and 22.6, corresponding to low- and high-viscosity ratio, the particles are almost in their equilibrium morphology at almost halfway of the extruder for the first blend. In contrast, more time (or screw space; almost ¾ of the barrel length) is necessary for the blend of higher viscosity ratio to reach a similar phase morphology.

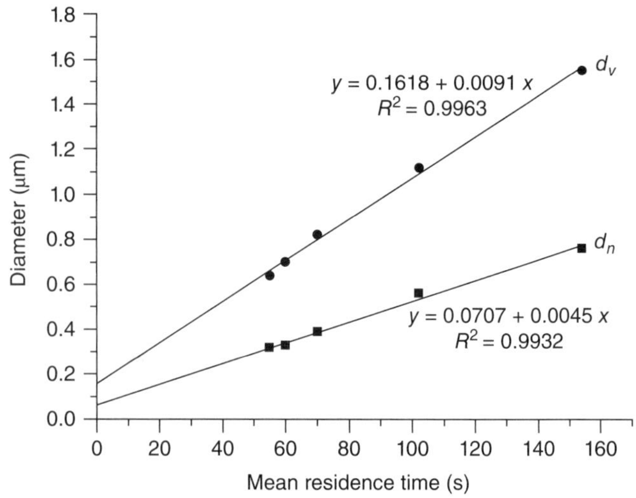

Figure 22.4 Number-average (dn) and volume-average (dv) diameters of EVA droplets in the PP/EVA 80/20 blends compounded in a Co-TSE as a function of the mean residence time (t), showing the coarsening kinetics during compounding in the Co-TSE. Canto [27]. Reproduced with permission of Elsevier.

22.3.2 Effects of Processing Parameters on Phase Morphology

22.3.2.1 Composition of the Blend The general tendency in uncompatiblized blends is that the droplet size increases as the composition increases up to the onset of co-continuity of the two phases. The driven mechanism accepted is that beyond the concentration of dilution, and particle–particle coalescence dominates the particle breakup process.

22.3.2.2 Effect of Shear Rate As one would *a priori* expect, the droplet size should decrease as the shear rate is increased, as *high shear implies high deformation*. Many researchers carried out experiments to verify this trend; but unfortunately they were not fully convincing as only a moderate size decrease was reported upon, increasing the shear rate [18, 19, 29]. Ghodgaokar and Sundararaj [30] found that the drop size exhibits a minimum at an intermediate shear rate. The observed trend can be due to the simultaneous effect the shear rate plays on enhancing the coalescence of particles, leading also to a size coarsening.

Few authors have investigated the role of the mixer type on phase morphology. We believe that the mixers, although designed differently, can mainly differ in the various types of flow or differing levels of shear. That means their effect is the same as the effect of the flow type and the shear rate they generate. Indeed, as observed by Sundararaj [24], no big differences, in terms of phase morphology, were found between an internal batch mixer [31] and a TSE [32].

22.3.2.3 Effect of Mixing Time As diffusion, flow, and deformation of polymer melts are kinetic processes, the

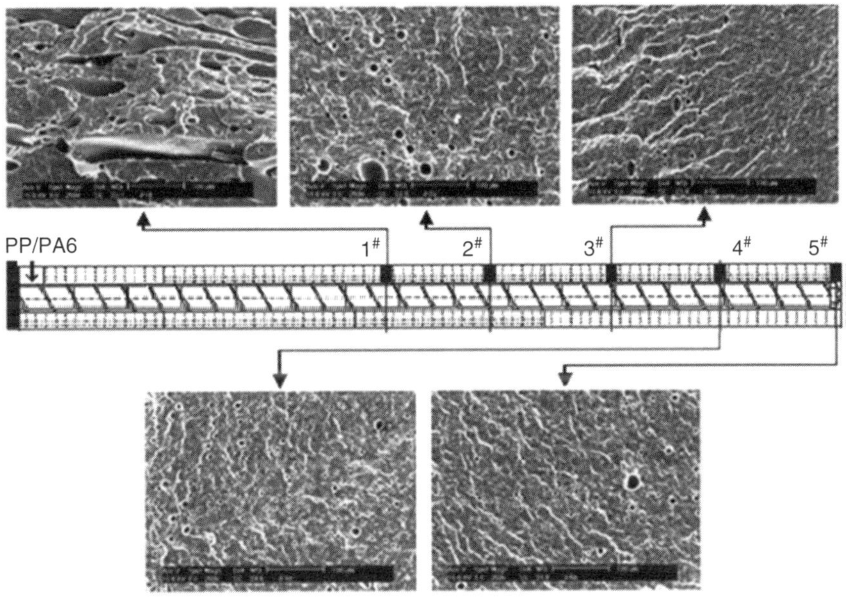

Figure 22.5 Typical SEM micrographs of PP/PA6-1 blend samples collected from five positions.

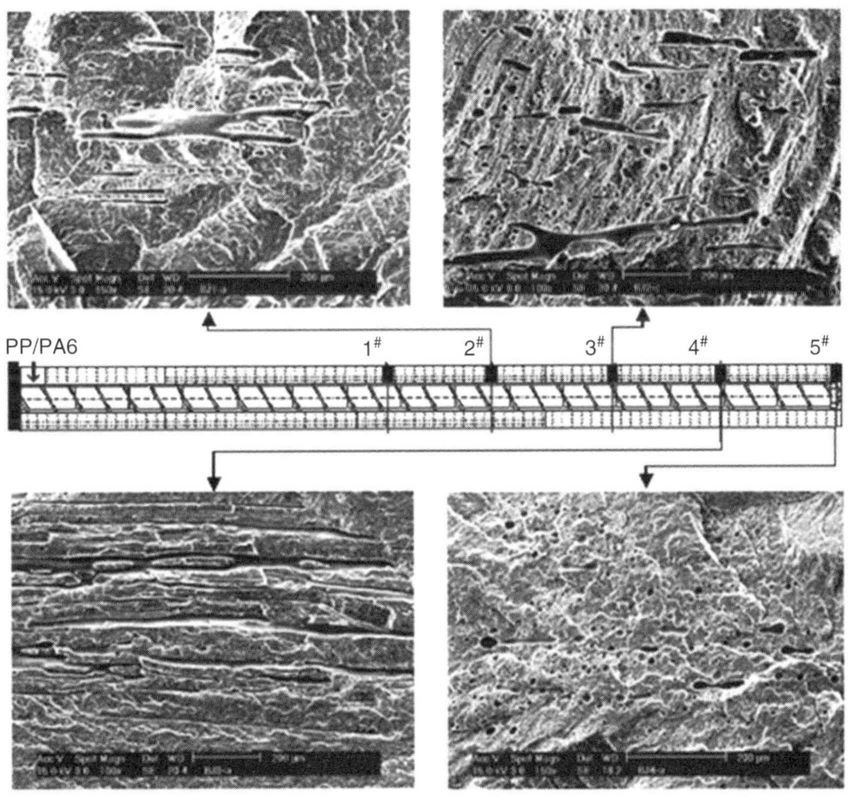

Figure 22.6 Typical SEM micrographs of PP/PA6-2 blend samples collected from four positions.

time of mixing two or more polymers is critical for the formation of phase morphology. As has been discussed, the phase morphology is developed during the first minute of mixing as long as the components of the blend are in the melt state. Maintaining a melt mixture under continuous shearing action can have the adverse effect of phase coarsening due to particle–particle coalescence [33]. As blend components exhibit high viscosity level and low interfacial tension, the size of the formed particles is rapidly reduced from millimeter- to micron-size scale. Once the micro scale, which is the stable droplet size, is reached, additional mixing does not bring any significant evolution to the morphology [34–36].

22.3.2.4 Effect of Surface Segregation on Phase Morphology In immiscible systems, interesting effects of segregation induced by the material used for compression molding of plastic film were reported by Bertoldo et al. [37]. The phase morphology of blends of low-density polyethylene (PE) with low-molecular-weight copolyamide (CPA) was investigated in films having 50- 100-μm thickness. Films were prepared by compression molding between two surfaces with different polarity, namely, Teflon and Aluminum sheets. The energy and the interactions at the surfaces produced nonhomogeneous phase distribution in the thickness direction (Fig. 22.7). The polar component CPA (the dispersed phase) enriches near the film side in contact with the aluminum sheet. An opposite effect is provided by Teflon, which has low surface energy and low interaction with the polar component of the blend. The shape, the dimension, and the distribution of the CPA domains depend on the treatment conditions including temperature, pressure, time, and the kind of contact surfaces used during the compression molding. Interestingly, a spherical shape was observed when two similar surfaces were used (either Teflon or Aluminum), whereas a lamellar shape was produced with asymmetric surfaces such as Teflon and Aluminum; each on either side of the film. The dispersed phase was mostly concentrated in the middle of the film obtained between two Teflon sheets, whereas a homogeneous distribution of the particles was observed in films compression molded between the two Aluminum sheets. A gradient distribution along the section was observed when two different surfaces (Aluminum and Teflon) were used with a higher copolyamide concentration near the Aluminum side. The elliptical copolyamide domains produced by asymmetric interactions evolved with the treatment time to a layered morphology, with the more polar component concentrated at the Aluminum surface as indicated by the increasing peeling strength.

22.4 CHEMISTRY INVOLVED IN POLYMER BLENDS

22.4.1 Effect of the Compatibilizer on Phase Morphology

The addition of a compatibilizer in a blend obviously changes its phase morphology development mainly by reducing the interfacial tension. Fortelny [38] tried to model it considering the decrease in interfacial tension, its effect on the droplet breakup and coalescence, and the associated change in interfacial area.

In the compatibilized system, the interfacial tension can be expressed as

$$\sigma = \sigma_0 - g\frac{Q^3}{27\phi_d^3}D^3 \tag{22.2}$$

where σ_0 is the blend interfacial tension without compatibilizer, g is a function of polymer molecular parameter, and Q is the total number of copolymer molecules on the interface.

The inverse of breakup time (t_b) can be expressed as a function of the difference between the initial capillary number and that of the compatibilized blend ($Ca - Ca_c$). Then the coalescence can be described by the derivative of the droplets number with respect to time:

$$\left(\frac{dn}{dt}\right)_c = -\frac{4}{\pi}\dot{\gamma}\phi_d n P_c \tag{22.3}$$

where P_c is the probability that particle collision is followed by fusion. P_c can be expressed as

$$P_c = \exp\left\{-\frac{9Ca^2D^2}{8h_c^2(1+(3C/p))}\right\} \tag{22.4}$$

where h_c is the critical distance at which the collapse of the matrix film between droplets occurs and C is a function of the mobility of the interface.

In steady-state condition, the following equation is valid:

$$\left(\frac{dn}{dt}\right)_{break-up} + \left(\frac{dn}{dt}\right)_c = 0 \tag{22.5}$$

The substitution of the previous expressions in 3 leads to

$$f\left(\frac{\eta_m\dot{\gamma}D}{\sigma} - Ca_c\right) = \frac{4}{\pi}\dot{\gamma}\phi_d \exp\left\{-\frac{9Ca^2D^2}{8h_c^2(1+(3C/p))}\right\} \tag{22.6}$$

This complex expression may be solved only when simplifying assumptions are made because of interdependence of the variables. Supposing, for example, a constant density of copolymer on the interface, σ, can be considered constant, so that the dependence of D on ϕ_d is the same as for blends without compatibilizer using the right value of σ and P_c.

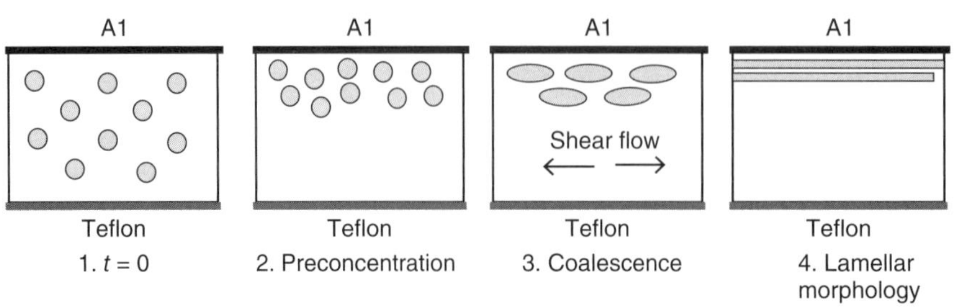

Figure 22.7 Simplified scheme of phase morphology development along the section of films molded between Aluminium (Al) and Teflon sheets. Bertoldo et al. [37]. Reproduced with permission of Elsevier.

22.4.2 Formation *in situ* of the Compatibilizer

The reactive compatibilization of a binary A/B immiscible polymer blend is usually ensured via the use of a chemical reaction during the melt-blending operation. The reaction leads to the formation of block or graft copolymers miscible, or at least sufficiently compatible, with both polymers A and B. Depending on its chemical composition and molecular architecture, the *in situ* formed copolymer is able to locate at the interface, improves the adhesion between the two phases, and constitutes a stabilizing barrier against coalescence (Fig. 22.8).

There are situations where a block or a graft copolymer bearing both A-compatible and B-compatible blocks can be previously synthesized and added to the A/B blend in the melt.

As on a thermodynamic basis, the effect of compatibilization should be the same with respect to a reactively compatibilized system, it was observed in several cases that the diffusion of the compatibilizing copolymer to the interfacial region is quite difficult for kinetic reasons. Because of the high viscosity of the molten polymer medium and the short time of blending usually adopted in industrial conditions, the premade copolymer does not quantitatively diffuse to the A/B interphase region. That is why in industrial processes the compatibilizers are preferentially *in situ* formed during the compounding step of blend development.

Different reactive routes were adopted to generate *in situ* the block copolymers. A criterion of classification of the different methods can be tentatively made by considering the different compatibilizer precursors added to polymer blends: a catalyst, a polymer bearing reactive groups, or a reactive additive. In the following, it is necessary to take into account that, often in industrial formulations, more than one compatibilizer precursor might be used.

22.4.2.1 Catalysts Polymer blends with ester or carbonate functionalities can be compatibilized by favoring catalyzed interchange reactions during the melt-blending process. This approach was described by many authors in blends of polyester/polyester [39–42], polyester/polycarbonate, [43, 44] and polyester with various polymers bearing ester groups [45–48]. Titanium-, zinc-, boron-, or tin-based catalysts with an electrophilic center were added during the melt extrusion of these polyester-based blends. The chemically modified blends exhibit a much improved compatibility compared to uncatalyzed ones. Although largely described in literature, this method suffers from two disadvantages including the relatively longer extrusion time required for the reaction to complete compared to time scales tolerated in industrial processes. Moreover, the presence of a low amount of humidity can result in polyester chain scission, as this side reaction is also catalyzed by the same catalysts [42, 47].

Recently, Liu et al. [49] investigated the effect of various catalysts including zinc borate, titanium pigment, and tetrabutyltitanate on the interchange reactions of PLA and PC. Attempts were made to elucidate the reaction mechanism under a flow field by preparing blends in a discontinuous mixer and adopting a long blending time (1000 s). The presence of PLA-PC copolymer was detected by a method based on selective extraction and spectroscopic characterization. A new T_g intermediate between those of pure PLA and PC was depicted via dynamic mechanical thermal analysis (DMTA). A similar result was claimed by Phuong et al. [27], employing both a trans-esterification catalyst (tetrabutyl ammonium tetraphenyl borate) and triacetin, but for only a short blending time of 60 s. In fact, triacetin favored the interchange reactions between PLA and PC probably because it can merely diffuse in the melt during extrusion, reacting with both components. In addition, a significant reduction of PLA molecular weight was also evidenced as a sign of molecular degradation, but that did not affect the tensile properties of the blend.

22.4.2.2 Polymers Bearing Reactive Groups Because of their spontaneous reactivity, the most frequently used reactive groups are anhydride and epoxide. Commercially, the polymer-bearing reactive groups are available as copolymers, synthesized by copolymerization, but also as functionalized polymers obtained by radical grafting post-modification of commercial polymers [50]. Note that this type of compatibilization approach is efficient if at least one polymer has reactive groups on its structure (in condensation polymers, usually the reactive groups are terminals).

The effect of copolymer-bearing reactive groups on compatibility was intensively investigated in polyamide/polyolefin and in polyolefin/polyester blends because of their industrial

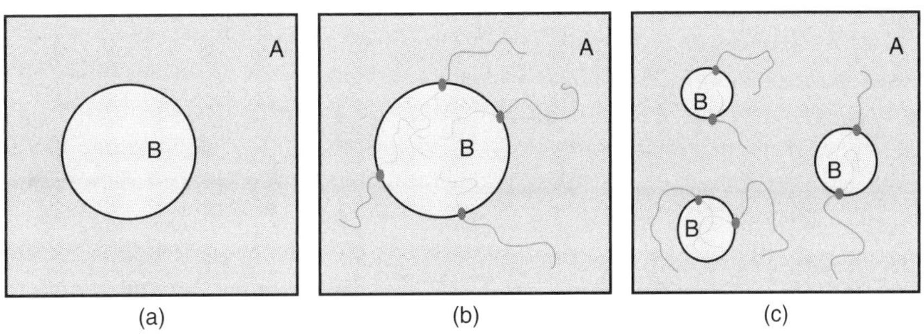

Figure 22.8 Scheme of reactive compatibilization of an A matrix/B dispersed phase immiscible polymer blend in the melt: (a) domain of B in the A matrix at the beginning; (b) formation of the compatibilizer thanks to a chemical reaction in the interphase region; (c) phase morphology rearrangement.

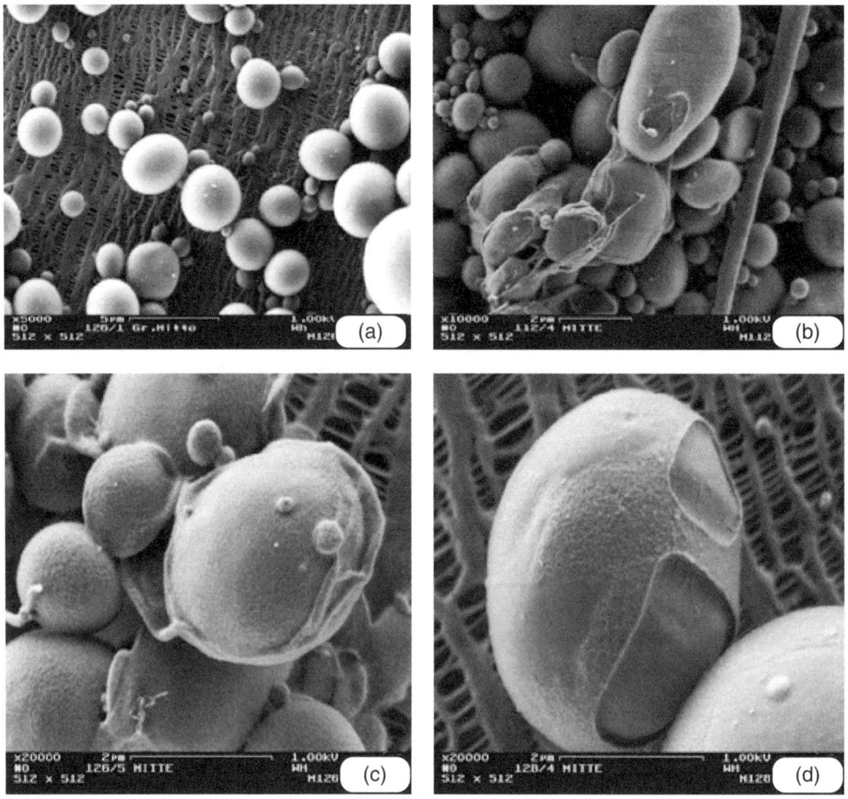

Figure 22.9 Visualization of compatibilizer location in immiscible blends: SEM micrographs of a blend of thermoplastic polyurethane (TPU) with 20 wt% polypropylene compatibilized with different ethylenic co- and terpolymers (blend melt mixed using ZSK 30 extruder). The TPU matrix was selectively dissolved in dimethyl formamide and the remaining PP particles without or with compatibilizer were separated on a membrane. Wallheinke et al. [58]. Reproduced with permission of Elsevier.

interests. In the former case, the addition of styrene-maleic anhydride [51] and poly(ethylene-*co*-vinyl alcohol-*co*-vinyl mercaptoacetate)[52] copolymers was reported to be effective for morphological modulation and enhancement of properties.

Copolymer-bearing epoxide groups such as ethylene-glycidyl methacrylate copolymer (E-GMA) or ethylene-ethyl acrylate-glycidyl methacrylate copolymer (E-EA-GMA) are reported to be very efficient compatibilizer precursors in high-density polyethylene (HDPE)/PET blends because of the high reactivity of the epoxy ring toward both hydroxyl and carboxyl PET terminal groups [53–55]. PET blends whose dispersed phase consists of an ethylene-propylene rubber (EPR) and a copolymer of ethylene and 8 wt% glycidyl methacrylate (E-GMA8) acting as a compatibilizing agent were studied. The epoxide-containing copolymer was more efficient than polymers functionalized with maleic anhydride (MAH) and glycidyl methacrylate (GMA) in attaining appreciable impact properties, probably because of the higher number of reactive groups in the copolymer. The morphological analysis of the blends allowed to make a direct correlation between the impact behavior and the generated phase morphology. Arostegui [56] obtained similar results when studying the mechanical properties of poly(ethylene-octene) copolymer/poly(butylene terephthalate) (PBT)/ethylene-glycidyl methacrylate copolymer (E-GMA) blends of different compositions. Moreover, an improvement of compatibility due to the use of E-EA-GMA copolymers was observed also in PET/ABS blends [57].

In situ-formed copolymer due to the chemical reaction in extrusion compounded blends of polyurethane containing 20 wt% PP modified with different reactive ethylenic co- and terpolymers was located by simple SEM observations. The polypropylene particles were separated on a membrane after dissolving the polyurethane matrix using dimethylformamide [58]. As illustrated in Figure 22.9, the isolated polypropylene particles recovered from reactively modified blends exhibit rough surfaces because they contain traces of copolymer. Note in Figure 22.9b and c that the PP particles are bridged by the compatibilizer. While the particles issued from uncompatibilized blends in Figure 22.9a are very smooth and free of any adherent, most of them are not agglomerated. These observations demonstrate that the copolymer has been formed reactively *in situ* during the melt-blending process of polyurethane and polypropylene in the presence of the reactive ethylenic co- and terpolymer containing acrylic acid and/or butylacrylate reactive groups.

22.4.2.3 Functionalized Polymers Some polyolefins, or olefin-based random or block copolymers can be post-modified by radical grafting in the melt by using an unsaturated functionalizing molecule and a radical initiator, usually a peroxide [59]. Such reactions are usually carried out on an industrial scale. Many interesting functionalized polymers are commercially available. The degree of modification is estimated by determining the functionalization degree (FD) using spectroscopic methods including IR or nuclear magnetic resonance analysis.[60]

The use of maleic-anhydride- or diethyl maleate–functionalized polyolefins was efficient in PA/polyolefin (PO) compatibilization. The formation of PA-PO graft copolymer was demonstrated to occur thanks to the formation of succinic amide rings.[61]The one-step process consisting in the functionalization of the polyolefin during the blending with the PA was successful [62], although some impact of side reaction occurring in the PA phase, as PA grafting, chain scission, and branching were observed [63].

The use of diethyl maleate–functionalized polyolefins with PET in the presence of transesterification catalysts, based on Ti [46] and Zn [47] derivatives, was efficient although a long blending time was necessary (1200 s). Interestingly, diethyl maleate–functionalized styrene-*b*-(ethylene-*co*-1-butene)-*b*-styrene triblock copolymer (SEBS) was able to compatibilize PET/PE blends , thanks to its preferential location in the interphase. In addition, PE and SEBS individually functionalized with DEM using the extrusion process led to two different kinds of phase morphologies. The DEM-grafted SEBS copolymer favored the development of co-continuous phase morphology structures [65].

Maleic anhydride–functionalized polyolefins were much more effective for PET/PE compatibilization [66]. This process was successfully applied in PET/PE recycling [67].

22.4.2.4 Addition of Reactive Moieties

The use of reactive additives was recently attempted, particularly when the polymers in the blend have similar functionalities as in polyester-based blends.

In an A/B polymer blend, the reactive additive can react with both polymers as a coupling agent. The advantages of this approach are the availability of many cheap reactive products and also the capability to diffuse in the melt as they possess a low molecular weight. The latter property makes the reaction kinetics suitable for extrusion. On the other hand, the disadvantage of a specific reactivity of the additives can lead to side reactions as the undesired polymer branching.

For example, Raffa et al. added diisocyanate [68] or molecules containing epoxide groups in PET/functionalized polyolefin blend and reported improved compatibility and enhancement of properties. It was difficult to ascribe the improvement of properties to only the copolymer formation because simultaneously some chain branching affects the viscosity of the blend.

Another example concerns the addition of a radical initiator, such as *2,5-dimethyl-2,5-di(tert-butylperoxy)hexane to PLA/PBAT blends* [70]. The peroxide decomposes during melt blending, abstracting a hydrogen atom from polymer P to induce the formation of a macroradical. Then the coupling of macroradicals of PLA with the macroradicals of PBAT leads to the formation of a copolymer in the interphase region. An interesting trend was observed as a function of peroxide concentration in the 75PLA/25PBAT blend. In fact, the process can be summarized by the following considerations:

- *At concentrations lower than 0.2 wt% of peroxide*: cross-linking of PBAT, slight degradation of PLA; the dispersed PBAT phase diameter decreases by increasing the content of peroxide.
- *At a concentration of about 0.2 wt% of peroxide*: the diameter of the dispersed phase reaches a minimum, which maximizes the formation of PLA-PBAT copolymer due to generation of larger interfacial area.
- *At a concentration higher than 0.2 wt%*: the cross-linking of PLA predominates, but also that of PBAT, takes place. The phase morphology features are complex, and the diameter of the dispersed phase increases by increasing the peroxide content.

In addition, it was observed that under nitrogen flow, the cross-linking is favored, as indicated by viscosity increase, but the formation of PLA-PBAT copolymer was adversely affected. Hence, the formation of PBAT-O-O or PLA-O-O should play a fundamental role in favoring the formation of PLA-PBAT copolymers.Sometimes a functionalizing molecule is added to polymer blends to form *in situ* a functionalized polymer, itself active in the copolymer formation as, for example, the use of maleic anhydride in starch/PLA blends [30]. This method represents the one-step version of the use of functionalized polymers.

A new method of compatibilizing has been described recently by Yousfi et al. [71] The innovation consists of using ionic liquids (ILs) based on tetraalkylphosphonium salts combined with different anions (phosphinate vs trifluoromethylsulfonylimide). Amounts of the ionic liquids comparable to those usually used in classical compatiblizers have been added to 80PP/20PA blend. The blends have been processed in melt by using a TSE. The influence of ILs on the different morphologies of these binary blends has been depicted by transmission electron microscopy (TEM). Although the location in the blend of the ILs was dependent on the balance of the chemical interactions the additive has with either of the polymer components, a significant PA6 particle size reduction by a factor of about 18 was observed upon addition of 1 wt% of the ionic liquid, which was able to locate preferentially at the PP/PA6 interphase. The modified blends exhibited thermal stability and excellent stiffness–toughness balance.

22.4.3 Case of Reactive Ternary Blends

The addition of a third reactive polymer to immiscible polymer blends can be useful for morphology modulation. In immiscible PET/PE blends, a DEM-functionalized very low density polyethyle (VLDPE) was added as compatibilizer precursor in

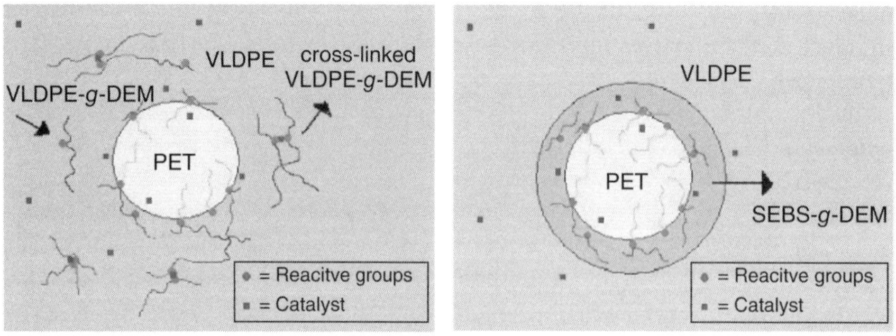

Figure 22.10 Phase morphology simplified scheme of VLDPE/VLDPE-*g*-DEM/PET blend and VLDPE/SEBS-*g*-DEM/PET. The latter scheme is extrapolated on the basis of literature survey of refs: [72–74].

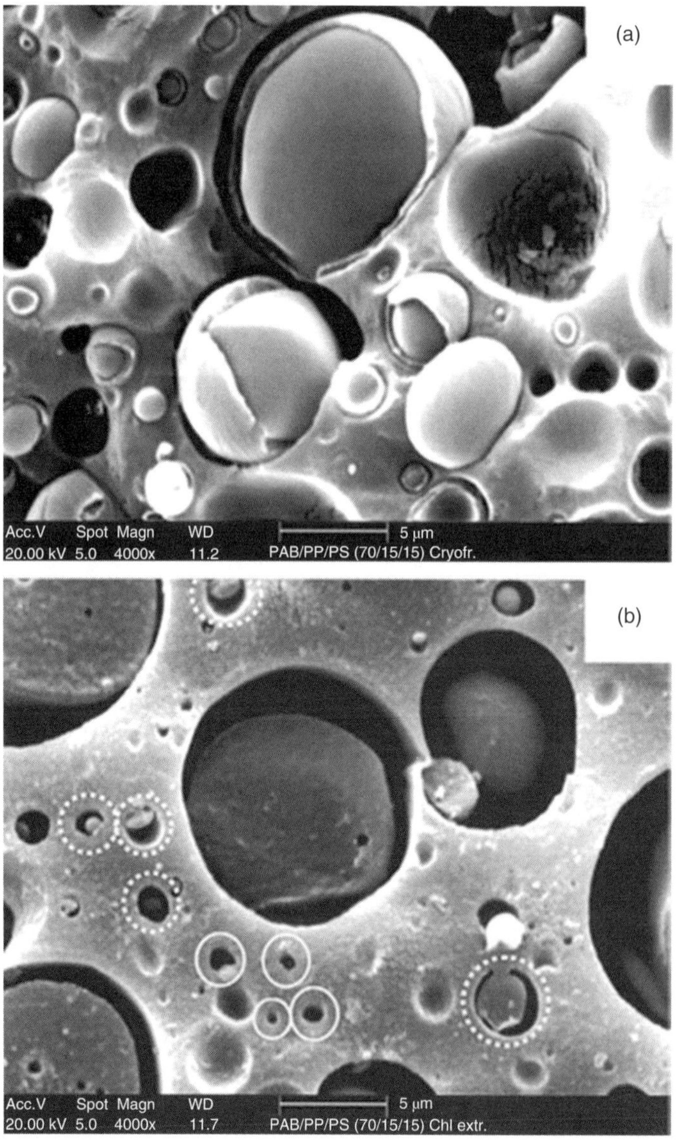

Figure 22.11 Morphology of the ternary 70PA6/15PP/15PS blends: (a) cryofractured, (b) chloroform extracted surfaces. Omonov et al. [77]. Reproduced with permission of Elsevier.

the presence of $Ti(OBu)_4$ as transesterification catalyst. During the blending, PET-PE copolymer was formed thanks to transesterification between the ester-functionalized PE and PET; however, the side reaction consisting in the cross-linking of functionalized VLDPE caused an adverse effect on the blend properties. To avoid this reaction, a compatibilizer precursor able to rapidly migrate to the interphase region between PET and PE was selected (Fig. 22.10).

This goal can be reached by using a nonreactive polymer *P* as a third component, which is able to develop an interfacial tension γ with the two immiscible polymers so as to fulfill Equation 22.7, written by applying Equation 22.1 to this specific case.

$$\lambda_{P\,PET} = [\sigma_{PET\,VLDPE} - \sigma_{P\,VLDPE} - \sigma_{PET\text{-}P}] > 0 \qquad (22.7)$$

Moreover, the viscosity ratio η_P/η_{PET} should be less than 1 to favor the encapsulation of PET with P nonreactive polymer [75]. The poly(styrene-*b*-ethylene-butylene-*b*-styrene) triblock copolymer (SEBS) shows these two characteristics in PET matrix hosting PE-dispersed phase.[76] If polyethylene is the matrix, only the former condition is respected. When the polymer *P* is reactive, the encapsulation of the dispersed phase was observed in different compositions of PA6/PP blends [72, 73] and of PET/PP blends.[74]

Ternary PS/PP/PA6 reactive blends were also the object of a report by Omonov et al. [77]. Depending on the composition, either dispersed encapsulated or co-continuous three-phase morphologies are developed. Uncompatibilized ternary 70PA6/15PP/15PS blends exhibited an encapsulated phase morphology having polypropylene cores and polystyrene shells in opposition to the prediction of Harkin's spreading coefficient (Eq 22.1) and free interfacial energy concept. This failure was ascribed to the substantial differences in viscosities between the dispersed components. The addition of the two compatibilizing reactive systems, that is, PP-*g*-MAH and St-*co*-MAH (SMA) affected the dispersed phase diameter, but still encapsulated structures were observed (Figs. 22.11 and 22.12).

In the ternary blends having almost equal amount of PA6, PP, and PS (i.e., composition of 40PA/30PP/30PS), a

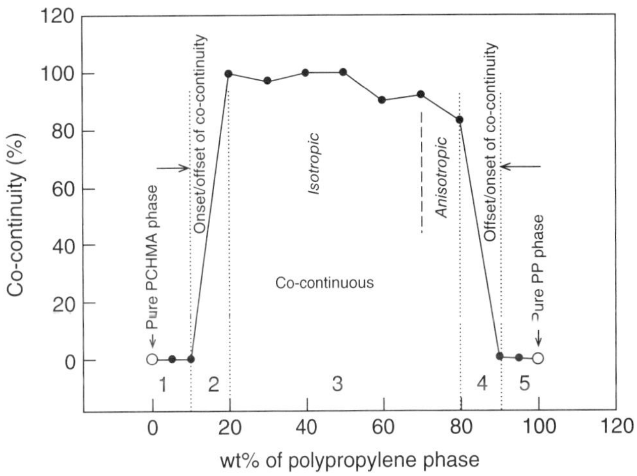

Figure 22.13 Percentage of co-continuity as a function of PP content in the uncompatibilized PP/PCHMA blends. Harrats et al. [78]. Reproduced with permission of Elsevier.

three-phase co-continuous morphology was developed. Their reactive compatibilization using the reactive precursors mentioned caused the PA6 phase to get dispersed preferentially in the polystyrene phase. Interestingly, a lower functionalization degree of PP-*g*-MAH (2 wt%) resulted in a lower dispersed phase diameter than PP-*g*-MAH containing 8 wt%. This result can be due to the competition between the reactions involving the polyamide 6 terminals, highly diluted in the melt system, which is the one with SMA and the one with PP-*g*-MAH. When the highly functionalized PP was employed, a higher yield of PP-PA6 resulted in a lower yield of PS-PA6 copolymer, thus leading to an increase of the PA6 droplet diameter in the PS phase.

22.4.4 Stability of Phase Morphology in Reactively Compatibilized Blends

The stability of phase morphology upon shearing, thermal annealing, and molding processes has been the subject of a huge volume of literature reports that is out of the scope of this review. Nevertheless, an interesting phenomenon

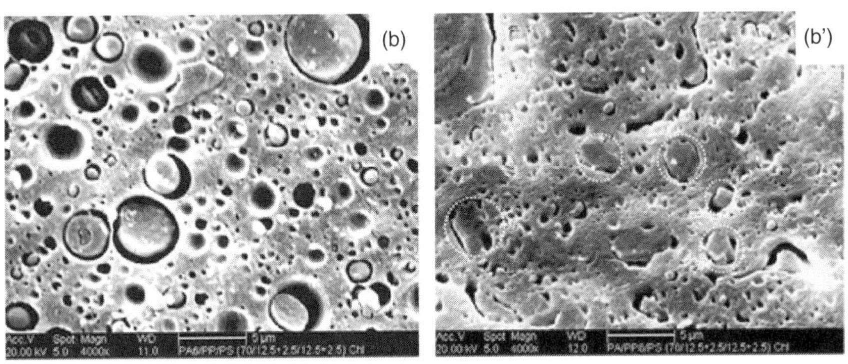

Figure 22.12 Morphology of the 70PA6/(12.5/2.5)(PP/PP–MA2)/(12.5/2.5)(PS/SMA2) (b) and 70PA6/(12.5/2.5)(PP/PP–MA8)/(12.5/2.5)(PS/SMA2) (b'); cryosmoothed surfaces. Omonov 2005, p. 12322–36. Reproduced with permission of Elsevier.

such as the isotropy of co-continuous phase morphology is worthy of being discussed. Harrats et al. [78] could show, in reactively melt-blended compatibilized polypropylene (PP)/poly(cyclohexylmethacrylate) (PCHMA), that a co-continuous phase morphology is not as simple as usually admitted but can be anisotropic depending on the direction in which one observes it. As shown in Figure 22.13, five distinct composition ranges corresponding to five different phase morphologies are identified in a plot of co-continuity versus blend composition. Below 20 and above 80 wt% PP (composition windows 1 and 5, respectively), the phase morphologies are constituted of PP droplets in PCHMA matrix and PCHMA droplets in PP matrix, respectively. In between these compositions, a co-continuous two-phase morphology prevails (window 3). The onset and offset of phase co-continuity take place over a composition window extending over about 10 wt%; from 10 to 20 wt% and from 80 to 90 wt% PP, respectively (windows 2 and 4). Within the delimited onset and offset range, the phase morphology might contain mixed structures as droplets, elongated structures, and even interconnected phases on a limited volume scale. The SEM pictures of the blends containing, 30, 40, 50, 70, and 80 wt% PP, situated in the composition window 3, are presented in Figure 22.14 for illustration. The pictures d and e apparently show a droplet-in-matrix phase morphology, whereas the solvent extraction results confirmed a total phase co-continuity. This behavior has not been reported before and was surprising to the authors. After performing morphological observations on surfaces of the same blends cut in the parallel direction with respect to the extrusion flow, a really oriented, elongated and infinite rodlike structure was found, as shown in Figure 22.15a and b, corresponding to the same blends as in Figure 22.14d and e, respectively. The question which arises from this morphology is: how is the continuity of these elongated rodlike structures of the PCHMA minor phase built up? A close observation of the

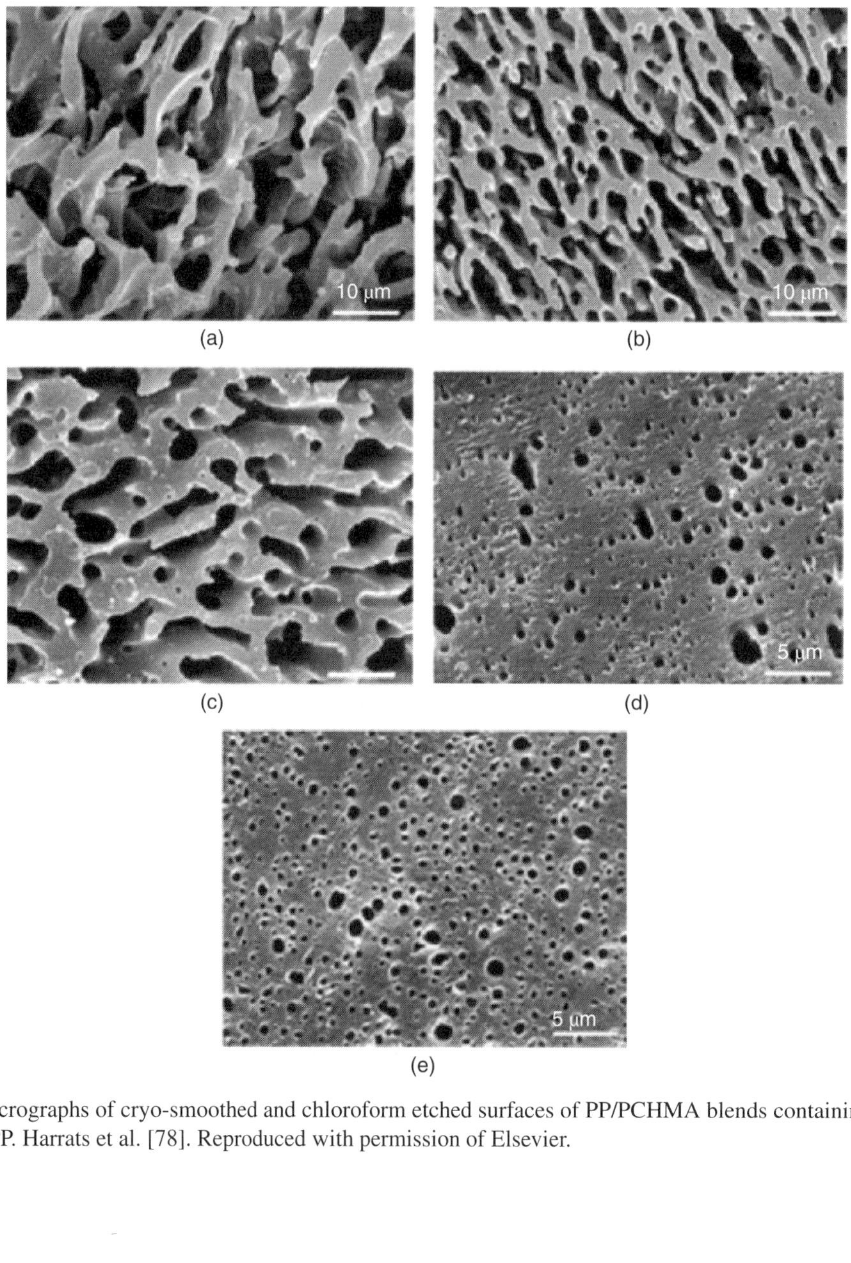

Figure 22.14 SEM micrographs of cryo-smoothed and chloroform etched surfaces of PP/PCHMA blends containing: (a) 30, (b) 40, (c) 50, (d) 70, and (e) 80 wt% PP. Harrats et al. [78]. Reproduced with permission of Elsevier.

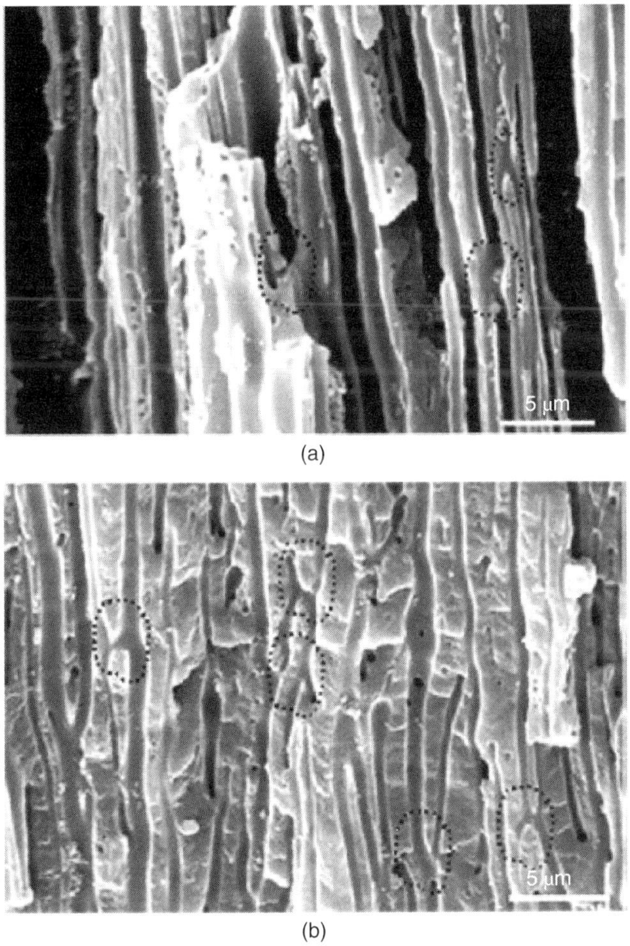

Figure 22.15 SEM micrographs of cryo-fractured and chloroform etched surfaces cut parallel to the extrusion direction: (a) blend containing 70 wt% PP; (b) blend containing 80 wt% PP. Harrats et al. [78]. Reproduced with permission of Elsevier.

pictures presented in Figure 22.15 reveals that these long rods are interconnected in two possible modes: the most likely is via multiple connections as indicated by the dashed lines on the SEM pictures, and also via small (<0.2 µm diameter) and short connecting rods acting as bridges between the long and continuous major rods (note the small holes on the SEM picture).This kind of phase morphology can be qualified as anisotropic co-continuous phase morphology.

More recently, the stability under shear of phase morphology was studied by many research groups. In particular, the structure of the copolymer generated at the interface during reactive blending was reported to deeply affect the stability. As an example, Pan et al. [60] investigated using TEM the PA/PE blends compatibilized by a terpolymer containing epoxide groups. Using polyamide of different molecular weight in the reactive blending process, they observed that the formation of micelles in the separate phases depends on the length of the polyamide chain grafted to the terpolymer. A copolymer with a long PA chain could be easily pulled out of the interface to form micelles in the polyamide phase. The pull out depends not only on the molecular architecture but also on the strength of shear fields, with micelle formation favored at high shear field. However, the structure of a premade block copolymer was reported to deeply affect the phase morphology development and its stability even under quiescent annealing conditions. The control of the copolymer molecular weight and its segmental balance is the correct strategy to attain fine and stable phase morphologies.[79]

In PLA/PBAT blended in the presence of $Ti(OBu)_4$ [41], transesterification reaction leading to the formation of PLA-PBAT copolymer was demonstrated by size-exclusion chromatography. Within a mixing time range of 10–40 minutes, the yield of copolymer formation was found to increase as a function of mixing time. The phase morphology was characterized by SEM after blending and also after a subsequent step of compression molding during two minutes. After 10 min of blending an increase in the diameter of the dispersed phase was observed after compression molding, whereas no significant change was observed after 40 min of blending (Fig. 22.16). The presence of higher amount of PLA-PBAT copolymer in the latter system stabilizes much of the morphology of the blend, certainly by forming a stable interfacial layer that prevents excessive particle–particle coalescence.

A similar effect was evidenced by Omonov et al. in PP/PS blends [80]. The phase morphology of uncompatibilized and reactively compatibilized blends using amino-terminated polystyrene and maleic anhydride–grafted polypropylene reactive precursors, was investigated. The structural instability of the phase morphology as a function of the annealing time at a temperature of 205 °C has been studied in 50PP/50PS blends. The *in situ* generation of 1 and 5 wt% compatibilizer based on maleic anhydride–functionalized PP (PP-*g*-MAH) having a content of maleic anhydride of 1% by weight did not affect the morphology stability significantly. Blends with 5 wt% compatibilizer based on PP-*g*-MA having a content of 8 wt% grafted MA caused a significant reduction in the characteristic phase size of the PS phase. Phase coarsening was drastically suppressed. Hence, the stability is a consequence of the yield of the interfacial reactions determining the formation of PP-PS copolymer. Interestingly, the specific interfacial area Q, that is, the average ratio between perimeter and area of observed particles decreased as a function of annealing time for compatibilized and uncompatibilized blends, with final Q values higher for the highly compatibilized system

22.4.5 Organoclay-Promoted Phase Morphology

The addition of a filler in multiphase polymer blends aimed at modifying the phase morphology is a new strategy progressively reported in blend literature. The initial objective was to distribute the filler preferentially in one of the two blend phases or, if the morphology is co-continuous, to have it at the interface. Electrical conductivity with a minimum filler loading was one of the applications searched for. For example, in a polystyrene/polyisoprene blend having co-continuous phase morphology as low as 0.2 vol% of carbon black particles was

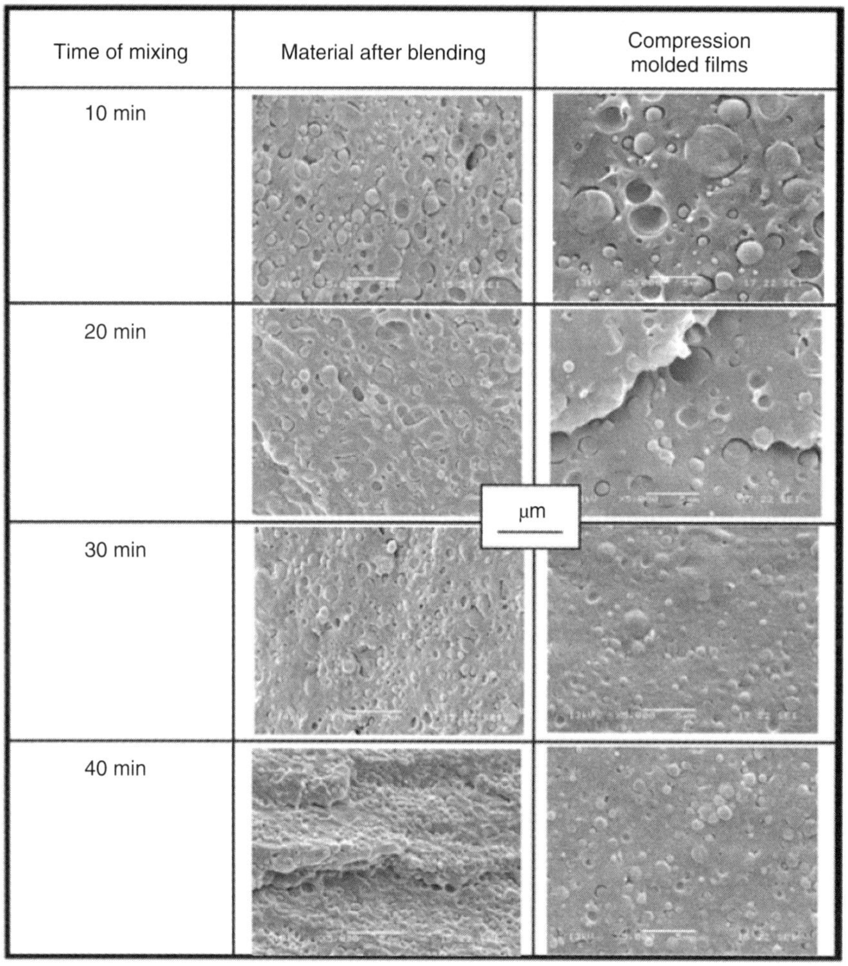

Figure 22.16 SEM micrographs of PLA/PBAT 75/25 blends obtained after different blending time. On the right column the corresponding micrographs obtained after an annealing step in a compression molding press are reported. Coltelli et al. [41]. Reproduced with permission of Elsevier.

sufficient to reach a percolated structure at the interface, resulting in a conductive polymer-based material [81]. Surprised by the effect that the inorganic filler had on the size of a dispersed polymer phase in a matrix of the second polymer, several groups have launched investigation of fillers, mainly silicate clay, as a new route to compatibilization of immiscible polymer blends [82–87]. A significant particle size reduction of the dispersed phase was reported in most of these studies. The authors attributed this effect to a reduction in the interfacial tension induced by the presence of the clay at the interface and to a significant modification of the viscosity ratio of the blend due to clay loading, particularly if it is exfoliated. Calcagno et al. [88, 89] investigated the effect of MMT clay on PP/PET/MAH-*g*-PP blends using melt-mixing process. The microscopy observations revealed that the clay was preferentially located at the interface and in the PET phase. As could be expected, hydrophilic MMT clay is attracted by the more polar phase of the blend.

Tiwari and Paul [90] recently published a consolidating work for the compatibilizing role of silicate clay in a model blend of immiscible PP/PS processed using a TSE. Maleated polypropylene was used, at a PP-*g*-MA/organoclay (MMT) ratio of 1, to preferentially promote dispersion of the organoclay in the PP matrix. The MMT content was fixed at 3 wt% based on the PP/PP-*g*-MA/MMT phase, while the PS content was varied within the full range of 0–100 wt% in the blend. TEM observations revealed that the organoclay resides in the PP phase and at the PP/PS interface (Fig. 22.17: note that L, M, and H are indices of low-, medium-, and high-viscosity PP grades). The dispersed PS particle size was significantly reduced by the presence of MMT, with a maximum decrease observed for the low-viscosity PP compared to its blends without MMT. The blends with MMT did not show any change in onset of co-continuity, although MMT shifts the phase inversion composition toward lower PS contents. The phase stability of the blend was significantly improved by the presence of MMT; for blends annealed at 210 °C for 2 h, the dispersed phase particle size increased by as much as 10× without MMT with only little change when clay was present in the blend.

Moghbelli et al. [91] revealed the same tendency of compatibilization induced by nanofillers in PA/styrene acrylonitrile copolymer (SAN). The addition of a few percentage

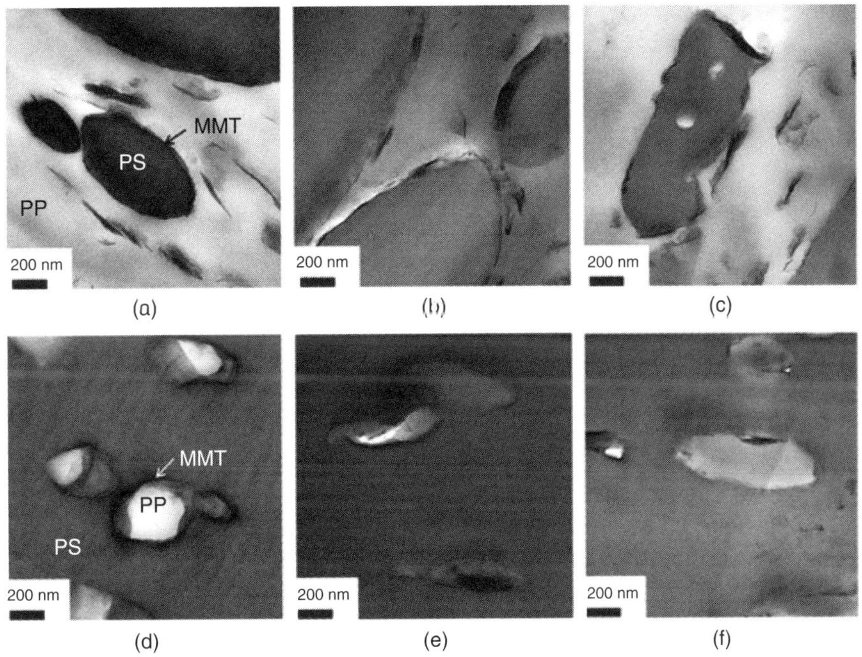

Figure 22.17 TEM micrographs showing location of the MMT particles in PP/PP-*g*-MA/MMT/PS blend having 30 wt% PS (top row) and 90 wt% PS (bottom row). The blends were prepared with different PP grades: L-PP (a and d); M-PP (b and e) and H-PP (c and f). Images were taken from the core and viewed perpendicular to the flow direction (FD). Tiwari and Paul [90]. Reproduced with permission of Elsevier.

of clay grade 1022C2 to PA/SAN blend, where the PA is the matrix, reduced the particle size of the SAN phase by a factor of 2 compared to the unmodified blend. No notable effect of the clay was observed in the opposite compositions where the SAN is the matrix. Furthermore, when the state of clay is considered, the exfoliation was more efficient than intercalation. The TEM observation confirmed the preferential location of the clay platelets in only the PA phase. They limited also the phase coarsening upon thermal annealing by hindering the coalescence process. The extent of reduction of the coalescence kinetics reported by the authors was comparable to that published by Khatua et al. [82] on clay in PA/EPR blends. In both studies, the phase size was increased by only 40% in the clay-modified blend but by 200% in the unmodified system

Very recently, Nuzzo et al. [92] compared the role played by organoclay (organomodified-MMT), organomodified-sepiolite and multiwalled carbon nanotubes (CNTs) in converting droplet-in-matrix to co-continuous phase morphology of 70PLA/30PA11 blends. The three types of fillers locate preferentially in the minor PA11 phase and they all induce phase co-continuity of the PLA and the PA11 phases provided a critical concentration of the filler is exceeded. Although it presents some limitations, the authors proposed a dimensionless group (Eq. 22.8) to estimate the conditions at which a conversion from droplet-in-matrix to co-continuous phase morphology takes place.

$$\propto = \frac{\tau_y P_c}{\sigma_{ij}^2} \qquad (22.8)$$

where σ is the interfacial tension, τ is the yield stress, and P_c is the filler critical buckling load (ability to resist bending). It was assumed that when $\propto$ exceeds a certain threshold (or its value is within 10^{-2}–10^{-1} range), the filled PA11 minor phase adopts elongated shapes that lead to the formation of co-continuous morphology with the PLA matrix. The modified sepiolites were more effective than the deformable CNTs but less effective than OMMT in promoting co-continuity of the minor phase where they are located.

Another role played by the addition of organoclay consists of freezing-in the phase morphology developed in poly(p-phenylene sulfide) (PPS)/polyamide66 (PA66)binary polymer blends [93]. A dispersed PA66 phase in PPS matrix transforms to PPS dispersed phase in PA66 continuous phase as a result of the addition of 1 wt% of clay. Substantial differences in phase morphology depending on whether the PA66 and PPS are mixed at 260 °C before the temperature is gradually increased to 300 °C or, whether they are melt mixed directly together at 300 °C. In the first processing case, the clay is intercalated/exfoliated in the molten PA66 phase before the PPS melts. The authors assume that the clay layers prevent the PPS droplets from coalescing, while in method 2 the clay is exfoliated after the PPS is melted and broken up into dispersed droplets.

22.4.6 Conclusions

Important aspects relating to the development of phase morphologies in polymer-based multicomponent systems were discussed. Focus is made on the co-continuous phase morphology in binary and ternary thermoplastics blends, both uncompatible and reactively compatibilized. The recent developments in the effects of the processing parameters used to generate these morphologies during the melt-compounding process are highlighted. They reveal that research aiming at understanding the interrelation between the processing conditions and the obtained morphologies still continues. New approaches of compatibilization as the use of organoclay are discussed, too. This can be a promising simple route for the generation of new materials at low cost away from complex chemistry. The chapter also shows that the effort of introducing more chemistry to promote interesting morphologies to polymer blends also continues mainly in research oriented toward industrial application.

REFERENCES

1. Harrats C. *Multiphase Polymer-Based Materials: An Atlas of Phase Morphology at the Nano and Micro Scale*. Boca Raton: CRC Press - Taylor & Francis; 2008.
2. Taylor GI. The viscosity of a fluid containing small drops of another fluid. Proc R Soc London A 1932;138:41.
3. Taylor GI. The formation of emulsions in definable fields of flow. Proc R Soc London A 1934;146:501.
4. Tomotika S. On the instability of a cylindrical thread of a viscous liquid surrounded by another viscous fluid. Proc R Soc London A 1935;A150:322.
5. Rayleigh L. On the capillary phenomena of jets. Proc R Soc London A 1879;29:71.
6. Van Oene H. Rheology of polymer blends and dispersion. In: Paul DR, Newman S, editors. *Polymer Blends Vol.1*. NY: Academic Press; 1978.
7. Elmendorp JJ; A study on polymer blending micro-rheology, PhD thesis, T.U Delft; 1986.
8. P.H.M. Elemans; PhD thesis, T.U. Eindhoven ; 1989.
9. Grace HP. Dispersion phenomena in high viscosity immiscible fluid systems and application of static mixers as dispersion devices in such systems. Chem Eng Commun 1982;14:225.
10. Harrats C, Mekhilef N. Co-continuous phase morphologies: Predictions, generation and practical applications. In: Harrats C, Thomas S, Groeninckx G, editors. *Micro- and Nanostructured Multiphase Polymer Blend Systems: Phase Morphology and Interfaces*. Boca Raton: CRC Taylor& Francis; 2008. p 91–132.
11. Utracki LA. *Polymer Alloys and Blends: Thermodynamics and Rheology. Part2*. Munich, Vienna, NY : Hanser Publisher; 1989.
12. Shokoohi S, Arefazar A. A review on ternary immiscible polymer blends: Morphology and effective parameters. Polym Adv Technol 2013;20:433–447.
13. Rastin H, Jafari SH, Saeb MR, Khonakdar HA, Wagenknecht U, Heinrich G. On the reliability of existing theoretical models in anticipating type of morphology and domain size in HDPE/PA-6/EVOH ternary blends. Eur Polym J 2014;53:1–12.
14. Ravati S, Favis BD. Morphological states for a ternary polymer blend demonstrating complete wetting. Polymer 2010;51:4547–4561.
15. Ravati S, Favis BD. Tunable morphologies for ternary blends with poly(butylene succinate): Partial and complete wetting phenomena. Polymer 2013;54:3271–3281.
16. Ravati S, Favis BD. Interfacial coarsening of ternary polymer blends with partial and complete wetting structures. Polymer 2013;54:6739–6751.
17. Wilkinson AN, Clemens ML, Harding VM. The effects of SEBS-g-maleic anhydride reaction on the morphology and properties of polypropylene/PA6/SEBS ternary blends. Polymer 2004;45:5239–5249.
18. Favis BD. The effect of processing parameters on the morphology of an immiscible binary blend. J Appl Polym Sci 1990;39:285.
19. Plochocki AP, Dagli SS, Curry JE, Starita J. The interface in binary-mixtures of polymers containing a corresponding block copolymer – Effects of industrial mixing processes and of coalescence. Polym Eng Sci 1990;30:741.
20. Scott CE, Macoskow CW. Model experiments concerning morphology development during the initial-stages of polymer blending. Polym Bull 1991;26:341.
21. Sundararaj U, Macosko CW, Nakayama A, Inoue T. Milligrams to kilograms - An evaluation of mixers for reactive polymer blending. Polym Eng Sci 1995;35:100.
22. Sundararaj U, Dori Y, Macosko CW. Sheet formation in immiscible polymer blends. Polymer 1995;36:1957.
23. Sundararaj U, Dori Y, Macosko CW. Visualization of polymer blending. SPE Tech Pap 1994;52:2448.
24. Sandararaj U. Phase morphology development in polymer blends: Processing and experimental aspects. In: Harrats C, Thomas S, Groeninckx G, editors. *Micro- and Nanostructured Multiphase Polymer Blend Systems: Phase Morphology And Interfaces*. Boca Raton: CRC Press; 2006.
25. Bourry D, Favis BD. Morphology development in a polyethylene/polystyrene binary blend during twin-screw extruder. Polymer 1998;39:1851–1856.
26. Canto LB, Mantovani GL, Covas JA, Hage E, Pessan LA. Phase morphology development during processing of compatibilized and uncompatibilized PBT/ABS blends. J Appl Polym Sci 2007;104:102–110.
27. Canto LB. On the coarsening of phase morphology of PP/EVA blends during compounding in a twin screw extruder. Polym Test 2014;34:175–182.
28. Huang H-X, Huang Y-F, Li X-J. Detecting blend morphology development during melt-blending along an extruder. Polym Test 2007;26:770–778.
29. Walczak ZK. Influence of shearing history on properties of polymer melt - 2. J Appl Polym Sci 1973;17:169.
30. Ghodgaokar P, Sundararaj U. Prediction of drop size in polymer clends. Polym Eng Sci 1996;36:1656.
31. Wu S. Formation of dispersed phase in incompatible polymer blends - interfacial and rheological effects. Polym Eng Sci 1987;27:335.
32. Favis BD, Chalifoux JP. The effect of viscosity ratio on the morphology of polypropylene polycarbonate blends during processing. Polym Eng Sci 1987;27:1591.
33. Thomas S, Groeninckx G. Nylon 6 ethylene propylene rubber (EPM) blends: Phase morphology development during processing and comparison with literature data. J Appl Polym Sci 1999;71:1405.
34. Huneault MA, Shi ZH, Utracki LA. Development of polymer blend morphology during compounding in a twin-screw extruder. 4. A new computational model with coalescence. Polym Eng Sci 1995;35:115.

35. Lee JK, Han CD. Evolution of a dispersed morphology from a co-continuous morphology immiscible polymer blends. Polymer 1999;40:6277.

36. Potente H, Bastian M, Gehring A, Stephan M, Potschke P. Experimental investigation of the morphology development of polyblends in corotating twin-screw extruders. J Appl Polym Sci 2000;76:708.

37. Bertoldo M, Coltelli MB, Miraglia L, Narducci P, Bronco S. Surface energy inducing asymmetric phase distribution in films of a binary polymeric blend. Polymer 2005;46:11311–11321.

38. Fortelny I, Zivny A. Competition between breakup and coalescence of droplets in polymer blends containing a compatibilizer. Macromol Symp 2000;149:157–164.

39. Kenwright AM, Peace SK, Richards RW, Bunn A, MacDonald WA. Transesterification in poly(ethylene terephthalate) and poly(ethylene naphthalene 2,6-dicarboxylate) blends; the influence of hydroxyl end groups. Polymer 1999;40:5851–5856.

40. Kim JH, Lyoo WS, Ha WS. Sequence analysis of poly(ethylene terephthalate)/poly(butylene terephthalate) copolymer prepared by ester-interchange reactions. J Appl Polym Sci 2001;82:159–168.

41. Coltelli MB, Toncelli C, Ciardelli F, Bronco S. Compatible blends of biorelated polyesters through catalytic transesterification in the melt. Polym Degrad Stab 2011;96:982–990.

42. Signori F, Coltelli MB, Bronco S. Thermal degradation of poly(lactic acid) (PLA) and poly(butylene adipate-co-terephthalate) (PBAT) and their blends upon melt processing. Polym Degrad Stab 2009;94:74–82.

43. Godard P, Dekoninck JM, Devlesaver V, Devaux J. Molten bisphenol-A polycarbonate-poly(ethylene terephthalate) blends. I. Identification of the reactions. J Polym Sci A 1986;24: 3301–3313.

44. Phuong VT, Coltelli MB, Cinelli P, CifelliM VS, Lazzeri A. Compatibilization and property enhancement of poly(lactic acid)/polycarbonate blends through triacetin-mediated interchange reactions in the melt. Polymer 2014;55(12):4498–4513.

45. Legros A, Carreau PJ, Favis BD, Michel A. Morphology modification by interfacial chemical reaction in a polyester/ethylene vinyl acetate/polyethylene blend. Polymer 1997;38:5085–5089.

46. Coltelli MB, Savi S, Della Maggiore I, Liuzzo V, Aglietto M, Ciardelli F. A model study of Ti(OBu)4 catalysed reactions during reactive blending of polyethylene (PE) and poly(ethylene terephthalate) (PET). Macromol Mater Eng 2004;289:400–412.

47. ColtelliMB BS, Aglietto M. Poly(ethylene terephthalate) (PET) degradation during the Zn catalysed transesterification with dibutyl maleate functionalized polyolefins. Polymer 2007;48:1276–1286.

48. Coltelli MB, Aglietto M, Ciardelli F. Influence of the transesterification catalyst structure on the reactive compatibilization and properties of poly(ethylene terephthalate) (PET)/dibutyl succinate functionalized poly(ethylene) blends. Eur Polym J 2008;44:1512–1524.

49. Liu C, Lin S, Zhou C, Yu W. Influence of catalyst on transesterification between poly(lactic acid) and polycarbonate under flow field. Polymer 2013;54:310–319.

50. Jerôme R, Pagnoulle C. Key role of structural features of compatibilizing polymer additives in reactive blending. In: Baker W, Scott C, Hu GH, editors. *Reactive Polymer Blending*. Munich: Hanser; 2001.

51. Triacca VJ, Ziaee S, Barlow J-W, Keskkula H, Paul DR. Reactive compatibilization of blends of nylon 6 and ABS materials. Polymer 1991;32:1401–1413.

52. Silva EF, Soares BG. Polyethylene polyamide-6 blends containing mercapto-modified Eva. J Appl Polym Sci 1996;60:1687–1694.

53. Loyens W, Groeninckx G. Ultimate mechanical properties of rubber toughened semicristalline PET at room temperature. Polymer 2002;43:5679–5691.

54. Loyens W, Groeninckx G. Rubber toughening semicristalline PET. Influence of the matrix properties and temperature. Polymer 2003;44:123–136.

55. Loyens W, Groeninckx G. Phase morphology development in reactively compatibilized polyethylene terephthalate elastomer blends. Polymer 2002;44:4929–4941.

56. Arostegui A, Nazabal J. New super-tough poly(butylene terephthalate) materials based on compatibilized blends with metallocenic poly(ethylene-octene) copolymer. Polym Adv Technol 2003;14:400–408.

57. Hale W, Keskkula H, Paul DR. Compatibilization of PBT/ABS blends by methyl methacrylate-glycidyl methacrylate-ethyl acrylate terpolymers. Polymer 1999;40:365–377.

58. Wallheinke KK, Heckmann W, Pötschke P, Stutz H. Localizing compatibilizers in immiscible blends by SEM. Polym Test 1998;17:247–255.

59. Ciardelli F, Aglietto M, Coltelli M-B, Passaglia E, Ruggeri G, Coiai S. Functionalization of polyolefins in the melt. In: Ciardelli F, Penczek S, editors. *Modification and Blending of Synthetic and Natural Macromolecules*. Kluwer; 2004. p 47–71.

60. Pan L, Chiba T, Inoue T. Reactive blending of polyamide with polyethylene: Pull-out of in-situ-formed graft copolymer. Polymer 2001;42:8825–8831.

61. Passaglia E, Aglietto M, Ruggeri G, Picchioni F. Formation and compatibilizing effect of the grafted copolymer in the reactive blending of 2-diethylsuccinate containing polyolefins with poly ε-caprolactam (Nylon-6). Polym Adv Technol 1998;9: 273–281.

62. Coltelli MB, Passaglia E, Ciardelli F. Formation of short and long chain branches during the free radical functionalization of polyamide 6 in the melt. Polymer 2006;47:85–97.

63. Coltelli MB, Passaglia E, Angiuli M, Castelvetro V, Ciardelli F. Formation of short and long chain branches during the free radical functionalization of polyamide 6 in the melt. Macromolecules 2006;39:2153–2161.

64. Coltelli MB, Della Maggiore I, Savi S, Aglietto M, Ciardelli F. Modified styrene–butadiene–styrene block copolymer as compatibiliser precursor in polyethylene/poly (ethylene terephthalate) blends. Polym Degrad Stab 2005;90(2):211–223 erratum 2006; 91:987.

65. Coltelli MB, Harrats C, Aglietto M, Groeninckx G. Influence of compatibilizer precursor structure on the phase distribution of low density poly(ethylene) in a poly(ethylene terephthalate) matrix. Polym Eng Sci 2008;48:1424–1433.

66. Coltelli M-B, Savi S, Aglietto M, Ciardelli F. A chemical view onto post consumer poly(ethylene terephthalate) valorization through reactive blending with functionalized polyolefins. Polym Sci A 2009;51(11):1–13.

67. Bartoli F, Bruni C, Coltelli MB, Castelvetro V, Ciardelli F. Conversion of post-industrial PET/PE scraps into compatibilized plastic blends for new applications. AIP

Conf Proc 2012;1459:160–162, ISSN: 0094-243X, doi: 10.1063/1.4738430.

68. Raffa P, ColtelliM-B SS, Bianchi S, Castelvetro V. Chain extension and branching of poly (ethylene terephthalate)(PET) with di-and multifunctional epoxy or isocyanate additives: An experimental and modelling study. React Funct Polym 2012;72:50–60.
69. Raffa P, Coltelli M-B, Castelvetro V. Expanding the application field of post-consumer poly(ethylene terephthalate) through structural modification by reactive blending. J Appl Polym Sci 2014;131:40881–40092.
70. Coltelli M-B, Bronco S, Chinea C. The effect of free radical reactions on structure and properties of poly(lactic acid) (PLA) based blends. Polym Degrad Stab 2010;95:332–341.
71. Yousfi M, Livi S, Duchet-Rumeau J. Ionic liquids: A new way of compatibilization of thermoplastic blend. Chem Eng J 2014;255:513–524.
72. Gonzalez-Montiel A, Keskkula H, Paul DR. Impact-modified nylon 6/polypropylene blends: 2. Effect of reactive functionality on morphology and mechanical properties. Polymer 1995;36(24):4587–4603.
73. Kim G-M, Michler GH, Rosch J, Mulhaupt R. Micromechanical deformation processes in toughened PP/PA/SEBS-g-MA blends prepared by reactive processing. Acta Polym 1998;49:88–95.
74. Heino M, Kirjava J, Hietoja P, Seppala J. Compatibilization of polyethylene terephthalate/polypropylene blends with styrene–ethylene/butylene–styrene (SEBS) block copolymers. J Appl Polym Sci 1997;65:241–249.
75. Reignier J, Favis BD, Heuzey M-C. Factors influencing encapsulation behavior in composite droplet-type polymer blends. Polymer 2003;44:49–59.
76. Ihm DJ, White JL. Interfacial tension of polyethylene/polyethylene terephthalate with various compatibilizing agents. J Appl Polym Sci 1996;60:1–7.
77. Omonov TS, Harrats C, Groeninckx G. Co-continuous and encapsulated three phase morphologies in uncompatibilized and reactively compatibilized polyamide6/polypropylene/polystyrene ternary blends using two reactive precursors. Polymer 2005;46:12322–12336.
78. Harrats C, Omonov T, Groeninckx G, Moldenaers P. Phase morphology development and stabilization inpolycyclohexylmethacrylate/polypropylene blends: Uncompatibilized and reactively compatibilized blends using two reactive precursors. Polymer 2004;45:8115–8126.
79. Harrats C, Fayt R, Jerome R. Effect of block copolymers of various molecular architecture on the phase morphology and tensile properties of LDPE rich (LDPE/PS) blends. Polymer 2002;43:863–873.
80. Omonov TS, Harrats C, Groeninckx G, Moldenaers P. Anisotropy and instability of the co-continuous phase morphology in uncompatibilized and reactively compatibilized polypropylene/polystyrene blends. Polymer 2007;48:5289–5302.
81. Soares BG, Gubbels F, Jérôme R, Teyssié P, Vanlathem E, Deltour R. Electrical conductivity in carbon black loaded polystyrene-polyisoprene blends selective localization of carbon black at the interface. Polym Bull 1995;35:223–228.
82. Khatua BB, Lee DJ, Kim HY, Kim JK. Effect of organoclay platelets onmorphology of nylon-6 and poly(ethylene-ran-propylene) rubber blends. Macromolecules 2004;37:2454.
83. Ray SS, Pouliot S, Bousmina M, Utracki LA. Role of organically modified layered silicate as an active interfacial modifier in immiscible polystyrene/polypropylene blends. Polymer 2004;45:8403–8413.
84. Hong JS, Namkung H, Ahn KH, Lee SJ, Kim C. The role of organically modified layered silicate in the breakup and coalescence of droplets in PBT/PE blends. Polymer 2006;47:3967–3975.
85. Si M, Araki T, Ade H, Kilcoyne ALD, Fisher R, Sokolov JC, Rafailovich MH. Compatibilizing bulk polymer blends by using organoclays. Macromolecules 2006;39:4793–4801.
86. Hong JS, Kim YK, Ahn KH, Lee SJ, Kim C. Interfacial tension reduction in PBT/PE/clay nanocomposite. Rheol Acta 2007;46:469–478.
87. Su Q, Feng M, Zhang S, Jiang J, Yang M. Melt blending of polypropyleneblend polyamide 6 blendorganoclay systems. Polym Int 2007;56:50–56.
88. Calcagno CIW, Mariani CM, Teixeira SR, Mauler RS. PP/PET nanocomposites. In: Proceedings of world polymer congress Macro 2006 – 41 international symposium on macromolecules; Rio de Janeiro, Brazil; 2006.
89. Calcagno CIW, Mariani CM, Teixeira SR, Mauler RS. The role of the MMT on the morphology and mechanical properties of the PP/PET blends. Compos Sci Technol 2008;68:2193–2200.
90. Tiwari RR, Paul DR. Effect of organoclay on the morphology, phase stability and mechanical properties of polypropylene/polystyrene blends. Polymer 2011;52:1141–1154.
91. Moghbelli E, Sue H-J, Jain S. Stabilization and control of phase morphology of PA/SAN blends via incorporation of exfoliated clay. Polymer 2010;51:4231–4237.
92. Nuzzo A, Bilotti E, Peijs T, Acierno D, Filippone G. Nanoparticle-induced co-continuity in immiscible polymer blends- A comparative study on bio-based PLA-PA11 blends filled with organoclay sepiolite and carbon nanotubes. Polymer 2014;55:4908–4919.
93. Zou H, Zhang Q, Tan H, Wang K, Du R, Fu Q. Clay locked phase morphology in the PPS/PA66/clay blends during compounding in an internal mixer. Polymer 2006;47:6–11.

INDEX

Polymer Morphology: Principles, Characterization, and Processing, First Edition. Edited by Qipeng Guo.